METHODS OF THEORETICAL PHYSICS

INTERNATIONAL SERIES IN PURE AND APPLIED PHYSICS

LEONARD I. SCHIFF, CONSULTING EDITOR

Allis and Herlin Thermodynamics and Statistical Mechanics
Becker Introduction to Theoretical Mechanics
Clark Applied X-rays
Collin Field Theory of Guided Waves
Evans The Atomic Nucleus
Finkelnburg Atomic Physics
Ginzton Microwave Measurements
Green Nuclear Physics
Gurney Introduction to Statistical Mechanics
Hall Introduction to Electron Microscopy
Hardy and Perrin The Principles of Optics
Harnwell Electricity and Electromagnetism
Harnwell and Livingood Experimental Atomic Physics
Harnwell and Stephens Atomic Physics
Henley and Thirring Elementary Quantum Field Theory
Houston Principles of Mathematical Physics
Hund High-frequency Measurements
Kennard Kinetic Theory of Gases
Lane Superfluid Physics
Leighton Principles of Modern Physics
Lindsay Mechanical Radiation
Livingston and Blewett Particle Accelerators
Middleton An Introduction to Statistical Communication Theory
Morse Vibration and Sound
Morse and Feshbach Methods of Theoretical Physics
Muskat Physical Principles of Oil Production
Present Kinetic Theory of Gases
Read Dislocations in Crystals
Richtmyer, Gennard, and Lauritsen Introduction to Modern Physics
Schiff Quantum Mechanics
Seitz The Modern Theory of Solids
Slater Introduction to Chemical Physics
Slater Quantum Theory of Matter
Slater Quantum Theory of Atomic Structure, Vol. I
Slater Quantum Theory of Atomic Structure, Vol. II
Slater and Frank Electromagnetism
Slater and Frank Introduction to Theoretical Physics
Slater and Frank Mechanics
Smythe Static and Dynamic Electricity
Stratton Electromagnetic Theory
Thorndike Mesons: A Summary of Experimental Facts
Townes and Schawlow Microwave Spectroscopy
White Introduction to Atomic Spectra

The late F. K. Richtmyer was Consulting Editor of the series from its inception in 1929 to his death in 1939. Lee A. DuBridge was Consulting Editor from 1939 to 1946; and G. P. Harnwell from 1947 to 1954.

METHODS OF THEORETICAL PHYSICS

Philip M. Morse
PROFESSOR OF PHYSICS
MASSACHUSETTS INSTITUTE OF TECHNOLOGY

Herman Feshbach
ASSOCIATE PROFESSOR OF PHYSICS
MASSACHUSETTS INSTITUTE OF TECHNOLOGY

PART I: CHAPTERS 1 TO 8

New York Toronto London
McGRAW-HILL BOOK COMPANY, INC.
1953

METHODS OF THEORETICAL PHYSICS

Library of Congress Catalog Card Number: 52-11515

IX

43317

THE MAPLE PRESS COMPANY, YORK, PA.

Preface

This treatise is the outgrowth of a course which has been given by one or the other of the authors for the past sixteen years. The book itself has been in the process of production for more than half this time, though with numerous interruptions, major and minor. Not the least difficult problem in its development has been to arrive at a general philosophy concerning the subject matter to be included and its order of presentation.

Theoretical physics today covers a vast area; to expound it fully would overflow a five-foot shelf and would be far beyond the authors' ability and interest. But not all of this area has recently been subjected to intensive exploration; the portions in which the most noticeable advances have been made in the past twenty years are mostly concerned with fields rather than with particles, with wave functions, fields of force, electromagnetic and acoustic potentials, all of which are solutions of partial differential equations and are specified by boundary conditions. The present treatise concentrates its attention on this general area. Fifty years ago it might have been entitled "Partial Differential Equations of Physics" or "Boundary Value Problems." Today, because of the spread of the field concept and techniques, it is perhaps not inappropriate to use a more general title.

Even this restricted field cannot be covered in two volumes of course. A discussion of the physical concepts and experimental procedures in all the branches of physics which use fields for their description would itself result in an overlong shelf, duplicating the subject matter of many excellent texts and, by its prolixity, disguising the fundamental unity of the subject. For the unity of field theory lies in its techniques of analysis, the mathematical tools it uses to obtain answers. These techniques are essentially the same, whether the field under study corresponds to a neutral meson, a radar signal, a sound wave, or a cloud of diffusing neutrons. The present treatise, therefore, is primarily concerned with an exposition of the mathematical tools which have proved most useful in the study of the many field constructs in physics, together with a series of examples, showing how the tools are used to solve various physical problems. Only enough of the underlying physics is given to make the examples understandable.

This is not to say that the work is a text on mathematics, however. The physicist, using mathematics as a tool, can also use his physical knowledge to supplement equations in a way in which pure mathematicians dare not (and should not) proceed. He can freely use the construct of the point charge, for example; the mathematician must struggle to clarify the analytic vagaries of the Dirac delta function. The physicist often starts with the solution of the partial differential equation already described and measured; the mathematician often must develop a very specialized network of theorems and lemmas to show exactly when a given equation has a unique solution. The derivations given in the present work will, we hope, be understandable and satisfactory to physicists and engineers, for whom the work is written; they will not often seem rigorous to the mathematician.

Within these twofold limits, on the amount of physics and the rigor of the mathematics, however, it is hoped that the treatise is reasonably complete and self-contained. The knowledge of physics assumed is that expected of a first-year graduate student in physics, and the mathematical background assumed is that attained by one who has taken a first course in differential equations or advanced calculus. The further mathematics needed, those parts of vector and tensor analysis, of the theory of linear differential equations and of integral equations, which are germane to the major subject, are all treated in the text.

The material is built up in a fairly self-contained manner, so that only seldom is it necessary to use the phrase "it can be shown," so frustrating to the reader. Even in the earlier discussion of the basic mathematical techniques an attempt is made to relate the equations and the procedures to the physical properties of the fields, which are the central subject of study. In many cases derivations are given twice, once in a semi-intuitive manner, to bring out the physical concepts, a second time with all the symbols and equations, to provide as much rigor as seems necessary. Often part of the derivation is repeated in a later chapter, from a different point of view, to obviate excessive back reference; this was considered desirable, though it has increased the size of the treatise.

An effort has been made to avoid trivial and special-case examples of solutions. As a result, of course, the examples included often require long and complicated explanations to bring out all the items of interest, but this treatise is supposed to explain how difficult problems can be solved; it cannot be illustrated by easy ones. The variational technique applied to diffraction problems, the iteration methods used in calculating the scattering of waves from irregular boundaries, the computation of convergent series for eigenstates perturbed by strong interaction potentials are all techniques which show their power only when used on problems not otherwise soluble.

Another general principle has also tended to lengthen the discussions: The authors have tried, as often as possible, to attack problems "head on," rather than by "backing into them." They have tried to show how to find the solution of a new and strange equation rather than to set down a list of results which someone has found to be solutions of interesting problems. A certain number of "backing-up" examples, where one pulls a solution out of the air, so to speak, and then proceeds to show it is indeed the solution, could not be avoided. Usually such examples have saved space and typographic complications; very many of them would have produced a state of frustration or of fatalism in the student.

It is hoped that the work will also prove to be reasonably self-contained in regard to numerical tables and lists of useful formulas. The tables at the end of each chapter summarize the major results of that chapter and collect, in an easily accessible spot, the properties of the functions most often used. Rather than scattering the references among the text, these also are collected at the end of each chapter in order to make them somewhat easier to find again. These only include titles of the books and articles which the authors feel will be useful to the reader in supplementing the material discussed in the chapter; they are not intended to indicate priority or the high points in historical development. The development of the subject matter of this book has been extensive and has involved the contributions of many famous persons. Techniques have been rediscovered and renamed nearly every time a new branch of physics has been opened up. A thorough treatment of the bibliography would require hundreds of pages, much of it dull reading. We have chosen our references to help the reader understand the subject, not to give each contributor in the field his "due credit." Frankly, we have put down those references we are familiar with and which we have found useful.

An attempt has been made to coordinate the choice of symbols for the various functions defined and used. Where usage is fairly consistent in the literature, as with Bessel functions, this has been followed. Where there have been several alternate symbols extant, the one which seemed to fit most logically with the rest and resulted in least duplication was chosen, as has been done with the Mathieu functions. In a few cases functions were renormalized to make them less awkward to use; these were given new symbols, as was done with the Gegenbauer polynomials. The relation between the notation used and any alternate notation appearing with reasonable frequency in physics literature is given in the Appendix, together with a general glossary of symbols used.

The numerical table in the Appendix should be adequate for the majority of the calculations related to the subject matter. It was considered preferable to include a number of tables of limited range and accuracy rather than to have a few tables each with a large number of entries and significant figures. Most of the functions used in actual physical problems are tabulated, though many of the auxiliary functions, such as the gamma and the elliptic functions, and some functions with too many independent parameters, such as the hypergeometric functions, are not represented. A few functions, such as the parabolic and the spheroidal wave functions, should have been included, but complete basic tables have not yet been published.

Several of the figures in this work, which have to do with three dimensions, are drawn for stereoscopic viewing. They may be viewed by any of the usual stereoscopic viewers or, without any paraphernalia, by relaxing one's eye-focusing muscles and allowing one eye to look at one drawing, the other at the other. Those who have neither equipment nor sufficient ocular decoupling may consider these figures as ordinary perspective drawings unnecessarily duplicated. If not benefited, they will be at least not worse off by the duplication.

The authors have been helped in their task by many persons. The hundreds of graduate students who have taken the related course since 1935 have, wittingly or unwittingly, helped to fix the order of presentation and the choice of appropriate examples. They have removed nearly all the proof errors from the offset edition of class notes on which this treatise is based; they have not yet had time to remove the inevitable errors from this edition. Any reader can assist in this by notifying the authors when such errors are discovered.

Assistance has also been more specific. The proof of Cauchy's theorem, given on page 364, was suggested by R. Boas. Parts of the manuscript and proof have been read by Professors J. A. Stratton and N. H. Frank; Doctors Harold Levine, K. U. Ingard, Walter Hauser, Robert and Jane Pease, S. Rubinow; and F. M. Young, M. C. Newstein, L. Sartori, J. Little, E. Lomon, and F. J. Corbató. They are to be thanked for numerous improvements and corrections; they should not be blamed for the errors and awkward phrasings which undoubtedly remain. We are also indebted to Professor Julian Schwinger for many stimulating discussions and suggestions.

Philip M. Morse
Herman Feshbach

May, 1953

Contents

CHAPTER 1

Types of Fields

Our task in this book is to discuss the mathematical techniques which are useful in the calculation and analysis of the various types of fields occurring in modern physical theory. Emphasis will be given primarily to the exposition of the interrelation between the equations and the physical properties of the fields, and at times details of mathematical rigor will be sacrificed when it might interfere with the clarification of the physical background. Mathematical rigor is important and cannot be neglected, but the theoretical physicist must first gain a thorough understanding of the physical implications of the symbolic tools he is using before formal rigor can be of help. Other volumes are available which provide the rigor; this book will have fulfilled its purpose if it provides physical insight into the manifold field equations which occur in modern theory, together with a comprehension of the physical meaning behind the various mathematical techniques employed for their solution.

This first chapter will discuss the general properties of various fields and how these fields can be expressed in terms of various coordinate systems. The second chapter discusses the various types of partial differential equations which govern fields, and the third chapter treats of the relation between these equations and the fundamental variational principles developed by Hamilton and others for classic dynamics. The following few chapters will discuss the general mathematical tools which are needed to solve these equations, and the remainder of the work will be concerned with the detailed solution of individual equations.

Practically all of modern physics deals with fields: potential fields, probability fields, electromagnetic fields, tensor fields, and spinor fields.

Mathematically speaking, a field is a set of functions of the coordinates of a point in space. From the point of view of this book a field is some convenient mathematical idealization of a physical situation in which *extension* is an essential element, *i.e.*, which cannot be analyzed in terms of the positions of a finite number of particles. The transverse displacement from equilibrium of a string under static forces is a very simple example of a one-dimensional field; the displacement y is different for different parts of the string, so that y can be considered as a function of the distance x along the string. The density, temperature, and pressure in a fluid transmitting sound waves can be considered as functions of the three coordinates and of time. Fields of this sort are obviously only approximate idealizations of the physical situation, for they take no account of the atomic properties of matter. We might call them *material fields*.

Other fields are constructs to enable us to analyze the problem of *action at a distance*, in which the relative motion and position of one body affects that of another. Potential and force fields, electromagnetic and gravitational fields are examples of this type. They are considered as being "caused" by some piece of matter, and their value at some point in space is a measure of the effect of this piece of matter on some test body at the point in question. It has recently become apparent that many of these fields are also only approximate idealizations of the actual physical situation, since they take no account of various quantum rules associated with matter. In some cases the theory can be modified so as to take the quantum rules into account in a more or less satisfactory way.

Finally, fields can be constructed to "explain" the quantum rules. Examples of these are the Schroedinger wave function and the spinor fields associated with the Dirac electron. In many cases the value of such a field at a point is closely related to a probability. For instance the square of the Schroedinger wave function is a measure of the probability that the elementary particle is present. Present quantum field theories suffer from many fundamental difficulties and so constitute one of the frontiers of theoretical physics.

In most cases considered in this book fields are solutions of partial differential equations, usually second-order, linear equations, either homogeneous or inhomogeneous. The actual physical situation has often to be simplified for this to be so, and the simplification can be justified on various pragmatic grounds. For instance, only the "smoothed-out density" of a gas is a solution of the wave equation, but this is usually sufficient for a study of sound waves, and the much more tedious calculation of the actual motions of the gas molecules would not add much to our knowledge of sound.

This Procrustean tendency to force the physical situation to fit the requirements of a partial differential equation results in a field which is both more regular and more irregular than the "actual" conditions. A solution of a differential equation is more smoothly continuous over most of space and time than is the corresponding physical situation, but it usually is also provided with a finite number of mathematical discontinuities which are considerably more "sharp" than the "actual" conditions exhibit. If the simplification has not been too drastic, most of the quantities which can be computed from the field will correspond fairly closely to the measured values. In each case, however, certain discrepancies between calculated and measured values will turn up, due either to the "oversmooth" behavior of the field over most of its extent or to the presence of mathematical discontinuities and infinities in the computed field, which are not present in "real life." Sometimes these discrepancies are trivial, in that the inclusion of additional complexities in the computation of the field to obtain a better correlation with experiment will involve no conceptual modification in the theory; sometimes the discrepancies are far from trivial, and a modification of the theory to improve the correlation involves fundamental changes in concept and definitions. An important task of the theoretical physicist lies in distinguishing between trivial and nontrivial discrepancies between theory and experiment.

One indication that fields are often simplifications of physical reality is that fields often can be defined in terms of a limiting ratio of some sort. The density field of a fluid which is transmitting a sound wave is defined in terms of the "density at a given point," which is really the limiting ratio between the mass of fluid in a volume surrounding the given point and the magnitude of the volume, as this volume is reduced to "zero." The electric intensity "at a point" is the limiting ratio between the force on a test charge at the point and the magnitude of the test charge as this magnitude goes to "zero." The value of the square of the Schroedinger wave function is the limiting ratio between the probability that the particle is in a given volume surrounding a point and the magnitude of the volume as the volume is shrunk to "zero," and so on. A careful definition of the displacement of a "point" of a vibrating string would also utilize the notion of limiting ratio.

These mathematical platitudes are stressed here because the technique of the limiting ratio must be used with caution when defining and calculating fields. In other words, the terms "zero" in the previous paragraph must be carefully defined in order to achieve results which correspond to "reality." For instance the volume which serves to define the density field for a fluid must be reduced several orders of magnitude smaller than the cube of the shortest wavelength of transmitted sound in order to arrive at a ratio which is a reasonably accurate solution of the wave equation. On the other hand, this volume must not be reduced to a size commensurate with atomic dimensions, or the resulting ratio will lose its desirable properties of smooth continuity and would not be a useful construct. As soon as this limitation is realized, it is not difficult to understand that the description of a sound wave in terms of a field which is a solution of the wave equation is likely to become inadequate if the "wavelength" becomes shorter than interatomic distances.

In a similar manner we define the electric field in terms of a test charge which is made small enough so that it will not affect the distribution of the charges "causing" the field. But if the magnitude of the test charge is reduced until it is the same order of magnitude as the electronic charge, we might expect the essential atomicity of charge to involve us in difficulties (although this is not necessarily so).

In some cases the limiting ratio can be carried to magnitudes as small as we wish. The probability fields of wave mechanics are as "fine-grained" as we can imagine at present.

1.1 *Scalar Fields*

When the field under consideration turns out to be a simple number, a single function of space and time, we call it a *scalar* field. The displacement of a string or a membrane from equilibrium is a scalar field. The density, pressure, and temperature of a fluid, given in terms of the sort of limiting ratio discussed previously, are scalar fields. As mentioned earlier, the limiting volume cannot be allowed to become as small as atomic dimensions in computing the ratio, for the concepts of density, pressure, etc., have little meaning for individual molecules. The ratios which define these fields must approach a "macroscopic limit" when the volume is small compared with the gross dimensions of the fluid but is still large compared with atomic size; otherwise there can be no physical meaning to the concept of scalar field here.

All these scalar fields have the property of *invariance* under a transformation of space coordinates (we shall discuss invariance under a space-time transformation later in this chapter). The *numerical* value of the field at a point is the same no matter how the coordinates of the point are expressed. The *form* of mathematical expression for the field may vary with the coordinates. For instance, a field expressed in rectangular coordinates may have the form $\psi = y$; in spherical coordinates it has the different form $\psi = r \sin \vartheta \cos \varphi$, but in *either* coordinate system, at the point $x = 10$, $y = 10$, $z = 0$ ($r = \sqrt{200}$, $\vartheta = 45°$, $\varphi = 0$) it has the value $\psi = 10$. This is to be contrasted to the behavior of the x component of the velocity of a fluid, where the direction of the x axis may change as the coordinates are changed. Therefore, the numerical value of the x component at a given point will change as the direction of the x axis is rotated.

This property of invariance of a scalar will be important in later discussions and is to be contrasted alternatively to the invariance of *form* of certain equations under certain coordinate transformations. For some of the scalar fields mentioned here, such as density or temperature or electric potential, the definition of the field has been such as to make the property of invariance almost tautological. This is not always the case with less familiar fields, however. In some cases, in fact, the property of invariance must be used as a touchstone to find the proper expression for the field.

Isotimic Surfaces. The surfaces defined by the equation ψ = constant, where ψ is the scalar field, may be called *isotimic surfaces* (from Greek *isotimos*, of equal worth). Isotimic surfaces are the obvious generalizations of the contour lines on a topographic map. In potential theory they are referred to as equipotentials; in heat conduction they are isothermals; etc. They form a family of nonintersecting surfaces,

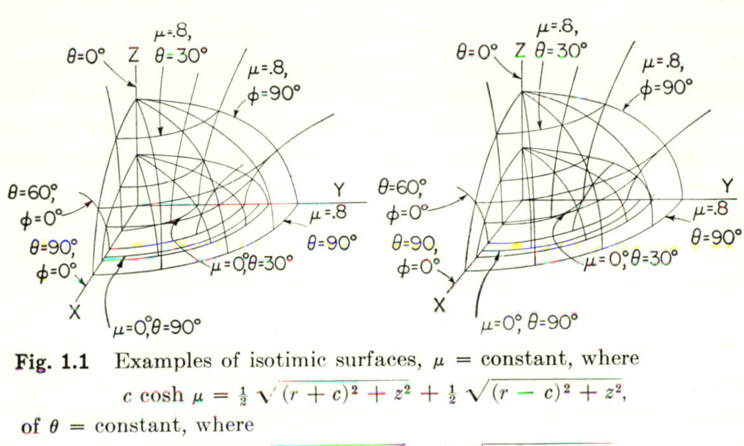

Fig. 1.1 Examples of isotimic surfaces, μ = constant, where

$$c \cosh \mu = \tfrac{1}{2}\sqrt{(r+c)^2 + z^2} + \tfrac{1}{2}\sqrt{(r-c)^2 + z^2},$$

of θ = constant, where

$$c \cos \theta = \tfrac{1}{2}\sqrt{(r+c)^2 + z^2} - \tfrac{1}{2}\sqrt{(r-c)^2 + z^2},$$

and of $\phi = C$, where $\tan \phi = y/x$.

which are often useful in forming part of a system of coordinates naturally suitable for the problem. For instance if the field is the well-known potential $\psi = (x^2 + y^2 + z^2)^{-\frac{1}{2}}$, the isotimic surfaces (in this case surfaces of constant potential) are concentric spheres of radius $r = \sqrt{x^2 + y^2 + z^2}$ = constant; and the natural coordinates for the problem are the spherical ones, r, ϑ, φ. Another set of surfaces is shown in Fig. 1.1, together with the corresponding coordinate system. The surfaces μ = constant correspond to the equipotential surfaces about a circular charged disk of radius c, lying in the x,y plane ($\mu = 0$).

The derivatives of the scalar ψ with respect to the rectangular coordinates x, y, z measure the rapidity with which the field changes as we change position. For instance the change in ψ from the point (x,y,z) to the point $(x + dx, y + dy, z + dz)$ is

$$d\psi = \left(\frac{\partial \psi}{\partial x}\right) dx + \left(\frac{\partial \psi}{\partial y}\right) dy + \left(\frac{\partial \psi}{\partial z}\right) dz \tag{1.1.1}$$

If the two points are in the same isotimic surface $d\psi = 0$; in fact the differential equation for these surfaces is

$$\left(\frac{\partial\psi}{\partial x}\right) dx + \left(\frac{\partial\psi}{\partial y}\right) dy + \left(\frac{\partial\psi}{\partial z}\right) dz = 0 \tag{1.1.2}$$

The displacement (dx,dy,dz) is perpendicular to the surface if the component displacements satisfy the equation.

$$\frac{dx}{\partial\psi/\partial x} = \frac{dy}{\partial\psi/\partial y} = \frac{dz}{\partial\psi/\partial z} \tag{1.1.3}$$

These relations are the differential equations for a family of lines, called the *normal lines*, which are everywhere perpendicular to the isotimic surfaces. Together with the isotimic surfaces they can be used to define the natural coordinate system for the field. For instance, for the field $\psi = (x^2 + y^2 + z^2)^{-\frac{1}{2}}$ the surfaces are spheres (as we noted before) and the normal lines are the radii, suggesting (but not completely defining) the spherical coordinates (r,ϑ,φ).

The normal lines are pointed along the direction of most rapid change of ψ. A little manipulation of Eqs. (1.1.1) and (1.1.3) will show that the change of ψ as one goes a distance ds along a normal line is

$$\sqrt{(\partial\psi/\partial x)^2 + (\partial\psi/\partial y)^2 + (\partial\psi/\partial z)^2}\, ds$$

The square-root factor is called the *magnitude of the gradient* of ψ. Its properties will be discussed in more detail later in the chapter.

The Laplacian. An extremely important property of a scalar field is expressed in terms of its second derivatives. In the simple one-dimensional case where ψ is the transverse displacement of a string from its straight-line equilibrium position, the second derivative $d^2\psi/dx^2$ is closely related to the difference between the value of ψ at x and its average values at neighboring points. Using the fundamental definition of the derivative

$$\begin{aligned}\lim \{\psi(x) - \tfrac{1}{2}[\psi(x - dx) + \psi(x + dx)]\} \\ = -\tfrac{1}{2} \lim \{[\psi(x + dx) - \psi(x)] - [\psi(x) - \psi(x - dx)]\} \\ = -\tfrac{1}{2}(d^2\psi/dx^2)(dx)^2\end{aligned}$$

Consequently if the second derivative is negative, ψ at x is larger than the average of ψ at $x + dx$ and $x - dx$ and the plot of ψ against x will have a downward curvature at that point. If the second derivative is zero, ψ will have no curvature.

It is not difficult to see that the equation for the shape of a stretched, flexible string acted on by a transverse force $F(x)$ per unit length of string, distributed along the string, must be expressed in terms of this second derivative. For the greater the transverse force at a point, the greater must be the curvature of the string there, in order that the tension T along the string may have greater transverse components to equalize the force. The equation turns out to be

$$T(d^2\psi/dx^2) = -F(x)$$

as a detailed analysis of the problem will show later.

We now seek a three-dimensional counterpart to this measure of curvature of ψ. The limiting value of the difference between ψ at x and the average value of ψ at neighboring points is $-\frac{1}{6}(dx\,dy\,dz)^2\nabla^2\psi$,

where
$$\nabla^2\psi = \frac{\partial^2\psi}{\partial x^2} + \frac{\partial^2\psi}{\partial y^2} + \frac{\partial^2\psi}{\partial z^2} \tag{1.1.4}$$

is the obvious generalization of the one-dimensional second-derivative operator. The mathematical operation given in Eq. (1.1.4) is labeled by the symbol ∇^2 (read *del squared*), which is called the *Laplace operator*. The result of the operation on the function ψ is called the *Laplacian* of ψ. If $\nabla^2\psi$ is negative at some point there is a tendency for ψ to *concentrate* at that point.

One immediate consequence of this definition is that a scalar function $\psi(x,y,z)$ *can have no maxima or minima in a region where* $\nabla^2\psi = 0$. This is a result of considerable importance.

The equation $\nabla^2\psi = 0$, called *Laplace's equation*, occurs in so many parts of physical theory that it is well to have a clear picture of its meaning. Accordingly we shall quote a number of facts concerning the solutions of Laplace's equation which will not be proved till later in the chapter.

Suppose a perfectly elastic membrane to be in equilibrium under a uniform tension applied around its boundary edge. If the edge lies entirely in a plane, then the membrane will lie entirely in the plane. If the boundary support is distorted so that it no longer lies all in a plane, the membrane will be distorted also. This distortion can be represented by $\psi(x,y)$, the displacement, normal to the plane, of the point (x,y) on the membrane. It turns out that this displacement satisfies Laplace's equation in two dimensions, $\nabla^2\psi = 0$. The equation simply corresponds to the statement that the tension straightens out all the "bulges" in the membrane, that the displacement at any point equals the average value of the displacement for neighboring points. It also corresponds (as we shall prove later) to the statement that the membrane has arranged itself so that its average slope is as small as it can be. Since the total stretching of the membrane is proportional to the average square of the slope, we see that the Laplace equation for the membrane corresponds to the requirement that the membrane assumes a shape involving the least stretching possible.

An additional loading force on the membrane, perpendicular to the equilibrium plane $\psi = 0$, produces a "bulging" of the membrane. As will be proved later, the Laplacian of ψ at a point on a loaded membrane is proportional to the load per unit area at that point. One can say that the two-dimensional Laplacian operator measures the "bulginess" of the shape function ψ.

The generalization of this discussion to three dimensions is somewhat harder to picture but not more complicated in principle. We might picture the scalar function ψ as corresponding to the concentration of a solute in a solvent. The three-dimensional analogue of "bulginess" might be termed "lumpiness"; if there is a tendency for the solute to "lump together" at any point, the Laplacian of the concentration will be

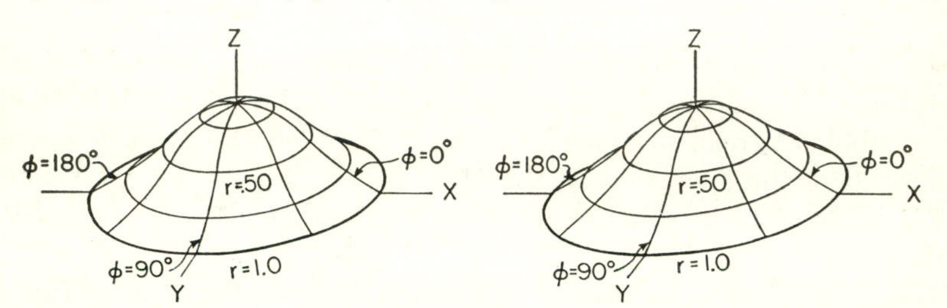

Fig. 1.2 Shape of circular membrane loaded uniformly ($\nabla^2\psi$ = constant) from $r = 0$ to $r = \frac{1}{2}$, unloaded ($\nabla^2\psi = 0$) from $r = \frac{1}{2}$ to $r = 1$, where it is clamped.

negative there. In a case where $\nabla^2\psi = 0$, the solute has no "lumpiness" at all, its density arranging itself so as to average out as smoothly as possible the differences imposed by the boundary conditions. As in the two-dimensional case, the Laplace equation corresponds to the requirement that ψ at every point be equal to the average value of ψ at neighboring points.

The presence of electric charge density ρ causes a (negative) concentration of the electric potential ψ, so that $\nabla^2\psi = -\rho/\epsilon$, where ϵ is a constant. The presence of a distributed source of heat Q in a solid causes a concentration of temperature T, so that $\nabla^2 T = KQ$, where K is a constant. In a great many cases the scalar field is affected by a *source* function $q(x,y,z)$ (which is itself another scalar field, obeying some other equations) according to the equation

$$\nabla^2\varphi = -q \tag{1.1.5}$$

This equation is called *Poisson's equation.* We shall discuss it in more detail later in this chapter and shall devote a considerable amount of space later in the book to its solution.

1.2 *Vector Fields*

We have discussed in a preliminary sort of way a number of fields which were characterized by a single magnitude at each point. These were called scalar fields. Many other fields require a *magnitude and direction* to be given at each point for complete characterization. These will be called *vector fields.* These fields also can often be defined in terms of limiting ratios, though the definitions are usually more complex than those for scalar fields. The force on a portion of fluid in a gravitational or electric field is a vector, having magnitude and direction. The limit of the ratio between this force and the volume occupied by the portion of fluid, as this volume is decreased in size, defines a vector at each point in space, which is the *force field.* As with scalar fields it is sometimes important not to let the volume decrease to atomic dimensions.

Sometimes the vector field is most easily defined by a scalar ratio, which has a direction inherent in it. For instance, for an electric conductor carrying current we can imagine an instrument which would measure the amount of current crossing an element of area dA centered at some point in the conductor. We should find that the amount of current measured would depend not only on the size of dA but also on the orientation of the element dA. In fact the measurement would turn out to correspond to the expression $J\,dA\cos\vartheta$, where ϑ is the angle between the normal to dA and some direction which is characteristic of the current distribution. The magnitude of the vector field at the point measured would then be J, and the direction would be that which defines the angle ϑ.

Vector fields in three dimensions are specified by giving *three* quantities at each point, the magnitude and two angles giving the direction, or three components of the vector along the three coordinate axes. Four-vectors will be considered later.

Boldface Latin capital letters (**A**,**F**,**X**) will be used in this book to denote vectors; the corresponding normal-weight letters (A,F,X) will denote the corresponding magnitudes (normal-weight letters will usually denote scalar quantities). The components of **A** along the three coordinate axes will be denoted A_x, A_y, and A_z. A vector of unit length in the direction of **A** is denoted by **a**; and in conformity with usual practice the unit vectors along the x, y, and z directions will be labeled **i**, **j**, and **k**, respectively. Unit vectors along curvilinear coordinate axes will usually be labeled **a**, with a subscript indicating the axis in question (for instance, in polar coordinates, the unit vector along r is $\mathbf{a}_r$ that along ϑ is $\mathbf{a}_\vartheta$, etc.). Unless otherwise noted the coordinate systems used will be right-handed ones: if a right-hand screw is placed along the z axis, then a rotation from x into y will move the screw in the positive z direction; or if the x, y plane is on a blackboard which is faced by the observer, the x direction can point to the observer's right, the y axis can point upward, and the z axis will then point toward the observer.

In the notation given above vectors **A** and **B** obey the following general equations:

$$\begin{aligned} \mathbf{A} &= A\mathbf{a} = A_x\mathbf{i} + A_y\mathbf{j} + A_z\mathbf{k} \\ \mathbf{A} + \mathbf{B} &= (A_x + B_x)\mathbf{i} + (A_y + B_y)\mathbf{j} + (A_z + B_z)\mathbf{k} \end{aligned} \tag{1.2.1}$$

which constitute a definition of vector components and of vector addition.

Vectors are not invariant to change of coordinates in the same sense as scalars are; for their components change as the coordinate directions change. The transformation properties of vectors will be discussed later.

Multiplication of Vectors. Two vectors can be multiplied together in two different ways: one resulting in a scalar and the other in vector. The *scalar* or *dot product* of two vectors **A** and **B** is equal to the product of the magnitude of one by the component of the other in the direction of the first:

$$\mathbf{A} \cdot \mathbf{B} = AB \cos \vartheta = A_xB_x + A_yB_y + A_zB_z \tag{1.2.2}$$

where ϑ is the angle between **A** and **B**. The expression $AB \cos \vartheta$ is independent of the coordinate system used to compute the components A_x, etc., so that the *value* of the dot product is independent of the coordinate system chosen. The dot product is therefore a true scalar, the simplest invariant which can be formed from two vectors.

The dot product is a useful form for expressing many physical quantities: The work done in moving a body equals the dot product of the force and the displacement; the electrical energy density in space is proportional to the dot product of the electric intensity and the electric displacement; and so on. The dot product of two unit vectors is equal to the direction cosine relating the two directions. The maximum value of the dot product of two vectors is obtained when the two vectors are parallel; it is zero when the two are perpendicular. In a sense the dot product is a measure of the coalignment of the two vectors.

The *vector* or *cross product* $\mathbf{A} \times \mathbf{B}$ of two vectors is a vector with magnitude equal to the area of the parallelogram defined by the two vectors and with direction perpendicular to this parallelogram. The choice as to which end of the perpendicular to put the arrowhead is decided arbitrarily by making the trio **A**, **B**, and $(\mathbf{A} \times \mathbf{B})$ a right-handed system: if a right-hand screw is placed perpendicular to **A** and **B**, a rotation of **A** into **B** will move the screw in the direction of $(\mathbf{A} \times \mathbf{B})$. In terms of right-hand rectangular coordinates

$$\begin{gathered} \mathbf{A} \times \mathbf{B} = (A_yB_z - B_yA_z)\mathbf{i} + (A_zB_x - B_zA_x)\mathbf{j} + (A_xB_y - B_xA_y)\mathbf{k} \\ \text{Magnitude of } (\mathbf{A} \times \mathbf{B}) = AB \sin \vartheta \end{gathered} \tag{1.2.3}$$

We note that vector multiplication is not commutative, for $\mathbf{A} \times \mathbf{B} = -\mathbf{B} \times \mathbf{A}$.

Axial Vectors. Although the cross product of two vectors is a vector with most of the transformation properties of a "true" vector (as we shall see later), there is one difference which is of importance. The cross product, as defined in Eq. (1.2.3), changes sign if we change from a right-handed to a left-handed system of coordinates. This is another aspect of the fact that the cross product has the directional properties of an element of area rather than of an arrow. The direction connected with an area element is uniquely determined, the direction normal to the element, except that there is no unequivocal rule as to which side of the element is the positive direction of the vector. The area fixes the shank of the arrow, so to speak, but does not say which end should bear the point. Which end does carry the point must be decided by some purely arbitrary rule such as the right-hand rule, mentioned above, which we shall use in this book.

Arealike vectors, with a fixed shank but interchangeable head, are called *axial vectors* (they are also sometimes called *pseudovectors*). We shall see later that the three components of an axial vector are actually the three components of a second-order antisymmetric tensor in three dimensions. Indeed only in three dimensions is it possible to represent an antisymmetric tensor by an axial vector.

As indicated above, the axial vector associated with an element of area $\mathbf{dA}$ can be written:

$$\mathbf{dA} = \mathbf{n}\, dA = \mathbf{dx} \times \mathbf{dy}$$

where $\mathbf{n}$ is the unit vector normal to the element and where $\mathbf{dx}$ and $\mathbf{dy}$ are the vectors associated with the component elements dx and dy. If the first two notations are used, it is necessary in addition to indicate which side of the element is positive; if the last notation is used, our arbitrary right-hand rule will automatically decide the question.

Other axial vectors can be represented in terms of cross products: the angular momentum of a particle about some point is equal to the cross product of the vector representing the particle's momentum and the radius vector from the origin to the particle; the torque equals the cross product of the force vector and the vector representing the lever arm; and so on. Rotation fixes a plane and an axis normal to the plane, the characteristics of an axial vector. According to our convention the direction of a rotational vector follows the motion of a right-handed screw turned by the rotation.

A useful example of a product of three vectors is the *scalar triple product*

$$\mathbf{A} \cdot (\mathbf{B} \times \mathbf{C}) = \mathbf{B} \cdot (\mathbf{C} \times \mathbf{A}) = \mathbf{C} \cdot (\mathbf{A} \times \mathbf{B}) = \begin{vmatrix} A_x & A_y & A_z \\ B_x & B_y & B_z \\ C_x & C_y & C_z \end{vmatrix} \qquad (1.2.4)$$

The magnitude of this quantity is equal to the volume (or to minus the volume) enclosed by the parallelepiped defined by the vectors **A**, **B**, and **C**. Such a scalar, a dot product of an axial vector and a true vector, which changes sign if we change from left- to right-handed coordinates or if we interchange the order of the vectors, is sometimes called a *pseudoscalar*.

We note that the dot product of two axial vectors (or of two true vectors) is a "true" scalar, with no uncertainty in sign.

Incidentally, the rules governing the products of the unit vectors are

$$\begin{aligned} \mathbf{i}^2 = \mathbf{i}\cdot\mathbf{i} = \mathbf{j}^2 = \mathbf{k}^2 = 1; \quad \mathbf{i}\cdot\mathbf{j} = \mathbf{i}\cdot\mathbf{k} = \mathbf{j}\cdot\mathbf{k} = 0 \\ \mathbf{i}\times\mathbf{i} = \cdots = 0; \quad \mathbf{i}\times\mathbf{j} = \mathbf{k}; \quad \mathbf{j}\times\mathbf{k} = \mathbf{i}; \quad \mathbf{k}\times\mathbf{i} = \mathbf{j} \end{aligned} \tag{1.2.5}$$

Lines of Flow. As we have said earlier, a vector field is defined by specifying a vector at each point in space, in other words, by specifying a vector which is a function of x, y, and z: $\mathbf{F}(x,y,z)$. In most cases of interest this vector is a continuous function of x, y, and z, except at isolated points, or *singularities*, or along isolated lines, or *singular lines*. Where the vector is continuous, we can define *lines of flow* of the field, which are lines at every point tangent to the vector at that point. The differential equation for the lines is obtained by requiring that the components dx, dy, dz of displacement along the line be proportional to the components F_x, F_y, and F_z of the vector field at that point:

$$\frac{dx}{F_x} = \frac{dy}{F_y} = \frac{dz}{F_z} \tag{1.2.6}$$

Compare this with Eq. (1.1.3).

In certain simple cases these equations can be integrated to obtain the algebraic equations for the family of lines of flow. For instance, if

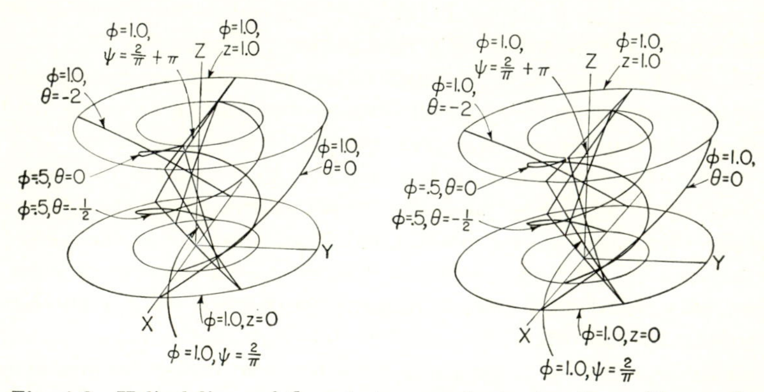

Fig. 1.3 Helical lines of flow θ, ϕ = constants, together with pseudopotential surfaces ψ = constant, as given on page 15.

$F_x = -ay$, $F_y = ax$, $F_z = b(x^2 + y^2)$, the lines of flow are helical. The equation $dx/F_x = dy/F_y$ becomes $x\,dx = -y\,dy$, which integrates to the equation for the circular cylinder, $x^2 + y^2 = \varphi^2$, with φ an arbitrary constant, denoting the particular line of flow chosen. The equation $dy/F_y = dz/F_z$ becomes $\varphi^2\,dy/\sqrt{\varphi^2 - y^2} = (a/b)\,dz$, after substitution for x from the equation relating x and y. This integrates to

$$z = (b\varphi^2/a)\sin^{-1}(y/\varphi) + \vartheta = (b/a)(x^2 + y^2)\tan^{-1}(y/x) + \vartheta$$

where ϑ is the other constant of integration which is needed to specify completely a particular line of flow. The two equations

$$\varphi = \sqrt{x^2 + y^2}, \quad \vartheta = z - (b/a)(x^2 + y^2) \tan^{-1} (y/x)$$

define a doubly infinite family of flow lines, one line for each pair of values chosen for φ and ϑ.

Another example is the case $F_x = x/r^3$, $F_y = y/r^3$, $F_z = z/r^3$, where $r^2 = (x^2 + y^2 + z^2)$. The equations for the flow lines reduce to $dx/x = dy/y = dz/z$. The first equation gives $\ln (x) = \ln (y) +$ constant, or $x/y =$ constant. Similarly we have either x/z or $y/z =$ constant or, combining the two, $(x^2 + y^2)/z^2 =$ constant. The most suitable forms for the constants of integration, analogous to the forms given in the previous paragraph, are

$$\varphi = \tan^{-1} (y/x); \quad \vartheta = \tan^{-1} (\sqrt{x^2 + y^2}/z)$$

Again, a choice of values for φ and ϑ picks out a particular flow line; in this case a straight line radiating from the origin.

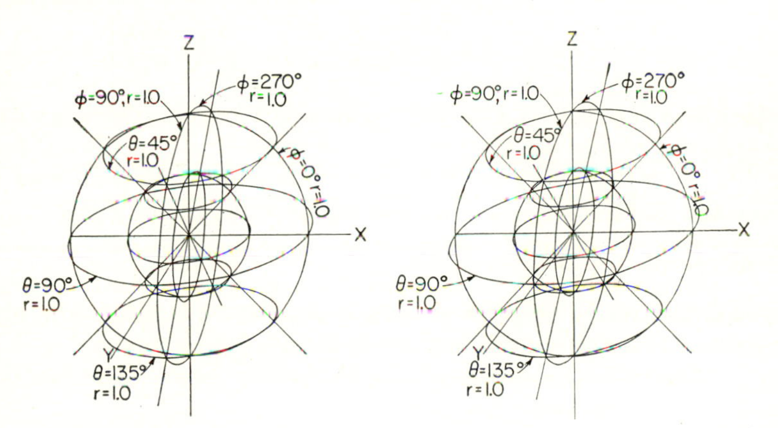

Fig. 1.4 Radial lines of flow and spherical potential surfaces for field about a source point.

From another point of view φ and ϑ can be considered as functions of x, y, and z and are called *flow functions*. The values of φ and ϑ at some point label the flow lines at that point. From still another point of view the two families of surfaces $\varphi =$ constant and $\vartheta =$ constant can be considered as two families of coordinate surfaces for a generalized system of coordinates. The intersection of two such surfaces, $\varphi = \varphi_0$ and $\vartheta = \vartheta_0$, is the flow line corresponding to the pair (φ_0, ϑ_0) and is a coordinate line in the new system.

Potential Surfaces. The lines of flow may also determine another family of surfaces, perpendicular to the lines (unless the lines are so "curled" that no such set of surfaces can exist). By analogy with Eq. (1.1.2) the equation for such a surface is

$$\mathbf{F}\cdot d\mathbf{s} = F_x\,dx + F_y\,dy + F_z\,dz = 0 \tag{1.2.7}$$

corresponding to the fact that any displacement vector in the surface must be perpendicular to **F**.

In certain cases this equation is integrable. If there is a function ψ such that $\mu(\partial\psi/dx) = F_x$, $\mu(\partial\psi/\partial y) = F_y$, $\mu(\partial\psi/\partial z) = F_z$, then the equation for the family of surfaces becomes ψ = constant. The quantity μ may be a function of x, y, and z and is called an *integrating factor*. The criterion for whether an equation for the surfaces exists in integral form is developed as follows: Assuming that there is a ψ as defined above, we substitute in the expression

$$\begin{aligned} F_x\left(\frac{\partial F_z}{\partial y} - \frac{\partial F_y}{\partial z}\right) + F_y\left(\frac{\partial F_x}{\partial z} - \frac{\partial F_z}{\partial x}\right) + F_z\left(\frac{\partial F_y}{\partial x} - \frac{\partial F_x}{\partial y}\right) \\ = \mu\frac{\partial\psi}{\partial x}\left[\frac{\partial}{\partial y}\left(\mu\frac{\partial\psi}{\partial z}\right) - \frac{\partial}{\partial z}\left(\mu\frac{\partial\psi}{\partial y}\right)\right] + \mu\frac{\partial\psi}{\partial y}\left[\frac{\partial}{\partial z}\left(\mu\frac{\partial\psi}{\partial x}\right) - \frac{\partial}{\partial x}\left(\mu\frac{\partial\psi}{\partial z}\right)\right] \\ + \mu\frac{\partial\psi}{\partial z}\left[\frac{\partial}{\partial x}\left(\mu\frac{\partial\psi}{\partial y}\right) - \frac{\partial}{\partial y}\left(\mu\frac{\partial\psi}{\partial x}\right)\right] = 0 \end{aligned} \tag{1.2.8}$$

If there is a function ψ, this last expression is zero. Therefore, if the expression involving the components of **F** turns out to be zero, then it is possible to integrate the differential equation for the surfaces perpendicular to the flow lines. In other words, if the vector whose x, y, and z components are $\left(\frac{\partial F_z}{\partial y} - \frac{\partial F_y}{\partial z}\right)$, $\left(\frac{\partial F_x}{\partial z} - \frac{\partial F_z}{\partial x}\right)$, and $\left(\frac{\partial F_y}{\partial x} - \frac{\partial F_x}{\partial y}\right)$ is perpendicular to the vector **F** *at every point*, then it is possible to obtain the equation for the normal surfaces in integral form, as $\psi(x,y,z)$ = constant. More will be said concerning this vector later. The function ψ is called a *pseudopotential function*.

In certain special cases μ is a constant and can be set equal to (-1), so that $F_x = -(\partial\psi/\partial x)$, $F_y = -(\partial\psi/\partial y)$, $F_z = -(\partial\psi/\partial z)$. The excuse for the negative sign will appear later. In these cases the scalar function is called the *potential function* for the vector field **F**, and the surfaces ψ = constant are called *equipotential surfaces*. For this to be true, the three quantities $\left(\frac{\partial F_z}{\partial y} - \frac{\partial F_y}{\partial z}\right)$, $\left(\frac{\partial F_x}{\partial z} - \frac{\partial F_z}{\partial x}\right)$, $\left(\frac{\partial F_y}{\partial x} - \frac{\partial F_x}{\partial y}\right)$ must each be zero, as can be seen by substituting the expressions assumed for the components of **F** in terms of ψ.

In other cases the equation for the surface is not integrable, either with or without the use of the integrating factor; then it is impossible to find a sensibly behaved family of surfaces everywhere perpendicular to the lines of flow. We shall return to this discussion on page 18.

In the first example given in the previous section, where the lines of flow are helices, the vector whose components are

$$\left(\frac{\partial F_z}{\partial y} - \frac{\partial F_y}{\partial z}\right) = 2by; \quad \left(\frac{\partial F_x}{\partial z} - \frac{\partial F_z}{\partial x}\right) = -2bx; \quad \left(\frac{\partial F_y}{\partial x} - \frac{\partial F_x}{\partial y}\right) = 2a$$

is perpendicular to the vector **F**. Therefore, the differential equation (1.2.7) for the family of surfaces is integrable. Dividing the equation by the integrating factor $\mu = x^2 + y^2$ results in a perfect differential. Setting $\psi = \tan^{-1}(y/x) + (b/a)z$, we find that

$$F_x = (x^2 + y^2)\frac{\partial \psi}{\partial x}; \quad F_y = (x^2 + y^2)\frac{\partial \psi}{\partial y}; \quad F_z = (x^2 + y^2)\frac{\partial \psi}{\partial z}$$

so that the integrated equation corresponding to (1.2.7) for this case is ψ = constant. The system of surfaces ψ = constant, ϑ = constant, and φ = constant form a set of generalized coordinate surfaces (in this case not mutually perpendicular) which are the natural ones for the vector field considered. The values of φ and ϑ for a point determine which flow line goes through the point, and the value of ψ determines the position of the point along the flow line.

In the second example considered above, the quantities $\left(\frac{\partial F_z}{\partial y} - \frac{\partial F_y}{\partial z}\right)$, $\left(\frac{\partial F_x}{\partial z} - \frac{\partial F_z}{\partial x}\right)$, and $\left(\frac{\partial F_y}{\partial x} - \frac{\partial F_x}{\partial y}\right)$ are all zero, so that Eq. (1.2.7) is integrable directly, without need of an integrating factor. The function

$$\psi = 1/\sqrt{x^2 + y^2 + z^2}$$

is, therefore, a potential function, and the spherical surfaces ψ = constant are equipotential surfaces. The components of **F** are related to ψ by the required equations:

$$\frac{\partial \psi}{\partial x} = -\frac{x}{r^3} = -F_x; \quad \frac{\partial \psi}{\partial y} = -\frac{y}{r^3} = -F_y; \quad \frac{\partial \psi}{\partial z} = -\frac{z}{r^3} = -F_z;$$
$$r^2 = x^2 + y^2 + z^2$$

The coordinate system corresponding to the flow lines and the potential surfaces is the spherical system

$$\varphi = \tan^{-1}(y/x); \quad \vartheta = \tan^{-1}(\sqrt{x^2 + y^2}/z); \quad r = \sqrt{x^2 + y^2 + z^2}$$

In this case the coordinate surfaces are mutually perpendicular.

When a family of equipotential surfaces is possible, the vector field **F** can then be represented in terms of the scalar potential field ψ, and it is usually much easier to compute the scalar field first, obtaining the vector field by differentiation.

Surface Integrals. There are a number of general properties of vector fields and their accompanying lines of flow which are important in our work. One is the "spreading out" or "net outflow" of lines of flow in a given region, whether lines originate or are lost in the region or simply pass through the region from one side to the other. Another interesting property is the amount by which the lines of force have "curled up" on themselves, whether the vector field is one with "whirlpools" in it or not.

The amount of divergence of flow lines out from a region can be measured by means of the *surface integral.* Suppose that we consider an element of surface area as being represented by the infinitesimal axial vector **dA**, of magnitude equal to the area of the surface element and having direction perpendicular to the surface. The scalar product $\mathbf{F} \cdot d\mathbf{A}$ then equals the product of the surface area with the component of the vector **F** normal to the surface. If the vector field $\mathbf{F}(x,y,z)$ represents the velocity vector for fluid motion, then $\mathbf{F} \cdot d\mathbf{A}$ will equal the volume flow of fluid across the surface element and the integral $\int \mathbf{F} \cdot d\mathbf{A}$ will equal the total volume flow over the surface covered by the integration. Its sign depends on the choice of direction of the axial vectors **dA**, that is, on whether they point away from one side of the surface or the other. Its magnitude is sometimes called *the number of flow lines of* **F** *crossing the surface* in question, a phrase which, in a sense, defines what we mean by "number of flow lines."

If the surface integrated over is a closed one, and if the vectors **dA** are chosen to point *away from* the region enclosed by the surface, then the integral will be written as $\oint \mathbf{F} \cdot d\mathbf{A}$ and will be called the *net outflow integral* of the vector **F** for the region enclosed by the surface. When **F** is the fluid velocity, then this integral equals the net volume outflow of fluid from the region in question. The enclosing surface need not be a single one, enclosing a simply connected region; it may be two surfaces, an inner and outer one; or it may be more than two surfaces. For that matter, the outer surface may be at infinity, so the region "enclosed" is all of space outside the inner surface or surfaces. In this case the vectors **dA** at the inner surface must point inward, away from the "enclosed" region.

The integral $\oint \mathbf{F} \cdot d\mathbf{A}$ is a measure of the number of flow lines originating within the enclosed region. If no flow lines originate there, all lines passing through from one side to the other, then the net outflow integral is zero.

Source Point. One simple case of vector field is particularly important in our later work and illustrates some interesting properties of net outflow integrals. This is the case where all flow lines originate at one point O, when the vector **F** at P has the magnitude Q/r^2 and is pointed outward along **r**. The quantity r is the distance from O to P, as shown in Fig. 1.5, the point O is called a *simple source* of flow lines, and the constant Q is called the *strength of the source.* The element of the surface integral for such a case is $\mathbf{F} \cdot \mathbf{dA} = (Q/r^2) \cos \vartheta \, dA$. However, $dA\,(\cos \vartheta / r^2)$ is equal to $d\Omega$, the element of solid angle subtended by the element of area dA when its normal is inclined at an angle ϑ to the radius. The net outflow integral thus reduces to $Q \oint d\Omega$, which is zero when O is outside the region enclosed by the integral and is equal to $4\pi Q$ when O is inside this region. Detailed analysis similar to the above shows that this is true for all shapes and varieties of enclosing surfaces.

The foregoing paragraph is a rather backhanded way of defining a simple source. A more straightforward way would be to say that a simple source of strength Q is a point singularity in a vector field such that a net outflow integral of the field over any surface enclosing the singularity is equal to $4\pi Q$.

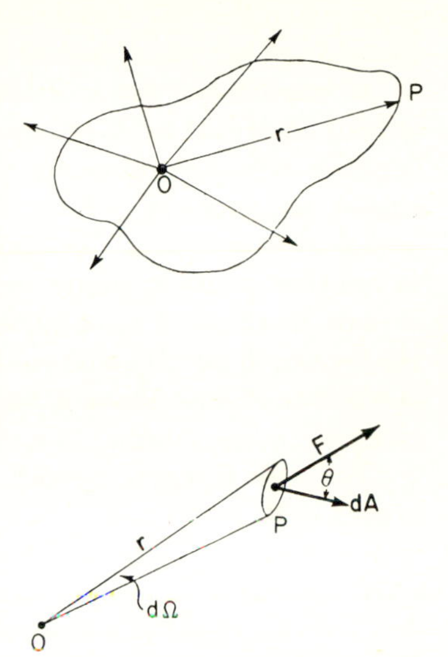

Fig. 1.5 Vector field about a source point. Element of net outflow integral.

The theorem which we have obtained can be stated by means of an equation

$$\oint \left(\frac{Q}{r^2}\right) \mathbf{a}_r \cdot \mathbf{dA} = \begin{cases} 0; & \text{source outside enclosed region} \\ 4\pi Q; & \text{source inside enclosed region} \end{cases} \qquad (1.2.9)$$

where $\mathbf{a}_r$ is the unit vector pointing outward along $\mathbf{r}$.

Sometimes the vector field is a combination due to several simple sources, one at O_1 with strength Q_1, one at O_2 with strength Q_2, one at O_n with strength Q_n, etc. In other words

$$\mathbf{F} = \Sigma (Q_n / r_n^2) \mathbf{a}_{rn}$$

where r_n is the distance from the point O_n to P and $\mathbf{a}_{rn}$ is the unit vector pointing along r_n. In this case the net outflow integral becomes

$$\oint \mathbf{F} \cdot \mathbf{dA} = \Sigma \oint (Q_n / r_n^2) \mathbf{a}_{rn} \cdot \mathbf{dA} = \Sigma' 4\pi Q_n \qquad (1.2.10)$$

where the unprimed summation is over all the sources but the primed summation is over only those sources *inside* the enclosed region.

Line Integrals. Instead of integrating the normal component of a vector over a surface, one can also integrate its component along a line. If $\mathbf{ds}$ is the vector element of length along some path, the integral $\int \mathbf{F} \cdot \mathbf{ds}$, taken along this path, is called the *line integral* of **F** along the path. When **F** is a force vector, its line integral is the work done along the path; when **F** is the electric intensity, the line integral is the emf between the ends of the path; and so on.

In general the value of the line integral between two points depends on the choice of path between the points. In some cases, however, it depends only on the position of the end points. This is the case discussed on page 14, where the components of **F** are derivatives of a potential function ψ. In such a case the line integral from a point O to another point P, along path A, is equal in magnitude and opposite in sign to the line integral in the reverse direction from P to O along another path B. Therefore, the integral around the closed path from O along A to P, then along B back to O, is zero for such a field. In general, however, a line integral of a vector field around a closed path is not zero.

The line integral around a closed path will be designated by the expression

$$\oint \mathbf{F} \cdot d\mathbf{s}$$

and will be called the *net circulation integral* for **F** around the path chosen. This integral is a measure of the tendency of the field's flow lines to "curl up." For instance if the flow lines are closed loops (as are the lines of magnetic intensity around a wire carrying current), a line integral of **F** around one of the loops will, of course, not be zero. The expression is called a *circulation integral*, because if **F** represents fluid velocity, then the integral $\oint \mathbf{F} \cdot d\mathbf{s}$ is a measure of the circulation of fluid around the path chosen.

We have seen above that, whenever the vector field has a potential function, the net circulation integral is zero. Therefore, we designate all fields having potential functions as circulation-free or *irrotational* fields. Fields having pseudopotentials do not necessarily have zero net circulation integrals. Indeed this will be true only if the gradient of μ (discussed on page 14) is everywhere parallel to **F**.

There are cases of vector fields which are irrotational everywhere except inside a limited region of space; in other words all the net circulation integrals enclosing this region differ from zero and all those not enclosing the region are zero. By analogy with the fluid-flow problem we call the region which "produces" the circulation a *vortex region*. Vortex regions must be of tubular shape, and the tube can never begin or end. It must either extend to infinity at both ends or close on itself to form a doughnut-shaped region. The only way for a vortex region to end above V (see Fig. 1.6) would be to have the circulation integral about path A different from zero, and to have the integrals about paths B, C, and D all equal to zero. However, a little close reasoning will show that, when the integral around A is not zero and, for instance, those around B and C are zero, then the integral around D *cannot be zero.* For by pairing off adjacent segments of the different paths one can show that the sum of the four line integrals around A, B, C, and D must be zero, since each pair consists of segments integrated over in opposite directions, which must cancel each other exactly. Therefore, if the integrals over B and C are zero, the integral over D must equal minus that over A, which by definition is not zero. Therefore, the vortex region cannot stop at V but must continue on in some such manner as is shown by the dotted region in Fig. 1.6. (Of course this reasoning tacitly assumes that the field is continuous at the vortex boundaries; anything may happen if discontinuities are present.)

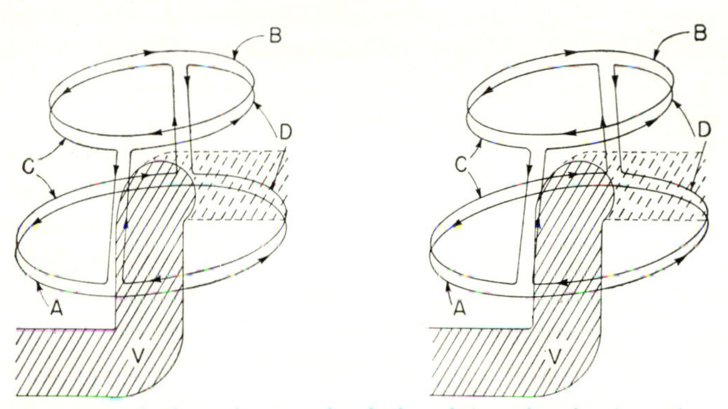

Fig. 1.6 Paths of net circulation integrals about vortex regions.

A vortex region may, of course, "fray out" into several tubes, some of which may return on themselves and others may go to infinity. It is possible to extend the theorem given above to include these more general cases; a differential form of the generalized theorem will be given on page 43.

Vortex Line. The simplest sort of vortex region is the *simple vortex line*, whose field is given by the equation

$$\mathbf{F} = (2/r)(\mathbf{Q} \times \mathbf{a}_r) = (2Q/r)\mathbf{a}_\varphi \tag{1.2.11}$$

where $\mathbf{Q}$, sometimes called the *vorticity vector*, is a vector of arbitrary length pointed along the vortex line L and $\mathbf{q}$ is a unit vector in the same direction. The vector $\mathbf{r}$ represents the line perpendicular to Q between L and the point P where the field is measured, $\mathbf{a}_r$ is the unit vector in the same direction, and $\mathbf{a}_\varphi = \mathbf{q} \times \mathbf{a}_r$ is a unit vector perpendicular to $\mathbf{a}_r$ and $\mathbf{Q}$. For a curve in a plane perpendicular to L, the line integral for this field is

$$\oint \mathbf{F} \cdot d\mathbf{s} = 2Q \oint (\cos \alpha / r)\, ds$$

where α is the angle between $\mathbf{ds}$ and $\mathbf{a}_\varphi$.

The second part of Fig. 1.7 shows that $(ds \cos \alpha/r) = d\varphi$, the element of angle of rotation of $\mathbf{r}$ about L. Therefore, the net circulation integral reduces to $2Q \oint d\varphi$ which can be integrated directly. For path B, enclosing L, it equals $4\pi Q$; for path A, not enclosing L, it equals zero.

This argument can be extended to paths not in a plane perpendicular to L, and the final result for the field of the field of the simple vortex line is

$$\oint \left(\frac{2Q}{r}\right) \mathbf{a}_\varphi \cdot d\mathbf{s} = \begin{cases} 0; & L \text{ not enclosed by path} \\ \pm 4\pi Q; & L \text{ enclosed by path} \end{cases} \qquad (1.2.12)$$

for all possible paths. The plus sign is used when the integration is clockwise about L when viewed in the direction of positive Q; the minus sign when the integration is in the opposite sense.

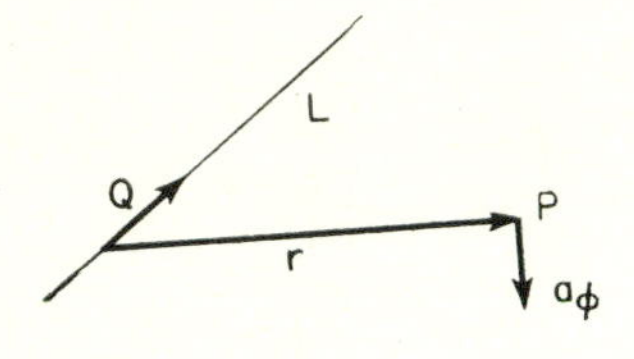

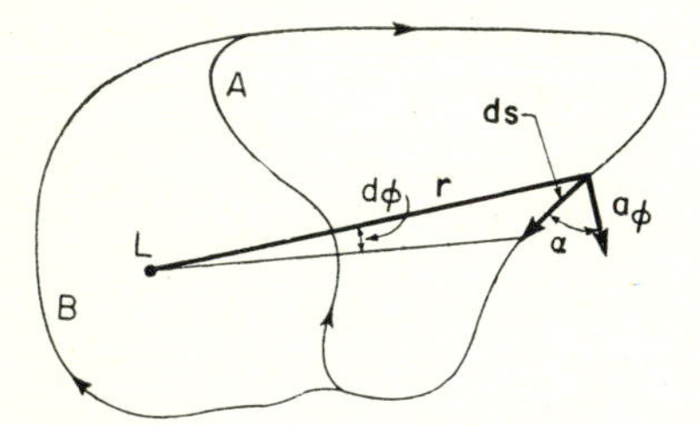

Fig. 1.7 Vector field about vortex line. Element of net circulation integral.

Singularities of Fields. It is interesting to notice the parallelism between the properties of the net outflow integral near a simple source field, discussed on page 17, and the net circulation integral near a vortex line, discussed immediately above. The source and the line are the simplest examples of *singularities* in vector fields. As our discussion progresses, we shall find that singularities like these two are usually the most important aspects of scalar and vector fields. The physical peculiarities of the problem at hand are usually closely related to the sort of singularities the field has. Likewise the mathematical properties of the solutions of differential equations are determined by the sorts of singularities which the equations and its solutions have. Much of our time will be spent in discussing the physical and mathematical properties of singularities in fields.

The field due to a simple source radiates out from a point, whereas that due to a vortex line rotates around the line. It is possible to spread sources out into lines or surfaces or even volume distributions, and it is possible to extend vortices into surfaces or volume distributions, but it is not possible to crowd a vortex into a point. This property is related to the fact that a rotation requires an axis, a line, to rotate about.

An interesting property which net circulation integrals and net outflow integrals, for any sort of field, have in common is their "additiveness." For instance, in Fig. 1.8, the net circulation integral about path C equals the sum of the integrals for paths A and B, because of the fact that the integration over the internal portion D is covered in opposite directions in paths A and B and, therefore, cancels out in the sum, leaving only the integral over path C. Similarly the net outflow integral for any region is equal to the sum of the net outflow integrals for all the subregions which together make up the original region. This again is due to the fact that the integrals over the surfaces internal to the original surface always occur in pairs, which cancel each other out in the sum, leaving only the integral over the outer surface of the original region.

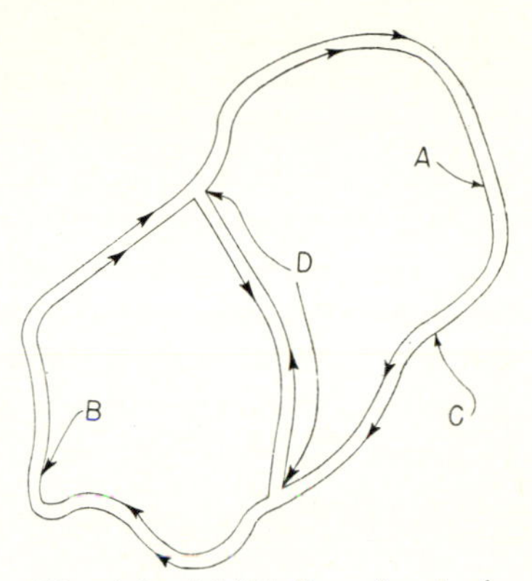

Fig. 1.8 Additivity of net circulation integrals.

1.3 *Curvilinear Coordinates*

Up to this point we have been tacitly assuming that the fields we have been discussing could always be expressed in terms of the three rectangular coordinates x, y, z (the fourth, or time, coordinate will be discussed later). Vector and scalar fields can in general be so expressed, but it often is much more convenient to express them in terms of other coordinate systems. We have already seen that it is sometimes possible to set up a system of coordinates "natural" to a vector field, using the lines of flow and the potential surfaces. In many cases the nature of the field is determined by specification of its behavior at a boundary surface or by specifying the nature and position of its singularities (or both), and the "natural" coordinate system for the field will bear some simple relationship to the boundary surface or to the distribution of singularities (or both). In many cases the expression for the field has a simple and tractable form in terms of these "natural" coordinates, whereas in terms of x, y, and z the expression is complex and the necessary calculations are intractable.

For all these reasons, and for others which will become apparent as we delve into the subject, we find it expedient to learn how to express our fields and the differential and integral operators which act on them in terms of generalized three-dimensional coordinates. We shall restrict ourselves to *orthogonal* coordinates, where the three families of coordinate surfaces are mutually perpendicular, because problems which require nonorthogonal coordinates practically never can be solved exactly. Approximate techniques for solving such problems usually involve the use of solutions for orthogonal coordinate systems.

A generalized coordinate system consists of a threefold family of surfaces whose equations in terms of the rectangular coordinates are $\xi_1(x,y,z)$ = constant, $\xi_2(x,y,z)$ = constant, $\xi_3(x,y,z)$ = constant (we assume that the reader is sufficiently acquainted with the properties of the rectangular coordinates x, y, z so that several pages of boring and obvious remarks can be omitted here). These equations give the functions ξ_1, ξ_2, and ξ_3 as functions of x, y, and z. In many cases it is more convenient to invert the equations and express x, y, and z in terms of ξ_1, ξ_2, and ξ_3.

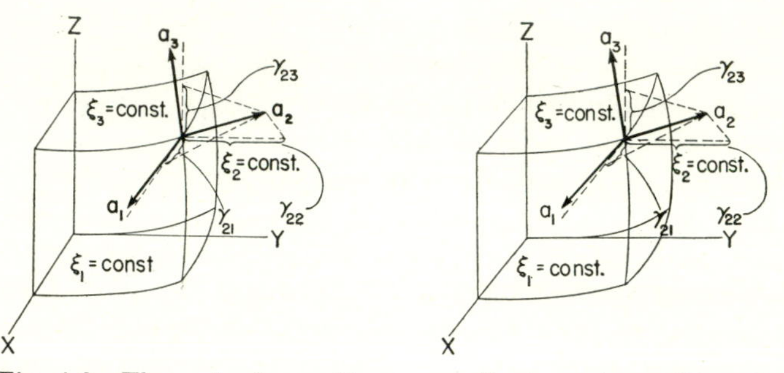

Fig. 1.9 Element of curvilinear coordinate system with unit vectors $\mathbf{a}_n$ and direction cosines γ_{nm}.

The lines of intersection of these surfaces constitute three families of lines, in general curved. At the point (x,y,z) or (ξ_1,ξ_2,ξ_3) we place three unit vectors $\mathbf{a}_1$, $\mathbf{a}_2$, and $\mathbf{a}_3$, each tangent to the corresponding coordinate line of the curvilinear system which goes through the point. These will be the new trio of unit vectors, in terms of which we can express the vector field $\mathbf{F}$. We note that the vectors $\mathbf{a}$ are of *unit* length, *i.e.*, 1 cm, or whatever the unit used, the same length as the vectors $\mathbf{i}$, $\mathbf{j}$, $\mathbf{k}$. For orthogonal systems the vectors $\mathbf{a}$ at a point are mutually perpendicular.

Direction Cosines. The direction cosines of the unit vector $\mathbf{a}_1$ with respect to the old axes are called $\alpha_1 = \mathbf{a}_1 \cdot \mathbf{i}$, $\beta_1 = \mathbf{a}_1 \cdot \mathbf{j}$, $\gamma_1 = \mathbf{a}_1 \cdot \mathbf{k}$; those of $\mathbf{a}_2$ are α_2, β_2, γ_2; and so on. In general these direction cosines will vary from point to point in space; *i.e.*, the α's, β's, and γ's will be functions of ξ_1, ξ_2, and ξ_3. From the properties of direction cosines we see that

$$\alpha_n^2 + \beta_n^2 + \gamma_n^2 = 1; \quad n = 1, 2, 3$$

for all values of the coordinates.

If the new unit vectors $\mathbf{a}$ are mutually perpendicular everywhere, the new coordinate system is *orthogonal*. In this case α_1, α_2, α_3 are the three direction cosines of $\mathbf{i}$ with respect to the $\mathbf{a}$'s, and the nine direction cosines are symmetrical with respect to the two systems. To emphasize this symmetry we can relabel these quantities:

$$\alpha_n = \gamma_{n1}; \quad \beta_n = \gamma_{n2}; \quad \gamma_n = \gamma_{n3},$$

so that $$\mathbf{a}_n = \gamma_{n1}\mathbf{i} + \gamma_{n2}\mathbf{j} + \gamma_{n3}\mathbf{k}; \quad \mathbf{i} = \sum_n \gamma_{n1}\mathbf{a}_n; \quad \text{etc.}$$

Since $\mathbf{i} \cdot \mathbf{i} = 1$, $\mathbf{i} \cdot \mathbf{j} = 0$, $\mathbf{a}_1 \cdot \mathbf{a}_1 = 1$, $\mathbf{a}_1 \cdot \mathbf{a}_3 = 0$, etc., the above relations between the direction cosines and the unit vectors result in a series of equations interrelating the values of the γ's:

$$\sum_s \gamma_{ms}\gamma_{ns} = \sum_s \gamma_{sm}\gamma_{sn} = \delta_{mn} \tag{1.3.1}$$

where δ_{mn} is the *Kronecker delta function*, which is zero when m is not equal to n, unity when m equals n.

Referring to Eq. (1.2.4) we note that, if the ξ coordinate system is right-handed (the x, y, z system is assumed to be right-handed), the determinant of the γ's, $|\gamma_{mn}|$, is equal to $+1$; if the ξ's are left-handed, then the determinant is equal to -1. Utilizing the second line of Eqs. (1.2.5) or solving Eq. (1.3.1) for one of the γ's in terms of the others, we obtain

$$\gamma_{mn} = \pm M_{mn} \tag{1.3.2}$$

where the plus sign is used if the ξ's form a right-handed system and the minus sign is used if the ξ's are left-handed. The quantity M_{mn} is the minor of γ_{mn} in the determinant $|\gamma_{mn}|$:

$$\begin{aligned} M_{11} &= \gamma_{22}\gamma_{33} - \gamma_{23}\gamma_{32} \\ M_{12} &= \gamma_{23}\gamma_{31} - \gamma_{21}\gamma_{33} \\ M_{31} &= \gamma_{12}\gamma_{23} - \gamma_{13}\gamma_{22}; \quad \text{etc.} \end{aligned}$$

It should, of course, be noted that the results expressed in Eqs. (1.3.1) and (1.3.2) hold, not only between any orthogonal system and the cartesian system, but also between *any* two orthogonal systems. The plus sign in Eq. (1.3.2) is used if both systems are right-handed or both left-handed; the minus sign if one system is right-handed and the other left-handed.

Since we have assumed that the curvilinear coordinate system is orthogonal, any vector $\mathbf{F}$ at (ξ_1,ξ_2,ξ_3) can be expressed in terms of its components along the new unit vectors:

$$\mathbf{F} = \sum_m F_m\mathbf{a}_m; \quad \text{where } F_m = \mathbf{F} \cdot \mathbf{a}_m$$

Utilizing the direction cosines γ_{mn} we can easily show that the relation between these components and the cartesian components of **F** are

$$F_m = \gamma_{m1}F_x + \gamma_{m2}F_y + \gamma_{m3}F_z = \alpha_m F_x + \beta_m F_y + \gamma_m F_z$$
$$F_x = \sum_m \gamma_{m1}F_m = \sum_m \alpha_m F_m;\quad \text{etc.} \tag{1.3.3}$$

Any set of three quantities defined with relation to a system of coordinates, so that choice of (x,y,z) gives a trio of functions of (x,y,z), a choice of (ξ_1,ξ_2,ξ_3) gives another trio of functions of (ξ_1,ξ_2,ξ_3), etc., can be considered as the components of a vector if, and only if, the two trios are related in the manner specified in Eq. (1.3.3).

Scale Factors. The above discussion does not help us much, however, unless we know the values of the direction cosines α, β, and γ for each point in space. Ordinarily we are given the equations for the new coordinate surfaces; we must translate these into expressions for the cosines. For instance, the usual definition for spherical coordinates $z = \xi_1 \cos \xi_2$, $x = \xi_1 \sin \xi_2 \cos \xi_3$, $y = \xi_1 \sin \xi_2 \sin \xi_3$ does not give us directly the α's, β's, and γ's in terms of ξ_1, ξ_2, and ξ_3 (ξ_1 = spherical coordinate r, ξ_2 = spherical coordinate ϑ, ξ_3 = spherical coordinate φ).

The needed connection is usually made by means of the line element. The length of the infinitesimal vector ds is given as

$$ds^2 = dx^2 + dy^2 + dz^2 = \sum_n h_n^2\, d\xi_n^2$$

if the new coordinate system is an orthogonal one. Simple substitution shows that the relation between the h's and ξ's is

$$h_n^2 = \left(\frac{\partial x}{\partial \xi_n}\right)^2 + \left(\frac{\partial y}{\partial \xi_n}\right)^2 + \left(\frac{\partial z}{\partial \xi_n}\right)^2 = \left[\left(\frac{\partial \xi_n}{\partial x}\right)^2 + \left(\frac{\partial \xi_n}{\partial y}\right)^2 + \left(\frac{\partial \xi_n}{\partial z}\right)^2\right]^{-1} \tag{1.3.4}$$

The quantity h_n is a *scale factor* for the coordinate ξ_n. A change $d\xi_n$ in this coordinate produces a displacement $h_n\, d\xi_n$ cm along the coordinate line. In general h_n varies from point to point in space.

We notice that, in order to obtain the h's in terms of the ξ's, it is necessary to express the old coordinates x, y, z in terms of the new ones, ξ_1, ξ_2, ξ_3, as was done for the spherical coordinates in the previous paragraph. This mode of writing the relation between the two coordinate systems is usually the most satisfactory.

Since $h_n\, d\xi_n$ is the length of the displacement corresponding to $d\xi_n$, the rate of displacement along ξ_n due to a displacement along the x axis will be $h_n(\partial \xi_n/\partial x)$. This quantity is, therefore, the direction cosine $\alpha_n = \gamma_{n1}$. Similarly, if x is expressed in terms of the ξ's, the rate of change of x due to the displacement $h_n\, d\xi_n$ will be $(1/h_n)(\partial x/\partial \xi_n)$, which is also equal to $\gamma_{n1} = \alpha_n$. Thus the direction cosines for the ξ_n axis with respect to the x, y, z axes can be written in terms of the derivatives relating x, y, z, and the ξ's in either of two ways:

$$\gamma_{n1} = \alpha_n = \frac{1}{h_n}\frac{\partial x}{\partial \xi_n} = h_n \frac{\partial \xi_n}{\partial x}; \quad \gamma_{n2} = \beta_n = \frac{1}{h_n}\frac{\partial y}{\partial \xi_n} = h_n \frac{\partial \xi_n}{\partial y}; \quad \gamma_{n3} = \gamma_n = \frac{1}{h_n}\frac{\partial z}{\partial \xi_n} = h_n \frac{\partial \xi_n}{\partial z} \tag{1.3.5}$$

depending on whether x, y, z are given in terms of the ξ's, or the ξ's in terms of x, y, z.

Equations (1.3.5) are useful but are not needed so often as might be expected. It is interesting to note here that all the differential expressions

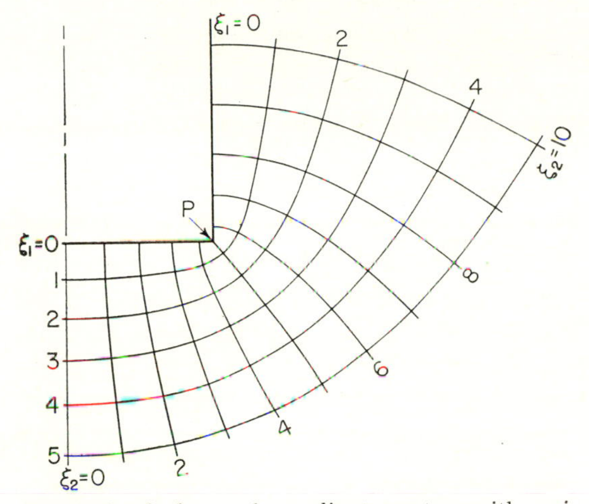

Fig. 1.10 Orthogonal coordinate system with variable scale factors h. In this case $h_1 = h_2$ everywhere and $h_1 = h_2 = 0$ at point P, a concentration point for the system.

we shall develop later in this chapter and use extensively later in the book need only the h's, and not the γ's, in order to be expressed in generalized coordinates. Evidently the scale of a new coordinate and the variation of scale from point to point determine the important properties of the coordinate. The direction of the coordinate with respect to x, y, z at a point is a property of relatively minor importance.

Curvature of Coordinate Lines. For instance, even the change in direction of the unit vectors $\mathbf{a}$ can be expressed in terms of the h's. The expressions for these changes are suggested in Fig. 1.11. In drawing (a), we see that the change in $\mathbf{a}_1$ due to change in ξ_2 is $\mathbf{a}_2\alpha$, where

$$\alpha = \frac{d\xi_1\, d\xi_2}{h_1\, d\xi_1}\frac{\partial h_2}{\partial \xi_1}$$

Therefore,

$$\frac{\partial \mathbf{a}_1}{\partial \xi_2} = \frac{\mathbf{a}_2}{h_1}\frac{\partial h_2}{\partial \xi_1}$$

Similarly, in drawing (*b*), the change in $\mathbf{a}_1$ due to change in ξ_1 has a ξ_2 component equal to

$$-\beta = -\frac{d\xi_1\,d\xi_2}{h_2\,d\xi_2}\frac{\partial h_1}{\partial \xi_2}$$

and a similar ξ_3 component. Therefore, the following formulas are suggested:

$$\begin{aligned}
\frac{\partial \mathbf{a}_1}{\partial \xi_1} &= -\frac{\mathbf{a}_2}{h_2}\frac{\partial h_1}{\partial \xi_2} - \frac{\mathbf{a}_3}{h_3}\frac{\partial h_1}{\partial \xi_3}; & \frac{\partial \mathbf{a}_1}{\partial \xi_2} &= \frac{\mathbf{a}_2}{h_1}\frac{\partial h_2}{\partial \xi_1}; & \frac{\partial \mathbf{a}_1}{\partial \xi_3} &= \frac{\mathbf{a}_3}{h_1}\frac{\partial h_3}{\partial \xi_1}\\
\frac{\partial \mathbf{a}_2}{\partial \xi_2} &= -\frac{\mathbf{a}_3}{h_3}\frac{\partial h_2}{\partial \xi_3} - \frac{\mathbf{a}_1}{h_1}\frac{\partial h_2}{\partial \xi_1}; & \frac{\partial \mathbf{a}_2}{\partial \xi_3} &= \frac{\mathbf{a}_3}{h_2}\frac{\partial h_3}{\partial \xi_2}; & \frac{\partial \mathbf{a}_2}{\partial \xi_1} &= \frac{\mathbf{a}_1}{h_2}\frac{\partial h_1}{\partial \xi_2}\\
\frac{\partial \mathbf{a}_3}{\partial \xi_3} &= -\frac{\mathbf{a}_1}{h_1}\frac{\partial h_3}{\partial \xi_1} - \frac{\mathbf{a}_2}{h_2}\frac{\partial h_3}{\partial \xi_2}; & \frac{\partial \mathbf{a}_3}{\partial \xi_1} &= \frac{\mathbf{a}_1}{h_3}\frac{\partial h_1}{\partial \xi_3}; & \frac{\partial \mathbf{a}_3}{\partial \xi_2} &= \frac{\mathbf{a}_2}{h_3}\frac{\partial h_2}{\partial \xi_3}
\end{aligned} \tag{1.3.6}$$

We can prove that these are the correct expressions by expressing

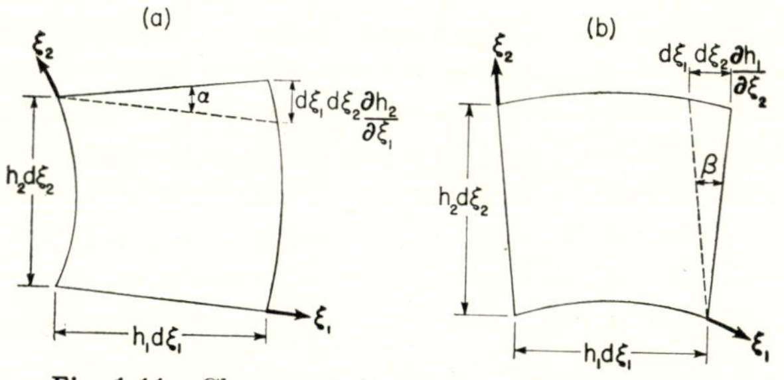

Fig. 1.11 Change of direction of unit vectors $\mathbf{a}_n$.

the **a**'s in terms of **i**, **j** and **k**, performing the differentiation, and utilizing the definitions of the h's given in Eqs. (1.3.4). For instance,

$$\begin{aligned}
\frac{\partial \mathbf{a}_1}{\partial \xi_2} &= \frac{\partial}{\partial \xi_2}\left[\frac{\mathbf{i}}{h_1}\frac{\partial x}{\partial \xi_1} + \frac{\mathbf{j}}{h_1}\frac{\partial y}{\partial \xi_1} + \frac{\mathbf{k}}{h_1}\frac{\partial z}{\partial \xi_1}\right]\\
&= h_1\mathbf{a}_1\frac{\partial}{\partial \xi_2}\left(\frac{1}{h_1}\right) + \frac{1}{h_1}\left[\mathbf{i}\frac{\partial^2 x}{\partial \xi_1\,\partial \xi_2} + \mathbf{j}\frac{\partial^2 y}{\partial \xi_1\,\partial \xi_2} + \mathbf{k}\frac{\partial^2 z}{\partial \xi_1\,\partial \xi_2}\right]\\
&= -\mathbf{a}_1\frac{\partial \ln h_1}{\partial \xi_2} + \frac{1}{h_1}\left\{\frac{\partial^2 x}{\partial \xi_1\,\partial \xi_2}\left[\frac{\mathbf{a}_1}{h_1}\frac{\partial x}{\partial \xi_1} + \frac{\mathbf{a}_2}{h_2}\frac{\partial x}{\partial \xi_2} + \frac{\mathbf{a}_3}{h_3}\frac{\partial x}{\partial \xi_3}\right]\right.\\
&\qquad + \frac{\partial^2 y}{\partial \xi_1\,\partial \xi_2}\left[\frac{\mathbf{a}_1}{h_1}\frac{\partial y}{\partial \xi_1} + \frac{\mathbf{a}_2}{h_2}\frac{\partial y}{\partial \xi_2} + \frac{\mathbf{a}_3}{h_3}\frac{\partial y}{\partial \xi_3}\right]\\
&\qquad \left. + \frac{\partial^2 z}{\partial \xi_1\,\partial \xi_2}\left[\frac{\mathbf{a}_1}{h_1}\frac{\partial z}{\partial \xi_1} + \frac{\mathbf{a}_2}{h_2}\frac{\partial z}{\partial \xi_2} + \frac{\mathbf{a}_3}{h_3}\frac{\partial z}{\partial \xi_3}\right]\right\}
\end{aligned}$$

or
$$\frac{\partial \mathbf{a}_1}{\partial \xi_2} = -\mathbf{a}_1\frac{\partial \ln h_1}{\partial \xi_2} + \frac{\mathbf{a}_1}{2h_1^2}\frac{\partial h_1^2}{\partial \xi_2} + \frac{\mathbf{a}_2}{2h_1h_2}\frac{\partial h_2^2}{\partial \xi_1} = \frac{\mathbf{a}_2}{h_1}\frac{\partial h_2}{\partial \xi_1}$$

which coincides with the expression given above.

Curvature and torsion of the coordinate surface ξ_1 = constant can easily be computed from Eqs. (1.3.6). The unit vector $\mathbf{a}_1$, perpendicular to the surface, changes in direction by an amount $(\partial\mathbf{a}_1/h_2\,\partial\xi_2)\,ds$ for a displacement ds in the ξ_2 direction. This change is in the ξ_2 direction and is a measure of the curvature of the ξ_1 surface in the direction of the ξ_2 coordinate line. In fact the magnitude $\mathbf{a}_2 \cdot (\partial\mathbf{a}_1/h_2\,\partial\xi_2)$ is just equal to the reciprocal of the radius of curvature of the ξ_2 coordinate line at the point (ξ_1,ξ_2,ξ_3); if the quantity is positive, it means that the curvature is convex in the direction of positive ξ_1; and if negative, the curve is concave in the positive ξ_1 direction.

It is not difficult to show that the total curvature of the surface ξ_1 = constant, at the point (ξ_1,ξ_2,ξ_3), is

$$C = -\left(\frac{\mathbf{a}_2}{h_2}\right)\cdot\left(\frac{\partial\mathbf{a}_1}{\partial\xi_2}\right) - \left(\frac{\mathbf{a}_3}{h_3}\right)\cdot\left(\frac{\partial\mathbf{a}_1}{\partial\xi_3}\right) = -\frac{1}{h_1h_2h_3}\frac{\partial}{\partial\xi_1}(h_2h_3)$$

where now the sign indicates the direction of the concavity of the curvature. This formula and the corresponding ones for the ξ_2 and ξ_3 surfaces will be of use later in calculating the restoring force produced by curved surfaces under tension.

As a simple example, for the spherical coordinates r, ϑ, ϕ, the curvature of the r surface is $-(2/r)$, the spherical surface being concave inward and being curved in both ϑ and ϕ directions (which accounts for the 2). The curvature of the conical ϑ = constant surface is $-(1/r)\cot\vartheta$, and the plane ϕ = constant surface has zero curvature.

In any case, once we know the equations giving x, y, and z in terms of the new coordinates, Eqs. (1.3.4) to (1.3.6) enable us to compute the scale of the new system, the components of a vector along these axes, their variation, and many other important expressions which will be discussed later.

The Volume Element and Other Formulas. Another quantity which will be of importance later will be the expression for the volume element in the new coordinate system. Since the elements $d\xi_1$, $d\xi_2$, $d\xi_3$, correspond to displacements $h_1\,d\xi_1$, $h_2\,d\xi_2$, $h_3\,d\xi_3$ along mutually perpendicular axes, the volume of the rectangular parallelepiped defined by these differentials is

$$dv = h_1h_2h_3\,d\xi_1\,d\xi_2\,d\xi_3 \tag{1.3.7}$$

This is the volume element in the new coordinate system. It is, of course, always positive in sign.

As an example of these properties of the scale factors, we consider the spherical coordinate system mentioned above: $x = \xi_1 \sin\xi_2 \cos\xi_3$ $y = \xi_1\sin\xi_2\sin\xi_3$, $z = \xi_1\cos\xi_2$. We find that the scale factors are

$$h_1 = 1;\quad h_2 = \xi_1;\quad h_3 = \xi_1\sin\xi_2$$

The direction cosines of the unit vectors along the spherical axes are therefore

$$\begin{aligned}
\alpha_1 &= \sin \xi_2 \cos \xi_3; \quad \alpha_2 = \cos \xi_2 \cos \xi_3; \quad \alpha_3 = -\sin \xi_3 \\
\beta_1 &= \sin \xi_2 \sin \xi_3; \quad \beta_2 = \cos \xi_2 \sin \xi_3; \quad \beta_3 = \cos \xi_3 \\
\gamma_1 &= \cos \xi_2; \quad \gamma_2 = -\sin \xi_2; \quad \gamma_3 = 0
\end{aligned}$$

These obey all the relations for orthogonality, etc., given on page 23.

The volume element in the new coordinate system is $dv = \xi_1^2 \sin \xi_2 \, d\xi_1 \, d\xi_2 \, d\xi_3$, and the components of a vector along the new coordinates are

$$\begin{aligned}
F_1 &= F_x \sin \xi_2 \cos \xi_3 + F_y \sin \xi_2 \sin \xi_3 + F_z \cos \xi_2 \\
F_2 &= F_x \cos \xi_2 \cos \xi_3 + F_y \cos \xi_2 \sin \xi_3 - F_z \sin \xi_2 \\
F_3 &= -F_x \sin \xi_3 + F_y \cos \xi_3
\end{aligned}$$

If the functions F_x, F_y, and F_z are expressed in terms of ξ_1, ξ_2, ξ_3, then the new components will be expressed in terms of the new coordinates and the transformation will be complete.

Rotation of Axes. Another example of coordinate transformation is the case where the new axes are also rectangular, being rotated with

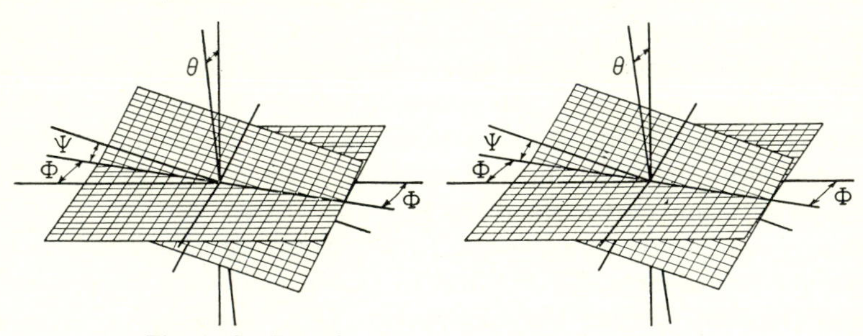

Fig. 1.12 Rotation of axes, showing Euler angles.

respect to the old set by the Euler angles θ, Φ, ψ, as shown in Fig. 1.12. The equations of transformation are

$$\begin{aligned}
x &= (\sin \psi \sin \Phi + \cos \psi \cos \Phi \cos \theta) \xi_1 + (\cos \psi \sin \Phi \\
&\qquad - \sin \psi \cos \Phi \cos \theta) \xi_2 + \sin \theta \cos \Phi \, \xi_3 \\
y &= (\sin \psi \cos \Phi - \cos \psi \sin \Phi \cos \theta) \xi_1 + (\cos \psi \cos \Phi \\
&\qquad + \sin \psi \sin \Phi \cos \theta) \xi_2 - \sin \theta \sin \Phi \, \xi_3 \\
z &= -\cos \psi \sin \theta \, \xi_1 + \sin \psi \sin \theta \, \xi_2 + \cos \theta \, \xi_3
\end{aligned} \tag{1.3.8}$$

The scale factors h_n for this case are all unity, as was to be expected, since the transformation does not involve a scale change. The direction cosines for the transformation are the coefficients of the linear equations given above:

$$\begin{aligned}
\alpha_1 &= \sin \psi \sin \Phi + \cos \psi \cos \Phi \cos \theta; \quad \text{etc.} \\
\beta_1 &= \sin \psi \cos \Phi - \cos \psi \sin \Phi \cos \theta; \quad \text{etc.}
\end{aligned}$$

From these direction cosines one can obtain the vector transformation formulas.

Law of Transformation of Vectors. We have seen that, in order for three functions of position to be considered as the three components of a vector, they must transform according to the rule given by Eqs. (1.3.3) and (1.3.5). If we transform the components from one curvilinear coordinate system ξ_1, ξ_2, ξ_3 with scale factors h_1, h_2, h_3 to another system ξ'_1, ξ'_2, ξ'_3 with scale factors h'_1, h'_2, h'_3, the components in the new system must be related to the components in the old system by the equations

$$F'_n = \sum_m \gamma_{nm} F_m$$

where

$$(h_m/h'_n)(\partial \xi_m/\partial \xi'_n) = \gamma_{nm} = (h'_n/h_m)(\partial \xi'_n/\partial \xi_m) \tag{1.3.9}$$

Since $h_m\, d\xi_m$ and $h'_n\, d\xi'_n$ are distances in centimeters, the new components F'_n have the same dimensions as the old. In developing a new theory, if we find three quantities which transform in a manner given by Eq. (1.3.9), we can be fairly sure we have found the components of a vector.

It is interesting at this point to investigate the transformation properties of the cross product $(\mathbf{A} \times \mathbf{B})$. We recall that on page 11 it was pointed out that $(\mathbf{A} \times \mathbf{B})$ is not a true vector. The present discussion will show why. Using Eq. (1.3.9), the component of the cross product of $\mathbf{A}$ and $\mathbf{B}$ along ξ'_1 is

$$\begin{aligned}(\mathbf{A}' \times \mathbf{B}')_1 = A'_2B'_3 - A'_3B'_2 &= \sum_{m,n} (\gamma_{m2}\gamma_{n3} - \gamma_{m3}\gamma_{n2})A_mB_n \\ &= \sum_{m>n} (\gamma_{m2}\gamma_{n3} - \gamma_{m3}\gamma_{n2})(A_mB_n - A_nB_m)\end{aligned} \tag{1.3.10}$$

Using the relations given in Eq. (1.3.2) (it is easily seen that these relations hold for the more general transformation discussed here as long as both systems are orthogonal) we finally obtain

$$(\mathbf{A}' \times \mathbf{B}')_1 = \pm \sum_n (\mathbf{A} \times \mathbf{B})_n \gamma_{1n} \tag{1.3.11}$$

where the plus sign holds if the coordinate systems are both right-handed or both left-handed, the negative sign holds if one is right-handed and one left-handed [moreover, Eq. (1.3.11) holds only if both systems are orthogonal].

It follows that $\mathbf{A} \times \mathbf{B}$ is an axial vector.

The equation also gives us a hint for a simple method of distinguishing true and axial vectors, for they will behave differently under a transformation which takes us from a right-handed to a left-handed system. A simple example of such a transformation is the *inversion* $\xi_1 = -x$, $\xi_2 = -y$, $\xi_3 = -z$. If the components of a true vector were A_x, A_y, A_z, then $A_1 = -A_x\ A_2 = -A_y$, etc.; *i.e.*, the components of a true vector change sign upon an inversion. On the other hand for an axial vector $\mathbf{A} \times \mathbf{B}$ there will be no change in sign, so that the components of an axial vector do not change sign upon an inversion.

Similarly, a true scalar or invariant, an example of which is $\mathbf{A} \cdot \mathbf{B}$, will not change sign upon an inversion. On the other hand a pseudoscalar, for example, $\mathbf{A} \cdot (\mathbf{B} \times \mathbf{C})$, does change sign upon an inversion.

The detailed use of transformation equation (1.3.9) for the purpose of discovering whether three quantities form the components a vector may be very tedious. Another method makes use of the knowledge of the invariants involving these quantities. For example, if $\Sigma A_i B_i$ form an invariant and B_i are the components of a true vector, then A_i must form the components of a true vector. Several examples of this method will occur later in this chapter.

Contravariant and Covariant Vectors. There are two other methods of writing vector components which have occasional use, obtained by using different "unit vectors" for the decomposition of $\mathbf{F}$ into components. Suppose the "unit vectors" are made to change scale with the coordinate system, so that the new vector $\hat{\mathbf{a}}_n = h_n \mathbf{a}_n$ corresponds to the size of a unit change of ξ_n rather than a unit length in centimeters (as does $\mathbf{a}_n$). The vector $\mathbf{F}$ can be written in terms of these new "unit vectors":

$$\mathbf{F} = \Sigma_n f^n \hat{\mathbf{a}}_n; \quad f^n = F_n/h_n$$

In this case the transformation properties of the new "components" are

$$(f^n)' = \sum_m f^m \frac{\partial \xi'_n}{\partial \xi_m} = \sum_m f^m \left(\frac{h_m}{h'_n}\right)^2 \frac{\partial \xi_m}{\partial \xi'_n} \tag{1.3.12}$$

The quantities f^n are called the contravariant components of a vector in the coordinate system ξ_1, ξ_2, ξ_3. They differ from the "actual" components by being divided by the scale factor; and they give an actual vector only when the components are multiplied by the "unit vector" $\hat{\mathbf{a}}_n = h_n \mathbf{a}_n$.

If the unit vectors along the coordinate lines have a scale inverse to the coordinate, $\hat{\mathbf{a}}^n = \mathbf{a}_n/h_n$, then the corresponding "components" are

$$\mathbf{F} = \sum_n f_n \hat{\mathbf{a}}^n; \quad f_n = h_n F_n$$

in which case the transformation properties for the f's are

$$(f_n)' = \sum_m f_m \frac{\partial \xi_m}{\partial \xi'_n} = \sum_m f_m \left(\frac{h'_n}{h_m}\right)^2 \frac{\partial \xi'_n}{\partial \xi_m} \tag{1.3.13}$$

The quantities f_m are called the *covariant* components of a vector in the coordinate system (ξ_1, ξ_2, ξ_3).

These transformations have a somewhat greater formal symmetry than does that given in Eqs. (1.3.9) for the usual vector components, for the h's have disappeared from the summation in at least one of the expressions for each case. This dubious advantage is often offset by the fact that the new components do not necessarily maintain their dimensionality when transforming from coordinate to coordinate. For instance, in spherical coordinates, if **F** has the dimensions of length, F_r, F_ϑ, and F_φ still have the dimensions of length after being transformed whereas f^ϑ and f^φ are dimensionless and f_ϑ and f_φ have the dimensions of area.

We shall return to this notation later, when we come to talk about tensors. There we shall find that the f's are useful in the preliminary exploration of a problem, where the formal symmetry of their transformation equations may aid in the manipulations. When it comes to the detailed calculations, however, we shall usually find it easier to use the "actual" components F_n, which always have the same dimensions as **F**, and the unit vectors $\mathbf{a}_n$ which are all of unit length.

1.4 *The Differential Operator* ∇

Now that we have discussed the fundamental principles of coordinate transformations and have indicated how we can recognize scalars and vectors by noting their transformation properties, it is time to study the general differential properties of vectors. In an earlier section we studied the gross, macroscopic properties of vector fields by use of surface and line integrals. Now we wish to study their detailed microscopic properties. By analogy to the differential operator d/dx which operates on a scalar function φ of x, changing it into a different function, the slope of φ, we have a differential operator, involving all three coordinates, which operates on scalar and vector fields, changing them into other fields. The resulting fields are measures of the rate at which the original fields change from point to point.

The Gradient. The rate of change of a scalar $\psi(x,y,z)$ is expressed in terms of a vector, the direction being that of the greatest rate of increase of ψ and the magnitude being the value of this maximum rate of increase. We have already indicated, in Eq. (1.1.1), that the change in ψ as we go from the point indicated by the vector $\mathbf{A} = x\mathbf{i} + y\mathbf{j} + z\mathbf{k}$ to the neighboring point $\mathbf{A} + \mathbf{ds}$ by the elementary displacement $\mathbf{ds} = \mathbf{i}\,dx + \mathbf{j}\,dy + \mathbf{k}\,dz$ is $d\psi = \mathbf{ds} \cdot \operatorname{grad} \psi$, where

$$\operatorname{grad} \psi = \nabla\psi = \frac{\partial\psi}{\partial x}\mathbf{i} + \frac{\partial\psi}{\partial y}\mathbf{j} + \frac{\partial\psi}{\partial z}\mathbf{k} \tag{1.4.1}$$

If $\mathbf{ds}$ is in the isotimic surface ψ = constant, $d\psi$ must be zero, so that the vector grad ψ must be perpendicular to the isotimic surface. The maximum value of $d\psi$ occurs when $\mathbf{ds}$ is also perpendicular to this surface, and in this case, as we have mentioned earlier,

$$d\psi = \sqrt{(\partial\psi/\partial x)^2 + (\partial\psi/\partial y)^2 + (\partial\psi/\partial z)^2}\, ds$$

The quantity labeled by the symbols grad ψ or $\nabla\psi$ (read *del* ψ) is, therefore, a measure of the rate of change of the scalar field ψ at the point (x,y,z). To show that it is definitely a vector, we must show that its transformation properties correspond to Eq. (1.3.9). This is not difficult, for it is obvious that the expression for the gradient in the curvilinear coordinates ξ_1, ξ_2, ξ_3 is

$$\operatorname{grad}\psi = \nabla\psi = \frac{\mathbf{a}_1}{h_1}\frac{\partial\psi}{\partial\xi_1} + \frac{\mathbf{a}_2}{h_2}\frac{\partial\psi}{\partial\xi_2} + \frac{\mathbf{a}_3}{h_3}\frac{\partial\psi}{\partial\xi_3} \tag{1.4.2}$$

When we now transform this expression to another set of coordinates ξ_1', ξ_2', ξ_3', as though it were a vector, it ends up having the same form in the new coordinates as it had in the old. Using Eq. (1.3.9) and the identities

$$\frac{\partial\psi}{\partial\xi_n'} = \sum_m \left(\frac{\partial\psi}{\partial\xi_m}\right)\left(\frac{\partial\xi_m}{\partial\xi_n'}\right)$$

we obtain

$$\nabla\psi = \sum_m \frac{\mathbf{a}_m}{h_m}\frac{\partial\psi}{\partial\xi_m} = \sum_n \mathbf{a}_n' \sum_m \left(\frac{1}{h_m}\frac{\partial\psi}{\partial\xi_m}\right)\left(\frac{h_m}{h_n'}\frac{\partial\xi_m}{\partial\xi_n'}\right) = \sum_n \frac{\mathbf{a}_n'}{h_n'}\frac{\partial\psi}{\partial\xi_n'}$$

as, of course, we must if grad ψ is to be a vector and have the form given in Eq. (1.4.2) for any orthogonal curvilinear coordinates.

Note that grad ψ is a true vector and not an axial vector. This is demonstrated by the fact that $d\psi = \operatorname{grad}(\psi)\cdot\mathbf{ds}$ is a true scalar invariant. Since $\mathbf{ds}$ is a true vector, grad ψ must also be a true vector. Note also that the circulation integral for grad ψ, $\oint \operatorname{grad}(\psi)\cdot\mathbf{ds} = \oint d\psi = 0$. The fact that the circulation integral of a gradient is always zero has already been pointed out on page 18.

This short discussion serves to emphasize that, if we express "physical" quantities in terms of the scale factors h for the coordinate ξ, then they must turn out to be expressible by the same *form* of expression in terms of the scale factors h' for another set of coordinates ξ'.

Directional Derivative. The quantity $\mathbf{B}\cdot\operatorname{grad}\psi$ will sometimes occur in our equations. When $\mathbf{B}$ is a unit vector, the quantity is called the *directional derivative* of ψ in the direction of the unit vector $\mathbf{B}$, and equals the rate of change of ψ in the direction of $\mathbf{B}$. Whether or not $\mathbf{B}$ is a unit vector, the value of $\mathbf{B}\cdot\operatorname{grad}\psi$ is

$$(\mathbf{B}\cdot\nabla)\psi = \mathbf{B}\cdot \operatorname{grad}\psi = B_x\frac{\partial\psi}{\partial x} + B_y\frac{\partial\psi}{\partial y} + B_z\frac{\partial\psi}{\partial z}$$
$$= \frac{B_1}{h_1}\frac{\partial\psi}{\partial\xi_1} + \frac{B_2}{h_2}\frac{\partial\psi}{\partial\xi_2} + \frac{B_3}{h_3}\frac{\partial\psi}{\partial\xi_3}$$

The scalar operator ($\mathbf{B}\cdot$ grad) can also be applied to a vector, giving

$$(\mathbf{B}\cdot\nabla)\mathbf{A} = (\mathbf{B}\cdot\operatorname{grad})\mathbf{A} = \mathbf{i}(\mathbf{B}\cdot\operatorname{grad}A_x) + \mathbf{j}(\mathbf{B}\cdot\operatorname{grad}A_y) + \mathbf{k}(\mathbf{B}\cdot\operatorname{grad}A_z)$$

In curvilinear coordinates this becomes somewhat more complicated, because the unit vectors $\mathbf{a}$ also change with position, as given in Eqs. (1.3.6). Using these equations, we can show that the ξ_1 component of the vector is

$$[(\mathbf{B}\cdot\operatorname{grad})\mathbf{A}]_1 = \frac{B_1}{h_1}\frac{\partial A_1}{\partial\xi_1} + \frac{B_2}{h_2}\frac{\partial A_1}{\partial\xi_2} + \frac{B_3}{h_3}\frac{\partial A_1}{\partial\xi_3} + \frac{A_2}{h_1h_2}\left[B_1\frac{\partial h_1}{\partial\xi_2} - B_2\frac{\partial h_2}{\partial\xi_1}\right] + \frac{A_3}{h_1h_3}\left[B_1\frac{\partial h_1}{\partial\xi_3} - B_3\frac{\partial h_3}{\partial\xi_1}\right] \quad (1.4.3)$$

and the other components can be obtained by cyclic permutations of the subscripts. The first three terms in this expression are just $\mathbf{B}\cdot\operatorname{grad}A_1$; the other terms are corrections due to the fact that the directions of the coordinate axes are not necessarily constant.

Infinitesimal Rotation. One type of coordinate transformation which will have special interest later is that caused by an infinitely small rotation of the rectangular coordinates about an axis through the origin. Suppose that the vector $\mathbf{d\omega}$ has magnitude equal to the angle of rotation in radians and has the direction of the axis of rotation such that a right-hand screw would move in the direction $\mathbf{d\omega}$ if rotated with the coordinates. The point designated by the vector $\mathbf{r}$ (in distance and direction from the origin) would, if it were fixed to the rotating coordinate system, be displaced by an amount represented by the vector $\mathbf{d\omega}\times\mathbf{r} = -\mathbf{r}\times\mathbf{d\omega}$. If the point is, instead, fixed in space, then its coordinates in the rotated system (which we shall label with a prime) will be related to its coordinates in the unrotated system by the equation $\mathbf{r}' = \mathbf{r} + \mathbf{r}\times\mathbf{d\omega}$, or

$$\begin{aligned} x' &= x + (y\,d\omega_z - z\,d\omega_y) \\ y' &= y + (z\,d\omega_x - x\,d\omega_z) \\ z' &= z + (x\,d\omega_y - y\,d\omega_x) \end{aligned}$$

We could also write $\mathbf{r} = \mathbf{r}' - \mathbf{r}'\times\mathbf{d\omega}$. [These equations only hold for very small rotations; otherwise we use the more general forms of Eq. (1.3.8), which reduce to (1.4.3) in the limit of θ and $(\Phi + \psi)$ small.]

Now suppose that a scalar field ψ is made to rotate slightly (the field could be the density of a solid which is rotated slightly). Because of this rotation, the value of the field ψ' at the point (x,y,z), fixed in space, after the rotation, is related to its value ψ before the rotation by the relation

$$\psi'(x,y,z) = \psi(x,y,z) + (\mathbf{r} \times \mathbf{d\omega}) \cdot \text{grad}\, \psi = \psi(x,y,z) - (\mathbf{r} \times \nabla\psi) \cdot \mathbf{d\omega} \tag{1.4.4}$$

for $(\mathbf{A} \times \mathbf{B}) \cdot \mathbf{C} = -(\mathbf{A} \times \mathbf{C}) \cdot \mathbf{B}$ for any vector trio. Therefore, the vector $(\mathbf{r} \times \nabla\psi)$ is a measure of the effect of any sort of infinitesimal rotation on ψ; to measure the effect of a particular rotation $\mathbf{d\omega}$ we take the dot product of the two. If the rotation axis is perpendicular to $(\mathbf{r} \times \nabla\psi)$, then a small rotation does not change ψ; if the axis is parallel to $(\mathbf{r} \times \nabla\psi)$, the effect of the rotation is the largest possible. Since any $f(r)$ is invariant against a rotation, then $\{\mathbf{r} \times \nabla[f(r)g(\vartheta,\varphi)]\} = f(r) \cdot \cdot [\mathbf{r} \times \nabla g(\vartheta,\varphi)]$.

The Divergence. There are two useful differential operations which can act on a vector field. One results in a scalar indicating the rate of increase of lines of flow, and the other in a vector indicating the rate of twisting of the lines of flow. One can be obtained from the limiting behavior of the net outflow integral for a vanishingly small enclosed

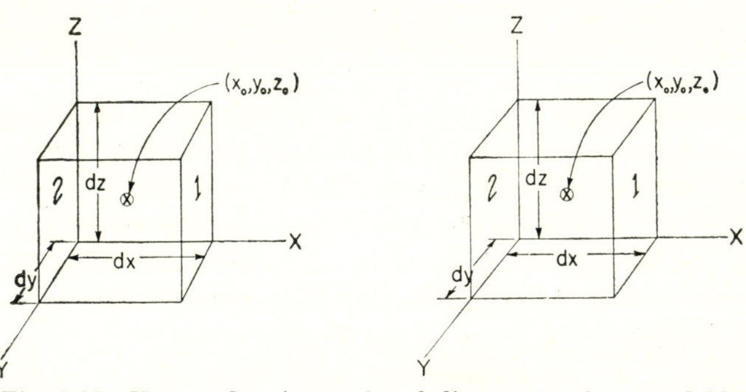

Fig. 1.13 Net outflow integral and divergence of vector field.

volume, and the other can be obtained by a similar limiting process on the net circulation integral.

To obtain the scalar we first compute the net outflow integral of a vector field $\mathbf{F}$ over the volume element $dx\, dy\, dz$ at the point (x_0,y_0,z_0). By an extension of Taylor's theorem, the x component of $\mathbf{F}$ near (x_0,y_0,z_0) is

$$F_x = F_x(x_0,y_0,z_0) + (\partial F_x/\partial x)(x - x_0) + (\partial F_x/\partial y)(y - y_0) + (\partial F_x/\partial z)(z - z_0) + \cdots$$

If the point (x_0,y_0,z_0) is taken at the center of the volume element, then the surface integral of the normal component of F over surface 1, shown in Fig. 1.13, is

$$\iint_1 F_x\,dA = dy\,dz\,[F_x(x_0,y_0,z_0) + \tfrac{1}{2}(\partial F_x/\partial x)\,dx]$$
$$+ \text{ higher orders of differentials}$$

The surface integral over surface 2 is

$$\iint_2 F_x\,dA = -dy\,dz\,[F_x(x_0,y_0,z_0) - \tfrac{1}{2}(\partial F_x/\partial x)\,dx] + \text{higher orders}$$

the negative sign outside the bracket coming in because the outward pointing component of **F** is being integrated, and the outward component of **F** for face 2 is $-F_x$. The sum of the surface integrals over these two faces is, therefore, simply $(\partial F_x/\partial x)\,dx\,dy\,dz$, to the order of approximation considered here. The contributions due to the other faces depend on F_y and F_z and can be computed in a similar manner. The net outflow integral for the volume element is therefore

$$\oint \mathbf{F}\cdot d\mathbf{A} = \left(\frac{\partial F_x}{\partial x} + \frac{\partial F_y}{\partial y} + \frac{\partial F_z}{\partial z}\right) dx\,dy\,dz$$

This is the three-dimensional vector analogue of the differential du of the scalar function $u(x)$ of one variable. The analogue of the derivative is the net outflow integral per unit volume at the point (x,y,z). This is a scalar quantity called the *divergence* of the vector **F** at the point (x,y,z) and is written

$$\text{div}\,\mathbf{F} = \lim_{\text{vol}\to 0}\left[\frac{\oint \mathbf{F}\cdot d\mathbf{A}}{\text{volume}}\right] = \frac{\partial F_x}{\partial x} + \frac{\partial F_y}{\partial y} + \frac{\partial F_z}{\partial z} = \nabla\cdot\mathbf{F} \qquad (1.4.5)$$

The divergence is equal to the rate of increase of lines of flow per volume.

According to the fundamental definition given in the last paragraph, the divergence of **F** at a point P is a property of the behavior of **F** in the neighborhood of P, and its value should not depend on the coordinate system we have chosen. It is to be expected that the expression for the divergence operator should have a different mathematical *form* for generalized curvilinear coordinates from what it has for rectangular coordinates; nevertheless the numerical value of div **F** at P should be the same for any sort of coordinate system. If the transformation of coordinates involves only a rotation and not a stretching (*i.e.*, if the h's all equal unity), then both form and value should be unchanged by the transformation. This is, after all, what we mean by a scalar invariant.

To find the expression for divergence in the generalized coordinates discussed on page 24, we return to the fundamental definition of div **F** and compute the net outflow integral for the volume element defined by the elementary displacements $h_n\,d\xi_n$ in the new system. The net outflow over face 1 in Fig. 1.14 is $d\xi_2\, d\xi_3\, \{F_1h_2h_3 + \frac{1}{2}[(\partial/\partial\xi_1)(F_1h_2h_3)]\, d\xi_1\}$. We have had to include the factors h_2h_3 inside the derivative in the second term for the following reason: The net outflow integral over face 1 differs from that over face 3, at the center of the element, both because F_1 changes as ξ_1 changes and because, in curvilinear coordinates, the area of face 1 differs from that of face 3; that is, h_2h_3 also depends on ξ_1. Therefore, both factors must be included in the differential term.

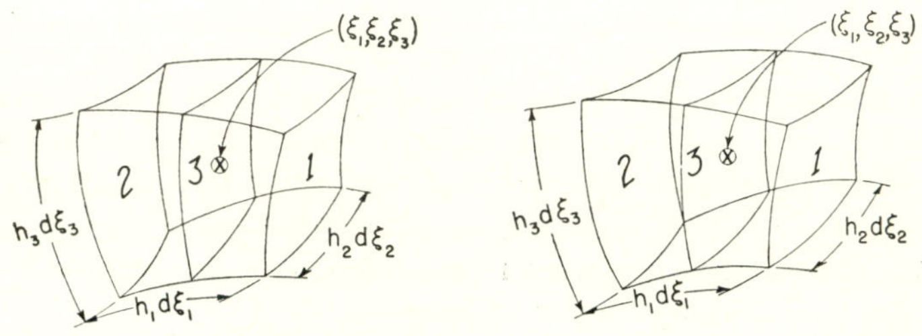

Fig. 1.14 Net outflow and divergence in curvilinear coordinates.

The outflow over face 2 is $d\xi_2\, d\xi_3\, \{-F_1h_2h_3 + [\frac{1}{2}(\partial/\partial\xi_1)(F_1h_2h_3)]\, d\xi_1\}$, and the net outflow over the two faces is, therefore, $d\xi_1\, d\xi_2\, d\xi_3\, (\partial/\partial\xi_1)(F_1h_2h_3)$. The divergence of **F** in generalized coordinates is, therefore,

$$\operatorname{div} \mathbf{F} = \frac{\oint \mathbf{F} \cdot \mathbf{dA}}{dV} = \frac{1}{h_1h_2h_3}\left[\frac{\partial}{\partial\xi_1}(F_1h_2h_3) + \frac{\partial}{\partial\xi_2}(h_1F_2h_3) + \frac{\partial}{\partial\xi_3}(h_1h_2F_3)\right] \tag{1.4.6}$$

where $dV = h_1h_2h_3\, d\xi_1\, d\xi_2\, d\xi_3$ is the volume of the elementary parallelepiped.

This expression has certainly a different mathematical form from the one in Eq. (1.4.5) for rectangular coordinates. In order to show that the two have the same value at a given point, we resort to direct transformation from one to the other, using the equations of transformation given on pages 23 to 25. Starting with the expression in rectangular coordinates, we express F_x, F_y, and F_z in terms of F_1, F_2, and F_3, using Eqs. (1.3.5). After expanding some of the differentials and rearranging terms, we obtain

$$\begin{aligned}\frac{\partial F_x}{\partial x} + \frac{\partial F_y}{\partial y} + \frac{\partial F_z}{\partial z} &= \frac{1}{h_1}\left[\frac{\partial x}{\partial\xi_1}\frac{\partial F_1}{\partial x} + \frac{\partial y}{\partial\xi_1}\frac{\partial F_1}{\partial y} + \frac{\partial z}{\partial\xi_1}\frac{\partial F_1}{\partial z}\right] \\ &+ \frac{1}{h_2}\left[\frac{\partial x}{\partial\xi_2}\frac{\partial F_2}{\partial x} + \frac{\partial y}{\partial\xi_2}\frac{\partial F_2}{\partial y} + \frac{\partial z}{\partial\xi_2}\frac{\partial F_2}{\partial z}\right] + \frac{1}{h_3}[\cdot\ \cdot\ \cdot] \\ &+ F_1\left[\frac{\partial x}{\partial\xi_1}\frac{\partial(1/h_1)}{\partial x} + \frac{\partial y}{\partial\xi_1}\frac{\partial(1/h_1)}{\partial y} + \frac{\partial z}{\partial\xi_1}\frac{\partial(1/h_1)}{\partial z}\right] + F_2[\cdot\ \cdot\ \cdot] + F_3[\cdot\ \cdot\ \cdot] \\ &+ \frac{F_1}{h_1}\left[\frac{\partial}{\partial x}\frac{\partial x}{\partial\xi_1} + \frac{\partial}{\partial y}\frac{\partial y}{\partial\xi_1} + \frac{\partial}{\partial z}\frac{\partial z}{\partial\xi_1}\right] + \frac{F_2}{h_2}[\cdot\ \cdot\ \cdot] + \frac{F_3}{h_3}[\cdot\ \cdot\ \cdot]\end{aligned}$$

The first and fourth brackets can be simplified by using the relation $\frac{\partial x}{\partial \xi_1} \cdot \frac{\partial}{\partial x} + \frac{\partial y}{\partial \xi_1} \cdot \frac{\partial}{\partial y} + \frac{\partial z}{\partial \xi_1} \cdot \frac{\partial}{\partial z} = \frac{\partial}{\partial \xi_1}$, so that they become

$$\frac{1}{h_1}\frac{\partial F_1}{\partial \xi_1} + F_1 \frac{\partial(1/h_1)}{\partial \xi_1}$$

The seventh bracket must be expanded further, by the use of

$$\frac{\partial}{\partial x} = \frac{\partial \xi_1}{\partial x}\frac{\partial}{\partial \xi_1} + \frac{\partial \xi_2}{\partial x}\frac{\partial}{\partial \xi_2} + \frac{\partial \xi_3}{\partial x}\frac{\partial}{\partial \xi_3}$$

and other similar expressions. Using also $\partial \xi_1/\partial x = (1/h_1^2)(\partial x/\partial \xi_1)$, and similar equations obtained from Eqs. (1.3.5) and rearranging terms, this seventh bracket becomes

$$\frac{F_1}{h_1}\left[\frac{1}{h_1^2}\left(\frac{\partial x}{\partial \xi_1}\frac{\partial^2 x}{\partial \xi_1^2} + \frac{\partial y}{\partial \xi_1}\frac{\partial^2 y}{\partial \xi_1^2} + \frac{\partial z}{\partial \xi_1}\frac{\partial^2 z}{\partial \xi_1^2}\right) + \frac{1}{h_2^2}(\cdot\ \cdot\ \cdot) + \frac{1}{h_3^2}(\cdot\ \cdot\ \cdot)\right]$$

But from Eqs. (1.3.4) defining the h's, we see that the first parenthesis in this bracket is $\frac{1}{2}(\partial h_1^2/\partial \xi_1)$, so that the whole bracket becomes

$$\frac{F_1}{2h_1}\left[\frac{1}{h_1^2}\frac{\partial h_1^2}{\partial \xi_1} + \frac{1}{h_2^2}\frac{\partial h_2^2}{\partial \xi_1} + \frac{1}{h_3^2}\frac{\partial h_3^2}{\partial \xi_1}\right] = \frac{F_1}{h_1^2 h_2 h_3}\frac{\partial(h_1 h_2 h_3)}{\partial \xi_1}$$

which is the first term in the expression for div **F** given in Eq. (1.4.6). The other six brackets likewise combine to give the other two terms, showing that div **F** in rectangular coordinates is equal numerically to div **F** at the same point expressed in any other orthogonal system of coordinates. We therefore can call div **F** an *invariant* to transformation of coordinates.

Incidentally, we have shown by this rather roundabout way that the net outflow integral for an infinitesimal surface depends only on the volume enclosed by the surface, and not on the shape of the surface, *i.e.*, not on whether the volume element is for rectangular or for curvilinear coordinates. If we had wished, we could have proved the invariance of div **F** by proving this last statement directly, instead of going through the tedious algebra of the previous page.

Gauss' Theorem. It is possible to combine our knowledge of the additive property of net outflow integrals and our definition of div **F** to obtain a very important and useful method of calculating the net outflow integral for any region of space. Because of the additive property, the net outflow integral for the whole region must equal the sum of the outflow integrals for all the elements of volume included in the region. According to Eq. (1.4.5) the integrals over the element of volume dv can be written div **F** dv, thus giving us the very important divergence theorem

$$\oint \mathbf{F} \cdot d\mathbf{A} = \iiint \operatorname{div} \mathbf{F}\, dv \qquad (1.4.7)$$

where the volume integral is taken over the whole region bound by the surface used for the surface integral. This is called *Gauss' theorem.*

This divergence theorem emphasizes the close relationship there must be between the behavior of a vector field along a closed surface and the field's behavior everywhere in the region inside this surface. It corresponds to the rather obvious property of flow lines; *i.e.*, their net flow out over a closed surface must equal the net number "created" inside the surface.

A Solution of Poisson's Equation. Furthermore, a juxtaposition of Gauss' theorem with the facts we have mentioned earlier concerning vector fields and source points [see Eq. (1.2.9)] enables us to work out a useful solution to Poisson's equation $\nabla^2\varphi = -q(x,y,z)$, [Eq. (1.1.5)], whenever q is a finite function of x, y, z which goes to zero at infinity and whenever the only requirement on the solution φ is that it also vanishes at infinity. The suggestive connection comes when we derive a vector field $\mathbf{F} = \text{grad}\,\varphi$ from the scalar-field solution φ. For $\mathbf{F}$ must satisfy Gauss' theorem so that

$$\oint(\text{grad}\,\varphi)\cdot d\mathbf{A} = \iiint(\nabla^2\varphi)\,dv$$

for any region c bounded by a closed surface S.

Another suggestion comes from the fact that the vector field $(Q/r^2)\mathbf{a}_r$ for a simple source turns out to be the gradient of the potential function $\varphi = -(Q/r)$.

Collecting all these hints we can guess that a solution of Poisson's equation $\nabla^2\varphi = -q(x,y,z)$ is the integral

$$\varphi(x,y,z) = \iiint \frac{q(x',y',z')}{4\pi R}\,dx'\,dy'\,dz' \tag{1.4.8}$$

where $R = \sqrt{(x-x')^2 + (y-y')^2 + (z-z')^2}$ is the distance from the point x, y, z to the point x', y', z'. The quantity φ goes to zero at infinity because q does.

To prove that φ is a solution, we form the vector field

$$\mathbf{F}(x,y,z) = \text{grad}\,\varphi = -\iiint \frac{q(x',y',z')}{4\pi R^2}\,\mathbf{a}_R\,dx'\,dy'\,dz'$$

where $\mathbf{a}_R$ is the unit vector pointed in the direction from point x', y', z' to point x, y, z, that is, along R in the direction from the primed to the unprimed end. Next we form the net outflow integral for F over a closed surface S which bounds some region c of space. Using Eq. (1.4.7), we obtain

$$\oint_S \mathbf{F}\cdot d\mathbf{A} = \iiint_c (\nabla^2\varphi)\,dx\,dy\,dz = -\oint_S dA \iiint \frac{q(x',y',z')}{4\pi R^2}\,\mathbf{a}_R\,dx'\,dy'\,dz'$$

where, of course, $\nabla^2\varphi = \text{div}\,(\text{grad}\,\varphi)$. The last integral is taken over all values of x', y', z' and over those values of x, y, z which are on the surface S. The order of integration may be reversed, and we shall consider first the integration over S. The integrand is $\mathbf{a}_R[q(x',y',z')/4\pi R^2]$ $dx'\,dy'\,dz'$. Referring to Eq. (1.2.9), we see that the net outflow integral over S for this integrand is just $[q(x',y',z')]\,dx'\,dy'\,dz'$ if the point x', y', z' is inside S and is zero if the point is outside S. Therefore the integral over $dx'\,dy'\,dz'$ is equal to the integral of q inside the region c, and the final result is

$$\iiint_c (\nabla^2\varphi)\,dx\,dy\,dz = -\iiint_c q(x',y',z')\,dx'\,dy'\,dz'$$

What has been shown is that the integral of $\nabla^2\varphi$ over any region c equals the integral of $-q$ over the same region, *no matter what shape or size the region has.*

It is not difficult to conclude from this that, if φ is the integral defined in Eq. (1.4.8), then φ is a solution of Poisson's equation $\nabla^2\varphi = -q$, when q is a reasonable sort of function going to zero at infinity. It is not the only solution, for we can add any amount of any solution of Laplace's equation $\nabla^2\psi = 0$ to φ and still have a solution of $\nabla^2\varphi = -q$ for the same q.

What amount and kind of solution ψ we add depends on the boundary conditions of the individual problem. If φ is to go to zero at infinity, no ψ needs to be added, for the φ given in Eq. (1.4.8) already goes to zero at infinity if q does. Of course, we might try to find a solution of Laplace's equation which is zero at infinity and not zero somewhere else, but we should be disappointed in our attempt. This is because no solution of Laplace's equation can have a maximum or minimum (see page 7), and a function which is zero at infinity and has no maximum or minimum anywhere must be zero everywhere. Consequently, the φ of Eq. (1.4.8) is the unique solution if the boundary condition requires vanishing at infinity.

For other boundary conditions the correct solution is φ plus some solution ψ of $\nabla^2\psi = 0$ such that $\varphi + \psi$ satisfy the conditions. This whole question will be discussed in much greater detail in Chap. 7.

The Curl. There now remains to discuss the differential operator which changes a vector into another vector. This operator is a measure of the "vorticity" of a vector field and is related to the net circulation integral discussed on page 18, just as the divergence operator is related to the net outflow integral. To find the vorticity of a vector field at a point P we compute the net circulation integral around an element of area at P and divide the result by the area of the element. However, it soon becomes apparent that the present limiting process is more complicated than that used to define the divergence, for the results obtained depend on the orientation of the element of area.

For instance, if the element is perpendicular to the x axis, the circulation integral for the path shown in Fig. 1.15 is

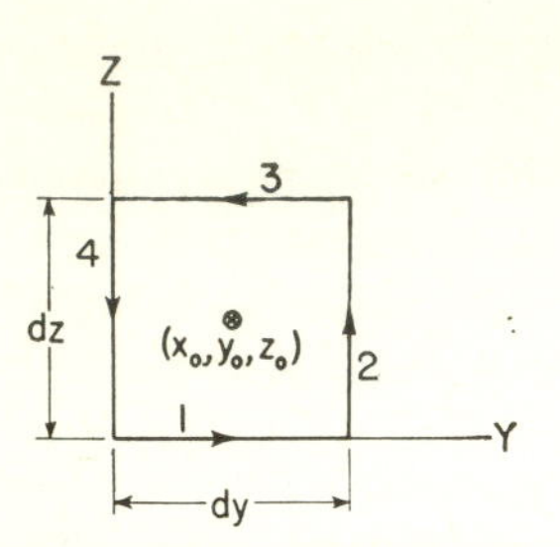

Fig. 1.15 Net circulation integral and curl of vector field.

$$\oint \mathbf{F}\cdot d\mathbf{s} = \int_1 F_y\,dy + \int_2 F_z\,dz - \int_3 F_y\,dy - \int_4 F_z\,dz$$

$$= F_y(x_0,y_0,z_0)\,dy - \frac{\partial F_y}{\partial z}\frac{dz}{2}\,dy$$

$$+ F_z(x_0,y_0,z_0)\,dz + \frac{\partial F_z}{dy}\frac{dy}{2}\,dz$$

$$-F_y(x_0,y_0,z_0)\,dy - \frac{\partial F_y}{\partial z}\frac{dz}{2}\,dy$$

$$- F_z(x_0,y_0,z_0)\,dz + \frac{\partial F_z}{\partial y}\frac{dy}{2}\,dz$$

$$= \left(\frac{\partial F_z}{\partial y} - \frac{\partial F_y}{\partial z}\right) dy\,dz$$

where we use the first-order terms in the Taylor's series expansion for both F_y and F_z.

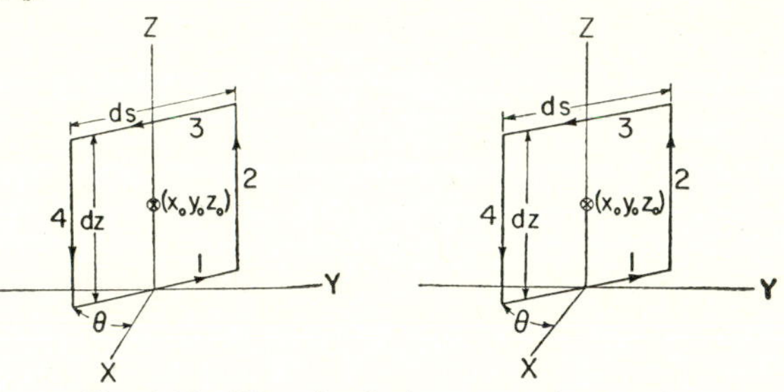

Fig. 1.16 Net circulation at angle to axis.

On the other hand, the circulation integral for the element perpendicular to the y axis is $\left[\left(\frac{\partial F_x}{\partial z}\right) - \left(\frac{\partial F_z}{\partial x}\right)\right] dx\,dz$, and so on. If the element is parallel to the z axis but at an angle θ to the x axis, as shown in Fig. 1.16, the computation is somewhat more complicated. For instance, the displacement ds along path 1 corresponds to a change $-ds \cos\theta$ along x and a change $ds \sin\theta$ along y. Similarly, at the mid-point of path 2, F_z has the value $F_z(x_0,y_0,z_0) - (ds/2)\cos\theta(\partial F_z/\partial x) + (ds/2)\cdot$ $\cdot \sin\theta(\partial F_z/\partial y)$. Taking all these aspects into account, we find for the circulation integral in this case

$$(F_y\,ds\sin\theta - F_x\,ds\cos\theta) - \frac{dz}{2}\left(ds\sin\theta\frac{\partial F_y}{\partial z} - ds\cos\theta\frac{\partial F_x}{\partial z}\right)$$

$$+ F_z\,dz + dz\left(\frac{ds}{2}\sin\theta\frac{\partial F_z}{\partial y} - \frac{ds}{2}\cos\theta\frac{\partial F_z}{\partial x}\right)$$

$$- (F_y\,ds\sin\theta - F_x\,ds\cos\theta) - \frac{dz}{2}\left(ds\sin\theta\frac{\partial F_y}{\partial z} - ds\cos\theta\frac{\partial F_x}{\partial z}\right)$$

$$- F_z\,dz + dz\left(\frac{ds}{2}\sin\theta\frac{\partial F_z}{\partial y} - \frac{ds}{2}\cos\theta\frac{\partial F_z}{\partial x}\right)$$

$$= \left[\left(\frac{\partial F_z}{\partial y} - \frac{\partial F_y}{\partial z}\right)\sin\theta + \left(\frac{\partial F_x}{\partial z} - \frac{\partial F_z}{\partial x}\right)\cos\theta\right] dz\,ds$$

The circulation integral for an element having an arbitrary orientation in space is a still more complicated matter.

However, the complication vanishes if we consider the three quantities

$$\left(\frac{\partial F_z}{\partial y} - \frac{\partial F_y}{\partial z}\right);\quad \left(\frac{\partial F_x}{\partial z} - \frac{\partial F_z}{\partial x}\right);\quad \left(\frac{\partial F_y}{\partial x} - \frac{\partial F_x}{\partial y}\right)$$

to be the x, y, and z components, respectively, of a vector. For it turns out that the circulation integral for any surface element dA is simply the component of this vector in the direction perpendicular to the surface element, multiplied by dA. For instance, the direction cosines for a line perpendicular to the element shown in Fig. 1.16 are $(\cos\theta, \sin\theta, 0)$, and the component in this direction of the vector under discussion, multiplied by $ds\,dz$, is just the result we have found.

The vector so defined is called the *curl* of **F**;

$$\text{curl}\,\mathbf{F} = \mathbf{i}\left(\frac{\partial F_z}{\partial y} - \frac{\partial F_y}{\partial z}\right) + \mathbf{j}\left(\frac{\partial F_x}{\partial z} - \frac{\partial F_z}{\partial x}\right) + \mathbf{k}\left(\frac{\partial F_y}{\partial x} - \frac{\partial F_x}{\partial y}\right) = \nabla \times \mathbf{F} \tag{1.4.9}$$

The net circulation integral around a surface element dA is, therefore $\mathbf{dA}\cdot\text{curl}\,\mathbf{F}$, where $\mathbf{dA}$ is the axial vector corresponding to the surface element. The vector curl **F** is a measure of the "vorticity" of the field at the point (x,y,z). When **F** represents a fluid velocity, the direction of curl **F** at point P is along the axis of rotation of the fluid close to P (pointed according to the right-hand rule) and the magnitude of curl **F** is equal to twice the angular velocity of this portion of the fluid.

The curl is an operator analogous to the vector product, just as the divergence is analogous to the scalar product. Note that curl **F** is an axial vector if **F** is a true vector, for the net circulation integral is a scalar and **dA** is an axial vector, so that curl **F** must be axial also.

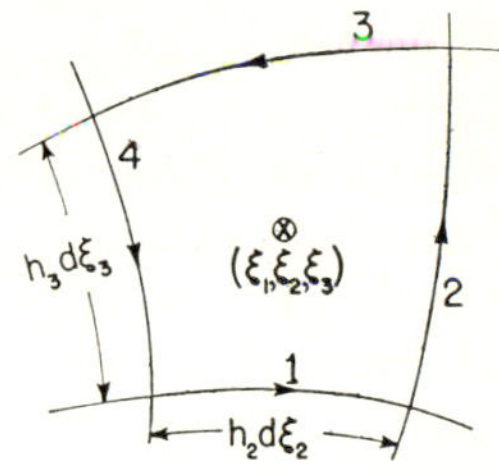

Fig. 1.17 Net circulation and curl in curvilinear coordinates.

To complete our discussion, we must now show that curl **F** acts like a vector, *i.e.*, transforms like one. In generalized coordinates the element of area perpendicular to the ξ_1 axis is shown in Fig. 1.17. By arguments similar to those used in obtaining Eq. (1.4.7), we see that the net circulation integral for the path is

$$\oint \mathbf{F}\cdot\mathbf{ds} = \left[h_2F_2 - \tfrac{1}{2}\,d\xi_3\frac{\partial}{\partial\xi_3}(h_2F_2)\right]d\xi_2 + \left[h_3F_3 + \tfrac{1}{2}\,d\xi_2\frac{\partial}{\partial\xi_2}(h_3F_3)\right]d\xi_3$$
$$- \left[h_2F_2 + \tfrac{1}{2}\,d\xi_3\frac{\partial}{\partial\xi_3}(h_2F_2)\right]d\xi_2 - \left[h_3F_3 - \tfrac{1}{2}\,d\xi_2\frac{\partial}{\partial\xi_2}(h_3F_3)\right]d\xi_3$$

Reducing this and dividing by the area $h_2h_3\,d\xi_2\,d\xi_3$ give us the ξ_1 component of the curl. Further calculation shows that the expression for the curl in generalized coordinates is

$$\operatorname{curl}\mathbf{F} = \frac{\mathbf{a}_1}{h_2h_3}\left[\frac{\partial(h_3F_3)}{\partial\xi_2} - \frac{\partial(h_2F_2)}{\partial\xi_3}\right] + \frac{\mathbf{a}_2}{h_1h_3}\left[\frac{\partial(h_1F_1)}{\partial\xi_3} - \frac{\partial(h_3F_3)}{\partial\xi_1}\right] + \frac{\mathbf{a}_3}{h_1h_2}\left[\frac{\partial(h_2F_2)}{\partial\xi_1} - \frac{\partial(h_1F_1)}{\partial\xi_2}\right] \qquad (1.4.10)$$

To show that this is the same vector as that given in Eq. (1.4.9) we consider the component of curl $\mathbf{F} = \mathbf{C}$ along ξ_1. By Eqs. (1.3.3) this equals

$$C_1 = \alpha_1C_x + \beta_1C_y + \gamma_1C_z$$
$$= h_1\left[\frac{\partial\xi_1}{\partial x}\left(\frac{\partial F_z}{\partial y} - \frac{\partial F_y}{\partial z}\right) + \frac{\partial\xi_1}{\partial y}\left(\frac{\partial F_x}{\partial z} - \frac{\partial F_z}{\partial x}\right) + \frac{\partial\xi_1}{\partial z}\left(\frac{\partial F_y}{\partial x} - \frac{\partial F_x}{\partial y}\right)\right]$$

But from Eq. (1.4.10), by the use of Eqs. (1.3.2) and (1.3.5), we obtain

$$C_1 = \frac{1}{h_2h_3}\left[\frac{\partial x}{\partial\xi_3}\frac{\partial F_x}{\partial\xi_2} - \frac{\partial x}{\partial\xi_2}\frac{\partial F_x}{\partial\xi_3} + \frac{\partial y}{\partial\xi_3}\frac{\partial F_y}{\partial\xi_2} - \frac{\partial y}{\partial\xi_2}\frac{\partial F_y}{\partial\xi_3} + \frac{\partial z}{\partial\xi_3}\frac{\partial F_z}{\partial\xi_2} - \frac{\partial z}{\partial\xi_2}\frac{\partial F_z}{\partial\xi_3}\right]$$
$$= \frac{1}{2h_2h_3}\left[\left(\frac{\partial y}{\partial\xi_3}\frac{\partial z}{\partial\xi_2} - \frac{\partial y}{\partial\xi_2}\frac{\partial z}{\partial\xi_3}\right)\left(\frac{\partial F_y}{\partial z} - \frac{\partial F_z}{\partial y}\right) + \left(\frac{\partial x}{\partial\xi_3}\frac{\partial z}{\partial\xi_2} - \frac{\partial x}{\partial\xi_2}\frac{\partial z}{\partial\xi_3}\right)\left(\frac{\partial F_x}{\partial z} - \frac{\partial F_z}{\partial x}\right) + \cdots\right]$$
$$= h_1\left[\frac{\partial\xi_1}{\partial x}\left(\frac{\partial F_z}{\partial y} - \frac{\partial F_y}{\partial z}\right) + \frac{\partial\xi_1}{\partial y}\left(\frac{\partial F_x}{\partial z} - \frac{\partial F_z}{\partial x}\right) + \frac{\partial\xi_1}{\partial z}\left(\frac{\partial F_y}{\partial x} - \frac{\partial F_x}{\partial y}\right)\right]$$

which is identical with the previous equation and shows that the expressions given in Eqs. (1.4.9) and (1.4.10) are the same vector. To show that curl $\mathbf{F}$ is an axial vector when $\mathbf{F}$ is a true vector, note that we were arbitrary in our choice of the direction of the line integral around the element of area. If we had reversed the direction, we should have changed the sign of the curl.

Vorticity Lines. The vector curl $\mathbf{F}$ defines another vector field, with new lines of flow. These lines might be called the *vorticity lines* for the field $\mathbf{F}$. For instance, for the field $F_x = -ay$, $F_y = ax$, $F_z = 0$, the vector curl $\mathbf{F}$ is directed along the positive z axis and has magnitude $2a$ everywhere. The flow lines of $\mathbf{F}$ are circles in planes perpendicular to the z axis. For the field $F_x = -ay$, $F_y = ax$, $F_z = b(x^2 + y^2)$, discussed on page 12, the vector curl $\mathbf{F}$ has components of $2by$, $-2bx$, $2a$. We have shown that the helical flow lines of this case are defined by the families of surfaces $\varphi = \sqrt{x^2 + y^2}$, $\vartheta = z - (b/a)(x^2 + y^2)\tan^{-1}(y/x)$.

By methods discussed on page 12, we find that the lines of vorticity are defined by the surfaces $\varphi = \sqrt{x^2 + y^2}$, $\psi = z + (a/b) \tan^{-1}(y/x)$.

In both of these examples the vorticity lines are everywhere perpendicular to the lines of flow. This is not always true, however. For instance, for the field $F_x = az$, $F_y = ax$, $F_z = ay$, the vector curl $\mathbf{F} = a\mathbf{i} + a\mathbf{j} + a\mathbf{k}$ is not perpendicular to $\mathbf{F}$.

One very interesting property is exhibited by vorticity lines for any field: *They never begin or end.* This is in line with the argument on page 18 about vortex tubes and can easily be shown from the properties of the curl operator. To say that a flow line never begins or ends is to say that the divergence of the corresponding vector is everywhere zero, for then by the divergence theorem, Eq. (1.4.8), every net outflow integral is zero. However, the divergence of every curl is zero, by its very definition, for

$$\operatorname{div}(\operatorname{curl}\mathbf{F}) = \frac{\partial^2 F_z}{\partial x\,\partial y} - \frac{\partial^2 F_y}{\partial x\,\partial z} + \frac{\partial^2 F_x}{\partial y\,\partial z} - \frac{\partial^2 F_z}{\partial y\,\partial x} + \frac{\partial^2 F_y}{\partial z\,\partial x} - \frac{\partial^2 F_x}{\partial z\,\partial y} = 0$$

In fact the easiest way of obtaining a divergenceless vector field is to use the curl of some other vector field, a dodge which is often useful in electromagnetic theory.

Stokes' Theorem. There is a curl theorem similar to the divergence theorem expressed in Eq. (1.4.7), which can be obtained from the fundamental definition of the curl and from the additive properties of the net circulation integral mentioned on page 21. We consider any surface S, whose boundary is the closed line (or lines) L, divide it into its elements of area dA, and add all the net circulation integrals for the elements. By our definition of the curl, this sum can be written $\int \operatorname{curl}\mathbf{F}\cdot\mathbf{dA}$, where $\mathbf{dA}$ is the vector corresponding to dA and where the integration is over the whole of S. According to the discussion on page 16 this defines the number of vorticity lines which cut surface S. However, because of the additive property of the circulation integral, the integral $\int \operatorname{curl}\mathbf{F}\cdot\mathbf{dA}$ must equal the net circulation integral around the boundary line (or lines) L:

$$\int \operatorname{curl}\mathbf{F}\cdot\mathbf{dA} = \oint \mathbf{F}\cdot\mathbf{ds} \tag{1.4.11}$$

This curl theorem is called *Stokes' theorem;* it enables one to compute the net circulation integral for any path. It is one other relation between the properties of a vector field at the boundary of a region and the behavior of the field in the region inside the boundary, this one corresponding to the requirement that the net circulation integral around any closed path must equal the number of vorticity lines enclosed by the path.

The Vector Operator ∇. Just as the divergence operator is an analogue of the dot product of vectors, so the curl operator is an analogue of the cross product. To make the analogy more complete, we can define a *vector operator*, called *del* and written ∇, with components given by the equation

$$\nabla = \mathbf{i}\frac{\partial}{\partial x} + \mathbf{j}\frac{\partial}{\partial y} + \mathbf{k}\frac{\partial}{\partial z} \tag{1.4.12}$$

In terms of this operator the three differential operators discussed in this section can be written symbolically:

$$\text{grad}\,\psi = \nabla\psi; \quad \text{div}\,\mathbf{F} = \nabla\cdot\mathbf{F}; \quad \text{curl}\,\mathbf{F} = \nabla\times\mathbf{F}$$

Some of the formulas involving the vector operator ∇ acting on a product of two quantities can be simplified. The formulas

$$\begin{aligned}
\text{grad}\,(\psi\Phi) &= \psi\,\text{grad}\,\Phi + \Phi\,\text{grad}\,\psi \\
\text{div}\,(a\mathbf{F}) &= a\,\text{div}\,\mathbf{F} + \mathbf{F}\cdot\text{grad}\,a \\
\text{div}\,(\mathbf{A}\times\mathbf{B}) &= \mathbf{B}\cdot\text{curl}\,\mathbf{A} - \mathbf{A}\cdot\text{curl}\,\mathbf{B} \\
\text{curl}\,(a\mathbf{B}) &= a\,\text{curl}\,\mathbf{B} + (\text{grad}\,a)\times\mathbf{B} \\
\text{curl}\,(\mathbf{A}\times\mathbf{B}) &= \mathbf{A}\,\text{div}\,\mathbf{B} - \mathbf{B}\,\text{div}\,\mathbf{A} + (\mathbf{B}\cdot\text{grad})\mathbf{A} - (\mathbf{A}\cdot\text{grad})\mathbf{B}
\end{aligned} \tag{1.4.13}$$

can all be obtained directly from the definitions of grad, div, and curl.

1.5 *Vector and Tensor Formalism*

The analogy between ∇ and a vector is only a symbolic one, however, for we cannot give the operator del a magnitude or direction and say that ∇ is perpendicular to $\mathbf{F}$ if div $\mathbf{F} = 0$ or that curl $\mathbf{F}$ is perpendicular to ∇ or even necessarily perpendicular to $\mathbf{F}$, as we could if ∇ were an actual vector. In fact the analogy is weaker still when we try to express ∇ in generalized coordinates, for the vector operator must have different forms for its different uses:

$$\nabla = \mathbf{a}_1\frac{1}{h_1}\frac{\partial}{\partial\xi_1} + \mathbf{a}_2\frac{1}{h_2}\frac{\partial}{\partial\xi_2} + \mathbf{a}_3\frac{1}{h_3}\frac{\partial}{\partial\xi_3}; \qquad \text{for the gradient}$$

$$= \frac{1}{h_1h_2h_3}\left[\mathbf{a}_1\frac{\partial}{\partial\xi_1}(h_2h_3) + \mathbf{a}_2\frac{\partial}{\partial\xi_2}(h_1h_3) + \mathbf{a}_3\frac{\partial}{\partial\xi_3}(h_1h_2)\right]; \qquad \text{for the divergence}$$

and no form which can be written down for the curl. In order to understand how these operators transform and to be able to set up more complex forms easily, we must delve into the formalism of tensor calculus.

Covariant and Contravariant Vectors. Tensor calculus is a formalism developed to handle efficiently problems in differential geometry, which has turned out to be decidedly useful in the study of general relativity. We shall touch on the subject briefly in this book, covering only enough to clarify the methods of calculating differential vector operators in curvilinear coordinates. We still consider only orthogonal coordinates in three dimensions, though tensor calculus in its full power can handle nonorthogonal coordinates of any number of dimensions.

In Eqs. (1.3.12) and (1.3.13) we defined the components of contravariant and covariant vectors and their law of transformation. If F_n are the components of an ordinary vector in a three-dimensional orthogonal system of coordinates with scale factors h_n, then the quantities $f_n = h_n F_n$ are said to be the *covariant* components of a vector in the same coordinate system and the quantities $f^n = F_n/h_n$ are called the contravariant components of a vector in the same system. Therefore, if f_n are the components of a covariant vector, then $f^n = f_n/h_n^2$ are the corresponding components of the contravariant vector in the same coordinate system.

As we have shown in Eqs. (1.3.12) and (1.3.13) the rules for transformation of these vectors from the coordinate system ξ_n to the system ξ'_n are

$$(f_m)' = \sum_n f_n \frac{\partial \xi_n}{\partial \xi'_m}; \quad (f^m)' = \sum_n f^n \frac{\partial \xi'_m}{\partial \xi_n} \tag{1.5.1}$$

As we have pointed out earlier, these transformations have the mathematical advantage of formal symmetry, but the new "vectors" have several disadvantages for the physicist. For one thing, the different components do not have the same dimensionality; if an ordinary vector has the dimensions of length, then the components of the corresponding contravariant vector have dimensions equal to the dimensions of the associated coordinate and the dimensions of the covariant components have still other dimensions.

The indices labeling the different components are written as superscripts for contravariant vectors in order to distinguish them from the covariant components. There is, of course, a chance of confusion between the components of real vectors and those of covariant vectors, for we use subscripts for both. The difficulty will not be great in the present book, for we shall not discuss covariant vectors very much and will mention it specifically whenever we do discuss them. A component F_n without such specific mention can be assumed to be for a real vector.

The quantities $\partial\psi/\partial\xi_1$, $\partial\psi/\partial\xi_2$, $\partial\psi/\partial\xi_3$ are the components of a covariant vector; they must be divided by h_1, h_2, h_3, respectively, to be components of a real vector, the gradient of ψ.

The quantities b_{ij}, b^{ij}, b^i_j are called the components of a *covariant* or *contravariant* or *mixed tensor*, respectively, of the second order in the coordinate system ξ_1 if they transform according to the formulas

$$b'_{ij} = \sum_{m,n} b_{mn} \frac{\partial \xi_m}{\partial \xi'_i} \frac{\partial \xi_n}{\partial \xi'_j}; \quad b'^{ij} = \sum_{m,n} b^{mn} \frac{\partial \xi'_i}{\partial \xi_m} \frac{\partial \xi'_j}{\partial \xi_n}; \quad b'^{i}_{j} = \sum_{m,n} b^m_n \frac{\partial \xi'_i}{\partial \xi_m} \frac{\partial \xi_n}{\partial \xi'_j} \tag{1.5.2}$$

The products of the components of two covariant vectors, taken in pairs, form the components of a covariant tensor. If the vectors are contravariant, the resulting tensor is contravariant. If A_i and B_j are the components of two ordinary vectors, then $(h_i/h_j)A_iB_j = c^j_i$ are the components of a mixed tensor.

The quantity $\sum_m b^m_m$, called a *contracted tensor*, formed from a mixed tensor, does not change its value when transformed, since

$$\sum_n b'^n_n = \sum_{m,k,n} b^m_k \frac{\partial \xi'_n}{\partial \xi_m} \frac{\partial \xi_k}{\partial \xi'_n} = \sum_{k,m} b^m_k \frac{\partial \xi_k}{\partial \xi_m} = \sum_m b^m_m$$

Such a quantity we have called a *scalar;* it is often called an *invariant.* The scalar product of two vectors is a contracted tensor, $\sum_n A_n B_n = \sum_n (h_n/h_n)A_nB_n$, and is, therefore, an invariant.

Axial Vectors. To discuss the properties of the cross product we must take into account the orthogonality and right-handedness of our generalized coordinates. An extension of the formulas derived on page 23 from the equations $\mathbf{a}_1 \times \mathbf{a}_2 = \mathbf{a}_3$, etc., analogous to Eqs. (1.2.5), shows that, for a transformation from one orthogonal system to another,

$$\left(\frac{\partial \xi_j}{\partial \xi'_\mu} \frac{\partial \xi_k}{\partial \xi'_\nu} - \frac{\partial \xi_k}{\partial \xi'_\mu} \frac{\partial \xi_j}{\partial \xi'_\nu}\right) = \frac{h'_1h'_2h'_3}{h_1h_2h_3} \frac{\partial \xi'_\lambda}{\partial \xi_i}$$

where if both systems are right-handed, the trios (i,j,k) and (λ,μ,ν) must be one or another of the trios (1,2,3), (2,3,1), or (3,1,2). By the use of this formula, we find that, for any tensor f_{ij}, the quantities

$$c^i = \frac{1}{h_1h_2h_3}(f_{jk} - f_{kj}); \quad i, j, k = 1, 2, 3 \text{ or } 2, 3, 1 \text{ or } 3, 1, 2$$

are the components of a contravariant vector, for by the equations for tensor transformation

$$\frac{f'_{12} - f'_{21}}{h'_1h'_2h'_3} = \sum_{mn} \frac{f_{mn}}{h'_1h'_2h'_3} \left[\frac{\partial \xi_m}{\partial \xi'_1} \frac{\partial \xi_n}{\partial \xi'_2} - \frac{\partial \xi_m}{\partial \xi'_2} \frac{\partial \xi_n}{\partial \xi'_1}\right] = \sum_n c^n \left(\frac{\partial \xi'_3}{\partial \xi_n}\right) = (c^3)'; \quad \text{etc.}$$

Similarly, $h_1'h_2'h_3'(f^{jk} - f^{kj}) = c_i$ are components of a covariant vector for orthogonal coordinates. We note that these vectors are axial vectors, as is shown by the arbitrary way we have to choose the order of the subscripts (1,2,3), etc. We also should mention that these definitions hold only in three dimensions.

Therefore, if A_m, B_n, C_k are components of ordinary vectors and a_m, b_n, c_k are components of the corresponding covariant vectors, the ith component of the cross product of A and B,

$$C_i = A_jB_k - A_kB_j = \frac{1}{h_kh_j}(a_jb_k - a_kb_j) = h_ic^i; \qquad i, j, k = 1, 2, 3 \text{ or } 2, 3, 1 \text{ or } 3, 1, 2 \tag{1.5.3}$$

is an ordinary vector. We note again that this holds only for three dimensions.

Christoffel Symbols. In order to discuss the properties of the divergence and curl we must define a set of useful quantities, called the *Christoffel symbols*, and discuss their properties. These symbols are defined as follows:

$$\left\{ {i \atop i\,i} \right\} = \frac{1}{h_i}\frac{\partial h_i}{\partial \xi_i}; \quad \left\{ {i \atop i\,j} \right\} = \left\{ {i \atop j\,i} \right\} = \frac{1}{h_i}\frac{\partial h_i}{\partial \xi_j}; \quad \left\{ {j \atop i\,i} \right\} = -\frac{h_i}{(h_j)^2}\frac{\partial h_i}{\partial \xi_j} \tag{1.5.4}$$
$$\left\{ {i \atop j\,k} \right\} = 0 \text{ for } i, j, k \text{ all different}$$

for orthogonal coordinates. These symbols are measures of the curvature of the coordinate axes. Referring to Eqs. (1.3.6), we see that the change in direction of the unit vectors $\mathbf{a}_i$ can be expressed in terms of the Christoffel symbols by the following simple expressions:

$$\frac{\partial}{\partial \xi_j}(h_i\mathbf{a}_i) = \sum_n h_n\mathbf{a}_n \left\{ {n \atop i\,j} \right\}; \quad \frac{\partial}{\partial \xi_j}\left(\frac{\mathbf{a}_i}{h_i}\right) = -\sum_n \left(\frac{\mathbf{a}_n}{h_n}\right)\left\{ {i \atop n\,j} \right\} \tag{1.5.5}$$

Now the unit vector $\mathbf{a}_i$ gives the direction of the ξ_i axis at the point P, (ξ_1,ξ_2,ξ_3); and h_i gives the scale of this coordinate, that is, h_1 equals the actual distance between the points (ξ_1,ξ_2,ξ_3) and $(\xi_1 + d\xi_1,\xi_2,\xi_3)$, divided by the change of coordinate $d\xi_1$. Therefore, the vector $h_i\mathbf{a}_i$ gives both direction and scale of the coordinate ξ_i at point P. The rate of change of this vector with respect to change in coordinate ξ_j is also a vector, the ith component giving the change of scale and the components perpendicular to $\mathbf{a}_i$ giving the change in direction as ξ_j changes. The nth component of this rate of change vector is h_n times the Christoffel symbol $\left\{ {n \atop i\,j} \right\}$.

We note that these symbols are symmetric in the two lower indices (i and j in the symbol just above). In other words, the vector representing the change of $h_i\mathbf{a}_i$ with respect to ξ_j is equal in magnitude and direction to the vector representing the change of $h_j\mathbf{a}_j$ with respect to ξ_i. This corresponds to the fact that, if coordinate ξ_i changes in scale as ξ_j is varied, then ξ_j will change in direction as ξ_i is varied, and vice versa. A study of Fig. 1.11 will show that this is true.

The Christoffel symbols are *not* tensors. It can be shown that their rule of transformation is

$$\sum_{m,n} \begin{Bmatrix} i \\ m\ n \end{Bmatrix}' \frac{\partial \xi'_m}{\partial \xi_k} \frac{\partial \xi'_n}{\partial \xi_s} = - \frac{\partial^2 \xi'_i}{\partial \xi_k \, \partial \xi_s} + \sum_n \begin{Bmatrix} n \\ k\ s \end{Bmatrix} \frac{\partial \xi'_i}{\partial \xi_n} \tag{1.5.6}$$

where a prime indicates quantities in terms of the new coordinates ξ'_i. Although Christoffel symbols are not tensors, they can be of great use in the formation of derivatives of vectors which have tensor form. The ordinary derivative $\partial f_i/\partial \xi_j$ is not a tensor, primarily because the coordinates are curvilinear, and the change in direction of the coordinate axes affects the vector components, adding a spurious variation to the real variation due to change in the vector itself. In other words, the components of the derivative of a vector are not just the derivative of the vector's components in curvilinear coordinates. To find the correct expression for the components of the derivative, we differentiate the whole vector at once and take components after differentiation.

Covariant Derivative. For instance, if f^i are the components of a contravariant vector, then the corresponding real vector is $\mathbf{F} = \sum_n \mathbf{a}_n h_n f^n$. The derivative of this vector with respect to ξ_i can be reduced to the form

$$\frac{\partial \mathbf{F}}{\partial \xi_j} = \sum_n \mathbf{a}_n h_n \frac{\partial f^n}{\partial \xi_j} + \sum_m f^m \frac{\partial}{\partial \xi_j} (\mathbf{a}_m h_m) = \sum_n \mathbf{a}_n h_n \left[\frac{\partial f^n}{\partial \xi_j} + \sum_m f^m \begin{Bmatrix} n \\ m\ j \end{Bmatrix} \right]$$

Therefore, the components of the contravariant vector which corresponds to the rate of change of the ordinary vector $\mathbf{F}$ with respect to ξ_j are the quantities

$$f^i{}_{,j} = \frac{\partial f^i}{\partial \xi_j} + \sum_m f^m \begin{Bmatrix} i \\ m\ j \end{Bmatrix} \tag{1.5.7}$$

where f^i are the components of the contravariant vector corresponding to $\mathbf{F}$. This derivative component has been corrected for the spurious effects of coordinate curvature, and the vector components correspond to the actual change in the vector with ξ_j. The comma before the subscript indicates the differentiation.

As a matter of fact, $f^i{}_{,j}$ are the components of a tensor, being covariant with respect to index j as well as contravariant with respect to index i. This can be shown by utilizing Eqs. (1.5.1) and (1.5.6):

$$
\begin{aligned}
(f^i{}_{,j})' &= \frac{\partial}{\partial \xi_j'} \sum_k f^k \left(\frac{\partial \xi_i'}{\partial \xi_k}\right) + \sum_{mk} f^k \frac{\partial \xi_m'}{\partial \xi_k} \begin{Bmatrix} i \\ m\, j \end{Bmatrix}' \\
&= \sum_{ks} \frac{\partial \xi_s}{\partial \xi_j'} \frac{\partial}{\partial \xi_s} \left(f^k \frac{\partial \xi_i'}{\partial \xi_k}\right) + \sum_{knms} f^k \frac{\partial \xi_s}{\partial \xi_j'} \frac{\partial \xi_n'}{\partial \xi_s} \frac{\partial \xi_m'}{\partial \xi_k} \begin{Bmatrix} i \\ m\, n \end{Bmatrix}' \\
&= \sum_{ns} \frac{\partial f^n}{\partial \xi_s} \frac{\partial \xi_i'}{\partial \xi_n} \frac{\partial \xi_s}{\partial \xi_j'} + \sum_{ks} f^k \frac{\partial \xi_s}{\partial \xi_j'} \left[\frac{\partial^2 \xi_i'}{\partial \xi_s\, \partial \xi_k} + \sum_{mn} \frac{\partial \xi_n'}{\partial \xi_s} \frac{\partial \xi_m'}{\partial \xi_k} \begin{Bmatrix} i \\ m\, n \end{Bmatrix}' \right] \\
&= \sum_{ns} \left[\frac{\partial f^n}{\partial \xi_s} + \sum_k f^k \begin{Bmatrix} n \\ k\, s \end{Bmatrix} \right] \left(\frac{\partial \xi_i'}{\partial \xi_n} \frac{\partial \xi_s}{\partial \xi_j'} \right) = \sum_{ns} f^n{}_{,s} \frac{\partial \xi_i'}{\partial \xi_n} \frac{\partial \xi_s}{\partial \xi_j'}
\end{aligned}
$$

Therefore, $f^i{}_{,j}$ are the components of a mixed tensor of the second order. The tensor is called the *covariant derivative* of the contravariant vector whose components are f^i.

Similarly, if f_i are the components of a covariant vector, the quantity $\mathbf{F} = \sum_n (\mathbf{a}_n/h_n) f_n$ is an ordinary vector, and

$$
\frac{\partial \mathbf{F}}{\partial \xi_j} = \sum_n \left(\frac{\mathbf{a}_n}{h_n}\right) \left[\frac{\partial f_n}{\partial \xi_j} - \sum_m f_m \begin{Bmatrix} m \\ n\, j \end{Bmatrix} \right]
$$

is also an ordinary vector. Therefore, the quantities

$$
f_{i,j} = \frac{\partial f_i}{\partial \xi_j} - \sum_m f_m \begin{Bmatrix} m \\ i\, j \end{Bmatrix} \tag{1.5.8}
$$

are components of the covariant tensor corresponding to the rate of change of $\mathbf{F}$ with respect to ξ_j. These quantities are the components of a covariant tensor of the second order which is called the covariant derivative of the covariant vector f_i.

The definition of covariant differentiation can be extended to tensors also:

$$
\begin{aligned}
f_{ij,k} &= \left(\frac{\partial f_{ij}}{\partial \xi_k}\right) - \sum_m f_{im} \begin{Bmatrix} m \\ j\, k \end{Bmatrix} - \sum_n f_{nj} \begin{Bmatrix} n \\ i\, k \end{Bmatrix} \\
f^{ij}{}_{,k} &= \left(\frac{\partial f^{ij}}{\partial \xi_k}\right) + \sum_m f^{im} \begin{Bmatrix} j \\ m\, k \end{Bmatrix} + \sum_n f^{nj} \begin{Bmatrix} i \\ n\, k \end{Bmatrix} \\
f^i{}_{j,k} &= \left(\frac{\partial f^i{}_j}{\partial \xi_k}\right) - \sum_m f^i{}_m \begin{Bmatrix} m \\ j\, k \end{Bmatrix} + \sum_n f^n{}_j \begin{Bmatrix} i \\ n\, k \end{Bmatrix}
\end{aligned} \tag{1.5.9}
$$

and so on. These quantities are the components of tensors of the *third order*, which transform according to obvious generalizations of Eqs.

(1.5.1). From these formulas it can be seen that covariant differentiation has the same rules of operation as ordinary differentiation. For instance, for the differential of a product, $(a_i b_j)_{,k} = a_i b_{j,k} + a_{i,k} b_j$, and so on.

Tensor Notation for Divergence and Curl. With these definitions made we can now express the differential operators div and curl in a symmetric form. The contracted tensor

$$\sum_n f^n{}_{,n} = \sum_n \frac{\partial f^n}{\partial \xi_n} + \sum_{n,m} f^n \begin{Bmatrix} m \\ n\ m \end{Bmatrix}$$

$$= \sum_n \frac{\partial f^n}{\partial \xi_n} + \sum_{m,n} f^n \frac{\partial}{\partial \xi_n} \ln h_m = \sum_n \frac{1}{h_1 h_2 h_3} \frac{\partial}{\partial \xi_n} (f^n h_1 h_2 h_3)$$

is a scalar invariant, according to the discussion above. If now $f^n = F_n/h_n$, where F_n are the components of an ordinary vector, then the contracted tensor is the *divergence* of $\mathbf{F}$:

$$\sum_n \left(\frac{F_n}{h_n}\right)_{,n} = \frac{1}{h_1 h_2 h_3} \sum_n \frac{\partial}{\partial \xi_n} \left(\frac{h_1 h_2 h_3}{h_n} F_n\right) = \operatorname{div} \mathbf{F} \qquad (1.5.10)$$

From this, the invariance of the divergence follows directly, from the general tensor rules.

Similarly, we have shown above that for orthogonal coordinates the quantities

$$c^i = -\frac{1}{h_1 h_2 h_3} (f_{j,k} - f_{k,j}); \quad ijk = 123,\ 231,\ 312$$

are components of a contravariant vector. If now $f_n = h_n F_n$, where F_n are the components of an ordinary vector, then the quantities $h_k c^k$ are components of an ordinary vector. Choosing one component and using the definitions of the Christoffel symbols, we see that

$$h_1 c^1 = \frac{1}{h_2 h_3} \left\{ \frac{\partial}{\partial \xi_2} (F_3 h_3) - (F_3 h_3) \frac{\partial}{\partial \xi_2} (\ln h_3) - (F_2 h_2) \frac{\partial}{\partial \xi_3} (\ln h_2) \right.$$
$$\left. - \frac{\partial}{\partial \xi_3} (F_2 h_2) + (F_2 h_2) \frac{\partial}{\partial \xi_3} (\ln h_2) + (F_3 h_3) \frac{\partial}{\partial \xi_2} (\ln h_3) \right\}$$

is, according to Eq. (1.4.10), the ξ_1 component of curl $\mathbf{F}$, an ordinary vector.

Other Differential Operators. Once we have become familiar with the technique of tensor calculus and with the definitions for covariant differentiation, we can set up the correct form for vector or scalar combinations of vectors, scalars, and operators with very few of the tedious complications we have met with earlier in this chapter while setting up such combinations. The formalism of tensor calculus has taken care of the complications once for all.

For instance, we can be sure that the combination $\sum_n b^n a_{i,n}$ is the ith component of a covariant vector, h_i times the corresponding component of the ordinary vector. Substituting $b^n = B_n/h_n$ and $a_i = h_i A_i$, where **A** and **B** are ordinary vectors, and dividing by h_i, we obtain the components of an ordinary vector, whose ξ_1 component is

$$F_1 = \frac{1}{h_1}\sum_n \frac{B_n}{h_n}(h_1A_1)_{,n} = \frac{1}{h_1}\sum_n \frac{B_n}{h_n}\left[\frac{\partial}{\partial \xi_n}(h_1A_1) - \sum_m h_m A_m \begin{Bmatrix} m \\ 1\ n \end{Bmatrix}\right]$$
$$= \left[\sum_n \frac{B_n}{h_n}\frac{\partial}{\partial \xi_n}(A_1) + \frac{A_2}{h_1h_2}\left(B_1\frac{\partial h_1}{\partial \xi_2} - B_2\frac{\partial h_2}{\partial \xi_1}\right)\right.$$
$$\left. + \frac{A_3}{h_1h_3}\left(B_1\frac{\partial h_1}{\partial \xi_3} - B_3\frac{\partial h_3}{\partial \xi_1}\right)\right]$$

and which is the vector (**B** grad)**A**, as reference to Eq. (1.4.3) shows. Thus the formalism of tensor calculus again enables us to obtain the components of a vector operator in any orthogonal system of coordinates.

The shorthand symbolism of tensor calculus enables one to express an equation in the same form in any coordinate system. Once one has set up a tensor equality and has made sure that the subscripts and superscripts on each side of the equality sign match, then one can be sure that the equation will hold in any coordinate system. This corresponds to the general aim of theoretical physics, which seeks to express laws in a form which will be independent of the coordinate system.

The Laplacian defined in Eq. (1.1.4) can also be obtained in its generalized form by means of the tensor formalism

$$\nabla^2\psi = \text{div}\,(\text{grad}\,\psi) = \sum_n \left(\frac{1}{h_n^2}\frac{\partial \psi}{\partial \xi_n}\right)_{,n} = \frac{1}{h_1h_2h_3}\sum_n \frac{\partial}{\partial \xi_n}\left[\frac{h_1h_2h_3}{h_n^2}\frac{\partial \psi}{\partial \xi_n}\right] \tag{1.5.11}$$

As we have mentioned on page 8, the Laplacian of ψ is a measure of the "lumpiness" of ψ.

The Laplacian operator can also be applied to a vector field **F**, resulting in a vector which can be considered as a measure of the lumpiness of either the direction or magnitude of the vector **F**. The x, y, and z components of this vector are obtained by taking the Laplacian of the x, y, and z components of **F**. To obtain components along general coordinates we use the relations

$$\nabla^2\mathbf{F} = \text{grad}\,(\text{div}\,\mathbf{F}) - \text{curl}\,(\text{curl}\,\mathbf{F})$$
$$(\nabla^2\mathbf{F})_i = h_i\sum_n \frac{1}{h_n^2}\left[\left(\frac{F_i}{h_i}\right)_{,n}\right]_{,n} = \frac{1}{h_i}\sum_n \frac{1}{h_n^2}[(h_iF_i)_{,n}]_{,n} \tag{1.5.12}$$

The first equation can readily be obtained in rectangular coordinates; it is, of course, true for any system. The second equation shows that the vector $\nabla^2\mathbf{F}$ is related to the contra- and covariant vectors formed by the double covariant differentiation of either F_i/h_i or h_iF_i and the contraction of the resulting third-order tensor. The resulting general form for $\nabla^2\mathbf{F}$, when written out, is a complicated one, but this form simplifies considerably for most of the systems we shall use.

The first equation in Eqs. (1.5.12) is of interest in itself, for it shows that $\nabla^2\mathbf{F}$ is related to the vorticity of $\mathbf{F}$ because of the last term, curl (curl $\mathbf{F}$). If $\mathbf{F}$ represents the velocity of an incompressible fluid, then div $\mathbf{F} = 0$ and $\nabla^2\mathbf{F} = -$ curl (curl $\mathbf{F}$). Therefore, in order that a divergenceless vector have a Laplacian which is not zero, not only must it have a vorticity, but its vorticity lines must themselves exhibit vorticity.

Other Second-order Operators. The other combinations of two operators ∇ are less important than the Laplacian ∇^2; however, they will sometimes enter into our equations, and it is well to discuss them briefly.

Several of these operators are zero. The relation

$$\text{curl (grad } \psi) \equiv \nabla \times (\nabla\psi) = 0 \tag{1.5.13}$$

has already been used in Sec. 1.2, where we showed that, if a vector was the gradient of a potential function, its curl had to be zero. The relation

$$\text{div (curl } \mathbf{F}) \equiv \nabla \cdot (\nabla \times \mathbf{F}) = 0 \tag{1.5.14}$$

has already been discussed on page 18, where we showed that lines of vorticity can neither begin nor end.

Equation (1.5.13) indicates a property of fields which was discussed on page 14, that if a field has zero curl everywhere, it can be represented as the gradient of a scalar, called the *potential function*. Equation (1.5.14) indicates a parallel property, which can be verified without much trouble, that if a field has a zero divergence, it can be represented as the curl of a vector, which may be called the *vector potential* for the zero-divergence field.

The operator grad (div $\;$) $= \nabla(\nabla\cdot\;\;)$ operates on a vector and results in a vector. It measures the possible change in the divergence of a vector field, and differs from the Laplacian of the same field, $\mathbf{F}$, by the quantity curl (curl $\mathbf{F}$), as given by Eq. (1.5.12). The operator curl (curl $\;$) $=$ $\nabla \times (\nabla \times \;\;)$, the last of the second-order operators, is thus defined in terms of two others. All these expressions can be given in tensor form.

Vector as a Sum of Gradient and Curl. We have now reached a degree of dexterity in our use of vector formulas to enable us to analyze further the statement made in the next to last paragraph. To be more specific, we shall prove that any vector field $\mathbf{F}$, which is finite, uniform, and continuous and which vanishes at infinity, may also be expressed as the sum of a gradient of a scalar φ and a curl of a zero-divergence vector **A**,

$$\mathbf{F} = \text{grad}\,\varphi + \text{curl}\,\mathbf{A};\quad \text{div}\,\mathbf{A} = 0 \tag{1.5.15}$$

The function φ is called the *scalar potential* of **F**, **A** is called its *vector potential*, and the theorem is called *Helmholtz's theorem.*

In order to prove this statement we need to show how φ and **A** can be computed from **F**, and in order to compute φ and **A** we shall have to use the solution of Poisson's equation $\nabla^2\varphi = -q$ which was given in Eq. (1.4.8),

$$\varphi = \iiint \frac{q(x',y',z')}{4\pi R}\,dx'\,dy'\,dz';\quad R^2 = (x-x')^2 + (y-y')^2 + (z-z')^2$$

By considering each of the three vector components in turn we can also show that the solution of the vector Poisson equation $\nabla^2\mathbf{F} = -\mathbf{q}$ has the same form as the scalar solution, with F and q changed to boldface letters. We shall show in Chap. 7 that both of these solutions are unique ones as long as the integral of **F** over all space is finite. If it is not, if, for example, $F = ax$, we can take F to be ax within a sphere of very large but finite radius and have it be zero outside the sphere. After the calculation is all finished, we could then let $R \to \infty$. If we are dealing with a field inside a finite volume, we can choose F to be the field inside the volume and be zero outside it. In all cases that F does not specifically go infinite, we can make $\int F\,dv$ finite.

To compute φ and **A** we compute first the vector function

$$\mathbf{W} = \iiint \frac{\mathbf{F}(x',y',z')}{4\pi R}\,dx'\,dy'\,dz' \tag{1.5.16}$$

which is a solution of the vector Poisson equation $\nabla^2\mathbf{W} = -\mathbf{F}$. From this we can see that div **W** may be set equal to $-\varphi$ and curl **W** may be set equal to **A**, for then [using the vector formula (1.5.12)]

$$\mathbf{F} = -\nabla^2\mathbf{W} = -\,\text{grad div}\,\mathbf{W} + \text{curl curl}\,\mathbf{W} = \text{grad}\,\varphi + \text{curl}\,\mathbf{A}$$

which is Eq. (1.5.15). Since **W** is determined uniquely by Eq. (1.5.16), φ and **A** are determined uniquely as long as the integral of **F** is finite (which can always be arranged by use of the trick mentioned above as long as **F** is not infinite somewhere a finite distance from the origin).

We can express φ and **A** in somewhat simpler terms if we use the symmetry of function $1/R$ with respect to (x,y,z) and (x',y',z'), for the gradient of $1/R$ with respect to (x,y,z) (written grad $1/R$) is the negative of the gradient of $1/R$ with respect to (x',y',z') (written grad$'$ $1/R$). Then using Gauss' theorem (1.4.7), we have

$$-\,\text{div}\,\mathbf{W} = \int \mathbf{F}\cdot\text{grad}'\left(\frac{1}{4\pi R}\right)dv' = \oint \frac{\mathbf{F}\cdot d\mathbf{A}'}{4\pi R} - \int \frac{\text{div}'\,\mathbf{F}}{4\pi R}\,dv'$$

or

$$\varphi = -\iiint [\text{div}'\,\mathbf{F}(x',y',z')/4\pi R]\,dx'\,dy'\,dz'$$

where the normal outflow integral of $\mathbf{F}/4\pi R$ is taken over a large, enclosing surface entirely in the region where $\mathbf{F}$ is (or has been made) zero.

Similarly, by using an equation related to Gauss' theorem,

$$\int \text{curl}\, \mathbf{B}\, dv = -\oint \mathbf{B} \times d\mathbf{A} \tag{1.5.17}$$

we can transform the expression for $\mathbf{A}$ to the simpler one

$$\text{curl}\, \mathbf{W} = \int \mathbf{F} \times \text{grad}' \left(\frac{1}{4\pi R}\right) dv' = \oint \frac{\mathbf{F} \times d\mathbf{A}}{4\pi R} + \int \frac{\text{curl}'\, \mathbf{F}}{4\pi R}\, dv'$$

or

$$\mathbf{A} = \iiint [\text{curl}'\, \mathbf{F}(x',y',z')/4\pi R]\, dx'\, dy'\, dz'$$

Consequently φ and $\mathbf{A}$ can be obtained directly from the divergence and the curl of $\mathbf{F}$, with $\mathbf{F}$ subject to the conditions outlined above.

This property of any vector field, of being uniquely separable into a divergenceless part, curl $\mathbf{A}$, and a curlless part, grad φ, is called *Helmholtz's theorem.* It will be of considerable use to us in this book, particularly in Chap. 13. It will be discussed from a different point of view in Sec. 2.3.

1.6 *Dyadics and Other Vector Operators*

We have already discussed the properties of vector fields and their correspondence to various physical phenomena in order to get a "physical feeling" for the concept of a vector field. We next must become familiar with the physical counterparts of the tensor forms defined in Eqs. (1.5.2). These forms, in three dimensions, have nine components, as contrasted to the three components of a vector. In a tensor field these nine components may vary from point to point; they transform under change of coordinate system in the manner given in Eqs. (1.5.2).

Dyadics. Just as we defined "real" vectors as distinguished from their contravariant and covariant forms, so here we shall define a dyadic as a set of nine components A_{ij}, each of them functions of the three coordinates, which transform from one coordinate system to the other according to the rule

$$\begin{aligned}(A_{ij})' &= \sum_{m,n} \frac{h_m h_n}{h'_i h'_j} \frac{\partial x_m}{\partial x'_i} \frac{\partial x_n}{\partial x'_j} A_{mn} \\ &= \sum_{m,n} \frac{h'_i h'_j}{h_m h_n} \frac{\partial x'_i}{\partial x_m} \frac{\partial x'_j}{\partial x_n} A_{mn} \\ &= \sum_{m,n} \frac{h'_i}{h_m} \frac{h_n}{h'_j} \frac{\partial x'_i}{\partial x_m} \frac{\partial x_n}{\partial x'_j} A_{mn} \\ &= \sum_{m,n} \gamma_{im}\, \gamma_{jn} A_{mn}\end{aligned} \tag{1.6.1}$$

The dyadic as a whole whose components obey Eq. (1.6.1) is represented by the German capital letter $\mathfrak{A}$. The relation between the dyadic components and the corresponding contravariant, covariant, and mixed tensor components can be derived from Eqs. (1.5.2):

$$a^{mn} = A_{mn}/h_m h_n; \quad a_{mn} = h_m h_n A_{mn}; \quad a_n^m = h_n A_{mn}/h_m \tag{1.6.2}$$

Two general properties of a dyadic $\mathfrak{A}$ can be written down immediately; the contracted form

$$|\mathfrak{A}| = \sum_m A_{mm} = \sum_m a_m^m \tag{1.6.3}$$

is a scalar invariant whose value at a point is independent of change of coordinates; and according to Eq. (1.5.3), the quantity

$$\langle\mathfrak{A}\rangle = \mathbf{a}_1[A_{23} - A_{32}] + \mathbf{a}_2[A_{31} - A_{13}] + \mathbf{a}_3[A_{12} - A_{21}] \tag{1.6.4}$$

is an axial vector, having the transformation properties of a vector. (The quantities $\mathbf{a}_m$ are, as before, unit vectors along the three directions of a right-handed system of coordinates.) The invariant $|\mathfrak{A}|$ can be called the *Spur* or the *expansion factor* of the dyadic, and the vector $\langle\mathfrak{A}\rangle$ will be called the *rotation vector* of $\mathfrak{A}$ for reasons which will be apparent shortly.

A dyadic can be combined with a vector to form a vector by contraction:

$$\mathfrak{A} \cdot \mathbf{B} = \sum_{mn} \mathbf{a}_m A_{mn} B_n; \quad \mathbf{B} \cdot \mathfrak{A} = \sum_{mn} B_m A_{mn} \mathbf{a}_n \tag{1.6.5}$$

Use of the definitions of transformation of vectors and dyadics will show that these quantities transform like a vector (a "true" vector). This result suggests that a formal mode of writing a dyadic in terms of its nine components along the axes (ξ_1,ξ_2,ξ_3) is

$$\mathfrak{A} = \mathbf{a}_1 A_{11}\mathbf{a}_1 + \mathbf{a}_1 A_{12}\mathbf{a}_2 + \mathbf{a}_1 A_{13}\mathbf{a}_3 + \mathbf{a}_2 A_{21}\mathbf{a}_1 + \mathbf{a}_2 A_{22}\mathbf{a}_2 + \mathbf{a}_2 A_{23}\mathbf{a}_3 + \mathbf{a}_3 A_{31}\mathbf{a}_1 + \mathbf{a}_3 A_{32}\mathbf{a}_2 + \mathbf{a}_3 A_{33}\mathbf{a}_3 \tag{1.6.6}$$

The quantities $\mathbf{a}_m\mathbf{a}_n$ are neither scalar nor vector products of the unit vectors but are operators such that the scalar product $(\mathbf{a}_m\mathbf{a}_n) \cdot \mathbf{B} = B_n\mathbf{a}_m$ is a vector pointed along ξ_m of magnitude equal to the component of $\mathbf{B}$ along ξ_n, etc.

We note that the vector $\mathbf{B} \cdot \mathfrak{A}$ is not usually equal to the vector $\mathfrak{A} \cdot \mathbf{B}$. The dyadic $\mathfrak{A}^*$ formed by reversing the order of the subscripts of the components $(A^*_{mn} = A_{nm})$ is called the *conjugate* of $\mathfrak{A}$. It is not difficult to see that $\mathfrak{A} \cdot \mathbf{B} = \mathbf{B} \cdot \mathfrak{A}^*$ and $\mathbf{B} \cdot \mathfrak{A} = \mathfrak{A}^* \cdot \mathbf{B}$.

Dyadics as Vector Operators. The last two paragraphs have suggested one of the most useful properties of dyadics: *They are operators which change a vector into another vector.* The new vector is related to the old one by a set of rules, represented by the nine components A_{ij}. The values of these components determine how the new vector differs from the old in magnitude and direction. The amount of difference, of course, also depends on the direction of the original vector. The vector operator represented by a dyadic is not the most general type of vector operator (we shall discuss others later), but it corresponds to so many physical phenomena that it merits detailed study.

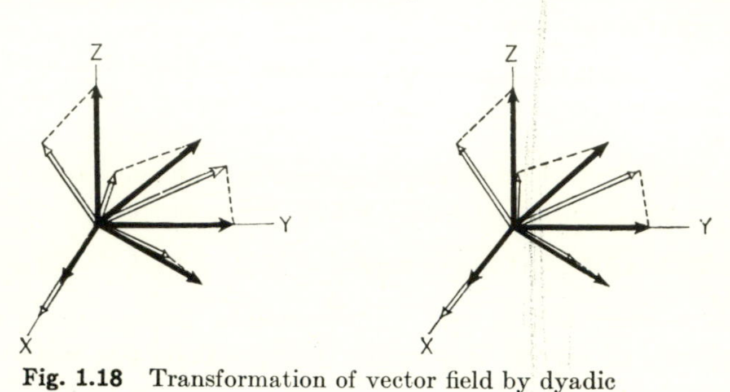

Fig. 1.18 Transformation of vector field by dyadic

$$\mathbf{i}(1.5\mathbf{i} + 0.2\mathbf{j}) + \mathbf{j}(\mathbf{j} - 0.4\mathbf{k}) + \mathbf{k}(0.5\mathbf{j} + 0.6\mathbf{k})$$

Black vectors are initial field; outlined vectors are transformed field.

Examples of phenomena which can be represented in terms of vector operators are found in many branches of physics. For instance, the relation between $\mathbf{M}$, the angular momentum of a rigid body, and its angular velocity $\boldsymbol{\omega}$ is $\mathbf{M} = \mathfrak{J} \cdot \boldsymbol{\omega}$, where $\mathfrak{J}$ is the moment of inertia dyadic. Also, in a nonisotropic porous medium through which fluid is being forced, the fluid velocity vector $\mathbf{v}$ is not in general in the same direction as the pressure gradient vector, but the two are related by a dyadic relationship, grad $p = \mathfrak{R} \cdot \mathbf{v}$, where $\mathfrak{R}$ is the dyadic resistance. Similarly the relation between the electric field and the electric polarization in a nonisotropic dielectric is a dyadic relationship. The most familiar example of dyadics is in the distortion of elastic bodies, which will be discussed shortly.

The concept of a dyadic as a vector operator, as well as consideration of the foregoing equations, leads us to the following rules of dyadic algebra:

$$\mathfrak{A} + \mathfrak{B} = \sum_{m,n} \mathbf{a}_m[A_{mn} + B_{mn}]\mathbf{a}_n = \mathfrak{B} + \mathfrak{A}$$

$$\mathfrak{A} \cdot \mathfrak{B} = \sum_{m,n} \mathbf{a}_m \left[\sum_i A_{mi}B_{in}\right] \mathbf{a}_n \neq \mathfrak{B} \cdot \mathfrak{A}$$

$$c\mathfrak{A} = \sum_{m,n} \mathbf{a}_m(cA_{mn})\mathbf{a}_n = \mathfrak{A}c$$

The first equation shows us that addition of dyadics is commutative and that a dyadic of general form can be built up as a sum of dyadics of simple form. The second equation shows that a dyadic times a dyadic is a dyadic and that dyadic multiplication is *not* commutative. The third equation defines multiplication by a scalar. Multiplication by a vector has already been defined. The scalar "double-dot" product $\mathfrak{A}:\mathfrak{B} = \sum_{m,n} A_{mn}B_{nm} = |\mathfrak{A} \cdot \mathfrak{B}|$ is, of course, the completely contracted form.

There is, of course, a zero dyadic $\mathfrak{O}$ and a unity dyadic $\mathfrak{J}$ called the *idemfactor:*

$$\mathfrak{O} \cdot \mathbf{F} = 0; \quad \mathfrak{J} \cdot \mathbf{F} = \mathbf{F}; \quad \mathfrak{J} = \mathbf{a}_1\mathbf{a}_1 + \mathbf{a}_2\mathbf{a}_2 + \mathbf{a}_3\mathbf{a}_3$$

where $\mathbf{F}$ is any vector.

One can also define $\mathfrak{A}^{-1}$, the *reciprocal* dyadic to $\mathfrak{A}$, as the dyadic, which, when multiplied by $\mathfrak{A}$, results in the idemfactor

$$(\mathfrak{A}^{-1}) \cdot \mathfrak{A} = \mathfrak{A} \cdot (\mathfrak{A}^{-1}) = \mathfrak{J}$$

The reciprocal of the zero dyadic is, of course, undefined. In terms of the nine components of $\mathfrak{A}$, as defined in Eq. (1.6.6), the components of the reciprocal are

$$(\mathfrak{A}^{-1})_{mn} = A'_{mn}/\Delta_A$$

where A'_{mn} is the minor of A_{mn} in the determinant

$$\Delta_A = \begin{vmatrix} A_{11} & A_{12} & A_{13} \\ A_{21} & A_{22} & A_{23} \\ A_{31} & A_{32} & A_{33} \end{vmatrix}$$

The definition of the multiplication of dyadics requires that the conjugate to the product $(\mathfrak{A} \cdot \mathfrak{B})$ is given in terms of the conjugates of $\mathfrak{A}$ and $\mathfrak{B}$ by the equation

$$(\mathfrak{A} \cdot \mathfrak{B})^* = (\mathfrak{B}^* \cdot \mathfrak{A}^*)$$

involving an inversion in the order of multiplication. Similarly the reciprocal of the product $(\mathfrak{A} \cdot \mathfrak{B})^{-1} = (\mathfrak{B}^{-1} \cdot \mathfrak{A}^{-1})$.

Since a dyadic at any point in space is determined by its nine components along orthogonal axes [which components change with coordinate rotation according to Eq. (1.6.1)], it can be built up by a proper combination of vectors having at least nine independently chosen constants. For instance, since a vector at a point is determined by three quantities, a dyadic can be formed in terms of three arbitrarily chosen vectors $\mathbf{A}_m$:

$$\mathfrak{A} = \mathbf{a}_1\mathbf{A}_1 + \mathbf{a}_2\mathbf{A}_2 + \mathbf{a}_3\mathbf{A}_3 = \mathbf{A}_1^*\mathbf{a}_1 + \mathbf{A}_2^*\mathbf{a}_2 + \mathbf{A}_3^*\mathbf{a}_3 \tag{1.6.7}$$

where the $\mathbf{a}$'s are unit vectors along three orthogonal, right-handed coordinate axes. The conjugate dyadic is

$$\mathfrak{A}^* = \mathbf{A}_1\mathbf{a}_1 + \mathbf{A}_2\mathbf{a}_2 + \mathbf{A}_3\mathbf{a}_3 = \mathbf{a}_1\mathbf{A}_1^* + \mathbf{a}_2\mathbf{A}_2^* + \mathbf{a}_3\mathbf{A}_3^*$$

These equations define the vectors $\mathbf{A}_n$ and $\mathbf{A}_n^*$. Their interrelation is given below. The vector $\mathbf{A}_m$ can be called the *component* vector with respect to the ξ_m axis. For an arbitrary dyadic this may be in any arbitrary direction and have any magnitude. A vector **B** pointing in the ξ_m direction is transformed by the operation $\mathbf{B} \cdot \mathfrak{A}$ into a vector pointing in the direction of $\mathbf{A}_m$ and is transformed by the operation $\mathfrak{A} \cdot \mathbf{B}$ into one pointing in the direction of $\mathbf{A}_m^*$.

It is not difficult to see that the component vectors are related to the nine components A_{mn} of $\mathfrak{A}$ in the ξ_1, ξ_2, ξ_3 axes [given in Eq. (1.6.6)] by the relations

$$\mathbf{A}_m = \sum_n A_{mn}\mathbf{a}_n; \quad \mathbf{A}_m^* = \sum_n \mathbf{a}_n A_{nm}$$

In the rectangular coordinates x, y, z the dyadic is represented as

$$\begin{aligned}
\mathfrak{A} &= \mathbf{i}\mathbf{A}_x + \mathbf{j}\mathbf{A}_y + \mathbf{k}\mathbf{A}_z = \mathbf{A}_x^*\mathbf{i} + \mathbf{A}_y^*\mathbf{j} + \mathbf{A}_z^*\mathbf{k} \\
\mathbf{A}_x &= \alpha_1\mathbf{A}_1 + \alpha_2\mathbf{A}_2 + \alpha_3\mathbf{A}_3; \quad \alpha_m = \mathbf{i} \cdot \mathbf{a}_m \\
\mathbf{A}_y &= \beta_1\mathbf{A}_1 + \beta_2\mathbf{A}_2 + \beta_3\mathbf{A}_3; \quad \beta_m = \mathbf{j} \cdot \mathbf{a}_m \\
\mathbf{A}_z &= \gamma_1\mathbf{A}_1 + \gamma_2\mathbf{A}_2 + \gamma_3\mathbf{A}_3; \quad \gamma_m = \mathbf{k} \cdot \mathbf{a}_m
\end{aligned}$$

More generally, a dyadic can be expressed in terms of a series of combinations of vectors

$$\mathfrak{A} = \sum_m \mathbf{A}_m\mathbf{B}_m$$

where there must be at least three terms in the sum in order to represent any arbitrary dyadic.

Symmetric and Antisymmetric Dyadics. The dyadic $A\mathbf{a}_1\mathbf{a}_1$ is a particularly simple vector operator; it converts any vector **F** into a vector of length $A(\mathbf{a}_1 \cdot \mathbf{F})$ pointed along $\mathbf{a}_1$. It is a symmetric dyadic, for its components in any rectangular system of coordinates are symmetric with respect to exchange of subscripts. For instance, in the rectangular coordinates x, y, z, the components are

$$\begin{aligned}
&A_{xx} = A\alpha_1^2; \quad A_{yy} = A\beta_1^2; \quad A_{zz} = A\gamma_1^2 \\
&A_{xy} = A_{yx} = A\alpha_1\beta_1; \quad A_{xz} = A_{zx} = A\alpha_1\gamma_1; \quad A_{yz} = A_{zy} = A\beta_1\gamma_1
\end{aligned}$$

where α_1, β_1, and γ_1 are the direction cosines of $\mathbf{a}_1$ with respect to x, y, z, etc.

The most general symmetric dyadic can be expressed in terms of three orthogonal unit vectors $\mathbf{a}_1$, $\mathbf{a}_2$, $\mathbf{a}_3$;

$$\mathfrak{A}_s = \mathbf{a}_1 A_1 \mathbf{a}_1 + \mathbf{a}_2 A_2 \mathbf{a}_2 + \mathbf{a}_3 A_3 \mathbf{a}_3 \tag{1.6.8}$$

Since a symmetric dyadic is specified by only six independent constants (three pairs of components are equal), specifying a definite symmetric dyadic uniquely specifies the three constants A_1, A_2, A_3 and uniquely specifies the directions in space of the three orthogonal unit vectors $\mathbf{a}_1$, $\mathbf{a}_2$, $\mathbf{a}_3$ [which are given by the Eulerian angles ψ, φ, ϑ, see Eq. (1.3.8)]. Referring to Eq. (1.6.7) we see that for a dyadic to be symmetric it must be possible to find a trio of unit vectors $\mathbf{a}$ such that the component vector $\mathbf{A}_1$ is parallel to $\mathbf{a}_1$, and so on.

Conversely, any symmetric dyadic can be expressed in the form of Eq. (1.6.8), and the values of the A's and the directions of the $\mathbf{a}$'s can be found, for Eq. (1.6.8) indicates that the symmetric operator $\mathfrak{A}_s$ operates on a vector in either of the three orthogonal directions $\mathbf{a}_1$, $\mathbf{a}_2$, or $\mathbf{a}_3$ by changing its length and not its direction whereas operating on a vector in any other direction, direction is changed as well as length. Those special directions in which the operation does not change the direction of the vector are called the *principal axes* of the dyadic.

The direction cosines of a principal axis $\mathbf{a}_1$ of the dyadic

$$\mathfrak{A}_s = A_x\mathbf{ii} + B_z\mathbf{ij} + B_y\mathbf{ik} + B_z\mathbf{ji} + A_y\mathbf{jj} + B_x\mathbf{jk} + B_y\mathbf{ki} + B_x\mathbf{kj} + A_z\mathbf{kk}$$

may be found by solving the equation

$$\mathfrak{A}_s \cdot \mathbf{a}_1 = A_1\mathbf{a}_1 \tag{1.6.9}$$

which is the mathematical statement of the definition of principal axis given above. Equation (1.6.9) is the first case of an *eigenvalue equation* that we have encountered, but it will not be long before we turn up other ones, in our discussions of "vector space" in quantum mechanics and wave theory and in many other aspects of field theory. The directions of the principal axes $\mathbf{a}_i$ are called *eigenvectors*, while the constants A_1, A_2, A_3 are known as the *eigenvalues*.

To solve Eq. (1.6.9) let $\mathbf{a}_1 = \alpha_1\mathbf{i} + \beta_1\mathbf{j} + \gamma_1\mathbf{k}$. Introducing this on both sides of the equation we obtain three linear homogeneous equations:

$$\begin{aligned}(A_x - A_1)\alpha_1 + B_z\beta_1 + B_y\gamma_1 &= 0\\ B_z\alpha_1 + (A_y - A_1)\beta_1 + B_x\gamma_1 &= 0\\ B_y\alpha_1 + B_x\beta_1 + (A_z - A_1)\gamma_1 &= 0\end{aligned}$$

This set of equations can be solved only if the determinant of the coefficients of α_1, β_1, γ_1 is zero:

$$\begin{vmatrix} A_x - A_1 & B_z & B_y \\ B_z & A_y - A_1 & B_x \\ B_y & B_x & A_z - A_1 \end{vmatrix} = 0$$

Solving this equation (a third-degree one) for A_1 yields three roots corresponding to the three numbers A_1, A_2, A_3. The determinant is known as a *secular determinant*. It turns up whenever one goes about solving an eigenvalue equation by means of a linear combination of vectors such as that used for $\mathbf{a}_1$.

Corresponding to the three numbers A_i there will be three sets of values for α_i, β_i, γ_i which are the direction cosines of the three principal axes. These axes are perpendicular, as shown by the following argument:

$$\mathfrak{A}_s \cdot \mathbf{a}_1 = A_1\mathbf{a}_1; \quad \mathfrak{A}_s \cdot \mathbf{a}_2 = A_2\mathbf{a}_2$$

Because of the symmetry of $\mathfrak{A}_s$

$$0 = \mathbf{a}_2 \cdot \mathfrak{A}_s \cdot \mathbf{a}_1 - \mathbf{a}_1 \cdot \mathfrak{A}_s \cdot \mathbf{a}_2 = (A_1 - A_2)\mathbf{a}_1 \cdot \mathbf{a}_2$$

Since A_1 and A_2 usually differ in value, this equation can hold only if $\mathbf{a}_1 \cdot \mathbf{a}_2 = 0$.

It may also be shown that the Spur or expansion factor is invariant,

$$A_x + A_y + A_z = A_1 + A_2 + A_3$$

as, of course, it must, since this expression is the scalar invariant $|\mathfrak{A}_s|$ of the dyadic. We can now see why this scalar is called the expansion factor of $\mathfrak{A}$, for it is three times the average fractional increase in length caused by dyadic $\mathfrak{A}_s$ on vectors pointed along its three principal axes. The other term "Spur" is the German word for trace or "identifying trail" (equivalent to the "spoor" of big game), a picturesque but understandable description.

We notice that the vector $\langle\mathfrak{A}_s\rangle$ formed from a symmetric dyadic is zero, and that for a symmetric dyadic $\mathfrak{A}_s \cdot \mathbf{F} = \mathbf{F} \cdot \mathfrak{A}_s$ for any vector $\mathbf{F}$. In other words, any symmetric dyadic is equal to its own conjugate.

An *antisymmetric* dyadic has its diagonal components A_{nn} zero, and its nondiagonal components change sign on interchange of subscripts; $A_{mn} = -A_{nm}$. The most general antisymmetric dyadic is specified by only three independent constants. It can always be expressed in the form

$$\mathfrak{A}_a = \mathbf{R} \times \mathfrak{J} = -\mathbf{ij}R_z + \mathbf{ik}R_y + \mathbf{ji}R_z - \mathbf{jk}R_x - \mathbf{ki}R_y + \mathbf{kj}R_x \quad (1.6.10)$$

where $\mathfrak{J} = (\mathbf{ii} + \mathbf{jj} + \mathbf{kk})$ is the idemfactor. Choice of the antisymmetric dyadic uniquely specifies the vector $\mathbf{R}$, which is minus half the rotation vector $\langle\mathfrak{A}_a\rangle$ of the antisymmetric dyadic $\mathfrak{A}_a$. We note that the expansion factor of an antisymmetric dyadic is zero. We also note that, for a vector $\mathbf{F}$, the operation $\mathfrak{A}_a \cdot \mathbf{F} = \mathbf{R} \times \mathbf{F} = -\frac{1}{2}\langle\mathfrak{A}_a\rangle \times \mathbf{F}$ yields a vector which is perpendicular to $\mathbf{F}$ and also perpendicular to the rotation vector of $\mathfrak{A}_a$.

We note that we may also set up an eigenvalue equation for an antisymmetric dyadic,

$$\mathfrak{A}_a \cdot \mathbf{a} = \mathbf{R} \times \mathbf{a} = \lambda\mathbf{a}$$

which may be formally solved to obtain the "principal axes" of $\mathfrak{A}_a$. Setting up the secular determinant shows that the three roots for λ are $A_1 = 0$, $A_2 = iR$, and $A_3 = -iR$, two of the roots being imaginary.

The unit vector along the principal axis corresponding to $\lambda = A_1 = 0$ is $\mathbf{a}_R$, parallel to $\mathbf{R}$; the other two unit vectors are not real. The sum $A_1 + A_2 + A_3 = 0$, as it should be, since the expansion factor for an antisymmetric dyadic is zero.

It is easy to see that *any* dyadic can be expressed in terms of a sum of a symmetric and an antisymmetric dyadic:

$$\mathfrak{A} = \mathfrak{A}_s + \mathfrak{A}_a$$
$$(\mathfrak{A}_s)_{mn} = \tfrac{1}{2}(A_{mn} + A_{nm}); \quad (\mathfrak{A}_a)_{mn} = \tfrac{1}{2}(A_{mn} - A_{nm})$$

By transforming to its principal axes (which are always real) we can always express the symmetric part $\mathfrak{A}_s$ in the form given in Eq. (1.6.8), and by proper choice of the vector $\mathbf{R}$ we can always express the antisymmetric part $\mathfrak{A}_a$ in the form given in Eq. (1.6.10).

Of course we can choose to find the principal axes of $\mathfrak{A}$ itself, by solving the eigenvalue equation

$$\mathfrak{A} \cdot \mathbf{e} = \lambda \mathbf{e}$$

directly (before separating into symmetric and antisymmetric parts) This results in a secular determinant equation

$$\begin{vmatrix} A_{xx} - \lambda & A_{xy} & A_{xz} \\ A_{yx} & A_{yy} - \lambda & A_{yz} \\ A_{zx} & A_{zy} & A_{zz} - \lambda \end{vmatrix} = 0$$

The roots of this cubic equation may be labeled $\lambda = A_1, A_2, A_3$, and the corresponding eigenvectors, along the principal axes, will be labeled

$$\mathbf{e}_1 = \alpha_1\mathbf{i} + \beta_1\mathbf{j} + \gamma_1\mathbf{k}$$
$$\mathbf{e}_2 = \alpha_2\mathbf{i} + \beta_2\mathbf{j} + \gamma_2\mathbf{k}$$
$$\mathbf{e}_3 = \alpha_3\mathbf{i} + \beta_3\mathbf{j} + \gamma_3\mathbf{k}$$

Theory of equations indicates that minus the coefficient of the λ^2 term $(A_{xx} + A_{yy} + A_{zz})$, is equal to the sum of the roots $(A_1 + A_2 + A_3)$. It also indicates that the roots are either all three real or else one is real and the other two are complex, one being the complex conjugate of the other (as long as all nine components A_{mn} of $\mathfrak{A}$ are real). When all the roots are real, the three eigenvectors $\mathbf{e}_m$ are real and mutually perpendicular, but when two of the roots are complex, two of the eigenvectors are correspondingly complex. It is to avoid this complication that we usually first separate off the antisymmetric part of $\mathfrak{A}$ and find the principal axes of $\mathfrak{A}_s$, for the eigenvalues and eigenvectors of a symmetric dyadic with real components are all real.

Rotation of Axes and Unitary Dyadics. A special type of vector operator corresponds to what might be called *rigid rotation*. Considering a number of vectors $\mathbf{F}$ as a sort of coordinate framework, the operation of rigid rotation is to rotate all the vectors $\mathbf{F}$ as a *rigid* framework, maintaining the magnitudes of all the **F**'s and the relative angles between them. If **F**, for instance, were the set of vectors giving the location of various parts of a rigid body with respect to some origin, then the rotation of the rigid body about this origin would correspond to the operation under consideration.

Suppose that we consider such a vector operation to be represented by the specialized dyadic $\mathfrak{G}$, with components γ_{xy}, etc. In order that the transformed vector $\mathfrak{G} \cdot \mathbf{F}$ will have the same magnitude as **F** for any **F**, we must have the following condition:

$$(\mathfrak{G} \cdot \mathbf{F}) \cdot (\mathfrak{G} \cdot \mathbf{F}) = \sum_{l,n} \left[\sum_m \gamma_{mn}\gamma_{ml} \right] F_n F_l = \mathbf{F} \cdot \mathbf{F} = \sum_n F_n^2; \quad l, m, n = x, y, z$$

or, in other words,

$$\sum_m \gamma_{mn}\gamma_{ml} = \delta_{nl} = \begin{cases} 1; & l = n \\ 0; & l \neq n \end{cases} \tag{1.6.11}$$

Dyadics with components satisfying this requirement are called *unitary dyadics*, for reasons shortly to be clear. Incidentally, when this requirement on the γ's is satisfied, it can easily be shown that, for any pair of vectors **A** and **B**, the dot product $\mathbf{A} \cdot \mathbf{B}$ is unaltered by a rotation corresponding to $\mathfrak{G}$,

$$(\mathfrak{G} \cdot \mathbf{A}) \cdot (\mathfrak{G} \cdot \mathbf{B}) = \mathbf{A} \cdot \mathbf{B}$$

and consequently (since magnitudes are unaltered by the rotation) angles between vectors are unaltered by the operator $\mathfrak{G}$. It is also easy to demonstrate that, if the dyadics $\mathfrak{G}$ and $\mathfrak{H}$ both represent rigid rotation operations, then the dyadic $(\mathfrak{G} \cdot \mathfrak{H})$ has components which also satisfy Eq. (1.6.11) and it consequently also represents a rotation. Products of unitary dyadics are unitary.

Incidentally, all the possible real values of components γ, corresponding to all real rotations about a center, may be obtained by referring to the Euler-angle rotation of axes given in Eq. (1.3.8). If the dyadic components of $\mathfrak{G}$ are

$$\begin{gathered} \gamma_{xx} = \sin\psi \sin\Phi + \cos\psi \cos\Phi \cos\theta; \\ \gamma_{xy} = \cos\psi \sin\Phi - \sin\psi \cos\Phi \cos\theta; \quad \gamma_{xz} = \sin\theta \cos\Phi \\ \gamma_{yx} = \sin\psi \cos\Phi - \cos\psi \sin\Phi \cos\theta; \\ \gamma_{yy} = \cos\psi \cos\Phi + \sin\psi \sin\Phi \cos\theta; \quad \gamma_{yz} = -\sin\theta \sin\Phi \\ \gamma_{zx} = -\cos\psi \sin\theta; \quad \gamma_{zy} = \sin\psi \sin\theta; \quad \gamma_{zz} = \cos\theta \end{gathered}$$

then reference to Fig. 1.12 shows that this transformation corresponds to the rotation of the rigid vector framework by an angle ψ about the z axis, a subsequent rotation by an angle θ about the y axis, and then a final rotation of the (already rotated) system once more about the z axis by an angle Φ.

A certain amount of algebra will suffice to show that the operation $\mathfrak{G}$, with components γ given above, satisfying Eq. (1.6.11), has all the properties of a rotation of axes. For one thing, the dyadic corresponding to a product $\mathfrak{A} \cdot \mathfrak{B}$, where $\mathfrak{A}$ and $\mathfrak{B}$ are both unitary, is also unitary and represents the rotation which results when the axes are first turned through the angles given by $\mathfrak{B}$ and then turned through the angles represented by $\mathfrak{A}$. As a matter of fact the statements made in the preceding paragraph correspond to the fact that $\mathfrak{G}$, with its components as given, is equal to the product $\mathfrak{C} \cdot \mathfrak{B} \cdot \mathfrak{A}$, where the three unitary dyadic factors are

$$\mathfrak{A} = \begin{pmatrix} \cos\psi & -\sin\psi & 0 \\ \sin\psi & \cos\psi & 0 \\ 0 & 0 & 1 \end{pmatrix}; \quad \mathfrak{B} = \begin{pmatrix} \cos\theta & 0 & \sin\theta \\ 0 & 1 & 0 \\ -\sin\theta & 0 & \cos\theta \end{pmatrix}$$

$$\mathfrak{C} = \begin{pmatrix} \cos\Phi & \sin\Phi & 0 \\ -\sin\Phi & \cos\Phi & 0 \\ 0 & 0 & 1 \end{pmatrix}$$

representing the elementary rotations by the three Euler angles. The elements of the product are computed using Eq. (1.6.6).

The reason these rotation dyadics are called unitary is that the determinant of their elements is equal to unity, as may be deduced from Eq. (1.6.11). But an even more useful property may be displayed by using the definition of the reciprocal dyadic on page 57, and that of the conjugate dyadic combining these with Eq. (1.6.11). The result is that, if $\mathfrak{G}$ is a unitary matrix [satisfying (1.6.11)], $\mathfrak{G}^{-1}$ is its inverse and, if $\mathfrak{G}^*$ is its conjugate, then

$$\mathfrak{G}^{-1} = \mathfrak{G}^* \quad \text{or} \quad \mathfrak{G}^* \cdot \mathfrak{G} = \mathfrak{J}; \quad \mathfrak{G} \text{ unitary} \tag{1.6.12}$$

Vice versa, if $\mathfrak{G}$ satisfies Eq. (1.6.12), then it is unitary and its components satisfy Eq. (1.6.11). Since $\mathfrak{G}^* \cdot \mathfrak{G}$ is vaguely analogous to the magnitude of a vector, we can say that the "amplitude" of a unitary dyadic is "unity."

Referring to Eq. (1.3.8) for the Euler-angle rotation, we see that, if the components of vector **F** are its components in the coordinate system x, y, z, then the components of $\mathfrak{G} \cdot \mathbf{F}$ are the components of **F** in the coordinate system ξ_1, ξ_2, ξ_3. Thus the unitary dyadic $\mathfrak{G}$ might be considered to represent the change of components of a vector caused by the rotation of the coordinate system. Instead of considering the coordinate system fixed and the vector changing, we can, in this case, consider the vector as unchanged in amplitude and direction and the coordinate axes as rotated, with the new components of vector **F** given by $\mathfrak{G} \cdot \mathbf{F}$. We note the correspondence of Eqs. (1.3.9) and (1.6.5) for a unitary dyadic.

If a unitary dyadic can represent the change in vector components caused by a rotation of axes, we might ask whether the change in components of a general dyadic $\mathfrak{A}$ caused by the same rotation may not be expressed in terms of the same unitary dyadic. This is answered in the affirmative by appeal to the last of Eqs. (1.6.1) or to the following: If $\mathfrak{G}$ is the unitary dyadic representing the rotation of axes and if $\mathfrak{A}$ is any dyadic, transforming vector **A** into vector **B** by the equation $\mathfrak{A} \cdot \mathbf{A} = \mathbf{B}$, then $\mathfrak{G} \cdot \mathbf{A}$ and $\mathfrak{G} \cdot \mathbf{B}$ give the components of **A** and **B**, respectively, in the new coordinates; and by juggling the equation relating **A** and **B**, we have

$$\mathfrak{G} \cdot \mathbf{B} = \mathfrak{G} \cdot \mathfrak{A} \cdot \mathbf{A} = (\mathfrak{G} \cdot \mathfrak{A} \cdot \mathfrak{G}^{-1}) \cdot (\mathfrak{G} \cdot \mathbf{A}) = (\mathfrak{G} \cdot \mathfrak{A} \cdot \mathfrak{G}^{*}) \cdot (\mathfrak{G} \cdot \mathbf{A})$$

Consequently the dyadic $\mathfrak{G} \cdot \mathfrak{A} \cdot \mathfrak{G}^*$, operating on the transform of **A**, gives the transform of **B**, which is, of course, *the definition of the transform of the dyadic* $\mathfrak{A}$. In other words the components of the dyadic $\mathfrak{G} \cdot \mathfrak{A} \cdot \mathfrak{G}^*$ are the components of $\mathfrak{A}$ in the new coordinates produced by the rotation represented by the unitary dyadic $\mathfrak{G}$.

In particular, if $\mathfrak{G}_A$ represents the rotation from the axes x, y, z to the principal axes of a symmetric dyadic $\mathfrak{A}_s$, then the transformed dyadic $(\mathfrak{G}_A \cdot \mathfrak{A}_s \cdot \mathfrak{G}_A^*)$ has the simple, diagonal form

$$(\mathfrak{G}_A \cdot \mathfrak{A}_s \cdot \mathfrak{G}_A^*) = \begin{pmatrix} A_1 & 0 & 0 \\ 0 & A_2 & 0 \\ 0 & 0 & A_3 \end{pmatrix}$$

No matter what sort of rotation is represented by $\mathfrak{G}$, as long as $\mathfrak{G}$ is unitary and its components are real, the transform $(\mathfrak{G} \cdot \mathfrak{A} \cdot \mathfrak{G}^{-1})$ of the dyadic $\mathfrak{A}$ is symmetric if $\mathfrak{A}$ is symmetric or is antisymmetric if $\mathfrak{A}$ is antisymmetric.

Dyadic Fields. So far we have been discussing the properties of a dyadic at a single point in space. A dyadic field is a collection of nine quantities, transforming in the manner given in Eq. (1.6.1), which are functions of x, y, z or ξ_1, ξ_2, ξ_3. At each point in space the dyadic represents an operator which changes a vector at that point into another vector, the amount of change varying from point to point. From still another point of view the expansion factor, the principal axes, and the rotation vector of the dyadic are all functions of position.

A dyadic field $\mathfrak{A}$ with components A_{nm} can be obtained from the covariant derivative of a vector field **F** [see Eq. (1.5.8)]

$$\begin{aligned} A_{nm} &= \left(\frac{1}{h_m h_n}\right) f_{m,n} = \left(\frac{1}{h_m h_n}\right) (h_m F_m)_{,n} \\ A_{mm} &= \frac{\partial}{\partial \xi_m}\left(\frac{F_m}{h_m}\right) + \frac{1}{h_m} \sum_n \frac{F_n}{h_n}\frac{\partial h_m}{\partial \xi_n} \\ A_{nm} &= \frac{1}{h_n}\frac{\partial F_m}{\partial \xi_n} - \frac{F_n}{h_m h_n}\frac{\partial h_n}{\partial \xi_m}; \quad m \neq n \end{aligned} \tag{1.6.13}$$

As shown in Eq. (1.5.10) *et seq.*, the expansion factor $|\mathfrak{A}|$ for this dyadic is the divergence of **F**, and the rotation vector $\langle\mathfrak{A}\rangle$ is the curl of **F**. This dyadic is, therefore, symmetric only when the vector **F** has zero curl.

The dyadic defined in Eq. (1.6.13) can be written symbolically as $\nabla\mathbf{F}$, with components along the x, y, z axes given by the formula

$$\begin{gathered}\mathfrak{A} = \mathbf{i}\frac{\partial\mathbf{F}}{\partial x} + \mathbf{j}\frac{\partial\mathbf{F}}{\partial y} + \mathbf{k}\frac{\partial\mathbf{F}}{\partial z} = \mathbf{A}_x^*\mathbf{i} + \mathbf{A}_y^*\mathbf{j} + \mathbf{A}_z^*\mathbf{k} \\ \mathbf{A}_x^* = \operatorname{grad} F_x;\quad \mathbf{A}_y^* = \operatorname{grad} F_y;\quad \mathbf{A}_z^* = \operatorname{grad} F_z\end{gathered} \tag{1.6.14}$$

The conjugate dyadic is, of course,

$$\mathfrak{A}^* = \mathbf{F}\nabla = \frac{\partial\mathbf{F}}{\partial x}\mathbf{i} + \frac{\partial\mathbf{F}}{\partial y}\mathbf{j} + \frac{\partial\mathbf{F}}{\partial z}\mathbf{k} = \mathbf{i}\mathbf{A}_x^* + \mathbf{j}\mathbf{A}_y^* + \mathbf{k}\mathbf{A}_z^*$$

The change in the vector **F** in a distance represented by the elementary vector $d\mathbf{r} = \mathbf{i}\,dx + \mathbf{j}\,dy + \mathbf{k}\,dz$ is obtained by operating on $d\mathbf{r}$ by $\nabla\mathbf{F}$.

$$d\mathbf{r}\cdot(\nabla\mathbf{F}) = \frac{\partial\mathbf{F}}{\partial x}dx + \frac{\partial\mathbf{F}}{\partial y}dy + \frac{\partial\mathbf{F}}{\partial z}dz = d\mathbf{F} \tag{1.6.15}$$

The symmetric dyadic corresponding to $\nabla\mathbf{F}$ is, of course, $\frac{1}{2}(\nabla\mathbf{F} + \mathbf{F}\nabla)$, having zero rotation vector.

The variation of a dyadic field from point to point can be computed by means of the differential operator ∇. For instance, $\nabla\cdot\mathfrak{A} = \mathbf{i}\cdot(\partial\mathfrak{A}/\partial x) + \mathbf{j}\cdot(\partial\mathfrak{A}/\partial y) + \mathbf{k}\cdot(\partial\mathfrak{A}/\partial z)$ is a vector formed by covariant differentiation of the corresponding mixed tensor and contracting the resulting third-order tensor.

$$\left(\frac{1}{h_n}\right)\sum_m a^m_{n,m} = \left(\frac{1}{h_n}\right)\sum_m\left(\frac{h_nA_{mn}}{h_m}\right)_{,m} = (\nabla\cdot\mathfrak{A})_n$$

In terms of the expansion in *component vectors*, given in Eq. (1.6.7), this vector is $(\partial\mathbf{A}_x/\partial x) + (\partial\mathbf{A}_y/\partial y) + (\partial\mathbf{A}_z/\partial z)$; whereas the conjugate vector

$$\mathfrak{A}\cdot\nabla = \nabla\cdot\mathfrak{A}^* = \mathbf{i}(\operatorname{div}\mathbf{A}_x) + \mathbf{j}(\operatorname{div}\mathbf{A}_y) + \mathbf{k}(\operatorname{div}\mathbf{A}_z) \tag{1.6.16}$$

The physical interpretation of this quantity will be discussed later.

There is also the dyadic formed by using the curl operator,

$$\begin{aligned}\nabla\times\mathfrak{A} &= \mathbf{i}\times\frac{\partial\mathfrak{A}}{\partial x} + \mathbf{j}\times\frac{\partial\mathfrak{A}}{\partial y} + \mathbf{k}\times\frac{\partial\mathfrak{A}}{\partial z} \\ &= \mathbf{i}\left(\frac{\partial\mathbf{A}_z}{\partial y} - \frac{\partial\mathbf{A}_y}{\partial z}\right) + \mathbf{j}\left(\frac{\partial\mathbf{A}_x}{\partial z} - \frac{\partial\mathbf{A}_z}{\partial x}\right) + \mathbf{k}\left(\frac{\partial\mathbf{A}_y}{\partial x} - \frac{\partial\mathbf{A}_x}{\partial y}\right) \\ &= (\operatorname{curl}\mathbf{A}_x^*)\mathbf{i} + (\operatorname{curl}\mathbf{A}_y^*)\mathbf{j} + (\operatorname{curl}\mathbf{A}_z^*)\mathbf{k}\end{aligned} \tag{1.6.17}$$

Related to these differential properties there are integral properties of dyadics analogous to Gauss' theorem [Eq. (1.4.8)] and to Stokes' theorem [Eq. (1.4.11)]. For a surface integral over a closed surface, for any dyadic $\mathfrak{B}$

$$\oint d\mathbf{A} \cdot \mathfrak{B} = \int \nabla \cdot \mathfrak{B} \, dv \tag{1.6.18}$$

where the volume integral is over the volume "inside" the surface and where the area element $d\mathbf{A}$ points outward, away from the "inside." The integrals, of course, result in vectors. For a line integral around some closed contour, for any dyadic $\mathfrak{B}$

$$\oint d\mathbf{s} \cdot \mathfrak{B} = \int d\mathbf{A} \cdot (\nabla \times \mathfrak{B}) \tag{1.6.19}$$

where the area integral is over a surface founded by the contour.

Deformation of Elastic Bodies. An important application of dyadic algebra is in the representation of the deformation of elastic bodies. A rigid body moves and rotates as a whole, but an elastic body can in addition change the relative position of its internal parts. For such a body the displacement of the part of the body originally at (x,y,z) is expressed in terms of three vectors:

$$\mathbf{D}(x,y,z) = \mathbf{T} + \mathbf{P}(x,y,z) + \mathbf{s}(x,y,z)$$

Here $\mathbf{T}$ is the constant vector representing the average translation of the body, $\mathbf{P}$ the part of the displacement due to the average rotation about the center of gravity, and $\mathbf{s}$ is the additional displacement due to the distortion of the body. By definition, therefore, $\mathbf{s}$ is zero at the center of gravity of the body and is zero everywhere if the body is perfectly rigid.

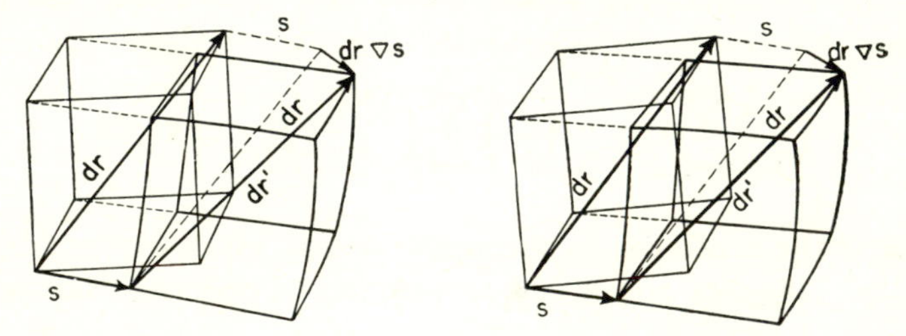

Fig. 1.19 Strain in elastic medium, producing displacement $\mathbf{s}$ and deformation represented by dyadic $\mathfrak{D} = \nabla\mathbf{s}$.

In general $\mathbf{s}$ is much smaller than either $\mathbf{T}$ or $\mathbf{P}$ may happen to be. In the present action we shall forget $\mathbf{T}$ and $\mathbf{P}$ and concentrate on $\mathbf{s}$, for we are not interested in the gross motions of the body, only its internal warps and strains.

Even the relative displacement $\mathbf{s}$ is not a good measure of the local strain in an elastic medium, for $\mathbf{s}$ is the total relative displacement of the point (x,y,z), which may be larger the farther (x,y,z) is from the center of gravity, even though the strain in any element of the medium is about the same everywhere. What is needed is a differential quantity which measures the strain *at the point* (x,y,z). This is obtained by computing the change in the vector $d\mathbf{r} = \mathbf{i}\,dx + \mathbf{j}\,dy + \mathbf{k}\,dz$ connecting the points (x,y,z) and $(x + dx,\ y + dy,\ z + dz)$ when the body is distorted. The point (x,y,z) is displaced by the amount given by the vector $\mathbf{s}(x,y,z)$, and the displacement of point $(x + dx,\ y + dy,\ z + dz)$ is given by the vector $\mathbf{s}(x + dx,\ y + dy,\ z + dz)$. The change in $d\mathbf{r}$ due to the strain is, to the first order,

$$dx\,\frac{\partial \mathbf{s}}{\partial x} + dy\,\frac{\partial \mathbf{s}}{\partial y} + dz\,\frac{\partial \mathbf{s}}{\partial z} = d\mathbf{r}\cdot\nabla\mathbf{s}$$

according to Eq. (1.6.15). Therefore, the vector $d\mathbf{r}$ representing the relative position of the points (x,y,z) and $(x + dx,\ y + dy,\ z + dz)$ is changed by the strain into the vector $d\mathbf{r}'$ representing the new relative position, where

$$d\mathbf{r}' = d\mathbf{r}\cdot(\mathfrak{I} + \mathfrak{D});\quad \mathfrak{D} = \nabla\mathbf{s} \tag{1.6.20}$$

The dyadic $\mathfrak{D}$ is a derivative operator, measuring the amount of strain *at the point* (x,y,z). As indicated on page 58, it may be divided into a symmetric and antisymmetric part;

$$\begin{aligned}
\mathfrak{D} &= \mathfrak{R} + \mathfrak{S};\quad \mathfrak{R} = -\tfrac{1}{2}(\text{curl }\mathbf{s})\times\mathfrak{I}\\
\mathfrak{S} &= \mathbf{i}e_{11}\mathbf{i} + \mathbf{j}e_{22}\mathbf{j} + \mathbf{k}e_{33}\mathbf{k} + e_{12}(\mathbf{ij} + \mathbf{ji}) + e_{13}(\mathbf{ik} + \mathbf{ki}) + e_{23}(\mathbf{jk} + \mathbf{kj})\\
&= \tfrac{1}{2}(\nabla\mathbf{s} + \mathbf{s}\nabla)
\end{aligned} \tag{1.6.21}$$

$$e_{11} = \frac{\partial s_x}{\partial x},\ \text{etc.};\quad e_{12} = \frac{1}{2}\left(\frac{\partial s_x}{\partial y} + \frac{\partial s_y}{\partial x}\right),\ \text{etc.}$$

The dyadic $\mathfrak{R}$ corresponds to the rotation of the element of volume around (x,y,z) due to the distortion of the medium. The axis of rotation is in the direction of curl $\mathbf{s}$, and the angle of rotation in radians is equal to the magnitude of curl $\mathbf{s}$. Note that this term is *not* due to the rotation of the body as a whole (for this part of the motion has been specifically excluded from our present study); it is due to the twisting of the material as it is being strained. This type of rotation is zero when curl $\mathbf{s}$ is zero.

The symmetric dyadic $\mathfrak{S}$ is called the *pure strain dyadic* for the point (x,y,z). When it is equal to zero, there is no true strain at this point.

As pointed out earlier, it is always possible to find three mutually perpendicular directions, the principal axes labeled by the unit vectors $\mathbf{a}_1$, $\mathbf{a}_2$, $\mathbf{a}_3$, in terms of which the symmetric tensor $\mathfrak{S}$ becomes

$$\mathbf{a}_1e_1\mathbf{a}_1 + \mathbf{a}_2e_2\mathbf{a}_2 + \mathbf{a}_3e_3\mathbf{a}_3 = \mathfrak{S} \tag{1.6.22}$$

The three quantities e_1, e_2, e_3 are called the *principal extensions* of the medium at (x,y,z). A rectangular parallelepiped of sides $d\xi_1$, $d\xi_2$, $d\xi_3$ oriented parallel to the principal axes is still a rectangular parallelepiped after the medium is strained (this would not be true if it were not parallel to the principal axes), but the length of its sides is now $(1 + e_1)\,d\xi,$

$(1 + e_2)\, d\xi_2$, $(1 + e_3)\, d\xi_3$. Therefore, the percentage *increase in volume* of the parallelepiped is

$$\theta = (e_1 + e_2 + e_3) = (e_{11} + e_{22} + e_{33}) = \mathrm{div}(\mathbf{s}) = |\mathfrak{D}| \quad (1.6.23)$$

The quantity θ, which is the expansion factor of the dyadic $\mathfrak{D}$ (in any coordinate system), is called the *dilation* of the medium at the point

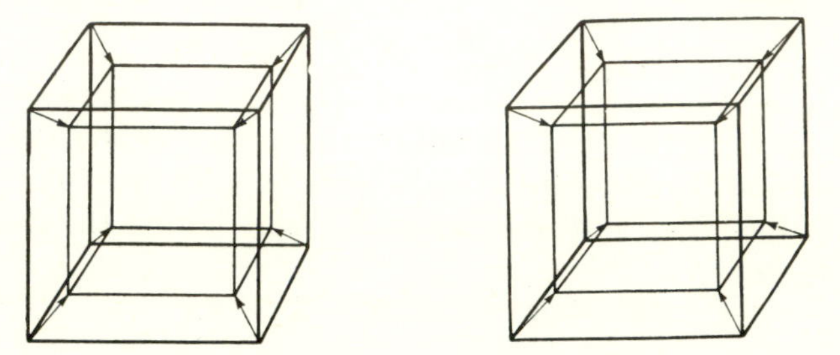

Fig. 1.20 Change of element of elastic medium in simple contraction.

(x,y,z). It also equals the percentage *decrease in density* (to the first order in the small quantities e) of the medium at (x,y,z).

Types of Strain. The simplest type of strain corresponds to a strain dyadic which is independent of position; this is called *homogeneous* strain. The simplest type of homogeneous strain is a *simple expansion*, corresponding to a distribution of displacement $\mathbf{s}$ and of strain dyadic $\mathfrak{S}$ as follows:

$$\mathbf{s} = e(x\mathbf{i} + y\mathbf{j} + z\mathbf{k});\quad \mathfrak{D} = \mathfrak{S} = e\mathfrak{I} \quad (1.6.24)$$

This type of strain is isotropic; any axes are principal axes; there is no rotation due to the strain. The dilation is $\theta = 3e$.

Another type of homogeneous strain, called *simple shear*, is obtained when the extension along one principal axis is equal and opposite to that along another, that along the third being zero:

$$\mathbf{s} = \tfrac{1}{2}e(x\mathbf{i} - y\mathbf{j});\quad \mathfrak{D} = \mathfrak{S} = \tfrac{1}{2}e(\mathbf{ii} - \mathbf{jj})$$

The dilation is zero, the extension in the x direction being just canceled

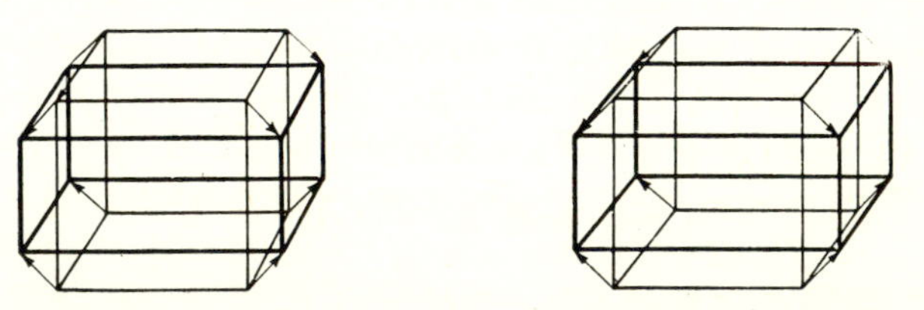

Fig. 1.21 Change of element of elastic medium in simple shear, as given by dyadic $e(\mathbf{ii} - \mathbf{jj})$.

by the contraction in the y direction. If the axes are rotated 45° about the z axis ($\sqrt{2}\,x = x' + y'$; $\sqrt{2}\,y = x' - y'$), the displacement and strain dyadic take on the form

$$\mathbf{s} = \tfrac{1}{2}e(x'\mathbf{j}' + y'\mathbf{i}'); \quad \mathfrak{D} = \mathfrak{S} = \tfrac{1}{2}e(\mathbf{i}'\mathbf{j}' + \mathbf{j}'\mathbf{i}')$$

This type of strain is called *pure shear*. If it is combined with a rigid rotation of the medium by $\frac{1}{2}e$ radians, corresponding to $\mathfrak{R} = -\frac{1}{2}e(\mathbf{i}'\mathbf{j}' - \mathbf{j}'\mathbf{i}')$, the resulting displacement and strain dyadic,

$$\mathbf{s} = ey'\mathbf{i}'; \quad \mathfrak{D} = e\mathbf{j}'\mathbf{i}' \tag{1.6.25}$$

correspond to what is usually called *shear in the y' direction*. The displacement is entirely in the x' direction, the parts of the medium sliding over each other as a pack of cards can be made to do.

Another type of homogeneous strain with zero dilation corresponds to a stretch in the x direction and a proportional shrinking in both y and z directions:

$$\mathbf{s} = e(x\mathbf{i} - \tfrac{1}{2}y\mathbf{j} - \tfrac{1}{2}z\mathbf{k}); \quad \mathfrak{D} = \mathfrak{S} = e(\mathbf{ii} - \tfrac{1}{2}\mathbf{jj} - \tfrac{1}{2}\mathbf{kk}) \tag{1.6.26}$$

This is the sort of strain set up in a material such as rubber when it is stretched in the x direction. It can be called a *dilationless stretch*.

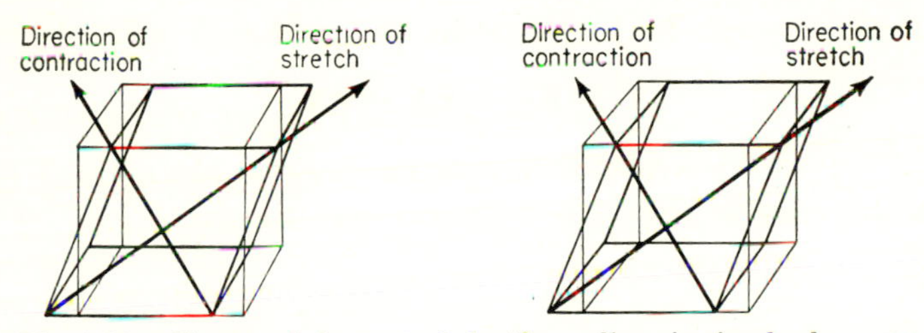

Fig. 1.22 Change of element of elastic medium in simple shear, as given in dyadic $e\mathbf{ji}$.

The most general homogeneous strain, referred to its principal axes, can be made up by combining various amounts of simple expansion and of shear and dilationless stretch in the three directions. Rotation of axes can then give the most general form to the expressions for $\mathbf{s}$ and $\mathfrak{S}$.

A simple type of nonhomogeneous strain is a helical twist along the x axis:

$$\begin{aligned}
\mathbf{s} &= ex(y\mathbf{k} - z\mathbf{j}) \\
\mathfrak{D} &= \mathfrak{R} + \mathfrak{S} = e[\mathbf{i}y\mathbf{k} - \mathbf{i}z\mathbf{j} + \mathbf{j}x\mathbf{k} - \mathbf{k}x\mathbf{j}] \\
\mathfrak{R} &= \tfrac{1}{2}e[2x(\mathbf{jk} - \mathbf{kj}) + y(\mathbf{ik} - \mathbf{ki}) + z(\mathbf{ji} - \mathbf{ij})] \\
&= -e[x\mathbf{i} - \tfrac{1}{2}y\mathbf{j} - \tfrac{1}{2}z\mathbf{k}] \times \mathfrak{J} \\
\mathfrak{S} &= e[y(\mathbf{ki} + \mathbf{ik}) - z(\mathbf{ij} + \mathbf{ji})]
\end{aligned} \tag{1.6.27}$$

This corresponds to a rotation of the element at (x,y,z) by an amount ex radians about the x axis [the term $ex\mathbf{jk} - ex\mathbf{kj}$ in $\mathfrak{D}$, see Eq. (1.6.10)] and a shear in the x direction of an amount proportional to the length of the vector $\mathbf{r} = y\mathbf{j} + z\mathbf{k}$ joining (x,y,z) with the x axis [the term $ey\mathbf{ik} - ez\mathbf{ij}$ in $\mathfrak{D}$, see Eq. (1.6.25)].

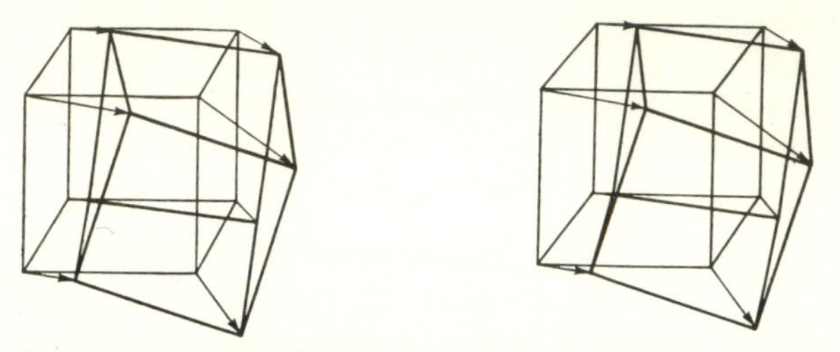

Fig. 1.23 Change of element of elastic medium in torsion, as given in Eqs. (1.6.27).

Stresses in an Elastic Medium. The forces inside an elastic medium which produce the strains are called *stresses*. They also are best represented by dyadics. The force across the element $dy\,dz$, perpendicular to the x axis is $\mathbf{F}_x\,dy\,dz$, for instance, where $\mathbf{F}_x$ is not necessarily parallel to the x axis. Similarly the forces across unit areas normal to the y and z axes can be called $\mathbf{F}_y$ and $\mathbf{F}_z$. It can easily be shown that the force across an area represented by the axial vector $d\mathbf{A}$ is $\mathfrak{T} \cdot d\mathbf{A}$, where

$$\mathfrak{T} = \mathbf{F}_x\mathbf{i} + \mathbf{F}_y\mathbf{j} + \mathbf{F}_z\mathbf{k}$$

More detailed consideration of the relation between the forces $\mathbf{F}$ and the areas $d\mathbf{A}$ will confirm that $\mathfrak{T}$ is a dyadic and transforms like any other dyadic.

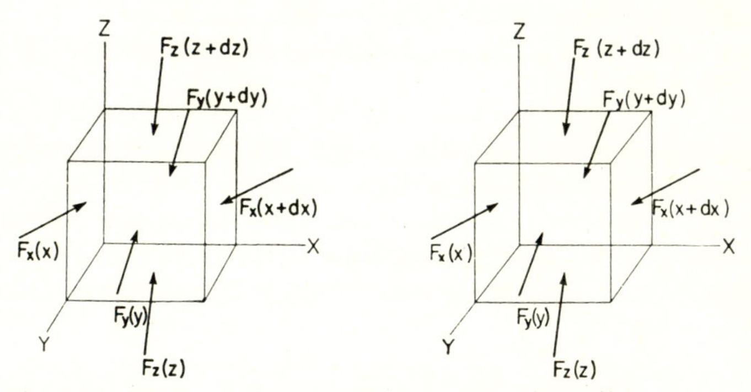

Fig. 1.24 Forces on faces of element of elastic medium corresponding to stress dyadic $\mathbf{F}_x\mathbf{i} + \mathbf{F}_y\mathbf{j} + \mathbf{F}_z\mathbf{k}$.

For static equilibrium these forces should not cause a rotation of any part of the medium. A consideration of the torques on an element of volume of the medium shows that

$$(\mathbf{F}_x)_y = (\mathbf{F}_y)_x; \quad \text{etc.}$$

Consequently the dyadic $\mathfrak{T}$ is symmetric and is equal to its conjugate $\mathfrak{T}^* = \mathbf{i}\mathbf{F}_x + \mathbf{j}\mathbf{F}_y + \mathbf{k}\mathbf{F}_z$. In terms of its principal axes and related orthogonal unit vectors $\mathbf{a}_n$, this can be written as

$$\mathfrak{T} = T_1\mathbf{a}_1\mathbf{a}_1 + T_2\mathbf{a}_2\mathbf{a}_2 + T_3\mathbf{a}_3\mathbf{a}_3$$

where the constant T_n is the *principal stress* along the nth principal axis. Various simple types of stress can be set up analogous to the strains given previously. For instance, where $T_2 = T_3 = 0$, the stress is called a *tension* in the $\mathbf{a}_1$ direction; when $T_2 = -T_1$ and $T_3 = 0$, it is called a *shearing stress;* and so on. The scalar $|\mathfrak{T}|$ is minus three times the pressure at the point.

Static Stress-Strain Relations for an Isotropic Elastic Body. When the elastic properties of a medium are independent of direction, it is said to be *isotropic.* If the medium can remain in equilibrium under a shearing stress, it is said to be *elastic.* When both of these requirements are satisfied, it turns out that the principal axes of strain are identical with the principal axes of stress everywhere and the strains due to the three principal stresses are independent and additive. For instance, the effect due to the principal stress T_1 will be a simple dilation plus an extension in the $\mathbf{a}_1$ direction. In other words the equations relating the principal strains and the principal stresses are

$$T_n = \lambda(e_1 + e_2 + e_3) + 2\mu e_n \qquad n = 1, 2, 3, \ldots$$

where the constants λ and μ are determined by the elastic properties of the medium.

Referred to the usual axes, x, y, z, both stress and strain dyadics take on a more general symmetrical form:

$$\begin{aligned} \mathfrak{T} &= T_{xx}\mathbf{ii} + \cdots + T_{xy}(\mathbf{ij} + \mathbf{ji}) + \cdots \\ \mathfrak{S} &= e_{xx}\mathbf{ii} + \cdots + e_{xy}(\mathbf{ij} + \mathbf{ji}) + \cdots \end{aligned}$$

The equations relating $\mathfrak{T}$ and $\mathfrak{S}$ and their components can be found by transformation from the principal axes:

$$\begin{aligned} \mathfrak{T} &= \lambda|\mathfrak{S}|\mathfrak{J} + 2\mu\mathfrak{S} \\ T_{xx} &= \lambda(e_{xx} + e_{yy} + e_{zz}) + 2\mu e_{xx}; \quad \text{etc.} \\ T_{xy} &= 2\mu e_{xy}; \quad \text{etc.} \end{aligned} \tag{1.6.28}$$

When the stress is an isotropic pressure, $\mathfrak{T} = -P\mathfrak{J}$, the strain is a simple compression, $\mathfrak{S} = -[P/(3\lambda + 2\mu)]\mathfrak{J}$. The constant $\frac{1}{3}(3\lambda + 2\mu) = P/\theta$ is, therefore, the *compression* or *bulk modulus* of the isotropic elastic medium. When the stress is a simple shearing stress $\mathfrak{T} = S(\mathbf{ii} - \mathbf{jj})$, the strain is a simple shear $\frac{1}{2}(S/\mu)(\mathbf{ii} - \mathbf{jj})$, so that the *shear modulus* of the medium is μ. When the stress is a simple tension in the x direction, $\mathfrak{T} = T\mathbf{ii}$, the strain is $\mathfrak{S} = [T/2\mu(3\lambda + 2\mu)][2(\lambda + \mu)\mathbf{ii} - \lambda(\mathbf{jj} + \mathbf{kk})]$, representing a stretch in the x direction and a shrinking in the y and z directions. The quantity $[\mu(3\lambda + 2\mu)/(\lambda + \mu)]$ is the ordinary *tension* or *Young's modulus* of the material. The ratio of lateral shrink to longitudinal stretch $[\lambda/2(\lambda + \mu)]$ is called *Poisson's ratio*.

Dyadic Operators. In order to discuss the relation between stress and strain dyadics for nonisotropic solids, we must introduce quantities which transform dyadics in a manner similar to the way dyadics transform vectors. The components must have four subscripts and should transform in a manner analogous to Eq. (1.6.1):

$$(G_{ijkl})' = \sum_{mnrs} \gamma_{im}\gamma_{jn}\gamma_{kr}\gamma_{ls}G_{mnrs}$$

These operators may be called *tetradics* and may be represented by the Hebrew characters in order to distinguish them from other entities. For instance the symbol representing the 81 components G_{ijkl} is ג (gimel), and the equation giving the nature of the transformation performed on a dyadic is

$$\mathfrak{B} = \text{ג} : \mathfrak{A} \quad \text{or} \quad B_{mn} = \sum_{rs} G_{mnrs}A_{rs} \tag{1.6.29}$$

A tetradic may be represented by juxtaposing two dyadics, just as a dyadic may be represented by juxtaposing two vectors. For instance a particularly simple tetradic is ע (ayin) = $\mathfrak{I}\mathfrak{I}$, which changes every dyadic into a constant times the idemfactor

$$\text{ע} : \mathfrak{B} = |\mathfrak{B}|\mathfrak{I}; \quad Y_{mnrs} = \delta_{mn}\delta_{rs}$$

There is, of course, the unity tetradic, י (yod), which reproduces every dyadic, and its conjugate,

$$\text{י} : \mathfrak{A} = \mathfrak{A}; \quad (\text{י}^*) : \mathfrak{A} = \mathfrak{A}^*; \quad I_{mnrs} = \delta_{mr}\delta_{ns}; \quad I^*_{mnrs} = \delta_{ms}\delta_{nr}$$

In this notation the stress dyadic is related to the strain dyadic, for isotropic solids, by the equation

$$\mathfrak{T} = [\lambda\text{ע} + \mu\text{י} + \mu\text{י}^*] : \mathfrak{S}$$

where the tetradic in the square brackets is of particularly simple form due to the isotropy of the medium. For nonisotropic media the relation is more complicated, and the elements of the tetradic ד (daleth)

$$\mathfrak{T} = \text{ד} : \mathfrak{S}; \quad T_{mn} = \sum_{rs} D_{mnrs}S_{rs}$$

are not mostly zero. Since both $\mathfrak{T}$ and $\mathfrak{S}$ are symmetric, we must have $D_{mnrs} = D_{nmrs}$ and $D_{nmrs} = D_{mnsr}$; we also have $D_{mnrs} = D_{rsmn}$. With all of these symmetries, the number of independent components of ד is reduced from 81 to 21. These components are called the *elastic components* of the nonisotropic solid.

One could carry out an analysis of tetradics, their "principal axes," and other properties in a manner quite analogous to the way we have analyzed the properties of dyadics. Lack of space and of importance for our work, however, forbids.

Complex Numbers and Quaternions as Operators. Before we go on to less familiar fields, it is well to review an example of vector operators which is so familiar that it is often overlooked. The use of complex numbers to represent vectors in two dimensions is commonplace; it is not as well realized that a complex number can also represent a linear vector operator for two dimensions.

The subject of complex numbers and of functions of a complex variable will be treated in considerable detail in Chap. 4, for we shall be using complex numbers as solutions to our problems all through this book. All that needs to be pointed out here is that the real unit 1 can be considered to represent a unit vector along the x axis and the imaginary unit $i = \sqrt{-1}$ can be considered to represent a unit vector along the y axis; then a vector in two dimensions with components x and y can be represented by the complex number $z = x + iy$. Such a quantity satisfies the usual rules for addition of vectors (that is, we add components) and of multiplication by a scalar (that is, $az = ax + iay$).

The vector which is the mirror image of z in the x axis is called the *complex conjugate* of z, $\bar{z} = x - iy$. The angle between z and the x axis is $\tan^{-1}(y/x)$, and the square of the length of z is $|z|^2 = z\bar{z} = x^2 + y^2 = \bar{z}z$. We note that multiplication of complex numbers does not correspond to the rules for multiplication of three-vectors. If $z = x + iy$ and $w = u + iv$, then $wz = zw = (ux - vy) + i(uy + vx)$ *is another vector* in the x, y plane. It is not the scalar product of the two vectors (the scalar product $ux + vy$ is the real part of $w\bar{z}$), nor is it the vector product (the vector product would have an amplitude equal to the imaginary part of $w\bar{z}$, but its direction would be perpendicular to both w and z, which requires the third dimension). Actually the product wz of two complex numbers corresponds more closely to the operation of a particular kind of dyadic w on a vector z.

The operation of multiplying by w changes the direction and magnitude of z. To put this in usual vector and dyadic form, the vector z would be $x\mathbf{i} + y\mathbf{j}$ and the dyadic w would be $u\mathbf{ii} - v\mathbf{ij} + v\mathbf{ji} + u\mathbf{jj}$, a combination of the antisymmetric dyadic $v(\mathbf{ji} - \mathbf{ij})$ and the symmetric dyadic $u(\mathbf{ii} + \mathbf{jj})$, with principal axes along the x and y axes. This dyadic is certainly not the most general two-dimensional linear operator we can devise, for it has only two independent constants rather than four. It represents the particularly simple type of operation which changes the direction of any vector by a constant angle and changes the magnitude of any vector by a constant factor, as we shall see in a moment.

The complex number represented symbolically by the exponential $e^{i\theta}$, where θ is a real number, is, by Euler's equation,

$$e^{i\theta} = \cos\theta + i\sin\theta$$

Viewed as an operator, this quantity *rotates any vector* $\mathbf{z}$ by an angle θ radians (counterclockwise) and *does not change its length.* Consequently, the operator $w = Ce^{i\theta}\,[C^2 = u^2 + v^2,\ \theta = \tan^{-1}(v/u)]$, when it multiplies any vector $z = x + iy$, rotates z by an angle θ and increases its length by a factor C. Many of our solutions will be representable as a complex number ψ multiplied by the time factor $e^{-i\omega t}$. This factor rotates the vector ψ with constant angular velocity ω, and if (as is often the case) the physical solution is the real part of the result, the solution will oscillate sinusoidally with time, with a frequency $\nu = \omega/2\pi$.

An extension of this same type of representation to three dimensions is not possible, which is why we have had to use the more complicated formalism of vectors and dyadics. It was shown by Hamilton, however, that vectors and operators in *four* dimensions can be represented by a rather obvious generalization of complex numbers, called *quaternions.*

We let the unit number 1 represent the unit vector in the fourth dimension and the unit vectors in the three space dimensions be three quantities i, j, k, with rules of multiplication analogous to those for $\sqrt{-1}$:

$$i^2 = j^2 = k^2 = -1;\quad ij = -ji = k;\quad jk = -kj = i;\quad ki = -ik = j$$

Then a space vector could be represented by the quantity $ix + jy + kz$, and a general quaternion $q = a + ib + jc + kd$ would represent a four-vector. The conjugate four-vector is $q^* = a - ib - jc - kd$, so that the square of the magnitude of f is $|q|^2 = q^*q = a^2 + b^2 + c^2 + d^2$, a simple extension of the rule for ordinary complex numbers.

As with complex numbers, the multiplication of quaternions pq can be considered as the operation which changes a four-vector represented by q into the four-vector represented by pq. If $q = a + ib + jc + kd$ and $p = \alpha + i\beta + j\gamma + k\delta$, then

$$\begin{aligned} pq = (\alpha a - \beta b - \gamma c - \delta d) + i(\alpha b + \beta a + \gamma d - \delta c) \\ + j(\alpha c - \beta d + \gamma a + \delta b) + k(\alpha d + \beta c - \gamma b + \delta a) \end{aligned} \quad (1.6.30)$$

is another quaternion, which can represent another four-vector. We note that quaternion multiplication is not commutative, that $pq \neq qp$. We cannot pursue this general discussion further (beyond pointing out that p cannot represent the most general four-dimensional dyadic), though the relationship between quaternions and space-time vectors in relativity will be mentioned later.

However, it is useful to consider a particular quaternion which is an interesting and suggestive generalization of the complex exponential. Utilizing the rules for multiplication of i, j, k and expanding the exponential, we can show that, when $\alpha^2 + \beta^2 + \gamma^2 = 1$,

$$e^{\theta(i\alpha+j\beta+k\gamma)} = \cos\theta + \sin\theta\,(i\alpha + j\beta + k\gamma)$$

which is the analogue of Euler's equation for the imaginary exponential. The exponent represents a space vector of length θ and of direction given by the direction cosines α, β, γ, whereas the exponential is of unit length. We note that *any* quaternion can be represented in the form $Qe^{\theta(i\alpha+j\beta+k\gamma)}$, where the number Q is the length of the four-vector and where the angle θ and the direction cosines fix the direction of the vector in four-space. We might expect, by analogy with the complex exponential, that this exponential operator (with $Q = 1$) is somehow related to a rotation operator, though the relationship cannot be so simple as for the complex plane.

The interesting and suggestive relationship is as follows: If f is a vector $ix + jy + kz$, in quaternion notation, then the quaternion

$$f' = x'i + y'j + z'k = e^{(\theta/2)(i\alpha+j\beta+k\gamma)}fe^{-(\theta/2)(i\alpha+j\beta+k\gamma)};\quad \text{where } \alpha^2 + \beta^2 + \gamma^2 = 1$$

represents a three-vector which is obtained by *rotating the vector f by an angle θ about an axis* having the direction cosines α, β, γ. Note that the angle of rotation about the axis is θ, not $\theta/2$. We can show this in general, but the algebra will be less cumbersome if we take the special case of rotation about the x axis. We have

$$\begin{aligned} f' &= e^{(\theta/2)i}(ix + jy + kz)e^{-(\theta/2)i} \\ &= ix + j(y\cos\theta - z\sin\theta) + k(y\sin\theta + z\cos\theta) \end{aligned}$$

which corresponds to the rotation of the y and z components of f by an angle θ in the y, z plane, *i.e.*, a rotation of the vector by an angle θ about the x axis. The proof of the general case only involves more algebra.

We can generalize this as follows: Suppose that we have a quaternion q expressed in terms of its magnitude Q and "direction operator" $e^{\theta(i\alpha+j\beta+k\gamma)}$; we can set up the formal "square-root quaternions" $\zeta = \sqrt{Q}\,e^{(\theta/2)(i\alpha+j\beta+k\gamma)}$ and $\zeta^* = \sqrt{Q}\,e^{-(\theta/2)(i\alpha+j\beta+k\gamma)}$; then the vector $f = ix + jy + kz$ is transformed into the vector f', with direction obtained from f by rotation by an angle θ about the axis with direction cosines α, β, γ and with magnitude equal to the magnitude $\sqrt{x^2 + y^2 + z^2}$ of f multiplied by the scalar factor Q, by means of the equation

$$f' = \zeta f \zeta^*$$

A subsequent operation, involving another rotation and change of magnitude, represented by the quaternions η, η^* is given by

$$f'' = \eta f' \eta^* = \eta \zeta f \zeta^* \eta^*$$

The geometric fact that two successive rotations give different end results depending on the order in which they are carried out is reflected in the fact that the quaternions η and ζ (and, of course, ζ^* and η^*) do not commute. The quantity Q is called the *tensor* (Latin for "stretcher") and the exponential is called the *versor* (Latin for "turner") of the operator ζ.

The most general rotation of a four-vector represented by $q = w + ix + jy + kz$ is given by the equation

$$q' = e^{\theta(i\alpha + j\beta + k\gamma)} q e^{-\varphi(i\lambda + j\mu + k\nu)}$$

where both $\alpha^2 + \beta^2 + \gamma^2$ and $\lambda^2 + \mu^2 + \nu^2$ equal unity. When both θ and φ are imaginary angles, the transformation corresponds to the Lorentz transformation discussed in the next section.

In later portions of this chapter and in Sec. 2.6, we shall have occasion to use generalizations of the versor $\mathfrak{Q} = e^{i\mathfrak{A}}$, where $\mathfrak{A}$ is a general dyadic operator. This function is always related to the rotation of the vector $\mathbf{F}$ on which the operator $\mathfrak{A}$ acts, and in many cases the rotation transformation will be represented by the formula

$$\mathbf{F}' = \mathfrak{Q} \cdot \mathbf{F} \cdot \mathfrak{Q}^*$$

as was the case with quaternions.

Abstract Vector Spaces. The three-dimensional concepts of vectors and dyadic operators discussed in the preceding sections may be generalized by considering vectors and dyadic operators in abstract spaces which may have many dimensions, often a denumerably infinite number of dimensions. This generalization has become one of the most powerful of modern mathematical tools, particularly because it permits a synthesis and a clearer understanding of a great many results taken from widely varying fields. We shall discuss this generalization briefly here, illustrating with some examples drawn from physics.

One of the simplest examples of the use of an abstract vector space occurs when normal coordinates are used to describe the motion of coupled oscillators. The number of normal coordinates, *i.e.*, the number of dimensions of the corresponding space, equals the number of degrees of freedom of the oscillator. The configuration or *state* of the system is described by a vector in this space. The principal axes of the space refer to particularly "elementary" states of motion, the most general motion being just a linear superposition of these "elementary" states.

These motions can be clarified by consideration of the system illustrated in Fig. 1.25. The "elementary" motions are two in number: (1) The masses oscillate in the same direction, *i.e.*, move together.

(2) The masses oscillate in opposite directions, *i.e.*, move toward each other and then away from each other. These motions are called elementary because there is a definite single frequency for each type of motion. The most general motion of the system is a linear superposition of the "elementary" states of motion and, as a result, has no definite frequency.

Let us now set up the two-dimensional space we require for the description of this system. We may plot, for example, x_1, the displacement of one of the masses along one of the axes, and x_2, the displacement

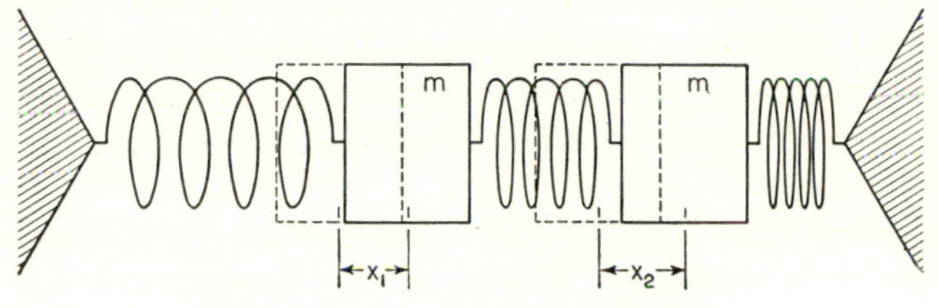

Fig. 1.25 Two coupled oscillators.

of the other along a perpendicular axis. If we let $\mathbf{e}_1$ be a unit vector in the one direction and $\mathbf{e}_2$ a unit vector in the two direction, then a general vector in this space is $\mathbf{e}_1 x_1 + \mathbf{e}_2 x_2 = \mathbf{r}$.

The equations of motion are

$$m(d^2x_1/dt^2) = -(k_1 + k_2)x_1 + k_2 x_2; \quad m(d^2x_2/dt^2) = -(k_1 + k_2)x_2 + k_2 x_1$$

or vectorwise

$$m(d^2\mathbf{r}/dt^2) = -\mathfrak{A} \cdot \mathbf{r}$$

where $\mathfrak{A}$ is the dyadic $\mathbf{e}_1 A_{11} \mathbf{e}_1 + \mathbf{e}_1 A_{12} \mathbf{e}_2 + \mathbf{e}_2 A_{21} \mathbf{e}_1 + \mathbf{e}_2 A_{22} \mathbf{e}_2$, and

$$A_{11} = (k_1 + k_2); \quad A_{12} = -k_2 = A_{21}; \quad A_{22} = (k_1 + k_2)$$

The "elementary" modes of motion $\mathbf{R}$ have a definite angular frequency ω; therefore they satisfy the relation $d^2\mathbf{R}/dt^2 = -\omega^2\mathbf{R}$, and the equation of motion for $\mathbf{R}$ becomes

$$\begin{vmatrix} A_{11} - m\omega^2 & A_{21} \\ A_{12} & A_{22} - m\omega^2 \end{vmatrix} = 0$$

Again, quoting the results given earlier, there are two elementary solutions $\mathbf{R}_1$ and $\mathbf{R}_2$ which are orthogonal to each other. Moreover, it is again clear that one may use these directions as new coordinate axes and that any vector in this two-dimensional space (*i.e.*, any possible motion) may be expressed as a linear superposition of the two elementary states of motion. The square of the cosine of the angle between a vector $\mathbf{F}$ and the $\mathbf{R}_1$ axis gives the relative fraction of the state designated by $\mathbf{F}$ which is of the $\mathbf{R}_1$ type, the remainder being of the $\mathbf{R}_2$ type, since the sum of the cosine square terms is unity. In other words, the square of this cosine gives the fraction of the total energy of the system which is of the $\mathbf{R}_1$ type.

Returning to the equations of motion, we now see that one may consider the motion of the system as a series of successive infinitesimal rotations produced by the operator $\mathfrak{A}$, the time scale being determined by the equation. The elementary solutions $\mathbf{R}_i$ have the important property of not rotating with time, for the operator $\mathfrak{A}$ operating on $\mathbf{R}_i$ just gives $\mathbf{R}_i$ back again. The $\mathbf{R}_i$'s thus are *stationary* states of motion.

It is now possible, of course, to generalize the above discussion for the case of N masses connected by a series of springs. Indeed such a system would serve as a one-dimensional model of a crystal. For this system, we require a space of N dimensions. There would be N elementary states of motion which define a set of fixed mutually orthogonal directions in abstract vector space.

Eigenvectors and Eigenvalues. The geometric character of these abstract spaces is completely determined by the operator $\mathfrak{A}$. The principal axes of the operator in the direction $\mathbf{e}_n$ (we shall use the symbol $\mathbf{e}_n$ to represent unit vectors in an abstract space as opposed to $\mathbf{a}_n$ for the unit vector in ordinary three-dimensional space) are determined by the *eigenvalue* equation

$$\mathfrak{A} \cdot \mathbf{e}_n = A_n \mathbf{e}_n \tag{1.6.31}$$

where A_n is a number called an *eigenvalue* of $\mathfrak{A}$. The vectors $\mathbf{e}_n$, called *eigenvectors* of $\mathfrak{A}$, are mutually orthogonal and serve to determine the coordinate axes of the space. Any vector in this space is then a linear combination of the eigenvectors. This suggests rather naturally that it might be worth while to classify the various types of operators which occur in physics and discuss the character of the corresponding abstract vector spaces. A fairly good coverage is obtained if we limit ourselves to the operators which occur in quantum mechanics.

Operators in Quantum Theory. The abstract formulation of quantum mechanics as given by Dirac and by von Neumann leans rather heavily on the concepts which arose in the discussion of coupled oscillators above. The state of a system is described by means of a vector in an abstract space, usually of an infinite number of dimensions. It is somewhat difficult to define the term state as used here; it connotes a relationship between the kind of system we are discussing (numbers of particles, kinds of forces, etc.), the initial conditions of position or velocity, etc., and the methods used to observe the system, which will become clear as the discussion progresses. It is one of the fundamental postulates of quantum theory that observation of a system disturbs (*i.e.*, changes) the state of the system. In abstract vector space this would mean that the vector representing the state of the system would be rotated by an observation of the position of a particle or its energy for instance. Since a rotation may be accomplished by means of a dyadic operator in abstract space, we are led to the conclusion that an *observation is to be represented by an operator*. Thus, the mechanical quantities energy, position, momentum, etc., are to be represented by operators. (One should say that observation of these quantities is to be represented by operators, but it is more convenient to use the more concise way of describing the operators used in the preceding sentence.)

Since the measurement of energy, etc., changes the state of a system, how are these mechanical quantities to be accurately determined? Our previous discussion suggests that only for certain special states will this be possible, *i.e.*, only in those states corresponding to the eigenvectors (principal axes in the ordinary space notation) of the operators in question. For example, the eigenvectors of the energy operator $\mathfrak{E}$ satisfy the equation

$$\mathfrak{E} \cdot \mathbf{e}_n = E_n \mathbf{e}_n$$

This equation states that for the eigenvectors $\mathbf{e}_n$ (*i.e.*, for certain particular states represented by directions $\mathbf{e}_n$ in the vector space) a measurement of the energy does *not* change the state of the system. Only then can one be certain that the observation results in an accurate value of the energy.

What then is the meaning of the constant E_n in this equation? It is usually assumed that it is possible to normalize operator $\mathfrak{E}$ in such a way that the eigenvalue E_n is actually the energy of the state represented by $\mathbf{e}_n$. This is, of course, automatic in ordinary three-dimensional cases discussed earlier. For example, the eigenvalues of the moment of inertia dyadic are just the three principal moments of inertia, and the eigenvalues of the stress dyadic are the principal stresses.

It is immediately clear that two quantities, such as energy and momentum, will be simultaneously measurable (or *observable* as Dirac puts it) if the eigenvectors for the energy operator are also the eigenvectors for the momentum operator. *The necessary and sufficient condition that two quantities be simultaneously measurable is that their corresponding operators commute.* The necessity follows from the fact that, if a vector $\mathbf{e}_n$ is an eigenvector for both $\mathfrak{E}$ and $\mathfrak{p}$, then $\mathfrak{E}\mathfrak{p}\mathbf{e}_n = \mathfrak{p}\mathfrak{E}\mathbf{e}_n$. Since any vector is a linear superposition (another assumption) of $\mathbf{e}_n$'s, we find that $\mathfrak{E}\mathfrak{p}\mathbf{e} = \mathfrak{p}\mathfrak{E}\mathbf{e}$ where $\mathbf{e}$ is any state vector.

The sufficiency of our statement in italics is somewhat more difficult to prove. Consider an eigenvector $\mathbf{e}_n$ of $\mathfrak{E}$. Assume commutability $\mathfrak{E}(\mathfrak{p}\mathbf{e}_n) = E_n(\mathfrak{p}\mathbf{e}_n)$. From this, it follows that $\mathfrak{p}\mathbf{e}_n$ is an eigenvector for $\mathfrak{E}$ with eigenvalue E_n. If there is only one eigenvector with eigenvalue E_n, it immediately follows that $\mathfrak{p} \cdot \mathbf{e}_n$ must be proportional to $\mathbf{e}_n$, yielding the required theorem. If on the other hand there are several eigenvectors $\mathbf{e}_{nm}$ with eigenvalue E_n (referred to as a degeneracy), it follows that

$\mathfrak{p} \cdot \mathbf{e}_{ni} = \sum_m p_{mi} \mathbf{e}_{nm}$. We may now, in this subspace for which all the eigenvectors have eigenvalue E_n, find the principal axes, *i.e.*, eigenvectors, of the operator $\mathfrak{p}$ and thus find states in which both $\mathfrak{p}$ and $\mathfrak{E}$ are simultaneously measurable.

It is an obvious but important corollary that, if two operators do not commute, they are not simultaneously measurable.

An important example of operators which do not all commute are the operators representing position $(\mathfrak{x},\mathfrak{y},\mathfrak{z})$ and the corresponding components of the momentum $(\mathfrak{p}_x,\mathfrak{p}_y,\mathfrak{p}_z)$, that is,

$$\begin{aligned} \mathfrak{x}\mathfrak{y} &= \mathfrak{y}\mathfrak{x}, \text{ etc.}; \quad & \mathfrak{p}_x\mathfrak{x} - \mathfrak{x}\mathfrak{p}_x &= \hbar/i \\ \mathfrak{p}_x\mathfrak{p}_y &= \mathfrak{p}_y\mathfrak{p}_x, \text{ etc.}; \quad & \mathfrak{p}_x\mathfrak{y} &= \mathfrak{y}\mathfrak{p}_x, \text{ etc.} \end{aligned} \tag{1.6.32}$$

where $\hbar$ is Planck's constant h divided by 2π. These are the fundamental equations of quantum theory. They will be discussed in more detail in Chap. 2.

Direction Cosines and Probabilities. What can be said about the state $\mathbf{e}$ which is not an eigenvector of the energy operator $\mathfrak{E}$ but rather is a linear combination of the $\mathbf{e}_n$ eigenvectors? Using the analogy with the abstract space of the coupled oscillator, one takes the square of the cosine of the angle between the state vector $\mathbf{e}$ and an eigenvector $\mathbf{e}_n$ as being that relative fraction of $\mathbf{e}$ which is "in the state $\mathbf{e}_n$." To say it in more physical language, if one measures the energy of a number of identical systems in state $\mathbf{e}$, the fractional number of measurements in which the energy will be E_n is equal to the square of the cosine of the angle between $\mathbf{e}$ and $\mathbf{e}_n$ in the abstract vector space for the system. For many quantum mechanical operators this cosine is a complex number (thus permitting interference between eigenvectors with arbitrary phases). In such a case the square of the absolute value of the cosine is used. The changes in the mathematics of abstract vector space which accompany the introduction of complex cosines will be discussed below.

Probabilities and Uncertainties. Referring the results of our discussion to a single measurement, we may interpret the square of the absolute value of the cosine as the probability that a state $\mathbf{e}$ will have an energy E_n. The average value of the energy for state $\mathbf{e}$ is then $\sum_n E_n(\mathbf{e}_n \cdot \mathbf{e})^2$ or $E_{\text{av}} = (\mathbf{e}\mathfrak{E}\mathbf{e})$. This is correct only if the cosines are real; the proper generalizations for complex cosines will be given below.

These results enable us to discuss the results of measurements of two quantities which are not simultaneously measurable. For example, suppose that $\mathbf{e}(x)$ were a state vector for which the position x was known precisely. What is the probability that this system will have a momentum p_x, the eigenvector for this momentum being $\mathbf{f}(p_x)$? This is given in the present formalism by $|\mathbf{e}(x) \cdot \mathbf{f}(p_x)|^2$. The average value of the momentum is $\mathbf{e}\mathfrak{p}\mathbf{e}$. One may then also express the mean-square deviation of a measurement of p_x from its average:

$$(\Delta p_x)^2_{\text{av}} = \mathbf{e}(\mathfrak{p}_x - p_x)^2\mathbf{e}$$

Only when $\mathbf{e}$ is an eigenvector for $\mathfrak{p}_x$ will p_x be known precisely for state $\mathbf{e}$, for then $\Delta p_x = 0$. The quantity Δp_x is known as the *uncertainty* in the measurement of p_x.

Complex Vector Space. We need now to generalize our notions of vector space so as to include the possibility of complex cosines. In such a case the length of a vector can no longer be specified by the sum of the squares of its components over the coordinate axes, for this square is no longer positive definite as is required by our usual understanding of a length. It is clear that the only way positive definite quantities can be obtained is by using the sum of the squares of the *absolute* values of the components for the definition of length. A simple way to make this result part of our formalism is to introduce a second space which is "complex conjugate" to the first and of the same number of dimensions. If the unit vectors in the first space are $\mathbf{e}_1$, $\mathbf{e}_2$, the corresponding unit vectors in the complex conjugate space will be labeled $\mathbf{e}_1^*$, $\mathbf{e}_2^*$, . . . To every vector $\mathbf{e} = \sum_i A_i\mathbf{e}_i$ having the components A_i in space 1, there will be a corresponding vector in the complex conjugate space $\mathbf{e}^* = \sum \mathbf{e}_i^*\bar{A}_i$ having $\bar{A}_i$ as components (where $\bar{A}_i$ is the complex conjugate of A_i). The scalar product (the essential operation in defining a length) is now redefined as follows:

$$(\mathbf{e}_i^* \cdot \mathbf{e}_j) = \delta_{ij} = \begin{cases} 0; & i \neq j \\ 1; & i = j \end{cases}$$

The dot product between $\mathbf{e}_i$ and $\mathbf{e}_j$ will not be defined, nor will it be needed. It is now clear that the length of vector $\mathbf{e}$ will now be taken as

$$(\mathbf{e}^* \cdot \mathbf{e}) = \Sigma|A_i|^2 > 0$$

a positive definite quantity. The "length" of $\mathbf{e}$ (written $|\mathbf{e}|$) will be taken as $\sqrt{\mathbf{e}^* \cdot \mathbf{e}}$.

Once we have made these definitions, a number of important consequences may be derived. It follows, for example, from them that, if $\mathbf{e}^* \cdot \mathbf{e} = 0$, then $|\mathbf{e}| = 0$. It may be easily verified that, if $\mathbf{e}$ and $\mathbf{f}$ are two vectors,

$$\mathbf{e}^* \cdot \mathbf{f} = \overline{\mathbf{f}^* \cdot \mathbf{e}}$$

Vectors in ordinary space have the property that their dot product is never larger than the product of their amplitudes:

$$AB \geq \mathbf{A} \cdot \mathbf{B}$$

This is called the *Schwarz inequality;* it must also be true in abstract vector space, in generalized form, since it follows from the fact that the square of the "length" of a vector is never less than zero, *i.e.*, is positive definite. For instance, if $\mathbf{e}$ and $\mathbf{f}$ are two vectors and a and b two complex numbers,

$$(\bar{a}\mathbf{e}^* - \bar{b}\mathbf{f}^*) \cdot (a\mathbf{e} - b\mathbf{f}) \geq 0$$

By setting $a = \sqrt{(\mathbf{f}^* \cdot \mathbf{f})(\mathbf{e}^* \cdot \mathbf{f})}$ and $b = \sqrt{(\mathbf{e}^* \cdot \mathbf{e})(\mathbf{f}^* \cdot \mathbf{e})}$ this can be reduced to

$$\sqrt{(\mathbf{e}^* \cdot \mathbf{e})(\mathbf{f}^* \cdot \mathbf{f})} \geq \sqrt{(\mathbf{f}^* \cdot \mathbf{e})(\mathbf{e}^* \cdot \mathbf{f})} \quad \text{or} \quad |\mathbf{e}| \cdot |\mathbf{f}| \geq |\mathbf{f}^* \cdot \mathbf{e}| \qquad (1.6.33)$$

which is the generalized form of the Schwarz inequality we have been seeking. This inequality will be used later in deriving the Heisenberg uncertainty principle.

Another inequality which has its analogue in ordinary vector space is the *Bessel inequality*, which states in essence that the sum of the lengths of two sides of a triangle is never less than the length of the third side:

$$|\mathbf{e}| + |\mathbf{f}| \geq |(\mathbf{e} + \mathbf{f})| \qquad (1.6.34)$$

Generalized Dyadics. We now turn to the discussion of operators in complex vector spaces. We consider first the linear transformation which rotates the axes of the space into another set of orthogonal axes:

$$\mathbf{e}'_i = \sum_n \mathbf{e}_n \gamma_{ni}$$

where $\mathbf{e}'_i$ is a unit vector along the ith axis of the new coordinate system and $\mathbf{e}_n$ are the unit vectors in the original system. The relation between $\mathbf{e}'_i$ and $(\mathbf{e}'_i)^*$ must be the same as that for any two vectors in the unprimed system so that

$$(\mathbf{e}'_i)^* = \sum_n \bar{\gamma}_{ni} \mathbf{e}^*_n$$

We can devise a dyadic operator which when operating on the vector $\mathbf{e}_n$ converts it into the new vector $\mathbf{e}'_i$. This dyadic is

$$\mathfrak{G} = \sum_{n,i} \mathbf{e}_n \cdot \gamma_{ni} \cdot \mathbf{e}^*_i \qquad (1.6.35)$$

so that

$$\mathfrak{G}\mathbf{e}_i = \sum_n \mathbf{e}_n \gamma_{ni} = \mathbf{e}'_i$$

The form of the dyadic implies the convention that we shall always place unstarred vectors to the *right* of an operator and starred vectors to the *left*. A useful property of the dyadic operator is that

$$(\mathbf{e}^* \mathfrak{G})\mathbf{f} = \mathbf{e}^*(\mathfrak{G}\mathbf{f})$$

so that no parentheses are actually needed and we shall usually write it $\mathbf{e}^*\mathfrak{G}\mathbf{f}$.

Our generalizations have been carried out so that, as with ordinary dyadics, products of operators are also operators:

$$(\mathfrak{G} \cdot \mathfrak{L}) = \left(\sum_{n,i} \mathbf{e}_n\gamma_{ni}\mathbf{e}_i^*\right) \cdot \left(\sum_{j,k} \mathbf{e}_j\lambda_{jk}\mathbf{e}_k^*\right) = \sum_{n,k} \mathbf{e}_n \left(\sum_j \gamma_{nj}\lambda_{jk}\right)\mathbf{e}_k^*$$

so that the (n,k) component of the operator $(\mathfrak{G} \cdot \mathfrak{L})$ is

$$(\mathfrak{G}\mathfrak{L})_{nk} = \left[\sum_j \gamma_{nj}\lambda_{jk}\right]$$

Returning now to the rotation operator $\mathfrak{G}$ defined in Eq. (1.6.35), we note that $\mathbf{e}_i^*\mathfrak{G}$ does not give $(\mathbf{e}_i')^*$. The proper operator to rotate $\mathbf{e}_i^*$ shall be written $\mathfrak{G}^*$ and defined so that

$$\mathbf{e}_i^*\mathfrak{G}^* = (\mathbf{e}_i')^*$$

The operator $\mathfrak{G}^*$ is therefore related to the operator $\mathfrak{G}$ by the equation

$$(\mathfrak{G}\mathbf{e}_i)^* = \mathbf{e}_i^*\mathfrak{G}^*$$

An operator $\mathfrak{G}^*$ which is related to an operator $\mathfrak{G}$ according to this equation is called the *Hermitian adjoint* of $\mathfrak{G}$. Writing out both sides of the equation we obtain

$$\sum_n \mathbf{e}_n^*\bar{\gamma}_{ni} = \sum_n (\gamma^*)_{in}\mathbf{e}_n^*$$

so that

$$\bar{\gamma}_{ni} = (\gamma^*)_{in} \tag{1.6.36}$$

This equation states that the (i,n) component of the Hermitian adjoint dyadic $\mathfrak{G}^*$ is obtained by taking the complex conjugate of the (n,i) component of $\mathfrak{G}$. The notion of adjoint is the generalization of the conjugate dyadic defined on page 55. The Hermitian adjoint of a product of two operators $\mathfrak{G}\mathfrak{L}$ is $(\mathfrak{G}\mathfrak{L})^* = \mathfrak{L}^*\mathfrak{G}^*$. The Hermitian adjoint of a complex number times an operator is the complex conjugate of the number times the adjoint of the operator.

Hermitian Operators. An operator which is self-adjoint, so that

$$\mathfrak{G} = \mathfrak{G}^* \quad \text{or} \quad \gamma_{nm} = \bar{\gamma}_{mn}$$

is called a *Hermitian* operator. All classical symmetric dyadic operators are Hermitian, since their components are all real numbers. The operators in quantum mechanics which correspond to measurable quantities must also be Hermitian, for their eigenvalues must be real (after all, the results of actual measurements are real numbers). To prove this we note that, if the eigenvalues a_n of an operator $\mathfrak{A}$ are real, then $\mathbf{e}^*\mathfrak{A}\mathbf{e}$ is real for any $\mathbf{e}$. For $\mathbf{e}$ can be expanded in terms of the eigenvectors

$\mathbf{e}_n$, giving a series of real numbers for $\mathbf{e}^*\mathfrak{A}\mathbf{e}$. Let $\mathbf{e} = \mathbf{f} + b\mathbf{g}$; then $\bar{b}(\mathbf{g}^*\mathfrak{A}\mathbf{f}) + b(\mathbf{f}^*\mathfrak{A}\mathbf{g})$ is real. But this is possible only if $(\mathbf{g}^*\mathfrak{A}\mathbf{f})$ is the complex conjugate of $(\mathbf{f}^*\mathfrak{A}\mathbf{g})$, that is, is $(\mathbf{g}^*\mathfrak{A}^*\mathbf{f})$. Consequently $\mathbf{g}^*(\mathfrak{A} - \mathfrak{A}^*)\mathbf{f} = 0$ for any $\mathbf{f}$ or $\mathbf{g}$; therefore $\mathfrak{A} = \mathfrak{A}^*$.

The rotation operator $\mathfrak{G}$ defined in Eq. (1.6.35) is still more narrowly specialized. The components γ_{mn} are direction cosines, so that one would expect the operator to have a "unit amplitude" somehow. Since a rotation of coordinates in vector space should not change the value of a scalar product, we should have

$$(\mathbf{e}^*\mathfrak{G}^*) \cdot (\mathfrak{G}\mathbf{f}) = \mathbf{e}^* \cdot \mathbf{f}$$

for $\mathbf{e}$ and $\mathbf{f}$ arbitrary state vectors. Therefore, it must be that

$$\mathfrak{G}^*\mathfrak{G} = \mathfrak{I}$$

where $\mathfrak{I}$ is the idemfactor $\sum_n \mathbf{e}_n\mathbf{e}_n^*$. This implies that the adjoint of $\mathfrak{G}$ is also its inverse:

$$\mathfrak{G}^* = \mathfrak{G}^{-1}$$

Such an operator, having "unit amplitude" as defined by these equations, is called a *unitary operator*. The rotation operator defined in Eq. (1.6.35) is a unitary operator [see also Eq. (1.6.12)].

Most of the operators encountered in quantum mechanics are Hermitian; their eigenvalues are real, but their effect on state vectors is both to rotate and to change in size. There are in addition several useful operators which are unitary; their eigenvalues will not all be real, but their effect on state vectors is to rotate without change of size.

Examples of Unitary Operators. Important examples of unitary operators occur in wave propagation, quantum mechanics, and kinetic theory. For example, the description of the manner in which a junction between two wave guides (say of different cross section) reflects waves traveling in the ducts may be made by using a *reflectance* dyadic. This operator rotates, in the related abstract space, the eigenvector corresponding to the incident wave into a vector corresponding to the reflected wave. The unitary condition corresponds essentially to the requirement that the process be a conservative one. The equation $\mathfrak{G}^* = \mathfrak{G}^{-1}$ may be related to the reciprocity theorem, *i.e.*, to the reversibility between source and detector. We shall, of course, go into these matters in greater detail in later chapters.

A unitary operator may be constructed from an ordinary Hermitian operator $\mathfrak{L}$ as follows:

$$\mathfrak{G} = (1 + i\mathfrak{L})/(1 - i\mathfrak{L})$$

The adjoint of $\mathfrak{G}$ is $(1 - i\mathfrak{L})/(1 + i\mathfrak{L})$ so that $\mathfrak{G}\mathfrak{G}^*$ is $\mathfrak{I}$ and $\mathfrak{G}$ is unitary. If $\mathfrak{G}$ is the reflectance dyadic, then the above equation yields the relation which exists between the reflection coefficients constituting $\mathfrak{S}$ and the *impedance* coefficients constituting the impedance dyadic $\mathfrak{Z}$.

Another construction using a Hermitian operator $\mathfrak{R}$ to obtain a unitary operator $\mathfrak{S}$ is

$$\mathfrak{S} = e^{i\mathfrak{R}} \tag{1.6.37}$$

where $e^{i\mathfrak{R}}$ is defined by its power series $1 + i\mathfrak{R} + \frac{1}{2}(i\mathfrak{R})^2 + \cdots$. For example, for $\mathfrak{S} = (1 + i\mathfrak{Z})/(1 - i\mathfrak{Z})$, we have $\mathfrak{S} = e^{i\mathfrak{R}}$, where $\mathfrak{R} = 2 \tan^{-1} \mathfrak{Z}$. In terms of the physical example above, this last transformation corresponds to using the shift in phase upon reflection to describe the reflected wave rather than the reflection coefficient itself.

Often a vector is a function of a parameter (we shall take time t as the typical parameter), the vector rotating as the parameter changes its value. The unitary operator corresponding to this rotation is useful in many problems. Its general form is not difficult to derive. We can call it $\mathfrak{T}(t)$. Then by our definition

$$\mathfrak{T}(t_1)\mathbf{e}(t_0) = \mathbf{e}(t_1 + t_0)$$

where t_1 and t_0 are specific values of the parameter t. Moreover,

$$\mathfrak{T}(t_2)\mathfrak{T}(t_1)\mathbf{e}(t_0) = \mathbf{e}(t_1 + t_0 + t_2) = \mathfrak{T}(t_1 + t_2)\mathbf{e}(t_0)$$

so that

$$\mathfrak{T}(t_2)\mathfrak{T}(t_1) = \mathfrak{T}(t_1 + t_2)$$

For this equation to hold for all t_2 and t_1, $\mathfrak{T}$ must be an exponential function of t. Since it must be unitary, it must be of the form

$$\mathfrak{T} = e^{i\mathfrak{H}t}$$

where $\mathfrak{H}$ is some unknown Hermitian operator.

To determine the equation of rotational motion for vector $\mathbf{e}$ we apply $\mathfrak{T}$ to $\mathbf{e}(t)$, changing t by an infinitesimal amount dt. Then $\mathfrak{T} = 1 + i\mathfrak{H}\,dt$ so that

$$(1 + i\mathfrak{H}\,dt)\mathbf{e}(t) = \mathbf{e}(t + dt) \quad \text{or} \quad \mathfrak{H}\mathbf{e}(t) = \frac{1}{i}\frac{\partial \mathbf{e}}{\partial t} \tag{1.6.38}$$

We must, of course, find $\mathfrak{H}$ from the physics of the problem. For example, in quantum mechanics and where t is the time, $\mathfrak{H}$ is proportional to the Hamiltonian function with the usual classical variables of momentum and position replaced by their corresponding operators.

Transformation of Operators. In the preceding discussion we have dealt with the rotation of a vector as a parameter changes upon which it depends. It is also possible to simulate the effects of the changes in the parameter by keeping the vector fixed and rotate "space," *i.e.*, to ascribe the appropriate parametric dependence to all the operators which occur. In other words, we change the meaning of the operators as the parameter changes, keeping the vectors fixed. We must find how this is to be done so as to give results equivalent to the first picture in which the vectors rotated but the operators were fixed. Let the rotation of the vectors be given by the unitary operator $\mathfrak{G}$:

$$\mathfrak{G}\mathbf{e} = \mathbf{e}'; \quad \mathfrak{G}\mathbf{f} = \mathbf{f}'; \quad \text{etc.}$$

The appropriate equivalent change in $\mathfrak{L}$ to $\mathfrak{L}'$ is found by requiring that the relation between $\mathfrak{L}\mathbf{e}'$ and $\mathbf{f}'$ (first picture $\mathbf{e}$ and $\mathbf{f}$ change, $\mathfrak{L}$ does not) be the same as the relation between $\mathfrak{L}'\mathbf{e}$ and $\mathbf{f}$ (second picture $\mathbf{e}$ and $\mathbf{f}$ do not change, $\mathfrak{L}$ transforms to $\mathfrak{L}'$). Analytically we write

$$\mathbf{f}'^* \cdot \mathfrak{L} \cdot \mathbf{e}' = \mathbf{f}^* \cdot \mathfrak{L}' \cdot \mathbf{e}$$

Inserting the relation between $\mathbf{f}'$ and $\mathbf{f}$ and recalling that $\mathfrak{G}^* = \mathfrak{G}^{-1}$, we find

$$\mathbf{f}'^* \cdot \mathfrak{L} \cdot \mathbf{e}' = \mathbf{f}^* \cdot (\mathfrak{G}^{-1}\mathfrak{L}\mathfrak{G}) \cdot \mathbf{e} \quad \text{or} \quad \mathfrak{L}' = \mathfrak{G}^{-1}\mathfrak{L}\mathfrak{G} \tag{1.6.39}$$

We now investigate the effects of this unitary (sometimes called *canonical*) transformation on the properties of $\mathfrak{L}$. For example, we shall first show that the eigenvalues of $\mathfrak{L}'$ and $\mathfrak{L}$ are the same. Let

$$\mathfrak{L}'\mathbf{e} = L'\mathbf{e}$$

Hence

$$(\mathfrak{G}^{-1}\mathfrak{L}\mathfrak{G}) \cdot \mathbf{e} = L'\mathbf{e}$$

Multiplying through by $\mathfrak{G}$ we obtain

$$\mathfrak{L}(\mathfrak{G}\mathbf{e}) = L'(\mathfrak{G}\mathbf{e})$$

In words if $\mathbf{e}$ is an eigenvector of $\mathfrak{L}'$ with eigenvalue L', then $\mathfrak{G}\mathbf{e}$ is an eigenvalue of $\mathfrak{L}$ with the same eigenvalue. This preservation of eigenvalue is sometimes a very useful property, for if $\mathfrak{L}$ is a difficult operator to manipulate, it may be possible to find a transformation $\mathfrak{G}$ which yields a simpler operator $\mathfrak{L}'$ whose eigenvalues are the same as those of $\mathfrak{L}$.

Because of the relation between an operator and its Hermitian adjoint, we can write

$$(\mathfrak{G}^{-1}\mathfrak{L}\mathfrak{G})^* = \mathfrak{G}^*\mathfrak{L}^*(\mathfrak{G}^{-1})^* = \mathfrak{G}^{-1}\mathfrak{L}^*\mathfrak{G}$$

Hence a Hermitian operator remains Hermitian after a unitary transformation.

The relationship between operators is not changed by a unitary transformation. For example, if

$$\mathfrak{L}\mathfrak{M} = \mathfrak{N}$$

then

$$\mathfrak{G}^{-1}\mathfrak{L}\mathfrak{G}\mathfrak{G}^{-1}\mathfrak{M}\mathfrak{G} = \mathfrak{G}^{-1}\mathfrak{N}\mathfrak{G} \quad \text{or} \quad \mathfrak{L}'\mathfrak{M}' = \mathfrak{N}'$$

Two unitary transformations applied, one after the other, correspond to a transformation which is also unitary. If the two unitary operators are $\mathfrak{F}\mathfrak{G}$, then

$$(\mathfrak{F}\mathfrak{G})^*\mathfrak{F}\mathfrak{G} = (\mathfrak{G}^*\mathfrak{F}^*)\mathfrak{F}\mathfrak{G} = \mathfrak{I}$$

proving the point in question.

Finally we may consider the changes in an operator $\mathfrak{L}$ under transformation by the unitary operator $\mathfrak{T}(t)$ defined in Eq. (1.6.38). This will enable us to calculate the change in an operator due to a change in a parameter t. The transformed operator is

$$e^{-i\mathfrak{H}t}\mathfrak{L}e^{i\mathfrak{H}t} = \mathfrak{L}'$$

It is clearly convenient to introduce the notation $\mathfrak{L}(t)$ for $\mathfrak{L}'$, calling $\mathfrak{L} = \mathfrak{L}(0)$, so that

$$\mathfrak{L}(t) = e^{-i\mathfrak{H}t}\mathfrak{L}(0)e^{i\mathfrak{H}t}$$

The rate of change of $\mathfrak{L}$ with t can be obtained by considering the difference $\mathfrak{L}(t + dt) - \mathfrak{L}(t) = e^{-i\mathfrak{H}dt}\mathfrak{L}e^{i\mathfrak{H}dt} - \mathfrak{L}$ so that

$$\frac{1}{i}\frac{\partial\mathfrak{L}}{\partial t} = \mathfrak{L}\mathfrak{H} - \mathfrak{H}\mathfrak{L} \tag{1.6.40}$$

Of course $\mathfrak{H}$ depends on the physics of the problem. For example, in many quantum mechanical problems $\mathfrak{H}$ is the Hamiltonian operator and t is time; the resulting equation is called the *Heisenberg equation* of motion. The equation has wide application; for an operator $\mathfrak{H}$ which is related to a rotation parameter t for state vectors of the form given in Eq. (1.6.38), the rate of change of any other operator $\mathfrak{L}$ related to the same system is proportional to the commutator $\mathfrak{H}\mathfrak{L} - \mathfrak{L}\mathfrak{H}$.

Quantum Mechanical Operators. Let us now apply some of these results to the operators occurring in quantum mechanics. We recall that the average result of measurements of a quantity (energy, momentum, etc.) represented by the operator $\mathfrak{p}$ on a system in a state represented by the state vector $\mathbf{e}$ is $p_{\text{av}} = (\mathbf{e}^*\mathfrak{p}\mathbf{e})$. Likewise, we point out again that the probability that a system in a state represented by the state vector $\mathbf{e}$ (*in* state $\mathbf{e}$, for short) turns out to be also in state $\mathbf{e}'$ is given by the square of the absolute magnitude of the cosine between the state vectors, $|(\mathbf{e}^* \cdot \mathbf{e}')|^2$.

Last section's discussion makes it possible to restate this last sentence in terms of unitary operators. Suppose that we find a unitary operator $\mathfrak{g}$ which transforms an eigenvector $\mathbf{e}(a_n)$ for operator $\mathfrak{A}$ with eigenvalue a_n to an eigenvector $\mathbf{e}'(b_m)$ for operator $\mathfrak{B}$ with eigenvalue b_m:

$$\mathfrak{g}(a,b)\mathbf{e}(a_n) = \mathbf{e}'(b_m)$$

Then, using Eq. (1.6.35), we see that *the probability that a measurement of $\mathfrak{B}$ gives the value b_m when $\mathfrak{A}$ has the value a_n* is

$$|\mathbf{e}^*(a_n) \cdot \mathbf{e}'(b_m)|^2 = |\mathbf{e}^*(a_n)\mathfrak{g}(a,b)\mathbf{e}(a_n)|^2 = |\gamma_{nn}(a,b)|^2 \tag{1.6.41}$$

At this point we can utilize the Schwarz inequality [Eq. (1.6.33)] to show the relation between the quantum equations (1.6.32) and the Heisenberg uncertainty principle. We have already shown that, if two operators do not commute, they cannot be simultaneously measurable; if one is measured accurately, the other cannot also be known accurately. From the physical point of view there is something inherent in the measurement of one which destroys our simultaneous accurate knowledge of the other. From the mathematical point of view, there is a reciprocal relationship between their uncertainties. As mentioned before, the uncertainty in measurement ΔA of an operator $\mathfrak{A}$ is the root-mean-square deviation defined by the equations

$$(\Delta A)^2 = \mathbf{e}^*[\mathfrak{A} - a]^2\mathbf{e}; \quad a = \mathbf{e}^*\mathfrak{A}\mathbf{e}$$

for the state denoted by the vector $\mathbf{e}$.

To apply these general statements to the quantum relations (1.6.32), we define the operators

$$\mathbf{\Delta}\mathfrak{p}_x = \mathfrak{p}_x - p_x; \quad p_x = (\mathbf{e}^*\mathfrak{p}_x\mathbf{e}); \quad \mathbf{\Delta}\mathfrak{x} = \mathfrak{x} - x; \quad x = (\mathbf{e}^*\mathfrak{x}\mathbf{e})$$

so that the rms value of the operator $\mathbf{\Delta}\mathfrak{x}$ is the uncertainty in measurement of x in state $\mathbf{e}$. By use of some of Eqs. (1.6.32), we can show that

$$\mathbf{\Delta}\mathfrak{p}_x\,\mathbf{\Delta}\mathfrak{x} - \mathbf{\Delta}\mathfrak{x}\,\mathbf{\Delta}\mathfrak{p} = \hbar/i$$

Taking the average value of this for state $\mathbf{e}$, we have

$$(\mathbf{\Delta}\mathfrak{p}_x\mathbf{e})^* \cdot (\mathbf{\Delta}\mathfrak{x}\mathbf{e}) - (\mathbf{\Delta}\mathfrak{x}\mathbf{e})^* \cdot (\mathbf{\Delta}\mathfrak{p}_x\mathbf{e}) = \hbar/i$$

The right-hand side of this equation is just twice the imaginary part of $(\mathbf{\Delta}\mathfrak{p}_x\mathbf{e})^* \cdot (\mathbf{\Delta}\mathfrak{x}\mathbf{e})$, so that we can conclude that

$$|(\mathbf{\Delta}\mathfrak{p}_x\mathbf{e})^* \cdot (\mathbf{\Delta}\mathfrak{x}\mathbf{e})| \geq \hbar/2$$

But the Schwarz inequality (1.6.33) says that the product of the amplitudes is never smaller than the amplitude of the dot product. Therefore, we finally obtain

$$(\Delta p_x)(\Delta x) \geq \hbar/2$$

which is the famous Heisenberg uncertainty principle. It states that simultaneous accurate measurement of position and momentum (in the same coordinate) is not possible, and that if simultaneous measurement is attempted, the resultant uncertainties in the results are limited by the Heisenberg principle.

Spin Operators. The statements made in the previous section were exceedingly general, and there is need for an example to clarify some of the concepts. An example which will be useful in later discussions is that of electron spin. It is experimentally observed that there are only two allowed values of the component of magnetic moment of the electron in any given direction, which leads us to expect that the angular momentum of the electron is similarly limited in allowed values. The angular momentum $\mathfrak{M}$ of a particle about the origin of coordinates is given in terms of its position from the origin and its momentum $\mathfrak{P}$ as follows:

$$\mathfrak{M}_x = \mathfrak{y}\mathfrak{P}_z - \mathfrak{z}\mathfrak{P}_y; \quad \mathfrak{M}_y = \mathfrak{z}\mathfrak{P}_x - \mathfrak{x}\mathfrak{P}_z; \quad \mathfrak{M}_z = \mathfrak{x}\mathfrak{P}_y - \mathfrak{y}\mathfrak{P}_x$$

We next compute the commutator for the components of $\mathfrak{M}$, utilizing the quantum equations (1.6.32):

$$\begin{aligned} \mathfrak{M}_x\mathfrak{M}_y - \mathfrak{M}_y\mathfrak{M}_x &= (\mathfrak{P}_z\mathfrak{z} - \mathfrak{z}\mathfrak{P}_z)(\mathfrak{x}\mathfrak{P}_y - \mathfrak{y}\mathfrak{P}_x) = i\hbar\mathfrak{M}_z; \\ \mathfrak{M}_y\mathfrak{M}_z - \mathfrak{M}_z\mathfrak{M}_y &= i\hbar\mathfrak{M}_x; \quad \mathfrak{M}_z\mathfrak{M}_x - \mathfrak{M}_x\mathfrak{M}_z = i\hbar\mathfrak{M}_y \end{aligned} \tag{1.6.42}$$

These equations indicate that we cannot know accurately all three components of the electron spin; in fact if we know the value of M_z accurately, we cannot know the values of M_x and M_y.

By utilizing these equations in pairs we can show that

$$\begin{aligned} &\mathfrak{M}_z(\mathfrak{M}_x \pm i\mathfrak{M}_y) - (\mathfrak{M}_x \pm i\mathfrak{M}_y)\mathfrak{M}_z = \pm\hbar(\mathfrak{M}_x \pm i\mathfrak{M}_y) \\ \text{or}\quad &\mathfrak{M}_z(\mathfrak{M}_x + i\mathfrak{M}_y) = (\mathfrak{M}_x + i\mathfrak{M}_y)(\mathfrak{M}_z + \hbar) \\ \text{and}\quad &\mathfrak{M}_z(\mathfrak{M}_x - i\mathfrak{M}_y) = (\mathfrak{M}_x - i\mathfrak{M}_y)(\mathfrak{M}_z - \hbar) \end{aligned} \tag{1.6.43}$$

Starting with a state vector $\mathbf{a}_m$ corresponding to a knowledge that the value of $\mathfrak{M}_z$ is $m\hbar$ (that is, $\mathbf{a}_m$ is an eigenvector for $\mathfrak{M}_z$ with eigenvalue $m\hbar$), we see that the vector $(\mathfrak{M}_x + i\mathfrak{M}_y)\mathbf{a}_m$ is an eigenvector for $\mathfrak{M}_z$ with eigenvalue $(m + 1)\hbar$ unless $(\mathfrak{M}_x + i\mathfrak{M}_y)\mathbf{a}_m$ is zero and that $(\mathfrak{M}_x - i\mathfrak{M}_y)\mathbf{a}_m$ is an eigenvector for $\mathfrak{M}_z$ with eigenvalue $(m - 1)\hbar$ unless $(\mathfrak{M}_x - i\mathfrak{M}_y)\mathbf{a}_m$ is zero, for from Eqs. (1.6.43),

$$\begin{aligned} \mathfrak{M}_z(\mathfrak{M}_x + i\mathfrak{M}_y)\mathbf{a}_m &= (\mathfrak{M}_x + i\mathfrak{M}_y)[(\mathfrak{M}_z + \hbar)\mathbf{a}_m] \\ &= (m + 1)\hbar[(\mathfrak{M}_x + i\mathfrak{M}_y)\mathbf{a}_m]; \quad \text{etc.} \end{aligned}$$

In the special case of electron spin, we will call the angular momentum $\mathfrak{S}$ instead of $\mathfrak{M}$. In order to have only two allowed values of m, differing by $\hbar$ (as the above equations require) and to have these values symmetric with respect to direction along x, we must have the allowed values of $\mathfrak{S}_z$ be $+\frac{1}{2}\hbar$ (with eigenvector $\mathbf{a}_+$) and $-\frac{1}{2}\hbar$ (with eigenvector $\mathbf{a}_-$), and that

$$(\mathfrak{S}_x + i\mathfrak{S}_y)\mathbf{a}_+ = 0 \quad \text{and} \quad (\mathfrak{S}_x - i\mathfrak{S}_y)\mathbf{a}_- = 0$$

Consequently the rules of operation of the spin operators on the two eigenvectors and the rules of multiplication of the spin operators are

$$\begin{aligned} &\mathfrak{S}_x\mathbf{a}_+ = (\hbar/2)\mathbf{a}_-; \quad \mathfrak{S}_y\mathbf{a}_+ = (i\hbar/2)\mathbf{a}_-; \quad \mathfrak{S}_z\mathbf{a}_+ = (\hbar/2)\mathbf{a}_+ \\ &\mathfrak{S}_x\mathbf{a}_- = (\hbar/2)\mathbf{a}_+; \quad \mathfrak{S}_y\mathbf{a}_- = -(i\hbar/2)\mathbf{a}_+; \quad \mathfrak{S}_z\mathbf{a}_- = -(\hbar/2)\mathbf{a}_- \\ &\mathfrak{S}_x\mathfrak{S}_y = (i\hbar/2)\mathfrak{S}_z = -\mathfrak{S}_y\mathfrak{S}_x; \quad \mathfrak{S}_y\mathfrak{S}_z = (i\hbar/2)\mathfrak{S}_x = -\mathfrak{S}_z\mathfrak{S}_y \\ &\mathfrak{S}_z\mathfrak{S}_x = (i\hbar/2)\mathfrak{S}_y = -\mathfrak{S}_x\mathfrak{S}_z; \quad (\mathfrak{S}_x)^2 = (\mathfrak{S}_y)^2 = (\mathfrak{S}_z)^2 = (\hbar/2)^2 \end{aligned} \tag{1.6.44}$$

We have here a fairly simple generalization of the concept of operators. The "spin space" is a two-dimensional space, one direction corresponding to a state in which the z component of spin is certainly $\hbar/2$ and the perpendicular direction corresponding to a state where $\mathfrak{S}_z$ is certainly $-(\hbar/2)$ (it should be clear by now that state space is a useful mathematical fiction which has nothing to do with "real" space). A unit state vector in an intermediate direction corresponds to a state where we will find $\mathfrak{S}_z$ sometimes positive and sometimes negative, the relative frequency of the two results depending on the square of the cosine of the angle between the unit vector and the two principal axes for $\mathfrak{S}_z$.

The operator $2\mathfrak{S}_x/\hbar$ reflects any state vector in the 45° line, *i.e.*, interchanges plus and minus components. Therefore, the eigenvectors for $\mathfrak{S}_x$ would be $(1/\sqrt{2})(\mathbf{a}_+ \pm \mathbf{a}_-)$, corresponding to a rotation by 45° of principal axes in spin space for a rotation of 90° of the quantizing direction in actual space (corresponding to changing the direction of the magnetic field which orients the electron from the z to the x axis, for instance). Therefore, if we know that $\mathfrak{S}_z$ is $\hbar/2$, there is an even chance that $\mathfrak{S}_x$ will have the value $\hbar/2$ or $-(\hbar/2)$.

On the other hand, the two eigenvectors for $\mathfrak{S}_y$ are $(1/\sqrt{2})(\mathbf{a}_+ \pm i\mathbf{a}_-)$, corresponding to an imaginary rotation in spin space (the square of the magnitude of a complex vector of this sort, we remember, is the scalar product of the complex conjugate of the vector with the vector itself).

Quaternions. Viewed in the abstract, the operators $\mathfrak{i} = 2\mathfrak{S}_x/i\hbar$, $\mathfrak{j} = 2\mathfrak{S}_y/i\hbar$, $\mathfrak{k} = 2\mathfrak{S}_z/i\hbar$, together with the unity operator $\mathfrak{J}\mathbf{a} = \mathbf{a}$ and a zero operator, exhibit a close formal relation to dyadic operators. They act on a vector (in this case a vector in two-dimensional spin space) to produce another vector. They can be added and multiplied to produce other operators of the same class. The multiplication table for these new operators is

$$\mathfrak{i}^2 = \mathfrak{j}^2 = \mathfrak{k}^2 = -1; \quad \mathfrak{i}\mathfrak{j} = \mathfrak{k} = -\mathfrak{j}\mathfrak{i}; \quad \mathfrak{j}\mathfrak{k} = \mathfrak{i} = -\mathfrak{k}\mathfrak{j}; \quad \mathfrak{k}\mathfrak{i} = \mathfrak{j} = -\mathfrak{i}\mathfrak{k} \tag{1.6.45}$$

Curiously enough, operators with just these properties were studied by Hamilton, long before the development of quantum mechanics, in his efforts to generalize complex numbers. As was discussed on page 73, the quantity $i = \sqrt{-1}$ can be considered to correspond to the operation of rotation by 90° in the complex plane. For this simple operator, the only multiplication table is $i^2 = -1$, where 1 is the unity operator. Corresponding to any point (x,y) on the complex plane is a complex number $z = x + iy$, where the square of the distance from the origin is $|z|^2 = z\bar{z} = (x - iy)(x + iy) = x^2 + y^2$. In order to reach the next comparable formal generalization one must use the three quantities $\mathfrak{i}$, $\mathfrak{j}$, and $\mathfrak{k}$ defined in Eqs. (1.6.45) to produce a *quaternion* (see page 74)

$$\mathfrak{p} = a + b\mathfrak{i} + c\mathfrak{j} + d\mathfrak{k}$$

The square of the magnitude of this quantity is obtained by multiplying $\mathfrak{p}$ by its conjugate

$$\mathfrak{p}^* = a - b\mathfrak{i} - c\mathfrak{j} - d\mathfrak{k}; \quad |\mathfrak{p}|^2 = \mathfrak{p}\mathfrak{p}^* = a^2 + b^2 + c^2 + d^2$$

Therefore, the only quaternion with zero magnitude is one where a, b, c, and d are all zero. In addition, the reciprocal to any quaternion is quickly found.

$$1/\mathfrak{p} = \mathfrak{p}^*/|\mathfrak{p}|^2$$

A quaternion can be related to the operation of rotation about an axis in three-dimensional space; the direction cosines of the axis are proportional to the constants b, c, and d, and the amount of the rotation is determined by the ratio of a^2 to $b^2 + c^2 + d^2$. Little further needs to be said about these quantities beyond what has been given on page 74, *et seq.*, for their interest is now chiefly historical. The spin operators are, however, closely related to them.

Rotation Operators. Unitary operators of the type defined in Eq. (1.6.37) are very closely related to the general angular momentum operators (indeed, they are proportional). Discussing them will enable us to use some more of the technique developed in the section on abstract spaces and at the same time shed some more light on the properties of angular momentum in quantum mechanics.

Suppose that a vector $\mathbf{e}$ in abstract space depends parametrically upon the orientation of a vector $\mathbf{r}$ in ordinary space. If we should now rotate $\mathbf{r}$ about the x axis through an angle θ_x, then $\mathbf{e}$ should rotate in abstract space. The operator which must be applied to $\mathbf{e}$ to yield the resultant vector is of the form given in Eq. (1.6.37). In the present case we write

$$\mathfrak{D}_x = e^{i(\mathfrak{M}_x/\hbar)\theta_x}$$

where now $\mathfrak{M}_x$ is an operator. Similarly, one may define a $\mathfrak{D}_y$ and a $\mathfrak{D}_z$ in terms of a θ_y and an θ_z. For most cases, since a rotation about the x axis of the vector $\mathbf{r}$ does not commute with a rotation about the y axis of this vector, $\mathfrak{D}_x\mathfrak{D}_y \neq \mathfrak{D}_y\mathfrak{D}_x$.

However, for an infinitesimal rotation of amounts $(d\theta)_x$, $(d\theta)_y$, $(d\theta)_z$, it is easy to see that the rotations do commute and that the corresponding operator in abstract space is $1 + (i/\hbar)(\mathfrak{M}_x\, d\theta_x + \mathfrak{M}_y\, d\theta_y + \mathfrak{M}_z\, d\theta_z)$. Since the effect of this infinitesimal operator on vector $\mathbf{e}$ cannot depend upon the orientation of the x, y, z axes it follows that $\mathfrak{M}_x\, d\theta_x + \mathfrak{M}_y\, d\theta_y + \mathfrak{M}_z\, d\theta_z$ must be invariant under a rotation. Since $(d\theta_x, d\theta_y, d\theta_z)$ form the three components of a space vector, it follows that $(\mathfrak{M}_x, \mathfrak{M}_y, \mathfrak{M}_z)$ must also form the three components of a vector $\mathfrak{M}$ and must transform, for example, as x, y, z in Eq. (1.3.8), for then the operator can be a simple scalar product, which is invariant under a rotation.

Since $\mathfrak{M}$ is a vector, it must transform like a vector. Specifically, if the space axes are rotated by an angle $d\theta_z$ about the z axis, the relation between the unprimed and the primed components are

$$\mathfrak{M}'_z = \mathfrak{M}_z; \quad \mathfrak{M}'_y = d\theta_z\,\mathfrak{M}_x + \mathfrak{M}_y; \quad \mathfrak{M}'_x = \mathfrak{M}_x - d\theta_z\,\mathfrak{M}_y \qquad (1.6.46)$$

However, $\mathfrak{M}_z$ is an operator related to a rotation in vector space, related to the parameter θ_z in a manner analogous to $\mathfrak{H}$ and t in Eq. (1.6.38). To be specific, it is related to the rate of change of a state vector $\mathbf{e}$ by the equation

$$\frac{1}{\hbar}\mathfrak{M}_z\mathbf{e} = \frac{1}{i}\frac{\partial\mathbf{e}}{\partial\theta_z}$$

Therefore, the rate of change of any other operator, related to the system, with respect to θ_z must be given by an equation of the form of (1.6.40).

For instance, the rate of change of the operator $\mathfrak{M}_x$ with respect to the parameter θ_z is given by the equation

$$\frac{1}{i}\frac{\partial\mathfrak{M}_x}{\partial\theta_z} = \frac{1}{\hbar}(\mathfrak{M}_z\mathfrak{M}_x - \mathfrak{M}_x\mathfrak{M}_z)$$

But from Eq. (1.6.46), $\partial\mathfrak{M}_x/\partial\theta_z = -\mathfrak{M}_y$, we have

$$i\hbar\mathfrak{M}_y = \mathfrak{M}_z\mathfrak{M}_x - \mathfrak{M}_x\mathfrak{M}_z$$

which is identical with the last of Eqs. (1.6.42) for the angular momentum operators.

The present derivation, however, has considered the operator vector $\mathfrak{M}$ to be the one related to an infinitesimal *rotation of axes* in ordinary space. The results show that, as far as abstract vector space is concerned, this operator is identical with the angular momentum operator defined in Eqs. (1.6.42). The reason for the identity is, of course, that a measurement of the angular momentum of a system usually results in a space rotation of the system (unless the state corresponds to an eigenvector for $\mathfrak{M}$) just as a measurement of linear momentum $\mathfrak{p}$ usually results in a change in position of the system.

In terms of the formalism of abstract vector space the operator

$$\mathfrak{g} = e^{i(\mathfrak{M}_z/\hbar)\theta_z}$$

performs the necessary reorientation of state vectors corresponding to a rotation of ordinary space by an amount θ_z about the z axis. Since, by Eqs. (1.6.43) the eigenvalues of $\mathfrak{M}_z$ are $m_z\hbar$, where m_z is either an integer or a half integer (depending on whether ordinary angular momentum or spin is under study), when the operator acts on an eigenvector of $\mathfrak{M}_z$ with eigenvalue $m_z\hbar$, it changes neither its direction nor its magnitude

$$e^{i(\mathfrak{M}_z/\hbar)\theta_z}\mathbf{e}(m_z) = e^{im_z\theta_z}\mathbf{e}(m_z)$$

This whole subject of quantum mechanics and operator calculus will be discussed in more detail in the next chapter.

1.7 *The Lorentz Transformation, Four-vectors, Spinors*

Up to this point we have discussed vectors and other quantities in three dimensions, and some of the formulas and statements we have made regarding axial vectors, the curl operator, and vector multiplication are only correct for three dimensions. Since a great deal of the material in this volume deals only with the three space dimensions, such specialization is allowable, and the results are of considerable value. In many cases, however, a fourth dimension intrudes, the time dimension. In classical mechanics this added no further complication, for it was assumed that no physically possible operation could rotate the time axis into the space axis or vice versa, so that the time direction was supposed to be the same for all observers. If this had turned out to be true, the only realizable transformations would be in the three dimensions, time would be added as an afterthought, and the three-dimensional analyses discussed heretofore would be the only ones applicable to physics.

Proper Time. Modern electrodynamics has indicated, however, and the theory of relativity has demonstrated that there are physically possible transformations which involve the time dimension and that, when the relative velocity of two observers is comparable to the speed of light, their time directions are measurably not parallel. This does not mean that time is just another space dimension, for, in the formulas, it differs from the three space dimensions by the imaginary factor $i = \sqrt{-1}$. It is found that, when an object is moved in space by amounts dx, dy, dz in a time dt, with respect to observer A, the time as measured by observer B, moving with the object, is $d\tau_B$, where

$$(d\tau_B)^2 = dt^2 - (1/c^2)(dx^2 + dy^2 + dz^2) \tag{1.7.1}$$

where c is the velocity of light and $d\tau_B$ is called the *proper time* for observer B ($dt = d\tau_A$ is, of course, the proper time for observer A). As long as the velocities dx/dt, etc., are small compared with the velocity c, the proper times $d\tau_A$, $d\tau_B$ differ little in value; but if the relative velocity nearly equals c, the time intervals may differ considerably.

Equation (1.7.1) is analogous to the three-dimensional equation for the total differential of distance along a path $ds^2 = dx^2 + dy^2 + dz^2$, and the analogy may be made more obvious if Eq. (1.7.1) is written in the form

$$(ic\, d\tau_B)^2 = dx^2 + dy^2 + dz^2 + (ic\, dt)^2$$

The path followed by an object in space-time is called its *world line*, and the distance along it measures its proper time. The equation suggests that the proper times for two observers moving with respect to each other are related by an *imaginary rotation* in space-time, the amount of rotation being related to the relative velocity of the observers.

The Lorentz Transformation. In order to begin the discussion as simply as possible, we shall assume that the relative velocity of the two observers is parallel to the x axis, which allows their y and z axes to remain parallel. We shall also assume that their relative velocity u is constant, so that their world lines are at a constant angle with respect to each other and the transformation from one space-time system to another is a simple (imaginary) rotation. Consideration of observers accelerating with respect to each other would involve us in the intricacies of general relativity, which will not be needed in the present book.

The transformation corresponding to a rotation in the (x,ict) plane by an imaginary angle $i\alpha$ is

$$\begin{aligned} x &= x' \cosh \alpha + ct' \sinh \alpha \\ y &= y'; \quad z = z' \\ ct &= x' \sinh \alpha + ct' \cosh \alpha \end{aligned} \tag{1.7.2}$$

where (x,y,z,t) are the space-time coordinates relative to observer A and (x',y',z',t') those relative to B, moving at a relative velocity u, parallel to the x axis. In order that the time axis for B moves with relative velocity u with respect to A (or vice versa), we must have

$$u = c \tanh \alpha; \quad \sinh \alpha = \frac{u/c}{\sqrt{1 - (u/c)^2}}; \quad \cosh \alpha = \frac{1}{\sqrt{1 - (u/c)^2}}$$

Consequently, we can write the transformation in the more usual form:

$$\begin{aligned} x &= \frac{x' + ut'}{\sqrt{1 - (u/c)^2}} \\ y &= y'; \quad z = z' \\ t &= \frac{(ux'/c^2) + t'}{\sqrt{1 - (u/c)^2}} \end{aligned} \tag{1.7.3}$$

Incidentally this transformation also requires that, if observer B has velocity $u = c \tanh \alpha$ with respect to A and observer C has velocity $v = c \tanh \beta$ with respect to B (in the same direction as u), then C has velocity $c \tanh (\alpha + \beta) = (u + v)/[1 + (uv/c^2)]$ with respect to A.

This set of equations, relating the space-time coordinates of two observers moving with constant relative velocity, is called a *Lorentz transformation.* It is a symmetrical transformation, in that the equations for x', t' in terms of x, t can be obtained from Eqs. (1.7.3) by interchange of primed and unprimed quantities and reversal of the sign of u. This can be seen by solving for x' and t' in terms of x and t. The Lorentz transformation is a very specialized sort of a change of coordinates, for it is a simple (imaginary) rotation in a space-time plane, but it will suffice for most of our investigations into space-time in the volume.

The equations for the general Lorentz transformation corresponding to relative velocity $u = c \tanh \alpha$ in a direction given by the spherical angles ϑ, φ with respect to the x axis are

$$\begin{aligned}
x &= [1 + \cos^2\varphi \sin^2\vartheta\,(\cosh\alpha - 1)]x' + \cos\varphi \sin\varphi \sin^2\vartheta\,(\cosh\alpha - 1)y' \\
&\qquad + \cos\varphi \cos\vartheta \sin\vartheta\,(\cosh\alpha - 1)z' + \cos\varphi \sin\vartheta\,(\sinh\alpha)ct' \\
y &= \cos\varphi \sin\varphi \sin^2\vartheta\,(\cosh\alpha - 1)x' + [1 + \sin^2\varphi \sin^2\vartheta\,(\cosh\alpha - 1)]y' \\
&\qquad + \sin\varphi \cos\vartheta \sin\vartheta\,(\cosh\alpha - 1)z' + \sin\varphi \sin\vartheta\,(\sinh\alpha)ct' \\
z &= \cos\varphi \cos\vartheta \sin\vartheta\,(\cosh\alpha - 1)x' + \sin\varphi \cos\vartheta \sin\vartheta\,(\cosh\alpha - 1)y' \\
&\qquad + [1 + \cos^2\vartheta\,(\cosh\alpha - 1)]z' + \cos\vartheta\,(\sinh\alpha)ct' \\
ct' &= \cos\varphi \sin\vartheta\,(\sinh\alpha)x' + \sin\varphi \sin\vartheta\,(\sinh\alpha)y' \\
&\qquad + \cos\vartheta\,(\sinh\alpha)z' + (\cosh\alpha)ct'
\end{aligned}$$

When $\varphi = 0°$ and $\vartheta = 90°$, this reduces to the simple form given in Eqs. (1.7.2).

The scale factors h are all unity for this transformation, since it is a rotation without change of scale. Of course, the scale factors here involve the four dimensions

$$(h_n)^2 = \sum_{m=1}^{4} \left(\frac{\partial x_n}{\partial x'_m}\right)^2; \quad x_1, x_2, x_3, x_4 = x, y, z, ict$$

Since the h's are all unity, there is no need to differentiate between contravariant, covariant, and "true" vector components.

Four-dimensional Invariants. Just as with three dimensions, we shall find it remunerative to search for quantities which do not change when a Lorentz transformation is applied, *i.e.*, which give the same result when measured by any observer traveling with any uniform velocity (less than that of light). These are analogous to the scalar quantities which we discussed earlier and should, in many cases, correspond to measurable quantities, for according to the theory of relativity many physical quantities should give the same result when measured by different observers traveling at different velocities. Such quantities are said to be *Lorentz invariant*.

The space-time length of a given portion of the world line of a nonaccelerated particle is a four-dimensional invariant. If its displacement in space to observer B is x' and its duration in time to the same observer is t', then the square of the *proper lengths* of its world line is

$$s^2 = (ct')^2 - (x')^2$$

To observer A, according to Eqs. (1.6.3), the square of the proper length is

$$\begin{aligned}
s^2 = (ct)^2 - x^2 = \frac{1}{1 - (u/c)^2}\Bigg[&(ct')^2 + 2ux't' + \left(\frac{ux'}{c}\right)^2 \\
&- (x')^2 - 2ux't' - (ut')^2\Bigg] = (ct')^2 - (x')^2
\end{aligned}$$

which is the same value. The square of the length of the same world line to an observer traveling with the particle would be his proper time, squared, multiplied by c^2, which would also have the same value.

Therefore, in relativity, the *space-length* of a line is *not* an invariant, nor is a time duration. The space-length of a line to observer A, moving with the line, is the distance between points on two world lines *measured at the same time* for observer A $(x_2 - x_1)$. To observer B moving at velocity u the space length of the same line is determined by measuring the distance between points on the world lines *at the same time for B*, that is, for $t'_2 = t'_1$. By Eqs. (1.7.3) we have $x_2 - x_1 = (x'_2 - x'_1)/\sqrt{1 - (u/c)^2}$, or the distance measured by observer B, moving with respect to the line, is $x'_2 - x'_1 = (x_2 - x_1)\sqrt{1 - (u/c)^2}$, shorter than the length measured by A by the factor $\sqrt{1 - (u/c)^2}$. Since the apparent size of objects changes with the relative velocity of the observer, the apparent density of matter is also not an invariant under a Lorentz transformation.

Many other quantities which were scalars in three dimensions (*i.e.*, for very small relative velocities) turn out not to be invariants in space-time. The mass of a body, for instance, is not a four-dimensional invariant; the correct invariant is a combination of the mass and the kinetic energy of the body, corresponding to the relativistic correspondence between mass and energy. This will be proved after we have discussed vectors in four dimensions.

Four-vectors. As might be expected from the foregoing discussion, we also encounter a few surprises when we expand our concept of a vector to four dimensions. As with a three-vector, a four-vector must have an invariant length, only now this length must be a proper length in space and time. The two points (x'_2,y'_2,z'_2,t'_2) and (x'_1,y'_1,z'_1,t'_1) as measured by observer B define a four-vector $F'_1 = x'_2 - x'_1, \ldots, F'_4 = c(t'_2 - t'_1)$. To observer A, traveling with velocity u (in the x direction) with respect to B, the components of this vector are

$$F_1 = \frac{F'_1 + (uF'_4/c)}{\sqrt{1 - (u/c)^2}}; \quad F_2 = F'_2; \quad F_3 = F'_3; \quad F_4 = \frac{(uF'_1/c) + F'_4}{\sqrt{1 - (u/c)^2}}$$

so that for this Lorentz transformation, the direction cosines are

$$\begin{gathered}\gamma_{11} = \cosh\alpha; \quad \gamma_{22} = \gamma_{33} = 1; \quad \gamma_{44} = \cosh\alpha; \quad \gamma_{14} = \gamma_{41} = \sinh\alpha \\ \gamma_{12} = \gamma_{21} = \gamma_{13} = \gamma_{31} = \gamma_{23} = \gamma_{32} = \gamma_{24} = \gamma_{42} = \gamma_{34} = \gamma_{43} = 0 \\ \tanh\alpha = u/c\end{gathered} \tag{1.7.4}$$

This transformation of components is typical of four-vectors. We note that the "sum" of the squares of the four components $F^2 = F_1^2 + F_2^2 + F_3^2 - F_4^2$ is invariant, as it must be.

A very important vector is the four-dimensional generalization of the momentum of a particle which is traveling at a constant velocity u in the x direction with respect to observer A. To an observer B, traveling with the particle, the mass of the particle is m_0 and the proper time is τ. With respect to observer A the particle travels a distance dx in time dt, where $(d\tau)^2 = (dt)^2 - (dx/c)^2$. The space component of the vector momentum cannot be $m_0(dx/dt)$, for it would be difficult to find a corresponding time component such that the square of the resulting magnitude would be invariant. On the other hand, if we choose it to be $m_0(dx/d\tau)$, which transforms like dx, so that

$$p_x = m_0 \frac{dx}{d\tau} = \frac{m_0 u}{\sqrt{1 - (u/c)^2}}; \quad p_t = m_0 c \frac{dt}{d\tau} = \frac{m_0 c}{\sqrt{1 - (u/c)^2}}; \quad u = \frac{dx}{dt}$$

where τ is the proper time for the particle under study (and, therefore, for observer B), then the square of the magnitude of p is $p_x^2 - p_t^2 = -(m_0 c)^2$, which is invariant.

With respect to observer C, traveling with velocity $v = c \tanh \beta$ in the x direction compared with A, the vector momentum transforms according to Eq. (1.7.4),

$$\begin{aligned} p'_x &= p_x \cosh \beta + p_t \sinh \beta = m_0 c \sinh (\alpha + \beta) = m_0 u' / \sqrt{1 - (u'/c)^2} \\ p'_t &= p_x \sinh \beta + p_t \cosh \beta = m_0 c \cosh (\alpha + \beta) = m_0 c / \sqrt{1 - (u'/c)^2} \end{aligned}$$

where $u = c \tanh \alpha$ is the velocity of the particle with respect to A and $u' = c \tanh (\alpha + \beta)$ is its velocity with respect to observer C. Thus the definition of the momentum four-vector is consistent with the rule of composition of velocities given earlier.

Therefore, the four-vector corresponding to the momentum as measured by observer A, for a particle moving with speed u with respect to A, is

$$\begin{aligned} p_x &= \frac{m_0(dx/dt)}{\sqrt{1 - (u/c)^2}}; \quad p_y = \frac{m_0(dy/dt)}{\sqrt{1 - (u/c)^2}}; \quad p_z = \frac{m_0(dz/dt)}{\sqrt{1 - (u/c)^2}} \\ p_t &= \frac{m_0 c}{\sqrt{1 - (u/c)^2}}; \quad u^2 = \left(\frac{dx}{dt}\right)^2 + \left(\frac{dy}{dt}\right)^2 + \left(\frac{dz}{dt}\right)^2 \end{aligned} \tag{1.7.5}$$

where x, y, z, t are the coordinates of the particle according to observer A.

The time component of the momentum is proportional to the total energy of the particle with respect to observer A,

$$E = c p_t = \frac{m_0 c^2}{\sqrt{1 - (u/c)^2}} = m_0 c^2 + \tfrac{1}{2} m_0 u^2 + \cdots$$

which is not invariant under a Lorentz transformation. This equation also shows that the total energy can be separated into a rest energy and a kinetic energy only when u is small compared with c.

Another four-vector is the space-time gradient of some scalar function of (x,y,z,t), $\square\psi$ (which may be called *quad* ψ), where

$$(\square\psi)_1 = \frac{\partial\psi}{\partial x}, \text{ etc.}; \quad (\square\psi)_4 = \frac{1}{c}\frac{\partial\psi}{\partial t}$$

Since the scale factors h are all unity and the Christoffel symbols therefore are all zero, these are also the components of the four-dimensional covariant derivative of ψ. Consequently, the contracted second-order derivative

$$\square^2\psi = \frac{\partial^2\psi}{\partial x^2} + \frac{\partial^2\psi}{\partial y^2} + \frac{\partial^2\psi}{\partial z^2} - \frac{1}{c^2}\frac{\partial^2\psi}{\partial t^2} \tag{1.7.6}$$

is Lorentz invariant. The operator $\square^2$ is called the *d'Alembertian*. It is analogous, in a formal sense, to the three-dimensional Laplacian operator ∇^2. However, because of the presence of the negative sign before the time term, we shall see later that solutions of the Laplace equation $\nabla^2\psi = 0$ differ markedly from solutions of the equation $\square^2\psi = 0$, which is the *wave equation* for waves traveling with the velocity of light, c.

Stress-Energy Tensor. By analogy with the three-dimensional case, dyadics or tensors may be set up in four dimensions, having transformation properties with respect to a Lorentz transformation which can easily be determined by extending the discussion of the previous section. An interesting and useful example of such a tensor is obtained by extending the stress dyadic defined in Eq. (1.6.28) for an elastic medium to four dimensions. The dimensions of the nine stress components T_{ij} are dynes per square centimeter or grams per centimeter per second per second, and they transform according to the rules for three dimensions. By analogy from the previous discussion of the momentum vector for a particle, we should expect that the time component of the four-tensor might be proportional to the energy density of the medium. In other words, this fourth component should be related to the total energy term ρc^2 (where ρ is the mass density of the medium), which has dimensions grams per centimeter per second per second.

Therefore, we assume that the stress-energy tensor P_{ij} at a point (x,y,z,ct) in a medium has the following components for an observer A at rest with that part of the medium which is at (x,y,z,ct):

$$\begin{aligned} P_{11} &= T_{xx}, \text{ etc.}; \quad P_{12} = T_{xy} = P_{21} = T_{yx}, \text{ etc.} \\ P_{14} &= P_{24} = P_{34} = P_{41} = P_{42} = P_{43} = 0 \\ P_{44} &= c^2\rho_0 \end{aligned} \tag{1.7.7}$$

where ρ_0 is the density of the medium at (x,y,z,t) as measured by observer A.

If these quantities are to be components of a true four-tensor, then an observer B traveling with velocity u in the x direction with respect to the medium at (x,y,z,ct) will measure the components of P to be those given by the general transformation rules:

$$P'_{mn} = \sum_{ij} \frac{\partial \xi'_m}{\partial \xi_i} \frac{\partial \xi'_n}{\partial \xi_j} P_{ij}; \quad \xi'_1 = \xi_1 \cosh \alpha + \xi_4 \sinh \alpha$$

$$\xi'_2 = \xi_2; \quad \xi'_3 = \xi_3; \quad \xi'_4 = \xi_1 \sinh \alpha + \xi_4 \cosh \alpha; \quad c \tanh \alpha = u$$

The results of the transformation are:

$$\begin{aligned}
P'_{11} &= T_{xx} \cosh^2 \alpha + \rho_0 c^2 \sinh^2 \alpha; \quad P'_{12} = P'_{21} = T_{xy} \cosh \alpha \\
P'_{13} &= P'_{31} = T_{xz} \cosh \alpha; \quad P'_{22} = T_{yy}; \quad P'_{33} = T_{zz} \\
P'_{23} &= P'_{32} = T_{yz}; \quad P'_{14} = P'_{41} = (T_{xx} + \rho_0 c^2) \cosh \alpha \sinh \alpha \qquad (1.7.8) \\
P'_{24} &= P'_{42} = T_{xy} \sinh \alpha; \quad P'_{34} = P'_{43} = T_{xz} \sinh \alpha \\
P'_{44} &= T_{xx} \sinh^2 \alpha + \rho_0 c^2 \cosh^2 \alpha
\end{aligned}$$

The space components (P'_{11}, P'_{23}, etc.) turn out to be the components of stress which observer B would measure in the medium if we take into account the finite velocity of light, and the component P'_{44} turns out to be c^2 times the effective density as measured by observer B. An examination of component P'_{14} shows that it can be considered to be proportional to the momentum flow density of the medium parallel to the x axis as measured by observer B. Correspondingly the components P'_{24}, P'_{34} must be proportional to momentum flows in the y and z directions as measured by observer B.

We therefore arrive at an interesting and important result, which can be verified by further analysis or by experiment, that relative motion transforms a stress into a momentum flow and vice versa. Moreover, since we can verify, in the system at rest with respect to the medium (observer A), that the contracted tensor equations

$$\sum_n \left(\frac{\partial P_{mn}}{\partial \xi_n} \right) = 0$$

or, in terms of the T's,

$$\frac{\partial T_{xx}}{\partial x} + \frac{\partial T_{xy}}{\partial y} + \frac{\partial T_{xz}}{\partial z} = 0, \text{ etc.}; \quad \frac{\partial}{\partial t} (c\rho_0) = 0$$

are true, then these equations should also be true for observers moving relative to the medium (or, what is the same thing, for a medium moving with respect to the observer). For instance, if we define the momentum vector M with respect to observer B as being the vector with space components

$$\begin{aligned}
M_x &= (1/c)(\rho_0 c^2 + T_{xx}) \cosh \alpha \sinh \alpha = P'_{41}/c \\
M_y &= P'_{42}/c; \quad M_z = P'_{43}/c
\end{aligned}$$

and the density with respect to observer B as being

$$\rho = \rho_0 \cosh^2 \alpha + (T_{xx}/c^2) \sinh^2 \alpha = P'_{44}/c^2$$

then one of the transformed equations would become

$$\sum_n \left(\frac{\partial P'_{4n}}{\partial \xi'_n}\right) = 0 \quad \text{or} \quad \frac{\partial M_x}{\partial x'} + \frac{\partial M_y}{\partial y'} + \frac{\partial M_z}{\partial z'} + \frac{\partial \rho}{\partial t'} = 0$$

where the primed coordinates are those suitable for observer B. This equation is, of course, the equation of continuity for the medium, relating momentum density (or mass flow) to change in mass density ρ. The other three transformed equations turn out to be the equations of motion for the medium.

Spin Space and Space-Time. One of the most interesting developments in modern theoretical physics is the demonstration that there is a close connection between the two-dimensional "state space" connected with the spin of the electron and the four-dimensional space-time describing the motion of the electron. In the previous section we have initiated the discussion of a spin space corresponding to the two possible states of spin of the electron and there pointed out that a change of direction of the spin in ordinary space by 180° (reversal of spin) corresponded to a change of the state vector in spin space by 90°. This is somewhat analogous to the relation between the graphical representation of (-1) and $\sqrt{-1}$ on the complex plane, and if one wished to become fanciful, one might consider that the relation between ordinary space and spin space was some sort of a "square-root" relation. Actually, it can be shown that the two-dimensional spin space is a "square-root space," not of three-dimensional space, but of four-dimensional space-time. More specifically, we will find that the four components of a dyadic in spin space could be identified with the components of a four-vector in space-time.

In order to show this we must consider the components of vectors in spin space, and even the unit vectors themselves, as complex quantities with complex conjugates (a and $\bar{a}$ for the components, which are numbers, and $\mathbf{e}$ and $\mathbf{e}^*$ for state vectors), so that $a\bar{a}$ and $\mathbf{e} \cdot \mathbf{e}^*$ are real positive numbers. We start out with the two mutually perpendicular state vectors $\mathbf{e}_1$ and $\mathbf{e}_2$ (with their complex conjugates $\mathbf{e}_1^*$ and $\mathbf{e}_2^*$) representing states where the electron spin is known to be pointed in a certain direction or known to be pointed in the opposite direction. For two complex vectors to be called unit, orthogonal vectors, we must have that $\mathbf{e}_1 \cdot \mathbf{e}_1^* = \mathbf{e}_1^* \cdot \mathbf{e}_1 = \mathbf{e}_2 \cdot \mathbf{e}_2^* = \mathbf{e}_2^* \cdot \mathbf{e}_2 = 1$ and that $\mathbf{e}_1 \cdot \mathbf{e}_2^* = \mathbf{e}_1^* \cdot \mathbf{e}_2 = \mathbf{e}_2 \cdot \mathbf{e}_1^* = \mathbf{e}_2^* \cdot \mathbf{e}_1 = 0$. The values of the complex quantities $\mathbf{e}_1 \cdot \mathbf{e}_1$, $\mathbf{e}_1 \cdot \mathbf{e}_2$, $\mathbf{e}_2^* \cdot \mathbf{e}_1^*$, etc., are not required.

Any vector in spin space can be expressed in terms of these two unit vectors:

$$\mathbf{s} = a_1\mathbf{e}_1 + a_2\mathbf{e}_2; \quad \mathbf{s}^* = \bar{a}_1\mathbf{e}_1^* + \bar{a}_2\mathbf{e}_2^*$$

and any dyadic in spin space can be expressed in the form

$$\mathfrak{S} = c_{11}\mathbf{e}_1\mathbf{e}_1^* + c_{12}\mathbf{e}_1\mathbf{e}_2^* + c_{21}\mathbf{e}_2\mathbf{e}_1^* + c_{22}\mathbf{e}_2\mathbf{e}_2^* \tag{1.7.9}$$

A dyadic in spin space is called a *spinor of second order*. It has the usual properties of dyadics in that it can operate on a state vector in spin space to produce another state vector:

$$\begin{aligned} \mathfrak{S}\cdot\mathbf{s} &= (c_{11} + c_{21})a_1\mathbf{e}_1 + (c_{12} + c_{22})a_2\mathbf{e}_2; \\ \mathbf{s}^*\cdot\mathfrak{S} &= \bar{a}_1(c_{11} + c_{12})\mathbf{e}_1^* + \bar{a}_2(c_{21} + c_{22})\mathbf{e}_2^* \end{aligned}$$

To give physical content to these definitions, we must relate the behavior of the spinor components a_i and the dyadic components c_{ij} under a Lorentz transformation, with their behavior under a rotation of axes in spin space. For example, a Lorentz invariant quantity should also be an invariant to rotation in spin space. Following our preceding remarks we shall require that the four second-order spinor components c_{ij} transform like the components of a four-vector in ordinary space-time. *A dyadic in spin space is a vector in space-time;* this is the consummation of our desire to have spin space be a "square-root space."

Spinors and Four-vectors. We still have not worked out the specific rules of transformation for spin space which will enable the c_{ij} components of the spinor to transform like a four-vector. The most general transformation is given as follows:

$$\begin{aligned} \mathbf{e}_1' &= \alpha_{11}\mathbf{e}_1 + \alpha_{12}\mathbf{e}_2; & \mathbf{e}_1^{*\prime} &= \bar{\alpha}_{11}\mathbf{e}_1^* + \bar{\alpha}_{12}\mathbf{e}_2^* \\ \mathbf{e}_2' &= \alpha_{21}\mathbf{e}_1 + \alpha_{22}\mathbf{e}_2; & \mathbf{e}_2^{*\prime} &= \bar{\alpha}_{21}\mathbf{e}_1^* + \bar{\alpha}_{22}\mathbf{e}_2^* \\ \mathbf{e}_1 &= \alpha_{22}\mathbf{e}_1' - \alpha_{12}\mathbf{e}_2'; & \mathbf{e}_1^* &= \bar{\alpha}_{22}\mathbf{e}_1^{*\prime} - \bar{\alpha}_{12}\mathbf{e}_2^{*\prime} \\ \mathbf{e}_2 &= -\alpha_{21}\mathbf{e}_1' + \alpha_{11}\mathbf{e}_2'; & \mathbf{e}_2^* &= -\bar{\alpha}_{21}\mathbf{e}_1^{*\prime} + \bar{\alpha}_{11}\mathbf{e}_2^{*\prime} \end{aligned} \tag{1.7.10}$$

where, in order that the scale be the same in the new coordinates as in the old, $\alpha_{11}\alpha_{22} - \alpha_{12}\alpha_{21} = 1$, $\bar{\alpha}_{11}\bar{\alpha}_{22} - \bar{\alpha}_{12}\bar{\alpha}_{21} = 1$.

Under this transformation in spin space the general spinor components undergo the following transformations:

$$c_{mn} = \sum_{ij} c'_{ij}\alpha_{im}\bar{\alpha}_{jn} \tag{1.7.11}$$

The safest way to carry on from here is to find a function of the c's which is invariant under transformation of the α's in spin space, which can then be made invariant under a Lorentz transformation. Utilizing the multiplication properties of the α's, we can soon show that one invariant is the quantity $c_{11}c_{22} - c_{12}c_{21}$ (it can be shown by substituting and multiplying out, utilizing the fact that $\alpha_{11}\alpha_{22} - \alpha_{12}\alpha_{21} = 1$, etc.). This quantity can also be made a Lorentz invariant if the c's are related to components F_n of a four-vector in such a way that $c_{11}c_{22} - c_{12}c_{21} = c^2F_4^2 - F_1^2 - F_2^2 - F_3^2$, for such a combination of the F's is invariant. There are a number of ways in which this can be accomplished, but the most straightforward way is as follows:

$$\begin{aligned} c_{11} &= cF_4 + F_1; & F_4 &= (1/2c)(c_{11} + c_{22}) \\ c_{22} &= cF_4 - F_1; & F_3 &= (1/2)(c_{11} - c_{22}) \\ c_{12} &= F_2 - iF_3; & F_2 &= (i/2)(c_{12} - c_{21}) \\ c_{21} &= F_2 + iF_3; & F_1 &= (1/2)(c_{12} + c_{21}) \end{aligned} \tag{1.7.12}$$

Lorentz Transformation of Spinors. For instance, for an observer B (primed coordinate) moving with velocity u in the x direction with respect to the electron, at rest with respect to A (unprimed coordinates), the transformation for the F's is $cF_4 = cF_4' \cosh\alpha + F_1' \sinh\alpha$, $F_1 = F_1' \cosh\alpha + cF_4' \sinh\alpha$, $F_2 = F_2'$, $F_3 = F_3'$, $u = c\tanh\alpha$. The transformation for the spinor components c is, therefore,

$$\begin{aligned} c_{11} &= c_{11}'e^{\alpha}; \quad c_{12} = c_{12}'; \quad c_{21} = c_{21}'; \\ c_{22} &= c_{22}'e^{-\alpha}; \quad e^{\alpha} = \sqrt{(c+u)/(c-u)} \end{aligned} \tag{1.7.13}$$

and that for the corresponding direction cosines for rotation of the unit vectors in spin space is

$$\alpha_{11} = e^{\alpha/2}; \quad \alpha_{22} = e^{-\alpha/2}; \quad \alpha_{12} = \alpha_{21} = 0 \tag{1.7.14}$$

Therefore, any state vector in spin space, with respect to observer A, $\mathbf{s} = a_1\mathbf{e}_1 + a_2\mathbf{e}_2$, becomes $\mathbf{s}' = a_1e^{\alpha/2}\mathbf{e}_1' + a_2e^{-\alpha/2}\mathbf{e}_2'$ with respect to observer B, moving with velocity $u = c\tanh\alpha$ with respect to A.

The transformation equations for the c's and α's for a more general Lorentz transformation will naturally be more complicated than those given here, but they can be worked out by the same methods. Any pair of complex quantities which satisfy the transformation requirements for the components of a state vector in spin space is called a *spinor of first order;* a quartet of quantities which satisfy the transformation requirements for the c's of Eqs. (1.7.9) and (1.7.11) is called a *spinor of the second order;* and so on. Equations (1.7.12) give the relationship between the spinor components and the components of a four-vector for this simple Lorentz transformation.

Space Rotation of Spinors. As a further example of the "square-root" relationship between spinor components and vector components, we consider the case where the time coordinate is not altered but the space coordinates are rotated in accordance with the Eulerian angles discussed on page 28. Under this transformation the time component of any four-vector will remain constant and, therefore, by Eq. (1.7.12), $c_{11} + c_{22}$ will stay constant. Transforming $c_{11} + c_{22}$ in terms of the

α's, we see that for a purely space rotation, where $c_{11} + c_{22} = c'_{11} + c'_{22}$, we must have

$$\bar{\alpha}_{1n}\alpha_{1m} + \bar{\alpha}_{2n}\alpha_{2m} = \bar{\alpha}_{n1}\alpha_{m1} + \bar{\alpha}_{n2}\alpha_{m2} = \delta_{nm} = \begin{Bmatrix} 0; & n \neq m \\ 1; & n = m \end{Bmatrix} \quad n, m = 1, 2$$

This result is not unexpected, for a consequence is that $\mathbf{e}'_n \cdot \mathbf{e}^{*\prime}_n = 1$, and we should expect that the "length" of a state-vector would be unchanged by a space rotation. Adding together all the limitations on the α's, we see that

$$\bar{\alpha}_{11} = \alpha_{22}; \quad \bar{\alpha}_{12} = -\alpha_{21}; \quad \text{etc.} \tag{1.7.15}$$

We write down the expressions for the transformation for the F's for a space rotation [modifying Eq. (1.3.8) and letting F_1 be F_z, F_2 be F_x, and F_3 be F_y].

$$\begin{aligned} F'_1 &= [\cos\Phi\cos\theta\cos\psi - \sin\Phi\sin\psi]F_1 \\ &\qquad + [\sin\Phi\cos\theta\cos\psi + \cos\Phi\sin\psi]F_2 - \sin\theta\cos\psi\, F_3 \\ F'_2 &= -[\cos\Phi\cos\theta\sin\psi + \sin\Phi\cos\psi]F_1 \\ &\qquad - [\sin\Phi\cos\theta\sin\psi - \cos\Phi\cos\psi]F_2 + \sin\theta\sin\psi\, F_3 \\ F'_3 &= \sin\theta\cos\Phi\, F_1 + \sin\theta\sin\Phi\, F_2 + \cos\theta\, F_3; \quad F'_4 = F_4 \end{aligned}$$

and we insert these into Eq. (1.7.12) both for the primed and unprimed components to determine the transformation equations for the c's:

$$\begin{aligned} c'_{12} = -\sin(\theta/2)\cos(\theta/2)e^{i\psi}(c_{11} - c_{22}) &+ \cos^2(\theta/2)e^{i(\psi+\Phi)}c_{12} \\ &- \sin^2(\theta/2)e^{i(\psi-\Phi)}c_{21}; \quad \text{etc.} \end{aligned}$$

In terms of the direction cosines α this component is related to the unprimed c's by the equation

$$c'_{12} = \alpha_{22}\alpha_{21}c_{11} + \alpha_{22}^2 c_{12} - \alpha_{21}^2 c_{21} - \alpha_{21}\alpha_{22}c_{22}$$

where we have inverted Eq. (1.7.11) and used Eq. (1.7.15).

Comparing these last two equations, we find that the direction cosines α for spin-space rotation corresponding to a space rotation given by the Euler angles Φ, θ, ψ, [see Eq. (1.3.8) and Fig. 1.6] are

$$\begin{aligned} \alpha_{11} &= \cos(\theta/2)e^{-i(\psi+\Phi)/2}; \quad \alpha_{21} = -\sin(\theta/2)e^{i(\psi-\Phi)/2} \\ \alpha_{12} &= \sin(\theta/2)e^{-i(\psi-\Phi)/2}; \quad \alpha_{22} = \cos(\theta/2)e^{i(\psi+\Phi)/2} \end{aligned} \tag{1.7.16}$$

where we have again used Eq. (1.7.15) to help untangle the equations. Therefore, under this rotation of the space coordinate system a state vector in spin space $\mathbf{s} = a_1\mathbf{e}'_1 + a_2\mathbf{e}'_2$ becomes

$$\begin{aligned} \mathbf{s} = [a_1\cos(\theta/2)e^{-i\Psi/2} - a_2\sin(\theta/2)e^{i\Psi/2}]e^{-i\Phi/2}\mathbf{e}_1 \\ + [a_1\sin(\theta/2)e^{-i\Psi/2} + a_2\cos(\theta/2)e^{i\Psi/2}]e^{i\Phi/2}\mathbf{e}_2 \end{aligned}$$

This last equation shows that rotation in spin space is governed by one-half the angles of rotation in ordinary space. A rotation of 180°

($\theta = \pi$, $\Phi = \psi = 0$) changes $\mathbf{s} = a_1\mathbf{e}'_1 + a_2\mathbf{e}'_2$ into $\mathbf{s} = a_2\mathbf{e}_1 - a_1\mathbf{e}_2$, which is a rotation of 90° in spin space.

The transformations given by Eqs. (1.7.14) and (1.7.16) are the cases of usual interest. Any other case can be handled in the same manner, by use of Eqs. (1.7.12). Although we began this discussion with a rather vague requirement to satisfy, we have developed the theory of a quantity which has definite transformation properties under a general rotation of coordinates (including a Lorentz rotation) and yet is not a tensor, according to our earlier discussions. This came as quite a surprise when spinors were first studied.

Spin Vectors and Tensors. A quartet of simple dyadics in spin space can be found which behave like four unit vectors in space-time:

$$\begin{aligned} \boldsymbol{\sigma}_4 &= \mathbf{e}_1\mathbf{e}_1^* + \mathbf{e}_2\mathbf{e}_2^* = \mathfrak{I} \\ \boldsymbol{\sigma}_1 &= \mathbf{e}_1\mathbf{e}_2^* + \mathbf{e}_2\mathbf{e}_1^* \\ \boldsymbol{\sigma}_2 &= i(\mathbf{e}_2\mathbf{e}_1^* - \mathbf{e}_1\mathbf{e}_2^*) \\ \boldsymbol{\sigma}_3 &= \mathbf{e}_1\mathbf{e}_1^* - \mathbf{e}_2\mathbf{e}_2^* \end{aligned} \tag{1.7.17}$$

These quantities operate on the spin vectors $\mathbf{e}$ as follows:

$$\boldsymbol{\sigma}_4 \cdot \mathbf{e}_n = \mathbf{e}_n; \quad \boldsymbol{\sigma}_1 \cdot \mathbf{e}_1 = \mathbf{e}_2; \quad \boldsymbol{\sigma}_2 \cdot \mathbf{e}_1 = i\mathbf{e}_2; \quad \boldsymbol{\sigma}_3 \cdot \mathbf{e}_1 = \mathbf{e}_1; \quad \text{etc.}$$

Comparison with Eqs. (1.6.44) shows that the quantities $\boldsymbol{\sigma}_1$, $\boldsymbol{\sigma}_2$, $\boldsymbol{\sigma}_3$ are $2/\hbar$ times the spin operators for the electron. They are called the *Pauli spin operators.* The quantity $\boldsymbol{\sigma}_4$ is, of course, the unity dyadic. We see also that $i\boldsymbol{\sigma}_1$, $i\boldsymbol{\sigma}_2$, $-i\boldsymbol{\sigma}_3$ are just the Hamilton quaternion operators.

We now can rewrite the spinor dyadic given in Eq. (1.7.9) as a four-vector, by using Eqs. (1.7.12) and (1.7.17):

$$\mathfrak{S} = cF_4\boldsymbol{\sigma}_4 + F_1\boldsymbol{\sigma}_1 + F_2\boldsymbol{\sigma}_2 + F_3\boldsymbol{\sigma}_3 \tag{1.7.18}$$

where the "unit vectors" $\boldsymbol{\sigma}$ are operators, operating on state vectors in spin space, but the components F are ordinary numbers transforming like components of an ordinary four-vector. Thus we finally see how operators in spin space can act like vectors in space-time. The extension of this discussion to the inversion transformation $\mathbf{r}' \to -\mathbf{r}$ and its correlation with spin space requires that we consider $\mathbf{e}^*$ and $\mathbf{e}$ as independent quantities, so that transformations between them are possible. We shall not go into this further, however, except to say that the vector $\boldsymbol{\sigma}$ transforms like an *axial* vector. (See Prob. 1.34.)

One can go on to form spinor forms which transform like dyadics in space time. For instance a spinor of fourth order,

$$\begin{aligned} \sum_{\mu,\nu} \boldsymbol{\sigma}_\mu\boldsymbol{\sigma}_\nu F_{\mu\nu} = {} & \boldsymbol{\sigma}_4(F_{11} + F_{22} + F_{33} + F_{44}) + \boldsymbol{\sigma}_1(F_{14} + F_{41} + iF_{23} - iF_{32}) \\ & + \boldsymbol{\sigma}_2(F_{24} + F_{42} + iF_{31} - iF_{13}) + \boldsymbol{\sigma}_3(F_{34} + F_{43} + iF_{12} - iF_{21}) \end{aligned}$$

has components $F_{\mu\nu}$ which behave like components of a dyadic in space-time. A particularly important form is the contracted tensor formed by multiplying one spinor vector by its conjugate:

$$(\boldsymbol{\sigma}_4 cF_4 + \boldsymbol{\sigma}_1 F_1 + \boldsymbol{\sigma}_2 F_2 + \boldsymbol{\sigma}_3 F_3)(\boldsymbol{\sigma}_4 cF_4 - \boldsymbol{\sigma}_1 F_1 - \boldsymbol{\sigma}_2 F_2 - \boldsymbol{\sigma}_3 F_3) = \boldsymbol{\sigma}_4(c^2F_4^2 - F_1^2 - F_2^2 - F_3^2) \quad (1.7.19)$$

giving the square of the magnitude of the four-vector. This relation will be of use when we come to discuss the Dirac theory of the electron.

Rotation Operator in Spinor Form. Reference to Hamilton's researches on quaternions (see page 75) has suggested a most interesting and useful spinor operator using the spin-vector direction cosines α as components:

$$\begin{aligned} \mathfrak{R} &= \alpha_{11}\mathbf{e}_1\mathbf{e}_1^* + \alpha_{21}\mathbf{e}_1\mathbf{e}_2^* + \alpha_{12}\mathbf{e}_2\mathbf{e}_1^* + \alpha_{22}\mathbf{e}_2\mathbf{e}_2^* \\ &= \sum_n R_n\boldsymbol{\sigma}_n; \quad R_1 = \frac{1}{2}(\alpha_{12} + \alpha_{21}); \\ R_2 &= \frac{1}{2i}(\alpha_{12} - \alpha_{21}); \quad R_3 = \frac{1}{2}(\alpha_{11} - \alpha_{22}); \quad R_4 = \frac{1}{2}(\alpha_{11} + \alpha_{22}) \end{aligned} \quad (1.7.20)$$

The α's, according to Eq. (1.7.10), are the direction cosines relating the primed and unprimed unit vectors $\mathbf{e}$ in spin space. If they have the values given in Eq. (1.7.16), they correspond to a rotation of space axes by the Euler angles θ, ψ, Φ. As we have shown above, a spinor operator of the form of $\mathfrak{R}$ has the transformation properties of a vector, and this is emphasized by writing it in terms of its components R "along" the unit spin vectors $\boldsymbol{\sigma}$.

However, vector (or spinor operator, whichever point of view you wish to emphasize) $\mathfrak{R}$ is a peculiar one in that its components

$$R_1 = i\sin(\theta/2)\sin[(\Phi - \psi)/2]; \quad R_2 = -i\sin(\theta/2)\cos[(\Phi - \psi)/2]$$
$$R_3 = i\cos(\theta/2)\sin[(\Phi + \psi)/2]; \quad R_4 = \cos(\theta/2)\cos[(\Phi + \psi)/2]$$

are themselves related to a particular transformation specified by the angles θ, ψ, Φ. (This does not mean that the vector $\mathfrak{R}$ cannot be expressed in terms of any rotated coordinates for any angles; it just means that it is especially related to one particular rotation for the angles θ, ψ, Φ.)

As might be expected, the vector has a particular symmetry for this rotation, for if the unit vectors $\mathbf{e}'$ are related to the vectors $\mathbf{e}$ through the same angles [see Eqs. (1.7.10) and (1.7.16)], then it turns out that $\mathfrak{R}$ has the same form in terms of the primed vectors as it does for the unprimed:

$$\begin{aligned} \mathfrak{R} &= \alpha_{11}\mathbf{e}_1\mathbf{e}_1^* + \alpha_{21}\mathbf{e}_1\mathbf{e}_2^* + \cdots \\ &= [\alpha_{11}\alpha_{22}\bar{\alpha}_{22} - \alpha_{21}\alpha_{22}\bar{\alpha}_{21} - \alpha_{12}\alpha_{21}\bar{\alpha}_{22} + \alpha_{22}\alpha_{21}\bar{\alpha}_{21}]\mathbf{e}_1'\mathbf{e}_1^{*\prime} \\ &\qquad - [\alpha_{11}\alpha_{22}\bar{\alpha}_{12} - \alpha_{21}\alpha_{22}\bar{\alpha}_{11} - \alpha_{12}\alpha_{21}\bar{\alpha}_{12} + \alpha_{22}\alpha_{21}\bar{\alpha}_{11}]\mathbf{e}_1'\mathbf{e}_2^{*\prime} \cdots \\ &= \bar{\alpha}_{22}\mathbf{e}_1'\mathbf{e}_1^{*\prime} - \bar{\alpha}_{12}\mathbf{e}_1'\mathbf{e}_2^{*\prime} \cdots \\ &= \alpha_{11}\mathbf{e}_1'\mathbf{e}_1^{*\prime} + \alpha_{21}\mathbf{e}_1'\mathbf{e}_2^{*\prime} + \cdots \end{aligned}$$

as one can prove by utilizing the multiplication properties of the α's.

However, $\mathfrak{R}$ is also a spin-vector operator. As a matter of fact it operates on any spin vector to rotate it by just the amount which the transformation $\mathbf{e} \to \mathbf{e}'$ produces. According to Eqs. (1.7.10),

$$\mathfrak{R} \cdot \mathbf{e}_n = \mathbf{e}'_n; \quad \mathfrak{R}^* \cdot \mathbf{e}^*_n = \mathbf{e}^{*\prime}_n \tag{1.7.21}$$

where $\mathfrak{R}^* = \bar{\alpha}_{11}\mathbf{e}^*_1\mathbf{e}_1 + \bar{\alpha}_{21}\mathbf{e}^*_1\mathbf{e}_2 + \bar{\alpha}_{12}\mathbf{e}^*_2\mathbf{e}_1 + \bar{\alpha}_{22}\mathbf{e}^*_2\mathbf{e}_2$

Another operator $\mathfrak{R}^{-1}$ performs the inverse operation $\mathbf{e}' \to \mathbf{e}$. Reference to Eqs. (1.7.10) indicates that the vector is

$$\begin{aligned} \mathfrak{R}^{-1} &= \alpha_{22}\mathbf{e}_1\mathbf{e}^*_1 - \alpha_{21}\mathbf{e}_1\mathbf{e}^*_2 - \alpha_{12}\mathbf{e}_2\mathbf{e}^*_1 + \alpha_{11}\mathbf{e}_2\mathbf{e}^*_2 \\ &= \alpha_{22}\mathbf{e}'_1\mathbf{e}^{*\prime}_1 - \alpha_{21}\mathbf{e}'_1\mathbf{e}^{*\prime}_2 - \alpha_{12}\mathbf{e}'_2\mathbf{e}^{*\prime}_1 + \alpha_{11}\mathbf{e}'_2\mathbf{e}^{*\prime}_2 \end{aligned}$$

so that $\mathfrak{R}^{-1} \cdot \mathbf{e}'_n = \mathbf{e}_n; \quad (\mathfrak{R}^{-1})^* \cdot \mathbf{e}^{*\prime}_n = \mathbf{e}^*_n$

But since $\bar{\alpha}_{22} = \alpha_{11}$, $\bar{\alpha}_{12} = -\alpha_{21}$, etc., we also can show that

$$\mathbf{e}_n \cdot (\mathfrak{R}^{-1})^* = \mathbf{e}'_n \quad \text{and} \quad \mathbf{e}^*_n \cdot \mathfrak{R}^{-1} = \mathbf{e}^{*\prime}_n$$

and that

$$\mathbf{e}'_n \cdot \mathfrak{R}^* = \mathbf{e}_n \quad \text{and} \quad \mathbf{e}^{*\prime}_n \cdot \mathfrak{R} = \mathbf{e}^*_n \tag{1.7.22}$$

which shows the close interrelation between the operator $\mathfrak{R}$ and its inverse, $\mathfrak{R}^{-1}$.

The particularly important property of the operators $\mathfrak{R}$ is that, in addition to causing a rotation of vectors in spin space, they can also cause a related rotation of four-vectors in ordinary space. For instance, the spinor

$$\mathfrak{S} = c_{11}\mathbf{e}_1\mathbf{e}^*_1 + c_{12}\mathbf{e}_1\mathbf{e}^*_2 + c_{21}\mathbf{e}_2\mathbf{e}^*_1 + c_{22}\mathbf{e}_2\mathbf{e}^*_2$$

(where the c's have any values) has the transformation properties of a four-vector [see Eq. (1.7.18)], with components F_n [see Eqs. (1.7.12)]. The vector formed by operating "fore-and-aft" on $\mathfrak{S}$ by $\mathfrak{R}$:

$$\begin{aligned} \mathfrak{R} \cdot \mathfrak{S} \cdot \mathfrak{R}^{-1} &= c_{11}\mathbf{e}'_1\mathbf{e}^{*\prime}_1 + c_{12}\mathbf{e}'_1\mathbf{e}^{*\prime}_2 + c_{21}\mathbf{e}'_2\mathbf{e}^{*\prime}_1 + c_{22}\mathbf{e}'_2\mathbf{e}^{*\prime}_2 \\ &= F_1\mathfrak{d}'_1 + F_2\mathfrak{d}'_2 + F_3\mathfrak{d}'_3 + F_4\mathfrak{d}'_4 \end{aligned} \tag{1.7.23}$$

is one with the same components F_n, but these components are now with respect to the primed unit vectors, rotated with respect to the unprimed ones. Therefore, the fore-and-aft operation by $\mathfrak{R}$ has effectively rotated the vector $\mathfrak{S}$ by an amount given by the angles θ, ψ and Φ. In keeping with the "square-root" relation between spin space and space-time, we operate on a spin vector once with $\mathfrak{R}$ to produce a rotation, but we must operate twice on a four-vector in order to rotate it by the related amount.

We note that we have here been dealing with rotations by finite-sized angles. If the rotation is an infinitesimal one, the Euler angles θ and $(\Phi + \psi)$ become small and the rotation can be represented by the infinitesimal vector $\Delta\boldsymbol{\omega}$, its direction giving the axis of rotation and its magnitude giving the angle of rotation in radians. Consideration of the properties of the cross product shows that the operation of changing an ordinary three-dimensional vector **A** into another **A′** by infinitesimal rotation is given by the equation

$$\mathbf{A}' = \mathbf{A} + \Delta\boldsymbol{\omega} \times \mathbf{A} \tag{1.7.24}$$

Inspection of the equations on page 103 for rotation in terms of the Euler angle shows that, when θ and $(\Phi + \psi)$ are small,

$$(\Delta\omega)_1 \to -(\Phi + \psi); \quad (\Delta\omega)_2 \to -\theta \sin \psi; \quad (\Delta\omega)_3 \to -\theta \cos \psi$$

Inspection of the equations for the components of $\mathfrak{R}$ results in a related set of equations:

$$R_4 = 1; \quad R_1 = -(i/2)(\Phi + \psi); \quad R_2 = -(i/2)\theta \sin \psi;$$
$$R_3 = -(i/2)\theta \cos \psi$$

when θ and $(\Phi + \psi)$ are small.

Consequently, for an infinitesimal rotation represented by the vector $\Delta\boldsymbol{\omega}$, the rotation spinor operator is

$$\mathfrak{R} \to \boldsymbol{\sigma}_4 + (i/2)[(\Delta\omega)_1\boldsymbol{\sigma}_1 + (\Delta\omega)_2\boldsymbol{\sigma}_2 + (\Delta\omega)_3\boldsymbol{\sigma}_3] \tag{1.7.25}$$

These equations are sometimes useful in testing an unknown operator to see whether its components satisfy the transformation rules for four-vectors.

Problems for Chapter 1

1.1 The surfaces given by the equation

$$(x^2 + y^2) \cos^2 \psi + z^2 \cot^2 \psi = a^2; \quad 0 < \psi < \pi$$

for ψ constant are equipotential surfaces. Express ψ in terms of x, y, z and compute the direction cosines of the normal to the ψ surface at the point x, y, z. Show that ψ is a solution of Laplace's equation. What is the shape of the surface ψ = constant? $\psi = 0$? $\psi = \pi$?

1.2 The surfaces given by the equation

$$[\sqrt{x^2 + y^2} - \psi]^2 + z^2 = \psi^2 - a^2; \quad a < \psi < \infty$$

for ψ constant, define a family of surfaces. What is the shape of the surface? What is the shape for the limiting cases $\psi = 0$, $\psi = \infty$? Express ψ in terms of x, y, z and compute the direction cosines of the normal to the ψ surface at x, y, z. Is ψ a solution of Laplace's equation?

1.3 The three components of a vector field are

$$F_x = 2zx; \quad F_y = 2zy; \quad F_z = a^2 + z^2 - x^2 - y^2$$

Show that the equations for the flow lines may be integrated to obtain the flow functions φ and μ, where

$$\frac{y}{x} = \tan\varphi; \qquad \frac{x^2 + y^2 + z^2 + a^2}{2a\sqrt{x^2 + y^2}} = \coth\mu$$

Show that a pseudopotential exists and is given by ψ, where

$$\frac{x^2 + y^2 + z^2 - a^2}{2az} = \cot\psi$$

Show that the surfaces φ, μ, ψ constant are mutually orthogonal.

1.4 The three components of a vector field are

$$F_x = 3xz; \quad F_y = 3yz; \quad F_z = 2z^2 - x^2 - y^2$$

Integrate the equations for the flow lines to obtain the flow functions

$$\varphi = \tan^{-1}\left(\frac{y}{x}\right); \quad \vartheta = \frac{x^2 + y^2}{(x^2 + y^2 + z^2)^{\frac{3}{2}}}$$

and show that the pseudopotential is

$$\psi = \frac{z}{(x^2 + y^2 + z^2)^{\frac{3}{2}}}$$

Is ψ a solution of Laplace's equation?

1.5 Compute the net outflow integral, for the force fields of Probs. 1.3 and 1.4, over a sphere of radius r with center at the origin and also over the two hemispheres, one for $z < 0$, the other for $z > 0$ (plus the plane surface at $z = 0$). Compute the net outflow integral, over the same three surfaces, for the vector field

$$F_x = \frac{x}{[x^2 + y^2 + (z - a)^2]^{\frac{3}{2}}}; \quad F_y = \frac{y}{[x^2 + y^2 + (z - a)^2]^{\frac{3}{2}}};$$
$$F_z = \frac{z - a}{[x^2 + y^2 + (z - a)^2]^{\frac{3}{2}}}$$

1.6 Compute the net circulation integral around the circle, in the x, y plane, of radius r, centered at the origin, for the field

$$F_x = \frac{(x - a)}{(x - a)^2 + y^2} - \frac{x}{x^2 + y^2}; \quad F_y = \frac{y}{x^2 + y^2} - \frac{y}{(x - a)^2 + y^2};$$
$$F_z = 0$$

Compute the net circulation integral for the field of Prob. 1.3 for the circle defined by the equations $\phi = 0$, $\mu =$ constant.

1.7 Parabolic coordinates are defined by the following equations:

$$\lambda = \sqrt{\sqrt{x^2 + y^2 + z^2} + z}; \quad \mu = \sqrt{\sqrt{x^2 + y^2 + z^2} - z}:$$
$$\phi = \tan^{-1}(y/x)$$

Describe (or sketch) the coordinate surfaces. Calculate the scale factors and the direction cosines for the system in terms of (x,y,z). Express x, y, z in terms of λ, μ, ϕ, and thence obtain the scale factors and direction cosines in terms of λ, μ, ϕ. Write out expressions for curl **F**, $\nabla^2\psi$. Calculate, in terms of λ, μ, ϕ, the λ, μ, ϕ components of the following vector field:

$$F_x = \frac{x/\sqrt{x^2+y^2+z^2}}{z+\sqrt{x^2+y^2+z^2}}; \quad F_y = \frac{y/\sqrt{x^2+y^2+z^2}}{z+\sqrt{x^2+y^2+z^2}};$$

$$F_z = \frac{1}{\sqrt{x^2+y^2+z^2}}$$

In terms of λ, μ, ϕ, calculate the divergence of **F**.

1.8 The flow functions ϕ, μ and the pseudopotential ψ, given in Prob. 1.3, form the toroidal coordinate system. Describe (or sketch) the surfaces. Calculate the scale factors as functions of x, y, z and also of μ, ψ, ϕ. Write out the expressions for curl **F**, div **F**, and $\nabla^2 U$. Express the vector **F** given in this problem in terms of components along the toroidal coordinates, and calculate the direction of its vorticity lines.

1.9 One family of coordinate surfaces, which may be used for a family of coordinates, is

$$\ln(x^2+y^2) - 2z = \xi$$

for ξ constant. Show that an appropriate additional pair of families, to make a three-dimensional system, is

$$\eta = \tfrac{1}{2}(x^2+y^2) + z; \quad \phi = \tan^{-1}(y/x)$$

i.e., show that they are mutually orthogonal. These may be termed exponential coordinates. Why? Compute the scale factors and direction cosines for transformation of vector components.

1.10 The bispherical coordinate system is defined by the equations

$$x = \frac{a\sin\vartheta\cos\phi}{\cosh\mu - \cos\vartheta}; \quad y = \frac{a\sin\vartheta\sin\phi}{\cosh\mu - \cos\vartheta}; \quad z = \frac{a\sinh\mu}{\cosh\mu - \cos\vartheta}$$

Describe (or sketch) the surfaces, and give the effective range of μ, ϑ, ϕ. Calculate the scale factors and direction cosines. Write out the expressions for the Laplacian and the gradient. Show that the curvature of the μ surfaces is a constant; *i.e.*, show that $(1/h_\vartheta)(\partial \mathbf{a}_\mu/\partial\vartheta) = (1/h_\phi) \cdot \cdot (\partial \mathbf{a}_\mu/\partial\phi)$ is independent of ϑ and ϕ and that therefore these surfaces are spheres.

1.11 Write out the expressions for the components of directional derivatives $(\mathbf{a}_\vartheta \cdot \nabla)\mathbf{A}$ and $(\mathbf{a}_\phi \cdot \nabla)\mathbf{B}$ in spherical coordinates and in the spheroidal coordinates

$$x = a\cosh\mu\cos\vartheta\cos\phi; \quad y = a\cosh\mu\cos\vartheta\sin\phi; \quad z = a\sinh\mu\sin\vartheta$$

1.12 A scalar function $\psi(\xi_1, \xi_2, \xi_3)$ in an orthogonal, curvilinear coordinate system ξ_1, ξ_2, ξ_3 may be made into a vector by multiplication by the unit vector $\mathbf{a}_1$, normal to the ξ_1 coordinate surfaces. Another vector may be obtained by taking the curl $\mathbf{A} = \text{curl}\,(\mathbf{a}_1\psi)$. Show that $\mathbf{A}$ is tangential to the ξ_1 surfaces. What equation must ψ satisfy, and what are the limitations on the scale factors h_n in order that $\mathbf{A}$ satisfy the equation

$$\nabla^2\mathbf{A} + k^2\mathbf{A} = 0$$

1.13 By the use of the tensor notation, find the expression for $\nabla \times (u\nabla v)$ in general orthogonal curvilinear coordinates.

1.14 We can define the curvature of the ξ_n coordinate surfaces in the ξ_m direction as the component along $\mathbf{a}_m$ of the rate of change of $\mathbf{a}_n$ with respect to distance in the $\mathbf{a}_m$ direction. Express the two curvatures of the ξ_n surface in terms of the Christoffel symbols.

1.15 Work out the expressions for the Christoffel symbols and for the covariant derivative of the components $f_i = h_iF_i$ for the bispherical coordinates given in Prob. 1.10 and the parabolic coordinates given by

$$x = \lambda\mu \cos\phi; \quad y = \lambda\mu \sin\phi; \quad z = \tfrac{1}{2}(\lambda^2 - \mu^2)$$

1.16 Give explicit expressions for the components of the symmetric dyadic $\frac{1}{2}(\nabla\mathbf{A} + \mathbf{A}\nabla)$ for the spheroidal coordinates given in Prob. 1.11 and for the elliptic cylinder coordinates given by

$$x = a \cosh\lambda \cos\phi; \quad y = a \sinh\lambda \sin\phi; \quad z = z$$

Also give expressions for the Laplacian of a vector in both of these systems.

1.17 Find the principal axis for the strain dyadic

$$\mathfrak{S} = -\mathbf{i}(\tfrac{1}{2} + \alpha y^2)\mathbf{i} - \mathbf{j}(\tfrac{1}{2} - \tfrac{1}{3}\alpha y^2)\mathbf{j} + \mathbf{kk} - \alpha xy(\mathbf{ij} + \mathbf{ji})$$

at the point (x,y,z). What are the elongations along each of these axes (principal extensions)?

1.18 Separate the dyadic

$$\mathfrak{A} = \mathbf{ii} - 2\mathbf{jj} + \mathbf{kk} + \sqrt{2\alpha^2 - 4}\,\mathbf{ij} + \sqrt{2}\,\alpha\mathbf{kj} + \beta(\mathbf{jk} - \mathbf{kj})$$

into its symmetric and antisymmetric parts. Compute the rotation vector of the antisymmetric dyadic and the direction of the principal axes of the symmetric dyadic. What is the form of the dyadic after transformation to these principal axes?

1.19 Transform the dyadic

$$\mathfrak{D} = z(\mathbf{ij} - \mathbf{ji}) - y\mathbf{ki} + x\mathbf{kj}$$

into components in the cylindrical coordinates $r = \sqrt{x^2 + y^2}$, $\phi = \tan^{-1}(y/x)$, z. Then separate the dyadic into symmetric and antisymmetric parts, and determine the rotation vector for the antisymmetric and the principal axes of the symmetric parts.

1.20 The vector displacement $\mathbf{s}$ can be represented by a combination of the gradient of a scalar potential ϕ plus the curl of a vector potential $\mathbf{A}$. Set up the expression, in general orthogonal coordinates ξ_1, ξ_2, ξ_3, of the strain dyadic

$$\mathfrak{S} = \tfrac{1}{2}(\nabla\mathbf{s} + \mathbf{s}\nabla); \quad \mathbf{s} = \text{grad}\,\phi + \text{curl}\,\mathbf{A}$$

in terms of ϕ and the components of $\mathbf{A}$ along ξ_1, ξ_2, ξ_3. Write out these expressions explicitly for cylindrical coordinates r, ϕ, z and for spherical coordinates r, ϑ, ϕ.

1.21 The displacement vector $\mathbf{s}$ for an elastic medium may, for some cases, be represented by the expression $\mathbf{s} = \text{curl}\,(\mathbf{a}_z\psi)$, where ψ is some function of the cylindrical coordinates r, ϕ, z. Show that, when $\psi = r^2 f(\phi,z) + g(r,z)$, the diagonal terms in the dyadic are all three equal to zero. Compute the strain dyadic and calculate the principal axes and the principal extensions for the case $\psi = zr^2 \cos\phi$.

1.22 Show that possible displacements of an elastic medium giving zero diagonal terms in the strain tensor $\mathfrak{S}$ expressed in the spherical coordinates r, ϑ, ϕ, may be expressed by a sum of two vectors, curl $[\mathbf{a}_\phi r \sin\vartheta\, g(\phi)]$ and curl $[\mathbf{a}_\vartheta r f(\vartheta)]$. Calculate the strain dyadic, the principal axes and extensions for the two cases $\mathbf{s} = \text{curl}\,[\mathbf{a}_\phi r \sin\vartheta \cos\phi]$ and $\mathbf{s} = \text{curl}\,[\mathbf{a}_\vartheta r \sin(2\vartheta)]$.

1.23 Three coupled oscillators satisfy the simultaneous equations

$$\frac{d^2y_1}{dt^2} + \omega^2 y_1 = \kappa^2 y_2; \quad \frac{d^2y_2}{dt^2} + \omega^2 y_2 = \kappa^2 y_1 + \kappa^2 y_3; \quad \frac{d^2y_3}{dt^2} + \omega^2 y_3 = \kappa^2 y_2$$

Express these equations in terms of abstract vector space and vector operators; find the principal axes of the operator and thus the natural frequencies of the system.

1.24 A system of $N - 1$ coupled oscillators has the following equation of motion, in abstract vector representation:

$$(d^2/dt^2)\mathbf{R} + \omega^2\mathbf{R} = \kappa^2\mathfrak{U}\cdot\mathbf{R}$$

where $\mathbf{R} = \sum_{n=0}^{N} y_n(t)\mathbf{e}_n$, with the boundary requirement that $y_0 = y_N = 0$, and where the operator $\mathfrak{U}$, operating on the unit vectors $\mathbf{e}_n$ (corresponding to the amplitude of vibration y_n of the nth oscillator), satisfies the following equation:

$$\mathfrak{U}\cdot\mathbf{e}_n = \tfrac{1}{2}\mathbf{e}_{n-1} + \tfrac{1}{2}\mathbf{e}_{n+1}$$

Show that the principal axes for $\mathfrak{U}$ are in the direction of the following eigenvectors:

$$\mathbf{a}_m = C_m \sum_{n=0}^{N} \sin\left(\frac{\pi nm}{N}\right)\mathbf{e}_n; \quad m = 1, 2, 3, \ldots, N-1$$

in other words, that $\mathfrak{U} \cdot \mathbf{a}_m = u_m \mathbf{a}_m$. Find the values of the eigenvalues u_m and thus the allowed frequencies of the system. Find the values of the constants C_m so that the new vectors $\mathbf{a}_m$ are unit vectors. Show that the vectors $\mathbf{a}_m$ are mutually orthogonal.

1.25 Prove the Bessel inequality $|\mathbf{e}| + |\mathbf{f}| \geq |\mathbf{e} + \mathbf{f}|$.

1.26 A Hermitian operator $\mathfrak{R}$ which satisfies the inequality $(\mathbf{e}^* \cdot \mathfrak{R} \cdot \mathbf{e}) \geq 0$ for all vectors $\mathbf{e}$ is called *positive-definite.* Show that, if $\mathfrak{R}$ is positive-definite,

$$|\mathbf{e}^* \cdot \mathfrak{R} \cdot \mathbf{f}| \leq \sqrt{(\mathbf{e}^* \cdot \mathfrak{R} \cdot \mathbf{e})(\mathbf{f}^* \cdot \mathfrak{R} \cdot \mathbf{f})}$$

1.27 *a.* Show that

$$\frac{\partial}{\partial \lambda} (e^{i\lambda\mathfrak{S}}\mathfrak{A}e^{-i\lambda\mathfrak{S}}) = i[\mathfrak{S}, e^{i\lambda\mathfrak{S}}\mathfrak{A}e^{-i\lambda\mathfrak{S}}]$$

and
$$\frac{\partial^2}{\partial \lambda^2} (e^{i\lambda\mathfrak{S}}\mathfrak{A}e^{-i\lambda\mathfrak{S}}) = (i)^2[\mathfrak{S},[\mathfrak{S}, e^{i\lambda\mathfrak{S}}\mathfrak{A}e^{-i\lambda\mathfrak{S}}]]$$

where $[\mathfrak{S},\mathfrak{T}] = [\mathfrak{S}\mathfrak{T} - \mathfrak{T}\mathfrak{S}]$ and $[\mathfrak{S},[\mathfrak{S},\mathfrak{T}]] = \mathfrak{S}[\mathfrak{S},\mathfrak{T}] - [\mathfrak{S},\mathfrak{T}]\mathfrak{S}$

b. From (*a*) derive the expansion

$$e^{i\lambda\mathfrak{S}}\mathfrak{A}e^{-i\lambda\mathfrak{S}} = \mathfrak{A} + i\lambda[\mathfrak{S},\mathfrak{A}] + \frac{(i\lambda)^2}{2!}[\mathfrak{S},[\mathfrak{S},\mathfrak{A}]] + \frac{(i\lambda)^3}{3!}[\mathfrak{S},[\mathfrak{S},[\mathfrak{S},\mathfrak{A}]]] + \cdots$$

c. If $\mathfrak{p}$ and $\mathfrak{q}$ are two operators such that $[\mathfrak{p},\mathfrak{q}] = i$, show that

$$e^{b\mathfrak{q}}e^{(a\mathfrak{p}+b\mathfrak{q})}e^{-b\mathfrak{q}} = e^{-iab}e^{a\mathfrak{p}+b\mathfrak{q}}$$

1.28 If $\mathbf{e}_n$ and $\mathbf{f}_n$ are two sets of orthogonal vectors (that is, $\mathbf{e}_n^* \cdot \mathbf{f}_p = 0$), then the projection operator $\mathfrak{P}$ on set $\mathbf{e}_n$ is defined by the equations $\mathfrak{P}\mathbf{e}_n = \mathbf{e}_n$; $\mathfrak{P}\mathbf{f}_p = 0$. Show that

a. $\mathfrak{P}^2 = \mathfrak{P}$

b. $\mathfrak{P}^* = \mathfrak{P}$

c. If $\mathfrak{P}^2 = \mathfrak{P}$, $\mathfrak{P}^* = \mathfrak{P}$, and there exists a set of vectors $\mathbf{e}_n$ such that $\mathfrak{P}\mathbf{e}_n = \mathbf{e}_n$, that $\mathfrak{P}$ is the projection operator on this set.

d. If $\mathfrak{P}_1$ and $\mathfrak{P}_2$ are projection operators on two different sets of vectors the necessary and sufficient condition that $\mathfrak{P}_1\mathfrak{P}_2$ be a projection operator is that $\mathfrak{P}_1\mathfrak{P}_2 - \mathfrak{P}_2\mathfrak{P}_1 = 0$.

1.29 A four-dimensional coordinate system analogous to spherical coordinates is $(\tau,\alpha,\vartheta,\phi)$, where

$$x_4(= ict) = c\tau \cosh \alpha; \quad x = ic\tau \sinh \alpha \cos \vartheta$$
$$y = ic\tau \sinh \alpha \sin \vartheta \cos \phi; \quad z = ic\tau \sinh \alpha \sin \vartheta \sin \phi$$

where a Lorentz transformation is any transformation which leaves the scale of τ invariant. Calculate the scale factors and the direction cosines, and show that the equation $\square^2\psi = 0$ becomes

$$\frac{1}{c^2\tau^2 \sinh^2 \alpha}\left[\frac{\partial}{\partial \alpha}\left(\sinh^2 \alpha \frac{\partial \psi}{\partial \alpha}\right) + \frac{1}{\sin \vartheta}\frac{\partial}{\partial \vartheta}\left(\sin \vartheta \frac{\partial \psi}{\partial \vartheta}\right) + \frac{1}{\sin^2 \vartheta}\frac{\partial^2 \psi}{\partial \phi^2}\right] - \frac{1}{c^2\tau^3}\frac{\partial}{\partial \tau}\left(\tau^3 \frac{\partial \psi}{\partial \tau}\right) = 0$$

Show that a solution of this equation is $\psi = (1/\tau^3) \cosh \alpha$. Give the x, y, z, t components of the four-vector formed by taking the four-gradient of ψ. Show that this is a true four-vector.

1.30 A particle of rest mass m_0, traveling with velocity v in the x direction, strikes another particle of eaual rest mass originally at rest (with respect to observer A). The two rebound, with no change in total energy momentum, the striking particle going off at an angle θ with respect to the x axis (with respect to the observer). Calculate the momentum energy four-vectors for both particles, before and after collision, both with respect to observer A (at rest with the struck particle before it is struck) and to observer B, at rest with respect to the center of gravity of the pair, and explain the differences.

1.31 A fluid is under a uniform isotropic pressure p according to observer A at rest with respect to it. Calculate the density, momentum density, and stress in the fluid, with respect to an observer B, moving at 0.8 the velocity of light with respect to the fluid.

1.32 Give the direction cosines α for the transformation of spin vectors for a combined Lorentz transformation (along the x axis) plus a space rotation.

1.33 An electron certainly has a spin in the positive x direction with respect to an observer at rest with respect to the electron. What are the probabilities of spin in plus and minus x directions for an observer B moving with velocity u in the x direction with respect to the electron? What is the probability that the electron has a spin in a direction at 45° with respect to the positive x axis, with respect to observer A? For observer B?

1.34 Let $\boldsymbol{\sigma}$ be a three-component vector-spin operator with components σ_1, σ_2, σ_3.

a. Show that, if $\mathbf{A}$ is a vector,

$$\begin{aligned}
(\boldsymbol{\sigma} \cdot \mathbf{A})\boldsymbol{\sigma} &= \mathbf{A} + i(\boldsymbol{\sigma} \times \mathbf{A}) \\
\boldsymbol{\sigma}(\boldsymbol{\sigma} \cdot \mathbf{A}) &= \mathbf{A} - i(\boldsymbol{\sigma} \times \mathbf{A}) \\
(\boldsymbol{\sigma} \times \boldsymbol{\sigma}) &= 2i\boldsymbol{\sigma} \\
\boldsymbol{\sigma} \times (\boldsymbol{\sigma} \times \mathbf{A}) &= i(\boldsymbol{\sigma} \times \mathbf{A}) - 2\mathbf{A}
\end{aligned}$$

b. Show that, if $\mathbf{a}$ is a unit vector and λ is a constant,

$$\frac{d^2}{d\lambda^2} \exp (i\lambda\boldsymbol{\sigma} \cdot \mathbf{a}) = - \exp (i\lambda\boldsymbol{\sigma} \cdot \mathbf{a})$$

and therefore $$\exp (i\lambda\boldsymbol{\sigma} \cdot \mathbf{a}) = \cos \lambda + i(\boldsymbol{\sigma} \cdot \mathbf{a}) \sin \lambda$$

Table of Useful Vector and Dyadic Equations

$$\mathbf{A}\cdot\mathbf{B} = A_xB_x + A_yB_y + A_zC_z;\quad \mathbf{A}\times\mathbf{B} = \mathbf{i}(A_yB_z - A_zB_y) + \mathbf{j}(A_zB_x - A_xB_z) + \mathbf{k}(A_xB_y - A_yB_x)$$

$$(\mathbf{A}\times\mathbf{B})\times\mathbf{C} = (\mathbf{A}\cdot\mathbf{C})\mathbf{B} - (\mathbf{B}\cdot\mathbf{C})\mathbf{A}$$

$$\mathbf{A}\times(\mathbf{B}\times\mathbf{C}) = (\mathbf{A}\cdot\mathbf{C})\mathbf{B} - (\mathbf{A}\cdot\mathbf{B})\mathbf{C}$$

$$\mathbf{A}\cdot(\mathbf{B}\times\mathbf{C}) = (\mathbf{A}\times\mathbf{B})\cdot\mathbf{C} = (\mathbf{C}\times\mathbf{A})\cdot\mathbf{B} = \mathbf{C}\cdot(\mathbf{A}\times\mathbf{B}) = \mathbf{B}\cdot(\mathbf{C}\times\mathbf{A}) = (\mathbf{B}\times\mathbf{C})\cdot\mathbf{A}$$

$$(\mathbf{A}\times\mathbf{B})\cdot(\mathbf{C}\times\mathbf{D}) = (\mathbf{A}\cdot\mathbf{C})(\mathbf{B}\cdot\mathbf{D}) - (\mathbf{A}\cdot\mathbf{D})(\mathbf{B}\cdot\mathbf{C})$$

$$(\mathbf{A}\times\mathbf{B})\times(\mathbf{C}\times\mathbf{D}) = [\mathbf{A}\cdot(\mathbf{B}\times\mathbf{D})]\mathbf{C} - [\mathbf{A}\cdot(\mathbf{B}\times\mathbf{C})]\mathbf{D} = [\mathbf{A}\cdot(\mathbf{C}\times\mathbf{D})]\mathbf{B} - [\mathbf{B}\cdot(\mathbf{C}\times\mathbf{D})]\mathbf{A}$$

$$\nabla u = \operatorname{grad} u;\quad \nabla\cdot\mathbf{F} = \operatorname{div}\mathbf{F};\quad \nabla\times\mathbf{F} = \operatorname{curl}\mathbf{F}$$

$$\nabla(uv) = u\nabla v + v\nabla u;\quad \nabla\cdot(u\mathbf{A}) = (\nabla u)\cdot\mathbf{A} + u\nabla\cdot\mathbf{A}$$

$$\nabla\times(u\mathbf{A}) = (\nabla u)\times\mathbf{A} + u\nabla\times\mathbf{A}$$

$$\nabla\cdot(\mathbf{A}\times\mathbf{B}) = \mathbf{B}\cdot(\nabla\times\mathbf{A}) - \mathbf{A}\cdot(\nabla\times\mathbf{B})$$

$$\nabla\cdot(\nabla\times\mathbf{F}) = 0;\quad \nabla\times(\nabla u) = 0;\quad \nabla\cdot(\nabla u) = \nabla^2 u$$

$$\nabla\times(\nabla\times\mathbf{F}) = \nabla(\nabla\cdot\mathbf{F}) - \nabla^2\mathbf{F}$$

$$\iiint(\nabla\cdot\mathbf{F})\,dv = \iint\mathbf{F}\cdot d\mathbf{A};\quad \iiint(\nabla\times\mathbf{F})\,dv = -\iint\mathbf{F}\times d\mathbf{A}$$

$$\iiint(\nabla\varphi)\cdot(\nabla\psi)\,dv = \iint\varphi(\nabla\psi)\cdot d\mathbf{A} - \iiint\varphi\nabla^2\psi\,dv$$

where the triple integrals are over all the volume inside the closed surface A and the double integrals are over the surface of A ($d\mathbf{A}$ *pointing outward*).

$$\iint(\nabla\times\mathbf{F})\cdot d\mathbf{A} = \int\mathbf{F}\cdot d\mathbf{r}$$

where the double integral is over an area bounded by a closed contour C, and the single integral is along C in a clockwise direction when looking in the direction of $d\mathbf{A}$.

A vector field $\mathbf{F}(x,y,z)$ can be expressed in terms of a scalar potential ψ and a vector potential $\mathbf{A}$,

$$\mathbf{F} = \operatorname{grad}\psi + \operatorname{curl}\mathbf{A};\quad \operatorname{div}\mathbf{A} = 0$$

When $\mathbf{F}$ goes to zero at infinity, the expressions for ψ and $\mathbf{A}$ in terms of $\mathbf{F}$ are

$$\psi = -\iiint\frac{\operatorname{div}\mathbf{F}(x',y',z')}{4\pi R}\,dx'\,dy'\,dz';\quad \mathbf{A} = \iiint\frac{\operatorname{curl}\mathbf{F}(x',y',z')}{4\pi R}\,dx'\,dy'\,dz'$$

where $R^2 = (x - x')^2 + (y - y')^2 + (z - z')^2$.

$$\mathfrak{A} = \mathbf{i}\mathbf{A}_x + \mathbf{j}\mathbf{A}_y + \mathbf{k}\mathbf{A}_z;\quad \mathfrak{A}^* = \mathbf{i}\mathbf{A}_x^* + \mathbf{j}\mathbf{A}_y^* + \mathbf{k}\mathbf{A}_z^* = \mathbf{A}_x\mathbf{i} + \mathbf{A}_y\mathbf{j} + \mathbf{A}_z\mathbf{k}$$

$$|\mathfrak{A}| = \mathbf{i}\cdot\mathbf{A}_x + \mathbf{j}\cdot\mathbf{A}_y + \mathbf{k}\cdot\mathbf{A}_z;\quad \langle\mathfrak{A}\rangle = \mathbf{i}\times\mathbf{A}_x + \mathbf{j}\times\mathbf{A}_y + \mathbf{k}\times\mathbf{A}_z$$

$$\mathfrak{A}\cdot\mathbf{B} = \mathbf{A}_x^*B_x + \mathbf{A}_y^*B_y + \mathbf{A}_z^*B_z = \mathbf{i}(\mathbf{A}_x\cdot\mathbf{B}) + \mathbf{j}(\mathbf{A}_y\cdot\mathbf{B}) + \mathbf{k}(\mathbf{A}_z\cdot\mathbf{B}) = \tfrac{1}{2}(\mathbf{A}_x + \mathbf{A}_x^*)B_x + \tfrac{1}{2}(\mathbf{A}_y + \mathbf{A}_y^*)B_y + \tfrac{1}{2}(\mathbf{A}_z + \mathbf{A}_z^*)B_z - \tfrac{1}{2}\langle\mathfrak{A}\rangle\times\mathbf{B}$$

$\mathfrak{A} \cdot \mathfrak{B} = \mathbf{A}_x^*\mathbf{B}_x + \mathbf{A}_y^*\mathbf{B}_y + \mathbf{A}_z^*\mathbf{B}_z$

$\mathfrak{A} : \mathfrak{B} = \mathbf{A}_x^* \cdot \mathbf{B}_x + \mathbf{A}_y^* \cdot \mathbf{B}_y + \mathbf{A}_z^* \cdot \mathbf{B}_z = |\mathfrak{A} \cdot \mathfrak{B}|$

$\nabla \mathbf{F} = \mathbf{i}\dfrac{\partial \mathbf{F}}{\partial x} + \mathbf{j}\dfrac{\partial \mathbf{F}}{\partial y} + \mathbf{k}\dfrac{\partial \mathbf{F}}{\partial z}; \quad \mathbf{F}\nabla = \dfrac{\partial \mathbf{F}}{\partial x}\mathbf{i} + \dfrac{\partial \mathbf{F}}{\partial y}\mathbf{j} + \dfrac{\partial \mathbf{F}}{\partial z}\mathbf{k} = (\nabla \mathbf{F})^*$

$\text{grad}\,(\mathbf{A} \cdot \mathbf{B}) = \mathbf{A} \cdot (\nabla \mathbf{B}) + \mathbf{B} \cdot (\nabla \mathbf{A}) + \mathbf{A} \times (\nabla \times \mathbf{B}) + \mathbf{B} \times (\nabla \times \mathbf{A})$

$\text{curl}\,(\mathbf{A} \times \mathbf{B}) = \mathbf{B} \cdot (\nabla \mathbf{A}) - \mathbf{A} \cdot (\nabla \mathbf{B}) + \mathbf{A}(\nabla \cdot \mathbf{B}) - \mathbf{B}(\nabla \cdot \mathbf{A})$

$\nabla \cdot \mathfrak{A} = (\partial \mathbf{A}_x/\partial x) + (\partial \mathbf{A}_y/\partial y) + (\partial \mathbf{A}_z/\partial z)$

$\qquad = \mathbf{i}\,\text{div}\,(\mathbf{A}_x^*) + \mathbf{j}\,\text{div}\,(\mathbf{A}_y^*) + \mathbf{k}\,\text{div}\,(\mathbf{A}_z^*)$

$\nabla \cdot (\nabla \mathbf{F}) = \nabla^2 \mathbf{F}; \quad \nabla \cdot (\mathbf{F}\nabla) = \nabla(\nabla \cdot \mathbf{F}) = \nabla^2 \mathbf{F} + \nabla \times \nabla \times \mathbf{F}$

$\nabla \cdot (\mathfrak{A} \cdot \mathbf{B}) = (\nabla \cdot \mathfrak{A}) \cdot \mathbf{B} + |\mathfrak{A} \cdot (\nabla \mathbf{B})|$

Table of Properties of Curvilinear Coordinates

For orthogonal, curvilinear coordinates ξ_1, ξ_2, ξ_3 with unit vectors $\mathbf{a}_1$, $\mathbf{a}_2$, $\mathbf{a}_3$, line element $ds^2 = \sum_n h_n^2(d\xi_n)^2$, and scale factors h_n, where

$$h_n^2 = \left[\left(\frac{\partial x}{\partial \xi_n}\right)^2 + \left(\frac{\partial y}{\partial \xi_n}\right)^2 + \left(\frac{\partial z}{\partial \xi_n}\right)^2\right] = \left[\left(\frac{\partial \xi_n}{\partial x}\right)^2 + \left(\frac{\partial \xi_n}{\partial y}\right)^2 + \left(\frac{\partial \xi_n}{\partial z}\right)^2\right]^{-1}$$

the differential operators become

$$\text{grad}\,\psi = \nabla\psi = \sum_n \mathbf{a}_n \frac{1}{h_n}\frac{\partial \psi}{\partial \xi_n}$$

$$\text{div}\,\mathbf{A} = \nabla \cdot \mathbf{A} = \frac{1}{h_1h_2h_3}\sum_n \frac{\partial}{\partial \xi_n}\left(h_1h_2h_3\frac{A_n}{h_n}\right)$$

$$\text{curl}\,\mathbf{A} = \nabla \times \mathbf{A} = \frac{1}{h_1h_2h_3}\sum_{l,m,n} h_l\mathbf{a}_l\left[\frac{\partial}{\partial \xi_m}(h_nA_n) - \frac{\partial}{\partial \xi_n}(h_mA_m)\right];$$

$l, m, n = 1, 2, 3$ or $2, 3, 1$ or $3, 1, 2$

$$\text{div grad}\,\psi = \nabla^2\psi = \frac{1}{h_1h_2h_3}\sum_n \frac{\partial}{\partial \xi_n}\left[\frac{h_1h_2h_3}{h_n^2}\frac{\partial \psi}{\partial \xi_n}\right]$$

$$\nabla\mathbf{A} = (\nabla\mathbf{A})_s + (\nabla\mathbf{A})_a; \quad (\nabla\mathbf{A})_a = \tfrac{1}{2}(\text{curl}\,\mathbf{A}) \times \mathfrak{J}$$

$$(\nabla\mathbf{A})_s = \tfrac{1}{2}(\nabla\mathbf{A} + \mathbf{A}\nabla) = \sum_m \left[\frac{\partial}{\partial \xi_m}\frac{A_m}{h_m} + \mathbf{A} \cdot \text{grad}\,(\ln h_m)\right]\mathbf{a}_m\mathbf{a}_m$$
$$+ \tfrac{1}{2}\sum_{m<n}\left[\frac{h_m}{h_n}\frac{\partial}{\partial \xi_n}\frac{A_m}{h_m} + \frac{h_n}{h_m}\frac{\partial}{\partial \xi_m}\frac{A_n}{h_n}\right](\mathbf{a}_m\mathbf{a}_n + \mathbf{a}_n\mathbf{a}_m)$$

and the volume element is $h_1h_2h_3\,d\xi_1\,d\xi_2\,d\xi_3 = dV$.

For cylindrical coordinates $\xi_1 = r$, $\xi_2 = \phi$, $\xi_3 = z$, we have $h_1 = 1$, $h_2 = r$, $h_3 = 1$, $dV = r\,dr\,d\phi\,dz$.

$$\text{grad}\,\psi = \mathbf{a}_r \frac{\partial \psi}{\partial r} + \mathbf{a}_\phi \frac{1}{r}\frac{\partial \psi}{\partial \phi} + \mathbf{a}_z \frac{\partial \psi}{\partial z}$$

$$\text{div}\,\mathbf{A} = \frac{1}{r}\frac{\partial}{\partial r}(rA_r) + \frac{1}{r}\frac{\partial A_\phi}{\partial \phi} + \frac{\partial A_z}{\partial z}$$

$$\text{curl}\,\mathbf{A} = \mathbf{a}_r\left(\frac{1}{r}\frac{\partial A_z}{\partial \phi} - \frac{\partial A_\phi}{\partial z}\right) + \mathbf{a}_\phi\left(\frac{\partial A_r}{\partial z} - \frac{\partial A_z}{\partial r}\right) + \mathbf{a}_z\left(\frac{1}{r}\frac{\partial}{\partial r} rA_\phi - \frac{1}{r}\frac{\partial A_r}{\partial \phi}\right)$$

$$\nabla^2\psi = \frac{1}{r}\frac{\partial}{\partial r}\left(r\frac{\partial \psi}{\partial r}\right) + \frac{1}{r^2}\frac{\partial^2\psi}{\partial \phi^2} + \frac{\partial^2\psi}{\partial z^2}$$

$$\nabla^2\mathbf{A} = \mathbf{a}_r\left[\nabla^2 A_r - \frac{A_r}{r^2} - \frac{2}{r^2}\frac{\partial A_\phi}{\partial \phi}\right] + \mathbf{a}_\phi\left[\nabla^2 A_\phi - \frac{A_\phi}{r^2} + \frac{2}{r^2}\frac{\partial A_r}{\partial \phi}\right] + \mathbf{a}_z \nabla^2 A_z$$

$$\tfrac{1}{2}(\nabla\mathbf{A} + \mathbf{A}\nabla) = \mathbf{a}_r \frac{\partial A_r}{\partial r}\mathbf{a}_r + \mathbf{a}_\phi\left[\frac{1}{r}\frac{\partial A_\phi}{\partial \phi} + \frac{A_r}{r}\right]\mathbf{a}_\phi + \mathbf{a}_z\frac{\partial A_z}{\partial z}\mathbf{a}_z$$
$$+ \tfrac{1}{2}\left[r\frac{\partial}{\partial r}\left(\frac{A_\phi}{r}\right) + \frac{1}{r}\frac{\partial A_r}{\partial \phi}\right](\mathbf{a}_r\mathbf{a}_\phi + \mathbf{a}_\phi\mathbf{a}_r) + \tfrac{1}{2}\left[\frac{\partial A_z}{\partial r} + \frac{\partial A_r}{\partial z}\right](\mathbf{a}_r\mathbf{a}_z + \mathbf{a}_z\mathbf{a}_r)$$
$$+ \tfrac{1}{2}\left[\frac{\partial A_\phi}{\partial z} + \frac{1}{r}\frac{\partial A_z}{\partial \phi}\right](\mathbf{a}_\phi\mathbf{a}_z + \mathbf{a}_z\mathbf{a}_\phi)$$

For spherical coordinates $\xi_1 = r$, $\xi_2 = \vartheta$, $\xi_3 = \varphi$, we have $h_1 = 1$, $h_2 = r$, $h_3 = r\sin\vartheta$. $dV = r^2\,dr\,\sin\vartheta\,d\vartheta\,d\varphi$.

$$\text{grad}\,\psi = \mathbf{a}_r\frac{\partial \psi}{\partial r} + \frac{\mathbf{a}_\vartheta}{r}\frac{\partial \psi}{\partial \vartheta} + \frac{\mathbf{a}_\varphi}{r\sin\vartheta}\frac{\partial \psi}{\partial \varphi}$$

$$\text{div}\,\mathbf{A} = \frac{1}{r^2}\frac{\partial}{\partial r}(r^2 A_r) + \frac{1}{r\sin\vartheta}\frac{\partial}{\partial \vartheta}(\sin\vartheta A_\vartheta) + \frac{1}{r\sin\vartheta}\frac{\partial A_\varphi}{\partial \varphi}$$

$$\text{curl}\,\mathbf{A} = \frac{\mathbf{a}_r}{r\sin\vartheta}\left[\frac{\partial}{\partial \vartheta}(\sin\vartheta A_\varphi) - \frac{\partial A_\vartheta}{\partial \varphi}\right] + \frac{\mathbf{a}_\vartheta}{r}\left[\frac{1}{\sin\vartheta}\frac{\partial A_r}{\partial \varphi} - \frac{\partial}{\partial r}(rA_\varphi)\right]$$
$$+ \frac{\mathbf{a}_\varphi}{r}\left[\frac{\partial}{\partial r}(rA_\vartheta) - \frac{\partial A_r}{\partial \vartheta}\right]$$

$$\nabla^2\psi = \frac{1}{r^2}\frac{\partial}{\partial r}\left(r^2\frac{\partial \psi}{\partial r}\right) + \frac{1}{r^2\sin\vartheta}\frac{\partial}{\partial \vartheta}\left(\sin\vartheta\frac{\partial \psi}{\partial \vartheta}\right) + \frac{1}{r^2\sin^2\vartheta}\frac{\partial^2\psi}{\partial \varphi^2}$$

$$\nabla^2\mathbf{A} = \mathbf{a}_r\left[\nabla^2 A_r - \frac{2}{r^2}A_r - \frac{2}{r^2\sin\vartheta}\frac{\partial}{\partial \vartheta}(\sin\vartheta A_\vartheta) - \frac{2}{r^2\sin\vartheta}\frac{\partial A_\varphi}{\partial \varphi}\right]$$
$$+ \mathbf{a}_\vartheta\left[\nabla^2 A_\vartheta - \frac{A_\vartheta}{r^2\sin^2\vartheta} + \frac{2}{r^2}\frac{\partial A_r}{\partial \vartheta} - \frac{2\cos\vartheta}{r^2\sin^2\vartheta}\frac{\partial A_\varphi}{\partial \varphi}\right]$$
$$+ \mathbf{a}_\varphi\left[\nabla^2 A_\varphi - \frac{A_\varphi}{r^2\sin^2\vartheta} + \frac{2}{r^2\sin\vartheta}\frac{\partial A_r}{\partial \varphi} + \frac{2\cos\vartheta}{r^2\sin^2\vartheta}\frac{\partial A_\vartheta}{\partial \varphi}\right]$$

$$\tfrac{1}{2}(\nabla\mathbf{A} + \mathbf{A}\nabla) = \mathbf{a}_r \frac{\partial A_r}{\partial r} \mathbf{a}_r + \mathbf{a}_\vartheta \left[\frac{1}{r}\frac{\partial A_\vartheta}{\partial \vartheta} + \frac{A_r}{r}\right] \mathbf{a}_\vartheta$$
$$+ \mathbf{a}_\varphi \left[\frac{1}{r \sin\vartheta}\frac{\partial A_\varphi}{\partial \varphi} + \frac{A_r}{r} + \frac{A_\vartheta}{r}\cot\vartheta\right] \mathbf{a}_\varphi$$
$$+ \tfrac{1}{2}\left[\frac{1}{r}\frac{\partial A_r}{\partial \vartheta} + r\frac{\partial}{\partial r}\left(\frac{A_\vartheta}{r}\right)\right](\mathbf{a}_r\mathbf{a}_\vartheta + \mathbf{a}_\vartheta\mathbf{a}_r)$$
$$+ \tfrac{1}{2}\left[\frac{1}{r\sin\vartheta}\frac{\partial A_r}{\partial \varphi} + r\frac{\partial}{\partial r}\left(\frac{A_\varphi}{r}\right)\right](\mathbf{a}_r\mathbf{a}_\varphi + \mathbf{a}_\varphi\mathbf{a}_r)$$
$$+ \tfrac{1}{2}\left[\frac{1}{r\sin\vartheta}\frac{\partial A_\vartheta}{\partial \varphi} + \frac{\sin\vartheta}{r}\frac{\partial}{\partial \vartheta}\left(\frac{A_\varphi}{\sin\vartheta}\right)\right](\mathbf{a}_\vartheta\mathbf{a}_\varphi + \mathbf{a}_\varphi\mathbf{a}_\vartheta)$$

Bibliography

As mentioned in the preface, this bibliography is not meant to be complete. Included are only those books and articles which the authors feel will be useful supplements to the text.

General references for material in this chapter:

Jeffreys, H. J., and B. S. Jeffreys: "Methods of Mathematical Physics," Cambridge, New York, 1946.

Joos, G.: "Theoretical Physics," Stechert, New York, 1944.

Margenau, H., and G. M. Murphy: "The Mathematics of Physics and Chemistry," Van Nostrand, New York, 1943.

Murnaghan, F. D.: "Introduction to Applied Mathematics," Wiley, New York, 1948.

Riemann-Weber: "Differential- und Integralgleichungen der Mechanik und Physik," ed. by P. Frank and R. von Mises, Vieweg, Brunswick, 1935; reprint, Rosenberg, New York, 1943.

Slater, J. C., and N. H. Frank: "Introduction to Theoretical Physics," McGraw-Hill, New York, 1933.

Webster, A. G.: "Partial Differential Equations of Mathematical Physics," Stechert, New York, 1933.

Additional material on vector and tensor analysis:

Craig, H. V.: "Vector and Tensor Analysis," McGraw-Hill, New York, 1943.

Frazer, B. A., Duncan, W. J., and Collar, A. R., "Elementary Matrices," Cambridge, New York, 1938.

Gibbs, J. W.: "Vector Analysis," ed. by E. B. Wilson, Scribner, New York, 1901.

Phillips, H. B.: "Vector Analysis," Wiley, New York, 1933.

Rutherford, D. E.: "Vector Methods Applied to Differential Geometry, etc.," Oliver & Boyd, Edinburgh, 1944.

Weatherburn, C. E.: "Elementary and Advanced Vector Analysis," 2 vols., G. Bell, London, 1928.

Books on elasticity:

Brillouin, L.: "Les tenseurs en méchanique et en élastique," Masson et Cie, Paris, 1938.

Love: "Mathematical Theory of Elasticity," Cambridge, New York, 1927, reprint, Dover, New York, 1945.

Sokolnikoff, I. S.: "Mathematical Theory of Elasticity," McGraw-Hill, New York, 1946.

Useful reference works on various aspects of the theory of abstract vector space:

Dirac, P. A. M.: "The Principles of Quantum Mechanics," Oxford, New York.

Laporte, O., and G. E. Uhlenbeck: Application of Spinor Analysis to Maxwell and Dirac Equations, *Phys. Rev.*, **37,** 1380 (1931).

Rojansky, V. B.: "Introductory Quantum Mechanics," Prentice-Hall, New York, 1938.

Stone, M. H.: "Linear Transformations in Hilbert Space," American Mathematical Society, New York, 1932.

Van der Waerden, B. L.: "Gruppentheoretische Methode in der Quantenmechanik," Springer, Berlin, 1932, reprint, Edwards Bros., Inc., Ann Arbor, 1943.

Von Neumann, J.: "Mathematische Grundlagen der Quantenmechanik," Springer, Berlin, 1932, reprint, Dover, New York, 1943.

Some texts having sections discussing special relativity:

Bergmann, P. G.: "Introduction to the Theory of Relativity," Prentice-Hall New York, 1942.

Corben, H. C., and P. Stehle: "Classical Mechanics," Chaps. 17 and 18, Wiley, New York, 1950.

Eddington, A. S.: "Mathematical Theory of Relativity," Cambridge, New York, 1930.

Goldstein, H.: "Classical Mechanics," Chap. 6, Addison-Wesley, Cambridge, 1950.

Tolman, R. C.: "Relativity, Thermodynamics and Cosmology," Oxford, New York, 1934.

CHAPTER 2

Equations Governing Fields

The physical phenomena which can be represented by fields are related from point to point and from time to time in ways which can usually be expressed in terms of partial differential equations. A change in the field at one point usually affects its value at nearby points, and these changes affect the values still farther away, and so on, a stepwise interrelation which is most naturally expressed in terms of space or time derivatives. The field which corresponds to a particular physical situation is, therefore, usually a solution of some partial differential equation, that particular solution which satisfies the particular set of "boundary conditions" appropriate to the situation.

The greater part of this book will concern itself with the finding of particular solutions of specific partial differential equations which correspond to given boundary and initial conditions. This chapter and the next one, however, will be devoted to a discussion of the ways in which differential equations are chosen to correspond to given physical situations. This process, of abstracting the most important interrelations from the phenomena under study and expressing them in differential form, is one of the more difficult tasks of the theoretical physicist.

We shall not try to discuss all the partial differential equations which have been found useful in physics; to do this would require a complete review of physics. Even with the equations which are here derived, we shall assume that the physical situation is fairly familiar and that the quantities mentioned do not need detailed explanation. Such explanations can be found in other texts. What is of interest here is the process of obtaining the differential equation from the physics.

We shall find that certain types of partial differential equations turn up again and again in a wide variety of situations and that therefore a detailed knowledge of the solutions of these relatively few equations will enable us to solve problems in an astonishingly large number of situations of physical interest.

2.1 *The Flexible String*

Before we take up more complicated situations, it is well to examine a one-dimensional example in some detail so as to bring out some of the procedures and concepts in their simplest form. The flexible string under tension is a good choice, since it is easy to picture and is familiar to most readers.

The physical prototype is a lower register piano string, which is a more or less uniformly loaded wire stretched between two fairly rigid supports. Such a string has stiffness, but experiment can show that the resistance to displacement of the string from its equilibrium shape is in the main due to the tension in the string rather than to its stiffness. Therefore, one simplification usually made in obtaining the equation governing the string's shape is that stiffness can be neglected (several books on vibration and sound analyze the effects of stiffness and show when it can safely be neglected). Other simplifying assumptions are that the mass of the string is uniformly distributed along its length, that the tension is likewise uniform, and that the displacement of a point of the string from equilibrium is always small compared with the distance of this point from the closer end support. These last two assumptions are not independent of each other.

The shape of such a string at any instant can be expressed in terms of its displacement from equilibrium. More specifically, each point of the string is labeled by its distance x from some origin point, measured when the string has its equilibrium shape (a straight line between the supports). The displacement $\psi(x)$ of point x from its equilibrium position is a function of x (and sometimes of time). If we consider only motion in one plane, the appropriate field for this example, ψ, is a scalar, one-dimensional field.

Forces on an Element of String. Study of Fig. 2.1 shows that, as long as the slope of the string $\partial\psi/\partial x$ is small, the net force $F_T(x)\,dx$ acting on that portion of the string between x and $x + dx$ due to the tension T in the string is

$$F_T(x)\,dx = T(\partial\psi/\partial x)_{x+dx} - T(\partial\psi/\partial x)_x \quad \text{or} \quad F_T(x) = T(\partial^2\psi/\partial x^2) \qquad (2.1.1)$$

having a direction approximately perpendicular to the equilibrium line. This net force due to tension on an element of the string at x, which tends to restore the string to equilibrium, is proportional to the rate of change of the string's slope at x. It tends to straighten each portion of the string; if the curvature is positive, it is upward; wherever the curvature is negative, it is downward. The force, therefore, depends only on the shape of that part of the string immediately adjacent to x, and not on the shape of the string as a whole. If the string is allowed to yield to this force, however, it will only come to equilibrium when *every* portion of the string is straight. Thus the differential equation for the string, though it deals only with the shape of each elementary portion of the string in turn, places in the end a definite restriction on the over-all shape.

Other forces also act on an actual piano string—the force due to the stiffness of the string (which we have already said we can ignore in many cases) and the frictional reaction of the air through which it moves, among others.

The frictional force is also relatively small when the string moves in air; and when we are interested in the motion over a short space of time

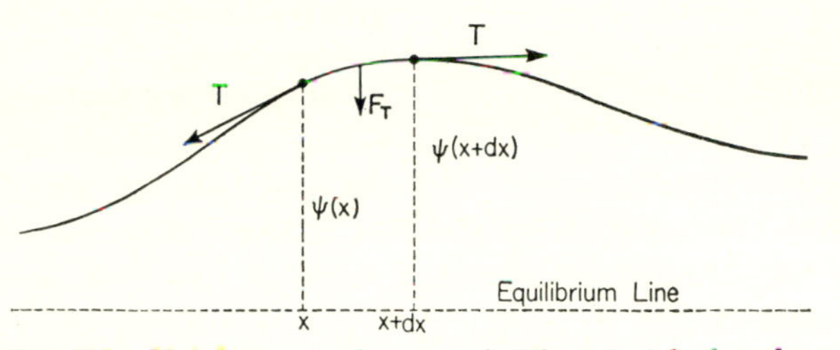

Fig. 2.1 Net force on element of string stretched under tension T.

or when we are interested in the shape when the string is not moving, this force may also be neglected. Other forces which may enter are the force of gravity on each portion of the string (if it is stretched horizontally) and the force due to a piano hammer or to a harpist's finger or to a violin bow. Which of these forces are important enough to include in the equation for the shape of the string depends on the particular case under study.

Poisson's Equation. For instance, the string may be subjected to steady, transverse forces distributed along its length, and we may be interested in calculating the resulting equilibrium shape of the string under the combined effect of this external force and that due to the tension. In this case, the time dependence does not enter and the differential equation for the shape is

$$d^2\psi/dx^2 = -f(x); \quad f = F(x)/T \tag{2.1.2}$$

where the transverse force applied to the element of string between x and $x + dx$ is $F(x)\,dx$. Here this applied transverse force is balanced at every point by the net transverse force due to the tension T. Equation (2.1.2) is a one-dimensional case of *Poisson's equation*.

As an example of the cases which this equation represents, we can consider the case of a horizontal string acted on by the force of gravity due to its own weight. If each centimeter length of the string weighs ρ gm, the force $F(x)$ is just $-\rho g$, where g is the acceleration of gravity. The general solution of the resulting equation $d^2\psi/dx^2 = \rho g/T$ is then

$\psi = a + bx + (\rho g/2T)x^2$, where a and b are determined by "boundary conditions." When the string supports at the two ends are rigid (*i.e.*, when their displacement can be neglected) and a distance L apart, these boundary conditions are that $\psi = 0$ when $x = 0$ when $x = L$. It is not difficult to see that the quadratic expression in x, which has a second-order term $(\rho g/2T)x^2$ and which goes to zero at $x = 0$ and $x = L$, is $\psi = (\rho g/2T)x(x - L)$. This, therefore, is the solution of the problem: The shape is parabolic with constant curvature $\rho g/T$ and with greatest displacement at the center of the string.

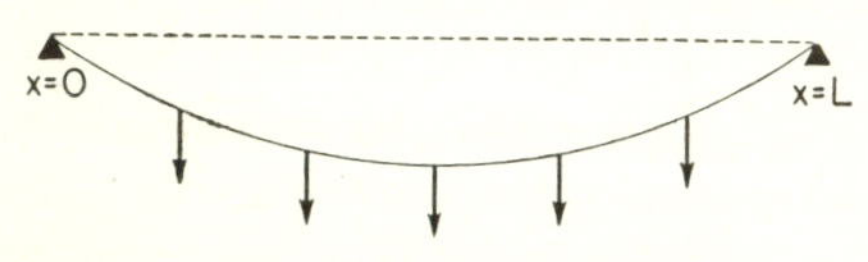

Fig. 2.2 Shape of string acted on transversely by force of gravity, longitudinally by tension.

Several interesting general properties of solutions of Eq. (2.1.2) may be deduced from the fact that ψ enters into the equation to the first power. For instance, if ψ is a solution of Eq. (2.1.2) for a specified function $f(x)$, then $a\psi$ is a solution of the equation $d^2\psi/dx^2 = -af(x)$. This new solution often also satisfies the same boundary conditions as ψ (it does for the string between fixed supports, for instance). Similarly if ψ_1 is a solution of $d^2\psi/dx^2 = -f_1$ and ψ_2 is a solution of $d^2\psi/dx^2 = -f_2$, then $\psi = \psi_1 + \psi_2$ is a solution of

$$d^2\psi/dx^2 = -f_1 - f_2 \tag{2.1.3}$$

Both of these properties will be utilized many times in this volume.

Concentrated Force, Delta Function. In many cases of practical interest the transverse force is applied to only a small portion of the string. This suggests a rather obvious idealization, a force applied "at a point" on the string. Mathematically speaking this idealization corresponds to a consideration of the limiting case of a force

$$F(x) = \begin{cases} 0; & x < \xi - (\Delta/2) \\ F/\Delta; & \xi - (\Delta/2) < x < \xi + (\Delta/2) \\ 0; & x > \xi + (\Delta/2) \end{cases}$$

when the length Δ of the portion of string acted on by the force is allowed to go to zero.

Similar idealizations of concentrated forces, electric charges, etc., will be of great utility in our subsequent discussions. They can all be expressed in terms of a "pathological function" called the *delta function*,

$$\delta(x) = \lim_{\Delta \to 0} \begin{cases} 0; & x < -\Delta/2 \\ 1/\Delta; & -(\Delta/2) < x < \Delta/2 \\ 0; & x > \Delta/2 \end{cases} \tag{2.1.4}$$

It is called a "pathological function" because it does not have the "physically normal" properties of continuity and differentiability at

$x = 0$. If we do not expect too much of the function, however, it will turn out to be of considerable aid in later analysis of many problems. Remembering the usual definition of integration as a limiting sum, one can derive the following integral rules for the delta function:

$$\int_{-\infty}^{\infty} f(\xi)\delta(\xi - x)\, d\xi = f(x) \tag{2.1.5}$$

A closely related function, one which illustrates the integral properties of the delta function, is the *unit step function:*

$$u(x) = \int_{-\infty}^{x} \delta(\xi)\, d\xi = \begin{cases} 0; & x < 0 \\ 1; & x > 0 \end{cases} \tag{2.1.6}$$

This function is also pathological in that differentiation should be attempted only with considerable caution.

Returning now to the problem of solving Eq. (2.1.2) for a force concentrated at the point $x = \xi$, we first work out the solution of

$$d^2\psi/dx^2 = -\delta(x - \xi)$$

The solution ψ satisfies the homogeneous equation $d^2\psi/dx^2 = 0$ at all points for which $x \neq \xi$. To obtain its behavior at $x = \xi$, we integrate both sides from $x = \xi - \epsilon$ to $x = \xi + \epsilon$, where ϵ is a vanishing small quantity. Using Eq. (2.1.6) we see that the solution must have a unit change of slope at $x = \xi$. If the supports are rigid, the shape of the string of length L for a force $F = T$ concentrated at $x = \xi$ must be

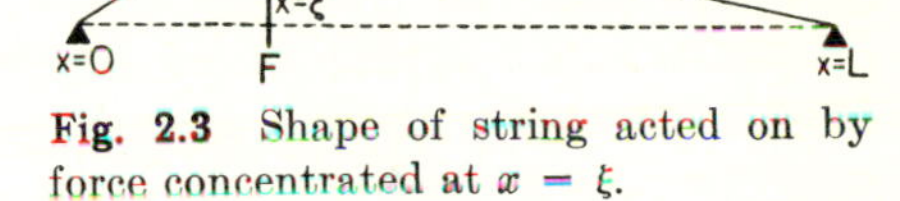

Fig. 2.3 Shape of string acted on by force concentrated at $x = \xi$.

$$\psi = G(x|\xi) = \begin{cases} x(L - \xi)/L; & 0 < x < \xi \\ \xi(L - x)/L; & \xi < x < L \end{cases} \tag{2.1.7}$$

This function is called the *Green's function* for Eq. (2.1.2) for the point $x = \xi$. We see that the solution for a string with force F concentrated at $x = \xi$ is, therefore, $(F/T)G(x|\xi)$ and that the solution for forces F_1 concentrated at ξ_1 and F_2 at ξ_2 is $(F_1/T)G(x|\xi_1) + (F_2/T)G(x|\xi_2)$.

Going from sum to integral and using Eq. (2.1.5) we see that the steady-state shape of a string under tension T between rigid supports a distance L apart, under the action of a steady transverse force $F(x)$, is

$$\psi = \int_0^L [F(\xi)/T]G(x|\xi)\, d\xi \tag{2.1.8}$$

Thus the Green's function, which is a solution for a concentrated force at $x = \xi$, can be used to obtain a solution of Poisson's equation for a force of arbitrary form, distributed along the string.

This technique, of obtaining a solution for the general equation in terms of an integral involving the Green's function, the solution of a particularly simple form of the equation, will be discussed in considerable detail later in the book.

The Wave Equation. In connection with other problems we may be less interested in the steady-state shape of the string under the influence of applied transverse forces than we are in the motion or the string after applied forces have been removed. The simplest case, and the one of greatest interest, occurs when all forces can be neglected except that due to the tension T of the string. When the mass of the string is uniformly distributed, being ρ gm per unit length, the equation for transverse motion is obtained by equating the mass times acceleration of each element of length of the string, $\rho\, dx(\partial^2\psi/\partial t^2)$, to the transverse force on that element due to the tension, $T\, dx(\partial^2\psi/\partial x^2)$, as given in Eq. (2.1.1). The resulting equation

$$\partial^2\psi/\partial x^2 = (1/c^2)(\partial^2\psi/\partial t^2);\quad c^2 = T/\rho \tag{2.1.9}$$

is called the *wave equation* for reasons which will shortly become apparent. It states that the transverse acceleration of any part of the string is proportional to the curvature of that part.

A wave may be roughly described as a configuration of the medium (transverse shape of string, distribution of density of fluid, etc.) which moves through the medium with a definite velocity. The velocity of propagation of the wave is not necessarily related to the velocity of any portion of the medium. In fact, for waves governed by the simple equation (2.1.9) the wave velocity is completely independent of the velocity of parts of the medium; or, in other words, as long as Eq. (2.1.9) is valid, the velocity of any wave on the string is the same, no matter what shape the wave has. The wave shape moves *along* the string with velocity c, whereas a point on the string moves back and forth with a velocity $\partial\psi/\partial t$ determined by the shape of the wave as it goes by.

A wave of this sort can be represented by stating that the displacement of the string from equilibrium is a function of $(x - ct)$, for a wave in the positive x direction, or of $(x + ct)$ for a wave in the negative x direction. To show that Eq. (2.1.9) requires such motion, we can transform coordinates from x and t to $\xi = x - ct$ and $\eta = x + ct$:

$$\frac{\partial}{\partial x} = \frac{\partial}{\partial \xi}\frac{\partial \xi}{\partial x} + \frac{\partial}{\partial \eta}\frac{\partial \eta}{\partial x} = \frac{\partial}{\partial \xi} + \frac{\partial}{\partial \eta}$$

$$\frac{\partial^2}{\partial x^2} = \frac{\partial^2}{\partial \xi^2} + \frac{2\partial^2}{\partial \xi\,\partial \eta} + \frac{\partial^2}{\partial \eta^2}$$

$$\frac{1}{c^2}\frac{\partial^2}{\partial t^2} = \frac{\partial^2}{\partial \xi^2} - \frac{2\partial^2}{\partial \xi\,\partial \eta} + \frac{\partial^2}{\partial \eta^2}$$

Therefore, Eq. (2.1.9) becomes

$$4\frac{\partial^2\psi}{\partial \xi\,\partial \eta} = 0$$

A solution of this equation is $\psi = f(\xi) + F(\eta)$, where f and F are any functions which satisfy the requirements of continuity and small amplitude that were assumed in the derivation of Eq. (2.1.9). As we shall show later, this is the most general solution of Eq. (2.1.9), so that the most general motion of the string always turns out to be a superposition of a wave traveling to the right and another to the left, each traveling with constant velocity and unchanged shape.

It is to be noted that, if c were the velocity of light, the wave equation would be invariant to a Lorentz transformation, for then the expression $(\partial^2\psi/\partial x^2) - (1/c^2)(\partial^2\psi/\partial^2 t)$ is the scalar obtained by contraction of the second-order tensor $\partial^2\psi/\partial x_n\, \partial x_m$ and therefore is invariant to space-time rotations of the sort discussed in the first chapter. The lines $\xi = x - ct$ and $\eta = x + ct$ represent the world lines of zero proper length $(c^2\, dt^2 - dx^2 = 0)$ which represent rays of light.

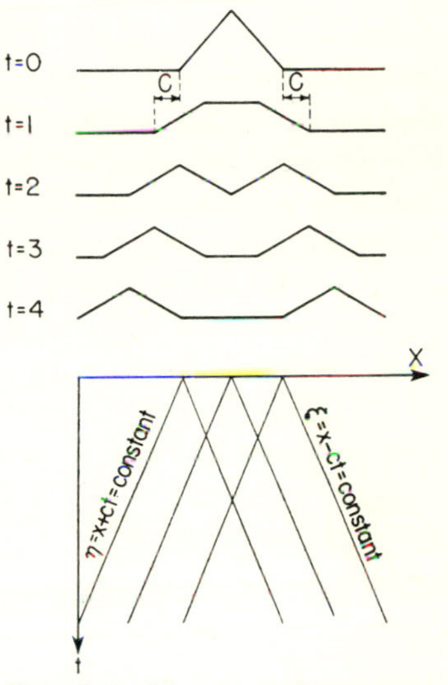

Fig. 2.4 Wave motion on a string, showing waves in opposite directions.

Simple Harmonic Motion, Helmholtz Equation. In some cases the wave motion is sinusoidal in its time dependence, so that one can factor out ("separate off" is the phrase often used) a term depending only on the time of the form $e^{-i\omega t}$. Since we have agreed to use only the real part of a complex solution, this factor ensures sinusoidal dependence on time. The constant ω is called the *angular velocity* of the oscillation, and the quantity $\nu = \omega/2\pi$ is called the *frequency* of oscillation of the wave.

Inserting the expression $\psi = \gamma(x)e^{-i\omega t}$ into the wave equation (2.1.9) results in an equation for the space part of ψ

$$(d^2\gamma/dx^2) + (\omega/c)^2\gamma = 0 \qquad (2.1.10)$$

which is called the *Helmholtz equation.*

We shall show later in this book (but it would be a useful exercise for the reader to show for himself now) that the Green's function [see discussion of Eq. (2.1.7)] for this equation for a string of infinite length, corresponding to a force $Te^{-i\omega t}$ concentrated at the point $x = 0$, is

$$G(x|0) = \begin{cases} \dfrac{i}{2}\dfrac{c}{\omega}\, e^{-i(\omega/c)x}; & x < 0 \\[2ex] \dfrac{i}{2}\dfrac{c}{\omega}\, e^{i(\omega/c)x}; & x > 0 \end{cases}$$

The general equation for any applied force having sinusoidal dependence on time, with frequency $\omega/2\pi$, is obtained by an integral analogous to Eq. (2.1.8).

Wave Energy. Since the waves in both directions are propagated with constant velocity and unvarying shape, it is reasonable to expect that the wave energy, once it is given to the string, is propagated without loss. The total energy of the string is made up of the sum of the kinetic energy of each element of the string,

$$\text{KE} = \tfrac{1}{2}\rho\int(\partial\psi/\partial t)^2\,dx$$

integrated over the length of the string, plus the total potential energy of the string. If the shape of the string at any time t is $\psi(x,t)$, the potential energy may be obtained by imagining this shape being produced by applying a transverse distributed force of the proper amount so as to move the string slowly from its equilibrium shape $\psi = 0$ to its final shape $\psi(x,t)$. The intermediate shape can be taken to be $\beta\psi$, where β varies between zero and unity as the string is moved. The applied force on an element dx of string in order to have reached the intermediate shape $\beta\psi$ is $-T\beta(\partial^2\psi/\partial x^2)\,dx$, and the work done by this force to move this element from $\beta\psi$ to $(\beta + d\beta)\psi$ is $-T\psi(\partial^2\psi/\partial x^2)\,dx\,\beta\,d\beta$. The total work done in moving the string from equilibrium ($\beta = 0$) to its final shape $\psi(\beta = 1)$ is, therefore,

$$\text{PE} = -T\int\psi(\partial^2\psi/\partial x^2)\,dx\int_0^1\beta\,d\beta = -\tfrac{1}{2}T\int\psi(\partial^2\psi/\partial x^2)\,dx$$

The total energy possessed by that part of the string from $x = a$ to $x = b$ is, therefore,

$$\begin{aligned} W(a,b) = \text{KE} + \text{PE} &= \tfrac{1}{2}\rho\int_a^b\left[\left(\frac{\partial\psi}{\partial t}\right)^2 - c^2\psi\left(\frac{\partial^2\psi}{\partial x^2}\right)\right]dx \\ &= \tfrac{1}{2}\rho\int_a^b\left[\left(\frac{\partial\psi}{\partial t}\right)^2 + c^2\left(\frac{\partial\psi}{\partial x}\right)^2\right]dx - \tfrac{1}{2}T\left[\psi\left(\frac{\partial\psi}{\partial x}\right)\right]_a^b \end{aligned}$$

where the final, symmetric form is obtained by integrating the potential energy term by parts.

When a and b represent the two ends of the string, which are held rigid, ψ is zero at a and b and the last term in the symmetric form is zero. The energy of the whole string is, therefore, proportional to the square of the velocity of each part plus c^2 times the square of the slope of each part, integrated over the length of the string.

This expression for the energy of only a portion of a string is not unique, for the question of the energy of the "ends" of the chosen portion cannot be uniquely determined. The only unique quantity is the energy of the entire string, including the supports, for only this energy is conserved. This fact can be most vividly illustrated by computing the potential energy of part of the string by another method and showing that this yields another answer. For example, because of the difference in configuration $\psi(x)$ from the equilibrium configuration, the string will be stretched. The corresponding stored energy due to stretching is just the potential energy of the string, for it is the work done by the principal component (the horizontal one) of the tension T. The length of a section dx of the string becomes $\sqrt{1 + (\partial\psi/\partial x)^2}\, dx$ because of the stretch. The potential energy due to the constant force T is, therefore,

$$\mathrm{PE} = +T \int_a^b \left\{ \sqrt{1 + \left(\frac{\partial\psi}{\partial x}\right)^2} - 1 \right\} dx$$

This to the second order, is

$$\mathrm{PE} = \tfrac{1}{2}T \int_a^b \left(\frac{\partial\psi}{\partial x}\right)^2 dx$$

so that the energy $W(a,b)$ is

$$W(a,b) = \tfrac{1}{2}\rho \int_a^b \left\{ \left(\frac{\partial\psi}{\partial t}\right)^2 + c^2 \left(\frac{\partial\psi}{\partial x}\right)^2 \right\} dx \qquad (2.1.11)$$

Comparing this with our previous answer we see that the two differ by the amount $-\frac{1}{2}T[\psi \partial\psi/\partial x]_a^b$, which involves only values at the two ends a and b. Each answer is equally good for the complete string, for when a and b are the ends of the string (which are rigidly or freely supported so that no energy is given to the supports), both results are the same. This is the only case which should give a unique answer. Since expression (2.1.11) is simpler, we shall use it in our subsequent discussions.

Energy Flow. The rate of change of the energy of the portion of string between a and b is obtained by differentiating $W(a,b)$ with respect to the time:

$$\begin{aligned} \frac{d}{dt} W(a,b) &= \rho \int_a^b \left[\frac{\partial\psi}{\partial t}\frac{\partial^2\psi}{\partial t^2} + \frac{T}{\rho}\frac{\partial\psi}{\partial x}\frac{\partial^2\psi}{\partial x\, \partial t} \right] dx \\ &= T \int_a^b \left[\frac{\partial\psi}{\partial t}\frac{\partial^2\psi}{\partial x^2} + \frac{\partial\psi}{\partial x}\frac{\partial^2\psi}{\partial x\, \partial t} \right] dx = T \int_a^b \frac{\partial}{\partial x}\left[\frac{\partial\psi}{\partial t}\frac{\partial\psi}{\partial x} \right] dx \end{aligned}$$

Therefore,

$$\frac{d}{dt} W(a,b) = T \left[\frac{\partial\psi}{\partial t}\frac{\partial\psi}{\partial x} \right]_a^b = T \left(\frac{\partial\psi}{\partial t}\frac{\partial\psi}{\partial x} \right)_{x=b} - T \left(\frac{\partial\psi}{\partial t}\frac{\partial\psi}{\partial x} \right)_{x=a} \qquad (2.1.12)$$

These two terms represent energy flow into or out of the length of string, across the two ends. If $-T(\partial\psi/\partial t)(\partial\psi/\partial x)$ represents the average energy flow in the positive x direction across the point x, then the first term on the right-hand side represents energy flow into the string from the far end of the portion ($b > a$) and the second term represents flow into the portion from the left-hand end.

It is not difficult to verify that $-T(\partial\psi/\partial t)(\partial\psi/\partial x)$ equals the flow of energy along the string in the positive x direction, for $-T(\partial\psi/\partial x)$ equals the transverse force which the part of the string to the left of x exerts on the part of the string to the right to make it move and $\partial\psi/\partial t$ is the transverse velocity of the point x of the string. Force times velocity, of course, equals power or rate of energy flow.

In this respect the term $-T(\partial\psi/\partial x)$ is analogous to a voltage across a transmission line at some point, and $\partial\psi/\partial t$ is analogous to the current past the same point; the product of the two equals the power transmitted.

Power and Wave Impedance. Moreover the analogy to a transmission line can be carried further. For alternating currents, the complex ratio of voltage to current is called the *impedance* of the line. Often this impedance[1] is a function of the a-c frequency, but sometimes, when the impedance is a pure resistance, it can be independent of frequency.

The analogue to the electrical impedance is the complex ratio of the transverse driving force to the transverse velocity, which can be called mechanical impedance. For the simple string, long enough so that waves are not reflected back from the far end (which is taken to be at $x = \infty$), the displacement of point x due to an alternating wave going in the direction of increasing x can be represented by the expression $\psi = A_+e^{i(\omega/c)(x-ct)}$. The force and velocity at point x are

$$-T(\partial\psi/\partial x) = -iT(\omega/c)\psi; \quad \partial\psi/\partial t = -i\omega\psi$$

Therefore, the power flow across x, the average product of the real parts of these two terms, is

$$\text{Power} = \tfrac{1}{2}\rho c\omega^2|A_+|^2 = \tfrac{1}{2}\rho c|U_+|^2; \quad T = \rho c^2$$

for a sinusoidal wave in the positive x direction. The quantity $U_+ = -i\omega A_+$ can be called the *velocity amplitude* of the string (not the wave) for a sinusoidal wave.

The impedance at point x for this simple wave is

$$Z_\omega = [-T(\partial\psi/\partial x)]/[\partial\psi/\partial t] = \rho c \tag{2.1.13}$$

This is called the *wave impedance* for the string. It is a constant, independent of x and of frequency for the simple string for waves in one direction. Indeed, we need not have dealt with so specialized a one-directional wave as we did to obtain this value for wave impedance; for any wave in the positive x direction, $f(x - ct)$, the transverse force is

[1] Since we shall be using the *negative* exponential for the time variation $e^{-i\omega t}$, the signs of the reactance terms (imaginary parts of the impedances) will have the opposite sign from that encountered in the usual electrical engineering notation. This is most easily done by making the minus sign explicit before the i. For instance, if $Z = R - iX$, then X will be the same reactance term as encountered in the electrical engineering notation. In fact, the impedance formulas in this book can be changed to the engineering notation simply by changing every $(-i)$ in the formulas to $(+j)$.

$-Tf'(x - ct)$ (where the prime indicates the derivative), the corresponding transverse velocity is $-cf'(x - ct)$, and the ratio of force to velocity is, therefore, $T/c = \rho c$, independent of x and t and of the shape of the wave.

Of course, if we have to deal with waves in both directions, the impedance does depend on frequency and position. If $\psi = [A_+e^{i\omega x/c} + A_-e^{-i\omega x/c}]e^{-i\omega t}$, then the average rate of flow of energy obtained by averaging the expression (2.1.11) over a cycle is

$$\text{Power} = \tfrac{1}{2}\rho c\omega^2[|A_+|^2 - |A_-|^2]$$

and the impedance is

$$Z(x) = \rho c \frac{A_+e^{i\omega x/c} - A_-e^{-i\omega x/c}}{A_+e^{i\omega x/c} + A_-e^{-i\omega x/c}}$$

There will be many cases encountered in this chapter where the analogy with voltage, current, power, and impedance can be usefully applied and generalized. In wave motion of all sorts, for instance, one can usually find two quantities derivable from the wave function such that their product equals the rate of flow of energy in the wave and their ratio can be taken as a generalized impedance. In most of these cases when the wave is in only one direction, the impedance is a real constant, independent of position and of frequency, in which case this constant value can be called the *wave impedance* for the wave under study. The more complicated expressions for the impedance, for more complicated forms of waves, are most easily given in terms of this wave impedance as a scale factor. For instance, for the string, the constant $\rho c = T/c = \sqrt{\rho T}$ is the scale factor in the general expression for the impedance.

Forced Motion of the String. As an example of the utility of the generalized concept of impedance, we can consider the motion of a string of length l held at $x = l$ under a tension T by a support which is not completely rigid and driven at $x = 0$ by a transverse force. The ratio between a sinusoidal transverse force on the support at $x = l$, represented by the real part of $F_l e^{-i\omega t}$ and the transverse velocity of the support $U_l e^{-i\omega t}$ which is produced by the force, is called the *transverse mechanical impedance* of the support, $Z_l = F_l/U_l$. This quantity usually depends on the frequency $\omega/2\pi$ but is independent of the amplitude of F_l or U_l within certain limits.

The shape of the string must be representable by a combination of a sinusoidal wave $A_+e^{i(\omega/c)x - i\omega t}$ going from source ($x = 0$) to support ($x = l$) and another wave $A_-e^{-i(\omega/c)x - i\omega t}$ reflected from the support back to the source:

$$\psi = [A_+e^{i(\omega/c)x} + A_-e^{-i(\omega/c)x}]e^{-i\omega t} = A \cosh [i(\omega x/c) + \pi\alpha_0 - i\pi\beta_0]e^{-i\omega t}$$

where $\qquad A_+ = Ae^{\pi(\alpha_0 - i\beta_0)}; \quad A_- = Ae^{-\pi(\alpha_0 - i\beta_0)}$

The transverse force exerted by the string on the support is, therefore,

$$\begin{aligned}
F_l e^{-i\omega t} &= -T(\partial\psi/\partial x) \text{ at } x = l \\
&= -i\omega\rho c[A_+ e^{i(\omega/c)l} - A_- e^{-i(\omega/c)l}]e^{-i\omega t} \\
&= -i\omega\rho c A \sinh [i(\omega l/c) + \pi\alpha_0 - i\pi\beta_0]e^{-i\omega t} \\
&= Z_l U_l e^{-i\omega t} = Z_l(-\partial\psi/\partial t) \text{ at } x = l \\
&= -i\omega Z_l[A_+ e^{i(\omega/c)l} + A_- e^{-i(\omega/c)l}]e^{-i\omega t} \\
&= -i\omega Z_l A \cosh [i(\omega l/c) + \pi\alpha_0 - i\pi\beta_0]e^{-i\omega t}
\end{aligned}$$

where we have used the definition of the transverse mechanical impedance of the support to obtain the last four forms.

From these equations, we can obtain the complex ratio between the wave amplitudes A_-, A_+, and also the constants α_0 and β_0 in terms[1] of the impedance Z_l:

$$\frac{A_-}{A_+} = \frac{\rho c - Z_l}{\rho c + Z_l} e^{2i(\omega/c)l}; \quad \alpha_0 - i\beta_0 = \frac{1}{\pi} \tanh^{-1}\left(\frac{Z_l}{\rho c}\right) - i\frac{2l}{\lambda} \tag{2.1.14}$$

where $\lambda = c/\nu = 2\pi c/\omega$ is the wavelength of the waves on the string The ratio A_-/A_+ is called the *standing-wave ratio* or, alternately, the *reflection coefficient.* If Z_l is a pure imaginary, *i.e.*, just reactive, $|A_-/A_+| = 1$, so that the amplitudes of the reflected and incident waves are equal, as they should be, though, of course, the *phase* of the reflected wave will be different from that of the incident one. The relation between the reflection coefficient and Z_l given in (2.1.14) is an example of the relation between the unitary reflection operator and the impedance operator discussed in the section on abstract vector spaces. From (2.1.14) we see that the boundary condition at $x = l$ fixes the relative phases and amplitudes of the incident and reflected waves. Once this is known, the ratio Z_0 between the applied force and the velocity of the driving point ($x = 0$), which is the *driving-point impedance* for the string, can be obtained at once:

$$Z_0 = \rho c \frac{1 - (A_-/A_+)}{1 + (A_-/A_+)} = \rho c \tanh [\pi(\alpha_0 - i\beta_0)] \tag{2.1.15}$$

In other words, if the force is known, the string velocity at $x = 0$ can be calculated and also the expression A_+, A_-, A, and ψ. For instance, if the driving force is $f(\omega)e^{-i\omega t}$, the expression for the wave is

$$\begin{aligned}
\psi(\omega,x,t) &= \frac{f(\omega)e^{-i\omega t}}{-i\omega Z_0} \frac{\cosh [i(\omega x/c) + \pi\alpha_0 - i\pi\beta_0]}{\cosh [\pi\alpha_0 - i\pi\beta_0]} \\
&= \frac{f(\omega)e^{-i\omega t}}{-i\omega\rho c}\left[\coth (\pi\alpha_0 - i\pi\beta_0) \cos\left(\frac{\omega x}{c}\right) + i \sin\left(\frac{\omega x}{c}\right)\right]
\end{aligned} \tag{2.1.16}$$

[1] See the footnote on page 128.

Transient Response, Fourier Integral. Just as with the Poisson equation discussed on page 121, a solution for several different forces acting simultaneously is the sum of the solutions for the forces acting separately. For instance, if forces of all frequencies are acting, as would be the case when we could express the total transverse force acting on the $x = 0$ end of the string in the form of an integral

$$F(t) = \int_{-\infty}^{\infty} f(\omega)e^{-i\omega t}\,d\omega \tag{2.1.17}$$

then the expression for the shape of the string as a function of x and t would be

$$\psi = \int_{-\infty}^{\infty} \psi(\omega,x,t)\,d\omega \tag{2.1.18}$$

where $\psi(\omega,x,t)$ is given in Eq. (2.1.16).

In Chap. 4 we shall show that a very wide variety of functions of t can be expressed in terms of an integral of the type given in Eq. (2.1.17) (which is called a *Fourier integral*), and we shall show there how to compute $f(\omega)$ if $F(t)$ is known. Therefore, the integral of Eq. (2.1.18) is a general solution for the motion of the string under the action of nearly any sort of physically realizable force applied transversely to its end. This technique of solution is analogous to the Green's function technique touched on in the discussion of Eq. (2.1.18) and will also suggest similar methods for solving other equations discussed later in this chapter. One finds a solution for a particularly simple form of "force," which involves a parameter (point of application for the Poisson equation, frequency for the wave equation). A very general form of force can then be built up by expressing it as an integral of the simple force over this parameter; the resulting solution is a similar integral of the simple solutions with respect to the same parameter. This is the general principle of the Green's function technique to be discussed in Chap. 7 and elsewhere in this book.

Operator Equations for the String. Before leaving the problem of wave motion in a simple string, it will be of interest to outline an alternative approach to the problem which is related to the discussions of operators in abstract vector space given in Chap. 1 and later in this chapter. We start out by considering the string to be an assemblage of equal mass points connected by equal lengths of weightless string. At first we consider that there are only a finite number N of these masses (obviously a poor approximation for a uniform string), and then we approach the actual string by letting N go to infinity. Thus we can show the relation between the coupled oscillators discussed on page 77 and the flexible string.

We approximate the string of uniform density stretched under tension T between rigid supports a distance l apart by N equally spaced mass points, each of mass $\rho l/N$ a distance $l/(N+1)$ apart. A glance at Fig. 2.5

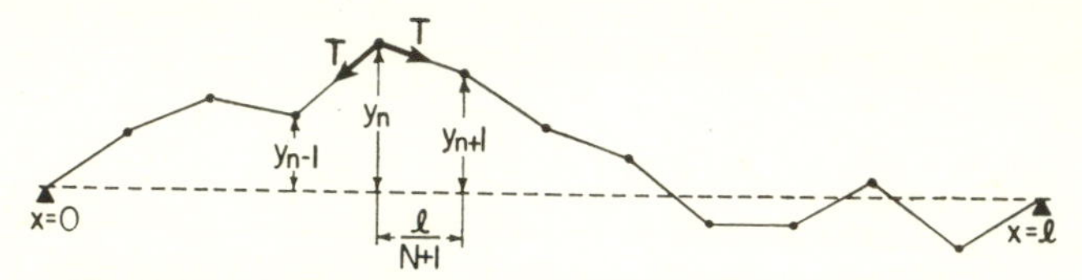

Fig. 2.5 Displacements of mass points on elastic string.

shows that, if the displacement from equilibrium of the nth mass is y_n, the transverse force on this mass due to the displacements of its neighbors is

$$(N+1)T\{[(y_{n+1}-y_n)/l]+[(y_{n-1}-y_n)/l]\} = (N+1)(T/l)(y_{n+1} + y_{n-1} - 2y_n)$$

(The last expression in parentheses is the analogue, for finite differences, of the second derivative.) Therefore, our set of simultaneous equations of motion for the N particles is

$$\begin{aligned}
\frac{d^2y_1}{dt^2} + 2\omega_0^2 y_1 &= \omega_0^2 y_2 \\
\frac{d^2y_2}{dt^2} + 2\omega_0^2 y_2 &= \omega_0^2 (y_1 + y_3) \\
&\cdots\cdots\cdots \\
\frac{d^2y_n}{dt^2} + 2\omega_0^2 y_n &= \omega_0^2 (y_{n-1} + y_{n+1}) \\
&\cdots\cdots\cdots \\
\frac{d^2y_N}{dt^2} + 2\omega_0^2 y_N &= \omega_0^2 y_{N-1}
\end{aligned} \tag{2.1.19}$$

where $\omega_0^2 = N(N+1)(T/\rho l^2)$.

We now consider the displacements y_n to be the components of a vector $\mathbf{y}$ in an abstract vector space of N dimensions with unit vectors $\mathbf{e}_n$ along the coordinate axes. The parts of the equations on the right-hand side represent the operation of a dyadic $\omega_0^2\mathfrak{U}$ which transforms the vector $\mathbf{e}_n$ into a vector with components along $\mathbf{e}_{n-1}$ and $\mathbf{e}_{n+1}$. The dyadic $\mathfrak{U}$ can be called the *unit shift operator*, for it shifts the index n by a unit up or down. It can be written in terms of the $\mathbf{e}$'s in the form

$$\mathfrak{U} = \mathbf{e}_1\mathbf{e}_2 + \mathbf{e}_2(\mathbf{e}_1 + \mathbf{e}_3) + \cdots + \mathbf{e}_n(\mathbf{e}_{n-1} + \mathbf{e}_{n+1}) + \cdots + \mathbf{e}_N\mathbf{e}_{N-1} \tag{2.1.20}$$

Therefore, the equation for the vector $\mathbf{y}$ which represents the displacements of all the particles,

$$\mathbf{y} = \sum_{n=1}^{N} y_n \mathbf{e}_n$$

can be written in the form

$$(d^2\mathbf{y}/dt^2) + 2\omega_0^2\mathbf{y} = \omega_0^2\mathfrak{U} \cdot \mathbf{y}$$

Eigenvectors for the Unit Shift Operator. The solution of the differential equation for $\mathbf{y}$ can be most easily effected by using the eigenvectors $\mathbf{u}_n$ of the operator $\mathfrak{U}$.

$$\mathfrak{U} \cdot \mathbf{u}_n = \eta_n \mathbf{u}_n$$

where $\mathbf{u}_n$ is a unit vector along a principal axis of the operator $\mathfrak{U}$. Introducing $\mathbf{u}_n$ for $\mathbf{y}$ into the equation for $\mathbf{y}$ we obtain the equation determining the time dependence of $\mathbf{u}_n$:

$$(d^2\mathbf{u}_n/dt^2) + \omega_0^2(2 - \eta_n)\mathbf{u}_n = 0$$

so that the time dependence of $\mathbf{u}_n$ is $e^{-i\omega_0\sqrt{2-\eta_n}\,t}$. The space ("vector space") dependence of $\mathbf{u}_n$ may be determined by solving the eigenvalue equation above.

Let $\mathbf{u}_n$ be expressed in terms of $\mathbf{e}_m$ by the expansion

$$\mathbf{u}_n = \sum_{m=1}^{N} \gamma_{nm}\mathbf{e}_m$$

the γ's being the direction cosines of the transformation. Then the γ's satisfy the equation

$$\gamma_{n,m-1} - \eta_n\gamma_{n,m} + \gamma_{n,m+1} = 0 \tag{2.1.21}$$

except for the first and last equations, for $m = 1$ and $m = N$, where the quantities γ_{n0} and $\gamma_{n,N+1}$ are naturally omitted. Even these two equations can be given this same form, however, if we just assume that the quantities γ_{n0} and $\gamma_{n,N+1}$ are always zero.

The solution of Eqs. (2.1.21) is obtained by the use of the trigonometric formula

$$\cos\alpha \sin(m\alpha) = \tfrac{1}{2}\sin[(m-1)\alpha] + \tfrac{1}{2}\sin[(m+1)\alpha]$$

for if we set $\gamma_{nm} = A\sin[m\alpha_n]$ (the time dependence being understood) and $\eta_n = 2\cos\alpha_n$, all of the equations are satisfied. One of the additional requirements, that $\gamma_{n0} = 0$, is likewise complied with, and the remaining requirement, that $\gamma_{n,N+1} = 0$, can be satisfied if we allow α_n to equal $[n\pi/(N+1)]$.

Since $$\sum_{m=1}^{N} \sin\left(\frac{mn\pi}{N+1}\right)\sin\left(\frac{mn'\pi}{N+1}\right) = \begin{cases} 0; & n' \neq n \\ \frac{1}{2}(N+1); & n' = n \end{cases}$$

we can choose the value of the constant A so that the γ's are properly normalized direction cosines, and the **u**'s will be unit vectors if the **e**'s are. The final results are

$$\mathbf{u}_n = \sqrt{\frac{2}{N+1}} \sum_{m=1}^{N} \mathbf{e}_m \sin\left(\frac{mn\pi}{N+1}\right) \exp\left[-2i\omega_0 t \sin\left(\frac{n\pi}{2(N+1)}\right)\right]$$

$$\mathfrak{U} \cdot \mathbf{u}_n = 2\cos[n\pi/(N+1)]\,\mathbf{u}_n \tag{2.1.22}$$

Thus we have discovered an alternative set of N mutually orthogonal unit vectors in abstract vector space which point along the principal axes of the operator $\mathfrak{U}$ (*i.e.*, which are eigenvectors for $\mathfrak{U}$).

In terms of this new coordinate system the solution of the equation of motion for the vector **y** representing the N particles is straightforward:

$$\sum_{m=1}^{N} y_m \mathbf{e}_m = \mathbf{y} = \sum_{n=1}^{N} U_n \mathbf{u}_n = \sqrt{\frac{2}{N+1}} \sum_{n,m=1}^{N} U_n \mathbf{e}_m \sin\left(\frac{mn\pi}{N+1}\right) \exp\left\{-2i\omega_0 t \sin\left[\frac{n\pi}{2(N+1)}\right]\right\}$$

Therefore,

$$y_m = \sqrt{\frac{2}{N+1}} \sum_{n=1}^{N} U_n \sin\left(\frac{mn\pi}{N+1}\right) \exp\left\{-2i\omega_0 t \sin\left[\frac{n\pi}{2(N+1)}\right]\right\} \tag{2.1.23}$$

The allowed frequencies are $\omega_n/2\pi$, where $\omega_n = 2\omega_0 \sin[n\pi/2(N+1)]$. The component motions $\mathbf{u}_n$ are called *normal modes* of the motion.

If the particles are initially displaced by the amounts y_n° and initially all have zero velocities, then the values of the U_n's can be obtained by use of the last equation on page 133:

$$(\mathbf{y} \cdot \mathbf{u}_n)_{t=0} = U_n = \sqrt{\frac{2}{N+1}} \sum_{m=1}^{N} y_m^\circ \sin\left(\frac{mn\pi}{N+1}\right) \tag{2.1.24}$$

Thus the coefficients of the series for the y's can be obtained in terms of the initial values of the y's and the direction cosines of the transformation.

Limiting Case of Continuous String. To go from a collection of N particles to a continuous string we increase N to infinity, so that each "point" on the string is labeled by a different n. If the string were actually continuous, this would require N to be nondenumerably infinite, which would mean that the corresponding abstract vector space would have a nondenumerable infinity of mutually perpendicular directions. Such a vector space is rather difficult to imagine, though we may console ourselves that such niceties in distinguishing of types of infinities are rather academic here, since any actual string is only approximately continuous, and our present solutions are not valid for details of atomic size or smaller. We can also reassure ourselves that only a small subspace of the "supervector" space corresponds to physics, for as the distance between successive points goes to zero, continuity requires that y_n approach y_{n+1} in value.

At any rate, for the continuous string we can discard the nondenumerable set of indices m and use the distance x of the point from one end as the label, *i.e.*, set $x = ml/(N + 1)$. Moreover since N is so large, the difference between N and $N + 1$ is negligible. The index n labeling the different allowed modes of motion need not become infinite or continous, however, since we are usually interested in the lowest few (the first hundred or so!) allowed frequencies. Therefore, n will be retained as an integer and n/N will be a small quantity. To be specific, the transition is as follows:

$$\omega_n \to n\pi c/l; \quad c = \sqrt{T/\rho}; \quad \mathbf{e}_m \to \mathbf{e}(x); \quad \mathbf{y} = \sum_x y(x)\mathbf{e}(x)$$

$$\mathbf{y} \to \sqrt{\frac{N}{2}} \sum_n Y_n \mathbf{u}_n = \sum_{n,x} Y_n \mathbf{e}(x) \sin\left(\frac{n\pi x}{l}\right) e^{-i\omega_n t}$$

$$y(x) = \sum_n Y_n \sin\left(\frac{n\pi x}{l}\right) e^{-i\omega_n t}$$

The last equation is the usual *Fourier series* for the free oscillations of the uniform string between rigid supports. The function $\sin(n\pi x/l)\, e^{-i\omega_n t}$, giving the shape of the nth normal mode, is the *transformation function*, changing the denumerably infinite set of eigenvectors $\mathbf{u}_n$ for the operator $\mathfrak{U}$ to the nondenumerably infinite set of unit vectors $\mathbf{e}(x)$, each of which corresponds to a different point along the continuous string. The summation over all the points is symbolized by the summation sign $\sum_x$, though it could also be expressed in terms of an integral over x. The limiting case of Eq. (2.1.24) is best expressed in terms of an integral, for instance. We have let Y_n be the limiting value of $U_n \sqrt{2/N + 1}$, so that the equation for y_n in terms of the initial values of displacement y_n^c (when the initial velocity is zero) is

$$Y_n = \lim_{N\to\infty} \left\{\frac{2}{N} \sum_{m=1}^{N} y_m^\circ \sin\left(\frac{mn\pi}{N}\right)\right\}; \quad \text{where } m \to \frac{xN}{l}$$

The number of terms in the sum over m between x and $x + dx$ is, therefore, $(N/l)\, dx$. Therefore, in the limit the sum for Y_n becomes the integral

$$Y_n = \frac{2}{l} \int_0^l y^\circ(x) \sin\left(\frac{n\pi x}{l}\right) dx$$

which is the usual integral for the Fourier series coefficients giving the amplitudes of the various normal modes.

Finally, if the distance between supports is increased without limit (we place the origin at the mid-point of the string), another series of limiting calculations, which will be explained in detail in Chap. 4, brings us to the general solution for the wave motion on an infinite string originally held at a displacement $y^\circ(x)$ and released at $t = 0$:

$$y(x, t) = \frac{1}{2\pi} \int_{-\infty}^{\infty} e^{i\alpha(x-ct)}\, d\alpha \int_{-\infty}^{\infty} y^\circ(\xi) e^{-i\alpha\xi}\, d\xi \qquad (2.1.25)$$

where the real part of this expression gives the actual displacement of point x at time t.

Finally it is of interest to see what limiting form of the operator $\mathfrak{U}$ takes on for the continuous string. To fit in with the equation for the vector $\mathbf{y}$, we ask for the limiting expression for the operator $\omega_0^2[\mathfrak{U} - 2]$ on the vector $\mathbf{y} = \sum_m y_m \mathbf{e}_m \to \sum_x y(x)\mathbf{e}(x)$. Before going to the limit, the operator has the following effect on the vector components y_m:

$$\omega_0^2[\mathfrak{U} - 2] \cdot \mathbf{y} = \frac{N(N+1)T}{\rho l^2} \sum_{m=1}^{\infty} [(y_{n+1} - y_n) - (y_n - y_{n-1})]\mathbf{e}_n$$

As the distance between particles gets smaller and smaller, the difference $(y_{n+1} - y_n)$ approaches the differential $dy(x)$, the distance between particles, $l/(N + 1)$, being dx. Therefore, $(N/l)(y_{n+1} - y_n)$ goes in the limit to $\partial y/\partial x$ and the expression above becomes

$$\omega_0^2[\mathfrak{U} - 2] \cdot \mathbf{y} \to c^2 \sum_x \frac{\partial^2 y(x)}{\partial x^2} \mathbf{e}(x)$$

so the equation of motion for $\mathbf{y}$ becomes in the limit

$$\frac{\partial^2 \mathbf{y}}{\partial t^2} = \sum_x \frac{\partial^2 y\,(x)}{\partial t^2} \mathbf{e}(x) = c^2 \sum_x \frac{\partial^2 y\,(x)}{\partial x^2} \mathbf{e}(x)$$

and therefore the equation for the transformation functions $y(x)$ is

$$\partial^2 y/\partial t^2 = c^2(\partial^2 y/\partial x^2)$$

which is just the wave equation (2.1.9).

Thus we have come back again to the partial differential equation for waves on the simple string via the roundabout route of abstract vector space. This excursion has been taken because similar ones will have to be taken later in this chapter for cases where alternative routes are not quite so direct or so simple as is the case of the string.

The Effect of Friction. So far we have neglected the reaction of the surrounding medium (air or water) on the motion of the string. For small-amplitude motions this reaction is opposed to the motion of each element of length and is proportional to its velocity. The force on the element between x and $x + dx$ is proportional to the velocity $\partial\psi/\partial t$ of this element and is opposed to this velocity. The proportionality constant R is usually dependent on the frequency of oscillation of the string, but when the viscosity of the medium is great enough, it is independent of frequency. This last case is the simplest and will be taken up first.

The equation of motion, when we take into account tension and friction of the medium but not stiffness or internal friction, is

$$\frac{\partial^2\psi}{\partial t^2} + 2k\frac{\partial\psi}{\partial t} - c^2\frac{\partial^2\psi}{\partial x^2} = 0; \quad k = \frac{R}{2\rho}; \quad c^2 = \frac{T}{\rho}$$

The effect of friction is, of course, to damp out the free vibrations. If the string is held between two rigid supports a distance l apart, the shapes of the normal modes are not affected by the friction, being still $\sin(\pi nx/l)$. However, the individual oscillations are damped out in time, for a solution of this equation is

$$\psi = \sum_n A_n \sin\left(\frac{\pi nx}{l}\right) e^{-kt-i\omega_n t}; \quad \omega_n^2 = \frac{\pi nc}{l} - k^2$$

If k depends on frequency, it will have the value corresponding to ω_n for the nth normal mode, so that the different modes will damp out at different rates.

On the other hand if the string is very long and is driven at one end by a sinusoidal force $Fe^{-i\omega t}$, then the waves will be damped in space rather than in time. A solution is

$$\psi = Ae^{i\alpha x - i\omega t}; \quad \alpha^2 = (\rho\omega^2/T) + i(R\omega/T)$$

Therefore, α has a positive imaginary part, which produces damping in the direction of wave motion.

Diffusion Equation. In one limiting case, the viscous forces may completely predominate over the inertial effects, so that the equation becomes

$$\frac{\partial^2\psi}{\partial x^2} = \kappa^2\frac{\partial\psi}{\partial t}; \quad \kappa^2 = \frac{R}{T} \tag{2.1.26}$$

This equation will be encountered many times in this book. Since it also represents the behavior of some solute diffusing through a solvent (where ψ is the density of the solute), it is usually called the *diffusion equation*.

As with the wave equation (2.1.9) the tendency is to straighten out the curvature; however, here the *velocity* of any part of the string is proportional to but opposite in sign to the curvature of the part, whereas in the wave equation it is the *acceleration* that is proportional and opposite in sign to the curvature. In short, we are essentially dealing with an equilibrium condition. In the wave equation a curved portion continually increases in velocity until it is straightened out and only then starts slowing down, thus ensuring oscillatory motion. But with the diffusion equation the velocity of any portion comes to zero when this portion is finally straightened out, so there is no oscillatory motion. One would expect this behavior of a string of no mass in a viscous fluid, for the damping is more than critical.

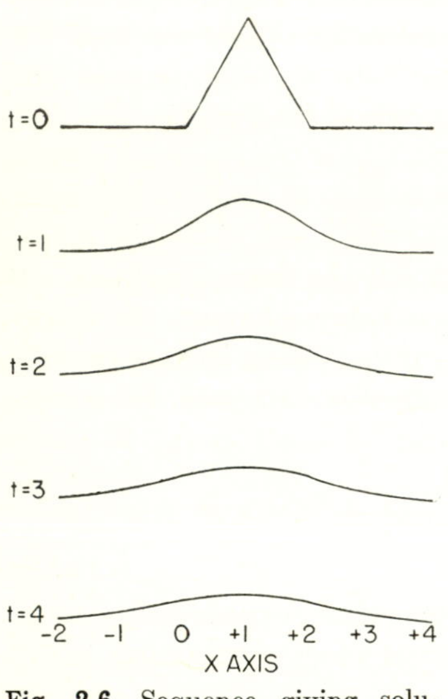

Fig. 2.6 Sequence giving solutions of diffusion equation after initial shape as shown at top.

In the case of the wave equation the general solution could be expressed as a superposition of two waves in opposite direction, $f(x + ct) + F(x - ct)$, due to the symmetrical relation between x and t in the equation. In the diffusion equation this symmetry is not present, and there is no simple form for the general solution. Here also there is a difference between the positive and negative time direction, due to the fact that the time derivative is a first derivative whereas there is a second derivative in the wave equation.

For instance, if the string has a sinusoidal shape $Ae^{i\omega x/c}$, then for the wave equation the time dependence is also sinusoidal, $e^{-i\omega t}$. But for the diffusion equation the time-dependent term is $e^{-(\omega/\kappa c)^2 t}$, which is not symmetrical with respect to time. For positive time the sinusoidal shape damps out exponentially, but, looking backward in time, we see that the wave amplitude increases without limit as t is made more and more negative. The smaller the wavelength of the fluctuations (*i.e.*, the larger ω is), the more rapidly are the functions damped out in positive time and the more rapidly do they increase in negative time. As we shall show in detail later, with the wave equation we can both predict future motion and reconstruct past motion from present conditions. With the diffusion equation prediction only is possible; attempts at reconstructing the past result only in divergent expressions.

Klein-Gordon Equation. A type of equation of some interest in quantum mechanics (it is used to describe a "scalar" meson) can also be exemplified by the flexible string with additional stiffness forces provided by the medium surrounding the string. If the string is embedded in a thin sheet of rubber, for instance (or if it is along the axis of a cylinder of rubber whose outside surface is held fixed), then in addition to the restoring force due to tension there will be a restoring force due to the rubber on each portion of string. If the displacement of the element dx of string at x is $\psi(x)$, this restoring force will be $-K\psi\,dx$, where K is a constant depending on the elastic properties and the geometrical distribution of the rubber.

Therefore, the equation of motion for the string is

$$\frac{1}{c^2}\frac{\partial^2\psi}{\partial t^2} = \frac{\partial^2\psi}{\partial x^2} - \mu^2\psi; \quad c^2 = \frac{T}{\rho}; \quad \mu^2 = \frac{K}{T} \tag{2.1.27}$$

where ρ is the linear density and T is the tension of the string. This equation is called the *Klein-Gordon equation* when it occurs in quantum mechanical problems. We note that, if c is the velocity of light, this equation is also invariant in form under a Lorentz transformation, as is the wave equation, so that solutions of the equation behave properly with regard to the space-time rotations of special relativity.

The reaction of this type of string to a unit, steady, transverse force, applied at point $x = \xi$, differs from the reaction of a string with only tension acting. The shape of the elastically braced string of infinite length which corresponds to Eq. (2.1.7) is

$$G(x|\xi) = \begin{cases} (1/2\mu)e^{\mu(x-\xi)}; & x < \xi \\ (1/2\mu)e^{\mu(\xi-x)}; & x > \xi \end{cases} \tag{2.1.28}$$

In the case of the usual string, with only tension, we had to consider the string to be a finite length, for the end supports were the only "anchors" to prevent the force from pushing the string by an indefinitely large amount. In the present case, however, the elastic medium in which the string is embedded absorbs nearly all of the thrust, and except for distances small compared with $1/\mu$ from either end, the exact position of the end supports are not important. Consequently we can write a Green's function which is independent of end points (*i.e.*, for an infinite string) for this case. The formula shows that a portion of the medium, spread out a distance of about $1/\mu$ from the point of application, supports the majority of the force, and the displacement of the string beyond this distance becomes quite small.

For applied transverse forces of arbitrary type distributed along the string, the corresponding shape of the string is obtained in the form of an integral over the Green's function of Eq. (2.1.28) of the general form given in Eq. (2.1.8).

If an elastically braced string is held under tension between rigid supports a distance L apart, it can vibrate with a sequence of normal modes of motion similar in shape but differing in frequency from the normal modes of the string without elastic support.

The Fourier series for the general free vibration is

$$\psi = \sum_{n=1}^{\infty} A_n \sin\left(\frac{n\pi x}{L}\right) e^{-i\omega_n t}; \quad \omega_n^2 = c^2\left[\left(\frac{n\pi}{L}\right)^2 + \mu^2\right]$$

The allowed frequencies are all larger than those for the usual string because of the quantity μ^2 which is proportional to the elastic constant K of the medium surrounding the string. This result is not surprising, for the added stiffness of the medium should increase the natural frequencies.

Forced Motion of the Elastically Braced String. A string embedded in rubber driven from one end by a transverse alternating force also exhibits certain characteristic differences of behavior compared with the ordinary string.

The solution for a wave traveling to the right only, suitable for the case of an infinitely long string, is

$$\psi = \begin{cases} A \exp\left[-x\sqrt{\mu^2 - (\omega/c)^2} - i\omega t\right]; & \omega^2 < \mu^2 c^2 = K/\rho \\ A \exp\left\{i(\omega/c)\left[x\sqrt{1 - (\mu c/\omega)^2} - ct\right]\right\}; & \omega^2 > K/\rho \end{cases} \tag{2.1.29}$$

At very high driving frequencies the wave motion is very similar to

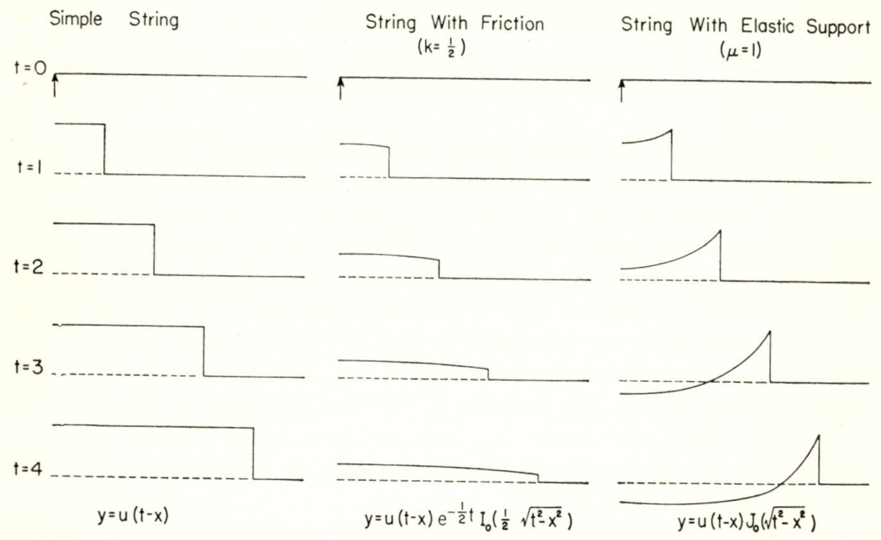

Fig. 2.7 Three sequences showing successive shapes of different strings when struck laterally at one end at $t = 0$.

the ordinary string, except that the wave velocity is always somewhat larger than c, by a factor $1/\sqrt{1 - (\mu c/\omega)^2}$. Here again the additional elastic forces tend to "speed up" the waves. The wave number $\lambda^{-1} = \sqrt{(\omega/c)^2 - \mu^2}$ is no longer a linear function of the frequency. Thus the string plus bracing will behave like a *dispersive medium.* A general wave shape composed of waves having different values of ω will no longer hold together but will diffuse from its original shape into a succession of shapes more and more dispersed over the entire length of the string. This behavior is, of course, strikingly different from the behavior of a wave form on an ordinary string which is not braced, for then the medium is not dispersive and the form will not change during propagation.

At low frequencies the effect of the elastic medium predominates over the inertial effect of the mass of the string, and for frequencies less than $(1/2\pi)\sqrt{K/\rho}$ there is no true wave motion at all, the string moving back and forth all in phase by an amount which is largest near the driving force and which decreases rapidly farther from the driving end.

The wave impedance for this type of string is obtained in a manner similar to that which obtained Eq. (2.1.13):

$$Z_\omega = \begin{cases} i(T/\omega)\sqrt{\mu^2 - (\omega/c)^2}; & \omega < \mu c \\ \rho c\sqrt{1 - (\mu c/\omega)^2}; & \omega > \mu c \end{cases} \tag{2.1.30}$$

At high frequencies the wave impedance is real and nearly equal to the value ρc for the simple string. As the frequency is diminished, however, the wave impedance decreases and the wave velocity increases until, at $\omega = \mu c = \sqrt{K/\rho}$, the wave impedance is zero and the wave velocity is infinite. This is the resonance frequency for the mass of the string and the elasticity of the medium. Below this frequency the impedance is imaginary, similar to a stiffness reactance, and there is no true wave motion.

Recapitulation. We have discussed the motions of a flexible string in detail for several reasons. In the first place our procedure in studying the string is a simple example of what will be our discussion of other equations for fields. We shall in each case discuss the various equations which result when one force after another is allowed to become predominant; thus by the study of the various limiting cases we shall arrive at a fairly complete understanding of the most general case. Second, the motions of different sorts of strings are easily picturable representations, in the simplest terms, of the solutions of a number of important partial differential equations, which turn up in many contexts and which represent many physical phenomena. Many of the other manifestations of the same equations correspond to physical situations which are much harder to visualize. In the third place, the techniques of solution which have been touched upon here and related to the physical problem have useful application in many other cases, and reference to the simple picture of the string will help our understanding of the others.

We shall now broaden the scope of our discussions and study a number of typical physical phenomena of considerable interest and importance in order to show what types of fields can be used for their picturization and what partial differential equations they must satisfy.

2.2 Waves in an Elastic Medium

For our first three-dimensional analysis of the interrelation between physical phenomena and differential equations for fields, we shall return to the problem of the behavior of an elastic medium (which was begun in Sec. 1.6) to take up the problem of wave motion in the medium. As in the case of the string, we assume that the displacements of the medium are small and that we are not concerned with translation or rotation of the medium as a whole. The displacement $\mathbf{s}(x,y,z;t)$ of an element $dx\,dy\,dz$ of the medium at x, y, z and at time t is small, and its rotation due to the strain is also small. The inertial reaction of the element to an acceleration of $\mathbf{s}$ is $(\partial^2\mathbf{s}/\partial t^2)\rho\,dx\,dy\,dz$, where ρ is the density of the medium.

In Sec 1.6 we defined the stress dyadic $\mathfrak{T} = \mathbf{F}_x\mathbf{i} + \mathbf{F}_y\mathbf{j} + \mathbf{F}_z\mathbf{k} = \mathbf{i}\mathbf{F}_x + \mathbf{j}\mathbf{F}_y + \mathbf{k}\mathbf{F}_z$ by saying that the force across a surface element $d\mathbf{A}$ of the medium is $\mathfrak{T} \cdot d\mathbf{A}$. For instance, the force across the face $dy\,dz$ of the element, perpendicular to the x axis, is $\mathbf{F}_x\,dy\,dz$. Therefore, the net force on the element $dx\,dy\,dz$ due to the difference in $\mathbf{F}_x$ from one face $dy\,dz$ to the opposite face of the element is $dx\,(\partial\mathbf{F}_x/\partial x)\,dy\,dz$, and the net force due to forces acting on all faces of the element is thus $\nabla \cdot \mathfrak{T}\,dx\,dy\,dz$.

But in Eq. (1.6.28) we showed that the stress dyadic is related to the strain dyadic $\mathfrak{S}$ by the relation $\mathfrak{T} = \lambda|\mathfrak{S}|\mathfrak{I} + 2\mu\mathfrak{S}$, where μ is the shear modulus of the medium and $(\lambda + \frac{2}{3}\mu)$ is its compression modulus (ratio of isotropic pressure to fractional rate of decrease of volume). From Eq. (1.6.21) we have the relation between $\mathfrak{S}$ and the displacement $\mathbf{s}$ given symbolically by $\mathfrak{S} = \frac{1}{2}(\nabla\mathbf{s} + \mathbf{s}\nabla)$. Putting all these equations together we finally arrive at the equation of motion for the medium under the influence of its own elastic restoring forces:

$$\rho\frac{\partial^2\mathbf{s}}{\partial t^2} = \nabla \cdot [\lambda\mathfrak{I}\,\mathrm{div}\,\mathbf{s} + \mu\nabla\mathbf{s} + \mu\mathbf{s}\nabla] = (\lambda + \mu)\,\mathrm{grad}\,\mathrm{div}\,\mathbf{s} + \mu\,\mathrm{div}\,\mathrm{grad}\,\mathbf{s}$$
$$= (\lambda + 2\mu)\,\mathrm{grad}\,\mathrm{div}\,\mathbf{s} - \mu\,\mathrm{curl}\,\mathrm{curl}\,\mathbf{s} \qquad (2.2.1)$$

where we have used Eq. (1.5.12) to rearrange the vector operators.

Longitudinal Waves. The form of the equation of motion suggests that at least part of the vector $\mathbf{s}$ may be expressed in terms of the gradient of a scalar potential ψ, since the equation then simplifies considerably: When $\mathbf{s} = \mathrm{grad}\,\psi$, the equation for ψ,

$$\mathrm{div}\,\mathrm{grad}\,\psi = \nabla^2\psi = \frac{1}{c_c^2}\frac{\partial^2\psi}{\partial t^2};\quad c_c^2 = \frac{\lambda + 2\mu}{\rho} \qquad (2.2.2)$$

is just the *wave equation* for the scalar wave potential ψ, the three-dimensional generalization of Eq. (2.1.9). The wave velocity c_c is greater the larger are the moduli λ and μ (*i.e.*, the stiffer is the medium) and is smaller the larger ρ is (*i.e.* the more dense is the medium).

When the solution of Eq. (2.2.1) is a gradient of a scalar, the dyadic operator $\mathfrak{D} = \nabla(\nabla\psi)$ is symmetric, rotation dyadic $\mathfrak{R}$ is zero and $\mathfrak{D} = \mathfrak{S}$, the pure strain dyadic. For such a solution there is no twisting of the medium, only stretching and squeezing. Waves of this sort are called *longitudinal* or compressional waves. They can be propagated even in liquid and gaseous media, where the shear modulus μ is zero.

But a gradient of a scalar potential is certainly not the most general vector field possible for the strain displacement **s**, as was shown on page 53. The most general vector field requires three scalar functions of position to specify, one for each component, whereas the gradient of a scalar is specified by a single function, the potential. Consequently, two more scalar functions of position are needed to specify the most general solution of Eq. (2.2.1).

Of course, we could set up equations for each of the rectangular components of **s**, but this would result in three equations, each containing the three components, which would have to be solved simultaneously—a cumbersome procedure.

Transverse Waves. It would be much better to utilize some of the properties of the vector operator ∇ to obtain the other solutions, as we did for the gradient of the potential: since the curl of a gradient is zero, one term in the equation dropped out and the wave equation for the scalar potential resulted. This result [plus the results of Eq. (1.5.15)] suggests that we try the curl of some vector, for the divergence of a curl is zero and therefore the divergence term would drop out. Accordingly we let another solution for **s** be curl **A**, and the resulting equation for **A** is

$$-\text{ curl curl } \mathbf{A} = c_s^2(\partial^2\mathbf{A}/\partial t^2); \quad c_s^2 = \mu/\rho \qquad (2.2.3)$$

which is also a wave equation, as we shall later demonstrate. The wave velocity c_s for this wave is smaller than the velocity for longitudinal waves, being proportional to the square root of the shear modulus μ instead of the combination $\lambda + 2\mu$. It suggests that this part of the solution is a *shear wave*, which indeed turns out to be the case. For with this type of displacement the dilation $\theta = \text{div } \mathbf{s}$ [see Eq. (1.6.23)] is zero, so there is no expansion or contraction, and therefore the strain must be a type of shear. We shall usually call this type of wave the *transverse wave*.

This separation of the general solution into a longitudinal part, which is the gradient of a scalar potential ψ, plus a transverse part, which is the curl of a vector potential **A** as suggested on page 53, is a neat one, for these two waves travel at different speeds and any other separation of the solution would result in waves of both velocities being part of both solutions, certainly a more clumsy procedure.

But at first sight the two solutions we have obtained appear to be redundant. We mentioned earlier that only three independent functions of position are needed to give the most general solution for the vector **s**, but here we appear to have four: one for the scalar potential and three for the components of the vector potential. This redundancy is only apparent, however, for we do not use all the degrees of freedom of the vector potential **A**. The process of taking the curl of **A** to obtain **s** discards a part of **A**, namely, the part which can be expressed as a gradient of a scalar, and uses only that part which has zero divergence. Therefore, the part of **A** which is used to contribute to the general solution for s involves only two independent functions of position, and these, with the scalar potential, make up the requisite three.

To put the whole argument in another form, any vector solution of Eq. (2.2.1) can be split into two parts: a *longitudinal part* having zero curl, which can always (see page 53) be represented as the gradient of a scalar potential, and a *transverse part* having zero divergence, which can always (see page 54) be represented as the curl of a vector potential. Equation (2.2.2) shows that, if the solution starts out as a longitudinal one, it will continue to be longitudinal or, if it starts out transverse, it will remain transverse as long as the quantities λ and μ have everywhere the same values. If λ or μ or both change abruptly at a boundary surface or change continuously in a region of space, then wave reflection will occur and the longitudinal and transverse waves may become intermingled.

In the present case the longitudinal part corresponds to wave motion of one velocity and the transverse part corresponds to wave motion of another, lesser velocity. Actually there are two independent transverse parts. One of these may be taken to be the curl of some solution **A** of Eq. (2.2.3) [incidentally the curl of a solution of Eq. (2.2.3) is also a solution of Eq. (2.2.3), as may be quickly verified]; this will be called the *first transverse solution*. The other transverse part may be taken to be the curl of the first solution (which is proportional to the zero-divergence part of **A** itself, as may be quickly verified); this will be called the *second transverse solution* of Eq. (2.2.1).

Wave Motion in Three Dimensions. Waves on a simple string are only of two general types: ones which travel to the right, represented by the general function $F(x - ct)$, and ones which travel to the left, represented by $f(x + ct)$. In three dimensions many more types of wave motion are possible. Confining ourselves for the moment to the scalar wave potential φ, we, of course, can have a simple generalization of the one-dimensional wave,

$$\varphi = f(\mathbf{a} \cdot \mathbf{r} - c_c t) \tag{2.2.4}$$

where $\mathbf{r} = x\mathbf{i} + y\mathbf{j} + z\mathbf{k}$ and $\mathbf{a}$ is a unit vector in some arbitrary direction given by the spherical angles θ and ϕ (see Fig. 2.8). The wave motion here is all in one direction, and the comments made in Sec. 2.1 concerning one-dimensional waves need no modification to apply here. Such waves are called *plane waves* for reasons which will shortly become apparent.

Inherent in our thinking of three-dimensional waves is the motion of a *wave front.* Crests and troughs of a wave often maintain their identity as they move along, which can be represented by surfaces everywhere perpendicular to the direction of wave motion and which move with the wave velocity c. These surfaces are called *surfaces of constant phase* or simply phase surfaces. For the simple plane wave form $f(\mathbf{a} \cdot \mathbf{r} - ct)$

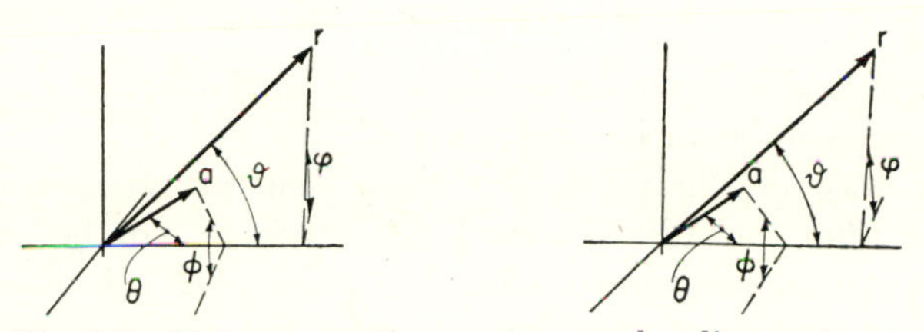

Fig. 2.8 Unit propagation vector **a** and radius vector **r**.

the surfaces are the planes $\mathbf{a} \cdot \mathbf{r} =$ constant, perpendicular to the unit vector **a**, which gives the direction of wave motion. If the wave is a sinusoidal one, represented by the complex exponential term $\psi = Ae^{i(\omega/c)(\mathbf{a}\cdot\mathbf{r}-ct)}$ for all points on one of the surfaces, the wave function has the same value of the phase angle of the complex exponential (which is why the surfaces are called phase surfaces).

We can ask whether there are other types of waves in three dimensions having crests and troughs which maintain their identity as the wave moves along. A bit of investigation will suffice to convince one that only plane waves of the form given in Eq. (2.2.4) maintain their shape and size completely unchanged as they travel. It is possible to have waves, other than plane, which keep their shape but not their size; these have the form

$$\psi = A(x,y,z)f[\varphi(x,y,z) - ct] \tag{2.2.5}$$

The function f provides for the motion of the wave, and the surfaces $\varphi =$ constant are the surfaces of constant phase; the factor A provides for the change in size of the wave from point to point.

Substituting this form into the wave equation $\nabla^2\psi = (1/c^2)(\partial^2\psi/\partial t^2)$ results in the equation

$$f\nabla^2 A + (f'/A) \operatorname{div} [A^2 \operatorname{grad} \varphi] + Af''[\operatorname{grad}^2 \varphi - 1] = 0$$

where the primes indicate differentiation of f with respect to its argument. If f is to be any arbitrary function of its argument $(\varphi - ct)$, the coefficients of f, f', and f'' must each be equal to zero:

$$\nabla^2 A = 0; \quad [\operatorname{grad} \varphi]^2 = 1; \quad \operatorname{div} [A^2 \operatorname{grad} \varphi] = 0 \tag{2.2.6}$$

The last equation is equivalent to stating that the vector $A^2 \operatorname{grad} \varphi$ is equal to the curl of some vector, and the second equation states that grad φ is a unit vector. These are stringent limitations on A and φ, and not many solutions can be found. In other words, not many types of waves maintain their shape as they move through space.

One solution of these equations has spherical wave fronts, $\varphi = r = \sqrt{x^2 + y^2 + z^2}$, and has an amplitude $A = B/r$ which varies inversely with the distance r from the center of the wave (B is a constant). The solution

$$\psi = (B/r)f(r - ct)$$

represents a spherical wave radiating out from the center $r = 0$. Naturally there can also be an ingoing wave $(D/r)F(r + ct)$.

Another way of analyzing the same general problem consists in determining what curvilinear coordinate surfaces can be wave fronts. If the wave equation in some coordinate system has solutions which are functions of only one of the three coordinates, then a traveling-wave solution can be formed from these solutions which will have one set of the coordinate surfaces as its surfaces of constant phase.

Suppose that we choose an orthogonal, curvilinear coordinate system ξ_1, ξ_2, ξ_3, with scale factors h_1, h_2, h_3 and unit vectors $\mathbf{a}_1, \mathbf{a}_2, \mathbf{a}_3$. According to Eq. (1.5.11) the wave equation for ψ in terms of these coordinates is

$$\frac{1}{h_1h_2h_3}\sum_n \frac{\partial}{\partial \xi_n}\left(\frac{h_1h_2h_3}{h_n^2}\frac{\partial \psi}{\partial \xi_n}\right) = \frac{1}{c^2}\frac{\partial^2 \psi}{\partial t^2} \tag{2.2.7}$$

To simplify matters we separate off the time dependence in the exponential factor $e^{-i\omega t}$. If, in addition, we can separate the space part of ψ into three factors, each dependent on only one coordinate, the equation is said to be *separable* and one or another of the three families of coordinate surfaces can be the family of phase surfaces of a wave. In other words, if the equation

$$\frac{1}{h_1h_2h_3}\frac{\partial}{\partial \xi_1}\left(\frac{h_2h_3}{h_1}\frac{\partial \psi}{\partial \xi_1}\right) + \left(\frac{\omega}{c}\right)^2 \psi = 0$$

will yield solutions which are functions of ξ_1 alone, then the wave equation is separable for the coordinate ξ_1.

If one solution of this equation can be found, two independent solutions $y(\xi_1)$ and $Y(\xi_1)$ can be obtained (this will be proved in Chap. 5), and the combination

$$y + iY = A(\xi_1)e^{i(\omega/c)\varphi(\xi_1)}$$

will serve to give us an expression for a simple harmonic wave having the coordinate surfaces ξ_1 = constant as the surfaces of constant phase of the wave:

$$\psi = (y + iY)e^{-i\omega t} = A(\xi_1)e^{i(\omega/c)[\varphi(\xi_1) - ct]} \tag{2.2.8}$$

This form of wave is more specialized than that given in Eq. (2.2.5), since we have contented ourselves here with sinusoidal dependence on time. In exchange for the simplification the limiting requirements on A and φ are not so stringent as in Eq. (2.2.6). For instance grad φ need not be a unit vector, which corresponds to the statement that the surfaces of constant phase do not travel with the velocity c everywhere. The functions A and φ may depend on ω, so that the shape of the wave may differ for differing frequencies. Nevertheless we again find that a form of wave front which allows wave propagation with reasonable permanence of wave form is not at all common; as we shall see in Chap. 5, only a few coordinate systems have a separable equation. The wave equation, it turns out, is rather particular about the shape of the wave fronts it allows.

Further discussion of these points is not profitable here. It has been sufficient to point out that there is a close connection between the property of a coordinate system of allowing separable solutions of the wave equation (solutions consisting of factors, each functions of only one coordinate) and the possibility for the corresponding coordinate surfaces to be surfaces of constant phase for some wave. In Chap. 5 we shall deal with the problem of separability in more detail.

Vector Waves. We must now return to the shear waves which cannot be represented by a scalar wave function but which can be represented by a divergenceless vector potential, satisfying the equation

$$\text{curl curl } \mathbf{A} + (1/c)^2(\partial^2\mathbf{A}/\partial t^2) = 0$$

These also can have plane wave solutions:

$$\mathbf{A} = (\mathbf{B} \times \mathbf{a})f(\mathbf{a} \cdot \mathbf{r} - ct) \tag{2.2.9}$$

where $\mathbf{B}$ is any constant vector and, therefore, $(\mathbf{B} \times \mathbf{a})$ is a constant vector perpendicular to the unit vector $\mathbf{a}$, which determines the direction of propagation of the wave. Since the magnitude of $\mathbf{A}$ is independent of position along a line in the direction of $\mathbf{A}$ (*i.e.*, since the gradient of f is perpendicular to $\mathbf{B} \times \mathbf{a}$), the divergence of $\mathbf{A}$ is zero, as was required. The curl of $\mathbf{A}$ is a vector perpendicular both to $\mathbf{A}$ and to $\mathbf{a}$,

$$\text{curl } \mathbf{A} = (\text{grad } f) \times (\mathbf{B} \times \mathbf{a}) = [\mathbf{B} - \mathbf{a}(\mathbf{a} \cdot \mathbf{B})]f'$$

and the curl of this vector is again parallel to $\mathbf{A}$,

$$\text{curl curl } \mathbf{A} = -(\mathbf{B} \times \mathbf{a})f'' = -(1/c^2)(\partial^2\mathbf{A}/\partial t^2)$$

as, of course, it must be in order to satisfy the vector wave equation (2.2.3). The directions of $\mathbf{A}$ and curl $\mathbf{A}$ are both perpendicular to the direction $\mathbf{a}$ of propagation of the wave, which is the reason for calling these waves transverse.

There is also a vector potential representing a spherical vector wave, analogous to the scalar spherical wave $(B/r)f(r - ct)$ mentioned earlier. If $\mathbf{a}_r$ is a unit vector pointed along the radius r, $\mathbf{a}_\varphi$ a unit vector perpendicular to $\mathbf{a}_r$ and to the axis of the spherical coordinates, and $\mathbf{a}_\vartheta = \mathbf{a}_\varphi \times \mathbf{a}_r$ another unit vector perpendicular to both, then the vector $\mathbf{A} = (\mathbf{a}_\vartheta/r)f(r \pm ct)$ is a solution of the vector wave equation which is satisfactory except along the spherical axis $\vartheta = 0$. For the outgoing wave, for instance,

$$\operatorname{curl} \mathbf{A} = (\mathbf{a}_\varphi/r)f'(r - ct)$$

and

$$\operatorname{curl}\operatorname{curl} \mathbf{A} = -(\mathbf{a}_\vartheta/r)f''(r - ct) = -(1/c^2)(\partial^2\mathbf{A}/\partial t^2)$$

It is obvious that the vector curl $\mathbf{A}$ is also a solution of the vector wave equation, so that the most general outgoing spherical vector wave is

$$(\mathbf{a}_\vartheta/r)f(r - ct) + (\mathbf{a}_\varphi/r)F(r - ct)$$

The more complex problem of the separability of the vector wave equation will be discussed later.

Integral Representations. More general types of waves can be constructed by adding up plane waves in different directions. As shown in Fig. 2.8 the vector $\mathbf{a}(\theta,\phi)$ is the unit propagation vector pointed in the direction defined by the spherical angles θ, ϕ and $\mathbf{r}$ is the radius vector of length r, with direction defined by the angles ϑ and φ. The most general sort of scalar wave can be represented by the integral

$$\psi = \int d\phi \int \sin\theta\, d\theta\, f[\phi, \theta; \mathbf{r} \cdot \mathbf{a}(\theta,\phi) - ct] \tag{2.2.10}$$

where f is a traveling wave of shape depending on the angles θ and ϕ. The limits of integration are usually from 0 to 2π for ϕ and from 0 to π for θ, but they may extend to imaginary or complex values [such as from 0 to $(\pi/2) + i\infty$ for θ].

The most general vector wave function can be formed in a similar manner:

$$\mathbf{A} = \int d\phi \int \sin\theta\, d\theta\, \mathbf{F}[\phi, \theta; \mathbf{r} \cdot \mathbf{a}(\theta,\phi) - ct] \tag{2.2.11}$$

where $\mathbf{F}(\phi,\theta;z)$ is a vector function of ϕ and θ and z which is pointed in a direction perpendicular to $\mathbf{a}(\theta,\phi)$. Since every element in the integrand is a transverse wave, the result must have zero divergence.

One can also express more specialized waves in this same manner. In the very important case of simple harmonic waves, for instance, with time factor $e^{-i\omega t}$, the expression for the scalar wave becomes

$$\psi = \int d\phi \int Y(\phi,\theta) e^{i(\omega/c)(r\cos\Omega - ct)} \sin\theta\, d\theta \tag{2.2.12}$$

where $r \cos \Omega = r[\cos\theta \cos\vartheta + \sin\theta \sin\vartheta \cos(\phi - \varphi)] = \mathbf{r} \cdot \mathbf{a}(\theta,\phi)$ and $Y(\phi,\theta)$ is some function of the spherical angles. For the vector solution $\mathbf{Y}$ is a vector perpendicular to $\mathbf{a}$ for every value of θ and ϕ. In future chapters we shall find it extremely valuable to express all the solutions of the wave equation in such an integral form.

Stress and Strain. To return to physics after this incursion into mathematics, it will be well to compute the stresses in the medium corresponding to the various types of wave motion. For a compressional (longitudinal) wave traveling to the right along the x axis, the scalar potential is $\psi = f(x - c_c t)$ and the displacement of the medium at point x, y, z at time t is

$$\mathbf{s} = \text{grad}\,\psi = \mathbf{i}f'(x - c_c t); \quad f'(\xi) = (d/d\xi)f(\xi) \qquad (2.2.13)$$

The strain dyadic is $\mathfrak{S} = \frac{1}{2}[\nabla\mathbf{s} + \mathbf{s}\nabla] = \mathbf{ii}f''(x - c_c t)$, and the stress dyadic is

$$\mathfrak{T} = \lambda\mathfrak{J}\,\text{div}\,\mathbf{s} + \mu(\nabla\mathbf{s} + \mathbf{s}\nabla) = [(\lambda + 2\mu)\mathbf{ii} + \lambda(\mathbf{jj} + \mathbf{kk})]f''(x - ct); \quad f''(\xi) = (d^2/d\xi^2)f(\xi) \qquad (2.2.14)$$

In other words the force across a unit area perpendicular to the x axis is in the x direction and of magnitude $(\lambda + 2\mu)f''$, whereas the force across a unit area parallel to the x axis is perpendicular to the area and equal to $\lambda f''$. The motion is entirely in the x direction, and the tractile forces are all normal; no shear is present.

For a shear (transverse) wave traveling to the right along the x axis, with motion parallel to the z axis, the vector potential is $\mathbf{A} = \mathbf{j}F(x - c_s t)$ and the displacement of the medium at point x, y, z at time t is

$$\mathbf{s} = \text{curl}\,\mathbf{A} = \mathbf{k}F'(x - c_s t); \quad F'(\xi) = (d/d\xi)F(\xi) \qquad (2.2.15)$$

The strain dyadic is $\mathfrak{S} = (\mathbf{ik} + \mathbf{ki})F''(x - c_s t)$, and the stress dyadic is

$$\mathfrak{T} = \mu(\nabla\mathbf{s} + \mathbf{s}\nabla) = \mu[\mathbf{ik} + \mathbf{ki}]F''(x - c_s t); \quad F''(\zeta) = (d^2/d\zeta^2)F(\zeta) \qquad (2.2.16)$$

since div $\mathbf{s}$ is zero. In this case the force across a unit area perpendicular to the x axis is in the z direction and of magnitude $\mu F''$; the force across one perpendicular to the z axis is in the x direction and also equal to $\mu F''$. There is no tractile force across a surface perpendicular to the y axis. This stress is, of course, a simple shear in the x, z plane.

Wave Energy and Impedance. To find the potential energy stored in a certain volume of medium when its strain dyadic is $\mathfrak{S} = \frac{1}{2}(\nabla\mathbf{s} + \mathbf{s}\nabla)$ and its stress dyadic is $\mathfrak{T} = \lambda\mathfrak{J}|\mathfrak{S}| + 2\mu\mathfrak{S}$, we first find the increase in potential energy when the displacement $\mathbf{s}$ of the medium at x, y, z is increased by the small amount $\delta\mathbf{s}$. The work done by the stress forces on the medium in the volume can be computed in terms of the scalar product of the tractive force $(\mathfrak{T} \cdot d\mathbf{A})$ across each element $d\mathbf{A}$ of the surface of the volume and the displacement $\delta\mathbf{s}$ of the element:

$$\delta w = \iint \delta\mathbf{s} \cdot (\mathfrak{T} \cdot d\mathbf{A}) = \iiint [\text{div}\,(\mathfrak{T} \cdot \delta\mathbf{s})]\,dv$$

where the first integral is over the surface bounding the volume in question and the second integral is over the volume itself. We have made use of Gauss's theorem, Eq. (1.4.7), to derive the second integral from the first.

However, a juggling of components shows that, for any vector **A** and dyadic $\mathfrak{B}$, the following formula holds:

$$\operatorname{div}(\mathfrak{B}\cdot\mathbf{A}) = (\nabla\cdot\mathfrak{B})\cdot\mathbf{A} + \mathfrak{B}:(\nabla\mathbf{A}); \quad \text{where } \mathfrak{B}:\mathfrak{D} = \sum_{mn} B_{mn}D_{nm} = |\mathfrak{B}\cdot\mathfrak{D}|$$

Therefore,

$$\iiint \operatorname{div}(\mathfrak{T}\cdot\delta\mathbf{s})\,dv = \iiint[(\nabla\cdot\mathfrak{T})\cdot\delta\mathbf{s} + \mathfrak{T}:(\nabla\delta\mathbf{s})]\,dv$$

Since $(\nabla\cdot\mathfrak{T})\,dv$ is the net force on the volume element dv, which is zero when the medium is in equilibrium (as it is when we measure potential energy), the integrand of the potential-energy integral becomes

$$\begin{aligned}\mathfrak{T}:(\nabla\delta\mathbf{s}) &= \mathfrak{T}:\delta[\tfrac{1}{2}(\nabla\mathbf{s} + \mathbf{s}\nabla)] = \mathfrak{T}:\delta\mathfrak{S} \\ &= [\lambda|\mathfrak{S}||\delta\mathfrak{S}| + 2\mu\mathfrak{S}:\delta\mathfrak{S}] = \delta[\tfrac{1}{2}\mathfrak{T}:\mathfrak{S}] = \delta[\tfrac{1}{2}|\mathfrak{T}\cdot\mathfrak{S}|]\end{aligned}$$

where, since $\mathfrak{T}$ is symmetric, $|\mathfrak{T}\cdot\nabla\mathbf{s}| = |\mathfrak{T}\cdot\mathbf{s}\nabla|$. This represents the increase of potential energy due to the increment of displacement $\delta\mathbf{s}$. It is clear, therefore, that the total potential energy due to the displacement field **s** is given by the volume integral

$$\begin{aligned}V &= \tfrac{1}{2}\iiint(\mathfrak{T}:\mathfrak{S})\,dv = \tfrac{1}{2}\iiint[\lambda|\mathfrak{S}|^2 + 2\mu\mathfrak{S}:\mathfrak{S}]\,dv \\ &= \tfrac{1}{2}\iiint\left\{\lambda(\operatorname{div}\mathbf{s})^2 + 2\mu\left[\left(\frac{\partial s_x}{\partial x}\right)^2 + \left(\frac{\partial s_y}{\partial y}\right)^2 + \left(\frac{\partial s_z}{\partial z}\right)^2\right]\right. \\ &\quad \left. + \mu\left[\left(\frac{\partial s_x}{\partial y} + \frac{\partial s_y}{\partial x}\right)^2 + \left(\frac{\partial s_x}{\partial z} + \frac{\partial s_z}{\partial x}\right)^2 + \left(\frac{\partial s_y}{\partial z} + \frac{\partial s_z}{\partial y}\right)^2\right]\right\}dv \qquad (2.2.17)\end{aligned}$$

The kinetic energy is, of course, the integral of $\frac{1}{2}\rho(\partial\mathbf{s}/\partial t)^2$ over the same volume. The total energy density in the medium is, therefore,

$$w = \tfrac{1}{2}\rho(\partial\mathbf{s}/\partial t)^2 + \tfrac{1}{2}|\mathfrak{T}\cdot\mathfrak{S}|; \quad W = \iiint w\,dv \qquad (2.2.18)$$

For the plane compressional and shear waves given in Eqs. (2.2.13) and (2.2.15) the energy densities turn out to be

$$\begin{aligned} w &= \tfrac{1}{2}\rho c_c^2[f'']^2 + \tfrac{1}{2}(\lambda + 2\mu)[f'']^2 = (\lambda + 2\mu)[f''(x - c_ct)]^2 \\ \text{and}\quad w &= \tfrac{1}{2}\rho c_s^2[F'']^2 + \tfrac{1}{2}\mu[F'']^2 = \mu[F''(x - c_st)]^2 \end{aligned} \qquad (2.2.19)$$

The flow of energy across any given closed surface may be obtained by finding the rate of change of the total energy inside the surface. Using Eq. (2.2.1) in the process, we find that

$$\frac{\partial W}{\partial t} = \iiint \left[\rho \left(\frac{\partial \mathbf{s}}{\partial t}\right) \cdot \left(\frac{\partial^2 \mathbf{s}}{\partial t^2}\right) + \left| \mathfrak{T} \cdot \frac{\partial \mathfrak{S}}{\partial t} \right| \right] dv$$

$$= \iiint \left[\left(\frac{\partial \mathbf{s}}{\partial t}\right) \cdot \nabla \cdot \mathfrak{T} + \left| \mathfrak{T} \cdot \nabla \left(\frac{\partial \mathbf{s}}{\partial t}\right) \right| \right] dv$$

$$= \iiint \operatorname{div} \left[\left(\frac{\partial \mathbf{s}}{\partial t}\right) \cdot \mathfrak{T} \right] dv = \iint \left[\left(\frac{\partial \mathbf{s}}{\partial t}\right) \cdot \mathfrak{T} \right] \cdot d\mathbf{A}$$

The last integral, being a surface integral, must equal the flow of energy in through the surface to cause the increase in W. With a minus sign in front of it, it is the net *outflow* of energy across the closed surface.

Therefore, the vector representing the energy flow density in a medium carrying elastic waves is

$$\mathbf{S} = -(\partial \mathbf{s}/\partial t) \cdot \mathfrak{T} \tag{2.2.20}$$

This is not a surprising result. The quantity $\partial \mathbf{s}/\partial t$ is the velocity of the particle of the medium at x, y, z. The tractile force across an element of surface $d\mathbf{A}_u$ perpendicular to $\partial \mathbf{s}/\partial t$ is $\mathfrak{T} \cdot d\mathbf{A}_u$, and the expression for power is force times velocity. Since the dimensions of $\mathfrak{T}$ are force per unit area, the dimensions of $\mathbf{S}$ are power per unit area.

For the plane longitudinal wave given in Eqs. (2.2.13) and (2.2.14) the transmitted power is

$$\mathbf{S} = \mathbf{i}(\lambda + 2\mu)c_c[f''(x - c_c t)]^2 \tag{2.2.21}$$

and for the transverse plane wave given in Eqs. (2.2.15) and (2.2.16) it is

$$\mathbf{S} = \mathbf{i}\mu c_s[F''(x - c_x t)]^2 \tag{2.2.22}$$

The density of energy flow for a plane elastic wave is usually called the *intensity* of the wave. We see in each case that the magnitude of the intensity is the energy density times the wave velocity. In a plane wave the energy moves along with the velocity of the wave.

In these cases we can consider the quantities $c_c f''$ and $c_s F''$, the amplitudes of velocity of the medium, as being analogous to an electric-current density and the quantities $(\lambda + 2\mu)f''$ and $\mu F''$, the amplitudes of the tractile forces, as being analogous to voltages. The product of the two gives power density. The ratio of the two would give a quantity which could be called the *impedance of the medium* for waves of the type considered. For compressional waves the impedance is $(\lambda + 2\mu)/c_c = \rho c_c$, and for shear waves it is $\mu/c_s = \rho c_s$.

2.3 *Motion of Fluids*

A fluid differs from an elastic solid in that it yields to a shearing stress. We cannot expect to relate the displacement of a fluid with the stress tensor, for if the displacement were kept constant, the shearing stress would vary with time, or if the shearing stress were kept constant, the displacement would vary with time. It requires a constant *rate* of shear to maintain a constant shearing force in a fluid.

This indicates (if it were not clear already!) that it is more convenient to express the behavior of a fluid in terms of velocities rather than displacements. Two types of description can be used; one which gives the velocity of each particle of the fluid at each instant of time and another which gives the fluid velocity at each point in space at each instant of time. In the first description the vector field follows the particles of fluid as they move around; in the second case the field is attached to a fixed coordinate system, the vector at a given point giving the velocity of that part of the fluid which is at that point at the time.

The two types of description of the motion of a fluid correspond in a distant way to the atomic and to the continuum picture of a fluid. An actual fluid, of course, is a collection of molecules, each moving under the influence of forces. Some of the forces are *internal*, due to other molecules nearby; the nature of these forces determines the compressibility of the fluid. Other forces are *external*, due to bodies at some distance, such as gravitational or electrical forces, which act more or less equally on all molecules in a given small region.

In a thoroughgoing analysis of the first type of description we would start by labeling each molecule by its position in space at $t = 0$. For a detailed analysis we should also have to know the initial velocity of each molecule before we could expect to determine in detail their subsequent motions. For many problems, however, it will suffice to know only the average position and velocity of the molecules in each element of volume (such as the one $dx\, dy\, dz$ at x_0, y_0, z_0) with dimensions large compared with molecular size but small compared with the total extent of the fluid considered. When these averages are obtained, the internal forces cancel out (except in determining the relation between pressure and density) and leave only the external forces acting on the portion of fluid in the element. By this averaging procedure we obtain equations for the gross motions of the fluid which disregard its detailed discontinuities and correspond to a continuous, nongranular approximation to the actual fluid. The discussion in Sec. 2.4 will show how this transition, from an overdetailed molecular picture to a smoothed-out, average picture for the fluid, is performed.

The second type of description usually starts immediately from the smoothed-out approximation. The average velocity of those fluid particles which are close to the fixed point x, y, z at time t is computed as a function of t, as though the fluid actually were continuous.

We shall choose the second method of representation, for it corresponds more closely to the types of fields studied in other parts of this chapter. The vector $\mathbf{v}(x,y,z,t)$ is the velocity of that portion of the fluid which happens to be at x, y, z at time t. The expression div $\mathbf{v}$ is the net outflow of fluid from the "region around x, y, z"; that is, $dx\,dy\,dz$ div $\mathbf{v}$ is the net outflow of fluid from the element $dx\,dy\,dz$. If div $\mathbf{v}$ is everywhere zero, the fluid is then said to be *incompressible*. The vector $\mathbf{w} = \frac{1}{2}$ curl $\mathbf{v}$ represents the circulation of fluid "around the point x, y, z"; it is called the *vorticity vector* of the fluid (see page 42). If $\mathbf{w}$ is everywhere zero, the flow of fluid is said to be *irrotational* (in which case the vector $\mathbf{v}$ can be expressed as the gradient of a scalar *velocity potential*).

This brings us back to the discussion of vector fields given in Sec. 1.2. As a matter of fact we used there the example of fluid flow to help us picture a vector field, and a number of terms, such as vorticity, flow lines, and net outflow, were chosen to further the analogy. We can now return to this point of view to obtain quantitative measures for the fluid motion.

For instance, the flow lines plot the average paths of the various particles of fluid. The differential equation for these lines is $dx/v_x = dy/v_y = dz/v_z$. The number of flow lines crossing a given surface, which is equal to the outflow integral $\int \mathbf{v} \cdot d\mathbf{A}$ across the surface, is also equal to the average flow of fluid across the surface, and so on. If there is no vorticity (*i.e.*, if curl $\mathbf{v} = 0$) and a velocity potential exists, the flow lines are everywhere perpendicular to the equipotential surfaces and constitute a natural coordinate system for the problem.

Equation of Continuity. Two general properties of the velocity field for a fluid should be mentioned before we go into details. One has to do with the relation between net outflow and change of density of fluid. If $\mathbf{v}$ is the fluid velocity and ρ is the fluid density at x, y, z, t, then $\rho\mathbf{v}$ is the vector representing the flow of mass per square centimeter and $dx\,dy\,dz$ div $(\rho\mathbf{v})$ is then the net outflow of mass from the volume element $dx\,dy\,dz$. Since matter is neither created nor destroyed in most of the cases considered, this net outflow of mass must equal the loss of mass $\rho\,dx\,dy\,dz$ of the fluid in the element. In other words

$$\partial\rho/\partial t = -\operatorname{div}(\rho\mathbf{v}) \tag{2.3.1}$$

This equation is called the *equation of continuity* for the fluid. From this equation it is obvious that for a fluid of constant density ρ (incompressible fluid) the net outflow div $\mathbf{v}$ must be zero.

In some problems it will be convenient to assume that fluid is being created (or destroyed) at some point or points. Such a point is called a *source* (or a *sink*) of fluid. Naturally the equation of continuity does not hold there.

The other general property of the velocity field is related to the fact that the coordinate system for the vector field does not move with the fluid. To find the rate of change of some property $F(\mathbf{r},t)$ of the fluid at or near a specified fluid particle (whose position is given by the radius vector $\mathbf{r}$ at time t) we cannot just compute the rate of change $\partial F/\partial t$ of F at the point x, y, z, for the particle does not usually stay at the one point. The change in F we are interested in is the difference between the value $F(\mathbf{r},t)$ at point x, y, z, where the fluid particle is supposed to be at time t, and the value $F(\mathbf{r} + \mathbf{v}\, dt,\ t + dt)$ at point $x + v_x\, dt,\ y + v_y\, dt,\ z + v_z\, dt$, which is where the particle is at time $t + dt$. This difference, when expanded out and the first-order terms kept, turns out to be $dF = [(\partial F/\partial t) + \mathbf{v} \cdot \nabla F]\, dt$. The rate of change of the property F *of the fluid*, which is denoted by the total derivative sign, is therefore given by the equation

$$\frac{dF}{dt} = \frac{\partial F}{\partial t} + \mathbf{v} \cdot \nabla F \tag{2.3.2}$$

in terms of the time rate of change of F at point x, y, z (given by the partial derivative of F) and the space dependence of F near x, y, z (given by the ∇F term).

For instance, the acceleration of the part of the fluid which is "at" x, y, z at time t is

$$\frac{d\mathbf{v}}{dt} = \frac{\partial \mathbf{v}}{\partial t} + \mathbf{v} \cdot \nabla \mathbf{v} = \frac{\partial \mathbf{v}}{\partial t} + \tfrac{1}{2}\nabla(\mathbf{v} \cdot \mathbf{v}) - \mathbf{v} \times \operatorname{curl} \mathbf{v}$$
$$= \frac{\partial \mathbf{v}}{\partial t} + \tfrac{1}{2}\nabla(v^2) - 2\mathbf{v} \times \mathbf{w} \tag{2.3.3}$$

The second form of this expression is obtained by a reshuffling of vector components, and the third form is obtained by substituting the vorticity vector $\mathbf{w}$ for $\frac{1}{2}$ curl $\mathbf{v}$. According to the discussion of page 41, the magnitude of $\mathbf{w}$ equals the angular velocity of the portion of fluid "near" x, y, z, and the direction of $\mathbf{w}$ is that along which a right-hand screw would move if it were turning with the fluid.

The rate of change of a scalar property of the fluid can also be computed. The rate of change of density of a given element of fluid, which happens to be "at" x, y, z at time t as it travels along, may also be calculated by the same method, with the following result:

$$\frac{d\rho}{dt} = \frac{\partial \rho}{\partial t} + \mathbf{v} \cdot \operatorname{grad} \rho \tag{2.3.4}$$

But the equation of continuity has that $\partial\rho/\partial t = -\operatorname{div}(\rho\mathbf{v})$, so that

$$(d\rho/dt) = -\operatorname{div}(\rho\mathbf{v}) + \mathbf{v} \cdot \operatorname{grad} \rho = -\rho \operatorname{div} \mathbf{v} \tag{2.3.5}$$

Solutions for Incompressible Fluids. When the fluid density ρ is everywhere constant, the equation determining $\mathbf{v}$ is just $\operatorname{div} \mathbf{v} = 0$. The most general solution of this can be expressed in terms of a scalar and vector potential (as shown on page 53).

$$\mathbf{v} = \operatorname{curl} \mathbf{A} + \operatorname{grad} \psi; \quad \nabla^2\psi = \operatorname{div} \operatorname{grad} \psi = 0 \tag{2.3.6}$$

The vector **A** can be any well-behaved vector field which satisfies the boundary conditions. The equation for the velocity potential ψ is called *Laplace's equation.* It will be discussed at great length later in the book. The flow lines, discussed on page 12, are, of course, perpendicular to the surfaces of constant velocity potential.

When there is no vorticity, $\mathbf{A} = 0$ and the velocity is completely determined by the scalar potential. If, in addition, the flow lines lie in

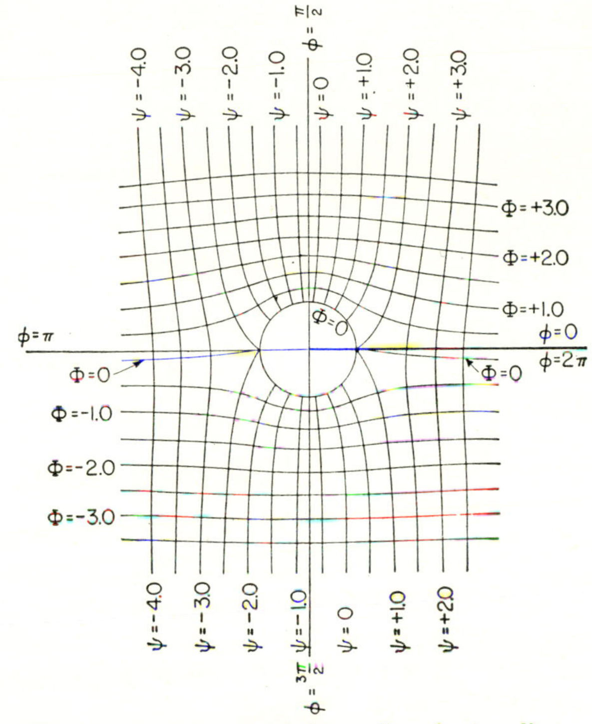

Fig. 2.9 Potential ψ and flow Φ lines for two-dimensional flow of incompressible fluid. Circulation is present, so there is a discontinuity in ψ at $\phi = 0$.

parallel planes, the velocity potential can be made a function of only two coordinates, and the motion is called two-dimensional flow. This special case has numerous important applications in aerodynamics. Here the flow lines and the potential lines make an orthogonal set of curvilinear coordinates in the two dimensions.

The equation for the flow lines is (see page 12)

$$dx/v_x = dy/v_y \quad \text{or} \quad -v_y\,dx + v_x\,dy = 0$$

Therefore if $v_y = -(\partial\Phi/\partial x)$ and $v_x = \partial\Phi/\partial y$, we have that

$$(\partial\Phi/\partial x)\,dx + (\partial\Phi/\partial y)\,dy = 0 \quad \text{or} \quad \Phi(x,y) = \text{constant}$$

along a stream line. The function Φ is called a stream function; it is related to the velocity potential ψ by the relations

$$\partial\Phi/\partial y = \partial\psi/\partial x, \quad \partial\Phi/\partial x = -(\partial\psi/\partial y)$$

which are called *Cauchy-Riemann equations* and which will be discussed in much greater detail in Chap. 4 in connection with functions of a complex variable.

We have mentioned earlier that the "density of flow lines" is a measure of the total flow and therefore of the velocity of the fluid. This can be quickly shown in the two-dimensional case, for the outflow integral $\int \mathbf{v} \cdot d\mathbf{A}$ between two stream lines $\Phi(x,y) = \Phi_2$ and $\Phi(x,y) = \Phi_1$ can be reduced to a line integral in the xy plane. The outflow integral concerned is between two planes parallel to the x, y plane a unit distance

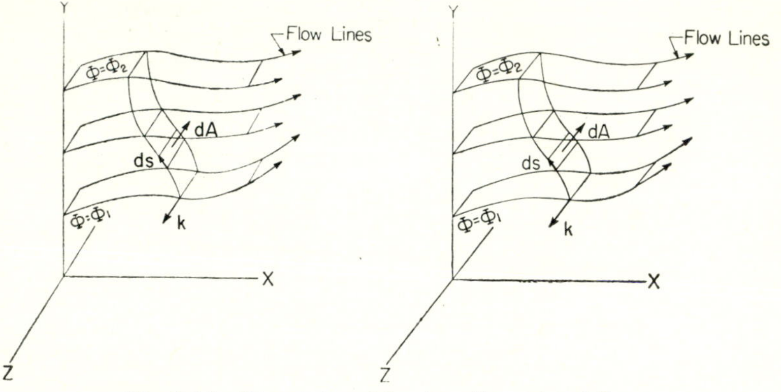

Fig. 2.10 Flow integral for two-dimensional flow.

apart, and the element of area $d\mathbf{A}$ can be a thin strip of unit length and of width equal to ds, where ds is the element of length along the path from Φ_1 to Φ_2 in the x, y plane.

The direction of $d\mathbf{A}$ is, of course, perpendicular to the direction of $d\mathbf{s}$; in fact $d\mathbf{A} = d\mathbf{s} \times \mathbf{k}$ where, of course, $d\mathbf{s}$ is always perpendicular to $\mathbf{k}$. The flow integral is then

$$\int^2 \mathbf{v} \cdot d\mathbf{A} = \int_1^2 \mathbf{v} \cdot (d\mathbf{s} \times \mathbf{k}) = \int_1^2 (\mathbf{v} \times d\mathbf{s}) \cdot \mathbf{k}$$
$$= \int_1^2 (v_x\, dy - v_y\, dx) = \int_1^2 d\Phi = \Phi_2 - \Phi_1$$

In other words the total flow of fluid along the region enclosed between the planes $z = 0$ and $z = 1$ and the surfaces defined by the flow lines 1 and 2 is just the difference between the values Φ_2 and Φ_1 of the flow function.

The usual boundary conditions in fluid flow are that the velocity is tangential to all bounding surfaces. When viscosity is important, we must require that the fluid immediately next to the surface move with the surface; *i.e.*, if the surface is at rest, even the tangential component of the velocity must go to zero at the boundary. If viscosity is not large, however, we may safely assume that the fluid may slip along the surface without appreciable drag, so that a finite tangential component next to the surface is allowed.

Examples. A few simple examples will perhaps clarify some of these statements and definitions. The scalar potential and related velocity field, given by the equations

$$\psi = -(Q/r); \quad \mathbf{v} = (Q/r^2)\mathbf{a}_r \tag{2.3.7}$$

have been shown on page 17 to be due to a point source of fluid at the origin ($r = 0$) in a fluid of infinite extent. As indicated in Eq. (1.2.9),

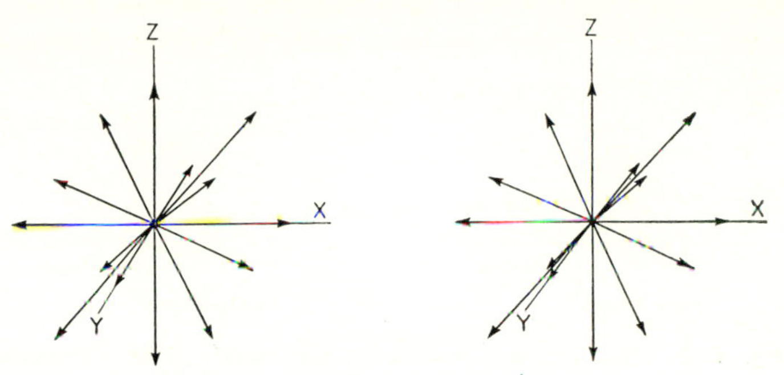

Fig. 2.11 Flow lines from a point source.

the total normal outflow from the source is $4\pi Q$, which is called the *strength of the source.* Since no vector potential enters here, there is no related vorticity vector $\mathbf{w} = \frac{1}{2}$ curl $\mathbf{v}$ (see page 153) and the flow is said to be *irrotational.*

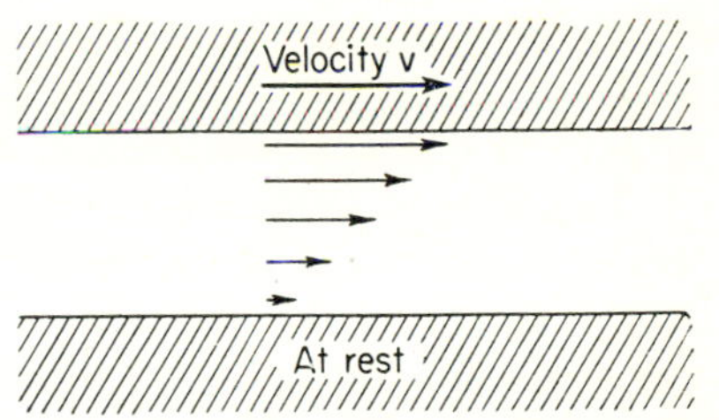

Fig. 2.12 Flow velocity for fluid in shear.

Another case represents the shearing flow which results when fluid is between two plane parallel surfaces ($z = 0$ and $z = 1$ for instance) one of which moves with respect to the other. If the surface at $z = 0$ is at rest and the surface at $z = 1$ is moving in the x direction with unit velocity, the fluid between is said to be subjected to a unit shearing rate. The velocity field, which is at rest with respect to both surfaces and uniformly distributed between, is derivable from a vector potential:

$$\mathbf{A} = -\tfrac{1}{2}z^2\mathbf{j}; \quad \mathbf{v} = z\mathbf{i} \tag{2.3.8}$$

There is no normal outflow (div $\mathbf{v} = 0$), but the vorticity vector $\mathbf{w} = \frac{1}{2}$ curl $\mathbf{v} = \frac{1}{2}\mathbf{j}$ is uniform over the region.

Another example of velocity field, exhibiting both vector and scalar potentials, is the following, expressed in cylindrical coordinates, r, ϕ, z:

$$\psi = \begin{cases} 0; & r < a \\ \omega a^2\phi; & r > a \end{cases}; \quad \mathbf{A} = \begin{cases} -\tfrac{1}{2}\omega r^2\mathbf{a}_z; & r < a \\ 0; & r > a \end{cases} \tag{2.3.9}$$

where ω is the angular velocity of the fluid inside the cylinder $r = a$. The velocity vector is then

$$\mathbf{v} = \begin{cases} \omega r\mathbf{a}_\phi; & r < a \\ (\omega a^2/r)\mathbf{a}_\phi; & r > a \end{cases}$$

The vorticity vector $\mathbf{w} = \frac{1}{2}$ curl $\mathbf{v}$ is $\omega\mathbf{a}_z$ (as is to be expected from its definition on page 41) for $r < a$ and is zero for $r > a$. We note that for $r > a$ the velocity field is that found outside a simple vortex line, as given in Eq. (1.2.11). Here we have made the vortex motion finite in

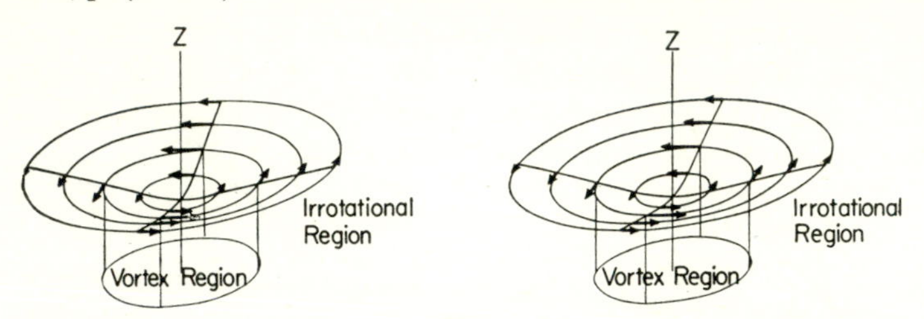

Fig. 2.13 Flow velocity, flow lines, and surface of zero pressure (free surface) for simple vortex.

extent ($r < a$) rather than concentrated in an infinitely narrow line, as was done in Chap. 1.

Stresses in a Fluid. Before we can go much further in this analysis, we must study the internal stresses in the fluid. There is, of course, the pressure, which may be due to gravitational or other forces on the fluid or may be due to a compression of the fluid or both.

In addition, there are the frictional stresses due to *rate of change of strain*, proportional to the velocity vector $\mathbf{v}$ instead of the displacement vector $\mathbf{s}$, as was the case with elastic solids. We saw on page 67 that the strain in an elastic solid could be represented by a symmetric dyadic $\mathfrak{S} = \frac{1}{2}(\nabla\mathbf{s} + \mathbf{s}\nabla)$. The rate of change of this strain is also a dyadic

$$\mathfrak{U} = \tfrac{1}{2}(\nabla\mathbf{v} + \mathbf{v}\nabla)$$

The expansion factor $|\mathfrak{U}| = \operatorname{div}\mathbf{v}$ is, by the equation of continuity, proportional to the rate of change of density of the fluid (which is zero if the fluid is incompressible). The "remainder" of $\mathfrak{U}$, which can be represented by the dyadic

$$\mathfrak{U}_s = \tfrac{1}{2}(\nabla\mathbf{v} + \mathbf{v}\nabla) - \tfrac{1}{3}\mathfrak{J}\operatorname{div}\mathbf{v}; \quad |\mathfrak{U}_s| = 0$$

corresponds to pure shearing rate and would represent the form of the rate of change of strain dyadic for incompressible fluids.

Now we must work out the form of the stress dyadic and how it depends on the rate of strain. When the fluid is not in motion, the only stress is the static pressure p, which is completely symmetrical;

$$\mathfrak{T} = -p\mathfrak{I}$$

so that the force across any element of area $d\mathbf{A}$ is $-p\,d\mathbf{A}$ (the negative sign indicating pressure, negative tension). When the fluid is expanding without shear ($\mathfrak{U}_s = 0$), it is possible that there is a frictional effect to pure expansion and that the pressure is altered by the rate of expansion (this turns out to be the case with all fluids except monatomic gases). In such a case the stress would be

$$\mathfrak{T} = (-p + \lambda \operatorname{div} \mathbf{v})\mathfrak{I}$$

where λ can be called the *coefficient of expansive friction.*

If, in addition, there is a rate of shear of the fluid, there will be a proportional shearing stress, $2\eta\mathfrak{U}_s$, where η is called the *coefficient of viscosity.* The total stress dyadic is therefore related to the pressure and to the rate of strain by the following equation:

$$\mathfrak{T} = -p\mathfrak{I} + \lambda\mathfrak{I}|\mathfrak{U}| + 2\eta\mathfrak{U}_s = -(p + \gamma \operatorname{div} \mathbf{v})\mathfrak{I} + \eta(\nabla\mathbf{v} + \mathbf{v}\nabla) \qquad (2.3.10)$$

where $\gamma = \frac{2}{3}\eta - \lambda$ can be called the second viscosity coefficient. This equation is similar to Eq. (1.6.28) for the stresses in an elastic solid, except that velocity $\mathbf{v}$ now enters where displacement $\mathbf{s}$ occurred before (and, of course, the pressure term has been added). This difference is not trivial, however, for a force proportional to a velocity is a dissipative force whereas the stresses in Eq. (1.6.28) are conservative.

One might, of course, have assumed that the constants γ and η were dyadics rather than scalars, but we have less reason to expect such complication here than we did in isotropic solids. We expect that a fluid is isotropic, and experimental results seem to bear this out.

Returning to our examples, we can use the expression for the dyadic $\frac{1}{2}(\nabla\mathbf{v} + \mathbf{v}\nabla)$ in spherical coordinates, given on page 117, to calculate the stress tensor,

$$\mathfrak{T} = \left(-p + \frac{2\eta Q}{r^3}\right)\mathfrak{I} - \frac{6\eta Q}{r^3}\,\mathbf{a}_r\mathbf{a}_r \qquad (2.3.11)$$

for the flow from a simple source given in Eq. (2.3.7). In other words, the force across a surface element perpendicular to a radius vector is a compressional one of magnitude $p + (4\eta Q/r^3)$, whereas the compressional force across any surface perpendicular to the former element is $p - (2\eta Q/r^3)$. When there is viscosity ($\eta > 0$), therefore, the force on an element of fluid is not isotropic, and for a large enough flow (Q large) or a small enough radius, the force "across" a radial flow line becomes a tension, whereas the force "along" the flow line is everywhere compressional. This sort of force is, of course, needed to change the shape of an element of fluid as it travels out radially from the source, for the element must spread out in all directions perpendicular to the radius and must correspondingly become thinner radially. If the fluid is viscous, it takes a nonisotropic force to produce such a deformation. The constant γ does not enter, for we are assuming that the fluid is incompressible and div $\mathbf{v} = 0$.

For the unit shear case given in Eq. (2.3.8), the stress tensor is

$$\mathfrak{T} = -p\mathfrak{I} + \eta(\mathbf{ki} + \mathbf{ik}) \tag{2.3.12}$$

Here the force on a unit area of stationary surface at $z = 0$ is just $\mathfrak{T} \cdot \mathbf{k} = -p\mathbf{k} + \eta\mathbf{i}$. The component $-p\mathbf{k}$ normal to the surface is, of course, the pressure (the minus sign indicating force *into* the surface). The component $\eta\mathbf{i}$ parallel to the motion of the upper surface (at $z = 1$) is that due to the viscosity of the fluid; in fact we have set up just the conditions corresponding to the fundamental definition of the *coefficient of viscosity* η of a fluid (η is the magnitude of the tangential force per unit area for a unit rate of shear).

In the last example, given in Eq. (2.3.9), we have for the stress

$$\mathfrak{T} = \begin{cases} -p\mathfrak{I}; & r < a \\ -p\mathfrak{I} - (4\eta\omega a^2/r^2)(\mathbf{a}_r\mathbf{a}_\varphi + \mathbf{a}_\varphi\mathbf{a}_r); & r > a \end{cases} \tag{2.3.13}$$

In the portion of fluid for $r < a$ the only stress is the isotropic pressure, which is not surprising, for this portion of the fluid is rotating as a rigid solid with angular velocity ω. Outside this vortex core, for $r > a$, there is shear of the fluid, and the force on unit area perpendicular to r has a tangential component $-(4\eta\omega a^2/r^2)\mathbf{a}_\varphi$, representing the drag of the fluid outside the cylinder of radius r on the fluid inside the cylinder (or vice versa).

The force on an element of fluid at x, y, z is, as we have stated, $(\nabla \cdot \mathfrak{T} + \mathbf{F})\, dx\, dy\, dz$. This must equal the acceleration of the element $d\mathbf{v}/dt$ times its mass $\rho\, dx\, dy\, dz$. The resulting equation (which is obtained by the use of the formulas on page 115)

$$\begin{aligned} \rho(\partial\mathbf{v}/\partial t) + \rho\mathbf{v} \cdot \nabla\mathbf{v} &= \mathbf{F} + \nabla \cdot [-(p + \gamma\nabla \cdot \mathbf{v})\mathfrak{I} + \eta(\nabla\mathbf{v} + \mathbf{v}\nabla)] \\ &= \mathbf{F} - \text{grad}\,[p - (\tfrac{1}{3}\eta + \lambda)\,\text{div}\,\mathbf{v}] + \eta\nabla^2\mathbf{v} \\ &= \mathbf{F} - \text{grad}\,[p - (\tfrac{4}{3}\eta + \lambda)\,\text{div}\,\mathbf{v}] - \eta\,\text{curl}\,\text{curl}\,\mathbf{v} \end{aligned} \tag{2.3.14}$$

where $\gamma = \frac{2}{3}\eta - \lambda$, serves to calculate the pressure if the velocity is known or enables the transient and oscillatory motions to be computed. This equation, together with the equation of continuity, Eq. (2.3.1), and the equation of state, relating the pressure and the compression of the fluid, is fundamental to all the multiform problems encountered in fluid dynamics. The various forms are obtained by considering one term after another in this equation negligibly small and combining the remaining terms.

Bernouilli's Equation. The simplest case is for the steady-state motion of an incompressible fluid, for then $\partial \mathbf{v}/\partial t$ and div $\mathbf{v}$ are both zero. In addition we assume that the external force $\mathbf{F}$ can be obtained from a potential energy V, $\mathbf{F} = -\,\text{grad}\,V$, and we use the vector relation

$$\tfrac{1}{2}\,\text{grad}\,v^2 = \mathbf{v}\cdot\nabla\mathbf{v} + \mathbf{v}\times\text{curl}\,\mathbf{v}$$

We finally obtain

$$\begin{aligned} 2\eta\,\text{curl}\,\mathbf{w} - 2\rho\mathbf{v}\times\mathbf{w} &= -\,\text{grad}\,U \\ U = V + p + \tfrac{1}{2}\rho v^2; \quad \mathbf{w} &= \tfrac{1}{2}\,\text{curl}\,\mathbf{v} \end{aligned} \tag{2.3.15}$$

The scalar quantity U can be considered to be the energy density of the moving fluid. The first term is the potential energy of position due to external forces; the second term is the kinetic energy density. If the fluid motion is *irrotational*, the vorticity vector $\mathbf{w}$ is zero and U is a constant everywhere for a given flow pattern. In this case we determine the fluid velocity, in terms of a velocity potential, from the boundary conditions and then calculate the pressure from the equation

$$p = U - V - \tfrac{1}{2}\rho v^2 \tag{2.3.16}$$

where U is a constant determined by the boundary conditions. This is called *Bernouilli's equation* for incompressible fluids (ρ = constant).

We note that it is possible for the solution to require a large enough velocity, in certain regions, so that the pressure, computed from this equation, would turn out to be negative. In principle this cannot happen, for cavitation would result and the boundary conditions would be modified.

A very large number of problems of practical interest can be computed with fair accuracy by assuming that the flow can be represented by a velocity potential (*i.e.*, irrotational flow) which is a solution of Laplace's equation. The pressure at any point can then be computed from Bernouilli's equation. Many problems even in aerodynamics can be calculated in this manner, although air is far from being incompressible. Only when the velocity of an important portion of the air approaches the speed of sound does the approximation become invalid. The more complicated case of *supersonic* flow will be touched later in this section.

As an example of irrotational, incompressional fluid motion, we return to the flow from a simple source, given in Eq. (2.3.7). If we neglect the gravitational potential, the pressure as a function of r is $p_\infty - (\rho Q^2/2r^4)$, where p_∞ is the pressure an infinite distance from the source. We see that, if the actual size of the source is too small (r too small), the pressure will be negative and cavitation will result.

Finally we consider the case given in Eq. (2.3.9) of a vertical vortex of radius a. This time we shall take into account the gravitational potential $V = \rho gz$. For $r > a$ the vorticity **w** is zero, so that U is a constant. Suppose that the fluid has a free surface ($p = 0$) at $z = 0$ when it is at rest. The constant value of U is therefore set so that $p = 0$ is at $z = 0$ and $r = \infty$; that is,

$$\rho gz + p + (\rho\omega^2 a^4/2r^2) = 0; \quad r > a$$

For $r < a$ the vorticity **w** is not zero but it has zero curl, so that grad $U = 2\rho\mathbf{v} \times \mathbf{w} = 2\rho\omega^2 r\mathbf{a}_r$. Integrating for U and adjusting the constant of integration so that the pressure is continuous at $r = a$, we have

$$\rho gz + p + (\rho\omega^2/2)(2a^2 - r^2) = 0; \quad r < a$$

The equation for the free surface is the equation for the surface $p = 0$:

$$z = \begin{cases} (\omega^2/2g)(r^2 - 2a^2); & r < a \\ -(\omega^2 a^4/2gr^2); & r > a \end{cases} \tag{2.3.17}$$

In both of these cases, the viscosity has had no effect on the pressure, because the only term involving viscosity in the equation for the pressure for steady-state motion of incompressible fluids is one involving the curl of the vorticity **w**, and the examples have been simple enough so that curl **w** was zero. Other examples can be worked out for which curl **w** is not zero and the viscosity does have an effect on the pressure, but the most frequently encountered examples of this sort are cases where **v** and p change with time.

The Wave Equation. The first examples to be considered of non-steady motion will be for small-amplitude vibrations. In this case all terms in Eq. (2.3.14) involving the squares of v can be neglected, and we obtain the simpler equation

$$\rho(\partial\mathbf{v}/\partial t) = -\operatorname{grad}(p + V) + (\tfrac{4}{3}\eta + \lambda)\operatorname{grad}\operatorname{div}\mathbf{v} - \eta\operatorname{curl}\operatorname{curl}\mathbf{v} \tag{2.3.18}$$

where we have again set $\mathbf{F} = -\operatorname{grad} V$ and where we do not now assume that the fluid is incompressible.

In order to get any further we must discuss the relation between the pressure and the state of compression of the fluid. Flow of material out of any volume element will reduce the pressure in a compressible fluid; in fact for any elastic fluid, as long as the compression is small, the rate of change of p is proportional to the divergence of **v**, $\partial p/\partial t = -\kappa \operatorname{div}\mathbf{v}$. The constant κ is called the *compressibility modulus* for the fluid under consideration. When the displacements are small, we can write this in terms of displacement **s**: $p = -\kappa \operatorname{div}\mathbf{s}$, $\partial\mathbf{s}/\partial t = \mathbf{v}$.

We have seen on page 53 that every vector field can be separated in a unique way into a part which is a gradient and a part which is a curl. Here we utilize this fact twice, once by setting the unknown velocity **v** equal to the gradient of a *velocity potential* ψ plus the curl of a vector potential **A**. Inserting this into Eq. (2.3.18), we equate the gradients and curls of each side separately. The equation for the curls is

$$\rho(\partial \mathbf{A}/\partial t) = -\eta \text{ curl curl } \mathbf{A} \tag{2.3.19}$$

This is not a vector wave equation, but a vector analogue of the diffusion equation mentioned on page 137 and to be discussed in Sec. 2.4. Since only the first time derivative of **A** occurs instead of the second derivative, solutions of this equation are not true waves, propagating with definite velocity and unchanged energy, but are critically damped disturbances, dying out in time and attenuated in spatial motion. They will be discussed more fully in Chap. 12. We note that the pressure is not affected by these waves. We note also that the equation for the vorticity $\mathbf{w} = \frac{1}{2}$ curl $\mathbf{v}$ is identical with that for **A**. Viscosity causes vorticity to diffuse away, a not unexpected result.

Collecting the gradient terms on both sides of the equation derived from Eq. (2.3.18) and differentiating both sides with respect to time finally give us the equation for the longitudinal wave:

$$\frac{\partial^2 \psi}{\partial t^2} = c^2 \nabla^2 \psi + \frac{1}{\rho}(\tfrac{4}{3}\eta + \lambda)\nabla^2\left(\frac{\partial \psi}{\partial t}\right); \quad c^2 = \frac{\kappa}{\rho} \tag{2.3.20}$$

When the compressional viscosity $\frac{4}{3}\eta + \lambda$ is small, ordinary compressional waves are transmitted through the fluid with velocity c, and all the remarks we have made concerning compressional waves in elastic media are applicable here. If this is not zero a damping term is introduced. For instance, for simple harmonic waves, with time dependence given by the exponential $e^{-i\omega t}$, the equation for the space dependence of ψ is

$$\nabla^2 \psi + \frac{\omega^2 \psi}{c^2 - i(\omega/3\rho)(4\eta + 3\lambda)} = 0$$

In other words the space dependence will have a complex exponential factor, representing a space damping of the wave.

On the other hand if a standing wave has been set up, with space part satisfying the equation $\nabla^2\psi + k^2\psi = 0$, the equation for the time dependence of ψ is

$$\frac{\partial^2 \psi}{\partial t^2} + \frac{1}{\rho}(\tfrac{4}{3}\eta + \lambda)k^2 \frac{\partial \psi}{\partial t} + c^2 k^2 \psi = 0$$

which is the equation for damped oscillators in time.

Irrotational Flow of a Compressible Fluid. Our next example of the different sorts of fluid motion represented by Eq. (2.3.14) is that of the steady, irrotational flow of a fluid which is compressible. This is the case of importance in aerodynamics when the fluid velocity approaches the speed of compressional waves, $c = \sqrt{\kappa/\rho}$, discussed in the previous subsection. Since the applications are nearly always to the flow of air, we may as well specialize our expression for the compressibility κ to the case of a gas.

We cannot go far enough afield here to discuss the thermodynamics of a perfect gas in detail; many texts on this subject are available for reference. We need only to write down two equations relating the pressure p, density ρ, and temperature T of a gas during adiabatic expansion (expansion without loss of heat contained in the gas):

$$p/p_0 = (\rho/\rho_0)^\gamma = (T/T_0)^{\gamma/(\gamma-1)} \tag{2.3.21}$$

where the indices zero designate the pressure, density, and temperature at standard conditions (for instance, where the fluid is at rest). Another way of writing this is to relate the pressure and density to the entropy S of the gas:

$$p/\rho^\gamma = Ae^{\alpha S}$$

An adiabatic expansion is one for constant entropy S. The constant γ is the ratio of specific heats at constant pressure and constant volume (its value for air is 1.405).

Taking the differential of this equation, for constant S, $dp/p = \gamma d\rho/\rho$, and using the equation of continuity, (2.3.5), we get $dp/p = -\gamma\, dt \operatorname{div} \mathbf{v}$. Comparing this result with the definition of the compressibility modulus κ we see that $\kappa = \gamma p$ and that the speed of sound (compressional waves) in a gas at pressure p and density ρ is

$$c = \sqrt{\gamma p/\rho} = \sqrt{dp/d\rho} \tag{2.3.22}$$

In flow of a compressible gas both pressure and density (and therefore the speed of sound) change from point to point in the fluid. The relation between them is obtained from Eqs. (2.3.15), where again we start by considering irrotational, steady-state flow ($\mathbf{w} = 0$, $\partial\mathbf{v}/\partial t = 0$ we also neglect the potential V). Since now ρ is not constant, the integration of grad $U = 0$ is a little less simple than before. Both pressure and density turn out to be functions of the air speed v at any point. The maximum pressure p_0 and related density ρ_0 and speed of sound c_0 are for those points where $v = 0$ (stagnation points). At any other point Eqs. (2.3.15) and (2.3.21) indicate that

$$v^2 = -2\int_{v=0}^{v} \frac{dp}{\rho} = 2\,\frac{p_0^{1/\gamma}}{\rho_0}\int_p^{p_0} p^{-1/\gamma}\, dp = \frac{2\gamma}{\gamma - 1}\,\frac{p_0}{\rho_0}\left[1 - \left(\frac{p}{p_0}\right)^{(\gamma-1)/\gamma}\right]$$

This indicates that there is a maximum fluid velocity

$$v_{\max} = \sqrt{2\gamma p_0/\rho_0(\gamma - 1)},$$

for which the pressure is zero. This would be the velocity of flow into a vacuum for instance. For air at 15°C at the stagnation points ($T_0 =$

288) this limiting velocity is 75,700 cm per sec. At this limiting velocity the speed of sound is zero (since the pressure is zero). At the stagnation point the air is at rest and the velocity of sound, $\sqrt{\gamma p_0/\rho_0}$, is maximum. Therefore as v goes from zero to $v_{\max}$, the dimensionless ratio $M = v/c$ goes from zero to infinity. This ratio is called the *Mach number* of the air flow at the point. If it is smaller than unity, the flow is *subsonic;* if it is larger than unity, the flow is *supersonic.*

The equations giving pressure, density, temperature, sound velocity, and Mach number at a point in terms of the fluid velocity v at the point and the pressure and density p_0 and ρ_0 at a stagnation point are

$$\begin{aligned} p &= p_0[1 - (v/v_{\max})^2]^{\gamma/(\gamma-1)} = p_0(c/c_0)^{2\gamma/(\gamma-1)} \\ \rho &= \rho_0[1 - (v/v_{\max})^2]^{1/(\gamma-1)} = \rho_0(c/c_0)^{2/(\gamma-1)} \\ T &= T_0[1 - (v/v_{\max})^2] = T_0(c/c_0)^2 \\ c &= \sqrt{\tfrac{1}{2}(\gamma - 1)(v_{\max}^2 - v^2)} \\ M &= \frac{v}{c} = \sqrt{\frac{2}{(\gamma - 1)}\left(\frac{v^2}{v_{\max}^2 - v^2}\right)} \\ v_{\max} &= \sqrt{2\gamma p_0/\rho_0(\gamma - 1)} = 75{,}700 \text{ cm per sec} \\ c_0 &= v_{\max}\sqrt{(\gamma - 1)/2} = \sqrt{\gamma p_0/\rho_0} = 34{,}100 \text{ cm per sec} \end{aligned} \tag{2.3.23}$$

The velocity v_s at which the Mach number M is unity (fluid speed equals sound speed) turns out to be equal to $\sqrt{(\gamma - 1)/(\gamma + 1)}\, v_{\max} = 31{,}100$ cm per sec for air at 15°C ($T_0 = 288$) at stagnation points. At this speed the pressure, density, etc., are $p_s = 0.528p_0$, $\rho = 0.685\rho_0$, $T = 0.832T_0 = 240°\text{K} = -33°\text{C}$, $c_s = v_s$.

Subsonic and Supersonic Flow. Several examples will show the importance of the region where $M = 1$ and will indicate that the phe-

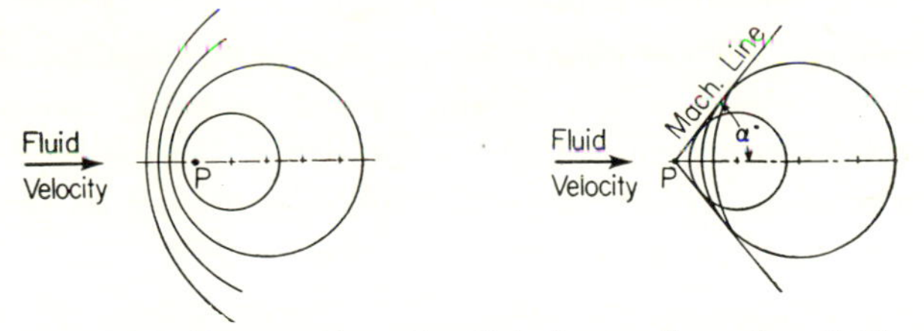

Fig. 2.14 Propagation of a disturbance through a fluid traveling past small obstruction at P with velocity smaller (*left*) and larger (*right*) than speed of sound.

nomena of gas flow for speeds above this (supersonic flow) are quite different from the phenomena for speeds below this (subsonic flow). As a very simple example, we suppose air to be flowing past a small object at rest at point P in Fig. 2.14. The presence of the object continuously decelerates the air in front of it, which continuously produces a sound wave in front of it. If the air velocity is less than that of sound, these waves can travel upstream from the obstruction at P and warn the fluid of its impending encounter, so to speak. But if the fluid is moving faster than sound, then no warning can be sent upstream and the expanding wave fronts are carried downstream as shown in the right-hand sketch of Fig. 2.14. The envelop of these waves is a "bow wave" of disturbance, which is called a *Mach line* or *Mach surface*. The first intimation of the presence of the obstruction at P occurs when the air strikes this line or surface. Incidentally, it is not hard to see that the angle of inclination of this line, the *Mach angle*, is given by the equation

$$\alpha = \sin^{-1} (1/M) = \sin^{-1} (c/v)$$

We shall come back to these Mach lines later.

As another example, consider air to flow along a tube of varying cross section $S(x)$, as in Fig. 2.15. In order that no air pile up anywhere (*i.e.*, that there be steady flow), the same mass Q of air must pass through

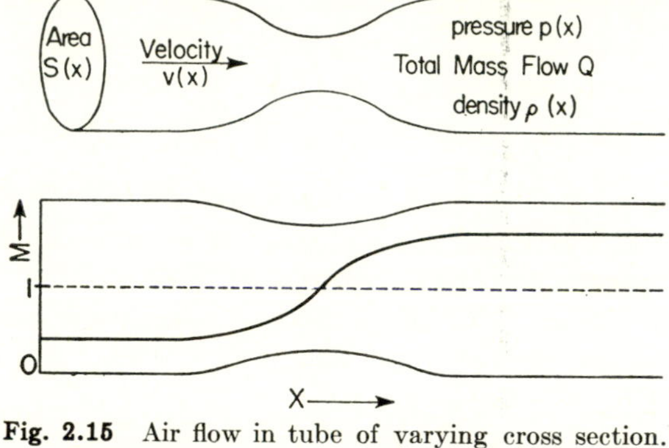

Fig. 2.15 Air flow in tube of varying cross section. Lower plot shows three curves of possible variation of $M = v/c$ along tube.

each cross section. If the tube does not change too rapidly with change of x, and if the inner surface of the tube is smooth, the density and velocity will be nearly uniform across each cross section, and with fairly good approximation, we can say that ρ and p and v are all functions of x alone. Then, to this approximation, for steady flow

$$Q = S(x)\rho(x)v(x) = (S\rho_0/v_{\max}{}^{2/(\gamma-1)})(v_{\max}^2 - v^2)^{1/(\gamma-1)}v$$

or $$\ln S = \frac{2}{\gamma - 1} \ln v_{\max} + \ln\left(\frac{Q}{\rho_0}\right) - \ln v - \frac{1}{\gamma - 1} \ln (v_{\max}^2 - v^2)$$

Differentiating this last equation with respect to x and using the equation for c given in Eq. (2.3.23), we obtain

$$\frac{1}{S}\frac{dS}{dx} = \frac{1}{v}\frac{dv}{dx}(M^2 - 1); \quad M = \frac{v}{c} \tag{2.3.24}$$

Therefore if the flow is everywhere subsonic ($M < 1$), wherever S *decreases* in size, the air speed v *increases*, and vice versa. On the other hand if the flow is everywhere supersonic ($M > 1$), wherever S *decreases*, the air speed v *decreases*, and vice versa. In each case wherever S has a maximum or minimum, there v has a maximum or minimum. These cases are shown in the lower plot of Fig. 2.15.

If, however, the pressure p_0, total flow Q, etc., are adjusted properly, the Mach number M can be made to equal unity at a minimum of S. In this case dv/dx need not be zero even though dS/dx is zero, and the velocity can increase from subsonic to supersonic as it passes through the constriction (or, of course, it could start supersonic and end up subsonic). This case is shown by the center curve in the lower part of Fig. 2.15.

Velocity Potential, Linear Approximation. We must now set up the equation which will enable us to compute the vector velocity field to satisfy any given boundary conditions. As with the irrotational flow of incompressible fluids, we assume that this field can be obtained from a scalar velocity potential field, $\mathbf{v} = \text{grad}\,\psi$. The equation for ψ comes from the equation of continuity, Eq. (2.3.1) for $\partial\rho/\partial t = 0$,

$$0 = \text{div}\,(\rho\mathbf{v}) = \text{div}\,[(\rho_0/v_{\max}{}^{1/\gamma-1})(v_{\max}^2 - v^2)^{1/\gamma-1}\mathbf{v}]$$

or

$$0 = \text{div}\,[(v_{\max}^2 - v^2)^{1/\gamma-1}\mathbf{v}]$$

Therefore, if $\mathbf{v} = \text{grad}\,\psi$,

$$\nabla^2\psi = J;\quad J = \sum_{m,n=1}^{3} \frac{1}{c^2}\frac{\partial\psi}{\partial x_m}\frac{\partial\psi}{\partial x_n}\frac{\partial^2\psi}{\partial x_m\,\partial x_n} \tag{2.3.25}$$

where $x_1 = x$, $x_2 = y$, $x_3 = z$, and $c^2 = \frac{1}{2}(\gamma - 1)(v_{\max}^2 - |\text{grad}\,\psi|^2)$.

For two-dimensional flow the equation becomes

$$\frac{\partial^2\psi}{\partial x^2}\left[1 - \frac{1}{c^2}\left(\frac{\partial\psi}{\partial x}\right)^2\right] + \frac{\partial^2\psi}{\partial y^2}\left[1 - \frac{1}{c^2}\left(\frac{\partial\psi}{\partial y}\right)^2\right] = \frac{2}{c^2}\frac{\partial^2\psi}{\partial x\,\partial y}\frac{\partial\psi}{\partial x}\frac{\partial\psi}{\partial y} \tag{2.3.26}$$

There is also a flow function Φ which defines the lines of flow and which measures the mass flow of air between two flow lines. We obtain this from the equation of continuity for steady flow, $\text{div}\,(\rho\mathbf{v}) = 0$, for we can set

$$v_x = \frac{\partial\psi}{\partial x} = \frac{\rho_0}{-\rho}\frac{\partial\Phi}{\partial y};\quad v_y = \frac{\partial\psi}{\partial y} = \frac{\rho_0}{\rho}\frac{\partial\Phi}{\partial x}$$

and then $\text{div}\,(\rho\mathbf{v})$ is automatically zero. Likewise, as we showed on page 156, the total mass flow between two flow lines (per unit of extent in z) is equal to ρ_0 times the difference between the values of Φ for the two flow lines. The equation for Φ is similar to that for ψ:

$$\frac{\partial^2\Phi}{\partial x^2}\left[1 - \left(\frac{\rho_0}{\rho c}\right)^2\left(\frac{\partial\Phi}{\partial y}\right)^2\right] + \frac{\partial^2\Phi}{\partial y^2}\left[1 - \left(\frac{\rho_0}{\rho c}\right)^2\left(\frac{\partial\Phi}{\partial x}\right)^2\right] = -2\left(\frac{\rho_0}{\rho c}\right)^2\frac{\partial^2\Phi}{\partial x\,\partial y}\frac{\partial\Phi}{\partial x}\frac{\partial\Phi}{\partial y} \tag{2.3.27}$$

Equation (2.3.25) is, of course, a nonlinear equation for ψ, one quite difficult to solve exactly. When M is small (subsonic flow), J may be neglected to the first approximation, and the equation reduces to the linear Laplace equation characteristic of incompressible fluids. When the solution of Laplace's equation, ψ_0, is found for the particular case of interest, J_0 may be calculated for each point from ψ_0 by use of the equation for J. Then a second approximation to the correct ψ may be obtained by solving the Poisson equation $\nabla^2\psi = J_0$, and so on.

If M is not small, however, such iterative methods cannot be used and other approximate methods must be used. One technique is useful when the flow is not greatly different from uniform flow, $\mathbf{v} = \mathbf{v}_u$, a constant. In this case the direction of the unperturbed flow can be taken along the x axis, and we can set

$$\mathbf{v} = v_u\mathbf{i} + \mathbf{v}_1; \quad \psi = v_u x + \psi_1$$

where $\mathbf{v}_1$ is small compared with $\mathbf{v}_u$, though $\mathbf{v}_u$ is not necessarily small compared with c. To the first order in the small quantity $\mathbf{v}_1/c$, we have

$$\frac{\partial^2\psi_1}{\partial x^2}(1 - M_u^2) + \frac{\partial^2\psi_1}{\partial y^2} + \frac{\partial^2\psi_1}{\partial z^2} \simeq 0 \tag{2.3.28}$$

where $M_u^2 = [2/(\gamma - 1)][v_u^2/(v_{\max}^2 - v_u^2)] = v_u^2/c_u^2$ is the square of the Mach number of the unperturbed flow.

This equation, being a linear one in ψ, can be solved to determine the steady flow around irregularities in the boundary surfaces as long as the irregularities do not produce large changes in air velocity near them.

Mach Lines and Shock Waves. Equation (2.3.28) again shows the essential difference in nature between subsonic and supersonic flow. The difference can be illustrated by a two-dimensional case, where the equation is

$$\frac{\partial^2\psi_1}{\partial x^2}(1 - M_u^2) + \frac{\partial^2\psi_1}{\partial y^2} = 0 \tag{2.3.29}$$

When M_u is less than unity, this equation can be changed into a Laplace equation for ψ_1 by changing the scale of y to $y' = y\sqrt{1 - M_u^2}$, $x' = x$. Therefore the flow lines and potential surfaces are similar to those for incompressible flow, except that the y axis is stretched by an amount $1/\sqrt{1 - M_u^2}$.

However, if M_u is larger than unity, we can no longer transform the equation into a Laplace equation, for the $\partial^2\psi/\partial x^2$ term changes sign and the equation is more analogous to a wave equation (see page 124) with x analogous to time and the "wave velocity" $c_u = 1/\sqrt{M_u^2 - 1}$. Solutions of the resulting equation are

$$\psi_1 = f(y - c_u x) + F(y + c_u x)$$

As mentioned on page 166, any irregularity in the boundary shape (which is, of course, here a plane parallel to the x, z plane) produces a "bow wave" which spreads out at an angle $\alpha = \tan^{-1}(c_u) = \sin^{-1}(1/M_u)$ to the x axis, the direction of unperturbed motion. This is the Mach angle mentioned on page 166.

In two dimensions we have also an approximate equation for the flow function Φ, discussed on page 167. We assume that $\Phi = (\rho/\rho_0)v_u y + \Phi_1$, and inserting in Eq. (2.3.27) and neglecting terms smaller than the first order in Φ_1 we have

$$\frac{\partial^2\Phi_1}{\partial x^2}(1 - M_u^2) + \frac{\partial^2\Phi_1}{\partial y^2} = 0 \tag{2.3.30}$$

which is similar to the approximate equation for the correction to the velocity potential ψ.

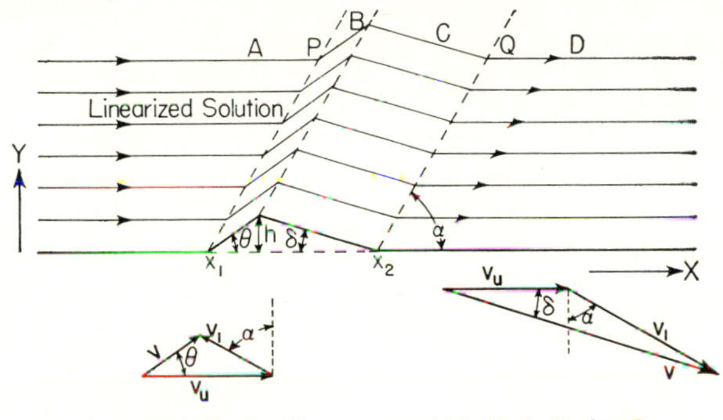

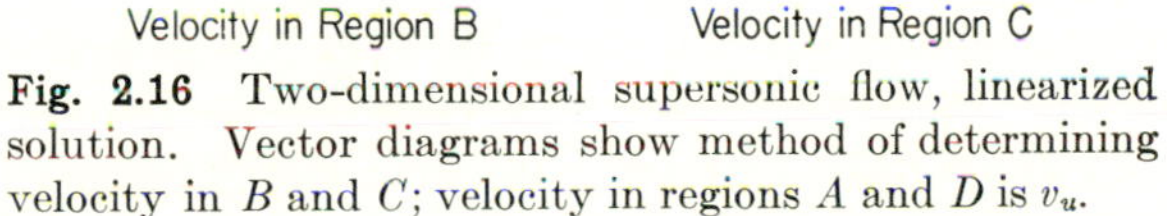

Velocity in Region B Velocity in Region C

Fig. 2.16 Two-dimensional supersonic flow, linearized solution. Vector diagrams show method of determining velocity in B and C; velocity in regions A and D is v_u.

As an example we can assume that air flows at supersonic velocity past a surface which is, in the main, just the x, z plane. At a certain point, however, a ridge of material, forming an irregularity parallel to the z axis, occurs in the boundary, as shown in Fig. 2.16. If the height h of the irregularity is small, the approximations of Eqs. (2.3.29) and (2.3.30) are valid and both the stream function Φ_1 and the velocity potential ψ_1 are functions of $y - c_u x$ or of $x - y\sqrt{M_u^2 - 1}$, which is the same thing. We do not need the function of $y + c_u x$ in this case (why?). Therefore the change in velocity v_1 is perpendicular to the Mach line, *i.e.*, is at an angle α to the y axis.

If the boundary surface is given by the equation $y = B(x)$, where $B(x)$ is zero for $x > x_2$ and for $x < x_1$, then the expression for the flow function is

$$\Phi = v_0 y - B(x - y\sqrt{M_u^2 - 1})$$

The flow lines $\Phi =$ constant are shown in the top part of Fig. 2.16. We note that the flow lines are undisturbed in region A to the left of the

Mach line (x_1,P), that they are again undisturbed in region D to the right of Mach line (x_2,Q), and that both these lines are inclined at the Mach angle $\alpha = \sin^{-1}(1/M_u)$ to the x axis.

The actual velocity $\mathbf{v}$ (to this approximation) in regions B and C of Fig. 2.16 can be obtained graphically by utilizing two obvious facts: first, that $\mathbf{v}$ must be parallel to the flow lines, *i.e.*, must be at an angle θ to the x axis, and second, that $\mathbf{v}_1$ is perpendicular to the Mach line, *i.e.*, must be at an angle α with respect to the y axis. The graphical construction is shown in Fig. 2.16. Since $\mathbf{v}_1$ is supposed to be much smaller than $\mathbf{v}_u$, we have for the air speed, density, Mach number, and mass flow in the two regions the approximate expressions.

Region B:
$$v \simeq v_u(1 - \theta \tan\alpha)$$
$$\rho \simeq \rho_u(1 + M_u^2\theta \tan\alpha)$$
$$\rho v \simeq \rho_u v_u[1 + (M_u^2 - 1)\theta \tan\alpha]$$
$$M \simeq M_u\left[1 - \frac{v_{\max}^2\theta \tan\alpha}{v_{\max}^2 - v_u^2}\right]$$

Region C:
$$v \simeq v_u(1 + \delta \tan\alpha)$$
$$\rho \simeq \rho_u(1 - M_u^2\delta \tan\alpha)$$
$$\rho v \simeq \rho_u v_u[1 - (M_u^2 - 1)\delta \tan\alpha]$$
$$M \simeq M_u\left[1 + \frac{v_{\max}^2\delta \tan\alpha}{v_{\max}^2 - v_u^2}\right]$$

The interesting point about the flow in region B is that, although the air velocity v *decreases*, the mass flow per unit area ρv *increases*. The drawing for the stream lines also shows this, for in Fig. 2.16 in region B these stream lines are closer together, a concomitant of increased mass flow, as the discussion on page 167 indicated. At the first Mach line the air suddenly slows down and compresses; at the second it speeds up and expands; at the third it slows down and compresses to its original condition.

If the boundary were actually distorted as much as Fig. 2.16 indicates (*i.e.*, if θ and δ were actually as large as is shown), v_1 would not be very small compared with v_u and the first approximation would not be sufficient to compute the motion. One difficulty which immediately arises is that the Mach angle for the air in region B is appreciably different from the Mach angle for air in region A whenever the speeds v differ appreciably in the two regions. The question then arises: What should be the angle between the x axis and the Mach line dividing regions A and B, the angle $\alpha_u = \sin^{-1}(1/M_u)$ appropriate for region A or the angle appropriate for the air in region B (which is greater than α_u)? Detailed study of an exact solution indicates that the angle between the x axis and the actual "shock front" is intermediate between the two discussed in the previous sentence and that the air as it flows across this front undergoes a practically instantaneous change of state to a new speed, density, and pressure appropriate to region B.

As for the demarcation between regions B and C in the exact solution, here the Mach lines appropriate for the two regions diverge (as shown by dotted lines Oa, Ob in Fig. 2.17), leaving a region between for the air to change *continuously* from the state appropriate to region B to that appropriate to region C. A plot of pressure along the flow line H is also shown in Fig. 2.17; it shows that change from a region of faster to one of slower speed involves a discontinuous *increase* in pressure,

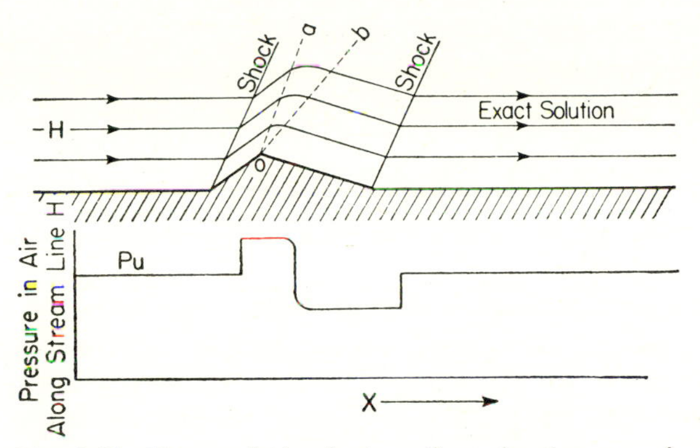

Fig. 2.17 Exact solution for two-dimensional supersonic flow; flow lines and pressure distribution along H.

whereas the reverse change can be accomplished in a more continuous manner.

A great deal more could be said here about supersonic aerodynamics, but it would carry us still further from our course in this chapter. We are here studying techniques of deriving field equations, not primarily discussing various parts of physics in an exhaustive manner.

2.4 *Diffusion and Other Percolative Fluid Motion*

In the previous section we have tacitly assumed that the fluid under consideration is the sole inhabitant of space, that within the rigid boundaries specified in the problem there is no other fluid or solid which hinders the fluid flow. Many problems of interest, however, involve the interpenetrative motion of a fluid through a porous solid (or another fluid) which interacts at every point with the diffusing fluid. Examples include the percolation of liquids through porous media and the motion of special fluids, such as free electrons through a gas or of neutrons through matter. A very important problem is the calculation of the flow of heat through matter. Heat is, of course, the internal energy of the matter itself, but it behaves in many respects as though it were a fluid, with a measurable "density" and rate of flow.

In all these cases we can still talk of an effective density ρ of the fluid under consideration, an average mass per unit volume (or heat per unit volume, etc.) at each point, even though each element of volume also contains another fluid or solid. In many cases we cannot speak of an average velocity (what is the velocity of the heat, for instance?) but we can always speak of a *mass flow* (or total heat flow, etc.) per square centimeter at each point. This quantity $\mathbf{J}$ is a vector and is equal to $\rho\mathbf{v}$ when there is such a thing as a fluid velocity $\mathbf{v}$.

In general the fluid under consideration is not evanescent (although when the fluid is a neutron gas some neutrons are lost by nuclear absorption), so that the equation of continuity usually holds:

$$\partial\rho/\partial t = -\operatorname{div}\mathbf{J} \tag{2.4.1}$$

From this point on, however, the discussion depends on the particular fluid which is studied.

Flow of Liquid through a Porous Solid. For instance, for a liquid seeping through a porous solid we can usually neglect fluid expansion and contraction, so we can set $\partial\rho/\partial t = 0$. We can also neglect viscosity in comparison with the friction of seepage, and we can neglect vorticity. Therefore the mass flow $\mathbf{J}$ can be taken as the gradient of a scalar potential ψ, which is a solution of Laplace's equation div [grad ψ] $= \nabla^2\psi = 0$, as with any irrotational flow of any incompressible fluid. The difference from other cases, considered in the previous section, comes in the equation of motion, which determines the pressure at any point. In the present case it takes a force to make the liquid flow through the pores; to the first approximation this force is proportional to the flow.

Referring to Eq. (2.3.14), we see that the force equation for the present case is

$$\frac{\partial\mathbf{J}}{\partial t} + R\mathbf{J} = \mathbf{F} - \operatorname{grad} p \tag{2.4.2}$$

where $\mathbf{F}$ is the external (gravitational, etc.) force per unit volume on the liquid, p is the pressure, and R is the *flow resistivity* of the porous material. When the porous material is equally resistant in all directions, R can be taken to be a scalar; for nonisotropic materials it would be a dyadic, operating on $\mathbf{J}$. For steady-state conditions, when $\mathbf{J} = \operatorname{grad}\psi$ and when $\mathbf{F}$ is also the gradient of a potential energy V, this equation reduces to

$$\operatorname{grad}(V - p) = R\operatorname{grad}\psi$$

and when R is constant it becomes simply $p = V - R\psi$, which serves to determine the pressure p at every point.

Diffusion. A more interesting problem occurs when the fluid is compressible and when we cannot talk about over-all "forces" on an element of the fluid. This is the case with heat flow, with the diffusion of one liquid through another, and with the diffusion of neutrons through matter. In the case of heat flow the concept of force on the fluid has no meaning; in the case of neutron diffusion the important forces are the internal ones causing collisions between the neutrons and the particles of matter.

In none of these cases does the equation of motion given in Eq. (2.4.2) hold. The flow is not caused by external forces or by pressure gradients but is simply due to concentration gradient. The fluid, for one reason or another, tends to flow from a place of higher density to a place of lesser density, the flow being proportional to the gradient of the density,

$$\mathbf{J} = -a^2 \operatorname{grad} \rho \tag{2.4.3}$$

where the constant a is called the *diffusion constant.* Combining this with the equation of continuity (2.4.1) gives us the *diffusion equation*

$$\partial\rho/\partial t = a^2\nabla^2\rho \tag{2.4.4}$$

which has already been referred to on pages 137 and 163.

In the case of heat flow ρ is the "amount of heat" per unit volume, which is proportional to the temperature, $\rho = CT$, where C is the heat capacity of the material *per unit volume.* Since the equation for heat flow is $\mathbf{J} = -K \operatorname{grad} T$, where K is the heat conductivity of the material, it will be seen that T must be a solution of the diffusion equation, with $a^2 = K/C$.

The proof that a neutron gas, for instance, satisfies approximately the flow equation (2.4.3) and the determination of the corresponding diffusion constant will require a detailed examination of the mechanism of neutron diffusion, which will be sketched later in this section. In such a case the diffusion is due to random motion of the particles of the fluid, and the diffusion constant is a measure of the hindrance offered to this motion by the other matter present.

Diffusion equation (2.4.4) differs from the wave equation (2.2.2) by having a first time derivative instead of a second. This corresponds to the fact that diffusion is an irreversible process analogous to friction, where energy is lost (or entropy gained), whereas wave motion is reversible and conservative. For one space dimension the equation

$$\frac{\partial\rho}{\partial t} = a^2 \frac{\partial^2\rho}{\partial x^2}$$

is the simplest form of the *parabolic partial differential equation.* As noted on page 138 the fluid density moves as though it were completely damped (as, of course, it is), tending always toward uniformity and never "overshooting the mark" to produce oscillations, as occurs with wave and vibrational motion.

Phase Space and the Distribution Function. Before we can go any further in the analysis of diffusion, we shall have to consider the detailed motions of the particles of diffusing fluid, which means that we must for the time being set aside the smoothed-out, continuum picture of a fluid and look into its discontinuous, atomic structure. It is well to carry out such an analysis once in this chapter, if only to show how it is possible to describe the behavior of large numbers of atoms by means of continuous fields satisfying partial differential equations and to show how one relates the properties of the individual atoms with the constants occurring in the field equations.

The connection is, of course, made by means of the techniques of kinetic theory. Our fluid is made up of N atoms (or molecules), each of mass m. The "state" of the nth atom at any instant is given by its position and velocity (for a molecule there are other internal motions which need not be considered here). The position can be specified in terms of the *radius vector* $\mathbf{r} = \mathbf{i}x + \mathbf{j}y + \mathbf{k}z$ to the atom from some origin; the velocity can be specified in terms of the *momentum* vector $\mathbf{p} = m\mathbf{v} = \mathbf{i}p_x + \mathbf{j}p_y + \mathbf{k}p_z$. Therefore the state of the atom (to the extent that is useful here) can be specified by giving its location in six-dimensional *phase space*, with coordinates x, y, z, p_x, p_y, p_z. The methods of kinetic theory provide a means whereby we can go from the motions in phase space of individual atoms, under the influence of external and internal (interatomic) forces, to the average motions of swarms of atoms.

The transition is effected by means of the *distribution function*. To define this function, we consider a particular fluid, consisting of N atoms, subject to given initial and boundary conditions of the usual sort. At some instant the atoms of this fluid can be represented as a swarm of N dots in phase space. At some regions in phase space there will be a concentration of dots; other regions will be very sparsely populated. If N is a large enough number, we shall find a tendency toward "smoothness" in the consistency of the cloud. We can write down an average density of dots at various regions of phase space, which density will vary more or less smoothly from point to point. Now suppose that we duplicate the initial and boundary conditions with a similar set of N atoms and look at the distribution of these N dots in phase space at the corresponding instant of time. Since the initial and boundary conditions are large-scale conditions, affecting only the average positions of atoms, the second swarm of dots will not exactly coincide, dot for dot, with the first swarm. The *average density* of dots, however, will be more or less equal in the two cases.

Suppose that we carry through the same experiment, not just twice, but a large number of times, starting out the N atoms under conditions which are identical from a large-scale point of view. We can then obtain a *probability density* $f(x,y,z,p_x,p_y,p_z,t) = f(\mathbf{r},\mathbf{p},t)$ for each point of phase space, such that $f\,dx\,dy\,dz\,dp_x\,dp_y\,dp_z$ gives the fraction of the experiments for which a dot is found at time t in the element of phase space, $dx\,dy\,dz\,dp_x\,dp_y\,dp_z$ at the point $\mathbf{r}$, $\mathbf{p}$. The function f is called the *distribution function*. If we have set up our experiment sensibly, we shall find that f is a reasonably continuous sort of function of $\mathbf{r}$ and $\mathbf{p}$ and t, and we can expect that it will satisfy some sort of differential equation, which will provide the link between small-scale interactions among the atoms of the fluid and large-scale motions of the fluid as a whole.

We can obtain the large-scale fluid properties which we have been discussing in the previous section in terms of integrals involving the distribution function. The volume element in phase space can be written $dV_x\,dV_p$, where $dV_x = dx\,dy\,dz$ and $dV_p = dp_x\,dp_y\,dp_z$. Sometimes spherical coordinates r, ϑ, φ in ordinary space and p, θ, ϕ in momentum space are useful; the corresponding forms for the volume element are then $dV_x = r^2\,dr\,\sin\vartheta\,d\vartheta\,d\varphi$ and $dV_p = p^2\,dp\,\sin\theta\,d\theta\,d\phi$.

In the first place the integral of f over all space inside the boundaries must just equal the number of atoms in the fluid:

$$\int\int\int\int\int\int f(\mathbf{r},\mathbf{p},t)\,dV_x\,dV_p = N \tag{2.4.5}$$

The average number of particles per unit volume of ordinary space must be the integral of f over momentum space; this number times m the mass of an individual particle must be the density

$$\rho(x,y,z,t) = m\int\int\int f(\mathbf{r},\mathbf{p},t)\,dV_p \tag{2.4.6}$$

mentioned earlier in this section. The total average momentum per cubic centimeter

$$\mathbf{J}(x,y,z,t) = \int\int\int \mathbf{p}\,f(\mathbf{r},\mathbf{p},t)\,dV_p \tag{2.4.7}$$

is the mass flow vector $\mathbf{J}$ mentioned earlier. The total kinetic energy of the fluid is

$$U = (1/2m)\int\int\int p^2 f(\mathbf{r},\mathbf{p},t)\,dV_x\,dV_p \tag{2.4.8}$$

If the forces between atoms are negligible, this is equal to the total internal energy of the gas and is proportional to its temperature. These integrals show that f must go to zero, for large values of p, decidedly enough so that the integral for U does not diverge.

One differential property of f is of general validity and is related to the equation of continuity (2.4.1). All the particles in a given element of momentum space dV_p are traveling with the same velocity $\mathbf{p}/m$. There are $f(\mathbf{r},\mathbf{p},t)\,dV_x\,dV_p$ of them in element dV_x at point $\mathbf{r}(x,y,z)$ at time t. At time $t + dt$ they are at point $\mathbf{r} + (\mathbf{p}/m)\,dt$. Therefore the probability density f at point $\mathbf{r}$ at time t must equal the density f at point $\mathbf{r} + (\mathbf{p}/m)\,dt$ at time $t + dt$, for the particles of momentum $\mathbf{p}$; $f(\mathbf{r} + (\mathbf{p}/m)\,dt,\ \mathbf{p},\ t + dt) = f(\mathbf{r},\mathbf{p},t)$. Expanding this in a series and retaining only terms in the first power of dt, we obtain the *equation of continuity* for the distribution function:

$$\frac{\partial}{\partial t} f(\mathbf{r},\mathbf{p},t) = -\frac{\mathbf{p}}{m} \cdot \operatorname{grad}\,[f(\mathbf{r},\mathbf{p},t)] \tag{2.4.9}$$

where the gradient operator operates on the space dependence of f. This equation, combined with Eqs. (2.4.6) and (2.4.7) gives immediately the usual form of the equation of continuity (2.4.1). It will be modified by other effects, to be discussed later.

Pressure and the Equation of State. A very simple example will indicate how we can use the distribution function to relate the smoothed-out pressure, discussed in the previous section, to the motions of the individual particles. Suppose that we have a container of volume V with N gas atoms uniformly distributed inside. By "uniformly distributed" we mean that f is independent of $\mathbf{r}$ inside V. For the gas to remain uniformly distributed, f must be independent of t and $\mathbf{J}$ must be zero everywhere. The simplest assumption that will satisfy all these requirements is that $f = (N/4\pi V)\psi(p)$, where ψ is a function only of the magnitude p of the particle momentum having the following properties:

$$\int_0^\infty \psi(p)p^2\,dp = 1; \quad \frac{1}{2m}\int_0^\infty \psi(p)p^4\,dp = \epsilon \tag{2.4.10}$$

The first integral determines the arbitrary constant giving the magnitude of ψ (*i.e., normalizes* ψ) so that $\psi p^2\,dp$ is the probability that a gas particle has momentum magnitude between p and $p + dp$. The quantity ϵ is then the average kinetic energy of a particle.

Substituting this expression for f into the integrals for the field quantities mentioned earlier, we obtain the equations

$$\rho = (Nm/V); \quad J = 0; \quad U = N\epsilon$$

It should be pointed out that the statement that f is $(N/4\pi V)\psi(p)$ thereby imposes certain restrictions on the nature of the boundary walls around the volume V. In the first place, these walls must bounce *every* impinging particle back into the volume V and not let any escape; otherwise N and f would not be independent of time. In the second place the walls must duplicate the momentum distribution given by $\psi(p)$ in their reflection of particles, so that a sample of particles just reflected from a wall would be indistinguishable, as far as velocity distribution, from a sample of unreflected particles. This is not to say that every particle must bounce off as if from a perfectly elastic boundary; it does say that for every particle reflected with reduced energy there is one reflected with increased energy. In any other case the distribution f

would change as the boundary surface was approached, and f could not be independent of $\mathbf{r}$.

Let us assume that all these slightly improbable requirements are complied with, and let us examine the average behavior of those particles close to a portion of the boundary wall. We shall assume that this portion is plane and arrange the axes so that they coincide with the y, z plane, the portion for negative x being inside the gas and the portion for positive x being inside the wall. Then for f at $x = 0$, all the particles having positive values of p_x have not yet struck the wall and all having negative values have just been reflected from the wall.

We are now in a position to ask what property of the gas particles produces a steady pressure on the boundary of the container. Obviously it is the repulsive interaction between the wall and the particles which strike the wall, the same interaction which reflects the particles back into the gas when they strike. Since action equals reaction, we can say that the force exerted by the gas per square centimeter on the surface is the same as the force exerted by a square centimeter of surface on the impinging particles to reflect them back into the gas. Since force equals rate of change of momentum, this reaction of the wall equals the average change of momentum of all particles which strike the wall in 1 sec. The number of particles in momentum element dV_p at momentum p which strike a unit area of y, z plane per second is equal to $v_x(N/4\pi V)\psi(p)dV_p = (N/4\pi Vm)\psi(p)p^3\,dp \;\cos\theta \;\sin\theta\,d\theta\,d\phi$, where, for these impinging particles, $0 < \theta < \pi/2$. The total average change in momentum for each of these particles is $2p_x = 2p\cos\theta$, so that the force exerted by the the square centimeter on the gas (and therefore the pressure) is normal to the surface and of average value

$$\frac{N}{4\pi Vm}\int_0^{2\pi} d\phi \int_0^{\pi/2} \cos^2\theta\,\sin\theta\,d\theta \int_0^\infty 2p^4\psi(p)\,dp = P = \frac{2N\epsilon}{3V} = \frac{2}{3}\frac{U}{V}$$

by using Eq. (2.4.10). We therefore obtain the equation relating pressure P, volume V, and internal kinetic energy U:

$$PV = \tfrac{2}{3}U \quad \text{or} \quad P = \tfrac{2}{3}E\rho \tag{2.4.11}$$

which is called the *equation of state* of the gas. The quantity $E = U/Nm$ is the kinetic energy per unit mass of gas, and $\rho = Nm/V$ is the average density of the gas.

We could go on to show that, when the volume of such a container with reflecting walls is reduced (this corresponds to an *adiabatic* compression, see page 164), then E changes, being proportional to the $(\gamma - 1)$th power of the density (where γ is a constant for the gas under study, being 1.4 for air); therefore the pressure for adiabatic compression is proportional to the γth power of ρ. This will take us too far afield, however, and we must get on with our discussion of the relation between internal forces in the fluid and the diffusive properties of the fluid.

Mean Free Path and Scattering Cross Section. In a gas the individual particles, atoms or molecules, are far enough apart that most of the time they move as free bodies, with constant momentum. Only occasionally does one come near enough to another that its motion is affected by the mutual force field. We can therefore separate the motion of each particle into two parts: a *free-flight* portion, when the particle is unaffected by other particles, and a shorter portion, when its momentum is being changed by momentary proximity to another particle, after which the particle sails off for another free flight. These momentary encounters with other particles, during which the momentum of each particle is radically changed, are, of course, called *collisions.* In the case of a neutron gas in a solid, the collision is between neutron and a nucleus of one of the atoms in the solid, but the same reasoning holds.

If thermodynamic equilibrium has been attained, these collisions will, on the average, be conservative of energy and therefore elastic. The law governing the average amount of deflection of path depends on the law of force between the particle and the "target," the nucleus of an atom for a neutron or another similar atom for a gas. The simplest assumption is that the law of force is similar to that between two billiard balls, zero force for a distance of separation larger than R and a very large repulsive force for distances smaller than R; and such a simple assumption corresponds closely to actuality in a gratifyingly large number of cases. Billiard balls, if they collide at all, rebound with equal frequency in all directions, and this result is observed in many actual cases, particularly when the relative velocities are not very large. When the mass of the target is the same order of magnitude as that of the impinging particle, this uniform distribution in angle of scattering on collision is with respect to coordinates moving with the center of gravity of the colliding pair, and the analysis becomes somewhat complicated.

To avoid complexity we shall first analyze cases where the target is much more massive than the particle, as is the case when neutrons are the particles and heavy nuclei the targets or when photons are the particles and fog particles the targets, for example. In these cases the targets are considered to be at rest and to remain at rest, n_t of them distributed at random throughout each cubic centimeter. Each target can be pictured as an elastic sphere of radius R, and the particles as mass points. Thus we can neglect the collisions between particles and concentrate on the collisions between a particle and a target.

The first question to answer is concerned with the relative frequency of collisions between a particle and the various randomly placed targets. The chance that the collision occurs after the particle has moved a distance x and before it has moved a distance $x + dx$ is proportional to the fraction of the area of the thin sheet of space of thickness dx, perpendicular to the path of the particle, which is filled up with targets. The area of *cross section* of a target is $Q_e = \pi R^2$, and there are $n_t\,dx$ targets in the volume of space unit area in extent and dx thick. The fraction of the unit area blocked out by the targets is, therefore, $\pi R^2 n_t\,dx = Q_e n_t\,dx$, and the probability of collision between x and $x + dx$ equals this factor times the probability $P(x)$ that the particle has gone a distance x without collision. We therefore have a differential equation for $P(x)$:

$$(d/dx)P(x) = -Q_e n_t P(x) \quad \text{or} \quad P(x) = e^{-Q_e n_t x} \tag{2.4.12}$$

where we have assumed (quite reasonably) that the probability $P(0)$ of going at least zero distance after the last collision without another collision is unity (certainty).

We have thus obtained an expression for the probability of free flight of length x between collisions, in terms of the density n_t of targets and of the *cross section* Q_e for collision between the particle and a target atom. Detailed calculations concerning the details of the force field between particle and target are thus needed only to determine the value of Q_e when it comes to computing mean lengths of free flight between collisions.

The average length of path between collisions is

$$\lambda = \int_0^\infty P(x)\,dx = \frac{1}{Q_e n_t} \tag{2.4.13}$$

where the length λ is called the *mean free path* of the particle among the cloud of targets.

In the case of an ordinary gas there is a mean free path of the particle among others of its like, for here the targets are other gas molecules. In the case of denser fluids, such as liquids, the mean free path is approximately the same size as the average distance R of approach of the particles, so here a particle is never long free from the influence of its neighbors, but even here the expression for the probability $P(x)$ is valid. One might say that the particles of fluid make contact with the atoms of matter (target atoms) through which they percolate only every mean free path on the average.

The possibility of collisions provides another way whereby the distribution function is modified from point to point or from time to time. For instance, during an instant dt the particles in the momentum element dV_p with momentum $\mathbf{p}$ travel a distance $dx = (p/m)\,dt$ and a fraction $(Q_e n_t p/m)\,dt$ of them suffer collisions. Those which collide change direction of momentum and therefore vanish from the momentum element dV_p. Therefore there is a rate of loss of f due to collisions according to the formula

$$df(\mathbf{r},\mathbf{p},t) = -(Q_e n_t p/m) f(\mathbf{r},\mathbf{p},t)\,dt \tag{2.4.14}$$

But there is also an increase of f due to collisions, for if some particles vanish from dV_p due to collisions, there are also particles, originally in other momentum elements, which are scattered *into* dV_p by collisions. Assuming (as we have been so far) that particles are scattered with equal probability in all directions and that there is no change in particle velocity on scattering of any group of particles undergoing collision, a fraction $d\omega/4\pi$ will be scattered *into* directions of motion within the solid angle $d\omega$. Referring to Fig. 2.18, if there are $f(\mathbf{r},\mathbf{p}',t)\,dV'_p$ particles in the momentum element $dV'_p = d\phi' \sin\theta'\,d\theta'\,(p')^2\,dp'$, then in time dt a number $(Q_e n_t p/4\pi m) f(\mathbf{r},\mathbf{p}',t)\,dV_p\,d\phi' \sin\theta'\,d\theta'\,dt$ are scattered *into* the momentum element $dV_p = d\phi \sin\theta\,d\theta\,p^2\,dp$ (p is equal to p', as we have been assuming so far). The total increase in $f(\mathbf{r},\mathbf{p},t)$ due to scattering

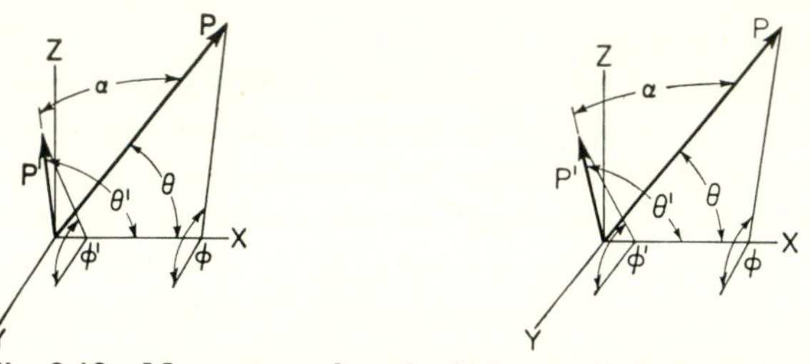

Fig. 2.18 Momenta and angles before and after collision between diffusing particle and target atom of medium.

into the final momentum element is the integral of this quantity over all initial directions of motion, given by θ' and ϕ';

$$df(r,\vartheta,\varphi,p,\theta,\phi,t) = (Q_e n_t p/4\pi m)\iint f(r,\vartheta,\varphi,p,\theta',\phi') \sin\theta'\,d\theta'\,d\phi'\,dt \quad (2.4.15)$$

Diffusion of Light, Integral Equation. A simple yet instructive example of the way these equations can be used concerns the diffusion of light through a slab of milky glass (or of fog), where the illumination is uniform over one surface (the y, z plane, for instance) of the slab, so that f is independent of y and z and only depends on x. This example was first studied by Milne in connection with the flow of light in a stellar atmosphere and is called *Milne's problem.* The "particles" are photons all having the same momentum, and from symmetry the function f depends on the angle θ between $\mathbf{p}$ and the x axis and is independent of ϕ. Therefore we can write f as $f(x,\theta,t)$. The rates of change of f given in Eq. (2.4.9) due to variation of f with x and in Eqs. (2.4.14) and (2.4.15) due to collisions all cancel when steady-state conditions are reached, and the resulting equation for f is

$$\cos\theta \frac{\partial}{\partial x} f(x,\theta) = -n_t Q_t f(x,\theta) + \tfrac{1}{2} n_t Q_e \int_0^\pi f(x,\theta') \sin\theta'\,d\theta'$$

where we have divided out the common factor p/m. Solution of this integrodifferential equation will enable one to determine any required property of the diffusing light. We have here included the possibility that a photon may be absorbed by one of the scattering centers (fog particles or whatever) as well as be scattered. For Q_t is the sum of the scattering cross section Q_e and the absorption cross section Q_a. Naturally Q_a does not enter into the integral term, for this represents photons scattered, after collision, *into* the direction denoted by θ, and photons which are absorbed do not survive the collision.

If we measure distance in mean free paths, $x = \xi/n_tQ_t = \lambda\xi$, this equation becomes

$$\cos\theta\frac{\partial}{\partial\xi}f(\xi,\theta) = -f(\xi,\theta) + \tfrac{1}{2}\kappa\int_0^\pi f(\xi,\theta')\sin\theta'\,d\theta' \qquad (2.4.16)$$

where $\kappa = Q_e/Q_t$ is the ratio of scattering to total cross section. The term on the left side of the equation represents the tendency of f to change due to motion of the particles. The first term on the right is the change of f due to absorption and scattering, and the second term is the change due to rescattering back into the original direction.

By change of normalization of f we can arrange it so that

$$c\int_0^\pi \cos\theta\, f(\xi,\theta)\sin\theta\,d\theta = \mathbf{J}(\xi) \qquad (2.4.17)$$

is the mean flow of light energy per unit area per second in the positive x direction at the point ξ. The constant c is the velocity of light. Then the integral

$$\int_0^\pi f(\xi,\theta)\sin\theta\,d\theta = \rho(\xi) \qquad (2.4.18)$$

is the mean density of light energy at ξ.

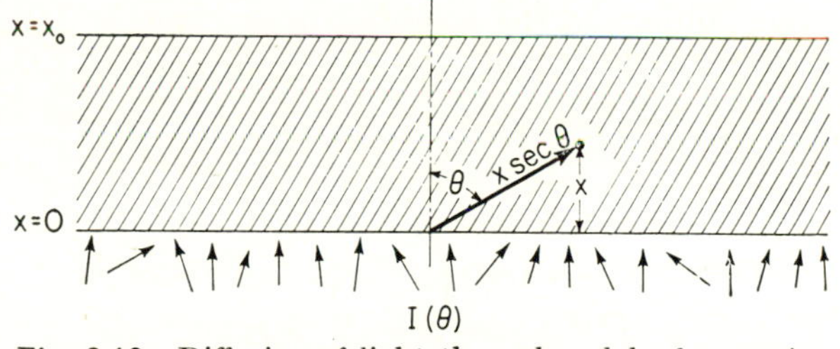

Fig. 2.19 Diffusion of light through a slab of scattering material. Incident intensity given by $I(\theta)$.

Now suppose that the slab of scattering material is between the planes $x = 0$ and $x = x_0$ and that a flux $I(\theta)$ is incident on the surface $x = 0$, as shown in Fig. 2.19. The function I can vary in any arbitrary manner with θ, in the range from 0 to $\pi/2$, but it must be zero for $(\pi/2) \le \theta \le \pi$, for this range of θ corresponds to flow *out* of the slab and could not correspond to incident flux. This flux distribution penetrates into the slab, gradually disappearing as its constituent photons strike a target and get absorbed or scattered. For that part of the flux at an angle θ with respect to the x axis, by the time it has penetrated a distance $x = \lambda\xi$ into the slab, it has traversed ξ sec θ mean free paths inside the material and, by Eq. (2.4.12) only $e^{-\xi \sec \theta}$ of the original amount remains. Therefore one part of $f(\xi,\theta)$ is $I(\theta)e^{-\xi \sec \theta}$.

Another part comes from photons scattered at least once. The number scattered at distance ξ' free paths from the front of the slab will be proportional to the density $\rho(\xi')$, given in Eq. (2.4.18), and the number of such photons arriving at depth ξ at angle θ will be proportional to $\rho(\xi')e^{-|\xi-\xi'| \sec \theta}$, where ξ' will be less than ξ if θ is less than $\pi/2$ and will be greater than ξ for θ larger than $\pi/2$ (backward scattering). Consequently it is reasonable to expect that the solution of Eq. (2.4.16) for the distribution function will have the general form

$$f(\xi,\theta) = \begin{cases} I(\theta)e^{-\xi \sec \theta} + \frac{1}{2}\kappa \sec \theta \int_0^\xi e^{(\xi'-\xi) \sec \theta} \rho(\xi')\, d\xi'; & 0 \le \theta < \frac{1}{2}\pi \\ \frac{1}{2}\kappa \sec \theta \int_{\xi_0}^{\xi} e^{(\xi'-\xi) \sec \theta} \rho(\xi')\, d\xi'; & \frac{1}{2}\pi < \theta \le \pi \end{cases} \tag{2.4.19}$$

Of course this is not yet a solution, for we have not yet calculated the density ρ. However, ρ is a simpler function than is f, for it depends only on ξ and not on θ.

To show that Eq. (2.4.19) is the correct form for f, we transform Eq. (2.4.16) into the following:

$$(\partial/\partial\xi)f(\xi,\theta) + \sec \theta\, f(\xi,\theta) = \tfrac{1}{2}\kappa\rho(\xi) \sec \theta$$

Assuming that ρ is known, we find the solution of the linear inhomogenous equation, subject to the condition that $f(0,\theta) = I(\theta)$, to be just that given in Eq. (2.4.19) (we of course take into account the difference between θ less than $\pi/2$ and θ greater than $\pi/2$). To find the equation which determines ρ we multiply Eq. (2.4.19) by $\sin \theta\, d\theta$ and integrate over θ:

$$\begin{aligned} \rho(\xi) &= \rho_0(\xi) + \tfrac{1}{2}\kappa \int_0^{\xi_0} E_1(|\xi - \xi'|)\rho(\xi')\, d\xi' \\ \rho_0(\xi) &= \int^{\pi/2} e^{-\xi \sec \theta} I(\theta) \sin \theta\, d\theta \\ E_1(|\xi - \xi'|) &= \int_1^\infty e^{-|\xi-\xi'|y}(dy/y); \quad y = \sec \theta \end{aligned} \tag{2.4.20}$$

This is an integral equation of standard type, which will be analyzed in some detail later in this book. When ξ_0 is infinite, the equation is said to be of the *Weiner-Hopf type*. The equation states that the value of ρ at ξ depends on the value of ρ within a mean free path of ξ.

Diffusion of Light, Differential Equation. However, we were setting out to obtain a differential equation for ρ and $\mathbf{J}$ to relate to the diffusion equation, so that further discussion of the integral equation will be postponed. Although it turns out that the integral equation is an exact expression and the differential equation is only approximate, the differential equation is enough easier to solve to make it advisable to use its solutions whenever they are valid.

A differential equation of the diffusion type is a good approximation whenever the fractional change in ρ or $\mathbf{J}$ per mean free path is small. As long as the absorption cross section Q_a is small compared with the elastic cross section Q_e and as long as we do not require too much detail concerning the behavior of ρ and $\mathbf{J}$ within a free path of the boundary, this is possible, for then the distribution function f is nearly independent of the angle of direction of momentum θ and the net flux $\mathbf{J}$ is considerably smaller than the rms magnitude of ρ.

This statement that flux $\mathbf{J}$ is small is equivalent to saying that the distribution function can be approximately expressed by the form

$$f(\xi,\theta) \simeq \tfrac{1}{2}\rho(\xi) + \tfrac{3}{2}\cos\theta\, J(\xi)/c \tag{2.4.21}$$

wherever J/c is considerably smaller than ρ. Referring to Eqs. (2.4.17) and (2.4.18), the flux and density are just equal to the quantities ρ and $\mathbf{J}$ in this expression. This is the simplest function of θ which makes it possible for values of ρ and $\mathbf{J}$ to be set independently.

Suppose we set up the equivalent to Eq. (2.4.16) with the time derivative still present. We take, as time scale, the mean free time for the photons, λ/c. Referring to Eqs. (2.4.9), (2.4.14), and (2.4.15), we have

$$\frac{\partial}{\partial\tau} f(\xi,\theta,\tau) = -\cos\theta \frac{\partial}{\partial\xi} f(\xi,\theta,\tau) - \tfrac{1}{2}\alpha\rho(\xi,\tau) + \tfrac{1}{2}\int_0^\pi [f(\xi,\theta',\tau) - f(\xi,\theta,\tau)]\sin\theta'\, d\theta' \tag{2.4.22}$$

where $\xi = (\eta_t Q_t x)$, $\tau = (\eta_t Q_t ct)$, $p/m = c$, $\alpha = Q_a/Q_t = 1 - \kappa$. Substituting the approximate expression (2.4.21) for f in terms of ρ and J, we obtain an integrodifferential equation of some complexity. The last term, the integral over θ', does simplify however. The parts of $[f(\xi,\theta',\tau) - f(\xi,\theta,\tau)]$ involving ρ cancel out, and the square bracket becomes $\tfrac{3}{2}[\cos\theta' - \cos\theta](1/c)J(\xi,\tau)$, which integrates out to $-(3/c)J(\xi,\tau)\cos\theta$.

We can separate the resulting equation into two equations by multiplying by $\sin\theta\, d\theta$ and integrating or else multiplying by $\cos\theta\sin\theta\, d\theta$ and integrating. These integrations remove the dependence on θ and result in two equations relating to ρ and J. The first,

$$\frac{\partial\rho}{\partial\tau} \simeq -\frac{1}{c}\frac{\partial J}{\partial\xi} - \alpha\rho \tag{2.4.23}$$

is just the equation of continuity (2.4.1) for the present case, as expressed in the dimensionless variables τ and ξ and including the loss of photons due to absorption (the term $\alpha\rho$). If there are, in addition, q photons per mean free time per cubic mean free path added to the distribution, the quantity q will be added to the right-hand side.

The second equation

$$\frac{1}{c}\left[\frac{\partial J}{\partial \tau} + J\right] \simeq -\frac{1}{3}\frac{\partial \rho}{\partial \xi}$$

is related to the diffusion gradient equation (2.4.3). It has an extra term $\partial J/\partial \tau$ giving the rate of change of J per mean free time. For the relatively slow changes involved in diffusion, this term is negligible compared with J, so that, to the approximation considered here, the second equation is

$$J \simeq -\frac{c}{3}\frac{\partial \rho}{\partial \xi} \tag{2.4.24}$$

For diffusing photons, a gradient of photon density produces a drift toward the region of lower density.

Combining Eqs. (2.4.23) and (2.4.24) results in the diffusion equation for the flux density:

$$\frac{\partial \rho}{\partial \tau} \simeq \frac{1}{3}\frac{\partial^2 \rho}{\partial \xi^2} - \alpha\rho + q \tag{2.4.25}$$

The diffusion constant for these units of distance and time is just $\sqrt{\frac{1}{3}}$ (see Eq. 2.4.4). Wherever the source function q is large, there the density ρ tends to increase rapidly; wherever the density is strongly concentrated ($\partial^2\rho/\partial\xi^2$ large and negative), there ρ tends to decrease rapidly. Since only the first derivative in τ enters, the solution is not reversible in time, as is the wave equation.

The distribution function, to the same approximation, is given by the following equation:

$$f(\xi,\theta) \simeq \tfrac{1}{2}\rho(\xi) - \tfrac{1}{2}(\partial\rho/\partial\xi)\cos\theta \tag{2.4.26}$$

in terms of the solution of Eq. (2.4.25). This is valid as long as $\partial\rho/\partial\xi$ is small compared with ρ.

For a typical solution of Eq. (2.4.25) we can return to the steady-state solution for a slab of scattering material. We assume that the beam incident on the surface $x = 0$ is of intensity I and is all directed in the positive x direction; that is, $I(\theta) = I\delta$, where $\delta = \delta(1 - \cos\theta)$ is the delta function discussed on page 122. We also assume that the slab is infinitely thick ($x_0 = \infty$). The part of the incident beam inside the slab which has not yet collided with a scattering target obviously can not be represented by the approximate formula (2.4.26), but we can handle this part separately [call it $f_i = (I/c)e^{-\xi}\delta$] and consider that Eq. (2.4.25) applies only to photons which have had at least one collision. As far as this part of the solution goes, the incident photons appear inside the slab at the point where they suffer their first collision, as though there were a source distribution of strength $q = (1 - \alpha)Ie^{-\xi}/c$

inside the material (the factor $1 - \alpha = \kappa$ appears because only the photons which are scattered, not absorbed, appear in the diffuse distribution). The diffuse density ρ_d is therefore a solution of

$$\frac{\partial^2 \rho_d}{\partial \xi^2} - 3\alpha\rho_d \simeq -\frac{3}{c}(1 - \alpha)Ie^{-\xi}$$

which results from substitution into Eq. (2.4.25), with the time derivative set equal to zero (for steady state).

A solution of this equation is

$$\rho_d \simeq \frac{3I}{c}\frac{1-\alpha}{1-3\alpha}\left\{\exp\left[-\sqrt{3\alpha}\,\xi + (1 - \sqrt{3\alpha})\,\Delta\right] - \exp(-\xi)\right\} \quad (2.4.27)$$

where the constant Δ is to be adjusted to fit the boundary condition at $\xi = 0$.

Since α, the ratio between absorption and scattering per collision, is supposed to be a small quantity, the first exponential diminishes with increasing ξ more slowly than the second exponential; at considerable distance inside the slab the density is proportional to $e^{-\sqrt{3\alpha}\,\xi}$. In other words, the attenuation deep inside the slab is due only to the relatively infrequent absorption of photons. The total distribution function is then

$$f(\xi,\theta) \simeq f_i + \frac{3I}{2c}\frac{1-\alpha}{1-3\alpha}\left\{[1 + \sqrt{3\alpha}\cos\theta]\exp\left[-\sqrt{3\alpha}\,\xi \right.\right.$$
$$\left.\left. + (1 - \sqrt{3\alpha})\,\Delta\right] - (1 + \cos\theta)\exp(-\xi)\right\} \quad (2.4.28)$$

Within a mean free path of the boundary $\xi = 0$ the diffuse part of the distribution function becomes, to the first power in the small quantities ξ, Δ, and α,

$$\frac{3I}{2c}\frac{1-\alpha}{1+\sqrt{3\alpha}}[(\xi + \Delta) - (1 - \xi)\cos\theta]$$

and the corresponding part of the density becomes $(3I/c)[(1 - \alpha)/(1 + \sqrt{3\alpha})](\xi + \Delta)$. These expansions show simply that any solution of Eq. (2.4.25) is invalid near the boundary $\xi = 0$ if the boundary conditions are such as to require that Δ be small, for when $(\xi + \Delta)$ is not large compared with unity, the $\cos\theta$ term of f is no longer small compared with the term independent of θ and the approximation on which Eq. (2.4.25) was built no longer holds.

Boundary Conditions. To see what value Δ should have and what form f and ρ should have close to a boundary we must go back to the exact equations for f and to the integral equation (2.4.20) for ρ, for formula (2.4.26) or (2.4.28) is obviously inaccurate for f at the boundary surface $\xi = 0$, where there is no scattering material to the left. At this point the only part of the distribution function having values of θ less than $\pi/2$ (corresponding to photons coming in to the material from outside) should be the incident beam. The diffuse part of the distribution should be zero for θ from zero to $\pi/2$. Formula (2.4.26), of course, cannot conform to this requirement, no matter what values ρ and $\partial\rho/\partial\xi$ have. All that can be done is to satisfy the requirement on the average, by making the average value of $f(0,\theta)\cos\theta$, over the range $0 \le \theta \le (\pi/2)$, to be zero. This requirement results in the approximate boundary condition [see also Eq. (2.4.34)]

$$\int_0^{\pi/2} \rho\cos\theta\sin\theta\,d\theta - \int_0^{\pi/2}\left(\frac{\partial\rho}{\partial\xi}\right)\cos^2\theta\sin\theta\,d\theta = 0 \quad \text{or}$$

$$\rho \simeq \Delta\frac{\partial\rho}{\partial\xi};\quad \text{at } \xi = 0,\ \Delta \simeq \tfrac{2}{3} \qquad (2.4.29)$$

which means that the constant Δ in Eqs. (2.4.27) and (2.4.28) should be set equal to $\frac{2}{3}$ in order to fit the boundary conditions as well as this

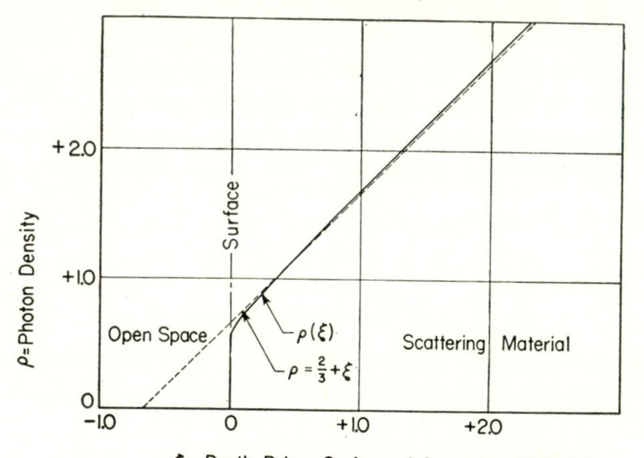

Fig. 2.20 Density of diffusion particles near surface of medium: solid line exact solution, dashed line approximate solution, corresponding to use of diffusion equation.

approximation can do so. Even this formula is none too accurate, for $\partial\rho/\partial\xi$ is certainly not then small compared with ρ near $\xi = 0$, so that any use of the diffusion equation to compute ρ near a boundary is quite a dubious proceeding.

Our only recourse is to go back to the integral equation (2.4.20) to check the validity of these approximations. Unfortunately we are not yet in a position to go through with the solution of Eq. (2.4.20), so we shall just quote results here; the techniques of solving integral equations of this sort will be taken up in Chaps. 8 and 12.

It will be more clear-cut if we compare results for a simpler case, the one considered by Milne for stellar atmospheres. Here photons are created well below the surface of the layer and diffuse upward through the outer scattering layer and radiate out from the surface. This outer layer is many mean free paths thick, so we can again consider x_0 to be extremely large. This time, however, there is no incident flux from above onto the outer surface $x = 0$, so that the integral equation is

$$\rho(\xi) = \tfrac{1}{2} \int_0^\infty E_1(|\xi - \xi'|)\rho(\xi')\, d\xi' \qquad (2.4.30)$$

which is called *Milne's equation*. Function E_1 is defined in Eq. (2.4.20). We have here assumed for simplicity that $\kappa \to 1$, in other words that there is no absorption in the material, only scattering.

The solution of the approximate diffusion equation (2.4.25) with α and q zero subject to the boundary condition given in Eq. (2.4.29) is just

$$\rho \simeq A(\xi + \Delta); \quad \Delta \simeq \tfrac{2}{3} \qquad (2.4.31)$$

The approximate expression for the distribution-in-angle of the outwardly radiated flux from the surface is

$$f \simeq \Delta - \cos\theta; \quad \pi/2 < \theta < \pi \qquad (2.4.32)$$

The solution of integral equation (2.4.30) results in the following expression for the density function:

$$\rho \simeq A\{\xi + 0.7104[1 - 0.3429E_2(\xi) + 0.3159E_3(\xi)]\};$$
$$E_n(z) = \int_1^\infty u^n e^{-uz}\, du \qquad (2.4.33)$$

which is never more than 0.3 per cent in error over the whole range of ξ. Since both E_2 and E_3 rapidly vanish as ξ is increased, we see that for ξ larger than unity (deeper than one free path below the surface) the approximate solution given in Eq. (2.4.31) is perfectly valid, except that Δ should be equal to 0.7104 instead of $\frac{2}{3}$. Within one free path of the surface the actual density falls below the approximate solution to some extent, as is shown in Fig. 2.20.

The exact solution for the angular distribution of flux escaping from the surface is too complicated an expression to be worth writing down here. It is plotted in Fig. 2.21 and there compared with the distribution given by approximate formula (2.4.32). The correspondence is not too bad.

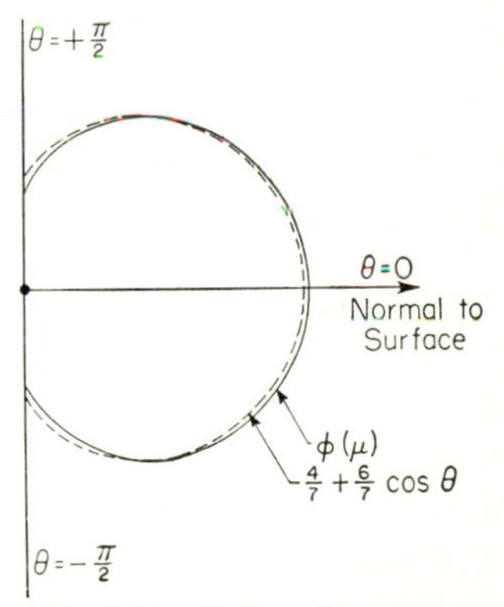

Fig. 2.21 Polar diagram of intensity of particles, emitted from surface of scattering medium, at different angles of emission θ. Solid line exact solution, dashed line diffusion equation approximation.

At any rate we see that the solution of the diffusion equation is a better approximation to the exact solution than we should have any right to expect *if we require it to satisfy the following boundary conditions at each free boundary.*

$$\text{Value of density at surface} \simeq 0.7104\ (\text{value of inward normal gradient of density at surface}) \qquad (2.4.34)$$

where gradient is given as the rate of change per mean free path. Only in the outer portion of the scattering material, within a free path of the surface, does the actual density depart much from the values computed according to these specifications. Inside the material the diffusion equation solution, with this boundary condition, gives quite accurate results, and even the angular distribution of radiation escaping from the surface is reasonably well portrayed by approximate formula (2.4.32).

It is of interest to note that many useful calculations concerning the diffusion of neutrons through solids may also be solved by using the diffusion equation and the boundary conditions of Eq. (2.4.34).

Effect of Nonuniform Scattering. We must next consider the case where the diffusing particle is not scattered uniformly in all directions, though we shall assume for a while yet that no energy is lost during the scattering. A particle, instead of having equal probability $(Q_e/4\pi)\,d\omega$ of being scattered into the solid angle element $d\omega$, no matter at what angle, actually has a probability $\sigma(\alpha)\,d\omega$ of being scattered by an angle α into the solid angle element $d\omega$. As shown in Fig. 2.18, the particle originally has momentum $\mathbf{p}'$ and after the collision has momentum $\mathbf{p}$. The relation between the function σ and the cross section for elastic scattering is

$$Q_e = \int_0^{2\pi} d\beta \int_0^{\pi} \sigma(\alpha) \sin\alpha\, d\alpha; \quad d\omega = \sin\alpha\, d\alpha\, d\beta \qquad (2.4.35)$$

There is also a related cross section Q_m, called the *momentum transfer cross section*, which is obtained from σ by the integral

$$Q_m = \int_0^{2\pi} d\beta \int_0^{\pi} \sigma(\alpha)(1 - \cos\alpha) \sin\alpha\, d\alpha \qquad (2.4.36)$$

If the scattering is uniform in all directions (σ constant), then $Q_m = Q_e$; if there is more scattering backward ($\alpha > 90°$) than forward ($\alpha < 90°$), then Q_m is larger than Q_e, and vice versa. Both Q_m, Q_e, and Q_a, the cross section for absorption of the particle, depend on the incident momentum.

We still assume that the magnitude of the particle momentum is unchanged by the collision and that there are no external forces acting on the particles. We shall assume that at each point in space the drift current $\mathbf{J}$ is in the direction of the gradient of the distribution function f. This is anticipating the solution to some extent and, as it will turn out, rather needlessly, for we could prove that this assumption is true. However, it is so much easier to make the assumption and then verify it that we may perhaps be forgiven the inversion this once. Together with the requirement that the angle-dependent part of f is quite small (which is necessary for a diffusion equation to hold), all this corresponds to the assumption that [see Eq. (2.4.21)].

$$f(\mathbf{r},\mathbf{p},t) \simeq \frac{1}{4\pi m}\,\rho(\mathbf{r},p,t) + \frac{3}{4\pi p^2}\,\mathbf{p}\cdot\mathbf{J}(\mathbf{r},p,t) \qquad (2.4.37)$$

where ρ and $\mathbf{J}$ are defined in Eqs. (2.4.6) and (2.4.7) and $\mathbf{J}$ is in the direction of grad ρ, which is in the direction of the polar axis in Fig. 2.18 (so that $\mathbf{p}\cdot\mathbf{J} = pJ\cos\theta$). We note here that ρ and $\mathbf{J}$ are still functions of the magnitude p of the momentum, so they are the density and flux of particles of a given speed. To obtain the mean quantities discussed on page 175, we average over p,

$$\rho(\mathbf{r},t) = \int_0^\infty p^2\rho(\mathbf{r},p,t)\,dp; \quad \mathbf{J}(\mathbf{r},t) = \int_0^\infty p^2\mathbf{J}(\mathbf{r},p,t)\,dp$$

we also can define other average values,

$$p_{\text{av}} = \int_0^\infty p^3\rho(\mathbf{r},p,t)\,dp; \quad E_{\text{av}} = \frac{1}{2m}\int_0^\infty p^4\rho(\mathbf{r},p,t)\,dp \qquad (2.4.38)$$

The rate of change of f due to continuity is still given by Eq. (2.4.9). The rate of loss of f due to absorption is equal to $(n_tQ_ap/m)f$. The rate of change of f due to elastic scattering is, from Eqs. (2.4.14) and (2.4.15),

$$-\frac{3n_tJ}{4\pi m}\int^{2\pi} d\phi' \int_0^\pi \sigma(\alpha)[\cos\theta - \cos\theta']\sin\theta'\,d\theta'$$

where we have chosen for the direction of J the direction of the polar axes shown in Fig. 2.18. However, it is better to use the final momentum as the polar axis, so we change $\cos\theta'$ to $[\cos\theta\cos\alpha + \sin\alpha\cos(\beta - \phi)]$, where θ and ϕ are constant and α and β are the scattering angles to be integrated over. Referring to Eq. (2.4.36) and the appropriate modification of Eq. (2.4.22) (σ depends on the angle of scattering α but not on the angle of orientation β about the polar axis $\mathbf{p}$) we eventually obtain a differential equation for ρ and for $\mathbf{J}$:

$$\frac{\partial\rho}{\partial t} + \frac{3m}{p^2}\,\mathbf{p}\cdot\frac{\partial\mathbf{J}}{\partial t} \simeq -\frac{1}{m}\,\mathbf{p}\cdot\operatorname{grad}\rho - \frac{3}{p^2}\,\mathbf{p}\cdot\operatorname{grad}(\mathbf{p}\cdot\mathbf{J}) - \frac{n_t}{m}Q_ap\rho$$
$$- \frac{3}{p}\,n_tQ_a\mathbf{p}\cdot\mathbf{J} - \frac{3}{p}\,n_tQ_m\mathbf{p}\cdot\mathbf{J} \qquad (2.4.39)$$

First-order Approximation, the Diffusion Equation. The term in $\partial\mathbf{J}/\partial t$ can be neglected when we are dealing with the relatively slow rates of change involved in diffusion, as we pointed out in connection with the derivation of Eq. (2.4.24). Some of the rest of the terms change sign when the direction of $\mathbf{p}$ is reversed, and some do not; all the terms in $\mathbf{p}\cdot\mathbf{J}$ (grad ρ is in the same direction as $\mathbf{J}$) do so, but the term in $\mathbf{p}\cdot$ grad $(\mathbf{p}\cdot\mathbf{J})$ does not change sign, nor do the terms in ρ. Therefore on the basis of symmetry we can separate Eq. (2.4.38) into two equations, which serve to determine ρ and $\mathbf{J}$ in terms of $\mathbf{r}(x,y,z)$, $\mathbf{p}$, and t.

The first equation can be written

$$\mathbf{J}(\mathbf{r},\mathbf{p},t) \simeq -(p/3n_tQm) \operatorname{grad} \rho(\mathbf{r},p,t); \quad Q = Q_m + Q_a \tag{2.4.40}$$

which, of course, justifies our assumption that $\mathbf{J}$ is in the same direction as grad ρ. The quantity $1/n_tQ$ is the mean free path of the particle, when both absorption and elastic collision are considered. We note that, in this more accurate analysis, the momentum transfer cross section Q_m [see Eq. (2.4.36)] enters rather than the elastic cross section Q_e. Equation (2.4.40) is closely related to Eq. (2.4.3) for average diffusion. The quantities Q, ρ, and $\mathbf{J}$ in the present equation are functions of the magnitude of the momentum $\mathbf{p}$ (*i.e.*, of the speed of the particles), and the equation must be averaged over p to reduce to Eq. (2.4.3). For particles of momentum p, the diffusion constant is therefore proportional to the square root of p times the mean free path. The equation for average density and flux is

$$\mathbf{J}(\mathbf{r},t) \simeq \frac{1}{3m} p_{\text{av}}\lambda_{\text{av}} \operatorname{grad} [\rho(\mathbf{r},t)]; \quad \lambda_{\text{av}} = \frac{1}{p_{\text{av}}} \int_0^\infty \left(\frac{p^3}{n_tQ}\right) \rho(\mathbf{r},p,t)\, dp$$

which is equivalent to Eq. (2.4.3). Therefore the quantity $\sqrt{p_{\text{av}}\lambda_{\text{av}}/3m}$ is equal to the diffusion constant for the average distribution, as will be seen below.

Next we consider these parts of Eq. (2.4.39) which do not change sign when $\mathbf{p}$ is reversed. All but one of the terms is independent of θ, the direction of $\mathbf{p}$, but the term in $\mathbf{p} \cdot \operatorname{grad} (\mathbf{p} \cdot \mathbf{J})$ is a function of θ. This, of course, is an indication that the assumption of Eq. (2.4.37) for a form for the distribution function f was only an approximate one. The relation can be approximately satisfied by averaging over all directions of $\mathbf{p}$. Such an averaging process will make no change in terms such as $m(\partial\rho/\partial t)$ or $n_tQ_ap\rho$, but terms such as $\mathbf{p} \cdot \operatorname{grad} \rho$ will have zero average value. The only complicated term can be reduced by using Eq. (2.4.40) and by expanding in components of $\mathbf{p}$:

$$\begin{aligned}
-\frac{3}{p^2}\mathbf{p} \cdot \operatorname{grad} [\mathbf{p} \cdot \mathbf{J}] &= \frac{1}{pn_tQm} \mathbf{p} \cdot \operatorname{grad} [\mathbf{p} \cdot \operatorname{grad} \rho] \\
&= \frac{1}{pn_tQm} \left\{ p_x \left[p_x \frac{\partial^2\rho}{\partial x^2} + p_y \frac{\partial^2\rho}{\partial x\, \partial y} + p_z \frac{\partial^2\rho}{\partial x\, \partial z} \right] \right. \\
&\quad + p_y \left[p_x \frac{\partial^2\rho}{\partial y\, \partial x} + p_y \frac{\partial^2\rho}{\partial y^2} + p_z \frac{\partial^2\rho}{\partial y\, \partial z} \right] \\
&\quad \left. + p_z \left[p_x \frac{\partial^2\rho}{\partial x\, \partial z} + p_y \frac{\partial^2\rho}{\partial y\, \partial z} + p_z \frac{\partial^2\rho}{\partial z^2} \right] \right\}
\end{aligned}$$

When these terms are averaged over all directions of $\mathbf{p}$, the cross terms, containing factors of the general type p_xp_y, p_xp_z, etc., all average to zero, whereas the terms in p_x^2, p_y^2, or p_z^2 each average to $\frac{1}{3}p^2$, so that the whole quantity becomes simply $(p/3n_tQm)\nabla^2\rho$.

The second equation arising from Eq. (2.4.39) is therefore

$$\frac{\partial}{\partial t}\rho(\mathbf{r},p,t) \simeq \frac{p}{3n_tQm}\nabla^2\rho(\mathbf{r},p,t) - n_tQ_a\frac{p}{m}\rho(\mathbf{r},p,t) + q \tag{2.4.41}$$

which serves to determine the density of particles of momentum magnitude p. The function q is the *source function*, giving the number of particles "created" per second per cubic centimeter of momentum p at r and t.

The diffusion equation for the average density is now not difficult to obtain. In terms of average values defined by Eqs. (2.4.38) and by

$$a^2 = \frac{1}{3m}\rho_{av}\lambda_{av};\quad q = \int_0^\infty p^2 q(p)\,dp;\quad \kappa = \frac{n_t}{\rho(\mathbf{r},t)}\int_0^\infty Q_a p^3\rho(\mathbf{r},p,t)\,dp$$

we finally obtain

$$(\partial/\partial t)\rho(r,t) \simeq a^2\nabla^2\rho(r,t) - \kappa\rho(r,t) + q \tag{2.4.42}$$

which is the diffusion equation with two additional terms, corresponding to annihilation and creation of particles. As we have seen, this equation is valid only for cases where mJ/p_{av} is much smaller than ρ (or, at least, we cannot be sure of its validity if mJ/p_{av} is not very small compared with ρ). The boundary conditions at a boundary surface, beyond which there are no more scattering targets, is that given by Eq. (2.4.34).

Unit Solutions. By analogy with our discussion of the wave equation and Poisson's equation, we search for a solution of Eq. (2.4.42) representing a single particle being produced at $t = 0$ at the origin $[q = \delta(t)\delta(x)\delta(y)\delta(z)]$. The solution should be sharply concentrated near the origin when t is small and should spread out more and more as time goes on. An error function dependence on space coordinates $[\rho = B(t)e^{-r^2D(t)}$, where $r^2 = x^2 + y^2 + z^2]$ may be tried for a solution in unbounded space, and eventually one can show that the required solution of Eq. (2.4.42) is (assuming κ constant)

$$\rho = G(x,y,z;t) = \begin{cases} 0; & t < 0 \\ (4\pi a^2 t)^{-\frac{3}{2}}e^{-(r^2/4a^2t)-(\kappa t)}; & t > 0 \end{cases} \tag{2.4.43}$$

The exponential is not valid for negative values of t; when t is positive but infinitesimally small, ρ is a delta function around the origin (as it should be), but as t increases beyond $t = 0$, the density spreads out more and more. (As a matter of fact, according to our comments on page 185, the diffusion equation, and therefore this solution, is not valid until times somewhat longer than a mean free time $1/n_tQ_{av}v_{av}$ after the introduction of the particle.) If absorption is present (κ is not zero), the whole solution decays exponentially, so that the integral of G over all space (which is the probability that the particle has not been absorbed yet) equals $e^{-(\kappa t)}$.

If particles are produced in a whole region, over a time duration, so that $q(x_0,y_0,z_0;t)\,dx_0\,dy_0\,dz_0\,dt$ of them are produced in the volume element $dx_0\,dy_0\,dz_0$ at the point x_0, y_0, z_0, between the times t and $t + dt$, then the same reasoning which arrived at Eq. (2.1.8) shows that the resulting density of particles in an unbounded space at any time t and point x, y, z is

$$\rho = \int_{-\infty}^{t} d\tau \int_{-\infty}^{\infty} dx_0 \int_{-\infty}^{\infty} dy_0 \int_{-\infty}^{\infty} dz_0\, q(x_0,y_0,z_0;\tau) \cdot \cdot G(x - x_0,\, y - y_0,\, z - z_0;\, t - \tau) \qquad (2.4.44)$$

This solution and others will be discussed in considerable detail later in this book.

When a steady state is reached and ρ no longer depends on the time, the resulting field is a solution of a Helmholtz equation (see page 125), for Eq. (2.4.42) then becomes

$$a^2\nabla^2\rho - \kappa\rho = -q$$

In this case the sign of the term in ρ is opposite to the sign of the corresponding term in the Helmholtz equation obtained from the wave equation, given on page 125.

Loss of Energy on Collision. So far we have been assuming that the atoms of the material through which the fluid particles diffuse (the targets) are rigidly fixed in space, so that the collisions are perfectly elastic and no energy is lost to the material. This is, of course, an idealization of the situation which is approached only by a few actual cases (such as the case of scattered photons, discussed previously). In many more cases of interest, the target atoms are also moving and are not infinitely heavy compared with the diffusing particles, so that these particles can lose (and also gain) kinetic energy by collision. The analysis of the most general case is possible but is tedious and would lead us further astray than we care to go. We shall go only far enough in our discussion here to show what new effects enter when some energy is lost by collisions.

Suppose that the target atoms have mass M, which is considerably greater than mass m of the fluid particles, and also suppose that the fluid particles have, on the average, a much greater kinetic energy than do the target atoms. In this case the targets can be considered to be at rest when hit by the particles, and also only a small fraction of the particle's kinetic energy is lost at each individual collision.

An elementary analysis of momentum and kinetic energy balance shows that, when the particle's initial momentum is p' and its final momentum, after being deflected by an angle α (see Fig. 2.18), is p, then to the first order in the small quantity m/M

$$p \simeq p'[1 - (m/M)(1 - \cos\alpha)]; \quad p' \simeq p[1 + (m/M)(1 - \cos\alpha)] \qquad (2.4.45)$$

Referring to Eqs. (2.4.35) and (2.4.36), we can say that to the first order in the small quantity m/M, for particles of initial momentum p',

The average loss of kinetic energy per elastic collision is $2mQ_m/MQ_e$ times its initial kinetic energy
The average loss of momentum magnitude per elastic collision is mQ_m/MQ_e times its initial momentum p'
The average reduction in forward component of momentum per elastic collision is Q_m/Q_e times its initial forward momentum p' (2.4.46)

where the average number of elastic collisions per second of a particle of momentum p is n_tQ_ep/m. The third statement means that the average value of the component of the final momentum along the direction of initial motion, $(\mathbf{p}\cdot\mathbf{p}'/p')$, equals $[1 - (Q_m/Q_e)]p'$. Incidentally these statements explain why we call Q_m the cross section for *momentum transfer*.

We must now return to our discussion of Eqs. (2.4.14) and (2.4.15) giving the rate of change of the distribution function $f(\mathbf{r},\mathbf{p},t)$ due to elastic collisions. The rate of loss of f due to particles being scattered *out* of momentum element $dV_p = p^2\,dp\,d\phi\,\sin\theta\,d\theta$ (see Fig. 2.18) is, as before,

$$\left(\frac{Q_en_tp}{m}\right) f(\mathbf{r},p,\theta,\phi,t)\,dV_p = \frac{n_tp^3}{m}\,dp\,\sin\theta\,d\theta\,d\phi \int_0^{2\pi} d\beta \int_0^{\pi} \sigma(p,\alpha)\sin\alpha\,d\alpha\, f(\mathbf{r},p,\theta,\phi,t)$$

The rate of increase of f due to particles being scattered *into* element dV_p which were originally in momentum element $dV'_p = (p')^2\,dp'\,d\phi'\,\sin\theta'\,d\theta'$ is

$$\frac{n_t}{m}\sin\theta\,d\theta\,d\phi \int_0^{2\pi} d\beta \int_0^{\pi} [(p')^3\sigma(p',\alpha)f(\mathbf{r},p',\theta',\phi',t)\,dp']\sin\alpha\,d\alpha$$

But from Eq. (2.4.45), p' differs from p by the small quantity $(pm/M)(1 - \cos\alpha)$. The expression in the bracket can be expanded in a Taylor series about p, giving to the first approximation

$$p^3\,dp\left\{\sigma(p,\alpha)f(\mathbf{r},p,\theta',\phi',t) + \frac{m}{Mp^3}(1 - \cos\alpha)\frac{\partial}{\partial p}[p^4\sigma(p,\alpha)f(\mathbf{r},p,\theta',\phi',t)]\right\}$$

We have been assuming that the particle distribution is nearly isotropic and that the drift motion is small compared with the random motion, according to Eq. (2.4.37), where mJ/p is small compared with ρ. We have also assumed that vector $\mathbf{J}$ points in the direction of the polar axis of θ and θ'. As long as these assumptions are valid, the two expressions can be combined and simplified to result in a net rate of change of f due to gain and loss of particles by scattering:

$$\frac{n_t p^3}{4\pi m} dp \sin\theta\, d\theta\, d\phi \int^{2\pi} d\beta \int_0^{\pi} \left\{ \frac{3}{p} \sigma(p,\alpha)[\cos\theta' - \cos\theta] J(\mathbf{r},\mathbf{p},t) + \frac{1}{Mp^3}(1 - \cos\alpha) \frac{\partial}{\partial p} [p^4 \sigma(p,\alpha) \rho(\mathbf{r},\mathbf{p},t)] \right\} \sin\alpha\, d\alpha$$
$$= \frac{n_t p}{4\pi m} dV_p \left[-\frac{3}{p^2} Q_m \mathbf{p} \cdot \mathbf{J} + \frac{1}{Mp^3} \frac{\partial}{\partial p} (p^4 Q_m \rho) \right] \quad (2.4.47)$$

The first term has been derived before [see Eq. (2.4.38) *et seq.*]; the second is a new term which represents the change in f due to loss of energy by collision.

Another rate of change which must be taken into account is that given in Eq. (2.4.9), corresponding to the equation of continuity. The change is

$$-\frac{dV_p}{4\pi m^2} \mathbf{p} \cdot \operatorname{grad} \rho - \frac{3}{4\pi p^2 m} dV_p \mathbf{p} \cdot \operatorname{grad} (\mathbf{p} \cdot \mathbf{J}) \quad (2.4.48)$$

when we insert the approximate expression for f given in Eq. (2.4.37).

Effect of External Force. When we start to take into account the effect on f of energy loss during collision, we also should take into account the possibility that the particle gains energy from an external field of force. If the particles are electrons, for instance, moving through a gas, an electric field could cause drift motion and alter the distribution function f. We suppose this force $\mathbf{F}$ to be uniform and in the direction of the drift current $\mathbf{J}$. (We shall not discuss the case where there is an external force and ρ has a gradient, so for our discussion in this section $\mathbf{J}$ *either* points along $\mathbf{F}$ *or* points along grad ρ, whichever is not zero.)

When a force $\mathbf{F}$ is applied, the momentum of each particle is changed; the particles having momentum $\mathbf{p}$ at time t will have momentum $\mathbf{p} + \mathbf{F}\,dt$ at time $t + dt$. Similarly to the argument resulting in the equation of continuity (2.4.9), we must have the density follow the accelerated particles, $f(\mathbf{r}, \mathbf{p} + \mathbf{F}\,dt, t + dt) = f(\mathbf{r},\mathbf{p},t)$, or the rate of change of $f\,dV_p$ due to the external field is

$$-\left[F_x \frac{\partial f}{\partial p_x} + F_y \frac{\partial f}{\partial p_y} + F_z \frac{\partial f}{\partial p_z} \right] dV_p$$
$$= -F \left[\cos\theta \frac{\partial f}{\partial p} + \frac{1}{p}(1 - \cos^2\theta) \frac{\partial f}{\partial \cos\theta} \right] dV_p$$
$$= -\frac{F}{4\pi} \left[\frac{\cos\theta}{m} \frac{\partial \rho}{\partial p} + 3\cos^2\theta \frac{\partial}{\partial p}\left(\frac{J}{p}\right) + \frac{3J}{p^2}(1 - \cos^2\theta) \right] dV_p \quad (2.4.49)$$

Finally, combining expressions (2.4.47) to (2.4.49) and inserting a term $-(n_t Q_a p f/m)$ to represent absorption of particles, we obtain an expression giving the rate of change of ρ and J due to collision, drift acceleration, and absorption:

$$\frac{\partial\rho}{\partial t} + \frac{3m}{p}\cos\theta\,\frac{\partial J}{\partial t} \simeq -\frac{1}{m}\,\mathbf{p}\cdot\operatorname{grad}\rho - \frac{3}{p^2}\,\mathbf{p}\cdot\operatorname{grad}(\mathbf{p}\cdot\mathbf{J})$$
$$- F\cos\theta\,\frac{\partial\rho}{\partial p} - 3mF\cos^2\theta\,\frac{\partial}{\partial p}\left(\frac{J}{p}\right) - \frac{3mFJ}{p^2}(1-\cos^2\theta)$$
$$- 3n_tQ_mJ\cos\theta + \frac{n_t}{Mp^2}\frac{\partial}{\partial p}(p^4Q_m\rho) - \frac{1}{m}n_tQ_ap\rho - 3n_tQ_aJ\cos\theta \qquad (2.4.50)$$

This can be separated into two equations according to the symmetry with respect to θ (or to the direction of $\mathbf{p}$), and solution of the two will determine $\rho(\mathbf{r},p,t)$ and $\mathbf{J}(\mathbf{r},p,t)$. Two further examples will be discussed, both for steady-state conditions, for cases where the loss-of-energy terms are crucial for the solution.

Uniform Drift Due to Force Field. For our first example consider a uniform distribution of particles is drifting through a medium under the influence of a uniform force $\mathbf{F}$. This is the case of electrons in a gas moving in response to an electric field. Here grad ρ is zero and the drift vector $\mathbf{J}$ must point along $\mathbf{F}$. For steady-state conditions $\partial f/\partial t$ is zero and we shall take Q_a to be zero. The remainder of Eq. (2.4.50) can be multiplied by $\sin\theta\,d\theta$ or by $\cos\theta\sin\theta\,d\theta$ and integrated over θ, resulting in the two equations

$$F\frac{\partial}{\partial p}(pJ) = \frac{n_t}{Mm}\frac{\partial}{\partial p}(p^4Q_m\rho);\quad \frac{\partial\rho}{\partial p} = -\frac{3n_tQ_m}{F}J \qquad (2.4.51)$$

The first equation, when integrated, yields

$$Fp^2J(p) = (n_t/Mm)p^5Q_m\rho(p) - Kp$$

The constant of integration K turns out to be zero according to the following reasoning: Integration of the equation (with $K = 0$) once more with respect to p gives us the energy-balance equation [see (Eq. 2.4.38)]

$$FJ = \int_0^\infty \left(\frac{p^2}{2m}\right)\left(\frac{2mQ_m}{MQ_e}\right)\left(\frac{n_tQ_ep}{m}\right)\rho(p)p^2\,dp$$

The left-hand side is just equal to the energy per cubic centimeter per second picked up from the force field F by the particle drift current density J. On the right-hand side, $(p^2/2m)(2mQ_m/MQ_e)$ has been shown in Eq. (2.4.46) to be the average kinetic energy lost per collision at momentum p; n_tQ_ep/m is the number of collisions per second per particle, and $\rho(p)p^2\,dp$ is the number of particles per cubic centimeter in the momentum range from p to $p + dp$; so that the integral is just the total energy lost per cubic centimeter per second because of collisions. If a steady state has been reached, energy lost by collisions should equal energy gained from force, and therefore this equation should hold. Consequently, K should be zero.

We can now return to the second part of Eqs. (2.4.51) and compute the dependence of ρ on p. Since $J(p) = (n_tQ_m/mMF)p^3\rho(p)$, we have, by integration,

$$\rho(p) = A \exp\left[-\frac{3m}{M}\int_0^p \left(\frac{n_tQ_mp}{mF}\right)^2 p\,dp\right] \qquad (2.4.52)$$

where A is chosen so that the integral of Eq. (2.4.38) gives ρ, the average density of particles. When Q_m is independent of velocity, this becomes

$$\rho(p) = Ae^{-h^4p^4}; \quad h^4 = \frac{3m}{M}\left(\frac{n_tQ_m}{2Fm}\right)^2; \quad A = 3.2642\rho h^3$$

From this distribution function we can compute the average kinetic energy ϵ_{av} of the particles and the mean drift velocity v_{av} down the field, in terms of the energy $\epsilon_e = F/n_tQ_m$ and velocity $v_e = \sqrt{2F/mn_tQ_m}$ which would be gained by the particle in falling from rest down the field for a mean free path. These quantities are

$$\epsilon_{av} = 0.4270(M/m)^{\frac{1}{2}}\epsilon_e; \quad v_{av} = 0.6345(m/M)^{\frac{1}{4}}v_e$$

These relations are true only if Q_m is independent of p. It is seen that the mean energy is larger the heavier are the atoms relative to the particles, because if M/m increases, less energy can be lost by the particle per collision. On the other hand, the average drift velocity diminishes if M/m is increased.

This example bears little relation to the diffusion equation, but it does show how the equations governing the distribution function f can be adjusted, in various cases, to correspond to a very wide variety of conditions and of phenomena. It indicates, for instance, that the distribution in velocity of electrons in a gas, under the influence of an electric field, is not the Maxwell distribution produced by thermal agitation but has a fourth power of the velocity (instead of the square) in the exponential. This is true, of course, only as long as the mean kinetic energy of the electrons is very much higher than the mean kinetic thermal energy of the gas atoms through which the electrons are drifting. Otherwise we cannot consider the gas atoms as relatively at rest, and the electron distribution would be modified by the Maxwell distribution of the atoms.

Slowing Down of Particles by Collisions. The diffusion equation turns up in the next (and last) example, which is of interest in the study of the slowing down of neutrons in a "moderator." We consider here solutions of Eq. (2.4.50) when $\mathbf{F}$ is zero, which are not independent of (x,y,z) and which take into account the slowing-down effect of collisions.

Suppose that particles are introduced into a portion of space all at an initial momentum p_0; they strike atoms and gradually slow down and spread out at the same time. Since this is a steady-state process, new particles being continually introduced to replace those slowed down, the distribution function is independent of time, but it does depend on the momentum **p** and on the space coordinates. As we shall see, **J** is pointed along grad ρ.

The momentum can be taken as a measure of the average "age" of an individual particle since it was introduced at momentum p_0. As this age increases, the distribution in space changes, so that one might expect to find an equation relating the space dependence of f with its dependence on particle age in the same manner as the diffusion equation relates the space dependence of f with its time dependence, in the non-steady-state case.

The average loss of momentum p per collision is shown in Eq. (2.4.46) to be given by

$$(dp/d\tau) \simeq -(mQ_m/MQ)p$$

where τ is the average number of collisions suffered by the particle; consequently we have for the relation between v and τ

$$\tau - \tau_0 = \int_p^{p_0} \left(\frac{MQ}{mQ_m}\right)\left(\frac{dp}{p}\right) \tag{2.4.53}$$

The quantity $\tau - \tau_0$ is called the *age* of the particle in the distribution; it takes the place of time when we consider the case of the steady-state solution with slowing down.

Referring again to Eq. (2.4.50), we omit the force terms and the time derivatives but include energy loss and absorption terms. A source term q, as in Eq. (2.4.41), is not included, for it will be brought in as an "initial condition" for $\tau = \tau_0$. When we multiply by $\cos\theta$ and integrate over all directions, we obtain as before [see Eq. (2.4.40)]

$$\mathbf{J} = -(p/3n_tQm)\ \text{grad}\ \rho; \quad Q = Q_m + Q_a$$

only now the equation is exact, for we now are dealing with a steady-state case, so that the time derivative of **J** is zero instead of being negligibly small.

Substituting this back into Eq. (2.4.50) and averaging over all directions, we obtain [see page 190 for the grad $(\mathbf{p}\cdot\mathbf{J})$ term]

$$-\frac{mn_t}{Mp^2}\frac{\partial}{\partial p}(p^4Q_m\rho) = \frac{p}{3n_tQ}\nabla^2\rho - n_tQ_ap\rho$$

Letting $\psi = (n_tQ_mp^4\rho/M)$ and substituting τ for p, according to Eq. (2.4.53) we finally obtain an equation for ψ having the form of the diffusion equation:

$$\frac{\partial\psi}{\partial\tau} = \frac{1}{3n_t^2Q^2}\nabla^2\psi - \alpha\psi; \quad \alpha = \frac{Q_a}{Q} \tag{2.4.54}$$

where, instead of time, we have $\tau - \tau_0$, the age of the particles having momentum p. The physical meaning of ψ will become apparent shortly.

Solutions of this equation are completely analogous to solutions of the usual diffusion equation. For instance, if the initial particles at momentum p_0 are introduced in a concentrated region of space, the distribution of slower particles is more spread out, the spreading being the greater the larger is τ (*i.e.*, the smaller is p/p_0). Once the steady state is reached, we can take a census of all the particles in a given cubic centimeter at a given instant to find the number present in a given range of momentum (or, rather, in a given interval $d\tau$). From the resulting density with respect to τ we can compute the function ψ at the point as a function of τ, or vice versa, if ψ is known, we can predict the results of the census.

To determine the "initial value" ψ_0, the value of ψ at $\tau = \tau_0$, we must relate the number of particles being produced per cubic centimeter per second at a given point to the total loss of momentum per cubic centimeter per second, *i.e.*, to the number of particles per cubic centimeter times the rate of loss of momentum of a particle per second. The loss of momentum per collision is, from Eq. (2.4.46), equal to mpQ_m/MQ_e on the average. The number of collisions per particle per second is n_tQ_ep/m, so that the total rate of loss of momentum of a particle per second is $dp/dt = n_tQ_mp^2/M$. Since the number of particles per cubic centimeter in the momentum range dp is, by Eq. (2.4.38), $p^2\rho\, dp$, the total loss of momentum per cubic centimeter in the time interval dt is $p^2\rho(dp/dt)\, dt = n_tQ_mp^4\rho dt/M$. Consequently, the quantity $n_tQ_mp^4\rho/M = \psi(\tau)$ is the total number of particles which, at a given time, had momenta greater than p and which, 1 sec later, have momenta less than p. It is the total rate of degradation of momentum.

For instance, if n_0 particles were introduced each second in each cubic centimeter of an enclosed space, ψ_0 would equal n_0 everywhere over the enclosed region (this assumes that the particles are introduced *isotropically*, with all directions of p_0 equally likely). If the boundaries of the region are perfect reflectors for particles, ψ would be independent of space coordinates for all values of τ, and the solution of Eq. (2.4.54) would be

$$\begin{aligned}\psi &= n_0e^{-(Q_a/Q)(\tau-\tau_0)}; && \text{for } \tau > \tau_0 \\ &= n_0(p/p_0)^{(mQ_a/MQ_m)}; && \text{if } Q_m/Q \text{ is independent of } p\end{aligned} \tag{2.4.55}$$

and the number of particles having momenta between p and $p + dp$ is

$$p^2\rho\, dp = (M\psi/n_tQ_mp^2)\, dp = (n_0M/n_tQ_mp^2)(p/p_0)^{(mQ_a/MQ_m)}\, dp$$

for $p < p_0$. The last form of solution is valid only if Q_m/Q does not vary with p. If there is no absorption, $(Q_a = 0)$ ψ is independent of τ and ρ is inversely proportional to p^4Q_m. This must be, for if there is no absorption, the rate of degradation of momenta ψ must be the same for all speeds. This solution, of course, predicts an infinite number of particles having infinitesimal speeds. If our analysis of energy lost per collision holds to the limit of zero speed, we must have this infinite population in order to reach a steady state. In reality, of course, the atoms of the medium are not completely at rest, and therefore, at low enough particle velocities the analysis made above, assuming that the particles always lose energy, becomes invalid. Therefore Eq. (2.4.55) does not hold for particles having kinetic energies of the same magnitude as the mean energy of the atoms or smaller.

As another example, we can imagine one particle of momentum p_0 introduced (isotropically) per second at the point x_0, y_0, z_0 in an unbounded space. Then ψ for $\tau = \tau_0$ would be a delta function $\delta(x - x_0) \cdot \delta(y - y_0)\delta(z - z_0)$, and using the same methods which resulted in Eq. (2.4.43), we obtain $\psi = G(x - x_0, y - y_0, z - z_0|\tau - \tau_0)$, where

$$G(x,y,z|\tau - \tau_0) = \begin{cases} 0; & \tau < \tau_0 \\ \left[\dfrac{3n_t^2Q^2}{4\pi(\tau - \tau_0)}\right]^{\frac{3}{2}} \exp\left[-\dfrac{3n_t^2Q^2r^2}{4(\tau - \tau_0)} - \dfrac{Q_a}{Q}(\tau - \tau_0)\right]; & \tau > \tau_0 \end{cases} \tag{2.4.56}$$

$$\tau = \int_p^\infty \left(\frac{MQ}{mQ_m}\right)\left(\frac{dp}{p}\right); \quad \rho(r,p) = \frac{M\psi}{n_tQ_mp^4}$$

Finally we can compute the result if $q(x_0,y_0,z_0|p_0)\,dp_0$ particles are introduced isotropically per second per cubic centimeter at the point x_0, y_0, z_0 in unbounded space, in the range of momentum between p_0 and $p_0 + dp_0$. The number of particles introduced per second between the "ages" τ_0 and $\tau_0 + d\tau_0$ is, therefore, $(p_0mQ_m/MQ)q(x_0,y_0,z_0|\tau_0)\,d\tau_0$, where p_0 is related to τ_0 in the same way that p and τ are related in Eq. (2.4.56). The resulting steady-state distribution of particles in space for different momenta p can be obtained from the solution for ψ:

$$\psi(x,y,z|\tau) = \frac{mQ_m}{MQ}\int_{-\infty}^{\tau} p_0\,d\tau_0 \int_{-\infty}^{\infty} dx_0 \int_{-\infty}^{\infty} dy_0 \int_{-\infty}^{\infty} dz_0 \cdot q(x_0,y_0,z_0|\tau_0)G(x - x_0, y - y_0, z - z_0|\tau - \tau_0) \tag{2.4.57}$$

where the number of particles per cubic centimeter in the range of momentum between p and $p + dp$ is $p^2\rho\,dp = (M\psi/n_tp^2Q_m)\,dp$. For some simple forms of q the integration can be performed and a closed analytic solution can be found for ψ.

Recapitulation. Many other applications of the diffusion equation can be found. The only requirement for its occurrence is that some quantity (density, partial pressure, heat, etc.) satisfy two requirements: first, that it obey the equation of continuity, that the rate of change of the quantity with time be equal to minus the divergence of the flow of the quantity and, second, that the flow of the quantity be proportional to the negative gradient of the quantity. Other solutions will be given in Chap. 12.

Since the time derivative enters to the first order whereas the space derivatives are to the second order, the solutions of the diffusion equation are irreversible in time. Nearly all other equations we shall discuss represent reversible phenomena in the thermodynamic sense, whereas solutions of the diffusion equation represent degradation of entropy (this will be discussed again in the next chapter). All of this is roughly equivalent to saying that phenomena represented by the diffusion equation are inherently statistical events.

2.5 *The Electromagnetic Field*

Another important branch of physics, where the concept of field turns out to be remunerative, is that of electricity. Some of the elementary particles of matter are electrically charged, and most, if not all, have magnetic moments. Electromagnetic theory has been elaborated to describe their interactions in bulk (the interactions between individual particles usually involve quantum phenomena, which will be described later in this chapter).

As with the fields encountered earlier in this chapter, the electric charge can often be considered to be a continuous fluid rather than a swarm of charged particles. Classical electromagnetic theory deals with the fields produced by various configurations of such a fluid and with the interaction of these fields with other parts of the fluid. Of course it might be possible to discuss the force on one portion of the fluid (or on one particle) due to another portion (or particle) without talking about a field at all. But it seems considerably easier, and perhaps also better, to break the problem into two parts: first, the "creation" of an electromagnetic field by a distribution of charge and current, second, the effect of this field on the distribution of charge and current.

The Electrostatic Field. The effect of one charged particle on another is quite similar to the interactive effect of gravity. The magnitude of the force on each particle is inversely proportional to the square of their distance of separation and directly proportional to the product of the "strength" of their charges; the direction of the force is along the line joining them (as long as the particles are relatively at rest). In the case of gravitation the force is always attractive and the "strength" of the gravitational charge is proportional to its mass;

between static electric charges the force is repulsive if the charges are the same sign and attractive if the charges are opposite in sign.

The force on any one particle or portion of charge is therefore proportional to the "strength" of its own charge. We can thus define a vector field, called the *electrostatic* (or the *gravitational*) *field*, **E**, at some point which is the ratio of the force on a test particle at the point to the strength of charge of the test particle. The vector **E** is called the electric (or gravitational) *intensity* at the point. This field, being the sum of forces with magnitudes inversely proportional to the squares of the distances from the various charges present, is the one which was discussed on page 17, due to a number of "source points." We showed in Eq. (1.2.10) that the net outflow integral for this sort of field, over any closed surface, is equal to 4π times the sum of the charges of all the particles inside the surface.

As long as we are considering only large-scale effects, we need not consider the microscopic irregularities of the field due to the fact that the charge is concentrated on discrete particles instead of being spread out smoothly; we need consider only average fields over elements of surface area large compared with interparticle distances but small compared with the total surface considered. When this can be done, the resulting averaged field is equivalent to one caused by a "smoothed-out," continuous charge distribution, and we do not need to bother ourselves about the exact positions of each individual particle. We can choose an element of volume $dx\,dy\,dz$ "around" the point x, y, z, containing a fairly large number of particles. The total charge inside the element is the average density of charge "at" x, y, z, times $dx\,dy\,dz$. This is proportional to the net outflow integral over the surface of the element, which is, by Eq. (1.4.5), equal to $dx\,dy\,dz$ times the divergence of the field **E**.

Consequently, for large-scale effects, we can replace the swarm of charged particles by a smooth distribution of charge of density $\rho(x,y,z,t)$. When this density is independent of time, Eq. (1.4.5) shows that the resulting static field **E** is related to ρ by the equation

$$\operatorname{div} \mathbf{E} = (4\pi/\epsilon)\rho$$

The factor of proportionality ϵ is characteristic of the medium and is called the *dielectric constant* of the medium. Whenever ϵ changes from point to point, it is better to compute a vector field related to **E**, called the *displacement* field **D**, where

$$\operatorname{div} \mathbf{D} = 4\pi\rho; \quad \mathbf{D} = \epsilon\mathbf{E} \tag{2.5.1}$$

We solve for **D** in terms of ρ; then, knowing ϵ, compute **E** and from **E** compute the force $\rho\mathbf{E}$ on a cubic centimeter of the electric fluid.

The vector **E** can always be expressed in terms of the curl of a vector and the gradient of a scalar potential, as was shown on page 53. But since the divergence of a curl is zero, the vector potential for **E** is not determined by Eq. (2.5.1) and does not enter into electrostatic calculations. The scalar potential φ must satisfy the following equation:

$$\operatorname{div}[\epsilon \operatorname{grad} \varphi] = \epsilon \nabla^2 \varphi + (\operatorname{grad} \epsilon) \cdot (\operatorname{grad} \varphi) = -4\pi\rho; \quad \mathbf{E} = -\operatorname{grad} \varphi \qquad (2.5.2)$$

When ϵ is constant, this becomes a Poisson equation for φ [see Eq. (2.1.2)].

In the case of gravity the quantity ϵ is everywhere constant and the equation for the intensity has a reversed sign, $\operatorname{div} \mathbf{E} = -(4\pi\rho/\epsilon)$, corresponding to the fact that the force is an attraction, not a repulsion. In this case also there is a scalar potential, which is everywhere a solution of Poisson's equation $\nabla^2\varphi = -(4\pi\rho/\epsilon)$.

The Magnetostatic Field. Ferromagnetic materials, having atoms with unneutralized magnetic moments, behave as though they were charged with a magnetic fluid[1] analogous to the electrical fluid we have just discussed. If there were a unit positive magnetic charge, it would be acted on by a force represented by a vector field **H**, analogous to the electric field **E**. Analogous to the dielectric constant ϵ is the permeability μ, and analogous to the displacement vector $\mathbf{D} = \epsilon\mathbf{E}$ is the *magnetic induction* $\mathbf{B} = \mu\mathbf{H}$.

The important difference between electricity and magnetism, however, is that there is no magnetic charge. The equation for the induction field **B**, instead of Eq. (2.5.1), is

$$\operatorname{div} \mathbf{B} = 0 \qquad (2.5.3)$$

One could, of course, express **B** as the gradient of a scalar potential which would always be a solution of Laplace's equation [Eq. (2.3.6)]. But it is better to utilize the fact that the divergence of the curl of any vector is zero and express **D** in terms of a *vector potential*, $\mathbf{B} = \operatorname{curl} \mathbf{A}$.

This would be about as far as one could go with magnetostatics were it not for the fact that the magnetic field turns out to be related to the flow of electric charge, the electric current. For instance, if a long straight wire of negligible diameter carries current **I**, where the direction of the vector is the direction of the current along the wire, then the magnetic field around the wire is given by the equation

$$\mathbf{H} = (2\mathbf{I} \times \mathbf{r})/r^2$$

[1] Since there are no magnetic charges, but only magnetic moments, it would be more logical to derive the magnetic equations by considering the torque on a magnetic dipole. This is not a text on electromagnetics, however, and we may be forgiven (we hope) for saving space by deriving the equations by analogy with electrostatics rather than using a few more pages to present the more logical derivation, which is given in detail in such texts as Frank, "Introduction to Electricity and Optics," McGraw-Hill, or Stratton, "Electromagnetic Theory," McGraw-Hill.

where $\mathbf{r}$ is the vector, perpendicular to the wire, from the wire to the point at which $\mathbf{B}$ is measured.

But this is just the field, discussed on page 19, due to a simple vortex. Reference to this and to the definition of the vorticity vector, given on page 41, leads to a general relation between $\mathbf{B}$ and a steady current. If the charge fluid is moving, the velocity $\mathbf{v}$ of the charge times the charge density ρ is expressible as a current density $\mathbf{J}$, charge per square centimeter per second, having the direction of the velocity of the charge at each point. This current is related to the vorticity vector for $\mathbf{H}$ by the simple equation

$$\text{curl } \mathbf{H} = 4\pi\mathbf{J} \tag{2.5.4}$$

(Incidentally, it is possible to have a current and yet not have free charge if the current is caused by the opposing motion of equal amounts of positive and negative charge.)

Reciprocally, there is a force on a current in a magnetic field. The force on a cubic centimeter of moving charge at point x, y, z is

$$\mathbf{F} = \rho\mathbf{v} \times \mathbf{B} = \mathbf{J} \times \mathbf{B}; \quad \mathbf{B} = \mu\mathbf{H} \tag{2.5.5}$$

For steady-state problems these equations suffice. The charge sets up the electrostatic field ($\mathbf{E}$,$\mathbf{D}$); the current sets up the magnetic field ($\mathbf{H}$,$\mathbf{B}$). The electric field in turn acts on the charge, and the magnetic field on the current. Electric field is caused by, and causes force on, static charges; magnetic field is caused by, and causes force on, moving charges. A rather far-fetched analogy can be made between the scalar potential, determining the electric field, and the scalar wave potential for purely compressional waves in elastic media and between the vector potential, determining the magnetic field, and the vector wave potential for shear waves. Thus far, however, there can be no wave motion, for we have considered only the steady state.

For unbounded space containing finite, steady-state charge and current distribution and with dielectric constant ϵ everywhere the same, the solution of Eq. (2.5.2) is, according to Eq. (1.4.8),

$$\varphi(x,y,z) = \frac{1}{\epsilon} \iiint_{-\infty}^{\infty} \frac{1}{R} \rho(x',y',z')\, dx'\, dy'\, dz' \tag{2.5.6}$$

where $R^2 = (x - x')^2 + (y - y')^2 + (z - z')^2$ and $\mathbf{E} = -\text{grad } \varphi$, $\mathbf{D} = -\epsilon \text{ grad } \varphi$.

The vector potential $\mathbf{A}$ can just as well be adjusted so that its divergence is zero (since we are interested only in that part of $\mathbf{A}$ which has a nonzero curl). Since $\nabla^2\mathbf{A} = \text{grad div } \mathbf{A} - \text{curl curl } \mathbf{A}$, we have, from Eq. (2.5.4),

$$\nabla^2\mathbf{A} = -4\pi\mathbf{J} \tag{2.5.7}$$

and, from Eq. (1.5.16),

$$\mathbf{A}(x,y,z) = \iiint_{-\infty}^{\infty} (1/R)\mathbf{J}(x',y',z')\, dx'\, dy'\, dz' \tag{2.5.8}$$

where $\mathbf{H} = \text{curl}\,\mathbf{A}$, $\mathbf{B} = \mu\,\text{curl}\,\mathbf{A}$.

Dependence on Time. So far we have been discussing steady-state conditions, where $\partial\rho/\partial t$ and div $\mathbf{J}$ are both zero [div $\mathbf{J}$ must be zero if $\partial\rho/\partial t$ is zero because of the equation of continuity (2.3.1)]. But if ρ and $\mathbf{J}$ vary with time, Eqs. (2.5.1) and (2.5.4) must become related by reason of the equation of continuity relating charge density and charge flow. Here we must relate the units of charge, current, and field in the two equations. If we use the mks system, the equations stand as written, with ϵ for vacuum being $\epsilon_0 \simeq (\frac{1}{9}) \times 10^{-9}$ and with μ for vacuum being $\mu_0 \simeq 10^{-7}$. We prefer, however, to use the mixed system of Gauss, measuring charge in statcoulombs and current in statamperes, magnetic field in electromagnetic units (μ for vacuum being 1) and electric field in electrostatic units (ϵ for vacuum being 1). Then Eq. (2.5.4) for the steady state becomes

$$-c\,\text{curl}\,\mathbf{H} = -4\pi\mathbf{J} \tag{2.5.9}$$

and Eq. (2.5.5) becomes $\mathbf{F} = (1/c)\,\mathbf{J} \times \mathbf{B}$, where $c = \sqrt{1/\mu_0\epsilon_0} \simeq 3 \times 10^8$ meters per sec $= 3 \times 10^{10}$ cm per sec.

Since the equation of continuity $\partial\rho/\partial t = -$ div $\mathbf{J}$ must hold, we should obtain an identity by taking the divergence of Eq. (2.5.9) and the time derivative of Eq. (2.5.1) and equating the two. The left-hand sides do not balance, however, for there is a term div$(\partial\mathbf{D}/\partial t)$ left over. This is not surprising, for these two equations were set up for steady-state fields, and any term in the time derivative of $\mathbf{D}$ should have vanished. The equation for the time-dependent magnetic field evidently should be

$$c\,\text{curl}\,\mathbf{H} - (\partial\mathbf{D}/\partial t) = 4\pi\mathbf{J} \tag{2.5.10}$$

This is confirmed by experiment. The equation is called the equation for *magnetic induction* or the Ampère circuital law for **H**. Not only does electric current produces a magnetic field; a change in the electric field also produces it.

Maxwell's Equations. We have nearly arrived at a symmetric pattern for the forms of the field equations; there are two equations dealing with the divergences of **B** and **D** [Eqs. (2.5.1) and (2.5.3)] and an equation dealing with the curl of **H** [Eq. (2.5.10)]. We need a fourth equation, dealing with the curl of **E**, in order to obtain a symmetric pattern. The fourth equation, however, cannot be completely symmetric to Eq. (2.5.10), for there is no magnetic current, any more than there is a magnetic charge. The nearest we can come to symmetry is an equation relating curl **E** and $\partial\mathbf{B}/\partial t$. This equation is also confirmed experimentally; it is called the Faraday law of *electric induction* and relates the change of magnetic field to the vorticity of the electric field. The experimental results show that a factor $-(1/c)$ must be included on the right side of the equation. We can therefore write down the four symmetric equations relating the fields to the currents,

$$\begin{aligned} &\operatorname{curl}\mathbf{H} = \frac{1}{c}\frac{\partial\mathbf{D}}{\partial t} + \frac{1}{c}4\pi\mathbf{J}; \quad && \operatorname{curl}\mathbf{E} = -\frac{1}{c}\frac{\partial\mathbf{B}}{\partial t} \\ &\operatorname{div}\mathbf{B} = 0; && \operatorname{div}\mathbf{D} = 4\pi\rho \\ &\mathbf{B} = \mu\mathbf{H}; && \mathbf{D} = \epsilon\mathbf{E} \end{aligned} \tag{2.5.11}$$

which are called *Maxwell's equations*. The force on a cubic centimeter of charge current is

$$\mathbf{F} = \rho\mathbf{E} + (1/c)\mathbf{J}\times\mathbf{B} \tag{2.5.12}$$

These are the fundamental equations defining the classical electromagnetic field, resulting from a "smoothed-out" charge and current density.

The equations for the scalar and vector potentials are also modified by the variation with time. We still set $\mathbf{B} = \operatorname{curl}\mathbf{A}$, for it automatically makes $\operatorname{div}\mathbf{B} = 0$. Placing this into the equation for curl **E** results in

$$\operatorname{curl}\mathbf{E} = -\frac{1}{c}\operatorname{curl}\left(\frac{\partial\mathbf{A}}{\partial t}\right) \quad\text{or}\quad \operatorname{curl}\left[\mathbf{E} + \frac{1}{c}\frac{\partial\mathbf{A}}{\partial t}\right] = 0$$

A vector whose curl is zero can be derived from a scalar potential function, so that $[\mathbf{E} + (1/c)(\partial\mathbf{A}/\partial t)]$ is the gradient of some scalar. In the steady-state case we had that $\mathbf{E} = -\operatorname{grad}\varphi$, where φ was the scalar potential, so we might as well define φ by the equation

$$\mathbf{E} = -\operatorname{grad}\varphi - \frac{1}{c}\frac{\partial\mathbf{A}}{\partial t} \tag{2.5.13}$$

Setting these equations for **A** and φ into the equation for curl **H** (and assuming that μ and ϵ are constant), we obtain

$$\frac{1}{\mu}\operatorname{curl}\operatorname{curl}\mathbf{A} = \frac{-\epsilon}{c}\left[\operatorname{grad}\frac{\partial\varphi}{\partial t} + \frac{1}{c}\frac{\partial^2\mathbf{A}}{\partial t^2}\right] + \frac{1}{c}4\pi\mathbf{J}$$

or

$$\nabla^2\mathbf{A} - \frac{\epsilon\mu}{c^2}\frac{\partial^2\mathbf{A}}{\partial t^2} = \operatorname{grad}\left[\operatorname{div}\mathbf{A} + \frac{\epsilon\mu}{c}\frac{\partial\varphi}{\partial t}\right] - \frac{\mu}{c}4\pi\mathbf{J}$$

This would be an equation just for **A**, a generalization of Eq. (2.5.7), if the quantity in square brackets were zero. Since only the part of **A** having nonzero curl is as yet defined, it is always possible to adjust the divergence of **A** so that

$$\operatorname{div}\mathbf{A} = -\frac{\epsilon\mu}{c}\frac{\partial\varphi}{\partial t} \tag{2.5.14}$$

Inserting Eqs. (2.5.13) and (2.5.14) into the last of Maxwell's equations, the one for div **D** (still assuming that μ and ϵ are constant), we have an equation for the scalar potential. Together with the final equation for **A**, we have

$$\nabla^2\varphi - \frac{\epsilon\mu}{c^2}\frac{\partial^2\varphi}{\partial t^2} = -\frac{4\pi\rho}{\epsilon};\quad \nabla^2\mathbf{A} - \frac{\epsilon\mu}{c^2}\frac{\partial^2\mathbf{A}}{\partial t^2} = -\frac{4\pi\mu\mathbf{J}}{c} \tag{2.5.15}$$

which serve to determine the scalar and vector potentials in terms of ρ and **J**. When ρ and **J** are zero, they are wave equations with wave velocity equal to $c\sqrt{1/\epsilon\mu}$ in Gaussian units (or $\sqrt{1/\mu\epsilon}$ in mks units). When ϵ and μ are both unity (in a vacuum), the speed of the waves is equal to the speed of light, 3×10^{10} cm per sec; in fact it *is* the speed of light. We note from these equations that relation (2.5.14) between φ and **A** is a consequence of the equation of continuity for ρ and **J**. By analogy with waves in elastic solids we call the waves in φ *longitudinal* waves and the waves in **A** *transverse* waves.

Retardation and Relaxation. By the same sort of reasoning which obtained the integral of Eq. (1.4.8) for a solution of Poisson's equation in infinite space, we can show that solutions of the equations for $\varphi(x,y,z,t)$ and $\mathbf{A}(x,y,z,t)$ in infinite space caused by a finite distribution of $\rho(x,y,z,t)$ and $\mathbf{J}(x,y,z,t)$ are

$$\begin{aligned}\varphi^\circ(x,y,z,t) &= \int_{-\infty}^{\infty} \frac{1}{R\epsilon}\,\rho\left(x',y',z',t - \frac{R}{c'}\right) dx'\,dy'\,dz' \\ \mathbf{A}^\circ(x,y,z,t) &= \int_{-\infty}^{\infty} \frac{\mu}{Rc}\,\mathbf{J}\left(x',y',z',t - \frac{R}{c'}\right) dx'\,dy'\,dz'\end{aligned} \tag{2.5.16}$$

where, as before, $R^2 = (x - x')^2 + (y - y')^2 + (z - z')^2$ and also $(c')^2 = c^2/\epsilon\mu$.

These integral solutions show that the effect on the potentials at the point x, y, z due to the charge and current at x', y', z' is felt a time R/c' later in time. The potential has been *retarded* in its effect by the time required for a wave with speed $c' = c/\sqrt{\epsilon\mu}$ to go from x', y', z' to x, y, z. Since the wave equation is symmetrical with respect to reversal of the direction of time, another solution can be obtained by replacing $t - (R/c')$ by $t + (R/c')$ in the integrands. Up until the present, however, such advanced potentials have not turned out to be of much practical use.

Specifying the potentials "overspecifies" the fields, or vice versa, more than one set of potentials has the same fields. The solutions given in Eq. (2.5.16) can be modified by adding to each a different function, so related that Eq. (2.5.14) still holds. In other words, we can take any function χ, satisfying appropriate boundary conditions, and form new solutions of Eqs. (2.5.15) by setting

$$\mathbf{A} = \mathbf{A}^\circ - \text{grad }\chi; \quad \varphi = \varphi^\circ + \frac{1}{c}\frac{\partial \chi}{\partial t} \tag{2.5.17}$$

Then $\text{div }\mathbf{A} + (\mu\epsilon/c)(\partial\varphi/\partial t) = -\nabla^2\chi + (\mu\epsilon/c^2)(\partial^2\chi/\partial t^2)$, so that the equations for the new $\mathbf{A}$ and φ may differ from the forms given in Eqs. (2.5.15). The electric and magnetic fields, however (which are the measurable quantities), are not affected by the choice of χ. In the process of taking the curl of $\mathbf{A}$ to obtain $\mathbf{B}$, the grad χ term vanishes, and in the process of computing $\mathbf{E}$ from Eq. (2.5.13) the terms involving grad $(\partial\chi/\partial t)$ cancel. This invariance of the actual fields to changes in the potentials, which leave Eq. (2.5.14) invariant, is called *gauge invariance*, discussed again later.

It can be used to simplify the solutions of Maxwell's equations in regions where there is no free charge. In this case φ itself is a solution of the wave equation, so that we can adjust χ to cancel φ and we need not consider the scalar potential at all. This means, according to Eq. (2.5.14), that $\mathbf{A}$ would have zero divergence, and the necessary equations become

$$\begin{gathered}\mathbf{H} = \text{curl }\mathbf{A}; \quad \mathbf{E} = -(1/c)(\partial\mathbf{A}/\partial t); \quad \text{div }\mathbf{A} = 0 \\ \nabla^2\mathbf{A} = -\text{ curl curl }\mathbf{A} = \frac{\mu\epsilon}{c^2}\frac{\partial^2\mathbf{A}}{\partial t^2} - 4\pi\mu\frac{\mathbf{J}}{c}\end{gathered} \tag{2.5.18}$$

Even when there is free charge, we can choose $\mathbf{A}$ so that $\text{div }\mathbf{A} = 0$. Then, from Eqs. (2.5.14) and (2.5.15), the scalar potential must be a solution of Poisson's equation $\nabla^2\varphi = -4\pi\rho/\epsilon$.

Inside metallic conductors there is no free charge, and the current density is proportional to $\mathbf{E}$, $\mathbf{J} = \sigma\mathbf{E}$, where σ is the volume *conductivity* of the metal. The equation for the vector potential then becomes the linear equation

$$\mu\epsilon\frac{\partial^2\mathbf{A}}{\partial t^2} + 4\pi\mu\sigma\frac{\partial\mathbf{A}}{\partial t} - c^2\nabla^2\mathbf{A} = 0 \tag{2.5.19}$$

This equation is very similar to Eq. (2.3.20) for damped compressional waves in a fluid. The term in $\partial\mathbf{A}/\partial t$ introduces a damping term in the solution, either in the space or in the time dependence or both, as with the compressional waves.

For instance, if the wave is simple harmonic, of frequency $\omega/2\pi$, the equation for space dependence is

$$\nabla^2\mathbf{A} + (1/c^2)[\mu\epsilon\omega^2 + 4\pi i\mu\sigma\omega]\mathbf{A} = 0$$

for the time factor $e^{-i\omega t}$. This, as on page 137, represents an attenuation of the wave in space. On the other hand, for standing waves, where

$\nabla^2\mathbf{A} + k^2\mathbf{A} = 0$, the equation for the time dependence becomes

$$\frac{\partial^2\mathbf{A}}{\partial t^2} + \frac{4\pi\sigma}{\epsilon}\frac{\partial\mathbf{A}}{\partial t} + \frac{c^2k^2}{\mu\epsilon}\mathbf{A} = 0$$

which corresponds to damped oscillations in time. Such free oscillations of current inside a conductor are called *relaxation* oscillations.

Lorentz Transformation. In space, where μ and ϵ are both unity, the wave velocity is c, the velocity of light. In this case we should expect that Eqs. (2.5.11) and (2.5.15), relating the fields and the potentials to the charge current, should bear some simple relationship to the Lorentz transformation discussed in Sec. 1.7. The operator ($\mu = \epsilon = 1$)

$$\square^2 = \nabla^2 - \frac{1}{c^2}\frac{\partial^2}{\partial t^2}$$

called the *d'Alembertian*, obviously [see (Eq. 1.7.6)] has Lorentz invariance of form (see page 98). By saying that the equations for $\mathbf{A}$ and φ are Lorentz invariant, we mean that they should have the same form for any observer, no matter what his relative velocity (as long as it is a constant velocity). This will be true if the quantity $(J_x,J_y,J_z,i\rho c) = \mathbf{I}$ is a four-vector, with the transformation properties given in Eq. (1.7.4). The easiest way to show that charge current is a four-vector is to show that its "scalar product" with a known four-vector is a Lorentz invariant. A known four-vector operator is the four-dimensional gradient, with components

$$\left(\frac{\partial}{\partial x}, \frac{\partial}{\partial y}, \frac{\partial}{\partial z}, \frac{1}{ic}\frac{\partial}{\partial t}\right) = \square$$

The "scalar product" of these two,

$$\square \cdot \mathbf{I} = \operatorname{div} \mathbf{J} + (\partial\rho/\partial t)$$

is just zero, by the equation of continuity [Eq. (2.3.1) and page 204]. Zero is certainly a Lorentz invariant, so we have proved that the quantity involving $\mathbf{J}$ and ρc is a four-vector. In other words, in terms of the coordinates

$$x_1 = x; \quad x_2 = y; \quad x_3 = z; \quad x_4 = ict$$

the components

$$I_1 = J_x; \quad I_2 = J_y; \quad I_3 = J_z; \quad I_4 = ic\rho$$

and

$$V_1 = A_x; \quad V_2 = A_y; \quad V_3 = A_z; \quad V_4 = i\varphi$$

are components of two four-vectors. The Lorentz-invariant equations

$$\square \cdot \mathbf{I} = \sum_n \frac{\partial I_n}{\partial x_n} = 0 \quad \text{and} \quad \square \cdot V = \sum_n \frac{\partial V_n}{\partial x_n} = 0$$

are just the equation of continuity and Eq. (2.5.14) relating $\mathbf{A}$ and φ. The set of four equations, for the four components of

$$\square^2\mathbf{V} = -(4\pi\mathbf{I}/c)$$

are just the Lorentz-invariant equations (2.5.15), the wave equations for the potentials.

The electric and magnetic fields are *not* four-vectors. According to Eq. (2.5.13) (and remembering that, when $\epsilon = \mu = 1$, then $\mathbf{B} = \mathbf{H}$ and $\mathbf{D} = \mathbf{E}$),

$$H_x = \frac{\partial A_z}{\partial y} - \frac{\partial A_y}{\partial z} = \left[\frac{\partial V_3}{\partial x_2} - \frac{\partial V_2}{\partial x_3}\right];$$

$$E_x = -\frac{\partial \varphi}{\partial x} - \frac{1}{c}\frac{\partial A_x}{\partial t} = i\left[\frac{\partial V_4}{\partial x_1} - \frac{\partial V_1}{\partial x_4}\right]$$

But the expressions in square brackets are components of an antisymmetric tensor (or what might better be called a four-dyadic) $\mathfrak{F}$, with components

$$f_{mn} = \frac{\partial V_n}{\partial x_m} - \frac{\partial V_m}{\partial x_n};\quad f_{11} = f_{22} = f_{33} = f_{44} = 0$$

$$f_{12} = -f_{21} = H_z;\quad f_{13} = -f_{31} = -H_y;\quad f_{23} = -f_{32} = H_x$$

$$f_{14} = -f_{41} = -iE_x;\quad f_{24} = -f_{42} = -iE_y;\quad f_{34} = -f_{43} = -iE_z$$

The Maxwell equations can then be put into Lorentz-invariant form. The components of the equation $\square \cdot \mathfrak{F} = (4\pi/c)\mathbf{I}$,

$$\sum_n \frac{\partial f_{mn}}{\partial x_n} = \frac{4\pi}{c} I_m \tag{2.5.20}$$

correspond to the equations

$$\operatorname{div}\mathbf{E} = 4\pi\rho;\quad \text{for } m = 4 \quad\text{and}\quad \operatorname{curl}\mathbf{H} - \frac{1}{c}\frac{\partial \mathbf{E}}{\partial t} = \frac{4\pi}{c}\mathbf{J};\quad \text{for } m = 1, 2, 3$$

The equations $\operatorname{div}\mathbf{H} = 0$ and $\operatorname{curl}\mathbf{E} + (1/c)(\partial\mathbf{H}/\partial t) = 0$, which correspond to the definition of $\mathbf{H}$ and $\mathbf{E}$ in terms of the potentials $\mathbf{A}$ and φ [Eqs. (2.5.13) and $\mathbf{H} = (1/\mu)\operatorname{curl}\mathbf{A}$], can be rewritten in the forms

$$\frac{\partial f_{23}}{\partial x_1} + \frac{\partial f_{31}}{\partial x_2} + \frac{\partial f_{12}}{\partial x_3} = 0;\quad \frac{\partial f_{34}}{\partial x_2} + \frac{\partial f_{42}}{\partial x_3} + \frac{\partial f_{23}}{\partial x_4} = 0$$

$$\frac{\partial f_{34}}{\partial x_1} + \frac{\partial f_{41}}{\partial x_3} + \frac{\partial f_{13}}{\partial x_4} = 0;\quad \frac{\partial f_{24}}{\partial x_1} + \frac{\partial f_{41}}{\partial x_2} + \frac{\partial f_{12}}{\partial x_4} = 0$$

which can be shown to be Lorentz-invariant also [*i.e.*, whereas in three dimensions Eq. (1.5.3) shows that $c_{23} - c_{23} = b_1$, etc., are components of a three-vector, in four dimensions $c_{123} + c_{231} + c_{312} = b_4$, etc., are components of a pseudovector].

As an example of the way the fields transform under a Lorentz transformation, we shall start with coordinates x_1', etc.; potentials A_1', etc.; and fields E_1', etc., for an observer B at rest with respect to the primed coordinates. With respect to an observer A moving with velocity $u = c \tanh \alpha$ along the x axis with respect to B, the coordinates, charge current, and potentials become [see Eqs. (1.7.2) and (1.7.4)]

$$\begin{gathered} x = x' \cosh \alpha + ct' \sinh \alpha; \quad y = y'; \quad z = z'; \\ ct = x' \sinh \alpha + ct' \cosh \alpha \\ J_x = J_x' \cosh \alpha + c\rho' \sinh \alpha; \quad J_y = J_y'; \quad J_z = J_z'; \\ c\rho = J_x' \sinh \alpha + c\rho' \cosh \alpha \\ A_x = A_x' \cosh \alpha + \varphi' \sinh \alpha; \quad A_y = A_y'; \quad A_z = A_z'; \\ \varphi = A_x' \sinh \alpha + \varphi' \cosh \alpha \\ \tanh \alpha = u/c; \quad \sinh \alpha = u/\sqrt{c^2 - u^2}; \quad \cosh \alpha = c/\sqrt{c^2 - u^2} \end{gathered} \tag{2.5.21}$$

We should expect that the new x component of the current would have a term in ρ', because the charge is now moving. But it might not be expected that the new charge density has a term depending on the x component of the current.

The fields, being tensor components, follow the corresponding transformation rules:

$$E_x = if_{14} = i \sum_{mn} \gamma_{1m}\gamma_{4n}f'_{mn} = i \cosh^2 \alpha f'_{14} + i \sinh^2 \alpha f'_{41} = if'_{14} = E'_x; \text{ etc.}$$

so that

$$\begin{aligned} E_x &= E_x'; \quad E_y = E_y' \cosh \alpha + H_z' \sinh \alpha; \quad E_z = E_z' \cosh \alpha - H_y' \sinh \alpha \\ H_x &= H_x'; \quad H_y = H_y' \cosh \alpha - E_z' \sinh \alpha; \quad H_z = H_z' \cosh \alpha + E_y' \sinh \alpha \end{aligned} \tag{2.5.22}$$

We see that electric and magnetic fields are intermingled by the motion, which is not surprising, since we saw earlier that stationary charge causes only an electric field but moving charge also causes a magnetic field.

Gauge Transformation. Although it is more convenient for our formal discussions to consider the vector and scalar potential to be components of a four-vector, they do not have to follow the Lorentz transformation in this simple manner. The electric and magnetic fields, of course, must transform as parts of the four-dyadic f_{nm}, for they are the physically measurable quantities and they must satisfy the Maxwell equations, which conform to the Lorentz transformation. But there is a "loose-jointedness" in the potentials, as has been mentioned on page 207. If we relate $\mathbf{A}$ and φ by the usual relation $\operatorname{div} \mathbf{A} + (1/c)(\partial\varphi/\partial t) = 0$, then potential $(\mathbf{A},i\varphi)$ is a four-vector. But if we perform a *gauge transformation* of the sort $\mathbf{A}' = \mathbf{A} - \operatorname{grad} \chi$, $\varphi' = \varphi + (1/c)(\partial\chi/\partial t)$, where χ is a solution of the wave equation, the new potentials $(\mathbf{A}',i\varphi')$ do not necessarily form a four-vector, though the fields computed by their means are the same as before the transformation.

The invariance of the fields under *gauge* transformations of the potentials allows us to choose a form for the potentials which transforms under a *Lorentz* transformation as a four-vector but which may not be particularly easy to compute, or else to choose a form which is easy to compute in a particular Lorentz system, even though it is more difficult to compute its behavior under a transformation of the Lorentz system.

For instance, if we choose, for a particular Lorentz system, the gauge corresponding to div $\mathbf{A} = 0$, then we shall have achieved a more complete separation between "longitudinal" and "transverse" fields (for that Lorentz system) than any other choice of gauge would allow. The equations for the new $\mathbf{A}$ and φ (which we shall label $\mathbf{A}^\circ$ and φ°) can be obtained by reworking the calculations between Eqs. (2.5.13) and (2.5.15), making the new assumption that div $\mathbf{A}^\circ = 0$. The equation for the scalar potential simplifies; it becomes

$$\nabla^2\varphi^\circ = -(4\pi\rho/\epsilon) \tag{2.5.23}$$

which is just the Poisson equation which always governs in the steady-state case. However, here we do not require that ρ be independent of time; the scalar potential φ° is to follow the fluctuations of ρ without any retardation, for there is no $(1/c^2)(\partial^2\varphi/\partial t^2)$ term to bring in retardation (this, by itself, makes it apparent that φ° cannot be the time component of a four-vector). Since it is usually easier to solve a Poisson equation than an inhomogeneous wave equation, Eq. (2.5.23) is usually easier to solve than the second of Eqs. (2.5.15). The resulting potential φ° is due to the free charge ρ.

Once Eq. (2.5.33) is solved, giving φ° in terms of the free charge, the equation for the corresponding $\mathbf{A}^\circ$,

$$\text{curl}\ (\text{curl}\ \mathbf{A}^\circ) + \frac{\mu\epsilon}{c^2}\left(\frac{\partial^2\mathbf{A}^\circ}{\partial t^2}\right) = \frac{4\pi\mu}{c}\mathbf{J} - \frac{\epsilon\mu}{c}\,\text{grad}\left(\frac{\partial\varphi^\circ}{\partial t}\right) \tag{2.5.24}$$

can be solved, considering φ° as already known. The resulting solution is due to the transverse, or no-free-charge, current only, for $\mathbf{J}$ may be split into two parts (see page 52): one with zero divergence, $\mathbf{J}_t$, which may be expressed as the curl of some vector, and one with zero curl, $\mathbf{J}_l$, which may be expressed as the gradient of some scalar:

$$\mathbf{J} = \mathbf{J}_l + \mathbf{J}_t;\quad \mathbf{J}_l = \text{grad}\ Q;\quad \mathbf{J}_t = \text{curl}\ \mathbf{C}$$

But using the equation of continuity relating $\mathbf{J}$ and ρ, we have div $\mathbf{J} =$ div $\mathbf{J}_l = -(\partial\rho/\partial t)$, which relates the *longitudinal* part of $\mathbf{J}$ to the rate of change of free charge. Comparing this equation with the time derivative of Eq. (2.5.23) (and remembering that $\nabla^2 =$ div grad), we find that

$$\mathbf{J}_l = \frac{\epsilon}{4\pi}\,\text{grad}\,\frac{\partial\varphi^\circ}{\partial t}$$

Therefore Eq. (2.5.24) can be rewritten in terms of the *transverse* current $\mathbf{J}_t$, which is not related to the rate of change of free charge:

$$\text{curl}\,(\text{curl}\,\mathbf{A}^\circ) + \frac{\mu\epsilon}{c^2}\left(\frac{\partial^2\mathbf{A}^\circ}{\partial t^2}\right) = \frac{4\pi\mu}{c}\,\mathbf{J}_t$$

Therefore whenever there is no free charge ρ and the current is all transverse, φ° can be made zero and the fields can be computed by means of the vector potential $\mathbf{A}^\circ$ alone. If there is free charge, φ° is computed from it but $\mathbf{A}^\circ$ is not affected by it. (Note that this gauge is not Lorentz invariant.)

Field of a Moving Charge. As an example of the use of the general transformations given above, suppose that we compute the field due to a point charge moving along the x axis with velocity $u = c \tanh \alpha$. We can do this by computing the retarded potentials from Eq. (2.5.16), or we can calculate the static field around a stationary charge and then transform to a moving system by the Lorentz transformations just given. This second way is the easier, as will be apparent shortly.

We start with a point charge Q, stationary at the origin in the primed coordinates. The potentials and fields at a point (x',y',z') are therefore

$$\varphi' = Q/r'; \quad \mathbf{A}' = 0; \quad \mathbf{E}' = Q(1/r')^3\mathbf{r}'; \quad \mathbf{H}' = 0$$

where $\mathbf{r}' = \mathbf{i}x' + \mathbf{j}y' + \mathbf{k}z'$. This is the simple source solution discussed on page 17.

Next we transform to a set of unprimed coordinates traveling with velocity $u = c \tanh \alpha$ in the x direction. The position of the charge at time t is a distance ut along the x axis from the origin. The relationship between the primed and unprimed coordinates is given by Eqs. (2.5.21) or by

$$x' = x\cosh\alpha - ct\sinh\alpha; \quad y' = y; \quad z' = z;$$
$$ct' = -x\sinh\alpha + ct\cosh\alpha$$
$$\cosh\alpha = 1/\sqrt{1-(u/c)^2}; \quad \sinh\alpha = (u/c)/\sqrt{1-(u/c)^2}$$

The quantity r' must also be expressed in terms of the unprimed coordinates:

$$(r')^2 = (x-ut)^2\cosh^2\alpha + y^2 + z^2 = s^2/[1-(u^2/c^2)] = s^2\cosh^2\alpha$$

where $$s^2 = (x-ut)^2 + (y^2+z^2)[1-(u^2/c^2)]$$

Referring to Eqs. (2.5.21) we see that the potentials for a moving charge can be [using the gauge of Eq. (2.5.14)]

$$\varphi = \varphi'\cosh\alpha = Q/s; \quad \mathbf{A} = \mathbf{i}\varphi'\sinh\alpha = (Q/sc)\mathbf{u} \qquad (2.5.25)$$

where $\mathbf{u} = \mathbf{i}u$ is the vector representing the constant velocity of the charge in the unprimed system of coordinates. By use of Eqs. (2.5.22) or else by performing the proper differentiations on the potentials, we can also obtain the electric and magnetic fields of the moving charge:

$$\mathbf{E} = Q\left(1 - \frac{u^2}{c^2}\right)\frac{\mathbf{r}}{s^3};\quad \mathbf{H} = \frac{1}{c}\,\mathbf{u} \times \mathbf{E} \tag{2.5.26}$$

where $\mathbf{r} = [x - (ut)]\mathbf{i} + y\mathbf{j} + z\mathbf{k}$ is the vector representing the distance from the charge to the point of measurement x, y, z at the time t when the measurement is made.

It is to be noticed that, when u^2 can be neglected in comparison with c^2, s becomes equal to r and the familiar expressions for the fields due to a slow-moving charge result:

$$\mathbf{E} \simeq Q\,\frac{\mathbf{r}}{r^3};\quad \mathbf{H} \simeq \frac{Q}{c}\,\frac{\mathbf{u} \times \mathbf{r}}{r^3}$$

The exact expressions, given in Eqs. (2.5.26), reveal a certain amount of distortion because of the relativistic imposition of a maximum relative speed c, which is not infinite. The equipotential surfaces for φ, for instance, are not spheres but oblate spheroids, with the short axis along the direction of motion of the charge and with the ratio of minor to major axes equal to $\sqrt{1 - (u^2/c^2)}$. In the limiting case of high particle velocity $(u \to c)$, the field is confined to a thin disk perpendicular to $\mathbf{u}$, with Q at its center, practically a plane wave pulse.

Presumably these potentials and fields should be *retarded; i.e.*, they should be expressible in terms of ordinary potentials due to the charge at some earlier position. A naive use of Eqs. (2.5.16) would have us expect that φ should be equal to Q divided by a distance r_r, the distance between x, y, z and the position of the charge at a time r_r/c earlier than t, the time when the potential is measured at x, y, z. Examination of Fig. 2.22 shows that the quantity s, which does enter into the expression for the potentials, is not equal to r_r, the length of the line PQ_r, but is equal instead to the length of the line PR, where QR is perpendicular to PR. It is true that s is equal to $r_r + (u/c)X_r = r_r - (1/c)(\mathbf{r}_r \cdot \mathbf{u}) = r_r[1 - (1/c)\mathbf{a}_r \cdot \mathbf{u}]$, where $\mathbf{a}_r$ is a unit vector pointed from Q_r toward P. Therefore the potentials can be expressed in terms of distances and directions related to the position of the charge at the time $t - (r_r/c)$, which is what is meant by the term retarded potential. If we can find out how the unexpected factor $[1 - (1/c)\mathbf{a}_r \cdot \mathbf{u}]$ arises when we compute φ from Eq. (2.5.16), we shall have completed the reconciliation between the two methods of calculating the potentials (and, as is usual with such reconciliations, have learned a bit more mathematical physics!).

The root of the difficulty lies in the difference, mentioned on page 152, in the two ways of describing the motion of fluids. Equations (2.5.16), for the retarded potentials, use factors in the integrand for charge and current density which are expressed in terms of a coordinate system at rest with respect to the observer. In the case of a moving particle, however, the distribution of charge density is constant when expressed in terms of a coordinate system at rest with respect to the particle (*i.e.*, moving with respect to the observer). If we neglect to transform coordinates before integrating, we end up without the needed factor $[1 - (1/c)\mathbf{a}_r \cdot \mathbf{u}]$.

This is shown in the second part of Fig. 2.22, where we have enlarged the picture sufficiently to indicate the particle's size and distribution of charge. It indicates that, no matter how small the particle is, its motion must be taken into account when we perform the integration. For that part of the charge a distance r_r from the observer, we use the position

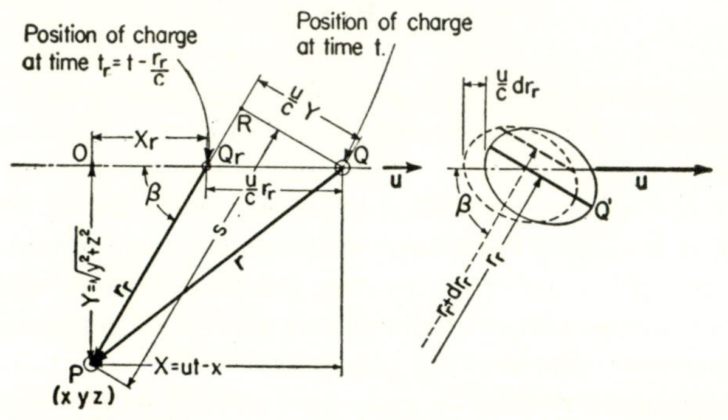

Fig. 2.22 Distances and angles involved in calculating the retarded potentials from a charge Q moving with velocity u with respect to observer at point P.

of the particle at time $t - (r_r/c)$, but for that part a distance $r_r + dr_r$ we use the position at time $t - (r_r/c) - (dr_r/c)$, involving a displacement of $u\,dr_r/c$ back along the x axis. A bit of cogitation will show that, in the integration, the amount of charge dq inside the volume element $dA\,dr_r$ is not $\rho\,dA\,dr_r$, as it would be if the charge were not moving, but is $[1 + (u/c)\cos\beta]\rho\,dA\,dr_r$. Therefore, $\rho\,dA\,dr_r$ is equal to $dq[1 + (u/c)\cos\beta]^{-1}$, and if the dimensions of the particle are small enough, the integral of Eq. (2.5.16) becomes

$$\frac{\int dq}{r_r[1 + (u/c)\cos\beta]} = \frac{Q}{r_r[1 - (1/c)\mathbf{a}_r \cdot \mathbf{u}]} = \frac{Q}{s}$$

which is just the value we obtained in Eq. (2.5.25) from the Lorentz transformation of the stationary potential field.

We thus see that the potential due to a moving charge at time t is the potential due to the charge at an earlier position, modified by a factor relating the velocity of the particle at the earlier position to the velocity of a light signal from particle to observer, *i.e.*, the factor $[1 - (\mathbf{u}/c) \cdot \mathbf{a}_r]$. The earlier position of the particle can be determined by imagining that the particle is sending light signals to the observer;

the position to be used in Eqs. (2.5.25) and (2.5.26) is the one from which the signal was sent which just reaches the observer at time t.

Incidentally Eqs. (2.5.25) for the potentials hold even if the velocity u of the particle is changing with time, if we interpret s as being $r_r[1 - (\mathbf{u}/c) \cdot \mathbf{a}_r]$. The expressions for $\mathbf{E}$ and $\mathbf{H}$ for an accelerating charge are of course, different from those given in Eq. (2.5.26), since $\mathbf{u}$ now depends on time and must be included in the differentiations.

Force and Energy. The three-vector $\mathbf{F} = \rho\mathbf{E} + (1/c)\mathbf{J} \times \mathbf{B}$ is the force on a cubic centimeter of charge current. It is the space part of a four-vector, with "time component" proportional to the work per cubic centimeter which the field performs per sec on the charge current, $\mathbf{J} \cdot \mathbf{E}$. Reference to Eq. (2.5.20) shows that the quantity $\mathbf{k}$, whose components are

$$k_m = \frac{1}{c} \sum_{n=1}^{4} f_{mn} I_n; \quad m = 1, 2, 3, 4 \tag{2.5.27}$$

is just this four-vector, having space components equal to the three components of $\mathbf{F}$ and having a time component $k_4 = i(\mathbf{E} \cdot \mathbf{J}/c)$.

The integral of ck_4/i over the inside of some closed surface is the total rate of work done by the field on the charge current within the volume, and this must equal the rate of reduction of internal energy U residing in the field inside the volume minus the power lost radiating out over the surface bounding the volume. Using Eqs. (2.5.11) and assuming that ϵ and μ do not change with time, we have

$$\begin{aligned}
\iiint \mathbf{E} \cdot \mathbf{J}\, dV &= \frac{c}{4\pi} \iiint \mathbf{E} \cdot \left[\text{curl}\, \mathbf{H} - \frac{1}{c} \frac{\partial \mathbf{D}}{\partial t} \right] dV \\
&= \frac{c}{4\pi} \iiint \left[\mathbf{E} \cdot \text{curl}\, \mathbf{H} - \frac{1}{2c} \frac{\partial}{\partial t} (\mathbf{E} \cdot \mathbf{D} + \mathbf{H} \cdot \mathbf{B}) - \mathbf{H} \cdot \text{curl}\, \mathbf{E} \right] dV \\
&= -\frac{\partial}{\partial t} \iiint \frac{1}{8\pi} (\mathbf{E} \cdot \mathbf{D} + \mathbf{H} \cdot \mathbf{B})\, dV - \frac{c}{4\pi} \iiint \text{div}\, (\mathbf{E} \times \mathbf{H})\, dV \\
&= -\frac{\partial}{\partial t} \iiint U\, dV - \oint \mathbf{S} \cdot d\mathbf{A}
\end{aligned}$$

$$\text{where} \quad U = (1/8\pi)(\mathbf{E} \cdot \mathbf{D} + \mathbf{H} \cdot \mathbf{B}) \quad \text{and} \quad \mathbf{S} = (c/4\pi)\mathbf{E} \times \mathbf{H} \tag{2.5.28}$$

As expected, we have a time rate of change of a volume integral minus a net outflow integral. The quantity U, which appears in the time rate of change of the volume integral, must be the energy density of the field, and the vector $\mathbf{S}$, which appears in the net outflow integral, must be the rate of energy flow in the field. The vector $\mathbf{S}$ is called the *Poynting vector*.

Returning to Eq. (2.5.27) for the force-energy four-vector $\mathbf{k}$ (and letting $\epsilon = \mu = 1$ again), we can substitute from the equations relating

I_m and the derivatives of f_{mn} to obtain

$$k_m = \frac{1}{4\pi} \sum_{rs} f_{mr} \frac{\partial f_{rs}}{\partial x_s}$$

It is of some interest to show that the four-vector k can be obtained by contraction of a "stress-energy" tensor $\mathfrak{T}$, where

$$T_{ms} = -\frac{1}{4\pi} \sum_r f_{mr} f_{sr} + \frac{1}{16\pi} \sum_{r,n} (f_{rn})^2 \delta_{ms} \tag{2.5.29}$$

The contracted form is

$$\sum_s \frac{\partial}{\partial x_s} T_{ms} = -\frac{1}{4\pi} \sum_{r,s} f_{mr} \frac{\partial f_{sr}}{\partial x_s} - \frac{1}{4\pi} \sum_{r,s} \frac{\partial f_{mr}}{\partial x_s} f_{sr} + \frac{1}{8\pi} \sum_{r,n} f_{rn} \frac{\partial f_{rn}}{\partial x_m}$$

The first sum is just k_m, for $f_{sr} = -f_{rs}$. In the third sum we utilize the equations $\partial f_{rn}/\partial x_m = -(\partial f_{mr}/\partial x_n) - (\partial f_{nm}/\partial x_r)$, given on page 209. It becomes, when we change the labels of the summation symbols,

$$-\frac{1}{8\pi} \sum_{r,n} f_{rn} \left(\frac{\partial f_{mr}}{\partial x_n} + \frac{\partial f_{nm}}{\partial x_r} \right) = \frac{1}{8\pi} \sum_{r,s} \frac{\partial f_{mr}}{\partial x_s} f_{sr} + \frac{1}{8\pi} \sum_{k,n} f_{kn} \frac{\partial f_{mn}}{\partial x_k}$$

$$= \frac{1}{4\pi} \sum_{r,s} \frac{\partial f_{mr}}{\partial x_s} f_{sr}$$

which just cancels the second summation. Therefore

$$k_m = \sum_s \frac{\partial}{\partial x_s} T_{ms}$$

where

$$\begin{aligned} T_{11} &= \frac{1}{8\pi} (E_x^2 - E_y^2 - E_z^2 + H_x^2 - H_y^2 - H_z^2); \quad \text{etc.} \\ T_{12} &= \frac{1}{4\pi} (E_x E_y + H_x H_y) = T_{21}; \quad \text{etc.} \\ T_{14} &= \frac{i}{4\pi} (\mathbf{E} \times \mathbf{H})_x = \frac{i}{c} S_x = T_{41}; \quad \text{etc.} \\ T_{44} &= \frac{1}{8\pi} (E_x^2 + E_y^2 + E_z^2 + H_x^2 + H_y^2 + H_z^2) = U \end{aligned} \tag{2.5.30}$$

The tensor $\mathfrak{T}$ is called the *Maxwell stress-energy tensor*. It is analogous to the stress-momentum tensor defined in Eq. (1.7.8). The space components can be considered to be a sort of electromagnetic stress; the space-time components, being proportional to the energy flow, can be considered to be proportional to the momentum of the field; and the term T_{44} is the energy density of the field, analogous to the mass density of the medium discussed earlier.

Surfaces of Conductors and Dielectrics. Two sorts of boundary conditions are of frequent interest in electromagnetic problems: those relating fields at the surface between two dielectrics, where there is an abrupt change in value of the dielectric constant ϵ (or, contrariwise, of permeability μ), and those at the free surfaces of a metallic conductor. In the case of the boundary between dielectrics one can usually neglect the conductivity of the media and assume that neither free charge nor current is present. The boundary conditions relating normal and tangential components of the electric field on both sides of the boundary can be obtained from Maxwell's equations (2.5.11).

Referring to Fig. 2.23 we can set up an elementary contour around which to compute a net circulation integral of $\mathbf{E}$. In the limit of small $l\delta$ this is equal to $l\delta|\text{curl }\mathbf{E}|$. If δ is small compared with l, the net circulation integral over the contour shown is $l(E_t^\circ - E_t^i)$, where E_t°, E_t^i

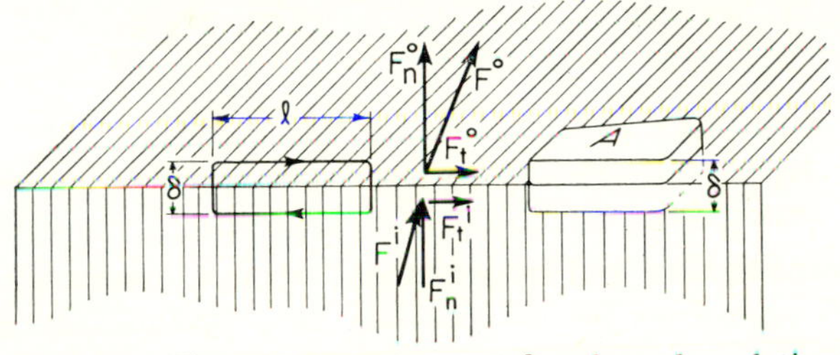

Fig. 2.23 Elementary contours and surfaces for relating inner and outer fields at boundary surfaces.

are the tangential components of the fields on both sides of the boundary. In the steady-state case curl $\mathbf{E}$ must be zero, so that E_t° and E_t^i must be equal.

The net outflow integral of $\mathbf{D}$ over the surface shown in Fig. 2.23 is $A(D_n^\circ - D_n^i)$ when δ is vanishingly small (where D_n°, D_n^i are the normal components of $\mathbf{D}$), and this by definition is proportional to div $\mathbf{D}$. With no free charge present, div $\mathbf{D}$ must be zero, so that the normal components of $\mathbf{D}$ are continuous across the boundary.

Similarly, for a boundary involving a change in permeability μ for the steady state, normal components of $\mathbf{B}$ and tangential components of $\mathbf{H}$ are continuous. The corresponding boundary conditions on the potentials are not difficult to work out.

A metallic conductor has a large value of conductivity σ and a consequently small value of relaxation time (see page 208). In many cases, involving fields with frequencies lower than this relaxation time, we can consider the metallic conductor to have an infinite conductivity and a zero relaxation time. Free charge can occur only on its surface, in amounts just sufficient to cancel the external electric field, for no electric field can exist inside a perfect conductor.

In the first place the tangential electric field E_t° just outside the conductor must be zero; the electric field must be everywhere normal to the surface. Using the elementary surface shown in Fig. 2.23 the net outflow integral for **D** must be just $A\mathbf{D}^\circ$, for $\mathbf{D}^\circ$ must be normal to the surface and $\mathbf{D}^i$ is zero. In order that this be so, there must be a surface charge on the conductor of surface density $\rho = D^\circ/4\pi$ [in order that Eq. (2.5.1) be satisfied]. Since the electric field is everywhere normal to the conductor surface, the surface must be an equipotential φ = constant (at least for the steady-state case).

The behavior of the magnetic fields near a conductor can best be shown by an example.

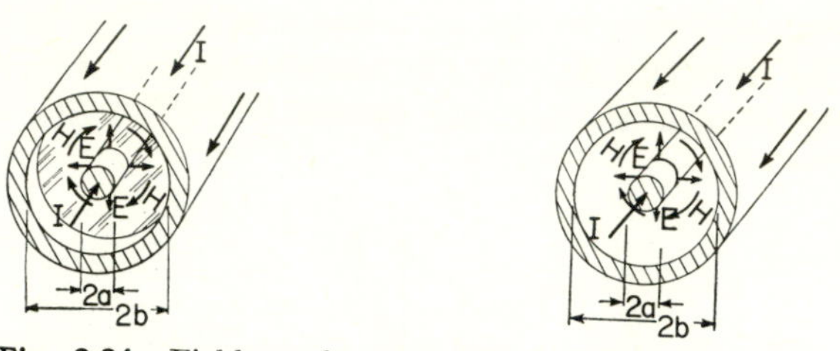

Fig. 2.24 Fields and currents in a concentric transmission line.

Wave Transmission and Impedance. As an example of the solutions of Maxwell's equations let us consider, in a preliminary sort of way, the fields inside a concentric transmission line consisting of two long coaxial cylinders of conductor, with a uniform medium of uniform dielectric constant ϵ, permeability μ, and conductivity σ filling the space between, as shown in Fig. 2.24. Suppose that the outer radius of the inner cylinder is a, the inner radius of the outer cylinder is b, and the coordinates are z, the distance along the axis from one end of the line, and r, the radial distance out from the axis.

We can assume that there is no free charge in the interspace between the conductors, so that only the vector potential needs to be used, and Eqs. (2.5.18) and (2.5.19) are applicable. The electric field must be normal to both conductor surfaces, so it must be radial at $r = a$ and $r = b$. The simplest way to do it would be to make **E** everywhere radial, and the simplest way to achieve this is to make the vector potential **A** everywhere radial. However, div **A** must be zero, and if **A** is radial ($\mathbf{A} = A\mathbf{a}_r$), the divergence of **A** is just $(1/r)(\partial/\partial r)(rA)$ as can be seen by referring to page 116. In order that div **A** be zero, its magnitude A must be of the form $1/r$ times a function of z and ϕ and t (for then rA is independent of r and its derivative with respect to r is zero). For the simplest case we shall assume that A is independent of ϕ:

$$\mathbf{A} = \frac{\mathbf{a}_r}{r}\psi(z,t) \tag{2.5.31}$$

Referring again to page 116, we see that the Laplacian of $\mathbf{A}$ is

$$\nabla^2\mathbf{A} = \mathbf{a}_r\left[\nabla^2\left(\frac{\psi}{r}\right) - \left(\frac{\psi}{r^3}\right)\right] = \frac{\mathbf{a}_r}{r}\frac{\partial^2\psi}{\partial z^2}$$

so that Eq. (2.5.19) turns out to result in a simple equation for ψ, the part of the amplitude of the vector potential which depends on z and t:

$$\frac{\partial^2\psi}{\partial z^2} = \frac{\mu\epsilon}{c^2}\frac{\partial^2\psi}{\partial t^2} + \frac{4\pi\mu\sigma}{c^2}\frac{\partial\psi}{\partial t} \tag{2.5.32}$$

This sort of equation has been encountered in our discussion of the motion of a string on page 137. Its solutions are damped waves, damped in either time or distance.

Suppose we assume that the transmission line is driven from the end $z = 0$ at a frequency $\omega/2\pi$. In this case the dependence on time would be simple harmonic, via the factor $e^{-i\omega t}$, and the attenuation will be in space, along z. Inserting this dependence on t into Eq. (2.5.32), we obtain

$$\frac{\partial^2\psi}{\partial z^2} + \left[\frac{\mu\epsilon\omega^2}{c^2} + \frac{4\pi i\mu\sigma\omega}{c^2}\right]\psi = 0$$

A solution of this can be given in exponential form:

$$\psi = \beta e^{-\kappa z + i(\omega/c)(nz - ct)} + \gamma e^{\kappa z - i(\omega/c)(nz-ct)} \tag{2.5.33}$$

where $\kappa = \frac{\omega}{c}\sqrt{\epsilon\mu}\sqrt{\frac{1}{2}\sqrt{1+\frac{16\pi^2\sigma^2}{\epsilon^2\omega^2}} - \frac{1}{2}} \simeq \frac{2\pi\sigma}{c}\sqrt{\frac{\mu}{\epsilon}};\quad \sigma \ll \epsilon\omega$

and $n = \sqrt{\epsilon\mu}\sqrt{\frac{1}{2}\sqrt{1+\frac{16\pi^2\sigma^2}{\epsilon^2\omega^2}} + \frac{1}{2}} \simeq \sqrt{\epsilon\mu};\quad \sigma \ll \epsilon\omega$

The first term represents a sinusoidal wave traveling in the positive z direction with velocity c/n and attenuation proportional to κ; the second term represents a sinusoidal wave traveling in the negative z direction with an equal velocity and attenuation. This is the most general solution which oscillates with sinusoidal time dependence of frequency $\omega/2\pi$. If the wave is generated from the end at $z = 0$, only the positive wave is present and γ is zero.

For this case, the fields turn out to be

$$\mathbf{E} = -\frac{1}{c}\frac{\partial\mathbf{A}}{\partial t} = \frac{i\beta\omega}{cr}\mathbf{a}_r e^{-\kappa z + i(\omega/c)(nz-ct)} = \left(\frac{i\omega}{c}\right)\mathbf{A} \tag{2.5.34}$$

$$\mathbf{H} = \text{curl}\,\mathbf{A} = \mathbf{a}_\phi\frac{i\beta}{r}\left(\frac{\omega n}{c} + i\kappa\right)e^{-\kappa z + i(\omega/c)(nz-ct)} = \left(n + \frac{i\kappa c}{\omega}\right)\mathbf{a}_z \times \mathbf{E}$$

The electric field is radial, as we initially required. The potential difference between inner and outer conductor is

$$V = \int_a^b E\,dr = \frac{i\beta\omega}{c}\ln(b/a)\,e^{-\kappa z + i(\omega/c)(nz-ct)}$$

and if the voltage across the end $z = 0$ is $V_0e^{-i\omega t}$, then we will have $\beta = [V_0c/i\omega \ln (b/a)]$, and the voltage between the conductors at z is

$$V(z) = V_0e^{-\kappa z + i(\omega/c)(nz - ct)}$$

A charge collects on the surfaces of both conductors, of surface density $D/4\pi = \epsilon E/4\pi$. The total charge on the surface of the inner conductor per unit length is $(2\pi a)(\epsilon E/4\pi)$ or

$$\frac{\epsilon V_0}{2 \ln (b/a)} e^{-\kappa z + i(\omega/c)(nz - ct)} = \frac{\epsilon V(z)}{2 \ln (b/a)}$$

which is, of course, also the charge per unit length on the inner surface of the outer conductor. Evidently the capacitance of the concentric line per unit length is $[\epsilon/2 \ln(b/a)]$ in these units.

The magnetic lines of force are concentric circles, which indicates that a current is transmitted down the inner conductor. Its value can be computed by using the circuital rule $\oint \mathbf{H} \cdot d\mathbf{s} = 4\pi I$, which is obtained from Eq. (2.5.4) or (1.2.12). Working out the algebra we obtain

$$I(z) = \frac{V_0}{2 \ln (b/a)} \left(n + \frac{i\kappa c}{\omega} \right) e^{-\kappa z + i(\omega/c)(nz - ct)}$$

The current attenuates as z increases because part of the current leaks across the partially conductive material between the perfectly conducting cylinders. If the medium were not conductive ($\sigma = 0$), then the wave would not attenuate (κ would be zero) and the current I would be in phase with the voltage V.

The ratio of the voltage to the current at any point

$$Z = \frac{2 \ln (b/a)}{n + i(\kappa c/\omega)} \tag{2.5.35}$$

is called the *characteristic impedance* of the transmission line. If the medium is nonconductive ($\kappa = 0$), then this impedance is real (purely resistive).

The parallelism between this discussion and the discussion on page 128 for waves on a string should be plain. It is, of course, possible to work out impedances for the transmission line when a reflected wave is present, as was done for the string, and it is possible to work out Fourier integrals for the transient response of the line.

Incidentally, we should notice that, since V is proportional to E and I to H, the impedance computed above is proportional to the ratio between E and H. In many cases of electromagnetic radiation, where it is not possible to compute a V or an I, it is still possible to compute the ratio between E and H at each point. This ratio is often called the *impedance* of the medium or of the wave. For instance, for a plane electromagnetic wave in the positive x direction, in open space ($\epsilon = \mu = 1$), the vector potential is $\mathbf{A}_0e^{i(\omega/c)(x-ct)}$, where $\mathbf{A}_0$ is perpendicular to the x axis. The fields are

$$\mathbf{E} = i\frac{\omega}{c}(\mathbf{A}_0)e^{i(\omega/c)(x-ct)};\quad \mathbf{H} = i\frac{\omega}{c}(\mathbf{i}\times\mathbf{A}_0)e^{i(\omega/c)(x-ct)} = \mathbf{i}\times\mathbf{E} \qquad (2.5.36)$$

In this case the impedance is unity (in these units). If ϵ and μ were not unity, the impedance would be $\sqrt{\mu/\epsilon}$.

Proca Equation. The electromagnetic equations can all be reduced to the Lorentz-invariant requirement that the d'Alembertian of the potentials, which is a four-vector, is proportional to the four-vector representing the charge-current density. This is the four-vector generalization of the inhomogeneous wave equation for the string

$$\frac{\partial^2\psi}{\partial x^2} - \frac{1}{c^2}\frac{\partial^2\psi}{\partial t^2} = f(x,t)$$

which we discussed in Sec. 2.1. By loose analogy, the electromagnetic potentials have an inertia and the sort of "restoring force" that the simple string under tension has, the sort which makes the field at a point tend to become the average of the field at surrounding points. The charge current is the analogue of the driving force f.

We saw in Sec. 2.1 that, if the string were embedded in an elastic medium, so that each part of the string tended to return to zero displacement, then an additional term proportional to ψ had to be added to the wave equation. One could set up a similar equation for the vector potentials, having an additional term proportional to the potentials,

$$\square^2\mathbf{V} - \alpha^2\mathbf{V} = \nabla^2\mathbf{V} - \frac{1}{c^2}\frac{\partial^2\mathbf{V}}{\partial t^2} - \alpha^2\mathbf{V} = -\frac{4\pi\mathbf{I}}{c};\quad \mathbf{V} = A_x,A_y,A_z,i\varphi;\qquad (2.5.37)$$
$$\mathbf{I} = J_x,J_y,J_z,ic\rho$$

which is analogous to Eq. (2.1.27) plus an inhomogeneous term for the force function. The equation for the scalar field, without the inhomogeneous term, is called the Klein-Gordon equation and is suitable to describe the behavior of a scalar meson. The analogous equation for the four-vector potential is called the *Proca equation;* it is of possible use in describing the behavior of a particle of unit spin (if there is such a particle!). The corresponding equations for the fields for the case where $\mu = \epsilon = 1$ (free space) and where the charge current is zero are

$$\begin{aligned}
&\operatorname{curl}\mathbf{H} = \frac{1}{c}\frac{\partial\mathbf{E}}{\partial t} - \alpha^2\mathbf{A}; && \operatorname{curl}\mathbf{E} = -\frac{1}{c}\frac{\partial\mathbf{H}}{\partial t}\\
&\operatorname{div}\mathbf{H} = 0; && \operatorname{div}\mathbf{E} = -\alpha^2 c\varphi\\
&\mathbf{H} = \operatorname{curl}\mathbf{A}; && \mathbf{E} = -\operatorname{grad}\varphi - \frac{1}{c}\frac{\partial\mathbf{A}}{\partial t} \qquad (2.5.38)\\
&\operatorname{div}\mathbf{A} = -\frac{1}{c}\frac{\partial\varphi}{\partial t}
\end{aligned}$$

[or another gauge, see page 211 and Eq. (3.4.21)].

The presence of this additional term has a marked effect on the solutions, as it did in the case of the string. The fields tend strongly to zero, except near a charge or current. The potential about a point charge, for instance, is $Qe^{-\mu r}/r$, which goes to zero much more rapidly than the usual Q/r. Such a potential function has been considered appropriate between nucleons (protons or neutrons) in an atomic nucleus. Therefore the Proca equations and also the Klein-Gordon equation (for the scalar case) may be of use in nuclear theory.

2.6 *Quantum Mechanics*

All through this chapter we have been citing cases where a "smoothed-out" continuous field can be substituted for an actually discontinuous distribution of mass or charge. As long as we are interested in large-scale events, we find we can substitute, for a random distribution of particles, a regular function of position having a magnitude proportional to the mean density of the particles. With the introduction of the electromagnetic field, however, we come to an essentially different linkage between continuous fields and discontinuous particles, a linkage which is basic to modern quantum theory.

There are two fundamental differences between the older relationships between field and particle and the newer quantum relationship. In the first place the magnitude of the classical field is proportional to the mean density of particles or, what is usually the same thing, the probability of the presence of a particle. In quantum theory it is the *square* of the magnitude of the field which is proportional to the probability of presence of the particle. This difference is more basic than might be apparent at first. It means, for instance, that classical fields must be *real numbers*, for they must equal densities or probabilities, which are real numbers. In many cases we shall carry out our calculations for classical fields with complex numbers, but it will always be with the convention that only the real part (or, in some cases, only the imaginary part) represents the actual field. In quantum mechanics, however, the field itself can be a complex quantity, since the square of its magnitude (which is always real) is proportional to the probability of presence of the particle. Indeed, in many cases, it is important that the complex conjugate of a quantum field be different from the field itself; *i.e.*, it is necessary for the fields to be complex.

In the second place the relationship between a classical field and its related distribution of particles is permissive but not, in principle, required. In principle one could always "go behind" the field and compute the individual motions of the particles themselves. Computation with the continuous fields is, of course, enormously easier than calculation of the motions of the individual particles, but the choice is one of convenience, not of necessity. On the other hand, the relationship between quantum fields and their related particles is required, not permissive. The fundamental tenets of quantum theory forbid us to "go behind" the fields to compute in detail the motions of the particles. They state that a complete description of the positions and motions of the related particles is impossible to obtain experimentally and that therefore such a description has no physical meaning. Only the probability density, which is proportional to the square of the magnitude of the field, can be measured and has physical meaning.

The meaning of these general statements can best be made by a specific example.

Photons and the Electromagnetic Field. For the moment let us set aside the question of the relation between the motions of the electrons and the charge-current four-vector; let us assume that somehow we have obtained values of J and ρ from which we can compute an electromagnetic field, subject to the boundary conditions imposed by the idiosyncrasies of the experiment. We shall concentrate on the effects of this field on other charged particles, electrons and ions.

When an electromagnetic wave strikes a photographic plate or is absorbed by a metallic surface, the amount of energy per square centimeter of wave front it can give up to a halide crystal or to move a photoelectron is equal to the *intensity* of the radiation, the magnitude of the Poynting vector $(c/4\pi)(\mathbf{E} \times \mathbf{H})$. This quantity is proportional to the square of the amplitudes of the potentials, and if the potentials are complex quantities, it is proportional to the square of their magnitudes. The field will act by giving energy to a number of the halide grains or electrons in its path, and the total energy lost will equal the total number of particles affected times the average energy given to each particle. Since the particles affected are not continuously distributed, one could not expect that the response would be absolutely uniform over the wave front. One could expect, however, that the more intense the radiation, the more particles will be strongly affected and the more uniform the resulting response will appear. This is actually what happens; the peculiarities of the quantum theory do not obtrude until we diminish the intensity (without changing the frequency) down to very small values.

The field itself is just as continuous, whether its magnitude is large or small. We might expect that the reactions of the photographic plate or the photoelectrons from the metal would try to remain as continuous as possible as the intensity is decreased, the reduction being taken up more or less equally by a reduction in the number of particles affected and in the energy absorbed per particle. However, as many experiments have clearly shown, this is *not* what happens. The energy absorbed per photoelectron does not reduce, only the number of photoelectrons produced decreases as the intensity is decreased. This state of affairs is most easily explained by assuming that the energy of the electromagnetic field is carried by discrete particles, each carrying a definite energy. When a light particle is absorbed by a photoelectron, it gives up the same amount of energy, no matter how intense the radiation is; variation of intensity changes only the *number* of the light particles present, not their individual energy.

As has been repeatedly demonstrated in many experiments, these light "particles," called *photons*, each carry an energy proportional to the frequency ν of the radiation field, $E = h\nu$, where h is Planck's constant. Their speed in free space is, of course, the speed of light, c. Their density of occurrence is determined by the magnitude of the intensity of the corresponding classical field. Thus the photon description of the electromagnetic field is not a competing "explanation" of the facts of the electromagnetic interaction with matter; it is a supplementary description of these facts, neither more nor less true than the classical field description. The photons have no "reality" in the classical sense, for we cannot compute the path of an individual photon or predict exactly its position and direction of motion at a given time. The best we can do is to compute the probability of the presence of a photon at a given place and time, and this probability is computed from the value of the square of the classical field at that point and time. The particle has no meaning separated from the field intensity, and the field can make itself felt only by means of the photons.

As an example of the way this works, we can imagine light falling on a metal surface provided with some mechanism to record the ejection of photoelectrons. Suppose the intensity at various points on the surface, as computed from the classical fields, is that given by the curve at the top of Fig. 2.25. If the intensity is very small and the exposure short, only a few photoelectrons would be emitted and it would be extremely difficult to reproduce the curve for intensity from the pattern of photoelectrons. As the intensity and exposure is increased, however, more and more photoelectrons are ejected and the relative values of density of dots can be measured with greater and greater accuracy. For very large intensities the distribution of dots is quite "smooth" and follows the variation of the classical field with great accuracy.

Even the static electric force between two stationary electric charges can be "explained" in terms of photons. The presence of the charges induces an interchange of photons, numbers of them being created at one charge and disappearing at the other. Their numbers are related to the energy of the field, and their combined effect "produces" the measured force on each charge. In this case, of course, it is easier to use the field aspect to compute these forces, just as it is easier to use the particle aspect to compute the phenomenon of the Compton effect.

The energy of the photon is proportional to the frequency of the field. But energy of a particle is the time component of a four-vector with space components proportional to the momentum of the particle, and frequency is the time component of a four-vector with space components proportional to the wave number (reciprocal of the wavelength) of the wave. Consequently our wave-particle combination requires a relation between energy and frequency, momentum and wave number,

$$\text{Energy} = h\nu; \quad \text{momentum} = h/\lambda \tag{2.6.1}$$

which involves the "quantum constant" h.

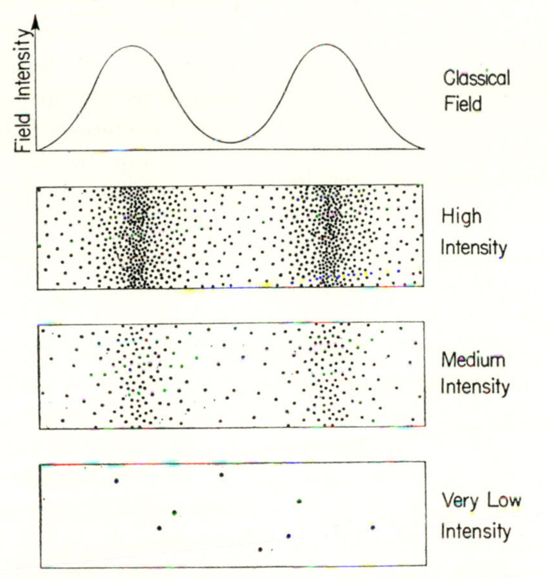

Fig. 2.25 Dots represent points of impact of individual photons. At high intensity the pattern has a density corresponding to the classical field intensity; at low intensities the random nature of the process is evident.

Photons and electromagnetic waves are not the only example of this new, inextricable duality between wave and particle. The experiments of Davisson and G. P. Thomson, and many subsequent ones, have demonstrated that the elementary particles of matters—electrons, protons, and neutrons—are governed by a "wave function," a continuous field satisfying a wave equation (the details of which will be discussed later in this section). The details of the trajectories of individual particles cannot be measured, even in principle, with arbitrarily great precision. All we can compute and predict is the probability of the presence of a particle at a given place and time, which probability is proportional to the square of the magnitude of the wave function. To complete the parallelism with the photon-field case, the energy and momentum of the particle are also related to the frequency and wave number of the wave function according to Eq. (2.6.1).

Uncertainty Principle. It is worth while to discuss further the second peculiarity of the quantum field—the fact that it is impossible to "go behind" the wave function to obtain details of the individual particle's motion. Suppose that we let a beam of monochromatic light (or a stream of electrons) fall on a screen with two slits in it, a distance apart as shown in Fig. 2.26. Some of the light (or electron beam) passes through the slits and can be detected on a plate P. According to the classical wave picture the intensity on the plate varies because of interference, the points of greatest intensity being at those angles θ where $\sin\theta = n\lambda/a$, the points of minimum intensity being where $\sin\theta = (n + \frac{1}{2})(\lambda/a)$, where n is an integer.

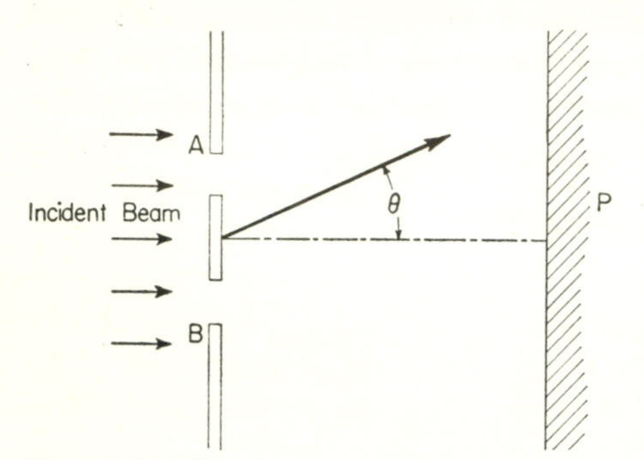

Fig. 2.26 Interference effects for light passing through two slits.

Our detecting plate gives evidence of discrete particles, photons or electrons, and for large intensities the density of such particles will follow closely the interference pattern required by the wave function. To see why we cannot "go behind" the probability density given by the wave function, let us try it once and see what happens. Let us try to follow a photon (or electron) as it goes through the slits and falls on the plate to be detected. Not all particles follow the same path, for they do not all end up in the same part of the plate. But does the photon whose path we are trying to follow go through one slit or the other or both? If it goes through only one slit, why is it so much more likely to fall at points corresponding to angles θ, where $\sin\theta = n\lambda/a$, than for intermediate angles? To say that the probability that the photon turns up at some given point on the plate depends on the distance a between the slit through which it has gone and the slit through which it has *not* gone is to voice an absurdity which serves only to demonstrate the impossibility of inquiring too closely into the paths of individual particles. As long as the occurrence of photons (or electrons) is governed by the square of the magnitude of a wave function, trying to picture the path of an individual particle will always lead us into such logical absurdities.

We might try to ensure that the photon goes through slit A by closing slit B. But then we shall not obtain the interference pattern on the plate. We have not resolved the paradox; we have just destroyed the experiment.

What we have been saying is that we cannot be sure what the transverse momentum of the photon is during its travel from slits to plate and that any attempt to measure it accurately destroys the experiment so that something else cannot be measured accurately. We could, for instance, try to measure the transverse momentum by measuring the momentum transferred to the screen as the particle passes through a slit and changes its direction of motion. But in order to measure the recoil momentum the screen must be of small mass and be suspended freely so it can move appreciably when it recoils, and if this is done, then we cannot be sure of the exact position of the screen when the particle goes through a slit.

The fact is that we never can measure position and momentum simultaneously with arbitrarily great precision. Figure 2.27 shows a related arrangement which will make this statement more definite. Here the particles are to go through but one slit, of width Δx. According to

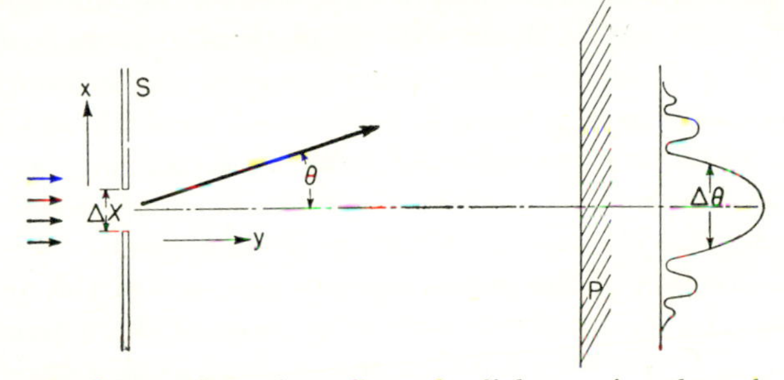

Fig. 2.27 Diffraction effects for light passing through single slit of width Δx. Curve to right of P is classical field intensity incident on P.

Frauenhofer diffraction theory, waves passing through a slit of width Δx will be diffracted, spreading into a divergent beam surrounded by diffraction bands, as shown in the smooth intensity plot at the right of the figure. Diffraction theory states that the angular width $\Delta\theta$ of the main beam is approximately equal to $\lambda/\Delta x$. But this angle is a measure of our uncertainty of the photon's transverse momentum after it has gone through the slit; this transverse momentum is most likely to be within the range between $\frac{1}{2}\Delta p \simeq \frac{1}{2}p\,\Delta\theta$ and $-\frac{1}{2}\Delta p \simeq -\frac{1}{2}p\,\Delta\theta$, where p is the momentum of the particle in the y direction, the initial direction of motion. But by Eq. (2.6.1), $p = h/\lambda$, so that $\Delta p \simeq p\,\Delta\theta \simeq (h/\lambda)(\lambda/\Delta x)$, or

$$\Delta p\,\Delta x \not< h \tag{2.6.2}$$

This equation is called the *Heisenberg uncertainty principle;* it is one of the fundamental equations of quantum physics.

The slit arrangement shown in Fig. 2.27 is a crude method of measuring the location of a particle in the x dimension. A stream of particles all traveling in the y direction is projected at the screen S. Before they hit the screen, the position of any one of the particles in the stream is unknown, but after one has gone through the slit, we can say that we know its x coordinate (or at least that we *knew* its x coordinate just after it went through) to within an uncertainty Δx. But since the behavior of the particle is governed by a wave equation, with related frequency and wavelength given in Eq. (2.6.1), the very fact that the particle went through the slit has produced an uncertainty in its subsequent direction of motion; the very fact that we have measured its x coordinate has introduced an uncertainty in its x component of momentum. The wave-particle relation has required a relationship between the uncertainty in our measurement of x and the uncertainty in corresponding momentum given by Eq. (2.6.2), which involves the ubiquitous constant h. The more accurately we measure x (the narrower the slit is), the less accurately do we know the momentum after the measurement.

Many other experiments could be imagined for measuring position and momentum, and analysis shows that all of them give the same result as long as we keep in mind the quantum, wave-particle interrelation. For instance, we could send a beam of light at an electron and measure its position by observing the reflected light by a microscope. However, light behaves like a particle when it reflects from an electron, and this reflection gives the electron a certain recoil momentum. We cannot be sure of the direction of this recoil, because the lens of the microscope is of finite size, and we cannot be sure what part of the lens the reflected photon went through before it was seen. We could reduce the uncertainty in direction of the recoil momentum by reducing the angular aperture of the microscope, but by the laws of diffraction this would reduce the resolving power of the microscope and therefore increase our uncertainty in locating where the electron was at the time of collision. When all the algebra is worked out, the relation between uncertainty in position of the electron at the time of reflection of the photon and uncertainty in subsequent momentum of the electron is again that given by Eq. (2.6.2).

The trouble (if we can call it trouble) is that physical entities cannot be subdivided indefinitely. We cannot shine an indefinitely small amount of light on the electron; the smallest amount possible still disturbs the electron by a nonvanishing amount. The very fact that matter (and electromagnetic radiation, and so on) comes only in finite packages produces an inherent clumsiness in all our experiments which makes the uncertainty principle inevitable. Whether we consider the wave-particle relationship, embodied in Eq. (2.6.1), as "explaining" the uncertainty principle, embodied in Eq. (2.6.2), or vice versa, both of them are different aspects of the consequences of the essential atomicity of interaction of field and matter, each aspect bearing the common trademark, the constant h.

Conjugate Variables and Poisson Brackets. Position x and momentum p_x are canonically conjugate variables in the jargon of classical dynamics. The uncertainty principle holds for any conjugate pair of variables: angle and angular momentum, energy and time, and so on. We shall discuss this matter of classical dynamical theory in somewhat greater detail in the next chapter; we shall only quote here enough aspects of the theory of contact transformations to elucidate this question of conjugate variables. Suppose we can completely express the configuration of some conservative dynamical system at a given time t in terms of some convenient set of coordinates $q_1, q_2, \ldots, q_n$. We can then express the *kinetic potential* $L = T - V$, the difference between kinetic and potential energies (sometimes called the *Lagrange function*), in terms of the q's and their time derivatives $\dot{q}_m = dq_m/dt$.

The momentum p_m conjugate to coordinate q_m is then defined by the equation

$$p_m = \partial L/\partial \dot{q}_m \tag{2.6.3}$$

for the dynamical system under consideration. The kinetic potential or the total energy or any other property of the system can be expressed in terms of the q's and the p's instead of the q's and $\dot{q}$'s. In many ways the q, p terminology is more satisfactory than the use of q and $\dot{q}$.

There are, of course, many other coordinate systems and conjugate momenta in terms of which the behavior of the system can be completely expressed. The transformation which goes from one set, $q_1, \ldots, q_n$; $p_1, \ldots, p_n$, describing a system to another set, $Q_1, \ldots, Q_n$; $P_1, \ldots, P_n$ which also can describe the same system, is called a *contact transformation*. These new coordinates Q may be chosen in any way which still suffices to describe completely the configuration at any instant. The kinetic potential L of the system may be calculated in terms of the Q's and $\dot{Q}$'s, and then the P's may be found by means of Eq. (2.6.3).

Another method of relating Q's and P's is by means of the *Poisson bracket* expressions. Starting with the original p's and q's, we express two functions of the system, u and v, in terms of the p's and q's. Then the *Poisson bracket* of u and v is defined by the following expression:

$$(u,v) = \sum_{m=1}^{n} \left[\frac{\partial u}{\partial p_m} \frac{\partial v}{\partial q_m} - \frac{\partial u}{\partial q_m} \frac{\partial v}{\partial p_m} \right] \tag{2.6.4}$$

These expressions will be discussed in more detail in Chap. 3.

The value of (u,v) is the same no matter how we choose the coordinates q as long as they completely describe the configuration of the system and as long as we define the conjugate momenta by Eq. (2.6.3). It is not difficult to see that

$$(u,p_m) = -(\partial u/\partial q_m) \quad \text{and} \quad (u,q_m) = \partial u/\partial p_m$$

It is an interesting and useful property of these expressions that, if u and v are conjugate variables, their Poisson bracket is unity. Therefore any complete set of coordinates Q_m and conjugate momenta P_m for the system under consideration will satisfy the equations

$$\begin{aligned} (Q_i,Q_j) &= 0; \quad (P_i,P_j) = 0; & i, j &= 1, \ldots, n \\ (P_i,Q_j) &= 0; & i, j &= 1, \ldots, n; \quad i \neq j \\ (P_m,Q_m) &= 1; & m &= 1, 2, \ldots, n \end{aligned} \tag{2.6.5}$$

We also find that, if $(u,v) = 1$ when u and v are expressed in the q's and p's, it also equals unity when u and v are expressed in any other coordinates and momenta, Q and P.

The most general form of the uncertainty principle is, therefore, for any two functions u and v of the dynamical variables p, q of a given system. If their Poisson bracket (u,v) is equal to a constant K (if it is K in one coordinate system, it is K in another), then the product of the uncertainties in simultaneous measurement of u and v is Kh;

$$\Delta u \, \Delta v \simeq |(u,v)h|_{\text{av}} \tag{2.6.6}$$

where the subscript "av" indicates that, if (u,v) does not turn out to be a constant, a quantum mechanical average (to be defined on page 232) is to be taken.

The wave-particle interconnection and the uncertainty relation, forced on us by experimental results, imply and require the formulation of a dynamical theory which will tell us what we need to know concerning the behavior of a system but which will *not* tell us what we should not know.

In summary of the results of the discussion of this subsection, we can say that the wave-particle dualism exhibited by both radiation and matter requires a fundamental change from the Newton-Maxwell description of physical phenomena, in which all the coordinates and their corresponding momenta could be measured precisely and thus the particle trajectory was completely known. Quantum mechanics asserts that no such exact knowledge is possible, that the more strenuous the attempt to obtain it, the more completely will the trajectory be distorted. This inherent indefiniteness arises from the fact that there is a wave of some kind associated with a particle; this wave may interfere with itself. The intensity of the wave (*i.e.*, the square of its magnitude) is related to the probability of finding the particle at a given point. The problem of a single particle is a central problem in quantum mechanics and will hold our main interest for some time. Once this is solved, there still remains the problem of an assembly of particles, free or interacting. This involves the quantization of fields, including, for example, the probability field mentioned above, in which case it is often referred to as second quantization.

The Fundamental Postulates of Quantum Theory. The quantitative formulation of these ideas involves a symbolic algebra of states, of the sort touched on in Chap. 1. The state of a system has a well-defined meaning in classical physics. For example, the trajectory or orbit of a particle describes its state; the direction of the polarization vector is required before one can describe the state of a plane electromagnetic wave and so on. In quantum physics, not so many things can be known about a system. Nevertheless, if we are willing to be in complete ignorance about one of a pair of conjugate variables, we can determine the other exactly. In general, there will be a number of such possible measurements which can be made simultaneously. The totality of the results of such measurements may be considered to describe the state of the system. Indeed, in quantum physics this is the most one can say about a system.

A measurement performed on a system changes the system. To represent this fact pictorially, it is most convenient to set up an abstract vector space in which a vector represents a given state of the system. This may, of course, be done in classical as well as in quantum physics, as, for example, the coupled oscillators discussed in the section on abstract spaces or the discussion of the vibration of a string in this chapter. As we have said, the effect of a measurement changes the state, *i.e.*, rotates the vector representing the state into another direction in the vector space. Thus a measurement may be represented by an *operator* in abstract space; to every dynamical variable there corresponds a definite operator.

In general for a given system, measurement of the value of a given dynamical variable will disturb the state so that a subsequent measurement of the same variable will give different results, and all we can obtain by repeated measurements is an average value and a mean deviation or uncertainty. Only a small subgroup of all possible states will be such that the uncertainty is zero, and each measurement of a given quantity yields the same value. If measurement of this one value is all that is needed to specify the state, then the state vector must be an *eigenvector* of the operator representing the variable (as defined on page 78). If values for more than one variable are needed to specify the state, a state vector for which the uncertainty of one of these variables is zero is not necessarily an eigenvector, but a set of eigenvectors can always be found, as can be demonstrated.

To recapitulate some of our discussion in Chap. 1, let $\mathfrak{P}$ be the operator corresponding to the momentum of part of the system, and let the states with zero uncertainty in the measurement of p (the eigenvectors of $\mathfrak{P}$) be $\mathbf{p}_n$. Then, according to Eq. (1.6.31),

$$\mathfrak{P} \cdot \mathbf{p}_n = p_n \mathbf{p}_n$$

where p_n is one of the measured values of momentum (we have adjusted the length of $\mathbf{p}_n$ so that p_n is the measured value). It is called an *eigenvalue* of $\mathfrak{P}$ for the state denoted by $\mathbf{p}_n$. Since p_n must be a real number, the operator $\mathfrak{P}$ must be *Hermitian,* as defined on page 83 (in other words, the operator $\mathfrak{P}$ is equal to its adjoint $\mathfrak{P}^*$).

When the system is an arbitrary state, corresponding to the state vector $\mathbf{e}$, the momentum will not in general be accurately measurable. All one can determine is an average value of a series of measurements. As we pointed out on page 80, the vector $\mathbf{e}$ can be expressed in terms of a sum of the eigenvectors $\mathbf{p}_m$, with the respective components of $\mathbf{e}$ in the $\mathbf{p}_m$ direction given by the scalar product $\mathbf{p}_m^* \cdot \mathbf{e}$. We have shown in Chap. 1 that the vectors $\mathbf{p}_m$ are mutually perpendicular and are so normalized that the scalar product $\mathbf{p}_m^* \cdot \mathbf{e}$ is the equivalent of the direction cosine, though in general it is a complex number. This means that the sum of the squares of the magnitudes of the direction cosines equals unity:

$$\sum_m |\mathbf{p}^* \cdot \mathbf{e}|^2 = 1; \quad \text{where } \mathbf{e} = \sum_m \mathbf{p}_m(\mathbf{p}_m^* \cdot \mathbf{e})$$

This equation suggests that we consider $|\mathbf{p}_m^* \cdot \mathbf{e}|^2$ to be a probability, namely, the probability that a measurement of p on the state $\mathbf{e}$ results in a value p_m (*i.e.,* the probability that the state $\mathbf{e}$ "is in" the state $\mathbf{p}_m$). Therefore the average value of p for the state e is

$$p_{\text{av}} = \sum_m p_m |\mathbf{p}_m^* \cdot \mathbf{e}|^2 = \left[\sum_n (\mathbf{e}^* \cdot \mathbf{p}_n)\mathbf{p}_n^*\right] \cdot \mathfrak{P} \cdot \left[\sum_m \mathbf{p}_m(\mathbf{p}_m^* \cdot \mathbf{e})\right] = \mathbf{e}^* \cdot \mathfrak{P} \cdot \mathbf{e}$$

This is the sort of average indicated in Eq. (2.6.6).

We have shown on page 80 that, if two dynamical variables are not simultaneously measurable, then they are not commutable; that is, $\mathfrak{A} \cdot \mathfrak{B} \neq \mathfrak{B} \cdot \mathfrak{A}$. We have spent some time in this section pointing out that a given momentum p and its conjugate coordinate q are *not* simultaneously measurable when we take into account the essential atomicity and wave-particle properties of physical interactions. Therefore the *commutator* $[\mathfrak{P} \cdot \mathfrak{Q} - \mathfrak{Q} \cdot \mathfrak{P}]$ (which will be written hereafter as $[\mathfrak{P},\mathfrak{Q}]$) is not zero. It is important to find out what value it has.

In the first place $[\mathfrak{P},\mathfrak{Q}]$ must be pure imaginary, for $(\mathbf{e}^* \cdot \mathfrak{P} \cdot \mathfrak{Q} \cdot \mathbf{e})$ is the complex conjugate of $(\mathbf{e}^* \cdot \mathfrak{Q} \cdot \mathfrak{P} \cdot \mathbf{e})$ for any vector $\mathbf{e}$ (for $\mathfrak{Q}^* = \mathfrak{Q}$ and $\mathfrak{P}^* = \mathfrak{P}$) and the difference between these two quantities (which is the mean value of $[\mathfrak{P},\mathfrak{Q}]$ in the state $\mathbf{e}$) must therefore be pure imaginary. In the second place we have shown on page 88 that the magnitude of the commutator $[\mathfrak{A},\mathfrak{B}]$ is proportional to the product of the uncertainties,

$(\Delta a\,\Delta b)$. Combining these leads (and deductions from experiment) we arrive at the equation $[\mathfrak{P},\mathfrak{Q}] = \hbar/i$ (where $\hbar = h/2\pi$) which has already been given [Eq. (1.6.32)].

Reference to Eq. (2.6.6) relating the quantum uncertainties with the classical Poisson bracket expression, we finally arrive at the *fundamental equation of quantum mechanics*, relating two operators, representing two dynamical variables of a system, with the Poisson bracket for the corresponding classical variables:

$$[\mathfrak{A},\mathfrak{B}] = \mathfrak{A}\cdot\mathfrak{B} - \mathfrak{B}\cdot\mathfrak{A} = (a,b)\frac{\hbar}{i} = \frac{\hbar}{i}\left\{\sum_m \left[\frac{\partial a}{\partial p_m}\frac{\partial b}{\partial q_m} - \frac{\partial a}{\partial q_m}\frac{\partial b}{\partial p_m}\right]\right\} \qquad (2.6.7)$$

where, unless (a,b) turns out to be a constant, we are to consider it to be a function of the operators $\mathfrak{P}_m$ and $\mathfrak{Q}_m$. A warning should be added, however, that the *order* of writing the terms (pqp instead of qp^2, etc.) is important for noncommuting operators and that Eq. (2.6.7) holds only when the correct order for the Poisson bracket expression is found (see page 234). This is a very interesting equation, relating as it does the quantum mechanical operators in abstract vector space with purely classical dynamical functions. It is not the only equation which might be devised to satisfy the general requirements we outlined on the previous page, but it is the simplest one, and its justification lies in the correctness of the results which flow therefrom.

We summarize the results of this section. In quantum physics a state is specified by giving the values of those quantities which are simultaneously measurable for this state. Their corresponding operators have the properties that they leave the state unchanged and that they commute with each other. The revelant operators are, of course, not functions of each other. One may distinguish these by finding an operator which commutes with one and not with another. When operators do not commute, they satisfy an uncertainty relation given by Eq. (2.6.6).

Independent Quantum Variables and Functions of Operators. There are, of course, many functions of the dynamical variables of a system, which have to be turned into operators. A function of one variable (of p_m or q_n, for instance) which can be defined classically in terms of a power series in the variable can be converted fairly easily. For any such function of $\mathfrak{P}_m$ (for instance) commutes with $\mathfrak{P}_m$ and its value can be measured accurately whenever $\mathfrak{P}_m$ can be so measured; the eigenvalue of the function can be computed from the eigenvalue of $\mathfrak{P}_m$. Finally, we can say that, if an operator $\mathfrak{F}$ and an operator $\mathfrak{P}$ commute, then $\mathfrak{F}$ is a function of $\mathfrak{P}$ if every operator which commutes with $\mathfrak{P}$ also commutes with $\mathfrak{F}$.

There are many cases, however, where two operators commute but one is not a function of the other. After all, many systems require the specification of several independent operators completely to specify the state. The operators $\mathfrak{P}_n$ and $\mathfrak{Q}_m$ ($n \neq m$) are independent, and they commute, according to Eq. (2.6.7). One is not a function of the other, however, because many operators can be found which commute with $\mathfrak{P}_n$ but which do not commute with $\mathfrak{Q}_m$ (for instance, any function of $\mathfrak{P}_m$). As a matter of fact if two operators commute and a third operator can be found which commutes with one and not with the other, then it is certain that the first two operators are independent of each other.

If we deal with functions of one pair of conjugate variables, then a use of Eq. (2.6.7) and a utilization of the invariance of the Poisson brackets under a contact transformation (*i.e.*, transformation from one set of conjugate variables to any other complete set) shows that

$$[\mathfrak{F}(q_n,p_n),\mathfrak{Q}_n] = \frac{\hbar}{i}\frac{\partial \mathfrak{F}}{\partial p_n}; \quad [\mathfrak{F}(q_n,p_n),\mathfrak{P}_n] = -\frac{\hbar}{i}\frac{\partial \mathfrak{F}}{\partial q_n} \tag{2.6.8}$$

where the derivative on the right-hand side is changed into an operator after the differentiation has been performed. Any function of conjugate variables must be carefully ordered before converting into an operator because of the noncommutative properties of the variables (for instance, the classical function p^2q could be $\mathfrak{P}\cdot\mathfrak{P}\cdot\mathfrak{Q}$ or $\mathfrak{P}\cdot\mathfrak{Q}\cdot\mathfrak{P}$ or $\mathfrak{Q}\cdot\mathfrak{P}\cdot\mathfrak{P}$, and each of these have different eigenvectors). Often the correct ordering can be determined only by trial and by check with experiment.

For example, suppose that one wanted to find the quantum mechanical operator corresponding to momentum conjugate to the spherical angle $\varphi = \tan^{-1}(y/x)$. Classically this is the z component of the angular momentum $= xp_y - yp_x$. Since $\mathfrak{x}$ and $\mathfrak{p}_y$ commute (so do $\mathfrak{y}$ and $\mathfrak{p}_x$), the quantum mechanical operator may be taken directly as $\mathfrak{x}\mathfrak{p}_y - \mathfrak{y}\mathfrak{p}_x$. It is easy to verify, using Eq. (2.6.8), that

$$[(\mathfrak{x}\mathfrak{p}_y - \mathfrak{y}\mathfrak{p}_x), \boldsymbol{\phi}] = \hbar/i \tag{2.6.9}$$

In a more complicated example, one asks for the quantum mechanical equivalent of the "radial momentum" p_r, conjugate to the radial coordinate, $r = \sqrt{x^2 + y^2}$. Classically the radial momentum

$$p_r = \cos\varphi\, p_x + \sin\varphi\, p_y = \frac{xp_x}{\sqrt{x^2+y^2}} + \frac{yp_y}{\sqrt{x^2+y^2}}$$

This has no unique quantum mechanical equivalent because $\mathfrak{x}$ and $\mathfrak{p}_x$, for example, do not commute. A form which has the correct commutator rule is just that stated above, *i.e.*,

$$\left[\frac{\mathfrak{x}}{\mathfrak{r}}\mathfrak{p}_x + \frac{\mathfrak{y}}{\mathfrak{r}}\mathfrak{p}_y, \mathfrak{r}\right] = \frac{\hbar}{i} \tag{2.6.10}$$

Since we require that the operator be Hermitian, the symmetric form must then be used.

Eigenvectors for Coordinates. So far we have discussed at some length the commutation rules which may be deduced from Eq. (2.6.7). We turn next to a discussion of the eigenvalues of these operators and the corresponding eigenvectors. First we show that the eigenvalues of the operator $\mathfrak{q}$ are continuous. To prove this, we show that, if an eigenvalue q of $\mathfrak{q}$ exists, an eigenvector with eigenvalue $q + dq$ may be formed.

Suppose that $\mathbf{e}(q)$ is the eigenvector for $\mathfrak{q}$, such that $\mathfrak{q} \cdot \mathbf{e}(q) = q\mathbf{e}(q)$. Operate on $\mathbf{e}(q)$ by the operator

$$\mathfrak{D} = \exp\left(-\frac{i}{\hbar}\mathfrak{p}\,dq\right) = 1 - \frac{i}{\hbar}\mathfrak{p}\,dq$$

The vector $\mathfrak{D} \cdot \mathbf{e}(q)$ is an eigenvector for $\mathfrak{q}$ with eigenvalue $q + dq$, for

$$\mathfrak{q} \cdot \mathfrak{D} \cdot \mathbf{e}(q) = \left[\mathfrak{q} - \frac{i}{\hbar}\mathfrak{q}\mathfrak{p}\,dq\right]\mathbf{e}(q) = \left[\mathfrak{q} + \frac{i}{\hbar}\left(\frac{\hbar}{i} - \mathfrak{p}\mathfrak{q}\right)dq\right]\mathbf{e}(q)$$
$$= (q + dq)\left(1 - \frac{i}{\hbar}\mathfrak{p}\,dq\right)\mathbf{e}(q) = (q + dq)\mathfrak{D} \cdot \mathbf{e}(q)$$

so that

$$\mathfrak{D} \cdot \mathbf{e}(q) = \mathbf{e}(q + dq)$$

We have used the commutation rule $[\mathfrak{p},\mathfrak{q}] = \hbar/i$ in obtaining this.

We note that, in our study of the vibrating string on page 136, we found another case having a continuous distribution of eigenvalues.

We also note that

$$\exp\left(-\frac{i}{\hbar}\mathfrak{p}q'\right) \cdot \mathbf{e}(q) = \mathbf{e}(q + q') \tag{2.6.11}$$

On page 134 we noted the possible complications of a sequence of eigenvectors for a nondenumerable set of eigenvalues. Each different eigenvector $\mathbf{e}(q)$, for each different value of q, is perpendicular to every other eigenvector of the sequence, so that $\mathbf{e}(q)$ is certainly not a continuous function of q. Therefore the result of the following manipulation

$$\left[1 - \frac{i}{\hbar}\mathfrak{p}\,\Delta q\right] \cdot \mathbf{e}(q) = \mathbf{e}(q + \Delta q) \quad \text{or}$$

$$\mathfrak{p} \cdot \mathbf{e}(q) = i\hbar \lim_{\Delta q \to 0}\left[\frac{\mathbf{e}(q + \Delta q) - \mathbf{e}(q)}{\Delta q}\right] \tag{2.6.12}$$

does not mean that the right-hand side is proportional to the derivative of $\mathbf{e}(q)$ with respect to q, because the difference $[\mathbf{e}(q + \Delta q) - \mathbf{e}(q)]$ does not approach zero in a continuous manner as Δq approaches zero. [We note that this result comes directly from Eq. (2.6.7) and could be written for any pair of canonically conjugate variables with continuous distribution of eigenvalues for one variable.]

The continuum of values of q also makes it a little confusing to decide what sort of a function the scalar product $\mathbf{e}^*(q) \cdot \mathbf{e}(q')$ is of q and q'.

Presumably it should be zero except when $q = q'$, which suggests a quite discontinuous function. It turns out to be most satisfactory to define it in terms of an integral. We extend the obvious formula $\sum_n \mathbf{e}_m^* \cdot \mathbf{e}_n = 1$, for discrete eigenvectors, to a continuum, obtaining $\int_{-\infty}^{\infty} \mathbf{e}^*(q) \cdot \mathbf{e}(q')\, dq' = 1$, which defines the magnitude of the **e**'s (see page 240).

Transformation Functions. However, we can set up integrals involving $\mathbf{e}(q)$, integrated over q, which are limits of sums of the integrand over every value of q within the limits of the integral. For instance, an arbitrary state vector **f** can be expressed in terms of a sum of all the different eigenvectors for q,

$$\mathbf{f} = \int_{-\infty}^{\infty} f(q)\mathbf{e}(q)\, dq \tag{2.6.13}$$

where $|f(q)|^2\, dq$ is the probability that the particle is between q and $q + dq$ *when the system is in state* **f**. The quantity $f(q)$, being proportional to the direction cosine between **f** and $\mathbf{e}(q)$, is a complex number, an ordinary function of the continuous variable q. The analogy between this integral and a Fourier integral will become increasingly apparent as this discussion progresses.

By this means, using integrals over the eigenvectors for the position operator $\mathfrak{q}$, we can correlate every state vector **f** with an ordinary function $f(q)$, which is called a *transformation function* since it describes the transformation between **f** and the **e**'s. An eigenvector for another operator can, of course, be expressed in terms of the $\mathbf{e}(q)$'s. For instance if $\mathfrak{A} \cdot \mathbf{e}(a) = a\mathbf{e}(a)$, then we can write

$$\mathbf{e}(a) = \int_{-\infty}^{\infty} \psi(a|q)\mathbf{e}(q)\, dq \tag{2.6.14}$$

where $|\psi(a|q)|^2\, dq$ is the probability that the coordinate of the system is between q and $q + dq$ when its state is such that the variable represented by $\mathfrak{A}$ is certain to have the value a. Such a transformation function was discussed on page 135.

If we do not wish to trust our skill with state-vector calculus, we can always fall back on calculation with the function $f(q)$. For instance, the average value of the dynamical quantity represented by the operator $\mathfrak{A}$ when the system is in state **f** is

$$\mathbf{f}^* \cdot \mathfrak{A} \cdot \mathbf{f} = \int_{-\infty}^{\infty} dq \int_{-\infty}^{\infty} \bar{f}(q)[\mathbf{e}^*(q) \cdot \mathfrak{A} \cdot \mathbf{e}(q')]f(q')\, dq'$$

so that, if we know the effect of an operator on the vectors $\mathbf{e}(q)$, then we can find the effect of an operator on any other state vector **f** in terms of the transformation function $f(q)$.

For instance, we could calculate the effect of the operator $\mathfrak{p}$, representing the canonical conjugate to the coordinate operator $\mathfrak{q}$, on an arbitrary state vector $\mathbf{f}$. Utilizing Eq. (2.6.12), we have

$$\mathfrak{p}\cdot\mathbf{f} = \int_{-\infty}^{\infty} f(q)\mathfrak{p}\cdot\mathbf{e}(q)\,dq = i\hbar\int_{-\infty}^{\infty}\lim\left[\frac{\mathbf{e}(q+\Delta q)-\mathbf{e}(q)}{\Delta q}\right]f(q)\,dq$$
$$= \frac{\hbar}{i}\int_{-\infty}^{\infty}\lim\left[\frac{f(q)-f(q-\Delta q)}{\Delta q}\right]\mathbf{e}(q)\,dq$$
$$= \int_{-\infty}^{\infty}\left[\frac{\hbar}{i}\frac{\partial f(q)}{\partial q}\right]\mathbf{e}(q)\,dq \quad (2.6.15)$$

In other words the effect of the operator $\mathfrak{p}$ acting on a state vector $\mathbf{f}$ corresponds to the analytic operation $(\hbar/i)(\partial/\partial q)$ acting on the transformation function $f(q)$. Since $f(q)$ is an ordinary function, the differential operator has its usual meaning. Repeating the operation, the operation $b(\mathfrak{p})^n$ corresponds to the operation $b(\hbar/i)^n(\partial^n/\partial q^n)$ acting on $f(q)$. Therefore if we have any operator function $\mathfrak{F}(\mathfrak{p})$ of $\mathfrak{p}$ which can be expanded in a power series, the corresponding operation on $f(q)$ is obtained by substituting for $\mathfrak{p}$ the differential operator $(\hbar/i)(\partial/\partial q)$.

It should not be difficult to utilize Eqs. (2.6.5) and (2.6.7) to generalize these remarks. We set up a complete set of coordinates $q_m(m = 1, \ldots, n)$ by which we can specify the configuration of a given system (in the classical sense). We find the momentum variables p_n conjugate to the q's, such that the set satisfies the Poisson bracket equations (2.6.5). Any dynamical variable of the system $B(p,q)$ can be expressed in terms of the p's and q's. The quantum mechanical operator $\mathfrak{B}$ corresponding to the classical quantity B can be formed by substituting the operators $\mathfrak{p}_m$ and $\mathfrak{q}_m$ for the quantities p_m and q_m in B. This is to operate on the abstract vector $\mathbf{f}$ representing the state of the system, to change its magnitude and/or direction. If, instead of discussing the effects of $\mathfrak{B}(p,q)$ on $\mathbf{f}$, we prefer to study the corresponding effect on the transformation function $f(q_1, \ldots, q_n)$, we can make use of the following generalization of Eq. (2.6.15):

$$\mathfrak{B}(p,q)\mathbf{f} = \int\cdots\int B\left(\frac{\hbar}{i}\frac{\partial}{\partial q}, q\right)f(q_1, \ldots, q_n)\,dq_1\cdots dq_n\,\mathbf{e}(q_1, \ldots, q_n) \quad (2.6.16)$$

where the differential operator in the integral, which acts on the transformation function $f(q, \ldots, q_n)$, is formed by changing each p_n in the classical function $B(p,q)$ into $(\hbar/i)(\partial/\partial q_n)$. Both in $\mathfrak{B}$ and in the corresponding differential operator for $f(q)$ we must watch the order of the p's and q's in each term of B, for the conjugate set of operators $\mathfrak{p}_n$ and $\mathfrak{q}_n$ do not commute, since they satisfy the equations

$$[\mathfrak{p}_m,\mathfrak{q}_n] = \left(\frac{\hbar}{i}\right)\delta_{mn};\quad \delta_{mn} = \begin{cases} 0; & m \neq n \\ 1; & m = n \end{cases}$$
$$[\mathfrak{p}_m,\mathfrak{p}_n] = 0 = [\mathfrak{q}_m,\mathfrak{q}_n]$$

which corresponds to the Poisson bracket equations (2.6.5). As defined earlier $[\mathfrak{A},\mathfrak{B}]$ is the commutator $(\mathfrak{A}\cdot\mathfrak{B} - \mathfrak{B}\cdot\mathfrak{A})$.

Operator Equations for Transformation Functions. For instance, instead of using the vector equation

$$\mathfrak{B}(p,q)\cdot\mathbf{b}_m = b_m\mathbf{b}_m$$

to determine the eigenvectors $\mathbf{b}_m$ and the eigenvalues b_m, we can use the corresponding differential equation

$$B\left(\frac{\hbar}{i}\frac{\partial}{\partial q}, q\right)\psi(b_m|q) = b_m\psi(b_m|q) \tag{2.6.17}$$

to compute the transformation functions $\psi(b_m|q)$ and their eigenvalues b_m of the operator $\mathfrak{B}$. The eigenvectors $\mathbf{b}_m$ can be calculated (if need be) by use of the formula

$$\begin{aligned} \mathbf{b}_m &= \int \cdots \int \psi(b_m|q)\mathbf{e}(q)\,dq_1 \cdots dq_n; \\ \mathbf{b}_m^* &= \int \cdots \int \bar{\psi}(b_m|q)\mathbf{e}^*(q)\,dq_1 \cdots dq_n \end{aligned} \tag{2.6.18}$$

The probability that the configuration of the system is between q_1 and $q_1 + dq_1, \ldots, q_n$ and $q_n + dq_n$ when B has the value b_n is $|\psi(b_m|q)|^2\,dq_1 \cdots dq_n$, and the average value of the dynamical variable $A(p,q)$ when B has value b_m is

$$(\mathbf{b}_m^*\cdot\mathfrak{A}\cdot\mathbf{b}_m) = \int \cdots \int \bar{\psi}(b_m|q)A\left(\frac{\hbar}{i}\frac{\partial}{\partial q}, q\right)\psi(b_m|q)\,dq_1 \cdots dq_n \tag{2.6.19}$$

Likewise, since the sequence of eigenvectors $\mathbf{b}_n$ can also be used to expand an arbitrary state vector, we have

$$\begin{aligned} \mathfrak{A}\cdot\mathbf{b}_m &= \sum_s (\mathbf{b}_s^*\cdot\mathfrak{A}\cdot\mathbf{b}_m)\mathbf{b}_s \\ (\mathbf{b}_s^*\cdot\mathfrak{A}\cdot\mathbf{b}_m) &= \int \cdots \int \bar{\psi}(b_s|q)\,A\left(\frac{\hbar}{i}\frac{\partial}{\partial q}, q\right)\psi(b_m|q)\,dq_1 \cdots dq_n \end{aligned} \tag{2.6.20}$$

for $$(\mathbf{b}_s^*\cdot\mathbf{b}_m) = \delta_{sm};\quad \delta_{sm} = \begin{cases} 0; & s \neq m \\ 1; & s = m \end{cases}$$

Thus, many of the operations of state-vector algebra are paralleled by integral and differential operations on the transformation functions. It should be apparent by now that these transformation functions are somehow related to the wave functions discussed in such a vague fashion at the beginning of this section. The square of their magnitude relates the probability that the system has a given configuration (*e.g.*, that the particle be at some given point) with the probability that the system be in a given state, characterized by the eigenvalue b_m. They are typical of the new analysis of quantum effects and represent the sort of facts we should expect to obtain about dynamical systems rather than the exact details of trajectories expected by classical dynamics. Of course when the energies and momenta involved are so large that the related period $1/\nu$ and wavelength, as given by Eq. (2.6.1), are very much smaller than the times and distances involved in the system, these transformation functions narrow down to rays which correspond closely to the trajectories of classical dynamics and the allowable states of the system correspond closely to the various initial conditions possible classically.

Transformation to Momentum Space. So far we have expressed our state vectors in terms of the basic set of vectors $\mathbf{e}(q)$ for the continuously variable coordinates $q_1 \cdot \cdot \cdot q_n$ of the system. We can, however, express everything in terms of the conjugate momenta $p_1 \cdot \cdot \cdot p_n$, since there is symmetry between p and q in both classical Poisson bracket and quantum commutator expressions. We have the set of eigenvectors $\mathbf{e}(p)$ for all the nondenumerable set of allowed values of $p_1 \cdot \cdot \cdot p_n$, and we can express any other state vector in terms of the integral

$$\mathbf{b}_m = \int \cdot \cdot \cdot \int \psi(b_m|p)\mathbf{e}(p)\,dp_1 \cdot \cdot \cdot dp_n \tag{2.6.21}$$

which is analogous to Eq. (2.6.18).

By use of the commutation relations between $\mathfrak{p}$ and $\mathfrak{q}$ we can show that operation by $\mathfrak{q}$ on the state vector $\mathbf{b}_m$ corresponds to the operation $-(\hbar/i)(\partial/\partial p)$ on the transformation function $\psi(b_m|p)$. Corresponding to Eqs. (2.6.17) and (2.6.20) are

$$B\left(p,\, -\frac{\hbar}{i}\frac{\partial}{\partial p}\right)\psi(b_m|p) = b_m\psi(b_m|p)$$

and

$$(\mathbf{b}_s^* \cdot \mathfrak{A} \cdot \mathbf{b}_m) = \int \cdot \cdot \cdot \int \bar{\psi}(b_s|p)\, A\left(p,\, -\frac{\hbar}{i}\frac{\partial}{\partial p}\right)\psi(b_m|p)\,dp_1 \cdot \cdot \cdot dp_n$$

The quantity $|\psi(b_n|p)|^2\,dp_1 \cdot \cdot \cdot dp_n$ is, of course, the probability that the system is in the momentum element $dp_1 \cdot \cdot \cdot dp_n$ when it is in the state given by b_m.

To complete the picture we need to know the relation between the eigenvectors $\mathbf{e}(q)$ for the coordinates and the eigenvectors $\mathbf{e}(p)$ for the momenta. This would also enable us to transform the function $\psi(b_m|q)$ into the function $\psi(b_m|p)$ and vice versa. As before, we have

$$\mathbf{e}(p) = \int_{-\infty}^{\infty} \psi(p|q)\mathbf{e}(q)\,dq \tag{2.6.22}$$

where we have considered only one coordinate and conjugate momentum for simplicity. The transformation function $\psi(p|q)$ relates the state where we know the position of the particle exactly (and have no knowledge of momentum) to the state where we know the momentum of the particle exactly and have no knowledge of its position (this last is the state of knowledge to the left of the screen in Figs. 2.26 and 2.27). To compute $\psi(p|q)$, we utilize the equation for the eigenvector for $\mathfrak{p}$, $\mathfrak{p}\mathbf{e}(p) = p\,\mathbf{e}(p)$ and transform it by Eq. (2.6.22) by the use of Eq. (2.6.15):

$$\int_{-\infty}^{\infty} \left[\frac{\hbar}{i}\frac{\partial}{\partial q}\psi(p|q) - p\psi(p|q)\right] \mathbf{e}(q)\,dq = 0$$

for all values of p and q. Therefore $\psi(p|q)$ is equal to $c\exp[(i/\hbar)pq]$, where c is the normalization constant.

The function is "normalized" by using the requirement (see page 236) that $\mathbf{e}^*(p)\cdot\mathbf{e}(p') = \delta(p - p')$, where δ is the Dirac delta function defined by the equations (see also page 122)

$$\delta(z) = \lim_{\Delta\to 0}\begin{cases} 0; & x < -\frac{1}{2}\Delta \\ 1/\Delta; & -\frac{1}{2}\Delta < x < \frac{1}{2}\Delta; \\ 0; & \frac{1}{2}\Delta < x \end{cases} \quad \int_{-\infty}^{\infty} f(z)\delta(x - z)\,dz = f(x) \tag{2.6.23}$$

We start out by expressing $\mathbf{e}^*(p)$ and $\mathbf{e}(p')$ in terms of the vectors $\mathbf{e}(q)$:

$$\mathbf{e}^*(p)\cdot\mathbf{e}(p') = c^2\int_{-\infty}^{\infty} dq \int_{-\infty}^{\infty} dq'\, e^{-(i/\hbar)pq+(i/\hbar)p'q'}[\mathbf{e}^*(q)\cdot\mathbf{e}(q')] = \delta(p - p')$$

But $[\mathbf{e}^*(q)\cdot\mathbf{e}(q')] = \delta(q - q')$, so that integration over q' results in

$$\delta(p - p') = c^2\int_{-\infty}^{\infty} dq\, e^{(i/\hbar)(p'-p)q}$$

This last integral is not particularly convergent, but the delta function is not a very well-behaved function either. It would be best if we utilized the definition of $\delta(p - p')$ and integrated both sides of the above equation over p' from $p - \frac{1}{2}\Delta$ to $p + \frac{1}{2}\Delta$ (it might have been safer if we had integrated over p' first and then integrated over q', rather than in the reverse order, but the results will be the same):

$$\int_{p-\frac{1}{2}\Delta}^{p+\frac{1}{2}\Delta} \delta(p - p')\,dp' = 1 = c^2\int_{-\frac{1}{2}\Delta}^{\frac{1}{2}\Delta} dz \int_{-\infty}^{\infty} dq\, e^{(i/\hbar)zq}$$
$$= 2c^2\hbar\int_{-\infty}^{\infty} \sin(q\Delta/2\hbar)\,(dq/q) = c^2\cdot 2\pi/\hbar$$

Therefore $c = 1/\sqrt{2\pi\hbar} = 1/\sqrt{h}$, and $\psi(p|q) = [1/\sqrt{2\pi\hbar}]\,e^{(i/\hbar)pq}$. Therefore the transformation functions $f(q)$ and $f(p)$ for the state vector $\mathbf{f}$ are related by the following equations

$$f(p) = \frac{1}{\sqrt{2\pi\hbar}} \int_{-\infty}^{\infty} f(q) e^{-(i/\hbar)pq}\, dq$$

and

$$f(q) = \frac{1}{\sqrt{2\pi\hbar}} \int_{-\infty}^{\infty} f(p) e^{(i/\hbar)pq}\, dp = \frac{1}{h} \int_{-\infty}^{\infty} dp \int_{-\infty}^{\infty} dq'\, f(q') e^{(2\pi i/h)(q-q')p} \tag{2.6.24}$$

This last equation has been encountered before [see Eq. (2.1.25)] and will be discussed in great detail in Chap. 4. It is one form of the Fourier integral.

The physical interpretation of the coordinate-momentum transformation function is worth mentioning. Since $|\psi(p|q)|^2$ is a constant, the probability $|\psi(p|q)|^2\, dq$, that the particle is between q and $q + dq$ when its momentum is p, is independent of q, and the probability $|\psi(p|q)|^2\, dp$, that the particle has momentum between p and $p + dp$ when it is at q, is independent of p. This is a natural consequence of the Heisenberg uncertainty principle, which says that, if p is precisely known ($\Delta p \to 0$), then the particle is equally likely to be anywhere ($\Delta q \to \infty$), and vice versa.

Second, we note that $e^{(i/\hbar)pq}$, viewed as a function of the coordinate q, could be the space factor of a traveling wave $\exp [(2\pi i/h)p(q - ut)]$ for a plane wave of length h/p and of frequency up/h traveling with some velocity u. We have not yet examined the time dependence of the state vectors and transformation functions, but we can see that the correlation $\lambda = (h/p)$ is the one we discussed [see Eq. (2.6.1)] when we first started to talk about waves and particles. To follow this relationship up, we should expect that the quantity ($h \times$ frequency) $= up$ would correspond to the energy of the particle, though we have not yet arrived at a point which would enable us to determine the wave velocity u.

The occurrence of a transformation function relating the state in which the particle has momentum p with its position q is a significant correlation with our preliminary discussions of waves and particles. For a state in which the particle may have either momentum p or momentum p', the eigenvector would be $\frac{1}{2}\sqrt{2}\,[\mathbf{e}(p) + e^{i\varphi}\,\mathbf{e}(p')]$ and the transformation function would be

$$\psi(p,q) = (1/\sqrt{2h})[e^{(2\pi i/\hbar)pq} + e^{(2\pi i/h)p'q+i\varphi}]$$

representing the space part of two linearly superposed waves of different wavelength. An arbitrary phase difference φ has been inserted, its value to be determined by a more exact specification of the state. The probability $|\psi|^2\, dq$ of finding the particle between q and $q + dq$ is not now everywhere the same, because the two parts of the wave will interfere, producing nodes and loops in the wave pattern. Since we are not now sure of the momentum, we can have some localization of the particle in space.

For a state **f** corresponding to a wide spread in momentum, as shown in Eq. (2.6.24), the interference between the partial waves may be such that $f(q)$ is only large near q_0 (*i.e.*, the particle is most likely to be near q_0) and the state may correspond to one in which the position of the particle is known with fair precision.

Hamiltonian Function and Schroedinger Equation. One of the most general "constants of the motion" in classical dynamics is the total energy of the system; in fact it is constant for all conservative systems. The classical analysis of the dynamics of such systems, developed by Hamilton, is based on the "Hamiltonian function" $H(p,q)$, see page 282, which is the total energy of the system, kinetic plus potential, expressed in terms of the coordinates and conjugate momenta (whereas the Lagrangian function L is usually the kinetic minus the potential energy, the Hamiltonian H is the sum of the two). The equations relating the time rate of change of the q's to the momenta are

$$\dot{q}_m = \frac{dq_m}{dt} = \frac{\partial H}{\partial p_m} \tag{2.6.25}$$

which corresponds to Eqs. (2.6.3). The equations of motion, relating the accelerations to the forces, are then

$$\dot{p}_m = \frac{dp_m}{dt} = -\frac{\partial H}{\partial q_m} \tag{2.6.26}$$

These equations will be discussed in more detail in Chap. 3.

In quantum mechanics the state vector $\mathbf{e}(E)$ for a given system, which represents the system having a definite energy E [$\mathbf{e}(E)$ is the eigenvector for the eigenvalue E of the energy], can be determined from the equation

$$\mathfrak{H}(p,q) \cdot \mathbf{e}(E) = E\mathbf{e}(E) \tag{2.6.27}$$

where the operator $\mathfrak{H}$ is obtained by changing the p's and q's in the classical Hamiltonian into operators. The correct order of the operators in the various terms is not completely defined by the classical $H(p,q)$, but this order can usually be determined by trial, as indicated on page 234.

Of course we may not wish to solve an equation involving operators and state vectors; it may be easier to solve the related differential equation for the transformation function $\psi(E|q)$, the square of which measures the probability density of various configurations of the system when we are sure that the energy is E. Referring to Eq. (2.6.17) we see that this equation is

$$H\left(\frac{\hbar}{i}\frac{\partial}{\partial q}, q\right)\psi(E|q) = E\psi(E|q) \tag{2.6.28}$$

where each p_m in the expression for H is changed to $(\hbar/i)(\partial/\partial q)$, operating on $\psi(E|q)$. This equation is called the *Schroedinger equation* for the system.

The Schroedinger equation often turns out to be of the general type of the Helmholtz equation (Eq. 2.1.10), an equation which results when the time dependence is separated out of a wave equation. To show this more clearly let us set up the Schroedinger equation for a particle of mass m in a potential field V (force $= -$ grad V). Appropriate coordinates q are the cartesian coordinates x, y, z. The kinetic energy is $\frac{1}{2}m(\dot{x}^2 + \dot{y}^2 + \dot{z}^2) = T$ and the kinetic potential $L = T - V$. The momentum p_x is, according to Eq. (2.6.3), equal to $\partial L/\partial \dot{x} = m\dot{x}$, etc. Expressing T and V in terms of the p's and q's, we finally obtain the classical Hamiltonian

$$H(p,q) = (1/2m)(p_x^2 + p_y^2 + p_z^2) + V(x,y,z)$$

The Schroedinger equation is obtained by substituting for the sum of the squares of the momenta $-(\hbar)^2$ times the sum of the second derivatives (the Laplacian):

$$\left[-H\left(\frac{\hbar}{i}\frac{\partial}{\partial q}, q\right) + E\right]\psi(E|q) = \left[\frac{\hbar^2}{2m}\nabla^2 + E - V\right]\psi(E|x,y,z) = 0$$

or

$$\nabla^2\psi + \frac{2m}{\hbar^2}[E - V(x,y,z)]\psi = 0$$

In those regions where $E > V$, ψ is oscillatory, and in those regions where $E < V$, the solution is exponential, either decreasing or increasing in size without limit. This is to be compared with Eq. (2.1.10).

Of course we could set up a similar equation for the transformation function from E to the momentum p. Here we should substitute for each q in $H(p|q)$ the differential operator $-(\hbar/i)(\partial/\partial p)$. Sometimes this is more difficult to do, for V is often a more complicated function of q than T is of p [how, for instance, do we interpret the operator formed from $1/\sqrt{x^2 + y^2 + z^2}$ when we substitute $-(\hbar/i)(\partial/\partial p_x)$ for x, etc.?]. We could, of course, solve first for $\psi(E|q)$ and obtain $\psi(E|p)$ by the Fourier transform given in Eq. (2.6.24). A more straightforward way, which can be demonstrated on the one-particle Hamiltonian, is as follows:

$$H = T(p) + V(q); \quad T = (1/2m)(p_x^2 + p_y^2 + p_z^2); \quad q = x, y, z$$

The operation corresponding to the kinetic energy is

$$\mathfrak{T}\mathbf{e}(E) = \iiint_{-\infty}^{\infty} \frac{1}{2m}(p_x^2 + p_y^2 + p_z^2)\psi(E|p)\,dp_x\,dp_y\,dp_z\,\mathbf{e}(p)$$

For the potential energy we can perform a second transformation to the $\mathbf{e}(q)$'s:

$$\mathfrak{V} \cdot \mathbf{e}(E) = \iiint_{-\infty}^{\infty} \mathfrak{V} \cdot \mathbf{e}(p')\psi(E|p')\, dp'_x\, dp'_y\, dp'_z, \text{ and by Eq. (2.6.20)}$$

$$\mathfrak{V} \cdot \mathbf{e}(p') = \iiint_{-\infty}^{\infty} [\mathbf{e}^*(p) \cdot \mathfrak{V} \cdot \mathbf{e}(p')]\, \mathbf{e}(p)\, dp_x\, dp_y\, dp_z$$

where

$$[\mathbf{e}^*(p) \cdot \mathfrak{V} \cdot \mathbf{e}(p')] = \iiint_{-\infty}^{\infty} \bar{\psi}(p|q)V(q)\psi(p'|q)\, dx\, dy\, dz$$

$$= \frac{1}{(2\pi\hbar)^3} \iiint_{-\infty}^{\infty} V(x,y,z)\, e^{(i/\hbar)[x(p_x'-p_x)+y(p_y'-p_y)+z(p_z'-p_z)]}\, dx\, dy\, dz = V_{pp'}$$

Therefore, instead of a differential equation for $\psi(E,p)$ we have an integral equation

$$(1/2m)(p_x^2 + p_y^2 + p_z^2)\psi(E|p) + \iiint_{-\infty}^{\infty} V_{pp'}\psi(E|p')\, dp'_x\, dp'_y\, dp'_z = E\psi(E|p)$$

which is, of course, completely equivalent to

$$H\left(p, -\frac{\hbar}{i}\frac{\partial}{\partial p}\right)\psi(E|p) = E\psi(E|p)$$

Which of the two equations is easier to solve depends on the form of H. If V can be expanded in simple polynomials of the q's, the differential equation is probably better; otherwise the integral equation is safer.

The Harmonic Oscillator. To illustrate the general principles we have adduced, it will be useful to discuss the quantum mechanics of a one-dimensional harmonic oscillator.

We first set up the operator corresponding to the Hamiltonian function H. For a particle of mass m and a spring of stiffness constant K, this Hamiltonian is

$$H(p,q) = (1/2m)p^2 + \tfrac{1}{2}Kq^2$$

so that the equation for an eigenvector of the energy operator is

$$\mathfrak{H} \cdot \mathbf{e} = [(\mathfrak{p}^2/2m) + \tfrac{1}{2}K\mathfrak{q}^2]\, \mathbf{e}(E) = E\, \mathbf{e}(E) \qquad (2.6.29)$$

We could, of course, set up the corresponding Schroedinger equation

$$\left[\frac{\hbar^2}{2m}\frac{\partial^2}{\partial q^2} + (E - \tfrac{1}{2}Kq^2)\right]\psi(E|q) = 0$$

and solve for ψ by methods to be discussed later in this book. But at present it will be more instructive to solve directly for $\mathbf{e}(E)$.

We note that the classical Hamiltonian can be factored:

$$H = \tfrac{1}{2}m[(p/m) - i\omega q][(p/m) + i\omega q]; \quad \omega^2 = K/m$$

The quantum mechanical operator cannot be factored quite so easily, but by watching the order of the $\mathfrak{p}$'s and $\mathfrak{q}$'s and by the use of the commutation relation $\mathfrak{p}\mathfrak{q} - \mathfrak{q}\mathfrak{p} = \hbar/i$, we obtain the following:

$$\tfrac{1}{2}m\mathfrak{G}_- \cdot \mathfrak{G}_+ = \mathfrak{H} + \tfrac{1}{2}\hbar\omega; \quad \tfrac{1}{2}m\mathfrak{G}_+ \cdot \mathfrak{G}_- = \mathfrak{H} - \tfrac{1}{2}\hbar\omega$$

where $\quad \mathfrak{G}_+ = (\mathfrak{p}/m) + i\omega\mathfrak{q} \quad$ and $\quad \mathfrak{G}_- = (\mathfrak{p}/m) - i\omega\mathfrak{q} = \mathfrak{G}_+^*$

Multiplying once more by $[(\mathfrak{p}/m) \pm i\omega\mathfrak{q}]$ and manipulating the results, we obtain

$$\mathfrak{H} \cdot \mathfrak{G}_+ = \mathfrak{G}_+ \cdot [\mathfrak{H} + \hbar\omega]; \quad \mathfrak{H} \cdot \mathfrak{G}_- = \mathfrak{G}_- \cdot [\mathfrak{H} - \hbar\omega] \tag{2.6.30}$$

These equations show that, if there is an eigenvector $\mathbf{e}(E)$ for $\mathfrak{H}$ with eigenvalue E, then the vector $\mathfrak{G}_+ \cdot \mathbf{e}(E)$ is an eigenvector for $\mathfrak{H}$ with eigenvalue $(E + \hbar\omega)$ [*i.e.*, it is $A\mathbf{e}(E + \hbar\omega)$ where A is some normalization constant] and the vector $\mathfrak{G}_- \cdot \mathbf{e}(E)$ is also an eigenvector for $\mathfrak{H}$ with eigenvalue $(E - \hbar\omega)$ [*i.e.*, it is $B\mathbf{e}(E - \hbar\omega)$]. This means that, if there is one eigenvalue E for $\mathfrak{H}$, then there is an infinite sequence of eigenvalues $(E + n\hbar\omega)$, where n is any integer, positive or negative.

This result is rather more than we had bargained on, for classically we should expect no allowed negative values of energy. A more careful scrutiny of Eqs. (2.6.30) and the related statements shows that we can be spared the unpleasant negative energies if we choose the value of E carefully. For we should have said that $\mathfrak{G}_- \cdot \mathbf{e}(E)$ is an eigenvector with eigenvalue $(E - \hbar\omega)$ *unless the vector* $\mathfrak{G}_- \cdot \mathbf{e}(E)$ is zero. Therefore if we are to have no allowed values of energy less than a certain minimum value $E_{\min}$, we must have

$$\mathfrak{G}_- \cdot \mathbf{e}(E_{\min}) = 0$$

or $\quad 0 = \mathfrak{G}_+ \cdot \mathfrak{G}_- \cdot \mathbf{e}(E_{\min}) = (\mathfrak{H} - \tfrac{1}{2}\hbar\omega)\mathbf{e}(E_{\min}) = (E_{\min} - \tfrac{1}{2}\hbar\omega)\mathbf{e}(E_{\min})$

Actually this choice is not merely most convenient; it is the only choice which makes physical sense. For if we do allow negative eigenvalues for the energy, the squares of the corresponding eigenvectors, $\mathbf{e}^* \cdot \mathbf{e}$, turn out to be *negative.* Since negative probabilities make no sense, the above choice is the *only* allowable one.

Consequently the minimum eigenvalue of the energy is equal to $\tfrac{1}{2}\hbar\omega$, and the sequence of allowed values is given by the formula

$$E_n = \hbar\omega(n + \tfrac{1}{2}); \quad n = 0, 1, 2$$

and the various eigenvectors can all be expressed in terms of the lowest, $\mathbf{e}_0 = \mathbf{e}(E_{\min})$, where $\mathfrak{H}\mathbf{e}_0 = \tfrac{1}{2}\hbar\omega\mathbf{e}_0$.

We must next normalize the eigenvectors so that $\mathbf{e}^* \cdot \mathbf{e} = 1$. We assume that $\mathbf{e}_0^* \cdot \mathbf{e}_0 = 1$. The next vector is $\mathbf{e}_1 = \mathbf{e}(E_{\min} + \hbar\omega) =$

$A_1\mathfrak{G}_+ \cdot \mathbf{e}_0$. To determine A_1 we set

$$1 = \mathbf{e}_1^* \cdot \mathbf{e}_1 = |A_1|^2\, \mathbf{e}_0^*\, \mathfrak{G}_- \cdot \mathfrak{G}_+\mathbf{e}_0 = \frac{2}{m}\,|A_1|^2\, \mathbf{e}_0^*(\mathfrak{H} + \tfrac{1}{2}\hbar\omega)\,\mathbf{e}_0 = \frac{2\hbar\omega}{m}\,|A_1|^2$$

Therefore $\quad \mathbf{e}_1 = i(m/2\hbar\omega)^{\frac{1}{2}}\mathfrak{G}_+ \cdot \mathbf{e}_0;\quad$ where $\mathfrak{H} \cdot \mathbf{e}_1 = \frac{3}{2}\hbar\omega\mathbf{e}_1$

Continuing in this manner we can show that

$$\mathbf{e}_n = \frac{1}{\sqrt{n!}}\left[i\sqrt{\frac{m}{2\hbar\omega}}\,\mathfrak{G}_+\right]^n \cdot \mathbf{e}_0;\quad \text{where } \mathfrak{H} \cdot \mathbf{e}_n = \hbar\omega(n + \tfrac{1}{2})\mathbf{e}_n \qquad (2.6.31)$$

We have therefore solved the problem of determining eigenvalues and eigenvectors for this Hamiltonian operator. Average values of other operator functions of $\mathfrak{p}$ and $\mathfrak{q}$ can be determined by playing about with the $\mathfrak{G}$ operators.

The equation for the lowest transformation function may be obtained by changing the equation $\mathfrak{G}_- \cdot \mathbf{e}_0 = 0$ into a differential equation for $\psi_0(q)$:

$$\frac{\hbar}{im}\frac{\partial\psi_0}{\partial q} - i\omega q\psi_0 = 0$$

which has for a solution

$$\psi_0 = \left[\frac{m\omega}{\pi\hbar}\right]^{\frac{1}{4}} e^{-\frac{1}{2}x^2};\quad x = \sqrt{\frac{m\omega}{\hbar}}\,q;\quad \int_{-\infty}^{\infty} \psi_0^2\, dq = 1$$

The operator $\mathfrak{G}_+$ has its differential counterpart

$$G_+ = \frac{1}{i}\sqrt{\frac{\hbar\omega}{m}}\left[\frac{\partial}{\partial x} - x\right]$$

By using Eq. (2.6.31) and translating it into an equation for the nth wave function, we have

$$\psi_n = \left[\frac{m\omega}{\pi\hbar 4^n(n!)^2}\right]^{\frac{1}{4}}\left[\frac{\partial}{\partial x} - x\right]^n e^{-\frac{1}{2}x^2};\quad \int_{-\infty}^{\infty}\psi_n^2\, dq = 1;\quad q = \sqrt{\frac{\hbar}{m\omega}}\,x$$

Thus we have also obtained an expression for the various wave functions with very little difficulty.

This example may perhaps have shown that dealing with the operators and eigenvectors directly is not difficult after all and that in some cases it may be actually easier than solving first for the wave functions.

Incidentally, the uncertainty principle could have told us that our state of minimum energy could not have been for zero energy, because zero energy implies that *both* p and q be zero, and the uncertainty principle states that it is impossible to know exact values of p and q simultaneously. The minimum value of the energy must be enough above zero so that the range of travel in q times the amplitude of oscillation of p

is not less than h. Putting in the right numbers we see that the minimum energy *cannot be smaller* than $\frac{1}{2}\hbar\omega$.

Dependence on Time. To complete our discussion of quantum mechanics, it is now necessary to introduce the time variable t in order to develop a kinematics and finally a dynamics so that we can state the quantum mechanical equations of motion.

Time enters into classical mechanics in two essentially different ways. In the case of conservative fields, time is only an implicit variable; *i.e.*, it is used as a *parameter* in terms of which the motion may be described. Indeed in two and three dimensions time may be eliminated completely and the motion described by giving the orbit. The space coordinate and the time coordinates do not play the same role. However, in relativistic mechanics, time and space must come into the theory in the same way, since by a Lorentz transformation one may mix space and time. A proper relativistic theory must therefore treat the time variable in a manner which is very different from a nonrelativistic theory. This difference maintains itself when we come to quantum mechanics.

Time also enters in as an *explicit* variable if the field of force is nonconservative or if the force field changes with time (*e.g.*, a field due to another system in motion) or finally in the statement of initial conditions. In all of these the time variable is needed to describe the force field acting on the system, so that the time coordinate and the space coordinates are employed in similar fashions.

This dichotomy appears in quantum mechanics also. For example, corresponding to $\Delta p_x \, \Delta x \simeq \hbar$ there is an uncertainty relation

$$\Delta E \, \Delta t \simeq \hbar \tag{2.6.32}$$

where ΔE measures the uncertainty in energy and Δt the uncertainty in time. One would expect this equation on the basis of relativistic requirements; indeed this is precisely the reasoning used by De Broglie when he guessed $\lambda = h/p$ from $\nu = E/h$. From Eq. (2.6.32) one would be tempted, in analogy with our deduction of the commutator equation $[\mathfrak{p}_x,\mathfrak{x}] = \hbar/i$ from the uncertainty relation (2.6.7), to set up a commutator between the operator corresponding to the energy and an operator corresponding to the time. This would not be a correct procedure for the reason that the uncertainty relation (2.6.32) applies *only when time occurs as an explicit variable.*

Stating this in another way, when time does not occur explicitly in the description of the force field, time is really a parameter. Its measurement by the observer (by looking at a clock, for instance) in no way affects the system under consideration. For example, it cannot affect the energy of such a system. However, if time occurs explicitly in the description of the force field, an uncertainty in time will lead to an uncertainty in the force and therefore in the energy.

A common example of the relation of ΔE and Δt considers the measurement of the time of travel between two points, 1 and 2, by a wave, say a light wave. Using a wave whose frequency (and therefore energy) is precisely known would require a wave infinite in extent. The time of travel would therefore be completely unknown ($\Delta t \to \infty$, $\Delta E \to 0$). The precision of the measurement of time of travel is increased by using a shutter at 1 which is tripped at some instant, say $t = 0$, and another shutter (in actual practice mirrors are used) at 2 which is to be tripped some time later. It is easy to see that the velocity being measured is that of a wave packet of width Δt in time, where Δt is the time difference between the action of the two shutters. However, such a wave packet must be a superposition of waves having different frequencies (a Fourier integral). These frequencies are spread approximately over a range $1/\Delta t$ so $\Delta \nu \, \Delta t \simeq 1$ or $\Delta E \, \Delta t \simeq h$. It is clear that the Hamiltonian describing this experiment is *time dependent;* the time variable is required to describe the interaction of matter (shutters) with the radiation. It is characteristic that the time dependence is due to the fact that all the interacting systems, *e.g.*, the operator of the shutter, have not been included in the discussion.

Recapitulating, we see that the uncertainty relation (2.6.32) applies only when the Hamiltonian depends explicitly on the time. When the time dependence is implicit, this equation does not apply, and indeed time may be considered just as a parameter. This Januslike behavior is reflected in the fact that the derivation of the time-dependent Schroedinger equation may differ, depending upon which situation applies. Fortunately for the relativistic generalizations of quantum mechanics it is possible to devise a single derivation which applies in both cases.

Time as a Parameter. In the interest of simplicity and clarity it is best to discuss the nonrelativistic case with conservative fields, *i.e.*, those cases in which the classical nonrelativistic Hamiltonian does not depend upon time, so that the time variable may be treated as a parameter. We have already shown in Chap. 1 (page 85) that the variation of the direction of a state vector with time may be obtained through the use of a unitary operator. We showed there that

$$e^{-(i/\hbar)\mathfrak{H}t'} \cdot \mathbf{e}(t) = \mathbf{e}(t + t')$$

where so far the operator $\mathfrak{H}$ is undefined. Indeed, in setting up a kinematics, one of the crucial steps will be that of choosing $\mathfrak{H}$. Here we have used $-(\mathfrak{H}/\hbar)$ instead of $\mathfrak{H}$, to bring the equation in line with Eq. (2.6.11) relating $\mathfrak{p}$ and $\mathfrak{q}$. In Chap. 1 we also showed that

$$\mathfrak{H} \cdot \mathbf{e} = i\hbar(\partial \mathbf{e}/\partial t) \tag{2.6.33}$$

where we have here also included the additional factor $-(1/\hbar)$. This equation is similar to Eq. (2.6.12) for the effect on $\mathfrak{p}$ on $\mathbf{e}(q)$. But there is a fundamental difference, which enables us to write $\partial \mathbf{e}/\partial t$ here but does not allow us to make this limiting transition in Eq. (2.6.12). In the earlier equation, giving the effect of $\mathfrak{p}$, we are dealing with eigenvectors $\mathbf{e}(q)$ for the operator $\mathfrak{q}$, so that each $\mathbf{e}$ is perpendicular to the other and there is no possibility of taking a limit. In the present case t is only a *parameter;* $\mathbf{e}$ is *not* its eigenvector, for t is not an operator. All eigenvectors for the system having $\mathfrak{H}$ as an operator are *continuous* functions of the parameter t, rotating in abstract space as t increases. Consequently we can here talk about a derivative of $\mathbf{e}$ with respect to t. The operator $(\mathfrak{H}/i\hbar)\,dt$ produces an infinitesimal rotation from its direction at time t to its direction at time $t + dt$, and this difference in direction becomes continuously smaller as dt approaches zero.

Kinematics in classical mechanics is concerned with the time variation of variables such as position q with time. In order to be able to draw the necessary analogies in quantum mechanics and so to determine $\mathfrak{H}$, it is necessary for us to consider the variation of operators with time. The development of quantum mechanics considered up to now in this chapter assumes that the meaning of the operator is independent of t as far as its operation on its eigenvector, so that the equation $\mathfrak{F} \cdot \mathbf{f} = f\mathbf{f}$ gives the same eigenvalue f for all values of time (as long as $\mathfrak{F}$ does not depend on t explicitly).

In many cases the *state vector* itself changes with time, in the manner we have just been discussing. However, we could just as well consider that the state vector is independent of time and blame all the variation with time on the operator. This formal change, of course, must not affect the measurable quantities, such as the eigenvalues f or the expansion coefficients defined in Eq. (2.6.20), for instance. In other words the operator $\mathfrak{A}(t)$, including the time parameter, can be obtained from the constant operator $\mathfrak{A}(0)$ by the use of Eq. (2.6.33) and the requirement that the quantity

$$\mathbf{f}^*(t) \cdot \mathfrak{A}(0) \cdot \mathbf{f}(t) = \mathbf{f}^*(0) \cdot \mathfrak{A}(t) \cdot \mathbf{f}(0) = \mathbf{f}^*(0) \cdot e^{(i/\hbar)\mathfrak{H}t}\mathfrak{A}(0)e^{-(i/\hbar)\mathfrak{H}t} \cdot \mathbf{f}(0)$$

be independent of t. Consequently

$$\mathfrak{A}(t) = e^{(i/\hbar)\mathfrak{H}t}\mathfrak{A}(0)e^{-(i/\hbar)\mathfrak{H}t} \tag{2.6.34}$$

gives the dependence of $\mathfrak{A}(t)$ on t if we are to consider the operator as time dependent.

Letting t become the infinitesimal dt, we can obtain an equation relating the time rate of change of the operator $\mathfrak{A}(t)$ to the as yet unknown operator $\mathfrak{H}$:

$$\mathfrak{A}(dt) = \left[1 + \frac{i}{\hbar}\mathfrak{H}\,dt\right]\mathfrak{A}(0)\left[1 - \frac{i}{\hbar}\mathfrak{H}\,dt\right]$$

or

$$\mathfrak{H} \cdot \mathfrak{A} - \mathfrak{A} \cdot \mathfrak{H} = [\mathfrak{H},\mathfrak{A}] = \frac{\hbar}{i}\frac{\mathfrak{A}(dt) - \mathfrak{A}(0)}{dt} = \frac{\hbar}{i}\frac{d}{dt}\mathfrak{A} \tag{2.6.35}$$

From the way we have derived this equation, the expression $d\mathfrak{A}/dt$ can be considered to be the time rate of change of the operator $\mathfrak{A}$ if we have let the state vectors be constant and put the time variation on $\mathfrak{A}$, or it can be the operator corresponding to the classical rate of change of the dynamical variable A when we let the operators be constant and let the state vectors vary.

For instance, the operator corresponding to the rate of change $\dot{q}_m$ of a coordinate can be computed from Eq. (2.6.35):

$$\dot{\mathfrak{q}}_m = (i/\hbar)[\mathfrak{H},\mathfrak{q}_m] = \partial\mathfrak{H}/\partial\mathfrak{p}_m \tag{2.6.36}$$

where we have used Eq. (2.6.8) to obtain the last expression. But this last expression is just the one needed to determine the nature of the operator $\mathfrak{H}$. In the limit of large energies and momenta this operator equation should reduce to the classical equations in ordinary variables. This will be true *if the operator* $\mathfrak{H}$ is the Hamiltonian for the system, with the p's and q's occurring in it changed to operators. In other words if $\mathfrak{H}$ is the Hamiltonian operator of Eq. (2.6.27), then Eq. (2.6.36) will correspond to the classical Eq. (2.6.25).

This can be double-checked, for if we set $\mathfrak{A} = \mathfrak{p}_m$ in Eq. (2.6.35) and use Eq. (2.6.8), again we obtain

$$\dot{\mathfrak{p}}_m = (i/\hbar)[\mathfrak{H},\mathfrak{p}_m] = -(\partial\mathfrak{H}/\partial\mathfrak{q}_m) \tag{2.6.37}$$

which corresponds to classical equation (2.6.26).

Thus we may conclude that the equations of motion of the operators in quantum mechanics have precisely the same form as their classical counterparts, with the classical quantities p and q replaced by their corresponding operators $\mathfrak{p}$ and $\mathfrak{q}$. For example, Newton's equation of motion becomes $m(d^2\mathfrak{q}/dt^2) = -(\partial\mathfrak{V}/\partial\mathfrak{q})$. By taking average values of any of these equations we see immediately that the classical Newtonian orbit is just the *average* of the possible quantum mechanical orbits. Stated in another way, the effect of the uncertainty principle is to introduce fluctuations away from the classical orbit. These average out. Of course, the average of the square of the fluctuations is not zero and is therefore observable, but in the limit of large energies the uncertainties become negligible, and quantum mechanics fades imperceptibly into classical mechanics. This statement is known as the *correspondence principle.* Its counterpart in equations is the statement that in the limit the commutator $(i/\hbar)[\mathfrak{A},\mathfrak{B}]$ goes into the classical (A,B). Because of the correspondence between commutator and the Poisson bracket, it follows that *any classical constant of the motion is also a quantum mechanical constant of the motion.*

Of course we may wish to work with transformation functions instead of eigenvectors. These functions also change with time, and corresponding to Eq. (2.6.33) we have the *time-dependent Schroedinger equation*

$$H\left(\frac{\hbar}{i}\frac{\partial}{\partial q}, q\right)\psi(t,q) = i\hbar\frac{\partial}{\partial t}\psi(t,q) \tag{2.6.38}$$

where $H(p,q)$ is the classical Hamiltonian and the time-dependent state vector is given by the integral

$$\mathbf{e}(t) = \int_{-\infty}^{\infty} \psi(t,q)\, \mathbf{e}(q)\, dq$$

Of course for stationary states

$$\mathfrak{H}\cdot\mathbf{e}(E) = i\hbar(\partial\mathbf{e}/\partial t) = E\mathbf{e}(E)$$

so that

$$\mathbf{e}(E,t) = \mathbf{e}(E,0)e^{-i(E/\hbar)t} \tag{2.6.39}$$

where E is an eigenvalue of the energy. Thus the time dependence for the stationary state is simple harmonic, with frequency equal to the value of the energy divided by h, so that the Planck equation $E = h\nu$, given in Eq. (2.6.1), is satisfied.

We have thus shown that the transformation function $\psi(E,q)$ is the "wave function" which we spoke about at the beginning of this section. The square of its magnitude gives the probability density of the various configurations of the system, and integrals of the form $\int \bar{\psi}B\left[\frac{\hbar}{i}\frac{\partial}{\partial q}, q\right]\psi\, dq$ give the average value of a sequence of measurements of the dynamical variable $B(p,q)$ when the system is in the state corresponding to ψ. This probability density and these average values are all that can be obtained experimentally from the system. For large systems, having considerable energy, the results correspond fairly closely to the precise predictions of classical dynamics, but for atomic systems the uncertainties are proportionally large and the results may differ considerably from the classical results.

We have also shown that these transformation functions have wave properties and exhibit interference effects which affect the probability density. The wave number of the wave in a given direction is equal to $1/h$ times the component of the momentum in that direction, and the frequency of the wave is equal to $1/h$ times the energy of the system, as pointed out in Eq. (2.6.1). Only by the use of the machinery of abstract vectors and operators and transformation functions is it possible to produce a theory of atomic dynamics which corresponds to the experimentally determined facts, such as the inherent uncertainties arising when atomic systems are observed.

Time-dependent Hamiltonian. Having discussed the case where time does not enter explicitly into the energy expression H, where time turns out to be a parameter rather than an operator, let us now consider the case when H does depend explicitly on the time t. In this case the time used in describing the change of energy should be considered as an operator (just as are the coordinates) rather than as a convenient parameter for keeping track of the system's motion.

The distinction is clearer in the quantum mechanical treatment than in the classical treatment, for we can distinguish between the operator corresponding to time and the continuous distribution of its eigenvalues. Classically we shall make the distinction by letting the explicit time be q_t, so that the total energy is a function of $q_t \cdots q_n$ and of the momenta $p_1 \cdots p_n$, which we can indicate formally by $H(q_t;p,q)$.

This function gives the proper classical equations of motion (2.6.25) and (2.6.26) for $\dot{q}_1 \cdots \dot{q}_n$ and $\dot{p}_1 \cdots \dot{p}_n$ but does not give a corresponding set of equations for q_t. As a matter of fact we have not yet considered the conjugate momentum p_t. It appears that we must modify the Hamiltonian function in the cases where H depends explicitly on time, so that the new Hamiltonian $H(p_t,q_t;p,q)$ satisfies the equation

$$\dot{q}_t = \frac{dq_t}{dt} = \frac{\partial H}{\partial p_t}$$

But before we decide on the form of H, we must decide on the meaning of $\dot{q}_t$. Since q_t is the explicit time, we should expect that in classical dynamics it should be proportional to the time parameter t and in fact that $dq_t/dt = 1$. Consequently we must have that the new Hamiltonian Θ is related to the total energy $H(q_t;p,q)$ and the new momentum p_t by the equation

$$\Theta(p_t,q_t;p,q) = H(q_t;p,q) + p_t \tag{2.6.40}$$

Then the equations of motion are

$$\frac{\partial \Theta}{\partial p_m} = \dot{q}_m; \quad \frac{\partial \Theta}{\partial q_m} = -\dot{p}_m; \quad m = t, 1, \ldots, n \tag{2.6.41}$$

The total rate of change of Θ with time (due to change of all the p's and q's with time) can be shown to be zero, for, using Eqs. (2.6.41), (2.6.25), and (2.6.26)

$$\frac{d\Theta}{dt} = \frac{\partial \Theta}{\partial q_t}\dot{q}_t + \frac{\partial \Theta}{\partial p_t}\dot{p}_t + \sum_m \left[\frac{\partial \Theta}{\partial q_m}\dot{q}_m + \frac{\partial \Theta}{\partial p_m}\dot{p}_m\right] = 0$$

Therefore the new Hamiltonian is constant even though the total energy H changes explicitly with time. We may as well add enough to make the constant zero; $\Theta = H + p_t = 0$. This means that the quantity p_t, the variable conjugate to the explicit time q_t, is just the negative of the value of the total energy, $p_t = -E$ (we write E because H is considered to be written out as an explicit function of q_t and the other p's and q's, whereas E is just a numerical value which changes with time). Therefore the explicit time is conjugate to minus the energy value.

The classical Poisson bracket expressions can also be generalized to include the new pair of variables,

$$(u,v) = \sum_m \left[\frac{\partial u}{\partial p_m}\frac{\partial v}{\partial q_m} - \frac{\partial u}{\partial q_m}\frac{\partial v}{\partial p_m}\right]; \quad m = t, 1, \ldots, n$$

The Poisson bracket including the Hamiltonian may be evaluated by the use of Eqs. (2.6.41):

$$(\Theta,v) = \sum_m \left[\frac{\partial \Theta}{\partial p_m}\frac{\partial v}{\partial q_m} - \frac{\partial \Theta}{\partial q_m}\frac{\partial v}{\partial p_m}\right]$$

$$= \sum_{m=1}^{n} \left[\frac{\partial v}{\partial q_m}\frac{dq_m}{dt} + \frac{\partial v}{\partial p_m}\frac{dp_m}{dt}\right] + \left[\frac{\partial v}{\partial q_t} + \frac{\partial \Theta}{\partial q_t}\frac{\partial v}{\partial p_t}\right] = \frac{dv}{dt} \tag{2.6.42}$$

since $\partial v/\partial q_t = \partial v/\partial t$ and $\partial\Theta/\partial q_t = 0$.

Introduction of the explicit time and its conjugate momentum into quantum mechanics is now fairly straightforward. We introduce the *operator* $\mathfrak{q}_t$, having a nondenumerable, continuous sequence of eigenvalues t, which *can* be used to specify the particular state which is of interest. The conjugate operator $\mathfrak{p}_t$ has eigenvalues equal to minus the allowed values of the energy. These operators are on a par with the operators for the different configuration coordinates and momenta. The commutator is

$$[\mathfrak{p}_t,\mathfrak{q}_t] = \hbar/i$$

so that the corresponding uncertainty relation is $\Delta E\, \Delta t \simeq \hbar$. The operators $\mathfrak{p}_t,\mathfrak{q}_t$ commute with all other $\mathfrak{p}$'s and $\mathfrak{q}$'s. Equations (2.6.8) and (2.6.11) also hold for this pair.

The Hamiltonian operator $\mathfrak{H}$ is now obtained by changing the p's and q's in the total energy function into the corresponding operators and the explicit time into the operator $\mathfrak{q}_t$; then

$$\mathfrak{H} = H(\mathfrak{q}_t;\mathfrak{p},\mathfrak{q}) + \mathfrak{p}_t \tag{2.6.43}$$

The unitary operator transforming a state vector at time t to the state vector at time t' is $\exp[(i/\hbar)\mathfrak{H}(t' - t)]$ [see Eq. (2.6.11)]. The equation of motion of a state vector $\mathbf{e}$ is

$$\mathfrak{H} \cdot \mathbf{e}(t) = i\hbar \lim \left[\frac{\mathbf{e}(t + dt) - \mathbf{e}(t)}{dt}\right] \tag{2.6.44}$$

as with Eq. (2.6.12), and the equation of motion for an operator $\mathfrak{A}$ is

$$[\mathfrak{H},\mathfrak{A}] = (\hbar/i)(d\mathfrak{A}/dt)$$

In particular

$$\dot{\mathfrak{q}}_t = \frac{i}{\hbar}[\mathfrak{H},\mathfrak{q}_t] = \frac{i}{\hbar}[\mathfrak{p}_t,\mathfrak{q}_t] = \mathfrak{I}$$

where $\mathfrak{J}$ is the constant operator which transforms every vector into itself (in Chap. 1 we called it the idemfactor).

We can now go to the properties of transformation functions for systems where the Hamiltonian varies explicitly with time. We still define the transformation function from q to E (often called the Schroedinger wave function) by the equation

$$\mathbf{e}(0) = \int \cdot \cdot \cdot \int \psi(0|q,t)\, dt\, dq_1 \cdot \cdot \cdot dq_n\, \mathbf{e}(q,t)$$

where t is the eigenvalue for $\mathfrak{q}_t$, q_m the eigenvalue for $\mathfrak{q}_m$, and 0 is the eigenvalue for $\mathfrak{H}$, the operator given in Eq. (2.6.43).

Just as was shown earlier that the operator $\mathfrak{p}_m$ operating on $\mathbf{e}$ corresponded to the differential operator $(\hbar/i)(\partial/\partial q_m)$ operating on the transformation function, so here the operator $\mathfrak{p}_t$ corresponds to the differential operator $(\hbar/i)(\partial/\partial t)$ acting on ψ. The differential equation for ψ corresponding to the vector equation $\mathfrak{H} \cdot \mathbf{e} = 0$ is

$$H\left(t; \frac{\hbar}{i}\frac{\partial}{\partial q}, q\right)\psi(|q,t) + \frac{\hbar}{i}\frac{\partial}{\partial t}\psi(|q,t) = 0 \qquad (2.6.45)$$

This is called the time-dependent Schroedinger equation and is to be compared with Eq. (2.6.38), where we considered time as simply a parameter. As we see, it is a consistent extension of the earlier procedure for changing the classical equations for time-dependent Hamiltonian to a quantum mechanical equation for the wave function ψ. The quantity $|\psi|^2$ is the probability density for a given configuration at a given time. The mean value of the particle current density at any point would seem to be proportional to

$$\mathbf{e}^* \cdot \mathfrak{p} \cdot \mathbf{e} = \int \cdot \cdot \cdot \int \bar{\psi} \frac{\hbar}{i} \operatorname{grad} \psi\, dq_1 \cdot \cdot \cdot dq_n$$

except that this is not necessarily a real quantity. However, we can now calculate what the current is.

Particle in Electromagnetic Field. For instance, for a particle of charge e (the electronic charge is $-e$) and mass m, moving through an electromagnetic field with potentials $\mathbf{A}$ and φ, the force on the particle [by Eq. (2.5.12) is $e[\mathbf{E} + (1/cm)\mathbf{p} \times \mathbf{H}]$ and the total energy (nonrelativistic) of the particle is

$$H(p,q) = \frac{1}{2m}\left[\left(\mathbf{p} - \frac{e}{c}\mathbf{A}\right) \cdot \left(\mathbf{p} - \frac{e}{c}\mathbf{A}\right)\right] + e\varphi$$

as will be shown in the next chapter (page 296). We substitute $(\hbar/i)(\partial/\partial q)$ for each p in H to obtain the differential equation for ψ. There is no ambiguity in order in the terms in $\mathbf{p} \cdot \mathbf{A}$; if div $\mathbf{A} = 0$, then proper order is $\mathbf{A} \cdot \mathbf{p}$. The resulting equation for ψ is

$$-\frac{\hbar^2}{2m}\nabla^2\psi - \frac{e\hbar}{imc}\mathbf{A} \cdot \operatorname{grad} \psi + \left[\frac{e^2A^2}{2mc^2} + e\varphi\right]\psi + \frac{\hbar}{i}\frac{\partial\psi}{\partial t} = 0 \qquad (2.6.46)$$

As with Eq. (2.6.45) the imaginary quantity i enters explicitly into the equation. This means that the equation for the complex conjugate $\bar{\psi}$ is

$$-\frac{\hbar^2}{2m}\nabla^2\bar{\psi} + \frac{e\hbar}{imc}\mathbf{A}\cdot\operatorname{grad}\bar{\psi} + \left(\frac{e^2A^2}{2mc^2} + e\varphi\right)\bar{\psi} - \frac{\hbar}{i}\frac{\partial\bar{\psi}}{\partial t} = 0$$

If $e\psi\bar{\psi}$ is to be the charge density ρ for the electromagnetic equations, then the current density $\mathbf{J}$ must be such as to satisfy the equation of continuity $(\partial\rho/\partial t) + \operatorname{div}\mathbf{J} = 0$. We use the equations for ψ and $\bar{\psi}$ to determine $\mathbf{J}$. Multiplying the equation for ψ by $\bar{\psi}$ and the one for $\bar{\psi}$ by ψ and subtracting, we obtain

$$\frac{\hbar^2}{2m}(\bar{\psi}\nabla^2\psi - \psi\nabla^2\bar{\psi}) - \frac{ie\hbar}{mc}\mathbf{A}\cdot\operatorname{grad}(\bar{\psi}\psi) + i\hbar\frac{\partial}{\partial t}(\bar{\psi}\psi) = 0$$

But from the rules of vector operation we can show that

$$\bar{\psi}\nabla^2\psi - \psi\nabla^2\bar{\psi} = \operatorname{div}(\bar{\psi}\operatorname{grad}\psi - \psi\operatorname{grad}\bar{\psi})$$

and if $\operatorname{div}\mathbf{A} = 0$, we have $\mathbf{A}\cdot\operatorname{grad}(\bar{\psi}\psi) = \operatorname{div}(\mathbf{A}\bar{\psi}\psi)$. Therefore

$$\frac{\partial}{\partial t}(e\bar{\psi}\psi) + \operatorname{div}\left[\frac{e\hbar}{2im}(\bar{\psi}\operatorname{grad}\psi - \psi\operatorname{grad}\bar{\psi}) - \frac{e^2}{mc}\mathbf{A}\bar{\psi}\psi\right] = 0$$

and if $\rho = e\bar{\psi}\psi$, the current density turns out to be

$$\mathbf{J} = \frac{e\hbar}{2im}(\bar{\psi}\operatorname{grad}\psi - \psi\operatorname{grad}\bar{\psi}) - \frac{e^2}{mc}\mathbf{A}\bar{\psi}\psi \qquad (2.6.47)$$

This expression is real, and since ρ and $\mathbf{J}$ satisfy the equation of continuity, presumably they are the expressions to insert in Maxwell's equations for the charge current. We note that these expressions are only probability densities, not "true densities" in the classical sense. This result is, however, in accord with our new conception of what is observable; since the "actual" positions and momenta of individual electrons cannot be known accurately, the only available expressions for the densities must come from the wave function ψ. As indicated at the beginning of this section, they involve the square of the magnitude of ψ, a characteristic of quantum densities and probabilities.

Relativity and Spin. The relationship between the four momentum operators for a single particle and the corresponding differential operators for use on the transformation function $\psi(\ |q,t)$ (where the blank before the vertical bar indicates that ψ could be for any eigenvector and eigenvalue),

$$p_m \to (\hbar/i)(\partial/\partial q_m); \quad q_m = x, y, z, t \qquad (2.6.48)$$

is a four-vector relationship which can satisfy the requirements of special relativity. The time-dependent Schroedinger equation (2.6.46) is not Lorentz invariant, however, even for the case of free flight, where $\mathbf{A}$ and φ are zero. The space operators occur in the second derivatives, and the time operator occurs in the first derivative, and no combination of p_x^2, p_y^2, p_z^2, and $p_t = -E$ can be Lorentz invariant.

The difficulty, of course, lies in the fact that the expression we used for $H(p,q)$ for a particle in an electromagnetic field was not relativistically invariant but was simply the first approximation to the correct relativistic Hamiltonian. This quantity can be obtained by combining the four-vector $p_x, p_y, p_z, -(i/c)H$ (see page 97) with the four-vector $A_x, A_y, A_z, i\varphi$ (see page 208) to form the Lorentz-invariant equation

$$-\left(p_x - \frac{e}{c}A_x\right)^2 - \left(p_y - \frac{e}{c}A_y\right)^2 - \left(p_z - \frac{e}{c}A_z\right)^2 + \left(\frac{1}{c}H + \frac{e}{c}\varphi\right)^2 = m^2c^2 \quad (2.6.49)$$

From this we can obtain the relativistic equations for the Hamiltonian:

$$H(p,q) = -e\varphi + c\sqrt{m^2c^2 + p^2 - (2e/c)\mathbf{A}\cdot\mathbf{p} + (e/c)^2A^2} \quad (2.6.50)$$

It is this function which should be converted into a differential operator to obtain the correct time-dependent Schroedinger equation.

This result, however, only poses a more serious problem: how do we interpret an operator involving the square root of a second derivative? Conceivably we could expand the radical in a series involving increasing powers of $1/m^2c^2$ (the Hamiltonian of page 254 is the first two terms of this expansion, with the constant mc^2 omitted), but this series will involve all the higher derivatives of ψ and would be an extremely "untidy" equation even if it did happen to converge. A possible solution is to use Eq. (2.6.49) as it stands, remembering that $-(1/c)H$ is the fourth component of the momentum vector and should be replaced by $(\hbar/ic)(\partial/\partial t)$. When the fields are zero, this results in the equation

$$\nabla^2\psi - \frac{1}{c^2}\frac{\partial^2\psi}{\partial t^2} - \left(\frac{mc}{\hbar}\right)^2\psi = 0 \quad (2.6.51)$$

which is the *Klein-Gordon equation* [see Eq. (2.1.27)].

This equation for the transformation function is relativistically invariant but has the disadvantage that, if $e|\psi|^2$ is the charge density, then the quantity given in Eq. (2.6.47) is not the current density. As a matter of fact the integral of $|\psi|^2$ over all space is no longer always a constant, as it is for a solution of Eq. (2.6.46), so that it is not clear that $e|\psi|^2$ is the charge density. We shall defer to the next chapter the determination of the correct expressions for ρ and J; it is only necessary to state here that the Klein-Gordon equation is not the correct one for electrons or for any particle with spin.

The time-dependent Schroedinger equation (2.6.46) is fairly satisfactory for particles moving slowly compared with the speed of light, but it does neglect two items: relativity and spin. We know that the electron has a spin and have discussed in Secs. 1.6 and 1.7 the properties of spin operators. These spin operators correspond to an additional degree of freedom for the electron, with, presumably, a new coordinate and momentum. We could, therefore, compute a transformation function including this new coordinate if we wished and obtain expressions for the spin operator in terms of differentiation with respect to this coordinate. Since the rules for operation of the spin operator are fairly simple, it is usually easier to keep the state vector.

The function used is therefore a mongrel one, consisting of a transformation function for the space and time components and the state vector **a** for the spin part. The total state vector is thus expanded as follows:

$$\mathbf{e}(E,s) = \int_{-\infty}^{\infty} \cdots \int_{-\infty}^{\infty} \psi(E|q_1 \cdots q_n)\mathbf{e}(q_1 \cdots q_n)\mathbf{a}(s)\, dq_1 \cdots dq_n$$

where s is one or the other of the two eigenvalues $\pm\hbar/2$ for the spin operator $\mathfrak{S}$ and **a** is one of the spin vectors defined in Eq. (1.6.44). Therefore if we have a Hamiltonian (nonrelativistic) which includes the spin operator $\mathfrak{S}$ as well as the p's and q's and time, the hybrid wave-function spin vector is $\Psi = \psi_+(|q,t)\mathbf{a}(\hbar/2) + \psi_-(|q,t)\mathbf{a}(-\hbar/2)$, and the equation is

$$H\left(t; \frac{\hbar}{i}\frac{\partial}{\partial q}, q; \mathfrak{S}\right)\Psi + \frac{\hbar}{i}\frac{\partial}{\partial t}\Psi = 0$$

which corresponds to Eq. (2.6.45). The average value of the quantity $B(p,q;\mathfrak{S})$ the state denoted by ψ is then

$$\int \cdots \int \bar{\Psi} B\left(\frac{\hbar}{i}\frac{\partial}{\partial q}, q; \mathfrak{S}\right)\Psi\, dq_1 \cdots dq_n$$

where the $\mathfrak{S}$ part of the operator works on the spin vectors **a** and the differential operators work on the wave functions ψ.

But this solution still does not give us a particle wave function which includes spin and also is relativistic. To attain this we turn to the spinor operators discussed in Sec. 1.7. The unit vectors $\boldsymbol{\sigma}_1 \cdots \boldsymbol{\sigma}_4$ defined in Eq. (1.7.17) provide operators which behave like components of a four-vector. They operate on state vectors **e** which have only two different directions, one corresponding to the z component of spin equal to $\hbar/2$ and the other corresponding to it being $-\hbar/2$ (the direction of the z axis is arbitrary). A suggestion for the Lorentz-invariant wave equation with spin is to form a scalar product of the four-vector p_x, p_y, p_z, $p_t/c = -E/c$ with the four-vector spin operator. Since a scalar product of two four-vectors is Lorentz invariant, we shall thus obtain a wave equation which has a first derivative with respect to time [as with Eq. (2.6.45) but not (2.6.51)] and which is also relativistic [as in Eq. (2.6.51) but not in (2.6.45)]. We should expect to set up an equation of continuity with such an equation so we can determine a charge and current density [as we did in Eq. (2.6.47) for solutions of Eq. (2.6.45)].

The simplest form for such an equation would be that the scalar product of the four-vectors $\boldsymbol{\sigma}$ and $\mathbf{p}$ operating on a spinor $\mathbf{e}$ would equal a constant times $\mathbf{e}$, or for a wave function $\boldsymbol{\Psi}$, consisting of two functions of position multiplied by the two spin vectors, as before, we should have

$$\frac{\hbar}{i}\left[\boldsymbol{\sigma}_1 \frac{\partial}{\partial x} + \boldsymbol{\sigma}_2 \frac{\partial}{\partial y} + \boldsymbol{\sigma}_3 \frac{\partial}{\partial z}\right]\boldsymbol{\Psi} = \left[\text{constant} - \frac{\hbar}{ic}\frac{\partial}{\partial t}\right]\boldsymbol{\Psi}$$

since $\boldsymbol{\sigma}_4 = \mathfrak{J}$ and E becomes $-(\hbar/i)\partial/\partial t)$ for the wave function. The only difficulty is that the vector $\boldsymbol{\sigma} = (\boldsymbol{\sigma}_1,\boldsymbol{\sigma}_2,\boldsymbol{\sigma}_3)$ is an axial vector (see page 104) whereas the gradient is a true vector, so that the quantity in brackets on the right-hand side should be a pseudoscalar (see page 11) changing sign for a reversal of sign of the coordinates. It is extremely difficult to see what fundamental constant we could find which would be a pseudoscalar, so difficult, in fact, that we are forced to look for a less simple form which will circumvent our need for a pseudoscalar.

Such a less simple form consists of a *pair* of equations

$$(\boldsymbol{\sigma}\cdot\mathbf{p})\mathbf{e} = \left[a + \frac{E}{c}\right]\mathbf{f};\quad (\boldsymbol{\sigma}\cdot\mathbf{p})\mathbf{f} = \left[b + \frac{E}{c}\right]\mathbf{e}$$

where $\mathbf{e}$ and $\mathbf{f}$ are *different* state vectors. By eliminating $\mathbf{f}$ and using Prob. 1.33, we discover that $b = -a$; then a can be a true scalar, not a pseudoscalar, and $\mathbf{e}$ is different from $\mathbf{f}$. As a matter of interest this pair of equations is analogous to the equations

$$\text{curl } \mathbf{H} = (1/c)(\partial\mathbf{E}/\partial t);\quad \text{curl } \mathbf{E} = -(1/c)(\partial\mathbf{H}/\partial t)$$

for the electromagnetic field in free space. There again, we could not have described the electromagnetic field in terms of just one vector (say $\mathbf{E}$) using a curl operator on one side and a time derivative on the other. For in the attempted equation

$$\text{curl } \mathbf{E} = a(\partial\mathbf{E}/\partial t)$$

the curl operator changes sign when we change from right- to left-handed systems, whereas $\partial/\partial t$ does not. Therefore a would have to be a pseudoscalar, which would be just as distressing for electromagnetics as it is with the wave equation for $\mathbf{e}$.

We can consider $\mathbf{e}$ and $\mathbf{f}$ to be vectors in the same two-dimensional spin space, related by the equations above, just as we consider $\mathbf{E}$ and $\mathbf{H}$ to be vectors in the same three space. But since $\mathbf{e}$ and $\mathbf{f}$ are independent vectors (in the same sense that $\mathbf{E}$ and $\mathbf{H}$ are) and since spin space is not quite so "physical" as three space, it is usual to consider $\mathbf{f}$ to be in *another spin space*, perpendicular to that for $\mathbf{e}$. In other words, we set up a *four*-dimensional spin space, with unit vectors $\mathbf{e}_1$, $\mathbf{e}_2$, $\mathbf{e}_3$, $\mathbf{e}_4$ all mutually perpendicular, and we ensure that $\mathbf{e}$ and $\mathbf{f}$ be mutually independent by making $\mathbf{e}$ a combination of $\mathbf{e}_1$ and $\mathbf{e}_2$ and $\mathbf{f}$ a combination of $\mathbf{e}_3$ and $\mathbf{e}_4$.

In this representation the change from $\mathbf{e}$ to $\mathbf{f}$ represents a rotation from one subspace to the other, which can be represented by an operator $\boldsymbol{\varrho}$ such that $\boldsymbol{\varrho}\cdot\mathbf{e} = \mathbf{f}$ and $\boldsymbol{\varrho}\cdot\mathbf{f} = \mathbf{e}$. Likewise the change from a to $-a$ in the earlier pair of equations may be expressed in operator form by $\boldsymbol{\varrho}_0$, such that $\boldsymbol{\varrho}_0\mathbf{e} = \mathbf{e}$ and $\boldsymbol{\varrho}_0\mathbf{f} = -\mathbf{f}$. In terms of this representation the two equations written down above can now be condensed into one equation:

$$\boldsymbol{\varrho}\cdot(\boldsymbol{\sigma}\cdot\mathbf{p})\mathbf{e} = [-\boldsymbol{\varrho}_0 a + (E/c)]\mathbf{e} \tag{2.6.52}$$

where $\mathbf{e}$ stands for either $\boldsymbol{\epsilon}$ or $\mathbf{f}$.

We must now extend our operator definitions of the spin operator $\boldsymbol{\sigma}$ to four-space, and these definitions, together with the detailed ones for $\boldsymbol{\varrho}$, $\boldsymbol{\varrho}_0$, and $\boldsymbol{\alpha} = \boldsymbol{\varrho}\cdot\boldsymbol{\sigma}$, are

$$\begin{array}{llll}
\boldsymbol{\sigma}_x\mathbf{e}_1 = \mathbf{e}_2; & \boldsymbol{\sigma}_x\mathbf{e}_2 = \mathbf{e}_1; & \boldsymbol{\sigma}_x\mathbf{e}_3 = \mathbf{e}_4; & \boldsymbol{\sigma}_x\mathbf{e}_4 = \mathbf{e}_3 \\
\boldsymbol{\sigma}_y\mathbf{e}_1 = i\mathbf{e}_2; & \boldsymbol{\sigma}_y\mathbf{e}_2 = -i\mathbf{e}_1; & \boldsymbol{\sigma}_y\mathbf{e}_3 = i\mathbf{e}_4; & \boldsymbol{\sigma}_y\mathbf{e}_4 = -i\mathbf{e}_3 \\
\boldsymbol{\sigma}_z\mathbf{e}_1 = \mathbf{e}_1; & \boldsymbol{\sigma}_z\mathbf{e}_2 = -\mathbf{e}_2; & \boldsymbol{\sigma}_z\mathbf{e}_3 = \mathbf{e}_3; & \boldsymbol{\sigma}_z\mathbf{e}_4 = -\mathbf{e}_4 \\
\boldsymbol{\varrho}\mathbf{e}_1 = \mathbf{e}_3; & \boldsymbol{\varrho}\mathbf{e}_2 = \mathbf{e}_4; & \boldsymbol{\varrho}\mathbf{e}_3 = \mathbf{e}_1; & \boldsymbol{\varrho}\mathbf{e}_4 = \mathbf{e}_2 \\
\boldsymbol{\varrho}_0\mathbf{e}_1 = \mathbf{e}_1; & \boldsymbol{\varrho}_0\mathbf{e}_2 = \mathbf{e}_2; & \boldsymbol{\varrho}_0\mathbf{e}_3 = -\mathbf{e}_3; & \boldsymbol{\varrho}_0\mathbf{e}_4 = -\mathbf{e}_4 \\
\boldsymbol{\alpha}_x\mathbf{e}_1 = \mathbf{e}_4; & \boldsymbol{\alpha}_x\mathbf{e}_2 = \mathbf{e}_3; & \boldsymbol{\alpha}_x\mathbf{e}_3 = \mathbf{e}_2; & \boldsymbol{\alpha}_x\mathbf{e}_4 = \mathbf{e}_1 \\
\boldsymbol{\alpha}_y\mathbf{e}_1 = i\mathbf{e}_4; & \boldsymbol{\alpha}_y\mathbf{e}_2 = -i\mathbf{e}_3; & \boldsymbol{\alpha}_y\mathbf{e}_3 = i\mathbf{e}_2; & \boldsymbol{\alpha}_y\mathbf{e}_4 = -i\mathbf{e}_1 \\
\boldsymbol{\alpha}_z\mathbf{e}_1 = \mathbf{e}_3; & \boldsymbol{\alpha}_z\mathbf{e}_2 = -\mathbf{e}_4; & \boldsymbol{\alpha}_z\mathbf{e}_3 = \mathbf{e}_1; & \boldsymbol{\alpha}_z\mathbf{e}_4 = -\mathbf{e}_2
\end{array} \tag{2.6.53}$$

We notice that the operator $\boldsymbol{\varrho}$ commutes with the operators $\boldsymbol{\sigma}_x$, $\boldsymbol{\sigma}_y$, $\boldsymbol{\sigma}_z$, $\boldsymbol{\alpha}_x$, $\boldsymbol{\alpha}_y$, $\boldsymbol{\alpha}_z$ but that $\boldsymbol{\varrho}\boldsymbol{\varrho}_0 + \boldsymbol{\varrho}_0\boldsymbol{\varrho} = 0$. The operator $\boldsymbol{\varrho}_0$ therefore commutes with the $\boldsymbol{\sigma}$'s but does not commute with the $\boldsymbol{\alpha}$'s. In tabular form these operators become

$$\boldsymbol{\sigma}_x = \begin{pmatrix} 0 & 1 & 0 & 0 \\ 1 & 0 & 0 & 0 \\ 0 & 0 & 0 & 1 \\ 0 & 0 & 1 & 0 \end{pmatrix}; \quad \boldsymbol{\sigma}_y = \begin{pmatrix} 0 & -i & 0 & 0 \\ i & 0 & 0 & 0 \\ 0 & 0 & 0 & -i \\ 0 & 0 & i & 0 \end{pmatrix};$$

$$\boldsymbol{\sigma}_z = \begin{pmatrix} 1 & 0 & 0 & 0 \\ 0 & -1 & 0 & 0 \\ 0 & 0 & 1 & 0 \\ 0 & 0 & 0 & -1 \end{pmatrix}; \quad \boldsymbol{\varrho} = \begin{pmatrix} 0 & 0 & 1 & 0 \\ 0 & 0 & 0 & 1 \\ 1 & 0 & 0 & 0 \\ 0 & 1 & 0 & 0 \end{pmatrix};$$

$$\boldsymbol{\varrho}_0 = \begin{pmatrix} 1 & 0 & 0 & 0 \\ 0 & 1 & 0 & 0 \\ 0 & 0 & -1 & 0 \\ 0 & 0 & 0 & -1 \end{pmatrix} = \boldsymbol{\alpha}_0; \quad \boldsymbol{\alpha}_x = \begin{pmatrix} 0 & 0 & 0 & 1 \\ 0 & 0 & 1 & 0 \\ 0 & 1 & 0 & 0 \\ 1 & 0 & 0 & 0 \end{pmatrix};$$

$$\boldsymbol{\alpha}_y = \begin{pmatrix} 0 & 0 & 0 & -i \\ 0 & 0 & i & 0 \\ 0 & -i & 0 & 0 \\ i & 0 & 0 & 0 \end{pmatrix}; \quad \boldsymbol{\alpha}_z = \begin{pmatrix} 0 & 0 & 1 & 0 \\ 0 & 0 & 0 & -1 \\ 1 & 0 & 0 & 0 \\ 0 & -1 & 0 & 0 \end{pmatrix}$$

where we shall use the symbols $\boldsymbol{\alpha}_0$ and $\boldsymbol{\varrho}_0$ interchangeably from now on.

In terms of these operators we can set up the operator equation (2.6.52) to operate on some vector $\mathbf{e}$, a combination of the four-unit vectors $\mathbf{e}_1$, $\mathbf{e}_2$, $\mathbf{e}_3$, $\mathbf{e}_4$, as follows:

$$[\boldsymbol{\varrho}(\boldsymbol{\sigma} \cdot \mathfrak{p}) + \varrho_0 a] \cdot \mathbf{e} = (E/c)\mathbf{e} \quad \text{or} \quad [\boldsymbol{\alpha}_x \mathfrak{p}_x + \boldsymbol{\alpha}_y \mathfrak{p}_y + \boldsymbol{\alpha}_z \mathfrak{p}_z + \varrho_0 a] \cdot \mathbf{e} = (E/c)\mathbf{e} \tag{2.6.54}$$

We must now "square" this equation to obtain a form analogous to Eq. (2.6.49).

The Dirac Equation. When the electromagnetic field is zero, Eq. (2.6.49) becomes

$$[p_x^2 + p_y^2 + p_z^2 + m^2c^2] = (p_t/c)^2$$

Taking Eq. (2.6.54) and squaring the operators on both sides (and remembering that $\boldsymbol{\alpha}_x\boldsymbol{\alpha}_y$ must not equal $\boldsymbol{\alpha}_y\boldsymbol{\alpha}_x$, etc.), we obtain

$$\{[\boldsymbol{\alpha}_x^2\mathfrak{p}_x^2 + \boldsymbol{\alpha}_y^2\mathfrak{p}_y^2 + \boldsymbol{\alpha}_z^2\mathfrak{p}_z^2 + \varrho_0^2 a^2] + [\boldsymbol{\alpha}_x\boldsymbol{\alpha}_y + \boldsymbol{\alpha}_y\boldsymbol{\alpha}_x]\mathfrak{p}_x\mathfrak{p}_y \\ + [\boldsymbol{\alpha}_z\varrho_0 + \varrho_0\boldsymbol{\alpha}_z]a\mathfrak{p}_z\} \cdot \mathbf{e} = (\mathfrak{p}_t/c)^2\mathbf{e}$$

To have this correspond to Eq. (2.6.49) as written above means that

$$\boldsymbol{\alpha}_x^2 = \boldsymbol{\alpha}_y^2 = \boldsymbol{\alpha}_z^2 = \varrho_0^2 = 1; \quad \boldsymbol{\alpha}_x\boldsymbol{\alpha}_y + \boldsymbol{\alpha}_y\boldsymbol{\alpha}_x = \cdots = \boldsymbol{\alpha}_z\varrho_0 + \varrho_0\boldsymbol{\alpha}_z = 0$$

and

$$a = mc \tag{2.6.55}$$

An examination of Eqs. (2.6.53) indicates that the operators defined there do obey the requirements of Eqs. (2.6.55), so that we have finally arrived at a relativistic equation for a single particle of mass m which has a first-order term in the operator for E (or $\mathfrak{p}_t$). To obtain it we have been forced to extend our "spin space" from two to four dimensions. Two of these spin states ($\mathbf{e}_1$,$\mathbf{e}_2$) correspond to a term $+ mc^2$ in the total energy, and the other two correspond to a term $- mc^2$, a negative mass energy. We know now that the negative energy states are related to the positron, a particle of opposite *charge* but the same mass as the electron.

The wave equation for a single particle of charge e and mass m in an electromagnetic field operates on a wave-function, spin-vector combination

$$\begin{aligned} \boldsymbol{\Psi} &= \psi_1\mathbf{e}_1 + \psi_2\mathbf{e}_2 + \psi_3\mathbf{e}_3 + \psi_4\mathbf{e}_4; \\ \bar{\boldsymbol{\Psi}} &= \bar{\psi}_1\mathbf{e}_1^* + \bar{\psi}_2\mathbf{e}_2^* + \bar{\psi}_3\mathbf{e}_3^* + \bar{\psi}_4\mathbf{e}_4^* \end{aligned} \tag{2.6.56}$$

where the ψ's are functions of x, y, z, and t and the $\mathbf{e}$'s are orthogonal unit vectors in spin space. The equation, which is called the *Dirac equation*, is

$$\left[\boldsymbol{\alpha}_0 mc + \boldsymbol{\alpha}_x\left(\frac{\hbar}{i}\frac{\partial}{\partial x} - \frac{e}{c}A_x\right) + \boldsymbol{\alpha}_y\left(\frac{\hbar}{i}\frac{\partial}{\partial y} - \frac{e}{c}A_y\right) \right. \\ \left. + \boldsymbol{\alpha}_z\left(\frac{\hbar}{i}\frac{\partial}{\partial z} - \frac{e}{c}A_z\right) + e\varphi\right]\boldsymbol{\Psi} = \left\{\boldsymbol{\alpha}_0 mc\boldsymbol{\Psi} + \boldsymbol{\alpha}\cdot\left[\frac{\hbar}{i}\operatorname{grad}\boldsymbol{\Psi} - \frac{e}{c}\mathbf{A}\boldsymbol{\Psi}\right]\right. \\ \left. + e\varphi\boldsymbol{\Psi}\right\} = -\frac{\hbar}{ic}\frac{\partial}{\partial t}\boldsymbol{\Psi} \tag{2.6.57}$$

where the operator $\boldsymbol{\alpha}$ is a vector of components $\boldsymbol{\alpha}_x$, $\boldsymbol{\alpha}_y$, $\boldsymbol{\alpha}_z$ and where $\boldsymbol{\alpha}_0 = \boldsymbol{\varrho}_0$. The $\boldsymbol{\alpha}$ operators obey the rules set forth in Eqs. (2.6.53) and (2.6.55). The equation for $\bar{\boldsymbol{\Psi}}$ is obtained by reversing the signs of all the i's in Eq. (2.6.57).

We must now see if all this work has resulted in an equation which will allow sensible expressions for charge-current density. It would seem reasonable that the charge density would be

$$\rho = e\bar{\boldsymbol{\Psi}}\boldsymbol{\Psi} = e[|\psi_1|^2 + |\psi_2|^2 + |\psi_3|^2 + |\psi_4|^2] \tag{2.6.58}$$

carrying on as before, multiplying the equation for $\boldsymbol{\Psi}$ by $\bar{\boldsymbol{\Psi}}$ and the equation for $\bar{\boldsymbol{\Psi}}$ by $\boldsymbol{\Psi}$ and subtracting, we obtain

$$-(\partial/\partial t)(\bar{\boldsymbol{\Psi}}\boldsymbol{\Psi}) = c[\bar{\boldsymbol{\Psi}}\boldsymbol{\alpha} \cdot \text{grad } \boldsymbol{\Psi} + \boldsymbol{\Psi}\boldsymbol{\alpha} \cdot \text{grad } \bar{\boldsymbol{\Psi}}] = c \text{ div } (\bar{\boldsymbol{\Psi}}\boldsymbol{\alpha}\boldsymbol{\Psi})$$

Therefore the vector whose components are

$$\begin{aligned} ce(\bar{\boldsymbol{\Psi}}\boldsymbol{\alpha}_x\boldsymbol{\Psi}) &= ce[\bar{\psi}_1\psi_4 + \bar{\psi}_2\psi_3 + \bar{\psi}_3\psi_2 + \bar{\psi}_4\psi_1] = J_x \\ ce(\bar{\boldsymbol{\Psi}}\boldsymbol{\alpha}_y\boldsymbol{\Psi}) &= -ice[\bar{\psi}_1\psi_4 - \bar{\psi}_2\psi_3 + \bar{\psi}_3\psi_2 - \bar{\psi}_4\psi_1] = J_y \\ ce(\bar{\boldsymbol{\Psi}}\boldsymbol{\alpha}_z\boldsymbol{\Psi}) &= ce[\bar{\psi}_1\psi_3 - \bar{\psi}_2\psi_4 + \bar{\psi}_3\psi_1 - \bar{\psi}_4\psi_2] = J_z \end{aligned} \tag{2.6.59}$$

is the current density vector $\mathbf{J}$.

It is interesting to note that, whereas the momentum density for the particle is $\bar{\boldsymbol{\Psi}}\mathfrak{p}\boldsymbol{\Psi} = (\hbar/i)\bar{\boldsymbol{\Psi}}$ grad $\boldsymbol{\Psi}$, the *velocity* density seems to be $c\bar{\boldsymbol{\Psi}}\boldsymbol{\alpha}\boldsymbol{\Psi}$. This can be shown in another way. We use the Hamiltonian equation $\partial H/\partial p = \dot{q} = u$ on Eq. (2.6.50) (we leave out the fields for simplicity and assume that the x axis is along $\mathbf{p}$ or $\mathbf{u}$). Then

$$\mathbf{u} = \frac{c\mathbf{p}}{\sqrt{p^2 + m^2c^2}} \quad \text{or} \quad \mathbf{p} = \frac{m\mathbf{u}}{\sqrt{1 - (u/c)^2}}$$

[see Eq. (1.7.5)] and

$$H = \frac{mc^2}{\sqrt{1 - (u/c)^2}} = \mathbf{u} \cdot \mathbf{p} + \sqrt{1 - (u/c)^2}\, mc^2$$

Comparing this classical equation for the total energy of the free particle (relativistic) with Eq. (2.6.54), we see that the vector operator $c\boldsymbol{\alpha}$ takes the place of the particle velocity u and the operator $\boldsymbol{\alpha}_0$ takes the place of $\sqrt{1 - (u/c)^2}$ when we change to the Dirac equation.

The transformations of the operators $\boldsymbol{\alpha}$, the spin vectors $\mathbf{e}$, and the wave function $\boldsymbol{\Psi}$ for a Lorentz rotation of space-time or for a space rotation can be worked out from the data given in Sec. 1.7. For instance, if the transformation corresponds to a relative velocity $u = c \tanh \theta$ along the x axis, the p's and $\mathbf{A}$'s transform as four-vectors:

$$p_x = p'_x \cosh \theta + (1/c)p'_t \sinh \theta; \quad p_y = p'_y; \quad p_z = p'_z;$$
$$p_t = p'_t \cosh \theta + cp'_x \sinh \theta$$

The spin vectors $\mathbf{e}$ are transformed according to the formula $\mathbf{e}' = \mathfrak{g} \cdot \mathbf{e}$, where

$$\mathfrak{g} = e^{\theta\boldsymbol{\alpha}_x/2} = \sum_{n=0}^{\infty} \frac{1}{n!}\left(\frac{\theta\boldsymbol{\alpha}_x}{2}\right)^n = \cosh\frac{\theta}{2} + \boldsymbol{\alpha}_x \sinh\frac{\theta}{2}$$

since $\boldsymbol{\alpha}_x^2 = 1$. The conjugate operator $\mathfrak{g}^*$, such that $\mathbf{e} = \mathfrak{g}^* \cdot \mathbf{e}'$, is equal to $\mathfrak{g}$ in this case. Therefore the new wave function is

$$\begin{aligned}\boldsymbol{\Psi}' = \mathfrak{g}\boldsymbol{\Psi} = \cosh(\theta/2)\ &[\psi_1\mathbf{e}_1 + \psi_2\mathbf{e}_2 + \psi_3\mathbf{e}_3 + \psi_4\mathbf{e}_4] \\ &+ \sinh(\theta/2)\ [\psi_4\mathbf{e}_1 + \psi_3\mathbf{e}_2 + \psi_2\mathbf{e}_3 + \psi_1\mathbf{e}_4]\end{aligned} \tag{2.6.60}$$

The operators $\boldsymbol{\alpha}$ are transformed according to the formula $\mathfrak{g}^* \cdot \boldsymbol{\alpha} \cdot \mathfrak{g} = \boldsymbol{\alpha}'$. For symmetry we set $\boldsymbol{\alpha}_t = \mathfrak{J}/c$, [see Eq. (1.7.17)] where $\mathfrak{J}$ is the idemfactor. We then have

$$\begin{aligned}\boldsymbol{\alpha}_t' &= \boldsymbol{\alpha}_t \cosh\theta + (1/c)\boldsymbol{\alpha}_x \sinh\theta \\ \boldsymbol{\alpha}_x' &= \boldsymbol{\alpha}_x \cosh\theta + c\boldsymbol{\alpha}_t \sinh\theta \\ \boldsymbol{\alpha}_y' &= e^{\theta\boldsymbol{\alpha}_x/2}\boldsymbol{\alpha}_y e^{\theta\boldsymbol{\alpha}_x/2} = e^{\theta\boldsymbol{\alpha}_x/2}e^{-\theta\boldsymbol{\alpha}_x/2}\boldsymbol{\alpha}_y = \boldsymbol{\alpha}_y \\ \boldsymbol{\alpha}_z' &= \boldsymbol{\alpha}_z; \quad \boldsymbol{\alpha}_0' = \boldsymbol{\alpha}_0\end{aligned}$$

so that the $\boldsymbol{\alpha}$'s $(\boldsymbol{\alpha}_1 \cdots \boldsymbol{\alpha}_4)$ transform like a four-vector. Therefore the scalar product of the $\boldsymbol{\alpha}$'s with the momentum four-vector is a Lorentz invariant, so that

$$\sum_{xyzt} \boldsymbol{\alpha}_x p_x = [\alpha_x p_x + \alpha_y p_y + \alpha_z p_z + \alpha_t p_t] = \sum_{xyzt} \mathfrak{g}^*\boldsymbol{\alpha}_x\mathfrak{g}p_x' = \sum_{xyzt} \boldsymbol{\alpha}_x' p_x'$$

Therefore the equation in unprimed coordinates can be changed as follows:

$$\begin{aligned}0 &= \left[\sum_{xyzt} \boldsymbol{\alpha}_x\left(p_x - \frac{e}{c}A_x\right) + \boldsymbol{\alpha}_0 mc\right]\boldsymbol{\Psi} \\ &= \left[\sum \mathfrak{g}^*\boldsymbol{\alpha}_x\mathfrak{g}\left(p_x' - \frac{e}{c}A_x'\right) + \mathfrak{g}^*\boldsymbol{\alpha}_0 mc\mathfrak{g}\right]\boldsymbol{\Psi} \\ &= \mathfrak{g}^*\left[\sum \boldsymbol{\alpha}_x\left(p_x' - \frac{e}{c}A_x'\right) + \boldsymbol{\alpha}_0 mc\right]\boldsymbol{\Psi}'\end{aligned}$$

which is the equation in the primed coordinates.

For a rotation in space by an angle θ about the x axis the rotation operator for the $\mathbf{e}$'s and $\boldsymbol{\alpha}$'s is, appropriately enough,

$$\mathfrak{g} = e^{-\theta\boldsymbol{\alpha}_y\boldsymbol{\alpha}_z/2}; \quad \mathfrak{g}^* = e^{\theta\boldsymbol{\alpha}_y\boldsymbol{\alpha}_z/2} \tag{2.6.61}$$

and the transformation equations are

$$\begin{aligned}p_t = p_t'; \quad p_x = p_x'; \quad p_y = p_y' \cos\theta + p_z' \sin\theta \\ p_z = -p_y' \sin\theta + p_z' \cos\theta; \quad \mathbf{e}' = \mathfrak{g}\mathbf{e}; \quad \boldsymbol{\alpha}' = \mathfrak{g}^*\boldsymbol{\alpha}\mathfrak{g}\end{aligned}$$

More complex rotations can always be made up by a series of rotations of the types discussed here; the corresponding rotation operators $\mathfrak{g}$ are just the products of the individual $\mathfrak{g}$'s for the component simple rotations, taken in the right order.

Total Angular Momentum. As an exercise in use of the operators $\boldsymbol{\alpha}$ and $\boldsymbol{\sigma}$ we shall show that, when there is no electromagnetic field, the total angular momentum of the particle is not just the mechanical momentum $\mathfrak{M}$ [see Eq. (1.6.42)] but is a combination of $\mathfrak{M}$ and the spin vector $\boldsymbol{\sigma}$. In other words we must include the spin of the particle in order to obtain the constant of the motion which we call the total angular momentum. Referring to Eq. (2.6.35) we see that for a constant of the motion represented by the operator $\mathfrak{A}$ we must have

$$\mathfrak{H}\mathfrak{A} - \mathfrak{A}\mathfrak{H} = 0$$

where $\mathfrak{H}$ is the Hamiltonian operator.

In the present case, with the potentials $\mathbf{A}$ and φ equal to zero, we have for the Hamiltonian, according to Eq. (2.6.57),

$$\mathfrak{H} = \boldsymbol{\alpha}_0 mc^2 + c(\boldsymbol{\alpha}_x\mathfrak{p}_x + \boldsymbol{\alpha}_y\mathfrak{p}_y + \boldsymbol{\alpha}_z\mathfrak{p}_z)$$

The operator for the z component of the mechanical angular momentum is $\mathfrak{M}_z = \mathfrak{x}\mathfrak{p}_y - \mathfrak{y}\mathfrak{p}_x$. We form the commutator $\mathfrak{H}\mathfrak{M}_z - \mathfrak{M}_z\mathfrak{H}$ to show that it is not zero. Now the $\boldsymbol{\alpha}$'s commute with the $\mathfrak{p}$'s and the coordinates, the $\mathfrak{p}$'s commute with each other, so that the only term in $\mathfrak{H}$ not commuting with the first term of $\mathfrak{M}_z$ is $c\boldsymbol{\alpha}_x\mathfrak{p}_x$ and the only term not commuting with the second is $c\boldsymbol{\alpha}_y\mathfrak{p}_y$, giving

$$\mathfrak{H}\mathfrak{M}_z - \mathfrak{M}_z\mathfrak{H} = -c\boldsymbol{\alpha}_y\mathfrak{p}_x(\mathfrak{p}_y\mathfrak{y} - \mathfrak{y}\mathfrak{p}_y) + c\boldsymbol{\alpha}_x\mathfrak{p}_y(\mathfrak{p}_x\mathfrak{x} - \mathfrak{x}\mathfrak{p}_x)$$

However, $\mathfrak{p}_x\mathfrak{x} - \mathfrak{x}\mathfrak{p}_x = (\hbar/i)$, etc., so that this expression comes out to be

$$\mathfrak{H}\mathfrak{M}_z - \mathfrak{M}_z\mathfrak{H} = -(\hbar c/i)(\mathfrak{p}_x\boldsymbol{\alpha}_y - \mathfrak{p}_y\boldsymbol{\alpha}_x)$$

which is certainly not zero, so that $\mathfrak{M}_z$ is not a constant of the motion.

By using the operator rules of Eqs. (2.6.53) we can show that [compare with Eqs. (1.6.43)]

$$\begin{aligned} (\boldsymbol{\sigma}_x)^2 = (\boldsymbol{\sigma}_y)^2 = (\boldsymbol{\sigma}_z)^2 = 1; \quad \boldsymbol{\sigma}_x\boldsymbol{\sigma}_y = -\boldsymbol{\sigma}_y\boldsymbol{\sigma}_x = i\boldsymbol{\sigma}_z; \\ \boldsymbol{\sigma}_x\boldsymbol{\sigma}_z = -\boldsymbol{\sigma}_z\boldsymbol{\sigma}_x = -i\boldsymbol{\sigma}_y; \quad \boldsymbol{\sigma}_y\boldsymbol{\sigma}_z - \boldsymbol{\sigma}_z\boldsymbol{\sigma}_y = i\boldsymbol{\sigma}_x \end{aligned} \tag{2.6.62}$$

Also, since the operator $\boldsymbol{\varrho}$ commutes with the $\boldsymbol{\sigma}$'s and since $\boldsymbol{\alpha} = \boldsymbol{\varrho}\boldsymbol{\sigma}$, we can obtain other equations, such as

$$\boldsymbol{\sigma}_x\boldsymbol{\alpha}_x = \boldsymbol{\alpha}_x\boldsymbol{\sigma}_x = \boldsymbol{\varrho}; \quad \boldsymbol{\alpha}_x\boldsymbol{\sigma}_z = -\boldsymbol{\sigma}_z\boldsymbol{\alpha}_x = i\boldsymbol{\alpha}_y; \quad \boldsymbol{\alpha}_y\boldsymbol{\sigma}_z = -\boldsymbol{\sigma}_z\boldsymbol{\alpha}_y = i\boldsymbol{\alpha}_x; \quad \text{etc.}$$

Therefore we can show that

$$\mathfrak{H}\boldsymbol{\sigma}_z - \boldsymbol{\sigma}_z\mathfrak{H} = (2c/i)(\mathfrak{p}_x\boldsymbol{\alpha}_y - \mathfrak{p}_y\boldsymbol{\alpha}_x)$$

Comparing the commutators for $\mathfrak{M}_z$ and $\boldsymbol{\sigma}_z$ we see that the combination $[\mathfrak{M}_z + (\hbar/2)\boldsymbol{\sigma}_z]$ does commute with $\mathfrak{H}$ and hence is a constant of the motion. This is also true of the x and y components.

Therefore the total angular momentum, which is a constant of the motion, is the mechanical angular momentum $\mathfrak{M}$ *plus* $\hbar/2$ times the spin vector $\boldsymbol{\sigma}$.

Field-free Wave Function. As another example we shall obtain the wave function when no field acts. For the field-free case the Dirac equation (2.6.57) takes on the form

$$\alpha_0 mc\Psi + \frac{\hbar}{i}\boldsymbol{\alpha}\cdot\operatorname{grad}\Psi = -\frac{\hbar}{ic}\frac{\partial\Psi}{\partial t}$$

A solution of this equation is

$$\Psi = [A_1\mathbf{e}_1 + A_2\mathbf{e}_2 + A_3\mathbf{e}_3 + A_4\mathbf{e}_4]e^{(i/\hbar)(\mathbf{p}\cdot\mathbf{r}-Et)} \tag{2.6.63}$$

where the A's are numerical coefficients, the radius vector $\mathbf{r} = x\mathbf{i} + y\mathbf{j} + z\mathbf{k}$, the vector $\mathbf{p} = p_x\mathbf{i} + p_y\mathbf{j} + p_z\mathbf{k}$ is a classical momentum vector with components p_x, p_y, p_z which are constants, not operators, and the number E is the magnitude of the electronic energy for the state designated by Ψ.

Inserting this in the Dirac equation, performing the differential operations and the spinor operations according to Eqs. (2.6.53), we finally obtain (we have set $p_x = p_y = 0$, $p_z = p$ with no loss of generality),

$$\begin{aligned}[(mc^2 - E)A_1 + cp\,A_3]\mathbf{e}_1 + [(mc^2 - E)A_2 - cp\,A_4]\mathbf{e}_2 \\ + [(-mc^2 - E)A_3 + cp\,A_1]\mathbf{e}_3 + [(-mc^2 - E)A_4 - cp\,A_1]\mathbf{e}_4 = 0\end{aligned}$$

This requires the four coefficients of the vectors $\mathbf{e}$ to be zero, which results in four homogeneous equations in the four coefficients A, two involving A_1 and A_3 and two involving A_2 and A_4. Both can be solved if the equation $E^2 = m^2c^4 + c^2p^2$ is satisfied. Therefore there are four solutions: two corresponding to the energy value

$$E = -mc^2\sqrt{1 + (p/mc)^2} \xrightarrow[p \ll mc]{} -[mc^2 + \tfrac{1}{2}(p^2/m)]$$

which are

$$\begin{aligned}\Psi_1 &= C[-\beta\mathbf{e}_1 + (1+\gamma)\mathbf{e}_3]e^{(i/\hbar)(px+mc^2\gamma t)} \\ \Psi_2 &= C[\beta\mathbf{e}_2 + (1+\gamma)\mathbf{e}_4]e^{(i/\hbar)(px+mc^2\gamma t)}\end{aligned} \tag{2.6.64}$$

and the other two corresponding to the energy value

$$E = +mc^2\gamma \xrightarrow[\beta\to 0]{} [mc^2 + \tfrac{1}{2}(p^2/m)]$$

which are

$$\begin{aligned}\Psi_3 &= C[(1+\gamma)\mathbf{e}_1 + \beta\mathbf{e}_3]e^{(i/\hbar)(px-mc^2\gamma t)}; \\ \Psi_4 &= C[(1+\gamma)\mathbf{e}_2 - \beta\mathbf{e}_4]e^{(i/\hbar)(px-mc^2\gamma t)}\end{aligned}$$

where $\beta = (p/mc)$ and $\gamma = \sqrt{1+\beta^2}$ and C is a normalization constant. The functions Ψ_1 and Ψ_3 correspond to a spin $\hbar\sigma_z/2$ equal to $+(\hbar/2)$

(because $\mathfrak{d}_z\mathbf{\Psi}_1 = \mathbf{\Psi}_1$, etc.), and $\mathbf{\Psi}_2$ and $\mathbf{\Psi}_4$ correspond to a spin of $-(\hbar/2)$.

Recapitulation. In this chapter we have endeavored to outline the basic connections between various phenomena in classical and quantum physics and the various sorts of fields discussed in Chap. 1. The connection usually can be represented, to a certain degree of approximation, by a differential equation specifying the point-to-point behavior of the field which is to describe a particular physical phenomenon. In the case of classical physics we found that we had to average over atomic discontinuities; in quantum physics we found that the uncertainty principle precluded our "going behind" the continuous "square-root-probability" wave function; the discontinuous details of the elementary particle trajectories turn out to be impossible to measure or predict. In either case we are left with a continuous field—scalar, vector, or dyadic—subject to a partial differential equation and specified uniquely by some set of boundary conditions (or initial conditions or both).

We have seen the same fields and the same differential equations turn up in connection with many and various physical phenomena. We find, for instance, that a scalar-field solution of Laplace's equation can represent either the electric field around a collection of charges or the density of a diffusing fluid under steady-state conditions or the velocity potential of a steadily flowing incompressible fluid or the gravitational potential around a collection of matter, and so on. From the point of view of this book, this lack of mathematical originality on the part of nature represents a great saving in effort and space. When we come to discuss the solutions of one equation, we shall be simultaneously solving several dozen problems in different parts of physics.

We have not gone in great detail into the physics of the various examples discussed in this chapter; this book is primarily concerned with working out solutions of the equations once they are derived. Other books, concentrating on various parts of physics, are available for detailed study of the physics involved. In the case of quantum mechanics, for instance, we have resisted the temptation to go beyond the bare outlines of the new point of view in dynamics. Only enough is given so that later, when solutions of Schroedinger's or Dirac's equations are studied, we shall be able to understand the physical implications of the solutions. More than this would make the section a text on quantum mechanics—a supererogative task at present.

It is true that more time was spent discussing the equations of quantum mechanics than was spent on the equations for classical fields. These newer equations are less familiar, and thus they have provided a chance to demonstrate what procedures must be used when carving out new field equations to describe new phenomena. The classical field equations have withstood the scrutiny of several generations of workers, and the logical structure of the relation to "reality" has become "second nature" to physicists. In quantum mechanics we have not quite completed the process of rationalization transforming an unfamiliar equation which "works" into a logically justified theory which is "obvious to anyone."

A new equation for the description of new phenomena is seldom first obtained by strictly logical reasoning from well-known physical facts; a pleasingly rigorous derivation of the equation usually is evolved only about the time the theory becomes "obvious." The first finding of the equation usually comes by the less deductive paths of analogy and "working backward" and by the constant use of a modern counterpart to Occam's razor. In the Dirac equation, for instance, we were held to a certain general form of equation because it was likely that the equation should be relativistically invariant, and we searched for the *simplest* equation which would result in "sensible" (*i.e.*, not redundantly complex) expressions for charge, current, and other measurable quantities. The result may not appear very simple at first sight, but a few days of prospecting by the reader (or of reading back issues of journals for the days when the Dirac equation was being put together) should suffice to demonstrate that it is much easier to construct equations of greater complication than it is to find one more simple.

Among the general principles which can be used to point the direction for search for a new equation, the requirement of invariance, particularly of Lorentz invariance, is one of the most valuable. But there are many others. One usually looks first for linear equations, for instance, and the Laplacian operator often turns up.

When an equation has been devised, all the related quantities must be investigated to see whether they can satisfactorily "correspond" to various physical quantities. There should usually be an energy density, for instance, and the quantity chosen to correspond should not turn out to have the annoying property of becoming negative somewhere or sometime. We used the choice of charge and current density expressions and the requirement that they satisfy the equation of continuity to guide us in our choice of the Dirac equation. A formal machinery to obtain these subsidiary quantities is the variational method, which will be discussed in the next chapter. Once these quantities are ground out, it is then possible to decide whether they are too complicated or not.

Another useful means of testing an equation which has just been devised is to find another aspect of physics where the same equation can be applied. The properties of the solutions of the Klein-Gordon equation could be tried out by devising a string-in-rubber system (see page 139) which satisfied the same equation and which could be pictured more easily than a wave function, because the motions of a string are familiar. Analogies of this sort occur throughout theoretical physics and represent a sort of cross-fertilization which is most useful. The early analysis of alternating electric currents was greatly aided by the analogy with more familiar mechanical oscillators. Now that "everyone knows about" alternating currents, we are prone to discuss other sorts of vibrational and wave motion (even mechanical oscillators) in terms of an analogy with alternating-current behavior, speaking of impedances, capacitances, and so on.

In the next chapter we discuss in detail an analogy between field behavior and the variational properties of classical dynamics, as developed by Hamilton. We shall find it a useful unifying agent for all the equations we have discussed in this chapter (as well as others).

Problems for Chapter 2

2.1 A membrane is stretched over one end of an airtight vessel so that both the tension T in the membrane and the excess pressure p of the air in the vessel act upon the membrane. If ψ gives the displacement of the membrane from equilibrium, show that

$$p = -(\rho c^2/V)\int\psi\,dA$$

where ρ, V, and c are the equilibrium values of the density, volume, and velocity of sound of the air in the vessel. Show that the equation of motion for the membrane is therefore

$$(1/v^2)(\partial^2\psi/\partial t^2) = \nabla^2\psi - (\rho c^2/VT)\int\psi\,dA$$

where $v^2 = T/\sigma$, where T is the tension and σ is the mass per unit area of the membrane. What assumptions have been made in obtaining this equation?

2.2 An infinite, elastic medium has piezoelectric properties for compression in the x direction and electric field in the y direction (for displacement s in the x direction, the electric intensity E and displacement vector D are in the y direction). The dielectric polarization P, also in the y direction, is related to D and E by the usual equation $D = E + 4\pi P$ and is related to the stress component $X = T_{xx}$ and E by the coupling equation $P = \delta X + \chi E$, where χ is the dielectric susceptibility and δ the piezoelectric constant. Alternately the strain $S_{xx} = u$ is related to the stress and to the electric intensity by the equations $u = \sigma X + \delta E$, where σ is the reciprocal of the elastic modulus. By use of the elastic equations and Maxwell's equations, set up the two simultaneous equations for compressional wave motion in the x direction. Show that two electroelastic waves are possible, one with velocity a little smaller than that for pure compressional waves (value if δ were zero) and the other a little larger than that for pure electromagnetic waves.

2.3 During the passage of a sound wave through matter, the temperature of the regions under compression is greater than the average temperature, while in the regions under expansion the temperature will be less than the average. These temperature differences will give rise to a flow of heat from one part of the material to another.

a. Show that the equations governing the heat flow and sound propagation are

$$\partial T/\partial t = (\partial T_0/\partial \rho_0)_S(\partial \rho/\partial t) + (\kappa/c_p\rho_0)\nabla^2 T$$
$$\partial^2\rho/\partial t^2 = (\partial p_0/\partial \rho_0)_T\nabla^2\rho + (\partial p_0/\partial T_0)_\rho\nabla^2 T$$

where the subscript zero is used to denote equilibrium values.

b. Assume that T and ρ propagate as plane waves

$$T = A \exp[i(kx - \omega t)]; \quad \rho = B \exp[i(kx - \omega t)]$$

Show that the relation between k and ω is given by

$$0 = i(\kappa/c_p\rho_0\omega)[(\partial p_0/\partial \rho_0)_T k^2 - \omega^2] - \{\omega^2 - k^2[(\partial p_0/\partial \rho_0)_T + (\partial T_0/\partial \rho_0)_S(\partial p_0/\partial T_0)_\rho]\}$$

Determine the velocity of propagation for $\kappa/c_p\rho_0\omega \ll 1$, for $\kappa/c_p\rho_0\omega \gg 1$. Discuss the propagation when $\kappa/c_p\rho_0\omega \simeq 1$.

2.4 A conducting fluid (electrical conductivity σ, permeability μ) in motion will generate a magnetic field, which in turn will affect the motion of the fluid. Show that the equations coupling the velocity $\mathbf{v}$ and the magnetic induction $\mathbf{B}$ are

$$\frac{\partial \mathbf{B}}{\partial t} = \operatorname{curl}(\mathbf{v} \times \mathbf{B}) + \left(\frac{c^2}{4\pi\mu\sigma}\right)\nabla^2\mathbf{B}$$

$$\rho\frac{\partial \mathbf{v}}{\partial t} + \rho(\mathbf{v}\cdot\nabla)\mathbf{v} = -\nabla p - \left(\frac{c}{4\pi\mu}\right)[\mathbf{B} \times (\nabla \times \mathbf{B})]$$

2.5 When dissolved in an appropriate solvent, many salts break up into positive and negative ions. Under the influence of an electric field these will diffuse. Show that the equations describing the motion of the positive ions, assuming that these move in a viscous medium with their terminal velocity, is

$$\partial c_1/\partial t = A_1^2\nabla^2 c_1 + B_1 Q \operatorname{div}(c_1 \operatorname{grad}\varphi)$$

where c_1 is the concentration, A_1^2 the diffusion constant, B_1 the ratio between the terminal velocity and the applied force, Q the ionic charge, and φ the electrostatic potential. Show that φ satisfies the equation

$$\nabla^2\varphi = -4\pi(F/e)(c_1 - c_2)$$

where F is Faraday's constant.

2.6 Particles of microscopic size are found to have a random motion, called Brownian motion, arising from molecular collisions. Let the number of particles at a time t_0 having a position between x_0 and $x_0 + dx_0$, a velocity between v_0 and $v_0 + dv_0$ be $f(x_0,v_0,t_0)\,dx_0\,dv_0$. Let the fraction of these particles which, in a time τ, are found in the region between x and $x + dx$ with a velocity $v + dv$ be $w(\Delta x,\Delta v,\tau|x_0,v_0,t_0)\,dx\,dv$, where $\Delta x = x - x_0$, $\Delta v = v - v_0$.

a. Show that

$$f(x,v,t) = \int_{-\infty}^{\infty}\int_{-\infty}^{\infty} w(\Delta x,\Delta v,\tau|x_0,v_0,t_0) f(x_0,v_0,t_0)\,dx_0\,dv_0$$

b. Show that, for small τ, Δx, and Δv,

$$\frac{\partial f(x,v,t_0)}{\partial t_0} = \frac{\partial}{\partial x}\left(\frac{f\,\overline{\Delta x}}{\tau}\right) - \frac{\partial}{\partial v}\left(\frac{f\,\overline{\Delta v}}{\tau}\right) + \frac{1}{2}\left\{\frac{\partial^2}{\partial x^2}\left(\frac{f\,\overline{\Delta x^2}}{\tau}\right) + 2\,\frac{\partial^2}{\partial x\,\partial v}\left(\frac{f\,\overline{\Delta x\,\Delta v}}{\tau}\right) + \frac{\partial^2}{\partial v^2}\left(\frac{f\,\overline{\Delta v^2}}{\tau}\right)\right\}$$

where $\overline{\Delta x} = \overline{\Delta x}(x,v,t,\tau) = \int_{-\infty}^{\infty}\int_{-\infty}^{\infty} \Delta x\, w(\Delta x,\Delta v,\tau|x,v,t_0)\,dx_0\,dv_0$ with corresponding definitions for the other average quantities.

c. If the particles move in a viscous medium and if the molecular collisions are random, $\overline{\Delta v} = -\alpha v\tau$ and $\overline{\Delta v^2} = A\tau$, where α and A are constants. Show that in the limit of small τ

$$\frac{\partial f}{\partial t} = -\frac{\partial}{\partial x}(vf) + \alpha\frac{\partial}{\partial v}(vf) + \tfrac{1}{2}A\frac{\partial^2 f}{\partial v^2}$$

d. Show that under steady-state conditions

$$\int_{-\infty}^{\infty} f(x,v)\,dx = F_0 e^{-(\alpha/A)v^2}$$

Show that the average of the nth power of the velocity defined by

$$\overline{v^n} = \int_{-\infty}^{\infty}\int_{-\infty}^{\infty} v^n f(x,v)\,dx\,dv$$

satisfies the differential equation

$$\frac{\overline{dv^n}}{dt} = -n\alpha\overline{v^n} + \tfrac{1}{2}An(n-1)\overline{v^{n-2}}$$

2.7 *a.* Two operators $\mathfrak{a}$ and $\mathfrak{a}^*$ obey the following commutation rule:

$$\mathfrak{a}\mathfrak{a}^* - \mathfrak{a}^*\mathfrak{a} = 1$$

Show that the eigenvalues of the operator $\mathfrak{a}^*\mathfrak{a}$ are 0, 1, 2, 3, ... If the corresponding states are $\mathbf{e}_n$, show that

$$\mathfrak{a}\mathbf{e}_n = \sqrt{n}\,\mathbf{e}_{n-1};\quad \mathfrak{a}^*\mathbf{e}_n = \sqrt{n+1}\,\mathbf{e}_{n+1}$$

b. Two operators $\mathfrak{a}$ and $\mathfrak{a}^*$ obey the following commutation rule:

$$\mathfrak{a}\mathfrak{a}^* + \mathfrak{a}^*\mathfrak{a} = 1$$

Also $\mathfrak{a}\mathfrak{a} = 0$, $\mathfrak{a}^*\mathfrak{a}^* = 0$. Show that the eigenvalues of $\mathfrak{a}^*\mathfrak{a}$ are only 0 or 1. If $\mathbf{e}_0$ and $\mathbf{e}_1$ are corresponding states, show that

$$\mathfrak{a}^*\mathbf{e}_0 = \mathbf{e}_1; \quad \mathfrak{a}^*\mathbf{e}_1 = 0; \quad \mathfrak{a}\mathbf{e}_0 = 0; \quad \mathfrak{a}\mathbf{e}_1 = \mathbf{e}_0$$

2.8 An electron moves in the coulomb field of a nucleus of charge Z.

a. Prove that if

$$r = [x^2 + y^2 + z^2]^{\frac{1}{2}}$$

then the appropriate conjugate momentum p_r is

$$\mathfrak{p}_r = (1/r)(\mathbf{r} \cdot \mathbf{p} - i\hbar)$$

b. Show that the Hamiltonian for the electron may be written

$$\mathfrak{H} = (1/2m)\mathfrak{p}_r^2 + (\mathfrak{L}^2/2mr^2) - (Ze^2/r)$$

where $\mathfrak{L}$ is the angular momentum operator.

c. Determine the energy values E_n for an electron having a given angular momentum l using the following method. Find an operator $\mathfrak{A}(r)$ such that

$$(\mathfrak{p}_r + i\mathfrak{A})(\mathfrak{p}_r - i\mathfrak{A})\mathbf{e}_{nl} = \left(2m\mathfrak{H}_l + \frac{m^2e^4Z^2}{l^2}\right)\mathbf{e}_{nl}$$

$$(\mathfrak{p}_r - i\mathfrak{A})(\mathfrak{p}_r + i\mathfrak{A})\mathbf{e}_{nl} = \left(2m\mathfrak{H}_{l-1} + \frac{m^2e^4Z^2}{l^2}\right)\mathbf{e}_{nl}$$

Hence show that E_{nl} is independent of l, that for a given E_{nl} there is a maximum l which we shall call $n - 1$. Express E_{nl} in terms of n.

2.9 Show that under the transformation

$$\mathbf{e} = \exp(-i\mathfrak{H}_0 t/\hbar)\mathbf{f}$$

The Schroedinger equation

$$(\mathfrak{H}_0 + \mathfrak{H}_1)\mathbf{e} = i\hbar(\partial\mathbf{e}/\partial t)$$

becomes

$$\mathfrak{H}_1(t)\mathbf{f} = i\hbar(\partial\mathbf{f}/\partial t)$$

where

$$\mathfrak{H}_1(t) = \exp(i\mathfrak{H}_0 t/\hbar)\mathfrak{H}_1 \exp(-i\mathfrak{H}_0 t/\hbar)$$

Show that

$$\mathbf{f} = \mathfrak{U}\mathbf{f}_0$$

where

$$\mathfrak{U} = 1 + (1/i\hbar)\int_{-\infty}^{t} \mathfrak{H}_1(t')\,dt' + (1/i\hbar)^2 \int_{-\infty}^{t} \mathfrak{H}_1(t')\,dt' \int_{-\infty}^{t'} \mathfrak{H}_1(t'')\,dt'' + \cdots$$

where $\mathbf{f}_0$ is time independent. Relate $\mathbf{f}_0$ to the solutions of

$$\mathfrak{H}_0\mathbf{e}_0 = i\hbar(\partial\mathbf{e}_0/\partial t)$$

2.10 Decompose the solution $\mathbf{e}$ of the Dirac wave equation as follows:

$$\mathbf{e} = \mathbf{f} + \mathbf{g}; \quad \mathbf{f} = \tfrac{1}{2}(1 + \boldsymbol{\alpha}_0)\mathbf{e}; \quad \mathbf{g} = \tfrac{1}{2}(1 - \boldsymbol{\alpha}_0)\mathbf{e}$$

Show that

$$\mathbf{f}^* \cdot \mathbf{g} = 0$$

and that

$$(E + eV + mc^2)\mathbf{f} = c[\boldsymbol{\alpha} \cdot (\mathbf{p} + e\mathbf{A}/c)]\mathbf{g};$$
$$(E + eV + mc^2)\mathbf{g} = -c[\boldsymbol{\alpha} \cdot (\mathbf{p} + e\mathbf{A}/c)]\mathbf{f}$$

Show, for positive energy states and $e\mathbf{A}$ and eV small compared with mc^2, that $\mathbf{g}^* \cdot \mathbf{g} \ll \mathbf{f}^* \cdot \mathbf{f}$.

2.11 Define a set of four states $\mathbf{e}_i$ which satisfy the Dirac equation for a particle at rest:

$$(\boldsymbol{\alpha}_0 mc^2)\mathbf{e}_i = E_0\mathbf{e}_i$$

Show that the four solutions of the Dirac equation for a particle of momentum $\mathbf{p}$ are

$$[c(\boldsymbol{\alpha} \cdot \mathbf{p}) + \boldsymbol{\alpha}_0(mc^2 + |E|)]\mathbf{e}_i$$

where $E^2 = c^2p^2 + m^2c^4$.

Standard Forms for Some of the Partial Differential Equations of Theoretical Physics

	EQ. NO.
Laplace Equation: $\nabla^2\psi = 0$	(1.1.4), (2.3.6)
Vector Form: $\operatorname{curl}(\operatorname{curl}\mathbf{A}) = 0$; $\operatorname{div}\mathbf{A} = 0$	
Poisson Equation: $\nabla^2\psi = -4\pi\rho$	(1.1.5), (2.1.2), (2.5.2)
Vector Form: $\operatorname{curl}(\operatorname{curl}\mathbf{A}) = 4\pi\mathbf{J}$; $\operatorname{div}\mathbf{A} = 0$	(2.5.7)
Helmholtz Equation: $\nabla^2\psi + k^2\psi = 0$	(2.1.10)
Vector Form: $\operatorname{curl}(\operatorname{curl}\mathbf{A}) - k^2\mathbf{A} = 0$; $\operatorname{div}\mathbf{A} = 0$	
Wave Equation: $\square^2\psi = \nabla^2\psi - \dfrac{1}{c^2}\dfrac{\partial^2\psi}{\partial t^2} = 0$	(2.1.9), (2.2.2)
Vector Form: $\operatorname{curl}(\operatorname{curl}\mathbf{A}) + \dfrac{1}{c^2}\dfrac{\partial^2\mathbf{A}}{\partial t^2} = 0$; $\operatorname{div}\mathbf{A} = 0$	(2.2.3), (2.5.15)
Diffusion Equation: $\nabla^2\psi = \dfrac{1}{a^2}\dfrac{\partial\psi}{\partial t}$	(2.4.4)
Vector Form: $\operatorname{curl}(\operatorname{curl}\mathbf{A}) + \dfrac{1}{a^2}\dfrac{\partial\mathbf{A}}{\partial t} = 0$; $\operatorname{div}\mathbf{A} = 0$	(2.3.19)

Klein-Gordon Equation: $\Box^2\psi = \mu^2\psi$ (2.1.27)

Vector Form: $\text{curl}\,(\text{curl}\,\mathbf{A}) + \frac{1}{c^2}\frac{\partial^2\mathbf{A}}{\partial t^2} + \mu^2\mathbf{A} = 0;$

$$\text{div}\,\mathbf{A} = 0 \quad \text{(Proca Equation)} \tag{2.5.37}$$

Maxwell's Equations: $\text{div}\,\mathbf{B} = 0; \quad \text{div}\,\mathbf{D} = 4\pi\rho$

$$\text{curl}\,\mathbf{H} = \frac{1}{c}\frac{\partial\mathbf{D}}{\partial t} + \frac{1}{c}4\pi\mathbf{J}; \quad \text{curl}\,\mathbf{E} = -\frac{1}{c}\frac{\partial\mathbf{B}}{\partial t} \tag{2.5.11}$$

$$\mathbf{B} = \mu\mathbf{H}; \quad \mathbf{D} = \epsilon\mathbf{E}$$

Electromagnetic Potential Equations:

$$\Box^2\varphi = -4\pi\rho/\epsilon; \quad \Box^2\mathbf{A} = -4\pi\mu\mathbf{J}/c \tag{2.5.15}$$

$$\mathbf{B} = \text{curl}\,\mathbf{A}; \quad \mathbf{E} = -\,\text{grad}\,\varphi - \frac{1}{c}\frac{\partial\mathbf{A}}{\partial t}; \quad \text{div}\,\mathbf{A} = -\frac{\epsilon\mu}{c}\frac{\partial\varphi}{\partial t}$$

(For the forms of these equations for other gauges, see pages 207 and 332.)

Elastic Wave Equation (isotropic media):

$$\rho\frac{\partial^2\mathbf{s}}{\partial t^2} = \nabla\cdot[\lambda\mathfrak{I}\,\text{div}\,\mathbf{s} + \mu(\nabla\mathbf{s}) + \mu(\mathbf{s}\nabla)]$$

$$= (\lambda + 2\mu)\,\text{grad div}\,\mathbf{s} - \mu\,\text{curl curl}\,\mathbf{s} \tag{2.2.1}$$

Viscous Fluid Equation:

$$\rho\frac{\partial\mathbf{v}}{\partial t} + \rho\mathbf{v}\cdot(\nabla\mathbf{v}) = \nabla\cdot[-(p + \gamma\,\text{div}\,\mathbf{v})\mathfrak{I} + \eta(\nabla\mathbf{v}) + \eta(\mathbf{v}\nabla)] \tag{2.3.14}$$

$$\rho\frac{\partial\mathbf{v}}{\partial t} = -\,\text{grad}\,[p - (\tfrac{4}{3}\eta + \lambda)\,\text{div}\,\mathbf{v} + \tfrac{1}{2}\rho v^2] - \eta\,\text{curl curl}\,\mathbf{v} + \rho\mathbf{v}\times\text{curl}\,\mathbf{v}$$

where $\lambda = \frac{2}{3}\eta - \gamma$.

Schroedinger Equation for single particle of mass m in potential V:

$$-\frac{\hbar^2}{2m}\nabla^2\psi + V\psi = i\hbar\frac{\partial\psi}{\partial t}; \quad |\psi|^2 \text{ is probability density} \tag{2.6.38}$$

Dirac Equation for electron in electromagnetic field:

$$\boldsymbol{\alpha}_0 mc\boldsymbol{\psi} + \boldsymbol{\alpha}\cdot\left(\frac{\hbar}{i}\,\text{grad}\,\boldsymbol{\psi} + \frac{e}{c}\mathbf{A}\boldsymbol{\psi}\right) + \left(\frac{\hbar}{ic}\frac{\partial\boldsymbol{\psi}}{\partial t} - e\varphi\boldsymbol{\psi}\right) = 0 \tag{2.6.57}$$

$$\boldsymbol{\psi} = \sum_{n=1}^{4}\mathbf{e}_n\psi_n; \quad \boldsymbol{\psi}^*\cdot\boldsymbol{\psi} \text{ is probability density}$$

Bibliography

General references for material in this chapter:

Joos, G.: "Theoretical Physics," Stechert, New York, 1944.
Landau, L. D., and E. Lifschitz: "Classical Theory of Fields," Addison-Wesley, Cambridge, 1951.
Lindsay, R. B., and H. Margenau: "Foundations of Physics," Wiley, New York, 1936.
Margenau, H., and G. M. Murphy: "Mathematics of Physics and Chemistry," Van Nostrand, New York, 1943.
Rayleigh, J. W. S.: "The Theory of Sound," Macmillan & Co., Ltd., London, 1896, reprinted, Dover, New York, 1945.
Riemann-Weber, "Differential- und Integralgleichungen der Mechanik und Physik," ed. by P. Frank and R. von Mises, Vieweg, Brunswick 1935, reprint Rosenberg, New York, 1943.
Slater, J. C., and N. H. Frank: "Introduction to Theoretical Physics," McGraw-Hill, New York, 1933.
Schaeffer, C.: "Einfuhrung in die Theoretische Physik," 3 vols., De Gruyter, Berlin, 1937.
Sommerfeld, A.: "Partial Differential Equations in Physics," Academic Press, New York, 1949.
Webster, A. G.: "Partial Differential Equations of Mathematical Physics," Stechert, New York, 1933.

Additional texts of particular interest in connection with the subject of vibration and sound:

Coulson, C. A.: "Waves, a Mathematical Account of the Common Types of Wave Motion," Oliver & Boyd, Edinburgh, 1941.
Lamb, H.: "The Dynamical Theory of Sound," E. Arnold & Co., London, 1925.
Morse, P. M.: "Vibration and Sound," McGraw-Hill, New York, 1948.

Books on elasticity and elastic vibrations:

Brillouin, L.: "Les tenseurs en méchanique et en élastique," Masson et Cie, Paris, 1938.
Love, A. E. H.: "Mathematical Theory of Elasticity," Cambridge, New York, 1927, reprint Dover, New York, 1945.
Sokolnikoff, I. S.: "Mathematical Theory of Elasticity," McGraw-Hill, New York, 1946.
Timoshenko, S.: "Theory of Elasticity," McGraw-Hill, New York, 1934.

Additional material on hydrodynamics and compressional wave motion:

Chapman, S., and T. G. Cowling: "Mathematical Theory of Non-uniform Gases," Cambridge, New York, 1939.
Hadamard, J. S.: "Leçons sur la propagation des ondes et les equations de l'hydrodynamique," Hermann & Cie, Paris, 1903.
Lamb, H.: "Hydrodynamics," Cambridge, New York, 1932, reprint Dover, New York, 1945.
Milne-Thomson, L. M.: "Theoretical Hydrodynamics," Macmillan & Co., Ltd., London, 1938.
Sauer, R.: "Introduction to Theoretical Gas Dynamics," Edwards Bros., Inc., Ann Arbor, Mich., 1947.

Books on diffusion, heat flow, and transport theory:

Chandrasekhar, S.: "Radiative Transfer," Oxford, New York, 1950.
Chapman, S., and T. G. Cowling: "Mathematical Theory of Non-uniform Gases," Cambridge, New York, 1939.
Fowler, R. H.: "Statistical Mechanics," Cambridge, New York, 1936.
Hopf, E.: "Mathematical Problems of Radiative Equilibrium," Cambridge, New York, 1934.
Lorentz, H. A.: "Theory of Electrons," B. G. Teubner, Leipzig, 1909.

Texts on electromagnetic theory, particularly on fundamental concepts:

Abraham, M., and R. Becker: "Classical Theory of Electricity and Magnetism," Blackie, Glasgow, 1932.
Stratton, J. A.: "Electromagnetic Theory," McGraw-Hill, New York, 1941.
Van Vleck, J. H.: "Theory of Electric and Magnetic Susceptibilities," Chaps. 1–4, Oxford, New York, 1932.

Discussions of the fundamental principles of quantum mechanics from various points of view:

Bohm, D.: "Quantum Theory," Prentice-Hall, New York, 1951.
De Broglie, L.: "L'Electron magnetique," Hermann & Cie, Paris, 1945.
Dirac, P. A. M.: "Principles of Quantum Mechanics," Oxford, New York, 1935.
Jordan, P.: "Anschauliche Quantentheorie," Springer, Berlin, 1936.
Kemble, E. C.: "Fundamental Principles of Quantum Mechanics," McGraw-Hill, New York, 1937.
Kramers, H. A.: "Grundlagen der Quantentheorie," Akademische Verlagsgesellschaft m.b.H., Leipzig, 1938.
Sommerfeld, A.: "Wellenmechanik," Vieweg, Brunswick, 1939, reprinted, Ungar, New York, 1946.
Van der Waerden, B. L.: "Gruppentheoretische Methode in der Quantenmechanik," Springer, Berlin, 1932.
Von Neumann, J.: "Mathematische Grundlagen der Quantenmechanik," Springer, Berlin, 1932.

CHAPTER 3

Fields and the Variational Principle

The use of superlatives enables one to express in concise form a general principle covering a wide variety of phenomena. The statements, for instance, that a straight line is the shortest distance between two points or that a circle is the shortest line which encloses a given area, are deceptively simple ways of defining geometrical entities. To say that electric current distributes itself in a network of resistors so that the least power is converted into heat is a description of direct-current flow which encompasses many individual cases without the use of mathematical complexities (though the complexities inevitably intrude when the general principle is applied to the individual case). The statement that a physical system so acts that some function of its behavior is least (or greatest) is often both the starting point for theoretical investigation and the ultimate distillation of all the relationships between facts in a large segment of physics.

The mathematical formulation of the superlative is usually that the integral of some function, typical of the system, has a smaller (or else larger) value for the actual performance of the system than it would have for any other imagined performance subject to the same very general requirements which serve to particularize the system under study. We can call the integrand L; it is a function of a number of independent variables of the system (coordinates, field amplitudes, or other quantities) and of the derivatives of these variables with respect to the parameters of integration (velocities or gradients of fields, etc.). If the variables are $\varphi_1, \ldots, \varphi_n$, the parameters $x_1, \ldots, x_m$ and the derivatives $\partial\varphi_r/\partial x_s = \varphi_{rs}$, then the integral which is to be minimized is

$$\mathfrak{L} = \int_{a_1}^{b_1} \cdots \int_{a_m}^{b_m} L\left(\varphi, \frac{\partial\varphi}{\partial x}, x\right) dx_1 \cdots dx_m \tag{3.1.1}$$

From the minimization of this function we can obtain the partial differential equations governing the φ's as functions of the x's and many other things. This process of obtaining the φ's is called the *variational method*.

In the present chapter we shall first indicate in a general way how the variational method can be used to obtain equations for the variables involved, then discuss in more detail the variational principles of classical dynamics, because they provide a well-explored example of this technique and its utility; and then we shall proceed to apply the method to the various types of fields which are to be studied in this book.

3.1 *The Variational Integral and the Euler Equations*

The integrand L of the integral to be minimized (or maximized) will be called the *Lagrange density* of the system. It is a *function of functions* of the basic parameters of the system. In the case of classical dynamics, for instance, the parameter is the time and the functions are the coordinates and velocities, at various times, of the configuration of the system as it moves in conformity with the applied forces and the initial conditions; in the case of fields the basic parameters are the coordinates locating every point where the field is to be measured, and the functions are the various components of the field and their gradients, which are determined, as functions of these coordinates, by the distribution of the various "sources" (or charges) in space and by the boundary conditions.

Thus when we require that the integral of L should be minimized (or maximized), we mean that the functions in terms of which L is expressed (the coordinates and velocities or fields and gradients) are to be adjusted, at every value of the parameters, so that the integral of L has the least possible value. We wish so to adjust the functions φ that the integral of L, a function of the φ's and $\partial\varphi/\partial x$'s, is as small as it can become, subject to the conditions imposed by the state of the system.

In order to solve the problem we must first make the step from a variational requirement on an integral of L to a set of equations which determine the best values of the functions φ.

The Euler Equations. Before we can make even this step we must be more specific about what we mean by "minimizing the integral" and "best values of the functions." To do this, suppose we arbitrarily choose a functional form for each of the functions $\varphi_1, \ldots, \varphi_n$, as functions of the parameters $x_1, \ldots, x_m$. This arbitrary choice will, of course, fix the form of the functions $\varphi_{rs} = \partial\varphi_r/\partial x_s$ and therefore will determine the value of the integral $\mathfrak{L}$ given in Eq. (3.1.1). Now let us change the φ's a little bit; for the function φ_r suppose the change to be expressed by the term $\epsilon_r\eta_r$, where η_r is an arbitrary function of the parameters and ϵ is a small quantity, independent of the parameters. The shorthand notation $\delta\varphi_r$ is often used instead of $\epsilon_r\eta_r$, where $\delta\varphi$ is considered to be an arbitrary small "variation" of the function φ. This modification of the φ's will also result in a change in the components φ_{rs} of the gradients. These are related to the changes in the φ's, for $\partial\epsilon_r\eta_r/\partial x_s = \epsilon_r\eta_{rs}$. In the shorthand variational notation, this relation is represented as $\delta\varphi_{rs} = \partial\delta\varphi_r/\partial x_s$.

By using a Taylor's series expansion of L we can show that the first-order change in the integral $\mathfrak{L}$ due to the small changes of the φ's can be written

$$\delta\mathfrak{L} = \int_{a_1}^{b_1} \cdots \int_{a_m}^{b_m} \sum_{r=1}^{n} \epsilon_r \left[\frac{\partial L}{\partial \varphi_r} \eta_r + \sum_{s=1}^{m} \frac{\partial L}{\partial \varphi_{rs}} \frac{\partial \eta_r}{\partial x_s} \right] dx_1 \cdots dx_m$$

We assume that the parameters are so chosen that the limits of integration can all be constants and that all the η's go to zero at these limits, which would be true, for instance, if the limits coincided with physical boundaries where certain boundary conditions are imposed on the φ's. This situation is usually the case, so we shall confine ourselves here to its study; the more general case, where limits are all varied, will be touched upon later.

The term $(\partial L/\partial\varphi_{rs})(\partial\eta_r/\partial x_s)$ can be integrated by parts over x_s, giving $\left[\frac{\partial L}{\partial\varphi_{rs}}\eta_r\right]_{a_s}^{b_s} - \int_{a_s}^{b_s} \frac{\partial}{\partial x_s}\left(\frac{\partial L}{\partial \varphi_{rs}}\right)\eta_r\, dx_s$. The first term is zero, since $\eta_r = 0$ at a_s and b_s. Therefore the first-order variation of $\mathfrak{L}$ is

$$\delta\mathfrak{L} = \int_{a_1}^{b_1} \cdots \int_{a_m}^{b_m} \sum_{r=1}^{n} \epsilon_r \left[\frac{\partial L}{\partial \varphi_r} - \sum_{s=1}^{m} \frac{\partial}{\partial x_s}\left(\frac{\partial L}{\partial \varphi_{rs}}\right) \right] \eta_r \, dx_1 \cdots dx_m \qquad (3.1.2)$$

If $\delta\mathfrak{L}$ is not zero, $\mathfrak{L}$ cannot be a maximum or minimum. When $\delta\mathfrak{L}$ is zero, no matter what (small) values the ϵ_r's have, then the functional forms of all the φ's have been chosen so that $\mathfrak{L}$, as a function of the ϵ_r's, has *either a minimum or a maximum or a point of inflection* for $\epsilon_r = 0$. Usually we can tell from the physical situation which of these cases is true; if we are not sure, it is always possible to calculate the second-order terms in the ϵ_r's in the Taylor's series expansion of $\mathfrak{L}$ to see whether they are positive, negative, or zero. To save space from here on we shall use the terms "minimize" and "minimum" when we mean "minimized or maximized or minimaxed" and "minimum or maximum or point of inflection."

Therefore in order that $\mathfrak{L}$ have its extreme value (maximum or minimum), the functional form of the φ's must be chosen so that the coefficient of each of the ϵ_r's in the integral for $\delta\mathfrak{L}$ is zero. This results in a set of equations for the desired behavior of the φ's:

$$\sum_{s=1}^{m} \frac{\partial}{\partial x_s}\left(\frac{\partial L}{\partial \varphi_{rs}}\right) = \frac{\partial L}{\partial \varphi_r}; \quad r = 1, \ldots, n \qquad (3.1.3)$$

where $\varphi_{rs} = \partial\varphi_r/\partial x_s$. These equations, which serve to determine the optimum functional form of the φ's, are called the *Euler equations.* We shall use these equations extensively later in this chapter.

Several general comments should be made concerning these results. In the first place, if the variational principle is to be of general validity, then $\mathfrak{L}$ should be an invariant, and the density L, or L divided by the scale factors coming into the element of integration, should be invariant to coordinate transformation of the parameters of integration. This will be of use later in finding new Lagrange densities.

A still more general comment is that the variational principle is generally useful in unifying a subject and consolidating a theory rather than in breaking ground for a new advance. It usually happens that the differential equations for a given phenomenon are worked out first, and only later is the Lagrange function found, from which the differential equations can be obtained. This is not to minimize the importance of finding the Lagrange density L, for it is of considerable utility to find what physical quantity must be minimized to obtain the differential equations for the phenomena, and the form of the variational equations often suggest fruitful analogies and generalizations.

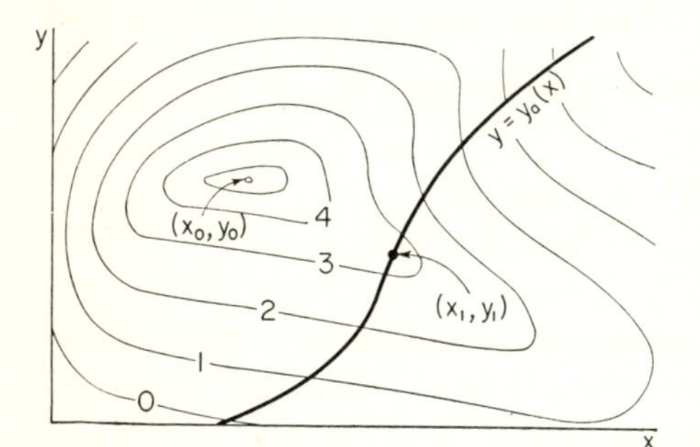

Fig. 3.1 Maximum point (x_0,y_0) for function $f(x,y,)$ represented by contour lines 0, 1, 2, ... Maximum point $(x_1,y_1,)$ along line $y = y_a(x)$.

Auxiliary Conditions. In many cases the Lagrange integral is to be minimized (or maximized) subject to some one or more additional requirements further restricting the independent variables and parameters. In this case we use the method of Lagrange multipliers to obtain the modified answer. Just how these multipliers work can best be shown by an example.

Suppose that the function $f(x,y)$ is to be maximized. If there are no auxiliary conditions, we solve the two equations

$$\partial f/\partial x = 0; \quad \partial f/\partial y = 0 \tag{3.1.4}$$

simultaneously. The resulting pair (or pairs) of values of x and y, (x_0,y_0), specify the point (or points) at which f has a maximum, minimum, or saddle point, and the value $f(x_0,y_0)$ is the value of f at this maximum or minimum. Here the function is of two variables, and the two equations (3.1.4) are needed to obtain a pair of values (x_0,y_0) for the maximum or minimum. A typical case is pictured in Fig. 3.1, where the function f is depicted in terms of contour lines.

But suppose that we wish to find the maximum of $f(x,y)$ along the line given by the *auxiliary equation* $y = y_a(x)$. This line does not usually run through the point (x_0,y_0), so the solution cannot be the same. There may be one or more points along the line, however, where $f(x,y)$ has a maximum (or minimum) value, such as the point (x_1,y_1) shown in Fig. 3.1. This may be computed by inserting the expression for y in terms of x into the form for f, which gives the value of f along the line as a function the *single* parameter x. We then differentiate with respect to x to find the maximum value,

$$\frac{d}{dx} f(x,y_a(x)) = \frac{\partial f}{\partial x} + \frac{\partial f}{\partial y}\frac{d}{dx}[y_a(x)] = 0 \tag{3.1.5}$$

The position of the maximum is then the solution x_1 of this equation and the related value $y_1 = y_a(x_1)$.

However, we can solve this same problem by a method which at first sight appears to be different from and more complicated than the one resulting in Eq. (3.1.5). Suppose the auxiliary equation is $g(x,y) = 0$. We first introduce a third unknown, λ, and then maximize the new function $f + \lambda g$, subject to the relation $g = 0$. In other words we are to solve the three equations

$$(\partial f/\partial x) + \lambda(\partial g/\partial x) = 0; \quad (\partial f/\partial y) + \lambda(\partial g/\partial y) = 0; \quad g = 0 \tag{3.1.6}$$

simultaneously to determine the proper values for x, y, *and* λ.

It is not immediately apparent that the solution of Eqs. (3.1.6) is identical with the solution of Eq. (3.1.5), but the connection becomes clearer when we write the auxiliary equation $g(x,y) = 0$ in the form used above, $y_a(x) - y = 0$. Then the first two of Eqs. (3.1.6) are

$$(\partial f/\partial x) + \lambda(dy_a/dx) = 0; \quad (\partial f/\partial y) - \lambda = 0$$

Substituting the second of these into the first gives us

$$(\partial f/\partial x) + (dy_a/dx)(\partial f/\partial y) = 0$$

which is just Eq. (3.1.5). Therefore in this simple case the method of Lagrange multipliers gives us the same result as the straightforward method. This is true in general. In this simple case it appears to be a more cumbersome method than the use of Eq. (3.1.5), but in more complex cases it turns out to be an easier method.

As applied to the variational integral of Eq. (3.1.1) the method of Lagrange multipliers can be stated as follows: Suppose that the Lagrange density is $L(\varphi_r,\varphi_{rs},x_s)$, $(s = 1, 2, \ldots, m)$, $(r = 1, 2, \ldots, n)$, and the auxiliary equations are

$$\int_{a_1}^{b_1} \cdots \int_{a_m}^{b_m} G_t\left(\varphi, \frac{\partial \varphi}{\partial x}, x\right) dx_1 \cdots dx_m = C_t \tag{3.1.7}$$

where $t = 1, 2, \ldots, k$ $(k < m)$ and where the C's are constants.

Then the modified variational integral is

$$\mathfrak{L}' = \int_{a_1}^{b_1} \cdots \int_{a_m}^{b_m} L'\left(\varphi, \frac{\partial\varphi}{\partial x}, x\right) dx_1 \cdots dx_m$$

where

$$L' = L(\varphi_r, \varphi_{rs}, x_s) + \sum_{t=1}^{k} \lambda_t G_t(\varphi_r, \varphi_{rs} x_s) \tag{3.1.8}$$

Then the m new Euler equations,

$$\sum_{s=1}^{n} \frac{\partial}{\partial x_s}\left(\frac{\partial L'}{\partial \varphi_{rs}}\right) = \frac{\partial L'}{\partial \varphi_s} \tag{3.1.9}$$

plus the k equations (3.1.7) serve to determine the φ's as well as the values of the λ's. In this case the Lagrange multiplier method is definitely the easier.

3.2 *Hamilton's Principle and Classical Dynamics*

In classical dynamics the parameter is time t and the functions φ in the Lagrange function are the coordinates q which specify the configuration of the system at each time. If the system has n degrees of freedom, we can choose n independent q's ($q_1, \ldots, q_n$) which will completely specify the configuration; the corresponding velocities will be $\dot{q}_r = dq_r/dt$. No matter how the q's are chosen, the kinetic energy of an inertial system always turns out to be a quadratic function of the $\dot{q}$'s:

$$T = \tfrac{1}{2} \sum_{r,s} a_{rs} \dot{q}_r \dot{q}_s \tag{3.2.1}$$

where the a's may be functions of the q's. If the system is *conservative* (*i.e.*, has a total mechanical energy which is constant in time), then the external force on the system can be represented in terms of the gradient of a scalar potential function V, so that the equivalent force along the q_r coordinate is

$$F_r = -(\partial V/\partial q_r) \tag{3.2.2}$$

The potential energy may depend on time explicitly, but it is not a function of the $\dot{q}$'s.

When the system is conservative, the variational principle determining the equations of motion, called *Hamilton's principle*, uses the *kinetic potential* (see page 229) $T - V$ as the Lagrange function and is

$$\delta \int_{t_0}^{t_1} (T - V)\, dt = 0 \tag{3.2.3}$$

This states that for any actual motion of the system, under the influence of the conservative forces, when it is started according to any reasonable initial conditions, the system will move so that the time average of the difference between kinetic and potential energies will be a minimum (or in a few cases, a maximum).

Lagrange's Equations. The Euler equations for the coordinates for this case,

$$\frac{d}{dt}\left[\frac{\partial(T - V)}{\partial \dot{q}_r}\right] = \frac{\partial(T - V)}{\partial q_r}; \quad r = 1, 2, \ldots, n \tag{3.2.4}$$

are called *Lagrange's equations of motion* for the system. The left-hand terms represent the accelerations of the system, and the right-hand terms are the corresponding forces, derived from the potential energy V, plus the "kinetic forces" (such as centrifugal force) due to the motions. When the forces are not conservative, so that there is no potential energy, the variational principle is

$$\int_{t_0}^{t_1}\left[\delta T + \sum_{r=1}^{n} F_r\,\delta q_r\right] dt = 0$$

and Lagrange's equations are

$$\frac{d}{dt}\left(\frac{\partial T}{\partial \dot{q}_r}\right) = \frac{\partial T}{\partial q_r} - F_r \tag{3.2.5}$$

Lagrange's equations are among the most generally useful equations of classical dynamics.

The kinetic and potential energies are scalar invariants to coordinate transformations, so that they can be expressed in terms of any complete set of coordinates, and Lagrange's equations will have the same form in each case. In each case the quantity

$$\frac{\partial}{\partial \dot{q}_r}(T - V) = \frac{\partial T}{\partial \dot{q}_r} = p_r$$

is called the *momentum* for the rth coordinate. Lagrange's equations can be written

$$\frac{dp_r}{dt} - \frac{\partial T}{\partial q_r} = \begin{cases} -(\partial V/\partial q_r; & \text{if the system is conservative} \\ F_r; & \text{in general} \end{cases}$$

and if we use rectangular coordinates for the q's, so that T does not depend on the q's but only on the $\dot{q}$'s, the equations reduce to the familiar Newton's equations

$$(d/dt)(\text{momentum}) = \text{force}$$

Consequently Hamilton's principle represents, in simple invariant form, all the equations of classical dynamics.

Energy and the Hamiltonian. When the system is conservative, a function of the coordinates and momenta which remains constant throughout the motion of the system is the total energy E, the sum of kinetic and potential energy.

When this is expressed in terms of the coordinates q and the momenta p, it is called the Hamiltonian function H for the system. Since the Lagrange function L is $T - V$ and the energy E is $T + V$, the two can be related by the equation $E = 2T - L$. Therefore we can express the variational principle as $\delta\int(2T - E)\,dt = 0$. From this we can obtain equations relating the Hamiltonian to the velocity and acceleration. Although these are just other forms of the same old equations of motion, their form turns out to be particularly well adapted for translation into quantum language.

First we must translate $(2T - E)$ into functions of q and p instead of q and $\dot{q}$. As shown earlier the momentum p_r is obtained by differentiating the kinetic energy T with respect to $\dot{q}_r$. Once we know p_r as a function of $\dot{q}_r$, it is not hard to eliminate the $\dot{q}$'s from L and T. From Eq. (3.2.1) we see that

$$p_r = \sum_{s=1}^{n} a_{rs}\dot{q}_s$$

$$2T = \sum_{r=1}^{n} p_r\dot{q}_r = L + H \qquad (3.2.6)$$

This equation enables us to answer a question which we have so far ignored: whether H can be expressed *only* in terms of the q's and the p's (as defined in Eq. 3.2.6) with no dependence on $\dot{q}$ left over. To demonstrate, we use the equation $H = \Sigma p\dot{q} - L$, where L, in general, cannot be expressed in terms of just p's and q's but can be expressed in terms of just q's and $\dot{q}$'s. We now make a small change in the q's, p's, and $\dot{q}$'s.

$$dH = \Sigma p\,d\dot{q} + \Sigma \dot{q}\,dp - \Sigma(\partial L/\partial\dot{q})\,d\dot{q} - \Sigma(\partial L/\partial q)\,dq$$

But by definition, $p = \partial L/\partial\dot{q}$, so that $dH = \Sigma\dot{q}\,dp - \Sigma(\partial L/\partial q)\,dq$. Therefore the total change in H is given in terms of changes of the q's and p's; so we have proved that H *can* be expressed as a function of the q's and p's alone (unless L depends on t explicitly, when $\partial H/\partial t = -\partial L/\partial t$ and H is a function of the q's, of the p's, and of t).

When the energy is expressed in terms of the p's and q's (and t, if necessary), we call it the Hamiltonian H. The variation of $\int L\,dt$ becomes

$$\delta\int_{t_0}^{t_1}(2T - H)\,dt = \int_{t_0}^{t_1}\sum_{r=1}^{n}\left[p_r\,\delta\dot{q}_r + \dot{q}_r\,\delta p_r - \frac{\partial H}{\partial q_r}\,\delta q_r - \frac{\partial H}{\partial p_r}\,\delta p_r\right]dt$$

where the δq's and δp's represent the variations of the quantities q and p from the values taken on along the actual path (corresponding to the quantities $\epsilon\eta$ used earlier). Integrating the terms $p_r\,\delta\dot{q}_r = p_r(d\,\delta q_r/dt)$ by parts, we can finally separate the variation of the integrand into a part due to the variation of the q's and another part due to the variation of the p's,

$$\delta\int L\,dt = \int_{t_0}^{t_1}\sum_{r=1}^{n}\left[\left(-\dot{p}_r - \frac{\partial H}{\partial q_r}\right)\delta q_r + \left(\dot{q}_r - \frac{\partial H}{\partial p_r}\right)\delta p_r\right]dt = 0$$

Since we are assuming that we can vary the p's independently of the q's, each of the parentheses must be zero, and we arrive at the alternate form of the equations of motion

$$\dot{q}_r = \partial H/\partial p_r; \quad \dot{p}_r = -(\partial H/\partial q_r) \tag{3.2.7}$$

which are called *Hamilton's canonical equations.* They have been used a number of times in the previous chapter (see pages 229 and 242). We shall return again to use them later in this chapter.

It is not difficult to see that the Hamiltonian function of the p's and q's is independent of time. For the time rate of change of H is

$$\frac{dH}{dt} = \frac{\partial H}{\partial q}\frac{dq}{dt} + \frac{\partial H}{\partial p}\frac{dp}{dt}$$

which is zero according to Eqs. (3.2.7) unless H depends explicitly on the time. Any Hamiltonian for which the canonical equations hold is thus a constant of the motion for the system. Note that this equation states that for conservative systems the *total* change of H with time is zero. In some cases H may depend *explicitly* on t; in these cases the variation of H with t due to changes of the q's and p's is still zero, and the total change dH/dt is equal to the explicit change $\partial H/\partial t$ which is, incidentally, equal to $-(\partial L/\partial t)$, as we have already seen.

Impedance. In Chap. 2 (page 128) we introduced the concept of mechanical impedance, the ratio between a simple harmonic driving force $F_0e^{-i\omega t}$ and the corresponding velocity. If the system is a linear one, this ratio is independent of the amplitude of motion and is a function of ω and of the constants of the system. As we shall see in the next chapter, the use of the impedance function enables us to transform a problem from one of studying response as a function of time to one of studying impedance as a function of frequency. Often the latter problem is a simpler one.

At any rate the canonical equations (3.2.7) enable us to approach the concept of impedance from a new direction. We first note that, if an external force F_r is applied to the coordinate q_r, then the canonical equations become

$$\dot{q}_r = \partial H/\partial p_r; \quad \dot{p}_r + (\partial H/\partial q_r) = F_r$$

We now see that the canonical equations have broken the second-order Lagrange equations into two simultaneous first-order equations of precisely the form required to compute the impedance (if the system is of such a form that an impedance function has meaning). For the ratio of applied force F_r to velocity $\dot{q}_r$, for the rth coordinate is just

$$Z_r = \frac{\dot{p}_r + (\partial H/\partial q_r)}{(\partial H/\partial p_r)}$$

From this new point of view we can imagine ourselves learning the fundamental properties of a system by "probing" it with oscillating forces. To each coordinate in turn we apply such a force, and we measure the ratio between force and coordinate velocity. If this ratio is independent of amplitude, then it can be used to reconstruct the system. Alternately, if we know the nature of the Lagrange function for the system, we can compute the impedance for each of its coordinates. We assume that $q_r = A_r e^{i\omega t}$ (in Chap. 2 and many times later in the book, we use the exponent $e^{-i\omega t}$ to represent simple harmonic motion; here and in Chap. 4, we shall be considering the behavior of Z for *all* values of ω, both positive and negative and imaginary, so we might as well compute it first for the positive exponential). The equation $p_r = \partial L/\partial \dot{q}_r$ enables us to compute p_r as a function of the $\dot{q}_s$ and therefore as a function of the exponential $e^{i\omega t}$ and of the amplitudes A_r. Therefore the derivatives $\partial H/\partial p_r$ and $\partial H/\partial q_r$ can be expressed in this form, and eventually the ratio Z_r. When this ratio turns out to be independent of the A's and of time, it is an impedance.

When the potential energy of the system has a minimum value for a certain set of values of the coordinates q, we can set the origin of coordinates at this minimum, and for sufficiently small values of displacement away from this origin, the potential energy can be expressed as a *quadratic function* of the q's:

$$V = \tfrac{1}{2} \sum_{m,n} b_{rs} q_r q_s + V_{\min} \tag{3.2.8}$$

analogous to Eq. (3.2.1). Sometimes this minimum is not an absolute one but depends on a "dynamical equilibrium." For instance we may find that one of the p's (say p_n) must be a constant; that is, $\partial H/\partial q_n = 0$. It is then possible to eliminate q_n from the equations and consider the constant p_n as a property of a system with one fewer coordinates, with possible additional terms in the potential energy which depend on p_n and are "caused" by the motion of constant momentum (they can be called *dynamic potential energies*). This new system may have equilibrium points where the "dynamic" forces are just balanced by the "true" forces, and then again the new potential energy takes on the form of Eq. (3.2.8) near the equilibrium, where some of the b's depend on the constant p_n (which is no longer considered as a momentum).

For displacements of the system from equilibrium (dynamical or otherwise) which are sufficiently small, therefore, the Hamiltonian function H is a quadratic function of the p's and q's. The expression equal to the applied force F_r is

$$\dot{p}_r + \frac{\partial H}{\partial q_r} = \sum_m [a_{rm}\ddot{q}_m + b_{rm}q_m]$$

This can be written in simpler form in abstract vector space. We say that the component of the displacement vector $\mathbf{q}$ along the rth coordinate is q_r and the corresponding component of the force vector $\mathbf{F}$ is F_r. The equation relating the force, displacement, and acceleration vectors is then

$$\mathbf{F} = \mathfrak{A} \cdot \ddot{\mathbf{q}} + \mathfrak{B} \cdot \mathbf{q}$$

where the components of the dyadics $\mathfrak{A}$ and $\mathfrak{B}$ are a_{mn} and b_{mn}, respectively. If, now, the vector $\mathbf{F}$ is simple harmonic with frequency $\omega/2\pi$, it can be represented by the vector $\mathbf{F}^\circ e^{i\omega t}$ and the steady-state velocities $\dot{\mathbf{q}}$ can be represented by the vector $\mathbf{U}e^{i\omega t}$, where the components U_r of $\mathbf{U}$ are complex numbers with magnitude equal to the amplitude of the velocity of the rth coordinate.

In this case we can write the previous equation in the form

$$\mathbf{F}^\circ = \mathfrak{Z} \cdot \mathbf{U} \tag{3.2.9}$$

where

$$\mathfrak{Z}(\omega) = i\omega\mathfrak{A} - (i/\omega)\mathfrak{B}$$

is called the *impedance dyadic* for the system near the equilibrium point under study. Therefore the impedance concept always has meaning sufficiently near to points of equilibrium of the system (if there are such). The diagonal element Z_{mm} is called the *input impedance* for the mth coordinate, and the nondiagonal element Z_{mn} is called the *transfer impedance* coupling the mth and nth coordinates. It is always possible (see page 59) to make a transformation to the principal axes of $\mathfrak{Z}$, to *normal coordinates* q_r°, so that all the transfer impedances are zero and the diagonal elements $Z_r^\circ(\omega)$ are the *principal values* of the impedance. This transformation may be different for different values of ω. One can also express the displacements $\mathbf{q} = \mathbf{A}e^{i\omega t}$ in terms of $\mathfrak{Z}$ and $\mathbf{F}^\circ$:

$$\mathbf{F} = i\omega\mathfrak{Z} \cdot \mathbf{A} = (-\omega^2\mathfrak{A} + \mathfrak{B}) \cdot \mathbf{A}$$

where $|A_m|$ is the amplitude of motion of q_m.

This transformation to normal coordinates is a special case of the *rotation* of axes in abstract vector space. The new coordinates $\mathbf{q}'$ are related to the old ones q by the equation

$$q'_r = \sum_{m=1}^{n} \gamma_{rm} q_m; \quad \text{where } \sum_{m=1}^{n} \gamma_{rm}\gamma_{sm} = \delta_{rs}$$

In other words the abstract vector operator, with elements γ_{mn}, is a unitary dyadic. The quantities γ_{rm} are then the generalizations of the direction cosines of page 22. For a rotational transformation they are independent of the q's. As we showed on page 61, the sum of the diagonal terms of the dyadic $\mathfrak{Z}$ is invariant under rotation:

$$|\mathfrak{Z}| = \sum_{m=1}^{n} Z_{mm} = \sum_{m=1}^{n} Z_m^\circ$$

as is also the determinant of the components of $\mathfrak{Z}$.

$$\Delta_z = |Z_{mn}| = Z_1^\circ(\omega) Z_2^\circ(\omega) \cdot \cdot \cdot Z_n^\circ(\omega)$$

It is also useful to compute the dyadic $\mathfrak{Y}$ reciprocal to $\mathfrak{Z}$, such that

$$\mathbf{U} = \mathfrak{Y} \cdot \mathbf{F}^\circ; \quad \mathbf{A} = (1/i\omega)\mathfrak{Y} \cdot \mathbf{F}^\circ; \quad \mathfrak{Y} \cdot \mathfrak{Z} = \mathfrak{Z} \cdot \mathfrak{Y} = \mathfrak{J}$$

where $\mathfrak{J}$ is the idemfactor. Reference to page 57 indicates that the relationship between the components Y_{mn} and Z_{mn} is given by the equation

$$Y_{mr} = Z'_{mr}/\Delta_z$$

where Z'_{mr} is the first minor of Z_{mr} in the determinant Δ_z. It should also be obvious that the principal axes for $\mathfrak{Y}$ are the same as the principal axes for $\mathfrak{Z}$ and that the principal values of $\mathfrak{Y}$ are

$$Y_m^\circ = 1/Z_m^\circ$$

The dyadic $\mathfrak{Y} = \mathfrak{Z}^{-1}$ is called the *admittance dyadic* for the system.

When the determinant Δ_z is zero, we naturally cannot compute $\mathfrak{Y}$. This occurs whenever the angular velocity ω has a value which makes one of the principal impedances Z_m° zero. A reference to the functional dependence of $\mathfrak{Z}$ on ω indicates that the determinant

$$(-i\omega)^n \Delta_z = |\omega^2 a_{mr} - b_{mr}| = (-i\omega)^n Z_1^\circ(\omega) Z_2^\circ(\omega) \cdot \cdot \cdot Z_n^\circ(\omega)$$

is an nth order polynomial in ω^2, which goes to zero for n different values of ω^2 (though two or more of these roots may be equal). Corresponding to the root of lowest value (which we shall call ω_1^2) one of the Z_r°'s becomes zero, and so on, for $\omega_2 \cdot \cdot \cdot \omega_n$. Since we have been arbitrary about the numbering of the principal axes, we may as well choose it so that $Z_r^\circ(\omega)$ goes to zero when $\omega = \pm\omega_r$. In other words Z_1° goes to zero for the lowest root ω_1, Z_2° goes to zero for the next root ω_2, and so on. Furthermore, an application of the theory of equations indicates that, with these definitions, we can write the principal impedances in the form

$$Z_r^\circ(\omega) = i\omega M_r - (i/\omega)K_r = (i/\omega)M_r(\omega^2 - \omega_r^2) = 1/Y_r^\circ(\omega) \quad (3.2.10)$$

where M_r and $K_r = \omega_r^2 M_r$ are constants determined by the values of a_{mr} and b_{mr}. We therefore see that the principal impedances for a conservative system are pure imaginary quantities which are odd functions of ω, that is, $Z(-\omega) = -Z(\omega)$.

When $\omega = \pm\omega_r$, the amplitude of the rth normal coordinate becomes infinite unless $F_r^\circ = 0$, so a steady-state solution cannot be obtained. These frequencies $\omega_r/2\pi$, roots of the determinant Δ_z, are called the *resonance frequencies* for the system.

Incidentally, we should note that the constants M_r and K_r and ω_r^2 are all positive, for otherwise the potential energy would not have an absolute minimum at $q = 0$.

Canonical Transformations. The general form of Eqs. (3.2.7) is of suggestive simplicity. In the first place the second-order Lagrange equations (3.2.4) are replaced with pairs of first-order equations, which are to be solved simultaneously to determine p and q. This separation of the description of the motion into two sets of independent variables, the p's and the q's, corresponds to the fundamental peculiarity of classical dynamics: that the *acceleration*, the second derivative, is proportional to the force and therefore that *both* the initial position and initial velocity can be chosen arbitrarily. The q's are the generalized components of position, and the p's are related to the corresponding velocities in such a way as to make the interrelations come out in symmetric form.

The fundamental equations, which at the same time relate the p's and the q's for a given system and also determine the behavior of the system, are the canonical equations (3.2.7). Choice of the Hamiltonian H, a function of the p's and q's, determines the system. A set of pairs of variables q and p, related according to Eqs. (3.2.7) by a given Hamiltonian H, are called *canonically conjugate variables* for the Hamiltonian H (or, more, simply, *conjugate variables*).

The same system can, of course, be described in terms of different coordinates (and conjugate momenta). Just as the study of fields was clarified by investigating the effects of coordinate transformations on field components, so here it will be illuminating to discuss the effect of transformations on conjugate variables p and q. We can take the long way around by transforming the q's into new coordinates Q, which also are capable of describing the configuration of the system under study, by expressing the Lagrange function L in terms of the Q's and the $\dot{Q}$'s, by finding the conjugate momenta P, by the equations $P_r = -(\partial L/\partial \dot{Q}_r)$, and finally by obtaining a new Hamiltonian $K = \Sigma P\dot{Q} - L$ in terms of the new conjugate variables P and Q. Or else we can develop new techniques of simultaneously transforming from the conjugate pairs p, q

to the new pairs P, Q by a transformation which keeps the canonical equations (3.2.7) invariant in form. Such transformations are called *canonical transformations.*

Canonical transformations are related to a family of transformations, called by mathematicians *contact transformations,* which transform *line elements* (*i.e., position and direction*) rather than points. Since we wish to transform both position (the q's) and momentum (the p's, related to the direction of the motion of the system), the connection is obvious. The basis of a contact transformation is a function S of both old and new coordinates.

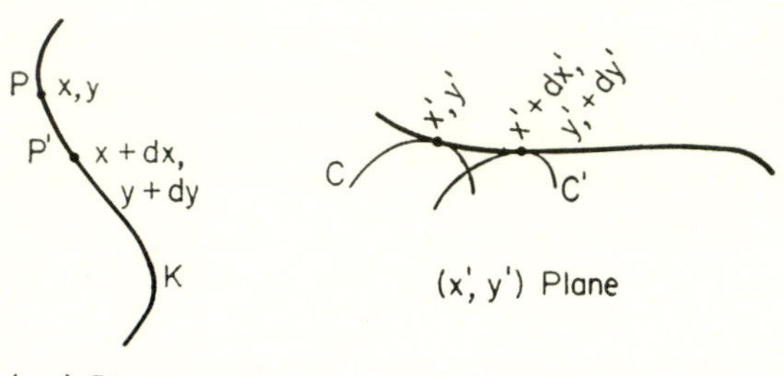

Fig. 3.2 Contact transformation in two dimensions.

As an example we consider a two-dimensional case shown in Fig. 3.2 where S is a function of x, y and of x', y'. Corresponding to each point P in (x,y) space (*i.e.*, to fixed values of x and y), the equation

$$S(x,y;x',y') = \text{constant}$$

defines a curve C in (x',y') space; and, vice versa, to every point on the (x',y') plane there is a related curve on the (x,y) plane. If we move the point in the (x,y) plane so it traces out a curve K, the related sequence of curves in the (x',y') plane may have an *envelope curve* E, which can be said to be the related curve traced out on the (x',y') plane. To each point in (x,y) there is a related curve in (x',y'), but to each curve in (x,y) there is a corresponding envelope curve in (x',y'). Therefore to each line element (*i.e.*, position plus direction) in (x,y) there is a corresponding line element in (x',y').

The correspondence between line elements can be shown by taking the two points (x,y) and $(x + dx, y + dy)$ defining a line element in (x,y) and working out the related line element in (x',y'). The curves in (x',y') are given by the two equations

$$S(x,y;x',y') = C;$$

$$S(x + dx,\, y + dy;\, x',y') = S(x,y;x',y') + \frac{\partial S}{\partial x}\, dx + \frac{\partial S}{\partial y}\, dy = C$$

If we set $dx = \dot{x}\,ds$ and $dy = \dot{y}\,ds$, where $\dot{x}/\dot{y}$ is the slope of the line element, then we arrive at two simultaneous equations:

$$S(x,y;x',y') = C; \quad \dot{x}\frac{\partial S}{\partial x} + \dot{y}\frac{\partial S}{\partial y} = 0$$

which we can solve to obtain the point (x',y') corresponding to the point (x,y). The direction of the envelope curve in (x',y') is obtained by differentiating the first equation with respect to the primed coordinates,

$$dx'\left(\frac{\partial S}{\partial x'}\right) + dy'\left(\frac{\partial S}{\partial y'}\right) = 0 \quad \text{or} \quad \dot{x}'\left(\frac{\partial S}{\partial x'}\right) + \dot{y}'\left(\frac{\partial S}{\partial y'}\right) = 0$$

if $dx' = \dot{x}'\,ds$ and $dy' = \dot{y}'\,ds$. The symmetry of the equations in the derivatives ensures that the transformation is symmetrical with respect to the two planes.

A simple example might make this clearer. Suppose the transformation function is $S = (x - x')^2 + (y - y')^2$ and the constant C is R^2. The point $x = a$, $y = b$ corresponds to the circle in the (x',y') plane of radius R and center (a,b). If the line element from (a,b) to $(a + dx, b)$ is used, the two equations to solve are (in this case $\dot{x} = 1$, $\dot{y} = 0$)

$$(x' - a)^2 + (y' - b)^2 = R^2; \quad 2(x' - a) = 0$$

The envelope of circles of radius R drawn with centers on the horizontal line $y = b$ is the horizontal line $y = b + R$ (or else the line $y = b - R$). Therefore the transformed line element goes from $(a, b \pm R)$ to $(a + dx, b \pm R)$.

Incidentally, this example indicates the close relationship between contact transformations and Huygen's principle.

For a dynamical system with a Hamiltonian function H which does not depend specifically on time, it turns out, as we shall show shortly, that the quantity $\Sigma p\,dq - \Sigma P\,dQ$, when expressed as a function of the p's and q's (or of the q's and Q's, etc.), is a complete differential whenever the transformation from the p's and q's to the P's and Q's is a canonical transformation. For then the transformation function S can be obtained by integrating the quantity

$$\Sigma p_r\,dq_r - \Sigma P_r\,dQ_r = dS$$

The function S can be expressed in terms of the q's and the Q's and is then the function defining the contact transformation.

Such a transformation leaves Hamilton's variational principle unchanged and therefore leaves the canonical equations (3.2.7) unchanged in form; for adding the quantity $(K - H)\,dt$ (which equals zero for K is the new Hamiltonian) to the equation for dS and integrating with respect to time, we have

$$\int_{t_0}^{t_1} \left[\sum p\dot{q} - H \right] dt = \int_{t_0}^{t_1} \left[\sum P\dot{Q} - K \right] dt + \int_0^1 dS$$

If we do not change the end points t_0 and t_1, the integral for S is unchanged when the intermediate path is varied, and so if $\delta\int[\Sigma p\dot{q} - H]\,dt$ is zero, then $\delta\int[\Sigma P\dot{Q} - K]\,dt$ is likewise zero.

Therefore the transformation defined by the function S, so obtained, is a canonical transformation and the P's, Q's, and K are related by the equations

$$\dot{Q}_r = \partial K/\partial P_r; \quad \dot{P}_r = -(\partial K/\partial Q_r)$$

which are the canonical equations (3.2.7).

Even when H (and therefore K) depends explicitly on the time, the transformation function S can be obtained by integrating the equation

$$\Sigma p_r\dot{q}_r - H - \Sigma P_r\dot{Q}_r + K = dS/dt$$

or the equation in its differential form

$$\Sigma p_r\,dq_r - \Sigma P_r\,dQ_r + (K - H)\,dt = dS$$

where dS is a complete differential. Since in any case

$$dS = \sum_{r=1}^{n} \left[\left(\frac{\partial S}{\partial q_r}\right) dq_r + \left(\frac{\partial S}{\partial Q_r}\right) dQ_r + \frac{\partial S}{\partial t}\,dt \right]$$

we have, by equating coefficients of the differentials,

$$p_r = \partial S/\partial q_r; \quad P_r = -(\partial S/\partial Q_r); \quad K - H = \partial S/\partial t \qquad (3.2.11)$$

which gives us the expressions for the momenta, in terms of the coordinates for the transformation.

Poisson Brackets. The study of the invariants for canonical transforms covers most of the basic aspects of classical dynamics. The energy function H is one such invariant (unless H depends explicitly on t). One whole class of invariants can be expressed most conveniently in terms of *Poisson bracket* expressions. These brackets, for two arbitrary functions u and v of the p's and q's, are defined as follows:

$$(u,v) = \sum_{r=1}^{n} \left[\frac{\partial u}{\partial p_r}\frac{\partial v}{\partial q_r} - \frac{\partial u}{\partial q_r}\frac{\partial v}{\partial p_r} \right]$$

as was given in Eq. (2.6.4).

The Poisson brackets have several interesting algebraic properties which are formally related to the properties of ordinary derivatives:

$$(u,c) = 0$$

where c is a constant independent of the p's and q's.

$$(u,v + w) = (u,v) + (u,w); \quad (u + v,w) = (u,w) + (v,w);$$
$$(uv,w) = u(v,w) + v(u,w); \quad \text{etc.}$$

The brackets are, of course, antisymmetric, so that $(u,v) = -(v,u)$. The reason the expressions are useful is that they are invariant with respect to a canonical transformation. If the set q, p are related to the set Q, P by the relation that $\Sigma p\, dq - \Sigma P\, dQ$ is a complete differential, then, for any pair of functions u, v of p, q or of P, Q,

$$(u,v) = -\sum_{r=1}^{n}\left[\frac{\partial u}{\partial q_r}\frac{\partial v}{\partial p_r} - \frac{\partial u}{\partial p_r}\frac{\partial v}{\partial q_r}\right] = -\sum_{r=1}^{n}\left[\frac{\partial u}{\partial Q_r}\frac{\partial v}{\partial P_r} - \frac{\partial u}{\partial P_r}\frac{\partial v}{\partial Q_r}\right]$$

Therefore dynamical equations set up in terms of Poisson brackets are invariant under canonical transformation. The canonical equations of motion (3.2.7), for instance, become

$$\dot{q}_r = (H,q_r); \quad \dot{p}_r = (H,p_r)$$

In fact, by using the canonical equations plus the definition of the Poisson brackets, we can show that for any function u of the q's and p's (or of the Q's and P's)

$$du/dt = (H,u); \quad \partial u/\partial q_r = (p_r,u); \quad \partial u/\partial p_r = (u,q_r)$$

We can also use the Poisson brackets to test for a contact transformation. Starting with the original set of n coordinates q and the conjugate momenta, another set of n coordinates Q and conjugate momenta are related to the original set by a contact transformation if, and only if, the following relations are satisfied:

$$(Q_r,Q_s) = 0; \quad (P_r,P_s) = 0; \quad (P_r,Q_s) = \delta_{rs}$$

where $\delta_{rs} = 0$ if $r \neq s$, $= 1$ if $r = s$.

The Action Integral. We note that the transformation function S has the dimensions of action and that, when the Q's are held constant and S is considered as a function of the q's, then the vector p, being the gradient of S, is normal to the surface S = constant. In other words, a choice of values for the Q's and K picks out a family of action surfaces S = constant and a family of trajectories for the system which are orthogonal to the action surfaces. If we pick new values of the Q's and K, we get new families of surfaces and trajectories. From one point of view the Q's can be considered as initial conditions, and the subsequent behavior of the system can be considered to be the unfolding of a contact transformation as time marches on.

A differential equation for the action function S, as function of initial and final coordinates, can be obtained by setting up the equation stating that the Hamiltonian is constant, $H(p,q) = E$, and substituting in for each p_r the expression $\partial S/\partial q_r$, from Eqs. (3.2.11). This results in a differential equation

$$H\left(\frac{\partial S}{\partial q}, q\right) = E \tag{3.2.12}$$

which is called the *Hamilton-Jacobi* equation. Its solution is a function of the n q's, of E, and of n constants of integration, which we can call $Q_2, Q_1, \ldots, Q_n, \alpha$ (we can always juggle the constants around so that α is simply an additive constant). If we let $E = Q_1$, then we can consider the other Q's to be a new set of coordinates for the system. According to Eqs. (3.2.11) the conjugate momenta are $P_r = -(\partial S/\partial Q_r)$, and the transformed coordinates and momenta satisfy the canonical equations

$$\dot{P}_1 = -\partial H/\partial Q_1 = -\partial H/\partial E = -1; \quad \dot{P}_r = -\partial H/\partial Q_r = 0; \quad r = 2, 3, \ldots, n$$

for in this case $K = H$, and H is independent of the initial condition constants Q_r. Therefore the equations of motion can be written

$$\partial S/\partial E = t + c_1; \quad \partial S/\partial Q_r = c_r; \quad r = 2, 3, \ldots, n$$

where the c's are another set of constants (corresponding, along with the Q's, to the initial conditions). It should be noted that P_1 is related to the quantity q_t of page 252.

The relations between the Hamilton-Jacobi equation and Schroedinger's equation (2.6.28), $H[(\hbar/i)(\partial/\partial q),q]\psi = E\psi$, for the quantum mechanical wave function ψ, are important and interesting but cannot be expanded on here. We need point out only that, if we set $\psi = e^{(i/\hbar)S}$, the Schroedinger equation reduces to the Hamilton-Jacobi, if S is so much larger than $\hbar$ that we can neglect $(i/\hbar)(\partial^2 S/\partial q^2)$ compared with $(i/\hbar)^2(\partial S/\partial q)^2$. In the limit of large values of action and energy, the surfaces of constant phase for the wave function ψ become the surfaces of constant action S for the corresponding classical system. Wave mechanics goes over to geometrical mechanics, just as wave optics goes over to geometrical optics for vanishingly small wavelengths.

We have displayed here a large part of classical dynamical theory, undiluted with any examples. Before we go on to the real business of this chapter, the application of the variational principle to fields, we had better consolidate our position thus far by discussing a few examples.

The Two-dimensional Oscillator. A useful example is that of the motion of a mass on the end of a stiff rod which can bend in both directions. For small oscillations the motion is in a plane, and the coordinates q_1, q_2 for the mass can be taken to be the principal axes for the spring, so that the force in the q_1 direction is proportional to q_1 only, and similarly with q_2. The kinetic energy of the mass is then $\frac{1}{2}m(\dot{q}_1^2 + \dot{q}_2^2)$. If the spring is equally stiff in both directions, the potential energy can be written $\frac{1}{2}m\omega^2(q_1^2 + q_2^2)$. The Lagrange equations (3.2.4) are just the Newton's equations

$$\ddot{q}_1 = -\omega^2 q_1; \quad \ddot{q}_2 = -\omega^2 q_2$$

The solution is, therefore, that q_1 and q_2 vary sinusoidally with time, with a frequency $\omega/2\pi$.

The related momenta are, of course, $p_1 = m\dot{q}_1$, $p_2 = m\dot{q}_2$, so that the Hamiltonian is

$$H(p,q) = (1/2m)[p_1^2 + p_2^2 + m^2\omega^2(q_1^2 + q_2^2)] \tag{3.2.13}$$

The solutions are already determined from the Lagrange equations, but we are now seeking to illustrate some of the concepts and quantities discussed above, so we shall try solving the problem by the use of a contact transformation. It would be handy if the new momenta were constant, and this could be accomplished by letting the new Hamiltonian K be independent of Q_1 and Q_2.

The easiest way is to let K be proportional to $(P_1 + P_2)$, for then $\dot{P} = -(\partial K/\partial Q) = 0$. Since $\dot{Q} = \partial K/\partial P$, the Q's will be proportional to time t. This suggests the following transformation:

$$q = A\sin(\omega t); \quad p = m\omega A\cos(\omega t); \quad Q \propto \omega t \quad \text{and} \quad P \propto A^2$$

or

$$Q_1 = \tan^{-1}(m\omega q_1/p_1); \quad Q_2 = \tan^{-1}(m\omega q_2/p_2)$$
$$P_1 = (1/2m\omega)(p_1^2 + m^2\omega^2 q_1^2); \quad P_2 = (1/2m\omega)(p_2^2 + m^2\omega^2 q_2^2)$$

The quantity $p_1\,dq_1 + p_2\,dq_2 - P_1\,dQ_1 - P_2\,dQ_2$, expressed in terms of the p's and q's, turns out to be $\frac{1}{2}(p_1\,dq_1 + q_1\,dp_1 + p_2\,dq_2 + q_2\,dq_2)$, which is a complete differential of the quantity

$$S = \tfrac{1}{2}(p_1q_1 + p_2q_2) = \tfrac{1}{2}m\omega[q_1^2\cot(Q_1) + q_2^2\cot(Q_2)] \tag{3.2.14}$$

which is the transformation function. Therefore this is a contact transformation. The inverse equations

$$\begin{aligned} q_1 &= \sqrt{2/m\omega}\,\sqrt{P_1}\sin Q_1; \quad q_2 = \sqrt{2/m\omega}\,\sqrt{P_2}\sin Q_2 \\ p_1 &= \sqrt{2m\omega}\,\sqrt{P_1}\cos Q_1; \quad p_2 = \sqrt{2m\omega}\,\sqrt{P_2}\cos Q_2 \end{aligned} \tag{3.2.15}$$

enable us to see that the new Hamiltonian is

$$K = \omega(P_1 + P_2)$$

Since this is a contact transformation, Hamilton's equations (3.2.7) still hold, and since $\dot{Q} = \partial K/\partial P$, we have

$$Q_1 = \omega t + \varphi_1; \quad Q_2 = \omega t + \varphi_2$$

Also, since $\dot{P} = -(\partial K/\partial Q) = 0$, we have that P_1 and P_2 are constant.

This completes the solution, for we can substitute these simple expressions back in Eqs. (3.2.15) to obtain formulas for the coordinates and momenta in terms of time and the energy of motion K.

We notice that the P's have the dimensions of action and the Q's are angles. In fact, if we integrate $p\,dq$ over a whole cycle of the oscillation and express this in terms of P and Q, the result is

$$\int p\,dq = 2P \int_0^{2\pi} \cos^2 Q\,dQ = 2\pi P$$

proportional to P itself. These canonically conjugate variables Q and P are called *angle* and *action variables.* All vibrational problems with sinusoidal solutions can be simplified and solved by performing a contact transformation to the appropriate action and angle variables.

But to return to the harmonic oscillator in two dimensions, we can also express the motion in terms of the polar coordinates r, φ instead of the rectangular coordinates q_1 and q_2. The contact transformation is given by the equations

$$\begin{gathered} r = \sqrt{q_1^2 + q_2^2}; \quad \varphi = \tan^{-1}(q_2/q_1) \\ p_r = (1/r)(p_1q_1 + p_2q_2); \quad p_\varphi = (p_2q_1 - p_1q_2) \end{gathered} \tag{3.2.16}$$

and the Hamiltonian is

$$K = (1/2m)[p_r^2 + (p_\varphi^2/r^2) + m^2\omega^2r^2] \tag{3.2.17}$$

Since $\partial K/\partial\varphi$ is zero, p_φ, the *angular momentum* of the system, is a constant. It is not difficult to see that this is true no matter what the potential energy is as long as it depends on r alone and not on φ. The rest of the solution, if we wish it, can be obtained from the solution in rectangular coordinates.

Charged Particle in Electromagnetic Field. There are times when it is not obvious what form the Lagrange function $L = T - V$ should have, for substitution in the variational integral (3.2.3). This is particularly true when forces due to fields enter. In many such cases we must combine our knowledge of scalar invariants of the system with investigations of simple, limiting cases to arrive at the correct answer.

For instance, in the case of a charged particle in an electromagnetic field, should we call the energy of interaction between the magnetic field and the motion of the particle part of the kinetic energy T (since it depends on the velocity of the particle) or part of the potential energy V (since it is due to an externally produced field)? We start by putting down all the scalar invariants (in three space) of the particle and of the field. The Lagrange function $L = T - V$ must be an invariant, for Hamilton's principle must hold in any space coordinate system. The kinetic energy of the particle alone, $\frac{1}{2}mv^2$, is such an invariant, for it is proportional to the dot product of the vector $\mathbf{v}$ with itself. The electric potential φ is also an invariant (in three space). So are the squares of the magnitudes of the fields, E^2 and H^2, and the square of the vector potential A^2. However, the fields are obtained from the potentials by differentiation and the forces on the particle are obtained from the Lagrange function by differentiation [see Eqs. (3.2.2) and (3.2.4)], so it would be natural to expect only the potentials $\mathbf{A}$ and φ to appear in the L for the particle. Another invariant which might enter is the dot product $\mathbf{v} \cdot \mathbf{A}$.

The forces on the particle are $e\mathbf{E} = -e \text{ grad } \varphi - (e/c)(\partial\mathbf{A}/\partial t)$ [see Eq. (2.5.13)] and $(e/c)\mathbf{v} \times \mathbf{B} = (e/c)\mathbf{v} \times \text{curl } \mathbf{A}$ [see (Eq. 2.5.5)], and these must come from the Lagrange equations (3.2.4) by differentiation of L. Since there is to be a time derivative of $\mathbf{A}$, we must have a term involving the product of $\mathbf{v}$ and $\mathbf{A}$, presumably the $\mathbf{v} \cdot \mathbf{A}$ term. In considering the equation for the particle only (we are not yet considering the equations for the fields), we have no term to the second order in $\mathbf{A}$. Therefore L for the particle must be a combination of v^2, $\mathbf{v} \cdot \mathbf{A}$, and φ.

The first term must obviously be the kinetic energy of the particle, $\frac{1}{2}mv^2$. The term in φ must be a potential energy, and if the charge on the particle is e, this term must be $-e\varphi$. The third term must give the term $-(e/c)(\partial\mathbf{A}/\partial t)$, for the rest of $e\mathbf{E}$, and also the term $(e/c)\mathbf{v} \times \text{curl } \mathbf{A}$. Since $\text{grad }(\mathbf{v} \cdot \mathbf{A}) = \mathbf{v} \times \text{curl } \mathbf{A} + \mathbf{v} \cdot (\nabla\mathbf{A})$ (see page 115), it appears that the term should be $(e/c)\mathbf{v} \cdot \mathbf{A}$. Therefore the Lagrange function for the charged particle in an electromagnetic field is

$$L = \tfrac{1}{2}mv^2 + (e/c)\mathbf{v} \cdot \mathbf{A} - e\varphi \tag{3.2.18}$$

Remembering that the coordinates in $\mathbf{A}$ and φ are the coordinates x, y, z of the particle at time t, we can express the three equations (3.2.4) for the three coordinates in terms of a single vector equation. Since

$$(\partial/\partial v_x)L = mv_x + (e/c)A_x = p_x$$

the vector equation is

$$\frac{d}{dt}\left(m\mathbf{v} + \frac{e}{c}\mathbf{A}\right) = \text{grad } L = -e \text{ grad } \varphi + \frac{e}{c}\mathbf{v} \times \text{curl } \mathbf{A} + \frac{e}{c}\mathbf{v} \cdot (\nabla\mathbf{A})$$

The term entering into the expression for $\mathbf{E}$ [Eq. (2.5.13)] has the partial time derivative $\partial\mathbf{A}/\partial t$, representing the change of $\mathbf{A}$ at a fixed *point in space*, whereas the quantity on the left-hand side is the total derivative $d\mathbf{A}/dt$, representing the total rate of change of $\mathbf{A}$ at the particle, which is moving. From Eq. (2.3.2) we have that the change in a field at a point moving with velocity $\mathbf{v}$ is

$$\frac{d\mathbf{A}}{dt} = \frac{\partial\mathbf{A}}{\partial t} + \mathbf{v} \cdot (\nabla\mathbf{A})$$

Therefore, the vector equation of motion of the particle reduces to

$$\frac{d}{dt}(m\mathbf{v}) = -e \text{ grad } \varphi - \frac{e}{c}\frac{\partial\mathbf{A}}{\partial t} + \frac{e}{c}\mathbf{v} \times \text{curl } \mathbf{A} = e\mathbf{E} + \frac{e}{c}\mathbf{v} \times \mathbf{H} \tag{3.2.19}$$

which corresponds to Eq. (2.5.12) for the effective force on the charged particle.

We are now in a position to set up the Hamiltonian for the particle. The "momentum" of the particle is the vector with x component $\partial L/\partial v_x$:

$$\mathbf{p} = m\mathbf{v} + (e/c)\mathbf{A}$$

In this case the action of the field is to change continuously the particle velocity, so that $m\mathbf{v}$ is no longer a quantity which is "conserved." If we are to have conservation of momentum, $\mathbf{p}$ cannot equal $m\mathbf{v}$. According to Eq. (3.2.6), the Hamiltonian is

$$H = \mathbf{p}\cdot\mathbf{v} - L = \left(m\mathbf{v} + \frac{e}{c}\mathbf{A}\right)\cdot\mathbf{v} - \tfrac{1}{2}mv^2 - \frac{e}{c}\mathbf{v}\cdot\mathbf{A} + e\varphi$$

$$= \frac{1}{2m}\left(\mathbf{p} - \frac{e}{c}\mathbf{A}\right)^2 + e\varphi \qquad (3.2.20)$$

This is the result quoted on pages 256 and 254.

This gives us a foretaste of the consequences of the introduction of fields and field interactions into the formalism of classical dynamics. The momenta are no longer simply proportional to the velocities, it is no longer quite so obvious what is kinetic energy or H or L, and we must depend more on the formal equations, such as Eqs. (3.2.4), (3.2.6), and (3.2.7), instead of an "intuition," to reach the correct results.

As a specific example we might consider the case of a particle of mass m and charge e in a constant magnetic field of magnitude $B = mc\omega/e$ in the z direction. The scalar potential is $\varphi = 0$, and the vector potential is

$$\mathbf{A} = (mc\omega/2e)(-y\mathbf{i} + x\mathbf{j})$$

The Lagrange function is

$$L = \tfrac{1}{2}m(\dot{x}^2 + \dot{y}^2) + \tfrac{1}{2}m\omega(-\dot{x}y + \dot{y}x); \quad \dot{x} = dx/dt; \quad \text{etc.}$$

and the two momenta are

$$p_x = m(\dot{x} - \tfrac{1}{2}\omega y); \quad p_y = m(\dot{y} + \tfrac{1}{2}\omega x)$$

The Lagrange equations are

$$\frac{d^2x}{dt^2} = \omega\frac{dy}{dt}; \quad \frac{d^2y}{dt^2} = -\omega\frac{dx}{dt}$$

and the solutions turn out to be

$$x = R\sin(\omega t + \alpha) + x_0; \quad y = R\cos(\omega t + \alpha) + y_0$$

representing a circular orbit of radius R with center at x_0, y_0. The Hamiltonian is, of course,

$$H = \frac{1}{2m}(p_x + \tfrac{1}{2}m\omega y)^2 + \frac{1}{2m}(p_y - \tfrac{1}{2}m\omega x)^2 = \tfrac{1}{2}m\omega^2R^2 = \tfrac{1}{2}mv^2$$

We note that the radius of the orbit is equal to v/ω, where $\omega = e/mc$ times the magnitude of the magnetic field B.

A contact transformation which simplifies the Hamiltonian considerably is

$$x = \sqrt{1/m\omega}\,[\sqrt{2P_1}\sin Q_1 + P_2]; \quad p_x = \tfrac{1}{2}\sqrt{m\omega}\,[\sqrt{2P_1}\cos Q_1 - Q_2]$$
$$y = \sqrt{1/m\omega}\,[\sqrt{2P_1}\cos Q_1 + Q_2]; \quad p_y = \tfrac{1}{2}\sqrt{m\omega}\,[-\sqrt{2P_1}\sin Q_1 + P_2]$$

Use of the Poisson bracket equations (2.6.4) in reverse (using the Q's and P's as independent variables) will confirm that this is a contact transformation. Substituting in for H, we find that the transformed Hamiltonian is just

$$K = \omega P_1$$

Therefore using Eqs. (3.2.8), we see that P_1, P_2, and Q_2 are constants and Q_1 is linearly dependent on time, with the proportionality constant equal to ω, the angular velocity of the particle in its circular orbit.

Relativistic Particle. Another example of the use of the classical dynamical equations, which will be needed later in the chapter, involves the behavior of a particle moving so fast that we can no longer neglect v^2 compared with c^2. We have pointed out earlier that L is not a Lorentz invariant. It should not be, for the variational integral $\int L\,dt$, integrated along the world line of the particle, should be the invariant. If the particle is moving with speed u with respect to the observer, the differential dt of the observer's time is related to the proper time $d\tau$ for the world line of the particle by the relation $d\tau = \sqrt{1 - (u/c)^2}\,dt$. Consequently if the integral $\int L\,dt = \int [L/\sqrt{1 - (u/c)^2}]\,d\tau$ is invariant and $d\tau$ is invariant, then the integrand $[L/\sqrt{1 - (u/c)^2}]$ is invariant and L must be some Lorentz invariant multiplied by $\sqrt{1 - (u/c)^2}$.

For instance, for a force-free particle the relativistic Lagrange function is

$$L = -m_0c^2\sqrt{1 - (u/c)^2} \simeq -m_0c^2 + \tfrac{1}{2}m_0u^2 \cdots; \quad u \ll c \qquad (3.2.21)$$

The term corresponds to the kinetic energy term minus the rest energy m_0c^2; if there were a potential energy V, it would be subtracted from the term shown here. The momentum is then obtained by differentiating L with respect to the components of u:

$$p = \frac{m_0 u}{\sqrt{1 - (u/c)^2}}$$

which corresponds to Eq. (1.7.5).

The Hamiltonian function is

$$H = pu - L = \frac{m_0c^2}{\sqrt{1 - (u/c)^2}} = m_0c^2\sqrt{1 + (p/m_0c)^2}$$
$$\simeq m_0c^2 + \left(\frac{1}{2m_0}\right)p^2 + \cdots; \quad p \ll m_0c \qquad (3.2.22)$$

which is the expression used on pages 256 and 260, in connection with the Dirac equation for the electron. This expression is, of course, the time component of a four-vector, the space components of which are the components of $c\mathbf{p}$. Any potential energy term which is to be added should thus also be the time component of a four-vector.

Dissipative Systems. Finally, before we turn to the application of Hamilton's principle to fields, we shall introduce a formalism which will enable us to carry on calculations for dissipative systems (*i.e.*, ones with friction nonnegligible) as though they were conservative systems (*i.e.*, ones with negligible friction). The dodge is to consider, simultaneously with the system having the usual friction, a "mirror-image" system with *negative* friction, into which the energy goes which is drained from the dissipative system. In this way the total energy is conserved, and we can have an invariant Lagrange function, at the sacrifice of a certain amount of "reality " in some of the incidental results.

For an example of what we mean, let us take the one-dimensional oscillator with friction, having the equation of motion

$$m\ddot{x} + R\dot{x} + Kx = 0 \tag{3.2.23}$$

We wish to obtain this equation from some Lagrange function by the usual variational technique. In order to do this we set up the purely formal expression

$$L = m(\dot{x}\dot{x}^*) - \tfrac{1}{2}R(x^*\dot{x} - x\dot{x}^*) - Kxx^* \tag{3.2.24}$$

This is to be considered as the Lagrange function for two coordinates, x and x^*. The coordinate x^* represents the "mirror-image" oscillator with negative friction. Applying our formalism, we obtain for the two "momenta"

$$p = m\dot{x}^* - \tfrac{1}{2}Rx^*; \quad p^* = m\dot{x} + \tfrac{1}{2}Rx$$

which have little to do with the actual momentum of the oscillator. Nevertheless a continuation of the formal machinery results in two Lagrange equations for the two systems:

$$m\ddot{x}^* - R\dot{x}^* + Kx^* = 0; \quad m\ddot{x} + R\dot{x} + Kx = 0$$

The equation for x is just Eq. (3.2.23), which we started out to get. The equation for x^* involves a negative frictional term, as we mentioned above.

The Hamiltonian is

$$\begin{aligned} H = p\dot{x} + p^*\dot{x}^* - L &= m\dot{x}\dot{x}^* + Kxx^* \\ &= (1/m)(p + \tfrac{1}{2}Rx^*)(p^* - \tfrac{1}{2}Rx) + Kxx^* \end{aligned} \tag{3.2.25}$$

Since x^* increases in amplitude as fast as x decreases, then H will stay constant.

By this arbitrary trick we are able to handle dissipative systems as though they were conservative. This is not very satisfactory if an alternate method of solution is known, but it will be necessary, in order to make any progress when we come to study dissipative fields, as in the diffusion equation. As an indication of the fact that such cases are far from typical we note that we had previously assumed that L was a quadratic function of the $\dot{q}$'s, whereas in the present case terms in $\dot{x}^*\dot{x}$ occur.

Impedance and Admittance for Dissipative Systems. At this point it is worth while to take up again the discussion of mechanical impedance from the point where we left it on page 286, to discuss the effect of resistive forces. For a system with dynamic or static equilibrium, as we have seen, we can transform coordinates to the set $x_1, x_2, \ldots, x_n$ which go to zero at the point of equilibrium. For displacements sufficiently near to this equilibrium the potential energy will be

$$V = \tfrac{1}{2}\sum_{m,r} b_{mr}x_m x_r + V_0 = \tfrac{1}{2}\mathbf{x}\cdot\mathfrak{B}\cdot\mathbf{x} + V_0$$

where $\mathfrak{B}$ is the dyadic with elements b_{mr} and $\mathbf{x}$ is the n-dimensional vector with components x_r. The kinetic energy may, as always, be given by the expression

$$T = \tfrac{1}{2}\sum_{m,r} a_{mr}\dot{x}_m\dot{x}_r = \tfrac{1}{2}\dot{\mathbf{x}}\cdot\mathfrak{A}\cdot\dot{\mathbf{x}}$$

where $\mathfrak{A}$ is the dyadic with elements a_{mr}.

In case there is friction the retarding force on the mth coordinate can be written in the form

$$\sum_r r_{mr}\dot{x}_r = (\mathfrak{R}\cdot\dot{\mathbf{x}})_m$$

where $\mathfrak{R}$ is the resistance dyadic, with components r_{mr}. Here, as with the potential and kinetic energies, we have included terms corresponding to the coupling between different displacements x_m, represented by the nondiagonal terms of $\mathfrak{R}$.

Following the pattern of Eq. (3.2.24) we write the Lagrange function

$$\begin{aligned} L &= \sum_{m,r}[a_{mr}\dot{x}_m^*\dot{x}_r - \tfrac{1}{2}r_{mr}(x_m^*\dot{x}_r - \dot{x}_m^*x_r) - b_{mr}x_m^*x_r] \\ &= \dot{\mathbf{x}}^*\cdot\mathfrak{A}\cdot\dot{\mathbf{x}} - \tfrac{1}{2}\mathbf{x}^*\cdot\mathfrak{R}\cdot\dot{\mathbf{x}} + \tfrac{1}{2}\dot{\mathbf{x}}^*\cdot\mathfrak{R}\cdot\mathbf{x} - \mathbf{x}^*\cdot\mathfrak{B}\cdot\mathbf{x} \qquad (3.2.26) \end{aligned}$$

where $\mathbf{x}^*$ is the vector conjugate to $\mathbf{x}$. The momentum vector and its conjugate are then

$$\mathbf{p} = \dot{\mathbf{x}}^*\cdot\mathfrak{A} - \tfrac{1}{2}\mathbf{x}^*\cdot\mathfrak{R};\quad \mathbf{p}^* = \mathfrak{A}\cdot\dot{\mathbf{x}} + \tfrac{1}{2}\mathfrak{R}\cdot\mathbf{x}$$

and the Hamiltonian is

$$H = \mathbf{p} \cdot \dot{\mathbf{x}} + \dot{\mathbf{x}}^* \cdot \mathbf{p}^* - L = \dot{\mathbf{x}}^* \cdot \mathfrak{A} \cdot \dot{\mathbf{x}} + \mathbf{x}^* \cdot \mathfrak{B} \cdot \mathbf{x}$$
$$= (\mathbf{p} + \tfrac{1}{2}\mathbf{x}^* \cdot \mathfrak{R}) \cdot (\mathfrak{A}^{-1}) \cdot (\mathbf{p}^* - \tfrac{1}{2}\mathfrak{R} \cdot \mathbf{x}) + \mathbf{x}^* \cdot \mathfrak{B} \cdot \mathbf{x} \quad (3.2.27)$$

where the dyadic $(\mathfrak{A}^{-1})$ is reciprocal to the dyadic $\mathfrak{A}$, such that $\mathfrak{A} \cdot \mathfrak{A}^{-1} = \mathfrak{A}^{-1} \cdot \mathfrak{A} = \mathfrak{J}$.

The generalized driving force vector, acting on the displacement x_m but not on the mirror-image displacements, is the one causing changes in the p^*'s. The Hamilton canonical equation for the component acting on the mth coordinate is

$$F_m = \dot{p}_m^* + (\partial H/\partial x_m^*)$$

or, in abstract vector space,

$$\mathbf{F} = \dot{\mathbf{p}}^* + \sum_{m=1}^{n} (\partial H/\partial x_m^*)\mathbf{e}_m = \dot{\mathbf{p}}^* + \tfrac{1}{2}\mathfrak{R} \cdot (\mathfrak{A}^{-1}) \cdot (\mathbf{p}^* - \tfrac{1}{2}\mathfrak{R} \cdot \mathbf{x}) + \mathfrak{B} \cdot \mathbf{x}$$
$$= \mathfrak{A} \cdot \ddot{\mathbf{x}} + \mathfrak{R} \cdot \dot{\mathbf{x}} + \mathfrak{B} \cdot \mathbf{x} \quad (3.2.28)$$

If now the driving force is oscillatory, $\mathbf{F} = \mathbf{F}°e^{i\omega t}$, each displacement (for steady-state motion) will be oscillatory with the same frequency with $\mathbf{x} = \mathbf{A}e^{i\omega t}$ (where a component of $\mathbf{A}$ is A_m, the amplitude of oscillation of the mth coordinate) and the velocity is $\dot{\mathbf{x}} = \mathbf{U}e^{i\omega t}$ (where the mth component of $\mathbf{U}$ is U_m). The relation between $\mathbf{F}°$ and $\mathbf{A}$ or $\mathbf{U}$ is

$$\mathbf{F} = \mathfrak{Z} \cdot \mathbf{U} = i\omega\mathfrak{Z} \cdot \mathbf{A}; \quad \text{where } Z_{mr} = i\omega a_{mr} + r_{mr} + (1/i\omega)b_{mr} \quad (3.2.29)$$

The impedance dyadic is now complex instead of pure imaginary. The real part of each term is called the *resistance*, and the imaginary part is called the *reactance*. The reciprocal dyadic $\mathfrak{Y} = \mathfrak{Z}^{-1}$ can be computed from the determinant $\Delta_z = |Z_{mr}|$:

$$Y_{mr} = Z'_{mr}/\Delta_z; \quad Z'_{mr} = \partial\Delta_z/\partial Z_{mr} \quad (3.2.30)$$

where Z'_{mr} is the first minor of Z_{mr} in the determinant Δ_z. This reciprocal is called the *admittance dyadic*, the real part of each term is called the *conductance*, and the imaginary part is called the *susceptance*.

Since $\mathfrak{Z}$ is a symmetric dyadic, it is possible to find the principal axes for it. In terms of these principal axes, or *normal coordinates*, the dyadic is diagonal, having diagonal elements Z_r, which are the principal *values* of the *impedance*. The determinant transforms into the product of these principal values, so that

$$(-i\omega)^n\Delta_z = |\omega^2 a_{mr} - i\omega r_{mr} - b_{mr}| = (-i\omega Z_1)(-i\omega Z_2) \cdots (-i\omega Z_n) \quad (3.2.31)$$

Since all the diagonal elements of $\mathfrak{A}$, $\mathfrak{R}$, and $\mathfrak{B}$ are positive, the determinant can be factored into n factors of the form $M\omega^2 - iR\omega - K$, where M, R, and K are all positive. The roots of these factors are $i(R/2M) \pm (1/2M)\sqrt{4KM - R^2}$, and the factors can be ordered in order of increasing real part of the roots. The factor $i\omega Z_1$ has the smallest value of $\sqrt{(K/M) - (R/2M)^2}$, and so on (unless one or more of the factors has roots which are pure imaginary, as when $R^2 > 4KM$, in which case these factors will be labeled in order of *decreasing* magnitude of the root). Therefore the rth principal value of the impedance has the form

$$Z_r = \frac{i}{\omega}[M_r\omega^2 - iR_r\omega - K_r] = \frac{iM_r}{\omega}[\omega^2 - 2ik_r\omega - \omega_r^2 - k_r^2]$$
$$= \frac{iM_r}{\omega}(\omega - ik_r - \omega_r)(\omega - ik_r + \omega_r) \quad (3.2.32)$$

where $k_r = (R_r/2M_r)$; $\omega_r = \sqrt{(K_r/M_r) - k_r^2}$ as long as K_r/M_r is larger than k_r^2. The constants M_r, R_r, K_r, k_r and ω_r all are determined by the values of a_{mr}, r_{mr}, and b_{mr}.

The dyadic $\mathfrak{Y}$ therefore becomes infinite for $2n$ different complex values of ω, n of them, $(\omega_r + ik_r)$, having both real and imaginary values positive and being ordered so that $\omega_{m-1} \leq \omega_m$; the other n of them, $(-\omega_n + ik_r)$, have the same imaginary parts but a real part having a reversed sign. In other words, if we plot the roots on the complex plane, they will all be above the real axis and the set for negative ω will be images of the set for positive ω with respect to the imaginary axis. (Unless k_r^2 is larger than K_r/M_r, in which case ω_r is imaginary and all such roots are on the imaginary axis, above the real axis.) These roots correspond to the free vibrations of the system; for the mth normal mode of oscillation the time dependence will be given by the term $e^{-k_rt+i\omega_rt}$ and the relative amplitudes of motion will be proportional to the components of the unit vector along the rth normal coordinate in abstract vector space. The time term shows that the free oscillations are damped out.

There are, of course, impedance and admittance dyadics for the conjugate coordinates, giving the relation between the forces $F_m^* = \dot{p}_m + (\partial H/\partial x_m)$ and the velocities $\dot{x}_m^*$. The elements of these dyadics turn out to differ from the elements of $\mathfrak{Z}$ and $\mathfrak{Y}$ simply by having negative resistance terms instead of positive. Put another way, the components of $[-i\omega\mathfrak{Z}^*]$ are the complex conjugates of the components of $[-i\omega\mathfrak{Z}]$, corresponding to the mirror-image relationship.

3.3 *Scalar Fields*

In the case of classical dynamics the problem is solved when we have obtained expressions for the coordinates of the system as functions of time; the variational integral from which the solution is obtained contains the Lagrange function, in terms of these coordinates and their time derivatives, integrated over time. The material fields discussed in the previous chapter (elastic displacement, diffusion density, fluid velocity potential, etc.) all represent "smoothed-out" averages of the behavior of some complex system containing many atoms. To solve such problems we can obtain the equations of motion of the particles and then average to obtain the equation for the field, as we did, in principle at least, in Chap. 2. Or we can average the Lagrange function for the whole system, before minimization, to obtain a variational integral for the *field*, which will be the approach used in this chapter.

In many cases the resulting field is a scalar function of time and of the coordinates, which are in this case parameters of integration only. Here the field is the quantity which is varied to find the minimum value of the integrated Lagrange function, and the Euler equations (3.1.3) turn out to be partial differential equations giving the dependence of the field on the coordinates and on time.

The Flexible String. The simple example of the flexible string under tension will illustrate several of these points and can serve as a guide in the analysis of more complex cases. We can start out with the Lagrange function for every one of the atoms in the string (of course we should start with the Schroedinger equation, but here we deal with gross motions of many millions of atoms, so that quantum effects are negligible and classical dynamics may be used).

The total kinetic energy is

$$T = \sum_{s=1}^{N} \tfrac{1}{2} m_s v_s^2$$

where we have assumed that there are N atoms present. The motion of each atom in an element of the string between x and $x + dx$ can be considered as the vector sum of the average motion of the element $\mathbf{j}(d\psi/dt)$ (we here assume for simplicity that the average motion is transverse, in one plane, and that the average displacement of point x on the string from the equilibrium line is ψ) and the fluctuating motion $\mathbf{w}_s$ of the individual atom away from this average. The total kinetic energy for the element dx is therefore

$$T = \tfrac{1}{2} \sum_{dx} m_s[(\dot{\psi})^2 + 2\dot{\psi}(\mathbf{j} \cdot \mathbf{w}_s) + w_s^2]; \quad \dot{\psi} = \frac{\partial \psi}{\partial t}$$

where the time average of the terms in $\mathbf{j} \cdot \mathbf{w}_s$ is zero. The sum is taken over all of the particles in the length dx of the string. We are not interested here in the fluctuating motions of the individual atoms, so we shall not bother about setting up the individual Lagrange equations for the coordinates corresponding to the velocities $\mathbf{w}_s$. Consequently the last term in the bracket will be dropped out, not because it is negligible (actually it represents the internal heat energy of the string and so is not small in the aggregate) but because such motion is not of interest to us here. The second term in the bracket can be neglected because the derivative of this with respect to $\dot{\psi}$ (which comes into the Lagrange equation for ψ) has a zero time average. The total kinetic energy which is of interest to us is therefore

$$T = \tfrac{1}{2}\rho \int (\dot{\psi})^2 \, dx \tag{3.3.1}$$

where $\rho \, dx$ is equal to the sum of all the masses m_s of all the particles in the element of length of the string between x and $x + dx$.

The potential energy of the string is a complex function of the coordinates of all the atoms in the string. It also can be broken up into a term representing the average increase in potential energy of the string when it is displaced by an amount $\psi(x)$ from equilibrium, plus some terms involving the individual displacements of the particles away from their average position, which may be omitted from consideration here. The average term can be obtained by measuring the amount of work required to stretch the string when it is displaced from equilibrium. If the string is under tension T, this work is T times the increase in length of the string as long as this increase is a small fraction of the length of the string. Expressed mathematically, the part of the potential energy which is of interest to us here is

$$V = T \int_0^l \left[\sqrt{1 + \left(\frac{\partial \psi}{\partial x}\right)^2} - 1 \right] dx \simeq \tfrac{1}{2} T \int_0^l \left(\frac{\partial \psi}{\partial x}\right)^2 dx \tag{3.3.2}$$

when $(\partial \psi / \partial x)^2 \ll 1$ and when the string is stretched between supports at $x = 0$ and $x = l$.

Consequently the Lagrange function for the average transverse motion of the string in a given plane is

$$\begin{aligned} \mathfrak{L} = T - V &= \tfrac{1}{2} \int_0^l \left[\rho \left(\frac{\partial \psi}{\partial t}\right)^2 - T \left(\frac{\partial \psi}{\partial x}\right)^2 \right] dx \\ &= \tfrac{1}{2}\rho \int_0^l \left[\left(\frac{\partial \psi}{\partial t}\right)^2 - c^2 \left(\frac{\partial \psi}{\partial x}\right)^2 \right] dx; \quad c^2 = \frac{T}{\rho} \end{aligned} \tag{3.3.3}$$

This over-all function is an integral of a *Lagrange density*

$$L = \tfrac{1}{2}\rho[(\partial \psi / \partial t)^2 - c^2 (\partial \psi / \partial x)^2]$$

over the whole length of the string.

As before, the time integral of the function $\mathfrak{L}$ is to be minimized. The Euler equation (3.1.3) corresponding to this minimization is

$$\frac{\partial}{\partial t}\left(\frac{\partial L}{\partial(\partial \psi / \partial t)}\right) + \frac{\partial}{\partial x}\left(\frac{\partial L}{\partial(\partial \psi / \partial x)}\right) = \frac{\partial L}{\partial \psi} = 0 \quad \text{or} \quad \left(\frac{\partial^2 \psi}{\partial t^2}\right) - c^2 \left(\frac{\partial^2 \psi}{\partial x^2}\right) = 0$$

which is the wave equation for the string [Eq. (2.1.9)].

The Wave Equation. Consequently the wave equation for the string corresponds to the requirement that *the difference between the total kinetic energy of the string and its potential energy be as small as possible,* on the average and subject to the initial and boundary conditions. This is a very suggestive and useful result, from which many other relations can be derived.

If, for instance, a transverse force $F(x)$ per unit length of string is applied, an additional term $-F\psi$ should be added to the potential energy; or if the string is embedded in an elastic medium (as discussed on page 139), the added term is $\frac{1}{2}K\psi^2$. The resulting Lagrange density is

$$L = \tfrac{1}{2}\rho\left(\frac{\partial\psi}{\partial t}\right)^2 - \tfrac{1}{2}T\left(\frac{\partial\psi}{\partial x}\right)^2 - \tfrac{1}{2}K\psi^2 + F\psi$$

and the equation of motion is

$$\rho\,\frac{\partial^2\psi}{\partial t^2} - T\,\frac{\partial^2\psi}{\partial x^2} = F - K\psi$$

which corresponds to Eq. (2.1.27).

The derivative of $\mathfrak{L}$ with respect to $\dot{\psi}$ should correspond to the momentum in particle dynamics. The corresponding density

$$p = \partial L/\partial\dot{\psi} = \rho(\partial\psi/\partial t) \tag{3.3.4}$$

will be called the *canonical momentum density* of the field at x. In the case of the string, as we see, it is the momentum of a unit length of string when moving with a velocity $[\partial\psi(x)/\partial t]$.

The Hamiltonian density is, according to Eq. (3.2.6),

$$\begin{aligned} H = p\dot{\psi} - L &= \tfrac{1}{2}\rho\left[\left(\frac{\partial\psi}{\partial t}\right)^2 + c^2\left(\frac{\partial\psi}{\partial x}\right)^2\right] + \tfrac{1}{2}K\psi^2 - F\psi \\ &= \frac{1}{2\rho}\,p^2 + \tfrac{1}{2}T\left(\frac{\partial\psi}{\partial x}\right)^2 + \tfrac{1}{2}K\psi^2 - F\psi \end{aligned} \tag{3.3.5}$$

which is to be compared with Eq. (2.1.11). The integral of this density over the string is the total energy, but we notice, in contrast to the case of classical dynamics, that H is dependent not only on p and ψ but also on $\partial\psi/\partial x$. Consequently we should expect that the equations corresponding to Hamilton's canonical equations (3.2.7) will be of more complex form. We obtain, by use of Hamilton's principle and by integration of two terms by parts,

$$\begin{aligned} 0 &= \delta\int dt\int dx[p\dot{\psi} - H(p,\psi,\psi')] \\ &= \int dt\int dx[\dot{\psi}\,\delta p + p\,\delta\dot{\psi} - (\partial H/\partial p)\,\delta p - (\partial H/\partial\psi)\,\delta\psi - (\partial H/\partial\psi')\,\delta\psi'] \\ &= \int dt\int dx\{[\dot{\psi} - (\partial H/\partial p)]\,\delta p - [\dot{p} + (\partial H/\partial\psi) - (\partial/\partial x)(\partial H/\partial\psi')]\,\delta\psi\} \end{aligned}$$

where $\quad \dot{\psi} = \partial\psi/\partial t;\quad \dot{p} = \partial p/\partial t;\quad \psi' = \partial\psi/\partial x$

so that

$$\frac{\partial\psi}{\partial t} = \frac{\partial H}{\partial p};\quad \frac{\partial p}{\partial t} = \frac{\partial}{\partial x}\left(\frac{\partial H}{\partial\psi'}\right) - \frac{\partial H}{\partial\psi} \tag{3.3.6}$$

which differs from the canonical equations by the term in $\partial H/\partial\psi'$ These equations, when combined, give the equation of motion for the string

$$\rho\frac{\partial^2\psi}{\partial t^2} = T\frac{\partial^2\psi}{\partial x^2} - K\psi + F$$

The situation here is more complicated than for dynamics, for the variable ψ, which now corresponds to the dynamical coordinate q, in addition to depending on the parameter t, also depends on a parameter x. This means that the interrelations between momenta, field, and field gradient must be more complicated than those given by the canonical equations (3.2.7).

This increased complexity is also apparent from the point of view of relativity. As we have seen on page 97, the energy of a particle is the time component of a four-vector, with space components proportional to the momentum. In the present case, however, the energy density

$$H = \frac{\partial L}{\partial\dot\psi}\dot\psi - L = \rho\frac{\partial\psi}{\partial t}\frac{\partial\psi}{\partial t} - L$$

appears to be the (4,4) component of a tensor, $\mathfrak{W}$, with components

$$\begin{aligned}
W_{11} &= \frac{\partial\psi}{\partial x}\frac{\partial L}{\partial\psi'} - L = -\tfrac{1}{2}\rho(\dot\psi)^2 - \tfrac{1}{2}T(\psi')^2 + \tfrac{1}{2}K\psi^2 - F\psi \\
W_{41} &= \frac{\partial\psi}{\partial t}\frac{\partial L}{\partial\psi'} = -T\dot\psi\psi'; \quad W_{14} = \frac{\partial\psi}{\partial x}\frac{\partial L}{\partial\dot\psi} = \rho\dot\psi\psi' \\
W_{44} &= \frac{\partial\psi}{\partial t}\frac{\partial L}{\partial\dot\psi} - L = H
\end{aligned} \tag{3.3.7}$$

These components satisfy the divergence equations

$$\begin{aligned}
\frac{\partial W_{11}}{\partial x} + \frac{\partial W_{14}}{\partial t} &= \frac{\partial\psi}{\partial x}\left[\rho\frac{\partial^2\psi}{\partial t^2} - T\frac{\partial^2\psi}{\partial x^2} + K\psi - F\right] = 0; \\
\frac{\partial W_{41}}{\partial x} + \frac{\partial W_{44}}{\partial t} &= \frac{\partial\psi}{\partial t}\left[\rho\frac{\partial^2\psi}{\partial t^2} - T\frac{\partial^2\psi}{\partial x^2} + K\psi - F\right] = 0
\end{aligned} \tag{3.3.8}$$

which have interesting physical interpretations. Taking the second equation first and integrating it over x from a to b, we see that

$$-\int_a^b\left(\frac{\partial W_{44}}{\partial t}\right)dx = -\frac{\partial}{\partial t}\int_a^b H\,dx = [W_{41}]_a^b$$

But W_{41} is, by Eq. (2.1.12), the rate of energy flow along the string, it should be naturally related to the change of energy H in the manner given in the equation. The second divergence equation is therefore the equation of continuity for energy flow along the string.

The first divergence equation relates the change in energy flow with time to the distribution of stress along the string, for $W_{14} = -(W_{41}/c^2)$

has the dimensions of momentum density, the momentum related to the energy flow of wave motion. The integral of the first divergence equation,

$$-\frac{\partial}{\partial t}\int_a^b W_{14}\,dx = [W_{11}]_a^b$$

indicates that, if W_{14} is the wave momentum, then W_{11} is a force, which might be called the *wave stress*. The equation states that the rate of change of wave momentum in a certain portion of the string is equal to the net wave stress across the ends of the portion.

The *wave momentum density* $P = W_{14}$ is more closely related to the wave motion than is the canonical momentum density $p = \rho(\partial\psi/\partial t)$. For p is the *transverse* momentum of the various parts of the string, whereas P is related to the flow of energy *along* the string due to wave motion.

Incidentally, Eqs. (3.3.6) can be used to show that the integral of the Hamiltonian density is a constant of the motion, independent of time, for

$$\frac{d}{dt}\int_0^l H(p,\psi,\psi')\,dx = \int_0^l \left[\frac{\partial H}{\partial p}\dot{p} + \frac{\partial H}{\partial \psi}\dot{\psi} + \frac{\partial H}{\partial \psi'}\frac{\partial^2 \psi}{\partial x\,\partial t}\right]dx$$
$$= \int_0^l \left[\frac{\partial \psi}{\partial t}\frac{\partial^2 H}{\partial x\,\partial \psi'} + \frac{\partial H}{\partial \psi'}\frac{\partial^2 \psi}{\partial x\,\partial t}\right]dx = \left[\dot{\psi}\frac{\partial H}{\partial \psi'}\right]_0^l = 0$$

since $\dot{\psi}$ or $\partial H/\partial\psi'$ is zero at both ends of the string.

Helmholtz Equation. When the string vibrates with simple harmonic motion, the dependence on time can be expressed as $\psi = Y(x)e^{-i\omega t}$, where the function Y must satisfy the *Helmholtz equation*

$$(d^2Y/dx^2) + k^2Y = 0; \quad k = \omega/c$$

where the value of the constant k must be determined by boundary conditions. This equation, too, can be derived from a variational principle.

The Lagrange density in this case is simply the potential energy term $-T(dY/dx)^2$. In order to obtain a nonzero result we must insist that the rms amplitude of Y be larger than zero. We can ensure this by requiring that

$$\int_0^l Y^2(x)\,dx = 1$$

This is, of course, a subsidiary requirement, which, together with the variational equation

$$\delta\int_0^l \left(\frac{dY}{dx}\right)^2 dx = 0$$

constitutes the requirement to be met by Y.

Reference to page 279 shows that the solution of such a variational problem can be obtained by the use of Lagrange multipliers. We include the auxiliary requirement by requiring that

$$\int_0^l \left[\lambda Y^2 - \left(\frac{dY}{dx}\right)^2\right] dx$$

be a minimum or maximum, where λ is the multiplier to be determined. The Euler equation for this is

$$(d^2Y/dx^2) + \lambda Y = 0$$

which is the Helmholtz equation again. The best values for λ must equal the required values for k^2.

Velocity Potential. Going next to a three-dimensional case, we can consider the motion of a fluid, discussed earlier in Sec. 2.3. When the motion is irrotational, the fluid velocity can be expressed as the gradient of a velocity potential ψ. The kinetic energy density of the fluid is then

$$T = \tfrac{1}{2}\rho\left[\left(\frac{\partial\psi}{\partial x}\right)^2 + \left(\frac{\partial\psi}{\partial y}\right)^2 + \left(\frac{\partial\psi}{\partial z}\right)^2\right] \tag{3.3.9}$$

If the fluid is incompressible, the potential energy is constant and the Lagrange function is just $L = T$. In this case the Euler equation (3.1.3) is just the Laplace equation

$$\nabla^2\psi = 0$$

Therefore the Laplace equation for the steady, irrotational motion of an incompressible fluid is equivalent to the requirement that the *total kinetic energy of all the fluid is as small as it can be*, subject to initial and boundary conditions.

If the fluid is compressible but has negligible viscosity, then it will have a potential energy density which can be expressed in terms of the velocity potential ψ. This energy is the work $p\,dV$ required to compress a unit portion of the fluid from standard conditions of density, etc., to the conditions of the problem. We shall assume here that the fractional change from standard conditions is small; if the density at standard conditions is ρ, that at the actual conditions can be written as $\rho(1 + s)$, where s is quite small compared with unity. At standard conditions 1 cc of fluid will occupy $(1 - s)$ cc when the density is $\rho(1 + s)$ (to the first order in the small quantity s).

To determine the potential energy we must know the relation between the pressure and the density as the state of the fluid is changed. For instance, we can consider the case of a gas, as discussed on page 163, where the ratio between pressure and density is given in Eq. (2.3.21). In the present case, we call p_0 the pressure at standard conditions and the pressure at actual conditions to be $p_0 + p$, so that p is the difference in pressure between actual and standard conditions. This definition is generally used in acoustics (note that in deference to custom we use p for the pressure in this section whereas everywhere else in this chapter, p stands for the canonical momentum). Rewriting Eq. (2.3.21) in our new notation, we have

$$1 + (p/p_0) = (1 + s)^\gamma \simeq 1 + \gamma s \quad \text{or} \quad p \simeq \rho c^2 s \tag{3.3.10}$$

where $c^2 = \gamma p_0/\rho$ for a gas. For other fluids the excess pressure is also proportional to the fractional increase in density [so that Eq. (3.3.10) still holds] but the constant c^2 depends on different properties of the material. In any case c is the velocity of sound in the fluid, as we shall shortly see.

As the fluid is being compressed from standard to actual conditions, an element of fluid will change volume from dV to $dV\,(1 - s) = dV\,[1 - (p/\rho c^2)]$. The work done in making this change,

$$\int_0^s p\,dV\,ds = \frac{dV}{\rho c^2}\int_0^p p\,dp = \left(\frac{1}{2\rho c^2}\right) p^2\,dV$$

is the potential energy of compression in an element of volume dV. The potential energy density is therefore $p^2/2\rho c^2$.

This still is not in form to set in the variational equation, for we must relate p to the velocity potential ψ (though it is perfectly possible to set up the variational equations in terms of the scalar p rather than ψ). The needed relationship can most easily be obtained from Eq. (2.3.14). When the viscosity coefficients η and λ are negligible, and when the external force F is zero, then $\rho(\partial \mathbf{v}/\partial t) = -\operatorname{grad} p$. If a velocity potential exists, $\mathbf{v} = \operatorname{grad} \psi$, and equating gradients, we see that

$$p = -\rho(\partial\psi/\partial t) + C_0 \tag{3.3.11}$$

where the constant of integration C_0 is usually set zero. Therefore the pressure is the time component of the four-vector which has, as space components, the velocity components of the fluid.

Compressional Waves. We are now ready to set up the Lagrange density for small vibrations of a nonviscous, compressible fluid:

$$L = T - V = \tfrac{1}{2}\rho\left\{|\operatorname{grad}\psi|^2 - \frac{1}{c^2}\left(\frac{\partial\psi}{\partial t}\right)^2\right\} \tag{3.3.12}$$

This is an invariant density, the space-time integral of which is to be minimized. The inversion here represented, where the velocity is given by space derivatives and the force by the time derivative, is due to the fact that the velocity must be a vector (a gradient) whereas the force (pressure) is a scalar.

The Euler equation (3.1.3) for this Lagrange density is just the wave equation in three space dimensions

$$\nabla^2\psi = (1/c^2)(\partial^2\psi/\partial t^2)$$

for a wave velocity c.

It is of interest that the role of space and time derivatives is the reverse of that for the string. Here the space derivatives of ψ correspond to the kinetic energy and the time derivative corresponds to the potential energy. Here the so-called "canonical momentum density" [see Eq. (3.3.4)] $\partial L/\partial\dot{\psi}$ is proportional to the pressure, not to the fluid velocity at all. It shows that the simple pattern of canonically conjugate variables q and p, suitable for particle dynamics, must be replaced by a more complex pattern. The useful quantity here is the four-dyadic $\mathfrak{W}$, having components

$$W_{ij} = L\delta_{ij} - \frac{\partial\psi}{\partial x_i}\frac{\partial L}{\partial(\partial\psi/\partial x_j)} \tag{3.3.13}$$

$$\delta_{ij} = \begin{cases} 1; & j = i \\ 0; & j \neq i \end{cases}; \quad x_1 = x;\, x_2 = y;\, x_3 = z;\, x_4 = ict$$

The component W_{44} is the energy density

$$W_{44} = \tfrac{1}{2}\rho\left[\frac{1}{c^2}\left(\frac{\partial\psi}{\partial t}\right)^2 + |\text{grad}\,\psi|^2\right] = \tfrac{1}{2}\rho v^2 + \frac{1}{2\rho c^2}p^2 = H \tag{3.3.14}$$

the integral of which is independent of time (as one can show by a procedure analogous to that on page 306 for the string).

The time-space components are proportional to a three-vector $\mathbf{S}$,

$$W_{k4} = W_{4k} = \frac{i\rho}{c}\frac{\partial\psi}{\partial t}\frac{\partial\psi}{\partial x_k} = \frac{1}{ic}pv_k = \frac{1}{ic}S_k; \quad k = 1, 2, 3 \tag{3.3.15}$$

which gives the direction and magnitude of the flow of energy due to wave motion. The quantity $\mathbf{P} = \mathbf{S}/c^2$ is a vector having the dimensions of momentum density, which may be called the *field momentum density*. On the other hand the vector $\rho\mathbf{S}/p = (\rho\mathbf{v})$ is the momentum density of the moving fluid.

We note also that the four-divergences of the vectors formed from $\mathfrak{W}$ are all zero. This can be shown as follows:

$$\sum_j \frac{\partial W_{ij}}{\partial x_j} = \frac{\partial L}{\partial x_i} - \sum_j\left[\frac{\partial^2\psi}{\partial x_i\,\partial x_j}\frac{\partial L}{\partial\psi_j} + \frac{\partial\psi}{\partial x_i}\frac{\partial}{\partial x_j}\left(\frac{\partial L}{\partial\psi_j}\right)\right]$$

where $\psi_j = \partial\psi/\partial x_j$. If L depends on x_i *only* through the function ψ and its derivatives, we have next,

$$\frac{\partial L}{\partial x_i} = \frac{\partial L}{\partial\psi}\frac{\partial\psi}{\partial x_i} + \sum_j \frac{\partial L}{\partial\psi_j}\frac{\partial^2\psi}{\partial x_j\,\partial x_i}$$

Finally, using the Euler equations (3.1.3), we have

$$\sum_j \frac{\partial W_{ij}}{\partial x_j} = \frac{\partial \psi}{\partial x_i}\left[\frac{\partial L}{\partial \psi} - \sum_j \frac{\partial}{\partial x_j}\frac{\partial L}{\partial \psi_j}\right] = 0$$

which proves the statement regarding the divergences if L does not depend explicitly on the x's. From it we can obtain the equation of continuity for $\mathbf{S}$ and H,

$$\operatorname{div} \mathbf{S} + (\partial H/\partial t) = 0$$

showing that, if H is the energy density, then $\mathbf{S}$ is the energy flow vector (which was pointed out earlier). Although the integral of H over all space is constant, H at any point can vary with time, for the energy can flow about within the boundaries.

Wave Impedance. Returning to the Lagrange-Euler equation for ψ, we see that, if a "force density" f is applied to ψ, the equation relating f to the changes of ψ will be

$$f = \frac{\partial}{\partial t}\left(\frac{\partial L}{\partial \dot{\psi}}\right) + \sum_i \frac{\partial}{\partial x_i}\left(\frac{\partial L}{\partial(\partial \psi/\partial x_i)}\right) - \frac{\partial L}{\partial \psi}$$

$$= -\left(\frac{\rho}{c^2}\right)\left(\frac{\partial^2 \psi}{\partial t^2}\right) + \rho \operatorname{div} (\operatorname{grad} \psi)$$

In the case of classical dynamical systems we can usually apply a force to one coordinate of the system and observe the relationship between the applied force and the displacement or velocity of the system, as was shown on page 284. In the case of fields, however, a force concentrated at a point usually produces an infinite displacement at the point of contact; the force must usually be applied over an area or a volume to produce a physically realizable reaction. In fact it is usually the case that an externally applied force is applied over some part of the surface bounding the field. In the present case, for instance, a vibrating loudspeaker diaphragm produces sound waves in the fluid in front of it and these waves react back on the diaphragm. A measure of this reaction is the *acoustic impedance* of the fluid in front of the diaphragm, which is characteristic of the kind of wave produced.

If the driving force is applied to the boundary surface, we must integrate the Lagrange-Euler equation over the volume enclosed by the boundary to find the total reaction. The result is

$$-\frac{\rho}{c^2}\iiint \frac{\partial^2 \psi}{\partial t^2}\, dv + \rho \iint (\operatorname{grad} \psi) \cdot d\mathbf{A}$$

where the second term has been changed to a surface integral by the use of Gauss' theorem, Eq. (1.4.7). Any force acting on ψ along the boundary surface is balanced by this surface term, so that, if $\mathbf{F}$ is the driving "force" on ψ per unit of surface area, then $\mathbf{F} = \rho \operatorname{grad} \psi$. If $\mathbf{F}$ is a simple harmonic driving force, $\mathbf{F} = \mathbf{F}_0 e^{-i\omega t}$, then the corresponding rate of change of ψ (analogous to the velocity) is $-i\omega\psi$, so that the ratio of surface force density to rate of change of ψ at the surface is

$$-\mathbf{F}/i\omega\psi = -(\rho/i\omega\psi) \operatorname{grad} \psi$$

Because of the reversal of the role between gradient and time derivative in this case, the quantity $(i\omega\psi)$ is proportional to the pressure and $\operatorname{grad} \psi$ is proportional to the fluid velocity. In acoustics we usually consider the pressure to be analogous to the driving force, instead of $\mathbf{F}$ (which is the "force" causing change of ψ), so that the ratio given above is more analogous to an admittance rather than an impedance.

The usual definition of *acoustic admittance* is given by the equation

$$\mathbf{Y} = \mathbf{v}/p = (1/i\omega\rho\psi) \operatorname{grad} \psi \tag{3.3.16}$$

where p is the pressure at some point on the boundary surface where a driving force is applied and $\mathbf{v}$ is the fluid velocity at the same point. To compute this admittance we must first solve the wave equation to find the field caused by the vibration of the given part of the boundary surface, after which we can compute the ratio $(1/i\omega\rho\psi) \operatorname{grad} \psi$ for the various parts of the driving surface and, if required, can integrate over this surface to find the acoustic admittance for the whole driving surface.

The admittance $\mathbf{Y}$ is a vector because $\mathbf{v}$ is a vector and p a scalar. It is usually sufficient to compute the *normal acoustic admittance*, which is the component of $\mathbf{Y}$ normal to the surface.

$$Y_n = (1/i\omega\rho\psi)(\partial\psi/\partial n)$$

The reciprocal of this,

$$Z_n = 1/Y_n = i\omega\rho\psi(\partial\psi/\partial n)^{-1}$$

is called the *normal acoustic impedance*. It is the ratio of pressure at the driving surface to normal velocity at the same point. Since the normal velocity of the fluid at the surface is equal to the velocity of the surface itself, this normal impedance is the most useful part of the impedance.

Plane-wave Solution. As an example of the various quantities we have spoken about so glibly in this section, we shall consider a particularly simple sort of wave motion, a *plane wave*. Such a wave is represented by the formula

$$\psi = Ce^{i\mathbf{k}\cdot\mathbf{r}-i\omega t}$$

where $C = |C|e^{i\phi}$ is a constant giving the amplitude $|C|$ and phase angle ϕ of the velocity potential and where $\mathbf{k}$ is a constant vector of magnitude

ω/c, pointing in the direction of the wave motion. The wave surfaces for this wave are planes perpendicular to **k**, traveling in the direction of **k** with a velocity c.

The expressions for the pressure and fluid velocity for the plane wave can be obtained from the velocity potential by means of the relations given earlier. They are the real parts of the following expressions:

$$p = -\rho(\partial\psi/\partial t) = i\omega\rho C e^{i\mathbf{k}\cdot\mathbf{r}-i\omega t}; \quad \mathbf{v} = \operatorname{grad}\psi = i\mathbf{k}Ce^{i\mathbf{k}\cdot\mathbf{r}-i\omega t}$$

In other words the actual value of the pressure at point x, y, z at time t is $-\omega\rho|C| \sin [(\omega/c)(\alpha x + \beta y + \gamma z - ct) + \phi]$ where α, β, γ are the direction cosines for **k** and $|\mathbf{k}| = \omega/c$. The fluid motion is in the direction of **k**, perpendicular to the wave fronts, and the velocity is in phase with the pressure for a plane wave.

In computing the stress-energy tensor we must use the real parts of the quantities given above, for the components are quadratic expressions involving ψ. Letting the symbol Ω stand for $[(\omega/c)(\alpha x + \beta y + \gamma z - ct) + \phi] = (\mathbf{k}\cdot\mathbf{r} - \omega t + \phi)$, we have

$$\psi = |C| \cos \Omega; \quad p = -\omega\rho|C| \sin \Omega; \quad \mathbf{v} = -\mathbf{k}|C| \sin \Omega;$$

$$W_{44} = H = \tfrac{1}{2}\rho v^2 + \frac{p^2}{2\rho c^2} = \frac{\rho\omega^2}{c^2}|C|^2 \sin^2 \Omega;$$

$$\mathbf{S} = p\mathbf{v} = \rho\omega\mathbf{k}|C|^2 \sin^2 \Omega = \mathbf{P}c^2;$$

$$W_{11} = -\frac{\rho\omega^2}{c^2}\alpha^2|C|^2 \sin^2 \Omega; \quad W_{12} = -\frac{\rho\omega^2}{c^2}\alpha\beta|C|^2 \sin^2 \Omega$$

where α, β, γ are the direction cosines for the *propagation vector* **k**. In matrix form the stress-energy tensor for the plane wave is, therefore,

$$\mathfrak{W} = -\frac{\rho\omega^2}{c^2}|C|^2 \sin^2 \Omega \begin{pmatrix} \alpha^2 & \alpha\beta & \alpha\gamma & -i\alpha \\ \beta\alpha & \beta^2 & \beta\gamma & -i\beta \\ \gamma\alpha & \gamma\beta & \gamma^2 & -i\gamma \\ i\alpha & i\beta & i\gamma & -1 \end{pmatrix}$$

It is not difficult to confirm the divergence equations for $\mathfrak{W}$,

$$\sum_{n=1}^{4}\left(\frac{\partial W_{mn}}{\partial x_n}\right) = 0$$

or that the space components transform like a dyadic.

The principal axes of the space part of $\mathfrak{W}$ are in the direction of **k** and in any two mutually perpendicular directions which are orthogonal to **k**. In terms of these coordinates, ξ_1, ξ_2, ξ_3, the matrix for $\mathfrak{W}$ becomes

$$\mathfrak{W} = \frac{\rho\omega^2}{c^2}|C|^2 \sin^2\left[\frac{\omega}{c}(\xi_1 - ct)\right]\begin{pmatrix} -1 & 0 & 0 & i \\ 0 & 0 & 0 & 0 \\ 0 & 0 & 0 & 0 \\ -i & 0 & 0 & 1 \end{pmatrix}$$

All the terms are proportional to the square of the frequency and to the square of the wave amplitude.

We can imagine the plane wave to be produced by a vibrating plane surface perpendicular to **k**, oscillating in the direction of **k** with velocity equal to grad $\psi = i\mathbf{k}Ce^{i\mathbf{k}\cdot\mathbf{r}-i\omega t}$. The acoustic admittance for this case, measuring the reaction of the wave back on the vibrating surface, is

$$\mathbf{Y} = \mathbf{v}/p = \mathbf{k}/\rho\omega = (1/\rho c)\mathbf{a}_k$$

where $\mathbf{a}_k$ is the unit vector in the direction of **k**. The acoustic impedance normal to the plane wave fronts is therefore (ρc), a real quantity. In other words a plane wave produces a resistive impedance, independent of frequency, on the driving surface. Impedances for other wave configurations will be worked out later in this book.

Diffusion Equation. When we come to dissipative equations, such as the fluid-flow case with viscosity or the case of diffusion, we must use the technique discussed on page 298 to bring the system within the formal framework. For instance, for the diffusion equation the Lagrange density is

$$L = -\,(\text{grad}\,\psi)\cdot(\text{grad}\,\psi^*) - \tfrac{1}{2}a^2\left(\psi^*\frac{\partial\psi}{\partial t} - \psi\frac{\partial\psi^*}{\partial t}\right) \tag{3.3.17}$$

where ψ is the density of the diffusing fluid, a^2 is the diffusion constant, and ψ^* refers to the mirror-image system where the fluid is "undiffusing" (or whatever it is that one calls the reverse of diffusing). The canonical momentum densities are

$$p = \partial L/\partial\dot\psi = -\tfrac{1}{2}a^2\psi^*;\quad p^* = +\tfrac{1}{2}a^2\psi$$

which has little to do with a physical momentum.

The Euler equations for this Lagrange density are

$$\nabla^2\psi = a^2(\partial\psi/\partial t);\quad \nabla^2\psi^* = -a^2(\partial\psi^*/\partial t) \tag{3.3.18}$$

The equation for ψ is the usual diffusion equation; that for ψ^* is for the mirror-image system, which gains as much energy as the first loses.

The Hamiltonian density is $(\text{grad}\,\psi)\cdot(\text{grad}\,\psi^*)$, the integral of which over the volume is independent of time. It is the 4,4 component of the tensor $\mathfrak{W}$ defined by the equations

$$W_{ij} = L\delta_{ij} - \frac{\partial\psi}{\partial x_i}\frac{\partial L}{\partial\psi_j} - \frac{\partial\psi^*}{\partial x_i}\frac{\partial L}{\partial\psi_j^*} \tag{3.3.19}$$

where $\psi_j = \partial\psi/\partial x_j$. The components W_{k4} contain the components of the vector, grad ψ, which gives the magnitude and direction of the diffusive flow.

The introduction of the mirror-image field ψ^*, in order to set up a Lagrange function from which to obtain the diffusion equation, is probably too artificial a procedure to expect to obtain much of physical significance from it. It is discussed here to show that the variational technique can also be applied to dissipative systems and also because similar introductions of ψ^* fields are necessary in some quantum equations and the diffusion case is a useful introduction to these more complicated cases.

A similar procedure can be used to obtain a Lagrange function for the dissipative case of fluid motion when viscosity is taken into account.

Schroedinger Equation. Somewhat the same procedure can be used to obtain the Schroedinger equation (2.6.38), though it is not a dissipative equation. The wave function ψ is a complex quantity, so that its real and imaginary parts can be considered as independent variables or, what is the same thing, ψ and its complex conjugate ψ^* can be considered as separate field variables, which can be varied independently. The product $\psi^*\psi$ is to be real and, for the best values of the variables, will equal the probability density for the presence of the particle, *i.e.*, for the configuration of the system specified by the coordinates x.

As an example, we consider a single particle of mass m, acting under the influence of a potential field $V(x,y,z)$. The Lagrange density turns out to be

$$L = -\frac{\hbar^2}{2m}(\text{grad}\,\psi^*)\cdot(\text{grad}\,\psi) - \frac{\hbar}{2i}\left(\psi^*\frac{\partial\psi}{\partial t} - \frac{\partial\psi^*}{\partial t}\psi\right) - \psi^*V\psi \tag{3.3.20}$$

and ψ^* and ψ are to be varied until $\mathfrak{L} = \iiiint L\,dv\,dt$ is a minimum.

The Lagrange-Euler equations are then

$$-\frac{\hbar^2}{2m}\nabla^2\psi^* - \frac{\hbar}{i}\frac{\partial\psi^*}{\partial t} = -V\psi^*; \quad -\frac{\hbar^2}{2m}\nabla^2\psi + \frac{\hbar}{i}\frac{\partial\psi}{\partial t} = -V\psi$$

or

$$-\frac{\hbar^2}{2m}\nabla^2\psi + V\psi = i\hbar\frac{\partial\psi}{\partial t}; \quad -\frac{\hbar^2}{2m}\nabla^2\psi^* + V\psi^* = -i\hbar\frac{\partial\psi^*}{\partial t} \tag{3.3.21}$$

It can easily be seen that these equations correspond to Eq. (2.6.38), when we write down the classical Hamiltonian for the particle, $H(p,q) = (1/2m)(p_x^2 + p_y^2 + p_z^2) + V$. Substituting $(\hbar/i)(\partial/\partial x)$ for p_x, etc., operating on ψ turns Eq. (2.6.38) into the first of Eqs. (3.3.21). The equation for the complex conjugate ψ^* is obtained by changing the sign of the i in the time derivative term.

The two canonical momenta are

$$p = \partial L/\partial\dot{\psi} = -(\hbar\psi^*/2i); \quad p^* = \hbar\psi/2i \tag{3.3.22}$$

They will have use when we wish to perform the "second quantization" often resorted to in modern quantum theory, but not treated in this book. The "stress-energy" tensor $\mathfrak{W}$ has components defined by the equations

$$W_{mn} = \psi_m^* \frac{\partial L}{\partial \psi_n^*} + \psi_m \frac{\partial L}{\partial \psi_n} - \delta_{mn} L \tag{3.3.23}$$

where $\psi_m = \frac{\partial \psi}{\partial x_m}; \quad x_m = (x,y,z,t); \quad \delta_{mn} = \begin{cases} 0; & m \neq n \\ 1; & m = n \end{cases}$

The energy density is the (4,4) component of $\mathfrak{W}$

$$H = W_{44} = (\hbar^2/2m)(\text{grad } \psi^*) \cdot (\text{grad } \psi) + \psi^* V \psi$$

It should be pointed out that in the present case, as with the diffusion equation, the time derivative terms (equivalent to the $\dot{q}$'s) enter as a *linear* function in L, rather than as a quadratic function of the form $\Sigma a_{rs}\dot{q}_r\dot{q}_s$, as had been assumed in classical dynamics. Whenever the $\dot{q}$'s occur as a linear function in L, the canonical momenta $\partial L/\partial \dot{q}$ will not be a function of $\dot{q}$'s but will be a function of just the q's, so that p and q are not independent variables. In this case the definition $H = \Sigma p\dot{q} - L$ will ensure that H is a function of the q's alone, without either p's or $\dot{q}$'s, and of course, the canonical equations will not have the same form as heretofore [Eq. (3.3.6)]. In both the diffusion equation and the Schroedinger equation p is a function of ψ^* and p^* a function of ψ, so that we cannot have one canonical equation for p and the other for q but must have one for ψ and the other for ψ^*.

As always, we have (where $\psi_2 = \partial\psi/\partial y$, $\psi_3^* = \partial\psi^*/\partial z$, etc.)

$$\begin{aligned}
\delta\mathfrak{L} = 0 = \delta \int dt \int dv \, [p\dot{\psi} + \dot{\psi}^* p^* - H(\psi,\psi^*,\psi_n,\psi_n^*)] \\
= \int dt \int dv \bigg[\delta p \, \dot{\psi} + p \, \delta\dot{\psi} + \dot{\psi}^* \, \delta p^* + \delta\dot{\psi}^* \, p^* \\
- \left(\frac{\partial H}{\partial \psi}\right) \delta\psi - \left(\frac{\partial H}{\partial \psi^*}\right) \delta\psi^* - \sum_n \left(\frac{\partial H}{\partial \psi_n}\right) \delta\psi_n - \sum_n \left(\frac{\partial H}{\partial \psi_n^*}\right) \delta\psi_n^* \bigg]
\end{aligned}$$

But now $\delta p = (dp/d\psi^*) \, \delta\psi^*$ and, integrating by parts, $\int dt \, p \, \delta\dot{\psi} = -\int dt \, \dot{p} \, \delta\psi = -\int dt \, \dot{\psi}^*(dp/d\psi^*) \, \delta\psi$, etc. As before we also have

$$-\int dv \sum_n \left(\frac{\partial H}{\partial \psi_n}\right) \delta\psi_n = \int dv \sum_n \frac{\partial}{\partial x_n} \left(\frac{\partial H}{\partial \psi_n}\right) \delta\psi, \text{ etc.}$$

Setting all these into the integral we find that $\delta\mathfrak{L}$ divides into an integral of a quantity times $\delta\psi$ plus another quantity times $\delta\psi^*$. Since $\delta\mathfrak{L}$ must be zero, no matter what values $\delta\psi$ and $\delta\psi^*$ have, the two quantities must be zero, giving us the two equations which are the new canonical equations:

$$\begin{aligned}
\dot{\psi} \left[\frac{dp}{d\psi^*} - \frac{dp^*}{d\psi}\right] = \frac{\partial H}{\partial \psi^*} - \sum_n \frac{\partial}{\partial x_n} \left(\frac{\partial H}{\partial \psi_n^*}\right); \\
\dot{\psi}^* \left[\frac{dp^*}{d\psi} - \frac{dp}{d\psi^*}\right] = \frac{\partial H}{\partial \psi} - \sum_n \frac{\partial}{\partial x_n} \left(\frac{\partial H}{\partial \psi_n}\right)
\end{aligned} \tag{3.3.24}$$

These equations, when applied to the Hamiltonian for the diffusion equation or for the Schroedinger equation, will again give the equations of motion (3.3.18) or (3.3.21). It is not certain, however, how useful Eqs. (3.3.24) will turn out to be, since they do not appear to tell us more than the Lagrange-Euler equations do.

The energy flow vector for the Schroedinger case is

$$\mathbf{S} = \mathbf{i}W_{41} + \mathbf{j}W_{42} + \mathbf{k}W_{43} = -\frac{\hbar^2}{2m}\left[\left(\frac{\partial\psi^*}{\partial t}\right)\operatorname{grad}\psi + \left(\frac{\partial\psi}{\partial t}\right)\operatorname{grad}\psi^*\right] \tag{3.3.25}$$

This satisfies the equation of continuity with the energy density W_{44}, $\operatorname{div}\mathbf{S} + (\partial H/\partial t) = 0$.

The field momentum density vector is

$$\mathbf{P} = \mathbf{i}W_{14} + \mathbf{j}W_{24} + \mathbf{k}W_{34} = -(\hbar/2i)[\psi^*\operatorname{grad}\psi - \psi\operatorname{grad}\psi^*] \tag{3.3.26}$$

Referring to page 255, we see that (when the magnetic field is zero) the current density corresponding to the wave function ψ is $\mathbf{J} = -(e/m)\mathbf{P}$, so that the field momentum vector $\mathbf{P}$ is related to the probable density of flow of the particle associated with the wave function ψ.

Klein-Gordon Equation. Another equation which can be dealt with in a manner similar to the preceding is the Klein-Gordon equation (2.6.51), a possible wave equation for a relativistic particle (though it is not the correct equation for electron or proton). Here again we use two independent field variables, ψ and ψ^*. The quantities $(\hbar/i)(\partial\psi/\partial x, \partial\psi/\partial y, \partial\psi/\partial z, \partial\psi/\partial ict)$ are the components of a four-vector, as are the similar derivatives of ψ^*. Combining these and the four-vector $(A_x,A_y,A_z,i\varphi)$ for the electromagnetic potential in a manner suggested by Eq. (2.6.49), we obtain a Lagrange density for a "particle" of charge e and mass m in an electromagnetic field:

$$L = -\frac{\hbar^2}{2m}\left[\left(\operatorname{grad}\psi^* + \frac{ie}{\hbar c}\mathbf{A}\psi^*\right)\cdot\left(\operatorname{grad}\psi - \frac{ie}{\hbar c}\mathbf{A}\psi\right) - \left(\frac{1}{c}\right)^2\left(\frac{\partial\psi^*}{\partial t} - \frac{ie}{\hbar}\varphi\psi^*\right)\left(\frac{\partial\psi}{\partial t} + \frac{ie}{\hbar}\varphi\psi\right) + \left(\frac{mc}{\hbar}\right)^2\psi^*\psi\right] \tag{3.3.27}$$

From this we can obtain the Lagrange-Euler equation for ψ,

$$\sum_{n=1}^{4}\frac{\partial}{\partial\zeta_n}\left(\frac{\partial L}{\partial\psi_n^*}\right) - \frac{\partial L}{\partial\psi^*} = 0;\ \zeta_1 = x;\ \zeta_2 = y;\ \zeta_3 = z;\ \zeta_4 = t;\ \psi_n^* = \frac{\partial\psi^*}{\partial\zeta_n},$$

resulting in the Klein-Gordon equation for the presence of an electromagnetic field:

$$\sum_{n=1}^{3}\left(\frac{\partial}{\partial\zeta_n} - \frac{ie}{\hbar c}A_n\right)^2\psi - \frac{1}{c^2}\left(\frac{\partial}{\partial t} + \frac{ie}{\hbar}\varphi\right)^2\psi = \left(\frac{mc}{\hbar}\right)^2\psi \tag{3.3.28}$$

The equation for ψ^* is of the same form. Here we have used the equation div $\mathbf{A} + (1/c)(\partial\varphi/\partial t) = 0$ several times to obtain the above result. This equation reduces to the simple form given in Eq. (2.6.51) when $\mathbf{A}$ and φ are zero.

To simplify the rest of the discussion we shall treat the case when $\mathbf{A}$ and φ are zero. The Lagrange function is then

$$L = -\frac{\hbar^2}{2m}(\text{grad}\,\psi^*)\cdot(\text{grad}\,\psi) + \frac{\hbar^2}{2mc^2}\left(\frac{\partial\psi^*}{\partial t}\right)\left(\frac{\partial\psi}{\partial t}\right) - \tfrac{1}{2}mc^2\psi^*\psi \quad (3.3.29)$$

and the canonical momenta are

$$p = (\partial L/\partial\dot{\psi}) = (\hbar^2/2mc^2)(\partial\psi^*/\partial t);\quad p^* = (\hbar^2/2mc^2)(\partial\psi/\partial t) \quad (3.3.30)$$

The 4,4 component of the stress-energy tensor

$$W_{mn} = \psi_m^*\frac{\partial L}{\partial\psi_n^*} + \psi_m\frac{\partial L}{\partial\psi_n} - \delta_{mn}L \quad (3.3.31)$$

is, of course, the energy density H. This can be expressed in terms of the canonical momenta p, p^*, the ψ's and their gradients:

$$\begin{aligned} W_{44} &= \frac{\hbar^2}{2m}(\text{grad}\,\psi^*)\cdot(\text{grad}\,\psi) + \frac{\hbar^2}{2mc^2}\left(\frac{\partial\psi^*}{\partial t}\right)\left(\frac{\partial\psi}{\partial t}\right) + \tfrac{1}{2}mc^2\psi^*\psi \\ &= \frac{2mc^2}{\hbar^2}(p^*p) + \frac{\hbar^2}{2m}(\text{grad}\,\psi^*)\cdot(\text{grad}\,\psi) + \tfrac{1}{2}mc^2\psi^*\psi = H \end{aligned} \quad (3.3.32)$$

From this, by using the canonical equations (3.3.6),

$$\frac{\partial\psi}{\partial t} = \frac{\partial H}{\partial p} = \frac{2mc^2}{\hbar^2}p^*$$

$$\frac{\partial p^*}{\partial t} = \frac{\hbar^2}{2mc^2}\frac{\partial^2\psi}{\partial t^2} = \sum_{n=1}^{3}\frac{\partial}{\partial\xi_n}\left(\frac{\partial H}{\partial\psi_n^*}\right) - \frac{\partial H}{\partial\psi^*} = \frac{\hbar^2}{2m}\nabla^2\psi - \tfrac{1}{2}mc^2\psi$$

plus two others for $\partial\psi^*/\partial t$ and $\partial p/\partial t$. These again give the Klein-Gordon equations for ψ and ψ^*.

The field momentum density vector is

$$\mathbf{P} = \mathbf{i}W_{14} + \mathbf{j}W_{24} + \mathbf{k}W_{34} = \frac{\hbar^2}{2mc^2}\left[\frac{\partial\psi^*}{\partial t}\,\text{grad}\,\psi + \frac{\partial\psi}{\partial t}\,\text{grad}\,\psi^*\right] \quad (3.3.33)$$

and the energy flow vector $\mathbf{S}$ is equal to $-c^2\mathbf{P}$.

The expressions for charge and current density for this equation may be obtained in several ways. One way, which will be useful later, is to refer ahead to Eq. (3.4.11) to note that the part of the Lagrange function which involves the interaction between electromagnetic potentials and charge current is the expression $(1/c)\mathbf{A}\cdot\mathbf{J} - \varphi\rho$. Therefore the part of the Lagrange density (3.3.27) which involves the ψ's and the potentials;

$$\frac{\hbar e}{2imc}\mathbf{A}\cdot[\psi^* \operatorname{grad}\psi - \psi \operatorname{grad}\psi^*] + \frac{\hbar e}{2imc^2}\varphi\left(\psi^*\frac{\partial\psi}{\partial t} - \psi\frac{\partial\psi^*}{\partial t}\right)$$

should have this form. This indicates that the current density vector should be

$$\mathbf{J} = (e\hbar/2im)[\psi^* \operatorname{grad}\psi - \psi \operatorname{grad}\psi^*] \qquad (3.3.34)$$

which is the same as the expression (2.6.47) for the Schroedinger equation, when $\mathbf{A}$ and φ are zero. The corresponding expression for the charge density where the potentials are zero is

$$\rho = -\frac{e\hbar}{2imc^2}\left[\psi^*\frac{\partial\psi}{\partial t} - \psi\frac{\partial\psi^*}{\partial t}\right] \qquad (3.3.35)$$

which is *not* the same as for the Schroedinger equation. In fact this expression for charge density is not necessarily everywhere positive (or not everywhere negative, depending on the sign of e), which is not particularly satisfactory for a wave function (unless we are willing to consider the possibility of the change of sign of the charge!).

Incidentally, these expressions for $\mathbf{J}$ and ρ can be obtained from the Klein-Gordon equation itself, by using the same methods as those given on page 255 to obtain ρ and $\mathbf{J}$ for the Schroedinger equation.

3.4 *Vector Fields*

When the field needed to describe some physical phenomenon has several components, the analysis becomes somewhat more complicated but the general principles are the same as those already discussed. The independent variables, which are to be varied until the integrated Lagrange function is a minimum, are the components $\psi_1, \ldots, \psi_n$, functions of the parameters x, y, z, t (or another set of four-dimensional coordinates). The Lagrange density L is an invariant function of the ψ's and their derivatives $\psi_{ij} = \partial\psi_i/\partial\xi_j$, ($\xi_1 = x$, $\xi_2 = y$, $\xi_3 = z$, $\xi_4 = t$), and the integral

$$\mathfrak{L} = \iiiint L(\psi_i,\psi_{ij})\,d\xi_1\,d\xi_2\,d\xi_3\,d\xi_4$$

is to be minimized. The Euler equations, the equations of motion of the field, are

$$\sum_{s=1}^{4}\frac{\partial}{\partial\xi_s}\left(\frac{\partial L}{\partial\psi_{is}}\right) = \frac{\partial L}{\partial\psi_i};\quad i = 1, 2, \ldots, n$$

or

$$\frac{\partial}{\partial t}\left(\frac{\partial L}{\partial\psi_{i4}}\right) = \frac{\partial L}{\partial\psi_i} - \sum_{s=1}^{3}\frac{\partial}{\partial\xi_s}\left(\frac{\partial L}{\partial\psi_{is}}\right) \qquad (3.4.1)$$

We note that the Lagrange integral $\mathfrak{L}$ and the corresponding Lagrange-Euler equations have a sort of "gauge invariance" (see page 211). Addition to the density L of the four-divergence of some four-vector function of the field variables or their derivatives, which goes to zero at the boundaries of the volume, will not change the value of $\mathfrak{L}$. For the fourfold integral of a four-divergence is equal to the four-dimensional equivalent of the net outflow integral of the vector function over the boundary surface, and this is zero if the vector function is zero at the boundary. Since $\mathfrak{L}$ is not changed by changing L to $L + \nabla \cdot \mathbf{F} = L'$, the new Lagrange density L' will also satisfy the Lagrange-Euler equations (3.4.1). Therefore $\mathfrak{L}$ and the Lagrange-Euler equations are invariant under such a change of L.

General Field Properties. The quantity $p_i = \partial L/\partial \psi_{i4}$ is the canonical momentum density for the ith component ψ_i, though we have seen that its relation to what is usually considered momentum is sometimes quite tenuous. Nevertheless, the quantity $\partial p_i/\partial t$ entering into the Euler equations is analogous to the mass times acceleration in a simpler system. The quantity

$$F_i = \frac{\partial L}{\partial \psi_i} - \sum_{s=1}^{3} \frac{\partial}{\partial \xi_s}\left(\frac{\partial L}{\partial \psi_{is}}\right)$$

which is equal to the time rate of change of p_i, is therefore analogous to a force component corresponding to the field component ψ_i. The first term $\partial L/\partial \psi_i$ usually has to do with the presence of external forces acting on the field. The second term often represents the effect of the rest of the field on the ith component at x, y, z, t.

The tensor $\mathfrak{W}$, having components

$$W_{ij} = \sum_{r=1}^{n} \frac{\partial \psi_r}{\partial \xi_i} \frac{\partial L}{\partial \psi_{rj}} - L\delta_{ij} \tag{3.4.2}$$

is the stress-energy tensor. Its time component W_{44} is the energy density H of the field, the integral of which is independent of time.

We can show, as we did earlier, that H can be expressed in terms of the ψ_r's, the canonical momenta p_r, and the gradients ψ_{rj}. We can proceed as we did on page 304 to obtain the Hamilton canonical equations from the variational principle. These equations turn out to be

$$\frac{\partial \psi_r}{\partial t} = \psi_{r4} = \frac{\partial H}{\partial p_r}; \quad \frac{\partial p_r}{\partial t} = \sum_{j=1}^{3} \frac{\partial}{\partial x_j}\left(\frac{\partial H}{\partial \psi_{rj}}\right) - \frac{\partial H}{\partial \psi_r}; \quad r = 1, 2, \ldots, n$$

From them also we can obtain the equations of motion (3.4.1).

The tensor $\mathfrak{W}$ is often not symmetric, which can be a serious matter inasmuch as we have come to expect stress dyadics to be symmetric. If it is desirable to use a symmetric tensor, we can usually do so by utilizing the "gauge invariance" of the function $\mathfrak{L}$ and of the Lagrange equations. We add to the density function L the divergence of some particular vector function of the ψ's and their derivatives and at the same time adjust the scales of the coordinates in a manner which will make the tensor $\mathfrak{W}$ symmetric and still have W_{44} the energy density. This uncertainty in the exact form of the stress-energy tensor is analogous to the uncertainty in form of the energy density of the string, as discussed on page 127. It is only a formal indeterminacy, however, for physically measurable quantities are not affected.

As we showed on page 309, the four-vector obtained by differentiation, having components

$$\sum_{j=1}^{4} \frac{\partial W_{ij}}{\partial \xi_j} = 0 \tag{3.4.3}$$

has zero magnitude.

We note, however, that the proof that these divergences are zero depends on the assumption that L and $\mathfrak{W}$ depend on the parameters ξ_j *only through the functions* ψ_r. If L (and therefore $\mathfrak{W}$) contains other terms (such as potentials or current densities) which are explicit functions of the ξ's, then Eqs. (3.4.3) will differ from zero by terms involving the derivatives of these extra terms with respect to the ξ's. Explicit dependence of L on the coordinates occurs only when the field is coupled to a set of particles or to a material medium causing the field (such as electric charge current). The Lorentz force on an electron, for instance, is given in terms of the field at a particular point in space, namely, the position of the electron. Interactions between the various parts of the field are expressed in terms of integrals over all space, and the dependence on the coordinates only enters through the ψ's.

At any rate, when L and $\mathfrak{W}$ depend on the ξ's *only* through the field variables ψ, Eqs. (3.4.3) hold, and in that case the three-vector

$$\mathbf{S} = \mathbf{i}W_{41} + \mathbf{j}W_{42} + \mathbf{k}W_{43} = \sum_{r=1}^{n} \frac{\partial \psi_r}{\partial t} \left[\mathbf{i} \frac{\partial L}{\partial \psi_{r1}} + \mathbf{j} \frac{\partial L}{\partial \psi_{r2}} + \mathbf{k} \frac{\partial L}{\partial \psi_{r3}} \right] \tag{3.4.4}$$

satisfies the equation of continuity, $\operatorname{div} \mathbf{S} + (\partial H/\partial t) = 0$, for the energy. Therefore it must represent the density of energy flow in the field. It can be called the *field intensity*.

The complementary vector

$$\mathbf{P} = \mathbf{i}W_{14} + \mathbf{j}W_{24} + \mathbf{k}W_{34} = \sum_{r=1}^{n} \frac{\partial L}{\partial \psi_{r4}} \operatorname{grad} \psi_r \tag{3.4.5}$$

has the dimensions of momentum per unit volume and can be called the *field momentum density*. If L has been modified so that $\mathfrak{W}$ is a symmetric tensor, then $\mathbf{P} = \mathbf{S}$; in any case $\mathbf{P}$ is closely related to $\mathbf{S}$.

The space part of the tensor $\mathfrak{W}$ is a three dyadic

$$\mathfrak{U} = \mathbf{i}\mathbf{W}_1 + \mathbf{j}\mathbf{W}_2 + \mathbf{k}\mathbf{W}_3$$

where

$$\mathbf{W}_1 = W_{11}\mathbf{i} + W_{12}\mathbf{j} + W_{13}\mathbf{k}$$
$$= \sum_{r=1}^{n} \frac{\partial \psi_r}{\partial x_1}\left[\mathbf{i}\frac{\partial L}{\partial \psi_{r1}} + \mathbf{j}\frac{\partial L}{\partial \psi_{r2}} + \mathbf{k}\frac{\partial L}{\partial \psi_{r3}}\right] - L\mathbf{i};\quad \text{etc.} \tag{3.4.6}$$

The other three divergence equations (3.4.3) are then given by the vector equation

$$\mathfrak{U} \cdot \nabla = \mathbf{i}\operatorname{div}\mathbf{W}_1 + \mathbf{j}\operatorname{div}\mathbf{W}_2 + \mathbf{k}\operatorname{div}\mathbf{W}_3 = -(\partial\mathbf{P}/\partial t)$$

indicating that, if $\mathbf{P}$ is a momentum, the derivative of $\mathfrak{U}$ is a force tensor, so that $\mathfrak{U}$ is related to the potential energy due to the field.

In tabular form, the stress energy tensor is

$$\mathfrak{W} = \begin{pmatrix} W_{11} & W_{12} & W_{13} & P_1 \\ W_{21} & W_{22} & W_{23} & P_2 \\ W_{31} & W_{32} & W_{33} & P_3 \\ S_1 & S_2 & S_3 & H \end{pmatrix}$$

with W_{ij} given by Eq. (3.4.2) and with $W_{n4} = P_n$ and $W_{4n} = S_n$.

An angular momentum density vector may also be generated (if it is needed) by taking the cross product of the radius vector $\mathbf{r}$ from some origin to the point (x,y,z) with the vector $\mathbf{P}$ at (x,y,z):

$$\mathbf{M} = \mathbf{r} \times \mathbf{P} = \sum_{r=1}^{n} \frac{\partial L}{\partial \psi_{r4}}[\mathbf{r} \times \operatorname{grad}\psi_r]$$

In quantum mechanics this property of the wave-function field turns out to be related to the probable angular momentum of the particles associated with the wave function. In the case of the nonviscous, compressible fluid, for instance, the angular momentum density would be

$$\mathbf{M} = (p/c^2)(\mathbf{r} \times \mathbf{v}) = \rho s(\mathbf{r} \times \mathbf{v})$$

according to Eqs. (3.3.15) and (3.3.10). This is the angular momentum of the excess density due to the motion of the fluid.

Therefore the change in energy density H with time requires an energy flow vector $\mathbf{S}$, and a change in momentum density $\mathbf{P}$ with time requires an internal stress dyadic $\mathfrak{U}$. For these reasons the tensor $\mathfrak{W}$ is called the *stress-energy tensor*, though it would be more accurate to call it the stress-momentum-energy tensor.

Isotropic Elastic Media. We are now in a position to apply these general formulas to a few interesting cases to see what some of this formalism means in physical terms. The first example is that of the motion of an elastic solid, discussed earlier in Secs. 1.6 and 2.2. From the latter section [Eqs. (2.2.17) and (2.2.18)] we see that the Lagrange density for an isotropic elastic medium is

$$\begin{aligned} L = T - V &= \tfrac{1}{2}\rho(\partial \mathbf{s}/\partial t)^2 - \tfrac{1}{2}|\mathfrak{T}\cdot\mathfrak{S}| \\ &= \tfrac{1}{2}\left\{\rho\left[\left(\frac{\partial s_x}{\partial t}\right)^2 + \left(\frac{\partial s_y}{\partial t}\right)^2 + \left(\frac{\partial s_z}{\partial t}\right)^2\right] - \lambda\left[\frac{\partial s_x}{\partial x} + \frac{\partial s_y}{\partial y} + \frac{\partial s_z}{\partial z}\right]^2 \right. \\ &\quad - 2\mu\left[\left(\frac{\partial s_x}{\partial x}\right)^2 + \left(\frac{\partial s_y}{\partial y}\right)^2 + \left(\frac{\partial s_z}{\partial z}\right)^2\right] \\ &\quad \left. - \mu\left[\left(\frac{\partial s_x}{\partial y} + \frac{\partial s_y}{\partial x}\right)^2 + \left(\frac{\partial s_x}{\partial z} + \frac{\partial s_z}{\partial x}\right)^2 + \left(\frac{\partial s_y}{\partial z} + \frac{\partial s_z}{\partial y}\right)^2\right]\right\} \end{aligned} \quad (3.4.7)$$

where the vector $\mathbf{s}$ is the displacement of the point x, y, z from undistorted equilibrium, ρ is the density of the medium, λ and μ its elastic constants, $\mathfrak{S}$ is the strain dyadic $\frac{1}{2}(\nabla\mathbf{s} + \mathbf{s}\nabla)$, and $\mathfrak{T}$ the stress dyadic

$$\mathfrak{T} = \lambda\mathfrak{I}|\mathfrak{S}| + 2\mu\mathfrak{S} = \lambda(\text{div }\mathbf{s})\mathfrak{I} + \mu(\nabla\mathbf{s} + \mathbf{s}\nabla)$$

The field variables ψ_n can be the three components of the displacement, s_x, s_y, s_z, which are to be varied until the total Lagrange function $\mathfrak{L} = \iiiint L\,dx\,dy\,dz\,dt$ is a minimum. The Lagrange-Euler equation (3.4.1) for s_x turns out to be

$$\rho\frac{\partial^2 s_x}{\partial t^2} = \lambda\frac{\partial}{\partial x}(\text{div }\mathbf{s}) + \mu\nabla^2 s_x + \mu\frac{\partial}{\partial x}(\text{div }\mathbf{s})$$

This is equivalent to the x component of the vector equation

$$\rho(\partial^2\mathbf{s}/\partial t^2) = (\lambda + \mu)\text{ grad }(\text{div }\mathbf{s}) + \mu\nabla^2\mathbf{s}$$

which is the equation of motion given already in Eq. (2.2.1).

The time part of the tensor $\mathfrak{W}$, defined in Eq. (3.4.2), is the energy density,

$$W_{44} = \tfrac{1}{2}\rho(\partial\mathbf{s}/\partial t)^2 + \tfrac{1}{2}|\mathfrak{T}\cdot\mathfrak{S}| = H$$

and the field intensity vector, defined by Eq. (3.4.4), is

$$\mathbf{S} = -(\partial\mathbf{s}/\partial t)\cdot\mathfrak{T}$$

which is the energy flow vector given in Eq. (2.2.20). It satisfies the equation of continuity for energy, $\text{div }\mathbf{S} + (\partial H/\partial t) = 0$, as shown in Eq. (3.4.3), and proved on page 309 (for in this case L depends on the coordinates *only* through the field variables s).

The tensor $\mathfrak{W}$ is not a symmetric one. The space part, corresponding to the force dyadic defined in Eqs. (3.4.6), is

$$\mathfrak{U} = -(\nabla\mathbf{s})\cdot\mathfrak{T} - L\mathfrak{I}$$

The field momentum density, defined in Eq. (3.4.5), is

$$\mathbf{P} = \rho(\nabla\mathbf{s}) \cdot (\partial\mathbf{s}/\partial t)$$

These two quantities satisfy the divergence equation $\mathfrak{U} \cdot \nabla + (\partial\mathbf{P}/\partial t) = 0$. If $\mathbf{P}$ is a momentum density, the dyadic $\mathfrak{U}$ is related to the stress density, as is verified by its definition in terms of $\mathfrak{T}$, the stress dyadic.

To illustrate the convenience and compactness of the dyadic and vector notation, we shall write out in full a few of the components of the stress-energy tensor $\mathfrak{W}$:

$$\begin{aligned} W_{11} = \tfrac{1}{2}(\lambda + 2\mu)\left[-\left(\frac{\partial s_x}{\partial x}\right)^2 + \left(\frac{\partial s_y}{\partial y}\right)^2 + \left(\frac{\partial s_z}{\partial z}\right)^2\right] - \tfrac{1}{2}\rho\left[\left(\frac{\partial s_x}{\partial t}\right)^2 \right.\\ \left. + \left(\frac{\partial s_y}{\partial t}\right)^2 + \left(\frac{\partial s_z}{\partial t}\right)^2\right] + \lambda\left(\frac{\partial s_y}{\partial y}\right)\left(\frac{\partial s_z}{\partial z}\right) \\ + \tfrac{1}{2}\mu\left[\left(\frac{\partial s_y}{\partial x}\right)^2 - \left(\frac{\partial s_x}{\partial y}\right)^2 + \left(\frac{\partial s_z}{\partial x}\right)^2 - \left(\frac{\partial s_x}{\partial z}\right)^2 + \left(\frac{\partial s_y}{\partial z} + \frac{\partial s_z}{\partial y}\right)^2\right] \end{aligned}$$

$$W_{12} = -\frac{\partial s_x}{\partial x}\left[\mu\left(\frac{\partial s_x}{\partial y} + \frac{\partial s_y}{\partial x}\right)\right] - \frac{\partial s_y}{\partial x}\left[(\lambda + 2\mu)\frac{\partial s_y}{\partial y} + \lambda\frac{\partial s_x}{\partial x} + \lambda\frac{\partial s_z}{\partial z}\right]; \quad \text{etc.}$$

There are not many additional comments which are appropriate here. Certainly the variational principle has collected most of the equations and formulas we so laboriously derived in Chap. 2 all in one compact package. Whether or not we can squeeze further physical meaning from the synthesis or can discover a use for the by-product quantities such as field momentum and force dyadic, we have at least developed a straightforward, as well as a suggestive, method for obtaining such important quantities as intensity, energy density, and equations of motion from the expression for a Lagrange density.

Plane-wave Solutions. To make more specific the formulas we have derived here, let us apply them to the plane wave, simple harmonic solutions of the equation of motion (2.2.1). Following Eq. (2.2.2), one solution is $\mathbf{s} = \text{grad}\,\psi$, where

$$\psi = Ce^{i\mathbf{k}\cdot\mathbf{r}-i\omega t}; \quad k = \omega/c_c; \quad c_c^2 = (\lambda + 2\mu)/\rho$$

The actual displacement is then obtained by taking the gradient of ψ,

$$\begin{aligned} \mathbf{s}_c &= i\mathbf{k}Ce^{i\mathbf{k}\cdot\mathbf{r}-i\omega t} = \mathbf{a}_k Ae^{i\mathbf{k}\cdot\mathbf{r}-i\omega t}; \quad A = ikC = |A|e^{i\varphi} \\ \mathbf{k} &= k\mathbf{a}_k; \quad \mathbf{a}_k = \alpha\mathbf{i} + \beta\mathbf{j} + \gamma\mathbf{k} \end{aligned}$$

where α, β, γ are the direction cosines for the propagation vector $\mathbf{k}$. Therefore, the displacement in this compressional wave is in the direction of propagation (as we mentioned before) with an amplitude $|A|$. The strain tensor is then

$$\begin{aligned} \mathfrak{S} &= \tfrac{1}{2}(\nabla\mathbf{s} + \mathbf{s}\nabla) = -\mathbf{kk}Ce^{i\mathbf{k}\cdot\mathbf{r}-i\omega t} = ik\mathbf{a}_k Ae^{i\mathbf{k}\cdot\mathbf{r}-i\omega t}; \\ \mathfrak{T} &= -[\lambda(\omega^2/c_c^2)\mathfrak{J} + 2\mu\mathbf{kk}]Ce^{i\mathbf{k}\cdot\mathbf{r}-i\omega t} \end{aligned}$$

where the dyadic $\mathbf{kk}$ is symmetric and its expansion factor $|\mathbf{kk}| = (\omega/c_c)^2$.

In order to compute the stress-energy tensor we must take the real parts of these expressions. The energy density, for instance, is

$$W_{44} = \rho\omega^2|A|^2 \sin^2 \Omega$$

where $\Omega = \mathbf{k} \cdot \mathbf{r} - \omega t + \varphi = (\omega/c_c)(\alpha x + \beta y + \gamma z - c_c t) + \varphi$. The energy flow vector is

$$\mathbf{S} = \mathbf{a}_k \rho c_c \omega^2 |A|^2 \sin^2 \Omega$$

and the wave momentum vector is $\mathbf{P} = \mathbf{S}/c_c^2$. The space part of the dyadic turns out to be

$$\mathfrak{U} = \mathbf{a}_k \mathbf{a}_k \rho\omega^2 |A|^2 \sin^2 \Omega$$

All of this, of course, is very nearly the same as the results given on page 312 for compressional waves in a fluid.

In the case of transverse or shear waves

$$\mathbf{s} = \mathbf{a}_p B e^{i\mathbf{k}\cdot\mathbf{r} - i\omega t}; \quad k = \omega/c_s; \quad c_s^2 = \mu/\rho; \quad B = |B|e^{i\varphi}$$

where $\mathbf{a}_p$ is a unit vector perpendicular to $\mathbf{k}$. The stress and strain dyadics are given by the equations

$$\mathfrak{T} = 2\mu\mathfrak{S} = i\rho c_s \omega B(\mathbf{a}_k \mathbf{a}_p + \mathbf{a}_p \mathbf{a}_k) e^{i\mathbf{k}\cdot\mathbf{r} - i\omega t}$$

The dyadic $(\mathbf{a}_k \mathbf{a}_p + \mathbf{a}_p \mathbf{a}_k)$ is symmetric but has zero expansion factor, so that $|\mathfrak{T}|$ and $|\mathfrak{S}|$ are both equal to zero.

The various parts of the stress-energy tensor are

$$W_{44} = \rho\omega^2|B|^2 \sin^2 \Omega; \quad \text{where } \Omega = (\omega/c_s)(\alpha x + \beta y + \gamma z - c_s t) + \varphi$$
$$\mathbf{S} = \mathbf{a}_k \rho c_s \omega^2 |B|^2 \sin^2 \Omega, \quad \mathbf{P} = \mathbf{S}/c_s^2; \quad \mathfrak{U} = \mathbf{a}_k \mathbf{a}_k \rho\omega^2 |B|^2 \sin^2 \Omega$$

which have the same form as the expressions for a compressional plane wave. In other words the energy flow vector and the wave momentum vector are pointed along the propagation vector $\mathbf{k}$, even though the displacement of the medium is perpendicular to $\mathbf{k}$.

Impedance. In the case of nonisotropic media the Lagrange density is [see Eq. (1.6.29)]

$$L = \tfrac{1}{2}\rho|\partial\mathbf{s}/\partial t|^2 - \tfrac{1}{2}(\nabla\mathbf{s}) :\gimel: (\nabla\mathbf{s})$$

where $\gimel$ (gimel) is a tetradic with elements g_{mnrs} determined by the nature of the medium. Due to the symmetry of the dyadics $\mathfrak{S}$ and $\mathfrak{T}$, there are certain symmetries of the tetradic which always hold: $g_{mnrs} = g_{rsmn} = g_{mnsr}$. In the case of an isotropic medium the elements are

$$g_{mnrs} = [\lambda\delta_{mn}\delta_{rs} + \mu\delta_{mr}\delta_{ns} + \mu\delta_{ms}\delta_{nr}]$$

or $\quad \gimel = \lambda ע + \mu ז + \mu ז^*; \quad ע : \mathfrak{A} = |\mathfrak{A}|\mathfrak{J}; \quad ז : \mathfrak{A} = \mathfrak{A}; \quad ז^* : \mathfrak{A} = \mathfrak{A}^*$

For the nonisotropic case, the equation of motion is

$$\rho(\partial^2\mathbf{s}/\partial t^2) = \nabla \cdot \gimel : (\nabla\mathbf{s})$$

which is a complicated second-order partial differential equation for the components of $\mathbf{s}$. Here it is not always possible to separate out pure compressional and pure transverse waves; also waves in different directions travel at different velocities.

The elements of the stress-energy tensor are

$$W_{44} = \tfrac{1}{2}\rho(\partial\mathbf{s}/\partial t)^2 + \tfrac{1}{2}(\nabla\mathbf{s}) : \mathfrak{T} : (\nabla\mathbf{s})$$
$$\mathbf{S} = -(\partial\mathbf{s}/\partial t)\cdot\mathfrak{T} : (\nabla\mathbf{s}); \quad \mathbf{P} = \rho(\nabla\mathbf{s})\cdot(\partial\mathbf{s}/\partial t)$$
$$\mathfrak{U} = -(\nabla\mathbf{s})\cdot[\mathfrak{T} : (\nabla\mathbf{s})] - L\mathfrak{J}$$

This same symbolism can be used to discuss the impedance for waves in an elastic medium. As we stated on page 310, the usual driving force is applied to the boundary surface of the medium and is equal to the volume integral of the inertial reaction $\rho(\partial\mathbf{s}/\partial t)$. But this is equal to a divergencelike expression, $\nabla\cdot\mathfrak{T} : (\nabla\mathbf{s})$, and the volume integral becomes equal to a surface integral of the surface force density dyadic

$$\mathfrak{F} = \mathfrak{T} : (\nabla\mathbf{s})$$

This expression is a dyadic (as are all stresses in an elastic medium) because the force is a vector which changes as the surface is changed in orientation. The force density on an element of boundary having the inward normal pointed along the unit vector $\mathbf{a}_n$ is $\mathbf{a}_n\cdot\mathfrak{T} : (\nabla\mathbf{s})$, which is a vector.

When the driving force is simple harmonic, the steady-state displacement vector also has a factor $e^{i\omega t}$ (or $e^{-i\omega t}$, in which case the impedance and admittance will be complex conjugates of the expressions for $e^{i\omega t}$). The force density across the part of the boundary surface which is vibrating with velocity $\mathbf{v} = \mathbf{V}e^{i\omega t} = i\omega\mathbf{s}$ is given by the equation

$$\mathbf{F} = \mathbf{a}_n\cdot\mathfrak{T} : (\nabla\mathbf{s}) = \mathfrak{Z}\cdot\mathbf{v} = i\omega\mathfrak{Z}\cdot\mathbf{s}$$

where $\mathbf{a}_n$ is a unit vector normal to the surface at the point where $\mathbf{F}$ is measured. The dyadic $\mathfrak{Z}$, which can be expressed in terms of the components g and the properties of the solution for $\mathbf{s}$, is the *impedance dyadic* which measures the reaction of the medium to a driving force.

For instance, for the isotropic case, $\mathfrak{T} = \lambda\mathfrak{J}\mathfrak{J} + \mu\mathfrak{K} + \mu\mathfrak{K}^*$, so that

$$\mathbf{a}_n\cdot\mathfrak{T} : (\nabla\mathbf{s}) = (\lambda\,\mathrm{div}\,\mathbf{s})\mathbf{a}_n + \mu\mathbf{a}_n\cdot(\nabla\mathbf{s} + \mathbf{s}\nabla)$$

For a plane compressional wave, with the driving surface perpendicular to the propagation vector (that is, $\mathbf{a}_n = \mathbf{a}_k$), then $(\nabla\mathbf{s}) = i\mathbf{a}_k\mathbf{k}Ae^{i\mathbf{k}\cdot\mathbf{r}-i\omega t} = (\mathbf{s}\nabla)$

$$\mathbf{a}_k\cdot\mathfrak{T} : (\nabla\mathbf{s}) = \mathbf{a}_k i\omega\rho c_c Ae^{i\mathbf{k}\cdot\mathbf{r}-i\omega t}; \quad \rho c_c^2 = \lambda + 2\mu$$

In this case the driving force is in the same direction as the velocity of the medium, $\mathbf{a}_k i\omega Ae^{i\mathbf{k}\cdot\mathbf{r}-i\omega t}$, so that the impedance dyadic is equal to the characteristic compressional impedance of the medium ρc_c times the idemfactor.

For a plane shear wave the velocity of the surface $i\omega \mathbf{a}_p A e^{-i\omega t}$ is is perpendicular to the propagation vector, and using the formulas on page 323,

$$\mathbf{a}_k \cdot \mathfrak{T}:(\nabla \mathbf{s}) = \mathbf{a}_p i\omega\rho c_s B e^{i\mathbf{k}\cdot\mathbf{r}-i\omega t}; \quad \rho c_s^2 = \mu$$

so that also in this case the driving force is parallel to the velocity and the impedance dyadic is the characteristic shear impedance ρc_s times the idemfactor, although here the driving force and velocity are perpendicular to the propagation vector.

The Electromagnetic Field. Next we come to a field which is expressed in terms of four-vectors, which, in fact, is the field for which the Lorentz transformation was devised, the electromagnetic field. A study of Sec. 2.5 suggests that the fundamental field quantities ψ_i should be the components of the potential four-vector given on page 208,

$$V_1 = A_x; \quad V_2 = A_y; \quad V_3 = A_z; \quad V_4 = i\varphi$$

where $\mathbf{A}$ is the vector potential and φ the scalar potential. In this case we may as well discard the coordinates ξ used in the previous example in favor of the Lorentz coordinates $x_1 = x$, $x_2 = y$, $x_3 = z$, $x_4 = ict$, as we did in Sec. 2.5. This choice will ensure Lorentz invariance but will require the factor ic to be added at times to retain the proper dimensions. The potential derivatives ψ_{ij} are therefore

$$V_{12} = (\partial A_x/\partial y), \text{ etc.}; \quad V_{14} = (1/ic)(\partial A_x/\partial t)$$
$$V_{41} = i(\partial\varphi/\partial x), \text{ etc.}; \quad V_{44} = (1/c)(\partial\varphi/\partial t)$$

In the present notation, then, the field vectors become

$$\begin{aligned} E_x &= i(V_{41} - V_{14}) = if_{14}; \quad E_y = i(V_{42} - V_{24}) = if_{24} \\ H_x &= (V_{32} - V_{23}) = f_{23}; \quad H_y = (V_{13} - V_{31}) = f_{31} \end{aligned} \tag{3.4.8}$$

if we assume that μ and ϵ are both unity. We have as a new complication, not encountered in the case of the elastic solid, the fact that the components of the potential are interrelated by means of an *auxiliary divergence condition*

$$\sum_{n=1}^{4} V_{nn} = \operatorname{div} \mathbf{A} + \left(\frac{1}{c}\right)\left(\frac{\partial\varphi}{\partial t}\right) = 0 \tag{3.4.9}$$

which is equivalent to Eq. (2.5.14), relating $\mathbf{A}$ and φ. This zero-value divergence can be added to or subtracted from various expressions to help simplify their form.

What we must now do is to set up a Lagrange density which will generate the equations of motion (see Eq. 2.5.20)

$$\sum_n \frac{\partial}{\partial x_n} f_{mn} = \sum_n \frac{\partial}{\partial x_n}(V_{nm} - V_{mn}) = \frac{\partial}{\partial x_m}\sum_n V_{nn} - \sum_n \frac{\partial^2 V_m}{\partial x_n^2} = \frac{4\pi}{c} I_m \tag{3.4.10}$$

equivalent to Maxwell's equations or to the wave equations (2.5.15) and which will produce, as the (4,4) component of the stress energy tensor, the energy density [see Eq. (2.5.28)]

$$W_{44} = \frac{1}{8\pi}(E^2 + H^2) = \frac{1}{8\pi}[f_{12}^2 + f_{23}^2 + f_{31}^2 - f_{14}^2 - f_{24}^2 - f_{34}^2]$$

when the four-vector **I** is zero. This vector **I** was defined on page 208, as the charge-current-density vector

$$I_1 = J_x; \quad I_2 = J_y; \quad I_3 = J_z; \quad I_4 = ic\rho$$

It is a little difficult to use the definition that L should be the difference between the kinetic energy density and the potential energy density, for it is not obvious which is kinetic and which potential energy. Examination of possible invariants suggests that part of the expression be $(1/8\pi)(E^2 - H^2)$; presumably the rest includes a scalar product of vector **I** with the potential vector **V**. Computing the Lagrange-Euler equation for such an L and comparing with Eq. (3.4.10) show that the proper expression for the Lagrange density is

$$\begin{aligned} L = &-\frac{1}{8\pi}\{(V_{41} - V_{14})^2 + (V_{42} - V_{24})^2 + (V_{43} - V_{34})^2 \\ &+ (V_{12} - V_{21})^2 + (V_{23} - V_{32})^2 + (V_{31} - V_{13})^2\} \\ &+ \frac{1}{c}\{I_1V_1 + I_2V_2 + I_3V_3 + I_4V_4\} \\ = &-\frac{1}{16\pi}\sum_{n,m} f_{nm}^2 + \frac{1}{c}\sum_n I_nV_n = \frac{1}{8\pi}(E^2 - H^2) + \frac{1}{c}\mathbf{J}\cdot\mathbf{A} - \rho\varphi \end{aligned} \tag{3.4.11}$$

Therefore Maxwell's equations for free space (**J** and ρ, zero) correspond to the requirement that E^2 be as nearly equal to H^2 as the boundary conditions allow.

The Lagrange-Euler equations (3.4.1) are just the Maxwell equations (3.4.10). The canonical momentum density vector **p**, having components $p_n = (1/ic)(\partial L/\partial V_{n4})$ $(n = 1, 2, 3)$, which turn out to be $(1/4\pi ic)(V_{4n} - V_{n4})$, is the vector $-(1/4\pi c)\mathbf{E}$. The "force vector" corresponding to the rate of change of this momentum with respect to time is then $(\mathbf{J}/c) - (1/4\pi)$ curl **H**. The time component of the canonical momentum density $(\partial L/\partial V_{44})$, is zero, and the time component of the Lagrange-Euler equations (3.4.10), div $\mathbf{E} = 4\pi\rho$ is a sort of equation of continuity for the canonical momentum density vector $\mathbf{p} = -(1/4\pi c)\mathbf{E}$.

Stress-energy Tensor. The time component of the momentum-energy tensor $\mathfrak{W}$ should be the Hamiltonian density:

$$W_{44} = \sum_{i=1}^{4} V_{i4} \frac{\partial L}{\partial \psi_{i4}} - L = -\frac{1}{8\pi} \sum_{m=1}^{3} (V_{4m} - V_{m4})^2$$
$$+ \frac{1}{4\pi} \sum_{m=1}^{3} V_{4m}(V_{4m} - V_{m4}) + \frac{1}{8\pi} [(V_{12} - V_{21})^2$$
$$+ (V_{23} - V_{32})^2 + (V_{31} - V_{13})^2] - \frac{1}{c} \sum_{m=1}^{4} I_n V_n$$
$$= -\frac{1}{4\pi} \sum_r f_{4r}^2 + \frac{1}{16\pi} \sum_{s,r} f_{sr}^2 - \frac{1}{c} \sum_n I_n V_n + \frac{1}{4\pi} \sum_m V_{4m} f_{4m}$$
$$= \frac{1}{8\pi} (E^2 + H^2) - \frac{1}{c} \mathbf{J} \cdot \mathbf{A} + \rho\varphi + \frac{1}{4\pi} \mathbf{E} \cdot \operatorname{grad} \varphi \tag{3.4.12}$$

which differs from the results of Eq. (2.5.28) by the terms $[\rho\varphi + (1/4\pi)\mathbf{E} \cdot \operatorname{grad} \varphi]$. However the expression $\mathbf{E} \cdot \operatorname{grad} \varphi$ is equal to $\operatorname{div} (\varphi \mathbf{E}) - \varphi \operatorname{div} \mathbf{E}$; and by remembering that $\operatorname{div} \mathbf{E} = 4\pi\rho$, we see that the extra terms are just equal to $(1/4\pi) \operatorname{div} (\varphi \mathbf{E})$. Since the integral of a divergence over all space equals the net outflow integral at infinity, which is zero, we see that the *average* value of W_{44} is equal to the *average* value of the Hamiltonian density

$$U = \frac{1}{8\pi} (E^2 + H^2) - \frac{1}{c} \mathbf{J} \cdot \mathbf{A} = T_{44} - \frac{1}{c} \mathbf{J} \cdot \mathbf{A} \tag{3.4.13}$$

where the tensor $\mathfrak{T}$ is defined in Eq. (2.5.30). This is an example of the gauge-invariant properties of the field mentioned on page 211 and also of the fact, noted on page 126, that energy density and energy flow are not uniquely determined, except in terms of the integral over all space.

On the other hand, to obtain the correct results for the Hamilton canonical equations given on page 319, we must use the full expression for W_{44}, with the canonical momentum $\mathbf{p}$ inserted for $-(1/4\pi c)\mathbf{E}$. The Hamiltonian then takes on the form

$$\mathfrak{H} = 2\pi c^2 p^2 + \frac{1}{8\pi} H^2 - c\mathbf{p} \cdot \operatorname{grad} \varphi - \frac{1}{c} \mathbf{J} \cdot \mathbf{A} + \rho\varphi$$
$$= \sum_{n=1}^{3} (2\pi c^2 p_n^2 + icV_{4n}p_n) - \frac{1}{c} \sum_{m=1}^{4} V_m I_m$$
$$+ \frac{1}{8\pi} [(V_{12} - V_{21})^2 + (V_{13} - V_{31})^2 + (V_{23} - V_{32})^2]$$

The equation $\partial\psi_n/\partial t = icV_{n4} = \partial H/\partial p_n$ becomes

$$ic(V_{n4} - V_{4n}) = 4\pi c^2 p_n \quad \text{or} \quad p_n = (1/4\pi ic)(V_{4n} - V_{n4})$$

which is the original definition of p_n. The equations

$$\frac{\partial p_n}{\partial t} = \sum_{r=1}^{3} \frac{\partial}{\partial x_r}\left(\frac{\partial H}{\partial V_{nr}}\right) - \frac{\partial H}{\partial V_r}$$

become the Maxwell equations

$$-\frac{1}{4\pi c}\frac{\partial \mathbf{E}}{\partial t} = -\frac{1}{4\pi}\operatorname{curl}\mathbf{H} + \frac{1}{c}\mathbf{J};\quad p_4 = 0;\quad -\frac{1}{4\pi}\operatorname{div}\mathbf{E} + \rho = 0$$

Therefore one can use the component W_{44} to calculate the Hamiltonian, but one should use U to compute the conventional energy density.

A similar sort of adjustment must be made to obtain the familiar forms for intensity vector and field momentum from the nondiagonal terms of the stress-energy tensor. We have

$$W_{mn} = \sum_{r=1}^{4} V_{rm}\left(\frac{\partial L}{\partial V_{rn}}\right) = -\frac{1}{4\pi}\sum_r V_{rm}(V_{rn} - V_{nr})$$

$$= -\frac{1}{4\pi}\sum_r (V_{rm} - V_{mr})(V_{rn} - V_{nr}) - \frac{1}{4\pi}\sum_r (V_{mr}V_{rn} - V_{mr}V_{nr}) \tag{3.4.14}$$

The second sum in the last expression can be modified by using the auxiliary condition (3.4.9) and also the wave equation for the potentials [Eqs. (2.5.15)] $\sum_r (\partial V_{nr}/\partial x_r) = -(4\pi I_n/c)$:

$$-\frac{1}{4\pi}\sum_r (V_{mr}V_{rn} - V_{mr}V_{nr}) = -\frac{1}{4\pi}\sum_r \frac{\partial}{\partial x_r}[V_m(V_{rn} - V_{nr})] + \frac{V_m}{4\pi}\left[\frac{\partial}{\partial x_n}\sum_r V_{rr} - \sum_r \frac{\partial}{\partial x_r} V_{nr}\right]$$

The first sum is a four-divergence, which is zero on the average. The second sum is zero because of Eq. (3.4.9) and the third is equal to $V_m I_n/c$.

Therefore the average value of W_{mn} $(m \neq n)$ is equal to the average value of the terms

$$T_{mn} + \left(\frac{V_m I_n}{c}\right) = -\frac{1}{4\pi}\sum_{r=1}^{4} f_{mr}f_{nr} + \left(\frac{V_m I_n}{c}\right) \tag{3.4.15}$$

In fact the average value of any of the terms W_{mn} of the tensor $\mathfrak{W}$ is equal to the average value of the tensor with terms

$$T_{mn} + \frac{1}{c} V_m I_n - \frac{1}{c} \delta_{mn} \sum_r V_r I_r,$$

where
$$T_{mn} = \frac{1}{4\pi} \left\{ \sum_r f_{mr} f_{rn} + \tfrac{1}{4} \delta_{mn} \sum_{r,s} f_{rs}^2 \right\} \tag{3.4.16}$$

The tensor $\mathfrak{T}$ has been discussed earlier, on page 216.

In those parts of space where the charge-current vector is zero the tensor $\mathfrak{T}$ is the stress-energy tensor. Expressed in terms of fields, the components are

$$\begin{aligned}
T_{11} &= \frac{1}{8\pi} [E_x^2 - E_y^2 - E_z^2 + H_x^2 - H_y^2 - H_z^2]; \quad \text{etc.} \\
T_{44} &= \frac{1}{8\pi} [E_x^2 + E_y^2 + E_z^2 + H_x^2 + H_y^2 + H_z^2] = U \\
T_{12} &= \frac{1}{4\pi} [E_x E_y + H_x H_y] = T_{21}; \quad \text{etc.} \\
T_{14} &= \frac{1}{4\pi i} [E_y H_z - E_z H_y] = \frac{1}{4\pi i} (\mathbf{E} \times \mathbf{H})_x = T_{41}; \quad \text{etc.}
\end{aligned} \tag{3.4.17}$$

Field Momentum. The fact that we have discarded the tensor $\mathfrak{W}$ for the tensor $\mathfrak{T}$ need not trouble us unduly, for $\mathfrak{W}$ does not satisfy the divergence conditions (3.4.3) unless $\mathbf{J}$ and ρ are zero, so $\mathfrak{W}$ would not be very useful anyway. The divergence relations for $\mathfrak{T}$ are not simple either because of the separation off of the four-divergence terms. We have

$$\sum_r \frac{\partial T_{mr}}{\partial x_r} = \frac{1}{4\pi} \sum_{r,s} f_{mr} \frac{\partial f_{rs}}{\partial x_s} = \frac{1}{c} \sum_r f_{mr} I_r = k_m \tag{3.4.18}$$

where k_m is the mth component of the force-density vector defined in Eq. (2.5.27). The space part

$$\rho \mathbf{E} + (1/c)\mathbf{J} \times \mathbf{H}$$

gives the magnitude and direction of the force on the charge-current distribution. It should equal the time rate of change of the momentum of the charge, which, together with the rate of change of the momentum of the field, should equal the net force on the field plus charge. Taking the integral of k_1, for instance, over a given portion of space, and calling Π_1 the x component of the momentum of the charge current in this region, we have

$$\begin{aligned}
\frac{d\Pi_1}{dt} = \iiint k_1 \, dv &= \iiint \frac{\partial T_{14}}{\partial x_4} dv + \iiint \operatorname{div} \mathbf{F}_1 \, dv \\
&= -\frac{1}{4\pi c} \frac{\partial}{\partial t} \iiint (\mathbf{E} \times \mathbf{H})_x \, dv + \iint \mathbf{F}_1 \cdot d\mathbf{A}
\end{aligned}$$

where $\mathbf{F}_1 = T_{11}\mathbf{i} + T_{12}\mathbf{j} + T_{13}\mathbf{k}$ is the net force acting on the x components of field and charge momenta, and where the last integral is a surface integral over the boundary of the given portion of space. If now

$$\mathbf{P} = \frac{1}{4\pi c}(\mathbf{E} \times \mathbf{H}) = \frac{i}{c}[T_{14}\mathbf{i} + T_{24}\mathbf{j} + T_{34}\mathbf{k}] \qquad (3.4.19)$$

is called the field momentum (see page 321), then the previous equation states that the net stress $\mathbf{T}$ acting over the surface of a portion of space equals the rate of change of the momentum Π of the charge current inside the surface *plus* the rate of change of the field momentum $\mathbf{P}$ inside the same surface.

The time component k_4 of Eq. (3.4.18) is the time rate of change of the kinetic energy T of the charge current. This equation also has physical significance, which becomes clearer if we define the rate of flow of energy by the usual vector (called here the *Poynting vector*)

$$\mathbf{S} = ic[T_{41}\mathbf{i} + T_{42}\mathbf{j} + T_{43}\mathbf{k}] = (c/4\pi)(\mathbf{E} \times \mathbf{H}) \qquad (3.4.20)$$

The component T_{44} is, of course, just the energy density U of the field. The $m = 4$ part of Eq. (3.4.18) is therefore

$$\operatorname{div} \mathbf{S} + (\partial U/\partial t) = -(\partial T/\partial t)$$

so that the equation of continuity for energy flow is that the net outflow integral of $\mathbf{S}$ over a closed boundary is equal to the negative rate of change of energy of charge current T and of the field U for the volume inside the boundary. Thus all the components of tensor $\mathfrak{T}$ have physical significance.

The density of angular momentum in a field with no charge current present is

$$\mathbf{M} = \mathbf{r} \times \mathbf{P} = \frac{1}{4\pi c}\mathbf{r} \times (\mathbf{E} \times \mathbf{H}) = \frac{1}{4\pi c}[(\mathbf{r} \cdot \mathbf{H})\mathbf{E} - (\mathbf{r} \cdot \mathbf{E})\mathbf{H}]$$

The total angular momentum of the field about the origin is obtained by integrating $\mathbf{M}$ over all the volume occupied by the field.

When an electromagnetic field, divorced from charge current, is confined inside a finite volume of space (a wave packet) which moves about as time is changed, we can show that the integral of the four quantities $(P_x,P_y,P_z,U) = (iT_{14}/c,\ iT_{24}/c,\ iT_{34}/c,\ T_{44})$ over the space occupied by the field (*i.e.*, integrated over the three space perpendicular to the time axis at any given instant) is a four-vector satisfying the Lorentz requirements for transformation of a four-vector. For in this case $\sum_r (\partial T_{mr}/\partial x_r) = 0$ so that, if C_m are the components of a constant four-vector, the four-divergence of the four-vector with components $B_r = \sum_m C_m T_{mr}$, $\sum_r (\partial B_r/\partial x_r)$ is zero, and the integral of the normal component of this vector over the surface of an arbitrary volume in four space is zero. We choose for the volume in four space the "four prism" with axis along the time dimension parallel to the motion of the wave packet and with space part perpendicular to this axis and large enough to contain the packet completely. The surface integral over the space part (along the sides of the four prism) is zero, for the field is zero outside the packet. Hence the integral of the time component of B, $\sum_m C_m T_{m4}$, over the packet at one end of the four prism must be equal to the same integral over the other end of the prism at an earlier time. Therefore, for this case, the integral of $\Sigma C_m T_{m4}$ over the packet is a Lorentz invariant and the components given by the integration of T_{m4} over the packet (over volumes perpendicular to the time axis) are components of a true four-vector. This is what we set out to prove.

This result indicates that, if we have such a thing as a wave packet of electromagnetic field, the vector, having as components the integrated field momentum $\mathbf{P}$ and the integrated field energy U, is a true momentum-energy vector behaving just as if the packet were a material particle. Its angular momentum can be obtained by integrating the $\mathbf{M}$ just obtained.

There are many other interesting properties of the electromagnetic field which can be obtained by means of the variational machinery we have set up.

Gauge Transformation. Many of the difficulties we have encountered in going from Lagrange density to energy density can be simplified by choosing the right gauge. If, instead of the gauge defined by the equation $\text{div}\,\mathbf{A} + (1/c)(\partial\varphi/\partial t) = 0$, we use the gauge defined by the equation $\varphi = 0$, the Maxwell equations reduce to

$$\begin{gathered}\text{curl}\,\mathbf{A} = \mathbf{B} = \mu\mathbf{H};\quad \mathbf{E} = -(1/c)(\partial\mathbf{A}/\partial t) = \mathbf{D}/\epsilon \\ \text{div}\,(\partial\mathbf{A}/\partial t) = -(4\pi\rho c/\epsilon) \\ \text{curl}\,(\text{curl}\,\mathbf{A}) + (\epsilon\mu/c^2)(\partial^2\mathbf{A}/\partial t^2) = (4\pi\mu/c)\mathbf{J}\end{gathered} \tag{3.4.21}$$

In other words, we use both longitudinal and transverse parts of $\mathbf{A}$, the longitudinal part being determined by the charge density and the transverse part being largely determined by the current density. This gauge is particularly useful for cases where there is no free charge ρ, though it is also useful at other times.

In this gauge, the Lagrange density is

$$L = \frac{\epsilon}{8\pi c^2}\left|\frac{\partial\mathbf{A}}{\partial t}\right|^2 - \frac{1}{8\pi\mu}|\text{curl}\,\mathbf{A}|^2 + \frac{1}{c}\mathbf{J}\cdot\mathbf{A} = \frac{1}{8\pi}(\mathbf{E}\cdot\mathbf{D} - \mathbf{H}\cdot\mathbf{B}) + \frac{1}{c}\mathbf{J}\cdot\mathbf{A} \tag{3.4.22}$$

The canonical momentum density is, therefore, $\mathbf{p} = \epsilon\dot{\mathbf{A}}/4\pi c^2 = -(\mathbf{D}/4\pi c)$. The Lagrange-Euler equations give us the last of Eqs. (3.4.21); the first two equations define the relation between the fields and the potential, and the third equation fixes the gauge.

The Hamiltonian density is then

$$W_{44} = \mathbf{p} \cdot \dot{\mathbf{A}} - L = \frac{1}{8\pi}(\mathbf{E} \cdot \mathbf{D} + \mathbf{H} \cdot \mathbf{B}) - \frac{1}{c}\mathbf{J} \cdot \mathbf{A}$$

$$= \frac{2\pi c^2}{\epsilon} p^2 + \frac{1}{8\pi\mu} |\mathrm{curl}\, \mathbf{A}|^2 - \frac{1}{c}\mathbf{J} \cdot \mathbf{A} = \mathfrak{H} \quad (3.4.23)$$

The second of the modified canonical equations (3.4.2) again corresponds to the last of Eqs. (3.4.21).

To find the rest of the stress energy tensor $\mathfrak{W}$ in this gauge we note that the dyadic with (x,y) component $\partial L/\partial(\partial A_x/\partial y)$ is $(\mathbf{H} \times \mathfrak{J}/4\pi) = -(\mathfrak{J} \times \mathbf{H}/4\pi)$. Utilizing this expression, we see that the energy flow vector is (for $\mu = \epsilon = 1$)

$$\mathbf{S} = -\dot{\mathbf{A}} \cdot (\mathfrak{J} \times \mathbf{H}/4\pi) = (c/4\pi)(\mathbf{E} \times \mathbf{H})$$

which is the same as the expression given in Eq. (3.4.20). Therefore this particular choice of gauge gives the standard form for the energy density and the Poynting vector without all the fussing with divergences which the usual choice of gauge requires and which we displayed in earlier pages. The field momentum vector, on the other hand, has a modified form,

$$\mathbf{P} = -\frac{1}{4\pi}(\nabla\mathbf{A}) \cdot \dot{\mathbf{A}} = \frac{c}{4\pi}[(\mathbf{D} \times \mathbf{B}) + \mathbf{D} \cdot (\nabla\mathbf{A})]$$

and, correspondingly, the space part of the stress-energy tensor is modified, becoming

$$\mathfrak{U} = (1/4\pi)(\nabla\mathbf{A}) \times \mathbf{H} - \mathfrak{J}L$$

These quantities are not quite so familiar to us as are the energy density and Poynting vector, so they may, perhaps, be allowed to take on these modified forms (or else the divergence argument may be applied to arrive at the more familiar form).

Impedance Dyadic. To determine the field impedance for the electromagnetic field it is most convenient to use this latest choice of gauge, which gives the "correct" form for energy density and energy flow density. We return to the Lagrange-Euler equations (or the canonical equations)

$$\dot{\mathbf{p}} = -\frac{1}{4\pi}\nabla \cdot (\mathfrak{J} \times \mathbf{H}) + \frac{1}{c}\mathbf{J}$$

The quantity on the right is the "force" which causes a rate of change of the momentum $\mathbf{p} = \epsilon\mathbf{A}/4\pi c^2 = -(\mathbf{D}/4\pi c)$. The part which can be applied at the boundary surface, according to the arguments of page 310, is the dyadic $(-1/4\pi)(\mathfrak{J} \times \mathbf{H})$, the divergence of which enters into the above expression for force density. If an electromagnetic wave is started at some part of the boundary surface, the "reaction" of the wave back on the element of area $d\mathbf{A}$ is, accordingly,

$$(1/4\pi)d\mathbf{A} \cdot (\mathfrak{I} \times \mathbf{H}) = (1/4\pi)(d\mathbf{A} \times \mathbf{H})$$

a vector perpendicular to $\mathbf{H}$ and to $d\mathbf{A}$ (*i.e.*, tangential to the boundary surface). Relating this to the circuital rule (see page 220) $c\oint \mathbf{H} \cdot d\mathbf{s} = 4\pi\mathbf{I}$ we see that if the wave is "caused" by a surface current in the boundary surface, then the vector $-(c/4\pi)(d\mathbf{A} \times \mathbf{H})$ is just equal, in amount and direction, to the part of this surface current which is contained in the element $d\mathbf{A}$. The integral of this vector over all the driving surface gives us just the total current sheet.

The "velocity" vector is $\dot{\mathbf{A}} = -c\mathbf{E}$, so that the quantity corresponding to the impedance for the potential $\mathbf{A}$, in the direction given by the unit vector $\mathbf{a}$, is the dyadic which changes the vector $-c\mathbf{E}$ into the vector $(\mathbf{a}/4\pi) \times \mathbf{H}$. However, the choice we have made as to which expression is to be "force" and which "velocity" is just the inverse of the usual definition of impedance, which is that Z is the ratio of voltage to current (H is proportional to current and E to voltage).

Consequently we define the *impedance dyadic* $\mathfrak{Z}$ of the electromagnetic field as the "ratio" between the electric field and $c/4\pi$ times the magnetic field, and the *admittance dyadic* $\mathfrak{Y}$ of the field as its inverse:

$$4\pi\mathbf{E} = -c\mathfrak{Z} \cdot \mathbf{H}; \quad c\mathbf{H} = -4\pi\mathfrak{Y} \cdot \mathbf{E}; \quad \mathfrak{Y} = \mathfrak{Z}^{-1} \tag{3.4.24}$$

The admittance of the field in the direction of the unit vector $\mathbf{a}$ is then $\mathbf{a} \times \mathfrak{Y}$, as can be seen by vector multiplication of the second equation by $\mathbf{a}$.

Incidentally we notice that, if $\mathbf{E}$ is analogous to a voltage and $(c/4\pi)(\mathbf{a} \times \mathbf{H})$ to a current, so that the "ratio" is an impedance, the "product" $(c/4\pi)(\mathbf{E} \times \mathbf{H})$ is the ratio of energy consumption, *i.e.*, the energy flow density [which Eq. (3.4.20) shows it to be]. Therefore our analogy is complete.

Plane-wave Solution. If there is no charge current and if $\epsilon = \mu = 1$, a simple solution of the equations of motion (3.4.21) is

$$\mathbf{A} = \mathbf{a}_p A e^{i\mathbf{k}\cdot\mathbf{r} - i\omega t}; \quad \mathbf{k} = (\omega/c)\mathbf{a}_k; \quad A = |A|e^{i\varphi}$$

where $\mathbf{a}_k$ and $\mathbf{a}_p$ are two mutually perpendicular unit vectors. The fields are therefore

$$\mathbf{E} = i(\omega/c)\mathbf{a}_p A e^{i\mathbf{k}\cdot\mathbf{r} - i\omega t}; \quad \mathbf{H} = i(\omega/c)(\mathbf{a}_k \times \mathbf{a}_p) A e^{i\mathbf{k}\cdot\mathbf{r} - i\omega t} = \mathbf{a}_k \times \mathbf{E}$$

so that the vectors $\mathbf{k}$, $\mathbf{E}$, and $\mathbf{H}$ form a right-handed, orthogonal trio of vectors. As usual with plane-wave solutions, the value of the Lagrange function is zero. The energy density and the Poynting vector are

$$U = \frac{E^2}{4\pi} = \frac{\omega^2|A|^2}{4\pi c^2} \sin^2 \Omega; \quad \mathbf{S} = \frac{\omega^2|A|^2}{4\pi c} \mathbf{a}_k \sin^2 \Omega$$

$$\Omega = k(\alpha x + \beta y + \gamma z - ct + \varphi)$$

where α, β, and γ are the direction cosines of $\mathbf{k}$ on the x, y, z axes. Dyadic $(\nabla\mathbf{A})$ is $i(\omega/c)\mathbf{a}_k\mathbf{a}_p A e^{i\mathbf{k}\cdot\mathbf{r}-i\omega t}$ so that the field momentum density is

$$\mathbf{P} = \frac{\omega^2|A|^2}{4\pi c}\,\mathbf{a}_k \sin^2\Omega = \mathbf{S}$$

and the space part of the stress-energy tensor is the symmetric dyadic

$$\mathfrak{U} = \frac{\omega^2|A|^2}{4\pi c^2}\,\mathbf{a}_k\mathbf{a}_k \sin^2\Omega$$

In matrix form the whole stress-energy tensor has the following symmetric form:

$$\mathfrak{W} = \frac{\omega^2|A|^2}{4\pi c^2}\sin^2(\mathbf{k}\cdot\mathbf{r} - \omega t + \varphi)\begin{pmatrix} \alpha^2 & \alpha\beta & \alpha\gamma & c\alpha \\ \beta\alpha & \beta^2 & \beta\gamma & c\beta \\ \gamma\alpha & \gamma\beta & \gamma^2 & c\gamma \\ c\alpha & c\beta & c\gamma & 1 \end{pmatrix}$$

Finally the impedance of the plane wave is the ratio between the vector $\mathbf{E}$ and the vector $-(c/4\pi)\mathbf{H}$, which is the dyadic

$$\mathfrak{Z} = (4\pi/c)\mathfrak{J} \times \mathbf{a}_k$$

and the admittance is

$$\mathfrak{Y} = (c/4\pi)\mathfrak{J} \times \mathbf{a}_k$$

The impedance of the wave in the direction of propagation is thus

$$(4\pi/c)\mathbf{a}_k \times \mathfrak{J} \times \mathbf{a}_k = -(4\pi/c)(\mathfrak{J} - \mathbf{a}_k\mathbf{a}_k)$$

In the Gaussian units, which we are using here, the "magnitude" of the impedance of an electromagnetic plane wave in vacuum is thus $4\pi/c$.

There are only a few other fields meriting attention in this chapter.

Dirac Equation. For instance, we should be able to set up a Lagrange density for the Dirac equation for the *electron*, Eq. (2.6.57). Here we have eight independent field functions, the four components of ψ, ψ_1, ψ_2, ψ_3, ψ_4 [given in Eq. (2.6.56)] along the four directions in spin space and the corresponding components of ψ^*. A little juggling of expressions will show that the Lagrange density is

$$L = \frac{\hbar c}{2i}[(\operatorname{grad}\psi^*)\cdot\boldsymbol{\alpha}\psi - \psi^*\boldsymbol{\alpha}\cdot(\operatorname{grad}\psi)] + \frac{\hbar}{2i}\left[\frac{\partial\psi^*}{\partial t}\psi - \psi^*\frac{\partial\psi}{\partial t}\right] - e\psi^*\boldsymbol{\alpha}\cdot\mathbf{A}\psi + ec\psi^*\varphi\psi - mc^2\psi^*\alpha_0\psi \qquad (3.4.25)$$

where $\mathbf{A}$ and φ are the electromagnetic potentials at the position of the electron, m and e the electronic mass and charge, where ψ and ψ^* represent all four components of each vector and where the operators $\boldsymbol{\alpha} = \alpha_x\mathbf{i} + \alpha_y\mathbf{j} + \alpha_z\mathbf{k}$ and α_0 are those defined in Eqs. (2.6.55).

The Lagrange-Euler equations may be obtained in the usual manner after substituting for ψ^*, ψ in terms of ψ_1^*, ψ_2^*, ψ_3^*, ψ_4^*, ψ_1, ψ_2, ψ_3, ψ_4 in

Eq. (3.4.25) and performing the necessary operations required by the operators α. For instance, the equation

$$\frac{\partial}{\partial x}\left(\frac{\partial L}{\partial \psi_{1x}^*}\right) + \frac{\partial}{\partial y}\left(\frac{\partial L}{\partial \psi_{1y}^*}\right) + \frac{\partial}{\partial z}\left(\frac{\partial L}{\partial \psi_{1z}^*}\right) + \frac{\partial}{\partial t}\left(\frac{\partial L}{\partial \psi_{1t}^*}\right) - \frac{\partial L}{\partial \psi_1^*} = 0$$

results in

$$c\left[\left(\frac{\hbar}{i}\frac{\partial \psi_4}{\partial x} + \frac{e}{c}A_x\psi_4\right) + i\left(\frac{\hbar}{i}\frac{\partial \psi_4}{\partial y} + \frac{e}{c}A_y\psi_4\right) + \left(\frac{\hbar}{i}\frac{\partial \psi_3}{\partial z} + \frac{e}{c}A_z\psi_3\right) + \left(\frac{\hbar}{ic}\frac{\partial \psi_1}{\partial t} - e\varphi\psi_1\right) + mc\psi_1\right] = 0 \quad (3.4.26)$$

which is one term of the Dirac equations (2.6.57). However, we can obtain the same result more easily by considering that only *two* field variables, ψ and ψ^*, are involved and performing the necessary partial derivatives formally as though they were simple functions instead of vectors in spin space. For instance, the Lagrange-Euler equation

$$\frac{\partial}{\partial x}\left(\frac{\partial L}{\partial \psi_x^*}\right) + \frac{\partial}{\partial y}\left(\frac{\partial L}{\partial \psi_y^*}\right) + \frac{\partial}{\partial z}\left(\frac{\partial L}{\partial \psi_z^*}\right) + \frac{\partial}{\partial t}\left(\frac{\partial L}{\partial \psi_t^*}\right) - \frac{\partial L}{\partial \psi^*} = 0$$

corresponds to the whole of Eq. (2.6.57),

$$c\left[\alpha_0 mc\psi + \boldsymbol{\alpha}\cdot\left(\frac{\hbar}{i}\,\mathrm{grad}\,\psi + \frac{e}{c}\mathbf{A}\psi\right) + \left(\frac{\hbar}{ic}\frac{\partial \psi}{\partial t} - e\varphi\psi\right)\right] = 0 \quad (3.4.27)$$

one part of which is Eq. (3.4.26). The corresponding equation for the spin vector ψ^* is

$$c\left[\psi^* mc\alpha_0 + \left(-\frac{\hbar}{i}\,\mathrm{grad}\,\psi^* + \frac{e}{c}\mathbf{A}\psi^*\right)\cdot\boldsymbol{\alpha} + \left(-\frac{\hbar}{ic}\frac{\partial \psi^*}{\partial t} - e\varphi\psi^*\right)\right] = 0$$

The energy is again the (4,4) component of the tensor $\mathfrak{W}$;

$$\begin{aligned} W_{44} &= \psi_t^*\frac{\partial L}{\partial \psi_t^*} + \psi_t\frac{\partial L}{\partial \psi_t} - L = H \\ &= mc^2(\psi^*\alpha_0\psi) + e\mathbf{A}\cdot(\psi^*\boldsymbol{\alpha}\psi) - ec(\psi^*\varphi\psi) \\ &\qquad + \frac{\hbar c}{2i}[\psi^*\boldsymbol{\alpha}\cdot\mathrm{grad}\,\psi - \mathrm{grad}\,\psi^*\cdot\boldsymbol{\alpha}\psi] \end{aligned} \quad (3.4.28)$$

and the "field intensity" vector is

$$\mathbf{S} = \mathbf{i}W_{41} + \mathbf{j}W_{42} + \mathbf{k}W_{43} = \frac{\hbar c}{2i}\left[\frac{\partial \psi^*}{\partial t}\boldsymbol{\alpha}\psi - \psi^*\boldsymbol{\alpha}\frac{\partial \psi}{\partial t}\right] \quad (3.4.29)$$

whereas the "field momentum" vector is

$$\mathbf{P} = \mathbf{i}W_{14} + \mathbf{j}W_{24} + \mathbf{k}W_{34} = \frac{\hbar}{2i}[(\mathrm{grad}\,\psi^*)\psi - \psi^*(\mathrm{grad}\,\psi)] \quad (3.4.30)$$

Neither of these vectors is proportional to the current density vector

$$\mathbf{J} = ce\psi^*\boldsymbol{\alpha}\psi$$

given in Eq. (2.6.59). As a matter of fact, since L is only a linear function of the time derivatives of the fields, the canonical moments are proportional to the fields themselves and the whole formalism of the Hamilton canonical equations must be modified in the manner described on page 315. What is more important is the expression for the Lagrangian, which is minimized, and the expression for the energy and momentum densities.

Problems for Chapter 3

3.1 *a.* Show that a generating function $S'(q,P,t)$ may be defined as follows:

$$S' = S(q,Q,t) + PQ$$

and that $\quad p = \partial S'/\partial q;\quad Q = \partial S/\partial P;\quad K = H + (\partial S'/\partial t)$

b. Show that $S' = qP$ is the identity transformation.

c. Show under an infinitesimal transformation

$$S' = qP + \epsilon T(q,P) = qp + \epsilon T(q,p);\quad \epsilon \ll 1$$

that

$$P - p = -\epsilon(\partial T/\partial q);\quad Q - q = \epsilon(\partial T/\partial P)$$

d. Show that $\Delta f = f(P,Q) - f(p,q)$ is given by

$$\Delta f = \epsilon[f,T]$$

(where $[f,T]$ is the Poisson bracket), and therefore show that the quantity T is a constant of the motion if the corresponding transformation leaves the Hamiltonian invariant.

e. Show that the proper T for an infinitesimal rotation about the z axis is $(\mathbf{r} \times \mathbf{p})_z = \mathbf{M}_z$.

3.2 Show that the Lagrange equations are not changed when a total time derivative is added to the Lagrangian. Hence show that the Lagrangian for a nonrelativistic particle moving in an electromagnetic field may be written

$$\mathfrak{L} = \tfrac{1}{2}mv^2 - e\varphi - (e/c)[(\partial\mathbf{A}/\partial t) + \mathbf{v}\cdot(\nabla\mathbf{A})]\cdot\mathbf{r}$$

where $\nabla\mathbf{A}$ is a dyadic (the gradient operating on $\mathbf{A}$ only). Show that the corresponding Hamiltonian is

$$\mathfrak{H} = (\tfrac{1}{2}m)|\mathbf{p} + (e/c)(\nabla\mathbf{A})\cdot\mathbf{r}|^2 + (e/c)\mathbf{r}\cdot(\partial\mathbf{A}/\partial t) + e\varphi \quad \text{(Richards)}$$

3.3 Show that the Lagrange-Euler equation for a generalized orthogonal coordinate system ξ_1, ξ_2, ξ_3 is

$$\frac{\partial L}{\partial \psi} - \left(\frac{1}{h_1h_2h_3}\right) \sum_i \frac{\partial}{\partial \xi_i} \left[h_1h_2h_3 \frac{\partial L}{\partial(\partial\psi/\partial\xi_i)} \right] - \frac{\partial}{\partial t}\left[\frac{\partial L}{\partial(\partial\psi/\partial t)}\right] = 0$$

Employing $(\nabla\psi)^2$ as the Lagrangian density, derive the result

$$\nabla^2\psi = \frac{1}{h_1h_2h_3} \sum_i \frac{\partial}{\partial \xi_i}\left[\frac{h_1h_2h_3}{h_i^2}\frac{\partial\psi}{\partial\xi_i}\right]$$

3.4 Show that the tensor of the third rank ($T\mu\nu = W\mu\nu$, see page 319)

$$M_{\mu\nu\lambda} = T_{\mu\nu}x_\lambda - T_{\mu\lambda}x_\nu$$

satisfies the continuity equation

$$\sum_\mu \frac{\partial M_{\mu\nu\lambda}}{\partial x_\mu} = 0$$

only if $T_{\mu\nu}$ is symmetric. Show that M_{4jk} is just the angular momentum density and that the continuity equation yields, upon integration, the principle of the conservation of angular momentum.

3.5 *a.* Show in an infinitesimal Lorentz transformation

$$x'_\mu = x_\mu + \sum_\sigma \omega_{\mu\sigma}x_\sigma$$

that $\omega_{\mu\sigma} = -\omega_{\sigma\mu}$.

b. From the fact that the Lagrangian density is an invariant against Lorentz transformation, show, in the electromagnetic case, where

$$L = -\tfrac{1}{4}\sum_{\mu\nu}\left[\left(\frac{\partial A_\nu}{\partial x_\mu}\right) - \left(\frac{\partial A_\mu}{\partial x_\nu}\right)\right]^2$$

that

$$\sum_{\nu\sigma} \omega_{\nu\sigma}\Gamma_{\nu\sigma} = 0$$

where

$$\Gamma_{\nu\sigma} = -T_{\nu\sigma} + \sum_\mu \frac{\partial}{\partial x_\mu}\left[\frac{\partial L}{\partial(\partial A_\nu/\partial x_\mu)} A_\sigma\right]$$

Show also that

$$\Gamma_{\nu\sigma} = \Gamma_{\sigma\nu}$$

3.6 When $T_{\mu\nu}$ is not symmetric, it is always possible to find a symmetric tensor $S_{\mu\nu}$ which is symmetric and which has all the physical properties of $T_{\mu\nu}$.

a. Show that $S_{\mu\nu}$ must satisfy the conditions

$$S_{\mu\nu} = S_{\nu\mu}; \quad \sum_\mu (\partial/\partial x_\mu) S_{\mu\nu} = 0; \quad \int S_{4\nu}\, dV = \int T_{4\nu}\, dV$$

b. Show that $S_{\mu\nu}$ must have the form

$$S_{\mu\nu} = T_{\mu\nu} - \sum_\lambda \left(\frac{\partial G_{\lambda\mu\nu}}{\partial x_\lambda}\right)$$

where $G_{\lambda\mu\nu} = -G_{\mu\lambda\nu}$ and $T_{\mu\nu} - T_{\nu\mu} = \sum_\lambda \frac{\partial}{\partial x_\lambda}[G_{\lambda\mu\nu} - G_{\lambda\nu\mu}]$

c. Using the results of Prob. 3.5, part *b*, show that

$$G_{\lambda\mu\nu} - G_{\lambda\nu\mu} = H_{\lambda\mu\nu} = \frac{\partial L}{\partial(\partial A_\mu/\partial x_\lambda)} A_\nu - \frac{\partial L}{\partial(\partial A_\nu/\partial x_\lambda)} A_\mu$$

Hence show that

$$G_{\nu\mu\lambda} = \tfrac{1}{2}(H_{\nu\mu\lambda} + H_{\mu\lambda\nu} + H_{\lambda\mu\nu})$$

d. Evaluate $S_{\mu\nu}$ for the electromagnetic case.

3.7 Show that the homogeneous integral equation

$$\psi(x) = \lambda \int_a^b K(x|x_0)\psi(x_0)\, dx_0$$

follows from the variational requirement

$$\delta \int_a^b \psi(x)\left[\psi(x) - \lambda \int_a^b K(x|x_0)\psi(x_0)\, dx_0\right] dx = 0$$

if $$K(x|x_0) = K(x_0|x)$$

Show that, if $K(x|x_0) \neq K(x_0|x)$,

$$\delta \int_a^b \tilde{\psi}(x)\left[\psi(x) - \lambda \int_a^b K(x|x_0)\psi(x_0)\, dx_0\right] dx = 0$$

where $\tilde{\psi}$ satisfies the integral equation

$$\tilde{\psi}(x) = \lambda \int_a^b K(x_0|x)\tilde{\psi}(x_0)\, dx_0$$

3.8 The equation of motion of a membrane stretched over one end of an airtight vessel is given in Prob. 2.1 as

$$(1/c^2)(\partial^2\psi/\partial t^2) = \nabla^2\psi - (\rho c^2/VT)\int \psi\, dS$$

Determine the corresponding Lagrangian and Hamiltonian densities.

3.9 The equation for the damped vibration of a string is of the form

$$(\partial^2\psi/\partial t^2) + 2k(\partial\psi/\partial t) = c^2(\partial^2\psi/\partial x^2)$$

Show that the proper Lagrangian density is

$$L = \{(\partial\tilde{\psi}/\partial t)(\partial\psi/\partial t) + k[\psi(\partial\tilde{\psi}/\partial t) - \tilde{\psi}(\partial\psi/\partial t)] - c^2(\partial\tilde{\psi}/\partial x)(\partial\psi/\partial x)\}$$

and determine the equation satisfied by $\tilde{\psi}$. Determine the momenta canonical to ψ and $\tilde{\psi}$, and find the Hamiltonian density. Discuss the physical significance of the results.

3.10 The steady-state transport equation for anisotropic scattering from very heavy scatterers (see Secs. 2.4 and 12.2) may be put into the form

$$\cos\theta\,(\partial f/\partial\xi) = -f(\xi,\theta) + (\kappa/4\pi)\int w(\mathbf{a} - \mathbf{a}_0)f(\xi,\theta_0)\,d\Omega_0$$

where κ is a constant, unit vectors $\mathbf{a}$ and $\mathbf{a}_0$ are two directions given by spherical angles θ, φ and θ_0, φ_0, respectively, and $d\Omega_0$ is the differential solid angle at $\mathbf{a}_0$. Show that this equation may be obtained from the variational principle

$$\delta\int_0^{\xi_0} d\xi \int d\Omega\,\tilde{f}(\xi,\theta)\Big[\cos\theta\,(\partial f/\partial\xi) + f - (\kappa/4\pi)\int w(\mathbf{a} - \mathbf{a}_0)f(\xi,\theta_0)\,d\Omega_0\Big] = 0$$

Show that the equation satisfied by $\tilde{f}$ is

$$-\cos\theta\,(\partial\tilde{f}/\partial\xi) = -\tilde{f}(\xi,\theta) + (\kappa/4\pi)\int w(\mathbf{a}_0 - \mathbf{a})\tilde{f}(\xi,\theta_0)\,d\Omega_0$$

Interpret these results.

3.11 The diffusion of charged particles under the influence of an external field $\mathbf{E}$ is given by

$$\partial c/\partial t = a^2\nabla^2 c + b(\nabla c\cdot\mathbf{E})$$

where the assumptions involved are given in Prob. 2.5. Show that the corresponding variational principle is

$$\delta\int\int dV\,dt\,\tilde{c}[(\partial c/\partial t) - a^2\nabla^2 c - b(\nabla c\cdot\mathbf{E})] = 0$$

Find the equation for $\tilde{c}$ and interpret.

3.12 A pair of integral equations which occur in the theory of the deuteron may be written

$$u(r) = \lambda\int_0^\infty G_0(r|r_0)[f(r_0)u + g(r_0)w]\,dr_0$$

$$w(r) = \lambda\int_0^\infty G_2(r|r_0)[g(r_0)u + h(r_0)w]\,dr_0$$

where both G_0 and G_2 are symmetric. Show that the variational integral is

$$\int_0^\infty [u^2f + 2uwg + w^2h]\,dr - \lambda\int_0^\infty\int_0^\infty \{[f(r)u(r) + g(r)w(r)]G_0(r|r_0)[f(r_0)u(r_0) + g(r_0)w(r_0)] + [g(r)u + h(r)w]G_2(r|r_0)[g(r_0)u(r_0) + h(r_0)w(r_0)]\}\,dr\,dr_0$$

3.13 The equations describing the coupling between mechanical motion and heat conduction in a sound wave are

$$\partial T/\partial t = \alpha(\partial\rho/\partial t) + \beta\nabla^2 T;\quad \partial^2\rho/\partial t^2 = \gamma\nabla^2\rho + \epsilon\nabla^2 T$$

where the constants α, β, γ, and ϵ are given in Prob. 2.3. Show that these equations follow from the variational integral:

$$\iint dV\,dt\{\epsilon\nabla^2\tilde{T}[(\partial T/\partial t) - \alpha(\partial\rho/\partial t) - \beta\nabla^2 T] - \alpha(\partial\tilde{\rho}/\partial t)[(\partial^2\rho/\partial t^2) - \gamma\nabla^2\rho - \epsilon\nabla^2 T]\}$$

Show that $\tilde{T}$, $\tilde{\rho}$ satisfy the time-reversed equations if appropriate initial conditions are employed.

3.14 An infinite piezoelectric medium has the properties relating electric field E, polarization P, stress, and strain given in Prob. 2.2. If the x, y, z axes are placed along the three major axes of the crystal, then the coupling relations between E, P, the stress dyadic $\mathfrak{S}$, and the strain dyadic $\mathfrak{T}$ may be expressed in three sets of three equations, three typical ones being

$$\begin{aligned} T_{xx} &= \lambda_x^x S_{xx} + \lambda_x^y S_{yy} + \lambda_x^z S_{zz}; \quad \text{etc.} \\ T_{xy} &= T_{yx} = \lambda_{xy} S_{xy} + \sigma_{xy} P_z; \quad \text{etc.} \\ E_z &= \kappa_z P_z + \sigma_{xy} S_{xy}; \quad \text{etc.} \end{aligned}$$

where the λ's are the elements of the elastic modulus tetradic reduced to its principal axes, the κ's are the reciprocals of the dielectric susceptibilities along the three axes, and the σ's are the elements of a nondiagonal "triadic" (changing a vector into a dyadic and vice versa) representing the coupling between strain and polarization. Combine these equations with Maxwell's equations and with the ones for elastic motion for the special case of a transverse wave moving in the z direction, elastic displacement being in the y direction. Show that the result is a pair of coupled wave equations corresponding to two possible shear-electric waves, one with velocity somewhat less than that of pure shear waves (value if the σ's were zero), the other with velocity somewhat greater than that of light in the medium. Compute the Lagrange density. For plane shear waves in the z direction (**E** in y direction) compute the momentum density and the stress-energy dyadic. What is the relative proportion of energy carried by electric to that carried by elastic field for the slow wave? For the fast wave?

Tabulation of Variational Method

The *Lagrange density* L is a function of field variables $\psi_i(i = 1, 2, \ldots, n)$ and their gradients $\psi_{is} = (\partial\psi_i/\partial\xi_s)$ (ξ_1, ξ_2, ξ_3 are space coordinates, $\xi_4 = t$). Sometimes L also depends explicitly on the ξ's (through potential functions or charge-current densities, for instance). The total *Lagrangian integral*

$$\mathfrak{L} = \int_{a_1}^{b_1} \cdots \int_{a_4}^{b_4} L\,d\xi_1\,d\xi_2\,d\xi_3\,d\xi_4 \qquad (3.1.1)$$

is an invariant. The requirement that $\mathfrak{L}$ be minimum or maximum (*i.e.*, that the first-order variation be zero) corresponds to the *Lagrange-Euler equations*

$$\sum_{s=1}^{4} \frac{\partial}{\partial \xi_s}\left(\frac{\partial L}{\partial \psi_{is}}\right) - \frac{\partial L}{\partial \psi_i} = 0 \tag{3.4.1}$$

for the field variables. If L is a quadratic function of the ψ_{i4}'s then the *canonical momentum density*

$$p_i = \partial L / \partial \psi_{i4}$$

is a linear function of ψ_{i4}. If L is a linear function of the ψ_{i4}'s, then p_i and the Hamiltonian are independent of ψ_{i4}. The *stress-energy tensor* $\mathfrak{W}$, having components

$$W_{ms} = \sum_{i=1}^{n} \psi_{im} \frac{\partial L}{\partial \psi_{is}} - \delta_{ms} L$$

contains most of the other important physical properties of the field. For instance, the (4,4) component is the energy density,

$$W_{44} = H = \sum_{i=1}^{n} p_i \psi_{i4} - L$$

If p_i depends on ψ_{i4}, then the terms ψ_{i4} can be eliminated from W_{44}, obtaining the *Hamiltonian density* H, a function of the p_i's, the ψ_i's and their space derivatives. In this case the equations of motion may also be written in *canonical* form,

$$\dot{\psi}_i = \psi_{i4} = \frac{\partial H}{\partial p_i}; \quad \dot{p}_i = \frac{\partial p_i}{\partial t} = \sum_{s=1}^{3} \frac{\partial}{\partial \xi_s}\left(\frac{\partial H}{\partial \psi_{is}}\right) - \frac{\partial H}{\partial \psi_i} \tag{3.4.2}$$

which equations only apply when L contains a quadratic function of the ψ_{i4}'s. If L contains a linear function of the ψ_{i4}'s, H is independent of the p's (see page 315). The field intensity vector $\mathbf{S}$ and the field momentum vector $\mathbf{P}$ are defined as follows:

$$\mathbf{S} = \sum_{s=1}^{3} W_{4s}\mathbf{a}_s = \sum_{i=1}^{n} \psi_{i4} \left[\sum_{s=1}^{3} \frac{\partial L}{\partial \psi_{is}} \mathbf{a}_s\right] \tag{3.4.4}$$

$$\mathbf{P} = \sum_{s=1}^{3} \mathbf{a}_s W_{s4} = \sum_{i=1}^{n} (\text{grad } \psi_i) \left(\frac{\partial L}{\partial \psi_{i4}}\right) \tag{3.4.5}$$

The rest of $\mathfrak{W}$ is the three-dyadic $\mathfrak{U} = \sum_{r,s=1}^{3} \mathbf{a}_r W_{rs} \mathbf{a}_s$, called the *stress dyadic.* The elements of $\mathfrak{W}$ satisfy the divergence equations

$$\sum_{s=1}^{4} \frac{\partial W_{ms}}{\partial \xi_s} = -\frac{\partial L}{\partial \xi_m} \tag{3.4.3}$$

where $\partial L/\partial \xi_m$ is the rate of change of L with ξ_m due to the *explicit* variation of L with ξ_m (through potentials or charge-currents, etc.). When L does not depend explicitly on the ξ's then $\partial L/\partial \xi$ is zero. In terms of the field intensity and momentum this then becomes

$$\nabla \cdot \mathbf{S} + (\partial \mathbf{H}/\partial t) = 0; \quad (\mathfrak{U} \cdot \nabla) + (\partial \mathbf{P}/\partial t) = 0$$

when L does not depend explicitly on the ξ's. The *field angular momentum* density about the origin is

$$\mathbf{M} = \mathbf{r} \times \mathbf{P} = \sum_{i=1}^{n} \frac{\partial L}{\partial \psi_{i4}} (\mathbf{r} \times \text{grad } \psi_i)$$

Flexible String or Membrane

Field variable ψ is transverse displacement.
Parameters ξ_s are x and t for string; x, y, and t for membrane.

Lagrange density: $L = \frac{1}{2}\rho \left[\left(\frac{\partial \psi}{\partial t} \right)^2 - c^2 \text{ grad}^2 \psi \right]; \quad c^2 = \frac{T}{\rho}.$

Lagrange-Euler equation:

$$c^2 \nabla^2 \psi - (\partial^2 \psi/\partial t^2) = 0 \text{ (scalar wave equation)}$$

Canonical momentum density: $p = \rho(\partial \psi/\partial t)$.
Hamiltonian: $H = (1/2\rho)p^2 + \frac{1}{2}T \text{ grad}^2 \psi$.
Field intensity: $\mathbf{S} = -T(\partial \psi/\partial t) \text{ grad } \psi$.
Field momentum: $\mathbf{P} = \rho(\partial \psi/\partial t) \text{ grad } \psi = -(1/c^2)\mathbf{S}$.

Compressible, Nonviscous Fluid

Field variable ψ is velocity potential;

$$\text{Field velocity} = \text{grad } \psi; \quad \text{excess pressure} = -\rho(\partial \psi/\partial t)$$

Parameters ξ_s are x, y, z and t.

Lagrange density: $L = -\frac{1}{2}\rho \left[(\text{grad } \psi)^2 - \frac{1}{c^2} \left(\frac{\partial \psi}{\partial t} \right)^2 \right]; \quad c^2 = \frac{p_0 \gamma}{\rho}.$

Lagrange-Euler equation: $\nabla^2 \psi - \frac{1}{c^2} \frac{\partial^2 \psi}{\partial t^2} = 0$ (scalar wave equation).

Canonical momentum density; $p = (\rho/c^2)\,\dot{\psi} = -(\text{excess pressure})/c^2$.
Hamiltonian: $H = \frac{1}{2}(1/\rho c^2)p^2 + \frac{1}{2}\rho(\text{grad } \psi)^2$.
Field intensity:

$$\mathbf{S} = -\rho(\partial \psi/\partial t) \text{ grad } \psi = (\text{excess pressure}) \text{ (fluid velocity)}.$$

Field momentum: $\mathbf{P} = (\rho/c^2)(\partial\psi/\partial t)$ grad $\psi = -(1/c^2)\mathbf{S}$.

Diffusion Equation

Field variables are temperature or concentration density ψ and its "conjugate" ψ^*.

Parameters ξ_s are x, y, z, and t.

Lagrange density: $L = -$ (grad ψ) · (grad ψ^*) $- \frac{1}{2}a^2\left(\psi^* \frac{\partial\psi}{\partial t} - \psi \frac{\partial\psi^*}{\partial t}\right)$.

Lagrange-Euler equation for ψ, $\nabla^2\psi = a^2(\partial\psi/\partial t)$ (diffusion equation).

Canonical momentum densities: $p = -\frac{1}{2}a^2\psi^*$; $p^* = \frac{1}{2}a^2\psi$.

Energy density: $U = W_{44} =$ (grad ψ) · (grad ψ^*).

Field intensity: $\mathbf{S} = -\dot{\psi}^*$(grad ψ) $-$ (grad ψ^*)$\dot{\psi}$.

Field momentum: $\mathbf{P} = \frac{1}{2}a^2$[(grad ψ^*)$\psi - \psi^*$(grad ψ)]

Schroedinger Equation

Field variables are the wave function ψ and its conjugate ψ^*. $\psi^*\psi$ is the probability density for presence of the particle.

Parameters ξ_s are x, y, z, and t.

Lagrange density

$$L = -\frac{\hbar^2}{2m}(\text{grad}\,\psi^*)\cdot(\text{grad}\,\psi) - \frac{\hbar}{2i}\left(\psi^*\frac{\partial\psi}{\partial t} - \frac{\partial\psi^*}{\partial t}\psi\right) - \psi^* V\psi.$$

$V(x,y,z)$ is potential energy of particle.

Lagrange-Euler equation for ψ,

$$-(\hbar^2/2m)\nabla^2\psi + V\psi = i\hbar(\partial\psi/\partial t) \text{ (Schroedinger equation).}$$

Canonical momentum densities: $p = -(\hbar/2i)\psi^*$; $p^* = (\hbar/2i)\psi$.

Energy density: $U = W_{44} = (\hbar^2/2m)$(grad ψ^*) · (grad ψ) $+ \psi^* V\psi$.

Field intensity: $\mathbf{S} = -(\hbar^2/2m)[(\partial\psi^*/\partial t)$ grad ψ + grad $\psi^*(\partial\psi/\partial t)]$.

Field momentum: $\mathbf{P} = -(\hbar/2i)[\psi^*$(grad ψ) $-$ (grad ψ^*)ψ].

Current density: $\mathbf{J} = (e\hbar/2im)[\psi^*$(grad ψ) $-$ (grad ψ^*)ψ] where e, m are the charge and mass of the particle.

Klein-Gordon Equation

Field variables are the wave function ψ and its conjugate ψ^*.

Charge density for particle is $\frac{\hbar e}{2imc^2}\left[\frac{\partial\psi^*}{\partial t}\psi - \psi^*\frac{\partial\psi}{\partial t}\right]$, where m is the particle mass.

Parameters ξ_s are x, y, z, and t.

Lagrange density $L = -\frac{\hbar^2}{2m}\left[(\text{grad}\,\psi^*)\cdot(\text{grad}\,\psi) - \frac{1}{c^2}\left(\frac{\partial\psi^*}{\partial t}\right)\left(\frac{\partial\psi}{\partial t}\right) + \left(\frac{mc^2}{\hbar}\right)\psi^*\psi\right]$ for the field-free case.

Lagrange-Euler equation for ψ, $\nabla^2\psi - \frac{1}{c^2}\left(\frac{\partial^2\psi}{\partial t^2}\right) = \left(\frac{mc}{\hbar}\right)^2 \psi$ (Klein-Gordon equation).

Canonical momentum densities $p = \frac{\hbar^2}{2mc^2}\left(\frac{\partial\psi^*}{\partial t}\right)$; $p^* = \frac{\hbar^2}{2mc^2}\left(\frac{\partial\psi}{\partial t}\right)$.

Hamiltonian:

$$H = (2mc^2/\hbar^2)p^*p + (\hbar^2/2m)(\text{grad } \psi^*) \cdot (\text{grad } \psi) + (mc^2/2)\psi^*\psi.$$

Field intensity: $\mathbf{S} = -\frac{\hbar^2}{2m}\left[\frac{\partial\psi^*}{\partial t}(\text{grad } \psi) + (\text{grad } \psi^*)\frac{\partial\psi}{\partial t}\right]$.

Field momentum: $\mathbf{P} = -(1/c^2)\mathbf{S}$.

Current density: $\mathbf{J} = (e\hbar/2im)[\psi^*(\text{grad } \psi) - (\text{grad } \psi^*)\psi]$ where e is the particle charge.

Elastic Wave Equation

Field variables ψ_n are the components of the displacement vector $\mathbf{s}$.

Parameters ξ_s are the coordinates x, y, z, and t.

Lagrange density: $L = \frac{1}{2}\rho\dot{\mathbf{s}}^2 - \frac{1}{2}\mathfrak{S}:\mathfrak{T}$, where $\mathfrak{S} = \frac{1}{2}(\nabla\mathbf{s} + \mathbf{s}\nabla)$ is the strain dyadic and $\mathfrak{T} = \lambda|\mathfrak{S}|\mathfrak{J} + 2\mu\mathfrak{S}$ is the stress dyadic for isotropic solids.

Lagrange-Euler equation: $\rho(\partial^2\mathbf{s}/\partial t^2) = (\lambda + \mu) \text{ grad } (\text{div } \mathbf{s}) + \mu\nabla^2\mathbf{s}$.

Canonical momentum density: $\mathbf{p} = \rho(\partial\mathbf{s}/\partial t)$.

Hamiltonian density: $H = W_{44} = (1/2\rho)p^2 + \frac{1}{2}|\mathfrak{T} \cdot \mathfrak{S}|$.

Field intensity: $\mathbf{S} = -(\partial\mathbf{s}/\partial t) \cdot \mathfrak{T}$.

Field momentum: $\mathbf{P} = \rho(\nabla\mathbf{s}) \cdot (\partial\mathbf{s}/\partial t)$.

For a nonisotropic solid the stress dyadic $\mathfrak{T} = \mathfrak{g}:\mathfrak{S}$, where $\mathfrak{g}$ is a tetradic with coefficients g_{mnrs} which are arbitrary except for the general symmetry requirements $g_{mnrs} = g_{nmrs} = g_{mnsr} = g_{rsmn}$.

The Lagrange-Euler equation is then

$$\rho\left(\frac{\partial^2\mathbf{s}}{\partial t^2}\right) = \nabla \cdot (\mathfrak{g}:\nabla\mathbf{s}); \quad \text{that is; } \rho\ddot{s}_n = \sum_{mrs} g_{nmrs}\frac{\partial^2 s_s}{\partial x_m\,\partial x_r}$$

The Hamiltonian density is

$$H = W_{44} = (1/2\rho)p^2 + (\nabla\mathbf{s}):\mathfrak{g}:(\nabla\mathbf{s}); \quad \mathbf{p} = \rho(\partial\mathbf{s}/\partial t)$$

The new expressions for $\mathbf{S}$, $\mathbf{P}$, etc., can be obtained by substituting the new expression for $\mathfrak{T}$ into the equation for $\mathbf{S}$, $\mathbf{P}$ given above.

Electromagnetic Equations

Field variables are the components of the vector potential $\mathbf{A}$ and the scalar potential φ. For simplicity we shall choose the gauge for which $\varphi = 0$, so that $\text{curl } \mathbf{A} = \mathbf{B} = \mu\mathbf{H}$, $\partial\mathbf{A}/\partial t = -c\mathbf{E} = -(c\mathbf{D}/\epsilon)$, and div $(\partial \mathbf{A}/\partial t) = -(4\pi\rho c/\epsilon)$, where ρ is the density of free charge. Parameters are x, y, z, and t.

Lagrange density: $L = \dfrac{\epsilon}{8\pi c^2}\left|\dfrac{\partial \mathbf{A}}{\partial t}\right|^2 - \dfrac{1}{8\pi\mu}\,|\,\mathrm{curl}\,\mathbf{A}|^2 + \dfrac{1}{c}\,\mathbf{J}\cdot\mathbf{A}$, where $\mathbf{J}$ is the current density.

Lagrange-Euler equation: $\mathrm{curl}\,(\mathrm{curl}\,\mathbf{A}) + (\mu\epsilon/c^2)(\partial^2\mathbf{A}/\partial t^2) = (4\pi\mu/c)\mathbf{J}$

Canonical momentum density: $\mathbf{p} = -(\mathbf{D}/4\pi c)$.

Hamiltonian density:

$$H = W_{44} = (2\pi c^2/\epsilon)p^2 + (1/8\pi\mu)|\,\mathrm{curl}\,\mathbf{A}|^2 - (1/c)\mathbf{J}\cdot\mathbf{A}.$$

Field intensity: $\mathbf{S} = (c/4\pi)(\mathbf{E} \times \mathbf{H})$.

Field momentum: $\mathbf{P} = -(\epsilon/4\pi)(\nabla\mathbf{A})\cdot(\partial\mathbf{A}/\partial t)$.

Dirac Equation

Field variables ψ_n^* and ψ_n $(n = 1, 2, 3, 4)$. Probability density for presence of electron is $\psi_1^*\psi_1 + \psi_2^*\psi_2 + \psi_3^*\psi_3 + \psi_4^*\psi_4 = \mathbf{\Psi}^*\mathbf{\Psi}$. Parameters are x, y, z, t. Wave functions $\mathbf{\Psi} = \Sigma \mathbf{e}_n\psi_n$ and $\mathbf{\Psi}^* = \Sigma\psi_n^*\mathbf{e}_n^*$ where the $\mathbf{e}$'s are unit vectors in spin space. Operators $\boldsymbol{\alpha}_x$, $\boldsymbol{\alpha}_y$, $\boldsymbol{\alpha}_z$, $\boldsymbol{\alpha}_0$ operate on the $\mathbf{e}$'s in a manner given in Eqs. (2.6.53). Lagrange density: $L = \dfrac{\hbar c}{2i}\,[(\mathrm{grad}\;\mathbf{\Psi}^*)\cdot\boldsymbol{\alpha}\mathbf{\Psi} - \mathbf{\Psi}^*\boldsymbol{\alpha}\cdot\mathrm{grad}\;\mathbf{\Psi})] + \dfrac{\hbar}{2i}\left[\left(\dfrac{\partial\mathbf{\Psi}^*}{\partial t}\right)\mathbf{\Psi} - \mathbf{\Psi}^*\left(\dfrac{\partial\mathbf{\Psi}}{\partial t}\right)\right] - e\mathbf{\Psi}^*(\boldsymbol{\alpha}\cdot\mathbf{A})\mathbf{\Psi} + ec\mathbf{\Psi}^*\varphi\mathbf{\Psi} - mc^2\mathbf{\Psi}^*\boldsymbol{\alpha}_0\mathbf{\Psi}$ where $\mathbf{A}$ and φ are the electromagnetic potentials and m the particle mass.

Lagrange-Euler equations:

$$\boldsymbol{\alpha}_0 mc\mathbf{\Psi} + \boldsymbol{\alpha}\cdot\left(\frac{\hbar}{i}\,\mathrm{grad}\;\mathbf{\Psi} + \frac{e}{c}\,\mathbf{A}\mathbf{\Psi}\right) + \left(\frac{\hbar}{ic}\frac{\partial\mathbf{\Psi}}{\partial t} - e\varphi\mathbf{\Psi}\right) = 0$$

$$mc\mathbf{\Psi}^*\boldsymbol{\alpha}_0 + \left(-\frac{\hbar}{i}\,\mathrm{grad}\;\mathbf{\Psi}^* + \frac{e}{c}\,\mathbf{A}\mathbf{\Psi}^*\right)\cdot\boldsymbol{\alpha} - \left(\frac{\hbar}{ic}\frac{\partial\mathbf{\Psi}^*}{\partial t} + e\varphi\mathbf{\Psi}^*\right) = 0$$

Canonical momentum density: $\mathbf{p} = -(\hbar/2i)\mathbf{\Psi}^*$; $\mathbf{p}^* = (\hbar/2i)\mathbf{\Psi}$.

Hamiltonian density: $H = \dfrac{\hbar c}{2i}\,[\mathbf{\Psi}^*\boldsymbol{\alpha}\cdot(\mathrm{grad}\;\mathbf{\Psi}) - (\mathrm{grad}\;\mathbf{\Psi}^*)\cdot\boldsymbol{\alpha}\mathbf{\Psi}] + e\mathbf{\Psi}^*\boldsymbol{\alpha}\cdot\mathbf{A}\mathbf{\Psi} - ec\mathbf{\Psi}^*\varphi\mathbf{\Psi} + mc^2\mathbf{\Psi}^*\alpha_0\mathbf{\Psi}$.

Field intensity: $\mathbf{S} = (\hbar c/2i)[(\partial\mathbf{\Psi}^*/\partial t)\boldsymbol{\alpha}\mathbf{\Psi} - \mathbf{\Psi}^*\boldsymbol{\alpha}(\partial\mathbf{\Psi}/\partial t)]$.

Field momentum: $\mathbf{P} = (\hbar/2i)[(\mathrm{grad}\;\mathbf{\Psi}^*)\mathbf{\Psi} - \mathbf{\Psi}^*(\mathrm{grad}\;\mathbf{\Psi})]$.

Current density: $\mathbf{J} = ce\mathbf{\Psi}^*\boldsymbol{\alpha}\mathbf{\Psi}$, where e is the particle charge.

Bibliography

Few books cover the central subject of this chapter in any satisfactory detail, though several deal with some aspects of the subject. The references dealing with calculus of variations in general:

Bliss, G. A.: "Calculus of Variations," Open Court, La Salle, 1925.

Courant, R., and D. Hilbert: "Methoden der mathematischen Physik," Vol. 1, pp. 165 *et seq.*, Springer, Berlin, 1937.

Rayleigh, J. W. S.: "The Theory of Sound," pp. 109 *et seq.*, Macmillan & Co., Ltd., London, 1896, reprinted Dover, New York, 1945.

Books on the transformation theory of dynamics, including a discussion of Hamilton's principle:

Born, M.: "Mechanics of the Atom," G. Bell, London, 1927.

Corben, H. C., and P. Stehle: "Classical Mechanics," Chaps. 10 to 15, Wiley, New York, 1950.

Goldstein, H.: "Classical Mechanics," Chaps. 7, 8 and 9, Addison-Wesley, Cambridge, 1950.

Lanczos, C.: "The Variational Principles of Dynamics," University of Toronto Press, 1949.

Webster, A. G.: "Dynamics," Chaps. 4 and 9, Stechert, New York, 1922.

Whittaker, E. T.: "Analytic Dynamics," Chaps. 9 to 12, Cambridge, New York, 1937.

Works discussing application of Hamilton's principle to fields from various points of view:

Fermi, E.: Quantum Theory of Radiation, *Rev. Modern Phys.*, **4,** 87 (1932).

Goldstein, H.: "Classical Mechanics," Chap. 11, Addison-Wesley, Cambridge, 1950.

Heitler, W.: "Quantum Theory of Radiation," Oxford, New York, 1936.

Landau, L. D., and E. Lifschitz: "Classical Theory of Fields," Addison-Wesley, Cambridge, 1951.

Pauli, W.: Relativistic Field Theories of Elementary Particles, *Rev. Modern Phys.*, **13,** 203 (1941).

Schiff, L. I.: "Quantum Mechanics," Chaps. 13 and 14, McGraw-Hill, New York, 1949.

Wentzel, G.: "Quantum Theory of Fields," Interscience, New York, 1949.

Weyl, H.: "Theory of Groups and Quantum Mechanics," Chap. 2, Methuen London, 1931.

CHAPTER 4

Functions of a Complex Variable

The past two chapters contain a discussion of the connection between physical phenomena and the partial differential equations for fields to represent these phenomena. The next several chapters must be devoted to a discussion of the general mathematical properties of the differential equations and their solutions. We have begun to familiarize ourselves with the possible physical interpretations of field quantities: tensors, divergences, line integrals, and the like. Now we must learn to recognize the different types of equations and their solutions. We must become familiar with the sort of tests which can be applied to tell how a given function will depend on its argument: where it goes to infinity or zero, where it can be integrated and differentiated, and so on. And we must be able to tell what sort of functions will be solutions of given differential equations, how the "singularities" of the equation are related to the singularities of the solutions, and the like. The general properties of functions will be treated in the present chapter, and the interrelation between equations and solutions in the next chapter.

To be more specific, we shall devote this chapter to a discussion of functions of the complex variable $z = x + iy$, where i is the square root of (-1). We have already (pages 73 and 74) shown that such a variable can be represented as a two-dimensional vector, with x the x component and y the y component of the vector; and we have indicated that z can also be considered as an operator, which rotates any other complex number vector by an angle $\tan^{-1}(y/x)$ and changes its length by a factor $\sqrt{x^2 + y^2}$. In this chapter we shall continually use the two-dimensional vector representation and occasionally use the concept of vector operator.

It could be asked why it is necessary to study complex numbers when many parts of physics are interested only in real solutions. One might expect that a study of the real functions of a real variable going from $-\infty$ to $+\infty$ would suffice to obtain knowledge of the physically interesting solutions in many cases. The answer is that it is desirable to extend our study to complex values of the variables and the solutions for reasons of *completeness* and *convenience*.

The set of real numbers is not even a sufficient basis for reproduction of the roots of algebraic equations. On the other hand *all* roots of *all* algebraic equations can be expressed as complex numbers. In addition, knowledge of the behavior of a function $f(z)$ for all complex values of z gives us a much more complete picture of the principal properties of f (even its properties for z real) than does a knowledge of its behavior for only real values of z. The location, on the complex plane for z, of the zeros and infinities of f [*i.e.*, the position of the roots of $f = 0$ and $(1/f) = 0$] will tell us a great deal about the behavior of f for all values of z. Often an integral of $f(z)$ over real values of z (along the real axis) may be modified into an integral along some elementary path for z in the complex plane, thereby considerably simplifying the integration. It is usually convenient to consider the solution of an equation as complex, to deal with it as a complex number up until the final answer is to be compared with a measured value, and only then to consider the real or the imaginary part of the solution as corresponding to the actual physical problem.

But the most important reason for the study of complex functions is the insight we shall obtain into the general properties of functions. For example, the various types of singularities a function might have may be classified. In general these singularities will be related to physical singularities, such as those caused by sources, point electric charges, etc. It turns out to be possible, simply from knowledge of the singularities of a function, to specify the function completely. The corresponding statement in electrostatics is that, once the size and distribution of all the electric charges are given, the electric field at any point can be determined. Because of the close connection between electrostatics and complex variables, it is not surprising that our study will reveal, in addition, a method for generating solutions for the Laplace equation (*i.e.*, will locate possible sets of equipotential lines). We recall from Chap. 1 that these equipotentials and their normals form the basis of an orthogonal coordinate system. We may therefore say that a method can be developed for generating new coordinate systems, systems which are appropriate to the geometry of the problem at hand.

4.1 *Complex Numbers and Variables*

Perhaps the first use that a student of physics makes of complex numbers is in the expression $Ae^{-i\omega t}$, for a vector rotating with constant angular velocity ω, where A gives the length of the vector. This representation is useful also in simple harmonic motion, for its real part is $A \cos \omega t$ while its imaginary part is $-A \sin \omega t$. We have already used this fact several times in the preceding chapters.

The connection between vectors and complex numbers is obtained by making a proper definition of the symbol i. *Here i is to be considered to be an operator which, when applied to a vector, rotates the vector through 90° counterclockwise.* The operator i^2, meaning the application of the operator twice in succession, results in the rotation of the vector through 180°. Since this yields a vector which is antiparallel to the original vector, we find

$$i^2 = -1 \tag{4.1.1}$$

in agreement with the more usual definition of i. The symbol of i^3 meaning the application of i three times results in a rotation of the original vector through 270° or $-90°$ so that $i^3 = -i$. Similarly $i^4 = 1$.

We may now differentiate between real and imaginary numbers. We shall plot all real numbers as vectors in the x direction. Thus, we multiply a real number by i to obtain a vector directed along the y axis. Vectors in the y direction are called *imaginary* numbers. Any vector[1] f may, of course, be decomposed into its two components u and v along the x and y axis so that we may write

$$f = u + iv \tag{4.1.2}$$

establishing a connection between complex numbers and vectors. The *magnitude* of f, written $|f|$, is equal to the absolute value of the complex number $u + iv = \sqrt{u^2 + v^2}$, while the direction of f, the angle φ it makes with the x axis, is just the *phase*, $\tan^{-1}(v/u)$, of $u + iv$. This angle is sometimes called the *argument of f*. The *conjugate* of $u + iv$

$$\bar{f} = u - iv$$

may be obtained from vector f by reflecting it in the x axis.

The Exponential Rotation Operator. To obtain the operator for a finite rotation of angle θ, it is necessary only to consider the results for an infinitesimal rotation $d\theta$. The operator for the infinitesimal rotation must yield the original vector f plus a vector at right angles to f of magnitude $f\,d\theta$. The new vector is $f + if\,d\theta = (1 + i\,d\theta)f$. Thus the change in f, df is

$$df = if\,d\theta$$

This equation may be integrated to yield f after a rotation through θ radians. Let the original $(\theta = 0)$ value of f be f_0. Then f for $\theta = \theta$ radians is

$$f_\theta = e^{i\theta} f_0 \tag{4.1.3}$$

The operator rotating a vector through θ radians is thus $e^{i\theta}$ (see page 74).

This operator when applied to a vector along the real axis, say of unit length, yields a vector in the direction θ. Decomposing this new vector into components and expressing the vector in complex notation,

[1] We shall not use the customary boldface type for these complex vectors.

we obtain *De Moivre's relation* $e^{i\theta} = \cos\theta + i\sin\theta$ mentioned on page 74. This agrees with the starting assumption that i yields a rotation of 90° (as may be seen by putting $\theta = \pi/2$). A unit vector rotating counterclockwise with angular velocity ω is simply $e^{i\omega t}$ where the unit being operated on is understood, as is customary. Any vector f may now be expressed in terms of its absolute magnitude $|f|$ and the required operator to rotate it, from the x axis to its actual direction, to wit,

$$f = |f|e^{i\varphi};\quad \bar{f} = |f|e^{-i\varphi}$$

The angle φ is called the *phase angle*, or *argument*, of f.

Vectors and Complex Numbers. Having established the one-to-one relation between complex numbers and vectors, we shall now explore the relations between various possible combinations of complex numbers and the corresponding combinations of vectors. The law of addition for two vectors, the parallelogram law, is the same as that for the addition of two complex numbers. However, when two complex numbers are multiplied together, the result, when expressed in vector notation, involves both the scalar and vector product. Consider $\bar{f}g$ where $f = u + iv$ and $g = s + it$:

$$\bar{f}g = (us + vt) + i(ut - vs)$$

In vector language:

$$\bar{f}g = \mathbf{f}\cdot\mathbf{g} + i|\mathbf{f}\times\mathbf{g}| \tag{4.1.4}$$

Thus if two vectors are orthogonal, the real part of their product is zero, while if they are parallel, the imaginary part of their product is zero. It should be noticed that rule (4.1.4) is the same in two dimensions as Eq. (1.6.30), for quaternion multiplication, is for three. This should occasion no surprise, since Hamilton originally wrote down the quaternion algebra in an attempt to generalize the method of complex variables to three dimensions and to three-dimensional vectors.

The differential properties of a vector field involve the operator ∇. Since we are limiting ourselves to two dimensions, the x, y plane, the operator ∇ can be written as

$$\nabla = \frac{\partial}{\partial x} + i\frac{\partial}{\partial y} \tag{4.1.5}$$

If now we should operate with $\bar{\nabla}$ on a vector g, then from Eq. (4.1.4) one obtains the relation

$$\bar{\nabla}g = \operatorname{div} g + i|\operatorname{curl} g| \tag{4.1.6}$$

so that $\bar{\nabla}$ immediately gives both the divergence and curl of a vector. Note that $\bar{\nabla}$ operating on a real function (g along x axis) yields directly from (4.1.6)

$$\bar{\nabla}g = \frac{\partial g}{\partial x} - i\frac{\partial g}{\partial y}$$

as it should. We thus see that one great advantage of the complex notation is that it affords a means by which one may condense several vector operations into just one.

Some further condensation is possible if we introduce two new variables in place of x and y:

$$z = x + iy; \quad \bar{z} = x - iy; \quad x = \tfrac{1}{2}(z + \bar{z}); \quad y = -\tfrac{1}{2}i(z - \bar{z}) \tag{4.1.7}$$

where z is just the radius vector to the point (x,y).

There is usually some confusion in the reader's mind at this point as to how it is possible to consider z and $\bar{z}$ as independent variables (no such confusion seems to come up for variables $x - y$, $x + y$) for it is often stated that, if z is known, so is $\bar{z}$. This is, however, *not* the case. Given a vector z as a line drawn from the origin to some point, $\bar{z}$ is not yet determined, for, in addition, the direction of the x axis must be given. Vice versa, if both z and $\bar{z}$ are known, the direction of the x axis is determined as a line bisecting the angle between the vectors z and $\bar{z}$, and then x and y can be found. In terms of these variables and using Eq. (4.1.5),

$$2\frac{\partial}{\partial \bar{z}} = 2\frac{\partial x}{\partial \bar{z}}\frac{\partial}{\partial x} + 2\frac{\partial y}{\partial \bar{z}}\frac{\partial}{\partial y} = \nabla; \quad \text{likewise } \bar{\nabla} = 2\frac{\partial}{\partial z} \tag{4.1.8}$$

The Two-dimensional Electrostatic Field. Suppose that we have an electrostatic field generated by line charges all perpendicular to the x, y plane. The electric vector $\mathbf{E}$ will everywhere lie in the x, y plane, and we need consider only two dimensions. Therefore the electric vector $\mathbf{E}$ can be represented by a complex number, say $u - iv$ (the reason for the minus sign will be apparent shortly), where u and v are functions of x and y determined by the distribution of the line charges. We first look at that part of the x, y plane which is free from charge density. In these places Maxwell's equations (2.5.11) state that

$$\text{div } \mathbf{E} = 0; \quad \text{curl } \mathbf{E} = 0 \tag{4.1.9}$$

Referring to Eqs. (4.1.6) and (4.1.8), we see that both of these requirements can be written down (in this two-dimensional case only) in the extremely simple form

$$\partial E/\partial z = 0$$

This states that the vector E is *not a function of* $z = x + iy$, *but only of* $\bar{z} = x - iy$. Contrariwise the conjugate vector $\bar{E} = u + iv$ *is a function only of* z *and not of* $\bar{z}$. Since we usually shall deal with functions of z, we may as well deal with $\bar{E}$, from which we can find E, the electric vector. We have just shown that $\bar{E}$ is a function of z and not of $\bar{z}$.

By using the equations for $\bar{E}$ analogous to Eq. (4.1.9) or by writing out the equation $2(\partial \bar{E}/\partial \bar{z}) = \nabla(u + iv) = 0$ in terms of the derivatives with respect to x and y and collecting real and imaginary parts, we find the interesting pair of cross connections between u and v:

$$\partial u/\partial x = \partial v/\partial y; \quad \partial u/\partial y = -(\partial v/\partial x) \tag{4.1.10}$$

which are called the *Cauchy-Riemann conditions.* We have derived them here for an electric vector (two-dimensional) in a region free from charge and current, but the way we have obtained them shows that they apply for any complex function $f = u + iv$ which is a function of z only (not $\bar{z}$). Any such function, with real and imaginary parts satisfying Eqs. (4.1.10), is called an *analytic function* of the complex variable $z = x + iy$.

Therefore any analytic function of z can represent a two-dimensional electrostatic field. Such a function may be created by taking any well-behaved function of a real variable and making it a function of $z = x + iy$ instead [for instance, $\sin(x + iy)$, $1/[(x + iy)^2 + a^2]$, $\log(x + iy)$ are all analytic functions for all values of z where the functions do not become infinite].

In a region free from charge current, an electric potential V exists such that $E = \nabla V = (\partial V/\partial x) + i(\partial V/\partial y)$, where V is a function of x and y. We may generalize somewhat and allow V also to become a complex function (its real part or imaginary part can be the actual potential). Then we can write $E = 2\partial V/\partial \bar{z}$, and since $\partial E/\partial z = 0$, we have

$$4\left(\frac{\partial^2 V}{\partial z\, \partial \bar{z}}\right) = \frac{\partial^2 V}{\partial x^2} + \frac{\partial^2 V}{\partial y^2} = 0 \tag{4.1.11}$$

which is the Laplace equation for two dimensions. Naturally both real and imaginary parts of V are separately solutions of Laplace's equation, and in fact, combining Eqs. (4.1.10) we see that the real or imaginary parts of *any* analytic function are solutions of Laplace's equation in two dimensions. Therefore either an analytic function can be used to generate an electric vector, or else its real or imaginary parts can be used as a potential function.

Contour Integrals. Integration of complex functions is a natural extension of the process of integration of real functions. The integrand is some analytic function $f(z)$; the variable of integration is, of course, z. But since z can move over the complex plane instead of just along the real axis, we shall have to specify along what line the integration is to be performed. This line of integration is called a *contour*, and if the contour is closed on itself, the integral is called a *contour integral* and is denoted $\oint f(z)\, dz = \oint f e^{i\varphi}\, ds$, where ds is the magnitude of dz and φ is its phase.

In this two-dimensional generalization of an integral we can no longer give enough details by writing in the upper and lower limits; we must describe, or draw out, the contour followed, as in Fig. 4.1. The expression is analogous to the two-dimensional form of the net outflow integral of Sec. 1.2 and also to the net circulation integral of that same section. As a matter of fact the complex contour integral is a compacted combination of both, as we can see by writing out the contour integral of the electric vector, and using Eq. (4.1.4)

$$\oint \bar{E}\,dz = \oint \mathbf{E}\cdot d\mathbf{s} + i\oint|\mathbf{E}\times d\mathbf{s}| = \oint E_t\,ds + i\oint E_n\,ds \quad (4.1.12)$$

where E_t is the component of E along ds and E_n the component normal to ds. Therefore the real part of the contour integral of $\bar{E}$ is the net circulation integral of **E** around the contour, and the imaginary part is the net outflow integral of **E** over the sides of a cylinder with axis perpendicular to the x, y plane, of cross section equal to the contour. (Since the field in this case is all parallel to the x, y plane, the total net outflow integral over the cylinder is equal to $\oint E_n\,ds$ times the length of the cylinder.)

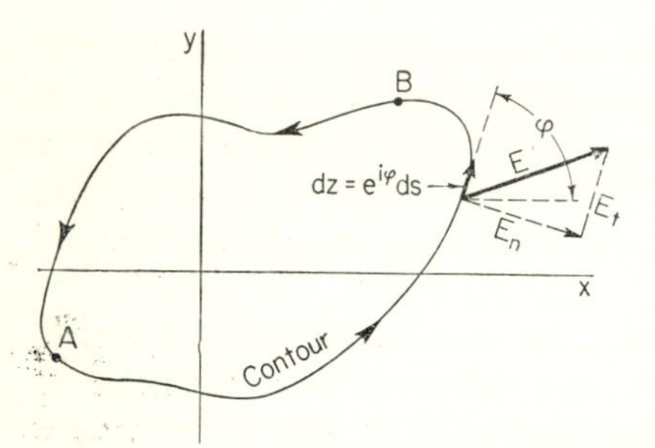

Fig. 4.1 Contour integration in the complex plane.

In the case of an electric vector in a region devoid of charge density, both outflow and circulation integrals are zero, so that for any contour in such regions

$$\oint \bar{E}\,dz = 0 \quad (4.1.13)$$

This equation is just *Cauchy's theorem*, which states that, if $f(z)$ is an analytic function of z at all points on and inside a closed contour, then $\oint f(z)\,dz$ around this contour is zero. Therefore the two-dimensional electric vector can be represented by an analytic function at all points where there is no charge or current.

By a specialization of Eq. (1.2.9) for the cylindrical surface we have set up on the contour, we see that, if the field **E** is due to a series of line charges each uniformly distributed along lines perpendicular to the x, y plane, the rth line having charge q_r per unit length, then, from Eq. (4.1.12)

$$\oint \bar{E}\,dz = 4\pi i \sum_r{}' q_r \quad (4.1.14)$$

where the sum is taken over all the lines r which cut the x, y plane *inside the contour*.

Suppose that we take the case where there is only one line, of charge density q_1, which cuts the x, y plane inside the contour at the point $z_1 = x_1 + iy_1$. The electric field, represented by $\bar{E}$, can then be split into two parts: $\bar{E}_s$, due to the source q_1 inside the contour; and $\bar{E}_0$, due to the sources outside. Elementary integration of the equations for electrostatic fields indicates that $\mathbf{E}_s = (2q_1/r)\mathbf{a}_r$, where $r^2 = (x - x_1)^2 + (y - y_1)^2 = |z - z_1|^2$ is the distance, in the x, y plane, from the source line and $\mathbf{a}_r$ is a unit vector in the x, y plane pointing away from the source line. In complex notation

$$E_s/q_1 = (2/r)\,[\cos\varphi + i\sin\varphi] = (2/r)e^{i\varphi}$$

where φ is the angle between $\mathbf{a}_r$ and the x axis. Therefore

$$\bar{E}_s/q_1 = (2/r)e^{-i\varphi} = (2/re^{i\varphi}) = 2/(z - z_1) \tag{4.1.15}$$

since $re^{i\varphi} = z - z_1$.

Adding $\bar{E}_0$ to $\bar{E}_s$ we can write

$$\bar{E} = [f(z)]/(z - z_1) \tag{4.1.16}$$

where $f(z) = \bar{E}_0(z - z_1) + 2q_1$ is an analytic function within and on the contour (why?). Thus we have, *for any analytic function* $f(z)$, the formula,

$$\oint [f(z)/(z - z_1)]\,dz = 2\pi i f(z_1) \tag{4.1.17}$$

which is a more general form of Cauchy's theorem.

Therefore Cauchy's theorem is just a restatement, in the notation of analytic functions, of Gauss' theorem for electrostatics.

Similarly one can use a function $\bar{H}$ of z to represent the magnetic field caused by line currents along lines perpendicular to the x, y plane. A current I along a line cutting the x, y plane at a point z_0 inside the contour will produce a field $H = 2(\mathbf{I} \times \mathbf{a}_r)/r$, which can be represented by the function $\bar{H} = 2I/i(z - z_0)$. If there are a number of line currents I_r, then, according to Eq. (1.2.11)

$$\oint \bar{H}\,dz = 4\pi \sum_r{}' I_r \tag{4.1.18}$$

where the summation is over all currents cutting inside the contour. Here we use the real part of the contour integral, but if we substitute for $\bar{H}$ its form in terms of $(z - z_0)$, we eventually obtain Cauchy's theorem again.

Returning to Fig. 4.1, we note that the integral from A to B along a contour (not closed) gives

$$\int_A^B \bar{E}\,dz = \int_A^B E_t\,ds + i\int_A^B E_n\,ds = W = V + iU \tag{4.1.19}$$

The real part V of this integral is just the electrostatic potential difference between points A and B. The imaginary part U measures the number of lines of force which cut across the contour between A and B.

We note that the family of curves U = constant is orthogonal to the family V = constant, so they can be used to set up orthogonal, two-dimensional coordinate systems. If a conducting cylinder is present, with axis perpendicular to the x, y plane, then the field must adjust itself so that it cuts the x, y plane along one of the equipotential lines V = constant. The lines of force are the orthogonal lines U = constant, and the surface charge density on the conductor, per unit length of cylinder, inside the region limited by the points A and B is $U(B) - U(A)$. The function U was called the *flow function* in Chap. 1 (see page 155).

In this section we have related complex variables and electrostatics and have given as examples some electrostatic interpretations of some familiar theorems in function theory. In the remainder of the chapter, we shall develop a more rigorous theory but shall use electrostatic interpretations to give some intuitive meaning to the theorems in much the same way as was done above for the Cauchy theorem and the integral formula.

4.2 *Analytic Functions*

The electrostatic analogue has furnished us with heuristic derivations of some of the fundamental theorems in function theory. In particular we have seen that analytic functions form a restricted class to which most functions do *not* belong. In this section we shall attempt to understand the nature of these limitations from the point of view of the geometer and analyst. This procedure will have the merit of furnishing more rigorous proofs of the aforementioned theorems. (Rigor in this particular subject is actually useful!)

We have taken as a rough definition of an analytic function that it be a function of z only, not a function of both z and $\bar{z}$. Thus the study of a function of a complex variable $f(z) = u + iv$, where u and v are *real* functions of x and y, is not so general as the study of functions of two variables, for u and v are related in a special way as given by the Cauchy-Riemann conditions (4.1.10). A more precise definition of an analytic function may be obtained by considering the behavior of the derivative of $f(z)$ with respect to z at a point a. The meaning of the derivative is fairly obvious. The function $f(z)$ is a vector. We ask how does this vector change, both in magnitude and direction, as z moves away from the point a in a direction specified by the vector dz. If $f(z)$ is a function of z only (for example, z^2), one would expect the derivative (for example, $2z$) to depend only upon the point at which it is evaluated. It is characteristic of *a single-valued function which is analytic at a point a that the derivative at point a is unique*, *i.e.*, independent of the direction dz along which the derivative is taken. No matter how one moves away from point a, the rate at which f will change with z will be the same. This is not true for an arbitrary complex function $u + iv$, where u and v are any sort of functions of x and y. It is true only when u and v satisfy Eqs. (4.1.10).

This definition of an analytic function has the advantage over the simpler one used earlier in being more precise. Using it, one may determine whether or not a function is analytic at a given point. We see once more how special an analytic function is. Most functions do not possess an "isotropic" derivative.

It may be shown that the Cauchy-Riemann equations form *a necessary condition only* that a function have a unique derivative. To show this, consider the change in $f(x) = u + iv$ as z changes from a to $a + \Delta z$:

$$\frac{\Delta f}{\Delta z} = \frac{f(a + \Delta z) - f(a)}{\Delta z} = \frac{[(\partial u/\partial x) + i(\partial v/\partial x)]\,\Delta x + [(\partial u/\partial y) + i(\partial v/\partial y)]\,\Delta y}{\Delta x + i\,\Delta y}$$

$$= \frac{(\partial u/\partial x) + i(\partial v/\partial x)}{1 + i(\Delta y/\Delta x)}\left\{1 + i\,\frac{\Delta y}{\Delta x}\left(\frac{(\partial v/\partial y) - i(\partial u/\partial y)}{(\partial u/\partial x) + i(\partial v/\partial x)}\right)\right\}$$

The last equation shows that, except for exceptional circumstances, the derivatives $df/dz = \lim\limits_{\Delta z \to 0} (\Delta f/\Delta z)$ will depend upon $\Delta y/\Delta x$, that is, upon the direction of Δz. For an analytic function, however, there cannot be any dependence on $\Delta y/\Delta x$. This may be achieved *only if*

$$\frac{\partial v}{\partial y} - i\,\frac{\partial u}{\partial y} = \frac{\partial u}{\partial x} + i\,\frac{\partial v}{\partial x}$$

or

$$\frac{\partial u}{\partial x} = \frac{\partial v}{\partial y}; \quad \frac{\partial u}{\partial y} = -\frac{\partial v}{\partial x} \tag{4.2.1}$$

which are the *Cauchy-Riemann* conditions. These conditions only are necessary. For *sufficiency* one must also include the requirement that the various derivatives involved be *continuous* at a. If this were not so, then $\partial u/\partial x$, etc., would have different values depending upon how the derivative were evaluated. This would again result in a nonunique first derivative. For example, suppose that

$$u = \frac{x^3 - y^3}{x^2 + y^2}; \quad \frac{\partial u}{\partial x} = \frac{x^4 + 3x^2y^2 + 2xy^3}{(x^2 + y^2)^2}$$

Then $\lim\limits_{x \to 0} \lim\limits_{y \to 0} (\partial u/\partial x) = 1$. However, $\lim\limits_{y \to 0} \lim\limits_{x \to 0} (\partial u/\partial x) = 0$.

The Cauchy-Riemann conditions show that, if the real (or imaginary) part of a complex function is given, then the imaginary (or real) part is, within an additive constant, determined. We shall later discuss special techniques for performing this determination. A simple example will suffice for the present. Let the real part be known, and let us try to determine the imaginary part.

$$v = \int dv = \int \left(\frac{\partial v}{\partial x} dx + \frac{\partial v}{\partial y} dy \right)$$

or

$$v = \int \left(-\frac{\partial u}{\partial y} dx + \frac{\partial u}{\partial x} dy \right) \qquad (4.2.2)$$

so that, if u is known, v can be determined by performing the integration indicated in Eq. (4.2.2). For example, let $u = \ln r = \frac{1}{2} \ln (x^2 + y^2)$. Then

$$v = \int \left(\frac{-y\,dx}{r^2} + \frac{x\,dy}{r^2} \right) = \tan^{-1} (y/x) + \text{constant}$$

so that $\ln r + i \tan^{-1} (y/x)$ is an analytic function. It may be more simply written $\ln z$.

The special nature of an analytic function is further demonstrated by the fact that, if u and v satisfy the Cauchy-Riemann conditions, so do the pair $\partial u/\partial x$ and $\partial v/\partial x$ and similarly $\partial u/\partial y$ and $\partial v/\partial y$. This seems to indicate that, if *$f(z)$ is analytic, so are all of its derivatives.* We cannot yet prove this theorem, for we have not shown that these higher derivatives exist; this will be done in the following section. It is a useful theorem to remember, for it gives a convenient test for analyticity which involves merely testing for the existence of higher derivatives.

Points at which functions are not analytic are called *singularities.* We have already encountered the singularity $1/(z - a)$ which represents the electric field due to a point charge located at a. Point a is a singularity because the derivative of this function at $z = a$ does not exist. Some functions are not analytic at any point, for example, $|z|^2$. This equals $z\bar{z}$ and is clearly not a function of z only. Another example is $z^{p/q}$, where p/q has been reduced to its lowest terms. For example, $z^{\frac{3}{2}}$ is not analytic at $z = 0$; its second derivative at $z = 0$ is infinite.

Conformal Representation. Any analytic function $f(z) = u + iv$, $z = x + iy$, can be represented geometrically as a transformation of two-dimensional coordinates. One can imagine two complex planes, one to represent the chosen values of z and the other to represent the resulting values of f. Any line drawn on the z plane has a resulting line on the f plane. Of course, many other pairs of functions u, v of x and y can be used to define such a general coordinate transformation. The transformations represented by the real and imaginary parts of analytic functions, however, have several unique and useful characteristics. The most important and obvious of these characteristics is that the transformation is "angle-preserving" or *conformal.*

When two lines drawn on the z plane cross, the corresponding lines on the f plane also cross. For a conformal transformation the angle between the lines on the f plane, where they cross, is *equal* to the angle between the lines on the z plane, where they cross. As shown in Fig. 4.2, for instance, the lines cross at $z = a$ on the z plane and at $f(a)$ on the f plane. An elementary length along line number 1 can be represented by $dz_1 = |dz_1|e^{i\varphi_1}$ and an elementary length along line 2 is given by $dz_2 = |dz_2|e^{i\varphi_2}$.

The corresponding elementary lengths on the f plane are $dz_1(df/dz)$ and $dz_2(df/dz)$. If the function f is analytic at $z = a$, then df/dz is independent of the direction of dz; that is, df/dz at $z = a$ is equal to $|df/dz|e^{i\alpha}$, independent of the angle φ of dz_1 or dz_2. Therefore the elementary length along line 1 in the f plane is $|dz_1(df/dz)|e^{i(\alpha+\varphi_1)}$ and the elementary length along line 2 is $|dz_2(df/dz)|e^{i(\alpha+\varphi_2)}$. The direction of

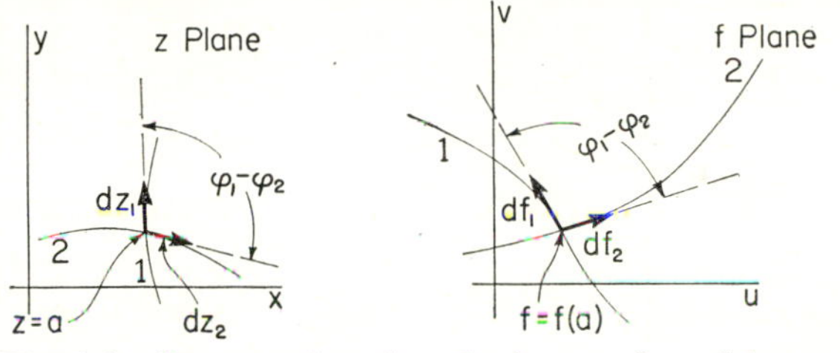

Fig. 4.2 Conservation of angles in a conformal transformation.

the two lines is rotated by the angle α from the lines on the z plane, but the *angle between the lines*, $(\alpha + \varphi_1) - (\alpha + \varphi_2) = \varphi_1 - \varphi_2$, is the same as the angle between the corresponding lines on the z plane. Therefore the transformation represented by the analytic function $f(z) = u + iv$ is a *conformal transform*, preserving angles. Likewise, as can be proved by retracing the steps of this discussion, if a two-dimensional transform from x, y to u, v is conformal, it can be represented by a function $f = u + iv$ which is an analytic function of $z = x + iy$, where u and v satisfy the Cauchy-Riemann equations (4.2.1).

We can likewise draw lines on the f plane and see where the corresponding lines fall on the z plane. For instance, the lines u = constant, v = constant, generate a rectangular coordinate system on the f plane; on the z plane the lines $u(x,y)$ = constant, $v(x,y)$ = constant constitute an orthogonal curvilinear coordinate system (orthogonal because right angles stay right angles in the transformation). We note from the discussion above or by using the definitions given in Sec. 1.3 [see (Eq. 1.3.4)] that the scale factors for these coordinates are equal,

$$h_u = \sqrt{(\partial u/\partial x)^2 + (\partial u/\partial y)^2} = h_v = \sqrt{(\partial v/\partial x)^2 + (\partial v/\partial y)^2} = |df/dz|$$

as long as the Cauchy-Riemann equations (4.2.1) are valid (*i.e.*, as long as f is analytic). Therefore any infinitesimal figure plotted on the f plane is transformed into a *similar* figure on the z plane, with a possible change of position and size *but with angles and proportions preserved,* wherever f is analytic. This is an alternative definition of a conformal transformation.

The simplest conformal transform is represented by the equation $f = ze^{i\theta} + c$, where the real angle θ and the complex quantity c are constants. Here the scale factor $h_u = h_v = |df/dz| = 1$, so that scale is preserved. The transform corresponds to a translation given by the complex constant c plus a rotation through the angle θ. In other cases, however, scale will be changed, and it will be changed differently in different regions so that the whole space will be distorted even though small regions will be similar. Since, as we have seen, any function of a

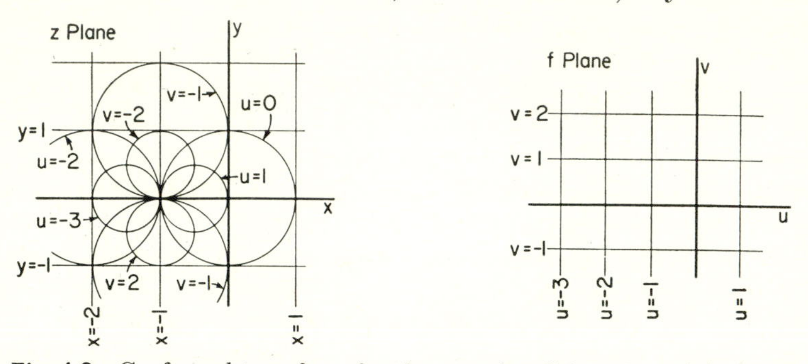

Fig. 4.3 Conformal transform for the equation $f(z) = (1 - z)/(1 + z)$.

complex variable may be regarded as an example of an electrostatic field, we may say that the effect of the field may be simulated by warping the space and replacing the field by a simple radial electrostatic field $\bar{E} = u + iv$. The motion of a charged particle may thus be explained as either due directly to the electrostatic field or to the warping of space by that field. This point of view is reminiscent of the procedure used by Einstein in discussing the effects of gravitation as given by the general theory of relativity.

Figure 4.3 shows the conformal transform corresponding to the equation

$$f = \frac{1 - z}{1 + z}; \quad u = \frac{1 - x^2 - y^2}{(1 + x)^2 + y^2}; \quad v = \frac{-2y}{(1 + x)^2 + y^2}; \quad z = \frac{1 - f}{1 + f}$$

The function is analytic except at $z = -1$, and the transformation is conformal except at this point. The curves u = constant, v = constant on the z plane constitute an orthogonal set of coordinates consisting of two families of tangential circles. Other cases will be pictured and discussed later (see Sec. 4.7).

We note that, at points where the scale factor $|df/dz|$ is zero, the transform also is not conformal. The region in the neighborhood of such points becomes greatly compressed when looked at in the f plane.

Inversely the corresponding region on the f plane is tremendously expanded. This suggests the possibility of a singularity of the inverse function to f, $z(f)$, at the point for which $f'(z) = 0$. An example will show that this is indeed the case. The simplest is $f(z) = z^2$ for which $f'(0) = 0$. The inverse $z = f^{\frac{1}{2}}$ has, as predicted, a singularity at this point. Thus the transformation cannot be conformal at this point. This can be shown directly, for if two line elements passing through $z = 0$ make an angle $\varphi_2 - \varphi_1$ with respect to each other, then the corresponding lines on the f plane make an angle of $2(\varphi_2 - \varphi_1)$ with respect to each other and thus mapping is not conformal at $z = 0$. It is also clear that, whenever $f'(a) = 0$, the mapping will not be conformal whether $f(z)$ behaves as $(z - a)^2$, as in this particular example, or as $(z - a)^n$ (n integer).

But for the region where f is analytic and where the scale factor $|df/dz|$ is *not* zero, the transform is reciprocally conformal. In mathematical language we can say: *Let $f(z)$ be analytic at $z = a$, and $f'(a) \neq 0$, then the inverse of $f(z)$ exists and is analytic in a sufficiently small region about $f(a)$ and its derivative is $1/f'(a)$.*

First as to the existence of the inverse function, we note that the inverse to the transformation $u = u(x,y)$ and $v = v(x,y)$ exists if the quantity $(\partial u/\partial x)^2 + (\partial u/\partial y)^2 \neq 0$. This is indeed the case if $f'(a) \neq 0$. We note that, if $f'(a)$ were equal to zero, the inverse function would not exist at this point, indicating the existence of a singularity in the inverse function at this point. It now remains to show that the inverse function is analytic, that is, $\partial x/\partial u = \partial y/\partial v$ and $\partial x/\partial v = -\partial y/\partial u$. Let us try to express $\partial x/\partial u$, etc., in terms of $\partial u/\partial x$, etc. To do this note that

$$dx = \frac{\partial x}{\partial u}\,du + \frac{\partial x}{\partial v}\,dv$$

so that

$$1 = \left(\frac{\partial x}{\partial u}\right)\left(\frac{\partial u}{\partial x}\right) + \left(\frac{\partial x}{\partial v}\right)\left(\frac{\partial v}{\partial x}\right) \quad \text{and} \quad 0 = \left(\frac{\partial x}{\partial u}\right)\left(\frac{\partial u}{\partial y}\right) + \left(\frac{\partial x}{\partial v}\right)\left(\frac{\partial v}{\partial y}\right)$$

or using Eq. (4.2.1)

$$1 = \left(\frac{\partial x}{\partial u}\right)\left(\frac{\partial u}{\partial x}\right) - \left(\frac{\partial x}{\partial v}\right)\left(\frac{\partial u}{\partial y}\right) \quad \text{and} \quad 0 = \left(\frac{\partial x}{\partial u}\right)\left(\frac{\partial u}{\partial y}\right) + \left(\frac{\partial x}{\partial v}\right)\left(\frac{\partial u}{\partial x}\right)$$

Similarly

$$1 = \left(\frac{\partial y}{\partial u}\right)\left(\frac{\partial u}{\partial y}\right) + \left(\frac{\partial y}{\partial v}\right)\left(\frac{\partial v}{\partial y}\right) \quad \text{and} \quad 0 = \left(\frac{\partial y}{\partial u}\right)\left(\frac{\partial u}{\partial x}\right) - \left(\frac{\partial y}{\partial v}\right)\left(\frac{\partial u}{\partial y}\right)$$

This gives us four equations with four unknowns, $\partial x/\partial u$, $\partial x/\partial v$, $\partial y/\partial u$, and $\partial y/\partial v$, which may be solved. One obtains the derivatives of x:

$$\frac{\partial x}{\partial u} = \frac{\partial u/\partial x}{(\partial u/\partial x)^2 + (\partial u/\partial y)^2}; \quad \frac{\partial x}{\partial v} = \frac{-\partial u/\partial y}{(\partial u/\partial x)^2 + (\partial u/\partial y)^2}$$

The derivatives of y turn out to be just as required by the Cauchy-Riemann conditions $\partial x/\partial u = \partial y/\partial v$, etc. It is also possible now to evaluate

$$\frac{dz}{df} = \frac{\partial x}{\partial u} + i\frac{\partial y}{\partial u} = \frac{\partial x}{\partial u} - i\frac{\partial x}{\partial v} = \frac{1}{(\partial u/\partial x) - i(\partial u/\partial y)} = \frac{1}{df/dz}$$

proving the final statement of the theorem. We shall devote a considerable portion of the chapter later to develop the subject of conformal representation further, for it is very useful in application.

Integration in the Complex Plane. The theory of integration in the complex plane is just the theory of the line integral. If C is a possible contour (to be discussed below), then from the analysis of page 354 [see material above Eq. (4.1.12)] it follows that

$$\int_C \bar{E}\,dz = \int_C E_t\,ds + i\int_C E_n\,ds; \quad ds = |dz|$$

where E_t is the component of the vector $\mathbf{E}$ along the path of integration while E_n is the normal component. Integrals of this kind appear frequently in physics. For example, if $\mathbf{E}$ is any force field, then the integral $\int_C E_t\,ds$ is just the work done against the force field in moving along the contour C. The second integral measures the total flux passing through the contour. If $\mathbf{E}$ were the velocity vector in hydrodynamics, then the second integral would be just the total fluid current through the contour.

In order for both of these integrals to make physical (and also mathematical) sense, it is necessary for the contour to be sufficiently smooth. Such a *smooth curve* is composed of arcs which join on continuously, each arc having a continuous tangent. This last requirement eliminates some pathological possibilities, such as a contour sufficiently irregular that is of infinite length. For purposes of convenience, we shall also insist that each arc have no multiple points, thus eliminating loops. However, loops may be easily included in the theory, for any contour containing a loop may be decomposed into a closed contour (the loop) plus a smooth curve, and the theorem to be derived can be applied to each. A *closed contour* is a closed smooth curve. A closed contour is described in a *positive* direction with respect to the domain enclosed by the contour if with respect to some point inside the domain the contour is traversed in a *counterclockwise* direction. The negative direction is then just the clockwise one. Integration along a closed contour will be symbolized by $\oint$.

One fairly obvious result we shall use often in our discussions: If $f(z)$ is an analytic function within and on the contour, and if df/dz is single-valued in the same region,

$$\oint (df/dz)\,dz = 0$$

This result is not necessarily true if df/dz is *not* single-valued.

Contours involving smooth curves may be combined to form new contours. Some examples are shown in Figure 4.4. Some of the contours so formed may no longer be smooth. For example, the boundary b' is not bounded by a smooth curve (for the inner circle and outer circle are not joined) so that this contour is not composed of arcs which join on continuously. Regions of this type are called *multiply connected,* whereas the remaining examples in the figure are *simply connected.* To test for connectivity of a region note that any closed contour drawn within a simply connected region can be shrunk to a point by continuous deformation without crossing the boundary of the region. In b' a

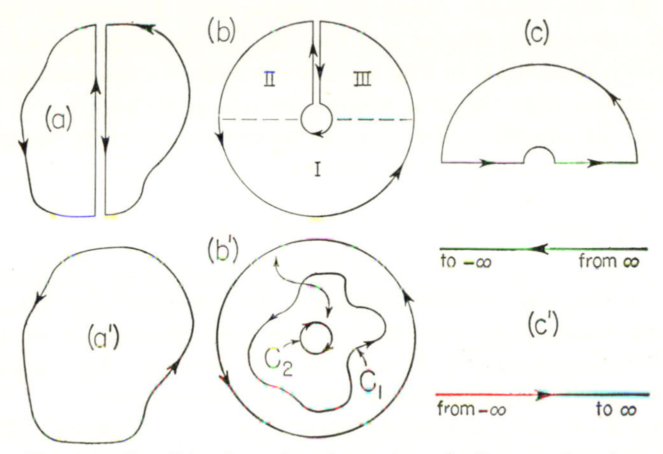

Fig. 4.4 Possible alterations in contours in the complex plane.

curve C_1 intermediate to the two boundary circles cannot be so deformed. The curve b illustrates the fact that any multiply connected surface may be made singly connected if the boundary is extended by means of crosscuts so that it is impossible to draw an irreducible contour. For example, the intermediate contour C_1 drawn in b' would not, if drawn in b, be entirely within the region as defined by the boundary lines. The necessity for the discussion of connectivity and its physical interpretation will become clear shortly.

Having disposed of these geometric matters, we are now able to state the central theorem of the theory of functions of a complex variable.

Cauchy's Theorem. *If a function $f(z)$ is an analytic function, continuous within and on a smooth closed contour C, then*

$$\oint f(z)\,dz = 0 \tag{4.2.3}$$

For a proof of Cauchy's theorem as stated above, the reader may be referred to several texts in which the Goursat proof is given. The simple proof given earlier assumes that $f'(z)$ not only exists at every point within C but is also continuous therein. It is useful to establish the theorem within a minimum number of assumptions about $f(z)$, for this extends the ease of its applicability. In this section we shall content ourselves with assuming that *C bounds a star-shaped region and that $f'(z)$ is bounded everywhere within and on C.*

The geometric concept of "star-shaped" requires some elucidation. A star-shaped region exists if a point O can be found such that every ray from O intersects the bounding curve in precisely one point. A simple example of such a region is the region bounded by a circle. A region which is not star-shaped is illustrated by any annular region. Restricting our proof to a star-shaped region is not a limitation on the theorem, for any simply connected region may be broken up into a number of star-shaped regions and the Cauchy theorem applied to each. This process is illustrated in Fig. 4.4c for the case of a semiannular region. Here the semiannular region is broken up into parts like II and III, each of which is star-shaped. The Cauchy theorem may then be applied to each along the indicated contours so that

$$\oint_{\text{II}} f\,dz + \oint_{\text{III}} f\,dz = 0$$

However, in the sum of these integrals, the integrals over the parts of the contour common to III and II cancel out completely so that the sum of the integrals over I, II, and III just becomes the integral along the solid lines, the boundary of the semiannular contour.

The proof of the Cauchy theorem may now be given. Take the point O of the star-shaped region to be the origin. Define $F(\lambda)$ by

$$F(\lambda) = \lambda \oint f(\lambda z)\,dz; \quad 0 \le \lambda \le 1 \tag{4.2.4}$$

The Cauchy theorem is that $F(1) = 0$. To prove it, we differentiate $F(\lambda)$:

$$F'(\lambda) = \oint f(\lambda z)\,dz + \lambda \oint z f'(\lambda z)\,dz$$

Integrate the second of these integrals by parts [which is possible only if $f'(z)$ is bounded]:

$$F'(\lambda) = \oint f(\lambda z)\,dz + \lambda\left\{\left[\frac{zf(\lambda z)}{\lambda}\right] - \frac{1}{\lambda}\oint f(\lambda z)\,dz\right\}$$

where the square bracket indicates that we take the difference of values at beginning and end of the contour of the quantity within the bracket. Since $zf(\lambda z)$ is a single-valued function, $[zf(\lambda z)/\lambda]$ vanishes for a closed contour so that

$$F'(\lambda) = 0 \quad \text{or} \quad F(\lambda) = \text{constant}$$

To evaluate the constant, let $\lambda = 0$ in Eq. (4.2.4), yielding $F(0) = 0 = F = F(\lambda)$. Therefore $F(1) = 0$, which proves the theorem. This proof, which appears so simple, in reality just transfers the onus to the question as to when an integral can be integrated by parts. The requirements, of course, involve just the ones of differentiability, continuity, and finiteness which characterize analytic functions.

Cauchy's theorem does *not apply to multiply connected regions,* for such regions are not bounded by a smooth contour. The physical reason for this restriction is easy to find. Recall from the discussion of page 354 that the Cauchy theorem, when applied to the electrostatic field, is equivalent to the statement that no charge is included within the region bounded by the contour C. Using Fig. 4.4b' as an example of a multiply connected region, we see that contours entirely within the region in question exist (for example, contour C_1 in Fig. 4.4b') to which Cauchy's

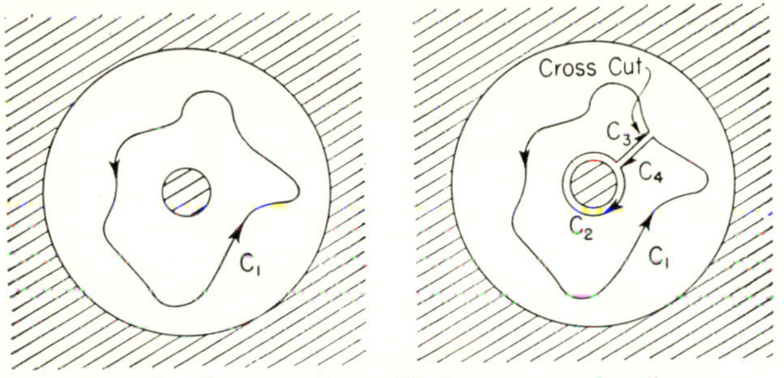

Fig. 4.5 Contours in multiply connected regions.

theorem obviously cannot apply because of the possible presence of charge outside the region in question, *e.g.*, charge within the smaller of the two boundary circles. The way to apply Cauchy's theorem with certainty would be to subtract the contour integral around the smaller circle; *i.e.*,

$$\oint_{C_1} f\,dz - \oint_{C_2} f\,dz = 0 \tag{4.2.5}$$

This may be also shown directly by using crosscuts to reduce the multiply connected domain to a single-connected one. From Fig. 4.5 we see that a contour in such a simply connected domain consists of the old contours C_1 and C_2 (C_1 described in a positive direction, C_2 in a negative direction) plus two additional sections C_3 and C_4. Cauchy's theorem may be applied to such a contour. The sections along C_3 and C_4 will cancel, yielding Eq. (4.2.5).

Some Useful Corollaries of Cauchy's Theorem. From Cauchy's theorem it follows that, *if $f(z)$ is an analytic function within a region bounded by closed contour C, then $\int_{z_1}^{z_2} f(z)\,dz$,* along any *contour within C depends only on z_1* and z_2. That is, $f(z)$ has not only a unique derivative but also a unique integral. The uniqueness requirement is often used as motivation for a discussion of the Cauchy theorem. To prove this, we compare the two integrals $\int_{C_1}$ and $\int_{C_2}$, in Fig. 4.6, where C_1 and C_2 are two different contours starting at z_1 and going to z_2. According to Cauchy's theorem $\int_{C_1} f(z)\,dz - \int_{C_2} f(z)\,dz = \oint f(z)\,dz$, is zero, proving the corollary.

It is a very important practical consequence of this corollary that *one may deform a contour without changing the value of the integral, provided that the contour crosses no singularity of the integrand during the deformation.* We shall have many occasions to use this theorem in the

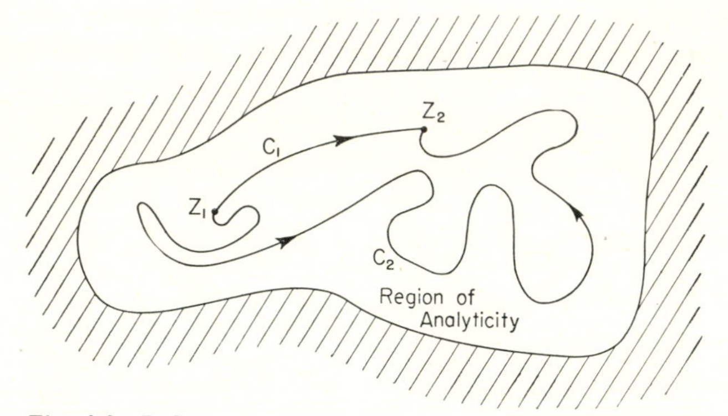

Fig. 4.6 Independence of integral value on choice of path within region of analyticity.

evaluation of contour integrals, for it thus becomes possible to choose a most convenient contour.

Because of the uniqueness of the integral $\int_{z_1}^{z_2} f\,dz$ it is possible to define an *indefinite integral of $f(z)$ by*

$$F(z) = \int_{z_1}^{z} f(z)\,dz$$

where the contour is, of course, within the region of analyticity of $f(z)$. It is an interesting theorem that, if $f(z)$ is analytic in a given region, then $F(z)$ is also analytic in the same region. Or, conversely, if $f(z)$ is singular at z_0, so is $F(z_0)$. To prove this result, we need but demonstrate the uniqueness of the derivative of $F(z)$, which can be shown by considering the identity

$$\frac{F(z) - F(\zeta)}{z - \zeta} - f(\zeta) = \frac{\int_{\zeta}^{z} [f(z) - f(\zeta)]\,dz}{z - \zeta}$$

Because of the continuity and single-valuedness of $f(z)$ the right-hand side of the above equation may be made as small as desired as z is made to approach ζ. Therefore in the limit

$$\lim_{z \to \zeta} \left[\frac{F(z) - F(\zeta)}{z - \zeta} \right] = f(\zeta)$$

Since the limit on the left is just the derivative $F'(\zeta)$, the theorem is proved.

We recall from Eqs. (4.1.19) *et seq.* that, if $f(z)$ is the conjugate of the electrostatic field, then the real part of $F(z)$ is the electrostatic potential while the imaginary part is constant along the electric lines of force and is therefore the stream function (see page 355). Therefore the two-dimensional electrostatic potential and the stream function form the real and imaginary parts of an analytic function of a complex variable.

Looking back through the proof of Cauchy's theorem, we see that we used only the requirements that $f(z)$ be continuous, one-valued, and that the integral be unique, with the result that we proved that $F(z)$ was *analytic*. We shall later show that, if a function is analytic in a region, so is its derivative [see Eq. (4.3.1)]. Drawing upon this information in advance of its proof, we see that, once we have found that $F(z)$ is analytic, we also know that $f(z)$ is analytic. This leads to the converse of Cauchy's theorem, known as *Morera's theorem:*

If $f(z)$ is continuous and single-valued within a closed contour C, and if $\oint f(z)\,dz = 0$ for any closed contour within C, then $f(z)$ is analytic within C.

This converse serves as a means for the identification of an analytic function and is thus the integral analogue of the differential requirement given by the Cauchy-Riemann conditions. Since the latter requires continuity in the derivative of f, the integral condition may sometimes be easier to apply.

The physical interpretation of Morera's theorem as given by the electrostatic analogue will strike the physicist as being rather obvious. It states that, if $f(z)$ is an electrostatic field and the net charge within any closed contour [evaluated with the aid of $f(z)$] within C is zero, then the charge density within that region is everywhere zero.

Cauchy's Integral Formula. This formula, a direct deduction from the Cauchy theorem, is the chief tool in the application of the theory of analytic functions to other branches of mathematics and also to physics. Its electrostatic analogue is known as Gauss' theorem, which states that the integral of the normal component of the electric field about a closed contour C equals the net charge within the contour. In electrostatics the proof essentially consists of separating off the field due to sources outside the contour from the field due to sources inside. The first, when integrated, must yield zero, while the second may be found by adding up the contribution due to each source. Cauchy's integral formula applies to the situation in which there is but one source inside C.

Consider the integral

$$J(a) = \oint [f(z)/(z-a)]\,dz \tag{4.2.6}$$

around a closed contour C within and on which $f(z)$ is analytic. The contour for this integral may be deformed into a small circle of radius ρ about the point a according to the corollary of Cauchy's theorem on the deformation of contours. Thus letting $z - a = \rho e^{i\varphi}$,

$$J = i \int_0^{2\pi} f(a + \rho e^{i\varphi})\, d\varphi = if(a) \int_0^{2\pi} d\varphi + i \int_0^{2\pi} [f(a + \rho e^{i\varphi}) - f(a)]\, d\varphi \tag{4.2.7}$$

Taking the limit as $\rho \to 0$, the second integral vanishes because of the continuity of $f(z)$. The Cauchy integral formula states, therefore, that, if $f(z)$ *is analytic inside and on a closed contour* C, *and if* a *is a point within* C, *then*

$$\oint [f(z)/(z - a)]\, dz = 2\pi i f(a) \tag{4.2.8}$$

If a is a point outside of C, then $\oint [f(z)/(z - a)]\, dz = 0$. If a is a point on C, the integral will have a Cauchy principal value[1] equal to $\pi i f(a)$ (just halfway between). The Cauchy principal value corresponds to putting half of the point source inside C and half outside. To summarize:

$$\oint \frac{f(z)}{z - a}\, dz = 2\pi i f(a) \begin{cases} 1; & \text{if } a \text{ within } C \\ \frac{1}{2}; & \text{if } a \text{ on } C \text{ (principal value)} \\ 0; & \text{if } a \text{ outside } C \end{cases} \tag{4.2.9}$$

Cauchy's formula is an *integral representation* of $f(z)$ which permits us to compute $f(z)$ anywhere in the interior of C, knowing only the value of $f(z)$ on C. Representations of this kind occur frequently in physics (particularly in the application of Green's or source functions) with the same sort of discontinuity as is expressed by Eq. (4.2.9). Thus if f is an electrostatic field, Eq. (4.2.8) tells us that the field within C may be computed in terms of the field along C. Similar theorems occur in the theory of wave propagation, where they are known collectively as Huygens' principle.

Cauchy's formula provides us with a very powerful tool for the investigation of the properties of analytic functions. It points up the strong correlation which exists between the values of an analytic function all over the complex plane. For example, using Eq. (4.2.7) we see that $f(a)$ is the *arithmetic average* of the values of f on any circle centered at a. *Therefore* $|f(a)| \leq M$ *where* M *is the maximum value of* $|f|$ *on the circle. Equality can occur only if* f *is constant on the contour,* in which case f *is constant within the contour.* This theorem may be easily extended to a region bounded by any contour C.

[1] The Cauchy principal value is defined as follows: Let $q(x) \to \infty$ as $x \to a$, then the principal value of

$\int_b^c q(x)\, dx$, written $\mathcal{P} \int_b^c q(x)\, dx$, $(b < a < c)$ is $\lim_{\delta \to 0} \left\{ \int_b^{a-\delta} q(x)\, dx + \int_{a+\delta}^c q(x)\, dx \right\}$

In terms of the electrostatic analogue, the largest values of an electrostatic field within a closed contour occur at the boundary. If $f(z)$ has no zeros within C, then $[1/f(z)]$ will be an analytic function inside C and therefore $|1/f(z)|$ will have no maximum within C, taking its maximum value on C. *Therefore $|f(z)|$ will not have a minimum within C but will have its minimum value on the contour C.* The proof and theorem do not hold if $f(z)$ has zeros within C. The absolute value of an analytic function can have neither a maximum nor a minimum within the region of analyticity. If the function assumes either the maximum or minimum value within C, the function is a constant. Points at which $f(z)$ has a zero derivative will therefore be saddle points, rather than true maxima or minima.

Applying these results to the electrostatic field, we see that the field will take on both its minimum and maximum values on the boundary curve.

These theorems apply not only to $|f(z)|$ but also to the real and imaginary parts of an analytic function and therefore to the electrostatic potential V. To see this result rewrite Eq. (4.2.7) as

$$2\pi i f(a) = 2\pi i(u + iv) = i \int_0^{2\pi} f(x + iy)\, d\varphi = i \int_0^{2\pi} (u + iv)\, d\varphi$$

Equating real parts of the second and fourth expressions in this sequence of equations one obtains

$$u = \frac{1}{2\pi} \int^{2\pi} u\, d\varphi \tag{4.2.10}$$

so that u at the center of the circle is the arithmetic average of the values of u on the boundary of the circle. We may now use precisely the same reasoning as was employed in the discussion above for $|f(z)|$ and conclude that u will take on its minimum and maximum value on the boundary curve of a region within which f is analytic.

We therefore have the theorem that the electrostatic potential within a source-free region can have neither a maximum nor a minimum within that region. This fact has already been established in Chap. 1 (page 7) in the discussion of the Laplace equation which the electrostatic potential satisfies. From the theorem it immediately follows that, if V is constant on a contour enclosing a source-free singly connected region, then V is a constant within that region. This is just the well-known electrostatic result which states that the electrostatic field within a perfect conductor forming a closed surface is zero.

From these examples the general usefulness of the Cauchy integral formula should be clear. In addition we have once more demonstrated the special nature of analytic functions. We shall return to a more thorough discussion of these properties later on in this chapter.

Real and Imaginary Parts of Analytic Functions. This last discussion indicates that it is possible to correlate and classify the component parts of an analytic function just as we have done for the entire analytic function in the preceding sections. Again we find that, because of the specialized nature of analytic functions, the behavior of its component parts in different regions of the complex plane is strongly correlated. In addition there is a close connection between the real and imaginary parts themselves. For example, we shall see that, if enough information about the real part is given, it is possible to *compute* the imaginary part (or vice versa).

We shall be particularly interested in this interconnection because of its great physical interest. In the electrostatic case, for instance, it yields a connection between the electrostatic potential and the total charge. In vibration theory, it yields a relation between the real and imaginary parts of the impedance, *i.e.*, a relation between the resistive and reactive parts. The complex variable in this case is the complex frequency.

We were able to obtain the beginnings of an understanding of analytic functions by using the integral representation obtained from the Cauchy integral formula. This representation is valid inside any domain bounded by a closed contour and relates the value of the analytic function f in the interior of the contour with its values on the contour. Such a simple relation for the real part of f (hereafter denoted as $\mathrm{Re}\, f$) inside a region in terms of $\mathrm{Re}\, f$ on the contour is easily exhibited for a circular contour. We shall show later it is possible to map, by a conformal transformation, the interior of a region onto the interior of a circle; as a result, it is possible, in principle, to relate the values of $\mathrm{Re}\, f$ inside a boundary to its values on the boundary, although in practice it is very often difficult to carry this out. However, because of the possibility of such a transformation the qualitative aspects of the theorems developed below will hold for any region. In all the discussion below it will be assumed that $f(z)$ *is analytic within the boundaries of the region being considered.*

Much information may be obtained by considering a circle of infinite radius forming, for convenience, the x axis, the interior of the circle becoming the upper half plane. Our problem then requires us to find the real and imaginary parts of $f = u + iv$ in the upper half plane when u is known on the x axis. The electrostatic analogue will suggest a solution, for the problem in this case is to find the potential in the upper half plane when the potential on the line $x = 0$ is known. The method of images (see Chap. 7) is the device used in electrostatics for solving such problems.

The Cauchy integral is our equivalent for a source function, the factor $1/(z - \zeta)$ representing a "source" at $z = \zeta = \xi + i\eta$. We set the point ζ in the upper half plane, so that the contour integral of $f(z)/(z - \zeta)$ around the boundaries of the upper half plane will be $2\pi i f(\zeta)$. The integral of $f(z)/(z - \bar{\zeta})$ will be zero because the image point $\bar{\zeta} = \xi - i\eta$ will be outside the contour. Consequently

$$f(\zeta) = \frac{1}{2\pi i} \oint f(z) \left[\frac{1}{z - \zeta} - \frac{1}{z - \bar{\zeta}} \right] dz \tag{4.2.11}$$

As long as f is analytic over the whole upper half plane, the integral around the semicircle at infinity is zero, so that the integral (4.2.11) becomes simply an integral along the real axis, yielding

$$f(\zeta) = \frac{\eta}{\pi} \int_{-\infty}^{\infty} \frac{f(x)}{(x - \xi)^2 + \eta^2} dx \tag{4.2.12}$$

where $\zeta = \xi + i\eta$. Since both sides of (4.2.12) contain only real coefficients, this equation holds also for both u and v. For instance,

$$u(\xi,\eta) = \frac{\eta}{\pi} \int_{-\infty}^{\infty} \frac{u(x,0)}{(x - \xi)^2 + \eta^2} dx \tag{4.2.13}$$

Thus if the potential is known at $y = 0$, *i.e.*, on the x axis, the integral (4.2.13) gives the potential u everywhere else in the upper half plane as long as f is analytic in the upper half plane.

To obtain a relation between v and u, the above derivation suggests adding the two source functions instead of subtracting and using the same contour as before:

$$f(\zeta) = \frac{1}{2\pi i} \oint f(z) \left[\frac{1}{z - \zeta} + \frac{1}{z - \bar{\zeta}} \right] dz$$

or

$$f(\zeta) = \frac{1}{\pi i} \int_{-\infty}^{\infty} f(x) \left[\frac{x - \xi}{(x - \xi)^2 + \eta^2} \right] dx \tag{4.2.14}$$

Equating real and imaginary parts of both sides of this equation yields the desired relations:

$$u(\xi,\eta) = \frac{1}{\pi} \int_{-\infty}^{\infty} \frac{(x - \xi)v(x,0)}{(x - \xi)^2 + \eta^2} dx \tag{4.2.15}$$

and

$$v(\xi,\eta) = -\frac{1}{\pi} \int_{-\infty}^{\infty} \frac{(x - \xi)u(x,0)}{(x - \xi)^2 + \eta^2} dx \tag{4.2.16}$$

Knowing the behavior of the real (or imaginary) parts of f on the x axis permits the calculation of both u and v in the upper half plane using (4.2.13) and (4.2.15). These equations represent the solutions of the electrostatic problems in which either the charge or potential distribution is given and the resultant fields required.

Equations (4.2.12) to (4.2.16) may also be used to determine the behavior of u and v on the boundary ($\eta = 0$) itself. However, because of the resultant singular integrals, one must be quite careful. Perhaps the simplest stratagem is to return to the Cauchy integral theorem, which for a point on the contour states [see (Eq. 4.2.9)]

$$f(\zeta) = \frac{1}{\pi i} \mathcal{P} \oint \frac{f(z)}{z - \zeta} dz$$

If the contour is the x axis, then

$$f(\xi,0) = \frac{1}{\pi i} \mathcal{P} \int_{-\infty}^{\infty} \frac{f(x,0)}{x - \xi} dx \tag{4.2.17}$$

or

$$u(\xi,0) = \frac{1}{\pi} \mathcal{P} \int_{-\infty}^{\infty} \frac{v(x,0)}{x - \xi} dx; \quad v(\xi,0) = -\frac{1}{\pi} \mathcal{P} \int_{-\infty}^{\infty} \frac{u(x,0)}{x - \xi} dx \tag{4.2.18}$$

These relations occur also in the theory of *Hilbert transforms*, u and v being the Hilbert transform of each other. We shall discuss these transforms in somewhat greater detail in later sections of this chapter. (It should be emphasized again that our formulas relate the real and imaginary parts of a function which is *analytic* in the upper half plane.) It is sometimes convenient to rewrite (4.2.17) so that the singularity is no longer present. Recalling that $\mathcal{P} \int_{-\infty}^{\infty} dx/(x - \xi) = 0$, we have

$$\begin{aligned} u(\xi,0) &= \frac{1}{\pi} \int_{-\infty}^{\infty} \frac{v(x,0) - v(\xi,0)}{x - \xi} dx; \\ v(\xi,0) &= -\frac{1}{\pi} \int_{-\infty}^{\infty} \frac{u(x,0) - u(\xi,0)}{x - \xi} dx \end{aligned} \tag{4.2.19}$$

Impedances. One of the most useful analytic functions we deal with is the *impedance* Z (see pages 283 and 324). An electrical impedance is defined as the ratio between voltage and current at a given point in a circuit, when both voltage and current are expressed in terms of the complex exponential of the time ($V = V_0 e^{i\omega t}$, etc.); a mechanical impedance is the complex ratio between force and velocity; the wave impedance of a string is defined on page 128; and so on.

The impedance is a function of the frequency ν (or, alternately, the angular velocity $\omega = 2\pi\nu$) and of the parameters of the system. When ω is real and positive (*i.e.*, when the driving force or voltage is $F_0 e^{i\omega t}$), then the real part of Z is called the *resistance* and is denoted by R and the imaginary part is called the reactance and is denoted by X. But we can imagine cases where ω is complex, when the driving force has a real exponential time factor, so that we can consider $Z(\omega) = R(\omega) + iX(\omega)$ to be an analytic function of the complex variable ω.

In most physically attainable cases Z is analytic in the upper half plane for ω (*i.e.*, when the imaginary part of ω is positive). The nature of the impedance is such that, if we reverse the sign of ω (*i.e.*, use $F_0e^{-i\omega t}$ as the force function), the resistance does not change but the reactance changes sign.

$$Z(-\omega) = \bar{Z}(\omega); \quad R(-\omega) = R(\omega); \quad X(-\omega) = -X(\omega) \qquad (4.2.20)$$

Specializing Eqs. (4.2.19) for such cases, we obtain

$$R(\omega) = \frac{2}{\pi}\int_0^\infty \frac{xX(x) - \omega X(\omega)}{x^2 - \omega^2}\,dx; \quad X(\omega) = \frac{2\omega}{\pi}\int_0^\infty \frac{R(x) - R(\omega)}{x^2 - \omega^2}\,dx \qquad (4.2.21)$$

which give the resistance for real values of ω in terms of the reactance for real values of ω, and vice versa. This indicates that, as long as Z is a well-behaved function of ω (analytic in the upper half plane), then resistance and reactance are related functions and cannot be chosen arbitrarily and independently.

Beside the direct applications mentioned here, formulas (4.2.19) are useful in evaluating integrals with infinite limits. This is done by applying (4.2.19) to functions whose real and imaginary parts are known. For example, suppose that $f(z) = e^{iz}$. Then $u(x,0) = \cos x$, $v(x,0) = \sin x$. One obtains

$$\cos\xi = \frac{1}{\pi}\int_{-\infty}^\infty \frac{\sin x - \sin\xi}{x - \xi}\,dx \qquad (4.2.22)$$

and, when $\xi = 0$,

$$\pi = \int_{-\infty}^\infty \frac{\sin x}{x}\,dx \qquad (4.2.23)$$

Many applications of this type may be made. Other examples will occur in this chapter and in the problems.

Poisson's Formula. Similar sets of relations may be obtained when the boundary curve is a circle. Here the image point to ζ is $a^2/\bar{\zeta}$, where a is the radius of a circle (if $|\zeta| < a$ then the image is outside the circle). Following the procedure used in Eqs. (4.2.11) *et seq.*, one writes

$$f(\zeta) = \frac{1}{2\pi i}\oint\left[\frac{f(z)}{z - \zeta} - \frac{f(z)}{z - (a^2/\bar{\zeta})}\right]dz$$

where the integral is taken along the circular contour. Converting to polar coordinates centered at the center of the circle we have $z = ae^{i\varphi}$, $\zeta = re^{i\vartheta}(r < a)$; one obtains, after a little manipulation, *Poisson's formula:*

$$f(re^{i\vartheta}) = \frac{1}{2\pi}\int_0^{2\pi} \frac{a^2 - r^2}{a^2 - 2ar\cos(\varphi - \vartheta) + r^2}\,f(ae^{i\varphi})\,d\varphi \qquad (4.2.24)$$

This formula is analogous to Eq. (4.2.13), enabling one to find the real (or imaginary) part of f *within* the circle in terms of its values on the circle.

Formulas similar to Eqs. (4.2.14) *et seq.* may be obtained by calculating

$$f(\zeta) = \frac{1}{2\pi i} \oint \left[\frac{f(z)}{z - \zeta} + \frac{f(z)}{z - a^2/\bar{\zeta}} \right] dz$$

One obtains, using Eq. (4.2.24),

$$f(re^{i\vartheta}) = f(0) - \frac{iar}{\pi} \int_0^{2\pi} \frac{\sin(\varphi - \vartheta)}{a^2 + r^2 - 2ar\cos(\varphi - \vartheta)} f(ae^{i\varphi})\, d\varphi \tag{4.2.25}$$

Equating real and imaginary parts yields

$$\begin{aligned} u(r,\vartheta) &= u(0) + \frac{ar}{\pi} \int_0^{2\pi} \frac{\sin(\varphi - \vartheta)}{a^2 + r^2 - 2ar\cos(\varphi - \vartheta)} v(a,\varphi)\, d\varphi \\ v(r,\vartheta) &= v(0) - \frac{ar}{\pi} \int_0^{2\pi} \frac{\sin(\varphi - \vartheta)}{a^2 + r^2 - 2ar\cos(\varphi - \vartheta)} u(a,\varphi)\, d\varphi \end{aligned} \tag{4.2.26}$$

where $f = u + iv$ and $u(0)$ is shorthand for u at $r = 0$.

Again values of u on the contour circle may be related to the totality of the values of v on the contour. One obtains

$$f(ae^{i\vartheta}) = f(0) - \frac{i}{2\pi} \mathcal{P} \int_0^{2\pi} f(ae^{i\varphi}) \cot \left[\frac{(\varphi - \vartheta)}{2} \right] d\varphi \tag{4.2.27}$$

Using the fact that $\mathcal{P} \int_0^{2\pi} \cot [(\varphi - \vartheta)/2]\, d\varphi = 0$ and equating real and imaginary parts in (4.2.27) yield

$$\begin{aligned} u(a,\vartheta) &= u(0) + \frac{1}{2\pi} \int_0^{2\pi} [v(a,\varphi) - v(a,\vartheta)] \cot \left[\frac{(\varphi - \vartheta)}{2} \right] d\varphi \\ v(a,\vartheta) &= v(0) - \frac{1}{2\pi} \int_0^{2\pi} [u(a,\varphi) - u(a,\vartheta)] \cot \left[\frac{(\varphi - \vartheta)}{2} \right] d\varphi \end{aligned} \tag{4.2.28}$$

the analogues of the Hilbert transforms for the circle.

The various formulas derived here demonstrate once more the close connection between u and v, which we have already seen in differential form in the Cauchy-Riemann equation (4.2.1).

4.3 *Derivatives of Analytic Functions, Taylor and Laurent Series*

One of the more remarkable properties of an analytic function is the fact that all its derivatives are also analytic, the region of analyticity being identical with that of $f(z)$. To prove the theorem, let us use the integral representation to evaluate the derivative, through the limiting process. The proof is able to proceed because the properties of the integral representation depend essentially on $1/(z - a)$, which is analytic when $z \neq a$.

$$
\begin{aligned}
2\pi i f'(a) &= 2\pi i \lim_{h\to 0} \left[\frac{f(a+h) - f(a)}{h}\right] \\
&= \lim_{h\to 0} \left\{\frac{1}{h} \oint f(z) \left[\frac{1}{z-a-h} - \frac{1}{z-a}\right] dz\right\} \\
&= \lim_{h\to 0} \oint \frac{f(z)}{(z-a)(z-a-h)}\, dz = \oint \frac{f(z)}{(z-a)^2}\, dz
\end{aligned}
$$

Taking the limit under the integration sign may be readily justified, for

$$
\begin{aligned}
&- \oint f(z) \left[\frac{1}{(z-a)^2} - \frac{1}{(z-a-h)(z-a)}\right] dz \\
&\qquad = h \oint \frac{f(z)\, dz}{(z-a)^2(z-a-h)} \leq \frac{hML}{b^2(b - |h|)}
\end{aligned}
$$

where M = maximum value of $|f(z)|$ on the contour, L is the length of the contour, and b = minimum value of $|z - a|$ on the contour. The right-hand side of this inequality approaches zero as $h \to 0$; therefore the left side also does.

We can continue with this same process to obtain higher derivatives, obtaining the important general formula for the nth derivative of f at $z = a$,

$$
f^{(n)}(a) = \frac{n!}{2\pi i} \oint \frac{f(z)}{(z-a)^{n+1}}\, dz \tag{4.3.1}
$$

Since all the derivatives exist, all the derivatives must be analytic within C. Note that we cannot claim existence of the derivatives *on the contour* from this proof, as is indicated by the discontinuity in the integral representation (4.2.9).

The Taylor Series. Using formula (4.3.1), it now becomes possible to derive the Taylor series and find its radius of convergence. The Taylor series is a power-series expansion of the function $f(a + h)$ in powers of h. Such a power series will in general be (1) *convergent within its radius of convergence and divergent outside it*, (2) *analytic within the radius of convergence, and* (3) *its radius of convergence will extend up to the singularity of $f(z)$ nearest a.* For the proof of statements 1 and 2 the reader is referred to any textbook which discusses the properties of power series in some detail. From the Cauchy integral formula (we make the assumptions involved in the Cauchy formula, such as a is inside a contour C and $f(z)$ is analytic inside and on C) we have

$$
f(a+h) = \frac{1}{2\pi i} \oint \frac{f(z)}{z-a-h}\, dz
$$

The contour is taken to be a circle about a, inasmuch as the region of convergence of the final series is circular.

We next use the identity

$$\left[1 + \frac{h}{z-a} + \frac{h^2}{(z-a)^2} + \cdots + \frac{h^{n-1}}{(z-a)^{n-1}}\right]\left(\frac{z-a-h}{z-a}\right) = 1 - \frac{h^n}{(z-a)^n}$$

to obtain the exact expression

$$\frac{1}{z-a-h} = \sum_{n=1}^{N}\left[\frac{h^{n-1}}{(z-a)^n}\right] + \frac{h^N}{(z-a-h)(z-a)^N}$$

Introducing this identity into the integral representation of $f(a+h)$ one obtains

$$f(a+h) = \sum_{n=0}^{N} \frac{h^n}{2\pi i}\oint \frac{f(z)}{(z-a)^{n+1}}\,dz + \frac{h^{N+1}}{2\pi i}\oint \frac{f(z)}{(z-a)^{N+1}(z-a-h)}\,dz$$

or

$$f(a+h) = \sum_{n=0}^{N} \frac{h^n}{n!} f^{(n)}(a) + R_N \tag{4.3.2}$$

where Eq. (4.3.1) has been used. The term R_N is just the remainder term; *i.e.*, it is the difference between $f(a+h)$ and the first $N+1$ terms of the Taylor series. To find the radius of convergence, we first note that the series certainly converges in the circular region of radius r centered at a and extending up to the next singularity. This follows from the following inequality:

$$R_N \leq \frac{|h|^{N+1}Mr}{r^{N+1}[r-|h|]}$$

where M is the maximum value of f on the contour. Within the radius r, $h < r$ so that, as $N \to \infty$, $R_N \to 0$. Therefore within the radius r

$$f(a+h) = \sum_{n=0}^{\infty} \frac{h^n}{n!} f^{(n)}(a) \tag{4.3.3}$$

is a convergent series, called the *Taylor series.* We also see that the radius of convergence is certainly as big as r. It cannot be larger than r, however, for we should not expect a power series to represent a singularity satisfactorily.

The fact that the radius of convergence of the Taylor series is determined by the distance to the nearest singularity explains some apparent paradoxes which occur if attention is paid only to values of the function and of the series along the real axis of z. A familiar example is the Taylor expansion $(1-z)^{-1} = 1 + z + z^2 + \cdots$. One would, of course, expect this series to "blow up" at $x = 1$; however, it is also divergent at $z = -1$ and indeed at $z = e^{i\varphi}$, that is, at any point on a unit circle surrounding the origin. The reason for this is clear from the point of view of the theorem expressed above. Another example is $f(x) = e^{-1/x^2}$. Every derivative of this function evaluated at $x = 0$ is zero, and if one puts this result blindly into the Taylor formula, one obtains apparent nonsense. The point here is that $z = 0$ is a singularity.

In fact it is often important to distinguish between the series representing a function and "the function itself" (whatever that phrase can be made to mean). A power series, such as a Taylor series, has only a limited range of representation. It can be made to correspond exactly to the function f in a given range. Beyond this range the function f "exists," but that particular series representation does not. In the example mentioned in the previous paragraph, the function $f = (1 - z)^{-1}$ "exists" and is analytic everywhere except at $z = 1$, but the power series $1 + z + z^2 + \cdots$ "exists" and represents f only inside the circle of unit radius centered at $z = 0$. Another series, $-\frac{1}{2} + \frac{1}{4}(z - 3) - \frac{1}{8}(z - 3)^3 + \cdots$, "exists" and represents f only inside the circle of radius 2 centered at $z = 3$, and so on.

It is as though power series (Taylor's or Laurent series or others) were analogous to pieces of a mold, by the use of which one would cast a copy of the function. Each piece of the mold reproduces the behavior of f perfectly as far as it goes but gives no inkling of the shape of f beyond its range; only when the whole mold is fitted together can one see the complete shape of f. This fitting together of pieces of the mold, to describe the whole of f, is called *analytic continuation* and will be discussed later in this chapter. As long as one has a closed form for f, such as $(z - 1)^{-1}$, for example, the process of fitting series representation is interesting but not vital, for we already know "the function itself." We have a recipe, so to speak, from which we can calculate f by a finite series of steps (in the present example, we subtract z from 1 and divide into 1); an infinite series, such as Taylor's series usually is, corresponds to an infinite number of steps in the recipe for calculation and is thus feasible only if it converges (if each successive step contributes less, so to speak, so we do not really have to take the infinity of steps to obtain a reasonable facsimile).

Unfortunately there are many functions for which we have no finite-length recipe for obtaining their value. Even the exponential and the trigonometric functions must be computed by the use of infinite series, for example. It is as though we had no complete casting of "the function itself," only the pieces of the mold to be fitted together. This will be the case for most of the functions studied in this treatise.

Integrals can also be used to represent functions (see particularly Secs. 4.8 and 5.3); often such "molds" are capable of representing the function over a much wider range of the variable than is a series. For example, some integral representations are valid within a band of constant width, extending across the complex plane, clear out to infinity, whereas series are convergent within a circle, a much more restricted area if the radius is finite. But integral representations also have their limitations of range, which it is necessary to keep in mind if we are to avoid mistakes. Indeed some of the questions of the range of convergence of integral representations are more subtle than those for series. We shall later encounter an integral representation of a constant, which is indeed a constant over half the complex z plane but which "blows up" for $\text{Im } z \geq 1$. If we know "the function itself" to be a constant everywhere, we can regard this integral representation simply as a peculiarly awkward way to represent a constant, the difficulty of the example is that we are given only the integral and must prove that the "function itself" is a constant.

The Laurent Series. If we wish to expand f about a singularity of f at a, we obviously cannot use Taylor's expansion. We may, however, obtain an expansion about a singular point which is valid as close to the singularity as we wish. The trick is to use the contour illustrated in Figure 4.7 such that its interior does not contain the point a. As is indicated the contour may be reduced to two circular contours C_1 and C_2 encircling $z = a$ positively and negatively, respectively. Applying Cauchy's theorem we have

$$\begin{aligned} f(a+h) &= \frac{1}{2\pi i} \oint_{C_1} \frac{f(z)}{z-a-h}\,dz + \frac{1}{2\pi i} \oint_{C_2} \frac{f(z)}{(z-a-h)}\,dz \\ &= \frac{1}{2\pi i} \left\{ \sum_0^\infty h^n \oint_{C_1} \frac{f(z)\,dz}{(z-a)^{n+1}} + \sum_1^\infty \frac{1}{h^n} \oint_{C_2} (z-a)^{n-1} f(z)\,dz \right\} \end{aligned}$$

$$f(a+h) = \sum_{n=-\infty}^{\infty} a_n h^n; \quad a_n = \frac{1}{2\pi i} \oint \frac{f(z)}{(z-a)^{n+1}}\,dz \tag{4.3.4}$$

where the contour for the coefficient a_n is C_1 for $n \geq 0$ and C_2 in the negative direction for $n < 0$ (although, since the integrand is analytic in the region between C_2 and C_1, we can just as well use C_1 as the contour for all values of n, reversing direction for negative n's).

This is the *Laurent series*. By using the same sort of reasoning as with the Taylor series, we can show that this series converges in an annulus whose larger radius (contour C_1) is determined by the singularity of $f(z)$ closest to a. The series containing the positive powers of h converges everywhere within the outer circle of convergence, while the series of negative powers converges anywhere outside the inner circle. Thus by using the Laurent series we divide $f(z)$ into two functions, one of which is analytic inside the outer circle of convergence; the other is analytic outside the inner circle of convergence. One may thus use the integrals over the bounding circles of an annulus (or any convenient contour into which they may be deformed in a given problem, *e.g.*, two lines parallel to the real axis) to resolve the series for function f into parts which are each analytic functions over different portions of the complex plane.

This series resolution has a simple physical interpretation which we shall discuss now. As was pointed out earlier, the function $1/h$ is proportional to the electrostatic field of a line charge located at $h = 0$.

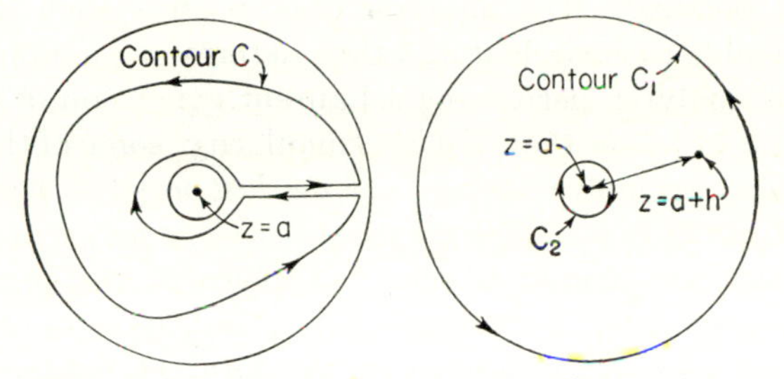

Fig. 4.7 Contour for Laurent series about a singularity at $z = a$.

What is the physical meaning of $1/h^2$, $(1/h^3)$, etc.? Since $1/h^2$ may be obtained from $1/h$ by differentiation, we might expect that $1/h^2$ represents the field of a *line doublet* (juxtaposed positive and negative lines). The orientation of the doublet is important and in this case is to be taken *along* the *x axis*. To show this, note that the field due to a doublet may be obtained by superposing the field due to a positive line charge q at $h = 0$, a negative line charge $-q$ at $h = \epsilon$ (ϵ real) and then taking the limit as ϵ goes to zero, $q\epsilon$ remaining finite and equal to the dipole moment p. Carrying this calculation out

$$-\lim\left[\frac{q}{h} - \frac{q}{h-\epsilon}\right] = \lim\left[\frac{q\epsilon}{h(h-\epsilon)}\right] = \frac{p}{h^2}; \quad p = q\epsilon$$

checking our expectation. In the same way $1/h^3$ represents the field due to a positive line charge q at $h = 0$, $-\frac{1}{2}q$ at both $\pm\epsilon$, so that the net charge is zero. This is just a *quadrupole* distribution, for it may be constructed of two doublets along the x axis of the type discussed above. In general h^{-n} will therefore represent a *multipole* of order 2^{n-1} whose total charge is zero.

That part of the Laurent series with negative powers of h, which is convergent outside the inner circle of convergence, therefore corresponds to the statement that the *field from the charge distribution within the inner circle of convergence may be expressed as a linear superposition of fields of a sequence of multipoles located at the point a.* A similar interpretation may be given of the series in positive powers of h. It states that the *field from the charge distribution outside the outer circle of convergence may be represented by fields from a linear superposition of multipoles located at infinity.* For an extension of this theorem to three dimensions, see Sec. 10.3.

It is no surprise that this expansion fails whenever a charge occurs within the region in question, for then it cannot be represented by multipoles located either at $h = 0$ or at infinity.

Isolated Singularities. The discussion in the last subsection leads to a classification of the types of isolated singularities a function of a complex variable may possess. If a function does possess such a singularity, it becomes possible to encircle it and thus determine an annulus in which the function is analytic, permitting a Laurent expansion of the function. From series (4.3.4) we see that, if a is a singularity, some of the coefficients a_n $(n < 0)$ must be nonvanishing. *If the last negative power coefficient which does not vanish is a_{-N}, then the point is said to have a pole of order N.* (Therefore poles are related to the multipoles of the preceding paragraph.) A pole of order N corresponds to a multipole of order 2^{N-1}.

If the series of negative powers in the Laurent expansion does not terminate, the point $z = a$ is called an essential singularity of f. A familiar example is $e^{1/z}$ near $z = 0$, having a Laurent expansion $\sum_{n=0}^{\infty}\left(\frac{1}{n!z^n}\right)$. If $f(z)$ has an essential singularity at $z = a$, so does $1/f(z)$. For if $1/f$ does *not* have an essential singularity at $z = a$, it will at most have a pole there (of order N, for instance) and can be expressed in terms of the series $\frac{1}{f} = \sum_{n=-N}^{\infty} b_n h^n$. Then f would be $h^N \Big/ \sum_{m=0}^{\infty} b_{m-N} h^m$, and since $[1/\Sigma b_{m-N} h^m]$ is an analytic function inside C_1, f can be expanded into a power series in h, starting with h^N, which is in contradiction to our assumption. Therefore wherever $f(z)$ has an essential singularity, $1/f$ also has one. This property of an essential singularity should be compared with the behavior of $1/f(z)$ at a pole. It is easy to see that $1/f(z)$ has a *zero* of order N if $f(z)$ has a pole of order N.

The function $e^{1/z}$ illustrates another important property of a function near an essential singularity. Consider its value as $z \to 0$ along different paths. For example, when z approaches zero from positive real values, the function goes to infinity, but if z approaches along the negative real axis, the limit is zero, and if the approach is along the imaginary axis, the magnitude of $e^{1/z}$ approaches unity. Near an essential singularity of f all values are taken on by the function f.

We thus see that the behavior of a function near an essential singularity is most complicated. Many of the functions which occur in the mathematical treatment of field problems have essential singularities. Their properties must therefore be carefully investigated in the neighborhood of such points.

The application of Laurent's expansion used here can be made only if the singularity is an isolated one. There are other singular points which are neither poles nor essential singularities. For example, neither $\sqrt{z}$ nor $\log z$ can be expanded in the neighborhood of $z = 0$ in a Laurent series. Both of these functions are discontinuous along an entire line (say the negative real axis) so that the singular point $z = 0$ is not isolated. We shall discuss singularities of this kind (called *branch points*) in Sec. 4.4 on multivalued functions.

Classification of Functions, Liouville's Theorem. Functions with isolated singularities may now be classified according to the position and nature of their singularities. The central theorem in this discussion is *Liouville's theorem: A function which is analytic for all finite values of z and is bounded everywhere is a constant.* The proof follows directly from an estimate of the derivative of $f(a)$:

$$f'(a) = \frac{1}{2\pi i} \oint \frac{f(z)}{(z-a)^2}\, dz$$

Let the contour be a circle of radius R centered at a, and let $|f(z)| \leq M$ by hypothesis. Then

$$|f'(a)| \leq \left(\frac{M}{2\pi R^2}\right)(2\pi R) = \frac{M}{R}$$

If now we let $R \to \infty$ (in other words, assume that f is bounded for all values of z), it follows that $f'(a) = 0$, so that $f(a)$ is a constant. As a corollary of this theorem we see that the slightest deviation of $f(z)$ from constancy anywhere implies the existence of a singularity somewhere else. Thus a function which is well behaved along the real axis will, if it is not exactly constant, have a singularity off the real axis. Again we note the strong correlation between the values of an analytic function everywhere on the complex plane.

By applying a similar technique it is possible to extend the above theorem to: *If $f(z)$ is analytic for all finite values of z, and if, as z approaches infinity, $|f|$ approaches infinity proportional to $|z|^k$, where k is an integer, then $f(z)$ is a polynomial of degree $\leq k$.* The nth derivative of f is $(n!/2\pi i)\oint [f(z)/(z-a)^{n+1}]\, dz$. Again we take the contour as being a circle of radius R. Then

$$|f^{(n)}(a)| < \frac{n!M}{R^n} = An!R^{k-n}$$

where A is a constant. As $R \to \infty$, all terms for which $k < n$ will vanish so that

$$f^{(n)}(a) = 0; \quad \text{for } n > k$$

which proves the theorem.

For a polynomial with $k > 0$ the point at infinity is a singularity.[1] Functions which are analytic in every finite region of the z plane are called *integral* or sometimes *entire* functions. A polynomial is an example. Many functions of importance in physical applications are in this class. For example e^z, $\cos z$, the Bessel function of order n, $J_n(z)$, etc. We shall later see that their analyticity, in finite regions, and the essential singularity at infinity are closely related to the geometric properties of the coordinate systems associated with these functions, *e.g.*, the Bessel function for circular cylindrical coordinates, etc.

Meromorphic Functions. We have now considered functions which are analytic everywhere (which are constants) and also functions which are analytic everywhere except at infinity (such as polynomials in z and e^z, etc.). Next in order of complication are functions all of whose singularities are poles. It is possible to show that such a function is a *rational function*, *i.e.*, the ratio of two polynomials.

To prove this, first note that the number of poles must be *finite;* for if they were infinite, there would be a point, either finite or at infinity, which would not be isolated and thus would not be a pole, contrary to hypothesis. Suppose that $f(z)$ has N poles, the nth pole being at a_n, and is of order i_n. Then the function

$$G(z) = \left[\prod_{n=1}^{N} (z - a_n)^{i_n} \right] f(z)$$

is analytic everywhere (except possibly at infinity). Whether there is a pole of $f(z)$ at $z \to \infty$ or not, it follows that $|G(z)|$ goes to infinity, as z goes to infinity, no faster than $|z|^k$, where k is finite; so from our previous theorem $G(z)$ is a polynomial. Therefore $f(z)$ is the ratio of two polynomials, G and $\prod_n (z - a_n)^{i_n}$, thus proving the theorem.

The electrostatic analogue yields a similar result, for the field caused by each pole may be expressed in terms of a finite series in inverse powers of $(z - a_n)$. Adding all these series together yields the desired result.

A generalization of the idea of a rational function may now be made. Suppose that all the singularities of a function in a given region of the complex plane are poles; then the function is said to be *meromorphic* in that region. A function which is meromorphic in the finite part of

[1] The point at infinity is investigated by first making the substitution $z = 1/\zeta$ and then considering the behavior of the result as $\zeta \to 0$. For example, z^k transforms to $(1/\zeta)^k$, which behaves as a pole of order k as $\zeta \to 0$.

the complex plane may be expanded in a partial fraction series just as the rational function may be expanded in such a series. The only difference is that now a singularity is permitted at infinity so that the partial fraction series is no longer finite. We shall now exhibit the expansion for the case where the poles are all first order.

Let $f(z)$ be the function in question, and let the poles be at points a_n, numbered in order of increasing distance from the origin at which point $f(z)$ is taken to be analytic. At pole a_n, $f(z) \to b_n/(z - a_n)$. Consider a circle R_p within which p poles may be found [$f(z)$ is not to have a pole on the circle]. Then the function

$$g_p(z) = f(z) - \sum_{n=1}^{p} \frac{b_n}{z - a_n} \tag{4.3.5}$$

will be analytic within this circle. Using a circular contour of radius R_p

$$g_p(z) = \frac{1}{2\pi i} \oint \frac{g_p(\zeta)}{\zeta - z} d\zeta = \frac{1}{2\pi i} \oint \frac{f(\zeta)}{\zeta - z} d\zeta - \frac{1}{2\pi i} \sum_{1}^{p} b_n \oint \frac{d\zeta}{(\zeta - z)(\zeta - a_n)}$$

It may be verified that the second term is zero so that

$$g_p(z) = \frac{1}{2\pi i} \oint \frac{f(\zeta)}{\zeta - z} d\zeta$$

Consider now a sequence of circles R_p as $R_p \to \infty$. This in turn defines a sequence of functions g_p each of which is analytic over a larger and larger region of the complex plane. It now remains to show that in the limit g_p is bounded, permitting application of Liouville's theorem. We can see that

$$|g_p(z)| \leq M_p R_p/(R_p - R)$$

where M_p is the maximum value of $|f|$ on the contour of radius R_p. It may happen that M_p does not increase indefinitely as R_p is increased [in other words it may be that $\lim M_p$ is bounded]. Therefore, $\lim_{p \to \infty} |g_p(z)|$ is bounded, so that by Liouville's theorem $g = \lim_{p \to \infty} g_p$ is a constant. Hence in this case,

$$f(z) = \text{constant} + \sum_{n=1}^{\infty} \frac{b_n}{z - a_n}$$

To determine the constant, let $z = 0$ on both sides of the above equation, inasmuch as $f(z)$ is analytic at $z = 0$. Then

$$f(z) = f(0) + \sum_{1}^{\infty} \left[\frac{b_n}{z - a_n} + \frac{b_n}{a_n} \right] \tag{4.3.6}$$

Even if $f(z)$ is not bounded, in the limit, on the circle of radius R_p, as p becomes infinite, we can still obtain an analogous expression. It is often possible, for instance, to find a power of z, say z^n, such that $f(z)/z^n$ *is* bounded on the circle of radius R_p as $p \to \infty$, so that the proper expression for $f(z)$ is z^n times the quantity given in Eq. (4.3.6).

Expansion (4.3.6) has, as pointed out above, a simple electrostatic analogy. It is also a common form for the admittance function for a dynamical system (see page 301). As such, it is a measure of the response of a system having many modes of motion (*e.g.*, a string) to a driving force. The denominators in (4.3.6), $z - a_n$ correspond to the resonance denominators which occur, the constants b_n to the effect of the space distribution of the driving force on the coupling to the nth mode. We shall, in Chap. 11, see an application of this fact to the discussion of the motion of fields.

An example of the application of (4.3.6) to the expansion of a function in a partial fraction series will now be given. Consider $f(z) = \tan z$. The poles of $\tan z$ occur at $a_n = \frac{1}{2}\pi(2n + 1)$. The value of b_n is determined from the formulas

$$\tan\,[(z - a_n) + \tfrac{1}{2}\pi(2n + 1)] = \tan\left[z - a_n + \frac{\pi}{2}\right]$$

$$= -\cot\,(z - a_n) \xrightarrow[z \to a_n]{} -\frac{1}{z - a_n}$$

Therefore $b_n = -1$. Next we set up a sequence of circles R_p. These are circles whose radius is $p\pi$ (p integer). On these circles, $\tan z$ is bounded for all values of p, satisfying the requirements of the theorem. Therefore

$$\tan z = -\sum_{-\infty}^{\infty}\left[\frac{1}{z - \frac{1}{2}(2n + 1)\pi} + \frac{1}{\frac{1}{2}(2n + 1)\pi}\right]$$

$$= -\sum_{0}^{\infty}\left[\frac{1}{z - \frac{1}{2}(2n + 1)\pi} + \frac{1}{\frac{1}{2}(2n + 1)\pi}\right]$$

$$-\sum_{0}^{\infty}\left[\frac{1}{z + \frac{1}{2}(2n + 1)\pi} - \frac{1}{\frac{1}{2}(2n + 1)\pi}\right]$$

$$\tan z = \sum_{n=0}^{\infty}\frac{2z}{[\frac{1}{2}(2n + 1)\pi]^2 - z^2} \tag{4.3.7}$$

Equation (4.3.6) may be also used to obtain an expansion of an *integral function* $f(z)$ into an infinite product, since the *logarithmic derivative of an integral function, $f'(z)/f(z)$, is a meromorphic function.* Its only singularities in the finite complex planes are poles at the zeros

a_n of $f(z)$. Suppose, again for simplicity, that these poles are simple poles, that is, $f(z) \xrightarrow[z \to a_n]{} \text{constant } (z - a_n)$

$$[f'(z)/f(z)] = [f'(0)/f(0)] + \sum_{n=1}^{\infty} \left[\frac{1}{z - a_n} + \frac{1}{a_n}\right] = \frac{d}{dz} \ln [f(z)]$$

or $$\ln [f(z)] = \ln [f(0)] + [f'(0)/f(0)]z + \sum_{n=1}^{\infty} \left[\ln\left(1 - \frac{z}{a_n}\right) + \frac{z}{a_n}\right]$$

or $$f(z) = f(0)e^{[f'(0)/f(0)]z} \prod_{n=1}^{\infty} \left(1 - \frac{z}{a_n}\right) e^{z/a_n} \tag{4.3.8}$$

For this formula to be valid it is required that $f(z)$ be an integral function, that its logarithmic derivative have simple poles, none of which are located at 0, and that it be bounded on a set of circles R_p etc.

Let us find the product representation of $\sin z$. We shall use it often in our future work. Now $\sin z$ satisfies all our requirements except for the fact that it has a zero at $a = 0$. We therefore consider $(\sin z)/z$. The logarithmic derivative is $\cot(z) - 1/z$, a function which satisfies all our requirements. The points a_n are $n\pi$, $n \neq 0$, so that

$$\frac{\sin z}{z} = \prod_{\substack{-\infty \\ (n \neq 0)}}^{\infty} \left[1 - \frac{z}{n\pi}\right] e^{z/n\pi} = \prod_{n=1}^{\infty} \left[1 - \left(\frac{z}{n\pi}\right)^2\right] \tag{4.3.9}$$

Similar expansions may be given for other trigonometric functions and for the Bessel functions $J_n(z)$.

Behavior of Power Series on the Circle of Convergence. In many problems it is impractical (or impossible) to obtain solutions in closed form, so that we are left with only power series representations of the desired solutions. Such series will generally have a finite radius of convergence, which may be determined if the general form of the series is known or may be found from other data available on the problem. The power series is not completely equivalent to the solution, of course; inside the radius of convergence it coincides exactly with the solution, but outside this radius there is no correspondence and another series must be found, which converges in the next region, and so on. As mentioned earlier, it is as though the solution had to be represented by a mold of many pieces, each series solution being one piece of the mold, giving the shape of a part of the solution over the area covered by the series but giving no hint as to the shape of the solution elsewhere. In order to ensure that the various pieces of the mold "join" properly, we must now inquire as to the relation between the solution and the power series on its circle of convergence.

A priori, it is clear that the investigation of the behavior of a power series on its circle of convergence will be an extremely delicate matter, involving rather subtle properties of analytic functions. Consequently, the proofs are involved and highly "epsilonic." Fortunately, the theorems are easy to comprehend, and the results, as stated in the theorems, are the relevant items for us in this book, not the details of mathematical machinery needed to prove the theorems. We shall, therefore, concentrate in this section upon a discussion of the aforementioned theorems, omitting most of the proofs. For these, the reader is referred to some of the texts mentioned at the end of the chapter.

Suppose, then, that the series representing the solution is written in the form

$$f = \sum_n a_n \zeta^n \tag{4.3.10}$$

The first question is to determine its *radius of convergence R*. If the general term a_n of the series is known, the *Cauchy test* yields the radius as

$$R = \lim_{n \to \infty} \left[\frac{a_n}{a_{n+1}} \right] \tag{4.3.11}$$

Finite radii of convergence occur only if the limit of the ratio $[a_n/a_{n+1}]$ is finite. In this case we can reduce the series to a standard form, with unit radius of convergence, by changing scale. Letting $\zeta = zR$, we have

$$f(z) = \Sigma b_n z^n; \quad b_n = a_n R^n \tag{4.3.12}$$

It is convenient to normalize the series to be discussed to this radius.

A great deal of care must be exercised in arguing from the behavior of b_n for n large to the behavior of $f(z)$ at a given point z on the circle of convergence. If series (4.3.12) diverges or converges at $z = e^{i\varphi}$, it is not correspondingly true that the function is, respectively, singular or analytic at $z = e^{i\varphi}$. For example, the series $\Sigma(-z)^n$, which represents $1/(1 + z)$ for $|z| < 1$, diverges at $z = 1$, but $1/(1 + z)$ is analytic there. On the other hand the function $-\int_0^z \ln(1 - w)\, dw = 1 + (1 - z)[\ln(1 - z) - 1]$ is singular (but finite) at $z = +1$, but the corresponding series, $\Sigma(z)^{n+1}/n(n + 1)$, converges at $z = +1$. Other series can be given where $b_n \to 0$ but which diverge at every point on the unit circle, and still others which converge at $z = 1$ but at no other point on the unit circle. With this warning to the reader to refrain from jumping to conclusions, we shall proceed to examine what it is possible to say about solutions and the corresponding series on their circles of convergence.

The first matter under discussion will be the available tests determining whether or not a point on the convergence circle is or is not a singular point. We first transform the point under examination to the position $z = 1$. Then there are two theorems which are of value. The first states that, if $f(z) = \Sigma b_n z^n$ and $g(z) = \Sigma \operatorname{Re} b_n z^n$ have radii of convergence equal to 1 and if $\operatorname{Re} b_n \geq 0$, then $z = 1$ is a singular point of $f(z)$. [$\operatorname{Re} f$ = real part of f.] In other words, if the phase angle of z on the circle of convergence is adjusted so that the real part of each term in the series is positive, this phase angle locates the position of the singularity on the circle of convergence.

For example, when this theorem is applied to $f = \Sigma z^{n+1}/n(n + 1)$, the series mentioned earlier, we find that $z = 1$ is a singular point of f in spite of the convergence of the series at $z = 1$. All the terms are positive, and the series for the derivative diverges, so we might expect complications at $z = 1$.

We note that, even if $f(z) = \Sigma b_n z^n$ does have a singularity at $z = 1$, according to this test, the series does not necessarily have a singularity for $z = e^{i\varphi}$, where φ is small but not zero; for we can write $z = Ze^{i\varphi}$ and $b_n = B_n e^{-in\varphi}$, so that $f(z) = \Sigma B_n Z^n$. In this case the real part of B_n has a term $\cos(n\varphi)$ in it, coming from the factor $e^{in\varphi}$.

A more powerful test is obtained by employing the following theorem. The theorem is: *If* $C_n = \sum_{m=0}^{n} \frac{n!}{m!(n-m)!} b_m$, *the necessary and sufficient condition that* $z = 1$ *be a singularity for* $\Sigma b_n z^n$ *is that the quantity* $(C_n)^{-1/n}$ *never gets smaller than* $\frac{1}{2}$ *as* n *goes to infinity.* For example, for the series $\Sigma(-z)^n$, $b_n = (-1)^n$, the point $z = 1$ is not a singular point. For the series Σz^n on the other hand, $C_n = 2^n$, $\lim (2^n)^{-1/n} = \frac{1}{2}$ so that $z = 1$ is a singular point. A more dramatic case is provided by the series $\Sigma(n + 1)(n + 2)(-z)^n$ [which represents the function $2/(1 + z)^3$] with $b_n = (-1)^n(n + 1)(n + 2)$; then for $n > 2$, $C_n = 0$, so that in spite of the strong divergence of the series, the function f is analytic at $z = 1$.

Having determined the radius of convergence and the positions of the singularities on the circle of convergence, it is obviously useful to be able to estimate the behavior of $f(z)$ on the whole circle of convergence, particularly the singular part. A tool for doing so is provided by the following theorem: *If* $f(z) = \Sigma b_n z^n$ *and* $g(z) = \Sigma C_n z^n$, *where* $\operatorname{Re} b_n$ and $\operatorname{Re} C_n \geq 0$ *and where* $\operatorname{Im} b_n$ *and* $\operatorname{Im} C_n \geq 0$, *and if as* $n \to \infty$, $b_n \to DC_n$ *(D constant), then as* $|z| \to 1$, $f(z) \to Dg(z)$. ($\operatorname{Im} f$ *is the imaginary part of* f.)

It should be noted that the condition $\operatorname{Re} b_n$, $\operatorname{Re} C_n \geq 0$, by a slight extension, may be relaxed to require that $\operatorname{Re} b_n$, $\operatorname{Re} C_n$ eventually maintain a constant sign; similarly for $\operatorname{Im} b_n$, $\operatorname{Im} C_n$.

This theorem simply states that, if two expansions have a similar behavior for large n, then the two functions have the same singularities. By using the theorem, some statement about the asymptotic behavior of the coefficient b_n may be made. For example, we may assert that, if $b_n \xrightarrow[n\to\infty]{} D\, n^{p-1}/(p-1)!$, then $f(z) \to D/(1-z)^p$ as $|z| \to 1$. To demonstrate this[1] we need only show that the coefficient C_n in the expansion $\Sigma C_n z^n = (1-z)^{-p}$ is asymptotically $n^{p-1}/(p-1)!$. We note that

$$(1-z)^{-p} = \sum \frac{(n+1)\cdots(n+p-1)}{(p-1)!} z^n$$

from which the required asymptotic behavior of C_n follows.

The hypergeometric function provides another example. This function, which we shall discuss at great length in Chap. 5, is given by

$$F(a,b|c|z) = 1 + \frac{ab}{c} z + \frac{(a)(a+1)(b)(b+1)}{(c)(c+1)} \frac{z^2}{2!} + \cdots$$

It is known to have a singularity at $z = 1$. The general coefficient b_n is

$$b_n = \frac{(a)(a+1)\cdots(a+n-1)(b)(b+1)\cdots(b+n-1)}{n!(c)(c+1)\cdots(c+n-1)}$$

Assuming a, b, c integers and $a + b > c \geq b$, the asymptotic behavior of b_n may be determined as follows:

$$\begin{aligned} b_n &= \frac{(c-1)!}{(b-1)!(a-1)!} \left[\frac{(a+n-1)!(b+n-1)!}{n!(c+n-1)!}\right] \\ &= \frac{(c-1)!}{(b-1)!(a-1)!} \left[\frac{(n+1)\cdots(a+n-1)}{(b+n)\cdots(c+n-1)}\right] \\ &\qquad\qquad \to \frac{(c-1)!}{(b-1)!(a-1)!} n^{a+b-c-1} \end{aligned}$$

Thus one may conclude that, when $c < a + b$,

$$F(a,b|c|z) \xrightarrow[z\to 1]{} \frac{(c-1)!(a+b-c-1)!}{(b-1)!(a-1)!} \frac{1}{(1-z)^{a+b-c}}$$

We noted earlier a case $f = \Sigma z^{n+1}/n(n+1)$, in which a series converged at a point $z = 1$ at which f was singular but finite. We might ask whether the convergent sum $s = \Sigma 1/n(n+1)$ is equal to f at $z = 1$. The answer to this query is given by a theorem due to Littlewood, which states: *If $f(z) \to s$ as $z \to 1$ along a smooth curve and* $\lim (na_n)$ *is bounded, then Σa_n converges to s.* In the example quoted above, $\Sigma 1/n(n+1) = \lim_{z\to 1} \{1 + (1-z)[\ln(1-z) - 1]\}$ which is, of course, unity, as can be verified directly from the series itself.

From the above theorems, it is clear that knowledge of the power

[1] The proof holds only for p integer. A suitable generalization for noninteger p in which the factorials are replaced by gamma functions (to be discussed later in this chapter) is possible.

series representation of a function yields fairly complete information about the function on the circle of convergence. Can the power series be used to determine properties of the function it represents outside the circle of convergence? Under certain fairly broad restrictions, the answer is yes. We shall discuss this point in the next subsection.

Analytic Continuation. It is often the case that knowledge of a function may be given in a form which is valid for only limited regions in the complex plane. A series with a finite radius of convergence yields no direct information about the function it represents outside this radius of convergence, as we have mentioned several times before. Another case, which often occurs, is in the representation of the function by an integral which does not converge for all values of the complex variable. The integral

$$\int_0^\infty e^{-zt}\,dt$$

represents $1/z$ only when $\mathrm{Re}\,z > 0$. It may be possible, however, by comparing the series or integral with some other form, to find values of the function outside of the region in which they are valid. Using, for example, the power series $f(z) = 1 + z + z^2 + \cdots$ which converges for $|z| < 1$, it is possible to find the values of $f(z)$ for $|z| < 1$ and to identify f with $1/(1 - z)$ which is then valid for $|z| > 1$.

The process by which the function is extended in this fashion is called *analytic continuation*. The resultant function may then (in most cases) be defined by sequential continuation over the entire complex plane, without reference to the original region of definition. In some cases, however, it is impossible to extend the function outside a finite region. In that event, the boundary of this region is called the *natural boundary* of the function, and the region, its *region of existence*.

Suppose, as an example, that a function f is given as a power series about $z = 0$, with a radius of convergence R, a singular point of f being on the circle of convergence. It is then possible to extend the function outside R. Note that at any point within the circle ($|z| < R$) it is possible to evaluate not only the value of the series but also all its derivatives at that point, since the derivatives will have the same region of analyticity and their series representation has the same radius of convergence. For example, all these derivatives may be evaluated at $z = z_0$. Using them, one can set up a Taylor series:

$$f = \sum_n \frac{f^{(n)}(z_0)}{n!}(z - z_0)^n \tag{4.3.13}$$

The radius of convergence of this series is the distance to the nearest singular point, say $z = z_s$. The resultant circle of convergence, with radius R_0, is indicated by the dashed circle in Fig. 4.8. One may continue this process using a new point, for example, $z = z_1$, not necessarily within the original circle of convergence, about which a new series like that of (4.3.13) can be set up. Continuing on in this way, it appears possible by means of such a series of overlapping circles to obtain values for f for every point in the complex plane excluding, of course, the singular points.

In the above discussion, we have assumed that there is only one singular point z_s on the circle of convergence. If, however, the singular points are everywhere dense on the circle or on some neighboring curve, it would be impossible to extend the function beyond this boundary. Such is the case for functions having a natural boundary.

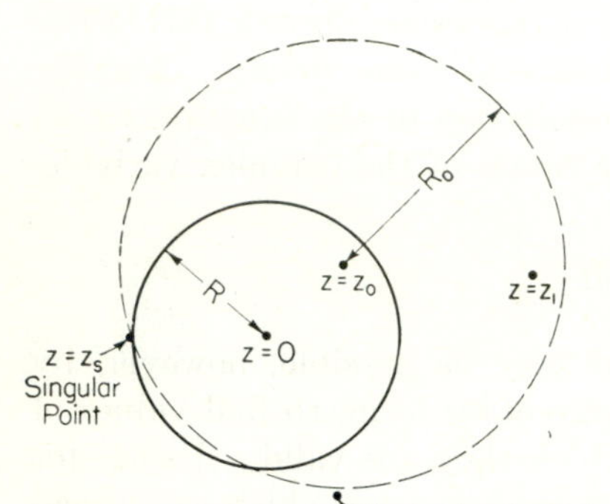

Fig. 4.8 Process of analytic continuation by sequence of series, first in z^n, then in $(z - z_0)^n$, and so on.

At this point the reader will justifiably raise the question of uniqueness. He will ask whether a function which is continued, as described above, along two different routes from one area to another, will have the same value and derivatives in the final area for one route as it does for the other. He will also ask whether, if the continuation is performed by two different techniques, the same answer will be obtained. We shall now attempt to answer such questions.

Fundamental Theorems. The theorem which will be found to be most useful in establishing these points states that, *if a function is analytic in a region and vanishes along any arc of a continuous curve in this region, then it must vanish identically in this region.* This follows from the fact that, under the hypothesis of the theorem, it is possible to evaluate all the derivatives of f at any point on the arc; all these derivatives will be zero. The resultant Taylor series will be identically zero, so that, within its circle of convergence, the function is zero. One can now take any arc within this circle, and keeping within the region of analyticity, we can repeat the process of finding a Taylor series, etc. In this manner the result may be extended to any point within the region of analyticity.

This remarkable theorem once more demonstrates the strong correlation between the behaviors of analytic functions in different parts of the complex plane. For example, if two functions agree in value over a small arc (as small as one might desire as long as it is not a point!), then they are identical in their common region of analyticity.

We may now prove the fundamental uniqueness theorem for analytic continuation. It states that, *if D_1 and D_2 are regions into which $f(z)$ has been continued from D, yielding the corresponding functions f_1 and f_2,* *and if D_3, a region common to D_1 and D_2, also overlaps D, then $f_1 = f_2$ throughout D_3.* This answers in the affirmative the first question raised above, subject, however (as may be seen from the theorem), to some restrictions. The proof is an immediate consequence of the preceding theorem, for the function $f_1 - f_2$ is analytic within D_3, it equals zero within that part of D_3 which is inside D and therefore is zero in the remainder of D_3. Thus identical values will be obtained at a point inside D_3 by use of two different routes for the analytic continuation.

However, this proof depends critically upon the fact that D_3 and D have a common region. If this is not the case, the uniqueness theorem may not hold. Instead one can say: *If a function f is continued along two different routes from z_0 to z_1 and two differing values are obtained for the function at z_1, then $f(z)$ must have a singularity between the two routes.*

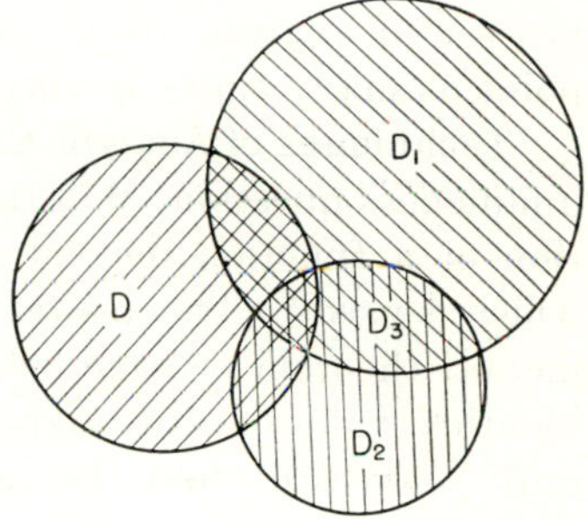

Fig. 4.9 Continuity of analytic continuation. If overlap D_3 between regions D_1 and D_2 also overlaps D there can be no singularity between the three regions and $f_1 = f_2 = f_3$.

This theorem is a fairly obvious consequence of the fact that the radius of convergence of a power series extends up to the next singularity of the function, for if there were no singularities between the two routes, then it would be possible, by means of the power-series method of continuation, completely to fill in the region between the two routes and obtain sufficient overlapping so that the uniqueness theorem is satisfied. In that event $f(z_1)$ for the two different routes would be identical, in contradiction to our hypothesis. There must therefore be a singularity between the two routes.

Branch Points. Note that this last theorem does *not* state that different values *must* be obtained if there is a singularity between the two routes. It must be a particular type of singularity to cause a discrepancy! We shall digress here for a few paragraphs to discuss this case. Suppose that the two routes of continuation do not yield the same values of $f(z)$ at z_1, what then? It clearly becomes useful to generalize our definition of an analytic function to include both of these values. We thus say that an analytic function will include not only the function as given in the original region of definition but also all the analytic continuations of the function from the original region, whether or not the results are unique. If they are unique, then the function is called *single-valued.* If they are not unique, it is called *many-valued.* From the previous theorem, these latter functions will have singularities such that continuation along two different routes surrounding the singular point will lead to nonunique results.

This type of singularity is called a *branch point,* and the various possible sets of values generated by the process of continuation are known as *branches.* We shall postpone to the next section (which will be completely devoted to many-valued functions) the more specific definition of branch points and the examples which would clarify the definition. Suffice it to say that the many-valued function arises naturally in the discussion of analytic continuation, that all the various possible values of a function at a given point may be obtained by the process of analytic continuation in which one winds about the branch point as many times as may be necessary.

Techniques of Analytic Continuation. We can now turn to the second uniqueness question: Will the results obtained for the analytic continuation of a function depend in any way upon the particular technique which is used? The answer is *no,* and it follows directly from the fundamental theorem of page 390. Any two methods must, of course, yield identical values in the original region of definition of the function. They must therefore yield the same functional values in any region which overlaps the original region, according to the theorem. By repeating this logic for each stage of the continuation process, the theorem is proved.

These three theorems on uniqueness are the fundamental theorems in the theory of analytic continuation. However, from the point of view of application the most relevant theorem would be one which would tell us if a function f_2 is the analytic continuation of a function f_1. In general, we shall not use the power-series method but rather any method which is natural for the problem at hand. Each of the techniques can be developed with the aid of the following theorem: *If two functions f_1 and f_2 are analytic, respectively, in the domains D_1 and D_2, and if these functions are identical in the region where the domains overlap, then f_2 is the continuation of f_1 into D_2 and vice versa.*

The theorem depends upon the fact that one could continue, from the common region into the remainder of D_1 and D_2 from the same power series in virtue of the analyticity of f_1 and f_2. Actually the overlapping involved in the theorem may be reduced to a line forming a common boundary of D_1 and D_2. We shall now show that, if f_1 *is analytic in* D_1 *and* f_2 *in* D_2, *if* f_1 *equals* f_2 *along their common boundary* C, *and if* f_1 *and* f_2 *are continuous along* C, *then* f_2 *is the continuation of* f_1 *in* D_2 *and vice versa.*

To prove this we shall show that the function $f = f_1$ in D_1 and f_2 in D_2 is analytic throughout $D_1 + D_2$ in Fig. 4.10. The fact that f_2 is a continuation of f_1 is then a consequence of this analyticity. Consider a point z_0 in D_2. Then from the Cauchy integral formula

$$f(z_0) = \frac{1}{2\pi i} \oint_{C_2+L} \frac{f(z)}{(z - z_0)}\, dz \quad \text{and} \quad 0 = \frac{1}{2\pi i} \oint_{C_1+L} \frac{f(z)}{(z - z_0)}\, dz$$

Adding these two equations, one obtains

$$f(z_0) = \frac{1}{2\pi i} \oint_{C_1+C_2} \frac{f(z)}{(z - z_0)} dz$$

since the integrals along L cancel by virtue of the equality of f_1 and f_2 along L. The contour $C_1 + C_2$ is a closed contour. One may now use the procedure employed in deriving Eq. (4.3.1) and evaluate the derivatives of f at z_0. Since the derivatives exist, the function is analytic at z_0, proving the theorem.

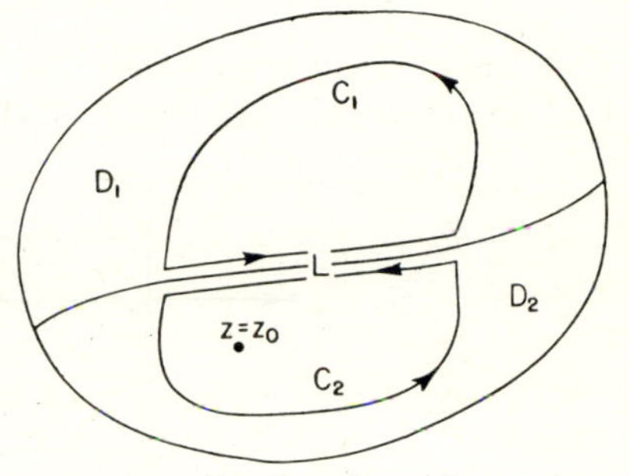

Fig. 4.10 Continuation of a contour integral about $z = z_{01}$.

We shall now go on to describe some possible methods of analytic continuation. We shall be principally concerned with two methods, the first making use of functional relationships, the second direct transformations of the power series. In the first category, we shall describe two methods, the first of which follows from the *Schwarz principle of reflection* which essentially makes use of the functional relation $f(\bar{z}) = \bar{f}(z)$. *If* $f(z)$ *is analytic within a region* D *intersected by the real axis and is real on the real axis, then one obtains for conjugate values of* z *conjugate values of* f. That is to say, $f(\bar{z}) = \overline{f(z)}$. To prove this, expand $f(z)$ in a Taylor series about a point a on the real axis. The coefficients of the Taylor series are real by virtue of the hypothesis that $f(z)$ is real on the real axis. We have then that, $f(z) = \Sigma a_n(z - a)^n$, a_n real. Then $\overline{f(z)} = \Sigma a_n(\bar{z} - a)^n = f(\bar{z})$, proving the theorem. The theorem thus holds for any point within the circle of convergence of the power series. By the methods of analytic continuation, it may be extended to include any nonsingular point conjugate to a point in D. One may thus see that the procedure of taking the conjugate of f immediately continues the function from the region above the real axis to the region below. The method may be generalized to a region where the real axis is replaced by any straight line. However, it is not necessary to state this theorem, for in general it is possible to rotate a given straight line so that it coincides with part of the real axis and then apply the theorem above.

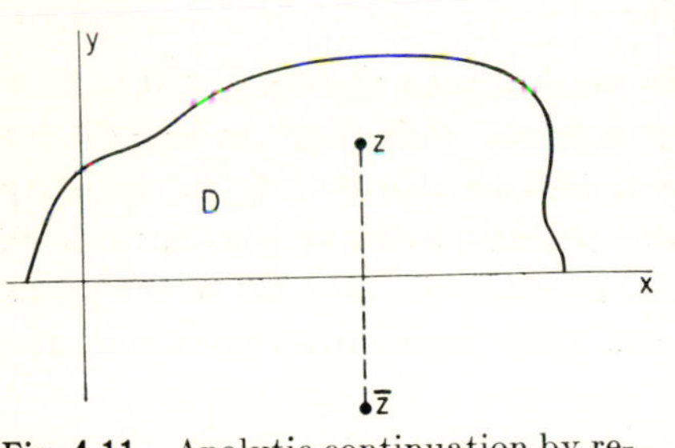

Fig. 4.11 Analytic continuation by reflection in the real axis.

A second method employs explicit functional relations such as addition formulas or recurrence relations. A simple example is provided by the addition formula $f(z + z_1) = f(z)f(z_1)$, satisfied, of course, by the exponential function. If f were known only in a given region, it would be continued outside that region to any point given by the addition of the coordinates of any two points inside the region. A less trivial example occurs in the theory of gamma functions. The gamma function is often defined by the integral

$$\Gamma(z) = \int_0^\infty e^{-t}t^{z-1}\,dt \tag{4.3.14}$$

This integral converges only for Re $z > 0$, so that the integral defines $\Gamma(z)$ for only the right half of the complex plane. From Eq. (4.3.14)

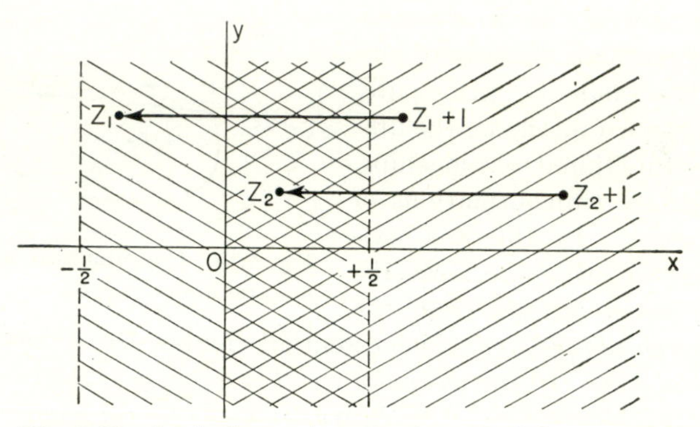

Fig. 4.12 Analytic continuation by use of the recurrence formula, for the gamma function.

one may readily derive (by integrating by parts) a functional relationship between $\Gamma(z)$ and $\Gamma(z + 1)$:

$$z\Gamma(z) = \Gamma(z + 1) \tag{4.3.15}$$

We may now use Eq. (4.3.15) to continue $\Gamma(z)$ into the Re $z < 0$ part of the complex plane. In Fig. 4.12 we assume that $\Gamma(z)$ is known for $x > 0$. Then using recurrence relation (4.3.15) the points in the strip $-\frac{1}{2} < x < \frac{1}{2}$ can be computed in terms of the values of $\Gamma(z)$ for $x > 0$. The function so defined and the original function have an overlapping region of convergence so that it is the analytic continuation into the negative x region.

We now consider methods of transforming the power series directly. The power-series method of continuation, although useful in the theoretical discussion of the properties of the process, is very cumbersome from a practical point of view and is thus to be shunned under all but the most extreme circumstances. If the method employing functional relationships does not apply, however, one may have to fall back upon a method due to Euler. Here the power series in z (divergent at $z = -1$) is transformed into a power series in the variable.

$$\zeta = z/(1 + z) \tag{4.3.16}$$

Thus a series which was slowly converging or diverging at $z = 1$ is converted into a power series which may converge at $\zeta = \frac{1}{2}$ fairly rapidly. Inasmuch as the two series will have common values in the region in which they are both convergent, the series in ζ is the analytic continuation of the series in z.

Let us now consider the transformation leading to a power series in ζ. Euler's transformation may be most conveniently applied to alternating series:

$$f(z) = \Sigma(-1)^n a_n z^n \tag{4.3.17}$$

Working within the radius of convergence of this series, multiply both sides by $(1 + z)$, and regrouping terms, one obtains

$$(1 + z)f(z) = a_0 + z\Sigma(a_n - a_{n+1})(-1)^n z^n$$

It is useful here to introduce a set of symbols which we shall use again when we come to discuss the calculus of finite differences. Let

$$\delta a_n = a_n - a_{n+1} \tag{4.3.18}$$

One may repeat the process:

$$\delta(\delta a_n) = \delta^2 a_n = a_n - 2a_{n+1} + a_{n+2}$$

In general
$$\delta^{(q)} a_n = \sum_{p=0}^{q} (-1)^p \binom{q}{p} a_{n+p} \tag{4.3.19}$$

where $\binom{q}{p}$ are the binomial coefficients $\binom{n}{m} = \frac{n!}{m!(n-m)!}$, so that $(1 + x)^n = \sum_m \binom{n}{m} x^m$. In terms of these symbols we have

$$f(z) = \frac{a_0}{1 + z} + \zeta \sum_{n=0}^{\infty} (\delta a_n)(-1)^n z^n \tag{4.3.20}$$

We may now apply the procedure used to obtain Eq. (4.3.20) to the coefficient of ζ, obtaining

$$f(z) = \frac{a_0}{1 + z} + \frac{\zeta \delta a_0}{1 + z} + \zeta^2 \sum_{n=0}^{\infty} (\delta^2 a_n)(-1)^n z^n$$

Continuing on in this way we obtain

$$f(z) = \frac{1}{1 + z} [a_0 + \zeta\, \delta a_0 + \zeta^2\, \delta^2 a_0 + \cdots] \tag{4.3.21}$$

To demonstrate the power of the transformation, let us apply it to some simple cases. As a first example let us take the series

$$\ln(1 + z) = z(1 - \tfrac{1}{2}z + \tfrac{1}{3}z^2 - \tfrac{1}{4}z^3 + \cdots) \tag{4.3.22}$$

which has a radius of convergence of 1. The series converges very slowly at z close to 1. We now compute the coefficients $\delta^{(n)}a_0$.

n	a_n	δa_n	$\delta^2 a_n$	$\delta^3 a_n$	—
0	1	$\frac{1}{2}$	$\frac{1}{3}$	$\frac{1}{4}$	—
1	$\frac{1}{2}$	$\frac{1}{6}$	$\frac{1}{12}$	$\frac{1}{20}$	—
2	$\frac{1}{3}$	$\frac{1}{12}$	$\frac{1}{30}$	—	—
3	$\frac{1}{4}$	$\frac{1}{20}$	—	—	—
—	—	—	—	—	—

The transformed series is

$$\ln(1 + z) = z/(1 + z)[1 + \tfrac{1}{2}\zeta + \tfrac{1}{3}\zeta^2 + \cdots] \tag{4.3.23}$$

which may be easily checked directly. The radius of convergence of the series in ζ is $|\zeta| \leq 1$, or $(x^2 + y^2)/[(1 + x)^2 + y^2] \leq 1$. This inequality will hold for $x > -\frac{1}{2}$ for any y, so that by means of the transformation we have continued the function into the half plane $x > -\frac{1}{2}$.

Another example is furnished by the series discussed on page 387, where $a_n = (n)(n + 1)$. From the table below

n	a_n	δa_n	$\delta^2 a_n$	$\delta^3 a_n$	—
0	0	−2	2	0	—
1	2	−4	2	0	—
2	6	−6	2	0	—
3	12	−8	2	0	—
—	—	—	—	—	—

the transformed series is finite, since all δ's higher than δ^2 are zero, so that

$$f(z) = \frac{1}{1 + z}[-2\zeta + 2\zeta^2] = -\frac{2z}{(1 + z)^3}$$

Note that in this case the original series diverged strongly at $z = 1$. Euler's transformation in this case yields a closed form for $f(z)$ and therefore succeeds in extending the function over the entire complex plane.

Various generalizations of Euler's transformation are possible. Perhaps the most useful one relates

$$g(z) = \sum_{n=0}^{\infty} b_n z^n \quad \text{and} \quad f(z) = \sum_{n=0}^{\infty} C_n b_n z^n \tag{4.3.24}$$

Assuming g as being a known function we now attempt to relate f and g. This may be accomplished by expressing the b_n in terms of derivatives of g as follows:

$$f = C_0 g(z) + \sum_1^{\infty} b_n(C_n - C_0)z^n$$

We have eliminated b_0. We may replace b_1 by noting that $g' = \sum_1^\infty n b_n z^{n-1}$.

Then

$$f = C_0 g + (C_1 - C_0)zg' + \sum_2^\infty b_n[(C_n - C_0) - (n)(C_1 - C_0)]z^n$$

Continuing this process one finally obtains

$$f = C_0 g - (\delta C_0)zg' + \frac{(\delta^2 C_0)z^2}{2!} g'' - \cdots \tag{4.3.25}$$

This reduces to Euler's transformation (4.3.21) if $g = 1/(1 + z)$. An important generalization of Eq. (4.3.21) is obtained if g is taken as $1/(1 + z)^p$. Then Eq. (4.3.25) becomes

$$f = \frac{1}{(1 + z)^p}\left[C_0 + p(\delta C_0)\zeta + \frac{(p)(p + 1)}{2!}(\delta^2 C_0)\zeta^2 + \cdots\right] \tag{4.3.26}$$

where $f = C_0 - C_1 pz + \dfrac{C_2(p)(p + 1)z^2}{2!} + \cdots ; \quad \zeta = \dfrac{z}{1 + z}$

Many important examples involving Eqs. (4.3.25) and (4.3.26) exist.

Note that Eq. (4.3.25) will yield a closed form if the series breaks off, which will occur if C_n is a polynomial in n. Second, considerable simplifications may be achieved if g satisfies some differential equation which relates all the higher derivatives to a finite number of derivatives. To illustrate let $g = \cos z = \sum \dfrac{(-1)^n z^{2n}}{(2n)!}$ (that is, *let* $b_{2n} = (-1)^n/(2n)!$, $b_{2n+1} = 0$) and let $C_n = n^2$. Then $f = \sum_n (-1)^n (2n)^2 \left[\dfrac{z^{2n}}{(2n)!}\right]$. From Eq. (4.3.25)

$$f = -z \sin z - z^2 \cos z$$

which may be easily verified directly.

A more complicated example, using Eq. (4.3.26), is a transformation which may be applied to the hypergeometric function of $-z$:

$$f = 1 - \frac{(a)(b)}{c} z + \frac{(a)(a + 1)(b)(b + 1)}{c(c + 1)} \frac{z^2}{2!} - \cdots \tag{4.3.27}$$

Let p in formula (4.3.26) be a. Then we can set $C_0 = 1$, $C_1 = b/c$, $C_2 = [(b)(b + 1)]/[(c)(c + 1)]$, etc., and from Eq. (4.3.26)

$$f = \frac{1}{(1 + z)^a}\left[1 + \frac{(a)(c - b)}{c}\zeta + \frac{(a)(a + 1)(c - b)(c - b + 1)}{(c)(c + 1)}\frac{\zeta^2}{2!} + \cdots\right] \tag{4.3.28}$$

This expression inside the brackets is another hypergeometric function of $z/(1+z)$. Its radius of convergence is $|\zeta| = 1$, which corresponds to a region in the z plane, $x > -\frac{1}{2}$. It thus may be used to compute the hypergeometric function for $z > 1$, originally impossible with Eq. (4.3.27).

A powerful generalization of the Euler method is due to Barnes, in which the sum is replaced by an integral in the complex plane of a function with poles adjusted so that the integral by Cauchy's integral formula yields the desired series. This technique exploits the properties of the gamma function, so we shall postpone a discussion of it to Sec. 4.6.

4.4 *Multivalued Functions*

The discussion of complex variables given here has, up to this point, been almost completely limited to single-valued functions which by definition are uniquely specified when z is given. When we come to consider multivalued functions, many theorems (principally Cauchy's theorem and Cauchy's formula) which have been discussed must be reconsidered. It is important to note that most of the functions we shall study later on in this volume either are multivalued or have inverses which are multivalued.

The necessary concepts are best delineated by discussing a specific example:

$$f(z) = z^{\frac{1}{2}} \tag{4.4.1}$$

Let $z = re^{i\phi}$ and $f(z) = Re^{i\theta}$. Then

$$R = r^{\frac{1}{2}}; \quad \theta = \phi/2 \tag{4.4.2}$$

This function is multivalued. For a point on the z plane specified by r and ϕ or alternatively by $(r, \phi + 2\pi)$ two values of $f(z)$ are obtained:

$$f_1 = r^{\frac{1}{2}}e^{\frac{1}{2}i\phi}; \quad f_2 = r^{\frac{1}{2}}e^{\frac{1}{2}i(\phi+2\pi)} = -r^{\frac{1}{2}}e^{\frac{1}{2}i\phi} \tag{4.4.3}$$

The reason for the multiple values obtained here is just the familiar fact that the square root of a number may have either a plus or a minus value. If the exponent were $\frac{1}{4}$ instead of $\frac{1}{2}$, there would be four possible values for f for a given z:

$$(z^{\frac{1}{4}})_1 = r^{\frac{1}{4}}e^{\frac{1}{4}i\phi}; \quad (z^{\frac{1}{4}})_2 = e^{\frac{1}{2}\pi i}(r^{\frac{1}{4}}e^{\frac{1}{4}i\phi}) = e^{\frac{1}{2}\pi i}(z^{\frac{1}{4}})_1; \quad (z^{\frac{1}{4}})_3 = e^{\pi i}(z^{\frac{1}{4}})_1;$$
$$(z^{\frac{1}{4}})_4 = e^{\frac{3}{2}\pi i}(z^{\frac{1}{4}})_1$$

On the other hand z raised to the π power (or any other transcendental number) will have an infinite number of values at $z = re^{i\phi}$, so that its phase at $re^{i\phi}$ may take on any value. This permits us, in a calculation in which the power of z is unspecified, to choose the phase of the function at will.

This multiplicity of values introduces discontinuities into f. For example, at the point $z = -r$, $z^{\frac{1}{2}}$ has the value $ir^{\frac{1}{2}}$ if $z = re^{\pi i}$ or the value $-ir^{\frac{1}{2}}$ if $z = re^{-\pi i}$. Such discontinuities can correspond to an actual physical situation, being a representation of a barrier placed along the line $\phi = \pm\pi$; the mathematics of such problems always involves multivalued functions.

Branch Points and Branch Lines. The discontinuity in value may also be exhibited in a graphical manner by considering the conformal transformation generated by $z^{\frac{1}{2}}$. In Fig. 4.13 we have drawn a circular contour $a \to b$ on the z plane; its transform on the $f = u + iv$ plane is a semicircle $A \to B$. The discontinuity is now clear, since the transforms

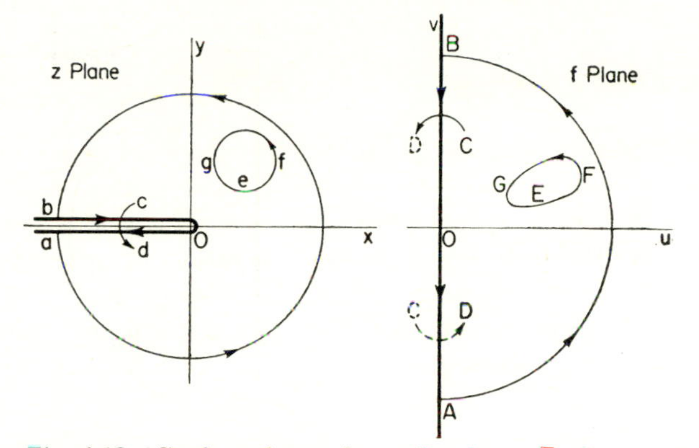

Fig. 4.13 Conformal transformation $f = \sqrt{z}$, showing multiple values and branch line aob on z^2 plane.

of the points a and b, A and B, respectively, are not equal. Since circle $a \to b$ may have any radius, we see that *all* the points on the z plane are correlated with only *half* of the points on the f plane, those for which $u > 0$. The remaining values of f are generated if a *second* circuit $a \to b$ is made. Then the values of f with $u < 0$ will be correlated with those values of z whose argument lies between π and 3π. Again associated with each value of z there are two values of f. However, we now notice that these values may be divided into two *independent* sets, those generated on the first tour of the z plane $-\pi < \phi < \pi$ and those generated on the second tour $\pi < \phi < 3\pi$. Upon making a third circuit $3\pi < \phi < 5\pi$, we find the resultant values of f to be the same as those obtained on the first circuit $-\pi < \phi < \pi$. These two independent sets of values for $z^{\frac{1}{2}}$ are called the *branches* of $z^{\frac{1}{2}}$. The line along which the discontinuities occur, $\phi = \pi$, is called the *branch line*. Branch lines, in future discussions, will be drawn as heavy, double lines.

The particular reason for singling out the branch line in this manner is that, upon crossing the branch line in the z plane, such as for contour cd, we cross over from one branch of $z^{\frac{1}{2}}$ to another, as may be seen by looking at the transform of cd, the solid CD. This latter is drawn on the assumption that C is on the branch of $z^{\frac{1}{2}}$ for which $u > 0$. If on the other hand C was in the other branch, the image of cd would be the dashed CD. In either case crossing the branch line on the z plane takes f from one of its branches to the other.

This fact may also be derived if we make use of the theorems developed in the section on analytic continuation. For example, consider circle efg in Fig. 4.13. If the function $z^{\frac{1}{2}}$ is continued along this path from e to g, then the value obtained at g should equal the value at e inasmuch as no singular point of the function has been enclosed. Pictorially this is exhibited by the transformed circle EFG. On the other hand if $z^{\frac{1}{2}}$

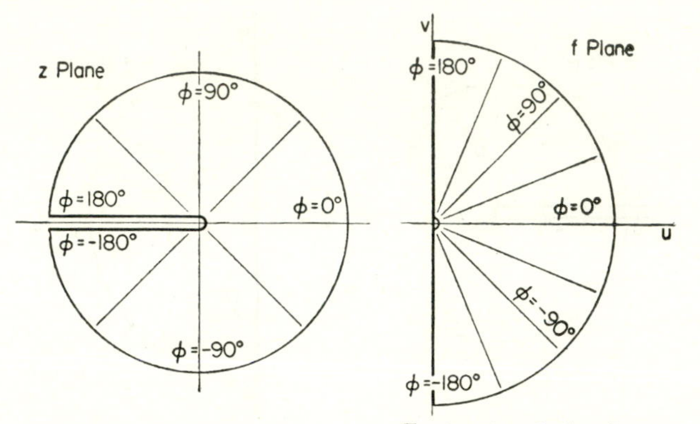

Fig. 4.14 Transformation $f = \sqrt{z}$, showing lack of conformality at $z = 0$.

is continued along the circle $a \to b$, we know that the value obtained at b will not equal the value at a. From our discussion of analytic continuation we know that there must be a singularity enclosed. Since circle $a \to b$ is of arbitrary radius, the singular point is $z = 0$ as may be verified by direct examination of the function z at $z = 0$. This type of singular point (see page 391) is called a *branch point*.

We note a few additional properties of branch points. The value of $f(z)$ at the branch point is common to all the branches of the function f. Most often it is the only point in common, though this is not always the case. Second, the transformation defined by $f(z)$ is not conformal at the branch point. This is illustrated in Fig. 4.14 where, for the function $z^{\frac{1}{2}}$, the angle between the transforms of two radial lines emanating from $z = 0$ is half of the angle between these two radii. For a function $z^{1/\alpha}$, the angle will be reduced by a factor of α. Finally, it is important to note that branch points always occur in *pairs* and that the *branch lines join the branch points*. For example, in the case of $z^{\frac{1}{2}}$, $z \to \infty$ is also a branch point This may be verified by making the substitution $z = 1/\zeta$, $z^{\frac{1}{2}} = \zeta^{-\frac{1}{2}}$, which is a multivalued function of ζ and has a branch point at $\zeta = 0$; that is, $z = \infty$.

The branch line we have used ran from the branch point at $z = 0$ to the branch point at $z = \infty$ along the negative real axis. *Any curve* joining the branch points 0 and ∞ would have done just as well. For example, we could have used the regions $0 < \phi < 2\pi$ and $2\pi < \phi < 4\pi$ as the defining regions for the first and second branch. On the f plane these two would correspond to $v > 0$ and $v < 0$, respectively. (The branch line in this case is thus the positive real axis of z.) This division into branches is just as valid as the one discussed on the previous page. We therefore may choose our branch line so that it is most convenient for the problem at hand.

Riemann Surfaces. The notion, for the case $z^{\frac{1}{2}}$, that the regions $-\pi < \phi < \pi$ and $\pi < \phi < 3\pi$ correspond to two different regions of

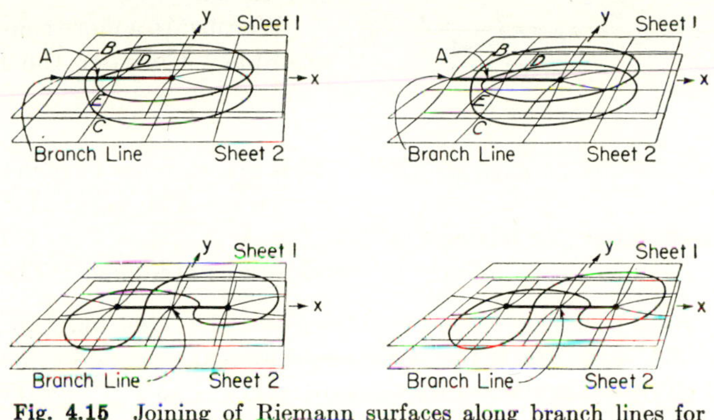

Fig. 4.15 Joining of Riemann surfaces along branch lines for the functions $f = \sqrt{z}$ and $f = \sqrt{z^2 - 1}$.

the f plane is an awkward one geometrically, since each of these two regions cover the z plane completely. To reestablish single-valuedness and continuity for the general purpose of permitting application of the various theorems developed in the preceding section, it is necessary to give separate geometric meanings to the two z plane regions. It is possible to do all this by use of the notion of *Riemann surfaces*. Imagine that the z plane, instead of being one single plane, is composed of several planes, sheets, or surfaces, arranged vertically. Each sheet, or ensemble of z values, is taken to correspond to a branch of the function, so that in the simple case of $z^{\frac{1}{2}}$ each sheet is in one-to-one correspondence with a part of the f plane. For the function $z^{\frac{1}{2}}$ only two sheets are needed, for there are only two branches, sheet 1 corresponding to $-\pi < \phi < \pi$, sheet 2 to $\pi < \phi < 3\pi$. These sheets must be joined in some fashion along the branch cut, for it should be recalled that it is possible to pass from one of the branches to the other by crossing the branch line.

The method is illustrated in Fig. 4.15. Each sheet is cut along a branch line, also often called a *branch cut* as a result. Then the lip of each side of the cut of a given sheet is connected with the opposite lip of the other sheet. A closed path $ABCDE$ on both sheets is traced out on the figure. Path AB is on sheet 1, but upon crossing the branch cut on sheet 1 at B we pass from sheet 1 to sheet 2 at C. Path CD is on sheet 2. Beyond D we cross the branch cut on sheet 2 and so pass on back through A to point E on sheet 1. It now becomes apparent what is meant by a closed contour. Path AB is not, while path $ABCDA$ is. The application of Cauchy's theorem and integral formula now becomes possible. We shall look into this matter in the next section. Right now let us continue to examine the behavior of multivalued functions.

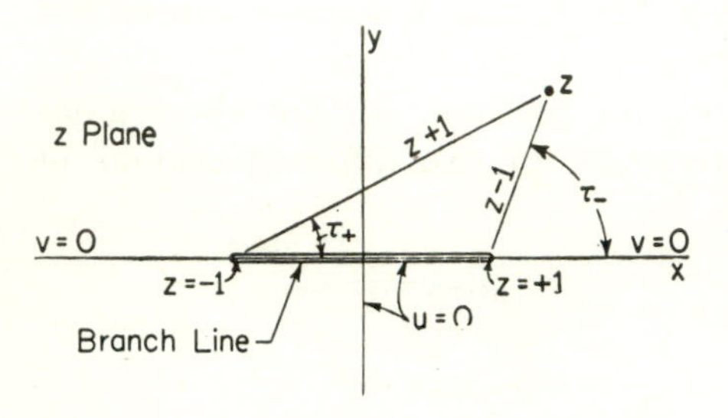

Fig. 4.16 Branch line for the transformation $f = \sqrt{z^2 - 1}$.

We turn to a more complicated example. Consider the function

$$f = \sqrt{z^2 - 1}$$

This function has branch points at $z = \pm 1$. The point at $z = \infty$ is not a branch point, for letting $z = 1/\zeta$, $f = \sqrt{(1/\zeta^2) - 1} = \sqrt{1 - \zeta^2}/\zeta \underset{\zeta \to 0}{\longrightarrow} 1/\zeta$. The point at infinity is therefore a simple pole. The branch line for f runs therefore from $z = 1$ to $z = -1$. One may go directly from -1 to 1 as in Fig. 4.15. One could also have gone between these points by going along the negative x axis from -1 to $x = -\infty$ and along the positive x axis from $x = \infty$ to $x = 1$. There are, of course, many other possibilities.

Along the real axis for $|x| > 1$, $\sqrt{z^2 - 1} = \sqrt{x^2 - 1}$ is real, so that $|x| > 1$ on the real axis corresponds to $\text{Im} f = v = 0$. As soon as $|x| < 1$, that is, along the branch line, the phase of f becomes uncertain; it may be either $\pi/2$ or $-\pi/2$. Resolving this uncertainty may be reduced into deciding where the branch lines of the constituent factors $\sqrt{z + 1}$, $\sqrt{z - 1}$ are to be placed. Different choices will lead to different branch lines for the product. To obtain the branch line of Fig. 4.15, we let the branch line $\sqrt{z + 1}$ extend from -1 to $-\infty$; for $\sqrt{z - 1}$ from $z = 1$ to $-\infty$. The phase of the product function $\sqrt{z^2 - 1}$ is then given by $\frac{1}{2}(\tau_+ + \tau_-)$, where τ_+ is the phase of $z + 1$ and τ_- the phase of $z - 1$ as shown in Fig. 4.16. Consider now a point just above the real axis between $+1$ and -1. Here $\tau_+ = 0$, $\tau_- = \pi$, so that the phase of f is $+\pi/2$. At a corresponding point just below the x axis, $\tau_+ = 0$, $\tau_- = -\pi$ so that the phase of $f = -\pi/2$. Therefore above the line $u = 0$, $v > 0$, while below the line $u = 0$, $v < 0$, again demonstrating the discontinuity which exists at a branch line. Now consider the y axis. For any point on the positive y axis, $\tau_+ + \tau_- = \pi$, so that the phase is $\pi/2$ and $u = 0$. Below the axis, the phase is $-\pi/2$.

With this information available we may now attempt to sketch the lines of constant u and constant v on the z plane. This is shown in Fig. 4.17. Note that these lines must be mutually orthogonal except at the branch points, where the transformation from the z plane to the f plane is not conformal. One example of this has already been found at points $z = \pm 1$ where the lines $u = 0$ and $v = 0$ meet. For large values of $z(|z| \gg 1)$ we may expect that $f \simeq z$, so that asymptotically the constant u and constant x lines are identical, and similarly for the constant v

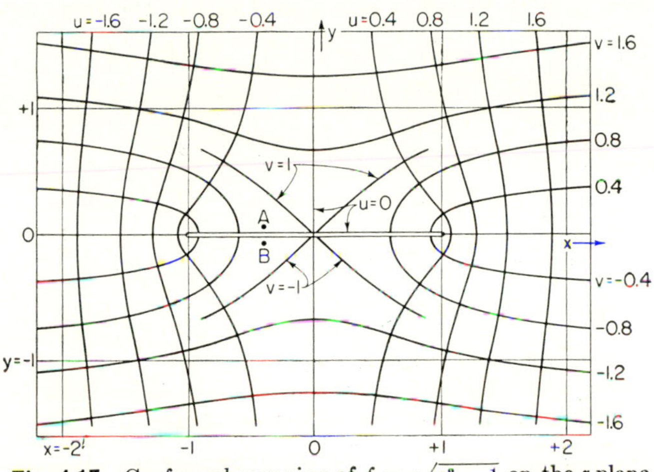

Fig. 4.17 Conformal mapping of $f = \sqrt{z^2 - 1}$ on the z plane. Contours of constant Re $f = a$ and constant Im $f = v$ are shown.

and constant y lines. We note in Fig. 4.17 that the contour $u = 0$ is the y axis, plus that part of the x axis between $x = -1$ and $x = +1$.

Between $z = -1$ and $z = 0$, v varies from 0 to 1 above the axis, while for points below the axis it varies from 0 to -1. For $|v| > 1$ we must turn to the y axis, for constant v lines for $v > 1$ (or $v < -1$) intersect the y axis at right angles, whereas for $|v| < 1$ v = constant lines intersect the x axis between 0 and -1. From this discussion, it becomes possible to sketch the v = constant lines. The u = constant lines may be drawn in from the requirement of orthogonality to the v = constant lines. We need to determine the sign of u to which they correspond. This may be done by asking for the phase of f along the x axis ($x < -1$). Here, just above the x axis, $\tau_+ = \tau_- = \pi$ so that the phase of f is π. For z just below, the same argument leads to a phase of $-\pi$, which, of course, is the same as π. We see then that the u = constant lines which are on the $x < 0$ half-plane are negative, those in the $x > 0$ half-plane are positive.

We have drawn in Fig. 4.17 the u and v contours corresponding to sheet 1 of the Riemann surface. The $v > 0$ lines, for example, must join with corresponding $v > 0$ lines in the lower half plane of the second sheet and vice versa. Aside from this change in phase the first and second sheets will be identical. To verify this, consider the phase at points A and B if we multiply the values of $\sqrt{1 + z}$ on its first sheet and $\sqrt{z - 1}$ on its second sheet. For A, $\tau_+ = 0$, $\tau_- = 3\pi$ so that the phase at A is $3\pi/2$ or $-\pi/2$. At B, $\tau_+ = 0$, $\tau_- = \pi$ so that the phase of f at B is $\pi/2$.

We could also have taken values of $\sqrt{z + 1}$ and $\sqrt{z - 1}$ from their second sheets only to find that this would lead to the same situation as

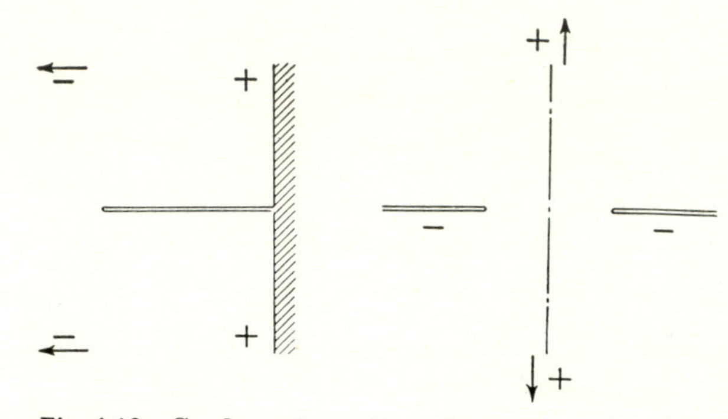

Fig. 4.18 Configurations of boundary surfaces for which the transformation of Fig. 4.17 is appropriate.

before. For example, at A, $\tau_+ = 2\pi$, $\tau_- = 3\pi$, and the phase of $f = \pi/2$. We now see in addition that there are only two sheets for the two branches of $\sqrt{z^2 - 1}$. These two sheets are joined through the branch line extending from -1 to 1.

As a final part of this discussion it is instructive to point to possible physical applications of the function $\sqrt{z^2 - 1}$ in order to observe the correlation between the mathematical branch cuts and the barriers of the corresponding physics problem. For example, the lines u = constant represent the equipotentials, the lines v = constant the lines of force for an infinitely conducting plate of shape shown in Fig. 4.18*a*, placed in a uniform electric field in the x direction. In hydrodynamics, u = constant would be the lines of flow about the same boundary. The lines u = constant may also be used as lines of force for another configuration illustrated in Fig. 4.18*b*, and the lines v = constant correspond to the equipotentials. The two plates are kept at a negative potential with respect to $|y| \gg 1$.

An Illustrative Example. As a final example of the behavior of multivalued functions, consider the function

$$z = (\tanh f)/f \tag{4.4.4}$$

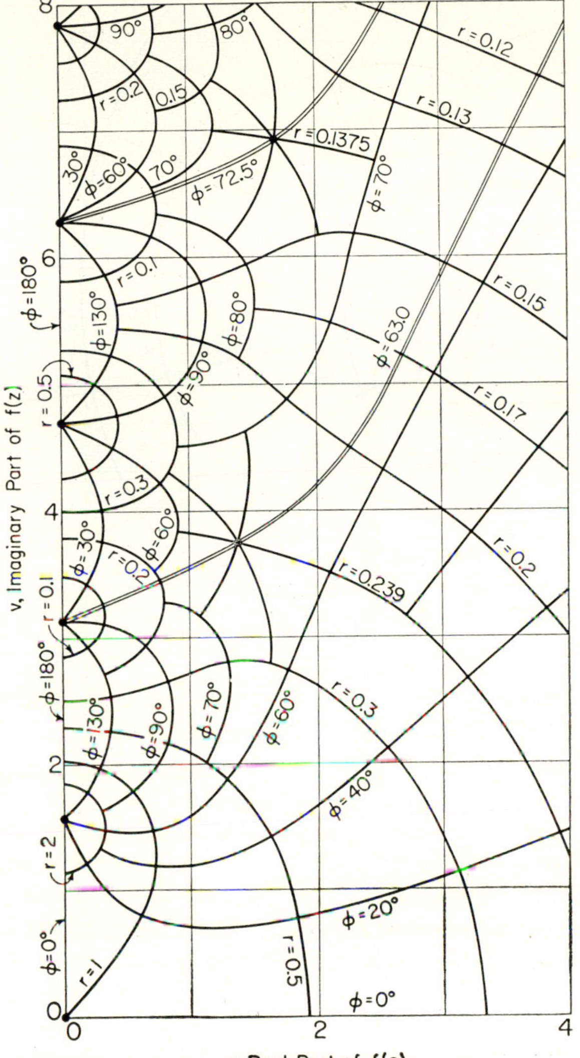

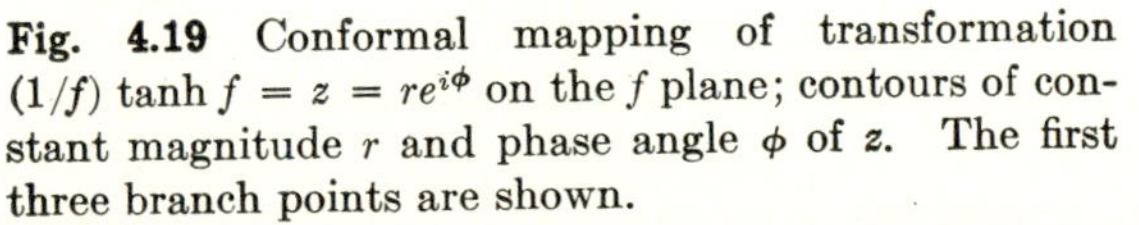

Fig. 4.19 Conformal mapping of transformation $(1/f)\tanh f = z = re^{i\phi}$ on the f plane; contours of constant magnitude r and phase angle ϕ of z. The first three branch points are shown.

This function occurs in the solution of the wave equation (see Chap. 11). The function z has simple poles at $f = [(2n + 1)/2]\pi i$ (n = integer). Its branch points occur at the points at which z has a zero derivative. At such a point, say a, the transformation defined by (4.4.4) is not conformal. On page 361 we pointed out that this results from the fact that $(z - a)$ has a multiple zero at a so that $f - f(a)$ in the neighborhood of

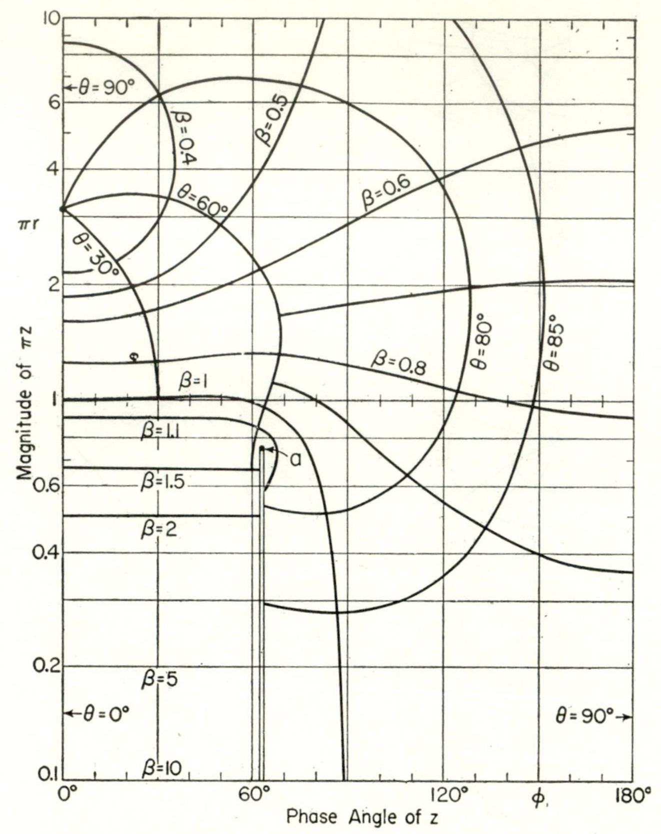

Fig. 4.20 Conformal mapping of the first sheet of the transformation $(1/f)\tanh f = z = e^w$ $(f = \pi\beta e^{i\theta})$ on the w plane. One branch point is at a.

$z = a$ is a fractional power of $(z - a)$. The branch points are therefore solutions of $[(\text{sech}^2 f)/f] - [(\tanh f)/f^2] = 0$ or

$$2f = \sinh 2f \tag{4.4.5}$$

A table of solutions of (4.4.5) is presented below (in addition to $z = a_0 = 1$).

Branch point	$f = u + iv$		$z = re^{i\phi}$	
	u	v	r	ϕ
a_1	1.3843	3.7488	0.23955	63.00°
a_2	1.6761	6.9500	0.13760	72.54°
a_3	1.8584	10.1193	0.09635	76.85°
a_4	1.9916	13.2773	0.07407	79.36°

The branch points for large values of v, $a_n(n \gg 1)$ are $a_n = u + iv$, $u \to \ln(4n + 1)\pi$, $v \to -(n + \frac{1}{4})\pi$, as may be ascertained by substitution in Eq. (4.4.5). We see that we have an infinite number of branch points and a corresponding infinite number of sheets. One simplification should be noted, however, that at each branch point f has the behavior $[z - a]^{\frac{1}{2}}$, for $z''[f(a)] \neq 0$. As a result the type of branch point is that

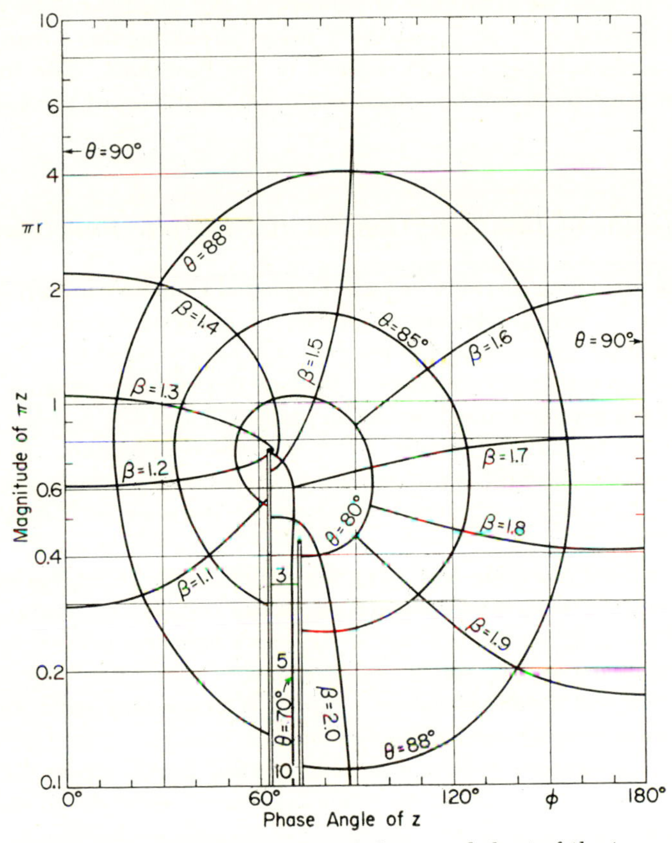

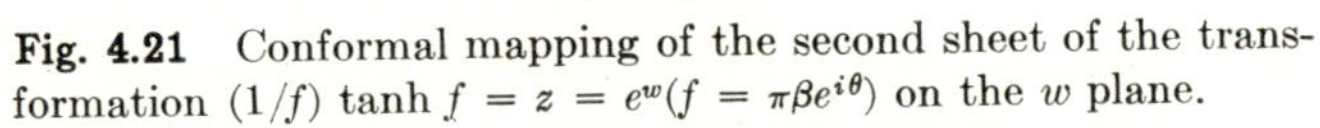

Fig. 4.21 Conformal mapping of the second sheet of the transformation $(1/f) \tanh f = z = e^w(f = \pi\beta e^{i\theta})$ on the w plane.

of the square-root function we have been discussing. We may at once say that only two branches of f will be in contact at $z = a$. With these details in mind, it now becomes possible to sketch the transformation (4.4.4). This is done in Figs. 4.20 and 4.21. To obtain a convenient plot on the z plane, we have performed a further transformation in which

$$z = re^{i\phi} = e^w; \quad \text{Im } w = \phi; \quad \text{Re } w = \ln r \tag{4.4.6}$$

Similarly for f:

$$u + iv = f = \pi\beta e^{i\theta} = e^F; \quad \text{Im } F = \theta; \quad \text{Re } F = \ln \pi\beta$$

As branch line emanating from point a, we have taken the line $\phi = 63°$. Note once more the apparent discontinuities of F at this line. Actually, of course, the constant θ lines upon crossing branch line $\phi = 63°$ go into the second sheet, which is sketched in Fig. 4.21. On this figure we have plotted both the $\phi = 63°$ branch line and the $\phi = 72.54°$ branch line joining the second and third sheets. Thus each sheet joins up with the next one and only two are in contact at any one time. In Fig. 4.19 we have plotted $w = w(f)$, on the f plane, breaking this plane up into sections corresponding to each branch of the function. The remaining details of these plots will be considered in the problems at the end of this chapter.

4.5 *Calculus of Residues; Gamma and Elliptic Functions*

In the preceding sections we have employed Cauchy's integral formula as a probe to reveal various properties of analytic functions. In the present section more mundane matters will be attended to. For example, Cauchy's integral formula may be used to evaluate integrals. Some of these have been discussed earlier on pages 373 and 374, but the applicability of the formula is broader than was there indicated. It may be used to obtain integral representations of functions which are discontinuous or have discontinuous derivatives, for instance, or it may be used to sum series.

Let us first consider the integrals which may be evaluated. The method employed is often referred to as the *calculus of residues*. The term *residue* is defined as follows: If $f(z)$ has a pole at a point a, it may be expanded in a Laurent series (4.3.4) $\Sigma a_n(z - a)^n$. The coefficient a_{-1} is defined as the *residue* of $f(z)$ at $z = a$. We may now state the fundamental theorem in the calculus of residues.

If $f(z)$ is analytic within a region bounded by C, except for a finite number of poles, the value of the contour integral $\oint f(z)\,dz$ around C will equal $2\pi i$ times the sum of the residues of f at the poles enclosed by the contour.

We shall leave the proof of this theorem to the reader. It follows directly from the Cauchy integral formula. Using the theorem it is possible to evaluate three types of integrals, listed below.

1. $\int_0^{2\pi} f(\cos\theta, \sin\theta)\,d\theta$, where f is a rational function of $\cos\theta$ and $\sin\theta$.

2. $\int_{-\infty}^{\infty} f(x)\,dx$ where $f(z)$ is analytic in the upper half plane except for poles which do not lie on the real axis. At infinity $|f(z)|$ should go to zero as $|Az|^{-m}$, where $m > 1$.

3. $\int_0^{\infty} x^{\mu-1}f(x)\,dx$ where $f(z)$ is a rational function, analytic at $z = 0$,

having no worse than simple poles on the positive, real axis and where $z^\mu f(z)$ approaches zero as z approaches zero or infinity.

Various generalizations of these three types are given in the problems.

Type 1 is evaluated by expressing $\sin\theta$ and $\cos\theta$ in terms of $z = e^{i\theta}$. The integrand now becomes a contour integral around the unit circle in the complex plane. The evaluation of the residues inside the circle completes the integration. As a simple example consider

$$I = \int_0^{2\pi} \frac{d\theta}{(1 - 2p\cos\theta + p^2)} \tag{4.5.1}$$

Let $z = e^{i\theta}$; then

$$I = \frac{1}{i}\oint \frac{dz}{(1 - pz)(z - p)} \tag{4.5.2}$$

where the contour is a unit circle. The evaluation of I now depends on the size of p. Suppose $|p| < 1$. Then the integrand has a pole at

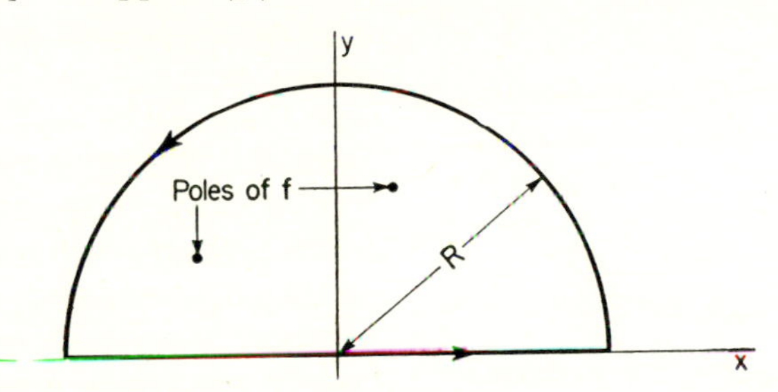

Fig. 4.22 Contour for evaluation of infinite integrals by the calculus of residues.

$z = p$. The residue there is $1/(1 - p^2)$ so that

$$I = \frac{2\pi}{1 - p^2}; \quad |p| < 1 \tag{4.5.3}$$

If $|p| > 1$, then the pole of the integrand occurs at $z = 1/p$. The integrand near $z = 1/p$ may be written $-1/[ip(z - 1/p)(z - p)]$ so that the residue is $-1/[i(1 - p^2)]$. The value of I is then

$$I = \frac{2\pi}{p^2 - 1}; \quad |p| > 1$$

Results (4.5.3) may be combined into one by writing $I = 2\pi/|p^2 - 1|$ for p real.

Turning next to integrals of type 2, we consider the integral of $f(z)$ along a closed contour consisting of the real axis from $-R$ to $+R$ and a semicircle in the upper half plane, as in Fig. 4.22. By virtue of the assumptions on $f(z)$, the integral along the semicircle will vanish in the limit $R \to \infty$. Then

$$\int_{-\infty}^{\infty} f(x)\,dx = 2\pi i \sum \text{residues of } f(z) \text{ in the upper half plane} \tag{4.5.4}$$

As a simple example, we compute

$$I = \int_{-\infty}^{\infty} \frac{dx}{1 + x^2}$$

which is known to equal π.

From Eq. (4.5.4), $I = (2\pi i)$ (residue at $z = i$) for $z = i$ is the only pole of $1/(1 + z^2) = 1/(z + i)(z - i)$ in the upper half plane. The residue is $1/2i$, and the value of I will therefore be π. Less simple examples will be found in the problems.

Integrals Involving Branch Points. Finally we consider type 3 with f having no poles on the positive real axis. Here we start with the contour integral

$$\oint (-z)^{\mu-1} f(z)\, dz$$

around a contour which does not include the branch point of the integrand at $z = 0$ and thus remains on one sheet of the associated Riemann surface. We choose the contour illustrated in Fig. 4.23, involving a small circle around $z = 0$ whose radius will eventually be made to approach zero, a large circle whose radius will eventually approach infinity, and two integrals along the positive real axis described in opposite directions and on opposite sides of the branch cut. Since the function is discontinuous along the branch cut, these two integrals will not cancel. Since μ is in general any number, the phase of $(-z)^{\mu-1}$ may be arbitrarily chosen at one point. We choose its phase at E to be *real*. At point D, achieved by rotating counterclockwise from E around $z = 0$, the phase factor is $(e^{\pi i})^{\mu-1}$ so that the integrand along DC is $e^{\pi i(\mu-1)}x^{\mu-1}f(x)$. To go from point E to A one must rotate clockwise so that the phase of the integrand at A is $-\pi i(\mu - 1)$. Along AB, the integrand is $e^{-\pi i(\mu-1)}x^{\mu-1}f(x)$. The integrals along the little and big circles vanish by virtue of the hypotheses for type 3 so that in the limit of an infinitely large radius for the big circle and zero radius for the little circle

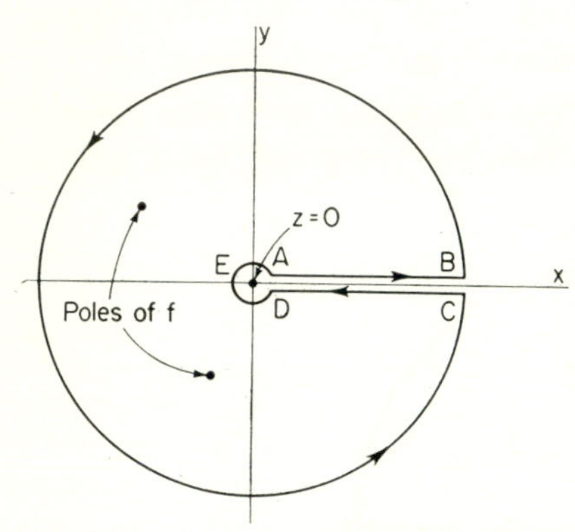

Fig. 4.23 Contour for evaluation of integrals involving a branch point at $z = 0$.

$$\oint (-z)^{\mu-1}f(z)\, dz = -\int_0^\infty e^{\pi i(\mu-1)}x^{\mu-1}f(x)\, dx + \int_0^\infty e^{-\pi i(\mu-1)}x^{\mu-1}f(x)\, dx$$

where the first of the integrals on the right is the contribution of the integral along DC while the second is the contribution from path AB.

Combining, we obtain

$$\oint (-z)^{\mu-1}f(z)\, dz = 2i \sin \pi\mu \int_0^\infty x^{\mu-1}f(x)\, dx$$

Applying Cauchy's integral formula to the contour integral we obtain

$$\int_0^\infty x^{\mu-1}f(x)\,dx = \pi\csc(\pi\mu)\sum \text{residues of } (-z)^{\mu-1}f(z) \text{ at all poles of } f \tag{4.5.5}$$

As a simple example consider $\int_0^\infty [x^{\mu-1}/(1+x)]\,dx$. From the theorem (4.5.5) this has the value $\pi \csc(\pi\mu)$ if $0 < \mu < 1$.

Finally we should point out that occasionally we encounter poles of the integrand of the second or higher order. Such cases can be calculated by using Eq. (4.3.1). Suppose that there is a pole of nth order in the integrand $f(z)$ at $z = a$. We set the integrand $f(z) = g(z)/(z-a)^n$ where $g(z)$ is analytic near $z = a$ (that is, of course, what we mean by a pole of nth order at $z = a$). Then by Eq. (4.3.1), the contribution to the integral due to this nth-order pole at $z = a$ is

$$\frac{1}{2\pi i(n-1)!}\left[\frac{d^{n-1}}{dz^{n-1}}g(z)\right]_{z=a}$$

The same result could be obtained by considering the nth-order pole to be the limiting case of n simple poles all pushed together at $z = a$.

From this result we can extend our discussion of the term *residue* at a pole. We say that the residue of a function $f(z)$ at the nth-order pole at $z = a$ [*i.e.*, if $f(z) = g(z)/(z-a)^n$ near $z = a$] is

$$\frac{1}{(n-1)!}\left[\frac{d^{n-1}}{dz^{n-1}}g(z)\right]_{z=a} \tag{4.5.6}$$

This extension allows us to use the prescriptions of Eqs. (4.5.3) and (4.5.4) for a wider variety of functions f.

Inversion of Series. The problem of inverting a function given in the form of a power series

$$w = f(z) = w_0 + \sum_{n=1}^{\infty} a_n(z-z_0)^n;\quad a_1 \neq 0 \tag{4.5.7}$$

is often encountered. We ask for the inverse function $z = z(w)$, such that $f(z) - w = 0$. From Eq. (4.5.7) we see that the inverse function may be expressed in terms of a power series

$$z(w) = z_0 + \sum_{n=1}^{\infty} b_n(w-w_0)^n \tag{4.5.8}$$

by virtue of the theorem on page 361. The coefficients b_n may be expressed in terms of a_n by introducing Eq. (4.5.8) directly into Eq. (4.5.7). However, it is possible to derive the coefficients more elegantly by employing Cauchy's formula. We shall first devise a contour integral whose residue is just the function $z(w)$. Using $f(z)$ as the independent variable, this integral is $(1/2\pi i)\oint[z\,df/(f-w)]$, for it has the required value when the contour is taken about the point $f = w$ and it includes no other zeros of $(f - w)$. In terms of z the contour integral can be written as

$$z(w) = \frac{1}{2\pi i}\oint \frac{zf'(z)\,dz}{f(z) - w} \tag{4.5.9}$$

Differentiating this with respect to w and then integrating by parts,

$$\frac{d}{dw}z(w) = \frac{1}{2\pi i}\oint \frac{dz}{f(z) - w} \tag{4.5.10}$$

In this form, it is possible to evaluate the integral. Writing

$$\frac{1}{f(z) - w} = \frac{1}{f(z) - w_0}\left[1 + \sum_{n=1}^{\infty}\frac{(w - w_0)^n}{[f(z) - w_0]^n}\right]$$

and referring to the series (4.5.8), differentiated with respect to w, we see that

$$nb_n = \frac{1}{2\pi i}\oint \frac{dz}{[f(z) - w_0]^n}$$

The value of the integral is

$$\frac{1}{(n-1)!}\frac{d^{n-1}}{dz^{n-1}}\left[\frac{(z - z_0)^n}{[f(z) - w_0]^n}\right]_{z=z_0}$$

or

$$b_n = \frac{1}{n!}\left\{\frac{d^{n-1}}{dz^{n-1}}[a_1 + a_2(z - z_0) + a_3(z - z_0)^2 + \cdots]^{-n}\right\}_{z=z_0}$$

$$b_n = \frac{1}{a_1^n n!}\left[\frac{d^{n-1}}{dx^{n-1}}\left(1 + \frac{a_2}{a_1}x + \frac{a_3}{a_1}x^2 + \cdots\right)^{-n}\right]_{x=0} \tag{4.5.11}$$

The derivative may be evaluated explicitly through use of the multinomial theorem

$$(1 + a + b + c + \cdots)^p = \sum_{r,s,t,\ldots}\left[\frac{p!}{r!s!t!\cdots}\right]a^s b^t \cdots;$$

where

$$r + s + t + \cdots = p$$

Introducing the multinomial expansion and performing the required differentiation yields

$$b_n = \frac{1}{na_1^n}\sum_{s,t,u,\ldots}(-1)^{s+t+u+\cdots}. \tag{4.5.12}$$

$$\cdot\frac{(n)(n+1)\cdots(n-1+s+t+u+\cdots)}{s!t!u!\cdots}\left(\frac{a_2}{a_1}\right)^s\left(\frac{a_3}{a_1}\right)^t\cdots$$

where

$$s + 2t + 3u + \cdots = n - 1$$

We list the first few b_n's, indicating their derivation from (4.5.12)

$$
\begin{aligned}
b_1 &= \frac{1}{a_1} \\
b_2 &= -\frac{1}{a_1^2}\frac{a_2}{a_1} = -\frac{a_2}{a_1^3} \\
b_3 &= \frac{1}{3a_1^3}\left[\frac{3\cdot 4}{2!}\left(\frac{a_2}{a_1}\right)^2 - \frac{3}{1!}\left(\frac{a_3}{a_1}\right)\right] = \frac{1}{a_1^3}\left[2\left(\frac{a_2}{a_1}\right)^2 - \left(\frac{a_3}{a_1}\right)\right] \qquad (4.5.13) \\
b_4 &= \frac{1}{4a_1^4}\left[-\frac{4\cdot 5\cdot 6}{3!}\left(\frac{a_2}{a_1}\right)^3 + \frac{4\cdot 5}{1!1!}\left(\frac{a_2}{a_1}\right)\left(\frac{a_3}{a_1}\right) - \frac{(4)}{1!}\left(\frac{a_4}{a_1}\right)\right] \\
&= \frac{1}{a_1^4}\left[-5\left(\frac{a_2}{a_1}\right)^3 + 5\left(\frac{a_2}{a_1}\right)\left(\frac{a_3}{a_1}\right) - \left(\frac{a_4}{a_1}\right)\right]
\end{aligned}
$$

. . . .

When this is inserted into Eq. (4.5.8) we have the required inverted series.

Summation of Series. The next application of the Cauchy formula to merit discussion is the summation of a series of the form $\sum_{n=-\infty}^{\infty} f(n)$. The device employed to sum the series first replaces the sum by a contour

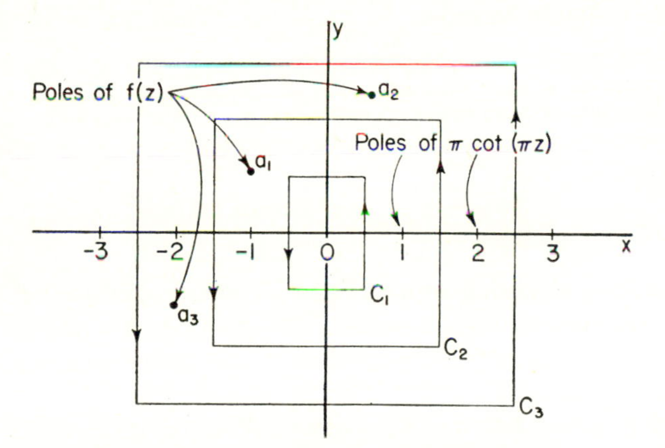

Fig. 4.24 Poles for integrand equivalent to series.

integral. To do this we require a function which has simple poles at $z = n$ and in addition is bounded at infinity. One such function is $\pi \cot(\pi z)$, which has simple poles at $n = 0, \pm 1, \pm 2 \pm \cdots$, each with residue 1. Moreover it is bounded at infinity except on the real axis. Another function of this type is $\pi \csc(\pi z)$ which has poles at $z = 0, \pm 1 \pm \cdots$ with residue $(-1)^n$.

The contour integral $\oint \pi f(z) \cot(\pi z)\, dz$ around the contour C_1 shown in Fig. 4.24 is just $2\pi i$ times the residue of the function $\pi f(z) \cot(\pi z)$ at

$z = 0$, which is $f(0)$. The integral about contour C_2 is

$$2\pi i\{f(0) + f(1) + f(-1) + \text{residue of } [\pi f(z) \cot(\pi z)] \text{ at } a_1\}$$

and so on. Finally, for a contour at infinity, the integral must be

$$2\pi i \left\{ \sum_{n=-\infty}^{\infty} f(n) + \text{residue of } [\pi f(z) \cot(\pi z)] \text{ at all the poles of } f(z) \right\}$$

if $f(z)$ has no branch points or essential singularities anywhere.

If in addition $|zf(z)| \to 0$ as $|z| \to \infty$, the infinite contour integral will be zero, so that in this case

$$\sum_{-\infty}^{\infty} f(n) = -\sum \text{residues of } \pi f(z) \cot(\pi z) \text{ at the poles of } f(z) \qquad (4.5.14)$$

If $\pi \csc(\pi z)$ is employed, one obtains

$$\sum_{-\infty}^{\infty} (-1)^n f(n) = -\sum \text{residues of } \pi f(z) \csc(\pi z) \text{ at the poles of } f(z) \qquad (4.5.15)$$

As a simple example consider $\sum_{-\infty}^{\infty} \dfrac{(-1)^n}{(a+n)^2}$. Then $f(z) = \dfrac{1}{(a+z)^2}$ with a double pole at $z = -a$. The residue of $\pi f(z) \csc(\pi z)$ at $z = -a$ is $-\pi^2 \csc(\pi a) \cot(\pi a)$ [see Eq. (4.5.5)] so that

$$\sum_{-\infty}^{\infty} \frac{(-1)^n}{(a+n)^2} = \pi^2 \csc(\pi a) \cot(\pi a)$$

This method for summing series also illustrates a method for obtaining an integral representation of a series. We shall employ this device when we wish to convert a power-series solution of a differential equation into an integral (see Sec. 5.3).

Integral Representation of Functions. We shall often find it useful to express a function as an integral, principally because very complex functions may be expressed as integrals of relatively simple functions. Moreover by changing the path of integration in accordance with Cauchy's theorem one may develop approximate expressions for the integral. In the chapter on differential equations we shall employ such techniques repeatedly.

As a first example consider the integral

$$u(x) = \frac{1}{2\pi i} \int_C \frac{e^{-ikx}}{k}\, dk \qquad (4.5.16)$$

The contour is given in Fig. 4.25. The function $u(x)$ occurs in the theory of the Heaviside operational calculus (see Sec. 11.1). It may be evaluated exactly, employing Cauchy's integral formula. If $x > 0$, we may close the contour by a semicircle of large (eventually infinite) radius in the lower half plane. The contribution along the semicircle vanishes. Therefore $u(x) = \text{Res}\,[e^{-ikx}/k]$ at $k = 0$ so that $u(x) = 1$ for $x > 0$. If $x < 0$, the contour may be closed by a semicircle of large radius in the upper half plane. Again the contribution along the semicircle vanishes.

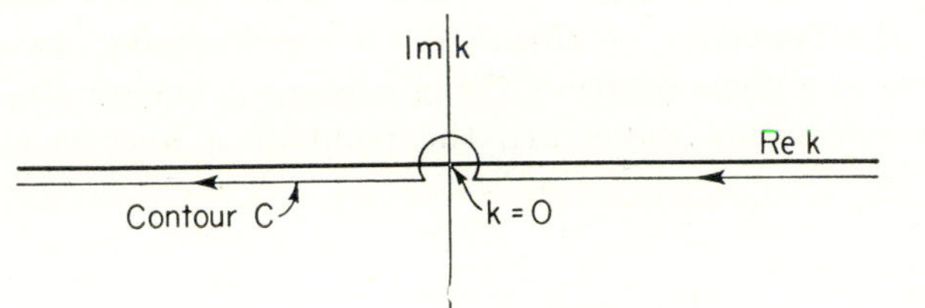

Fig. 4.25 Contour for unit step function $u(x)$.

Since $[e^{-ikx}/k]$ has no poles within this contour, $u(x) = 0$, $x < 0$. Therefore we find

$$u(x) = \begin{cases} 1; & x > 0 \\ 0; & x < 0 \end{cases} \tag{4.5.17}$$

Integral (4.5.16) thus represents a discontinuous function. It is often more useful in the integral form than in the more explicit form (4.5.17).

As a second example, we discuss an integral representation for the Green's function for motion of an infinite string discussed earlier [see also derivations of Eqs. (7.2.32) and (11.2.8)]. Consider the integral

$$G_K(x - x') = \frac{1}{2\pi} \int_C \frac{e^{ik(x-x')}}{k^2 - K^2}\, dk \tag{4.5.18}$$

The integral is undefined until we specify the contour relative to the singularities of the integrand at $k = \pm K$. We try two possible contours drawn as full and broken lines, respectively, as illustrated in Fig. 4.26, taking contour 1 first. When $x - x' > 0$, we may close the contour by a circle of large radius in the upper half plane. As we permit the radius of the semicircle to become infinite, the contribution of the integral along the semicircle tends to zero. The value of the integral is then $(2\pi i)$ (residue of the integrand at $k = K$). Therefore

$$G_k(x - x') = (i/2K)e^{iK(x-x')}; \quad x - x' > 0; \quad \text{contour } C_1 \tag{4.5.19}$$

If $x - x' < 0$, the contour may be closed from below. The value of the integral is now equal to $(-2\pi i)$ (residue at $k = -K$).

$$G_k(x - x') = (i/2K)e^{-iK(x-x')}; \quad x - x' < 0; \quad \text{contour } C_1 \tag{4.5.20}$$

Combining (4.5.19) and (4.5.20) we obtain

$$G_k(x - x') = (i/2K)e^{iK|x-x'|}; \quad \text{contour } C_1 \tag{4.5.21}$$

agreeing with page 125.

On the other hand if contour C_2 is used,

$$G_k(x - x') = (-i/2K)e^{-iK|x-x'|}; \quad \text{contour } C_2 \tag{4.5.22}$$

This result is no surprise, since contour 2 is just the reflection of contour 1 with respect to the real axis. Contour 1 gives us the Green's function satisfying the boundary condition requiring diverging waves; *i.e.*, the point x' acts as a point source. Using contour 2, one obtains the expression for a converging wave; *i.e.*, the point at x' acts as a sink. The

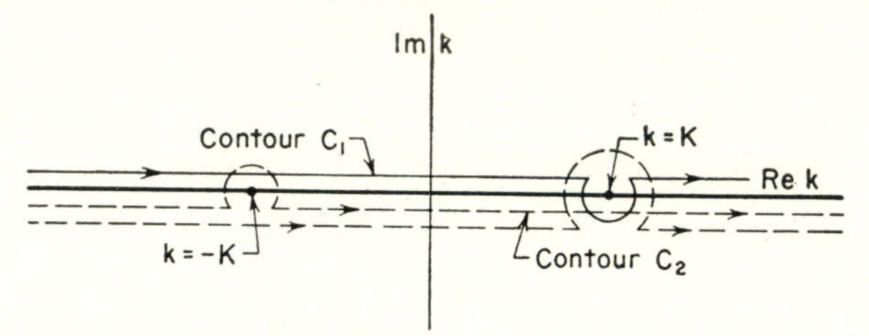

Fig. 4.26 Alternate contours for source function for string.

functions represented are continuous but have discontinuous first derivatives.

By proper manipulation it is possible for each type of contour to express $G_k(x - x')$ directly in terms of the step function $u(x' - x)$. For example, for contour 1

$$G_k = \frac{1}{4\pi K}\left\{e^{iK(x-x')}\int_C \frac{e^{-i(k-K)(x'-x)}}{k-K}\,dk + e^{-iK(x-x')}\int_C \frac{e^{-i(k+K)(x'-x)}}{k+K}\,dk\right\}$$

The first of these integrals has a singularity at $k = K$ only, so that no detour around $k = -K$ is necessary. Similarly in the second integral only the detour around $k = -K$ is required in the contour. Comparing with (4.5.16) and Fig. 4.25 one obtains

$$G_k(x - x') = (i/2K)\{e^{iK(x-x')}[1 - u(x' - x)] + e^{-iK(x-x')}u(x' - x)\}$$

which agrees with (4.5.21).

Integrals Related to the Error Function. So far we have dealt with integrals which could be evaluated, and thus expressed directly, in terms of the elementary transcendental functions. Let us now consider a case in which the integrals cannot be so expressed. An integral solution of the differential equation $(d^2\psi_\lambda/dz^2) - 2z(d\psi_\lambda/dz) + 2\lambda\psi_\lambda = 0$ is

$$\psi_\lambda = \frac{1}{2\pi i}\int_C \frac{e^{-t^2+2tz}}{t^{\lambda+1}}\,dt \tag{4.5.23}$$

where the contour is shown in Fig. 4.27*a*. We have chosen the branch line for the integrand to be the positive real axis. We now evaluate the integral for various special values of λ and for z small and z large. First, for λ an integer n, the origin is no longer a branch point, so that the contour may be deformed into a circle about the origin. Then by Eq. (4.3.1)

$$\psi_n = \frac{1}{n!}\left[\frac{d^n}{dt^n}(e^{-t^2+2tz})\right]_{t=0}$$

or

$$\psi_n = \frac{(-1)^n}{n!} e^{z^2}\left[\frac{d^n}{dz^n} e^{-z^2}\right]; \quad \lambda = n \tag{4.5.24}$$

The resulting polynomials are proportional to the Hermitian polynomials (see table at the end of Chap. 6).

For $\lambda < 0$, the contour integral may be replaced by a real integral as follows: The contour may be broken up into a line integral from ∞ to 0, a circle about zero, and an integral from 0 to ∞. For $\lambda < 0$, the value of the integral about the circle approaches zero as the radius approaches zero, so that we need consider only the two integrals extending from 0 to ∞. By virtue of the branch point at $t = 0$, we must specify what branch of the multivalued integrand is being considered. This is done by choosing a phase for $t^{-(\lambda+1)}$. Since λ is arbitrary, the phase (see page 398) may be chosen to suit our convenience, which is best served by making it zero at point A. Then the integral from infinity to A is

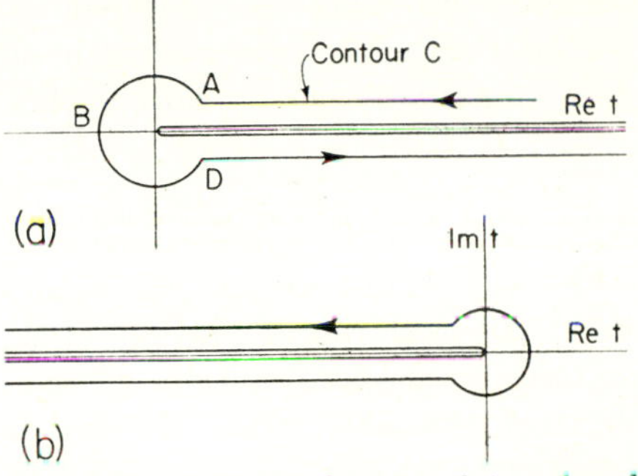

Fig. 4.27 Contours for error integral and gamma function.

$$\frac{1}{2\pi i}\int_\infty^0 \frac{e^{-t^2+2tz}}{t^{\lambda+1}}\,dt; \quad \text{where } t \text{ is real}$$

The phase of the integrand at D is obtained from that at A by rotating counterclockwise from A to D by 2π. Thus the phase is $-2\pi i(\lambda + 1)$. The integral from D to infinity is thus

$$\frac{e^{-2\pi i(\lambda+1)}}{2\pi i}\int_0^\infty \frac{e^{-t^2+2tz}}{t^{\lambda+1}}\,dt; \quad t \text{ real}$$

We thus obtain, for the integral (4.5.23), when $\lambda < 0$, the new form

$$\psi_\lambda = \frac{1}{\pi} e^{-\pi i(\lambda+1)} \sin \pi(\lambda + 1) \int_0^\infty \frac{e^{-t^2+2tz}}{t^{\lambda+1}}\,dt \tag{4.5.25}$$

Since the infinite integral is real and does not vanish (or have a pole) when λ is a negative integer, the function ψ_λ must vanish for these values of λ. This is not surprising, since then the value of the integral along the upper half of the contour is equal and opposite to the value along the lower half of the contour. We also see that it would have been more satisfactory (perhaps!) for us to have chosen the integrand to be real at point B on the contour, for then the exponential factor would not have remained in Eq. (4.5.25).

To investigate the behavior of ψ_λ as $z \to 0$, we expand e^{2tz} in a power series:

$$\psi_\lambda = \sum_{n=0}^{\infty} \frac{(2z)^n}{n!} \left\{ \frac{1}{2\pi i} \int_c e^{-t^2} t^{n-\lambda-1}\, dt \right\} \tag{4.5.26}$$

The integral may be expressed in terms of a gamma function. (The integral representation required is derived in the following section.) We shall therefore relegate the evaluation of (4.5.26) to a problem and shall content ourselves with noting one interesting point. If λ is an integer $p > 0$, the contour integral will vanish for all $n > p$. Thus ψ_p is a finite polynomial of order p. This is, of course, confirmed by further examination of the explicit formula for these cases, given in Eq. (4.5.24).

Finally we investigate the values of ψ_λ for z large. (The advantage of an integral representation of ψ_λ should be by now manifest. It contains *all* the information, all the details of the function's behavior in its most tractable form. It enables us to establish a correspondence between its values for $z \to 0$ and those for $z \to \infty$.) For $|z| \to \infty$ and for Re $z < 0$ the simplest procedure involves the substitution $2t|z| = u$.

$$\psi_\lambda = \frac{(2|z|)^\lambda}{2\pi i} \int_c \frac{e^{-(u/2|z|)^2} e^{-u}}{u^{\lambda+1}}\, du$$

We now expand the exponential in a power series:

$$\psi_\lambda = (2|z|)^\lambda \sum \frac{(-1)^n}{(2|z|)^{2n}} \frac{1}{2\pi i} \int_c u^{2n-\lambda-1} e^{-u}\, du; \quad \text{Re } z > 0 \tag{4.5.27}$$

Again the integral may be expressed in terms of gamma functions [see Eq. (4.5.36)]. For very large values of z, only the first term is important and $\psi_\lambda \to A|z|^\lambda$.

When Re $z > 0$ this procedure must be changed. Let $u = -2t|z|$

$$\psi_\lambda = \frac{(2|z|)^\lambda}{2\pi i} \int_c \frac{e^{-(u/2|z|)^2} e^{-u}}{(-u)^{\lambda+1}}\, du$$

The path of integration is obtained by performing the same transformation for the path, with the result illustrated in Fig. 4.27*b*. We may now expand $e^{-(u/2|z|)^2}$:

$$\psi_\lambda = (2|z|)^\lambda \sum (2|z|)^{-2n} \frac{1}{2\pi i} \int_c e^{-u} (-u)^{n-\lambda-1}\, du \tag{4.5.28}$$

As we shall see in the section immediately following, the contour integral is just a gamma function [see Eq. (4.5.36)].

Gamma Functions. Earlier [Eq. (4.3.14)] the gamma function was defined by the infinite integral

$$\Gamma(z) = \int_0^\infty e^{-t}t^{z-1}\,dt; \quad \Gamma(n) = (n-1)! \tag{4.5.29}$$

In order that this integral converge, the real part of z must be positive. When z is not an integer, $t = 0$ is a branch point for the integrand. We take the positive real axis for the branch line. A number of equivalent forms obtained by transformation will prove useful:

$$\begin{aligned}\Gamma(z) &= 2\int_0^\infty e^{-t^2}t^{2z-1}\,dt; && \text{Re } z > 0\\ \Gamma(z) &= 2\int_0^1 [\ln (1/t)]^{z-1}\,dt; && \text{Re } z > 0\end{aligned} \tag{4.5.30}$$

As discussed in Sec. 4.3, it is possible to extend definition (4.5.29) to negative values of real z by means of the recurrence relation

$$z\Gamma(z) = \Gamma(z+1) \tag{4.5.31}$$

Indeed this relation immediately yields considerable information about the singularities of $\Gamma(z)$.

Since the integral converges for $\text{Re } z > 0$, $\Gamma(z)$ is always finite for $\text{Re } z > 0$. Moreover since the derivative $\Gamma'(z)$ may be evaluated directly by differentiating under the integral sign, it follows that $\Gamma(z)$ is *analytic* for $\text{Re } z > 0$. To explore the behavior of $\Gamma(z)$ for $\text{Re } z < 0$, we find an integer n, for any given value of z, which will be sufficiently large so that $\text{Re } (z + n + 1) > 0$. Let n be the smallest possible integer of this sort. Then, of course,

$$\Gamma(z) = \frac{\Gamma(z+n+1)}{(z+n)(z+n-1)(z+n-2)\cdots(z)} \tag{4.5.32}$$

so that $\Gamma(z)$ is now defined in terms of $\Gamma(z + n)$ where $\text{Re } (z + n) > 0$. We note that $\Gamma(z)$ is generally finite, with a defined derivative for z, *except when* z is zero or a negative integer.

Near these points we can set $z = -n + \epsilon$, where $|\epsilon| \ll 1$, and use Eq. (4.5.32) to give us

$$\Gamma(-n+\epsilon) = \frac{(-1)^n\Gamma(1+\epsilon)}{(n-\epsilon)(n-\epsilon-1)\cdots(1-\epsilon)\epsilon}$$

which has a simple pole at $\epsilon \to 0$. Consequently the function $\Gamma(z)$ is analytic over the finite part of the z plane, with the exception of the points $z = 0, -1, -2, -3, \ldots$, where it has simple poles. The residue at the pole at $z = -n$ is $(-1)^n/n!$ [since $\Gamma(1) = 1$].

The regular spacing of these poles is reminiscent of the regular spacing of the poles of the trigonometric functions $\csc (\pi z)$ or $\cot (\pi z)$. However, the latter also has poles along the positive real axis for the positive integers. We may, of course, find a function which has poles there, namely, $\Gamma(-z)$. Thus the product $\Gamma(z)\Gamma(1 - z)$ has poles at $z = n$, where n is an integer (positive or negative). It has poles nowhere else. Thus $\sin(\pi z)\Gamma(z)\Gamma(1 - z)$ is analytic everywhere in the finite z plane. [We cannot use $\cot(\pi z)$ in place of $\csc(\pi z)$ because of its zeros at $z = \frac{1}{2}(2p + 1)\pi$, p integral.] In fact we shall now show that

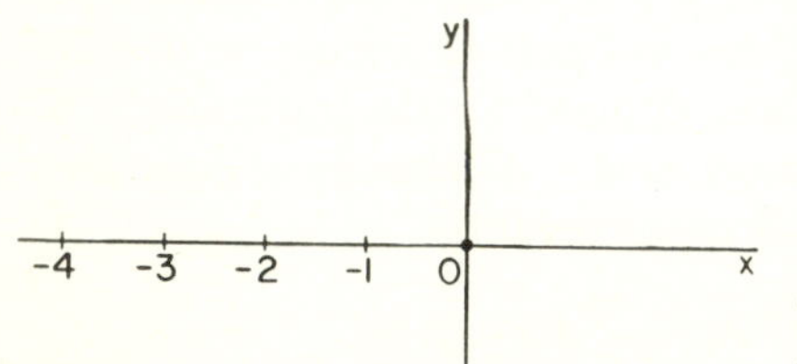

Fig. 4.28 Position of poles of gamma function.

$$\Gamma(z)\Gamma(1 - z) = \pi \csc(\pi z) \qquad (4.5.33)$$

We will show it first for z real and between 0 and 1; by analytic continuation (see theorem on page 390) we can show that $\sin(\pi z)\Gamma(z)\Gamma(1 - z) = \pi$ everywhere.

To prove Eq. (4.5.33) we use the first of Eqs. (4.5.30), obtaining the double integral

$$\Gamma(a)\Gamma(1 - a) = 4 \int_0^\infty \int^\infty e^{-(x^2+y^2)} x^{2a-1} y^{-(2a-1)}\, dx\, dy$$

We shift now to polar coordinates $x = r \cos\theta$, $y = r \sin\theta$ so that

$$\Gamma(a)\Gamma(1 - a) = 4 \int_0^{\pi/2} (\cot\theta)^{2a-1}\, d\theta \int_0^\infty re^{-r^2}\, dr = 2 \int_0^{\pi/2} (\cot\theta)^{2a-1}\, d\theta$$

To evaluate the final integral, use the substitution $\cot\theta = s$, then $\Gamma(a)\Gamma(1 - a) = 2 \int_0^\infty (s^{2a-1})/(1 + s^2)\, ds$. It will be recognized that for $0 < a < 1$, this integral is in a form which falls under formula (4.5.5). Then $\Gamma(a)\Gamma(1 - a) = 2 \csc(2\pi a)\{\Sigma$ residues of $[(-s)^{2a-1}]/[1 + s^2]$ at $s = \pm i\}$ if $0 < a < 1$. A little manipulation yields Eq. (4.5.33). As mentioned, analytic continuation will extend the range of the formula.

We shall employ this result to establish two results. First by letting $z = \frac{1}{2}$, we find $\Gamma^2(\frac{1}{2}) = \pi$ or $\Gamma(\frac{1}{2}) = \sqrt{\pi}$. Second, we shall establish *that* $[\Gamma(z)]^{-1}$ *is an integral function.* Since the singularities of $\Gamma(z)$ are poles, it is only necessary to show that $\Gamma(z)$ has no zeros in the finite complex plane. This is a consequence of Eq. (4.5.33), for if $\Gamma(z)$ is zero at some point, $\Gamma(1 - z)$ must be infinite there. However, the infinities of $\Gamma(1 - z)$ are known and $\Gamma(z)$ is not zero at these points, so it cannot be zero anywhere.

Contour Integrals for Gamma Functions. We now have a good idea of the general behavior of $\Gamma(z)$, and we may now proceed to find closed expressions for it which are valid over a greater range in z than (4.5.29). One such representation is obtained when the direct solution of the difference equation (4.5.31) is attempted. Care must be exercised, however, to choose a solution which reduces to (4.5.29) when Re $(z) > 0$, for any solution of (4.5.31) when multiplied by a periodic function with period 1 yields another solution of (4.5.31). The form of (4.5.29) suggests that it may be rewritten as a contour integral

$$\Gamma(z) = \int_C v(t)t^{z-1}\,dt$$

where the contour C and the function $v(t)$ are still to be determined. Substituting in (4.5.31) yields

$$\int_C v(t)t^z\,dt = \int_C v(t)zt^{z-1}\,dt = \int v(t)\frac{d}{dt}(t^z)\,dt$$

Integrating the second of these by parts yields

$$\int_C \left[v(t) + \frac{dv}{dt}\right] t^z\,dt = [v(t)t^z]$$

where the expression on the right is to be evaluated at the ends of the contour. We shall pick the contour so that $[v(t)t^z]$ vanishes. Hence v may be taken as the solution of

$$v(t) + \frac{dv}{dt} = 0; \quad v = (\text{constant})e^{-t}$$

so that

$$\Gamma(z) = A\int_C (e^{-t})(-t)^{z-1}\,dt \tag{4.5.34}$$

where A is a constant. The contour chosen is illustrated in Fig. 4.27a. Next we evaluate the integral for Re $z > 0$, and then choose A so that Eqs. (4.5.34) and (4.5.29) are identical. Choose B as the point at which the phase of $(-t)^{z-1}$ is zero. Then for Re $z > 0$

$$\begin{aligned}\Gamma(z) &= A\left\{\int_\infty^0 e^{-t}e^{-\pi i(z-1)}(t)^{z-1}\,dt + \int_0^\infty e^{-t}e^{\pi i(z-1)}t^{z-1}\,dt\right\}\\ &= 2iA\sin[\pi(z-1)]\int_0^\infty e^{-t}t^{z-1}\,dt\end{aligned}$$

Therefore $A = -1/(2i\sin\pi z)$ and

$$\Gamma(z) = -\frac{1}{2i\sin\pi z}\int_C e^{-t}(-t)^{z-1}\,dt \tag{4.5.35}$$

This representation is valid for all z. Combining it with the relation (4.5.33) yields another integral representation of $\Gamma(z)$:

$$\frac{1}{\Gamma(z)} = -\frac{1}{2\pi i}\int_C e^{-t}(-t)^{-z}\,dt \tag{4.5.36}$$

The behavior of $\Gamma(z)$ when z is an integer may be found directly from Eq. (4.5.35) or (4.5.36).

Infinite Product Representation for Gamma Functions. Another useful representation of $\Gamma(z)$ may be obtained by applying the formula developed for the product representation of an entire function. The function $[\Gamma(z+1)]^{-1}$ is an entire function, with zeros at $z = -1, -2, \ldots$. Using Eq. (4.3.8), one obtains

$$\frac{1}{\Gamma(z)} = ze^{\gamma z} \prod_{n=1}^{\infty} \left(1 + \frac{z}{n}\right) e^{-z/n} \tag{4.5.37}$$

The constant $\gamma = -\Gamma'(1)/\Gamma(1)$ equals 0.5772157 and is known as the *Euler-Mascheroni constant.* The value of γ may be obtained by placing $z = 1$ and taking logarithms of both sides of Eq. (4.5.37), yielding

$$\gamma = \sum_{n=1}^{\infty} \left[\frac{1}{n} - \log\left(1 + \frac{1}{n}\right)\right] \tag{4.5.38}$$

This series may be easily evaluated.

Other useful forms for γ may be directly derived from (4.5.38):

$$\gamma = \lim_{M \to \infty} \left(1 + \frac{1}{2} + \frac{1}{3} + \cdots + \frac{1}{M} - \ln M\right) \tag{4.5.39}$$

and
$$\gamma = \lim_{M \to \infty} \left\{\sum_{n=1}^{M} \int_0^\infty e^{-nq}\, dq - \int_0^\infty \frac{e^{-q} - e^{-Mq}}{q}\, dq\right\}$$

or
$$\gamma = \int_0^\infty e^{-q} \left[\frac{1}{1 - e^{-q}} - \frac{1}{q}\right] dq \tag{4.5.40}$$

Derivatives of the Gamma Function. Of considerable importance to subsequent applications is the logarithmic derivative of the gamma function

$$\psi_1(z) = \frac{d \ln \Gamma(z)}{dz} = \frac{\Gamma'(z)}{\Gamma(z)} \tag{4.5.41}$$

From (4.5.37)

$$\psi_1(z) = -\gamma - \frac{1}{z} + \sum_{n=1}^{\infty} \left[\frac{1}{n} - \frac{1}{n+z}\right] \tag{4.5.42}$$

When z is an integer N,

$$\psi_1(N) = -\gamma + \sum_{n=1}^{N-1} \frac{1}{n}; \quad \psi_1(1) = -\gamma \tag{4.5.43}$$

The derivative of ψ_1 is

$$\psi_2(z) = \sum_{n=0}^{\infty} \frac{1}{(n+z)^2} \tag{4.5.44}$$

In general

$$\psi_p(z) = \frac{d}{dz}\psi_{p-1}(z) = (-1)^p p! \sum_{n=0}^{\infty} \frac{1}{(n+z)^p} \tag{4.5.45}$$

These functions have been tabulated for real values of z, permitting the ready evaluation of sums, each term of which is a rational function of the summation index. For example, the sum

$$S = \sum_{n=0}^{\infty} \frac{1}{(n+1)^2(n+a)^2}$$

may be written

$$S = \frac{1}{(a-1)^2} \sum_{n=0}^{\infty} \left\{ \left[\frac{1}{(n+1)^2} + \frac{1}{(n+a)^2} \right] - \frac{2}{a-1} \left[\frac{1}{n+1} - \frac{1}{n+a} \right] \right\}$$

so that $S = \dfrac{1}{(a-1)^2} \left\{ [\psi_2(1) + \psi_2(a)] - \dfrac{2}{a-1} [-\psi_1(1) + \psi_1(a)] \right\}$

The polygamma functions, as the ψ_p functions are called, satisfy a simple recursion formula with respect to their argument.

$$\psi_p(z+1) = (-1)^p p! \sum_{n=0}^{\infty} \frac{1}{(n+1+z)^p} = (-1)^p p! \sum_{n=1}^{\infty} \frac{1}{(n+z)^p}$$

so that

$$\psi_p(z+1) = \psi_p(z) - [(-1)^p p!]/z^p \tag{4.5.46}$$

Integral representations for the polygamma function may be found by employing the device which led to Eq. (4.5.40) for γ. From Eq. (4.5.42) one has

$$\psi_1(z) = -\gamma + \int_0^{\infty} \left[\frac{e^{-q}}{1-e^{-q}} - \frac{e^{-zq}}{1-e^{-q}} \right] dq$$

Introducing Eq. (4.5.40) for γ, we find that

$$\psi_1(z) = \int_0^{\infty} \left[\frac{e^{-q}}{q} - \frac{e^{-zq}}{1-e^{-q}} \right] dq \tag{4.5.47}$$

Differentiating with respect to z yields the other polygamma functions:

$$\psi_p(z) = (-1)^p \int_0^{\infty} \frac{q^{p-1}e^{-zq}}{1-e^{-q}} dq \tag{4.5.48}$$

These formulas may be used to obtain the behavior of $\psi_p(z)$ as $z \to \infty$; thence, by integration with respect to z of $\psi_1 = [d \ln \Gamma(z)]/dz$ and, finally, by another integration, the behavior of $\Gamma(z)$ for large z. The determination of the constants of integration is rather involved so that we shall postpone discussion of the behavior of $\Gamma(z)$ for large values of z until the next section. For reference, the behavior is given by the Stirling approximation

$$\ln [\Gamma(z)] \xrightarrow[z\to\infty]{} -z + (z - \tfrac{1}{2}) \ln z + \tfrac{1}{2} \ln (2\pi) + 0(1/z) \quad (4.5.49)$$

when z is in the first quadrant, or

$$\Gamma(z) \to e^{-z}z^{z-\frac{1}{2}} \sqrt{2\pi}\,(1 + (1/12z) + \cdots); \quad z \to \infty \quad (4.5.50)$$

The Duplication Formula. Another application of the infinite product (4.5.37) is in the derivation of the *duplication formula*

$$2^{2z-1}\Gamma(z)\Gamma(z + \tfrac{1}{2}) = \sqrt{\pi}\,\Gamma(2z) \quad (4.5.51)$$

To prove this consider the ratio of $\Gamma(z)$ to $\Gamma(2z)$;

$$\frac{\Gamma(z)}{\Gamma(2z)} = 2e^{\gamma z} \frac{\prod_{n=1}^{\infty} [1 + (2z/n)]e^{-2z/n}}{\prod_{n=1}^{\infty} [1 + (z/n)]e^{-z/n}}$$

or $$\frac{\Gamma(z)}{\Gamma(2z)} = 2e^{\gamma z}(1 + 2z)e^{-2z} \prod_{p=1}^{\infty} \left(1 + \frac{2z}{2p + 1}\right) e^{-2z/(2p+1)}$$

$$= 2e^{\gamma z}(1 + 2z)e^{-2z} \frac{\prod_{p=1}^{\infty} [1 + (z + \frac{1}{2})/p]e^{-(z+\frac{1}{2})/p}}{\prod_{p=1}^{\infty} [1 + (1/2p)]e^{-\frac{1}{2}p}} \cdot \prod_{p=1}^{\infty} \exp\left[z\left(\frac{1}{p} - \frac{1}{p + \frac{1}{2}}\right)\right]$$

Utilizing the product representation of the gamma function (4.5.37), we may write

$$\frac{\Gamma(z)}{\Gamma(2z)} = \frac{2\Gamma(\frac{1}{2})e^{-2z}}{\Gamma(z + \frac{1}{2})} \exp\left[z \sum_{p=1}^{\infty} \left(\frac{1}{p} - \frac{1}{p + \frac{1}{2}}\right)\right] \quad (4.5.52)$$

To evaluate the exponential, let $z = 1$ in (4.5.52). Then one obtains

$$\exp\left[\sum_{p=1}^{\infty} \left(\frac{1}{p} - \frac{1}{p + \frac{1}{2}}\right)\right] = \frac{e^2}{4} \quad (4.5.53)$$

Introducing (4.5.53) into (4.5.52) yields the duplication formula (4.5.51).

Beta Functions. We conclude our discussion of the gamma function with an investigation of the *beta function* defined as

$$B(p,q) = \int_0^1 t^{p-1}(1-t)^{q-1}\,dt \tag{4.5.54}$$

or alternately by

$$B(p,q) = \int_0^\infty \frac{t^{p-1}}{(1+t)^{p+q}}\,dt; \quad B(p,q) = 2\int_0^{\pi/2} \sin^{2p-1}\theta\,\cos^{2q-1}\theta\,d\theta \tag{4.5.55}$$

We shall content ourselves with establishing the formula

$$B(p,q) = \frac{\Gamma(p)\Gamma(q)}{\Gamma(p+q)} \tag{4.5.56}$$

Using Eq. (4.5.30)

$$\Gamma(p)\Gamma(q) = 4\int_0^\infty e^{-x^2}x^{2p-1}\,dx \int_0^\infty e^{-y^2}y^{2q-1}\,dy \tag{4.5.57}$$

and transforming to polar coordinates, $x = r\cos\theta$, $y = r\sin\theta$, we find

$$\Gamma(p)\Gamma(q) = 4\int_0^\infty e^{-r^2}r^{2p+2q-1}\,dr \int_0^{\pi/2} \sin^{2p-1}\theta\,\cos^{2q-1}\theta\,d\theta$$

Employing (4.3.30) again, and (4.5.55), we eventually obtain Eq. (4.5.56).

Periodic Functions. There are many times when we shall encounter or need functions with the property of *periodicity*, of repeating themselves over and over again. Mathematically this is expressed by saying that, for the periodic function $f(z)$, there is a certain complex number a such that

$$f(z+a) = f(z)$$

for all values of z for which f is analytic. The constant a is called a *period* of f. After z goes from z_0 to $z_0 + a$, the function then starts in again and repeats its former behavior over the next period, from $z_0 + a$ to $z_0 + 2a$.

Of course, if a is a period for f, $2a$ is also a period and any integer (positive or negative) times a is likewise. Sometimes $\frac{1}{2}a$ is a period, or $\frac{1}{3}a$, etc. But the subdivision cannot go on forever, for a bit of the simple but close reasoning one uses in modern point-set theory will persuade one that there cannot be an infinitesimally small period (unless the periodic function is a constant!) In fact, if one arranges all the periods of a given periodic function in order, one soon finds that they all can be represented as integral multiples of a *fundamental period* τ, characteristic of the function f and such that the equation

$$f(z+\tau) = f(z) \tag{4.5.58}$$

holds, but no similar equation holds for f for any period smaller than τ.

The functions $\sin z$, $\sec z$ are periodic functions with fundamental period 2π; the function e^z is a periodic function with fundamental period

$2\pi i$; and so on. These functions may be characterized by series or by infinite product expressions, which display the fact that their poles (or zeros) are equally spaced along a line (or lines) in the z plane, with spacing equal to τ, such as the following

$$\frac{1}{\sin z} = \frac{1}{z} + \sum_{n=1}^{\infty} \frac{2(-1)^n z}{z^2 - (n\pi)^2}; \quad \sin z = z \prod_{n=1}^{\infty} \left[1 - \left(\frac{z}{n\pi} \right)^2 \right] \tag{4.5.59}$$

which have been discussed earlier. From these expressions, by algebraic manipulation, one can demonstrate the periodicity of the sine function.

We can, however, characterize the functions by their differential equations. For instance, the equation

$$(dy/dz)^2 = 1 - y^2 \tag{4.5.60}$$

or the second-order equation, formed from this by differentiation,

$$(d^2y/dz^2) = -y \tag{4.5.61}$$

may be used to characterize the periodic function $y = \sin z$.

To show that the function is periodic we use Eq. (4.5.61) to follow the behavior of y as z is increased from zero along the real axis. We start off by saying that y is the solution of Eq. (4.5.60) which is zero at $z = 0$; the equation then shows that its slope is unity at $z = 0$ (it could be ± 1, but we choose $+1$). As z is increased, y increases until $y = 1$, when dy/dx is zero, but d^2y/dx^2 is -1, as Eq. (4.5.61) shows. Consequently the maximum value of y along the real axis is $+1$. As z is increased further, y diminishes, dy/dz gets more and more negative, until, when y is zero again, dy/dz is -1. Increasing z still further results in an excursion of y to -1 and then back once more to zero, when it has a slope of $+1$, as at the beginning. At this value of z, y has gone through a full cycle and is ready to start again. We note that due to the symmetry of the equation this full period is divided into four equal parts, the first covering the distance over which y goes from zero to 1, the second over which y goes from 1 to 0, and so on.

Going back to Eq. (4.5.60) we see that the solution which goes to zero at $z = 0$ can be formally written

$$z = \int_0^y \frac{du}{\sqrt{1 - u^2}}; \quad \text{that is, } z = \sin^{-1} y$$

This gives us an inverse sort of solution, that is, z as a function of y, but it enables us to compute the period easily. For from the discussion of the previous paragraph we see that one-quarter of the total period τ is

$$\tfrac{1}{4}\tau = \int_0^1 \frac{du}{\sqrt{1 - u^2}}$$

Calculation of this integral shows it to be equal to $\frac{1}{2}\pi$ so that the period of the function y we are discussing is $\tau = 2\pi$.

In books on analysis the periodic properties of the trigonometric function are discussed in more detail and with much more rigor than we have space for here, particularly since we are using this only as an introduction to a more general type of periodic function.

For we might ask whether there exist functions of the complex variable z which have *two* periods, a_1 and a_2, such that $y(z + a_1) = y(z)$ and also $y(z + a_2) = y(z)$. This pair of statements would not introduce anything particularly new if a_1 and a_2 were in the same direction in the complex plane (*i.e.*, if a_1/a_2 were a real number); for with simply periodic functions, if a_1 is a period, $2a_1$ or $3a_1$ is also a period. Of course we might ask whether τ_1 and τ_2 could be in the same direction but be incommensurate. The same close reasoning which states that there cannot be periods of infinitesimal size (except when the function is a constant) shows that this cannot be. [If τ_2 were equal to $\lambda\tau_1$, where λ is real but not a rational fraction, then among the set $(m + \lambda n)\tau_1$ we would find values as close together as we wish and as small as we wish, which is impossible.] But if a_1 *is in a different direction* in the complex plane (*i.e.*, if a_1/a_2 is *not* a real number) then we have a situation specifically different from and more complicated than for the simply periodic functions.

For one thing, instead of a one-dimensional sequence of periods, there will be a two-dimensional lattice of parallelograms, with the function repeating, in each parallelogram, its behavior in every other parallelogram.

The smallest unit within which the function goes through all its behavior is called the *unit cell* for the function; each side of a unit cell is one of the fundamental periods for the function, τ_1 or τ_2. The behavior of f in the (m,n)th cell, at the point $z + m\tau_1 + n\tau_2$ $(m, n = 0, \pm 1, \pm 2, \ldots)$ is the same as the behavior of f in the central cell, at the point z. The points z and $z + m\tau_1 + n\tau_2$ are called *congruent points* for f.

We might ask if there are functions of the complex variable z which have more than two independent periods. The reasoning which proves that this is impossible is similar to that which proves that there cannot be two independent periods having a real ratio and depends basically on the fact that the complex plane is two-dimensional. Consequently, we can have singly and doubly periodic functions of z, but not triply or n-tuply periodic ones.

Fundamental Properties of Doubly Periodic Functions. Before we ever set down a formula for a doubly periodic function, we can use contour integration to prove certain of their properties. We make the contours go around the boundary of a unit cell, so that the periodicity ensures that the integral is zero. The doubly periodic function $f(z)$, with fundamental periods τ_1 and τ_2, may have poles and zeros. If it does, it must have the same number in each cell (we assume that we have drawn our unit cell so none of the poles or zeros is *on* the boundary). If it has branch points, they must come in pairs in each cell, so no branch lines need cut the cell boundaries.

Suppose that we concentrate on those doubly periodic functions $f(z)$ which have no branch points, only poles and zeros of integral order [where the function is of the form $(z - z_i)^{n_i}g(z)$ where n_i is an integer (positive or negative) the *order* of the zero, and g is analytic at $z = z_i$]. Such functions are called *elliptic functions.* According to page 408, the contour integral $(1/2\pi i)\oint f(z)\,dz$ for such an f is equal to the sum of the residues of f inside the closed contour. But if the contour is the boundary of a unit cell, such a contour integral must be zero, so we have

The sum of the residues of all the poles of an elliptic function, in one unit cell, is zero. (4.5.62)

If the elliptic function has only simple poles, it cannot therefore have *one* simple pole per cell; it must have at least two, so the residues can cancel out. The elliptic function could, of course, have one pole of order two per cell, if the residue of this double pole were zero. (Of course the function may have *no* poles per cell, in which case it is a constant. Why?)

We also see that, *if $f(z)$ is an elliptic function, $f + c$, $1/f$, f', f^n, f'/f are all elliptic functions.* If $f(z)$ is an elliptic function, then $[1/f(z)]$ is likewise an elliptic function, and a consideration of the integral $\oint(1/f)\,dz$ around the boundary of a unit cell shows that the sum of the residues of all the poles of $[1/f(z)]$ at all the zeros of $f(z)$ in one unit cell is zero. Therefore simple zeros of f must also come in pairs. Since, by definition, an elliptic function cannot have an essential singularity in the finite part of the complex plane, it cannot have an infinite number of poles in a unit cell, for this would result in an essential singularity. Consequently an elliptic function can have only a finite number of poles in each unit cell and, correspondingly, only a finite number of zeros.

We next look at the contour integral $\oint[f'(z)/f(z)]\,dz$, where $f' = df/dz$ and f has a finite number of simple poles and zeros inside the contour. Near the ith pole, which is of the n_ith order, the function $f \to (z - z_i)^{-n_i}g_i(z)$, where g_i is analytic (and not zero) at $z = z_i$. Therefore the function f'/f has a simple pole at z_i, with residue $-n_i$. Near the jth zero, which is of the m_jth order, the function $f \to (z - z_j)^{m_j}g_j(z)$, where g_j is analytic (and not zero) at $z = z_j$. Therefore the function f'/f has a simple pole at z_j, with residue $+m_j$. Therefore the contour integral

$$\frac{1}{2\pi i}\oint\frac{f'(z)}{f(z)}\,dz = \sum_j m_j - \sum_i n_i \tag{4.5.63}$$

In other words, this contour integral equals the sum of the orders of all the zeros, minus the sum of the orders of all the poles of f inside the contour. If the poles and zeros are all first order, then this integral equals the difference between the numbers of zeros and the numbers of poles.

Applying this to the contour around the unit cell for an elliptic function, we see that:

The sum of the orders of all the poles in one cell of an elliptic function equals the sum of the orders of all the zeros in the cell. If the poles and zeros are all simple, then the number of poles equals the number of zeros. (4.5.64)

By considering the integral $\oint[f'/(f - C)]\,dz$ around the boundary of a unit cell we see that

The number of roots of the equation $f(z) - C$ in a unit cell, where f is an elliptic function and C is any constant, is equal to the sum of the orders of all the poles of f in a unit cell. (4.5.65)

The sum of the order of the poles of f in a unit cell (which equals the number of poles in a unit cell if all the poles are first-order) is called the *order* of the elliptic function. By (4.5.62) we have seen that there cannot be an elliptic function of first order. An elliptic function of zero order is a constant, and the simplest elliptic function with nontrivial properties is one of second order.

Elliptic Functions of Second Order. There is an elliptic function with one second-order pole and two zeros per unit cell; it is called the *Weierstrass* elliptic function. Of greater use, however, are the *Jacobi elliptic functions,* which have two simple poles and two simple zeros in each cell. Therefore, by (4.5.65), *each of these functions takes on all possible values just twice in each cell.*

We can set up such functions by means of series or infinite products expressions or by means of inverse integrals, by analogy with the trigonometric functions. For the first such function we choose to have a sequence of zeros along the real axis (like the sine function), each a distance $\frac{1}{2}\tau_1$ apart, where τ_1 (real) is the first period. We shall take the second period, τ_2, to be pure imaginary and have a simple pole a distance $\frac{1}{2}|\tau_2|$ above each zero. Above these again, along the top of the first row of cells, is another row of zeros.

By analogy with the series of Eq. (4.5.59) we could set up a series which would account for all the zeros of this function. Suppose we assume that it starts out with unit slope at the origin, just as does the sine function; in fact let us assume that it behaves like a sine function at each of its zeros and that, when $\tau_2 \to \infty$ and the function becomes singly periodic, it reduces directly to $\sin z$. To emphasize this property we will call the function $\operatorname{sn} z$.

The poles of $(1/\operatorname{sn} z)$ (*i.e.*, the zeros of sn) along the real axis are obtained by putting in a term $(2\pi/\tau_1)\csc(2\pi z/\tau_1)$ in the series. This will have the desired behavior near each of the zeros of sn along the real axis and will reduce to $(1/\sin z)$ if τ_1 is made to become 2π when $\tau_2 \to \infty$. The next row of zeros of sn z is at the points $z = \tau_2 \pm \frac{1}{2}m\tau_1$, where τ_2 is pure imaginary.

Here we should have another row of poles of $(1/\text{sn } z)$. A term of the sort

$$\frac{\cos(2\pi\tau_2/\tau_1)}{\sin(2\pi z/\tau_1) - \sin(2\pi\tau_2/\tau_1)} = \frac{1}{\sin(2\pi\xi/\tau_1) - 2\tan(2\pi\tau_2/\tau_1)\sin^2(\pi\xi/\tau_1)}$$

where $\xi = z - \tau_2$, has poles which go to infinity at $\xi = \frac{1}{2}(n\tau_1)$ just as does the function $[1/\sin(2\pi z/\tau_1)]$ at $z = \frac{1}{2}n\tau_1$. Combining this with a term giving poles at the points $z = -\tau_2 \pm m\tau_1$, we have

$$\frac{2\cos(2\pi\tau_2/\tau_1)\sin(2\pi z/\tau_1)}{\sin^2(2\pi z/\tau_1) - \sin^2(2\pi\tau_2/\tau_1)} = \frac{4\sin(2\pi z/\tau_1)\cos(2\pi\tau_1/\tau_2)}{\cos(4\pi\tau_2/\tau_1) - \cos(4\pi z/\tau_2)}$$

Since there are rows of zeros of sn z at $iy = \pm n\tau_2$, we finally obtain a series

$$\frac{1}{\text{sn } z} = \frac{2\pi}{\tau_1}\left\{\csc\left(\frac{2\pi z}{\tau_1}\right) + 4\sum_{n=1}^{\infty}\frac{\sin(2\pi z/\tau_1)\cos(2\pi n\tau_1/\tau_2)}{\cos(4\pi n\tau_2/\tau_1) - \cos(4\pi z/\tau_2)}\right\} \tag{4.5.66}$$

which has the right distribution of zeros for sn z over the whole complex plane. It is not too difficult to show that the zeros of this series (the poles of sn) are midway between its poles (*i.e.*, at $z = \frac{1}{2}m\tau_1 + (n + \frac{1}{2})\tau_2$), but it is much more difficult to calculate the residues at these poles of sn.

These elliptic functions may also be approached from another direction. The series

$$F(z) = \sum_{n=-\infty}^{\infty} e^{\pi i n^2\gamma + 2inu} = 1 + 2\sum_{n=1}^{\infty} q^{n^2}\cos(2nu) \tag{4.5.67}$$

(where $\text{Im}\,\gamma > 0$ and where $q = e^{\pi i\gamma}$) is periodic in u with period π. It is convergent as long as the imaginary part of γ is positive, so that $|q| < 1$. Interestingly enough the series is also *pseudoperiodic* in u, with period $\pi\gamma$, for

$$F(u + \pi\gamma) = \sum_{n=-\infty}^{\infty} e^{(\pi i\gamma)(n^2+2n)+2inu} = e^{-\pi i\gamma - 2iu}\sum_{n=-\infty}^{\infty} e^{(\pi i\gamma)(n+1)^2+2i(n+1)u}$$
$$= q^{-1}e^{-2iu}F(u)$$

There are four such series, called the *theta functions:*

$$\vartheta_1(u,q) = 2\sum_{n=0}^{\infty}(-1)^n q^{(n+\frac{1}{2})^2}\sin[(2n+1)u]; \quad \vartheta_1(u+\pi,q) = -\vartheta_1(u,q);$$
$$\vartheta_1(u+\pi\gamma,q) = -N\vartheta_1(u,q) \tag{4.5.68}$$

$$\vartheta_2(u,q) = 2\sum_{n=0}^{\infty} q^{(n+\frac{1}{2})^2}\cos[(2n+1)u]; \quad \vartheta_2(u+\pi,q) = -\vartheta_2(u,q);$$

$$\vartheta_2(u+\pi\gamma,q) = N\vartheta_2(u,q)$$

$$\vartheta_3(u,q) = 1 + 2\sum_{n=1}^{\infty} q^{n^2}\cos(2nu);\quad \vartheta_3(u+\pi,q) = \vartheta_3(u,q);$$
$$\vartheta_3(u+\pi\gamma,q) = N\vartheta_3(u,q) \tag{4.5.68}$$
$$\vartheta_4(u,q) = 1 + 2\sum_{n=1}^{\infty}(-1)^n q^{n^2}\cos(2nu);\quad \vartheta_4(u+\pi,q) = \vartheta_4(u,q);$$
$$\vartheta_4(u+\pi\gamma,q) = -N\vartheta_4(u,q)$$
$$\vartheta_2(u,q) = \vartheta_1(u+\tfrac{1}{2}\pi,q);\quad \vartheta_3(u,q) = \vartheta_4(u+\tfrac{1}{2}\pi,q);\quad \text{etc.}$$

where $q = e^{\pi i\gamma}$ and where $N = q^{-1}e^{-2iu}$. These are, of course, not true elliptic functions because of the presence of the factor N.

By suitable manipulation of these series we can show that the quantities

$$\frac{\vartheta_1^2(u) + a\vartheta_4^2(u)}{\vartheta_2^2(u)} \quad\text{and}\quad \frac{\vartheta_1^2(u) + b\vartheta_4^2(u)}{\vartheta_3^2(u)}$$

are true elliptic functions and that, by suitable choice of a and b, we can arrange it so that they have at most only one simple pole in the unit cell, of side π and $\pi\gamma$. But, from Eq. (4.5.62), this means that these ratios, for these particular values of a and b, will be constants. Therefore it is possible to express ϑ_2^2 or ϑ_3^2 in terms of ϑ_1^2 and ϑ_4^2. The constants can be determined by setting $u = 0$ and $\frac{1}{2}\pi$; and we finally obtain

$$\begin{aligned}
\vartheta_2^2(u)\vartheta_4^2(0) &= \vartheta_4^2(u)\vartheta_2^2(0) - \vartheta_1^2(u)\vartheta_3^2(0)\\
\vartheta_3^2(u)\vartheta_4^2(0) &= \vartheta_4^2(u)\vartheta_3^2(0) - \vartheta_1^2(u)\vartheta_2^2(0)\\
\vartheta_1^2(u)\vartheta_4^2(0) &= \vartheta_3^2(u)\vartheta_2^2(0) - \vartheta_2^2(u)\vartheta_3^2(0)\\
\vartheta_4^2(u)\vartheta_4^2(0) &= \vartheta_3^2(u)\vartheta_3^2(0) - \vartheta_2^2(u)\vartheta_2^2(0)
\end{aligned} \tag{4.5.69}$$

where the second argument, q, has been omitted for brevity.

From the theta functions, as we have seen, we can build true elliptic functions, by taking proper ratios so that the multiplicative factor drops out. For instance, the function

$$\operatorname{sn} z = \frac{\vartheta_3(0,q)}{\vartheta_2(0,q)}\frac{\vartheta_1(u,q)}{\vartheta_4(u,q)};\quad u = \frac{z}{[\vartheta_3(0,q)]^2} \tag{4.5.70}$$

has the same distribution of zeros and the same behavior at each zero that the function, defined in Eq. (4.5.66) has, provided $\pi[\vartheta_3(0,q)]^2$ is equal to τ_1 and $\pi\gamma[\vartheta_3(0,q)]^2$ is equal to τ_2, the second period. Moreover, treatises on elliptic functions show that from the series of Eq. (4.5.68), we can calculate, with a great deal of difficulty, a differential equation for the theta functions and finally for sn z. We find, for instance, that

$$\frac{d}{du}\left[\frac{\vartheta_1(u,q)}{\vartheta_4(u,q)}\right] = [\vartheta_4(0,q)]^2\frac{\vartheta_2(u,q)}{\vartheta_4(u,q)}\frac{\vartheta_3(u,q)}{\vartheta_4(u,q)}$$

and that if $\eta = \vartheta_1(u,q)/\vartheta_4(u,q)$, we have, after further manipulation,

$$(d\eta/du)^2 = [\vartheta_2^2(0,q) - \eta^2\vartheta_3^2(0,q)][\vartheta_3^2(0,q) - \eta^2\vartheta_2^2(0,q)]$$

Finally, if we set $y = [\vartheta_3(0,q)/\vartheta_2(0,q)]\eta$ and $z = u[\vartheta_3(0,q)]^2$, we have for y

$$(dy/dz)^2 = (1 - y^2)(1 - k^2y^2) \tag{4.5.71}$$

where $\sqrt{k} = \vartheta_2(0,q)/\vartheta_3(0,q)$. All the properties of the elliptic function $\operatorname{sn} z = y$ may then be found in terms of the series (4.5.68), which converge rapidly. However, to discuss some of the properties it is useful to consider Eq. (4.5.71) as the basic definition.

Integral Representations for Elliptic Functions. This brings us to still a third way of defining elliptic functions of second order, which is the most useful of all (except, perhaps, for numerical computation of tables). We see that a formal solution of Eq. (4.5.71) is

$$z = \int_0^y \frac{du}{\sqrt{(1 - u^2)(1 - k^2u^2)}} = \operatorname{sn}^{-1} y \tag{4.5.72}$$

which defines the function $\operatorname{sn}^{-1} y$, inverse to the elliptic function $\operatorname{sn} z$. Referring to the discussion of $\sin^{-1} y$, on page 426, we see that the present function $y = \operatorname{sn} z$ is periodic in z with period $\tau_1 = 4K$, where

$$\begin{aligned} K &= \int_0^1 \frac{du}{\sqrt{(1 - u^2)(1 - k^2u^2)}} = \int_0^{\frac{1}{2}\pi} \frac{d\varphi}{\sqrt{1 - k^2 \sin^2 \varphi}} \\ &= \tfrac{1}{2} \int_0^1 \frac{dv}{\sqrt{v(1 - v)(1 - vk^2)}} = \frac{\pi}{2} F(\tfrac{1}{2},\tfrac{1}{2}|1|k^2) \end{aligned} \tag{4.5.73}$$

according to Eq. (5.3.16). As a matter of fact we can write $\operatorname{sn}(z + 2K) = -\operatorname{sn} z$. Equations (4.5.71) and (4.5.72) show that, for real values of z, $\operatorname{sn} z$ is never larger than unity.

Examination of the series expansion for the integral near $y = 0$ shows that $\operatorname{sn} z \to z - \frac{1}{6}(1 + k^2)z^3 + \cdots$, so that sn is an odd function of z and the residue of the simple pole of $(1/\operatorname{sn} z)$, at $z = 0$, is 1. Suppose we next investigate the behavior of $\operatorname{sn} z$ near the point $z = K$, where it has the value unity. Going along the real axis away from $z = K$ reduces the value of sn, so going at right angles to the real axis will increase the value of sn (why?). Going upward from the real axis gives us

$$z = \operatorname{sn}^{-1} y = K + iv;\quad v = \int_1^y \frac{du}{\sqrt{(u^2 - 1)(1 - k^2u^2)}};\quad y = \operatorname{sn}(K + iv)$$

Analysis of the equation for v shows that y is periodic in v with period $2K'$, where

$$K' = \int_1^{1/k} \frac{du}{\sqrt{(u^2 - 1)(1 - k^2u^2)}} = \frac{\pi}{2} F(\tfrac{1}{2},\tfrac{1}{2}|1|k'^2) \tag{4.5.74}$$

where $k'^2 = 1 - k^2$. Consequently $\operatorname{sn} z$ is periodic in z with period $\tau_2 = 2iK'$, the value of $\operatorname{sn} K$ being unity and the value of $\operatorname{sn}(K + iK')$ being $1/k$, and so on.

Finally, if we make y imaginary, z is imaginary, and we can have y go to infinity for z finite. This means that there is a pole of $\operatorname{sn} z = y$

on the imaginary axis, at the point ia, where

$$a = \int_0^\infty \frac{dw}{\sqrt{(1 + w^2)(1 + k^2w^2)}} = \int_1^{1/k} \frac{du}{\sqrt{(u^2 - 1)(1 - k^2u^2)}} = K';$$

$$u = \left[\frac{1 + w^2}{1 + k^2w^2}\right]^{\frac{1}{2}}$$

As a result of this type of exploration we find that the function sn z is an elliptic function with one period $\tau_1 = 4K$ which is real and one period $\tau_2 = 2iK'$ which is pure imaginary. The function has simple zeros at $z = 0$ and $z = 2K$ and at corresponding places in the other unit cells. It has simple poles at $z = iK'$ and $2K + iK'$ (and corresponding points), with the residue at the pole at $z = iK'$ being $1/k$ and consequently [from (4.5.62)] the residue at $z = 2K + iK'$ is $-(1/k)$. The parameter k is called the *modulus* of the function; whenever its value is to be emphasized, the elliptic function is written sn (z,k). Its value in terms of the quantity q is given in connection with Eq. (4.5.71): usually, however, k is treated as the basic parameter, in which case K and K' are found from Eqs. (4.5.73) and (4.5.74) (K' is the same function of $k' = \sqrt{1 - k^2}$ as K is of k) and then $q = e^{-\pi(K'/K)}$. We can then work back and show that the sn (z,k) defined by Eq. (4.5.72) is the same function as that defined by Eqs. (4.5.70) and (4.5.66).

Other Jacobi elliptic functions are defined in similar ways:

$$\begin{aligned}
\text{sn}\,(z,k) &= \frac{\vartheta_3(0,q)}{\vartheta_2(0,q)}\frac{\vartheta_1(u,q)}{\vartheta_4(u,q)}; \\
&\qquad z = \int_0^y \frac{dt}{\sqrt{(1 - t^2)(1 - k^2t^2)}} = \text{sn}^{-1}\,(y,k) \\
\text{cn}\,(z,k) &= \frac{\vartheta_4(0,q)}{\vartheta_2(0,q)}\frac{\vartheta_2(u,q)}{\vartheta_4(u,q)}; \\
&\qquad z = \int_y^1 \frac{dt}{\sqrt{(1 - t^2)(1 - k^2 + k^2t^2)}} = \text{cn}^{-1}\,(y,k) \\
\text{dn}\,(z,k) &= \frac{\vartheta_4(0,q)}{\vartheta_3(0,q)}\frac{\vartheta_3(u,q)}{\vartheta_4(u,q)}; \\
&\qquad z = \int_y^1 \frac{dt}{\sqrt{(1 - t^2)(t^2 + k^2 - 1)}} = \text{dn}^{-1}\,(y,k)
\end{aligned} \tag{4.5.75}$$

where $\quad u = z[\vartheta_3(0,q)]^{-2} = \pi z/2K; \quad q = e^{-\pi(K'/K)}$

and the constants K, K' are given by Eqs. (4.5.73) and (4.5.74).

Utilizing Eqs. (4.5.69) and our previous knowledge of the function sn, we can show that

$$\begin{aligned}
\text{sn}^2\,(z,k) + \text{cn}^2\,(z,k) = 1; \quad \text{cn}\,(0,k) = \text{dn}\,(0,k) = 1 \\
k^2\,\text{sn}^2\,(z,k) + \text{dn}^2\,(z,k) = 1; \quad (d/dz)\,\text{sn}\,(z,k) = \text{cn}\,(z,k)\,\text{dn}\,(z,k)
\end{aligned} \tag{4.5.76}$$

Many other properties of these functions may now be computed. Some of them are listed on page 486, and others are dealt with in the problems. We shall find use for the functions in a number of cases later in the book.

4.6 *Asymptotic Series; Method of Steepest Descent*

To study the behavior of the various functions we shall encounter, for large values of $|z|$, it will often be convenient to expand them in inverse powers of z:

$$f(z) = \varphi(z)\left[A_0 + \frac{A_1}{z} + \frac{A_2}{z^2} + \cdots\right] \tag{4.6.1}$$

where $\varphi(z)$ is a function whose behavior for large values of z is known. The expansion for $\Gamma(z)$ given in the preceding section [Eq. (4.5.50)] is of this type. If $f(z)/\varphi(z)$ has an essential singularity at $|z| \to \infty$, the series in Eq. (4.6.1) will diverge. Nevertheless, the series may be useful, not only for a qualitative understanding of the function, but even in its computation for large values of z.

The circumstance needed to make this possible is that the difference between $f(z)/\varphi(z)$ and the first $(n + 1)$ terms of the series be of the order of $1/z^{n+1}$, so that for sufficiently large z, this difference may be made quite small. More precisely, the series is said to represent $f(z)/\varphi(z)$ *asymptotically*, *i.e.*,

$$f(z) \simeq \varphi(z) \sum_{p=0}^{\infty} [A_p/z^p] \tag{4.6.2}$$

if

$$\lim_{|z|\to\infty} \left\{z^n\left[\frac{f(z)}{\varphi(z)} - \sum_{p=0}^{n} \frac{A_p}{z^p}\right]\right\} \to 0 \tag{4.6.3}$$

This equation states that, for a given n, the first n terms of the series may be made as close as may be desired to the ratio $f(z)/\varphi(z)$ by making z large enough. For each value of z and n there will be an error of the order of $1/z^{n+1}$. Since the series actually diverges, there will be an optimal number of terms of the series to be used to represent $[f(z)/\varphi(z)]$ for a given z. Associated with this there will be an unavoidable error. As z increases, the optimal number of terms increases and the error decreases.

An Example. Let us make these ideas more concrete by considering a simple example, the exponential integral

$$-Ei(-x) = \int_x^\infty [e^{-t}/t]\,dt$$

the asymptotic series for the exponential integral is obtained by a series of partial integrations. For example,

$$-Ei(-x) = \frac{e^{-x}}{x} - \int_x^\infty \left[\frac{e^{-t}}{t^2}\right] dt$$

Continuing this procedure yields

$$-Ei(-x) = \frac{e^{-x}}{x}\left[1 - \frac{1}{x} + \frac{2!}{x^2} - \frac{3!}{x^3} + \cdots + \frac{(-1)^n n!}{x^n}\right] + (-1)^{n+1}(n+1)! \int_0^\infty \frac{e^{-t}}{t^{n+2}}\,dt$$

The infinite series obtained by permitting $n \to \infty$ diverges, for the Cauchy convergence test yields

$$\lim_{n\to\infty} \left|\frac{u_{n+1}}{u_n}\right| = \lim_{n\to\infty} \left[\frac{n}{x}\right] \to \infty$$

Note that this indicates that two successive terms become equal in magnitude for $n = x$, indicating that the optimum number of terms for a given x is roughly the integer nearest x. To prove that the series is asymptotic, we must show that

$$x^{n+1}e^x(n+1)!(-1)^{n+1} \int_x^\infty \frac{e^{-t}}{t^{n+2}}\,dt \xrightarrow[x\to\infty]{} 0$$

This immediately follows, since

$$\int_x^\infty \left[\frac{e^{-t}}{t^{n+2}}\right] dt < \frac{1}{x^{n+2}} \int_x^\infty e^{-t}\,dt = \frac{e^{-x}}{x^{n+2}}$$

The error involved in using the first n terms is less than $[(n+1)!e^{-x}/x^{n+2}]$, which is exactly the next term in the series. We see that, as n increases, this estimate of the error first decreases and then increases without limit. Table 4.6.1 demonstrates how this works out for $-4e^4Ei(-4) = 0.82533$.

Table 4.6.1

n	Value of nth term	Sum including nth term	Upper bound to error
0	1.00000	1.00000	0.25000
1	−0.25000	0.75000	0.12500
2	0.12500	0.87500	0.09375
3	−0.09375	0.78125	0.09375
4	0.09375	0.87500	0.11719
5	−0.11719	0.75781	0.17579
6	0.17579	0.93360	0.31013
7	−0.31013	0.62347	0.62026
8	0.62026	1.24373	1.39559
9	−1.39559	−0.15186	

The exact value is approached to within 5 per cent by the values 0.87500 and 0.78125.

Averaging Successive Terms. The fact that these two values are on either side of the correct one suggests that the exact value might be more closely approached if the sequence defined by the average of two successive values is employed. In the present case, this average is 0.828125, within $\frac{1}{2}$ per cent of the correct value. We formulate this notion as follows:

$$\text{Let} \qquad S_n = \sum_{m=0}^{n} u_m; \quad T_{n+1} = \frac{S_n + S_{n+1}}{2} = S_n + \tfrac{1}{2}u_{n+1}$$

$$\text{then} \qquad T_{n+1} = \sum_{m=0}^{n+1} U_{m-1}; \quad \text{where } U_n = \frac{u_n + u_{n+1}}{2}; \quad U_{-1} = \tfrac{1}{2}u_0 \qquad (4.6.4)$$

The sequence U_n does not form an asymptotic series in the narrow sense, as defined by Eq. (4.6.1). We may, however, broaden our definition to include U_n as follows:

$$[f(z)/\varphi(z)] \simeq \sum_{p=0}^{\infty} U_p(z)$$

$$\text{if} \qquad \lim_{|z|\to\infty} \left\{ z^n \left[\frac{f(z)}{\varphi(z)} - \sum_{p=0}^{n} U_p(z) \right] \right\} \to 0 \qquad (4.6.5)$$

It is clear that the statements following (4.6.3) which apply to the asymptotic series (4.6.1) apply to the more general case defined by (4.6.5). For U_p as defined by (4.6.4)

$$\sum_{p=-1}^{n} U_p(z) = \sum_{p=0}^{n+1} u_p(z) - \tfrac{1}{2}u_{n+1}(z)$$

If $u_p(z)$ forms an asymptotic series, *i.e.*,

$$\lim_{|z|\to\infty} \left\{ z^n \left[(f/\varphi) - \sum_{p=0}^{n} u_p(z) \right] \right\} \to 0$$

then $U_p(z)$ forms an asymptotic series. The method of averaging given by (4.6.4) is particularly useful when the series is an alternating one. We illustrate this by applying such a rearrangement to the case tabulated in Table 4.6.1 (see Table 4.6.2).

There are a number of questions we must answer with regard to asymptotic series before we will feel free to manipulate them. First let us note that an asymptotic series is not *unique*. For example, the two functions

$$[f(z)/\varphi(z)] \quad \text{and} \quad \psi(z) = [f(z)/\varphi(z)] + e^{-z}$$

have the same asymptotic expansion when $\operatorname{Re} z > 0$. Moreover it is clear that the asymptotic expansion for $\psi(z)$ will change markedly in moving from $\operatorname{Re} z > 0$ to $\operatorname{Re} z < 0$. Examination of the asymptotic form of $\psi(z)$ as the phase of $z = |z|e^{i\vartheta}$ changes would thus show a discontinuity at $\vartheta = \pi/2,\ 3\pi/2$. These discontinuities are only apparent and essentially are a result of the fact that asymptotic series are not unique. Therefore in our frequent dealings with asymptotic series we shall need to keep them under close scrutiny always to be aware of the range in the phase of z for which they are valid. The apparent discontinuity in the series will manifest itself often, indeed in the chapter on differential equations, it will be referred to as the *Stokes' phenomenon* (see Sec. 5.3).

Table 4.6.2. Summing Asymptotic Series by Averaging Terms [see Eqs. (4.6.4) and Table 4.6.1]

n	U_p	T_p
−1	0.50000	0.50000
0	0.37500	0.87500
1	−0.06250	0.81250
2	0.015625	0.82813
3	0.00000	0.82813
4	−0.01720	0.810913
5	0.02930	0.84023
6	−0.06717	0.77306
7	0.15508	0.92813
8	0.38767	0.54046

The following general properties of these series should be noted: *Asymptotic series may be added term by term; asymptotic series may also be multiplied.* If $\chi(z) \simeq \sum_{p=0}^{\infty} (A_p/z^p)$ and $\psi(z) \simeq \sum_{p=0}^{\infty} \left(\frac{B_p}{z^p}\right)$;

then
$$\chi(z)\psi(z) \simeq \sum_{n=0}^{\infty} \left(\frac{C_n}{z^n}\right); \quad \text{where } C_n = \sum_{p=0}^{n} A_p B_{n-p}$$

Asymptotic series may be integrated:

$$\int \chi(z)\, dz = A_0 z + A_1 \ln z - \sum_{p=1}^{\infty} \left(\frac{A_{p+1}}{p z^p}\right)$$

On the other hand, asymptotic series may be differentiated to obtain an asymptotic expansion for the derivative function only if it is known through some other means that the derivative function has an asymptotic expansion.

Integral Representations and Asymptotic Series. We shall often have occasion to compute the asymptotic behavior of an integral representation of some function—by a procedure known alternately as the *method of steepest descent* and as the *saddle-point method.* This technique will now be discussed and applied to the gamma function.

First we should emphasize that the technique is successful only when the function under examination can be represented by an integral of a rather particular form:

$$J(z) = \int_C e^{zf(t)}\,dt \tag{4.6.6}$$

where the contour C is such that the integrand goes to zero at the ends of the contour. This form is closely related to the Laplace transform (see Sec. 5.3). We may expect many of the integrals of interest in this book to be of form (4.6.6), inasmuch as solutions of the scalar Helmholtz equation $(\nabla^2 + k^2)\psi = 0$ may be represented as a general superposition of plane waves, $e^{i\mathbf{k}\cdot\mathbf{r}}$; that is, $\psi = \int e^{i\mathbf{k}\cdot\mathbf{r}} f(\mathbf{k})\,d\Omega_n$ where $d\Omega_n$ is the differential solid angle in the direction of vector $\mathbf{k}$. In addition, it is often possible to transform an integral, not of this form, into one by some simple ruse. For example, the gamma function $\Gamma(z + 1)$ which for $\mathrm{Re}\,z > -1$ is represented by $\int_0^\infty e^{-\tau}\tau^z\,d\tau$ becomes, upon making the substitution $\tau = tz$,

$$\Gamma(z + 1) = z^{z+1} \int_0^\infty e^{-tz} t^z\,dt = z^{z+1} \int_0^\infty e^{z(\ln t - t)}\,dt \tag{4.6.7}$$

The function $\Gamma(z + 1)/z^{z+1}$ is therefore of form (4.6.6).

Let us now examine the behavior of $J(z)$ as $|z| \to \infty$, for a given phase ϕ of z. Large values of $|z|$ will usually result in some very rapid fluctuations in the value of the integrand. If z is complex, or if $f(t)$ is complex along portions of the contour C, the imaginary part of $[zf(t)]$ will generally increase as $|z|$ increases. Consequently, the factor $\exp\{i\,\mathrm{Im}\,[zf(t)]\}$ will oscillate rapidly, the oscillations increasing in frequency as $|z| \to \infty$. The existence of such oscillations makes it difficult to evaluate the integral without some further transformations because of the resultant cancellations of the integrand; in many cases large positive values are almost completely canceled by large negative values elsewhere on the contour. To obtain the relevant residue would require a forbiddingly accurate evaluation of the integral at every point. Under these conditions it is manifestly desirable to deform the contour so as to minimize such effects. In the following it will be assumed that the requisite deformations are possible, that, if singularities intervene, it is possible in some manner to evaluate their effect.

Choosing the Contour. In general the contour taken by the integral must run through regions where the real part of $zf(t)$ is positive and other regions where it is negative. The former regions are more important, since here the integrand is larger, and in these regions, where $\mathrm{Re}\,[zf(t)]$ is largest, it is most important to reduce oscillations. What we do then is to search for a contour along which the *imaginary part* of $[zf(t)]$ *is constant* in the region (or regions) where its real part is largest. Thus, in the region which contributes most to the final value, the integral may be written

$$J(z) = \int_C e^{zf(t)}\,dt = e^{i\mathrm{Im}[zf(t)]} \int_C e^{\mathrm{Re}[zf(t)]}\,dt \tag{4.6.8}$$

Then the path, in the regions where Re $[zf(t)]$ is least, may be chosen so that Im $[zf(t)]$ varies if this turns out to be necessary to complete the contour. In this way we have ensured that the oscillations of the integrand make the least trouble.

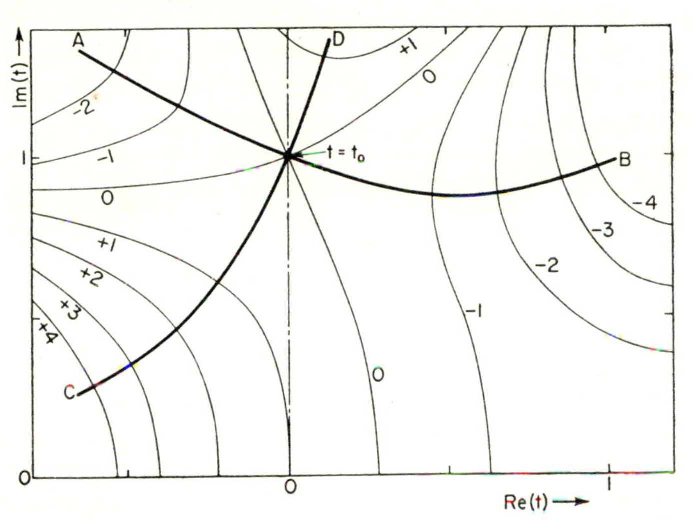

Fig. 4.29 Contour plot of real part of function $f(t)$ near a saddle point $t = t_0$.

The point at which Re $[zf(t)]$ is largest is the maximum point for $f(t)$, at $t = t_0$ where

$$f'(t_0) = 0 \tag{4.6.9}$$

Near this point it is particularly important that oscillations be suppressed, so we require that near t_0 the contour be chosen so that

$$\mathrm{Im}\,[zf(t)] = \mathrm{Im}\,[zf(t_0)] \tag{4.6.10}$$

A little geometry will make the meaning of these conditions clear. Set up a three-dimensional cartesian coordinate system for which the x coordinate is Re t, the y coordinate Im t, and an arbitrary vertical coordinate. To see the behavior of the integrand in (4.6.8) we plot the surface; vertical coordinate = Re $[f(t)]$; at every point t in the x, y plane the surface is at a height above the horizontal given by Re $[f(t)]$ at that point. The first question to answer concerns the behavior of the surface at $t = t_0$. Because of Eq. (4.6.9) the surface must be flat in the neighborhood of t_0. However, it cannot have a maximum or a minimum there; it cannot be a mountain top or a valley because, by the theorem of page 369, the real part of a function of a complex variable cannot have a maximum or minimum. The point t_0 must therefore be a *minimax* or *a saddle point* or, to use a topographical analogy, a mountain pass. This is illustrated in Fig. 4.29 by a drawing of a section of the surface near $t = t_0$. For curve AB, the point $t = t_0$ is a maximum, whereas for curve CD on the same surface, $t = t_0$ is a minimum. Curve CD runs from one mountain top to another, whereas curve AB runs from one valley to another between the mountain tops (Fig. 4.30).

What is the significance of the particular path Im $[zf(t)]$ = constant? We recall (page 371) that along such a path Re $[zf(t)]$ changes as rapidly as possible. Actually there are two such paths, as is evident from Fig.

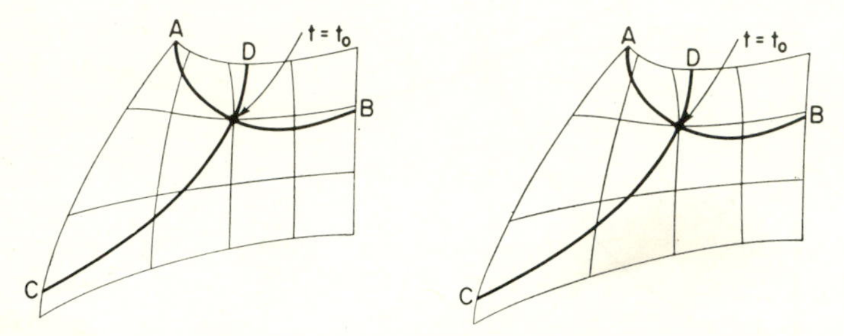

Fig. 4.30 Choice of path through saddle point.

4.29. On one of them, CD, Re $[zf(t)]$ *increases* as rapidly as possible; on the other, AB, Re $[zf(t)]$ *decreases* as rapidly as possible. Path AB is chosen, since this contour can, and CD cannot, correspond to the original contour of the convergent integral under discussion. Thus along the chosen path $e^{zf(t)}$ goes to its end point values by the steepest route possible and passes through the mountain pass t_0. Most of the value of the integral will come from the neighborhood of this point, since it is the maximum value of the integrand along this contour. This becomes increasingly true as $z \to \infty$, for the maximum becomes sharper and the descent to end-point value more rapid.

First Term in the Expansion. To illustrate the above discussion, let us use this information to derive a general formula, giving the first term in the asymptotic expansion of $J(z)$. In the neighborhood of t_0

$$f(t) = f(t_0) + [(t - t_0)^2/2]f''(t_0) + \cdots \tag{4.6.11}$$

[Note that, if it should happen that $f''(t_0) = 0$, then the analysis would require revision.] The path of steepest descent must be such that the integrand is a decreasing exponential. Therefore let

$$\tau = \sqrt{e^{i(\pi+\phi)}f''(t_0)}\,(t - t_0); \quad \text{where } z = |z|e^{i\phi} \tag{4.6.12}$$

The integral becomes

$$J(z) \simeq \frac{e^{zf(t_0)}}{\sqrt{e^{i(\pi+\phi)}f''(t_0)}} \int_C e^{-\frac{1}{2}|z|\tau^2}\, d\tau$$

We see that, as $|z|$ increases, the exponential becomes steeper and less and less of the contour becomes important. For sufficiently large $|z|$ the integrand will become effectively zero outside the range in which approximation (4.6.11) is valid. We may therefore replace the contour integral by a real integral covering the range from $-\infty$ to $+\infty$. The direction of integration is determined by the original contour and by Eq. (4.6.12). We assume that the direction of travel along the contour is such that the integration on τ is from $-\infty$ to $+\infty$. Then as $|z| \to \infty$,

$$J(z) \simeq e^{zf(t_0)} \sqrt{2\pi/ze^{\pi i}f''(t_0)} \tag{4.6.13}$$

In the case of the gamma function [see Eq. (4.6.7)] $f(t) = \ln t - t$. Then $f'(t) = (1/t) - 1$, so that $t_0 = 1$. We find that $f(t_0) = -1$ and $f''(t_0) = -1$. The variable t runs from 0 to $+\infty$. Transformation (4.6.12) changes to $\tau = (t - 1)e^{i\phi/2}$ so that the upper limit for τ is $+\infty$. Equation (4.6.13) applies, and

$$\Gamma(z + 1) \xrightarrow[z \to \infty]{} \sqrt{2\pi}\, z^{z+\frac{1}{2}}e^{-z}$$

in agreement with the first term of the asymptotic expansion for $\Gamma(z)$ given in Eq. (4.5.50). See also the discussion of Eq. (5.3.77), where another example is discussed.

The Rest of the Series. Equation (4.6.13) is the leading term in an asymptotic series. We shall now proceed to generalize the above discussion so as to obtain the next terms in the series. To simplify the discussion let the phase of z be included in $f(t)$ so that in the following it will be possible to regard z as real.

It is necessary to go back in our discussion to Eq. (4.6.11) and replace it by an exact substitution:

$$f(t) = f(t_0) - w^2 \tag{4.6.14}$$

Note that *w is real* in virtue of the choice $\text{Im}\,[f(t)] = \text{Im}\,[f(t_0)]$. Introducing (4.6.14) into $J(z)$ yields

$$J(z) = e^{zf(t_0)} \int_C e^{-zw^2}\,dt \quad \text{or} \quad J(z) = e^{zf(t_0)} \int_C e^{-zw^2} \left(\frac{dt}{dw}\right) dw$$

We once more assume that the original sense of the integration was such that the integration direction for w is from $-\infty$ to $+\infty$, so that

$$J(z) = e^{zf(t_0)} \int_{-\infty}^{\infty} e^{-zw^2} \left(\frac{dt}{dw}\right) dw \tag{4.6.15}$$

To complete the story it is necessary to compute dt/dw, that is, to invert Eq. (4.6.14) and obtain dt/dw as a function of w. This should be done in the form of a power series, namely,

$$\frac{dt}{dw} = \sum_{n=0}^{\infty} a_n w^n \tag{4.6.16}$$

where, according to Eq. (4.6.14), only even powers of w will enter. Substituting form (4.6.16) into (4.6.15) will then yield as the asymptotic series

$$J(z) \xrightarrow[z \to \infty]{} e^{zf(t_0)} \sqrt{\frac{\pi a_0^2}{z}} \sum_{n=0}^{\infty} \left(\frac{a_{2n}}{a_0}\right) \frac{\Gamma(n+\frac{1}{2})}{\Gamma(\frac{1}{2})} \left(\frac{1}{z}\right)^n \tag{4.6.17}$$

The coefficients a_n may be determined by the procedure described in the preceding section (see page 411). There it was shown that, if

$$w = \sum_{n=1}^{\infty} \frac{a_{n-1}}{n} (t - t_0)^n = g(t)$$

then

$$a_n = \frac{1}{n!} \frac{d^n}{dt^n} \left[\frac{t - t_0}{g(t)}\right]^{n+1} \tag{4.6.18}$$

In the present case $g(t) = \sqrt{f(t_0) - f(t)}$. As an aid, we expand in a series:

$$\left[\frac{f(t_0) - f(t)}{(t - t_0)^2}\right] = \sum_p A_p (t - t_0)^p$$

Then Eq. (4.6.18) becomes

$$a_n = \frac{1}{n!} \left\{\frac{d^n}{dx^n} \left[\sum_p A_p x^p\right]^{-\frac{1}{2}n-\frac{1}{2}}\right\}_{x=0}$$

or a_n *is the coefficient of the nth power of x in a power-series expansion of* $(\Sigma A_p x^p)^{-\frac{1}{2}n-\frac{1}{2}}$.

We tabulate below the first three relevant coefficients in terms of A_p:

$$\begin{aligned}
a_0 &= 1/\sqrt{A_0} \\
a_2/a_0 &= \tfrac{15}{8} A_1^2 A_0^{-3} - \tfrac{3}{2} A_2 A_0^{-2} \\
a_4/a_0 &= \frac{5 \cdot 7 \cdot 9 \cdot 11}{2^7 \cdot 3} A_1^4 A_0^{-6} - \frac{5 \cdot 7 \cdot 9}{2^4} A_0^{-5} A_1^2 A_2 \\
&\quad + \frac{5 \cdot 7}{2^3} A_0^{-4}(A_2^2 + 2A_1 A_3) - \tfrac{5}{2} A_0^{-3} A_4
\end{aligned} \tag{4.6.19}$$

To illustrate the method let us again consider the gamma function. We have $f(t) = \ln(t) - t$, $t_0 = 1$. The function $[f(t_0) - f(t)]/(t - t_0)^2$ is

$$\left[\frac{(t-1) - \ln t}{(t-1)^2}\right] = \frac{1}{2} - \frac{(t-1)}{3} + \frac{(t-1)^2}{4} - \cdots$$

so that $A_n = [(-1)^n/(n+2)]$. The values of a_n, by substitution into Eq. (4.6.19), are equal to

$$a_0 = 1/\sqrt{2}; \quad \frac{a_2}{a_0} = \tfrac{1}{6}; \quad \frac{a_4}{a_0} = \tfrac{1}{216}$$

Therefore, by Eq. (4.6.17), the asymptotic expansion for $\Gamma(z + 1)$ becomes

$$\Gamma(z + 1) = \sqrt{2\pi}\, z^{z+\frac{1}{2}} e^{-z} \left[1 + \frac{1}{12z} + \frac{1}{288z^2} + \cdots \right]$$

We shall use this saddle-point method several times in our investigations; in particular we shall study in detail the asymptotic behavior of Bessel functions and related functions by means of this technique in Sec. 5.3.

4.7 *Conformal Mapping*

Most of the interesting and important geometrical applications of the theory of analytic functions of a complex variable to problems in physics may be grouped under the heading of conformal mapping. We employ the electrostatic field as an example. A common problem is that of a point charge q (line charge in three dimensions) enclosed by a metallic conductor forming a closed surface C and kept at zero potential (see Fig. 4.31). The electric field E will diverge from the source and strike the bounding surface orthogonally. We have sketched the equipotentials and lines of force in the figure. Of course, lines of force and the equipotentials form an orthogonal grid. We may therefore expect that some function of a complex variable, $w(z) = u + iv$, may be found for which the lines determined by the real part of w, $u(x,y) =$ constant, yield the lines of force and the contour lines for the imaginary part of w, $v(x,y) =$ constant, yield the equipotentials. The correct function must be singular, at the point $z = z_q$, where the charge is located. In fact, at z_q the function should have a logarithmic singularity, and as we shall see in Chap. 10, it should have the form $w = -(2iq) \ln (z - z_q)$ plus some function analytic at z_q. The variables u and v may be used to define a conformal transformation in which $u =$ constant lines are parallel to the imaginary axis, while the $v =$ constant lines are parallel to the real axis. Since $v = 0$ on C, the bounding surface C transforms to the real axis in the w plane. The transform of Fig. 4.31 is illustrated in Fig. 4.32. The charge q is located at infinity ($w = i\infty$) in the w plane. The constant u lines constitute just the parallel lines of force generated by the charge at infinity.

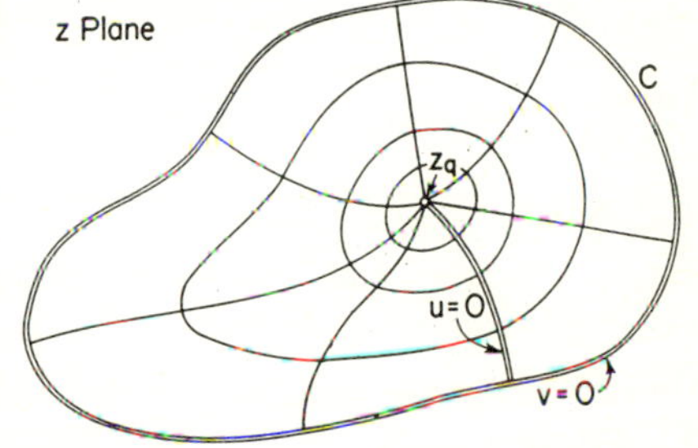

Fig. 4.31 Equipotentials and lines of force for point source inside grounded enclosure.

General Properties of the Transformation. The distortion of space resulting from the transformation can be given a more dramatic description, a description which is often useful in obtaining a "feeling" for the transformation. Essentially the transformation cuts C at some point and straightens C out, stretching it so that the point at which the cut is made goes to infinity. This procedure straightens out the potential lines; to straighten out the lines of force we must also move the charge q upward to infinity.

One result of paramount importance should be noted. *The function $w(z)$ transforms the interior of the region bounded by C into the upper half plane.* To perform this transformation it is only necessary to obtain

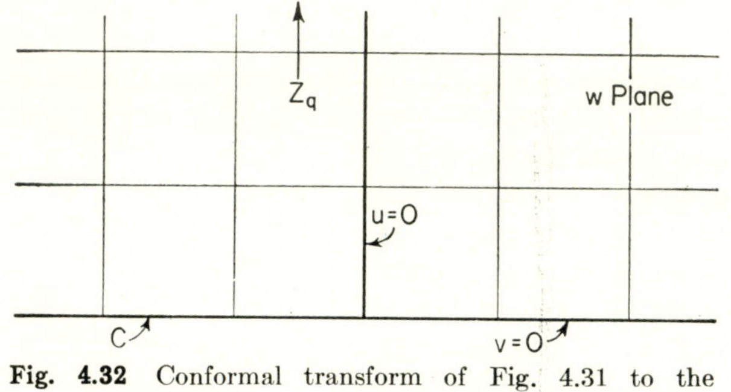

Fig. 4.32 Conformal transform of Fig. 4.31 to the $w = u + iv$ plane.

the point source function whose imaginary part satisfies the condition $V = 0$ on C and behaves as $-(q/2\pi) \ln |z - z_q|$ near the charge at z_q. It will be clear from the physics of the situation that such a point source function exists, with the consequence that *a transformation taking the interior of C into the upper half plane exists.* It is also clear that there are many such transformations, since the position of the charge can be anywhere in the interior of C.

Assuming the existence of the transformation function $f(z)$ we may now go on to show its great value in the solution of various electrostatic problems which involve the interior of C. The essential point is that by means of the transformation it is possible to transform the problem from the complicated geometry of Fig. 4.31 to the simple geometry of Fig. 4.32 for which solutions may be much more easily obtained. For example, suppose that the potential varies along C, there being no charge in the interior of C. Then the transformed problem is one in which the potential V is specified along the real axis ($v = 0$) and it is desired to find V in the upper half plane. We know that V is the imaginary part of a function of a complex variable which is analytic in the upper half plane. It is therefore permissible to employ Eq. (4.2.13):

$$V(u,v) = \frac{v}{\pi} \int_{-\infty}^{\infty} \left[\frac{V(u', 0)}{(u' - u)^2 + v^2} \right] du'$$

It should be apparent that, once the transformation function is known, it becomes possible to solve any electrostatic problem associated with the interior of C.

Schwarz-Christoffel Transformation. We now turn to the practical question of finding the transformation function. The most general contour C which may be discussed in great detail is that of the polygon. Special cases of importance which are also known are the ellipse and in

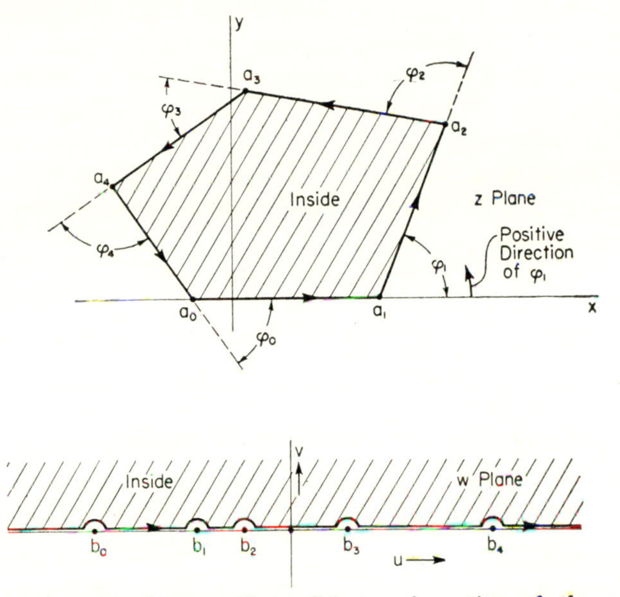

Fig. 4.33 Schwarz-Christoffel transformation of the inside of a polygon onto the upper half of the w plane.

particular the circle. Let us consider the polygon case; the transformation is then referred to as the *Schwarz-Christoffel transformation.* The polygon is illustrated in Fig. 4.33. The vertices of the polygon are labeled $a_0, a_1, \ldots$, the corresponding exterior angles $\varphi_0, \varphi_1, \ldots$, and the transforms of the vertices to the w plane $b_0, b_1, \ldots$. Note that

$$\varphi_0 + \varphi_1 + \cdots = 2\pi \tag{4.7.1}$$

The reader should pay particular attention to the manner in which the angles φ_i are defined and to the fact that the interior of the polygon is to the left as one goes around the perimeter in the direction of the angles. In this direction, φ's which correspond to turning farther to the left are positive, those for reentrant angles, where the turn is to the right, are negative. In some "degenerate" cases, φ may be π or $-\pi$; which angle is the right one is usually determined by appeal to Eq. (4.7.1).

Since the angles φ_i are not preserved by the transformation, it is evident that $w(z)$ must be singular at the points a_i. As a consequence we shall "round" these corners off as the contour C is described. The corresponding contour in the w plane is modified as indicated by means of the indicated semicircles. These are chosen so as to exclude the singular points b_i from the upper half plane inasmuch as $w(z)$ is to be analytic in this region.

Consider the behavior of dz and dw as the polygon is traversed in a counterclockwise direction as indicated by the arrows in the figure. The phase of dw in the region to the left of b_0 is zero (dw is real), whereas the phase of dz is given by the direction of the line a_4a_0. At the point a_0 the phase of dz suffers a discontinuity of amount φ_0, whereas the phase of dw remains real. What must be the behavior of dz/dw in the neighborhood of the point b_0? It must be real for $w > b_0$ and must suffer a change of phase of φ_0 while $w - b_0$ is changing phase by $-\pi$, as illustrated in Fig. 4.33. The functional dependence thus described leads uniquely to

$$dz/dw \xrightarrow[w \to b_0]{} A(w - b_0)^\alpha$$

In order to obtain the correct phase change at b_0, $\alpha = -\varphi_0/\pi$, so that

$$dz/dw \xrightarrow[w \to b_0]{} A(w - b_0)^{-\varphi_0/\pi} \tag{4.7.2}$$

Performing this analysis at each point b_i we obtain

$$dz/dw = A(w - b_0)^{-\varphi_0/\pi}(w - b_1)^{-\varphi_1/\pi}(w - b_2)^{-\varphi_2/\pi} \cdots \tag{4.7.3}$$

A must be a constant, for the only singularities and zeros of z in the upper half plane are at $b_0, b_1, \ldots$. By Schwarz's principle of reflection (page 393) the function z may be continued across the real axis to the lower half plane, so that the only singularities of z in the entire plane occur at b_i. Integrating (4.7.3) yields the *Schwarz-Christoffel transformation.*

$$z = z_0 + A\int(w - b_0)^{-\varphi_0/\pi}(w - b_1)^{-\varphi_1/\pi}(w - b_2)^{-\varphi_2/\pi} \cdots dw \tag{4.7.4}$$

Recapitulating, Eq. (4.7.3) is a transformation of the interior of a polygon with exterior angles $\varphi_0 \ldots$, into the upper half plane of w. The arbitrary constants z_0, $|A|$, arg A must be adjusted so as to yield the correct origin, scale, and orientation of the polygon on the z plane. The other constants $b_0, b_1, \ldots$ must be chosen so that they correspond to $a_0, a_1, \ldots$. Because of the three arbitrary constants available, *it is possible to choose three of these points at will; i.e.*, three of the b's may be placed arbitrarily on the w plane. The remaining b's must be determined by first integrating (4.7.3), determining z_0 and A, and then solving for each of the b's by noting that, when $w = b_i$, $z = a_i$. The transformation function $z = z(w)$ will generally have branch points at the vertices b_i.

Usually one of the transformed points is placed at infinity. Suppose it is b_0. Then the transformation (4.7.4) becomes

$$z = z_0 + A\int(w - b_1)^{-\varphi_1/\pi}(w - b_2)^{-\varphi_2/\pi} \cdots dw \tag{4.7.5}$$

Transformations (4.7.4) and (4.7.5) transform the interior of the region bounded by the polygon into the upper half plane. It often occurs that the region of interest is the *exterior of the polygon.* In that case the polygon must be described in the clockwise direction so the interior is to the right as one goes around the perimeter. The appropriate angles are now the negative of the angles φ_i employed in (4.7.4). There is, however, a complication at the point p in the w plane corresponding to the point in the z plane at infinity. It is necessary that the transformation z have a pole at the point $w = p$ (note that p cannot be on the real axis of w). Therefore the pole at p might imply an additional multiplicative factor in the integrands of either (4.7.4) or (4.7.5) of the form $1/(w - p)$. This, however, does not suffice because its phase changes as w traverses the real axis. To obtain the proper phase relation between the sides of the polygon and the real w axis, without distortion, the additional multiplicative factors must not add in any variable phase, or in other words, the multiplicative function must be real when w is real. Hence the multiplicative factor must be $1/[(w - p)(w - \bar{p})]^2$ so that (4.7.5) must be changed to

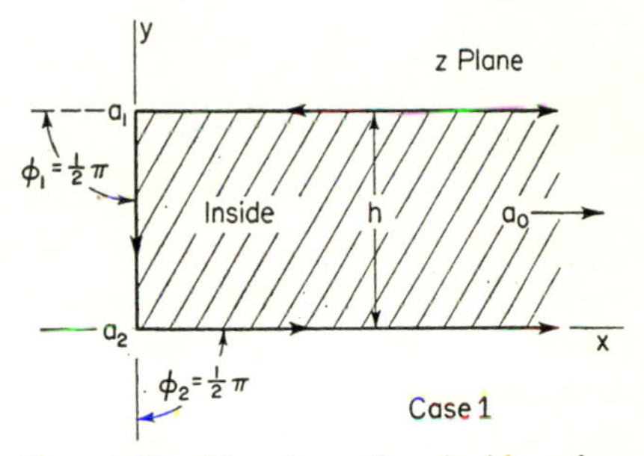

Fig. 4.34 Mapping the inside of a degenerate triangle onto the upper half of the w plane. The angles are $\varphi_0 = \pi$, $\varphi_1 = \frac{1}{2}\pi = \varphi_2$.

$$z = z_0 + A \int \frac{dw}{(w - p)^2(w - \bar{p})^2} \{(w - b_0)^{-\varphi_0/\pi}(w - b_1)^{-\varphi_1/\pi} \cdots\} \quad (4.7.6)$$

in order to transform the exterior of the polygon onto half the w plane. The angles φ are the negatives of the angles shown in Fig. 4.33.

Let us now consider some examples, principally to see how the theory behaves in practice. We shall restrict our cases to those for which the integrations may be evaluated in terms of the elementary transcendental functions. This will be the case when the figures are degenerate triangles and rectangles. If it is a triangle, Eq. (4.7.5) becomes

$$z = z_0 + A \int (w - b_1)^{-\varphi_1/\pi}(w - b_2)^{-\varphi_2/\pi}\, dw$$

For case 1, consider the region illustrated in Fig. 4.34. The angles φ_i are given on the figure. The angle at infinity (a_0 at infinity) may be evaluated by considering the triangle of Fig. 4.34 to be the limit of a triangle with the vertex at a_0 going to infinity.

The transformation for this case is then

$$z = z_0 + A \int (w - b_1)^{-\frac{1}{2}}(w - b_2)^{-\frac{1}{2}}\, dw$$

The symmetric position of a_1 and a_2 suggests placing $b_1 = -1$, $b_2 = +1$ so that

$$z = z_0 + A \int \frac{dw}{\sqrt{(w^2 - 1)}} = z_0 + A \cosh^{-1} w$$

To evaluate z_0 and A, let $z = 0$, then $w = 1$

$$0 = z_0 + A \cosh^{-1}(1) = z_0 \quad \text{or} \quad z_0 = 0$$

Now let $z = hi$, then $w = -1$

$$hi = A \cosh^{-1}(-1) = Ai\pi \quad \text{or} \quad A = h/\pi$$

The transformation is therefore

$$z = (h/\pi) \cosh^{-1} w \quad \text{or} \quad w = \cosh(\pi z/h) \tag{4.7.7}$$

We may check that this transformation maps the interior of the region of Fig. 4.34 into the upper half plane by testing with some simple point; *e.g.*, let $z = p + \frac{1}{2}ih$. Then $\cosh(\pi z/h) = \cosh[(\pi i/2) + (\pi p/h)] = i \sinh(\pi p/h)$, so that the line parallel to the lines $a_1 \to a_0$ and $a_2 \to a_0$ and halfway between becomes the imaginary axis of the w plane. We may view the transformation as one in which the lines a_0a_1 and a_1a_2 are folded back on to the real axis, the new origin chosen so that the points b_1 and b_2 are symmetrically placed about the origin. The line "halfway between" transforms naturally to the imaginary axis.

We might also desire to transform the exterior of the Fig. 4.34 into the upper half plane. Then $\varphi_0 = -\pi$, $\varphi_1 = -\pi/2$, $\varphi_2 = -\pi/2$. Since, in this case, we can set the p of (4.7.6) equal to $i\infty$, we have

$$z = z_0 + A\int(w^2 - 1)^{\frac{1}{2}}\, dw \quad \text{or} \quad z = z_0 + \tfrac{1}{2}A\{w\sqrt{w^2 - 1} - \cosh^{-1} w\}$$

When $z = 0$, $w = 1$, so that $0 = z_0$. When $z = hi$, $w = -1$, so that $hi = \frac{1}{2}A[-\pi i]$ or $A = -(2h/\pi)$. Therefore

$$z = +(h/\pi)[\cosh^{-1} w - w\sqrt{w^2 - 1}] \tag{4.7.8}$$

To define (4.7.8) more precisely it is necessary to state the branches upon which we must evaluate the two functions $\cosh^{-1} w$, $\sqrt{w^2 - 1}$. These are most easily chosen by requiring that the conditions set up by the transformation desired be satisfied. For example, if $w = ip$, then we choose the branches such that $\cosh^{-1} w = (\pi/2)i + \sinh^{-1} p$, and $\sqrt{w^2 - 1} = -i\sqrt{p^2 + 1}$ so that for

$$w = ip; \quad z = \frac{h}{\pi}\left[\frac{\pi}{2} i + \sinh^{-1} p - p\sqrt{1 + p^2}\right]$$

The function $\sinh^{-1} p$ is always less than $p\sqrt{1 + p^2}$ so that the real part of z is negative. It is then clear that the points on the imaginary axis in the upper half w plane corresponds to the half line parallel to the real axis of the z plane extending from $\frac{1}{2}ih$ to $\frac{1}{2}ih - \infty$. Thus it is safe to conclude that the exterior of Fig. 4.34 has been transformed into the upper half plane of w.

Some Examples. For illustrative purpose we now tabulate a number of transformations together with enough detail so that the reader may follow their derivation. The geometry of case 2 (Fig. 4.35) provides a simple example of a "fringing" field. If $a_0a_2a_1$ is at one potential, say 0,

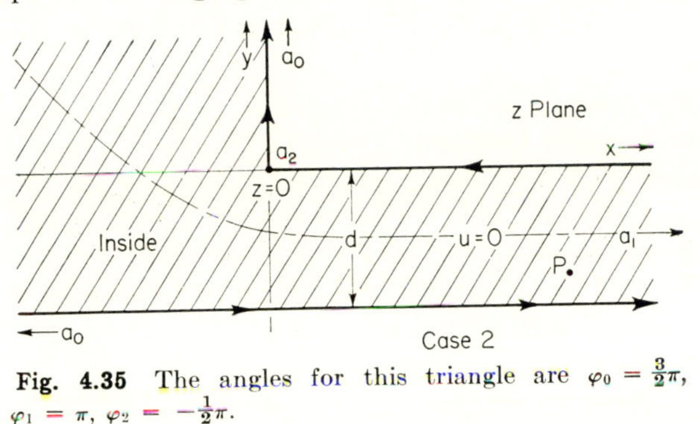

Fig. 4.35 The angles for this triangle are $\varphi_0 = \frac{3}{2}\pi$, $\varphi_1 = \pi$, $\varphi_2 = -\frac{1}{2}\pi$.

and the line a_0a_1 is at another, say V_0, then the field at point P, for $x \gg d$, is uniform, whereas near a_0 the lines of force are circular. Another feature is the effect of the sharp corner at a_2. It is expected that the field intensity should be infinite here; the order of infinity is of interest.

For this case, with $b_1 = 0$, $b_2 = 1$

$$z = z_0 + A \int \frac{(w-1)^{\frac{1}{2}}}{w} \, dw;$$

$$z = z_0 + 2A \left\{ \sqrt{w-1} - \frac{1}{2i} \ln \left[\frac{1 + i\sqrt{w-1}}{1 - i\sqrt{w-1}} \right] \right\}$$

To specify this transformation, it is necessary to give the branch lines for z as well as the w plane. The transformation naturally has its singularities at $w = 0$ and $w = 1$, *i.e.*, at b_i as well as at ∞. The branch line of most convenience is the Re $w = 0$ axis, for Re $w < 1$. It follows, from the requirement that $z = 0$ when $w = 1$, that $z_0 = 0$. We approach the point $w = 0$, ($z = a_1$) along two different routes. One, corresponding to $a_0 \to a_2 \to a_1$, is along the positive real w axis Re $w > 0$. The phase of the logarithm is zero for $a_2 \to a_1$ so that $z = 2A[-i\infty]$. Constant A must be a pure imaginary. The second route $a_0 \to a_1$ corresponds to approaching $w = 0$ along Re $w < 0$, so that the phase of the logarithm is π. Then $z = 2A[-i\infty - (1/2i)(\pi i)] = -A\pi + \infty = -di + \infty$; thus $A = di/\pi$. Therefore

$$z = \frac{2di}{\pi} \left\{ \sqrt{w-1} - \left(\frac{1}{2i}\right) \ln \left[\frac{1 + i\sqrt{w-1}}{1 - i\sqrt{w-1}} \right] \right\}$$

To verify our picture of the transformation, consider $w = iv$. Then when $v \to \infty$, $z \to (i - 1)\infty$. When $v \to 0$, $z \to -\frac{1}{4}di + \infty$. Thus the imaginary axis on the w plane corresponds to the dotted line on the z plane.

The original physical problem transforms into one in which half of the real axis $\text{Re}\, w > 0$ is at potential 0 while $\text{Re}\, w < 0$ is at potential V_0. The solution for the potential V is

$$V = \text{Im}\left[\frac{V_0}{\pi} \ln w\right] = \left(\frac{V_0}{\pi}\right)\theta$$

where θ is the polar angle on the w plane.

Let $W = U + iV$, then $w = e^{\pi W/V_0}$.

The quantity of most direct physical interest is the electric field E.

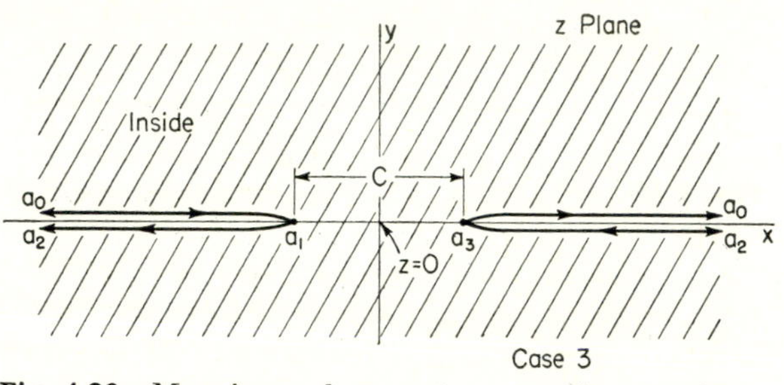

Fig. 4.36 Mapping a degenerate rectangle onto the w plane. "Inside" includes all the z plane except the regions between the lines $a_0a_1 - a_1a_2$ and between the lines $a_0a_3 - a_3a_2$. The angles are $\varphi_0 = 2\pi$, $\varphi_1 = -\pi$, $\varphi_2 = 2\pi$, $\varphi_3 = -\pi$.

This is very simply related to dW/dz for $dW/dz = (\partial U/\partial x) + i(\partial V/\partial x)$. By the Cauchy-Riemann conditions (4.2.1)

$$\frac{dW}{dz} = \frac{\partial V}{\partial y} + i\frac{\partial V}{\partial x} = -(iE_x + E_y)$$

$$E_x - iE_y = i(dW/dz)$$

We may express $E_x - iE_y$ in terms of w by means of $dW/dz = (dW/dw)\cdot$ $\cdot(dw/dz)$ or

$$E_x - iE_y = (V_0/d)(w - 1)^{-\frac{1}{2}}$$

In the neighborhood of $w = 0$, $E_x - iE_y$ is a pure imaginary, as it should be, so that $E_x = 0$ and E_y is a constant as it should be. For $\text{Re}\, w \gg 1$, $E_x - iE_y$ is real so that $E_y = 0$, as it should along $a_2 \to a_0$. For $|w| \gg 1$, $z = (2di/\pi)\sqrt{w}$ so that $E_x - iE_y \to (V_0/d\sqrt{w}) \to 2i(V_0/\pi z)$. The lines of force are circular, extending from a_0 on C to a_0 on D. Finally we examine the field near a_2, that is, near $w = 1$. The field becomes infinite as expected. The dependence of the field on z near this point may be found by solving for $\sqrt{w-1}$ in terms of z for $w \simeq 1$: $z \simeq (2di/\pi)(w-1)^{\frac{3}{2}}$ so that $(w-1)^{\frac{1}{2}} \simeq z^{\frac{1}{3}}$ and $E_x - iE_y \simeq 1/z^{\frac{1}{3}}$.

In case 3 consider the electric field generated by two infinite planes separated by a distance C, one at potential 0, the other at V_0. We shall limit ourselves to the transformation itself. The method employed to solve the physical problem is entirely similar to that of case 2. The slits form a degenerate rectangle with vertices a_0, a_1, a_2, a_3 as indicated in Fig. 4.36.

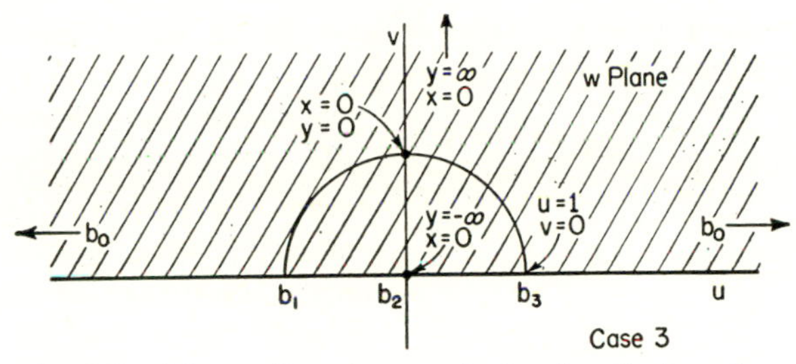

Fig. 4.37 The w plane for case 3, showing the transform of the x axis and the location of the points corresponding to $z = \infty$.

The two points b_1 and b_3 must be symmetrically situated with respect to b_2, which permits setting the values of both b_1 and b_3. Then

$$z = z_0 + A \int \frac{(w-1)(w+1)}{w^2}\,dw; \quad z = z_0 + A\left(w + \frac{1}{w}\right)$$

When $w = 1$, $z = c/2$, $c/2 = z_0 + 2A$. When $w = -1$, $z = -(c/2)$ so that $-(c/2) = z_0 - 2A$. From these two equations one obtains $z_0 = 0$ and $A = c/4$ so that

$$z = \frac{c}{4}\left[w + \frac{1}{w}\right]$$

To understand this transformation more completely the manner in which the x and y axis map on to the w plane is illustrated in Fig. 4.37.

The x axis between $-(c/2)$ and $+(c/2)$ becomes a semicircle of unit radius on the w plane. The y axis is telescoped into the $v > 0$ axis. One might think of the transformation as one in which the lines $a_3 \to a_0$ and $a_0 \to a_1$ are rotated from their original positions in the z plane through the lower half of the z plane joining finally after a 180° rotation. The lower half of the z plane is swept into the region inside the semicircle. The x axis bulges here in order to make room for the lower half of the z plane.

The Method of Inversion. The Schwarz-Christoffel method may be extended to regions bounded by circular arcs by the *method of inversion.* The inversion transformation arises naturally in the discussion of the images of a charge with respect to a circular boundary. This matter has already been discussed (see *Poisson's formula*, page 373) where it was stated that the image w of a point z, with respect to a circle of radius a whose center is at the origin is given by

$$w = a^2/\bar{z} \tag{4.7.9}$$

Equation (4.7.9) defines a transformation from the z to the w plane in which the interior points of the circle become exterior points and vice versa.

The inversion preserves the magnitude of the angle between two curves but reverses the sense of the angle. This may easily be seen, for (4.7.9) may be considered as the result of two successive transformations, the first a^2/z, the second a reflection with respect to the real axis. The first of these is conformal; the second maintains angle size but reverses the sense of the angle.

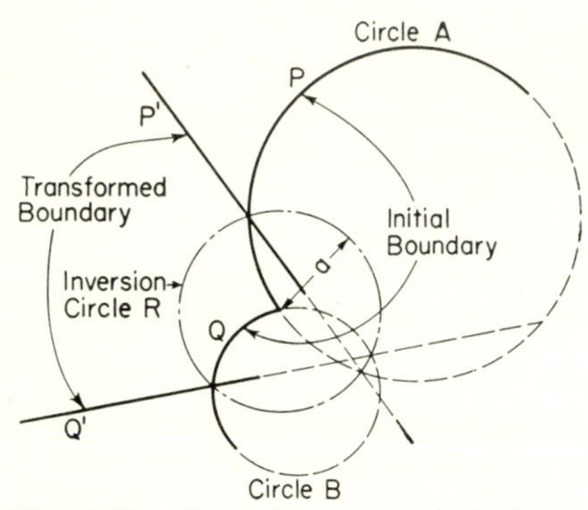

Fig. 4.38. Inversion of arcs PQ of two circles to form two straight lines $P'Q'$, by use of inversion circle R.

For the purposes of this section, we need to investigate the inversion of a circle which passes through the center of an inversion circle. We shall show that the transform of the circle is a straight line which must necessarily pass through the intersection of the circle and the inversion circle. On the z plane, the points of a circle of radius z_0 (no generality is lost if z_0 is taken as real) satisfy the equation

$$|z - z_0| = z_0$$

or

$$\bar{z}z - (z z_0 + \bar{z} z_0) = 0 \tag{4.7.10}$$

To find the corresponding points on the w plane, z and $\bar{z}$ must be replaced by $\bar{w}$ and w according to the defining equation (4.7.9). Equation (4.7.10) becomes

$$\frac{a^4}{w\bar{w}} - z_0\left(\frac{a^2}{w} + \frac{a^2}{\bar{w}}\right) = 0$$

From this it follows that

$$u = a^2/2z_0; \quad \text{where } w = u + iv$$

Therefore the transform is a straight line.

The role that inversion plays in extending the Schwarz-Christoffel transformation should now be clear. Two intersecting circular arcs such as P and Q in Fig. 4.38 may be transformed by an inversion with respect to circle R of radius a whose center is the intersection of the two circular arcs. The point at the intersection is transformed into the point at infinity, the arcs themselves being transformed into the solid portions of the lines P' and Q'. The Schwarz-Christoffel transformation may now be applied to these two intersecting straight lines.

A Schwarz-Christoffel transformation may also be applied to a more general polygon composed of circular arcs. However, this would carry us too far afield. We shall therefore restrict ourselves to the cases which involve the intersection of at most two circular arcs and refer the more inquisitive reader to other treatises, which deal with the more general case. The transformed problem for Fig. 4.39 has already been discussed as case 3 above.

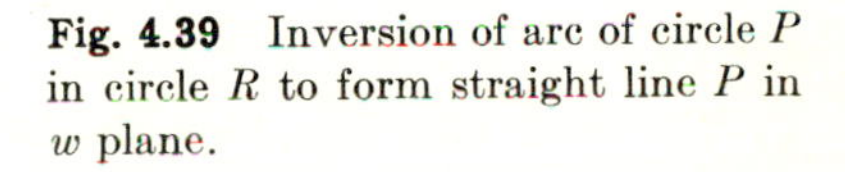

Fig. 4.39 Inversion of arc of circle P in circle R to form straight line P in w plane.

4.8 *Fourier Transforms*

We shall have many occasions in the future to employ the Fourier as well as the related Laplace and Mellin transforms. In this section, we shall study those properties of these transforms which will prove useful in our later work.

The *Fourier transform* $F(k)$ of a function $f(x)$ is defined by

$$F(k) = \frac{1}{\sqrt{2\pi}} \int_{-\infty}^{\infty} e^{ikx} f(x)\, dx \tag{4.8.1}$$

As indicated, we shall employ the capital letter to denote a transform. We shall employ this same notation to indicate any type of transform, for it is generally not necessary to distinguish. However, if it should be necessary, we shall employ a subscript f, l, or m, to denote the Fourier, Laplace, and Mellin transforms, respectively. For example, F_f is the Fourier transform of f. We shall also occasionally use script letters $\mathfrak{F}$, $\mathfrak{L}$, $\mathfrak{M}$ so that $\mathfrak{F}(f)$ will be the Fourier transform of f.

Perhaps the most important theorem in our discussion is the *Fourier integral theorem*:

$$f(x) = \frac{1}{2\pi} \int_{-\infty}^{\infty} dk \int_{-\infty}^{\infty} d\xi\, e^{ik(\xi - x)} f(\xi) \tag{4.8.2}$$

Introducing $F(k)$, Eq. (4.8.2) may be written as

$$f(x) = \frac{1}{\sqrt{2\pi}} \int_{-\infty}^{\infty} dk\, e^{-ikx} F(k)$$

In this form Eq. (4.8.2) is often referred to as *Fourier's inversion formula*.

Relation to Fourier Series. Before considering in some detail the conditions under which the Fourier integral theorem holds, it is instructive to show the connection between the theorem and Fourier series. Consider the Fourier sine series for a function $h(x)$ in the region $0 \leq x \leq l$:

$$h(x) = \sum_n A_n \sin\left(\frac{\pi n x}{l}\right)$$

Expansions of this kind have been discussed in Chap. 2 where it occurs as an essential part of the discussion on vibrating things; they will be discussed again in Chap. 6. The coefficients A_n follow from

$$\int_0^l \sin\left(\frac{n\pi x}{l}\right) \sin\left(\frac{m\pi x}{l}\right) dx = \tfrac{1}{2} l \delta_{nm}$$

so that

$$A_n = \frac{2}{l} \int_0^l h(x) \sin\left(\frac{n\pi x}{l}\right) dx$$

Introducing this value of A_n in the expansion for $h(x)$ we obtain

$$h(x) = \frac{2}{l} \sum_{n=0}^{\infty} \left[\int_0^l h(\zeta) \sin\left(\frac{n\pi\zeta}{l}\right) d\zeta \right] \sin\left(\frac{n\pi x}{l}\right)$$

We shall now consider the behavior of this series as $l \to \infty$. For this purpose introduce the variable k, which at discrete values k_n takes on the value $n\pi/l$. Moreover $\Delta k = k_{n+1} - k_n = \pi/l$. Then

$$h(x) = \frac{2}{\pi} \sum_{n=0}^{\infty} (\Delta k) \left[\int_0^l h(\zeta) \sin(k_n \zeta)\, d\zeta \right] \sin(k_n x)$$

It should be apparent that in the limit $l \to \infty$, that is, $\Delta k \to 0$, the sum may be replaced by the integral

$$h(x) = \frac{2}{\pi} \int_0^\infty dk \int_0^\infty d\zeta\, h(\zeta) \sin(k\zeta) \sin(kx) \tag{4.8.3}$$

This formula holds for odd functions of x only. For even functions it is more appropriate to employ the Fourier cosine series which leads to

$$g(x) = \frac{2}{\pi} \int_0^\infty dk \int_0^\infty d\zeta\, g(\zeta) \cos(k\zeta) \cos(kx) \tag{4.8.4}$$

It is clear that in both (4.8.3) and (4.8.4) the integrations on ζ and k may be extended to $-\infty$ so that Eq. (4.8.3) becomes

$$h(x) = \frac{1}{2\pi} \int_{-\infty}^\infty dk \int_{-\infty}^\infty h(\zeta) \sin(k\zeta) \sin(kx)\, d\zeta$$

To obtain an expression valid for a function which is neither odd nor even, note that a function may be divided into its even and odd parts, namely,

$$f(x) = \tfrac{1}{2}[f(x) + f(-x)] + \tfrac{1}{2}[f(x) - f(-x)]$$

where the first bracket is even and the second is odd. From the related properties

$$\int_{-\infty}^{\infty} h(x) \cos(kx)\, dx = 0 \quad \text{and} \quad \int_{-\infty}^{\infty} g(x) \sin(kx)\, dx = 0$$

we may write

$$\begin{aligned} f(x) &= (1/4\pi) \int_{-\infty}^{\infty} dk \int_{-\infty}^{\infty} d\zeta\, [f(\zeta) - f(-\zeta)] \sin(k\zeta) \sin(kx) \\ &\qquad + (1/4\pi) \int_{-\infty}^{\infty} dk \int_{-\infty}^{\infty} d\zeta\, [f(\zeta) + f(-\zeta)] \cos(k\zeta) \cos(kx) \\ &= (1/4\pi) \int_{-\infty}^{\infty} dk \int_{-\infty}^{\infty} d\zeta\, \{f(\zeta) \cos[k(\zeta - x)] + f(-\zeta) \cos[k(\zeta + x)]\} \end{aligned}$$

$$f(x) = (1/2\pi) \int_{-\infty}^{\infty} dk \int_{-\infty}^{\infty} d\zeta\, f(\zeta) \cos[k(\zeta - x)] \tag{4.8.5}$$

To obtain Eq. (4.8.2) we note that, since $\int_{-\infty}^{\infty} d\zeta\, f(\zeta) \sin[k(\zeta - x)]$ is an odd function of k,

$$0 = i \int_{-\infty}^{\infty} dk \int_{-\infty}^{\infty} d\zeta\, f(\zeta) \sin[k(\zeta - x)] \tag{4.8.6}$$

if this integral converges. Under this assumption, adding (4.8.5) and (4.8.6) yields the Fourier integral theorem, Eq. (4.8.2).

From this discussion we observe that the Fourier integral theorem is the generalization of Fourier series for the interval $-\infty < x < \infty$. In the theory of vibrating strings, the Fourier integral is appropriate for motion on strings stretching from $-\infty$ to $+\infty$. On the other hand, Eq. (4.8.3) is more appropriate for strings extending from 0 to $+\infty$, the end at $x = 0$ being fixed. Indeed Eqs. (4.8.3) and (4.8.4) are often dignified by the names Fourier's sine and cosine integral formulas. Associated with these integrals are the corresponding Fourier sine and cosine transforms, denoted by F_s and F_c, respectively, and defined by

$$F_s(k) = \sqrt{2/\pi} \int_0^{\infty} \sin(kx)\, f(x)\, dx \tag{4.8.7}$$

$$F_c(k) = \sqrt{2/\pi} \int_0^{\infty} \cos(kx)\, f(x)\, dx \tag{4.8.8}$$

The integral formulas become inversion formulas:

$$f(x) = \sqrt{2/\pi} \int_0^{\infty} \sin(kx)\, F_s(k)\, dk \tag{4.8.9}$$

$$f(x) = \sqrt{2/\pi} \int_0^{\infty} \cos(kx)\, F_c(k)\, dk \tag{4.8.10}$$

Some Theorems on Integration. The derivation given above of the Fourier integral theorem is strictly heuristic and does not contain any criteria for the range of its validity. We must now turn to a more rigorous procedure. To obtain a theorem of the widest applicability it is necessary to employ the Lebesgue theory of integration. Space does not permit our entering into the subtleties of Lebesgue measure and the corresponding definition of an integral. However, it is possible to point out the practical effect of replacing the more usual Riemann definition by the Lebesgue. The difference enters in integrals over functions with discontinuities. The Riemann integral is defined if and only if the number of ordinary discontinuities[1] is denumerable, so that, if the discontinuities are nondenumerable, the Riemann integral will not exist. On the other hand, the Lebesgue integral will exist in either event, being identical with the Riemann integral when the latter can be defined. In more picturesque language, it is possible to define a Lebesgue integral for a function which in some region oscillates with infinite rapidity. An example which illustrates this point is given by the following integrand:

$$f(x) = 1; \quad \text{at all rational points in the interval } (0,1)$$
$$f(x) = 0; \quad \text{at all irrational points in the interval } (0,1)$$

The Lebesgue integral is defined for this integrand, but the Riemann integral is not.

When Lebesgue integrals exist, they are classified according to their *Lebesgue class.* A function is in the Lebesgue class L^p in the interval (a,b) if $|f(x)|^p$ $(p > 0)$ is integrable over (a,b).

As a preliminary to proving the Fourier integral theorem, we shall first prove *Parseval's formula.* If $f(x)$ belongs to Lebesgue class L^2 in the interval $(-\infty,\infty)$, then, according to Parseval's formula,

$$\int_{-\infty}^{\infty} |F(k)|^2\,dk = \int_{-\infty}^{\infty} |f(x)|^2\,dx \tag{4.8.11}$$

It will be seen later that the Fourier integral theorem is an almost immediate consequence of the Parseval formula. To prove this formula, we consider the integral

$$I = \int_{-\infty}^{\infty} e^{-\frac{1}{2}\delta^2k^2}|F(k)|^2\,dk$$

Substitute the integral for $F(k)$ in I:

$$I = \frac{1}{2\pi}\int_{-\infty}^{\infty} e^{-\frac{1}{2}\delta^2k^2}\,dk \int_{-\infty}^{\infty} f(x)e^{ikx}\,dx \int_{-\infty}^{\infty} \bar{f}(\zeta)e^{-ik\zeta}\,d\zeta$$
$$= \frac{1}{2\pi}\int_{-\infty}^{\infty} f(x)\,dx \int_{-\infty}^{\infty} \bar{f}(\zeta)\,d\zeta \int_{-\infty}^{\infty} e^{-\frac{1}{2}\delta^2k^2+ik(x-\zeta)}\,dk$$

[1] A function $f(x)$ is said to have an ordinary discontinuity at $x = t$ if $\lim_{x\to t^+}[f(x)] \neq \lim_{x\to t^-}[f(x)]$ and both limits exist. In the first limit x approaches t from the right, while in the second it is from the left.

The integration on k may now be performed:

$$I = \frac{1}{\sqrt{2\pi}\,\delta}\int_{-\infty}^{\infty} f(x)\,dx \int_{-\infty}^{\infty} \bar{f}(\zeta)e^{-\frac{1}{2}(x-\zeta)^2/\delta^2}\,d\zeta \tag{4.8.12}$$

We may now show that $F(x)$ belongs to L^2 in the interval $(-\infty,\infty)$. We write

$$I = \frac{1}{\sqrt{2\pi}\,\delta}\int_{-\infty}^{\infty}\int_{-\infty}^{\infty} f(x)e^{-\frac{1}{4}(x-\zeta)^2/\delta^2}\,\bar{f}(\zeta)e^{-\frac{1}{4}(x-\zeta)^2/\delta^2}\,dx\,d\zeta$$

and by Schwarz's inequality[1]

$$I \le \frac{1}{\sqrt{2\pi}\,\delta}\left[\iint_{-\infty}^{\infty} |f(x)|^2 e^{-\frac{1}{2}(x-\zeta)^2/\delta^2}\,dx\,d\zeta \iint_{-\infty}^{\infty} |f(\zeta)|^2 e^{-\frac{1}{2}(x-\zeta)^2/\delta^2}\,dx\,d\zeta\right]^{\frac{1}{2}}$$

$$\le \frac{1}{\sqrt{2\pi}\,\delta}\iint_{-\infty}^{\infty} |f(x)|^2 e^{-\frac{1}{2}(x-\zeta)^2/\delta^2}\,dx\,d\zeta = \int_{-\infty}^{\infty} |f(x)|^2\,dx \tag{4.8.13}$$

Hence in the limit $\delta \to 0$

$$\int_{-\infty}^{\infty} |F(k)|^2\,dk \le \int_{-\infty}^{\infty} |f(x)|^2\,dx$$

From this inequality we may deduce that $F(k)$ belongs to L^2 in the interval $(-\infty,\infty)$ if $f(x)$ does.

We may now proceed to prove that the equality holds. Return to Eq. (4.8.12), inverting the order of integration:

$$I = \frac{1}{\sqrt{2\pi}\,\delta}\int_{-\infty}^{\infty} e^{-y^2/2\delta^2}\,dy \int_{-\infty}^{\infty} \bar{f}(x+y)f(x)\,dx \tag{4.8.14}$$

We see intuitively at any rate that the function $(e^{-\frac{1}{2}y^2/\delta^2}/\sqrt{2\pi}\,\delta)$ becomes more and more peaked at $y = 0$ as $\delta \to 0$ but that its integral over y remains equal to unity for all values of δ. Then, if the integral

$$h(y) = \int_{-\infty}^{\infty} \bar{f}(x+y)f(x)\,dx$$

has a sufficiently smooth behavior near $y = 0$, we might expect that I approaches $h(0)$ as $\delta \to 0$. This follows from the integrability of the absolute square of the function; *i.e.*, as a function of y it belongs to L^2 in $(-\infty,\infty)$. It then becomes possible to justify placing $y = 0$ in $h(y)$ as it appears in (4.8.14). Hence

$$I \xrightarrow[\delta\to 0]{} \frac{1}{\sqrt{2\pi}\,\delta}\int_{-\infty}^{\infty} e^{-\frac{1}{2}y^2/\delta^2}\,dy \int_{-\infty}^{\infty} |f(x)|^2\,dx$$

or

$$\int_{-\infty}^{\infty} |F(k)|^2\,dk = \int_{-\infty}^{\infty} |f(x)|^2\,dx \tag{4.8.15}$$

[1] A form of this inequality was derived in Chap. 1. In the present application it reads: Let $u(x)$ and $v(x)$ belong to L^2; then

$$\left|\iint u(x,\zeta)v(x,\zeta)\,dx\,d\zeta\right|^2 \le \left[\iint |u(x,\zeta)|^2\,dx\,d\zeta\right]\left[\iint |v(x,\zeta)|^2\,dx\,d\zeta\right]$$

which is the Parseval formula. It should be emphasized that this proof holds for complex as well as for real functions f.

The Fourier Integral Theorem. We now go on to prove the Fourier integral theorem as it is stated by Plancherel:

Let $f(x)$ belong to L^2 in the interval $(-\infty,\infty)$ and let

$$F(k,a) = \frac{1}{\sqrt{2\pi}} \int_{-a}^{a} f(x)e^{ikx}\, dx$$

Then as $a \to \infty$, $F(k,a)$ converges in the mean in the interval $(-\infty,\infty)$ to the function $F(k)$, and if

$$f(x,a) = \frac{1}{\sqrt{2\pi}} \int_{-a}^{a} F(k)e^{-ikx}\, dk$$

then $f(x,a)$ converges in the mean to $f(x)$.

Convergence in the mean for $F(k,a)$ in an interval $(-\infty,\infty)$ is defined as follows:

$$\lim_{a\to\infty} \left\{ \int_{-a}^{a} |F(k,a) - F(k)|^2\, dk \right\} \to 0 \tag{4.8.16}$$

In other words, the mean-square error of $F(k,a)$ in representing $F(k)$ goes to zero as a goes to infinity.

To prove this theorem, we proceed as follows: In Eq. (4.8.15) replace F by $F + G$ and f by $f + g$. Then

$$\int_{-\infty}^{\infty} |F + G|^2\, dk = \int_{-\infty}^{\infty} [|F|^2 + |G|^2 + 2\,\mathrm{Re}\,(F\bar{G})]\, dk$$
$$= \int_{-\infty}^{\infty} [|f|^2 + |g|^2 + 2\,\mathrm{Re}\,(f\bar{g})]\, dx$$

Employing Parseval's formula,

$$\int_{-\infty}^{\infty} \mathrm{Re}\,(F\bar{G})\, dk = \int_{-\infty}^{\infty} \mathrm{Re}\,(f\bar{g})\, dx$$

Replacing F in (4.8.15) by $F + iG$, we obtain

$$\int_{-\infty}^{\infty} \mathrm{Im}\,(F\bar{G})\, dk = \int_{-\infty}^{\infty} \mathrm{Im}\,(f\bar{g})\, dx$$

so that, following from Parseval's formula, we have

$$\int_{-\infty}^{\infty} F\bar{G}\, dk = \int_{-\infty}^{\infty} f\bar{g}\, dx \tag{4.8.17}$$

from which innocent little formula we can prove nearly all the theorems we desire concerning Fourier transforms.

Let $g(x) = 1$ in the interval $(0,\zeta)$ and zero elsewhere. Its transform is

$$G(k) = \frac{1}{\sqrt{2\pi}} \int_{0}^{\zeta} e^{ikx}\, dx = \frac{1}{\sqrt{2\pi}} \left[\frac{e^{ik\zeta} - 1}{ik}\right]$$

Hence, from the previous paragraph

$$\frac{1}{\sqrt{2\pi}} \int_{-\infty}^{\infty} F(k) \left[\frac{1 - e^{-ik\zeta}}{ik}\right] dk = \int_0^{\zeta} f\, dx$$

Differentiating both sides with respect to ζ we obtain

$$\frac{1}{\sqrt{2\pi}} \int_{-\infty}^{\infty} F(k)e^{-ik\zeta}\, dk = f(\zeta)$$

which is the usual formula for the inverse transform.

We still must show that $F(k)$ exists, *i.e.*, that in the mean,

$$\lim_{a\to\infty} [F(k,a)] = F(k)$$

$F(k,a)$ is the transform of the function which equals zero for $|x| > a$ and equals $f(x)$ for $|x| < a$, since

$$F(k,a) = \frac{1}{\sqrt{2\pi}} \int_{-a}^{a} f(x)\, e^{ikx}\, dk$$

Now consider $H(k,a) = F(k,a) - F(k)$. The transform of this is a function which is zero in the range $|x| < a$ and equals $f(x)$ in the range $|x| > a$. Applying Parseval's formula, we have

$$\int_{-\infty}^{\infty} |H(k,a)|^2\, dk = \int_{-\infty}^{\infty} |F(k,a) - F(k)|^2\, dk$$
$$= \int_{-\infty}^{-a} |f(x)|^2\, dx + \int_{a}^{\infty} |f(x)|^2\, dx$$

Taking the limit $a \to \infty$ we see that $\int|F(k,a) - F(k)|^2\, dk \to 0$, proving the theorem expressed in Eq. (4.8.16).

Properties of the Fourier Transform. Having once established the Fourier integral theorem, we may now turn to a consideration of the properties of the Fourier transform in the complex plane. The theorem we shall find of particular value is as follows:

Let $f(z)$, $(z = x + iy)$ be analytic in the strip

$$y_- < y < y_+$$

where $y_+ > 0$ and $y_- < 0$. If, for any strip within this strip

$$|f(z)| \to \begin{cases} Ae^{\tau_- x}; & \text{as } x \to \infty; \quad \tau_- < 0 \\ Be^{\tau_+ x}; & \text{as } x \to -\infty; \quad \tau_+ > 0 \end{cases}$$

Then $F(k)$, $(k = \sigma + i\tau)$ will be analytic everywhere in the strip

$$\tau_- < \tau < \tau_+$$

and in any strip within this strip

$$|F(k)| \to \begin{cases} Ce^{-y_+ \sigma}; & \text{as } \sigma \to +\infty \\ De^{-y_- \sigma}; & \text{as } \sigma \to -\infty \end{cases}$$

where A, B, C, D are real constants. The theorem is proved by noticing that the analyticity of $F(k)$ is determined completely by the convergence of the defining integral:

$$F(k) = \frac{1}{\sqrt{2\pi}} \int_{-\infty}^{\infty} f(x)e^{ikx}\,dx = \frac{1}{\sqrt{2\pi}} \int_{-\infty}^{\infty} f(x)e^{-\tau x}e^{i\sigma x}\,dx$$

Uniform convergence at the upper limit of integration requires that $e^{\tau_- x - \tau x}$ decays exponentially with x so that $\tau_- < \tau$, and vice versa at the lower limit of integration. Turning to the behavior of $F(k)$ in the strip $\tau_- < \tau < \tau_+$, consider the convergence of the integral for $F(k)$. Since $f(z)$ is analytic, we may take the path of integration along a line parallel to the x axis as long as $y_- < y < y_+$. Then

$$F(k) = \frac{1}{\sqrt{2\pi}} \int_{-\infty}^{\infty} f(z)e^{ikz}\,dz = \frac{e^{-\sigma y}}{\sqrt{2\pi}} \int_{-\infty}^{\infty} f(z)e^{i(\sigma x - \tau y)}e^{-\tau x}\,dz$$

Recalling the region of analyticity for $f(z)$ we see that

$$|F(k)| \to Ee^{-\sigma y}; \quad \text{as } |\sigma| \to \infty$$

For y near y_+ convergence occurs only for $\sigma > 0$, while for y near y_-, $\sigma < 0$, proving the theorem.

In the event that conditions at infinity, *i.e.*, belonging to L^2 in the interval $(-\infty, \infty)$, are not satisfied by $f(x)$ (say at $x = +\infty$), it is possible that the function

$$g(x) = f(x)e^{-\tau_0 x}$$

will belong to L^2 in the interval. Then the inversion formula may be applied to $g(x)$:

$$g(x) = f(x)e^{-\tau_0 x} = \frac{1}{\sqrt{2\pi}} \int_{-\infty}^{\infty} G(k)e^{-ikx}\,dk$$

$$\text{or}\quad f(x) = \frac{1}{\sqrt{2\pi}} \int_{-\infty}^{\infty} G(k)e^{-i(k+i\tau_0)x}\,dk = \frac{1}{\sqrt{2\pi}} \int_{-\infty+i\tau_0}^{\infty+i\tau_0} G(\zeta - i\tau_0)e^{-i\zeta x}\,d\zeta$$

The function $G(\zeta - i\tau_0)$ is simply related to $F(k)$, for

$$G(k) = \frac{1}{\sqrt{2\pi}} \int_{-\infty}^{\infty} f(x)e^{-\tau_0 x}e^{ikx}\,dx$$

$$G(\zeta - i\tau_0) = \frac{1}{\sqrt{2\pi}} \int_{-\infty}^{\infty} f(x)e^{i\zeta x}\,dx = F(\zeta)$$

We thus obtain the simple inversion formula

$$f(x) = \frac{1}{\sqrt{2\pi}} \int_{-\infty+i\tau_0}^{\infty+i\tau_0} F(k)e^{-ikx}\,dk \tag{4.8.18}$$

It is often the case, however, that one convergence factor $e^{-\tau_0 x}$ will not suffice for the entire range in x. Then we consider

$$f_+(x) = \begin{cases} f(x)e^{-\tau_0 x}; & \begin{cases} x > 0 \\ \tau_0 > 0 \end{cases} \\ 0; & x < 0 \end{cases} ; \quad f_-(x) = \begin{cases} 0; & x > 0 \\ f(x)e^{-\tau_1 x}; & \begin{cases} x < 0 \\ \tau_1 < 0 \end{cases} \end{cases}$$

Here τ_0 is the minimum value possible, while τ_1 is the maximum, to make f_+ and f_- converge. The corresponding transforms are

$$F_+(k) = \frac{1}{\sqrt{2\pi}} \int_0^\infty f(x)e^{ikx}\,dx = \frac{1}{\sqrt{2\pi}} \int_0^\infty f(x)e^{-\tau x}e^{i\sigma x}\,dx; \quad \tau > \tau_0$$

$$F_-(k) = \frac{1}{\sqrt{2\pi}} \int_{-\infty}^0 f(x)e^{ikx}\,dx = \frac{1}{\sqrt{2\pi}} \int_{-\infty}^0 f(x)e^{-\tau x}e^{i\sigma x}\,dx; \quad \tau < \tau_1$$

where $k = \sigma + i\tau$.

Applying the Fourier integral formula:

$$f_+(x) = \frac{1}{\sqrt{2\pi}} \int_{-\infty}^\infty F_+(\sigma + i\tau)e^{-ikx}\,d\sigma$$

Hence

$$\left.\begin{matrix} f(x) & (x > 0) \\ 0 & (x < 0) \end{matrix}\right\} = \frac{1}{\sqrt{2\pi}} \int_{-\infty}^\infty F_+(\sigma + i\tau_0)e^{-i(\sigma+i\tau_0)x}\,d\sigma$$

$$= \frac{1}{\sqrt{2\pi}} \int_{-\infty+i\tau_0}^{\infty+i\tau_0} F_+(k)e^{-ikx}\,dk$$

Similarly

$$\left.\begin{matrix} 0 & (x > 0) \\ f(x) & (x < 0) \end{matrix}\right\} = \frac{1}{\sqrt{2\pi}} \int_{-\infty+i\tau_1}^{\infty+i\tau_1} F_-(k)e^{ikx}\,dk$$

Adding, we obtain the inversion formula:

$$f(x) = \frac{1}{\sqrt{2\pi}} \int_{-\infty+i\tau_0}^{\infty+i\tau_0} F_+(k)e^{-ikx}\,dk + \int_{-\infty+i\tau_1}^{\infty+i\tau_1} F_-(k)e^{-ikx}\,dk \quad (4.8.19)$$

From the preceding theorem, we may also discuss the analyticity of F_+ and F_-. From the definition of F_+, convergence of the integral occurs for $\tau > \tau_0$, so that F_+ is analytic in the region in the upper half plane of $k = \sigma + i\tau$, above $i\tau_0$. Similarly, the function F_- is analytic in the lower half plane $\tau < \tau_1$.

For example, consider the function $f(x) = e^{|x|}$. Then $\tau_0 = 1^+$, and $\tau_1 = -1^-$ where $1^+ = 1 + \epsilon$, where ϵ is small. Then F_+ should be analytic in the region $\tau > 1^+$ while F_- should be analytic in the region $\tau < -1^-$. We now evaluate F_+ and F_- to check:

$$F_+ = \frac{1}{\sqrt{2\pi}} \int_0^\infty e^x\, e^{ikx}\,dx = -\frac{1}{\sqrt{2\pi}}\left(\frac{1}{1 + ik}\right); \quad \tau > 1$$

$$F_- = \frac{1}{\sqrt{2\pi}} \int_{-\infty}^0 e^{-x}\, e^{ikx}\,dx = \frac{1}{\sqrt{2\pi}}\left(\frac{1}{-1 + ik}\right); \quad \tau < -1$$

F_+ has a singularity at $k = i$ while F_- has a singularity at $k = -i$, which limits their range of analyticity. Function F_+ is analytic above the singularity, for example.

Asymptotic Values of the Transform. This is a convenient point at which to examine the asymptotic behavior of F_+ and f_+ and similarly for F_- and f_-. Assume that it is possible to expand $F_+(k)$ in a series in inverse powers of k:

$$F_+(k) = \frac{1}{\sqrt{2\pi}} \sum_{n=1}^{\infty} \frac{a_n}{(-ik)^n}; \quad \tau > \tau_0$$

Introducing this expansion into the inversion formula (4.8.19)

$$f_+(x) = \frac{1}{\sqrt{2\pi}} \int_{-\infty+i\tau_0}^{\infty+i\tau_0} F_+(k)e^{-ikx}\,dk$$

we obtain
$$f_+(x) = \frac{1}{2\pi} \sum_n a_n \int_{-\infty+i\tau_0}^{\infty+i\tau_0} \frac{e^{-ikx}}{(-ik)^n}\,dk$$

We may employ the transform

$$-\frac{1}{2\pi} \int_{-\infty+i\tau_0}^{\infty+i\tau_0} \left(\frac{e^{-ikx}}{ik}\right) dk = \begin{cases} 1; & x > 0 \\ 0; & x < 0 \end{cases}$$

which may be obtained by application of the inversion formula, since direct integration shows that the transform of the unit function on the right side of the above equation is $(1/\sqrt{2\pi})(1/ik)$. Integrating both sides of this equation with respect to x gives the general formula

$$\frac{1}{2\pi} \int_{-\infty+i\tau_0}^{\infty+i\tau_0} \left[\frac{e^{-ikx}}{(-ik)^n}\right] dk = \begin{cases} \dfrac{x^{n-1}}{(n-1)!}; & x > 0 \\ 0; & x < 0 \end{cases}$$

Hence
$$f_+(x) = \sum_n \frac{a_n x^{n-1}}{(n-1)!} \quad \text{or} \quad a_n = f_+^{(n-1)}(0)$$

and
$$F_+(k) = \frac{1}{\sqrt{2\pi}} \sum_n \frac{f_+^{(n-1)}(0)}{(-ik)^n}$$

Thus the behavior of $F_+(k)$ for *large values of* k in the upper half plane is connected with the behavior of f_+ near the origin.

General Formulation. We may consider the inverse of Eq. (4.8.19). Suppose we are given that

$$f(x) = \frac{1}{\sqrt{2\pi}} \left\{ \int_{-\infty+i\tau_0}^{\infty+i\tau_0} G(k)e^{-ikx}\,dk + \int_{-\infty+i\tau_1}^{\infty+i\tau_1} H(k)e^{-ikx}\,dk \right\}$$

When can we identify G and H with F_+ and F_-? From the preceding discussion we see that one of the functions would need to be analytic in the upper half plane, the other in the lower half plane. Moreover if these regions overlap, it would be necessary for $F_+ = F_-$ and therefore $G = K$. This follows from the definition of f_+ and f_-, for if the regions of analyticity overlap, both τ_0 and τ_1 could be placed equal to zero. Therefore F_+ would be equal to F_- in a common region in which they are both analytic. Hence, by the theory of analytic continuation, F_+ provides the continuation of F_- into the upper half plane and vice versa. We may therefore conclude that, if G is not equal to H, G and H may be identified with F_+ and F_- only if there is no region in which they are both analytic.

In the event that G and H are analytic in nonoverlapping strips rather than in nonoverlapping semi-infinite half planes, we also cannot identify

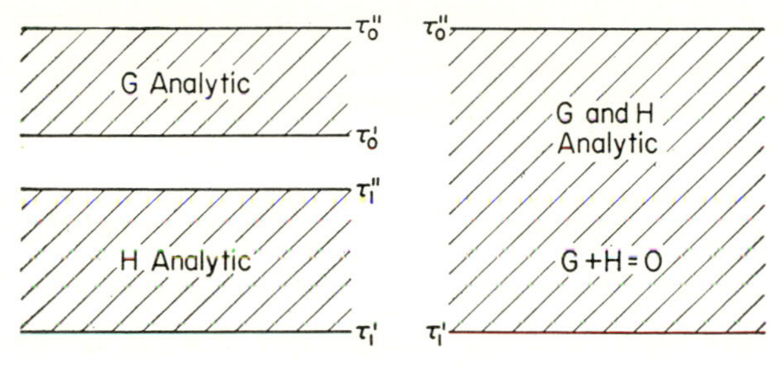

Fig. 4.40 Regions of analyticity of functions G and H.

these functions as F_+ and F_-. However, an important statement may be made if $f = 0$, that is, if

$$\int_{-\infty+i\tau_0}^{\infty+i\tau_0} G(k)e^{-ikz}\,dk + \int_{-\infty+i\tau_1}^{\infty+i\tau_1} H(k)e^{-ikz}\,dk = 0 \qquad (4.8.20)$$

To recapitulate, if Eq. (4.8.20) holds and if

1. G is analytic in the strip $\tau_0' \le \tau \le \tau_0''$, τ_0 within this strip
2. H is analytic in the strip $\tau_1' \le \tau \le \tau_1''$, τ_1 within this strip
3. $\tau_1'' < \tau_0'$
4. Both G and H are integrable L^1 in the interval $(-\infty, \infty)$
5. Both G and H approach 0 as $|\sigma| \to \infty$ in the region in which they are analytic (where $k = \sigma + i\tau$)

then we may show that

1. Both G and H are analytic in the strip $\tau_1' < \tau < \tau_0''$
2. $G + H = 0$ in that strip

The assumptions and theorem are illustrated in Fig. 4.40. The assumed regions of analyticity are shown in the left-hand part, the theorem on the right.

To prove this theorem, we multiply Eq. (4.8.20) by $e^{i\zeta z}$ and integrate with respect to z. Because of the absolute convergence of the integral in (4.8.20) we may invert the order of integration to obtain

$$\int_{-\infty+i\tau_0}^{\infty+i\tau_0} \frac{G(k)}{k-\zeta}\,dk + \int_{-\infty+i\tau_1}^{\infty+i\tau_1} \frac{H(k)}{k-\zeta}\,dk = 0 \tag{4.8.21}$$

We choose Im ζ to be between τ_0 and τ_0''. From Cauchy's integral formula we have, from the contour indicated in Fig. 4.41, page 466,

$$\int_{-\infty+i\tau_0}^{\infty+i\tau_0} \frac{G(k)}{k-\zeta}\,dk - \int_{-\infty+i\tau_0''}^{\infty+i\tau_0''} \frac{G(k)}{k-\zeta}\,dk = 2\pi i\,G(\zeta)$$

Hence
$$\int_{-\infty+i\tau_0''}^{\infty+i\tau_0''} \frac{G(k)}{k-\zeta}\,dk + \int_{-\infty+i\tau_1}^{\infty+i\tau_1} \frac{H(k)}{k-\zeta}\,dk = -2\pi i\,G(\zeta) \tag{4.8.22}$$

where $\tau_1'' > \tau_1 > \tau_1'$. In the left-hand side of the equation Im ζ may take on all values between τ_0'' and τ_1 so that Eq. (4.8.22) provides the analytic continuation for $G(k)$ outside of its original region of analyticity.

By the same procedure

$$\int_{-\infty+i\tau_0}^{\infty+i\tau_0} \frac{G(k)}{k-\zeta}\,dk + \int_{-\infty+i\tau_1'}^{\infty+i\tau_1'} \frac{H(k)}{k-\zeta}\,dk = 2\pi i\,H(\zeta) \tag{4.8.23}$$

where $\tau_1' < \text{Im}\,\zeta < \tau_0$.

We see that representations (4.8.23) and (4.8.22) are both valid for $\tau_1 < \text{Im}\,\zeta < \tau_0$. Pick ζ to be in this region, and subtract (4.8.22) from (4.8.23):

$$\begin{aligned}2\pi i[H(\zeta) + G(\zeta)] &= \int_{-\infty+i\tau_0}^{\infty+i\tau_0} \frac{G(k)}{k-\zeta}\,dk - \int_{-\infty+i\tau_0''}^{\infty+i\tau_0''} \frac{G(k)}{k-\zeta}\,dk \\ &\quad + \int_{-\infty+i\tau_1'}^{\infty+i\tau_1'} \frac{H(k)}{k-\zeta}\,dk - \int_{-\infty+i\tau_1}^{\infty+i\tau_1} \frac{H(k)}{k-\zeta}\,dk \\ &= \oint \frac{G(k)}{k-\zeta}\,dk + \oint \frac{H(k)}{k-\zeta}\,dk\end{aligned}$$

Since ζ is not within either contour, the right-hand side of the above equation is zero, thus proving the theorem.

This is as far as we shall need to take the general theory. We have considered the Fourier integral theorem; we have discussed the analytic properties of Fourier transforms and obtained transforms and inversion formulas which apply when the functions do not satisfy the requirements of the Fourier integral theorem at both $+\infty$ and $-\infty$. We now consider some applications.

Faltung. The *faltung* of two functions f and h is defined by the integral

$$\frac{1}{\sqrt{2\pi}} \int_{-\infty}^{\infty} f(y)h(x-y)\,dy \tag{4.8.24}$$

It is called faltung (German for *folding*) because the argument of h is "folded" $(x - y)$ with respect to y. We shall show, when both f and h

are L^2 integrable in $(-\infty, \infty)$, that the Fourier transform of this integral is $F(k)H(k)$, that is, just the product of the two transforms. This relation is exceedingly useful in the solution of integral equations as discussed in Sec. 8.4.

The theorem follows directly from Parseval's formula as given in Eq. (4.8.17). For if $\bar{g}(y) = h(x - y)$, then

$$G = \frac{1}{\sqrt{2\pi}} \int_{-\infty}^{\infty} e^{+iky}\bar{h}(x - y)\, dy \quad \text{and} \quad \bar{G} = \frac{1}{\sqrt{2\pi}} \int_{-\infty}^{\infty} e^{-iky}h(x - y)\, dy$$

Now let $x - y = \xi$ so that

$$\bar{G} = \frac{e^{-ikx}}{\sqrt{2\pi}} \int_{-\infty}^{\infty} e^{ik\xi}h(\xi)\, d\xi = e^{-ikx}\, H(k)$$

Hence, by Eq. (4.8.17),

$$\frac{1}{\sqrt{2\pi}} \int_{-\infty}^{\infty} f(y)h(x - y)\, dy = \frac{1}{\sqrt{2\pi}} \int_{-\infty}^{\infty} F(k)H(k)e^{-ikx}\, dk \qquad (4.8.25)$$

which is just the theorem to be proved. In other words, the Fourier transform of FH is just the integral of Eq. (4.8.24). Because of the reciprocal relationship existing between functions and their Fourier transforms, we can give another form to Eq. (4.8.25):

$$\frac{1}{\sqrt{2\pi}} \int_{-\infty}^{\infty} f(x)h(x)e^{ikx}\, dx = \frac{1}{\sqrt{2\pi}} \int_{-\infty}^{\infty} F(l)H(k - l)\, dl$$

so that the faltung of the transforms FH is the transform of the product fh.

This theorem may be generalized by regarding the integral on the right-hand side of Eq. (4.8.25) as a function of x. By multiplying by $p(z - x)$ and integrating over x one obtains a function of z:

$$\iint_{-\infty}^{\infty} p(z - x)h(x - y)f(y)\, dy\, dx = \int_{-\infty}^{\infty} P(k)H(k)F(k)e^{-ikz}\, dk$$

To illustrate the value of Eq. (4.8.25), consider the simple integral equation

$$g(x) = \int_{-\infty}^{\infty} f(y)h(x - y)\, dy$$

where g and h are known and it is desired to find f. Take the Fourier transform of both sides of the equation. We have

$$G(k) = \sqrt{2\pi}\, F(k)H(k)$$

Hence $F(k) = \left[\dfrac{1}{\sqrt{2\pi}} \dfrac{G(k)}{H(k)}\right]$ or $f(x) = \dfrac{1}{2\pi} \displaystyle\int_{-\infty}^{\infty} \left[\dfrac{G(k)}{H(k)}\right] e^{-ikx}\, dk$

This is, of course, only a particular solution of the integral equation and, of course, is only valid when G/H is integrable L^2 in the region $(-\infty, \infty)$. A fuller discussion of this type of problem will be considered in Sec. 8.4.

Poisson Sum Formula. The Fourier integral theorems also help us to evaluate sums. For example, the very general sort of series $S = \sum_{n=-\infty}^{\infty} f(\alpha n)$ can be evaluated by the following process: Let f be $L^{(2)}$ in the region $(-\infty, \infty)$. If we define

$$f_+(x) = \begin{cases} 0; & x < 0 \\ f; & x > 0 \end{cases}; \quad f_-(x) = \begin{cases} f; & x < 0 \\ 0; & x > 0 \end{cases}$$

Then
$$f_+(x) = \frac{1}{\sqrt{2\pi}} \int_{-\infty + i\tau_0}^{\infty + i\tau_0} F_+(k) e^{-ikx}\, dx \tag{4.8.26}$$

Because of the integrability of f, constant τ_0 can be less than zero. The

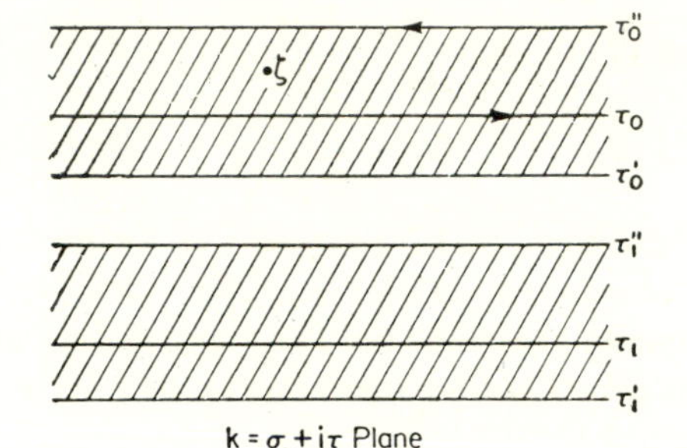

Fig. 4.41 Contours for analytic continuation of functions G and H.

transform $F_+(k)$ is analytic for $\tau > \tau_0'$, $\tau_0 > \tau_0'$, as shown in Fig. 4.41. We also have

$$f_-(x) = \frac{1}{\sqrt{2\pi}} \int_{-\infty + i\tau_1}^{\infty + i\tau_1} F_-(k) e^{-ikx}\, dx \tag{4.8.27}$$

where τ_1 can be greater than zero and $F_-(k)$ is analytic for $\tau < \tau_1''$, $\tau_1 < \tau_1''$.

With these definitions in mind, it becomes convenient to break S up into two parts:

$$S_+ = \sum_{n=0}^{\infty} f_+(\alpha n); \quad S_- = \sum_{n=-\infty}^{-1} f_-(\alpha n)$$

Consider S_+ first. Introduce Eq. (4.8.26) into S_+:

$$S_+ = \frac{1}{\sqrt{2\pi}} \sum \int_{-\infty + i\tau_0}^{\infty + i\tau_0} F_+(k) e^{-ik\alpha n}\, dk$$

Because of the absolute convergence of F we can exchange sum and integral:

$$S_+ = \frac{1}{\sqrt{2\pi}} \int_{-\infty + i\tau_0}^{\infty + i\tau_0} F_+(k) \sum_{n=0}^{\infty} e^{-ik\alpha n}\, dk = \frac{1}{\sqrt{2\pi}} \int_{-\infty + i\tau_0}^{\infty + i\tau_0} \left[\frac{F_+(k)}{1 - e^{-ik\alpha}} \right] dk$$

Since F_+ is analytic in the upper half plane, we can add a semicircle in the upper half plane to the line of integration $-\infty + i\tau_0$ to $\infty + i\tau_0$. Since τ_0 may be less than zero, we can evaluate, by Cauchy's integral formula, the poles occurring at the zeros of the denominator, *i.e.*, at $k = 2m\pi/\alpha$ where m is any positive or negative integer. Hence

$$S_+ = \frac{\sqrt{2\pi}}{\alpha} \sum_{m=-\infty}^{\infty} F_+\left(\frac{2m\pi}{\alpha}\right)$$

Similarly,

$$S_- = \frac{\sqrt{2\pi}}{\alpha} \sum_{m=-\infty}^{\infty} F_-\left(\frac{2m\pi}{\alpha}\right)$$

Hence

$$S_+ + S_- = \frac{\sqrt{2\pi}}{\alpha} \sum_{m=-\infty}^{\infty} (F_+ + F_-) = S$$

or

$$\sum_{n=-\infty}^{\infty} f(\alpha n) = \frac{\sqrt{2\pi}}{\alpha} \sum_{m=-\infty}^{\infty} F\left(\frac{2m\pi}{\alpha}\right) \tag{4.8.28}$$

where F is the Fourier transform of f. This is the *Poisson sum formula.*

As a simple example let $f(n) = 1/(1 + n^2)$. The Fourier transform of f can be evaluated by the calculus of residues:

$$F(k) = \sqrt{\pi/2}\, e^{-|k|}$$

Applying Eq. (4.8.28), we have

$$\sum_{n=-\infty}^{\infty} \left(\frac{1}{1 + \alpha^2 n^2}\right) = \frac{\pi}{\alpha} \left\{\sum_{m=0}^{\infty} e^{-(2m\pi/\alpha)} + \sum_{m=-\infty}^{-1} e^{2m\pi/\alpha}\right\}$$

$$= \frac{\pi}{\alpha} \left\{\frac{1}{1 - e^{-2\pi/\alpha}} + \frac{e^{-2\pi/\alpha}}{1 - e^{-2\pi/\alpha}}\right\}$$

or

$$\sum_{n=-\infty}^{\infty} \left(\frac{1}{1 + \alpha^2 n^2}\right) = \left(\frac{\pi}{\alpha}\right) \coth\left(\frac{\pi}{\alpha}\right) \tag{4.8.29}$$

We shall employ Poisson's formula often in later chapters [for one application see Eq. (7.2.30)].

The Laplace Transform. The Laplace transform of $f(x)$ is defined by

$$F_l(p) = \int_0^\infty f(x) e^{-px}\, dx \tag{4.8.30}$$

The discussion of the properties of the Laplace transform may be referred back to that for the Fourier transform, for, as we shall see, the Laplace transform is a special case of the latter. Consider the Fourier transform of

$$f_+(x) = \begin{cases} f(x); & x > 0 \\ 0; & x < 0 \end{cases}$$

This Fourier transform is

$$F_+(k) = \frac{1}{\sqrt{2\pi}} \int_0^\infty f(x) e^{ikx}\, dx$$

Comparing with the definition of $F_l(p)$ [Eq. (4.8.30)], we find that

$$F_l(p) = \sqrt{2\pi}\, F_+(ip) \tag{4.8.31}$$

Armed with this equation the known theorems on F_+ can be employed to obtain the corresponding theorems for $F_l(p)$. For example, from the inversion formula (4.8.19), we have

$$f_+(x) = \frac{1}{\sqrt{2\pi}} \int_{-\infty+i\tau_0}^{\infty+i\tau_0} F_+(k) e^{-ikx}\, dk$$

Let $k = ip$. Then we obtain, for the inverse Laplace transform,

$$f_+(x) = \frac{1}{2\pi i} \int_{-i\infty+\tau_0}^{i\infty+\tau_0} F_l(p) e^{px}\, dp; \quad \text{Re } x > 0 \tag{4.8.32}$$

As an example consider $f(x) = e^{iqx}$. Then

$$F_l(p) = 1/(p - iq)$$

Introducing $F_l(p)$ into (4.8.32)

$$f_+(x) = \frac{1}{2\pi i} \int_{-i\infty+\tau_0}^{i\infty+\tau_0} \frac{e^{px}}{p - iq}\, dp$$

This integral may be evaluated by the calculus of residues. For $x > 0$, we may close the contour by a semicircle in the left half of the plane, so that $f_+ = e^{iqx}$. For x negative, the semicircle must be placed in the right half plane. But in this region the integrand is analytic. Hence $f_+ = 0$ for $x < 0$ as it should be.

The faltung theorem has a central position in the application of Laplace transforms. Consider

$$\int_{-\infty}^{\infty} f_+(y) h_+(x - y)\, dy = \int^x f(y) h(x - y)\, dy$$

since $f_+(y) = 0$ for $y < 0$ and $h_+(x - y) = 0$ if $x < y$. Substituting in Eq. (4.8.25), the faltung theorem for Laplace transforms is obtained:

$$\int_0^x f(y) h(x - y)\, dy = \int_{-\infty+i\tau_0}^{\infty+i\tau_0} F_+(k) H_+(k) e^{-ikx}\, dk$$

$$\int_0^x f(x) h(x - y)\, dy = \frac{1}{2\pi i} \int_{-i\infty+\tau_0}^{i\infty+\tau_0} F_l(p) H_l(p) e^{px}\, dp \tag{4.8.33}$$

In words, the Laplace transform of $\int_0^x f(x)h(x-y)\,dy$ is the product of the Laplace transforms of f and h.

As an example of the usefulness of the faltung theorem, we apply it to the solution of the Volterra integral equation which arises in the discussion of initial-value problems.

$$f(t) = \int_0^t k(t-\tau)g(\tau)\,d\tau$$

Here g is unknown while f is known. The solution of this will be discussed in Chap. 8. Here we shall content ourselves with indicating, without proof, the manner in which a solution may be obtained. Take the Laplace transform of both sides of the above equation, obtaining

$$F(p) = K(p)G(p) \quad \text{or} \quad G(p) = [F(p)/K(p)]$$

and

$$g(t) = \frac{1}{2\pi i}\int_{-i\infty+\tau_0}^{i\infty+\tau_0} \left[\frac{F(p)}{K(p)}\right] e^{pt}\,dp$$

Further discussion and application of the Laplace transform will be found in Chaps. 5, 8, 11, and 12.

The Mellin Transform. Another transform of considerable value, also closely related to the Fourier transform, is given by

$$F_m(s) = \int_0^\infty f(x)x^{s-1}\,dx \tag{4.8.34}$$

A simple example is given by the Mellin transform of e^{-x} which is just $\Gamma(s)$. To obtain the relation of (4.8.34) to the Fourier transform, let

$$x = e^z$$

Then $F_m(s) = \int_{-\infty}^\infty f(e^z)e^{sz}\,dz = \int_{-\infty}^\infty g(z)e^{sz}\,dz; \quad f(e^z) = g(z)$

Comparing this with

$$G(k) = \frac{1}{\sqrt{2\pi}}\int_{-\infty}^\infty g(z)e^{ikz}\,dz$$

we see that

$$F_m(s) = \sqrt{2\pi}\,G(-is) \tag{4.8.35}$$

where $G(k)$ is the Fourier transform of $g(z)$.

We can now rewrite the theorems given for the Fourier transform theory so as to apply to Mellin transforms. First let us consider the integrability conditions. The condition that the Lebesgue integral

$$\int_{-\infty}^\infty |g(z)|^2\,dz$$

exists is equivalent to requiring the existence of

$$\int_0^\infty |f(x)|^2 \left(\frac{dx}{x}\right) \tag{4.8.36}$$

Second let us consider the inversion formula; for $g(z)$ it reads

$$g(z) = \frac{1}{\sqrt{2\pi}} \int_{-\infty}^{\infty} e^{-ikz} G(k)\, dk = \left(\frac{1}{2\pi i}\right) \int_{-i\infty}^{i\infty} e^{-sz} F_m(s)\, ds$$

Hence

$$f(x) = \left(\frac{1}{2\pi i}\right) \int_{-i\infty}^{i\infty} \frac{F_m(s)}{x^s}\, ds \tag{4.8.37}$$

In the event that condition (4.8.36) is not satisfied, Eq. (4.8.37) will be integrable only for certain values of Re s, say Re $s > \sigma_0$. In that event, integral (4.8.34) will define $F_m(s)$ for only the region to the right of $s = \sigma_0$, and the remainder of its dependence must be obtained by analytic continuation. An example of this procedure was discussed in Sec. 4.5 (the Γ function). For these circumstances inversion formula (4.8.37) is no longer correct. However, if we consider the function $x^{\sigma_0'}f(x)$ $(\sigma_0' > \sigma_0)$, then the above discussion again becomes valid. The transform of $x^{\sigma_0'}f(x)$ is just $F_m(s + \sigma_0')$. Applying Eq. (4.8.37) we find that

$$x^{\sigma_0'}f(x) = \left(\frac{1}{2\pi i}\right) \int_{-i\infty}^{i\infty} \frac{F_m(s + \sigma_0')}{x^s}\, ds$$

Let $s + \sigma_0' = \xi$. Then

$$f(x) = \left(\frac{1}{2\pi i}\right) \int_{-i\infty+\sigma_0'}^{i\infty+\sigma_0'} \frac{F_m(\xi)}{x^\xi}\, d\xi \tag{4.8.38}$$

If it is impossible to obtain one value of Re (s) which will allow convergence for the entire range of x in (4.8.34), it then becomes necessary to break up the range of integration and define two functions with different convergence factors as was done for the Fourier case [see Eq. (4.8.18) *et seq.*]. This is a straightforward generalization and will be included as a problem.

Finally we turn our attention to the faltung theorem for the Mellin transform. Starting from Eq. (4.8.25) and making the following substitutions:

$$e^y = \eta;\quad f(y) = v(\eta)$$
$$e^x = \xi;\quad h(x - y) = w(e^{x-y}) = w(\xi/\eta)$$

we obtain

$$\int_0^\infty v(\eta) w\left(\frac{\xi}{\eta}\right)\left(\frac{d\eta}{\eta}\right) = \frac{1}{2\pi i} \int_{-i\infty+\sigma_0'}^{i\infty+\sigma_0'} V_m(s) W_m(s) \left(\frac{ds}{\xi^s}\right) \tag{4.8.39}$$

Another form can also be extracted:

$$\int_0^\infty v(\xi) w(\xi) \xi^{s-1}\, d\xi = \frac{1}{2\pi i} \int_{-i\infty+\tau_0'}^{i\infty+\tau_0'} V_m(\rho) W_m(s - \rho)\, d\rho \tag{4.8.40}$$

As an example of the utility of Eq. (4.8.40) consider the evaluation of

$$I = \frac{1}{2\pi i} \int_{\sigma_0'-i\infty}^{\sigma_0'+i\infty} \Gamma(a + \rho)\Gamma(s - \rho)\, d\rho;\quad \text{where } -a < \sigma_0' < s$$

Recall that, if $F_m(s) = \Gamma(a+s)$, then $f(x) = e^{-x}x^a$. We can now apply Eq. (4.8.40):

$$I = \int_0^\infty (e^{-x}x^a)(e^{-x})(x^{s-1})\,dx = \int_0^\infty e^{-2x}\,x^{a+s-1}\,dx = \frac{\Gamma(a+s)}{2^{a+s}}$$

In I let $s = a$, and take the integration along the imaginary axis:

$$\int_0^\infty |\Gamma(a+i\tau)|^2\,d\tau = \left[\frac{\pi\Gamma(2a)}{2^{2a}}\right]$$

an interesting relation between values of $\Gamma(z)$ off the real axis to those for real argument $(z = a)$.

Having now discussed the behavior of functions of a complex variable z, we shall next investigate their relation to differential equations.

Problems for Chapter 4

4.1 Prove that

$$\int_0^{2\pi} \frac{\sin^2\vartheta\,d\vartheta}{a+b\cos\vartheta} = \frac{2\pi}{b^2}[a-\sqrt{(a^2-b^2)}]; \quad a > b > 0$$

4.2 Prove that

$$\int_0^{2\pi} \frac{e^{q\cos\vartheta}}{1-2p\sin\vartheta+p^2}\left\{\begin{matrix}\cos\\ \sin\end{matrix}(q\sin\vartheta)\ \begin{matrix}-\\+\end{matrix}\ p\ \begin{matrix}\sin\\ \cos\end{matrix}(q\sin\vartheta+\vartheta)\right\}d\vartheta = 2\pi\ \begin{matrix}\cos\\ \sin\end{matrix}(pq)$$

4.3 Show that

$$\int_{-\infty}^\infty \frac{\cos(x)\,dx}{(x^2+a^2)(x^2+b^2)} = \frac{\pi}{a^2-b^2}\left[\frac{e^{-b}}{b} - \frac{e^{-a}}{a}\right]; \quad \operatorname{Re} a > \operatorname{Re} b > 0$$

What is the value of the integral when $a = b$; $\operatorname{Re} a > 0$?

4.4 Show that

$$\int_0^{2\pi} e^{\cos\vartheta}\cos(n\vartheta - \sin\vartheta)\,d\vartheta = \frac{2\pi}{n!}$$

4.5 Prove that

$$\int_0^\infty \frac{x^{2a-1}}{1+x^2}\,dx = (\tfrac{1}{2}\pi)\csc(\pi a); \quad 0 < a < 1$$

4.6 Calculate

$$\int_{-\infty}^\infty \frac{x^2\,dx}{(1+x^2)(1-2x\cos\vartheta+x^2)}$$

4.7 Prove that

$$\int^{\infty} \frac{x^a}{(1+x^2)^2}\,dx = \frac{\pi(1-a)}{4\cos(\frac{1}{2}\pi a)}; \qquad -1 < a < 3$$

4.8 Prove that

$$\int_0^{\infty} \left[\frac{\ln^2 z}{1+z^2}\right] dz = \frac{\pi^3}{8}$$

4.9 Discuss the real integral $\int_a^b (b-x)^\mu (x-a)^{n-\mu-1} F(x)\,dx$, where a and b are real, $b > a$, $\mu > -1$, n is an integer greater than μ, and $z^n F(z)$ approaches zero as $|z|$ approaches infinity. The function $F(z)$ is analytic over the whole complex plane except for a finite number of poles, none of which are on the real axis to the left of or at $z = b$. Show that the integral equals

$$(1/2i \sin \pi\mu) \int_C (z-a)^{n-\mu-1}(z-b)^\mu F(z)\,dz$$

where the integral goes from $z = b + \epsilon$ to $-\infty$ just below the real axis, goes in a large circle counterclockwise to just above the real axis at $-\infty$, and returns to $z = b + \epsilon$ by a path just above the real axis. Hence show that the integral equals

$$\pi \csc(\pi\mu)\,\Sigma[\text{residues of } (z-b)^\mu (z-a)^{n-\mu-1} F(z) \text{ at poles of } F]$$

Show that, for $0 < \mu < 1$, $k > 1$,

$$\int_{-1}^{1} \left(\frac{1-x}{1+x}\right)^\mu \frac{dx}{(x-k)^2} = 2\pi\mu \csc(\pi\mu)\,(k-1)^{\mu-1}(k+1)^{-\mu-1}$$

4.10 Prove that

$$\int_0^{\infty} \frac{x^{-a}\,dx}{1 + 2x\cos\vartheta + x^2} = \frac{\pi}{\sin \pi a}\left(\frac{\sin a\vartheta}{\sin \vartheta}\right); \qquad -1 < a < 1,\ -\pi < \vartheta < \pi$$

4.11 Show from Eq. (4.2.19) that

$$\int_{-\infty}^{\infty} \frac{\sin(kx)\sin(k'x)}{x^2 - \xi^2}\,dx = \left(\frac{\pi}{\xi}\right)\begin{cases} \sin(k\xi)\cos(k'\xi); & k \le k' \\ \cos(k\xi)\sin(k'\xi); & k \ge k' \end{cases}$$

4.12 Show from Eq. (4.2.19) that

$$\int_{-\infty}^{\infty} \frac{\cos(px) - \cos(qx)}{x^2}\,dx = \pi(p-q)$$

4.13 Show from Eq. (4.2.19) that

$$\int_{-\infty}^{\infty} \frac{-(x^2 - ab)\sin x + (a+b)x\cos x}{x(x^2+a^2)(x^2+b^2)}\,dx = \frac{\pi}{ab}$$

4.14 Show that

$$\left(\frac{1}{2\pi i}\right)\int_{-\infty-i\beta}^{\infty-i\beta} \ln\left[\left(\frac{z^2+1}{z^2}\right)\left(1-\frac{\tan^{-1}z}{z}\right)\right]\frac{dz}{z} = -\tfrac{1}{2}\ln 3$$

4.15 Consider the integral

$$I(z) = \int_0^\infty \frac{f(t)}{t-z}\,dt$$

where the contour goes along the real axis except for a semicircle described positively about the point z, where z is real. Show that

$$I = \pi i\, f(z) + \mathcal{P}\int^\infty \frac{f(t)}{t-z}\,dt$$

where $\mathcal{P}$ denotes the principal value of the integral.

4.16 A function $f(z)$ is analytic in the strip $|\text{Im } z| < \alpha$. Show that

$$f(z) = f_-(z) - f_+(z)$$

where $\quad f_- = \dfrac{1}{2\pi i}\displaystyle\int_{-\infty+i\beta}^{\infty+i\beta}\frac{f(t)}{t-z}\,dt;\quad f_+ = \dfrac{1}{2\pi i}\displaystyle\int_{-\infty-i\beta}^{\infty-i\beta}\frac{f(t)}{t-z}\,dt$

with $\beta < \alpha$. Show that f_- is analytic for $\text{Im } z < \beta$ while f_+ is analytic for $\text{Im } z > -\beta$.

4.17 Function $\psi(x)$ is defined by the integral

$$\psi(x) = \frac{1}{2\pi i}\left\{\int_{-\infty+i\sigma}^{\infty+i\sigma}\frac{(1+z^2)e^{-izx}}{(z^2-z_0^2)(z+i\alpha)}\,dz - \int_{-\infty+i\tau}^{\infty+i\tau}\frac{(1+z^2)e^{-izx}}{(z^2-z_0^2)(z-i\alpha)}\,dz\right\}$$

where $\tau < 1$ and $\sigma > -1$. Show that

$$\psi(x) = \left\{\left(\frac{\alpha^2-1}{\alpha^2+z_0^2}\right)e^{\alpha|x|} + \left(\frac{z_0^2+1}{z_0\sqrt{\alpha^2+z_0^2}}\right)\cos\left[z_0|x| - \tan^{-1}\left(\frac{\alpha}{z_0}\right)\right]\right\}$$

4.18 Show that $z = 1$ is a singularity of the function represented by the following series, within its radius of convergence

$$f(z) = \sum_{n=0}^\infty \frac{\Gamma(\frac{1}{2})(-z)^n}{\Gamma(1+n)\Gamma(\frac{1}{2}-n)}$$

4.19 The generalized hypergeometric function $F(a_0,a_1, \ldots ,a_s|c_1,c_2, \ldots ,c_s|z)$ is defined by the power series

$$1 + \frac{a_0a_1\cdots a_s}{c_1\cdots c_s}z + \frac{a_0(a_0+1)a_1(a_1+1)\cdots a_s(a_s+1)}{c_1(c_1+1)\cdots c_s(c_s+1)}\left(\frac{z^2}{2!}\right) + \cdots$$

Show that F is singular at $z = 1$ and that for $a_i < c_i$

$$F \to \frac{\left[\prod_{n=1}^s \Gamma(c_n)\right]\Gamma(p)}{\left[\prod_{m=0}^s \Gamma(a_m)\right](1-z)^p};\quad \text{as } z \to 1$$

where $p = \sum_{n=1}^{s} (a_n - c_n) + a_0$.

4.20 Prove that

$$\sum_{n=-\infty}^{\infty} \frac{1}{(n^4 + a^4)} = \left(\frac{\pi}{\sqrt{2}\, a^3}\right) \left[\frac{\sinh(\pi a \sqrt{2}) + \sin(\pi a \sqrt{2})}{\cosh(\pi a \sqrt{2}) - \cos(\pi a \sqrt{2})}\right]$$

4.21 Prove that

$$\sum_{n=-\infty}^{\infty} \frac{1}{(n^2 + a^2)(n^2 + b^2)} = \frac{\pi}{(b^2 - a^2)} \left[\frac{\coth(\pi a)}{a} - \frac{\coth(\pi b)}{b}\right]$$

4.22 Show that an even integral function may be expressed as

$$f(z) = f(0) \prod_{n=1}^{\infty} \left[1 - \left(\frac{z^2}{a_n^2}\right)\right]$$

where $f(a_n) = 0$ and where we consider only one of the two roots a_n, $-a_n$ in the product. Assume the restrictions on $F(z)$ given on page 385. Show that, if $f(0) = 1$,

$$f''(0) = -2 \sum_{n=1}^{\infty} \left(\frac{1}{a_n^2}\right); \quad f^{(iv)}(0) = 3[f''(0)]^2 - 12 \sum_{n=1}^{\infty} \left(\frac{1}{a_n^4}\right)$$

Employing product representation (4.3.9) for $(1/z) \sin z$, show that

$$\sum_{n=1}^{\infty} \left(\frac{1}{n^2}\right) = \left(\frac{\pi^2}{6}\right); \quad \sum_{n=1}^{\infty} \left(\frac{1}{n^4}\right) = \left(\frac{\pi^4}{90}\right)$$

4.23 Prove that

$$e^{az} - e^{bz} = (a - b) z e^{\frac{1}{2}(a+b)z} \prod_{n=1}^{\infty} \left[1 + \frac{(a - b)^2 z^2}{4n^2\pi^2}\right]$$

4.24 Prove that

$$\sum_{m=-\infty}^{\infty} \sum_{n=-\infty}^{\infty} \frac{1}{(m^2 + a^2)(n^2 + a^2)} = \frac{\pi^2}{ab} \coth(\pi a) \coth(\pi b)$$

4.25 Employing the generalized Euler transformation, show that $F(a_0,a_1, \ldots | b_1,b_2, \ldots | z) = F(a_1 \cdots | b_2 \cdots | z)$

$$+ z\left(\frac{b_1 - a_0}{b_1}\right)\left(\frac{a_1a_2 \cdots}{b_2b_3 \cdots}\right) F(a_1 + 1, a_2 + 1, \ldots | b_2 + 1 \cdots | z)$$
$$+ \left(\frac{z^2}{2!}\right) \frac{(b_1 - a_0)(b_1 - a_0 + 1)a_1(a_1 + 1)a_2(a_2 + 1) \cdots}{b_1(b_1 + 1)b_2(b_2 + 1)b_3(b_3 + 1) \cdots} \cdot$$
$$\cdot F(a_1 + 2, a_2 + 2, \ldots | b_2 + 2 \cdots | z) + \cdots$$

4.26 By integrating $e^{-kz} z^{n-1}$ around the sector enclosed by the real axis, the line $\varphi = \alpha$ $(z = re^{i\varphi})$, the small arc $r = \epsilon$ and the large arc $r = R$, prove that

$$\int^{\infty} x^{a-1} e^{-px \cos \vartheta} \begin{matrix}\cos\\ \sin\end{matrix} (px \sin \vartheta)\, dx = p^{-m}\Gamma(a) \begin{matrix}\cos\\ \sin\end{matrix} (a\vartheta)$$

where $\vartheta < \frac{1}{2}\pi$, a and p are real and greater than zero.

4.27 By integrating $z^{a-1}e^z$ about the contour starting from a point just to the right of the origin $(z = \epsilon)$ to $-\infty$ just below the real axis, to $-\infty - i\infty$, to $b - i\infty$ to $b + i\infty$, to $-\infty + i\infty$, to $-\infty$, and thence to $z = \epsilon$ by a path just above the real axis, prove that, for $0 < a < 1$ and $b > 0$,

$$\int_{-\infty}^{\infty} e^{iy}(b + iy)^{a-1}\, dy = 2e^{-b} \sin(\pi a)\, \Gamma(a)$$

and therefore that

$$\int_0^{\frac{1}{2}\pi} \cos[\tan \vartheta - (1 - a)\vartheta] \sec^{a+1}(\vartheta)\, d\vartheta = \left(\frac{1}{e}\right) \sin(\pi a)\, \Gamma(a)$$

4.28 Show that

$$J = \iiint x^{l-1}y^{m-1}z^{n-1}\, dx\, dy\, dz$$

integrated over an octant of the volume bounded by the surface $(x/a)^p + (y/b)^q + (z/c)^r = 1$ is

$$\left(\frac{a^lb^mc^n}{pqr}\right) \frac{\Gamma(l/p)\Gamma(m/q)\Gamma(n/r)}{\Gamma[(l/p) + (m/q) + (n/r) + 1]}$$

4.29 Prove that

$$\Gamma(z)\Gamma\left(z + \frac{1}{n}\right)\Gamma\left(z + \frac{2}{n}\right) \cdots \Gamma\left(z + \frac{n-1}{n}\right) = (2\pi)^{\frac{1}{2}(n-1)} n^{\frac{1}{2} - nz}\Gamma(nz)$$

4.30 Express the integral

$$\int_C e^{-t^2} t^{n-\lambda-1}\, dt$$

occurring in Eq. (4.5.26), in terms of a gamma function.

4.31 Show that

$$1 - \frac{\cos(\pi\alpha)}{\cosh(\pi z)} = \frac{2\pi\Gamma(\frac{1}{2} - iz)\Gamma(\frac{1}{2} + iz)}{\Gamma\left(\frac{\alpha + iz}{2}\right)\Gamma\left(\frac{\alpha - iz}{2}\right)\Gamma\left(1 - \frac{\alpha + iz}{2}\right)\Gamma\left(1 - \frac{\alpha - iz}{2}\right)}$$

4.32 Show that

$$|\Gamma(iy)|^2 = \pi/(y \sinh \pi y)$$

4.33 Show that

$$|\Gamma(z+1)|^2 \to 2\pi r^{2x+1} e^{-2(y\varphi+x)} \left[1 + \frac{x}{6r^2} + \frac{1}{72r^2} + \cdots\right]$$

where $z = x + iy$, $r^2 = x^2 + y^2$, and $\varphi = \tan^{-1}(y/x)$.

Show that the phase of $\Gamma(z)$ is given asymptotically by

$$\tfrac{1}{2}y \ln(x^2+y^2) + (x - \tfrac{1}{2}) \tan^{-1}\left(\frac{y}{x}\right) \\ - y\left[1 + \frac{1}{12(x^2+y^2)} - \frac{3x^2-y^2}{360(x^2+y^2)^3} + \cdots\right]$$

4.34 Consider the integral

$$f(t) = \frac{1}{2\pi i}\int_c \frac{e^{zt-\lambda\sqrt{z}}}{z(\kappa+\sqrt{z})}\,dz$$

where the branch line for the integral is taken along the negative real axis and the contour starts at $(-\infty)$ below the branch line, encircles the branch point at the origin in a positive fashion, and goes out to $(-\infty)$ above the branch line. Break up the integral into three parts, consisting of the two integrals extending from $(-\infty)$ to a small quantity ϵ_0 and an integral over a circle of radius ϵ_0 about the origin. Show that, in the limit of ϵ_0 equal to zero, the third integral is $1/\kappa$ while the first two may be combined into

$$-\frac{1}{\pi}\sum_{n=0}^{\infty}\left(\frac{1}{\kappa^{n+1}}\right)\frac{d^n}{d\lambda^n}\left[\int_0^\infty e^{-ut}\sin(\lambda\sqrt{u})\,\frac{du}{u}\right]$$

Show that

$$\int_0^\infty e^{-ut}\sin(\lambda\sqrt{u})\,\frac{du}{u} = \sqrt{\pi}\int_0^{\lambda/\sqrt{t}} e^{-\xi^2}\,d\xi$$

and hence that

$$f(t) = \left(\frac{1}{\kappa}\right) - \sqrt{\pi}\sum_{n=0}^{\infty}\left(\frac{1}{\kappa^{n+1}}\right)\frac{d^n}{d\lambda^n}\left[\int_0^{\lambda/\sqrt{t}} e^{-\xi^2}\,d\xi\right]$$

4.35 Consider the integral

$$g(z) = \int_{-i\infty}^{i\infty} G(t)\Gamma(-t)(-z)^t\,dt$$

where $G(t)$ is regular in the half plane $\text{Re}\, t > 0$, $[G(t)z^t/\Gamma(t+1)]$ approaches zero for large values of $\text{Re}\, t$. The contour is along the imaginary axis except for a small semicircle going clockwise about the origin. Show by closing the contour with a semicircle of large radius in the half plane

Re $t > 0$ that

$$g(z) = 2\pi i \sum_{n=0}^{\infty} G(n) \left(\frac{z^n}{n!}\right)$$

Hence show that

$$F(a,b|c|z) = \frac{\Gamma(c)}{\Gamma(a)\Gamma(b)} \sum_{n=0}^{\infty} \frac{\Gamma(a+n)\Gamma(b+n)}{\Gamma(c+n)} \left(\frac{z^n}{n!}\right)$$
$$= \frac{\Gamma(c)}{2\pi i\Gamma(a)\Gamma(b)} \int_{-i\infty}^{i\infty} \frac{\Gamma(a+t)\Gamma(b+t)}{\Gamma(c+t)} \Gamma(-t)(-z)^t\, dt$$

for $\mathrm{Re}(a + b - c) < 0$, a and b not negative integers. Show that the contour may also be closed by a circle in the negative real plane. Deduce that

$$\frac{\Gamma(a)\Gamma(b)}{\Gamma(c)} F(a,b|c|z) = \frac{\Gamma(a)\Gamma(a-b)}{\Gamma(a-c)} (-z)^{-a} F\left(a,1-c+a|1-b+a|\frac{1}{z}\right)$$
$$+ \frac{\Gamma(b)\Gamma(b-a)}{\Gamma(b-c)} (-z)^{-b} F\left(b,1-c+b|1-a+b|\frac{1}{z}\right)$$

[see Eq. (5.2.49)].

4.36 Prove that

$$\int_{-\frac{1}{2}\pi}^{\frac{1}{2}\pi} (\cos\vartheta)^{a-2}\, e^{it\vartheta}\, d\vartheta = \frac{\pi\Gamma(a-1)}{2^{a-2}\Gamma[\frac{1}{2}(a+t)]\Gamma[\frac{1}{2}(a-t)]} \qquad a > 1$$

by calculating $\int [z + (1/z)]^{a-2} z^{t-1}\, dz$ around a contour running along the imaginary axis from $+i$ to $-i$ and returning to $+i$ via the right-hand half of the unit circle.

4.37 Using the method of steepest descents, show from the definition

$$H_\nu^{(1)}(z) = \left(\frac{e^{-i\nu\pi/2}}{\pi}\right) \int_{-\frac{1}{2}\pi+i\infty}^{\frac{1}{2}\pi-i\infty} e^{i(z\cos\varphi+\nu\varphi)}\, d\varphi$$

that

$$H_\nu^{(1)}(\nu\sec\alpha) \simeq \frac{e^{i[\nu(\tan\alpha-\alpha)+(\pi/4)]}}{\sqrt{\frac{1}{2}\nu\pi\tan\alpha}} \left[1 - i\,\frac{1 + (\frac{5}{3})\cot^2\alpha}{8\tan\alpha} + \cdots\right]$$

4.38 Show that the path of integration along the real axis for the integral

$$I = \int_{-\infty}^{\infty} e^{i(t^3-qt)}\, dt$$

may be deformed into the contour coming from $-\infty$ to 0 along a line of constant phase $\varphi = \frac{5}{6}\pi$ $(z = re^{i\varphi})$ and from the origin to $+\infty$ along a line of constant phase $\varphi = \frac{1}{6}\pi$, that consequently

$$I = \tfrac{2}{3} \sum_{m=0} \cos\,[(\tfrac{2}{3}m + \tfrac{1}{6})\pi]\Gamma[\tfrac{1}{3}(m+1)] \left[\frac{(1-q)^m}{m!}\right]$$

Why is q restricted to the real axis?

4.39 The method of steepest descent must be modified when $f''(t_0)$ is small. Expand the $f(t)$ in the integral described in Eq. (4.6.6)

$$J = \int_c e^{zf(t)}\, dt$$

in powers of $(t - t_0)$, where $f'(t_0) = 0$. Show that the appropriate transformation for f'' small is

$$t - t_0 = \alpha + \beta s$$

where $\alpha = -[f''(t_0)/f'''(t_0)]$ and $\beta = (6i/zf''')^{\frac{1}{3}}$

Show that

$$J \simeq \exp\left\{zf + \frac{1}{3}\frac{z[f''(t_0)]^3}{[f'''(t_0)]^2}\right\} \int_{c'} e^{i(s^3 - qs)}\, ds$$

where $q = \left(\dfrac{z}{2}\right)\dfrac{(f'')^2}{f'''}\left(\dfrac{i6}{zf'''}\right)^{\frac{1}{3}}$

4.40 Show that

$$H_p^{(1)}(p) = (e^{-\frac{ip\pi}{2}}/\pi) \int_{-\frac{1}{2}\pi + i\infty}^{\frac{1}{2}\pi - i\infty} e^{ip(\cos\varphi - \varphi)}\, d\varphi \to \left(\frac{1}{\pi}\right) e^{-\frac{1}{3}\pi i} \sqrt{3}\left(\frac{6}{p}\right)^{\frac{1}{3}} \Gamma(\tfrac{4}{3});$$

as $p \to \infty$

4.41 Show that the conformal transformation

$$z = x + iy = w + \sqrt{w^2 - 1}; \quad w = u + iv$$

transforms the real w axis into the part of the real z axis for $|x| \geq 1$, plus the unit circle $z = e^{i\varphi}$. Plot the lines on the z plane corresponding to the six lines $u = 0, \pm 1$, $v = 0, \pm 1$.

4.42 Show that the transformation

$$z = w + iae^{i\varphi}\tan\psi + \frac{a^2 e^{2i\varphi}}{w + iae^{i\varphi}\tan\psi}$$

(a, φ, ψ constants) transforms the circle of radius $a \sec\psi$ centered at the origin on the w plane into an arc of a circle the chord of which is of length $4a$ and is inclined at an angle φ with respect to the x axis. Plot the lines on the z plane corresponding to $u = 0$, $|v| \geq a \sec\psi$, to $v = 0$, $|u| \geq a \sec\psi$, for $a = 1$, $\varphi = 30°$, $\psi = 15°$. What is the angular extent of the arc on the z plane corresponding to the circle $a \sec\psi\, e^{i\vartheta}$ on the w plane?

4.43 Show that the Schwarz-Christoffel transformation transforming the interior of the four-sided "polygon" formed by the lines $x = 0$; $y \geq 0$; $y = 0$, $x \geq 0$; $y = -a$, $x \geq -b$; $x = -b$, $y \geq -a$ onto the upper half of the w plane is

$$z = \frac{2a}{\pi}\tanh^{-1}\sqrt{\frac{a^2 - wb^2}{a^2(1 + w)}} - \frac{2b}{\pi}\tan^{-1}\sqrt{\frac{a^2 - wb^2}{b^2(1 + w)}}$$

Where on the w plane are the points $z = 0$, $z = -b - ia$, $z = -b$, $z = -ib$, $z = -b - ia$?

4.44 Show that the conformal transformation

$$w = \tan\left(\tfrac{1}{2}\cos^{-1} z^2\right)$$

transforms the crossed lines $y = 0$, $-1 \leq x \leq 1$; $x = 0$, $-1 \leq y \leq 1$ into the real w axis. Where are the points $z = 1, i, -1, -i$ on the w plane? What physical problems could be solved by the use of this transformation?

4.45 A thin metal fin, kept at temperature T_0, is perpendicular to a metal plate at zero temperature so that the edge of the fin is parallel to and a distance a from the plate. The whole is immersed in a thermal conductor. By use of the Schwarz-Christoffel transformation show that the steady-state temperature distribution in the water is given by the transformation

$$z = a \sinh\left[(\pi/2T_0)(U + iT)\right]$$

where the lines T = constant are the isothermals. Show that the strip of fin from the edge to b cm in from the edge, of length L, loses

$$(4LT_0\kappa/\pi)\cosh^{-1}\left[(a + b)/a\right] \quad \text{cal per sec}$$

where κ is the heat conductivity of the medium.

4.46 Use the Schwarz-Christoffel transformation to show that the region of the z plane above the line $y = -a$ and "outside" the two parts of the x axis, one for $x \geq 0$, the other for $x \leq -b$ (two parallel planes a distance a apart, the upper plane broken by a slit of width b) is transformed to the upper half of the w plane by the equation

$$z = -\frac{a}{\pi}\left[\frac{(e^\beta - 1)(w - 1)}{w - e^\beta} + \ln w\right]$$

where $b = (2a/\pi)(\beta + \sinh \beta)$. For $b = a$ ($\beta = 0.7493$) give the positions on the w plane (to two significant figures) of the points $z = -a$; $z = -\frac{1}{2}a$; $z = -\frac{1}{2}a - ia$; $z = -\frac{1}{2}a + ia$.

4.47 Show that if

$$f(x) = \sum_{n=-\infty}^{\infty} A_n e^{i(n\pi x/l)}; \quad -l \leq x \leq l$$

then

$$A_n = \frac{1}{2l}\int_{-l}^{l} e^{-in\pi x/l} f(x)\, dx$$

Insert the expression for A_n into the series for $f(x)$ and obtain the Fourier integral theorem by taking the appropriate limit of the result as l goes to infinity.

4.48 The generalized Mellin transforms $F_-(s)$ and $F_+(s)$ are defined by

$$F_-(s) = \int_0^1 f(x)x^{s-1}\, dx; \quad F_+(s) = \int_1^\infty f(x)x^{s-1}\, dx$$

Show that F_- will be analytic for Re $s > \sigma_0$ while F_+ will be analytic for Re $s < \sigma_1$. Show that $\sigma_0 < \sigma_1$ if the Mellin transform exists. Show that

$$f(x) = \frac{1}{2\pi i}\left\{\int_{-i\infty+\sigma}^{i\infty+\sigma} F_-(s)\left(\frac{ds}{x^s}\right) + \int_{-i\infty+\tau}^{i\infty+\tau} F_+(s)\left(\frac{ds}{x^s}\right)\right\}$$

where $\sigma > \sigma_0$ and $\tau < \sigma_1$.

4.49 Find the Fourier transforms F_+ and F_- of $\cos(ax)$ for a complex, and determine the region in which each is analytic.

4.50 Find the Fourier transforms F_+ and F_- of $x^n e^{-x}$, and determine the region in which each is analytic.

4.51 From the Poisson sum formula show that

$$\vartheta_3(u, e^{-\frac{1}{2}\alpha^2}) = \frac{\sqrt{2\pi}}{\alpha} \sum_{m=-\infty}^{\infty} e^{-(2/\alpha^2)(u+m\pi)^2}$$

(see page 431 for definition of ϑ_3).

Tabulation of Properties of Functions of Complex Variable

A function $f = u + iv$ of the complex variable $z = x + iy$ is *analytic* in a given region R of the z plane when:

a. Its derivative df/dz at a point $z = a$ in the region R is independent of the direction of dz; this derivative is also continuous in the region R.

b. $\partial u/\partial x = \partial v/\partial y$; $\partial u/\partial y = -\partial v/\partial x$, these derivatives being continuous in R.

c. $\oint f(z)\,dz = 0$ for any closed contour within region R.

Any one of these three statements is a necessary and sufficient condition that f be analytic in R; the other two statements are then consequences of the first chosen. When any and all are true in R, then *all derivatives* of f with respect to z *are analytic* in R.

In addition, for contour integrals entirely within a singly connected region of analyticity of f,

$$\oint \frac{f(z)\,dz}{(z-a)} = 2\pi i f(a); \quad \oint \frac{f(z)\,dz}{(z-a)^{n+1}} = \frac{2\pi i}{n!} f^{(n)}(a)$$

when a is inside the contour; when a is outside the contour, the integrals are zero (see Sec. 4.3).

Points where $f(z)$ is not analytic are called *singularities* of f (see page 380). A singularity at $z = a$ is called a *simple pole* when

$$f(z) \to g(z)/(z-a); \quad z \to a$$

where $g(z)$ is analytic at $z = a$ and $g(a) \neq 0$. When

$$f(z) \to g(z)/(z - a)^n; \quad z \to a; \quad n = 1, 2, 3, \ldots$$

the singularity is called a *pole of order n*. If $f(z)(z - a)^n \to \infty \quad (z \to a)$ for all finite values of n, point a is called an *essential singularity* for f.

If, on the other hand, the singularity at $z = a$ is such that $f(a + \epsilon e^{i\varphi})$ is not periodic in φ with period 2π (ϵ chosen so that $a + \epsilon e^{i\varphi}$ is in the region of analyticity of f for all φ), then the point $z = a$ is a *branch point* for f. Examples of branch points are $f = (z - a)^\nu g(z)$, where ν is not an integer (ν may be negative), or $f = \ln (z - a)g(z)$, where $g(z)$ is analytic at $z = a$. Poles and essential singularities are *isolated singularities* of f; branch points (which never occur singly) are not isolated, for f is nonanalytic along a line ending at the branch point.

When $f(1/w)$ has a singularity at $w = 0$, $f(z)$ is said to have a singularity at infinity.

Functions $f(z)$ may be classified according to the nature and position of their singularities (see also Sec. 4.3):

1. If $f(z)$ has no singularities anywhere it is a *constant.*

2. If $f(z)$ has only a pole of order n at infinity it is a polynomial in z of order n.

3. If $f(z)$ has only an isolated singularity at infinity, it is called an *integral function* of z. (Case 3 includes case 2.)

4. If $f(z)$ has only poles (there can be only a finite number of poles), then f is the *ratio of two polynomials* in z. A function the only singularities of which are poles in a given region is said to be *meromorphic* in that region. A *rational function* is meromorphic in the entire z plane.

5. If $f(z)$ is meromorphic in the finite part of the z plane (only poles for z finite, possibly an essential singularity at infinity), it is called a *meromorphic function* of z. (Case 5 includes case 4.)

6. Functions with branch points are *multivalued functions* (cases 1 to 5 are single-valued functions).

A *rational function* may be expanded in a series of partial fractions, each term of which corresponds to one of the poles in the finite part of the plane. For example, if all the poles in the finite part of the z plane are simple poles (the nth one at $z = a_n$), and if we choose the origin $z = 0$ at a point which is *not* a pole, then [see Eq. (4.3.6) and restrictions on page 383]

$$f(z) = f(0) + \sum_n \left[\frac{b_n}{z - a_n} + \frac{b_n}{a_n}\right]$$

where the constant b_n, determined by the behavior of f near $z = a_n$, is called the *residue* of f at the simple pole $z = a_n$.

An *integral function* may be expressed as a product of terms related to its zeros in the finite part of the plane. For example, if f has only simple zeros ($1/f$ has simple poles) at the points $z = a_1, a_2, a_3, \ldots$, none of the a's being 0, then

$$f(z) = f(0)e^{[f'(0)/f(0)]z} \prod_n \left(1 - \frac{z}{a_n}\right) e^{z/a_n}$$

[see Eq. (4.3.8) and restrictions on page 385].

Euler's Algorithm Relating Series. Let g be the known series

$$g(z) = \sum_{n=0}^{\infty} b_n z^n \quad \text{and let} \quad f(z) = \sum_{n=0}^{\infty} C_n b_n z^n$$

be a series related to g via the multiplicative constants C_n. Then

$$f(z) = C_0 g(z) - (\delta C_0) z g'(z) + (\delta^2 C_0)(z^2/2!) g''(z) \cdots$$

where
$$\delta C_0 = C_0 - C_1; \quad \delta^2 C_0 = C_0 - 2C_1 + C_2;$$
$$\delta^3 C_0 = C_0 - 3C_1 + 3C_2 - C_3; \ldots$$

In particular, if $g(z) = 1/(1 + z) = \Sigma(-1)^n z^n$, then

$$f(z) = \sum (-1)^n a_n z^n = \frac{1}{1+z} [a_0 + (\delta a_0)\zeta + (\delta^2 a_0)\zeta^2 + \cdots]$$

where $\zeta = [z/(1 + z)]$. If C_n or a_n is an Nth order polynomial in n, all differences δ^n higher than $n = N$ will be zero and a finite expression for f has been found in terms of the known comparison function $g(z)$.

Asymptotic Series for Integral Representation. Often the function $\psi(z)$ is given in terms of an integral representation

$$\psi(z) = \kappa \int_C e^{zf(t)}\, dt$$

where C is some contour (either closed or going to infinity) in the t plane. To determine ψ for Re $z \gg 1$, we adjust the contour C, within its allowed limits, so that Re f is everywhere as small as possible. Then the largest value of Re f throughout the contour will be at a saddle point for $f(t)$, where $df/dt = 0$. Near this point ($t = t_0$) the series

$$f(t) = f(t_0) + \tfrac{1}{2}(t - t_0)^2 f''(t_0) + \tfrac{1}{6}(t - t_0)^3 f'''(t_0) + \cdots$$

will hold. We adjust the contour in this region so that Im f is constant, equal to the imaginary part of f at $t = t_0$. There are two such paths; we pick the one for which $f(t_0)$ is a maximum, not a minimum. Integrating over this path, the parts of the integral near $t = t_0$ contribute a preponderating fraction of the value of the integral. The asymptotic series for Re z large is then

$$\psi(z) \simeq \kappa \frac{e^{zf(t_0) - \frac{1}{2}\pi i}}{\sqrt{\frac{1}{2} z f''(t_0)}} \sum_{n=0}^{\infty} B_n \Gamma(n + \tfrac{1}{2}) \left(\frac{1}{z^n}\right)$$

where $B_0 = 1;\quad B_1 = \dfrac{-1}{12[f''(t_0)]^3}\{5[f'''(t_0)]^2 - 3f''(t_0)f^{IV}(t_0)\};\ \ldots$
See Eqs. (4.6.14) to (4.6.19) for further details.

Fourier Transforms. If $f(z)$ is such a function of z that $\int_{-\infty}^{\infty} |f(z)|^2\,dz$ is finite and if the function

$$F(k) = \frac{1}{\sqrt{2\pi}} \int_{-\infty}^{\infty} f(z)e^{ikz}\,dz$$

then $F(k)$ is called the *Fourier transform* of $f(z)$ and

$$f(z) = \frac{1}{\sqrt{2\pi}} \int_{-\infty}^{\infty} F(k)e^{-ikz}\,dk;\quad \int_{-\infty}^{\infty} |F(k)|^2\,dk = \int_{-\infty}^{\infty} |f(z)|^2\,dz$$

Furthermore if the expansion for $f_+(z)$ in the neighborhood of $z = 0$ is

$$f_+(z) = \sum_{n=0}^{\infty} \left(\frac{z^n}{n!}\right) f^{(n)}(0);\quad z > 0;\quad f_+ = 0;\quad z < 0$$

then the asymptotic behavior of F for large k is

$$F_+(k) = -\frac{1}{\sqrt{2\pi}} \sum_{n=1}^{\infty} \left(\frac{i}{k}\right)^n f^{(n-1)}(0)$$

If $F(k)$ is the Fourier transform of $f(z)$ and $G(k)$ the Fourier transform of $g(z)$, then the Fourier transform of

$$\frac{1}{\sqrt{2\pi}} \int_{-\infty}^{\infty} f(y)g(z-y)\,dy \quad \text{is} \quad F(k)G(k);\quad \text{faltung theorem}$$

and the Fourier transform of $f(z)g(z)$ is $\dfrac{1}{\sqrt{2\pi}} \displaystyle\int_{-\infty}^{\infty} F(l)G(k-l)\,dl.$ Also $\int_{-\infty}^{\infty} F(k)\bar{G}(k)\,dk = \int_{-\infty}^{\infty} f(z)\bar{g}(z)\,dz.$

If $F(k)$ is the Fourier transform of $f(z)$, then

$$\sum_{n=-\infty}^{\infty} f(\alpha n) = \frac{\sqrt{2\pi}}{\alpha} \sum_{m=-\infty}^{\infty} F\left(\frac{2\pi m}{\alpha}\right);\quad \text{Poisson sum formula}$$

Even if $\int_{-\infty}^{\infty} |f(z)|^2\,dz$ is not finite, if $\int_{-\infty}^{\infty} |f(z)|^2 e^{-2\tau_0 z}\,dz$ is finite and if $G(k)$ is the Fourier transform of $f(z)e^{-\tau_0 z}$, then the Fourier transform of $f(z)$ is $G(k - i\tau_0) = F(k)$, where

$$f(z) = \frac{1}{\sqrt{2\pi}} \int_{-\infty+i\tau_0}^{\infty+i\tau_0} F(k)e^{-ikz}\,dk$$

For other conditions of convergence see Eqs. (4.8.19) *et seq.*

Function $f(z)$	Fourier transform $F(k)$
$\lambda f(z)$	$\lambda F(k)$
$f(az)$	$(1/a)F(k/a)$
$izf(z)$	$\frac{d}{dk}F(k)$
$\frac{d}{dz}f(z)$	$-ikF(k)$
$e^{izk_0}f(z)$	$F(k+k_0)$
$f(z+z_0)$	$e^{-ikz_0}F(k)$
1	$\delta(k)$
$(z-iz_0)^{-1}$ (Re $z_0 > 0$)	$i\sqrt{2\pi}\,e^{-kz_0}$ (Re $k > 0$)
$[(z-iz_0)(z+iz_1)]^{-1}$ (Re z_0 and Re $z_1 > 0$)	$\frac{\sqrt{2\pi}}{(z_0+z_1)}\begin{cases} e^{-z_0k} & (\text{Re } k > 0) \\ e^{z_1k} & (\text{Re } k < 0)\end{cases}$
$\operatorname{sech}(k_0z)$	$(1/k_0)\sqrt{\pi/2}\operatorname{sech}(\pi k/2k_0)$
$\tanh(k_0z)$	$(i/k_0)\sqrt{\pi/2}\operatorname{csch}(\pi k/2k_0)$
$z^{-\alpha-1}e^{i/z}$	$i\sqrt{2\pi}\,e^{\frac{1}{2}\pi i\alpha}k^{\frac{1}{2}\alpha}J_\alpha(2\sqrt{k})$
$e^{-\frac{1}{2}z^2}$	$e^{-\frac{1}{2}k^2}$
$\sqrt{z}\,J_{-\frac{1}{4}}(\frac{1}{2}z^2)$	$\sqrt{k}\,J_{-\frac{1}{4}}(\frac{1}{2}k^2)$

Laplace Transforms. If $f(x)$ is zero for $x < 0$, if $\int_0^\infty |f(x)|^2 e^{-2\tau x}\,dx$ is finite for $\tau > \tau_0$, and if

$$F_l(p) = \int_0^\infty f(x)e^{-px}\,dx$$

then $F_l(p)$ is called the *Laplace transform of* $f(x)$ and

$$f(x) = \frac{1}{2\pi i}\int_{-i\infty+\tau}^{i\infty+\tau} F_l(p)e^{px}\,dp; \quad \tau > \tau_0; \quad x > 0$$

The faltung theorem becomes

$$\int_0^x f(y)h(x-y)\,dy = \frac{1}{2\pi i}\int_{-i\infty+\tau}^{i\infty+\tau} F_l(p)H_l(p)e^{px}\,dp$$

For further discussion of the Laplace transform see tables at end of Chap. 11.

Mellin Transforms. If $f(x)$ is defined for $(0 \le x \le \infty)$, if

$$\int_0^\infty |f(x)|^2 x^{-2\sigma-1}\,dx$$

is finite for $\sigma > \sigma_0$, and if

$$F_m(s) = \int_0^\infty f(x) x^{s-1}\,dx$$

then $F_m(s)$ is called the *Mellin transform* of $f(x)$ and

$$f(x) = \frac{1}{2\pi i}\int_{-i\infty+\sigma}^{i\infty+\sigma} F_m(s)\left(\frac{ds}{x^s}\right); \quad \sigma > \sigma_0; \quad x > 0$$

The faltung theorem takes on the following forms:

$$\int_0^\infty v(y) w\left(\frac{x}{y}\right)\left(\frac{dy}{y}\right) = \frac{1}{2\pi i}\int_{-i\infty+\sigma}^{i\infty+\sigma} V_m(s) W_m(s)\left(\frac{ds}{x^s}\right)$$

or

$$\int_0^\infty v(x) w(x) x^{s-1}\,dx = \frac{1}{2\pi i}\int_{-i\infty+\sigma}^{i\infty+\sigma} V_m(\rho) W_m(s-\rho)\,d\rho$$

Function $f(x)$	Mellin transform $F_m(s)$
$\lambda f(x)$	$\lambda F_m(s)$
$f(ax)$	$a^{-s}F_m(s)$
$x^\alpha f(x)$	$F_m(s+\alpha)$
$\frac{d}{dx}f(x)$	$-(s-1)F_m(s-1)$
$\ln(x) f(x)$	$\frac{d}{ds}F_m(s)$
e^{-x}	$\Gamma(s)$
$F(a,b\|c\|-x)$	$\frac{\Gamma(c)}{\Gamma(a)\Gamma(b)}\left[\frac{\Gamma(s)\Gamma(a-s)\Gamma(b-s)}{\Gamma(c-s)}\right]$
$(1+x)^{-a}$	$[\Gamma(s)\Gamma(a-s)/\Gamma(a)]$
$(1/x)\ln(1+x)$	$[\pi/(1-s)\sin(\pi x)]$
$\tanh^{-1}(x)$	$(\pi/2s)\tan(\pi s/2)$
$(1+x)^{-m}P_{m-1}\left(\frac{1-x}{1+x}\right)$	$[\Gamma(s)/\Gamma(1-s)][\Gamma(m-s)/\Gamma(m)]^2$
$F(a\|c\|-x)$	$[\Gamma(s)\Gamma(a-s)\Gamma(c)/\Gamma(a)\Gamma(c-s)]$
$J_\nu(x)$	$\left[2^{s-1}\Gamma\left(\frac{\nu+s}{2}\right)\Big/\Gamma\left(\frac{\nu-s}{2}+1\right)\right]$
$xj_\nu(x)$	$\left[2^{s-\frac{1}{2}}\Gamma\left(\frac{s+\nu+1}{2}\right)\Big/\Gamma\left(1+\frac{\nu-s}{2}\right)\right]$
$\sin(x)$	$\Gamma(s)\sin(\frac{1}{2}\pi s)$
$\cos(x)$	$\Gamma(s)\cos(\frac{1}{2}\pi s)$
$N_\nu(x)$	$-\frac{2^{\nu-1}}{\pi}\Gamma\left(\frac{s+\nu}{2}\right)\Gamma\left(\frac{s-\nu}{2}\right)\cos[\frac{1}{2}\pi(s-\nu)]$

where the functions J_ν, j_ν, F, P_m, and N_ν are defined in the tables at the end of Chaps. 5, 10, and 11.

Tables of Special Functions of General Use

(For other tables see the ends of Chaps. 5, 6, 10, 11, and 12)

The Gamma Function (see page 419)

$$\Gamma(z) = \int_0^\infty e^{-t}t^{z-1}\,dt (\text{Re } z > 0); \quad \Gamma(z+1) = z\Gamma(z)$$

$$\Gamma(n+1) = n!; \quad \Gamma(1) = 1; \quad \Gamma(\tfrac{1}{2}) = \sqrt{\pi}$$

$$\Gamma(z)\Gamma(1-z) = \frac{\pi}{\sin(\pi z)}; \quad \Gamma(2z) = \frac{2^{2z-1}}{\sqrt{\pi}}\Gamma(z)\Gamma(z+\tfrac{1}{2})$$

$$\frac{1}{\Gamma(z)} = ze^{\gamma z}\prod_{n=1}^{\infty}\left(1+\frac{z}{n}\right)e^{-z/n}; \quad \gamma = 0.577215 \ldots$$

$$\psi(z) = \frac{d}{dz}\ln[\Gamma(z)] = \frac{1}{\Gamma(z)}\frac{d}{dz}\Gamma(z)$$

$$= -\gamma + \sum_{n=0}^{\infty}\left[\frac{1}{n+1} - \frac{1}{z+n}\right]; \quad \psi(z+1) = \frac{1}{z} + \psi(z)$$

$$\psi(n+1) = -\gamma + 1 + \frac{1}{2} + \frac{1}{3} + \cdots + \frac{1}{n}; \quad \psi(\tfrac{1}{2}) = -\gamma - 2\ln 2$$

$$\Gamma(z) \to \sqrt{2\pi}\, z^{z-\frac{1}{2}}e^{-z}; \quad \text{for } z \gg 1$$

$$B(x,y) = \int_0^1 t^{x-1}(1-t)^{y-1}\,dt = \frac{\Gamma(x)\Gamma(y)}{\Gamma(x+y)}; \quad \text{Re } x, \quad \text{Re } y > 0$$

$$= 2\int_0^{\frac{1}{2}\pi} \sin^{2x-1}\phi\, \cos^{2y-1}\phi\, d\phi$$

Elliptic Functions (see page 428)

$$\int_0^x \frac{dx}{\sqrt{(1-x^2)(1-k^2x^2)}} = \text{sn}^{-1}(x,k);$$

$$\int_x^1 \frac{dx}{\sqrt{(1-x^2)(1-k^2+k^2x^2)}} = \text{cn}^{-1}(x,k)$$

$$\int_x^1 \frac{dx}{\sqrt{(1-x^2)(x^2+k^2-1)}} = \text{dn}^{-1}(x,k);$$

$$\int_0^x \frac{dx}{\sqrt{(1+x^2)(1+k'^2x^2)}} = \text{tn}^{-1}(x,k)$$

$$K = \int_0^1 \frac{dx}{\sqrt{(1-x^2)(1-k^2x^2)}}; \quad K' = \int_0^1 \frac{dx}{\sqrt{(1-x^2)(1-k'^2x^2)}};$$

$$k = \sin\alpha; \quad k' = \cos\alpha$$

$$\int_0^x \frac{dx}{\sqrt{x(1-x)(1-k^2x)}} = 2\,\text{sn}^{-1}(\sqrt{x},k);$$

$$\int_x^1 \frac{dx}{\sqrt{x(1-x)(k'^2+k^2x)}} = 2\,\mathrm{cn}^{-1}\,(\sqrt{x},k)$$

$$\int_x^1 \frac{dx}{\sqrt{x(1-x)(x-k'^2)}} = 2\,\mathrm{dn}^{-1}\,(\sqrt{x},k);$$

$$\int_0^x \frac{dx}{\sqrt{x(1+x)(1+k'^2x)}} = 2\,\mathrm{tn}^{-1}\,(\sqrt{x},k)$$

$\mathrm{cn}^2\,(u,k) = 1 - \mathrm{sn}^2\,(u,k);\quad \mathrm{dn}^2\,(u,k) = 1 - k^2\,\mathrm{sn}^2\,(u,k);$

$$\mathrm{tn}\,(u,k) = \frac{\mathrm{sn}\,(u,k)}{\mathrm{cn}\,(u,k)}$$

$\mathrm{sn}\,(0,k) = 0;\quad \mathrm{cn}\,(0,k) = 1;\quad \mathrm{dn}\,(0,k) = 1$

$\mathrm{sn}\,(-u,k) = -\,\mathrm{sn}\,(u,k);\quad \mathrm{cn}\,(-u,k) = \mathrm{cn}\,(u,k);\quad \mathrm{dn}\,(-u,k) = \mathrm{dn}\,(u,k)$

Periodic Properties of Elliptic Functions

	$\mathrm{sn}(-,k)$	$\mathrm{cn}(-,k)$	$\mathrm{dn}(-,k)$	$\mathrm{tn}(-,k)$
iu	$i\,\mathrm{tn}(u,k')$	$\dfrac{1}{\mathrm{cn}(u,k')}$	$\dfrac{\mathrm{dn}(u,k')}{\mathrm{cn}(u,k')}$	$i\,\mathrm{sn}(u,k)$
$u + K$	$\dfrac{\mathrm{cn}(u,k)}{\mathrm{dn}(u,k)}$	$-\,k'\dfrac{\mathrm{sn}(u,k)}{\mathrm{dn}(u,k)}$	$\dfrac{k'}{\mathrm{dn}(u,k)}$	$-\dfrac{1}{k\,\mathrm{tn}(u,k)}$
$u + 2K$	$-\,\mathrm{sn}(u,k)$	$-\,\mathrm{cn}(u,k)$	$\mathrm{dn}(u,k)$	$\mathrm{tn}(u,k)$
$u + iK'$	$\dfrac{1}{k\,\mathrm{sn}(u,k)}$	$-\dfrac{i\,\mathrm{dn}(u,k)}{k\,\mathrm{sn}(u,k)}$	$-\dfrac{i}{\mathrm{tn}(u,k)}$	$\dfrac{i}{\mathrm{dn}(u,k)}$
$u + K + iK'$	$\dfrac{\mathrm{dn}(u,k)}{k\,\mathrm{cn}(u,k)}$	$-\dfrac{ik'}{k\,\mathrm{cn}(u,k)}$	$ik'\,\mathrm{tn}(u,k)$	$\dfrac{i}{k'}\,\mathrm{dn}(u,k)$
$u + 2iK'$	$\mathrm{sn}(u,k)$	$-\,\mathrm{cn}(u,k)$	$-\,\mathrm{dn}(u,k)$	$-\,\mathrm{tn}(u,k)$
$u + 2K + 2iK'$	$-\mathrm{sn}(u,k)$	$\mathrm{cn}(u,k)$	$-\,\mathrm{dn}(u,k)$	$-\,\mathrm{tn}(u,k)$

The function (given in the top row) of the argument given in the left-hand column equals the quantity given in the appropriate space. For instance,

$$\mathrm{cn}(u + 2K,k) = -\mathrm{cn}(u,k);\quad \mathrm{dn}(u + K + iK') = ik'\,\mathrm{tn}(u,k);\quad \text{etc.}$$

Relations between Parameters of Elliptic Functions

K/K'	0	0.1	0.2	0.3	0.4	0.5	0.6	0.7	0.8	0.9	1.0	K/K'
K	1.571	1.571	1.571	1.571	1.573	1.583	1.604	1.643	1.699	1.768	1.854	K
K'	∞	15.71	7.855	5.237	3.933	3.166	2.673	2.347	2.124	1.966	1.854	K'
k	0	......	0.00156	0.0213	0.0784	0.171	0.265	0.407	0.520	0.622	0.707	k
k'	1.000	1.000	1.000	1.000	0.998	0.985	0.965	0.913	0.853	0.784	0.707	k'
α	0	......	5.4′	1°11.7′	4°30′	9°50′	15°22′	24°0′	31°23′	38°30′	45°	α
q	0	......	....	....	0.0004	0.0019	0.0053	0.0114	0.0197	0.0307	0.0432	q

where $q = e^{-\pi(K'/K)}$. For sufficiently small values of $k(k < 0.1)$ the equation $k \simeq 4e^{-(\pi K'/2K)} = 4\sqrt{q}$ is a good approximation. For values of K/K' larger than 1.0, interchange K and K', k and k', α and $90° - \alpha$.

In the following we shall omit the second part of the argument, k. It is assumed to be the same in all terms.

$$\mathrm{sn}(u+v) = \frac{\mathrm{sn}\,u\,\mathrm{cn}\,v\,\mathrm{dn}\,v + \mathrm{cn}\,u\,\mathrm{sn}\,v\,\mathrm{dn}\,u}{1 - k^2\,\mathrm{sn}^2\,u\,\mathrm{sn}^2\,v}$$

$$\mathrm{cn}(u+v) = \frac{\mathrm{cn}\,u\,\mathrm{cn}\,v - \mathrm{sn}\,u\,\mathrm{sn}\,v\,\mathrm{dn}\,u\,\mathrm{dn}\,v}{1 - k^2\,\mathrm{sn}^2\,u\,\mathrm{sn}^2\,v}$$

$$\mathrm{dn}(u+v) = \frac{\mathrm{dn}\,u\,\mathrm{dn}\,v - k^2\,\mathrm{sn}\,u\,\mathrm{sn}\,v\,\mathrm{cn}\,u\,\mathrm{cn}\,v}{1 - k^2\,\mathrm{sn}^2\,u\,\mathrm{sn}^2\,v}$$

$$\mathrm{tn}(u+v) = \frac{\mathrm{tn}\,u\,\mathrm{dn}\,v + \mathrm{tn}\,v\,\mathrm{dn}\,u}{1 - \mathrm{tn}\,u\,\mathrm{tn}\,v\,\mathrm{dn}\,u\,\mathrm{dn}\,v}$$

The functions sn, cn, dn have simple poles at the positions $[2mK + (2n+1)iK']$, where m and n are integers, positive or negative. The residues at these poles are given in the table

Residues for	iK'	$-iK'$	$2K + iK'$	$2K - iK'$
sn	$1/k$	$1/k$	$-(1/k)$	$-(1/k)$
cn	$-(i/k)$	i/k	i/k	$-(i/k)$
dn	$-i$	i	$-i$	i

The zeros of sn (u,k) are at $u = 2mK + 2inK'(m, n = 0, \pm 1, \pm 2, \ldots)$ also $\mathrm{cn}\,[(2m+1)K + 2inK', k] = 0$, $\mathrm{dn}\,[(2m+1)K + i(2n+1)K', k] = 0$, $\mathrm{sn}\,(u,k) \underset{u\to 0}{\longrightarrow} u + 0(u^3)$; the behavior of the other functions at their various zeros can be obtained from their periodic properties.

$$\frac{1}{\mathrm{sn}\,(u,k)} = \frac{\pi}{2K}\csc\left(\frac{\pi u}{2K}\right) + \frac{2\pi}{K}\sum_{n=1}^{\infty}\frac{\sin(\pi u/2K)\cosh(n\pi K'/K)}{\cosh(2\pi nK'/K) - \cos(\pi u/K)}$$

$$\frac{\mathrm{cn}\,(u,k)}{\mathrm{sn}\,(u,k)} = \frac{\pi}{2K}\cot\left(\frac{\pi u}{2K}\right) + \frac{\pi}{2K}\sum_{n=0}^{\infty}\frac{\sin(\pi u/K)}{\cosh[4\pi nK'/K] - \cos(\pi u/K)} - \frac{\pi}{2K}\sum_{n=0}^{\infty}\frac{\sin(\pi u/K)}{\cosh[\pi(4n+2)K'/K] - \cos(\pi u/K)}$$

$$\frac{\mathrm{dn}\,(u,k)}{\mathrm{sn}\,(u,k)} = \frac{\pi}{2K}\csc\left(\frac{\pi u}{2K}\right) + \frac{2\pi}{K}\sum_{n=1}^{\infty}(-1)^n\frac{\sin(\pi u/2K)\cosh(n\pi K'/K)}{\cosh(2\pi nK'/K) - \cos(\pi u/K)}$$

If $q = e^{-\pi K'/K}$, then

$$\mathrm{sn}\,(u,k) = \frac{2K}{\pi}\sin\left(\frac{\pi u}{2K}\right)\prod_{n=1}^{\infty}\left[\frac{(1-q^{2n-1})}{(1+q^{2n})}\right]^2\left[\frac{1 - 2q^{2n}\cos(\pi u/K) + q^{4n}}{1 - 2q^{2n-1}\cos(\pi u/K) + q^{4n-2}}\right]$$

$$\mathrm{cn}\,(u,k) = \cos\left(\frac{\pi u}{2K}\right)\prod_{n=1}^{\infty}\left[\frac{(1-q^{2n-1})}{(1+q^{2n})}\right]^2\left[\frac{1+2q^{2n}\cos\,(\pi u/K)+q^{4n}}{1-2q^{2n-1}\cos\,(\pi u/K)+q^{4n-2}}\right]$$

$$\mathrm{dn}\,(u,k) = \prod_{n=1}^{\infty}\left[\frac{(1-q^{2n-1})}{(1+q^{2n-1})}\right]^2\left[\frac{1+2q^{2n-1}\cos\,(\pi u/K)+q^{4n-2}}{1-2q^{2n-1}\cos\,(\pi u/K)+q^{4n-2}}\right]$$

$$\mathrm{tn}\,(u,k) = \frac{2K}{\pi}\tan\left(\frac{\pi u}{2K}\right)\prod_{n=1}^{\infty}\left[\frac{(1+q^{2n})}{(1-q^{2n})}\right]^2\left[\frac{1-2q^{2n}\cos\,(\pi u/K)+q^{4n}}{1+2q^{2n}\cos\,(\pi u/K)+q^{4n}}\right]$$

$(d/du)\ \mathrm{sn}(u,k) = \mathrm{cn}(u,k)\ \mathrm{dn}(u,k)$;
$(d/du)\ \mathrm{dn}(u,k) = -\ \mathrm{sn}(u,k)\ \mathrm{dn}(u,k)$
$(d/du)\ \mathrm{dn}(u,k) = -k^2\ \mathrm{sn}(u,k)\ \mathrm{cn}(u,k)$
$\int \mathrm{dn}(x,k)\,dx = \sin^{-1}[\mathrm{sn}\,(x,k)] = \mathrm{am}(x)$
$\mathrm{sn}\,(x,k) = \sin\,[\mathrm{am}(x)]$; $\quad \mathrm{cn}\,(x,k) = \cos\,[\mathrm{am}(x)]$
$\int \mathrm{sn}(x,k)\,dx = (1/k)\cosh^{-1}[\mathrm{dn}(x,k)/k']$
$\int \mathrm{cn}(x,k)\,dx = (1/k)\cos^{-1}[\mathrm{dn}(x,k)]$

$$\int\frac{dx}{\mathrm{sn}\,(x,k)} = \ln\left[\frac{\mathrm{sn}\,(x,k)}{\mathrm{cn}\,(x,k)+\mathrm{dn}\,(x,k)}\right]$$

$$\int\frac{dx}{\mathrm{cn}\,(x,k)} = \frac{1}{k'}\ln\left[\frac{k'\,\mathrm{sn}\,(x,k)+\mathrm{dn}\,(x,k)}{\mathrm{cn}\,(x,k)}\right]$$

$$\int\frac{dx}{\mathrm{dn}\,(x,k)} = \frac{1}{k'}\tan^{-1}\left[\frac{k'\,\mathrm{sn}\,(x,k)-\mathrm{cn}\,(x,k)}{k'\,\mathrm{sn}\,(x,k)+\mathrm{cn}\,(x,k)}\right]$$

Theta Functions. Defined as follows:

$$\vartheta_1(u,iv) = 2\sum_{n=0}^{\infty}(-1)^n e^{-\pi v(n+\frac{1}{2})^2}\sin\,[\pi(2n+1)u] = -\vartheta_2(u+\tfrac{1}{2},iv)$$

$$\vartheta_2(u,iv) = 2\sum_{n=0}^{\infty}e^{-\pi v(n+\frac{1}{2})^2}\cos\,[\pi(2n+1)u] = \vartheta_1(u+\tfrac{1}{2},iv)$$

$$\vartheta_3(u,iv) = 1+2\sum_{n=1}^{\infty}e^{-\pi v n^2}\cos\,(2\pi n u) = \vartheta_0(u+\tfrac{1}{2},iv)$$

$$\vartheta_0(u,iv) = 1+2\sum_{n=1}^{\infty}(-1)^n e^{-\pi v n^2}\cos\,(2\pi n u) = \vartheta_3(u+\tfrac{1}{2},iv)$$

$\vartheta_1(u+\tfrac{1}{2}iv,iv) = ie^{\frac{1}{4}\pi v - i\pi u}\vartheta_0(u,iv)$; $\quad \vartheta_2(u+\tfrac{1}{2}iv,iv) = e^{\frac{1}{4}\pi v - i\pi u}\vartheta_3(u,iv)$
$\vartheta_3(u+\tfrac{1}{2}iv,iv) = e^{\frac{1}{4}\pi v - i\pi u}\vartheta_2(u,iv)$; $\quad \vartheta_0(u+\tfrac{1}{2}iv,iv) = ie^{\frac{1}{4}\pi v - i\pi u}\vartheta_1(u,iv)$

$$\vartheta_1(u,iv+1) = e^{\frac{1}{4}i\pi}\vartheta_1(u,iv);\quad \vartheta_1(u,iv) = \frac{i}{\sqrt{v}}e^{-\pi u^2/v}\vartheta_1\left(\frac{u}{iv},\frac{i}{v}\right)$$

$$\vartheta_2(u,iv+1) = e^{\frac{1}{4}i\pi}\vartheta_2(u,iv);\quad \vartheta_2(u,iv) = \frac{1}{\sqrt{v}}e^{-\pi u^2/v}\vartheta_0\left(\frac{u}{iv},\frac{i}{v}\right)$$

$$\vartheta_3(u,iv+1) = \vartheta_0(u,iv);\quad \vartheta_3(u,iv) = \frac{1}{\sqrt{v}}e^{-\pi u^2/v}\vartheta_3\left(\frac{u}{iv},\frac{i}{v}\right)$$

$$\vartheta_0(u,iv+1) = \vartheta_3(u,iv); \qquad \vartheta_0(u,iv) = \frac{1}{\sqrt{v}} e^{-\pi u^2/v} \vartheta_2\left(\frac{u}{iv}, \frac{i}{v}\right)$$

$$\frac{\partial^2\vartheta}{\partial u^2} = 4\pi \frac{\partial\vartheta}{\partial v}$$

Zeros of	ϑ_1	ϑ_2	ϑ_3	ϑ_0
are at $u =$	$n + imv$	$n + \frac{1}{2} + imv$	$n + \frac{1}{2} + iv(m + \frac{1}{2})$	$n + iv(m + \frac{1}{2})$

(n,m,integers)

Bibliography

There are, of course, a large number of texts dealing with various aspects of the theory of functions of a complex variable. Perusal of some of the following will be rewarding:

Copson, E. T.: "Theory of Functions of a Complex Variable," Oxford, New York, 1935.
Hurwitz, A., and R. Courant: "Algemeine Funktionentheorie," Springer, Berlin, 1939, reprint Interscience, New York, 1944.
McLachlan, N. W.: "Complex Variables and Operational Calculus," Cambridge, New York, 1939.
Polya, G., and G. Szego: "Aufgabe und Lehrsatze aus der Analysis," Springer, Berlin, 1925.
Riemann-Weber: "Differential- und Integralgleichungen der Mechanik und Physik," ed. by P. Frank and R. von Mises, Chap. 3, Friedrich Vieweg & Sohn, Brunswick, 1935, reprint Rosenberg, New York, 1943.
Titchmarsh, E. C.: "Theory of Functions," Oxford, New York, 1939.
Whittaker, E. T., and G. N. Watson: "A Course in Modern Analysis," Chaps. 1 to 9, Cambridge, 1940.

Texts concerned with series and asymptotic developments:

Hadamard, J., and S. Mandelbrojt: "La Serie de Taylor et son prolongement analytique," Gauthier-Villars, Paris, 1926.
Landau, E.: "Darstellung and Begrundung einiger neuerer Ergebnisse der Funktionentheorie," Springer, Berlin, 1929.
Watson, G. N.: "Theory of Bessel Functions," Chaps. 7 and 8, Cambridge, 1944.

References of interest in connection with multivalued functions and with special functions:

Forsythe, A. R.: "Theory of Functions of a Complex Variable," Cambridge, 1893.
Neville, E. H.: "Jacobian Elliptic Functions," Oxford, New York, 1944.
Nielsen, N.: "Theorie der Gammafunktion," B. G. Teubner, Leipzig, 1906.

Books containing descriptions of various conformal transformations of interest:

Caratheodory, C.: "Conformal Representation," Cambridge, 1932.
Milne-Thomson, L. M.: "Theoretical Hydrodynamics," Macmillan & Co., Ltd., London, 1938.

Ramsey, A. S.: "Treatise on Hydromechanics," Part 2, "Hydrodynamics," Chap. 6, G. Bell, London, 1920.

Rothe, R., F. Ollendorff, and K. Pohlhausen: "Theory of Functions," Technology Press, Cambridge, 1933.

Books on the theory of the Fourier, Laplace, and other transforms:

Campbell, G. A., and R. M. Foster: "Fourier Integrals for Practical Applications," Bell System Technical Publication, Monograph B-584, 1942. (Table of Fourier transforms.)

Carslaw, H. S.: "Theory of Fourier Series and Integrals," Macmillan, New York, 1930.

Churchill, R. V.: "Fourier Series and Boundary Value Problems," McGraw-Hill, New York, 1941.

Doetsch, G.: "Theorie und Anwendung der Laplace-Transformation," Springer, Berlin, 1937, reprinted Dover, New York, 1943.

Gardner, M. F., and J. L. Barnes: "Transients in Linear Systems," Wiley, New York, 1942.

Paley, R. E. A. C., and N. Wiener: "Fourier Transforms in the Complex Plane," American Mathematical Society, New York, 1934.

Sneddon, I. N.: "Fourier Transforms," McGraw-Hill, New York, 1951.

Titchmarsh, E. C.: "Introduction to the Theory of Fourier Integrals," Oxford, New York, 1937.

CHAPTER 5

Ordinary Differential Equations

Having now reviewed the sorts of fields encountered in physics and the partial differential equations which govern them, and having discussed the analytic behavior of various kinds of solutions of these equations, we must now commence the central part of our work, the finding of a solution of the particular field equation under study at the moment. This usually requires two major steps; the first, which will be discussed in this chapter, is to find all (or nearly all) the possible solutions of the equation and the second, which will be discussed in later chapters, is to pick from these solutions the particular combination which satisfies the boundary conditions of the case under consideration.

The equations which we are going to study in this book are those listed in the table on page 271. As we have already indicated, they are far from being the only field equations encountered in physics, but as we have also indicated, they represent a surprisingly large fraction of the field equations which are important in modern physics. If we know how to find solutions to these equations, we are equipped to study many of the theoretical problems in all the major areas of present interest.

All the equations listed in the table have a number of important properties in common, which will allow us to narrow the field of study in this and future sections. The equations can all be written in the form

$$\mathfrak{H}\psi = F$$

where ψ, the field, is either a scalar or a vector (in real space or in abstract vector space or both) function of the coordinates and time. The operator $\mathfrak{H}$ is a combination of functions of the coordinates and time, of the partial derivatives with respect to space and time, and, sometimes, of vector operator functions in real space or in abstract vector space. The quantity F is a scalar or vector function of space and time. The quantities $\mathfrak{H}$ and F are known; the field ψ is the unknown.

The first common property to be noticed is that such equations are *linear* in the unknown ψ; none of the terms contains ψ^2, or products of two components of ψ, or higher powers. Linear equations are usually much easier to solve than nonlinear ones, and we shall use this property of linearity to help us many times in our discussions.

When F is zero, the equation is said to be homogeneous, for each of its terms contains a ψ. Homogeneous, linear equations possess the following extremely useful property: If $\psi_1, \psi_2, \ldots$ are a set of solutions of the equation, then *any linear combination* $\Sigma a_n\psi_n$ *of the solutions is also a solution.* A proof of this statement should be easy for the reader to provide.

When F is not zero, the equation is said to be *inhomogeneous*, for some of the terms do not include ψ. Inhomogeneous, linear equations also have useful properties: If ψ_i is a solution of the inhomogeneous equation $\mathfrak{H}\psi_i = F$ and ψ_h is any solution of the corresponding homogeneous equation $\mathfrak{H}\psi_h = 0$, then *the sum of the two solutions* $\psi_i + \psi_h$ *is a solution of the inhomogeneous equation;* also if ψ_1 is a solution of the equation $\mathfrak{H}\psi_1 = F_1$, if ψ_2 is a solution of $\mathfrak{H}\psi_2 = F_2$, and so on, then *the sum* $\Psi = \Sigma a_n\psi_n$ *is a solution of the inhomogeneous equation* $\mathfrak{H}\Psi = \Sigma a_nF_n$. Both of these results, which are not difficult to prove, will turn out to be of great utility in our future studies.

Another property of the equations listed in the table on page 271 is that *none of them are of higher order than second order; i.e.*, none of them contain higher derivatives than the second. There are a few field equations in physics which are of higher order than the second (the equations for the transverse motions of stiff plates, for instance), but the equations which are important enough to be considered in this book are either first or second order. This also will allow us to save time and space by specializing our general analysis to these cases.

Partial differential equations, involving more than one independent variable, are much more difficult to solve than are ordinary differential equations, which involve just one independent variable. Aside from the few cases where solutions are guessed and then verified to be solutions, only two generally practicable methods of solution are known, the *integral solution* and the *separated solution*. An example of the integral solution is given, on page 38, for the Poisson equation $\nabla^2\psi = -q(x,y,z)$. A solution can be written, as shown earlier,

$$\psi(x,y,z) = \iiint [q(x',y',z')/4\pi R]dx'\,dy'\,dz'$$

where R is the distance between the point (x,y,z) and the point (x',y',z'). This is typical of the integral solution. The inhomogeneous part of the equation, q, is in the integral, and the rest of the integrand, $1/4\pi R$, called a *Green's function*, is the same for any function q. Solutions of a similar type can also be obtained for homogeneous equations, where the function q is determined by the boundary conditions and where the Green's function is determined by the type of equation to be solved.

Integral solutions have the advantage of generality, for usually the integral is invariant under coordinate transformation, and once the Green's function is found, as shown in Chap. 7, any desired solution of the homogeneous or inhomogeneous equation may be found, in principle. However, as the phrase "in principle" indicates, the integral solution is not always the most satisfactory solution, for in many cases the integral cannot be integrated in closed form and numerical values are then extremely difficult to obtain.

The alternate method of solving linear partial differential equations is the method of factoring or separation, whereby the original equation, in several independent variables, is broken up, or *separated,* into a set of ordinary differential equations, each involving just one independent variable. This technique does not have the universality of the integral solution technique, for the separation must be different for each different coordinate system chosen and the separation can be carried out for only a few systems. But when the method can be used, it is usually much more satisfactory, for solutions of ordinary differential equations are much easier to obtain than are solutions of partial differential equations.

In this chapter we shall study the separation method. We shall indicate in what coordinate systems an equation will separate and then go on to study the resulting ordinary differential equations, showing how they can be solved and showing how the analytic properties of the solutions can be related to the characteristics of the equations and, ultimately, to the geometry of the coordinate system chosen.

5.1 *Separable Coordinates*

In order not to be confined to remarks which are so general as to be difficult to understand and apply, we shall center our comments on one equation of the general form $\mathfrak{H}\psi = F$, which will serve as an example with which the other equations may be compared. The differential operator $\mathfrak{H} = (\nabla^2 + k^2)$, corresponding to the homogeneous, second-order, linear equation for the scalar field ψ of the form

$$\nabla^2\psi + k^2\psi = 0 \tag{5.1.1}$$

where ∇^2 is the Laplace operator (see page 7), is a sufficiently typical form to be of general interest. When $k = 0$, the equation is Laplace's equation, for static potential fields; when k is a real constant, we have the wave equation for sinusoidal time dependence (Helmholtz equation) or the diffusion equation for exponential time dependence; and when k^2 is a function of the coordinates, we have Schroedinger's equation for a particle with constant energy E (k^2 is then proportional to the difference between E and the potential energy of the particle).

There are an infinite number of different solutions of an equation of the type of (5.1.1), an embarrassing variety of choices. Infinities and zeros can be placed anywhere; solutions can be found having any reasonably continuous distribution of chosen values (and some rather unreasonable distributions also) on any arbitrarily shaped surface we choose. The question is not to find a solution but to find the particular solution or solutions corresponding to the particular problem we wish to solve.

These various particular problems usually differ from each other in the nature of the *boundary conditions* applied; either the shape of the boundary varies or the prescribed behavior of the field at the boundary is different. (Initial conditions are, of course, boundary conditions for the surface $t = 0$). This suggests that we classify the solutions of a particular partial differential equation according to the shapes of boundary surfaces which "generate" the solution and also according to the nature of the boundary condition at the surface. Such a classification is a tedious and complicated one; we shall discuss it here only to the extent needed for our present purposes. Several sections will be devoted to the subject later in the book.

Boundary Surfaces and Coordinate Systems. In the first place two general classes of boundary surfaces should be distinguished: *open* and *closed surfaces.* A *closed* boundary surface is one which surrounds the field everywhere, confining it to a finite volume of space. In this case the boundary surface is definitely outside the field, and all energy that escapes from the field is absorbed by the boundary. An *open surface* is one which does not completely enclose the field but lets it extend to infinity in at least one direction, so that the field extends over an infinite volume and energy can escape "to infinity" as well as be absorbed by the boundary. This definition of open surface is rather different from the usual loose meaning of the phrase; for instance, a sphere is a closed surface to a field inside it, but it is an open surface for a field outside it.

Let us first examine open boundaries. The boundary conditions usually placed on the field are the specification of the value of the field at every point on the boundary surface or the specification of the normal gradient to the surface at the surface, or both. *Dirichlet boundary conditions* fix the *value* of ψ on the surface; *Neumann boundary conditions* fix the value of $\partial\psi/\partial n$ there; and *Cauchy conditions* fix both value and normal gradient there. Each is appropriate for different types of equations and different boundary surfaces. For instance, Dirichlet conditions on a closed surface uniquely specify a solution of Laplace's equation inside the closed surface, whereas Cauchy conditions would be "overspecifying" the same field. These matters will be dealt with in detail in Chap. 6.

For an open surface it would be expected that both value and slope at the surface (*i.e.*, Cauchy conditions) would have to be specified in order to result in a unique solution of a second-order differential equation outside the surface. This is true, though there are other ways of unique specification of the solution. In order to simplify the discussion it is best, if possible, to "erect" on the boundary a coordinate system suitable to the boundary. By this we mean that we choose a coordinate system ξ_1, ξ_2, ξ_3 such that the boundary surface is one of the coordinate surfaces, say $\xi_1 = X$, a constant. (There are modifications of this which add complication but do not change the essential principles, such as making a part of the surface $\xi_1 = X$, another part $\xi_2 = Y$, etc.) If the boundary surface is not too "pathological" in shape, we can find at least one suitable orthogonal system (see page 6). In many cases we can find more than one suitable system; this need not disturb us; we choose one of them and stick to it.

The partial differential equation [for instance Eq. (5.1.1), which we have chosen as an example] can be expressed in terms of these new coordinates, and the solutions will come out as functions of these ξ's. All possible solutions of the equation can be expressed, of course, as functions of the ξ's, but we are now in a position to classify the solutions according to the boundary conditions which they satisfy on the surface $\xi_1 = X$. There are the solutions, for instance, which go to zero at $\xi_1 = X$; these solutions must contain, as a factor, $(\xi_1 - X)$. There are also solutions with zero normal gradient on the surface; and, third, there are solutions with both value and normal gradient constant over all the surface.

As we mentioned above and will prove later, the third set achieves a one-to-one correspondence between solution and boundary condition; for each pair of values of ψ, $\partial\psi/\partial n$ on the surface there is just one corresponding solution. Some of these are just multiples of others and should not be considered as "different" solutions. For instance, the solution corresponding to $\psi = \alpha a$, $\partial\psi/\partial n = \alpha b$ on the surface $\xi_1 = X$ is just α times the solution for $\psi = a$, $\partial\psi/\partial n = b$. These should not be considered as independent solutions, any more than the vector **A** should be considered independent of the unit vector **a** pointed along **A**. We shall obtain different solutions, however, if we change the ratio of ψ and $\partial\psi/\partial n$.

Suppose that we arrange these solutions in order of the value of the ratio $[\psi_X/(\partial\psi/\partial n)_X] = P$. The solution for $P = 0$ (there are an infinite number of solutions for $P = 0$, but they differ only by a constant factor, so they are counted just as one solution) corresponds to the boundary condition $\psi = 0$ at $\xi_1 = X$. If we now look at the solutions as P is made to decrease in value below zero, we shall find that the surface $\psi = 0$ moves away from the surface $\xi_1 = X$. In general the family of surfaces $\psi = 0$ thus generated *does not* coincide with the family of surfaces $\xi_1 =$ constant, but *in a few cases they do coincide.* In these special cases the solution ψ must have a factor $F_1(\xi_1)$, the functional form of which depends on P. Whenever $F_1(\xi_1)$ equals zero, the corresponding surface $\psi = 0$ coincides with one of the coordinate surfaces $\xi_1 =$ constant.

The surfaces $\psi = 0$ are called *nodal surfaces* (or nodes) for ψ. The family generated by variation of P may not include all the nodal surfaces. In general these additional surfaces will not be orthogonal to the surfaces $\xi_1 =$ constant, so they will not coincide with the other coordinate surfaces.

So far we have been talking of a restricted set of solutions ψ, those with constant value or constant normal gradient (or both) over the surface $\xi_1 = X$. Most solutions have values and gradients at the boundary surface which vary over the surface, *i.e.*, are functions of the coordinates ξ_2 and ξ_3. Their nodal surfaces, in general, bear no simple relation to the ξ_1 coordinate surfaces. In a few cases, however, for a few simple boundary surfaces, we can separate out of the mass of different solutions a set having *all* their nodes either coincident with the ξ_1 coordinate surfaces or else orthogonal to these coordinate surfaces. Such solutions must be in the form of a product of two factors

$$\psi = F_1(\xi_1)\Phi(\xi_2,\xi_3)$$

one of them, F_1, a function of ξ_1 alone, the zeros of which are responsible for the nodes coincident with the ξ_1 surfaces and the other, Φ, a function of ξ_2 and ξ_3 but not ξ_1, responsible for the nodes orthogonal to the ξ_1 surfaces.

In a still more limited set of coordinate systems one can find a set of ψ's with nodal surfaces which all coincide with the three families of coordinate surfaces and which can be expressed as a product of three factors

$$\psi = F_1(\xi_1)F_2(\xi_2)F_3(\xi_3)$$

Such solutions are *separated* into factors, each dependent on only one coordinate. In a large number of coordinate systems a few solutions can be found having a few nodal surfaces coincident with the coordinate surfaces, but in only a few coordinates can one find a whole family of solutions with all their nodes behaving in this manner.

Of course, even in these special coordinates the family of separated solutions forms a small subset of the set of all solutions of Eq. (5.1.1). The interesting and important property of this subset, however, is that *all solutions of the partial differential equation can be built up out of linear combinations of the members of the family of separated solutions.* Once we have computed the separated solutions, we can obtain the rest.

Such coordinate systems, which allow families of separated solutions of a given equation from which all solutions of the equation can be built up, are called *separable coordinate systems for the equation in question.* Only for these systems is it possible to obtain solutions in more usable form than unwieldly multivariable series or integrals.

We have been dogmatic in the statements made in the past few pages in order to present the picture as a whole. It will take several chapters to analyze and prove all that we have said. The rest of this section will be devoted to finding separable systems of coordinates and showing how the separated solutions can be obtained.

Two-dimensional Separable Coordinates. Let us start with two dimensions, where our sample equation (5.1.1) has the form

$$\frac{\partial^2\psi}{\partial x^2} + \frac{\partial^2\psi}{\partial y^2} + k^2\psi = 0 \tag{5.1.2}$$

This equation is separable in the rectangular coordinates x, y if k is a constant, for if we set $\psi = X(x)Y(y)$ into Eq. (5.1.2), after a bit of juggling we obtain

$$\frac{1}{X}\frac{d^2X}{dx^2} + \frac{1}{Y}\frac{d^2Y}{dy^2} + k^2 = 0$$

The first term in this equation is a function of x only, the second term a function of y only, and the third term is a constant. The only way that it can be satisfied *for all values of x and y is for each term to be a constant* and for the sum of the three constants to be zero. That is

$$(1/X)(d^2X/dx^2) = -\alpha^2; \quad (1/Y)(d^2Y/dy^2) = -\beta^2; \quad \alpha^2 + \beta^2 = k^2$$

or

$$(d^2X/dx^2) + \alpha^2X = 0; \quad (d^2Y/dy^2) + (k^2 - \alpha^2)Y = 0 \tag{5.1.3}$$

We have therefore split the partial differential equation in two variables into two ordinary differential equations, each for a single independent variable. The constant α^2 is called the *separation constant.* The family of separated solutions of Eq. (5.1.2) consists of the products of solutions of Eqs. (5.1.3) for all values of the parameter α. A general solution of Eq. (5.1.2) can be expressed in terms of a linear combination of the separated solutions for various values of the parameter α. In other words, since solutions of Eqs. (5.1.3) are $e^{i\alpha x + y\sqrt{\alpha^2 - k^2}}$, the general solutions can be expressed as an integral

$$\psi = \int_{-\infty}^{\infty} f(\alpha)e^{i\alpha x + y\sqrt{\alpha^2 - k^2}}\,d\alpha \tag{5.1.4}$$

which is a form of Fourier integral (see page 453). The integral is sometimes a contour integral over complex values of α.

If k^2 is a function of the coordinates [as it would be if Eq. (5.1.2) were a Schroedinger equation], it must have the general form $k^2 = \epsilon^2 + f(x) + g(y)$ in order to be able to separate the equation, where f and g are arbitrary functions of a single variable. Then the separated equations become

$$(d^2X/dx^2) + [\alpha^2 + f(x)]X = 0; \quad (d^2Y/dy^2) + [\epsilon^2 - \alpha^2 + g(y)]Y = 0$$

with α^2 the separation constant again.

To discuss other coordinates we can utilize the techniques of the theory of analytic functions of the complex variable $z = x + iy$ in order to simplify the analysis, for Eq. (5.1.2) can be expressed in terms of derivatives with respect to z and its complex conjugate $\bar{z} = x - iy$ (which are independent variables). According to Eq. (4.1.11) we can transform Eq. (5.1.2) into

$$4(\partial^2\psi/\partial z\, \partial\bar{z}) + k^2\psi = 0$$

Now let us make a conformal transformation of coordinates from x, y to ξ_1, ξ_2. If the transformation is conformal, the function $w = \xi_1 + i\xi_2$ must be an analytic function of $z = x + iy$ (and the conjugate function $\bar{w} = \xi_1 - i\xi_2$ must be a function of $\bar{z} = x - iy$). The new coordinate lines are given by the equations $\xi_1(x,y) =$ constant, $\xi_2(x,y) =$ constant. The differentials $\partial/\partial z = (dw/dz)(\partial/\partial w)$ and $(\partial/\partial\bar{z}) = (d\bar{w}/d\bar{z})\cdot(\partial/\partial\bar{w})$ (since $dw/d\bar{z}$ and $d\bar{w}/dz$ are zero), so that the transformed equation is

$$4\,\frac{\partial^2\psi}{\partial w\, \partial\bar{w}} = \frac{\partial^2\psi}{\partial\xi_1^2} + \frac{\partial^2\psi}{\partial\xi_2^2} = -\,\frac{k^2\psi}{|dw/dz|^2} = -\,\frac{dz}{dw}\frac{d\bar{z}}{d\bar{w}}\,k^2\psi \qquad (5.1.5)$$

where we now express z as a function of w (and $\bar{z}$ as a function of $\bar{w}$).

Separable Coordinates for Laplace's Equation in Two Dimensions. We notice, first of all, that for Laplace's equation, where $k = 0$, *all coordinates obtained by conformal transformation from the rectangular coordinates x, y are separable coordinates for Laplace's equation in two dimensions.* The separated solutions are of the general form

$$\psi = Ae^{\pm i\alpha\xi_1 \pm \alpha\xi_2} \quad \text{or} \quad ae^{\beta w} + be^{\gamma\bar{w}}$$

where the separation constant α may have any value, real, imaginary, or complex. Integrals or sums of these elementary solutions, for different values of α, or β or γ can represent any solution of the two-dimensional Laplace equation. This will be treated in detail in Sec. 10.2.

We should notice that there is a close correlation between the geometrical properties of the coordinates and the behavior of the solutions. The geometrical properties of the coordinates ξ_1, ξ_2 are expressed most clearly by the scale factors $h_1 = \sqrt{(\partial x/\partial\xi_1)^2 + (\partial y/\partial\xi_1)^2}$ and h_2 (see page 24). But for conformal transforms the scale factors are equal, and by use of the Cauchy conditions, we can show that the scale factors are equal to $|dz/dw|$. Wherever $w(z)$ has a pole, there the scale factor $|dz/dw|$ has a zero and the ξ coordinate system has a concentration point. Turned around, we see that, wherever the coordinate system has a concentration point, $w(z)$ has a pole, and the solution $\psi = e^{\pm\alpha w}$ has an essential singularity.

Separation of the Wave Equation. For the two-dimensional Helmholtz equation (wave equation with sinusoidal time factor) k is a constant, equal to ω/c. In this case Eq. (5.1.5) will not separate unless $(dz/dw)(d\bar{z}/d\bar{w})$, which is a function of ξ_1 and ξ_2, turns out to be the *sum* of a function of ξ_1 alone and a function of ξ_2 alone. In other words,

$$k^2|dz/dw|^2 = f(\xi_1) + g(\xi_2)$$

or, in differential form,

$$\frac{\partial^2}{\partial \xi_1 \, \partial \xi_2}\left(\left|\frac{dz}{dw}\right|^2\right) = 0 \tag{5.1.6}$$

which must be solved to find the forms $|dz/dw|^2$ can take on for separable coordinates.

Since the quantity

$$\left|\frac{dz}{dw}\right| = \sqrt{\left(\frac{\partial x}{\partial \xi_1}\right)^2 + \left(\frac{\partial y}{\partial \xi_1}\right)^2} = \sqrt{\left(\frac{\partial x}{\partial \xi_2}\right)^2 + \left(\frac{\partial y}{\partial \xi_2}\right)^2}$$

[using the Cauchy conditions (4.2.1)] is the *scale* factor $h_1 = h_2$ for the two coordinates ξ_1, ξ_2 (for conformal transformations the scale factors are always equal, see page 359), what we have done is to obtain an equation for the kind of scale factor a coordinate system must have to be separable for Eq. (5.1.5). It is not surprising to find the problem turning out this way; we said on page 25 that, when we know the scale factors, we know all the important properties of the corresponding coordinate system.

The equation for h_1h_2 is still not in shape for solution. The quantity dz/dw is a function of w and $d\bar{z}/d\bar{w}$ a function of $\bar{w}$, so the differential operator needs transforming to the variables w, $\bar{w}$. We have

$$\frac{\partial^2}{\partial \xi_1 \, \partial \xi_2} = i\frac{\partial^2}{\partial w^2} - i\frac{\partial^2}{\partial \bar{w}^2}$$

and since dz/dw does not depend on $\bar{w}$, and vice versa, Eq. (5.1.6) becomes

$$\left(\frac{d\bar{z}}{d\bar{w}}\right)\frac{d^2}{dw^2}\left(\frac{dz}{dw}\right) = \left(\frac{dz}{dw}\right)\frac{d^2}{d\bar{w}^2}\left(\frac{d\bar{z}}{d\bar{w}}\right)$$

or

$$\frac{1}{dz/dw}\frac{d^2}{dw^2}\left(\frac{dz}{dw}\right) = \frac{1}{d\bar{z}/d\bar{w}}\frac{d^2}{d\bar{w}^2}\left(\frac{d\bar{z}}{d\bar{w}}\right)$$

The left-hand side of the last equation is a function of w only. In order that it equal the function on the right (a function of $\bar{w}$ alone) for all values of w and $\bar{w}$, both sides must equal the same constant, which we can call λ. Therefore, we finally obtain

$$\frac{d^2}{dw^2}\left(\frac{dz}{dw}\right) = \lambda\left(\frac{dz}{dw}\right);\quad \frac{d^2}{d\bar{w}^2}\left(\frac{d\bar{z}}{d\bar{w}}\right) = \lambda\left(\frac{d\bar{z}}{d\bar{w}}\right) \tag{5.1.7}$$

Only if dz/dw satisfies the first equation (and $d\bar{z}/d\bar{w}$ the conjugate equation) can $|dz/dw|$ be the scale factor for a separable coordinate system.

We must solve these equations to determine what coordinates are separable for Eq. (5.1.1).

Rectangular and Parabolic Coordinates. The simplest case, of course, is for $\lambda = 0$. The solution for dz/dw is then $\beta + \gamma w$, and if γ happens to be zero, we have ($\alpha = a + ib$, $\beta = c + id$)

$$z = \alpha + \beta w; \quad x = a + c\xi_1 - d\xi_2; \quad y = b + c\xi_2 + d\xi_1 \tag{5.1.8}$$

This corresponds to a simple rotation, change of scale, and translation, leaving the coordinates still rectangular. The new scale factor is $|dz/dw| = |\beta| = \sqrt{c^2 + d^2}$, and the equation is the same as Eq. (5.1.2) with $k^2|\beta|^2$ replacing k^2.

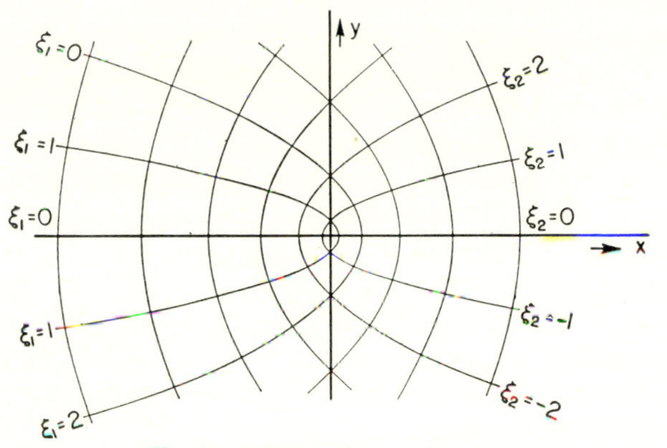

Fig. 5.1 Parabolic coordinates.

When γ is not zero, we may as well concentrate on this term, since the terms in α and β simply rotate and translate. Simple change in scale makes no essential change, so we choose $\gamma = 1$ (and omit the integration constant) for simplicity. Therefore we have

$$\begin{gathered} z = \tfrac{1}{2}w^2; \quad x = \tfrac{1}{2}(\xi_1^2 - \xi_2^2); \quad y = \xi_1\xi_2 \\ |dz/dw| = h_1 = h_2 = |w| = \sqrt{\xi_1^2 + \xi_2^2} \end{gathered} \tag{5.1.9}$$

These are parabolic coordinates, the coordinate lines being two orthogonal families of confocal parabolas, with axes along the x axis. These lines are given by $\xi_1 = [\sqrt{x^2 + y^2} + x]^{\frac{1}{2}} = \text{constant}$ and $\xi_2 = \pm[\sqrt{x^2 + y^2} - x]^{\frac{1}{2}} = \text{constant}$, as shown in Fig. 5.1. Such coordinates are suitable for a boundary consisting of the negative half of the x axis, for instance (which corresponds to $\xi_1 = 0$).

Equation (5.1.5) then becomes

$$\frac{\partial^2\psi}{\partial\xi_1^2} + \frac{\partial^2\psi}{\partial\xi_2^2} + (\xi_1^2 + \xi_2^2)k^2\psi = 0 \tag{5.1.10}$$

which is of separable form, as it should be. Applying the usual machinery, we see that the factored solution is $\psi = F(\xi_1)G(\xi_2)$, with the ordinary differential equations for the factors

$$(d^2F/d\xi_1^2) + (\kappa^2 + k^2\xi_1^2)F = 0; \quad (d^2G/d\xi_2^2) + (-\kappa^2 + k^2\xi_2^2)G = 0 \quad (5.1.11)$$

We shall discuss solutions of these equations later in this chapter.

Several points should be made here. One is that, while the function $z = \frac{1}{2}w^2$ is analytic everywhere except at infinity, nevertheless $w = 0$ is a branch point for the conformal transformation for the scale factor h goes to zero there. We take care of the double values by requiring that ξ_1 have only positive values whereas ξ_2 is allowed to range from $-\infty$ to $+\infty$. The line $\xi_1 = 0$, which is the negative half of the x axis, is thus made the cut, positive values of ξ_2 being above this line and negative values below it. If it is possible, we shall exclude this cut from the region under consideration by making the boundary coincide with one of the lines $\xi_1 =$ constant, having the field outside the boundary. This can be done for most cases with open boundaries, but for closed boundaries, consisting of the lines $\xi_1 =$ constant, $\xi_2 =$ constant, with the field inside the boundaries, part of the cut is in the region of interest. In such a case we avoid discontinuities in the solution along the cut by requiring that G be either an even or an odd function of ξ_2 and then choosing the corresponding $F(\xi_1)$ so that, if G is even, $dF/d\xi_1 = 0$ at $\xi_1 = 0$ or, if G is odd, $F(0) = 0$ (discussed in detail in Sec. 11.2).

Corresponding to the only singularity in w^2, that at infinity, we shall later show that the solutions F and G also have their only singularities for infinite values of ξ_1 and ξ_2.

For the Schroedinger equation in parabolic coordinates, we must have the function k^2 with the form $\epsilon^2 + [f(\xi_1) + g(\xi_2)]/(\xi_1^2 + \xi_2^2)$. In this case the separated equations are

$$(d^2F/d\xi_1^2) + [\epsilon^2\xi_1^2 + f(\xi_1)]F = 0; \quad (d^2G/d\xi_2^2) + [\epsilon^2\xi_2^2 + g(\xi_2)]G = 0$$

where f and g are any self-respecting functions of a single variable. The functions F and G will have singularities at $\xi = \infty$ and at any singularity of f or g.

Polar and Elliptic Coordinates. Returning to Eqs. (5.1.7) we next investigate the cases where λ is not zero. We take first the positive values of λ, and choose $\lambda = 1$ because a variation in numerical value of λ only changes scale and does not give a new set of coordinates (as long as λ is larger than zero, it might as well be unity). We have

$$dz/dw = e^{\pm w} \quad \text{or} \quad z = ae^w + be^{-w}$$

We first take the limiting case of $b = 0$ (and we let $a = 1$, since change of a only changes scale). Then the transformation is to the coordinates given by

$$\begin{aligned} z &= e^w = e^{\xi_1 + i\xi_2}; \quad x = e^{\xi_1}\cos\xi_2; \quad y = e^{\xi_1}\sin\xi_2 \\ r &= e^{\xi_1} = \sqrt{x^2 + y^2}; \quad \varphi = \xi_2 = \tan^{-1}(y/x) \\ |dz/dw| &= h_1 = h_2 = e^{\xi_1} = r \end{aligned} \quad (5.1.12)$$

which are polar coordinates, the family ξ_1 = constant (or r = constant) being concentric circles and the set $\xi_2 = \varphi$ = constant are the radial lines, as shown in Fig. 5.2 (we note that the scale factor h_1 is for the coordinate $\xi_1 = \ln r$, *not* for the coordinate r, which has a scale factor $h_r = 1$). These coordinates are useful for circular boundaries or ones consisting of two lines meeting at an angle.

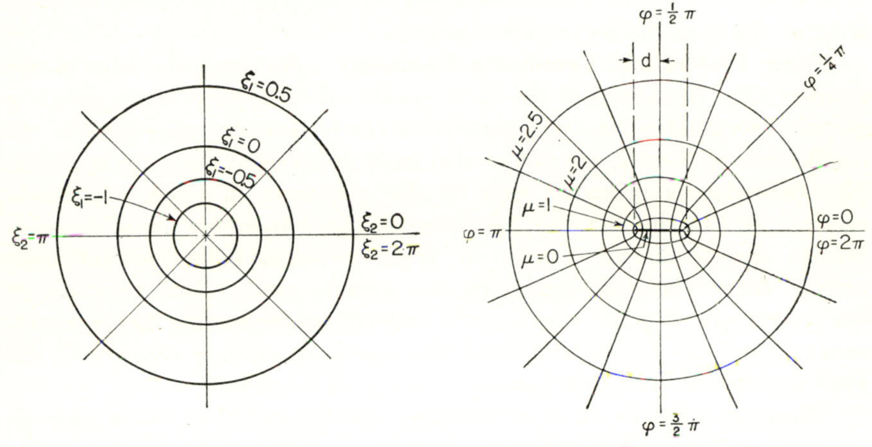

Fig. 5.2 Polar coordinates.

Fig. 5.3 Elliptic coordinates.

The form taken by Eq. (5.1.5) is

$$(\partial^2\psi/\partial\xi_1^2) + (\partial^2\psi/\partial\xi_2^2) + e^{2\xi_1}k^2\psi = 0 \tag{5.1.13}$$

which separates into the equations

$$\psi = F(\xi_1)G(\xi_2); \quad (d^2F/d\xi_1^2) + (k^2e^{2\xi_1} - \alpha^2)F = 0; \quad (d^2G/d\xi_2^2) + \alpha^2 G = 0 \tag{5.1.14}$$

where the constant α^2 is again the separation constant.

For the general case $z = ae^w + be^{-w}$, we can obtain the results in more convenient form by setting

$$a = \tfrac{1}{2}de^{-\beta} = \tfrac{1}{2}e^{\alpha-\beta}; \quad b = \tfrac{1}{2}de^{\beta} = \tfrac{1}{2}e^{\alpha+\beta}; \quad d = e^\alpha = \sqrt{4ab}; \quad e^\beta = \sqrt{b/a}$$

so that, if d, α, and β are real quantities,

$$\begin{gathered} z = d\cosh(w - \beta) = e^\alpha \cosh(\xi_1 + i\xi_2 - \beta) \\ x = d\cosh(\xi_1 - \beta)\cos(\xi_2); \quad y = d\sinh(\xi_1 - \beta)\sin(\xi_2) \\ |dz/dw| = h_1 = h_2 = d\sqrt{\cosh^2(\xi_1 - \beta) - \cos^2\xi_2} \end{gathered} \tag{5.1.15}$$

These are the elliptic coordinates, consisting of confocal ellipses and hyperbolas, with the foci at $x = \pm d$ and $y = 0$, as shown in Fig. 5.3. The constant β is usually set equal to zero for convenience. (However, it should be noted that, if we set $\alpha = \beta + \ln 2$ and then allow β to go to negative infinity, the elliptic foci merge together into the origin and the coordinate system changes to polar coordinates in the limit.) The forms of the partial differential equation in these coordinates and of the ordinary differential equations resulting from separation are

$$\frac{\partial^2\psi}{\partial\xi_1^2} + \frac{\partial^2\psi}{\partial\xi_2^2} + d^2k^2[\cosh^2(\xi_1 - \beta) - \cos^2\xi_2]\psi = 0$$

$$\psi = F(\xi_1)G(\xi_2);\quad (d^2F/d\xi_1^2) + [d^2k^2\cosh^2(\xi_1 - \beta) - \alpha^2]F = 0 \quad (5.1.16)$$

$$(d^2G/d\xi_2^2) - [d^2k^2\cos^2(\xi_2) - \alpha^2]G = 0$$

with α^2 again as the separation constant.

Scale Factors and Coordinate Geometry. We shall see later in this chapter that equations of the form produced by separation in polar or elliptic coordinates, having exponential (or hyperbolic) functions in the coefficients of F or G, are not in the most convenient form for analysis. The difficulty is primarily due to the geometry of these coordinate systems and to the fact that we have obtained them by use of the conformal transformation. Both of these systems have concentration points, where the scale factor $|dz/dw|$ goes to zero. Near such points the coordinate lines are very closely spaced, corresponding to the smallness of the scale factor, and since the transformation is conformal, the scale factors for both coordinates vanish at these points.

However, once we have determined the geometry of the separable system, we can modify the scale factors of each coordinate separately in such a manner as to retain the coordinate geometry but change the separated equations to forms which are more amenable to analysis. For instance, since algebraic functions for the coefficients of F and G are more desirable than exponential ones, in the separated equations, we could set e^{ξ_1} to be a new coordinate r for the polar coordinates and $\cosh(\xi_1 - \beta)$ to be μ for elliptic coordinates. The transformation would not then be conformal, but the shape of the coordinate lines would be unchanged and the coordinates would still be separable.

Referring back to Eq. (5.1.6), either for the wave equation or the Schroedinger equation, for separation we must have that $k^2|dz/dw|^2 = k^2h^2 = f(\xi_1) + g(\xi_2)$. In this case the separated equations are

$$(d^2F/d\xi_1^2) + f_1(\xi_1)F = 0;\quad (d^2G/d\xi_2^2) + g(\xi_2)G = 0;\quad \psi = F(\xi_1)G(\xi_2) \quad (5.1.17)$$

In line with the discussion of the previous paragraph, we choose some function $\mu(\xi_1)$ of the variable ξ_1, such that the function $f(\xi_1)$, expressed as $f(\mu)$, is a simple algebraic function of μ. Since $f + g$ is proportional to the scale factor $|dz/dw|^2$, the chances are that both f and g will go to zero at the concentration points of the coordinate system. We can choose our new coordinate μ so the concentration point is for some standard value of μ, such as 0 or 1 (or, perhaps, infinity). The scale factor for μ is related to the scale factor h for ξ_1 and ξ_2 by the following relation:

$$h_\mu = \sqrt{(\partial x/\partial\mu)^2 + (\partial y/\partial\mu)^2} = h\Phi_\mu;\quad \Phi_\mu = (d\xi_1/d\mu) \quad (5.1.18)$$

since $\partial x/\partial\mu = (\partial x/\partial\xi_1)(d\xi_1/d\mu)$. If h_μ does not go to zero at the concentration point of the coordinate system (where h is zero), then $\Phi(\mu) = h_\mu/h$ must become infinite at these points.

To transform the differential equation (5.1.17) for ξ_1 to one in μ, we also need the formulas

$$\frac{d}{d\mu} = \frac{d\xi_1}{d\mu}\frac{d}{d\xi_1};\quad \frac{d}{d\xi_1} = \frac{1}{\Phi_\mu}\frac{d}{d\mu};\quad \frac{d^2}{d\xi_1^2} = \frac{1}{\Phi_\mu^2}\frac{d^2}{d\mu^2} - \frac{\Phi'_\mu}{\Phi_\mu^3}\frac{d}{d\mu};\quad \Phi'_\mu = \frac{d^2\xi_1}{d\mu^2}$$

We can, if necessary, do the same for ξ_2, changing to a new function $\eta(\xi_2)$ and obtaining a new scale factor $h_\eta = h\Phi_\eta$, and so on. The new transformation and resulting separated equations then become

$$\begin{gathered} w(z) = \xi_1(\mu) + i\xi_2(\eta);\quad h_\mu = \sqrt{(\partial x/\partial\mu)^2 + (\partial y/\partial\mu)^2} = \Phi_\mu h;\quad h_\eta = \Phi_\eta h \\ \frac{\partial^2\psi}{\partial\xi_1^2} + \frac{\partial^2\psi}{\partial\xi_2^2} = -h^2k^2\psi = -[p(\xi_1) + q(\xi_2)]\psi;\quad \psi = F(\mu)G(\eta) \\ \frac{d^2F}{d\mu^2} - \frac{\Phi'_\mu}{\Phi_\mu}\frac{dF}{d\mu} + \Phi_\mu^2 f(\mu)F = 0;\quad \frac{d^2G}{d\eta^2} - \frac{\Phi'_\eta}{\Phi_\eta}\frac{dG}{d\eta} + \Phi_\eta^2 g(\eta)G = 0 \end{gathered} \tag{5.1.19}$$

These last two ordinary differential equations look more complicated than Eqs. (5.1.17), but if Φ'_μ/Φ_μ and $\Phi_\mu^2 f(\mu)$ are algebraic functions of μ instead of transcendental or other functions, they are easier to analyze and solve.

As we pointed out above, the function Φ_μ^2 or the function Φ'_μ/Φ_μ or both will probably become infinite at the values of μ corresponding to the concentration points of the coordinate system, so that the singularities of the coefficients of $dF/d\mu$ and F in the equation for F are closely related to the geometry of the corresponding coordinate system. This interrelation will be referred to later in the chapter.

To make specific these remarks we apply them to the polar coordinates of Eqs. (5.1.12). We wish to change the scale of the radial coordinate so that the function $e^{2\xi_1}$ becomes an algebraic function. An obvious choice is $\mu = r = e^{\xi_1}$ or $\xi_1 = \ln\mu$, where μ (or r) is the usual distance. The origin, the only concentration point for this coordinate system, is then at $r = 0$. The scale factor and resulting equation for F are then

$$\xi_1 = \ln r;\quad \Phi_r = \frac{1}{r};\quad \Phi'_r = -\left(\frac{1}{r^2}\right);\quad h_r = 1;\quad \frac{d^2F}{dr^2} + \frac{1}{r}\frac{dF}{dr} + \frac{1}{r^2}f(r)F = 0 \tag{5.1.20}$$

where, for the wave equation, $f(r) = r^2k^2 - \alpha^2$. We see, therefore, that the coefficients both of dF/dr and of F have singularities at the polar center, $r = 0$.

The scale of the ξ_2 coordinate does not need to be changed, since the last of Eqs. (5.1.14) has no transcendental coefficients. Nevertheless ξ_2 is an angle, so that the solution G is a periodic function and w is a multivalued function of z. To remove the "multivaluedness" it is sometimes useful to make the transformation $\eta = \cos \xi_2$, where the useful range of η is between -1 and $+1$. The related equations are

$$\xi_2 = \cos^{-1} \eta; \quad \Phi_\eta = -(1 - \eta^2)^{-\frac{1}{2}}; \quad \Phi'_\eta = -\eta(1 - \eta^2)^{-\frac{3}{2}}$$
$$h_\eta = -r/\sqrt{1 - \eta^2}; \quad \frac{d^2G}{d\eta^2} - \frac{\eta}{1 - \eta^2}\frac{dG}{d\eta} + \frac{g(\eta)}{1 - \eta^2} G = 0 \qquad (5.1.21)$$

where, for the wave equation, $g(\eta) = \alpha^2$, the separation constant. The singularities of the coefficients are here at $\eta = \pm 1$, the two ends of the useful range for η.

It should be obvious by now that the simplest transformation for the elliptic coordinate case is $\cosh(\xi_1 - \beta) = \mu$ and $\cos \xi_2 = \eta$, where the interesting range of μ is from 1 to ∞ and the useful range of η is from $+1$ to -1. The transformation and modified equations are (for $\beta = 0$)

$$x = d\mu\eta; \quad y = d\sqrt{(\mu^2 - 1)(1 - \eta^2)}; \quad \xi_1 = \cosh^{-1} \mu; \quad \xi_2 = \cos^{-1} \eta$$
$$\Phi_\mu = \frac{1}{\sqrt{\mu^2 - 1}}; \quad h_\mu = d\sqrt{\frac{\mu^2 - \eta^2}{\mu^2 - 1}}; \quad h_\eta = d\sqrt{\frac{\mu^2 - \eta^2}{1 - \eta^2}}$$
$$\frac{d^2F}{d\mu^2} + \frac{\mu}{\mu^2 - 1}\frac{dF}{d\mu} + \frac{f(\mu)}{\mu^2 - 1} F = 0; \quad \frac{d^2G}{d\eta^2} - \frac{\eta}{1 - \eta^2}\frac{dG}{d\eta} + \frac{g(\eta)}{1 - \eta^2} G = 0 \qquad (5.1.22)$$

where, for the wave equation, $f(\mu) = d^2k^2\mu^2 - \alpha^2$ and $g(\eta) = -d^2k^2\eta^2 + \alpha^2$. Here again, the coefficients in these equations have singularities only where the coordinates have concentration points ($\mu = 1$, $\eta = \pm 1$).

We should also note that $\mu + \eta = (1/d)\sqrt{(x + d)^2 + y^2} = r_1/d$ and $\mu - \eta = (1/d)\sqrt{(x - d)^2 + y^2} = r_2/d$, or

$$\mu = (r_1 + r_2)/2d; \quad \eta = (r_2 - r_2)/2d \qquad (5.1.23)$$

where r_1, r_2 are the distances from the point (x,y) to the two foci of the coordinates ($x = \mp d$, $y = 0$). Therefore the line μ = constant corresponds to the locus of points such that the sum of the distances of the point from the two foci are constant, a standard definition of an ellipse. Similarly the line η = constant involves the difference between the distances, a standard definition of an hyperbola.

Separation Constants and Boundary Conditions. If the coordinates suitable for the boundary surface of a given problem are separable coordinates, then one can, in principle, satisfy reasonable boundary conditions by the right combination of the factored solutions (what boundary conditions are "reasonable" and how to find the "right combination" will be discussed in Chaps. 6 and 7). For instance, the boundary may be the line ξ_1 = constant in one of the separable two-dimensional coordinates we have been discussing. This boundary may be finite in length (as for a closed boundary), in which case the factors $X_2(\xi_2)$ must be continuous as we vary ξ_2, going along the line ξ_1 = constant from a given point clear around to the same point.

For instance, for the polar coordinates r, φ, the line r = constant is a circle of finite size which is completely traversed by letting the angle φ go from 0 to 2π. The equation for the φ factor in the separated solution is

$$(d^2\Phi/d\varphi^2) + \alpha^2\Phi = 0; \quad \text{where } \alpha \text{ is the separation constant}$$

with solutions either $\cos(\alpha\varphi)$ or $\sin(\alpha\varphi)$ or a linear combination thereof. In order to have Φ continuous along the boundary r = constant, the solution Φ must have the same value at $\varphi = 2\pi$ as it does at $\varphi = 0$; in other words Φ must be *periodic* in φ with period 2π. Such a periodicity requirement imposes restrictions on the allowable values of the separation constant α. In our example, in order that $\cos(\alpha\varphi)$ or $\sin(\alpha\varphi)$ be periodic in φ with period 2π, the separation constant α must be an integer $m = 0, 1, 2, 3, \ldots$. In other cases where ξ_2 is a periodic coordinate, there are similar restrictions on the values of the separation constant for solutions to be continuous around the circuit of the boundary ξ_1 = constant. We can always order these allowed values, letting the lowest value be α_1 and so on, so that $\alpha_{n+1} > \alpha_n$, and we can label the corresponding factors in the solutions $X_2^1(\xi_2)$, $X_2^2(\xi_2)$, ... The factor for ξ_1 also depends on α, so that the complete solution corresponding to the allowed value α_n of the separation constant is $X_1^n(\xi_1)X_2^n(\xi_2)$.

It will be shown in Chap. 6 that any "reasonable" function of the periodic coordinate ξ_2 can always be expressed in terms of a series of these allowed functions

$$f(\xi_2) = \sum_{n=1}^{\infty} A_n X_2^n(\xi_2)$$

The rules for computing the coefficients A_n will also be given later. Therefore if the solution of interest $\psi(\xi_1,\xi_2)$ is to be one which satisfies the boundary condition $\psi(c,\xi_2) = f(\xi_2)$ along the boundary $\xi_1 = c$, then the solution may be expressed in terms of the separated solutions for the allowed values of the separation constant:

$$\psi(\xi_1,\xi_2) = \sum A_n \left[\frac{X_1^n(\xi_1)}{X_1^n(c)}\right] X_2^n(\xi_2) \tag{5.1.24}$$

Other, more complicated boundary conditions may be similarly satisfied.

We notice that the final solution, $\psi(\xi_1,\xi_2)$, is *not* separable but that it can be expressed in terms of a series of separable solutions. In each case the conditions of periodicity pick out a sequence of allowed values of the separation constant, and the general solutions are given in terms of a series over these allowed values. Even for open boundaries, generalizations of the periodicity requirements allowed us again to express a solution satisfying specified boundary conditions in terms of a series (or an integral) of separated solutions over the allowed values of the separation constant.

But this discussion has carried us as far afield as is needful here; we must next investigate the separability of three-dimensional partial differential equations.

Separation in Three Dimensions. Separation of solutions in two dimensions is particularly simple for several reasons. In the first place there is only one separation constant, so that the factored solutions form a one-parameter family, which makes the fitting of boundary conditions by series relatively straightforward. In the second place, the conditions for separation are simple: in the equation

$$\frac{\partial^2\psi}{\partial u^2} + \frac{\partial^2\psi}{\partial v^2} + k^2\left|\frac{dz}{dw}\right|^2 \psi = 0$$

the term $k^2|dz/dw|^2$ must consist of a simple sum of a function of u alone and a function of v alone, or else the equation would not separate. And third, the only cases giving rise to families of nodal surfaces which coincide with the coordinate surfaces are cases where the solutions have the simple factored form $X_1(\xi_1)X_2(\xi_2)$.

In all three of these respects the three-dimensional separation problem is more complex. Since there are three separated equations, there are two separation constants instead of one. Each of the three equations may contain both constants, in which case each of the three factors in the separated solution depend on both separation constants in a complicated manner and the satisfying of boundary conditions even by a series of separated solutions is a tedious and difficult task. For some coordinate systems, however, the results are such that one (or two) of the separated equations contains only one separation constant; in these cases the general series solution takes on a simpler and more amenable form.

Taking up the third item next, it turns out that in the three-dimensional Laplace equation $\nabla^2\psi = 0$, there are some coordinate systems in which the solution takes on the more complex form $R(\xi_1,\xi_2,\xi_3)X_1(\xi_1)\cdot$ $\cdot X_2(\xi_2)X_3(\xi_3)$, where the additional factor R (which might be called a *modulation factor*) is independent of the separation constants. For these systems there is a measure of separation, and boundary conditions can be satisfied, for the common factor R can be taken outside the summation over allowed values of the separation constant and the sum has the same general form as for the cases without a modulation factor.

Returning to the second item enumerated above, the term in the three-dimensional partial differential equation corresponding to the term $k^2|dz/dw|^2$ for two dimensions does not need to be simply a sum of functions each of only one coordinate; separation may be attained for more complicated cases than this.

The Stäckel Determinant. The general technique for separating our standard three-dimensional partial differential equation

$$\nabla^2\psi + k_1^2\psi = 0$$

depends on the properties of three-row determinants. The relation between such a determinant S and its elements Φ_{mn} is given by the equation

$$S = |\Phi_{mn}| = \begin{vmatrix} \Phi_{11} & \Phi_{12} & \Phi_{13} \\ \Phi_{21} & \Phi_{22} & \Phi_{23} \\ \Phi_{31} & \Phi_{32} & \Phi_{33} \end{vmatrix} = \Phi_{11}\Phi_{22}\Phi_{33} + \Phi_{12}\Phi_{23}\Phi_{31} + \Phi_{13}\Phi_{21}\Phi_{32} - \Phi_{13}\Phi_{22}\Phi_{31} - \Phi_{11}\Phi_{23}\Phi_{32} - \Phi_{12}\Phi_{21}\Phi_{33} \quad (5.1.25)$$

The *first minor* of the determinant for the element Φ_{mn} is the factor which multiplies the element Φ_{mn} in the expression for the determinant. For instance, the first minors for the elements in the first column, Φ_{11}, Φ_{21}, Φ_{31}, are

$$\begin{aligned} M_1 &= \partial S/\partial\Phi_{11} = \Phi_{22}\Phi_{33} - \Phi_{23}\Phi_{32} \\ M_2 &= \partial S/\partial\Phi_{21} = \Phi_{13}\Phi_{32} - \Phi_{12}\Phi_{33} \\ M_3 &= \partial S/\partial\Phi_{31} = \Phi_{12}\Phi_{23} - \Phi_{13}\Phi_{22} \end{aligned} \quad (5.1.26)$$

Since we need here only minors for the first column, we need not give the M's two subscripts.

The important property of determinants, which we shall use in separating three-dimensional equations, is the orthogonality property relating elements and first minors. For instance, for the minors defined in Eq. (5.1.26) we have

$$\sum_{n=1}^{3} M_n\Phi_{n1} = S; \quad \sum_{n=1}^{3} M_n\Phi_{nm} = 0; \quad m = 2, 3 \quad (5.1.27)$$

as though there were vectors with components Φ_{n2} or Φ_{n3} which were perpendicular to a vector with components M_n.

Therefore if the separated equations for the three-dimensional case were

$$\frac{1}{f_n}\frac{\partial}{\partial\xi_n}\left[f_n\frac{\partial X_n}{\partial\xi_n}\right] + [k_1^2\Phi_{n1} + k_2^2\Phi_{n2} + k_3^2\Phi_{n3}]X_n = 0 \quad (5.1.28)$$

we could combine the three equations in such a manner as to eliminate the separation constants k_2 and k_3. For by multiplying the equation for X_1 by $(M_1/S)X_2X_3$, and so on, and adding, we obtain

$$\sum_n \frac{M_n}{Sf_n}\frac{\partial}{\partial\xi_n}\left[f_n\frac{\partial\psi}{\partial\xi_n}\right] + k_1^2\psi = 0; \quad \psi = X_1X_2X_3 \quad (5.1.29)$$

This equation would correspond to our standard equation $\nabla^2\psi + k_1^2\psi = 0$

if the expression for the Laplacian in the ξ coordinates,

$$\nabla^2\psi = \sum_n \frac{1}{h_1h_2h_3} \frac{\partial}{\partial \xi_n} \left[\frac{h_1h_2h_3}{h_n^2} \frac{\partial \psi}{\partial \xi_n} \right] \tag{5.1.30}$$

were equal to the first expression of Eq. (5.1.29).

In order to have this happen, several restrictions must be placed on the scale factors h and on the elements Φ_{nm} of the determinant S. In the first place, since Eq. (5.1.28) is supposed to be a separated equation, the functions f_n, Φ_{n1}, Φ_{n2}, and Φ_{n3} must all be functions of ξ_n alone. A determinant with elements Φ_{1m} in the top row which are functions of ξ_1, with elements Φ_{2m} in the second row functions of ξ_2 and with elements Φ_{3m} in the bottom row functions of ξ_3 only, is called a *Stäckel determinant.* It is basic for the study of separation in three dimensions. We also note that, if Φ_{1m} is a function of ξ_1, etc., the first minor M_1 is a function of ξ_2 and ξ_3, but not of ξ_1, etc.

Next, we must have the quantity $h_1h_2h_3/h_n^2$ be equal to the product of the function f_n of ξ_n alone times some function g_n of the other ξ's. Then, for instance, the term in the Laplacian for ξ_1 becomes

$$\frac{1}{h_1h_2h_3} \frac{\partial}{\partial \xi_1} \left[\frac{h_1h_2h_3}{h_1^2} \frac{\partial \psi}{\partial \xi_1} \right] = \frac{g_1(\xi_2,\xi_3)}{h_1h_2h_3} \frac{\partial}{\partial \xi_1} \left[f_1(\xi_1) \frac{\partial \psi}{\partial \xi_1} \right] = \frac{1}{h_1^2 f_1} \frac{\partial}{\partial \xi_1} \left[f_1 \frac{\partial \psi}{\partial \xi_1} \right]$$

which has the general form of the term $(M_1/Sf_1)[\partial(f_1\partial\psi/\partial\xi_1)/\partial\xi_1]$ in Eq. (5.1.29). In fact in order to have the terms identical, we must have

$$1/h_n^2 = M_n/S \tag{5.1.31}$$

and this, together with the original restriction on $h_1h_2h_3/h_n^2$, leads to the *Robertson condition*

$$h_1h_2h_3/S = f_1(\xi_1)f_2(\xi_2)f_3(\xi_3) \tag{5.1.32}$$

which at the same time determines the Stäckel determinant and limits the kinds of coordinate systems which will allow separation. When it holds, the quantity $h_1h_2h_3/h_1^2$ is equal to f_1, a function of ξ_1, times $M_1f_2f_3$, a function of ξ_2 and ξ_3 but not of ξ_1, thus fulfilling the requirement stated earlier in this paragraph.

These interrelated requirements on the scale factors severely limit the number of independent coordinate systems which answer the specifications. The detailed analysis necessary to determine which systems do satisfy the requirements is much more tedious than was the corresponding analysis, given on page 500, for the two-dimensional case. There we showed that the separable coordinate systems (for the wave equation) consisted of *confocal conic sections* (ellipses and hyperbolas) or their degenerate forms (circles and radii, confocal parabolas, or parallel lines). The detailed analysis for the three-dimensional case reveals a corresponding limitation; for Euclidean space, the separable coordinates for the wave equation are *confocal quadric surfaces* or their degenerate forms.

Confocal Quadric Surfaces. The equation

$$\frac{x^2}{\xi^2 - a^2} + \frac{y^2}{\xi^2 - b^2} + \frac{z^2}{\xi^2 - c^2} = 1; \quad a \geq b \geq c \geq 0 \qquad (5.1.33)$$

for different values of the parameter ξ, represents three families of confocal quadric surfaces. For $\xi > a$ we have a complete family of confocal ellipsoids, with traces on the y, z plane which are ellipses with foci at $y = 0$, $z = \pm \sqrt{b^2 - c^2}$, traces on the x, z plane which are ellipses with foci at $x = 0$, $z = \pm \sqrt{a^2 - c^2}$ and traces on the x, y plane which are ellipses with foci at $x = 0$, $y = \pm \sqrt{a^2 - b^2}$. The limiting surface of this family is for $\xi \to a$, which is the part of the y, z plane *inside* an ellipse having major axis $2\sqrt{a^2 - c^2}$ along the z axis and minor axis $2\sqrt{a^2 - b^2}$ along the y axis.

For $a > \xi > b$ we have a complete set of confocal hyperboloids of one sheet, with traces on the y, z plane which are ellipses with foci at $y = 0$, $z = \pm \sqrt{b^2 - c^2}$, with traces on the x, z plane which are hyperbolas with foci at $x = 0$, $z = \pm \sqrt{a^2 - c^2}$ and with traces on the x, y plane which are hyperbolas with foci at $x = 0$, $y = \pm \sqrt{a^2 - b^2}$. The limiting surfaces are for $\xi \to a$, which is the part of the y, z plane *outside* the ellipse having major axis $2\sqrt{a^2 - c^2}$ along the z axis and minor axis $2\sqrt{a^2 - b^2}$ along the y axis, and for $\xi \to b$, which is the part of the x, z plane outside the hyperbolas having transverse axis $2\sqrt{b^2 - c^2}$ along z and conjugate axis $2\sqrt{a^2 - b^2}$ along x.

Finally for $b > \xi > c$ we have a complete family of confocal hyperboloids of two sheets, with traces on the y, z plane which are hyperbolas with foci at $y = 0$, $z = \pm \sqrt{b^2 - c^2}$, with traces on the x, z plane which are hyperbolas with foci at $x = 0$, $z = \pm \sqrt{a^2 - c^2}$ and with no traces on the x, y plane. The limiting surfaces here are for $\xi \to b$, which is the part of the x, z plane inside the hyperbolas having transverse axis $2\sqrt{b^2 - c^2}$ along z and conjugate axis $2\sqrt{a^2 - b^2}$ along the x axis, and for $\xi \to c$, which is the x, y plane. Incidentally, we can set $c = 0$ without any loss of generality.

Since the three families of surfaces are mutually orthogonal, we can consider the three ranges for the parameter ξ to correspond to three families of coordinate surfaces, with $\xi_1(\xi_1 > a)$ corresponding to the ellipsoids, $\xi_2(a > \xi_2 > b)$ to the hyperboloids of one sheet, and $\xi_3(b > \xi_3 > c)$ to the hyperboloids of two sheets. It can easily be verified that the relation between the coordinates x, y, z and the *ellipsoidal coordinates* ξ_1, ξ_2, ξ_3 (for $c = 0$) with their scale factors are

$$x = \sqrt{\frac{(\xi_1^2 - a^2)(\xi_2^2 - a^2)(\xi_3^2 - a^2)}{a^2(a^2 - b^2)}};\quad \xi_1 > a > \xi_2 > b > \xi_3 > 0$$

$$y = \sqrt{\frac{(\xi_1^2 - b^2)(\xi_2^2 - b^2)(\xi_3^2 - b^2)}{b^2(b^2 - a^2)}};\quad z = \frac{\xi_1\xi_2\xi_3}{ab} \tag{5.1.34}$$

$$h_1 = \sqrt{\frac{(\xi_1^2 - \xi_2^2)(\xi_1^2 - \xi_3^2)}{(\xi_1^2 - a^2)(\xi_1^2 - b^2)}};\qquad h_2 = \sqrt{\frac{(\xi_2^2 - \xi_1^2)(\xi_2^2 - \xi_3^2)}{(\xi_2^2 - a^2)(\xi_2^2 - b^2)}};\quad \text{etc.}$$

Following our earlier discussion, we find that $h_1h_2h_3/h_1^2$ equals

$$\sqrt{(\xi_1^2 - a^2)(\xi_1^2 - b^2)}$$

times the function $\sqrt{(\xi_3^2 - \xi_2^2)^2/(\xi_2^2 - a^2)(\xi_2^2 - b^2)(\xi_3^2 - a^2)(\xi_3^2 - b^2)}$, which does not contain ξ_1. Consequently the function

$$f_n(\xi_n) = \sqrt{(\xi_n^2 - a^2)(\xi_n^2 - b^2)} \tag{5.1.35}$$

is the one defined in Eq. (5.1.32). This in turn fixes the Stäckel determinant:

$$S = \frac{h_1h_2h_3}{f_1f_2f_3} = \frac{(\xi_1^2 - \xi_2^2)(\xi_2^2 - \xi_3^2)(\xi_3^2 - \xi_1^2)}{(\xi_1^2 - a^2)(\xi_2^2 - a^2)(\xi_3^2 - a^2)(\xi_1^2 - b^2)(\xi_2^2 - b^2)(\xi_3^2 - b^2)}$$

and this, together with Eqs. (5.1.31),

$$M_1 = \frac{S}{h_1^2} = -\frac{\xi_2^2 - \xi_3^2}{(\xi_2^2 - a^2)(\xi_3^2 - a^2)(\xi_2^2 - b^2)(\xi_3^2 - b^2)};\quad \text{etc.}$$

enables us to compute the elements of the Stäckel determinant. They are

$$\Phi_{n1}(\xi_n) = 1;\quad \Phi_{n2}(\xi_n) = \frac{1}{\xi_n^2 - a^2};\quad \Phi_{n3}(\xi_n) = \frac{1}{(\xi_n^2 - b^2)(a^2 - b^2)} \tag{5.1.36}$$

The Helmholtz equation and the resulting separated ordinary differential equations for these coordinates are therefore

$$\sum_n \frac{G_nf_n}{(\xi_1^2 - \xi_2^2)(\xi_2^2 - \xi_3^2)(\xi_3^2 - \xi_1^2)}\frac{\partial}{\partial\xi_n}\left[f_n\frac{\partial\psi}{\partial\xi_n}\right] + k_1^2\psi = 0;$$

$$G_1 = (\xi_2^2 - \xi_3^2);\quad G_2 = (\xi_3^2 - \xi_1^2);\quad G_3 = (\xi_1^2 - \xi_2^2)$$

and
$$\frac{1}{\sqrt{(\xi_n^2 - a^2)(\xi_n^2 - b^2)}}\frac{d}{d\xi_n}\left[\sqrt{(\xi_n^2 - a^2)(\xi_n^2 - b^2)}\frac{dX_n}{d\xi_n}\right] + \left[k_1^2 + \frac{k_2^2}{(\xi_n^2 - a^2)} + \frac{k_3^2}{(\xi_n^2 - b^2)(a^2 - b^2)}\right]X_n = 0 \tag{5.1.37}$$

This is the sort of ordinary differential equation we shall encounter in our future studies.

We note that in the three-dimensional Schroedinger equation for a particle the term k^2 (which we have been calling k_1^2) is not a constant but is the difference between a constant k_1^2 (the total energy of the particle) and the potential energy of the particle as a function of the ξ's. To have separability the potential energy of the particle must be such as to subtract some function $\mu_n(\xi_n)$ of ξ_n alone from the coefficient of X_n in Eq. (5.1.37). This means that the allowed form for the potential energy is

$$V = \sum_{n=1}^{3} \frac{\mu_n(\xi_n)}{h_n^2} \tag{5.1.38}$$

We note also that for ellipsoidal coordinates each of the three separated ordinary differential equations contains k_1 and both separation constants k_2 and k_3. Recalling our discussion of quantum mechanics in Chap. 2, we can consider the process of separation a process of rotation, in abstract vector space, from a vector set specified by the coordinates (x, y, z or ξ_1, ξ_2, ξ_3) to one specified by the parameters k_1, k_2, k_3. The factored solutions are the transformation functions (direction cosines) going from the eigenvalues for coordinate position to the eigenvalues for the k's. What we have found is that this transformation, for ellipsoidal coordinates, yields transformation functions which are separated for the coordinates (into factors) but *which are not separated* for the parameters k. For some of the degenerate forms of ellipsoidal coordinates the factored solutions are also separated with regard to the parameters, which adds considerably to the ease of manipulating the solutions.

It is these degenerate forms of the ellipsoidal coordinates, obtained by setting a, b, c equal or zero or infinity, which are of more use and interest than the general form. There are 10 such forms which are recognized as "different" coordinate systems and given special names. These 11 systems (the general ellipsoidal system plus the 10 degenerate forms) are the only systems which allow separation of the wave equation or the Schroedinger equation in three dimensions [even with these the Schroedinger equation separates only if the potential energy has a certain functional form, see Eq. (5.1.38)]. These forms, with the related forms for scale factors h, the determinant S, and so on, are given in the table at the end of this chapter.

Degenerate Forms of Ellipsoidal Coordinates. Starting from the transformation

$$x' = \sqrt{\frac{(x_1^2 - \alpha^2)(x_2^2 - \alpha^2)(x_3^2 - \alpha^2)}{\alpha^2(\alpha^2 - \beta^2)}}$$

$$y' = \sqrt{\frac{(x_1^2 - \beta^2)(x_2^2 - \beta^2)(x_3^2 - \beta^2)}{\beta^2(\beta^2 - \alpha^2)}}; \quad z' = \frac{x_1 x_2 x_3}{\alpha\beta}$$

for the generalized ellipsoidal coordinates, we can obtain all the ten degenerate forms listed at the end of the chapter by stretching, compressing, and translating. For instance, stretching all focal distances indefinitely gives, at the center of the ellipsoids,

I. Rectangular Coordinates. We set x_1^2 in the equations above equal to $\alpha^2 + \xi_1^2$, $x_2^2 = \beta^2 + \xi_2^2$, and $x_3 = \xi_3$; we let $\beta = \alpha \sin \varphi$, where φ can be arbitrary, and then allow α to become infinite. This will give the coordinates defined under I at the end of the chapter. On the other hand, letting β go to zero symmetrizes the ellipsoids into flattened figures of revolution:

IX. Oblate Spheroidal Coordinates. Setting $\alpha = a$, $x_1^2 = a^2 + \xi_1^2$, $x_2^2 = a^2 - a^2\xi_2^2$, $x_3 = \beta\xi_3$ and letting β go to zero, we change from ellipsoids to oblate (flattened) spheroids, from hyperboloids of one sheet to hyperboloids of revolution (still of one sheet) and from hyperboloids of two sheets to pairs of planes going through the axis of revolution. In order to have the form given in the table at the end of the chapter we must set $x' = z$, $y' = y$, $z' = x$, which makes z the axis of revolution. Making the ellipsoid into a figure of revolution about the longer axis gives us

VIII. Prolate Spheroidal Coordinates. This is done by letting $\beta \to \alpha$. according to the following formulas, $\alpha = a$, $\beta^2 = a^2 - \epsilon$, $x_1 = \xi_1$, $x_2^2 = a^2 - \epsilon\xi_3^2$, $x_2^2 = a^2\xi_2^2$, $\epsilon \to 0$. If then we stretch the long axis out indefinitely, we arrive at

II. Circular Cylinder Coordinates. We set $\alpha = a$, $\beta^2 = a^2 - \epsilon$, $x_1^2 = a^2 + \xi_1^2$, $x_2^2 = a^2 - \epsilon\xi_2^2$, $x_3 = \xi_3$, let $\epsilon \to 0$ and then $a \to \infty$, giving the simplest rotational coordinates. If we had stretched the long axis out indefinitely before symmetrizing, we would have obtained

III. Elliptic Cylinder Coordinates. Setting $\beta^2 = \alpha^2 + a^2, x_1^2 = \alpha^2 + \xi_1^2$, $x_2^2 = \alpha^2 + a^2\xi_2^2$, $x_3 = \xi_3$ and letting $\alpha \to \infty$ but keeping a finite, results in this system. On the other hand if we shorten the long axis of the prolate spheroidal coordinates, instead of stretching it, we arrive at the completely symmetric

V. Spherical Coordinates. Setting $\alpha = a, \beta^2 = a^2 - \epsilon, x_1 = \xi_1$, $x_2^2 = a^2 - \epsilon\xi_2^2$, $x_3 = a\xi_3$, we first let $\epsilon \to 0$, then let $a \to 0$, which produces complete symmetry. Finally, if we let β be proportional to α and have both go to zero at the same time, we obtain

VI. Conical Coordinates. These have spheres and cones of elliptic cross section as coordinate surfaces. They are obtained by setting $\alpha = ka$, $\beta = k'a$, $k^2 + k'^2 = 1$, $x_1^2 = \xi_1^2/(k^2 - k'^2)$, $x_2^2 = a^2[2k^2k'^2 + (k^2 - k'^2)\xi_2^2]$, $x_3^2 = a^2[2k^2k'^2 - (k^2 - k'^2)\xi_3^2]$ and then letting $a \to 0$.

The parabolic systems are obtained by changing the position of their origin to the "edge" of the ellipsoid before stretching. The most general case is

XI. Paraboloidal Coordinates. Here we set $\alpha^2 = d^2 + a^2d$, $\beta^2 = d^2 + b^2d$ and set the new origin at $z' = d$, so that $x = x'$, $y = y'$, and $z = z' - d$. We let $x_1^2 = d^2 + \eta_1^2d$, $x_2^2 = d^2 + \eta_2^2d$, and $x_3^2 = d^2 + \eta_3^2d$ and then finally allow $d \to \infty$. The new coordinates are

$$x = \sqrt{\frac{(\eta_1^2 - a^2)(\eta_2^2 - a^2)(\eta_3^2 - a^2)}{a^2 - b^2}};\quad y = \sqrt{\frac{(\eta_1^2 - b^2)(\eta_2^2 - b^2)(\eta_3^2 - b^2)}{b^2 - a^2}}$$

$$z = \tfrac{1}{2}(\eta_1^2 + \eta_2^2 + \eta_3^2 - a^2 - b^2)$$

which correspond to the surfaces

$$\frac{x^2}{\eta^2 - a^2} + \frac{y^2}{\eta^2 - b^2} = \eta^2 - 2z$$

For $\eta = \eta_1 > a$ they are a family of elliptic paraboloids with traces which are parabolas on the x, z and y, z planes but which are ellipses on the x, y plane. For $\eta = \eta_2$ (where $a > \eta_2 > b$) the surfaces are hyperbolic paraboloids, with traces which are parabolas in the x, z and y, z planes but which are hyperbolas in the x, y plane. Finally for $\eta = \eta_3 < b$ (we must let η_3^2 become negative to cover all of this family) the surfaces are again elliptic paraboloids, pointed in the opposite direction with respect to the z axis.

The limiting surfaces are for $\eta_1 \to a$, which is that part of the y, z plane which is inside the parabola with vertex at $z = \frac{1}{2}a^2$ and focus at $z = \frac{1}{2}b^2$; for $\eta_2 \to a$, which is the rest of the y, z plane; and for $\eta_2 \to b$, which is the part of the x, z plane outside the parabola with vertex at $z = \frac{1}{2}b^2$ and focus at $\frac{1}{2}a^2$; while the limiting surface for $\eta_3 \to b$ is the rest of the x, z plane. The scale factors and related functions for these coordinates are given on page 664.

As before, other coordinate systems can be obtained by modification of interfocal distances. For instance, if we set $a = b$, we obtain the rotational

VII. Parabolic Coordinates. We set $b^2 = a^2 - \epsilon$, $\eta_1^2 = \xi_1^2 + a^2$, $\eta_2^2 = a^2 - \epsilon\xi_3^2$, $\eta_3^2 = b^2 - \xi_2^2$ and then let $\epsilon \to 0$, giving the simpler system. On the other hand, if we pull out the longer elliptic axis, we eventually obtain

IV. Parabolic Cylinder Coordinates. Here we set $\eta_1^2 = a^2 + \xi_3^2$, $\eta_2^2 = b^2 + \xi_1^2$, $\eta_3^2 = b^2 - \xi_2^2$ and $x = z' - \frac{1}{2}b^2$, $y = y'$, $z = x'/a$ and then let $a \to \infty$.

This completes all the different degenerate systems which can be obtained from ellipsoidal coordinates. It is of interest to examine the Stäckel determinants and the finally separated equations for these cases to see whether there are systematic trends.

Confluence of Singularities. We have arranged the scale of all these coordinates so that the functions f_n and Φ_{nm} are algebraic functions of the coordinate ξ_n and so that, when the separated equation is put in the form

$$\frac{d^2X_n}{d\xi_n^2} + p_n(\xi_n)\frac{dX_n}{d\xi_n} + q_n(\xi_n)X_n = 0; \quad p_n = \frac{d}{d\xi_n}[\ln f_n(\xi_n)];$$
$$q_n = \sum_{m=1}^{3} k_m^2\Phi_{nm}(\xi_n) \qquad (5.1.39)$$

the functions p and q have singularities at the concentration points for the corresponding coordinates. For instance, for the ellipsoidal coordinates,

p and q have poles at $\xi = \pm a, \pm b$, and at infinity (*i.e.*, if we change to the variable $u = 1/\xi$ the corresponding functions p and q have poles at $u = 0$, or $\xi = \infty$). We obtain the degenerate forms of the coordinate systems by merging two or more of these singularities. A point which is a singularity for p or q is called a *singular point* of the corresponding equation, and a merging of singular points is called a *confluence* of singular points.

In the case of prolate spheroidal coordinates, for instance, there is confluence between the singular points at a and b and at $-a$ and $-b$ which, together with a change of scale, puts the singular points for the ξ_1 equation at $\pm a$ and infinity, for the ξ_2 and ξ_3 equations at ± 1 and ∞. In the spherical coordinates we let a go to zero, so that the singular points for the ξ_1 equation are at 0 and ∞, and so on.

Wherever there is a singular point of the differential equation, there is a singularity (pole, branch point, or essential singularity) of the general solution of the equation. Consequently we can say that the factored solution $\psi = X_1X_2X_3$ usually has a singularity at all the concentration points of the corresponding coordinate system. We can also say that all the ordinary differential equations obtained by separating the equation $\nabla^2\psi + k^2\psi = 0$ (which includes a large part of the equations we shall study) are obtained from the general equation with five singular points by confluence of the five singular points to form four, three, or two. Just as the specification of zeros and singularities specifies a function of a complex variable, so a specification of the position and nature of the singular points of a differential equation specifies the equation and its solutions, as we shall see later in this chapter. This is, of course, just another way of saying that the geometry of the coordinate system determines the nature of the solutions of the separated equations, a not surprising result.

Separation Constants. Examination of the Stäckel determinants of the 11 coordinate systems tabulated at the end of the chapter shows that there are quite a number with some of their elements equal to zero. For instance all the rotational coordinates have $\Phi_{31} = \Phi_{32} = 0$. This means that the factors $X_3(\xi_3)$, for rotational coordinates involve only the separation constant k_3. Since ξ_3 for rotational coordinates corresponds to the angle about the rotational axis, it is not surprising that this factor is particularly simple. We also see that all the cylindrical coordinates have two of the three elements $\Phi_{n1}(\xi_n)$ equal to zero, which means that only one of the factors X_n depends on k_1.

Therefore for some of the degenerate forms of the ellipsoidal coordinates, the factored solutions achieve a certain amount of separation of the parameters k.

Another way of stating this is in terms of the actual process of separating the equation, as it is usually done. One takes the equation

$$\nabla^2\psi + k_1^2\psi = \sum_{n=1}^{3} \frac{1}{h_1h_2h_3} \frac{\partial}{\partial\xi_n}\left[\frac{h_1h_2h_3}{h_n^2}\frac{\partial\psi}{\partial\xi_n}\right] + k_1^2\psi = 0$$

and divides by $\psi = X_1X_2X_3$. If the coordinates are separable, so that Eqs. (5.1.29) and (5.1.31) hold, we have

$$\sum_{n=1}^{3} \frac{1}{h_n^2 f_n X_n} \frac{d}{d\xi_n}\left[f_n \frac{dX_n}{d\xi_n}\right] + k_1^2 = 0 \qquad (5.1.40)$$

In some cases it will be possible to multiply this equation by some function of the ξ's so that *at least one* of the four terms in the resulting equation depends only on one coordinate and the other terms do not depend on this coordinate. We can then set this term equal to a constant α (for a function of ξ_n alone to be equal to a function of the other ξ's can only be if the function is independent of all ξ's constant), and in such a case the corresponding factor X depends only on the one constant α (which is a separation constant, either k_2^2 or k_3^2).

The possibility of separating the equation in this simple manner depends on the nature of the scale factors h_n, as Eq. (5.1.40) clearly indicates (all other factors in the nth term are functions of ξ_n alone, so that, if h_n were a constant or a function of ξ_n alone, this term would be all ready for separation without further manipulation). Three cases may be distinguished:

A. Solution Completely Separable for Separation Constants. In this case one can find a multiplier $\mu(\xi_1,\xi_2,\xi_3)$ such that *two* of the terms are functions of one coordinate each (suppose that they are for ξ_2 and ξ_3). Then the term for ξ_2, $(\mu/h_2^2f_2X_2)[d(f_2dX_2/d\xi_2)/d\xi_2]$ may be set equal to the constant k_2^2 and the corresponding term for X_3 may be set equal to the constant k_3^2. Consequently the remaining equation for X_1 is

$$\frac{\mu}{h_1^2 f_1 X_1} \frac{d}{d\xi_1}\left(f_1 \frac{dX_1}{d\xi_1}\right) + k_2^2 + k_3^2 + \mu k_1^2 = 0$$

and the factored solution takes on the form

$$(A) \qquad \psi = X_1(\xi_1;k_1,k_2,k_3)X_2(\xi_2;k_2)X_3(\xi_3;k_3) \qquad (5.1.41)$$

where two of the factors depend only on one parameter k as well as on just one coordinate ξ. Comparison with the Stäckel determinant method of separation indicates that the solution can have the form of Eq. (5.1.41) only if two rows of the Stäckel determinant each have two zeros, and reference to the table at the end of this chapter indicates that only solutions for rectangular and circular cylinder coordinates have this simple behavior. [Spherical coordinates give a solution of the form $X_1(\xi_1;k_1,k_2)X_2(\xi_2,k_2,k_3)X_3(\xi_3,k_3)$, which is as simple as Eq. (5.1.41) for the Laplace equation, when $k_1 = 0$.] This type of separation requires a high degree of symmetry in the coordinate system.

B. Solution Partially Separable for Separation Constants. In this case only one term (the one for ξ_3, for example) can be separated the first time round; the remaining equation

$$\frac{\mu}{h_1^2 f_1 X_1} \frac{d}{d\xi_1}\left(f_1 \frac{dX_1}{d\xi_1}\right) + \frac{\mu}{h_2^2 f_2 X_2} \frac{d}{d\xi_2}\left(f_2 \frac{dX_2}{d\xi_2}\right) + \mu k_1^2 + k_3^2 = 0$$

must be multiplied by still another factor, $\nu(\xi_1,\xi_2)$, in order that a second term can be split off. The factored solutions, therefore, have one of the following forms:

$$\begin{aligned} &(B_1) \qquad \psi = X_1(\xi_1;k_2,k_3)X_2(\xi_2;k_2k_3)X_3(\xi_3;k_1,k_3) \\ &(B_2) \qquad \psi = X_1(\xi_1;k_1,k_2,k_3)X_2(\xi_2;k_1,k_2,k_3)X_3(\xi_3;k_3) \end{aligned} \tag{5.1.42}$$

Reference to the table shows that parabolic cylinder coordinates correspond to (B_1) and elliptic cylinder, parabolic, oblate, and prolate spherical coordinates (all the rest of the cylindrical and rotational coordinates) correspond to (B_2). Here only the last row of the Stäckel determinant has two zeros.

C. Solution Nonseparable for Separation Constants. In this case no Stäckel element in the second or third column is equal to zero and the full machinery of the Stäckel determinant must be used to achieve separation. The possible forms are

$$\begin{aligned} &(C_1) \qquad \psi = X_1(\xi_1;k_1,k_2,k_3)X_2(\xi_2;k_2,k_3)X_3(\xi_3;k_2,k_3) \\ &(C_2) \qquad \psi = X_1(\xi_1;k_1,k_2,k_3)X_2(\xi_2;k_1,k_2,k_3)X_3(\xi_3;k_1,k_2,k_3) \end{aligned} \tag{5.1.43}$$

Only conical coordinates give form (C_1). Ellipsoidal and paraboloidal coordinates result in form (C_1), where no Stäckel element is zero.

It should be obvious that forms (A) are comparatively simple to apply to a given problem, forms (B) are more difficult, and forms (C) are very much more difficult to use.

Laplace Equation in Three Dimensions, Modulation Factor. It is obvious that the Laplace equation, $\nabla^2\psi = 0$, to which our standard equation reduces when $k_1 = 0$, is separable in all the 11 coordinate systems enumerated in the table. But since the Laplace equation in two dimensions separates in more systems than did the two-dimensional wave equation, we may well inquire whether this is also true for three dimensions. Investigations show that there are no more coordinate systems for which solutions of the Laplace equation can take on the form $X_1(\xi_1)X_2(\xi_2)X_3(\xi_3)$ of type (A), (B), or (C). However, it turns out that other systems can be found for which a series of solutions of the Laplace equation can be found having the more general form

$$\psi = X_1(\xi_1)X_2(\xi_2)X_3(\xi_3)/R(\xi_1,\xi_2,\xi_3) \tag{5.1.44}$$

where R is independent of the separation constants k_2 and k_3 (see page 509). The same investigation shows that the wave equation is not amenable to this same generalization, so that the additional coordinate systems are separable only for the Laplace equation.

The factor R can be called a *modulation factor;* it modifies all the family of factored solutions in the same way. Its presence modifies the formalism of the Stäckel determinant to some extent. For instance, we now set [instead of Eq. (5.1.32)]

$$h_1h_2h_3/S = f_1(\xi_1)f_2(\xi_2)f_3(\xi_3)R^2u \tag{5.1.45}$$

where u is a function of ξ_1, ξ_2, ξ_3. We also require that [instead of Eq. (5.1.31)]

$$1/h_n^2 = M_n/Su \tag{5.1.46}$$

where, in these two equations, the Stäckel determinant S and its first minors M_n satisfy the same requirements as before (the elements Φ_{nm} of S are functions of ξ_n only and therefore the minors M_n do not depend on ξ_n). Inserting all this into Laplace equation we first obtain the equation

$$\sum_n \frac{1}{h_n^2 f_n X_n} \frac{d}{d\xi_n}\left[f_n \frac{dX_n}{d\xi_n}\right] = \sum_n \frac{1}{h_n^2 f_n R} \frac{\partial}{\partial \xi_n}\left[f_n \frac{\partial R}{\partial \xi_n}\right]$$

This separation of terms for X and terms for R is the reason for the insertion of R in both Eqs. (5.1.44) and (5.1.45). If now we can find a form for R which satisfies the equation

$$\sum_n \frac{1}{h_n^2 f_n} \frac{\partial}{\partial \xi_n}\left[f_n \frac{\partial R}{\partial \xi_n}\right] + \frac{k_1^2 R}{u} = 0 \tag{5.1.47}$$

then, using Eq. (5.1.46), we finally obtain

$$\sum_n \frac{M_n}{Sf_n X_n} \frac{d}{d\xi_n}\left[f_n \frac{dX_n}{d\xi_n}\right] + k_1^2 = 0 \tag{5.1.48}$$

which separates, like the wave equation, into the ordinary differential equations

$$\frac{1}{f_n}\frac{d}{d\xi_n} f_n \frac{dX_n}{d\xi_n} + [k_1^2\Phi_{n1} + k_2^2\Phi_{n2} + k_3^2\Phi_{n3}]X_n = 0 \tag{5.1.49}$$

from which the factors X_n can be determined.

Confocal Cyclides. The fourth-order surfaces most nearly analogous to the quadric surfaces (ellipsoids, hyperboloids, paraboloids) are the cyclides. One interesting property of such a surface is that its inversion in a sphere is also a cyclide. The equation for the surface can most simply be expressed in terms of *homogeneous coordinates* λ, μ, ν, ρ

$$x = \lambda/\rho; \quad y = \mu/\rho; \quad z = \nu/\rho \tag{5.1.50}$$

or in terms of "pentaspherical coordinates"

$$\begin{aligned} x_1 &= i(\lambda^2 + \mu^2 + \nu^2 + \rho^2) = i\rho^2(x^2 + y^2 + z^2 + 1) \\ x_2 &= (\lambda^2 + \mu^2 + \nu^2 - \rho^2) = \rho^2(x^2 + y^2 + z^2 - 1) \\ x_3 &= 2\rho\lambda = 2\rho^2 x; \quad x_4 = 2\rho\mu = 2\rho^2 y; \quad x_5 = 2\rho\nu = 2\rho^2 z \end{aligned} \tag{5.1.51}$$

The surface defined by the equation

$$\sum_{n=1}^{5} \frac{x_n^2}{\xi - a_n} = 0; \quad a_{n+1} \geq a_n \tag{5.1.52}$$

where ξ and the a's are constant, is a *cyclide*. The surfaces obtained by taking different values of ξ, keeping the a's constant, is a family of surfaces which may be called a *family of confocal cyclides.*

One complete family is generated by taking all values of ξ between a_2 and a_3, another by taking values between a_3 and a_4, and a third corresponds to the range between a_4 and a_5. Furthermore these families are mutually orthogonal, so they can be used for coordinate surfaces. We take for ξ_1 the range between a_2 and a_3, for ξ_2 the range a_3 to a_4, and for ξ_3 the range a_4 to a_5.

The Laplace equation $\nabla^2\psi = 0$ is equivalent to the equation

$$\frac{\partial^2\psi}{\partial x_3^2} + \frac{\partial^2\psi}{\partial x_4^2} + \frac{\partial^2\psi}{\partial x_5^2} = 0$$

since $x = x_3/2\rho^2$, $y = x_4/2\rho^2$, $z = x_5/2\rho^2$, $\rho = -(x_2 + ix_1)/2\rho$ when ψ depends on x_3, x_4, x_5 alone. But also, since x_1 and x_2 enter in the combination $x_2 + ix_1$, we can say that, when ψ also depends on x_1 and x_2 (or on ρ),

$$\frac{\partial^2\psi}{\partial x_1^2} + \frac{\partial^2\psi}{\partial x_2^2} = 0$$

The result can be generalized, to say that if $x_1, \ldots, x_5$ are pentaspherical coordinates, related to the four homogeneous coordinates λ, μ, ν, ρ through second-order equations in such a manner that

$$\sum_{n=1}^{5} x_n^2 = 0 \tag{5.1.53}$$

then a solution of Laplace's equation is also a solution of the equation

$$\sum_{n=1}^{5} \frac{\partial^2\psi}{\partial x_n^2} = 0 \tag{5.1.54}$$

To finish our discussion, we must transform from pentaspherical coordinates to the confocal cyclidal coordinates ξ. To do this, we first consider the x_n coordinates as ordinary orthogonal coordinates in five dimensions, without attempting to satisfy Eq. (5.1.53) as yet. We then transform to another set of five coordinates defined by the five equations

$$x_1^2 = \xi_5 \frac{(\xi_1 - a_1)(\xi_2 - a_1)(\xi_3 - a_1)(\xi_4 - a_1)}{(a_1 - a_2)(a_1 - a_3)(a_1 - a_4)(a_1 - a_5)}; \quad \text{etc.} \tag{5.1.55}$$

or else by the equations

$$\sum_{n=1}^{5} \frac{x_n^2}{\xi_m - a_n} = 0; \quad m = 1, \ldots, 5$$

which are equivalent. We note that $\xi_5 = \sum_{n=1}^{5} x_n^2$, which will eventually go to zero.

The equation corresponding to Eq. (5.1.54) is obtained after a great deal of algebraic manipulation,

$$\frac{\sqrt{f(\xi_1)}}{(\xi_2 - \xi_1)(\xi_3 - \xi_1)(\xi_4 - \xi_1)} \frac{\partial}{\partial \xi_1}\left[\sqrt{f(\xi_1)}\, \frac{\partial \psi}{\partial \xi_1}\right] + \cdots$$
$$+ \frac{\sqrt{f(\xi_4)}}{(\xi_1 - \xi_4)(\xi_2 - \xi_4)(\xi_3 - \xi_4)} \frac{\partial}{\partial \xi_4}\left[\sqrt{f(\xi_4)}\, \frac{\partial \psi}{\partial \xi_4}\right]$$
$$+ \frac{1}{\sqrt{\xi_5}} \frac{\partial}{\partial \xi_5}\left[(\xi_5)^{\frac{3}{2}} \frac{\partial \psi}{\partial \xi_5}\right] = 0 \tag{5.1.56}$$

where $f(x) = (x - a_1)(x - a_2)(x - a_3)(x - a_4)(x - a_5)$. This is the equivalent of a five-dimensional Laplace equation in the new coordinates.

But we are not concerned with a five-dimensional equation; what we want to arrive at eventually is a three-dimensional equation in $\xi_1\xi_2\xi_3$. The coordinates are closely related, however, for we can solve the equations

$$\sum_{n=1}^{5} \frac{x_n^2}{\xi_m - a_n} = 0; \quad m = 1, 2, 3; \quad \sum_{n=1}^{5} x_n^2 = 0$$

to obtain the equations

$$x_1^2 = -\left[\sum_{n=1}^{5} a_n x_n^2\right] \frac{(\xi_1 - a_1)(\xi_2 - a_1)(\xi_3 - a_1)}{(a_1 - a_2)(a_1 - a_3)(a_1 - a_4)(a_1 - a_5)}; \quad \text{etc.} \tag{5.1.57}$$

which give the x_n's in terms of the three-dimensional coordinates ξ_1, ξ_2, ξ_3 and, eventually, give x, y, z in terms of these cyclidal coordinates.

We notice that the transformation of Eq. (5.1.57) is related to that of Eq. (5.1.55). In the latter case only three coordinates are in evidence, so we can say that the transformation of Eq. (5.1.57) represents a three-dimensional subspace of the transformation given in Eq. (5.1.55). We have known this right along, since we were aware that

$\xi_5 = \Sigma x_n^2$ would have to be set equal to zero, and the coordinate ξ_4 will also have to be set equal to some constant in order that we might finally obtain a three-dimensional equation. Comparison between Eqs. (5.1.57) and (5.1.55) indicates that, if we let ξ_5 go to zero and ξ_4 simultaneously go to infinity such that $\xi_4\xi_5 \to -\Sigma a_n x_n^2$, we should arrive at the correct cyclidal coordinates.

Presumably ψ will be a function of ξ_4 and ξ_5. If it is a function of $(\xi_4\xi_5)$, then it will not be complicated by additional zeros or infinities. In particular, if ψ contains $(\xi_4\xi_5)^\alpha$ as a factor, then in the limit the factor will become $[\Sigma a_n x_n^2]^\alpha$. If we wish the potential to go to zero at large distances as $1/r$, then the power α can be $-\frac{1}{4}$, and we can set

$$\psi = (\xi_4\xi_5)^{-\frac{1}{4}}\varphi(\xi_1,\xi_2,\xi_3) = \left[\sum_{n=1}^{5} a_n x_n^2\right]^{-\frac{1}{4}} \varphi(\xi_1,\xi_2,\xi_3) \qquad (5.1.58)$$

Setting this form for the solution into Eq. (5.1.56) and letting ξ_4 go to infinity results finally in a partial differential equation for φ. We multiply through by $(\xi_4)^{\frac{3}{2}}$ and expand in powers of $1/\xi_4$. The terms in the first power of ξ_4 cancel automatically. Those in the zeroth power of ξ_4 give the equation for φ:

$$\begin{aligned}\frac{\sqrt{f(\xi_1)}}{(\xi_1-\xi_2)(\xi_1-\xi_3)}\frac{\partial}{\partial\xi_1}\left[\sqrt{f(\xi_1)}\,\frac{\partial\varphi}{\partial\xi_1}\right] &+ \frac{\sqrt{f(\xi_2)}}{(\xi_2-\xi_1)(\xi_2-\xi_3)}\frac{\partial}{\partial\xi_2}\left[\sqrt{f(\xi_2)}\,\frac{\partial\varphi}{\partial\xi_2}\right] \\ &+ \frac{\sqrt{f(\xi_3)}}{(\xi_3-\xi_1)(\xi_3-\xi_2)}\frac{\partial}{\partial\xi_3}\left[\sqrt{f(\xi_3)}\,\frac{\partial\varphi}{\partial\xi_3}\right] \\ &+ \left[\tfrac{5}{16}(\xi_1+\xi_2+\xi_3) - \tfrac{3}{16}\sum_{n=1}^{5} a_n\right]\varphi = 0 \qquad (5.1.59)\end{aligned}$$

where $\qquad f(x) = (x-a_1)(x-a_2)(x-a_3)(x-a_4)(x-a_5)$

This equation can be separated by means of a Stäckel determinant, just as was the equation for ellipsoidal coordinates. In this case, however, the complete solution ψ is the product of a nonseparable part $[\Sigma a_n x_n^2]^{-\frac{1}{4}} = 1/R$ times a separable part φ, as we have already stated.

Various degenerate forms of cyclides can be produced by allowing one or more of the constants a_n to become equal or go to infinity. They include the ellipsoidal coordinates and all their degenerate forms. In addition they include the toroidal coordinates defined by the following equations:

$$\frac{(x^2+y^2+z^2+a^2)^2}{\xi_1^2} - \frac{4a^2(x^2+y^2)}{\xi_1^2-1} = 0;$$

$$\frac{(x^2+y^2+z^2-a^2)^2}{\xi_2^2} + \frac{4a^2z^2}{\xi_2^2-1} = 0 \qquad (5.1.60)$$

These are degenerate forms of Eqs. (5.1.52) (where we have used ξ_n^2 and a^2 instead of ξ_n and a_n); they correspond (for ξ_1 = constant) to toroids formed by rotating the circle of radius $a/\sqrt{\xi_1^2 - 1}$ with center at $z = 0$, $x = a\xi_1/\sqrt{\xi_1^2 - 1}$, about the z axis, and (for ξ_2 = constant) to spheres of radius $a/\sqrt{1 - \xi_2^2}$ with center at $x = y = 0$, $z = a\xi_2/\sqrt{1 - \xi_2^2}$ (all these spheres pass through the circle $z = 0$, $x^2 + y^2 = a^2$, which corresponds to the limiting toroid $\xi_1 = \infty$).

The equations giving x, y, z in terms of the ξ's, together with the forms for the scale factors h, the modulation factor R, and the Stäckel determinant S for this useful special case are all given on pages 665 and 666 of this chapter. Another useful special case, bispherical coordinates, is also given in these pages.

It appears that the general cyclidal coordinates, including all the degenerate forms, contain all the coordinate systems for which the Laplace equation separates, with or without a modulation factor R; just as the ellipsoidal coordinates contain all the systems for which the wave equation separates. We must now turn to the examination of the ordinary differential equations resulting from the separation.

5.2 *General Properties, Series Solutions*

We must now study the ordinary differential equations obtained by separating the partial differential equation $\nabla^2\psi + k^2\psi = 0$ in the various separable coordinates. As we showed on page 509, all the separated equations have the general form

$$\mathfrak{L}(\psi) = \frac{d^2\psi}{dz^2} + p(z)\frac{d\psi}{dz} + q(z)\psi = 0 = \frac{1}{f_n}\frac{d}{dz}\left(f_n\frac{d\psi}{dz}\right) + q\psi \quad (5.2.1)$$

where
$$p = \frac{d}{dz}\ln f_n = \left(\frac{1}{f_n}\right)\left(\frac{df_n}{dz}\right);\quad q = \sum_m k_m^2\Phi_{nm}$$

where f_n and Φ_{nm} are functions of ξ_n (here labeled z). As we have chosen the scale of the coordinates, the functions p and q are simple algebraic functions of z, with a finite number of poles, corresponding to the concentration points of the coordinate system.

Equation (5.2.1) is a second-order, linear, homogeneous differential equation. As we indicated on page 493, such equations may have several solutions. If $\psi_1, \psi_2, \ldots, \psi_n$ are solutions of $\mathfrak{L}(\psi) = 0$, then $\sum_m A_m\psi_m$ (where the A_m are arbitrary constant coefficients) is also a solution. The corresponding inhomogeneous equation $\mathfrak{L}(\psi) = r(z)$ sometimes occurs. We have pointed out earlier that, if Ψ_n is a solution of

$\mathcal{L}(\psi) = r_n$, then $\Psi_n + \sum_m A_m\psi_m$ is also a solution of $\mathcal{L}(\psi) = r_n$ and $\left[\sum_n \Psi_n + \sum_m B_m\psi_m\right]$ is a solution of $\mathcal{L}(\psi) = \sum_n r_n$.

Consequently we can have an infinite number of different solutions of Eq. (5.2.1), obtained by different choice of the constants A_n. Actually a great number of these differ only by a constant factor. The situation is perhaps clearer if we consider the matter from the point of view of abstract vector space. Any function $y(z)$ can be considered as corresponding to a vector $\mathbf{Y}$ in a nondenumerably infinite-dimensional space, where each $y(z)$ for each value of z is the component of $\mathbf{Y}$ along the direction corresponding to that value of z (see, for example, page 135). The differential operator $\mathcal{L}$ corresponds to a vector operation which, in general, transforms any vector $\mathbf{A}(z)$ into another vector. If y is a solution of $\mathcal{L}(y) = 0$, however, the corresponding vector $\mathbf{Y}$ gives zero when operated upon by the vector operator corresponding to $\mathcal{L}$. Any solution $u(z)$ which is simply a constant times $y(z)$ has a vector $\mathbf{U}$ which is of different *length* from $\mathbf{Y}$ but is *in the same direction* as $\mathbf{Y}$. The question we are trying to ask is: In how many *different directions* can we have vectors corresponding to solutions of Eq. (5.2.1)?

The Wronskian. If two solutions y_1 and y_2 correspond to vectors in the same direction, then $y_2 = ay_1$ and $y_2' = ay_1'$ (in this chapter we shall use the shorthand notation $y' = dy/dz$ and $y'' = d^2y/dz^2$) and the expression

$$\Delta(y_1,y_2) = y_1y_2' - y_2y_1' \tag{5.2.2}$$

will be zero for all values of z. On the other hand, if $\mathbf{Y}_2$ is in a different direction from $\mathbf{Y}_1$, then $\Delta(y_1,y_2)$ *will not be zero* everywhere. The expression $\Delta(y_1,y_2)$ defined in Eq. (5.2.2) (where y_1 and y_2 are both solutions of $\mathcal{L}(y) = 0$) is called the *Wronskian* of y_1 and y_2 for the given homogeneous equation. If the Wronskian is zero, then y_2 equals a constant times y_1. If the Wronskian differs from zero for any range of values of z, then $\mathbf{Y}_2$ is in a different direction from $\mathbf{Y}_1$ and the two solutions y_1 and y_2 are said to be *independent.*

That there can be at least one solution which is independent of y_1 can be shown fairly directly by use of the properties of the Wronskian. Suppose that we know the solution y_1 and from $z = z_0$ as a starting point we try to build out another solution y_2 which is to be related to y_1 at z_0 by the equations $y_2 = \alpha y_1$, $y_2' = \beta y_1'$ $(\alpha \neq \beta)$. The Wronskian $\Delta(y_1,y_2) = (\beta - \alpha)y_1y_1'$ will differ from zero at $z = z_0$ (we assume that neither y_1 nor y_1' is zero at $z = z_0$). Let us see what value Δ has for other values of z when y_2 is supposed to be a solution of $\mathcal{L}(y) = 0$. Taking the derivative of Δ with respect to z and utilizing Eq. (5.2.1), we have

$$\frac{d\Delta}{dz} = y_1 y_2'' - y_2 y_1'' = -y_1(py_2' + qy_2) + y_2(py_1' + qy_1)$$
$$= -p\Delta = -\Delta \frac{d}{dz} (\ln f)$$

This equation for Δ can be integrated, giving

$$\Delta(z) = \Delta(z_0) e^{-\int_{z_0}^{z} p\,dz} = \Delta(z_0) \frac{f(z_0)}{f(z)} \tag{5.2.3}$$

Therefore, unless we were unlucky enough to pick a z_0 where f is zero, if Δ differs from zero at $z = z_0$, it will differ from zero for other values of z; so it appears to be possible to have a second solution, independent of y_1.

Since $\quad \Delta(z) = y_1 y_2' - y_2 y_1' = y_1^2(z)(d/dz)[y_2(z)/y_1(z)]$

we can immediately find that

$$y_2(z) = y_1(z) \int_{z_0}^{z} \frac{\Delta(u)}{y_1^2(u)}\, du = \Delta(z_0) y_1(z) \int_{z_0}^{z} \frac{e^{-\int_{z_0}^{u} p(w)\,dw}}{y_1^2(u)}\, du$$
$$= \Delta(z_0) f(z_0) y_1(z) \int_{z_0}^{z} \frac{du}{f(u) y_1^2(u)} \tag{5.2.4}$$

It is not difficult to verify that Eq. (5.2.4) is a solution of Eq. (5.2.1), for if we set $y = uv$, then Eq. (5.2.1) becomes

$$v\mathcal{L}(u) + uv'' + puv' + 2u'v' = 0; \quad y = uv; \quad \mathcal{L}(y) = 0 \tag{5.2.5}$$

Letting $u = y_1$, $\mathcal{L}(u) = 0$, and $v = \int(\Delta/y_1^2)\, dz$, we soon see that Eq. (5.2.5) is satisfied. Therefore y_2, as defined in Eq. (5.2.4), is a solution of $\mathcal{L}(y) = 0$ which is independent of y_1; its vector $\mathbf{Y}_2$ in abstract vector space is in a different direction from $\mathbf{Y}_1$. If we have two independent solutions, since Eq. (5.2.1) is linear, any combination $Ay_1 + By_2$ is also a solution. Hence *any vector in the plane defined by* $\mathbf{Y}_1$ and $\mathbf{Y}_2$ corresponds to a solution of the corresponding equation.

Independent Solutions. We have just seen that, if one solution, $y_1(z)$, of $\mathcal{L}(y) = 0$ is known, a second solution can be obtained by integration,

$$y_2(z) = By_1(z) \int e^{-\int p\,dz} \frac{dz}{y_1^2(z)} \tag{5.2.6}$$

This solution is independent of $y_1(z)$, for the Wronskian $\Delta(y_1,y_2)$ is equal to $Be^{-\int p\,dz}$. The next question is whether *all possible* solutions of Eq. (5.2.1) can be expressed in terms of the combination $Ay_1 + By_2$ or we could have still another solution with related vector which does not lie in the plane defined by $\mathbf{Y}_1$ and $\mathbf{Y}_2$. Suppose that we pick a solution y_3 and see if we can always express it in terms of y_1 and y_2. Starting at the point $z = z_0$, we can always find values of A and B such that

$$y_3(z_0) = Ay_1(z_0) + By_2(z_0); \quad y_3'(z_0) = Ay_1'(z_0) + By_2'(z_0) \tag{5.2.7}$$

This can be done as long as the Wronskian $\Delta(y_1,y_2)$ is not zero, in other words as long as y_1 and y_2 are independent solutions.

Therefore it is always possible to find a combination of y_1 and y_2 such that the Wronskian of y_1 and the combination are zero at $z = z_0$. But to be sure that the function y_3 is really equal to $Ay_1 + By_2$ we should be sure that the higher derivatives at $z = z_0$ are equal. By use of Eqs. (5.2.1) and (5.2.7) we can see that

$$\begin{aligned} y_3''(z_0) = -py_3' - qy_3 = -p[Ay_1' + By_2'] - q[Ay_1 + By_2] \\ = Ay_1''(z_0) + By_2''(z_0) \end{aligned}$$

Continued differentiation and use of Eq. (5.2.1) show that, if Eqs. (5.2.7) hold, then the nth derivative of y, at z_0 is equal to the same combination of nth derivatives of y_1 and y_2. Consequently a Taylor series about $z = z_0$ gives

$$\begin{aligned} y_3(z) &= [y_3(z_0) + (z - z_0)y_3'(z_0) + \tfrac{1}{2}(z - z_0)^2y_3''(z) + \cdots] \\ &= A[y_1(z_0) + (z - z_0)y_1'(z_0) + \cdots] \\ &\qquad + B[y_2(z_0) + (z - z_0)y_2'(z_0) + \cdots] \\ &= Ay_1(z) + By_2(z) \end{aligned}$$

so that y_3 is actually represented by the combination $Ay_1 + By_2$ over the whole region of z where the Taylor series converges.

In this sense, therefore, *any solution* y_3 of $\mathfrak{L}(y) = 0$ *can be represented in terms of a linear combination of two independent solutions* y_1 and y_2. From the point of view of abstract vector space, the vectors representing the solutions $\mathfrak{L}(y) = 0$ *all lie in a plane.* It is not hard to see that, if we were dealing with third-order equations, we should need three independent solutions to form a basic set in terms of which we could express all solutions; in other words the vectors corresponding to the solutions would all lie within a three-dimensional subspace of abstract vector space; and so on, for higher order differential equations, the number of dimensions of the subspace being equal to the order of the linear differential equation which generates the solutions.

Integration Factors and Adjoint Equations. The solution of an unfamiliar differential equation in terms of known algebraic or transcendental functions is usually a more difficult task than the integration of an unfamiliar function; indeed the determination of the functional form of $u = \int v\, dz$ corresponds to the finding of the solution of the very simple, first-order, differential equation $(du/dz) - v(z) = 0$. What is usually done to perform an integration in closed form is to try a number of possible solutions u to see if the derivative u' of any of them is equal to v. The results of such exploration are compiled in tables of integrals.

If the required integral is not in the tables, one is usually forced to resort to a series expansion (with consequent limitation of the region of applicability of the series to its region of convergence) or else to numerical computation (with similar or more stringent limitations).

Our chief task in this chapter will be to classify equations (*i.e.*, make a list, similar to a table of integrals) so that one can recognize the variant forms of equations having known and tabulated solutions and to discuss various general techniques of solving such equations, so that we can find the general behavior of solutions of unfamiliar equations. In many cases we shall consider the solution found if we can express it in terms of one or more integrals, even if the integration can only be performed by expanding in series (or computing numerically). This is called "reducing to quadratures," a phrase by which the mathematician symbolically washes his hands of the remainder of the task of finding the solution.

For instance, the simplest linear differential equation

$$(dy/dz) + p(z)y = 0 \tag{5.2.8}$$

can be reduced to quadratures by rearrangement of terms

$$\int (dy/y) = -\int p(z)\, dz; \quad \ln y = -\int p\, dz + C; \quad y = Ae^{-\int p\, dz} \tag{5.2.9}$$

or by the use of an *integration factor*. We notice that, if we multiply Eq. (5.2.8) by the factor $e^{\int p\, dz}$, the result is a complete differential, which may be integrated immediately:

$$e^{\int p\, dz} y' + e^{\int p\, dz} py = (d/dz)[ye^{\int p\, dz}] = 0; \quad ye^{\int p\, dz} = A; \quad y = Ae^{-\int p\, dz}$$

An extensive literature has developed, reporting useful techniques for finding integration factors for more complicated first-order equations.

Second-order equations of the form of Eq. (5.2.1) may sometimes be reduced to quadrature by the use of an integration factor. Utilizing the identities

$$y''v - v''y = (d/dz)(y'v - v'y); \quad (d/dz)(pyv) = vpy' + y(pv)'$$

we obtain the following equation, for any reasonable functions y and v of z,

$$v[y'' + py' + qy] - y[v'' - (pv)' + qv] = (d/dz)[vy' - v'y + vpy]$$

which may be represented symbolically as

$$v\mathfrak{L}(y) - y\tilde{\mathfrak{L}}(v) = (d/dz)P(v,y) \tag{5.2.10}$$

where the original differential equation is $\mathfrak{L}(y) = y'' + py' + qy = 0$. The differential operator which operates on v,

$$\tilde{\mathfrak{L}}(v) = \frac{d^2v}{dz^2} - \frac{d}{dz}(pv) + qv = v'' - pv' + (q - p')v \tag{5.2.11}$$

is said to be *adjoint* to the operator $\mathfrak{L}(y)$ which operates on y, and the quantity which is differentiated by z,

$$P = vy' - v'y + vpy = vy\left[\frac{y'}{y} - \frac{v'}{v} + p\right] \tag{5.2.12}$$

is called the *bilinear concomitant*, a function of the functions v and y and of the variable z.

If, now, it is easy to solve the adjoint equation $\tilde{\mathfrak{L}}(v) = 0$, then the solution of the original equation, $\mathfrak{L}(y) = 0$, is equivalent to solving the first-order equation

$$P = vy[y'/y) - (v'/v) + p] = \text{constant}$$

Since any solution will suffice, we choose the easiest one by setting the constant equal to zero, so that, if v is a solution of $\tilde{\mathfrak{L}}(v) = 0$, then

$$\frac{dy}{y} = \frac{dv}{v} - p\,dz; \quad y_1 = ve^{-\int p\,dz} \tag{5.2.13}$$

(we pay no attention to arbitrary constants, since we can introduce them later). A second solution independent of y_1 can be found by using Eq. (5.2.6):

$$y_2 = ve^{-\int p\,dz} \int e^{\int p\,dz} \frac{dz}{v^2} \tag{5.2.14}$$

where we have to perform another integration. Then the general solution is $\psi = Ay_1 + By_2$.

A general type of second-order differential equation which may be solved in this manner is the case where $q = dp/dz$, for then the adjoint equation (5.2.11) for v takes on the simple form

$$v'' - pv' = 0; \quad v' = e^{\int p\,dz}; \quad v = \int e^{\int p\,dz}\,dz$$

and two independent solutions for y are

$$\begin{aligned} y_1 &= e^{-\int p\,dz} \int e^{\int p\,dz}\,dz = \frac{v}{v'} \\ y_2 &= -e^{-\int p\,dz}\left[\int e^{\int p\,dz}\,dz\right] \int \frac{e^{\int p\,dz}\,dz}{[\int e^{\int p\,dz}\,dz]^2} = \frac{1}{v'} \end{aligned} \tag{5.2.15}$$

for proper choice of the constants of integration.

In some of our separated equations the factor f_n has the form $(\xi - a)^\alpha$ so that the corresponding expression p, for the equation, is $(d \ln f/d\xi) = \alpha/(\xi - a)$. If now the elements of the Stäckel determinant are such as to make $q = -\alpha/(\xi - a)^2$, then the equation $\mathfrak{L}(y) = 0$ has the form

$$y'' + (\alpha/z)y' - (\alpha/z^2)y = 0; \quad z = \xi - a$$

the requirement $q = p'$ is satisfied, and we can use Eqs. (5.2.15). In this case $\int p\,dz = \alpha \ln z$, $v' = z^\alpha$, $v = z^{\alpha+1}/(\alpha + 1)$, so that the general solution of $\mathfrak{L}(y) = 0$ is

$$\psi = A'y_1 + B'y_2 = Az + (B/z^\alpha) \tag{5.2.16}$$

which has a branch point (or pole, if α is an integer) of order α at $z = 0$ ($\xi = a$, a concentration point of the coordinates) and a simple pole at $z = \infty$ (*i.e.*, inserting $z = 1/w$, ψ has a simple pole at $w = 0$).

This equation is adequate *unless* $\alpha = -1$, when the term $Bz^{-\alpha}$ is not independent of the first solution Az. When this is the case, $p = -(1/z)$ and the second solution may be obtained directly from Eq. (5.2.6):

$$y_2 = y_1 \int \frac{e^{-\int p\,dz}}{y_1^2}\,dz = z \int \frac{z\,dz}{z^2} = z \ln z$$

so that the general solution is then

$$\psi = z(A + B \ln z) \tag{5.2.17}$$

again having a branch point at $z = 0$.

Solution of Inhomogeneous Equation. Once we know two independent solutions y_1 and y_2 of the homogeneous equation $\mathfrak{L}(y) = 0$, we can find the general solution of the inhomogeneous equation $\mathfrak{L}(\psi) = r(z)$ in terms of an additional integration. Analogously to Eq. (5.2.5), we set $\psi = uv$ in $\mathfrak{L}(\psi) = r$, obtaining

$$v\mathfrak{L}(u) + uv'' + (up + 2u')v' = r$$

If now we set u equal to one of the solutions y_1 of the homogeneous equation, $\mathfrak{L}(y) = 0$, we obtain

$$v'' + [p + 2(y_1'/y_1)]v' = r/y_1 \tag{5.2.18}$$

However, a reference to the discussion of page 525 about the Wronskian shows that, for a second solution and Wronskian related by the equation $(y_2/y_1)' = \Delta/y_1^2$, we have

$$\left(\frac{y_2}{y_1}\right)'' = \left(\frac{\Delta}{y_1^2}\right)' = \frac{\Delta'}{y_1^2} - 2\frac{y_1'\Delta}{y_1^3} = -p\frac{\Delta}{y_1^2} - 2\frac{y_1'}{y_1}\frac{\Delta}{y_1^2}$$

so that

$$(y_2/y_1)'' + [p + 2(y_1'/y_1)](y_2/y_1)' = 0$$

Multiplying this equation by v' and multiplying Eq. (5.2.18) by $(y_2/y_1)'$ and subtracting, we obtain

$$\left(\frac{y_2}{y_1}\right)' v'' - v'\left(\frac{y_2}{y_1}\right)'' = \left[\left(\frac{y_2}{y_1}\right)'\right]^2 \frac{d}{dz}\left[\frac{v'}{(y_2/y_1)'}\right] = \left(\frac{r}{y_1}\right)\left(\frac{y_2}{y_1}\right)' = \left(\frac{r}{y_1}\right)\left(\frac{\Delta}{y_1^2}\right)$$

By this manipulation we have reduced our original second-order inhomogeneous equation $\mathfrak{L}(\psi) = r$ to a simple first-order inhomogeneous equation

$$\frac{d}{dz}\left[\frac{v'}{(y_2/y_1)'}\right] = \frac{ry_1}{\Delta}$$

where y_1, y_2, and $\Delta = y_1y_2' - y_2y_1'$ are obtained from the homogeneous equation $\mathfrak{L}(y) = 0$ and are supposed to be known functions of z. Integrating this first-order equation without more ado, we obtain

$$v' = \left[\frac{d}{dz}\left(\frac{y_2}{y_1}\right)\right]\int \frac{ry_1}{\Delta}\,dz \quad \text{or} \quad v' + \frac{ry_2}{\Delta} = \frac{d}{dz}\left[\left(\frac{y_2}{y_1}\right)\int \frac{ry_1}{\Delta}\,dz\right]$$

Since $v = \psi/y_1$, we therefore finally arrive at a formal solution of the inhomogeneous equation $\mathfrak{L}(\psi) = r$:

$$\psi = vy_1 = y_1\left[c_1 - \int \frac{ry_2\,dz}{\Delta(y_1,y_2)}\right] + y_2\left[c_2 + \int \frac{ry_1\,dz}{\Delta(y_1,y_2)}\right] \tag{5.2.19}$$

where the integrals are indefinite integrals and the constants c may be adjusted to fit the boundary conditions. In conformity with the discussion on page 524, this solution consists of the particular solution,

$$\int^z r(w)\,dw\left[\frac{y_1(w)y_2(z) - y_2(w)y_1(z)}{y_1(w)y_2'(w) - y_2(w)y_1'(w)}\right]$$

plus an arbitrary amount of the solution of the homogeneous equation $\mathfrak{L}(y) = 0$.

Series Solutions about Ordinary Points. As we pointed out several times before and shall prove shortly, the general solution of $\mathfrak{L}(y) = 0$ has its singularities at the points where the functions p and q have their poles. All other values of z, where p and q are analytic functions, are called *ordinary points for the equation*. The points where p or q (or both) have singularities are called *singular points for the equation*.

To show that the general solution is analytic at an ordinary point and also to illustrate one of the techniques for solving differential equations, we shall compute the series expansion for the solution of Eq. (5.2.1) at an ordinary point $z = a$. Since $z = a$ is an ordinary point, both p and q are analytic and can be expanded into Taylor's series about $z = a$;

$$p(z) = p(a) + (z - a)p'(a) + \tfrac{1}{2}(z - a)^2p''(a) + \cdots$$
$$q(z) = q(a) + (z - a)q'(a) + \tfrac{1}{2}(z - a)^2q''(a) + \cdots$$

The solution y (if it is analytic) can also be expressed as a series

$$y = a_0 + a_1(z - a) + a_2(z - a)^2 + \cdots$$

Substituting this into Eq. (5.2.1), we obtain

$$0 = [2a_2 + a_1p(a) + a_0q(a)] + [6a_3 + 2a_2p(a) + a_1p'(a) + a_0q'(a) + a_1q(a)](z - a) + \cdots$$

Equating coefficients of each power of $(z - a)$ to zero, we have a series of equations for determination of the coefficients a_n in the series for the solution. The first equation gives a_2 in terms of a_0 and a_1 [and the known quantities $p(a)$ and $q(a)$]. The second gives a_3 in terms of a_2, a_1, and a_0, and therefore in terms of a_0 and a_1, and so on. These equations can be solved, so the series for y is valid.

The series expansion for y can thus be written

$$\begin{aligned} y &= a_0 y_1 + a_1 y_2 \\ y_1 &= 1 - \tfrac{1}{2}q(a)(z-a)^2 + \tfrac{1}{6}[q(a)p(a) - q'(a)](z-a)^3 + \cdots \\ y_2 &= (z-a) - \tfrac{1}{2}p(a)(z-a)^2 \\ &\qquad + \tfrac{1}{6}[p^2(a) - p'(a) - q(a)](z-a)^3 + \cdots \end{aligned} \tag{5.2.20}$$

where y_1 and y_2 are independent solutions. They are, in fact, the particularly useful pair of solutions, one having unit value and zero slope at $z = a$ and the other having zero value and unit slope, which are called the *basic set* of solutions for the ordinary point $z = a$. Any boundary condition of the type $y(a) = A$, $y'(a) = B$ can easily be satisfied by setting $y = Ay_1 + By_2$. This series solution will hold within a circle of convergence of radius equal to the distance between a and the nearest singular point of the differential equation. For instance, the basic set of solutions of the differential equation $y'' + y = 0$ at the ordinary point $z = 0$ is $y_1 = \cos z$, $y_2 = \sin z$. Since the nearest singular point is at infinity, the series expansions for cosine and sine are valid over the entire finite part of the z plane.

It is of interest to note that, when p has a pole at $z = a$ but q is analytic at a, *one solution* of the equation is analytic, though the second solution has a singularity. Setting $p = F(z)/(z - a)^n$, where F is an analytic function which is not zero at $z = a$ and then carrying out the same series expansion as above, we find in general that $a_1 = a_2 = \cdots = a_n = 0$ but that a_{n+1}, a_{n+2}, etc., can be expressed in terms of a_0. For instance, if $n = 2$ in the expression for p, the series corresponding to the equation $\mathcal{L}(y) = 0$ is

$$\begin{aligned} 0 = \frac{1}{(z-a)^2} a_1 F(a) + \frac{1}{(z-a)} [2a_2 F(a) + a_1 F'(a)] \\ + [2a_2 + q(a)a_0 + 3a_3 F(a) + 2a_2 F'(a) + \tfrac{1}{2}a_1 F''(a)] + \cdots \end{aligned}$$

so that one solution of the equation is (letting $a_0 = 1$)

$$y_1 = 1 - \frac{q(a)}{3F(a)}(z-a)^3 + \left[\frac{q(a)}{2F^2(a)} + \frac{q(a)F'(a)}{4F^2(a)} - \frac{q'(a)}{4F(a)}\right](z-a)^4 + \cdots$$

The second solution may be calculated by using Eq. (5.2.6). Since

$$\int p\,dz = -\frac{F(a)}{(z-a)} + F'(a)\ln(z-a) + \tfrac{1}{2}F''(a)(z-a) + \cdots, \text{ the}$$

integral $\int dz\, e^{-\int p\,dz}/y_1^2$ will have an essential singularity of general form $(z - a)e^{F(a)/(z-a)}$. Therefore the general solution $Ay_1 + By_2$ has an essential singularity at $z = a$.

On the other hand if p has a simple pole at $z = a$ (for example, $n = 1$), then $\int p\,dz = F(a) \ln (z - a) + F'(a)(z - a) + \cdots$ and the second solution will have a branch point of the form $(z - a)^{-F-1}$ times an analytic function.

Singular Points, Indicial Equation. We have just seen that, near those singular points where q is analytic but p has a pole, one solution is analytic and the second solution has a branch point if p has a simple pole or has an essential singularity if p has a pole of higher order. This distinction is of general utility; we can define a *regular singular point* as one where the general solution has a pole or branch point, an *irregular singular point* as one where the general solution has an essential singularity.

To find out what behavior of p and q produces which type of singular point, we shall separate off the singularity in y by setting $y_1 = vu$, where u is supposed to be analytic. Using Eq. (5.2.5) we have

$$u'' + Hu' + Ju = 0; \quad H = p + 2(v'/v); \quad J = q + (v''/v) + p(v'/v) \tag{5.2.21}$$

Now let us assume that a singular point for $\mathfrak{L}(y) = 0$ is at $z = a$, which means that $p = F(z)/(z - a)^m$, $q = G(z)/(z - a)^n$, where F and G are analytic at $z = a$. In order that u be analytic, we must have that the coefficient J be analytic; u can then be the analytic solution of the equation $u'' + Hu' + Ju = 0$ with J analytic and H having a pole.

First we ask what are the restrictions placed on p and q (*i.e.*, on m and n) in order that $z = a$ be regular singular point. In this case, by definition, $v = (z - a)^s$, $v'/v = s/(z - a)$; $v''/v = s(s - 1)/(z - a)^2$. In order that J be analytic, the pole for q must be second order or less ($n = 2$) and the pole for p first order or less ($m = 1$). Consequently *if p has the form $F(z)/(z - a)$* and *q the form $G(z)/(z - a)^2$*, where F and G are analytic, *then the point $z = a$ is a regular singular point* for the equation $\mathfrak{L}(y) = 0$ and *the general solution has a branch point there.* The equation determining s comes from Eq. (5.2.21),

$$s^2 + [F(a) - 1]s + G(a) = 0 \tag{5.2.22}$$

and is called the *indicial equation* for the solution. The two roots s_1 and $s_2 (s_1 > s_2)$ correspond to the two solutions $y_1 = (z - a)^{s_1} u_1$, $y_2 = (z - a)^{s_2} u_2$, where u_1 and u_2 are functions analytic at a.

When $s_1 = s_2$ and, in most cases, when $(s_1 - s_2)$ is an integer, u_2 turns out to be $(z - a)^{s_1 - s_2} u_1$, so that the indicial equation and the series solution for u give one solution but not a second one. In this case we use Eq. (5.2.6) to obtain y_2. Since $e^{-\int p\,dz} = (z - a)^{-F(a)}$ times an analytic function and since $1 - F(a) = s_1 + s_2$, the integrand of Eq.

(5.2.6) is then

$$\frac{(z-a)^{s_1+s_2-1}\,(\text{analytic function})}{(z-a)^{2s_1}u_1^2} = \frac{\text{analytic function}}{(z-a)^{s_1-s_2+1}}$$

If $s_1 - s_2$ is an integer, this integrand has a pole at $z = a$ rather than a branch point, and when the analytic function is expanded in a series $[b_0 + b_1(z - a) + \cdots]$, the series for the second solution will be

$$(z-a)^{s_2}u_1\left[\frac{b_0}{s_1-s_2} + \frac{b_1(z-a)}{s_1-s_2-1} + \cdots \right.$$
$$\left. + b_{s_1-s_2}(z-a)^{s_1-s_2}\ln(z-a) + \cdots\right]$$

which has the distinguishing term $u_1(z - a)^{s_1} \ln (z - a)$.

This logarithmic term always appears in the second solution when $s_1 = s_2$ (see page 529) and nearly always appears when $(s_1 - s_2)$ is an integer. Therefore we can say that the general solution of Eq. (5.2.1) has a branch point at a regular singular point of the equation, for when s_1 and s_2 are both integers, so that we might expect no branch point to occur, in just these cases a logarithmic term will enter the second solution, bringing with it a branch point of its own. One of the exceptions to this rule is given in Eq. (5.2.16).

When q has a pole of higher order than second or when p has a pole of order higher than first, or both, one or both solutions must have an essential singularity and we have an irregular singular point. When q has a pole of order one higher than p plus a term in $(z - a)^{-2}$, only one of the solutions has an essential singularity.

A simple way of showing this is to insert the series

$$y = (z-a)^s \sum_{i=0}^{\infty} c_i(z-a)^i;\quad \text{that is, } v = (z-a)^s;\quad u = \sum c_i(z-a)^i$$

into Eq. (5.2.1). Then, if p and q can be represented by the Laurent series

$$\begin{aligned} p &= a_{-m}(z-a)^{-m} + a_{-m+1}(z-a)^{-m+1} + \cdots \\ q &= b_{-n}(z-a)^{-n} + b_{-n+1}(z-a)^{-n+1} + \cdots;\quad m, n \text{ integers} \end{aligned}$$

the following series:

$$\begin{aligned} &c_0s(s-1)(z-a)^{s-2} + c_1s(s+1)(z-a)^{s-1} + \cdots \\ &\quad + c_0a_{-m}s(z-a)^{s-m-1} + [c_0a_{-m+1}s + c_1a_{-m}(s+1)](z-a)^{s-m} + \cdots \\ &\qquad + c_0b_{-n}(z-a)^{s-n} + [c_0b_{-n+1} + c_1b_{-n}](z-a)^{s-n+1} + \cdots \end{aligned}$$

must be zero in the range of z where the series for u converges. Consequently the coefficient of each power of $(z - a)$ must be zero in turn. If the resulting infinite set of equations cannot be solved, then our original assumption as to the form of y is not adequate and y must have an essential singularity at $z = a$.

It turns out that, if the coefficient of the lowest power of $(z - a)$ can be made zero, then all the other equations can be satisfied by proper choice of the coefficients c_i. This crucial equation, for the lowest power of $(z - a)$, is just the indicial equation (5.2.22) if m and n have such values that s enters this equation at all. We see that, if $m \leq 1$ and $n \leq 2$, this equation is a quadratic in s; for example, for $m = 1$ and $n = 2$ we have

$$c_0[s^2 + (a_{-1} - 1)s + b_{-2}] = 0$$

which allows two roots for s; consequently *both* independent solutions y_1 and y_2 have the assumed form (branch point at $z = a$) and the point $z = a$ is by definition a *regular singular point.* (If the roots are equal, we have already seen that the second solution has a logarithmic term.)

If $1 < m \geq n - 1$, the equation for the lowest power of $(z - a)$ is linear in s, so that *only one* solution y can have the assumed form, and if $n > m + 1 \geq 2$, there is no indicial equation and *neither* solution can have the assumed form. Such a point is an *irregular singular point;* one or both solutions must have essential singularities at $z = a$. There is a definite hierarchy of irregular singular points, depending on what sort of essential singularity the solution has. For example, the form

$$y = [c_0(z - a)^s + c_1(z - a)^{s+1} + \cdots] \cdot \\ \cdot \exp [A_0(z - a)^{-k} + A_1(z - a)^{-k+1} + \cdots]$$

is a solution at some irregular singular points; the nature of the essential singularity of y at $z = a$ will be determined by the value of k required. One can order the singular points into different *species* according to the value of k required. When p or q has branch points or essential singularities, the singularities of y are even more "pathological." Luckily we need not penetrate this jungle of complications to solve the equations we are to work with.

Classification of Equations, Standard Forms. It should be apparent by now that the first thing to do when attempting to solve a linear differential equation of unfamiliar form is to locate the position of all the singularities of the functions p and q. If all these singularities are poles, we can go on to the next step; if some of them are branch points or essential singularities, we try to change the independent variable z so they all become poles (if this cannot be done, we are reduced to numerical integration; this sometimes happens!). We then distinguish the regular singular points from the irregular ones, and by solving the indicial equations for all the regular points, we find the values of the *indices* s, which determine the nature of the branch points in the solutions at these places. If there are irregular singular points, we also determine the nature of the essential singularity there by the methods discussed above.

The nature of the point at infinity is determined by setting $z = 1/w$; the resulting equation is

$$\frac{d^2y}{dw^2} + P(w)\frac{dy}{dw} + Q(w)y = 0; \quad P(w) = \frac{2}{w} - \frac{1}{w^2}p\left(\frac{1}{w}\right); \quad Q = \frac{1}{w^4}q\left(\frac{1}{w}\right) \tag{5.2.23}$$

The nature of the poles of P and Q for $w = 0$ tells us the nature of the singular point at infinity. For instance, if a solution of the indicial equation for $w = 0$ is s_1, then the solution has the form $w^{s_1}F(w) = (1/z)^{s_1}F(1/z)$, where $F(w)$ is analytic at $w = 0$.

The singular points of all the equations arising from the separation of the equation $\nabla^2\psi + k^2\psi = 0$ are listed in the table at the end of this chapter.

It is usually best next to transform the equation into a standard form, with standard positions for the singular points and with indices s for the solutions which are as simple as can be. For instance, if there are but two singular points, it is usually best to place them at zero and infinity; if there are three, the usual location is zero, unity, and infinity. This is done by changing the independent variable. If the points are originally at a, b, and c, the transformation

$$z = (w - a)\gamma/(w - c); \quad w = (\gamma a - cz)/(\gamma - z); \quad \gamma = (b - c)/(b - a) \tag{5.2.24}$$

will change the position of the singular points but will not change their nature or the indicial equations at each point. The related equations are

$$\begin{gathered}\frac{d^2\psi}{dw^2} + P(w)\frac{d\psi}{dw} + Q(w)\psi = 0; \quad \frac{d^2\psi}{dz^2} + p(z)\frac{d\psi}{dz} + q(z)\psi = 0 \\ p(z) = \frac{\gamma(a - c)}{(\gamma - z)^2}P\left(\frac{\gamma a - cz}{\gamma - z}\right) - \frac{2}{\gamma - z} \\ q(z) = \frac{\gamma^2(a - c)^2}{(\gamma - z)^4}Q\left(\frac{\gamma a - cz}{\gamma - z}\right)\end{gathered} \tag{5.2.25}$$

Here, and for the rest of this chapter, we use ψ for the general solution, y for specific solutions.

It is impossible to devise a transformation which will change four arbitrary points to four standard positions and not change the indices, so there is no single standard form for equations with more than three singular points. We usually set the irregular singular point at infinity (if there is only one), for the factor v discussed above then has the comparatively simple form $\exp\,(a_kz^k + a_{k-1}z^{k-1} + \cdots + a_{k-s}z^{k-s})$.

If the singular point at zero is regular, it is usually advisable next to change the dependent variable, by setting $\psi = vu$ and using Eq. (5.2.21), in order to obtain an equation for u which is comparatively easy to solve. Usually this involves letting v have a branch point corresponding to the lowest root of the indicial equation for the point at zero, so that one solution of the equation for u is analytic at $z = 0$. But the detailed considerations guiding the choice of transformations for both independent and dependent variable are best shown in terms of examples.

For these and other reasons it is useful to discuss some of the less complicated equations in a more or less systematic way so we can become familiar with their form and can learn something of the behavior of their solutions. We shall take the cases up in order of increasing number and complexity of singular points. In each case we discuss the general form first, then transform to a standard form, for which we will define standard solutions. The simplest case is, of course, that for *one regular singular point.* If this point is at $z = a$, the equation is

$$\psi'' + [2/(z - a)]\psi' = 0 \tag{5.2.26}$$

where the coefficient of the term $\psi'/(z - a)$ must be 2 in order that there be no singular point at infinity [see Eq. (5.2.23)]. The general solution is $\psi = A + B/(z - a)$. The standard form for the equation could be with the singular point at infinity; the equation and solutions are then

$$\psi'' = 0; \quad \psi = A + Bz \tag{5.2.27}$$

Two Regular Singular Points. Here the general form of the equation is

$$\frac{d^2\psi}{dw^2} + \frac{2w + c(\lambda + \mu - 1) - a(\lambda + \mu + 1)}{(w - a)(w - c)}\frac{d\psi}{dw} + \frac{\lambda\mu(a - c)^2}{(w - a)^2(w - c)^2}\psi = 0 \tag{5.2.28}$$

where the form of $P(w)$ is closely specified (particularly the $2w$ in the numerator) in order that the point at infinity be an ordinary point. The indicial equation (5.2.22) for $w = a$ is $s^2 - (\lambda + \mu)s + \lambda\mu = 0$, with roots λ and μ, so that the solutions are $y_1 = (w - a)^\lambda u_1^a$, $y_2 = (w - a)^\mu u_2^a$. The indicial equation for $w = c$ is $\alpha^2 + (\lambda + \mu)\alpha + \lambda\mu = 0$, with roots $-\lambda$ and $-\mu$, so that the solutions are $y_1 = (w - c)^{-\lambda} u_1^c$, $y_2 = (w - c)^{-\mu} u_2^c$. Direct substitution (or series solution) shows that $u_1^a = (w - c)^{-\lambda}$, $u_1^c = (w - a)^\lambda$, $u_2^a = (w - c)^{-\mu}$, $u_2^c = (w - a)^\mu$, so that the general solution is

$$\psi = A\left(\frac{w - a}{w - c}\right)^\lambda + B\left(\frac{w - a}{w - c}\right)^\mu \tag{5.2.29}$$

The task of solution would have been easier, however, if we had made the transformation $z = (w - a)/(w - c)$, placing one singular point at zero and one at infinity. The transformed equation is then

$$\frac{d^2\psi}{dz^2} - \left(\frac{\lambda + \mu - 1}{z}\right)\frac{d\psi}{dz} + \left(\frac{\lambda\mu}{z^2}\right)\psi = 0 \tag{5.2.30}$$

From this standard form it is not difficult to show that the solution is $\psi = Az^\lambda + Bz^\mu$. With only two singular points and these regular, the solution is forced to have a particularly simple form, and the indices at one of the two singular points must be the negatives of the indices at the other.

The one exceptional case here is when $\lambda = \mu$. In this case the solution, obtained from Eq. (5.2.6), is

$$\psi = z^\lambda(A + B \ln z) = \left(\frac{w - a}{w - c}\right)^\lambda \left[A + B \ln\left(\frac{w - a}{w - c}\right)\right] \quad (5.2.31)$$

One Irregular Singular Point. The equation for a single irregular point at $w = a$ turns out to be

$$\frac{d^2\psi}{dw^2} + \frac{2}{w - a}\frac{d\psi}{dw} - \frac{k^2}{(w - a)^4}\psi = 0 \quad (5.2.32)$$

The term in $(w - a)^{-4}$ is responsible for the irregular singular point, and the term $2/(w - a)$ prevents a singular point from occurring at $w = \infty$. The solution of this equation is

$$\psi = Ae^{k/(w-a)} + Be^{-k/(w-a)} \quad (5.2.33)$$

as can be found by direct substitution or else by transformation to $z = 1/(w - a)$, for the standard form of this type of equation is obtained by setting the irregular singular point at infinity:

$$(d^2\psi/dz^2) - k^2\psi = 0 \quad (5.2.34)$$

We should note that Eq. (5.2.34) is not the only equation having one irregular singular point at ∞, for there can be different *species* of irregular singular points. For instance, the equation coming from the wave equation in parabolic cylinder coordinates,

$$y'' + (a^2 + b^2z^2)y = 0$$

has one irregular singular point, at ∞, but its solutions are not the simple exponentials.

An interesting interrelationship between regular and irregular singular points is illustrated if we see what the case for two regular singular points becomes when we merge the singular points, at the same time keeping the term multiplying ψ finite. In Eq. (5.2.28) we set $c = a + \epsilon$, $\lambda = -\mu = k/\epsilon$, and then let ϵ go to zero; we obtain Eq. (5.2.32). The solution

$$\psi = \lim_{\epsilon \to 0} \left[A\left(1 + \frac{\epsilon}{\omega - c}\right)^{k/\epsilon} + B\left(1 + \frac{\epsilon}{\omega - c}\right)^{-k/\epsilon}\right]$$

corresponds to one of the familiar definitions of the exponential:

$$e^x = \lim_{\epsilon \to 0} (1 + \epsilon x)^{1/\epsilon}$$

Such a merging of two singular points, with corresponding modification of indices, is called a *confluence* of singular points.

Three Regular Singular Points. Equations with more complexity than the former have solutions which cannot be expressed in terms of elementary functions. Reference to the table at the end of this chapter indicates, however, that a majority of the separated equations listed there have either three regular singular points or one regular and one irregular singular point, so it behooves us to analyze these cases in detail.

As a means of further familiarizing ourselves with the techniques of analysis of singular points of equations, let us construct the equation (in the variable w) for three regular singular points at the positions $w = a, b, c$. This means that the function $p(w)$ must have simple poles at $w = a, b, c$ but nowhere else. This means that $p(w)$ must have either of the two equivalent forms

$$p(w) = \frac{Aw^3 + Bw^2 + Cw + D}{(w - a)(w - b)(w - c)} = \frac{\alpha}{w - a} + \frac{\beta}{w - b} + \frac{\gamma}{w - c} + \delta$$

We shall use the second form because it is easier to work with. Referring to Eq. (5.2.23) we see that, if infinity is to be an ordinary point, the function $(w = 1/u)$

$$\frac{2}{u} - \frac{1}{u^2} p\left(\frac{1}{u}\right) = \frac{2}{u} - \frac{\alpha/u}{1 - au} - \frac{\beta/u}{1 - bu} - \frac{\gamma/u}{1 - cu} - \frac{\delta}{u^2}$$

must be analytic at $u = 0$. This means that $\delta = 0$ and that $\alpha + \beta + \gamma = 2$.

In a similar manner, since q can have poles of order no higher than second order at a, b, c and since $q(1/u)/u^4$ must be analytic at $u = 0$, its form must be

$$q(w) = \frac{1}{(w - a)(w - b)(w - c)} \left[\frac{d}{w - a} + \frac{e}{w - b} + \frac{f}{w - c}\right]$$

To find the relation between the indices of the solution at $w = a$ and the constants $\alpha, \beta, \gamma, d, e, f$, we refer to Eq. (5.2.22). Near $w = a$,

$$p = \frac{F(w)}{w - a}; \quad F(a) = \alpha; \quad q = \frac{G(w)}{(w - a)^2}; \quad G(a) = \frac{d}{(a - b)(a - c)}$$

If the two indices at $w = a$ are to be $s = \lambda$ and $s = \lambda'$ [*i.e.*, if the two solutions at $w = a$ are $y_1 = (w - a)^\lambda u_1(w)$ and $y_2 = (w - a)^{\lambda'} u_2(w)$ with u_1 and u_2 analytic], then in the equation $s^2 + [F(a) - 1]s + G(a) = 0$ the term $1 - F(a)$ must equal the sum of the roots $\lambda + \lambda'$ and the term $G(a)$ must equal the product of the roots $\lambda\lambda'$. Therefore $\alpha = 1 - \lambda - \lambda'$ and $d = \lambda\lambda'(a - b)(a - c)$. If the indices at b are μ and μ' and those at c are ν and ν', then the rest of the constants are also given specifically in terms of these indices. But, as we have shown, in order that the point at infinity be an ordinary point, we must have $\alpha + \beta + \gamma = 2$ or

$$\lambda + \lambda' + \mu + \mu' + \nu + \nu' = 1 \tag{5.2.35}$$

With this restriction understood, we can now write the most general equation for three regular singularities,

$$\frac{d^2\psi}{dw^2} - \left[\frac{\lambda + \lambda' - 1}{w - a} + \frac{\mu + \mu' - 1}{w - b} + \frac{\nu + \nu' - 1}{w - c}\right]\frac{d\psi}{dw}$$
$$+ \left[\frac{\lambda\lambda'(a - b)(a - c)}{(w - a)^2(w - b)(w - c)} + \frac{\mu\mu'(b - a)(b - c)}{(w - a)(w - b)^2(w - c)} + \frac{\nu\nu'(c - a)(c - b)}{(w - a)(w - b)(w - c)^2}\right]\psi = 0 \quad (5.2.36)$$

This is called the *equation of Papperitz*. As we have seen, this equation (and therefore its solutions) is completely determined by specifying the position of its three singular points and the values of the two indices at each of the three points [or, rather, five out of the six indices, the sixth being fixed by Eq. (5.2.35)]. In other words the symbolic tabular display

$$\psi = P\begin{Bmatrix} a & b & c & \\ \lambda & \mu & \nu & w \\ \lambda' & \mu' & \nu' & \end{Bmatrix} \quad (5.2.37)$$

is completely equivalent (if the sum of the second and third rows is unity) to a statement that ψ is a solution of Eq. (5.2.36). This symbol, due to Riemann, will occasionally be a space saver.

A series solution of Eq. (5.2.36) would be a very tedious affair, so we shall next alter the singular points to their standard positions. Letting $a = 0$, $b = 1$, and $c \to \infty$, we obtain (letting $w \to z$)

$$\frac{d^2\psi}{dz^2} - \left[\frac{\lambda + \lambda' - 1}{z} + \frac{\mu + \mu' - 1}{z - 1}\right]\frac{d\psi}{dz}$$
$$- \left[\frac{\lambda\lambda'}{z} - \frac{\mu\mu'}{z - 1} + \nu(\lambda + \lambda' + \mu + \mu' + \nu - 1)\right]\frac{\psi}{z(z - 1)} = 0$$

or

$$\psi = P\begin{Bmatrix} 0 & 1 & \infty & \\ \lambda & \mu & \nu & z \\ \lambda' & \mu' & 1 - \lambda - \lambda' - \mu - \mu' - \nu & \end{Bmatrix} \quad (5.2.38)$$

Recursion Formulas. We have now learned that we can express ψ in terms of the following expansions about the singular points

$$\psi = z^{\lambda}\sum_{n=0}^{\infty} a_n^0 z^n + z^{\lambda'}\sum_{n=0}^{\infty} b_n^0 z^n$$
$$= (z - 1)^{\mu}\sum_{n=0}^{\infty} a_n^1 (z - 1)^n + (z - 1)^{\mu'}\sum_{n=0}^{\infty} b_n^1 (z - 1)^n$$
$$= \left(\frac{1}{z}\right)^{\nu}\sum_{n=0}^{\infty} a_n^{\infty}\left(\frac{1}{z}\right)^n + \left(\frac{1}{z}\right)^{1-\lambda-\lambda'-\mu-\mu'-\nu}\sum_{n=0}^{\infty} b_n^{\infty}\left(\frac{1}{z}\right)^n \quad (5.2.39)$$

or else in terms of a fundamental set of solutions about any ordinary point (see page 531). What we have to do next is to determine the relationship between the series coefficients a_n and b_n for different n's so we can compute a_n in terms of a_0 and b_n in terms of b_0 and therefore obtain the series for the solution. Insertion of any one of the six possible series into the equation and equating coefficients of powers of z, $(z - 1)$, or $1/z$ to zero will give us a whole series of equations of the general form

$$D_n(a_n) = \Gamma_{n1}a_{n-k+1} + \Gamma_{n2}a_{n-k+2} + \cdots + \Gamma_{nk}a_n = 0 \quad (5.2.40)$$

which are to be solved to obtain a_n in terms of a_0 (or b_n in terms of b_0). These equations are called *k-term recursion formulas.* The coefficients Γ are functions of the constants in the equation, are different for the different singular point and index chosen, and also depend on their subscripts n and j ($j = 1, 2, \ldots, k$). Since the series (5.2.39) are the correct forms, we shall find that $\Gamma_{0k} = 0$, which ensures that there need not be any a's with negative subscripts.

For Eq. (5.2.38), for instance, for $z = 0$ and for the index λ, the recursion formula is

$$\begin{aligned}D_n(a_n) = [n(n + \lambda - \lambda' - \mu - \mu' + 1) \\ - (\nu + \lambda)(\lambda' + \mu + \mu' + \nu - 1)]a_n \\ - [(n + 1)(2n + 2\lambda - 2\lambda' - \mu - \mu' + 3) - \mu\mu' \\ - (\nu + \lambda)(\lambda' + \mu + \mu' + \nu - 1)]a_{n+1} \\ + (n + 2)(n + \lambda - \lambda' + 2)a_{n+2} = 0 \quad (5.2.41)\end{aligned}$$

With such a three-term recursion formula, an explicit formula for a_n/a_0 is an extremely complicated thing, however.

At any rate we have reduced the problem of finding the solution of the differential equation to one of solving an infinite sequence of algebraic equations determining the successive coefficients of the series solution. In other words, we have made yet another transformation, from the continuous variable z to the sequence of integer values of the subscripts of the coefficients of the power series, the n of a_n or b_n; correspondingly the equation has changed from the differential equation $L_z(\psi) = 0$ to the *difference equation* $D_n(a_n) = 0$. The coefficients in the difference equation correspond to and are determined by the coefficients p and q in the differential equation. The interrelation between differential equation and recursion formula is given its most suggestive form when we express D_n in terms of the *difference operators:*

$$\begin{aligned}\delta(a_n) &= a_n - a_{n+1}; \quad \delta(na_n) = na_n - (n + 1)a_{n+1}; \quad \text{etc.} \\ \delta^2(a_n) &= \delta(\delta a_n) = a_n - 2a_{n+1} + a_{n+2}; \quad \text{etc.}\end{aligned}$$

For instance, the recursion formula for the series solution of Eq. (5.2.38), given in Eq. (5.2.41), can be written

$$D_n(a_n) = (n + \lambda - \lambda' + 2)\delta^2(na_n) - (\mu + \mu' + 3)\delta(na_n) + (\lambda + \nu)(\lambda + \nu')\delta(a_n) - \mu\mu'\delta(a_n) + (\mu\mu' + 2n)a_n = 0$$
$$(\nu' = 1 - \lambda - \lambda' - \mu - \mu' - \nu)$$

As long as we are limiting ourselves to the use of power series for the solutions, it behooves us to transform the independent and dependent variables of the differential equation into the form which will give the simplest recursion formula possible. The three-term recursion formula of Eq. (5.2.41), for instance, corresponds to a second-order difference equation; perhaps a transformation of dependent variable will change this to a two-term recursion formula, which would correspond to a first-order difference equation and which would be very much simpler to solve.

The Hypergeometric Equation. Having transformed the independent variable to shift the positions of the singular points to their standard position, we next transform the dependent variable so that the indices at the singular points in the finite part of the plane are as simple as possible. Such a change of ψ can change only the sum of the indices at the singular point; it cannot change their difference. We can, however, divide out by $z^\lambda(z - 1)^\mu$, so that one solution at each point is analytic, the other having the index $\lambda' - \lambda$ or $\mu' - \mu$. In other words, our solution is $\psi = z^\lambda(z - 1)^\mu y$, where the Riemann symbol for y is

$$y = P\left\{\begin{matrix} 0 & 1 & \infty & \\ 0 & 0 & \nu + \lambda + \mu & z \\ \lambda' - \lambda & \mu' - \mu & 1 - \lambda' - \mu' - \nu & \end{matrix}\right\}$$

The corresponding equation for y [obtained by substituting $\psi = z^\lambda(z - 1)^\mu y$ into Eq. (5.2.38) or by modifying Eq. (5.2.38) to correspond to the new P symbol] is

$$y'' + \left[\frac{\lambda - \lambda' + 1}{z} + \frac{\mu - \mu' + 1}{z - 1}\right]y' + \frac{(\nu + \lambda + \mu)(1 - \lambda' - \mu' - \nu)}{z(z - 1)}y = 0$$

But we have now too many constants for the quantities we have to fix. Only three constants are needed to fix the indices; there are four indices which have not been set zero (one at 0, one at 1, and two at ∞), but the sum of these four must equal unity, so three constants are enough. It will be found most convenient to set the two indices at ∞ equal to a and b, the one at 0 equal to $1 - c$, so that the one at 1 equals $c - a - b$. The Riemann P symbol is then

$$F = P\left\{\begin{matrix} 0 & 1 & \infty & \\ 0 & 0 & a & z \\ 1 - c & c - a - b & b & \end{matrix}\right\}$$

and the corresponding differential equation

$$z(z - 1)F'' + [(a + b + 1)z - c]F' + abF = 0 \qquad (5.2.42)$$

called the *hypergeometric equation*, is the standard equation for three regular singular points.

The analytic solution of this equation about $z = 0$ is called the *hypergeometric function*. To obtain its series expansion we set $F = \Sigma a_n z^n$ and insert in Eq. (5.2.42). The coefficient of z^n gives the recursion formula for the series

$$\begin{aligned} D_n(a_n) &= (n + a)(n + b)a_n - (n + 1)(n + c)a_{n+1} \\ &= (n + c)(n + 1)\delta(a_n) + [n(a + b - 1) - (n + 1)c + ab]a_n = 0 \end{aligned} \qquad (5.2.43)$$

This is a *two-term recursion formula*, a first-order difference equation, with the simple solution

$$a_0 = 1; \quad a_n = \frac{a(a + 1) \cdots (a + n - 1)b(b + 1) \cdots (b + n - 1)}{[1 \cdot 2 \cdots n]\, c(c + 1) \cdots (c + n - 1)}$$

The corresponding series

$$F(a,b|c|z) = 1 + \frac{ab}{c}z + \frac{a(a + 1)b(b + 1)}{2!c(c + 1)}z^2 + \cdots \qquad (5.2.44)$$

is called the *hypergeometric series* (see page 388). It is the analytic solution of Eq. (5.2.42) at $z = 0$. It converges as long as $|z| < 1$, for the next singular point is at $z = 1$. All the solutions of Eq. (5.2.42) about all its singularities can be expressed in terms of such series.

For instance, if we insert the function $z^{1-c}F_2$ (where F_2 can be analytic, for the second index at $z = 0$ is $1 - c$) into Eq. (5.2.42), we obtain

$$\begin{aligned} z(z - 1)F_2'' + [(a + b - 2c + 3)z - 2 + c]F_2' \\ + (a - c + 1)(b - c + 1)F_2 = 0 \end{aligned}$$

which is another hypergeometric equation with analytic solution the hypergeometric series $F_2 = F(b - c + 1, a - c + 1| 2 - c| z)$. The general solution of Eq. (5.2.42) is therefore

$$AF(a,b|c|z) + Bz^{1-c}F(b - c + 1, a - c + 1| 2 - c| z) \qquad (5.2.45)$$

Working backward now, the general solution of Eq. (5.2.38) at $z = 0$ is

$$\begin{aligned} Az^{\lambda}(z - 1)^{\mu}F(\lambda + \mu + \nu, 1 - \nu - \lambda' - \mu'| \lambda - \lambda' + 1| z) \\ + Bz^{\lambda'}(z - 1)^{\mu}F(\lambda' + \mu + \nu, 1 - \nu - \lambda - \mu'| \lambda' - \lambda + 1| z) \end{aligned}$$

and the general solution of Papperitz equation (5.2.36) at $w = a$ is

$$A\left(\frac{w-a}{w-c}\right)^\lambda \left(\frac{w-b}{w-c}\right)^\mu F\left(\lambda+\mu+\nu,\right.$$
$$\left. 1-\nu-\lambda'-\mu' \middle| \lambda-\lambda'+1 \middle| \frac{w-a}{w-c}\frac{b-c}{b-a}\right)$$
$$+ B\left(\frac{w-a}{w-c}\right)^{\lambda'} \left(\frac{w-b}{w-c}\right)^\mu F\left(\lambda'+\mu+\nu,\right.$$
$$\left. 1-\nu-\lambda-\mu' \middle| \lambda'-\lambda+1 \middle| \frac{w-a}{w-c}\frac{b-c}{b-a}\right) \qquad (5.2.46)$$

Solutions of Papperitz equation at $w = b$ or $w = c$ can be obtained by interchanging λ, λ' with μ, μ', etc., since the equation is symmetrical for interchange of singular points and corresponding indices (if we remember that $\nu' = 1 - \lambda - \lambda' - \mu - \mu' - \nu$).

The formula (5.2.45) expresses the general solution unless $|1 - c|$ is zero or an integer, in which case the series for one or the other of the solutions will have a zero factor in the denominator of all terms above a certain term in the series. For instance, if $c = 3$, the series for the second solution will have its first two terms finite but all the higher terms will be infinite because of the term $(2 - c + 1)$ in the denominator. This is an example of the special case, mentioned on page 529, where the indices for a given singular point differ by an integer. As mentioned on page 532, we must then find the second solution by means of Eq. (5.2.6).

Here $y_1 = F(a,b|c|z)$ and $p = (c/z) + [(a + b + 1 - c)/(z - 1)]$ so that $e^{-\int p\,dz} = z^{-c}(1 - z)^{c-a-b-1}$. The second solution is

$$y_2 = F(a,b|c|z)\int[F(a,b|c|z)]^{-2}z^{-c}(1 - z)^{c-a-b-1}\,dz$$

As long as $|z| < 1$, we can expand $(1 - z)^{c-a-b-1}[F(a,b|c|z)]^{-2}$ in a power series in z, which we can write $g_0 + g_1z + \cdots$. The series for the integrand is then

$$\frac{g_0}{z^c} + \frac{g_1}{z^{c-1}} + \cdots + \frac{g_{c-1}}{z} + g_c + g_{c+1}z + \cdots$$

where we have assumed that c is a positive integer, so that g_c is multiplied by the zero power of z and g_{c-1} is divided by the first power of z, and so on, with no fractional powers entering. In this case g_{c-1}/z will integrate to a logarithmic term, and the second solution will be

$$y_2 = F(a,b|c|z)g_{c-1}\ln z$$
$$+ (1/z^{c-1})(h_0 + h_1z + \cdots + h_{c-2}z^{c-2} + h_cz^c + \cdots)$$

where the coefficients h_n are functions of a, b, c, and n.

Functions Expressible by Hypergeometric Series. The series of Eq. (5.2.44) can express a great variety of functions. For instance, some of the simpler cases are

$$(1+z)^n = F(-n,b|b|-z); \quad \ln(1+z) = zF(1,1|2|-z)$$

The separated equation for ξ_2 for circular cylinder coordinates and for ξ_3 for spherical, parabolic, prolate, and oblate spheroidal coordinates is

$$\frac{1}{\sqrt{1-\xi^2}}\frac{d}{d\xi}\left[\sqrt{1-\xi^2}\frac{dX}{d\xi}\right] + \frac{k_3^2}{1-\xi^2}X = 0$$

or

$$\frac{d^2X}{d\xi^2} + \left[\frac{\frac{1}{2}}{\xi+1} + \frac{\frac{1}{2}}{\xi-1}\right]\frac{dX}{d\xi} - \frac{k_3^2}{\xi^2-1}X = 0$$

This has the form of Eq. (5.2.36) with $a = +1$, $b = -1$, $c = \infty$, $\lambda + \lambda' - 1 = -\frac{1}{2} = \mu + \mu' - 1$, $\lambda\lambda' = 0 = \mu\mu'$, $\nu\nu' = -k_3^2$ [and, of course, the usual Eq. (5.2.35), $\lambda + \lambda' + \mu + \mu' + \nu + \nu' = 1$]. The Riemann symbol for the solution is therefore

$$X = P\begin{Bmatrix} 1 & -1 & \infty & \\ \frac{1}{2} & \frac{1}{2} & k_3 & \xi \\ 0 & 0 & -k_3 & \end{Bmatrix}$$

and the general solution about $\xi = +1$ is (from Eq. 5.2.46)

$$X = A\sqrt{1-\xi^2}\,F\left(1+k_3, 1-k_3\middle|\tfrac{3}{2}\middle|\frac{1-\xi}{2}\right) + B\sqrt{1+\xi}\,F\left(\tfrac{1}{2}+k_3, \tfrac{1}{2}-k_3\middle|\tfrac{1}{2}\middle|\frac{1-\xi}{2}\right)$$

These solutions are called *Tschebyscheff functions.* They turn out [see Eq. (5.2.54)] to be proportional to sin $(k\varphi)$ and cos $(k\varphi)$, respectively, where $\xi = \cos\varphi$, as a transformation of the differential equation above will show. These functions are discussed in Sec. 5.3.

The separated equation for ξ_2 for spherical coordinates and also for spheroidal coordinates when $k_1 = 0$ is

$$\frac{d^2X}{d\xi^2} + \left[\frac{1}{\xi-1} + \frac{1}{\xi+1}\right]\frac{dX}{d\xi} - \left[\frac{k_2^2}{\xi^2-1} + \frac{k_3^2/2}{(\xi^2-1)(\xi+1)} - \frac{k_3^2/2}{(\xi^2-1)(\xi-1)}\right]X = 0$$

which is called the *Legendre equation.* It corresponds to Eq. (5.2.36) if $a = +1$, $b = -1$, $c = \infty$, $\lambda = -\lambda' = m/2 = \mu = -\mu'$, $\nu = -n$, $\nu' = n+1$, where we have set $k_3 = m$ and $k^2 = n(n+1)$ in order to make the results easier to write out. Therefore the solution corresponds to the Riemann symbol

$$X = P\begin{Bmatrix} 1 & -1 & \infty & \\ m/2 & m/2 & -n & \xi \\ -m/2 & -m/2 & n+1 & \end{Bmatrix}$$

and the general solution about $\xi = 1$ is [see also Eq. (5.2.52)]

$$X = A(1 - \xi^2)^{m/2} F\left(m - n, m + n + 1 | 1 + m | \frac{1 - \xi}{2}\right) + B\left(\frac{1 + \xi}{1 - \xi}\right)^{m/2} F\left(-n, n + 1 | 1 - m | \frac{1 - \xi}{2}\right) \quad (5.2.47)$$

The first solution is known as the *Legendre function of the first kind*, and the second solution is called the *Legendre function of the second kind*. If m is a positive integer, this second solution, as written, will "blow up" and a second solution will have to be constructed by use of Eq. (5.2.6) (see page 532). These Legendre functions and the related Gegenbauer functions will be discussed again in a few pages and in very considerable detail in Secs. 5.3 and 10.3, for they are of very great importance in our future discussions.

Analytic Continuation of Hypergeometric Series. As an exercise in dealing with these solutions we can derive an expression for the behavior of the hypergeometric series as z approaches unity. By referring to the Riemann symbol above Eq. (5.2.42), we see that the indices for the singular point $z = 1$ are 0 and $c - a - b$. Using Eq. (5.2.46) with $a = 1$, $b = 0$, $c = \infty$, $\lambda = 0$, $\lambda' = c - a - b$, $\mu = 0$, $\mu' = 1 - c$, $\nu = a$, $\nu' = b$ (in other words, exchanging the two singular points 0 and 1), we have that the general solution of the hypergeometric equation (5.2.42) about $z = 1$ is

$$AF(a,b|a + b - c + 1|1 - z) + B(1 - z)^{c-a-b}F(c - b, c - a| c - a - b + 1| 1 - z)$$

Now the hypergeometric series $F(a,b|c|z)$ is a solution of Eq. (5.2.42), and by analytic continuation, it must be some combination of the two solutions about $z = 1$. In other words we can be sure that

$$F(a,b|c|z) = \alpha F(a,b|a + b - c + 1|1 - z) + \beta(1 - z)^{c-a-b}F(c - b, c - a| c - a - b + 1| 1 - z) \quad (5.2.48)$$

If we could somehow determine the values of the coefficients α and β, we should have a means of computing the exact behavior of $F(a,b|c|z)$ at $z = 1$ and even beyond, almost out to $z = 2$.

On page 388 we showed that, when $b < c < a + b$,

$$F(a,b|c|z) \underset{z \to 1}{\longrightarrow} \frac{\Gamma(c)\Gamma(a + b - c)}{\Gamma(a)\Gamma(b)} (1 - z)^{c-a-b}$$

where we have substituted gamma functions for the factorials, so that a, b, and c can have nonintegral values. All that this equation shows is that, when $c < a + b$, the highest order infinity at $z = 1$ is of the form $(1 - z)^{c-a-b}$, and it gives the coefficient of this term. There are, of course, other terms of the form $(1 - z)^{c-a-b+1}$, $(1 - z)^{c-a-b+2}$, etc., some of which may also go to infinity at $z = 1$, but very close to $z = 1$ they are "drowned out" by the $(1 - z)^{c-a-b}$ term. The rest of the terms can be supplied, however, by use of Eq. (5.2.48).

Since $F(c - b, c - a | c - a - b + 1 | 1 - z) \xrightarrow[z\to 1]{} 1$, we have that the principal term on the right-hand side of Eq. (5.2.48), as $z \to 1$, is $\beta(1 - z)^{c-a-b}$ as long as $c < a + b$. Comparing this with the results referred to above we see that

$$\beta = [\Gamma(c)\Gamma(a + b - c)]/[\Gamma(a)\Gamma(b)]$$

which fixes the coefficients of all the terms in negative powers of $(1 - z)$.

But we can go further than this in the case that $c < 1$. We can use the limiting formula above in reverse by seeing what happens to Eq. (5.2.48) when z is allowed to go to zero. In this case the left-hand side goes to unity, of course. Using the limiting formula of page 388 on the right-hand side, we have eventually

$$1 \xrightarrow[z\to 0]{} \alpha \frac{\Gamma(a + b - c + 1)\Gamma(c - 1)}{\Gamma(a)\Gamma(b)} z^{1-c} + \beta \frac{\Gamma(c - a - b + 1)\Gamma(c - 1)}{\Gamma(c - a)\Gamma(c - b)} z^{1-c}$$

If $c < 1$, the right-hand side will become infinite at $z = 0$ unless the two terms just cancel each other. This gives a relation between α and β which enables us to solve the problem. Inserting the value of β already obtained and utilizing the property of the gamma function, $\Gamma(u + 1) = u\Gamma(u)$, we obtain

$$1 \xrightarrow[z\to 0]{} (a + b - c) \frac{\Gamma(a + b - c)\Gamma(c - 1)}{\Gamma(a)\Gamma(b)} \left[\alpha - \frac{\Gamma(c - a - b)\Gamma(c)}{\Gamma(c - a)\Gamma(c - b)}\right] z^{1-c} + \text{finite terms}$$

In order that this be true for $c < 1$, we must have the quantity in square brackets equal to zero, which gives an equation for α. Consequently, at least for $c < 1$, $c < a + b$, we have the useful formula

$$F(a,b|c|z) = \frac{\Gamma(c)\Gamma(c - a - b)}{\Gamma(c - a)\Gamma(c - b)} F(a, b| a + b - c + 1| 1 - z) + \frac{\Gamma(c)\Gamma(a + b - c)}{\Gamma(a)\Gamma(b)} (1 - z)^{c-a-b} F(c - a, c - b| c - a - b + 1| 1 - z) \tag{5.2.49}$$

which enables us to extend the solution through the singular point at $z = 1$. In the next section we shall show that the equation is valid over a much wider range of values of c than our present derivation would lead us to believe (for c larger than $a + b$, for instance). In fact, except for the places where the gamma functions in the numerators go to infinity, this formula is universally valid.

Another set of useful formulas relating hypergeometric series may be obtained by use of this "joining equation" plus a certain amount of manipulation of the hypergeometric equation (5.2.42). In this equation we set $a = 2\alpha$, $b = 2\beta$, $c = \alpha + \beta + \frac{1}{2}$ and change the independent variable from z to $u = 4z(1 - z)$, giving the equation

$$u(u - 1)\frac{d^2F}{du^2} + [(\alpha + \beta + 1)u - (\alpha + \beta + \tfrac{1}{2})]\frac{dF}{du} + \alpha\beta F = 0$$

which is the hypergeometric equation again, with new parameters α, β and $(\alpha + \beta + \frac{1}{2})$ instead of a, b, and c. Therefore we have shown that the function $F(\alpha, \beta| \alpha + \beta + \frac{1}{2}| 4z - 4z^2)$ is a solution of Eq. (5.2.42) for $a = 2\alpha$, $b = 2\beta$, $c = \alpha + \beta + \frac{1}{2}$ and must be expressible in terms of the two solutions about $z = 0$ [see Eq. (5.2.45)]:

$$AF(2\alpha, 2\beta| \alpha + \beta + \tfrac{1}{2}| z) + Bz^{\frac{1}{2}-\alpha-\beta}F(\beta - \alpha + \tfrac{1}{2}, \alpha - \beta + \tfrac{1}{2}| \tfrac{3}{2} - \alpha - \beta| z)$$

However, $F(\alpha, \beta| \alpha + \beta + \frac{1}{2}| 4z - 4z^2)$ is an analytic function of z at $z = 0$; therefore the second solution, having a branch point, cannot enter and B is zero. Also, since $F = 1$ for $z = 0$, A must equal unity, We therefore have

$$F(2\alpha, 2\beta| \alpha + \beta + \tfrac{1}{2}| z) = F(\alpha, \beta| \alpha + \beta + \tfrac{1}{2}| 4z - 4z^2) \qquad (5.2.50)$$

This formula, relating an F for 2α, 2β, and z to an F for α, β, and z^2 might be called a *duplication formula* for the hypergeometric function.

Gegenbauer Functions. We mentioned earlier that it is sometimes desirable to have the canonical form for three regular singular points so that two of them are at ± 1 rather than 0 and 1. This is particularly true when the indices at $+1$ are the same as the indices for -1. This suggests the Riemann symbol

$$\psi = P\begin{Bmatrix} -1 & 1 & \infty & \\ 0 & 0 & -\alpha & z \\ -\beta & -\beta & \alpha + 2\beta + 1 & \end{Bmatrix}$$

The corresponding equation

$$(z^2 - 1)\psi'' + 2(\beta + 1)z\psi' - \alpha(\alpha + 2\beta + 1)\psi = 0 \qquad (5.2.51)$$

is called *Gegenbauer's equation* [compared with the Legendre equation on page 544]. It is a satisfactory form for the equations for ξ_2 for circular cylinder and spherical coordinates. The solutions can, of course, be expressed in terms of hypergeometric functions:

$$AF\left(-\alpha, \alpha + 2\beta + 1 \middle| 1 + \beta \middle| \frac{1+z}{2}\right) + B(1+z)^{-\beta} F\left(-\alpha - \beta, \alpha + \beta + 1 \middle| 1 - \beta \middle| \frac{1+z}{2}\right)$$

or $$aF\left(-\alpha, \alpha + 2\beta + 1 \middle| 1 + \beta \middle| \frac{1-z}{2}\right) + b(1-z)^{-\beta} F\left(-\alpha - \beta, \alpha + \beta + 1 \middle| 1 - \beta \middle| \frac{1-z}{2}\right)$$

The solution which will be most useful to us has the various forms

$$T_\alpha^\beta(z) = \frac{\Gamma(\alpha + 2\beta + 1)}{2^\beta \Gamma(\alpha + 1)\Gamma(\beta + 1)} F(\alpha + 2\beta + 1, -\alpha | 1 + \beta | \tfrac{1}{2} - \tfrac{1}{2}z) \tag{5.2.52}$$

$$T_\alpha^\beta(z) = \frac{\Gamma(\alpha + 2\beta + 1)}{2^\beta \Gamma(\alpha + 1)\Gamma(\beta + 1)} \frac{\sin [\pi(\alpha + \beta)]}{\sin (\pi\beta)} \cdot F(\alpha + 2\beta + 1, -\alpha | 1 + \beta | \tfrac{1}{2} + \tfrac{1}{2}z) - \frac{(1+z)^{-\beta}}{\Gamma(1-\beta)} \frac{\sin (\pi\alpha)}{\sin (\pi\beta)} F(-\alpha - \beta, \alpha + \beta + 1 | 1 - \beta | \tfrac{1}{2} + \tfrac{1}{2}z)$$

where we have used Eq. (5.2.49) and the relation $\sin(\pi u)\Gamma(u)\Gamma(1-u) = \pi$ in order to obtain the second form. The function $T_\alpha^\beta(z)$ is called a Gegenbauer function. If α is not an integer, T has a branch point at $z = -1$ (unless β is an integer). When α is zero or a positive integer, T is a finite polynomial in z and, of course, is analytic at $z = \pm 1$.

The second solution about $z = 1$ is also sometimes used. As a matter of fact $(1 - z^2)^{\frac{1}{2}\beta}$ times this second solution,

$$P_{\alpha+\beta}^\beta(z) = \frac{1}{\Gamma(1-\beta)} \left(\frac{1+z}{1-z}\right)^{\frac{1}{2}\beta} F(-\alpha - \beta, \alpha + \beta + 1 | 1 - \beta | \tfrac{1}{2} - \tfrac{1}{2}z)$$
$$= \frac{(\alpha + 2\beta + 1)(1 - z^2)^{\frac{1}{2}\beta}}{2^\beta \Gamma(\alpha + 1)\Gamma(\beta + 1)} \left\{ \frac{\sin [\pi(\alpha + \beta)]}{\sin (\pi\alpha)} \cdot F(\alpha + 2\beta + 1, -\alpha | 1 + \beta | \tfrac{1}{2} - \tfrac{1}{2}z) - \frac{\sin (\pi\beta)}{\sin (\pi\alpha)} F(\alpha + 2\beta + 1, -\alpha | 1 + \beta | \tfrac{1}{2} + \tfrac{1}{2}z) \right\}$$

is often called the *generalized Legendre function of z*.

When α is an integer $n = 0, 1, 2, \ldots$, it can be shown by expansion of the polynomial and term-by-term comparison that

$$F(n + 2\beta + 1, -n | 1 + \beta | \tfrac{1}{2} + \tfrac{1}{2}z) = (-1)^n F(n + 2\beta + 1, -n | 1 + \beta | \tfrac{1}{2} - \tfrac{1}{2}z) = \frac{\Gamma(\beta + 1)}{2^n \Gamma(n + \beta + 1)(1 - z^2)^\beta} \frac{d^n}{dz^n} (1 - z^2)^{n+\beta}$$

and, consequently,

$$T_n(z) = \frac{(-1)^n \Gamma(n + 2\beta + 1)}{2^{n+\beta} n! \Gamma(n + \beta + 1)} (1 - z^2)^{-\beta} \frac{d^n}{dz^n} (1 - z^2)^{n+\beta} \tag{5.2.53}$$

which can be called a *Gegenbauer polynomial* (it differs, by a numerical factor, from the polynomial $C_n^{\beta+\frac{1}{2}}$ often called a Gegenbauer polynomial).

When β is an integer $m = 0, 1, 2, \ldots$ but α is not, it will be shown in the next section that T_α^m has logarithmic branch points at $z = \pm 1$ but that $(1 - z^2)^{-\frac{1}{2}m} P_{\alpha+m}^m$ is analytic over the whole range $-1 \leq z \leq 1$. Finally, when *both* α and β are positive integers (or zero), $T_n^m(z) = (1 - z^2)^{-\frac{1}{2}m} P_n^m(z)$ and both functions are analytic over this range of z. Since

$$\frac{1}{(1 - z^2)^m} \frac{d^n}{dz^n} (1 - z^2)^{n+m} = (-1)^n \frac{n!}{(n + 2m)!} \frac{d^{n+2m}}{dz^{n+2m}} (z^2 - 1)^{n+m}$$

we have

$$T_n^m(z) = \frac{1}{2^{n+m}(n + m)!} \frac{d^{n+2m}}{dz^{n+2m}} (z^2 - 1)^{n+m}$$

which is sometimes called a *tesseral polynomial*. The case of $m = 0$ is of particular importance, enough importance to merit giving the polynomial a special symbol and name. The function

$$P_n(z) = T_n^0(z) = \frac{1}{2^n n!} \frac{d^n}{dz^n} (z^2 - 1)^n$$

is called a *Legendre polynomial* [see Eq. (5.2.47)]. We shall meet it often in the rest of the book. We see that

$$T_n^m(z) = \frac{d^m}{dz^m} P_{n+m}(z)$$

We shall have much more to say about these functions in the next section.

In the special case $\beta = \pm\frac{1}{2}$, $\alpha = n = 1, 2, 3, \ldots$, the polynomials T can be shown (by direct expansion) to have the following forms:

$$\begin{aligned} T_n^{-\frac{1}{2}}(z) &= \frac{1}{n} \sqrt{\frac{2}{\pi}} \cosh\, [n \cosh^{-1}(z)] \\ T_n^{\frac{1}{2}}(z) &= \sqrt{\frac{2/\pi}{z^2 - 1}} \sinh\, [(n + 1) \cosh^{-1}(z)] \end{aligned} \tag{5.2.54}$$

called *Tschebyscheff polynomials*.

Incidentally, the general solution of the separated equation for ξ_2 in circular cylinder coordinates and for ξ_3 in parabolic, prolate, and oblate spheroidal coordinates,

$$(z^2 - 1)\psi'' + z\psi' - \alpha^2\psi = 0$$

has for general solution

$$\psi = A T_\alpha^{-\frac{1}{2}}(z) + B \sqrt{1 - z^2}\, T_{\alpha-1}^{\frac{1}{2}}(z)$$

as a comparison with the equation for T will show.

One Regular and One Irregular Singular Point. For this equation it is usual to put the regular singular point at $z = 0$ and the irregular one at $z = \infty$. Equations of this sort occur for the following separable coordinates:

1. For the wave equation in circular cylinder ($z = k_1\xi_1$, $\psi = X_1$), spherical ($z = k_1\xi_1$, $\psi = \sqrt{z}\, X_1$), and conical ($z = k_1\xi_1$, $\psi = \sqrt{z}\, X_1$) coordinates the equation is the *Bessel equation*

$$\frac{d^2\psi}{dz^2} + \frac{1}{z}\frac{d\psi}{dz} + \left(1 - \frac{n^2}{z^2}\right)\psi = 0$$

2. For the wave equation in parabolic cylinder coordinates ($z = \frac{1}{2}\xi_1^2, \frac{1}{2}\xi_2^2$, $\psi = X_1, X_2$)

$$\frac{d^2\psi}{dz^2} + \frac{1}{z}\frac{d\psi}{dz} + \left(k^2 + \frac{2\alpha}{z}\right)\psi = 0$$

3. For the wave equation in parabolic coordinates ($z = \xi_1^2, \xi_2^2$, $\psi = X_1, X_2$)

$$\frac{d^2\psi}{dz^2} + \frac{1}{z}\frac{d\psi}{dz} + \left(k^2 + \frac{2\alpha}{z} - \frac{m^2}{z^2}\right)\psi$$

which includes the parabolic cylinder function as a special case.

4. For the Schroedinger equation for one particle in a coulomb $1/r$ field, in spherical coordinates, for the radial factor ($z = \xi_1$, $\psi = X_1$)

$$\frac{d^2\psi}{dz^2} + \frac{2}{z}\frac{d\psi}{dz} + \left[-E + \frac{2\alpha}{z} - \frac{n(n+1)}{z^2}\right]\psi = 0$$

This suggests that we study the general equation

$$\frac{d^2\psi}{dz^2} + p(z)\frac{d\psi}{dz} + q(z)\psi = 0; \quad p = \frac{(1 - \lambda - \lambda')}{z}; \qquad q = -k^2 + \left(\frac{2\alpha}{z}\right) + \left(\frac{\lambda\lambda'}{z^2}\right) \tag{5.2.55}$$

This is not the most general equation with a regular singular point at $z = 0$ and an irregular one at infinity. The function p could have a constant, and both p and q could have terms like az or bz, etc., added to the expressions shown in Eq. (5.2.55). But these additional terms would make the essential singularity in the solution at $z = \infty$ more "singular," and since none of the equations coming from the separation of the wave equation exhibit such additional terms, we shall not include them here. A few more complicated cases will be included in the problems, however!

Returning to Eq. (5.2.55), we see that the indicial equation indicates that the expansions about the regular singular point at $z = 0$ are $z^\lambda u_1(z)$ and $z^{\lambda'} u_2(z)$, where u_1 and u_2 are analytic functions at $z = 0$. As before, we arrange it so that one solution is analytic, by setting $\psi = z^\lambda f(z)$, so that $f(z) = u_1(z) + z^{\lambda'-\lambda}u_2(z)$. The equation for f is then

$$f'' + [(1 + \lambda - \lambda')/z]f' + [(2\alpha/z) - k^2]f = 0 \qquad (5.2.56)$$

Setting $z = 1/w$, we next examine the singular point at infinity. The equation, in terms of w, becomes

$$\frac{d^2f}{dw^2} + \left[\frac{1 + \lambda' - \lambda}{w}\right]\frac{df}{dw} + \left[\frac{2\alpha}{w^3} - \frac{k^2}{w^4}\right]f = 0$$

which indicates an irregular singular point at $w = 0$, due to the terms $2\alpha/w^3$ and k^2/w^4. This equation has no indicial equation, so both solutions have essential singularities at $w = 0$. However, setting $f = e^{-k/w}F$, we can arrive at an equation

$$F'' + \left[\frac{2k}{w^2} + \frac{1 - \lambda + \lambda'}{w}\right]F' - \left[\frac{k(1 + \lambda - \lambda') - 2\alpha}{w^3}\right]F = 0$$

which does have an indicial equation, with one root,

$$2ks - [k(1 + \lambda - \lambda') - 2\alpha] = 0$$

Therefore a solution for F is $w^\beta v_1(w)$, where v_1 is an analytic function at $w = 0$ and $\beta = \frac{1}{2}(1 + \lambda - \lambda') - (\alpha/k)$. Another solution for f is $e^{k/w}w^{\beta'}v_2(w)$, where $\beta' = \frac{1}{2}(1 + \lambda - \lambda') + (\alpha/k)$ and v_2 is analytic at $w = 0$. Thus the essential singularity at $w = 0$ ($z \to \infty$) is of the form $e^{+k/w} = e^{+kz}$.

We can now return to Eq. (5.2.56) in z and set $f = e^{-kz}F(z)$ (or $\psi = z^\lambda e^{-kz}F$) and be sure that one solution for F is analytic at $z = 0$ and has a branch point only at $z \to \infty$. The equation for F in terms of z is

$$F'' + \left[\frac{1 + \lambda - \lambda'}{z} - 2k\right]F' - \left[\frac{k(1 + \lambda - \lambda') - 2\alpha}{z}\right]F = 0$$

which reduces to its simplest form when we set $z = x/2k$, $c = 1 + \lambda - \lambda'$, $a = \frac{1}{2}(1 + \lambda - \lambda') - (\alpha/k)$,

$$x\frac{d^2F}{dx^2} + (c - x)\frac{dF}{dx} - aF = 0 \qquad (5.2.57)$$

which is called the *confluent hypergeometric equation*. This name is chosen because Eq. (5.2.57) comes from the hypergeometric equation (5.2.42) by suitable confluence of the singular points at $z = 1$ and $z = \infty$.

The confluence may be more easily seen by starting with the Papperitz equation (5.2.36) for the case $a = 0$, $c = \infty$,

$$\psi'' + \left[\frac{1 - \lambda - \lambda'}{z} + \frac{1 - \mu - \mu'}{z - b}\right]\psi' + \left[\frac{-\lambda\lambda' b}{z^2(z - b)} + \frac{\mu\mu' b}{z(z - b)^2} + \frac{\nu(1 - \lambda - \lambda' - \mu - \mu' - \nu)}{z(z - b)}\right]\psi = 0$$

We now let $\lambda = 0$, $1 - \lambda' = c$ (using c now for another constant, not the position of the third singular point which went to infinity), $\mu' = -b$, and $\mu = a - \nu$ (not the a which went to zero). We finally arrive at the confluent hypergeometric equation (5.2.57) by allowing the second singular point to merge with the third at infinity, that is, we let $b \to \infty$. At the same time that the singular points merge, one of the indices at each of these points (that is, μ' and ν') also tends to infinity.

The one solution of Eq. (5.2.57) which is analytic at $x = 0$ may be found by substituting the general series form $F = \Sigma a_n x^n$ into the equation. The coefficient of the power x^n gives the recursion formula

$$(n + 1)(n + c)a_{n+1} - (n + a)a_n = 0 \tag{5.2.58}$$

which is again a two-term recursion formula, with a simple solution. Therefore the solution which is analytic at $z = 0$ is given by the series

$$F(a|c|x) = 1 + \frac{a}{c}x + \frac{a(a+1)}{2!c(c+1)}x^2 + \frac{a(a+1)(a+2)}{3!c(c+1)(c+2)}x^3 + \cdots \tag{5.2.59}$$

which is called the *confluent hypergeometric series.* This series converges in the range $-\infty < x < \infty$. Using the methods discussed on page 388 we can obtain an indication of the asymptotic behavior of this series. For large values of z the terms in the higher powers of z preponderate. But to the first approximation in z/n and when a and c are integers, the term in the series with z^n (n large) becomes

$$\frac{(c-1)!(a+n)!}{n!(a-1)!(c+n)!}z^n \simeq \frac{(c-1)!}{(a-1)!}\frac{n^{a-c}}{n!}z^n \simeq \frac{(c-1)!}{(a-1)!}\frac{z^n}{(n-a+c)!}$$

Therefore when z is large enough, the series approaches the series

$$\frac{(c-1)!}{(a-1)!}\sum_n \frac{z^n}{(n-a+c)!} \simeq \frac{(c-1)!}{(a-1)!}z^{a-c}\sum_m \frac{z^m}{m!} = \frac{(c-1)!}{(a-1)!}z^{a-c}e^z$$

Inserting gamma functions instead of factorials we finally arrive at the indication that

$$F(a|c|z) \xrightarrow[z\to\infty]{} \frac{\Gamma(c)}{\Gamma(a)}z^{a-c}e^z \tag{5.2.60}$$

In the next section we shall show that this asymptotic formula is valid over a wider range of values of a and c and z than the present derivation would concede.

By reversing the procedure by which we obtained F from ψ, we see that the general solution of Eq. (5.2.55) about $z = 0$ is

$$\psi = Ae^{-kz}z^\lambda F\left(\frac{1+\lambda-\lambda'}{2} - \frac{\alpha}{k}\Big|1+\lambda-\lambda'\Big|2kz\right) + Be^{-kz}z^{\lambda'}F\left(\frac{1-\lambda+\lambda'}{2} - \frac{\alpha}{k}\Big|1-\lambda+\lambda'\Big|2kz\right) \tag{5.2.61}$$

unless $(\lambda - \lambda')$ is an integer, in which case the second solution contains a logarithmic term and must be obtained by use of Eq. (5.2.6).

From this general solution, we can also see that a second solution of the confluent hypergeometric equation is

$$x^{1-c}F(a - c + 1|2 - c|x)$$

However, we can also see that, if we set $z = -x$ in Eq. (5.2.57) and set $F = e^{-\xi}F_2$, the equation for F_2 is also that for a confluent hypergeometric function. In other words, another solution of Eq. (5.2.57) is $e^xF(c - a|c| - x)$. This is not a third independent solution, however, for series expansion and series multiplication show that

$$e^xF(c - a|c| - x) = F(a|c|x) \tag{5.2.62}$$

Incidentally, comparison with Eq. (5.2.60) indicates that an asymptotic formula which might be satisfactory for z large and positive might not be satisfactory at all for z large and negative. In fact if Eq. (5.2.62) is correct and Eq. (5.2.60) is valid for Re $z \to \infty$, then

$$F(a|c|z) \xrightarrow[z\to -\infty]{} \frac{\Gamma(c)}{\Gamma(c - a)} (-z)^{-a} \tag{5.2.63}$$

However, we should postpone further discussion of this matter until next section, when we shall be much better equipped for the discussion.

Comparison of the equations on page 550 with Eq. (5.2.61) shows that the general solution of the Bessel equation is

$$\psi = e^{-iz}[Az^n F(\tfrac{1}{2} + n|1 + 2n|2iz) + Bz^{-n} F(\tfrac{1}{2} - n|1-2n|2iz)]$$

or, using Eq. (5.2.62),

$$\psi = Ae^{-iz} z^n F(\tfrac{1}{2} + n|1 + 2n|2iz) + Be^{iz} z^{-n} F(\tfrac{1}{2} - n|1 - 2n|-2iz)$$

The general solution for the wave equation in parabolic coordinates, which includes that for the parabolic cylinder, is

$$\psi = Ae^{-ikz} z^m F\left(\frac{1}{2} + m + \frac{i\alpha}{k}|1 + 2m|2ikz\right) + Be^{ikz} z^{-m} F\left(\frac{1}{2} - m - \frac{i\alpha}{k}|1 - 2m|-2ikz\right)$$

and the general solution of the Schroedinger equation for a particle in a coulomb potential field is, for $E = -k^2$,

$$\psi = Ae^{-ikz} z^n F\left(n + 1 + \frac{i\alpha}{k}|2n + 2|2ikz\right) + Be^{ikz} z^{-n-1} F\left(-n - \frac{i\alpha}{k}|-2n|-2ikz\right)$$

Many other functions (such as the error function and the incomplete gamma function) can be expressed in terms of the confluent hypergeometric function.

Asymptotic Series. Although we shall postpone a complete discussion of the behavior of the confluent hypergeometric function about $z = \infty$, we should investigate the series expansion of the solutions suitable about the singular point at $z = \infty$. Making the transformation $w = 1/x$ in Eq. (5.2.57), we see that the confluent hypergeometric equation has the form

$$\frac{d^2F}{dw^2} + \left[\frac{2 - c}{w} + \frac{1}{w^2}\right]\frac{dF}{dw} - \frac{a}{w^3}F = 0 \qquad (5.2.64)$$

about the point at infinity. Although this is an irregular singular point, there is one solution of the indicial equation, $s = a$. Inserting $F = \Sigma a_n w^{a+n}$ into Eq. (5.2.64) we find again a two-term recursion formula for the a's

$$(n + 1)a_{n+1} + (n + a)(n + a - c + 1)a_n = 0$$

Therefore a series expansion for F about $w = 0$ ($z = \infty$) is

$$F_1 = Az^{-a}\left[1 - \frac{a(a - c + 1)}{1!}\frac{1}{z} + \frac{a(a + 1)(a - c + 1)(a - c + 2)}{2!} \cdot \frac{1}{z^2} - \cdots\right] \qquad (5.2.65)$$

where we have inserted $1/z$ for w again. This should be compared with Eq. (5.2.63). Comparison with Eq. (5.2.60) suggests that a second solution should be of the form $w^{c-a}e^{1/w}\Sigma b_n w^n$. Solving for the b's results in a two-term recursion formula similar to the one above. The second series solution about $z = \infty$ is then

$$F_2 = Bz^{a-c}e^z\left[1 + \frac{(1 - a)(c - a)}{1!}\frac{1}{z} + \frac{(1 - a)(2 - a)(c - a)(c - a + 1)}{2!}\frac{1}{z^2} + \cdots\right]$$

The chief trouble with these two series solutions is that *they do not converge*, except at $z = \infty$. The divergence is of a peculiar type, however, for when z is large but finite, the series *first converges* and then, as we take more and more terms, *eventually diverges*. To be specific, we find that the difference $\Delta_n(z)$ between F_1 and the first n terms of the series in Eq. (5.2.65) first diminishes as n is increased and then increases without limit. For small values of z the minimum value for Δ_n comes at a relatively small value of n, and this minimum value is relatively large. As z is increased, the minimum for Δ_n comes for larger and larger n and the value of the minimum gets smaller and smaller. For any finite value of z, therefore, it is possible to obtain a fairly accurate value of F by taking a finite number of terms, whereas many more terms would result in a less accurate answer. As long as z is finite, there is a certain irreducible error in the computed result even if the optimum number of terms is used, but this error rapidly diminishes as z increases. In many cases of interest, this irreducible error is already smaller than 0.1 by the time z is as large as 10. In many such cases the first term only in the expansion is a satisfactory approximation for $z > 20$.

Such series, which are divergent but which can be used to compute a value that never exactly equals the "correct" value but rapidly approaches this "correct" value as z is increased, are called *asymptotic series.* They were discussed in some detail in Sec. 4.6. In some respects they turn out to be more useful than convergent series as long as they are handled with tact and understanding. Some of the requisite understanding will be supplied in next section; the tact must be left to the user.

Two Regular, One Irregular Singular Point. The equation for ξ_1 and ξ_2 for elliptic cylinder coordinates has the form ($\psi = X_1, X_2$, $z = \xi_1/d, \xi_2$)

$$(z^2 - 1)\psi'' + z\psi' + (h^2z^2 - b)\psi = 0$$

and the equations for ξ_1 and ξ_2 for prolate and oblate spheroidal coordinates have the form $[X = (z^2 - 1)^{\frac{1}{2}a}\psi]$

$$(z^2 - 1)\psi'' + 2(a + 1)z\psi' + (h^2z^2 - b)\psi = 0 \qquad (5.2.66)$$

The first equation is but a special case of the second equation ($a = -\frac{1}{2}$). To show the singular points we rewrite the second equation as

$$\frac{d^2\psi}{dz^2} + \left[\frac{a + 1}{z - 1} + \frac{a + 1}{z + 1}\right]\frac{d\psi}{dz} + \left[h^2 + \frac{h^2 - b}{z^2 - 1}\right]\psi = 0$$

It has regular singular points at $z = \pm 1$, with indices 0 and $-a$ at both and an irregular singular point at $z = \infty$. The equation is not the most general one having two regular and one irregular singular points, but it is the equation which is encountered in our work. The singular points are at standard positions (we could change the regular ones to 0 and 1, but ± 1 is more convenient), and one solution at each regular point is analytic, as we have previously required for a canonical form; therefore we shall consider Eq. (5.2.66) as the canonical form for this type of equation.

If we insert a power series for z in Eq. (5.2.66), we obtain the three-term recursion formula

$$(n + 1)(n + 2)a_{n+2} + [b - n(n + 2a + 1)]a_n + ha_{n-2} = 0$$

from which we can obtain the two fundamental solutions about the regular points $z = 0$. The series expansions about the singular points, in powers of $1 - z$ or $1 + z$, produce four-term recursion formulas, which are even more difficult to compute and analyze.

In such cases we try expansions in series of appropriate functions rather than in series of powers of $(1 \pm z)$. For instance, by transforming the independent variable in the elliptic cylinder case ($a = -\frac{1}{2}$) we can obtain

$$z = \cos\phi; \quad \frac{d^2\psi}{d\phi^2} + (b - h^2\cos^2\phi)\psi = 0$$

$$z = \tfrac{1}{2}\left(x + \frac{1}{x}\right); \quad x = e^{i\phi} \tag{5.2.67}$$

$$x^2\frac{d^2\psi}{dx^2} + x\frac{d\psi}{dx} + \left(\frac{h^2}{4}x^2 + \frac{h^2}{2} - b + \frac{h^2}{4}\frac{1}{x^2}\right)\psi = 0$$

The first of these equations is called *Mathieu's equation.* The second is quite interesting, for it is algebraic in form but the transformation from z to x has changed the singular points so it now has *two irregular singular points*, one at 0 and the other at ∞.

The first form of the equation can be used to bring out an interesting property of the solutions. Since $\cos^2\phi$ is periodic in ϕ with period π, if $\psi(\phi)$ is a solution of the equation, then $\psi(\phi + \pi)$ is also a solution. For instance, if ψ_1 and ψ_2 are two independent solutions, we shall have, in general,

$$\psi_1(\phi + \pi) = \alpha_{11}\psi_1(\phi) + \alpha_{12}\psi_2(\phi)$$

and

$$\psi_2(\phi + \pi) = \alpha_{21}\psi_1(\phi) + \alpha_{22}\psi_2(\phi)$$

where the α's are constants determined by the parameters b and h and by the particular set of solutions ψ_1 and ψ_2.

From this fact it can be shown that it is possible to find a solution of Eq. (5.2.67) which has the form $e^{is\phi}$ times a function which is periodic in ϕ. This solution (call it Ψ) would be, of course, some combination of ψ_1 and ψ_2, $\Psi = A\psi_1(\phi) + B\psi_2(\phi) = e^{is\phi}F(\phi)$, where F is periodic in ϕ with period π; that is, $F(\phi + \pi) = F(\phi)$. Using the properties of the ψ's we have that

$$\begin{aligned} e^{is(\phi+\pi)}F(\phi + \pi) &= A\psi_1(\phi + \pi) + B\psi_2(\phi + \pi) \\ &= (A\alpha_{11} + B\alpha_{21})\psi_1(\phi) + (A\alpha_{12} + B\alpha_{22})\psi_2(\phi) \\ &= e^{\pi is}\,e^{is\phi}F(\phi) \\ &= e^{\pi is}[A\psi_1(\phi) + B\psi_2(\phi)] \end{aligned}$$

Equating the coefficients of $\psi_1(\phi)$ and $\psi_2(\phi)$ we have two simultaneous equations for A and B and $e^{\pi is}$:

$$A(\alpha_{11} - e^{\pi is}) + B\alpha_{21} = 0; \quad A\alpha_{12} + B(\alpha_{22} - e^{\pi is}) = 0$$

For this to be soluble, the determinant of the coefficients must be zero:

$$\begin{vmatrix} (\alpha_{11} - e^{\pi is}) & \alpha_{21} \\ \alpha_{12} & (\alpha_{22} - e^{\pi is}) \end{vmatrix} = 0$$

This is a quadratic equation in $e^{\pi i s}$ with two roots, corresponding to two independent solutions of Eq. (5.2.66). The possibility of such solutions of Eq. (5.2.67) is called *Floquet's theorem.*

Referring back to the second form of Eq. (5.2.67), Floquet's theorem states that it should be possible to set up two independent solutions of the form $x^s(x = e^{i\phi})$ times a Laurent series in x^2, for a Laurent series in x^2 is a series of positive and negative integral powers of x, that is, a Fourier series in 2ϕ. Such a series is periodic in ϕ, with period π, as we have just shown it must be, and it can represent the function around both irregular singular points $x = 0$ and $x = \infty$. We therefore choose as a solution

$$\psi = \sum_{n=-\infty}^{\infty} a_n x^{s+2n} = e^{is\phi} \sum_{n=-\infty}^{\infty} a_n e^{2in\phi} \tag{5.2.68}$$

Setting this into the second Eq. (5.2.67) we arrive at the basic recursion formula

$$\begin{aligned} h^2 a_{n+1} + [2h^2 - 4b + 16(n + \tfrac{1}{2}s)^2]a_n + h^2 a_{n-1} &= 0 \\ \delta^2(a_n) + \frac{4}{h^2}[h^2 - b + 4(n + \tfrac{1}{2}s)^2]a_n &= 0 \end{aligned} \tag{5.2.69}$$

This is a three-term recursion formula for the unknowns a_n and s. It would be much more desirable to make other transformations of dependent or independent variable to obtain a two-term recursion formula. Unfortunately, as will be shown later in this section, such a pleasant outcome is not possible for equations of this complexity, so we are forced to undertake the analysis of three-term recursion formulas.

If we start with arbitrarily chosen values of a_0, a_1, and s, we can compute all the other a's for positive and negative n's. But in such a case the a's will not necessarily become smaller as n increases, so the series will not, in general, converge. Only for certain values of s and of a_1/a_0 will the series converge, and we must find a way to compute these certain values.

Continued Fractions. First, of course, we must make sure that the series *can* converge for some values of a_1/a_0 and of s. To do this we compute the value of a_n/a_{n-1} for large positive n and of a_n/a_{n+1} for large negative n. If these approach zero as $n \to \pm\infty$, we can be sure the series converges over the range $0 < x < \infty$.

$$\begin{aligned} \frac{a_n}{a_{n-1}} &= \frac{-h^2}{16(n + \frac{1}{2}s)^2 + 2h^2 - 4b + h^2(a_{n+1}/a_n)} \\ \frac{a_n}{a_{n+1}} &= \frac{-h^2}{16(n + \frac{1}{2}s)^2 + 2h^2 - 4b + h^2(a_{n-1}/a_n)} \end{aligned} \tag{5.2.70}$$

The first of these equations shows that, when n is large and positive, a_n/a_{n-1} is small and approaches zero proportional to $1/n^2$ *if the next ratio* a_{n+1}/a_n *also approaches zero* for large n, for then for large enough n the term $16(n + \frac{1}{2}s)^2$ is much larger than all the other terms in the denominator and $a_n/a_{n-1} \to -(h^2/16n^2)$. Use of the methods of page 388 indicates that for large values of x (ϕ large and along the negative imaginary axis) ψ is approximately proportional to $x^s \cos(hx/4)$.

Likewise the second of Eqs. (5.2.70) shows that, when a_{n-1}/a_n is small for large negative values of n, then $a_n/a_{n+1} \to -(h^2/16n^2)$. Therefore for very small values of x (ϕ large and along the positive imaginary axis) ψ is approximately proportional to $x^s \cos(h/4x)$. However, we still must find out how to compute the ratios a_n/a_{n-1} for small values of n and also how to determine the correct value of s.

Equations (5.2.70) can give us the hint which sets us on the right track, however. If we cannot start from a_0, a_1 and work outward, perhaps we can start from very large values of n and work inward. Suppose that we start from a value of n that is so large that a_{n+1}/a_n is very nearly $-h^2/16(n + \frac{1}{2}s)^2$. Substituting in the first equation gives a fairly accurate expression for a_n/a_{n-1}; then substituting this in again into the equation for a_{n-1}/a_{n-2} gives us a still more accurate expression for a_{n-1}/a_{n-2}; and so on, until we arrive at a value for a_1/a_0

$$\frac{a_1}{a_0} = \cfrac{-h^2}{16(1 + \frac{1}{2}s)^2 + 2h^2 - 4b - \cfrac{h^4}{16(2 + \frac{1}{2}s)^2 + 2h^2 - 4b - \cfrac{h^4}{16(3 + \frac{1}{2}s)^2 + 2h^2 - 4b - \cdots}}}$$

Similar use of the second equation for negative n's gives us

$$\frac{a_{-1}}{a_0} = \cfrac{-h^2}{16(1 - \frac{1}{2}s)^2 + 2h^2 - 4b - \cfrac{h^4}{16(2 - \frac{1}{2}s)^2 + 2h^2 - 4b - \cfrac{h^4}{16(3 - \frac{1}{2}s)^2 + 2h^2 - 4b - \cdots}}}$$

These expressions are called *continued fractions.* Questions of convergence can be answered by adaptation of rules for series.

Using Eq. (5.2.69) for $n = 0$ gives us a formula relating a_1/a_0 and a_{-1}/a_0:

$$s^2 = b - \frac{h^2}{4}\left[2 + \frac{a_1}{a_0} + \frac{a_{-1}}{a_0}\right] \tag{5.2.71}$$

This, together with the continued fraction equations above, results in an equation from which s can be determined in terms of b and h. We compute the continued fractions for assumed values of s and then check it by means of Eq. (5.2.71). If it does not check exactly, we use the square root of the right-hand side of the equation for a new value of s to insert in the continued fraction. Unless we have made too bad a guess at first, this process of reinsertion converges rapidly and a value correct to five or six places can usually be obtained in less than a dozen runs.

Of course, if h is small, we can perform this iterative process analytically. To the first order in h^2, we have $s = \sqrt{b} - (h^2/4\sqrt{b})$. Inserting this into the two continued fractions (and omitting the h^4 terms in them), we have

$$s^2 \simeq b - \tfrac{1}{2}h^2 + \frac{(h^4/64)}{1+\sqrt{b}} + \frac{(h^4/64)}{1-\sqrt{b}}$$

or

$$s \simeq \sqrt{b}\left[1 - \frac{h^2}{4b} - \frac{h^4}{64b^2}\left(\frac{2-3b}{1-b}\right)\right]$$

unless b is nearly equal to 1, in which case higher powers of h^2 must be included.

We note, from the symmetry of Eq. (5.2.71) and the continued fractions [or from Eq. (5.2.68)], that, if s is a solution then $-s$ is also a solution; indeed $\pm s \pm 2m$, where m is any integer, is also a solution.

When s is computed, a_1 and a_{-1} can be calculated in terms of a_0 (which can be set equal to 1) and the other a's can be computed by the auxiliary continued fractions

$$\frac{a_n}{a_{n-1}} = \frac{-h^2}{16(\frac{1}{2}s+n)^2 + 2h^2 - 4b - \dfrac{h^4}{16(\frac{1}{2}s+n+1)^2 + 2h^2 - 4b - \cdots}};\quad \text{etc.}$$

$$\frac{a_{-n}}{a_{-n+1}} = \frac{-h^2}{16(\frac{1}{2}s-n)^2 + 2h^2 - 4b - \dfrac{h^4}{16(\frac{1}{2}s-n-1)^2 + 2h^2 - 4b - \cdots}};\quad \text{etc.}$$

and therefore the whole series can be computed. The corresponding function

$$\mathcal{S}(b,h,e^{i\phi}) = e^{is\phi}\sum_{n=-\infty}^{\infty} a_n e^{2in\phi} \tag{5.2.72}$$

using these values of s and the a's is one solution of Eq. (5.2.66). The other solution is for the opposite sign of s, where a_n and a_{-n} change places:

$$\mathcal{S}(b,h,e^{-i\phi}) = e^{-is\phi}\sum_{n=-\infty}^{\infty} a_n e^{-2in\phi}$$

which is the complex conjugate of the first series. The corresponding real functions are obtained by addition or subtraction:

$$\begin{aligned} Se(b,h;z) &= \sum_{n=-\infty}^{\infty} a_n \cos[(s+2n)\phi]; \quad z = \cos\phi; \\ So(b,h;z) &= \sum_{n=-\infty}^{\infty} a_n \sin[(s+2n)\phi] \end{aligned} \tag{5.2.73}$$

These functions are even or odd about $\phi = 0$ but are not periodic in ϕ with period π or 2π unless s is an integer or zero.

For certain ranges of values of b (for negative values of b, for instance, or for b near 1 or near 4 or near 9, etc.) s turns out to be a complex number. In this case the real solutions Se and So have somewhat more complicated forms. The larger h is, the larger the ranges of b where s is complex. These ranges are called the *ranges of instability* of the solutions, so named because the real exponential factor, which then is present in the solutions, becomes extremely large for large values of z, positive or negative.

The Hill Determinant. Before we continue with our discussion of the solution of Mathieu's equation, we shall indicate a quite different method of computing s and the coefficients a_n, which is successful because of the particular symmetry of the recursion formulas (5.2.69). These equations are, of course, a set of homogeneous simultaneous differential equations for the a's (an infinite number because the a's are infinite). In order that they be solved to find the a_n's in terms of a_0, for instance, the determinant of the coefficients must be zero. This is an infinite determinant, so that we must watch its convergence. However, we can divide the nth recursion formula by $2h^2 - 4b + 16n^2$ before we form the determinant, which will improve convergence. The resulting determinant

$$\Delta(s) = \begin{vmatrix} \cdot & \cdot & \cdot & \cdot & \cdot & \cdot \\ \cdot & \dfrac{(\sigma+2)-\alpha^2}{4-\alpha^2} & \dfrac{\beta^2}{4-\alpha^2} & 0 & 0 & \cdot \\ \cdot & \dfrac{\beta^2}{1-\alpha^2} & \dfrac{(\sigma+1)^2-\alpha^2}{1-\alpha^2} & \dfrac{\beta^2}{1-\alpha^2} & 0 & \cdot \\ \cdot & 0 & \dfrac{\beta^2}{-\alpha^2} & \dfrac{\sigma^2-\alpha^2}{-\alpha^2} & \dfrac{\beta^2}{-\alpha^2} & \cdot \\ \cdot & 0 & 0 & -\dfrac{\beta^2}{1-\alpha^2} & \dfrac{(\sigma-1)^2-\alpha^2}{1-\alpha^2} & \cdot \\ \cdot & \cdot & \cdot & \cdot & \cdot & \cdot \end{vmatrix} \tag{5.2.74}$$

where $\sigma = \frac{1}{2}s$, $\alpha^2 = \frac{1}{4}b - \frac{1}{8}h^2$, and $\beta = h/4$, is called *Hill's determinant.* All we have to do is to solve it for σ!

Astonishingly enough such solution is possible because of the periodic nature of the dependence on α, which relates Δ to the trigonometric functions. First of all we simplify our determinant by multiplying the nth row (measured from the row $n = 0$) by $[n^2 - \alpha^2]/[(\sigma + n)^2 - \alpha^2]$, and so on, obtaining a new determinant $D(\sigma)$, where

$$\Delta(s) = D(\sigma) \prod_{n=-\infty}^{\infty} \frac{(\sigma+n)^2 - \alpha^2}{n^2 - \alpha^2}$$

$$= D\,\frac{\sigma^2 - \alpha^2}{-\alpha^2} \prod_{n=1}^{\infty} \frac{[1 - \{(\sigma+\alpha)/n\}^2][1 - \{(\sigma-\alpha)/n\}^2]}{[1 - (\alpha/n)^2]^2}$$

The determinant $D(\sigma)$ has a sequence of unities along the main diagonal, flanked by a sequence of the sort $\beta^2/[(\sigma + n)^2 - \alpha^2]$ on both neighboring diagonals, with all other terms zero. Referring to Eq. (4.3.9) we see that

$$\Delta(s) = -D(\sigma)\frac{\sin \pi(\sigma + \alpha) \sin \pi(\sigma - \alpha)}{\sin^2(\pi\alpha)} = D(\sigma)\frac{\sin^2(\pi\alpha) - \sin^2(\pi\sigma)}{\sin^2(\pi\alpha)} \tag{5.2.75}$$

which displays some of the periodic dependence on α and σ.

But determinant $D(\sigma)$ also has definite periodicity in σ, having simple poles at $\sigma = \pm n \pm \alpha$, due to the elements $\beta^2/[(\sigma + n)^2 - \alpha^2]$. In fact the only poles of $D(\sigma)$ are at these points, and the function also is bounded at infinity ($D \to 1$ as $\sigma \to \infty$), so it is an obvious function on which to practice the legerdemain of pages 381 to 385. We first subtract off all the poles:

$$K(\sigma) = D(\sigma) - C \sum_{n=-\infty}^{\infty} \frac{1}{(\sigma + n)^2 - \alpha^2}$$
$$= D(\sigma) - \frac{C}{2\alpha} \sum_{n=-\infty}^{\infty} \left[\frac{1}{\sigma + \alpha + n} - \frac{1}{\sigma - \alpha + n}\right]$$

where C equals the residue at each of the poles of D. However analysis of the sort resulting in Eq. (4.3.7) (see Prob. 4.21, page 474) gives

$$\pi \cot(\pi x) = \frac{1}{x} + \frac{2x}{x^2 - 1} + \frac{2x}{x^2 - 4} + \cdots = \sum_{n=-\infty}^{\infty} \frac{1}{x + n}$$

so that the function $K(\sigma)$ can be rewritten. This function

$$K(\sigma) = D(\sigma) - \frac{\pi C}{2\alpha} [\cot \pi(\sigma + \alpha) - \cot \pi(\sigma - \alpha)]$$

has no poles for any value of s and is bounded at $s \to \infty$. By Liouville's theorem (page 381) it must be a constant, and by letting $\sigma \to \infty$, we see the constant K is unity. Therefore, amazingly enough,

$$D(\sigma) = 1 + \frac{\pi C}{2\alpha} [\cot \pi(\sigma + \alpha) - \cot \pi(\sigma - \alpha)]$$

Returning to Eq. (5.2.75) we see that

$$\Delta(s) = 1 - \frac{\sin^2(\pi\sigma)}{\sin^2(\pi\alpha)} + \frac{\pi C}{\alpha} \cot(\pi\alpha)$$

where the only constant as yet undetermined is the constant C, the residue at the poles of $D(\sigma)$. This can be computed by setting $\sigma = 0$;

$(\pi C/\alpha)\cot(\pi\alpha) = \Delta(0) - 1$. Therefore, returning to the original notation, we have that the original determinant has the relatively simple functional dependence on s:

$$\Delta(s) = \Delta(0) - \frac{\sin^2(\pi s/2)}{\sin^2(\frac{1}{2}\pi\sqrt{b - \frac{1}{2}h^2})} \tag{5.2.76}$$

where

$$\Delta(0) = \begin{vmatrix} \cdot & \cdot & \cdot & \cdot & \cdot & \cdot \\ \cdot & 1 & \frac{h^2}{144+2h^2-4b} & 0 & 0 & \cdot \\ \cdot & \frac{h^2}{64+2h^2-4b} & 1 & \frac{h^2}{64+2h^2-4b} & 0 & \cdot \\ \cdot & 0 & \frac{h^2}{16+2h^2-4b} & 1 & \frac{h^2}{16+2h^2-4b} & \cdot \\ \cdot & 0 & 0 & \frac{h^2}{2h^2-4b} & 1 & \cdot \\ \cdot & 0 & 0 & 0 & \frac{h^2}{16+2h^2-4b} & \cdot \\ \cdot & \cdot & \cdot & \cdot & \cdot & \cdot \end{vmatrix}$$

is a convergent determinant which is independent of s. Since $\Delta(s)$ is to be zero, the equivalent of Eq. (5.2.71) for determining s is

$$\sin^2(\pi s/2) = \Delta(0)\sin^2(\tfrac{1}{2}\pi\sqrt{b - \tfrac{1}{2}h^2})$$

Mathieu Functions. We are now ready to return to our previous discussion of the allowed values of s and the solutions S. The quantity $\Delta(0)\sin^2(\frac{1}{2}\pi\sqrt{b - \frac{1}{2}h^2})$ is a periodic function of $\alpha = \frac{1}{2}\sqrt{b - \frac{1}{2}h^2}$ with period 1. For $h = 0$, $\Delta = 1$ and $s = \pm 2\alpha = \pm\sqrt{b}$; this is the limiting case where Eq. (5.2.66) reduces to $(d^2\psi/d\phi^2) + b\psi = 0$. When $b - \frac{1}{2}h^2$ is large enough negative, $\Delta(0)\sin^2(\pi\alpha)$ is negative and s is pure imaginary. The whole range of values of h and b shown shaded, to the left of line 0 in Fig. 5.4, corresponds to unstable solutions, with a real exponential factor.

For some value of b, depending on the value of h and plotted as curve 0 in Fig. 5.4, $\Delta(0)\sin^2(\pi\alpha)$ is zero, so that s is zero. In this case the solution $\mathcal{S}(b,h;e^{i\phi})$ is real, symmetrical in ϕ, and *periodic in ϕ with period π* [for $a_n = a_{-n}$ and $\mathcal{S}$ is a Fourier series in $\cos(2n\phi)$]. This function is called the *Mathieu function of zero order* and is given the special symbol

$$Se_0(h,z) = \sum_{n=\infty}^{\infty} B_{2n}\cos(2n\phi); \quad z = \cos\phi$$

where the B's are proportional to the a's, but adjusted so that $Se_0(h,1) = 1$. When $s = 0$, the two solutions $\mathcal{S}(b,h;e^{i\phi})$ and $\mathcal{S}(b,h;e^{-i\phi})$ are equal and the second solution, independent of Se_0, must be obtained by means of Eq. (5.2.6). It has a logarithmic term in it and hence is not periodic in ϕ.

Over the range of b and h indicated in Fig. 5.4 by the unshaded area between the line marked 0 and the line 1_0 expression $\Delta(0)\sin^2(\pi\alpha)$ is less than unity and thus there is a solution for s which is real and less than unity. *Se* and *So* as given in Eq. (5.2.73) are independent solutions, and the best way to compute s and the coefficients a_n is by means of the continued fraction of Eq. (5.2.71).

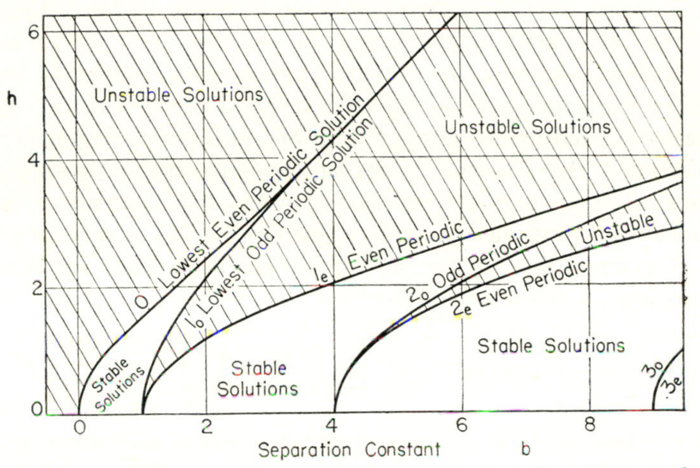

Fig. 5.4 Values of separation constants for periodic solutions of the Mathieu equation.

For the set of values of b and h given by the curve marked as 1_0, $\Delta(0)\sin^2(\pi\alpha)$ and s are unity. It turns out that $a_n = -a_{-n-1}$ so that both $\mathcal{S}(b,h,e^{i\phi})$ and $\mathcal{S}(b,h,e^{-i\phi})$ are proportional to

$$So_1(h,z) = \sum_{n=0}^{\infty} B_{2n+1}\sin(2n+1)\phi$$

which is called the *odd Mathieu function of the first order*. Above this, in the next shaded region, $\Delta(0)\sin^2(\pi\alpha)$ is larger than unity, s is complex, having the value $1 + \epsilon i$, and consequently the solution is unstable. At the upper edge of this region of instability s is again unity, but this time $a_n = a_{-n-1}$, so that both $\mathcal{S}$ solutions are proportional to

$$Se_1(h,z) = \sum_{n=0}^{\infty} B_{2n+1}\cos(2n+1)\phi$$

the *even Mathieu function of the first order*. The second solution again has a logarithmic term.

This behavior continues for increasing values of b: alternating regions of stability and instability, the boundary lines between, corresponding to the special case where s is an integer and where one solution is periodic, either an even function (using cosines in the Fourier series) or an odd function (using sines), and the other solution is nonperiodic, containing a logarithm. For the rest of the range of b and h, exclusive of the boundary lines, the solution is nonperiodic but oscillating, with solutions of the form given in Eq. (5.2.73), or else nonperiodic and unstable, with solutions containing a complex exponential factor times a Fourier series.

In many cases of physical interest the coordinate corresponding to ϕ is a periodic one, returning on itself as ϕ increases by 2π. This being the case the only useful solutions are the periodic ones, which we have called *Mathieu functions*, for integral values of s (one odd function So_m and one even function Se_m for each integral value m of s). When h goes to zero Se_m reduces to $\cos(m\phi)$ and So_m reduces to $\sin(m\phi)$.

To compute the allowed values of the separation constant b corresponding to these periodic functions we can use, instead of the Hill determinant, the continued fraction equation (5.2.71) in reverse to find b when s is given. We let $s = m$, an integer, and find, by successive approximations, a consistent solution of

$$b = m^2 + \tfrac{1}{4}h^2 \left[2 + \frac{a_1}{a_0} + \frac{a_{-1}}{a_0} \right]$$

where the ratios a_1/a_0, a_{-1}/a_0 are given in terms of continued fractions on page 559. Except for $m = 0$, there are two different solutions for each value of m, one giving rise to a sine series and one giving a cosine series (that is, $a_n = \pm a_{2m-n}$).

When $s = 0$, $a_1/a_0 = a_{-1}/a_0$, and we solve the equation

$$2b = h^2 - \cfrac{h^4}{16 + 2h^2 - 4b - \cfrac{h^4}{16 \cdot 4 + 2h^2 - 4b - \cdots}}; \quad \text{etc.}$$

then $$Se_0(h,z) = \sum_{n=0}^{\infty} B_{2n} \cos(2n\phi); \quad B_{2n} = a_n \Big/ \left[\sum_{n=0}^{\infty} a_n \right]$$

where the coefficients D are normalized so that $Se_0 = 1$ for $\phi = 0$. The allowed value of b for this case can be labeled $be_0(h)$.

However if we are interested only in the Mathieu functions, the periodic solutions, we can simplify the calculations considerably by making use of the fact that the solutions are Fourier series. We transform the first of Eqs. (5.2.67) into

$$(d^2\psi/d\phi^2) + (b - \tfrac{1}{2}h^2 - \tfrac{1}{2}h^2 \cos 2\phi)\psi = 0$$

As we have been showing, the periodic solutions of this equation are of four different types:

I. Even solutions of period π, s = even integer $= 2m$, allowed values of $b = be_{2m}$

$$Se_{2m}(h, \cos\phi) = \sum_{n=0}^{\infty} B_{2n} \cos(2n\phi)$$

II. Even solutions of period 2π, s = odd integer $= 2m + 1$, for $b = be_{2m+1}$

$$Se_{2m+1}(h, \cos\phi) = \sum_{n=0}^{\infty} B_{2n+1} \cos(2n+1)\phi$$

III. Odd solutions of period π, s = even integer $= 2m$, for $b = bo_{2m}$

$$So_{2m}(b, \cos\phi) = \sum_{n=1}^{\infty} B_{2n} \sin(2n\phi)$$

IV. Odd solutions of period 2π, s = odd integer $= 2m + 1$, for $b = bo_{2m+1}$

$$So_{2m+1}(h, \cos\phi) = \sum_{n=0}^{\infty} B_{2n+1} \sin(2n+1)\phi$$

where the coefficients B depend on h, on m (*i.e.*, on the value of s and the related value of b) and, of course, on n.

Inserting the Fourier series of type I into the differential equation and utilizing the identity $\cos a \cos b = \frac{1}{2}\cos(a + b) + \frac{1}{2}\cos(a - b)$, we have

$$B_2 = k_0 B_0;\quad B_4 = k_2 B_2 - 2B_0;\quad k_{2n} B_{2n} = B_{2n+2} + B_{2n-2}$$

where

$$k_m = h^{-2}(4b - 2h^2 - 4m^2)$$

From these equations, by rearranging into continued fractions, we can compute the ratio of coefficients and also the allowed value of b, which is be_{2m}. Letting the ratio be represented by

$$G_m = B_m/B_{m-2};\quad (1/G_0) = 0$$

we have two alternative sets of equations for the G's:

$$G_2 = \frac{2}{k_2 -}\,\frac{1}{k_4 -}\,\frac{1}{k_6 -}\cdots;\qquad G_{2n} = \frac{1}{k_{2n} -}\,\frac{1}{k_{2n+2} -}\,\frac{1}{k_{2n+4} -}\cdots;\quad n > 1 \tag{5.2.77}$$

or $G_2 = k_0;\quad G_4 = k_2 - (2/k_0)$

$$G_{2n} = k_{2n-2} - \frac{1}{k_{2n-4} -}\,\frac{1}{k_{2n-6} -}\cdots\frac{1}{k_2 - (2/k_0)};\quad n > 2 \tag{5.2.78}$$

either of which can be used, depending on the relative ease of calculation and speed of convergence.

Equating the two expressions for G_2 gives a continued fraction equation for determination of the allowed value of b. Setting $\alpha = b - \frac{1}{2}h^2$, $\theta = \frac{1}{4}h^2$, we have

$$\alpha = \cfrac{2\theta^2}{\alpha - 4 - \cfrac{\theta^2}{\alpha - 16 - \cfrac{\theta^2}{\alpha - 36 - \cdots}}}$$

which is equivalent to the equation of page 564. An infinite sequence of solutions of α as functions of θ can be found, from which values of be_{2m} can be determined. Some of the values are given in the tables. Values are also shown graphically in Fig. 5.4.

For solutions of type II, by the same methods, we can arrive at the following equations for the coefficient ratios G and the separation constant $be_{2m+1} = \alpha + \frac{1}{2}h^2$:

$$G_3 = \frac{B_3}{B_1} = k_1 - 1; \quad G_{2n+1} = k_{2n-1} - \cfrac{1}{k_{2n-3} - \cfrac{1}{k_{2n-5} - \cdots - \cfrac{1}{k_1 - 1}}}; \quad n > 1$$

$$G_{2n-1} = \cfrac{1}{k_{2n-1} - \cfrac{1}{k_{2n+1} - \cfrac{1}{k_{2n+3} - \cdots}}}; \quad n > 0$$

$$\alpha = 1 + \theta + \cfrac{\theta^2}{\alpha - 9 - \cfrac{\theta^2}{\alpha - 25 - \cfrac{\theta^2}{\alpha - 49 - \cdots}}}$$

In both of these cases it is convenient to normalize the function so that $Se_m = 1$ at $\phi = 0$. This means that $\Sigma B_n = 1$.

For type III solutions we have $B_0 = 0$, and the equations for the G's and bo's are

$$\frac{1}{G_2} = 0; \quad G_4 = k_2; \quad G_{2n} = k_{2n-2} - \cfrac{1}{k_{2n-4} - \cdots - \cfrac{1}{k_2}}; \quad n > 2$$

$$G_{2n} = \cfrac{1}{k_{2n} - \cfrac{1}{k_{2n+2} - \cdots}}; \qquad k_m = \frac{\alpha - m^2}{\theta}$$

$$\alpha = 4 + \cfrac{\theta^2}{\alpha - 16 - \cfrac{\theta^2}{\alpha - 36 - \cdots}};$$

Finally for type IV solutions we have

$$G_3 = k_1 + 1; \quad G_{2n+1} = k_{2n-1} - \cfrac{1}{k_{2n+1} - \cdots - \cfrac{1}{k_1 + 1}}; \quad n > 1$$

$$G_{2n-1} = \cfrac{1}{k_{2n-1} - \cfrac{1}{k_{2n+1} - \cdots}}; \qquad n > 0$$

$$\alpha = 1 - \theta - \cfrac{\theta^2}{\alpha - 9 - \cfrac{\theta^2}{\alpha - 25 - \cdots}}$$

For both sine series it is convenient to normalize so that the slope of the function, $dSo_m/d\phi$, is unity at $\phi = 0$. This means that $\Sigma nB_n = 1$.

When computing values of α (and thus of b) we can juggle the continued fractions to ease the work. For instance, for the type I solutions for the value of α near 16, we can invert Eq. (5.2.77) twice, to obtain

$$\alpha = 16 + \frac{\theta^2}{\alpha - 4 - (2\theta^2/\alpha)} + \frac{\theta^2}{\alpha - 36 - \dfrac{\theta^2}{\alpha - 64 - \cdots}}$$

It turns out that (unless h is quite large) the values of B_n are largest for $n \simeq m$. Consequently $G_n = B_n/B_{n-2}$ is small for $n > m$ and is large for $n < m$. Experience will show that the finite continued fractions, like Eq. (5.2.77), are best to use when computing G_n for $n < m$ and the infinite fractions, like Eq. (3.2.78), are best for the values of n larger than m.

Mathieu Functions of the Second Kind. As we have mentioned on page 559, when s is an integer *both* $\mathcal{S}(b,h,e^{i\phi})$ *and* $\mathcal{S}(b,h,e^{-i\phi})$ are proportional to either Se_m or So_m, the Mathieu function of the first kind, corresponding to $b = be_m$ or bo_m, respectively. For these particular values of b the second solution has a logarithmic singularity in $z = e^{i\phi}$ (in other words, it is not periodic in ϕ), and we must set up special solutions for these special cases. We indicate here the method for obtaining the second solutions corresponding to the even functions $Se_{2m}(h, \cos\phi)$.

Since $\phi = 0$ is an ordinary point for the Mathieu equation

$$\psi'' + (b - \tfrac{1}{2}h^2 - \tfrac{1}{2}h^2 \cos 2\phi)\psi = 0$$

one should be able to set up a fundamental set of solutions, one with unit value and zero slope and the other with zero value and unit slope at $\phi = 0$. For $b = be_{2m}$, the first solution is $Se_{2m}(h, \cos\phi)$, which has unit value and zero slope. The second solution should be of the general form

$$Fe_{2m}(h, \cos\phi) = \gamma_{2m}\left[\phi\, Se_{2m}(h, \cos\phi) + \sum_{n=1}^{\infty} D_{2n}\sin(2n\phi)\right] \quad (5.2.79)$$

which has a logarithmic singularity in z at $\phi = \cos^{-1}(z) = 0$. It is also not periodic in ϕ. Inserting this into the Mathieu equation and remembering that $Se_{2m} = \Sigma B_{2n}\cos(2n\phi)$ is a solution of the same equation for the same value of b, we finally obtain

$$\gamma_{2m}\sum_{n=1}^{\infty}[-4nB_{2n} - (2n)^2 D_{2n} + \tfrac{1}{4}h^2 D_{2n-2} \\ + (be_{2n} - \tfrac{1}{2}h^2)D_{2n} + \tfrac{1}{4}h^2 D_{2n+2}]\sin(2n\phi) = 0$$

with the term in D_{2n-2} being absent when $n = 1$. This gives rise to a set of simultaneous equations

$$[be_2 - \tfrac{1}{2}h^2 - 4]D_2 + \tfrac{1}{4}h^2 D_4 = 4B_2$$
$$\tfrac{1}{4}h^2 D_2 + [be_{2n} - \tfrac{1}{2}h^2 - 16]D_4 + \tfrac{1}{4}h^3 D_6 = 8B_4; \quad \text{etc.}$$

from which the D's can be obtained in terms of the B's (this is not simple, but we can find a solution for which the series converges). We fix the value of the constant γ_{2m} by requiring that the slope of Fe at $\phi = 0$, which is $\gamma_{2m}[1 + \Sigma 2nD_{2n}]$, be unity,

$$\gamma_{2m} = \left[1 + \sum_{n=1}^{\infty} 2nD_{2n}\right]^{-1}$$

Therefore we have the fundamental set for ϕ,

$$\left.\begin{array}{ll} Se_{2m} = 1; & (d/d\phi)Se_{2m} = 0 \\ Fe_{2m} = 0; & (d/d\phi)Fe_{2m} = 1 \end{array}\right\} \phi = 0$$

The Wronskian $\Delta(Se,Fe)$, with respect to ϕ, is a constant, and therefore

$$Se_{2m}(h, \cos\phi)\frac{d}{d\phi}Fe_{2m}(h, \cos\phi) - Fe_{2m}(h, \cos\phi)\frac{d}{d\phi}Se_{2m}(h, \cos\phi) = 1$$

for all values of ϕ. See also Eq. (5.3.91).

The second solutions for the other Mathieu functions are obtained in a similar manner. For instance, for $b = bo_{2m+1}$, the second solution is

$$Fo_{2m+1}(h, \cos\phi) = \delta_{2m+1}\left[\phi So_{2m+1}(h, \cos\phi) + \sum_{n=0}^{\infty} D_{2n+1}\cos(2n+1)\phi\right]$$

with equations for the D's similar to those written above. In this case we normalize so that $Fo_{2m+1} = 1$ at $\phi = 0$ (it has a zero slope there), which results in an equation for

$$\delta_{2m+1} = \left[\sum_{n=0}^{\infty} D_{2n+1}\right]^{-1}$$

and the Wronskian for this pair of solutions equals -1.

Therefore we have (at least) indicated the form of the second solutions for those values of b for which $\mathcal{S}(b,h,e^{-i\phi})$ is not independent of $\mathcal{S}(b,h,e^{i\phi})$. For all other values of b the two $\mathcal{S}$ functions are independent and constitute a satisfactory pair.

More on Recursion Formulas. We are now in a position to be a little more sophisticated about series solutions of differential equations and their related recursion formulas. Suppose that we have a differential equation $\mathcal{L}(\psi) \equiv \psi'' + p\psi' + q\psi = 0$, for which we wish to obtain a series solution about one of its singular points. For ease in computation we shall make the origin coincide with the singular point in question, which can be done without disturbing the other singular points. Then p or q or both have poles at $z = 0$. If p has no more than a simple pole and q no more than a pole of second order at $z = 0$, then the singular point there is regular, and we can, if we wish, expand the solution directly in a double power series in z. Each series is of the form $z^s \Sigma a_n z^n$, where s is one of the two roots of the indicial equation

$$s^2 + (P - 1)s + Q = 0$$

where $P = \lim_{z\to 0} zp(z)$ and $Q = \lim_{z\to 0} z^2 q(z)$. The coefficient of z^{n+s} in the series resulting when $\mathfrak{L}(\psi)$ is applied to the series $\Sigma a_n z^{n+s}$ is the *recursion formula* $D_n(a_n)$ for the power series about the singularity at $z = 0$. It, with its fellows, for other values of n, constitutes an infinite sequence of linear equations for the unknown coefficients a_n. If either p or q, or both, is a function of z which requires an infinite series to represent it, then each recursion formula D_n involves all the a's from a_0 to a_n (or perhaps even higher).

In principle these simultaneous equations can always be solved, to obtain the ratios between a_n and a_0. But if the recursion formulas have more than two terms apiece (*i.e.*, involve more than a pair of adjacent a's), then the task of computing the series and testing it for convergence, for asymptotic behavior, and so on, becomes very much more difficult. Let us see what we can say about the possibility of obtaining a set of two-term formulas.

Short recursion formulas can be obtained only when p and q are rational functions of z, *i.e.*, are ratios of polynomials in z (see page 382). If they are not rational functions, we can try to transform the independent variable so that the new p and q are rational functions: if this can be done, we can then proceed; if not, we are left to struggle with infinite recursion formulas.

The denominators of p and q, if they are rational functions, are determined by the position of the singular points of the equation. At least one of the two denominators has z as a factor, for one of the two functions has a pole at $z = 0$. If there are other singular points for finite values of z (say at $z = z_i$, $i = 1, 2, \ldots, N$), then there must be factors of the form $(z - z_i)$ in the denominators of either p or q or both. At any rate, when we clear the equation $\mathfrak{L}(\psi) = 0$ of fractions, it will have the form

$$\prod_{i=0}^{N} (z - z_i)^{n_i}\psi'' + F(z)\psi' + G(z)\psi = 0; \quad M = \sum_{i=0}^{N} n_i; \quad z_0 = 0$$

where F and G are polynomials in z. Parenthetically, if there is to be no singular point at infinity, $G(z)$ must be a polynomial in z of order $M - 4$ or less and F must be a polynomial of order $M - 1$, with its highest order term equal to $2z^{M-1}$(why?).

It is not difficult to see that, in general, such an equation will have an M-term recursion formula. If the point at infinity is a regular point, this can be reduced by one or two terms by transforming the independent variable [by the transformation $w = z/(z - z_j)$] so that the jth singular point is placed at infinity. When this is done (if it can be done) and there is a singular point at zero and at infinity, the equation will still have the form given above but the order of the polynomials multiplying ψ'', ψ', and ψ will be the smallest possible for the particular equation chosen.

We now can see that usually *it is possible to have a two-term recursion formula only if there is but one other singular point* beside the ones at zero and infinity, for the coefficient of ψ'' must have the form $z^{n_0}(z - z_1)$, F must have the form as $az^{n_0} + bz^{n_0-1}$, and G the form $\alpha z^{n_0-1} + \beta z^{n_0-2}$ in order that the powers of z arrange themselves in the series for $\mathfrak{L}(\psi) = 0$ so that two-term recursion formulas will result. One other, rather unlikely, case is for there to be *two* other singular points, symmetrically placed (that is, $z_2 = -z_1$) so that the coefficient of ψ'' is $z^{n_0}(z^2 - z_1^2)$. If then $F = az^{n_0+1} + bz^{n_0-1}$ and G has the unlikely form $\alpha z^{n_0} + \beta z^{n_0-2}$, a two-term recursion formula results, relating a_n and a_{n+2} (rather than a_n and a_{n+1}).

Even if there is only one other singular point, the functions F and G may not have the requisite simple forms. In this case we can sometimes help by transforming the dependent variable, letting $\psi = u(z)f(z)$, where u is some power of z times some power of $(z - z_1)$. Usually the proper powers are one of the indices s at each of the singular points, so that the new dependent variable f has one analytic solution both at $z = 0$ and $z = z_1$. This often reduces the order of the polynomial G and produces a two-term recursion formula. It is the trick which was used successfully on the Papperitz equation, to result in the hypergeometric equation.

When more than one singular point is irregular, F or G does not have the right form for a two-term recursion formula to result. As we have seen, the best we can do for the case of two irregular as well as the case of two regular and one irregular points is to obtain a three-term formula. Larger numbers of singular points or a higher species of irregularity will result in still more complex formulas. Luckily such cases have not as yet turned out to be of great practical importance, so we shall dismiss them without further ado, beyond remarking that, if they become important, further research will be needed to establish techniques for handling these more complicated recursion formulas.

Functional Series. But after all, it is not necessary to confine ourselves to a series of powers of z; a series of terms containing a set of functions f_n:

$$\psi = \Sigma A_n f_n(z)$$

can also be used. To see how this generalization may be carried out, we return to the power series method and ask why the set of functions

$$f_n = z^n$$

was so useful. One obvious reply is that f_n, for this case, satisfies eminently simple *recurrence relations:*

$$zf_n = f_{n+1}; \quad f'_n = nf_{n-1}$$

Using these relations it is possible to reduce a differential operator $\mathfrak{L}$ to a form in which only different powers of z^n are involved. In order to use another set of f_n's to represent the solution, the new set of functions must also satisfy recurrence relations.

Another important and useful property of power series is its property of *completeness.* By completeness we mean that, subject to certain conditions, a linear combination of the powers of z may be used to represent any function. This statement is a consequence of Laurent's theorem [Eq. (4.3.4)] and is subject to the conditions involved in Laurent's theorem. Before we can use other sets of functions, we must also show what functions may be represented by their means and what others cannot be so represented. Later on, in the chapter on eigenfunctions, we shall devote considerable space to a discussion of this question. However, it is worth while discussing whatever can be done at this point with the techniques we have already developed. We shall give a few examples and then return to the original question of the solving of equations by the functional series.

The method generally employed involves relating the functional series to be used to the power series. Then, by using the known properties of the power series, it is possible to obtain information about the set f_n. As a first example, we can establish the completeness of the Fourier series in $e^{in\theta}$ directly from the Laurent series. From Eq. (4.3.4)

$$\psi(z) = \sum_{-\infty}^{\infty} a_n z^n$$

Consider now the values of $\psi(z)$ on the unit circle, $z = e^{i\theta}$.

Then
$$\psi(e^{i\theta}) = \sum_{-\infty}^{\infty} a_n e^{in\theta}$$

From the completeness property of the power series, we may now conclude that the functional set $e^{in\theta}$ will represent any reasonable periodic function of θ of period 2π. The necessity for the periodicity is a consequence of the fact that we are representing by the Fourier series the values of ψ on a circle and ψ repeats itself as we complete the circuit.

By considering even or odd functions of θ, we immediately generate the sine or cosine Fourier series.

As a second example, let us examine the first solution of Legendre's equation (page 544) for $m = 0$ and integer values of n.

$$X = F(-n, n + 1 | 1 | \tfrac{1}{2} - \tfrac{1}{2}z)$$

For integer values of n, X is a polynomial in n, the *Legendre polynomials* P_n to be discussed in more detail later on page 595. Let us tabulate the first few of these functions:

$$P_0 = 1 \qquad P_3 = (5z^3 - 3z)/2$$
$$P_1 = z \qquad P_4 = (35z^4 - 30z^2 + 3)/8$$
$$P_2 = (3z^2 - 1)/2 \qquad P_5 = (63x^5 - 70z^3 + 15z)/8; \quad \text{etc.}$$

From this sequence, we may prove that any z^n may be expressed as a linear combination of P_n. For 1 and z it is obvious. We write out the answer for the next few powers:

$$\begin{aligned}
1 &= P_0 \\
z &= P_1 \\
z^2 &= \tfrac{1}{3}(2P_2 + P_0) \\
z^3 &= \tfrac{1}{5}(2P_3 + 3P_1) \\
z^4 &= \tfrac{1}{35}(8P_4 + 20P_2 + 7P_0) \\
z^5 &= \tfrac{1}{63}(8P_5 + 28P_3 + 27P_1); \quad \text{etc.}
\end{aligned}$$

From the fact that the power series in positive powers of z is complete for functions which are not singular, we may conclude that these functions may equally well be expressed in terms of the Legendre polynomials. In order to include singular functions, it would be necessary to include the second solution of the Legendre equation in our discussion (corresponding to the negative powers of z).

For present purposes it is not necessary to determine coefficients in the above expressions explicitly. It is enough to demonstrate the possibility of such a representation. For example, in the case of the solution of Bessel's equation for integer n (5.2.62) we have one set of solutions

$$J_n(z) = e^{-iz}z^nF(\tfrac{1}{2} + n|1 + 2n|2iz) \xrightarrow[z\to 0]{} z^n(1 + \cdots)$$

[see Eqs. (5.2.63) *et seq.*]. We may expect that by suitable combinations of these functions it would be possible to represent z^n. Similarly for z^{-n}, the second solutions, the Neumann functions (see page 625) would be useful. Again one sees that these functions are as complete as the power series. This statement is somewhat more difficult to verify than the similar one for the Legendre functions because the Bessel functions are not polynomials but rather infinite series. However, it is relatively easy to demonstrate that *in principle* a Bessel function representation is possible. We now are able to turn to some examples of the use of functional series, from which we shall be able to deduce the sort of maneuvers which must be usually employed. As a first example, consider the Mathieu equation discussed earlier (page 556):

$$\psi'' + \left[b - \frac{h^2}{2} - \frac{h^2}{2}\cos 2\phi\right]\psi = 0$$

This equation resembles to some extent the equation satisfied by the exponential functions

$$f_n = e^{i2(n+s)\phi};\quad f_n'' + [4(n+s)^2]f_n = 0$$

(the $\cos 2\phi$ term is not present). We substitute $\sum_{-\infty}^{\infty} A_n f_n$ into the Mathieu equation:

$$\sum_{-\infty}^{\infty} A_n \left\{\left[-4(n+s)^2 + \left(b - \frac{h^2}{2}\right)\right] e^{2i(n+s)\theta} - \frac{h^2}{4} e^{2i(n+s+1)\theta} - \frac{h^2}{4} e^{2i(n+s-1)\theta}\right\} = 0$$

Rearranging terms so that $e^{2i(n+s)\theta}$ is a common factor, we obtain

$$\sum_{-\infty}^{\infty} \left\{-\frac{h^2}{4}A_{n-1} - \frac{h^2}{4}A_{n+1} + A_n\left[b - \frac{h^2}{2} - 4(n+s)^2\right]\right\} e^{2i(n+s)\theta} = 0$$

From the completeness of $e^{i(n+s)\theta}$, the coefficient of each term must be zero. (Where we are using the result that, if a power series is identically zero, the coefficient of each power must be zero.) We thus obtain the three-term recursion formula:

$$\frac{h^2}{4}A_{n+1} + A_n\left[4(n+s)^2 - b + \frac{h^2}{2}\right] + \frac{h^2}{4}A_{n-1} = 0$$

This is identical with (5.2.69) (of course).

As a second example, consider a special case of the equation resulting from the separation of spheroidal coordinates:

$$(z^2 - 1)\psi'' + 2z\psi' + (h^2z^2 - b)\psi = 0$$

Compare this with the equation satisfied by the Legendre polynomials P_n:

$$(z^2 - 1)P_n'' + 2zP_n' - n(n+1)P_n = 0$$

If we insist that ψ be nonsingular at ± 1, the singular points of its differential equation, the use of $P_n = f_n$ is suggested; *i.e.*, let

$$\psi = \sum_0^{\infty} A_n P_n$$

The following recurrence relation, which will be derived later, will be useful here:

$$z^2P_n = \frac{n(n-1)}{4n^2-1}P_{n-2} + \left[\frac{n^2}{4n^2-1} + \frac{(n+1)^2}{(2n+1)(2n+3)}\right]P_n + \frac{(n+1)(n+2)}{(2n+1)(2n+3)}P_{n+2}$$

Substituting in the differential equation yields

$$\sum_0^\infty P_n \left\{A_{n+2}h^2 \frac{(n+2)(n+1)}{(2n+3)(2n+5)} + A_n\left[(n)(n+1) - b + h^2\left(\frac{n^2}{4n^2-1} + \frac{(n+1)^2}{(2n+1)(2n+3)}\right)\right] + A_{n-2}\frac{h^2(n-1)(n)}{(2n-1)(2n-3)}\right\} = 0$$

This is a three-term recursion formula which must now be solved for A_n, subject to the conditions $A_{-1} = A_{-2} = 0$.

What we have done here is to find some complete set of functions $f_n(z)$, in terms of which we choose to expand our solution. In practice we choose f's such that their differential equation $\mathfrak{M}_n(f_n) = 0$ is not very different from the equation we wish to solve, $\mathfrak{L}(\psi) = 0$. We then use the relations between successive f_n's in order to express the difference between $\mathfrak{L}(f_n)$ and $\mathfrak{M}_n(f_n)$ in terms of a series of f_n's.

$$[\mathfrak{L} - \mathfrak{M}_m]f_m = \sum_n \gamma_{mn}f_n \tag{5.2.80}$$

For instance, for the Legendre polynomial series discussed above, $[\mathfrak{L} - \mathfrak{M}_m]P_n = [h^2z^2 - b + n(n+1)]P_n$, which can be set into the series form of Eq. (5.2.80), with only three terms in the series (for P_{n-2}, P_n, and P_{n+2}).

If we have chosen well, the series in f_n will be finite, with a few terms. We call these equations, relating simple operators operating on f_n to simple series of f_m's, *recurrence* formulas in order to distinguish them from the *recursion formulas* of Eq. (5.2.40). Setting our series into the operator $\mathfrak{L}$, we have

$$\mathfrak{L}(\Sigma a_m f_m) = \left[\sum_m (\mathfrak{L} - \mathfrak{M}_m)a_m f_m + \sum_m a_m \mathfrak{M}_m(f_m)\right] = \sum_m a_m\left[\sum_n \gamma_{mn}f_n\right] = \sum_n\left[\sum_m a_m\gamma_{mn}\right]f_n = 0$$

where $\mathfrak{M}_m(f_m) = 0$ by definition. If the set f_n is complete, we can equate the coefficients of f_n in the last series to zero separately:

$$\sum_m a_m\gamma_{mn} = 0$$

which are the *recursion* formulas for the coefficients a_m. If these can be solved, we have obtained a solution for the equation $\mathfrak{L}(\psi) = 0$.

The general usefulness of the expansion in question depends, of course, on its rate of convergence, which in turn depends on the behavior of a_n as $n \to \infty$. To obtain this, consider the above equation in the limit $n \to \infty$:

$$A_{n+2}\frac{h^2}{4} + A_n\left[n^2 - b + \frac{h^2}{2}\right] + h^2\frac{A_{n-2}}{4} = 0; \quad \text{for } n \to \infty$$

This is just the recursion relation (5.2.69) derived for Mathieu functions if the substitution

$$s = 0; \quad A_{n+2} = \tfrac{1}{2}C_{n+2}; \quad n = 2\beta$$

is made. It may be recalled that the Mathieu recursion relation has a convergent result for a_n as $n \to \infty$ if, for a given s, b takes on only certain special values. These values must be determined from the recursion formulas for a_n by the methods outlined in the section on continued fractions.

Many other cases of series expansion of a function in terms of other functions will be found later in this book. Series of Legendre functions (hypergeometric functions, see page 593) and of Bessel functions (confluent hypergeometric functions, see page 619) will be particularly useful.

In general, what we can try to do with such series is to use solutions of equations with a given set of singular points to express solutions of equations with one more singular point (or with more complicated singular points). For instance, according to Eq. (5.2.30) a power of z is a solution of a differential equation with two regular singular points, at 0 and ∞. Therefore solutions of an equation with three regular points (hypergeometric function) or of one with one regular and one irregular point (confluent hypergeometric function) can be expressed in terms of a power series in z with comparative ease. On the other hand solutions of an equation with two regular points and one irregular point (spheroidal functions) can most easily be expressed in terms of a series of hypergeometric functions (Gegenbauer functions) or of confluent hypergeometric functions (Bessel functions). We defer the Bessel function series and any discussion of series of more complex functions until later in the book [see Eqs. (5.3.83) and (11.3.87)].

As a final remark, note that the series $\Sigma a_n f_n(z)$ may be generalized into an integral as the Fourier series was generalized into a Fourier integral. For example, instead of $\psi(z) = \sum_n a_n f_n(z)$ we could write

$$\psi(z) = \int K(z,t)v(t)\,dt$$

It is clear that the variable integer n has been replaced by the continuous variable t, that the functions $f_n(z)$ have gone over into $K(z,t)$, and that the coefficients a_n have become $v(t)$. We may, by comparison with the procedure for the series form, outline the manner in which we obtain the *integral representation* of ψ (as the above integral is called).

First we apply the operator $\mathfrak{L}$, where now, to indicate that $\mathfrak{L}$ operates on the z variable only, we write $\mathfrak{L}_z$ for $\mathfrak{L}$. Then

$$\mathfrak{L}_z\psi = \int \mathfrak{L}_z[K(z,t)]v(t)\,dt = 0$$

In our treatment of $\mathfrak{L}_z[f_n(z)]$, we made use of the recurrence relations of f_n to change the differential operator into a set of difference operators, by the recurrence relations

$$\mathfrak{L}_z[f_n] = \sum_p \gamma_{np} f_p$$

where γ_{np} are numerical coefficients. This amounted to replacing the operation on z *by an operation on the subscript* n. In the case of the integral representation, this means that we could arrange to have $\mathfrak{L}_z[K(z,t)] = \mathfrak{M}_t[K(z,t)]$, where $\mathfrak{M}_t$ is a differential operator in t, so that

$$0 = \int \mathfrak{M}_t[K(z,t)]v(t)\,dt$$

The next step in the series representation in terms of f_n was to rearrange the series so that f_n was a common factor; the coefficient of f_n, involving several a_n's, was then put equal to zero, yielding a *recursion relation* for a_n. Thus the operation on f_n was transformed to an operation on a_n. Similarly here the operation $\mathfrak{M}_t$ must now be transformed over to an operation on v. This may be accomplished by integrating by parts or equivalently by means of the adjoint to $\mathfrak{M}_t$ defined earlier (page 528). We recall that

$$v\mathfrak{M}_t[u] - u\mathfrak{M}_t^*[v] = \frac{d}{dt}P(u,v)$$

Therefore

$$0 = \int K(z,t)\mathfrak{M}_t^*(v)\,dt + [P(u,v)]$$

with the second term evaluated at the limits of integration. We now choose the limits or contour of integration so that the $P(u,v)$ term is zero; then the original differential equation is satisfied if the "amplitude" $v(t)$ in the integral representation satisfies the equation

$$\mathfrak{M}_t^*(v) = 0$$

This equation is the analogue of the recursion formulas for a_n. If we can solve the differential equation for $v(t)$, we shall then have a solution of the original differential equation for $\psi(z)$, which has some points of advantage over a series representation. But this subject is a voluminous one, and we had better devote a separate section to it.

5.3 *Integral Representations*

We have now proceeded far enough to see the way series solutions go. Expansion about an ordinary point is straightforward. The unusual cases are about the singular points of the differential equation, where the general solution has a singularity. We have indicated how we can obtain series solutions for two independent solutions about each singular point, which will converge out to the next singular point (or else are asymptotic series, from which reasonably accurate values of the solution can be computed over a more restricted range). In other words, we have worked out a means of calculating the behavior of any solution of the second-order, linear, differential equation in the immediate neighborhood of any point in the complex plane. In particular we can set up the series expansion for the particular solution which satisfies any reasonable boundary condition at any given point (what constitutes "reasonableness" in boundary conditions will be discussed in Chap. 6).

Quite often these boundary conditions are applied at the singular points of the differential equation. We have seen that such singular points correspond to the geometrical "concentration points" of the corresponding coordinate system. Often the shape of the physical boundary can be idealized to fit the simple shape corresponding to the singular point of one of the dimensions (for instance, the origin corresponding to $r = 0$ in spherical coordinates, the disk corresponding to $\mu = 0$ for oblate spheroidal coordinates, the slit corresponding to $\phi = 0$, π for elliptic cylinder coordinates, etc.). Often only one of the solutions, for only one of the indices (if the singular point is regular), can fit the boundary conditions, so one of the solutions discussed in the previous section will be satisfactory.

If we wish values of the allowed solution nearby, the series expansion is satisfactory and, indeed, is the only way whereby these intermediate values can be computed. But quite often we wish to calculate values and slopes near the next singular point, where the series which is satisfactory for the first singular point either converges extremely slowly or else diverges. For instance, we often must satisfy boundary conditions at both ends of the range for a coordinate corresponding to two consecutive singular points. What is needed is a joining factor, relating one of the series about one singular point with the two solutions about the other point, for then we should not need to stretch the convergence of the series. Suppose u_1, v_1 are the two independent series expanded about the singular point $z = a_1$ and u_2, v_2 are the series about $z = a_2$. If we can find relations of the sort, $u_1 = \gamma_{11}u_2 + \gamma_{12}v_2$, etc., we can then use u_1, v_1 when we wish to put boundary conditions at a_1 and u_2, v_2 for putting conditions at a_2. We shall have no convergence problems if we can express each solution at one end in terms of the solutions at the other.

For the simplest sorts of differential equation this joining of the behavior about one singular point to the behavior about another is easy. As long as the solutions are either rational functions (Eq. 5.2.29) or elementary transcendental functions [Eq. (5.2.33)], we know the behavior of the solutions at both ends of the range; the "joining" has already been done for us. For the next more complex equations this joining is not so simple. Equations (5.2.51) and (5.2.60) are examples of such joining equations, but our derivation of these formulas from the series solutions was not rigorous, nor was it even valid for much of the ranges of the parameters. Series solutions, as Stokes put it, "have the advantage of being generally applicable, but are wholly devoid of elegance." What we should prefer would be to express the solutions in terms of the rational functions or the elementary transcendentals in some finite way, which would be just as convergent at one singular point as at another. This can be done for some equations by the use of integrals, instead of series.

The expression

$$\psi(z) = \int K(z,t)v(t)\,dt \tag{5.3.1}$$

is general enough to express nearly any solution. If the functions K and v turn out to be rational functions or elementary transcendentals, then we have a "closed" form which can be used to calculate the solution anywhere on the complex plane. All we need is to determine how we are to find the correct forms for K and v for a given equation.

Some Simple Examples. Our acquaintance with the techniques of contour integration enables us to work out a few simple cases to illustrate the relationship between integral representation and series solution. For instance, since

$$\pi\cot(\pi t) = \sum_{n=-\infty}^{\infty} \frac{1}{t-n} \quad \text{or} \quad \pi\coth(\pi t) = \sum_{n=-\infty}^{\infty} \frac{1}{t-in}$$

(see page 561), we can utilize the residues of this function to produce a Fourier series. For instance, the integral

$$I(z) = \oint_C \coth(\pi t)\,F(t)e^{zt}\,dt$$

can be developed into a series if F is a rational function of t (see page 413) with all its poles to the right of the imaginary axis, which behaves at infinity such that $tF(t)$ stays finite or goes to zero as $t \to \infty$. The contour C includes a line just to the right of the imaginary axis, going from $\epsilon - i\infty$ to $\epsilon + i\infty$, and then returns to $-i\infty$ by going along the semicircle of infinite radius inscribed about the negative half of the t plane. Since $tF(t)$ does not become infinite at $|t| \to \infty$, the contour integral about the infinite semicircle is zero [since $\coth(\pi t) \to 1$ when $|t| \to \infty$ if $\mathrm{Re}\, t > 0$] as long as $\mathrm{Re}\, z > 0$.

The poles inside the contour are therefore those of $\coth(\pi t)$, which are $\pm in$. Inside the contour we have required that F be analytic everywhere. Along the imaginary axis F has a symmetric and unsymmetric part.

$$F(t) = s(t) + u(t); \quad s(t) = \tfrac{1}{2}F(t) + \tfrac{1}{2}\bar{F}(t) = \mathrm{Re}\, F;$$
$$u(t) = \tfrac{1}{2}F(t) - \tfrac{1}{2}\bar{F}(t) = i\,\mathrm{Im}\,(F)$$

The contour integral is $2\pi i$ times the sum of the residues for all the poles of $\coth(\pi z)$:

$$\int_{-i\infty+\epsilon}^{i\infty+\epsilon} \coth(\pi t)\, F(t) e^{zt}\, dt = 4i \sum_{n=0}^{\infty} [\mathrm{Re}\, F_n \cos(nz) - \mathrm{Im}\, F_n \sin(nz)] \tag{5.3.2}$$

where $F_n = F(in)$.

A more immediately useful integral representation can be obtained from the properties of the gamma function $\Gamma(-t)$. The function

$$\Gamma(-t) = \frac{-\pi}{\Gamma(t+1)\sin(\pi t)} \xrightarrow[t\to n]{} -(-1)^n \frac{1}{(t-n)\Gamma(t+1)}; \qquad n = 0, 1, 2, \ldots$$

has simple poles at $t = 0, 1, 2, \ldots$. If we include in the integrand a factor z^t, and if the whole integrand converges, then the sum of the residues will turn out to be a series in integral powers of z. As was shown on page 486, the asymptotic expression for $\Gamma(t + 1)$ is

$$\Gamma(t+1) \xrightarrow[t\to\infty]{} \sqrt{2\pi}\, e^{-t-1} t^{t+\frac{1}{2}}$$

Therefore if $G(t)$ is such a function that $[G(t)z^t/\Gamma(t + 1)] \to 0$ when $t \to \infty$ for $\mathrm{Re}\, t > 1$, and if all the singularities of G are to the left of the imaginary axis, then the integral

$$\int_{-i\infty}^{i\infty} G(t)\Gamma(-t)(-z)^t\, dt = 2\pi i \sum_{n=0}^{\infty} G(n) z^n/n! \tag{5.3.3}$$

where the contour passes to the left of the pole at $t = 0$ and is completed along the semicircle of infinite radius surrounding the positive half of the t plane, going from $+i\infty$ back around to $-i\infty$. Thus we have a fairly direct means of going from a series to an integral. If $G(n)$ is a "closed function" of n (*i.e.*, if the successive coefficients are related by a simple formula such as a two-term recursion formula), then the integrand is a closed form.

The uses of such a formula and the precautions necessary to ensure convergence are well illustrated by applying it to the hypergeometric series [see Eq. (5.2.44)]

$$F(a,b|c|z) = \frac{\Gamma(c)}{\Gamma(a)\Gamma(b)} \sum_{n=0}^{\infty} \frac{\Gamma(a+n)\Gamma(b+n)}{\Gamma(c+n)n!} z^n$$

It appears that the function G should be $\dfrac{\Gamma(c)/2\pi i}{\Gamma(a)\Gamma(b)} \dfrac{\Gamma(a+t)\Gamma(b+t)}{\Gamma(c+t)}$ if this function should have its poles to the left of the imaginary axis and if $Gz^t/\Gamma(t+1)\sin(\pi t)$ converges. Using the asymptotic formula for the gamma function, we find that, if $t = Re^{i\theta} = R\cos\theta + iR\sin\theta$, for R large enough

$$\frac{\Gamma(a+t)\Gamma(b+t)}{\Gamma(c+t)\Gamma(t+1)} \simeq e^{1+c-a-b} R^{a+b-c-1} e^{i(a+b-c-1)\theta}$$

Also if $z = re^{i\phi}$ and $(-z) = re^{i\phi - i\pi} = e^{\ln r + i(\phi - \pi)}$

$$(-z)^t = \exp\{R[(\ln r)\cos\theta + (\pi - \phi)\sin\theta] + iR[(\phi - \pi)\cos\theta + (\ln r)\sin\theta]\}$$

and $$\frac{1}{\sin(\pi t)} = \begin{cases} 2i\exp[i\pi R\cos\theta - \pi R\sin\theta]; & 0 < \theta < \pi \\ 2i\exp[i\pi R\cos\theta + \pi R\sin\theta]; & 0 > \theta > -\pi \end{cases}$$

Therefore, discarding the imaginary exponentials by taking the magnitude of the integrand

$$\left|\frac{\Gamma(a+1)\Gamma(b+t)}{\Gamma(c+t)\Gamma(t+1)} \frac{(-z)^t}{\sin(\pi t)}\right| \simeq 2\left(\frac{R}{e}\right)^{a+b-c-1} e^{R\ln r\cos\theta} \begin{cases} e^{-R\phi\sin\theta}; & 0 < \theta < (\pi/2) \\ e^{R(2\pi-\phi)\sin\theta}; & 0 > \theta > -(\pi/2) \end{cases}$$

Therefore as long as the magnitude of z is less than unity (*i.e.*, $\ln r < 0$) and the argument of z larger than zero and less than $2\pi (0 < \phi < 2\pi)$, the integrand will vanish along the semicircle part of the contour $[R \to \infty, -(\pi/2) < \theta < (\pi/2)]$. As long as these conditions hold, an integral representation for the hypergeometric series is

$$F(a,b|c|z) = \frac{\Gamma(c)}{\Gamma(a)\Gamma(b)} \int_{-i\infty}^{i\infty} \frac{\Gamma(a+t)\Gamma(b+t)}{2\pi i\Gamma(c+t)} \Gamma(-t)(-z)^t\,dt \quad (5.3.4)$$

if we can draw the contour between $-i\infty$ and $+i\infty$ so that all the poles of $\Gamma(-t)$ are to the right and the poles of $\Gamma(a+t)\Gamma(b+t)$ are to the left. Figure (5.5) shows that this may be done, even if Re a and Re b are negative, *as long as neither a nor b are negative integers.* If either a or b is a negative integer, Eq. (5.2.44) shows that F is a finite polynomial, not an infinite series (it also indicates that c should not be a negative integer).

This integral representation does not seem to be a useful or elegant result; in fact at first sight it appears less elegant than the expression for the series. It is not difficult to show, however, that the possibilities for rearranging the contour without changing the value of the integral make for considerable flexibility in expanding the result, just the flexibility needed to relate the expansion about one singularity to expansions about another singularity.

For instance, further examination of the asymptotic behavior of the integrand of Eq. (5.3.4) shows that it goes to zero when $R \to \infty$ for the range $(\pi/2) < \theta < (3\pi/2)$, $(t = Re^{i\theta})$, *i.e.*, for the infinite semicircle enclosing the *left-hand* part of the t plane. Therefore the contour enclosing all the poles of $\Gamma(-t)$ (which results in the hypergeometric

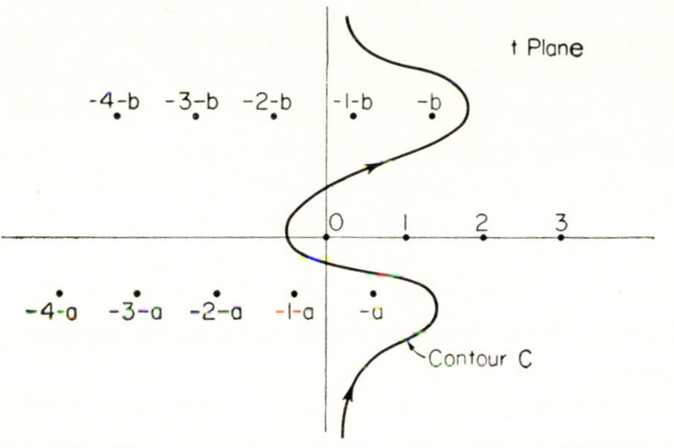

Fig. 5.5 Contour for integral representation of hypergeometric function by Mellin transform.

series above) is equal to the contour enclosing all the poles of $\Gamma(a + t)\Gamma(b + t)$. Using the equations $\Gamma(a + t) = [\pi/\Gamma(1 - a - t)\sin\pi(a + t)]$, etc., and evaluating residues at the pairs of poles, we have

$$\frac{\Gamma(a)\Gamma(b)}{\Gamma(c)} F(a,b|c|z) = \sum_{n=0}^{\infty} \frac{\Gamma(a + n)\Gamma(1 - c + a + n)}{\Gamma(1 + n)\Gamma(1 + a - b + n)} \cdot \frac{\sin\pi(c - a - n)}{\cos(n\pi)\sin\pi(b - a - n)} (-z)^{-a-n}$$

$$+ \sum_{n=0}^{\infty} \frac{\Gamma(b + n)\Gamma(1 - c + b + n)}{\Gamma(1 + n)\Gamma(1 - a + b + n)} \frac{\sin\pi(c - b - n)}{\cos(n\pi)\sin\pi(a - b - n)} (-z)^{-b-n}$$

$$= \frac{\Gamma(a)\Gamma(b - a)}{\Gamma(c - a)} (-z)^{-a} F\left(a, 1 - c + a\middle| 1 - b + a\middle| \frac{1}{z}\right)$$

$$+ \frac{\Gamma(b)\Gamma(a - b)}{\Gamma(c - b)} (-z)^{-b} F\left(b, 1 - c + b\middle| 1 - a + b\middle| \frac{1}{z}\right) \quad (5.3.5)$$

If $|a - b|$ is an integer or zero, one of these series "blows up," for the second solution should contain a logarithmic term.

The series on the left-hand side converges when $|z| < 1$, whereas the series on the right converges when $|z| > 1$. Strictly speaking, the two series should not be equated; what we should say is that the integral representation given in Eq. (5.3.4) has a series expansion in terms of $F(a,b|c|z)$ which is valid for $|z| < 1$; it also has another series expansion, given by the right-hand side of Eq. (5.3.5), which is valid for $|z| > 1$. The integral representation has validity for all (or nearly all) values of z and can be considered as the "real solution" of Eq. (5.2.42). The series expansions can be considered as partial representations of this "real solution," valid over limited ranges of z. By means of the integral representation we can perform the analytic continuation of the solution from one region of convergence around one singularity to another region of convergence about another.

Therefore we might call the integral on the right-hand side of Eq. (5.3.4) *the hypergeometric function*, $F(a,b|c|z)$, one solution of Eq. (5.2.42). When $|z| < 1$, we can compute this function by using the series representation given in Eq. (5.2.44), *the hypergeometric series* in z. For other ranges of z the hypergeometric function can be computed by using hypergeometric series in $1/z$, as given in Eq. (5.3.5), or in $(1 - z)$, as given later, and so on. The function itself, exhibiting different facets from different points of view, is the integral representation.

General Equations for the Integrand. The characteristics exhibited by the representation for the hypergeometric function are typical of integral representations in general. They make analytic continuation of a solution almost a tautology. When such a representation can be found, it is usually not difficult to fit boundary conditions at both ends of a range of z. When a representation cannot be found, such fitting is difficult and "untidy."

The methods used to obtain the integral representations of the previous subsection were far from straightforward; what is needed is a technique of proceeding directly from the differential equation to the form for the integrand. We shall first use the form given in Eq. (5.3.1), with a *kernel* $K(z,t)$ and a *modulation factor* $v(t)$. We choose a form for K which seems appropriate and then find out what differential equation v satisfies. If this equation is simple enough (enough simpler than the equation for ψ), then a closed form for v can be obtained and then the integral representation is manageable.

The differential equation to be solved is a second-order equation, with the independent variable transformed so that the coefficients are algebraic. Rather than using the form given in Eq. (5.2.1), it is better to clear of fractions, using the form

$$\mathfrak{L}_z(\psi) = f(z)\frac{d^2\psi}{dz^2} + g(z)\frac{d\psi}{dz} + h(z)\psi = 0 \tag{5.3.6}$$

where f, g, and h are then polynomials in z. When the differential operator $\mathfrak{L}$ is applied to the integral form of Eq. (5.3.1), the operator can be taken inside the integral, to operate on the z part of the kernel K,

$$\mathfrak{L}(\int Kv\,dt) = \int \mathfrak{L}_z(K(z,t))v(t)\,dt$$

if the integral is a reasonably convergent one. Operation by $\mathfrak{L}_z$ on $K(z,t)$ produces another function of z and t. If we have picked a satisfactory form for K, this new function of z and t is also equal to some operator function of t, operating on $K(z,t)$,

$$\mathfrak{L}_z(K) = \mathfrak{M}_t(K)$$

[In some cases it is sufficient to have $\mathfrak{L}_z(K)$ equal to an operator $\mathfrak{M}_t$, operating on some *other* kernel $K'(z,t)$.]

It is this equivalence between an operator in z and an operator in t, when operating on the kernel K, which at the same time makes possible the calculation of $v(t)$ and also severely limits the choice of forms for $K(z,t)$. Not many functions of z and t possess a simple reciprocal relation such as that exhibited by the exponential kernel e^{zt} [used in Eq. (5.3.2)]

$$(d/dz)e^{zt} = te^{zt}; \quad ze^{zt} = (d/dt)e^{zt}$$

where a differential operator involving derivatives in z and powers of z is transformed into an operator involving corresponding powers of t and derivatives in t. Nevertheless a number of other kernels have been found useful in various cases: z^t(used in Eq. 5.3.3), $(z - t)^\mu$, various functions of the product (zt), etc. In many cases several different kernels can be used, giving rise to several alternative integral representations of the same solution. Which kernel is likely to produce the most useful representation depends on the relation between the singularities of the kernel and the singular points of the differential equation. For instance, for the hypergeometric equation, with three regular singular points, one might expect a kernel of the form $(z - t)^\mu$ to be better than e^{zt}, which has an essential singularity at infinity. On the other hand, the confluent hypergeometric equation, having an irregular singular point at infinity, would appear to be suitable for the kernel e^{zt}.

But let us return to the immediate subject, which is the setting up of an equation for $v(t)$. We have reached the point where the integrand consists of a function v of t multiplied by a differential operator $\mathfrak{M}_t$ operating on the t part of the kernel $K(z,t)$. Referring to Eq. (5.2.10), we see that we can next transform this to K times the *adjoint* (see page 527) differential operator $\tilde{\mathfrak{M}}$ operating on v, plus the derivative of a bilinear concomitant P. Symbolically it goes as follows:

$$\mathfrak{L}_z(y) = f(z)\frac{d^2y}{dz^2} + g(z)\frac{dy}{dz} + h(z)y = 0; \quad y = \int K(z,t)v(t)\,dt$$

$$\mathfrak{L}_z(y) = \int \mathfrak{L}_z(K)\,v\,dt = \int \mathfrak{M}_t(K)\,v\,dt \tag{5.3.7}$$

$$= \int \left[K\tilde{\mathfrak{M}}_t(v) + \frac{d}{dt}P(v,K)\right]dt = \int K(z,t)\tilde{\mathfrak{M}}_t(v)\,dt + [P(v,K)]$$

where, if $\mathfrak{M}(K) = \alpha(t)(d^2K/dt^2) + \beta(t)(dK/dt) + \gamma(t)K$, the adjoint operator is

$$\tilde{\mathfrak{M}}(v) = \frac{d^2}{dt^2}(\alpha v) - \frac{d}{dt}(\beta v) + \gamma v$$

and the corresponding bilinear concomitant is

$$P(v,K) = \alpha v\frac{d}{dt}K - K\frac{d}{dt}(\alpha v) + \beta vK$$

If now the limits of integration and the contour along which the integral is taken are such *that P returns to its initial value at the end of the contour*, the integral of dP/dt is zero and

$$\mathfrak{L}_z(y) = \int K\tilde{\mathfrak{M}}_t(v)\,dt$$

Therefore if $v(t)$ is a solution of the differential equation $\tilde{\mathfrak{M}}_t(v) = 0$, the integral $y = \int Kv\,dt$ is a solution of the differential equation $\mathfrak{L}_z(y) = 0$, which is the equation we wish to solve. If we have been lucky in our choice of kernel K, the equation $\tilde{\mathfrak{M}}_t(v) = 0$ will be simpler than $\mathfrak{L}_z(y) = 0$ and v will be a simple function of t. Usually there are several different possible integration paths and limits, any one of which makes $\int(dP/dt)\,dt = 0$. These different integrals correspond to different independent solutions of $\mathfrak{L}_z(y) = 0$.

From another point of view, what we are doing here is investigating integral transforms, of the sort of the Fourier transform

$$f(\nu) = \int_{-\infty}^{\infty} F(\mu)e^{i\mu\nu}\,d\mu; \quad F(\mu) = \frac{1}{2\pi}\int_{-\infty}^{\infty} f(\nu)e^{-i\mu\nu}\,d\nu$$

discussed in Sec. 4.8. We transform from function $\psi(z)$ to function $v(t)$ by means of the kernel $K(z,t)$,

$$\psi(z) = \int v(t)K(z,t)\,dt$$

and try to find a type of transformation for which v, the transform, is a simpler function than ψ.

For instance, when the kernel is e^{zt} the relation is called the *Laplace transform*. This can be quickly obtained from the Fourier transform above; by setting $\mu = -it$, $\nu = z$, $F(-it) = -iv(t)$, and $f(\nu) = \psi(z)$, we have

$$\psi(z) = \int_{-i\infty}^{i\infty} v(t)e^{zt}\,dt; \quad v(t) = \frac{i}{2\pi}\int_{-\infty}^{\infty} \psi(z)e^{-zt}\,dz \qquad (5.3.8)$$

And when the kernel is z^t, the relation is called the *Mellin transform*. It also can be obtained from the Fourier transform by setting $\mu = -it$, $\nu = \ln z$, $f(\ln z) = \psi(z)$, and $F(-it) = -iv(t)$, giving

$$\psi(z) = \int_{-i\infty}^{i\infty} v(t)z^t\,dt; \quad v(t) = \frac{i}{2\pi}\int_0^{\infty} \psi(z)z^{-t-1}\,dz \qquad (5.3.9)$$

Still other transforms can be devised, not all of them as closely related to the Fourier transform as these two. The transform $\int v(t)(z - t)^\mu\,dt$, for instance, is called the *Euler transform* (see also Sec. 4.8).

The rest of this section will be devoted to the study of a number of examples of integral representations, both to illustrate the techniques of obtaining such solutions and the methods of using the results and also with an aim to make more familiar certain functions which will be generally useful later in this volume. Two types of kernel will be studied in considerable detail:

$(z - t)^\mu$; the Euler transform
e^{zt}; the Laplace transform

Other types of transforms, of less general utility, will be discussed in less detail.

The Euler Transform. As mentioned previously, we should expect that the kernel $(z - t)^\mu$ would be a satisfactory representation for solutions of equations having only regular singular points, such as the Papperitz equation (5.2.36) or the corresponding canonical form, the hypergeometric equation (5.2.42). This restriction of the form of equation for which the Euler transform is applicable evidences itself in a somewhat unfamiliar way when we come to apply the differential operator $\mathfrak{L}$ to the kernel $(z - t)^\mu$.

The form for $\mathfrak{L}$ is that given in Eq. (5.3.6), the form in which the hypergeometric equation (5.2.42) is already given. When $\mathfrak{L}$ is applied to the kernel $(z - t)^\mu$, there results a complicated algebraic function of z and t:

$$\mathfrak{L}_z((z - t)^\mu) = \mu(\mu - 1)f(z)(z - t)^{\mu-2} + \mu g(z)(z - t)^{\mu-1} + h(z)(z - t)^\mu$$

This is now to be converted into some differential operator $\mathfrak{M}_t$, operating on some power of $(z - t)$. The form of $\mathfrak{M}$ could be worked out painfully by dividing out $(z - t)^{\mu-2}$, expanding the rest in a bilinear form of z and t and then trying to work out the form for $\mathfrak{M}$ which would give the result. A more orderly procedure results when we expand the functions

f, g, h in a Taylor's series about $z = t$ (since f, g, and h are polynomials in z this can always be done). For instance, $f(z) = f(t) + (z - t)f'(t) + \frac{1}{2}(z - t)^2 f''(t) + \cdots$, and so on. The result is

$$\begin{aligned}\mathfrak{L}_z((z - t)^\mu) = \mu(\mu - 1)f(t)(z - t)^{\mu-2} \\ + [\mu(\mu - 1)f'(t) + \mu g(t)](z - t)^{\mu-1} \\ + [\tfrac{1}{2}\mu(\mu - 1)f''(t) + \mu g'(t) + h(t)](z - t)^\mu + \cdots\end{aligned}$$

where the fourth term in the series involves the third derivative of f, the second derivative of g, the first derivative of h, and so on.

In order to have this form represent a second-order operator, operating on $(z - t)^\mu$, we must arrange that the fourth and all higher terms in this series be zero. There are many ways of adjusting the functions f, g, h so that this is so, but the simplest way (and sufficient for our needs here) is to require that all derivatives of f higher than the second, of g higher than the first, and of h higher than the zeroth be zero. In other words, if $f(z)$ is a second-order polynomial in z, $g(z)$ a first-order polynomial, and $h(z)$ is a constant, then the expression above will *have only the three terms which are written out;* all higher terms will vanish.

We see that this automatically restricts us to an equation with three regular singular points; for f, being a quadratic function, will have two zeros, and when the equation is written in the form of Eq. (5.2.1)

$$\frac{d^2\psi}{dt^2} + \frac{g(z)}{f(z)}\frac{d\psi}{dz} + \frac{h(z)}{f(z)}\psi = 0$$

we can soon verify that, in general, there are three regular singular points; two at the roots of $f(z) = 0$ and one at infinity. The hypergeometric equation is of just this form (as, of course, it must be). Referring to Eq. (5.2.42) we see that $f = z(z - 1)$, $g = [(a + b + 1)z - c]$, and $h = ab$.

But having gone this far, we can go further toward simplifying the equation, for we have the liberty of choosing the value of μ. The coefficient of $(z - t)^\mu$ is now independent of z and t, for f'', g', and h are constants. Consequently if we set

$$\tfrac{1}{2}\mu(\mu - 1)f'' + \mu g' + h = 0 \tag{5.3.10}$$

we obtain two solutions for μ, either of which can be used in the integral representation.

The differential form $\mathfrak{M}_t$ is then

$$\mathfrak{M}_t = \alpha(t)\frac{d^2}{dt^2} - \beta(t)\frac{d}{dt} + \gamma$$

where $\alpha = f(t)$; $\beta = (\mu - 1)f'(t) + g(t)$; $\gamma = \frac{1}{2}\mu(\mu - 1)f'' + \mu g' + h$ (5.3.11) and the adjoint equation and the bilinear concomitant are

$$\mathfrak{M}_t(v) = \frac{d^2}{dt^2}[f(t)v] + \frac{d}{dt}[\beta(t)v] + \gamma v = 0$$

$$P(v,K) = fv\frac{d}{dt}(z-t)^\mu - (z-t)^\mu \frac{d}{dt}(fv) - \beta v(z-t)^\mu \tag{5.3.12}$$

$$= -\mu f v(z-t)^{\mu-1} - (\mu f'v + fv' + gv)(z-t)^\mu$$

If μ is adjusted so that Eq. (5.3.10) holds, then $\gamma = 0$ and the equation for v can quickly be solved:

$$y = fv;\quad \frac{d^2y}{dt^2} = -\frac{d}{dt}\left(\frac{\beta}{f}y\right);\quad \frac{dy}{dt} = -\frac{\beta}{f}y;\quad \ln y = -\int \frac{\beta}{f}dt + \ln A$$

or $$v(t) = \frac{A}{f}\exp\left\{-\int\left[(\mu-1)\frac{f'}{f} + \frac{g}{f}\right]dt\right\} = Af^{-\mu}e^{-\int (g/f)\,dt}$$

and $$P(v,K) = -\mu f(t)v(t)(z-t)^{\mu-1};\quad \psi = \int (z-t)^\mu v(t)\,dt \tag{5.3.13}$$

where the integral for ψ is such that P has the same value at the beginning and end of the path of integration.

We note that it is possible to make the differential equation, which is to be solved by the Euler transform, have *four* regular singular points, one at infinity, by letting f be a third-order polynomial, g a second-, and h a first-order polynomial and then arranging that that coefficient of $(z-t)^{\mu+1}$ in the series for $\mathfrak{L}_z((z-t)^\mu)$ be zero:

$$\tfrac{1}{6}\mu(\mu-1)f''' + \tfrac{1}{2}\mu g'' + h' = 0$$

Since f''', g'', and h' are constants, this equation can serve to determine μ, instead of Eq. (5.3.8), and we shall then have to solve Eq. (5.3.12) for v, when γ is *not* zero. The fly in the ointment is that in this case the equation $\mathfrak{M}(v)$ *is just as complicated as equation* $\mathfrak{L}(\psi)$, having just as many singular points. Consequently the Euler transform does not save us any work, as it does in the case of the equation with three regular points, when γ can be set equal to zero. The solution v does not have a simple form of the sort given in Eq. (5.3.13); it still must be expanded in an infinite series.

We can therefore say that the Euler transform is of especial utility for equations of the Papperitz form [Eq. (5.2.36)] and, in particular, for the hypergeometric function and its special cases, the Legendre functions [Eq. (5.2.47)] and the Gegenbauer functions [Eq. (5.2.52)].

Euler Transform for the Hypergeometric Function. The hypergeometric equation

$$\mathfrak{L}_z(\psi) = (z^2 - z)\psi'' + [(a+b+1)z - c]\psi' + ab\psi = 0$$

lends itself to solution by means of the Euler transform. The expressions for the coefficients in the operator $\mathfrak{M}_t$ are

$$\alpha = f(t) = t(t-1);\quad \beta = (a+b+2\mu-1)t - (c+\mu-1);$$
$$\gamma = \mu(\mu-1) + \mu(a+b+1) + ab$$

and the roots of the equation $\gamma = 0$ are $\mu = -a$, $\mu = -b$. We will choose $\mu = -a$; substitution in Eq. (5.3.13) [for $g = (a + b + 1)z - c$] results in

$$v = At^{a-c}(t - 1)^{c-b-1}; \quad P = aAt^{a-c+1}(t - 1)^{c-b}(z - t)^{-a-1}$$

Therefore a solution of the hypergeometric equation is

$$\psi = A \int (t - z)^{-a} t^{a-c} (t - 1)^{c-b-1} \, dt$$

taken between limits and over a contour which makes the corresponding integral of dP/dt equal to zero. The integrand in this case has branch points (unless a, b, and c are integers) at $t = 0, 1, z$, and ∞. If P vanishes at two of these, we can make these the limits of integration; otherwise we can make the integral over a closed contour such that P returns to its original value at the end of the circuit. For instance, if $c > b > 0$, P goes to zero at $t = 1$ and $t = \infty$.

To obtain a power series expansion for the integral to compare with the hypergeometric series, we expand $(t - z)^{-a}$ in the integrand

$$(t - z)^{-a} = t^{-a} \left[1 + a \left(\frac{z}{t}\right) + \frac{a(a + 1)}{2!} \left(\frac{z}{t}\right)^2 + \cdots \right]$$

and utilize the formula [see Eq. (4.5.54)]

$$\frac{\Gamma(m)\Gamma(n)}{\Gamma(m + n)} = \int_0^1 x^{m-1}(1 - x)^{n-1} \, dx = \int_1^\infty t^{-m-n}(t - 1)^{n-1} \, dt$$

Comparison with the definition of the hypergeometric series given in Eq. (5.2.44) shows that

$$F(a,b|c|z) = \frac{\Gamma(c)}{\Gamma(b)\Gamma(c - b)} \int_1^\infty (t - z)^{-a} t^{a-c} (t - 1)^{c-b-1} \, dt \qquad (5.3.14)$$

which is valid as long as Re $c >$ Re $b > 0$ and as long as z is not a real number larger than unity. For cases where Re $b \leq 0$, we can interchange a and b in this formula and obtain another, equally valid representation, since $F(a,b|c|z) = F(b,a|c|z)$.

The integral of Eq. (5.3.14) can then be considered to be the "real" hypergeometric function, from which one obtains series expansions about any desired point. It is interesting to note that this integral representation has a very different appearance from the one given in Eq. (5.3.4), which is equally valid. The difference is only superficial, however, for there is a close connection between gamma functions and the Euler transform, as will be indicated in some of the problems.

An interesting extension of this formula is the expression for one of the solutions of the Papperitz equation (5.2.36), corresponding to Eq. (5.2.46). Setting $t = [(u - a)(b - c)/(u - c)(b - a)]$ (where a, b, and c are now the positions of the singular points, not the indices as above) we have

$$\left(\frac{z-a}{z-c}\frac{b-c}{b-a}\right)^\lambda \left(\frac{z-b}{z-c}\frac{a-c}{a-b}\right)^\mu \cdot$$
$$\cdot F\left(\lambda+\mu+\nu, \lambda+\mu+\nu' \middle| \lambda-\lambda'+1 \middle| \frac{z-a}{z-c}\frac{b-c}{b-a}\right)$$
$$= \frac{(-1)^{\mu+\lambda}\Gamma(\lambda-\lambda'+1)}{\Gamma(\lambda+\mu+\nu')\Gamma(\lambda+\mu'+\nu)} \cdot$$
$$\cdot (z-a)^\lambda (z-b)^\mu (z-c)^\nu (a-c)^{\mu'} (c-b)^{\lambda'} (b-a)^{\nu'} \cdot$$
$$\cdot \int_b^c (u-z)^{-\lambda-\mu-\nu}(u-a)^{-\lambda-\mu'-\nu'}(u-b)^{-\lambda'-\mu-\nu'}(u-c)^{-\lambda'-\mu'-\nu}\,du \tag{5.3.15}$$

where $\lambda+\lambda'+\mu+\mu'+\nu+\nu'=1$. This has a suggestive symmetry with respect to singular points and to indices. From it, by setting a, b, c

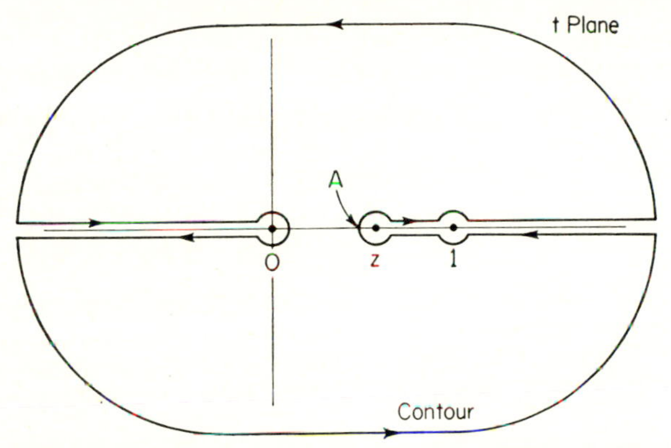

Fig. 5.6 Limiting contour for obtaining joining relations for hypergeometric function. Integrand is real at point A when z is real.

equal to 0, 1, ∞ in different orders, one can write out a whole series of useful integral representations for different solutions of the hypergeometric equation about each of the singular points.

The integral of Eq. (5.3.14) may be modified and manipulated to give a large number of useful and interesting relationships. For instance, Eq. (5.2.49) joining the solution about $z = 0$ with the solutions about $z = 1$ can be derived in a much more satisfactory manner than was given on page 546, with different restrictions from these which seemed to limit the validity of the result there. We start with a contour integral

$$\oint (z-t)^{-a} t^{a-c} (1-t)^{c-b-1}\,dt = 0$$

where the contour is the one shown in Fig. 5.6, carefully omitting all the singularities of the integrand. We make the integrand as shown real at the point A, for t on the real axis a little smaller than z but larger than zero (we assume here that z is real, which is not necessary but makes the arrangements somewhat less complicated to describe; all that really is necessary is that z not lie on that portion of the real axis between $+1$ and $+\infty$).

Now if $\text{Re}\, b \le \text{Re}\, c \le 1 + \text{Re}\, a \le 2$, we can use the methods described on page 410 to show that the equation above is equivalent to

$$\sin(\pi a) \int_z^1 (t - z)^{-a} t^{a-c} (1 - t)^{c-b-1}\, dt \\ - \sin \pi(a + b - c) \int_1^\infty (t - z)^{-a} t^{a-c} (t - 1)^{c-b-1}\, dt \\ - \sin \pi(a - c) \int_0^{-\infty} (z - t)^{-a} (-t)^{a-c} (1 - t)^{c-b-1}\, dt = 0$$

We now make use of the formula $\Gamma(z)\Gamma(1 - z) = \pi/\sin(\pi z)$ to convert $\dfrac{\sin \pi(c - a)}{\sin \pi(c - a - b)}$ into $\dfrac{\Gamma(c - a - b)\Gamma(a + b - c + 1)}{\Gamma(c - a)\Gamma(a - c + 1)}$ and $\dfrac{\sin(\pi a)}{\sin \pi(a + b - c)}$ into $\dfrac{\Gamma(a + b - c)\Gamma(c - a - b + 1)}{\Gamma(a)\Gamma(1 - a)}$. In the first integral we set $t = (w + z - 1)/w$, and in the third integral we set $t = 1 - u$; this eventually gives us

$$\int_1^\infty (t - z)^{-a} t^{a-c} (t - 1)^{c-b-1}\, dt = \frac{\Gamma(c - a - b)\Gamma(a + b - c + 1)}{\Gamma(c - a)\Gamma(a - c + 1)} \cdot \\ \cdot \int_1^\infty (u - 1 + z)^{-a} u^{c-b-1} (u - 1)^{a-c}\, dt \\ + \frac{\Gamma(a + b - c)\Gamma(c - a - b + 1)}{\Gamma(a)\Gamma(1 - a)} (1 - z)^{c-a-b} \cdot \\ \cdot \int_1^\infty (w - 1 + z)^{a-c} w^{b-1} (w - 1)^{-a}\, dw$$

which can quickly be converted into Eq. (5.2.49) by using Eq. (5.3.14) in reverse. In this case our restrictions are that $\text{Re}\, b \le \text{Re}\, c \le 1 + \text{Re}\, a \le 2$. Since the result is obtained by rearrangement of contours of integration and substitution of gamma functions and other finite procedures, it has, perhaps, a more dependable "feel" to it than does the juggling with infinite series required in the earlier derivation.

Between the two derivations and by use of analytic continuation, we can extend this joining formula to cover a still wider range of a, b, and c. For instance, our previous derivation is valid for $\text{Re}\, b > \text{Re}\, c - \text{Re}\, a$, whereas the present one is valid for $\text{Re}\, b \le \text{Re}\, c$ so that, between the two, the relation is valid over the whole range of b for which the functions are analytic. As long as z is not equal to unity, the function $F(a,b|c|z)$ is an analytic function of a, b, and c, except for a and $b = \infty$ and for c zero or a negative integer. The right-hand side of Eq. (5.2.49) is an analytic function of a, b, and c except for c zero or a negative integer or for $a + b - c$ a positive or negative integer or zero. For any one of these parameters we can find a region for which one or the other of the restrictions on the derivation of Eq. (5.2.49) is complied with and which at the same time overlaps the wider range over which F is analytic. Consequently we can extend the equation, by analytic continuation, eventually to cover the whole range of values of a, b, and c for which the left- and right-hand sides are analytic. The other joining formula (5.3.5) may likewise be extended. Between the two, it is possible to express any solution of the hypergeometric function around one of the three singular points in terms of the solutions about any of the other two points. Thus we have completely solved the problem of the joining factors, mentioned on page 577, for the case of equations with three regular singular points.

Analytic Continuation of the Hypergeometric Series. Another set of formulas, useful in later calculations, may be obtained by further contortions of the integral expression (5.3.14). Let $t = 1/u$ and $u = 1 - w$,

$$\begin{aligned} F(a,b|c|z) &= \frac{\Gamma(c)}{\Gamma(b)\Gamma(c-b)} \int_0^1 u^{b-1}(1-u)^{c-b-1}(1-uz)^{-a}\,du \\ &= \frac{\Gamma(c)(1-z)^{-a}}{\Gamma(b)\Gamma(c-b)} \int_0^1 w^{c-b-1}(1-w)^{b-1}\left(1-\frac{wz}{z-1}\right)^{-a} dw \\ &= (1-z)^{-a}\,F\left(a,\, c-b|\, c|\, \frac{z}{z-1}\right) \end{aligned} \tag{5.3.16}$$

Finally, by applying this equation to the right-hand side of Eq. (5.2.50), we can obtain the further specialized relation

$$F\left(a,b\left|\frac{a+b+1}{2}\right|\frac{1-z}{2}\right) = z^{-a}F\left(\frac{a}{2}, \frac{a+1}{2}\left|\frac{a+b+1}{2}\right| 1-\frac{1}{z^2}\right) \tag{5.3.17}$$

The integral representation for F can also be used to derive various recurrence relations between contiguous functions. For instance, since

$$(z-t)^{-a}t^{a-c}(t-1)^{c-b-1} = z(z-t)^{-a-1}t^{a-c}(t-1)^{c-b-1} - (z-t)^{-a-1}t^{a-c+1}(t-1)^{c-b-1}$$

therefore

$$-F(a,b|c|z) = (bz/c)F(a+1,\, b+1|\, c+1|\, z) - F(a+1,\, b|\, c|\, z)$$

Also, since $\quad (d/dz)(t-z)^{-a}t^{a-c}(t-1)^{c-b-1} = a(t-z)^{-a-1}t^{a-c}(t-1)^{c-b-1}$

we have
$$\frac{d}{dz}F(a,b|c|z) = \frac{ab}{c}F(a+1,\, b+1|\, c+1|\, z)$$

both of which, of course, can be easily obtained by manipulation of the hypergeometric series.

Before we finish our discussion of the hypergeometric function, it would be well to verify a statement made on page 584 that the different independent solutions for a differential equation may be obtained by changing the allowed limits of integration of the integral representation without changing the form of the integrand. In the case of the hypergeometric equation, the second solution about $z = 0$ is, according to Eq. (5.2.45),

$$y_2 = z^{1-c} F(b - c + 1, a - c + 1 | 2 - c | z)$$

when the first solution is $F(a,b|c|z)$. Use of Eq. (5.3.14) to obtain an integral representation for the new F gives us

$$Ay_2 = \frac{A\Gamma(2-c)}{\Gamma(a-c+1)\Gamma(1-a)} z^{1-c} \int_1^\infty (u-z)^{c-b-1} u^{b-1} (u-1)^{-a} \, du$$

which has a different integrand from the one for the first solution given in Eq. (5.3.14). However, by letting $u = z/t$ and

$$A = \frac{\Gamma(c)\Gamma(a-c+1)\Gamma(1-a)}{\Gamma(b)\Gamma(c-b)\Gamma(2-c)}$$

we finally obtain an integral representation for the second solution

$$\begin{aligned} Ay_2 &= \frac{\Gamma(c)}{\Gamma(b)\Gamma(c-b)} \int_0^z (z-t)^{-a} t^{a-c} (1-t)^{c-b-1} \, dt \\ &= \frac{\Gamma(c)\Gamma(a-c+1)\Gamma(1-a)}{\Gamma(b)\Gamma(c-b)\Gamma(2-c)} z^{1-c} F(b-c+1, a-c+1 | 2-c | z) \end{aligned} \tag{5.3.18}$$

which is valid for $\text{Re}\, c < \text{Re}\, a + 1 < 2$. This formula for the second solution differs from the formula (5.3.14) for the first solution only in the difference in limits of integration, thus verifying our earlier statement, at least as far as the hypergeometric equation goes.

Changing $2 - c$ to c, $b - c + 1$ to a, and $a - c + 1$ to b we find that

$$F(a,b|c|z) = \frac{\Gamma(c)\, z^{1-c}}{\Gamma(b)\Gamma(c-b)} \int_0^z (z-t)^{c-b-1} t^{b-1} (1-t)^{-a} \, dt \tag{5.3.19}$$

thus obtaining still another representation for the first solution, for $\text{Re}\, c > \text{Re}\, b > 0$ as before. This integral can be changed into a contour integral about 0 and z if we can arrange the contour so as to return the integral to its initial value at the end of the circuit. This requires a double circuit, as shown in Fig. (5.7), going around both points once in each direction. Denoting the integral in Eq. (5.3.19) by J, we have

$$\begin{aligned} \oint_C (t-z)^{c-b-1} t^{b-1} (1-t)^{-a} \, dt &= e^{i\pi b}[e^{i\pi(c-2b)} - e^{i\pi c} + e^{-i\pi(c-2b)} - e^{-i\pi c}]J \\ &= 4e^{i\pi b} \sin(\pi b) \sin \pi(c-b) J = \frac{4\pi^2 e^{i\pi b} J}{\Gamma(b)\Gamma(1-b)\Gamma(c-b)\Gamma(1+b-c)} \end{aligned}$$

Therefore a contour integral for F is

$$F(a,b|c|z) = \frac{e^{-i\pi b} z^{1-c}}{4\pi^2} \Gamma(c)\Gamma(1+b-c)\Gamma(1-b) \cdot \\ \cdot \oint_C (t-z)^{c-b-1} t^{b-1} (1-t)^{-a} \, dt \tag{5.3.20}$$

This equation may now be extended by analytic continuation to the whole range for a, b, and c except that which makes the gamma functions go to infinity.

Finally, by making the substitution $t = (u - z)/(1 - u)$ in Eq. (5.3.19) we can obtain

$$F(a,b|c|z) = \frac{\Gamma(c)z^{1-c}(1 - z)^{c-a-b}}{\Gamma(b)\Gamma(c - b)} \cdot \int_0^z (z - u)^{b-1}u^{c-b-1}(1 - u)^{a-c}\,du \quad (5.3.21)$$

which may also be converted into a contour integral similar to Eq. (5.3.20).

We have therefore found integral representations for both solutions of the hypergeometric equation which are valid over rather wide ranges of values of the parameters a, b, c. We could obtain other representations which are valid over other ranges by juggling the interrelations between F's. This is usually not necessary, however, for the integral representations are chiefly useful in obtaining other formulas, such as recursion relations, series expansions, and the like. Once these formulas are obtained by using the integral representations, they can be extended for other ranges of the parameters beyond the range of validity of the representation, by analytic continuation, if need be and if the formulas themselves allow it.

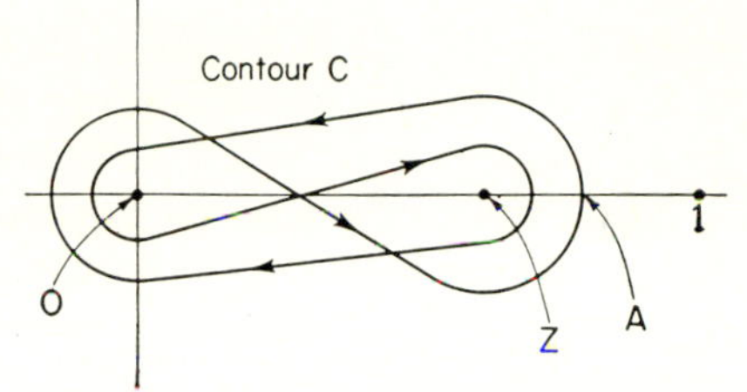

Fig. 5.7 Contour C for integral representation of hypergeometric function by Euler transform. Integrand is real at point A when z is real.

Rather than go into further detail in the properties of the general hypergeometric function, it will be more useful to study some of the special cases of particular interest later in this book.

Legendre Functions. The Gegenbauer functions, defined on page 547, represent a specialization of the hypergeometric function, since there are only two parameters, α and β, instead of the three, a, b, and c. However, these functions include the great majority of the functions of hypergeometric type actually encountered in mathematical physics, so it is important to discuss their special properties in some detail. The equation is

$$(z^2 - 1)\psi'' + 2(\beta + 1)z\psi' - \alpha(\alpha + 2\beta + 1)\psi = 0 \quad (5.3.22)$$

having three regular singular points, at -1, $+1$, and ∞, with indices $(0,-\beta)$, $(0,-\beta)$, and $(-\alpha,\ \alpha + 2\beta + 1)$, respectively.

It is interesting and useful to note that direct differentiation of the equation yields

$$(z^2 - 1)\psi''' + 2(\beta + 2)z\psi'' - (\alpha - 1)(\alpha + 2\beta + 2)\psi' = 0$$

which shows that, if $T_\alpha^\beta(z)$ is a solution of Eq. (5.3.22), $T_{\alpha-1}^{\beta+1}(z)$ is a solution of the equation directly above. This indicates that solutions for integral values of β can be obtained from the functions $T_\alpha^0(z)$ by differentiation. We also can see that, if $T_\alpha^\beta(z)$ is a solution of Eq. (5.3.22), $T^\beta_{-\alpha-2\beta-1}(z)$ is likewise a solution of the same equation.

The equation for $T_\alpha^0(z)$,

$$(z^2 - 1)\psi'' + 2z\psi' - \alpha(\alpha + 1)\psi = 0 \tag{5.3.23}$$

is called *Legendre's equation*. If α is a positive integer, it can be obtained from the equation

$$(z^2 - 1)(dV/dz) - 2\alpha z V = 0$$

where $V = (z^2 - 1)^\alpha$, by differentiation $(\alpha + 1)$ times, the final equation having $d^\alpha V/dz^\alpha$ for ψ. Therefore one solution of Legendre's equation is proportional to $d^\alpha(z^2 - 1)^\alpha/dz^\alpha$ when α is a positive integer.

With these simple properties pointed out, let us use the machinery of the present section to determine solutions of Eq. (5.3.22) for all values of α and β and expansions for these functions about the three singular points.

Reverting to page 586 for the general procedure in setting up integral representations for equations with three regular singular points, we set down, for this equation, $f(t) = (t^2 - 1)$, $g(t) = 2(\beta + 1)t$, $h = -\alpha(\alpha + 2\beta + 1)$. The equation corresponding to (5.3.10), determining μ, is

$$\mu^2 + (2\beta + 1)\mu - \alpha(\alpha + 2\beta + 1) = 0$$

so that $\mu = \alpha$ or $-\alpha - 2\beta - 1$. From Eq. (5.3.13) we then have the two alternative integral expressions for the solution and for the bilinear concomitant:

$$\psi = A \int \frac{(z - t)^\alpha}{(t^2 - 1)^{\alpha+\beta+1}}\, dt; \quad P = -\alpha \frac{(z - t)^{\alpha-1}}{(t^2 - 1)^{\alpha+\beta}}; \quad \mu = \alpha$$

$$\psi = B \int \frac{(t^2 - 1)^{\alpha+\beta}}{(z - t)^{\alpha+2\beta+1}}\, dt; \quad P = (\alpha + 2\beta + 1) \frac{(t^2 - 1)^{\alpha+\beta+1}}{(z - t)^{\alpha+2\beta+2}}$$

$$\mu = -\alpha - 2\beta - 1$$

as is usual with such representations, one integral can be changed into the other by suitable transformation of the integration variable and the limits of integration. We might as well consider that the real parts of α and β are positive to start with, for we can obtain the negative cases by analytic continuation later. When Re α and Re $\beta > 0$, proper limits for a line integral for the first alternative are $t = z$ and $t = \infty$, or the integral can be a contour integral about -1, $+1$ and z in the proper order so that P returns to its original value after the circuit.

Starting with the case $\beta = 0$, we wish first to find the solution which is proportional to the αth derivative of $(z^2 - 1)^\alpha$ when α is an integer.

Referring to Eq. (4.3.1) we see that a contour integral, using the second form, enclosing the point $t = z$ and (for instance) the point $t = 1$, is the simplest means. We accordingly set

$$T^0_\alpha(z) = P_\alpha(z) = A \oint_D \frac{(t^2-1)^\alpha}{(t-z)^{\alpha+1}}\,dt = \frac{2\pi iA}{n!}\frac{d^n}{dz^n}(z^2-1)^n$$

when $\alpha = n = 0, 1, 2, \ldots$, where the contour goes counterclockwise about both $t = 1$ and $t = z$, as shown in Fig. 5.8. The integrand is supposed to be real at point A when z is on the real axis between $+1$ and -1. We note that for z on the real axis to the left of -1 the integral has a

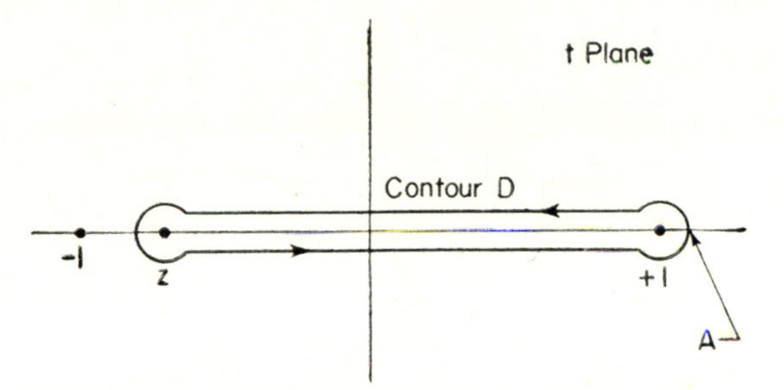

Fig. 5.8 Contour D for integral representation of Legendre function of first kind.

different value (when α is not an integer), depending on whether the contour is passed above or below the point $t = -1$; therefore we make a cut along the negative real axis from -1 to $-\infty$ to keep the function $T^0_\alpha(z)$ single-valued.

For $-1 < \mathrm{Re}\,\alpha < 0$ we can change the contour integral into a line integral from $t = z$ to $t = 1$. Then by letting $t = 1 - 2u$ and by use of Eq. (5.3.19), we obtain a series expansion for P_α:

$$\begin{aligned}
T^0_\alpha(z) &= -2iA\sin(\pi\alpha)\int_z^1 (t-z)^{-\alpha-1}(1-t)^\alpha(1+t)^\alpha\,dt \\
&= \frac{2\pi iA2^\alpha}{\Gamma(\alpha+1)\Gamma(-\alpha)}\int_0^{(1-z)/2}\left(\frac{1-z}{2}-u\right)^{-\alpha-1} u^\alpha(1-u)^\alpha\,du \\
&= 2\pi iA2^\alpha F[-\alpha,\alpha+1|1|(1-z)/2]
\end{aligned}$$

which final result may then be extended by analytic continuation to all ranges of α for which the hypergeometric series is analytic. Since it is convenient to have $T^0_\alpha(1) = 1$, we set $A = 1/2^{\alpha+1}\pi i$, so that

$$P_\alpha(z) = T^0_\alpha(z) = \frac{1}{2^{\alpha+1}\pi i}\oint_D \frac{(t^2-1)^\alpha}{(t-z)^{\alpha+1}}\,dt = F\left(-\alpha,\alpha+1|1|\frac{1-z}{2}\right) \tag{5.3.24}$$

which can be taken to be the *fundamental definition of the Legendre function* $P_\alpha(z)$. We note, because of the symmetry of F, that $P_{-\alpha-1}(z) = P_\alpha(z)$.

This formula also allows us to calculate the behavior of $P_\alpha(z)$ for very large values of z. Referring to Eq. (5.3.17),

$$P_\alpha(z) = F\left(-\alpha, \alpha + 1|1|\frac{1-z}{2}\right) = z^\alpha F\left(-\frac{\alpha}{2}, \frac{1-\alpha}{2}|1|1 - \frac{1}{z^2}\right)$$

By using Eq. (5.2.49) and the equations $\Gamma(x)\Gamma(1-x) = \pi/\sin(\pi x)$ and $\sqrt{\pi}\,\Gamma(2x) = 2^{2x-1}\Gamma(x)\Gamma(x+\frac{1}{2})$, we finally arrive at

$$\begin{aligned} P_\alpha(z) &= \frac{\Gamma(\alpha+\frac{1}{2})}{\sqrt{\pi}\,\Gamma(\alpha+1)}(2z)^\alpha F\left(-\frac{\alpha}{2}, \frac{1-\alpha}{2}\middle|\frac{1-2\alpha}{2}\middle|\frac{1}{z^2}\right) \\ &\quad + \frac{\Gamma(\alpha+1)}{\sqrt{\pi}\,\Gamma(\alpha+\frac{3}{2})}\frac{\tan(\pi\alpha)}{(2z)^{\alpha+1}}F\left(\frac{2+\alpha}{2}, \frac{1+\alpha}{2}\middle|\frac{2\alpha+3}{2}\middle|\frac{1}{z^2}\right) \end{aligned} \tag{5.3.25}$$

so that, when α is positive, $P_\alpha \propto z^\alpha$ when z is very large.

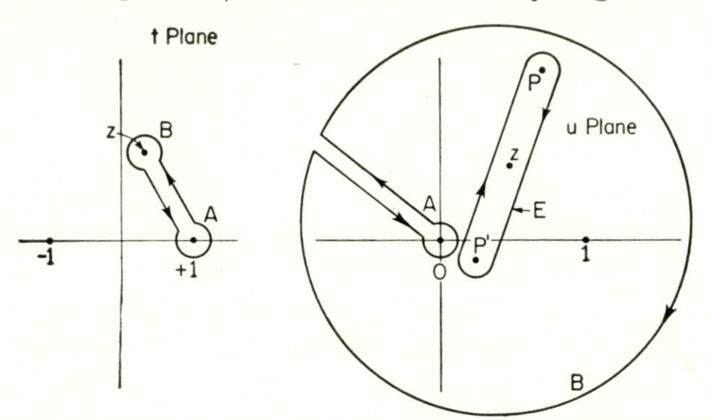

Fig. 5.9 Contour E for integral representation of Legendre function of first kind.

Another useful integral representation can be obtained from Eq. (5.3.24) by setting $u = \frac{1}{2}(t^2-1)/(t-z)$ or $t = u + \sqrt{u^2 - 2uz + 1}$, where t is real for z and u real, $|z| < 1$ and $u > 1$. The transformed integral is

$$P_\alpha(z) = T_\alpha^0(z) = \frac{1}{2\pi i}\oint_E \frac{u^\alpha\,du}{\sqrt{1-2uz+u^2}}$$

where the contour E is the one shown in Fig. 5.9; the part A, corresponding to the circuit about $t = 1$ going into a circuit about $u = 0$, and the part B, which was the circuit about $t = z$ going into a circuit a very large distance about $u = 0$, outside $u = 1$ or $u = z$. The two zeros of $\sqrt{1 - 2uz + u^2}$ are at $u = z \pm \sqrt{z^2 - 1}$ (the points P and P' in Fig. 5.9), so the contour reduces to the one marked E, about these two points.

Since $P_\alpha(z) = P_{-\alpha-1}(z)$, we also have

$$P_\alpha(z) = T_\alpha^0(z) = \frac{1}{2\pi i}\oint_E \frac{u^{-\alpha-1}\,du}{\sqrt{u^2-2uz+1}} \tag{5.3.26}$$

When α is a positive integer, the integral about the large circle B goes to zero and E reduces to the circle A about $u = 0$, so by Eq. (4.3.1)

$$P_n(z) = \frac{1}{n!}\left[\frac{d^n}{du^n}\frac{1}{\sqrt{u^2 - 2uz + 1}}\right]_{u=0}$$

and, by use of Taylor's series (4.3.3), we have

$$\frac{1}{\sqrt{1 - 2hz + h^2}} = \sum_{n=0}^{\infty} h^n P_n(z) \tag{5.3.27}$$

where $|h|$ must be smaller than the distance from the origin to the points P or P' of Fig. 5.9 and $|z| \leq 1$. Also, by use of the Laurent series and the integral with u^α, we have

$$\frac{1}{\sqrt{1 - 2hz + h^2}} = \sum_{n=0}^{\infty} \frac{1}{h^{n+1}} P_n(z)$$

where $|h|$ must be larger than either $|z \pm \sqrt{z^2 - 1}|$ or $z \leq 1$. Therefore we obtain the useful general formula

$$\frac{1}{\sqrt{r_1^2 - 2r_1r_2\cos\theta + r_2^2}} = \sum_{n=0}^{\infty}\left(\frac{r_2^n}{r_1^{n+1}}\right) P_n(\cos\theta); \quad r_2 < r_1 \tag{5.3.28}$$

Legendre Functions of the Second Kind. The second solution of Legendre's equation must be found by using a different contour for the integral expression (5.3.24). We cannot simply use the second solution of the hypergeometric functions given in Eq. (5.3.18), for with $c = 1$, $y_2 = y_1$. In the present case, since the bilinear concomitant for (5.3.24) goes to zero at $t = \pm 1$, we can use the integral

$$\int_{-1}^{1} (1 - t^2)^\alpha (z - t)^{-\alpha-1}\, dt$$

to form a second solution. Consequently we define the function

$$Q_\alpha(z) = \frac{1}{2^{\alpha+1}} \int_{-1}^{1} \frac{(1 - t^2)^\alpha}{(z - t)^{\alpha+1}}\, dt \tag{5.3.29}$$

as the *Legendre function of the second kind.* For this form $\mathrm{Re}\,\alpha > -1$ and z cannot be a real number between -1 and $+1$.

For negative values of $\alpha(\mathrm{Re}\,\alpha < -1)$, we must take a contour integral about $+1$ and -1. In order to bring the bilinear concomitant back to its original value, we make it a figure-eight contour, going around $t = -1$ in a positive direction and around $t = +1$ in a negative direction.

Hence, working out the details

$$Q_\alpha(z) = \frac{-1}{2^{\alpha+2}i\sin(\pi\alpha)} \oint_C \frac{(t^2-1)^\alpha}{(z-t)^{\alpha+1}}\,dt$$

unless α is an integer or zero. The remaining case, α a negative integer, can be taken care of by setting $Q_{-n}(\alpha) = Q_{n-1}(\alpha)$, which is allowable according to our discussion on page 595. Both integral formulas make it clear that, in order that Q be single-valued, a cut must be made between $z = +1$ and $z = -1$.

When α is a positive integer or zero, Eq. (5.3.29) may be integrated directly, giving

$$Q_0(z) = \tfrac{1}{2}\ln[(z+1)/(z-1)]; \quad Q_1(z) = \tfrac{1}{2}z\ln[(z+1)/(z-1)] - 1; \quad \text{etc.}$$

Even when α is not an integer, $Q_\alpha(z)$ has logarithmic singularities at $z = \pm 1$. For large values of z, we can develop a series in powers of $1/z$ as follows:

$$Q_\alpha(z) = \frac{1}{(2z)^{\alpha+1}} \int_{-1}^{1} dt \left\{ \sum_{m=0}^{\infty} \frac{\Gamma(\alpha+m+1)}{\Gamma(\alpha+1)m!} \left(\frac{t}{z}\right)^m (1-t^2)^\alpha \right\}$$

The integrals for m odd are zero, so that we can set $u = t^2$ and obtain

$$Q_\alpha(z) = \frac{1}{(2z)^{\alpha+1}} \sum_{n=0}^{\infty} \frac{\Gamma(\alpha+2n+1)}{\Gamma(\alpha+1)\Gamma(2n+1)} \left(\frac{1}{z^{2n}}\right) \int_0^1 u^{n-\frac{1}{2}}(1-u)^\alpha\,du;$$

$$m = 2n$$

Using Eq. 4.5.54 for evaluating the integrals and using the formula $\sqrt{\pi}\,\Gamma(2x) = 2^{2x-1}\Gamma(x)\Gamma(x+\frac{1}{2})$ several times, we finally arrive at an expression for Q_α which is useful for large values of z.

$$Q_\alpha(z) = \frac{\sqrt{\pi}}{(2z)^{\alpha+1}} \frac{\Gamma(\alpha+1)}{\Gamma(\alpha+\frac{3}{2})} F\left(\frac{2+\alpha}{2}, \frac{1+\alpha}{2} \middle| \alpha+\tfrac{3}{2} \middle| \frac{1}{z^2}\right) \tag{5.3.30}$$

as long as α is not a negative integer (in which case $Q_{-n} = Q_{n-1}$, a special case). We see, therefore, that $Q_\alpha(z) \xrightarrow[z\to\infty]{} 0$ whenever $\operatorname{Re}\alpha > -1$. Compare this with the corresponding Eq. (5.3.25) for $P_\alpha(z)$.

There are several interesting interrelations between the Legendre functions of the first and second kind. One may be obtained by manipulating the contour for $P_\alpha(-z)$ as shown in Fig. 5.10. We first set $t = -u$ in the contour integral for $P_\alpha(-z)$ and then change the contours as shown:

$$\begin{aligned} P_\alpha(-z) &= \frac{-1}{2^{\alpha+1}\pi i} \oint_A \frac{(u^2-1)^\alpha}{(z-u)^{\alpha+1}}\,du \\ &= \frac{-1}{2^{\alpha+1}\pi i} \oint_C \frac{(u^2-1)^\alpha}{(z-u)^{\alpha+1}}\,du - \frac{1}{2^{\alpha+1}\pi i} \oint_B \frac{(u^2-1)^\alpha}{(z-u)^{\alpha+1}}\,du \end{aligned}$$

But the contour C is just that for the function Q_α and contour B is that for P_α. In addition $(z - u)$ in the second integral must be changed to $(u - z)$ in order to compare with that in Eq. (5.3.24). If $\text{Im}\, z > 0$, as shown in the figure, $(z - u) = e^{i\pi}(u - z)$; if $\text{Im}\, z < 0$, then $(z - u) = e^{-i\pi}(u - z)$. Therefore we obtain

$$P_\alpha(-z) = -[2\sin(\pi\alpha)/\pi]\, Q_\alpha(z) + e^{\mp i\pi\alpha} P_\alpha(z) \tag{5.3.31}$$

with the negative exponential used when $\text{Im}\, z > 0$ and the positive when $\text{Im}\, z < 0$. This equation shows the nature of the singularity of $P_\alpha(z)$. Unless α is an integer or zero, $P_\alpha(z)$ (which is unity at $z = 1$) has a logarithmic singularity at $z = -1$; if n is an integer, $P_n(-1) = (-1)^n P_n(1) = (-1)^n$.

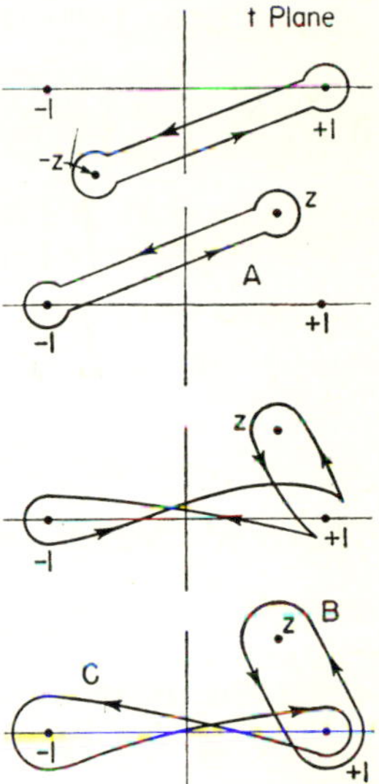

Fig. 5.10 Modification of contour to relate $P_\alpha(-z)$ with $P_\alpha(z)$ and $Q_\alpha(z)$.

This formula also allows us to obtain an expression for $Q_{-\alpha-1}$ in terms of Q_α and P_α for any values of α. Since for all α we have $P_\alpha(z) = P_{-\alpha-1}(z)$, the above equation can be rearranged to obtain

$$Q_{-\alpha-1}(z) = Q_\alpha(z) - \pi\cot(\pi\alpha)\, P_\alpha(z) \tag{5.3.32}$$

which is valid for all but integral values of α. For $\alpha = n = 0, 1, 2, \ldots$, we simply use the formula $Q_{-n-1}(z) = Q_n(z)$.

Since P and Q are independent solutions, their Wronskian $P_\alpha Q'_\alpha - P'_\alpha Q_\alpha$ should not be zero. Referring to Eq. (5.2.3), since $p = 2z/(z^2 - 1) = d\ln(z^2 - 1)/dz$, we obtain

$$\Delta(P_\alpha, Q_\alpha) = C/(z^2 - 1)$$

where the constant C can be obtained by calculating the value for some specific value of z. We choose the point at infinity, where we can use Eqs. (5.3.25) and (5.2.30). Since, for z very large,

$$P_\alpha(z) \to 2^\alpha\Gamma(\alpha + \tfrac{1}{2})z^\alpha/\sqrt{\pi}\,\Gamma(\alpha + 1)$$

and

$$Q_\alpha(z) \to \sqrt{\pi}\,\Gamma(\alpha + 1)/2^{\alpha+1}\Gamma(\alpha + \tfrac{3}{2})z^{\alpha+1}$$

we calculate that

$$[P_\alpha(z)Q'_\alpha(z) - P'_\alpha(z)Q_\alpha(z)] \to (-\alpha - 1 - \alpha)/[(2\alpha + 1)z^2]; \quad z \to \infty$$

so that $C = -1$ and we have, for all values of z,

$$\Delta(P_\alpha, Q_\alpha) = [P_\alpha(z)Q'_\alpha(z) - Q_\alpha(z)P'_\alpha(z)] = 1/(1 - z^2) \tag{5.3.33}$$

Consequently, from Eq. (5.2.4),

$$Q_\alpha(z) = P_\alpha(z) \int_z^\infty \frac{du}{[P_\alpha(u)]^2}\,(u^2 - 1)$$

Finally, we can utilize the integral representations (5.3.24) or (5.3.29) for P or Q to obtain recurrence formulas for the Legendre functions. Let $p_\alpha(z)$ be $P_\alpha(z)$ or $Q_\alpha(z)$ or any linear combination of these two functions (when the coefficients of P and Q are independent of α). Then

$$p_\alpha(z) = \frac{K}{2^\alpha} \int \frac{(t^2-1)^\alpha}{(t-z)^{\alpha+1}}\,dt$$

where the integral is over any of a number of allowed contours, depending on the linear combination involved. In any case the contour is such that $\int d[(t^2-1)^{\alpha+1}/(t-z)^{\alpha+1}]$ is zero. Therefore

$$\begin{aligned}
0 &= \frac{K}{2^{\alpha+1}} \int \frac{d}{dt}\left[\frac{(t^2-1)^{\alpha+1}}{(t-z)^{\alpha+1}}\right] dt \\
&= \frac{K(\alpha+1)}{2^{\alpha+1}} \int \left\{\frac{2t(t^2-1)^\alpha}{(t-z)^{\alpha+1}} - \frac{(t^2-1)^{\alpha+1}}{(t-z)^{\alpha+2}}\right\} dt \\
&= \frac{K(\alpha+1)}{2^\alpha} \int \frac{(t^2-1)^\alpha}{(t-z)^\alpha}\,dt + z\,\frac{K(\alpha+1)}{2^\alpha} \int \frac{(t^2-1)^\alpha}{(t-z)^{\alpha+1}}\,dt \\
&\qquad - \frac{K(\alpha+1)}{2^{\alpha+1}} \int \frac{(t^2-1)^{\alpha+1}}{(t-z)^{\alpha+2}}\,dt
\end{aligned}$$

Differentiating by z and dividing out by $(\alpha + 1)$, we have

$$p'_{\alpha+1}(z) - zp'_\alpha(z) = (\alpha+1)p_\alpha(z)$$

where the prime, as usual, indicates differentiation by z.

For another equation, we expand $\int d[t(t^2-1)^\alpha/(t-z)^\alpha] = 0$, obtaining

$$(\alpha+1)p_{\alpha+1}(z) - (2\alpha+1)zp_\alpha(z) + \alpha p_{\alpha-1}(z) = 0$$

From these, by combination and differentiation, we obtain

$$\begin{aligned}
zp_\alpha(z) &= [1/(2\alpha+1)][(\alpha+1)p_{\alpha+1}(z) + \alpha p_{\alpha-1}(z)] \\
p'_{\alpha+1}(z) - p'_{\alpha-1}(z) &= (2\alpha+1)p_\alpha(z) \\
p'_\alpha(z) &= [\alpha/(z^2-1)][zp_\alpha(z) - p_{\alpha-1}(z)]
\end{aligned} \tag{5.3.34}$$

where p_α is P_α or Q_α or a linear combination of both. Finally, using Eq. (5.3.33), we have another relation between P and Q:

$$\alpha[P_\alpha(z)Q_{\alpha-1}(z) - P_{\alpha-1}(z)Q_\alpha(z)] = 1$$

Gegenbauer Polynomials. It is now fairly easy to extend our calculations to the more general functions defined in Eqs. (5.2.52) *et seq.* We could use, as primary solution, either the function T^β_α, which is finite when α is an integer, or the function $(1-z^2)^{-\frac{1}{2}m}P^\beta_{\alpha+\beta}$, which simplifies when β is an integer (the two functions are equal when *both* α and β are integers). We prefer to make the first choice, since we are more often interested in the case α an integer β not an integer than we are in the inverse case. We accordingly define (see Eq. 5.3.21)

$$T_\alpha^\beta(z) = \frac{\Gamma(\alpha + 2\beta + 1)}{2^\beta\Gamma(\alpha + 1)\Gamma(\beta + 1)} F(\alpha + 2\beta + 1, -\alpha| 1 + \beta| \tfrac{1}{2} - \tfrac{1}{2}z)$$
$$= \frac{e^{i\pi(\alpha+\beta)}\Gamma(\alpha + 2\beta + 1)}{2^{\alpha+\beta+2}\pi\Gamma(\alpha + \beta + 1) \sin [\pi(\alpha + \beta)]} \oint_C \frac{(t^2 - 1)^{\alpha+\beta}}{(t - z)^{\alpha+1}} dt \quad (5.3.35)$$

where the contour C is the one given in Fig. 5.7, going around $t = 1$ and $t = z$ in both positive and negative directions. When α is an integer, the integral may be changed into the differential form given in Eq. (5.2.53).

On the other hand, if we had preferred the other solution, we could define

$$(1 - z^2)^{-\frac{1}{2}\beta}P_{\alpha+\beta}^\beta(z) = \frac{(z - 1)^{-\beta}}{\Gamma(1 - \beta)} F(-\alpha -\beta, \alpha + \beta + 1| 1 - \beta| \tfrac{1}{2} - \tfrac{1}{2}z)$$
$$= \frac{e^{-i\pi(\alpha+\beta)}\Gamma(\alpha + 2\beta + 1)}{2^{\alpha+\beta+2}\pi\Gamma(\alpha + \beta + 1) \sin [\pi(\alpha + \beta)]} \oint_C \frac{(t^2 - 1)^{\alpha+\beta}}{(t - z)^{\alpha+2\beta+1}} dt \quad (5.3.36)$$

which, when β is a positive integer m, is equal to $T_\alpha^m(z)$, which, in this case, equals the mth derivative of the Legendre function $P_{m+\alpha}(z)$. When *both* α and β are integers,

$$T_n^m(z) = (1 - z^2)^{-\frac{1}{2}m}P_{n+m}^m(z) = \frac{d^m}{dz^m} T_{m+n}^0(z)$$
$$= \frac{(n + 2m)!}{2^{n+m}n!(n + m)!} \frac{1}{(z^2 - 1)^m} \frac{d^n}{dz^n} (z^2 - 1)^{n+m}$$
$$= \frac{1}{2^{n+m}(n + m)!} \frac{d^{n+2m}}{dz^{n+2m}} (z^2 - 1)^{n+m} \quad (5.3.37)$$

as indicated in Eq. (5.2.53). These polynomials are called *associated Legendre polynomials.*

From the symmetry of the hypergeometric functions we have

$$T_{-\alpha-2\beta-1}^\beta(z) = \left[\frac{\sin \pi(\alpha + 2\beta)}{\sin(\pi\alpha)}\right] T_\alpha^\beta(z); \quad P_{\alpha+\beta}^\beta(z) = P_{-\alpha-\beta-1}^\beta(z)$$

and for $\beta = m$, an integer, we use the relation $F(a,b|c|w) = (1 - w)^{c-a-b} \cdot F(c - b, c - a| c| w)$ to show that

$$T_\alpha^{-m}(z) = (1 - z^2)^{\frac{1}{2}m}P_{\alpha-m}^{-m}(z)$$
$$= \frac{(z - 1)^m}{\Gamma(1 + m)} F(m - \alpha, \alpha - m + 1| 1 + m| \tfrac{1}{2} - \tfrac{1}{2}z)$$
$$= \frac{(z^2 - 1)^m}{2^m m!} F(2m - \alpha, 1 + \alpha| 1 + m| \tfrac{1}{2} - \tfrac{1}{2}z)$$
$$= \frac{\Gamma(\alpha - 2m + 1)}{\Gamma(\alpha + 1)} (z^2 - 1)^m T_{\alpha-2m}^m(z)$$
$$(-1)^m P_\gamma^{-m}(z) = \frac{\Gamma(\gamma - m + 1)}{\Gamma(\gamma + m + 1)} P_\gamma^m(z) \quad (5.3.38)$$

An analogous set of formulas to Eqs. (5.3.26) and (5.3.27) may also be found, showing that the generating function for the Gegenbauer polynomials is

$$\frac{2^\beta\Gamma(\beta+\frac{1}{2})/\sqrt{\pi}}{(1+h^2-2hz)^{\beta+\frac{1}{2}}} = \sum_{n=0}^{\infty} h^n T_n^\beta(z); \quad |h| < 1 \tag{5.3.39}$$

By using Eq. (5.3.5) and the relation

$$F\left(a,b|2b|\frac{2}{1-z}\right) = \left(\frac{z-1}{z}\right)^a F\left(\tfrac{1}{2}a,\ \tfrac{1}{2}a+\tfrac{1}{2}|\,b+\tfrac{1}{2}|\,\frac{1}{z^2}\right)$$

which is closely related to Eq. (5.3.17), we obtain the expansion for these solutions for large values of z:

$$\begin{aligned} T_\alpha^\beta(z) = {} & \frac{\Gamma(\alpha+2\beta+1)\sin(\pi\alpha)\, z^{-\alpha-2\beta-1}}{\sqrt{\pi}\, 2^{\alpha+\beta+1}\cos[\pi(\alpha+\beta)]\,\Gamma(\alpha+\beta+\frac{3}{2})} \cdot \\ & \cdot F\left(\frac{\alpha+2\beta+1}{2}, \frac{\alpha+2\beta+2}{2}\,|\,\alpha+\beta+\tfrac{3}{2}|\,\frac{1}{z^2}\right) \\ & + \frac{2^{\alpha+\beta}\Gamma(\alpha+\beta+\frac{1}{2})}{\sqrt{\pi}\,\Gamma(\alpha+1)} z^\alpha F\left(-\tfrac{1}{2}\alpha, \tfrac{1}{2}-\tfrac{1}{2}\alpha|\,\tfrac{1}{2}-\alpha-\beta|\,\frac{1}{z^2}\right) \end{aligned} \tag{5.3.40}$$

$$\begin{aligned} (1-z^2)^{-\frac{1}{2}\beta}P_{\alpha+\beta}^\beta(z) = {} & \frac{\Gamma(\alpha+2\beta+1)\sin[\pi(\alpha+2\beta)]\, z^{-\alpha-2\beta-1}}{\sqrt{\pi}\, 2^{\alpha+\beta+1}\cos[\pi(\alpha+\beta)]\,\Gamma(\alpha+\beta+\frac{3}{2})} \cdot \\ & \cdot F\left(\frac{\alpha+2\beta+1}{2}, \frac{\alpha+2\beta+2}{2}\,|\,\alpha+\beta+\tfrac{3}{2}|\,\frac{1}{z^2}\right) \\ & + \frac{2^{\alpha+\beta}\Gamma(\alpha+\beta+\frac{1}{2})}{\sqrt{\pi}\,\Gamma(\alpha+1)} z^\alpha F\left(-\tfrac{1}{2}\alpha, \tfrac{1}{2}-\tfrac{1}{2}\alpha|\,\tfrac{1}{2}-\alpha-\beta|\,\frac{1}{z^2}\right) \end{aligned}$$

which shows that both of these solutions go to infinity at $z \to \infty$ unless $0 > \alpha > -2\beta - 1$.

If we consider the function $T_\alpha^\beta(z)$ to be the first solution of Eq. (5.2.51), we now have collected an embarrassing richness of second solutions. The function $T_{-\alpha-2\beta-1}^\beta(z)$ is, of course, proportional to $T_\alpha^\beta(z)$ and thus is not independent, but the function variously labeled

$$\begin{aligned} (1-z^2)^{-\frac{1}{2}\beta}P_{\alpha+\beta}^\beta(z) &= (z^2-1)^{-\beta}\frac{\Gamma(\alpha+2\beta+1)}{\Gamma(\alpha+1)}\frac{\sin[\pi(\alpha+2\beta)]}{\sin(\pi\alpha)} T_{-\alpha-1}^{-\beta}(z) \\ &= (1-z^2)^{-\frac{1}{2}\beta}P_{-\alpha-\beta-1}^\beta(z) = (z^2-1)^{-\beta}\frac{\Gamma(\alpha+2\beta+1)}{\Gamma(\alpha+1)} T_{\alpha+2\beta}^{-\beta}(z) \end{aligned}$$

is independent of $T_\alpha^\beta(z)$ (unless β is an integer) as are likewise $T_\alpha^\beta(-z)$ and $(1-z^2)^{-\frac{1}{2}\beta}P_{\alpha+\beta}^\beta(-z)$. In addition, there are the functions with the sign of the argument reversed. Equation (5.2.49) can be used to show that

$$\sin(\pi\beta)\, T_\alpha^\beta(z) = \sin[\pi(\alpha+\beta)]\, T_\alpha^\beta(-z) - \sin(\pi\alpha)(1-z^2)^{-\frac{1}{2}\beta}\, P_{\alpha+\beta}^\beta(-z)$$

which also shows the special relationships occurring when either β or α is an integer.

All these functions, however, go to infinity at $z \to \infty$ (except when $0 > \alpha > -2\beta - 1$). We shall often prefer to have a second solution which goes to zero as $z \to \infty$ when $\alpha > 0$, as do the Q functions defined in Eq. (5.3.29). Also we should prefer a solution which stays independent of $T_\alpha^\beta(z)$ even when β is an integer.

Such a function may be obtained from Eq. (5.3.39). We set

$$\begin{aligned}
V_\alpha^\beta(z) &= \frac{e^{-i\pi(\alpha+\beta)}\Gamma(\alpha + 2\beta + 1)}{2^{\alpha+\beta+2}i\Gamma(\alpha + \beta + 1)\sin[\pi(\alpha+\beta)]} \oint_Q \frac{(t^2 - 1)^{\alpha+\beta}}{(t - z)^{\alpha+2\beta+1}}\,dt \\
&= \frac{e^{i\pi\beta}2^\beta\sqrt{\pi}\,\Gamma(\alpha + 2\beta + 1)}{\Gamma(\alpha + \beta + \frac{3}{2})(2z)^{\alpha+2\beta+1}} \cdot \\
&\qquad \cdot F\left(\frac{\alpha + 2\beta + 1}{2}, \frac{\alpha + 2\beta + 2}{2} \middle| \alpha + \beta + \tfrac{3}{2} \middle| \frac{1}{z^2}\right) \\
&= \frac{-\pi e^{i\pi\beta}}{2\sin(\pi\beta)}\left[T_\alpha^\beta(z) - \frac{\Gamma(\alpha + 2\beta + 1)}{\Gamma(\alpha + 1)(z^2 - 1)^\beta} T_{-\alpha+2\beta}^{-\beta}(z)\right] \\
&= \frac{\pi}{2\sin[\pi(\alpha + 2\beta)]}[e^{\mp i\pi(\beta+\alpha)}T_\alpha^\beta(z) - e^{\pm i\pi\beta}T_\alpha^\beta(-z)] \\
&= e^{2\pi i\beta}\frac{\Gamma(\alpha + 2\beta + 1)}{\Gamma(\alpha + 1)z^2 - 1)^\beta} V_{\alpha+2\beta}^{-\beta}(z) = e^{i\pi\beta}(z^2 - 1)^{-\frac{1}{2}\beta}Q_{\alpha+\beta}^\beta(z)
\end{aligned} \tag{5.3.41}$$

The contour for the integral is the figure eight, about $+1$ and -1, which was used for $Q_\alpha(z)$. The second line shows that V goes to zero when $z \to \infty$ as long as $\mathrm{Re}\,(\alpha + 2\beta + 1) > 0$, as was desired. The third line shows the relationship between V and the two independent T solutions. The fourth line, to be compared with Eq. (5.3.31), relates the solutions for $+z$ with those for $-z$; the upper sign is to be used when $\mathrm{Im}\,z > 0$, the lower when $\mathrm{Im}\,z < 0$. The last line shows that this second solution is adjusted so that change of sign of superscript does not produce an independent function, in contrast to T_α^β, where Eq. (5.3.38) holds only for β an integer. On the other hand, $V_{-\alpha-2\beta-1}^\beta$ is not proportional to V_α^β but

$$V_\alpha^\beta(z) = V_{-\alpha-2\beta-1}^\beta(z) + [\pi e^{i\pi\beta}\cos\pi(\alpha + \beta)/\sin(\pi\alpha)]T_\alpha^\beta(z)$$

When $\beta = m$, an integer, the third relation in Eqs. (5.3.41) goes to a finite limit, but in this case the following, simpler relations hold:

$$V_\alpha^m(z) = \frac{d^m}{dz^m}Q_{\alpha+m}(z); \quad V_{-\alpha-2m-1}^m = \frac{\Gamma(\alpha + 2m + 1)}{\Gamma(\alpha + 1)(z^2 - 1)^m} V_{-\alpha-1}^{-m} \tag{5.3.42}$$

The case of $\beta = \pm\frac{1}{2}$ is of some interest, both because of the resulting polynomials and also because of the special properties of the hypergeometric functions which are displayed:

$$T_\alpha^{-\frac{1}{2}}(z) = \sqrt{\frac{2}{\pi}}\frac{\cosh[\alpha\cosh^{-1}z]}{\alpha} = \frac{1}{\alpha}\frac{\sqrt{z^2-1}}{(1-z^2)^{\frac{1}{4}}}P_{\alpha-\frac{1}{2}}^{\frac{1}{2}}(z)$$
$$T_\alpha^{\frac{1}{2}}(z) = \sqrt{\frac{2}{\pi}}\frac{\sinh[(\alpha+1)\cosh^{-1}z]}{\sqrt{z^2-1}} = \alpha\frac{(1-z^2)^{\frac{1}{4}}}{\sqrt{z^2-1}}P_{\alpha-\frac{1}{2}}^{-\frac{1}{2}}(z) \quad (5.3.43)$$
$$V_\alpha^{-\frac{1}{2}}(z) = \sqrt{2\pi}\left(\frac{i}{\alpha}\right)\exp[-\alpha\cosh^{-1}z] = -\frac{1}{\alpha}\sqrt{z^2-1}\,V_{\alpha-1}^{\frac{1}{2}}(z)$$

which, for α an integer, are proportional to the Tschebyscheff polynomials.

The Confluent Hypergeometric Function. The hypergeometric equation for $F(a, b + c|\,c|\,z/b)$ is

$$z\left(1-\frac{z}{b}\right)F'' + \left[c - (a+b+c+1)\frac{z}{b}\right]F' - a\left(1+\frac{c}{b}\right)F = 0$$

This equation has regular singular points at 0, b, and ∞ with indices $(0, 1 - c)$, $(0, -a - b)$, and $(a, b + c)$, respectively. If we let b go to infinity, we shall at the same time merge two singular points (the ones originally at b and ∞) and make one index about each of the merged points go to infinity (the index $-a - b$ at the point b and the index $b + c$ at the point ∞). This double-barreled process is called a *confluence* of singular points; it has been mentioned before on page 537. The resulting equation

$$zF'' + (c - z)F' - aF = 0 \quad (5.3.44)$$

is called the *confluent hypergeometric equation* [see Eq. (5.2.57)]. The solution analytic at $z = 0$ is given by the limiting form of $F(a, b + c|\,c|\,z/b)$ when b goes to infinity;

$$F(a|c|z) = 1 + \frac{a}{c}z + \frac{a(a+1)}{2!c(c+1)}z^2 + \frac{a(a+1)(a+2)}{3!c(c+1)(c+2)}z^3 + \cdots \quad (5.3.45)$$

is called the *confluent hypergeometric series.*

To see how the confluence affects the integral representation, we start with a form derivable from Eq. (5.3.14) by changing t into $1/t$:

$$F\left(a, b+c\,|\,c\,|\,\frac{z}{b}\right) = F\left(b+c, a\,|\,c\,|\,\frac{z}{b}\right)$$
$$= \frac{\Gamma(c)}{\Gamma(a)\Gamma(c-a)}\int_0^1\left(1-\frac{zt}{b}\right)^{-b-c}t^{a-1}(1-t)^{c-a-1}\,dt$$

Letting b go to infinity changes the nature of the kernel $[1 - (zt/b)]^{-b-c}$ from an algebraic to an exponential function, for

$$\left[1-\frac{x}{b}\right]^{-b} \xrightarrow[b\to\infty]{} e^x$$

The singularity changes from a branch point at $x = b$ to an essential singularity at $x = \infty$, and the resulting integral representation for the confluent hypergeometric function is then

$$F(a|c|z) = \frac{\Gamma(c)}{\Gamma(a)\Gamma(c-a)} \int_0^1 e^{zt}\, t^{a-1}(1-t)^{c-a-1}\, dt \qquad (5.3.46)$$

as long as Re $c >$ Re $a > 0$. For other ranges of a and c we can devise a corresponding contour integral which will be valid except when c is a negative integer or zero, in which case even the series expansion "blows up." In these cases, which will be discussed later, the solution has a logarithmic branch point at $z = 0$.

The second solution of Eq. (5.3.44) may be found by performing the confluence on the second solution of the hypergeometric equation $z^{1-c} F(a - c + 1, b + 1|\, 2 - c|\, z/b)$ or may be found by remembering that $z^{1-c} f(z)$ is a solution and, by inserting this in Eq. (5.3.44), showing that the equation for f is also of the form of (5.3.44). In either case the second solution is found to be

$$z^{1-c} F(a - c + 1|\, 2 - c|\, z)$$

which is valid unless c is 2, 3, 4, ... If $c = 1$, this series is convergent but it is identical with the first solution $F(a|1|z)$, so that the above expression does not represent a second solution whenever c is a positive integer. The second solution in these special cases will be obtained later. There is no need to investigate the case of $c < 1$ separately, for if $c = 2 - c' < 1$, we can multiply our solutions by $z^{c-1} = z^{1-c'}$ and call the original second solution the first solution, and vice versa, and then c', which is the new c, is greater than 1.

An interesting interrelation, first pointed out in Eq. (5.2.62), may be derived from the integral representation, for since

$$\int_0^1 e^{zt}\, t^{a-1}(1-t)^{c-a-1}\, dt = e^z \int_0^1 e^{-zu}\, u^{c-a-1}\, (1-u)^{a-1}\, du$$

we must have $F(a|c|z) = e^z F(c - a|c|-z)$. Similarly another form of the second solution is $z^{1-c} e^z F(1 - a|2 - c|-z)$.

The Laplace Transform. But before we go further with a discussion of the solution, we should examine more carefully the new form of the integral representation, for we have now changed from an Euler transform to a *Laplace transform*, mentioned on page 584. The new kernel is e^{zt}, not $(t - z)^\mu$. Referring to page 583, where we discussed integral representations in general, we see that the exponential kernel has certain advantages. For instance, since $de^{zt}/dz = te^{zt}$ and $ze^{zt} = de^{zt}/dt$, if our original equation is

$$0 = \mathfrak{L}_z(\psi) = f\frac{d^2\psi}{dz^2} + g\frac{d\psi}{dz} + h\psi = \sum_{m,n} A_{mn}\, z^m \frac{d^n\psi}{dz^n}$$

$$f = \sum_m A_{m2} z^m; \quad g = \sum_m A_{m1} z^m; \quad h = \sum_m A_{m0} z^m$$

then the corresponding transformed expression will be

$$\mathfrak{M}_t(e^{zt}) = \sum_{m,n} A_{mn} t^n \frac{d^m}{dt^m} e^{zt}$$

and the adjoint to this will give the equation

$$\tilde{\mathfrak{M}}_t(v) = \sum_{m,n} (-1)^m A_{mn} \frac{d^m}{dt^m} (t^n v) = 0$$

If this is a simpler equation to solve than $\mathfrak{L}_z(\psi) = 0$, then the Laplace transform will be a suitable means for integral representation.

For instance, if in the original equation $\mathfrak{L}(\psi) \equiv f\psi'' + g\psi' + h\psi = 0$ the polynomials f, g, and h are none higher than the first order in z, then in the differential operator $\mathfrak{M}_t$ there will be no derivatives in t higher than the first and $v(t)$ can be easily found. The most general type of this sort of equation is

$$(z - \delta)y'' + (2\epsilon z - \xi)y' + (\eta z - \zeta)y = 0$$

where δ can be eliminated by proper choice of origin for z. This sort of equation for y may be obtained from the following equation for ψ, where $\psi = z^\alpha y$:

$$\psi'' + \left(2\epsilon - \frac{2\alpha + \xi}{z}\right)\psi' + \left[\eta - \frac{2\alpha\epsilon + \zeta}{z} + \frac{\alpha(\alpha + \xi + 1)}{z^2}\right]\psi = 0$$

which is of the general form of Eq. (5.2.55). This equation has a regular singular point at $z = 0$ with indices α and $\alpha + \xi + 1$ and an irregular point at $z = \infty$. The substitution $y = F \exp[(-\epsilon + \sqrt{\epsilon - \eta})z]$ results in a confluent hypergeometric equation for F. Consequently the confluent hypergeometric equation is particularly amenable to the use of the Laplace transform for its solution.

Other equations, having f, g, or h polynomials of higher order than the first, may be solved by a Laplace transform, but the resulting equation for v will be a second-order (or higher) one, which would have to be particularly simple for its solution to have a closed form.

At any rate we shall commence our study of the Laplace integral representation by applying it to the confluent hypergeometric equation. Referring to the above equation, if we set $F = \int e^{zt}v(t)\,dt$, the expression $\mathfrak{M}_t(e^{zt})$ corresponding to Eq. (5.3.44), is

$$(t^2 - t)(d/dt)e^{zt} + (ct - a)e^{zt}$$

and the adjoint equation is

$$\tilde{\mathfrak{M}}(v) \equiv -(d/dt)(t^2 - t)v + (ct - a)v = 0$$

with solution and bilinear concomitant

$$v = At^{a-1}(1 - t)^{c-a-1}; \quad P = -At^a(1 - t)^{c-a}\, e^{zt}$$

The possible paths of integration are therefore from 0 to 1 for Re $c >$ Re $a > 0$, as was given in Eq. (5.3.46), or from $-\infty$ to 0 for Re $a > 0$ and Re $z > 0$, or else a double contour about 0 and 1 which brings P back to its original value. Each of these integrals will, of course, represent a different solution. The one representing the confluent hypergeometric series is that given in Eq. (5.3.46) or by the contour integral

$$F(a|c|z) = \frac{e^{-i\pi a}}{4\pi^2}\,\Gamma(c)\Gamma(1-a)\Gamma(a-c+1)\oint_C e^{zt}\,t^{a-1}(t-1)^{c-a-1}\,dt \qquad (5.3.47)$$

where the integrand is real for z real and t on the real axis to the right of $t = 1$ and the contour goes counterclockwise about $t = 1$ and $t = 0$ and then clockwise about both, as in Fig. 5.7. This representation is valid except for the points where the gamma functions are not analytic (*i.e.*, when $1 - c$, a, or $a - c$ are negative integers).

Asymptotic Expansion. Let us next investigate the behavior of $F(a|c|z)$ for very large values of z, using Eq. (5.3.46) for the suitable representation. On page 552 we have already discussed the complications inherent in asymptotic expansions about irregular singular points and have indicated that, when the real part of z is large and positive,

$$F(a|c|z) \xrightarrow[\mathrm{Re}\, z \to \infty]{} \frac{\Gamma(c)}{\Gamma(a)}\, z^{a-c}e^{z}$$

We have also indicated that this expression is not correct for Re z large and negative but that the correct form for this case is probably

$$F(a|c|z) \xrightarrow[\mathrm{Re}\, z \to -\infty]{} \frac{\Gamma(c)}{\Gamma(c-a)}\,(-z)^{-a}$$

We must now verify these tentative results and try to understand what properties of the asymptotic expansion are responsible for this curious, indecisive behavior of the function which makes it have one form for z large and positive and another for z large and negative.

Referring to Eq. (5.3.46) we see that, if z is real, positive, and large, the most important part of the integrand is near $t = 1$ and that it does not change the integral appreciably if we extend the integration from 0 to $-\infty$ for t. To be more precise, we can write

$$F(a|c|z) = \frac{\Gamma(c)}{\Gamma(a)\Gamma(c-a)}\cdot \cdot\left[\int_{-\infty}^{1} e^{zt}\,t^{a-1}(1-t)^{c-a-1}\,dt - \int_{-\infty}^{0} e^{zt}\,t^{a-1}(1-t)^{c-a-1}\,dt\right]$$

which is valid as long as Re z is positive. Setting $t = 1 - (u/z)$ in the first integral and $t = -(w/z)$ in the second integral, we have

$$F(a|c|z) = \frac{\Gamma(c)}{\Gamma(a)\Gamma(c-a)} \left\{ z^{a-c}e^z \int_0^\infty e^{-u}u^{c-a-1}\left(1-\frac{u}{z}\right)^{a-1} du \right.$$
$$\left. + (-z)^{-a} \int_0^\infty e^{-w}\, w^{a-1}\left(1+\frac{w}{z}\right)^{c-a-1} dw \right\} \quad (5.3.48)$$

The two integrals are now of the same form and will be shown to be approximately equal to $\Gamma(c-a)$ and $\Gamma(a)$, respectively. Therefore the first term, due to the factor e^z, is vastly larger than the second term whenever Re z is large and positive, and the second term may then be neglected (though we must keep in mind the fact we have neglected it, for it is the essential clue to the indecisive behavior of F mentioned above).

An adaptation of the argument used on page 376 for the Taylor's series expansion shows that

$$\left(1-\frac{u}{z}\right)^{a-1} = \sum_{m=0}^{n-1} \frac{\Gamma(a)}{\Gamma(m+1)\Gamma(a-m)}\left(-\frac{u}{z}\right)^m + D_n$$

where $D_n \leq \dfrac{\Gamma(a)}{\Gamma(n+1)\Gamma(a-n)}\left|-\dfrac{u}{z}\right|^{n+1}$; when $n > \text{Re } a - 1$

Inserting this in the integral form we have

$$\int_0^\infty e^{-u}u^{c-a-1}\left(1-\frac{u}{z}\right)^{a-1} du$$
$$= \Gamma(a) \sum_{m=0}^{n} \frac{\Gamma(c-a+m)}{\Gamma(m+1)\Gamma(a-m)}\left(-\frac{1}{z}\right)^m + R_n$$
$$= \Gamma(c-a)\left\{1 + \frac{(c-a)(1-a)}{1!}\frac{1}{z} + \frac{(c-a)(c-a+1)(1-a)(2-a)}{2!}\frac{1}{z^2} + \cdots \right.$$
$$\left. + \frac{(c-a)\cdots(c+n-a-1)(1-a)\cdots(n-a)}{n!}\frac{1}{z^n}\right\} + R_n \quad (5.3.49)$$

where $R_n \leq \left|\dfrac{(1-a)\cdots(n-a)\Gamma(c+n-a+1)}{n!z^{n+1}}\right|$

We see that for finite values of z the series diverges but that the sum of the first n terms of the series approaches the "true" value as n is held constant and z is increased without limit. For instance, for $z = 10$ and $|c-a| < 2$ and $|1-a| < 1$ the first four terms of the series give a value for the integral which is correct to about one-tenth of a per cent whereas the first hundred terms for the same values of z and the parameters have a value which is far from the correct value. This property is typical of *asymptotic series*, as discussed on page 434. From the first few terms we get a fairly accurate value, but the error can be reduced to zero only when z is made infinite.

Usually we are interested in values for asymptotic series for values of z very much larger than unity ($z > 1{,}000$, for instance). In this case the first term in the series is sufficient for our purpose. In the present case we can write, therefore, that $F(a|c|z) \to [\Gamma(c)/\Gamma(a)]z^{a-c}e^z$, for z real, positive, and large. This corresponds to Eq. (5.2.60). We have not included the term $[\Gamma(c)/\Gamma(c-a)]z^{-a}$ coming from the second integral because this is, in general, smaller than the error inherent in the first asymptotic series and it thus would be senseless to include it.

The situation is quite different when z is real, *negative*, and large. In this case the integrand in Eq. (5.3.46) is largest for t near zero and the second integral in Eq. (5.3.48) is much larger than the first. In this case, starting from Eq. (5.3.46) and saving only the dominant term, we have (where $z = -|z|$ and setting $t = u/|z|$)

$$F(a|c|z) \to \frac{\Gamma(c)|z|^{-a}}{\Gamma(a)\Gamma(c-a)} \int_0^\infty e^{-u}u^{a-1}\left(1 - \frac{u}{|z|}\right)^{c-a-1} du \to \frac{\Gamma(c)}{\Gamma(c-a)} |z|^{-a}$$

for z real, negative, and large. This corresponds to Eq. (5.2.63). Here we have omitted the term in $e^z = e^{-|z|}$ coming from the first integral of Eq. (5.3.48) because it is, in general, smaller than the error inherent in the asymptotic expansion of the second integral.

Stokes' Phenomenon. We should have been able to see, by simple examination of Eq. (5.2.60), that this asymptotic form for F could not be an accurate expression for all positions of z along the circle at infinity. If we set $z = |z|e^{i\phi}$, where $|z|$ is very large, into the expression $[\Gamma(c)/\Gamma(a)]z^{a-c}e^z$, we see that the expression does not return to its original value as ϕ is increased from 0 to 2π. But since $F(a|c|z)$ is analytic over the whole finite part of the z plane, a correct formula for large z must predict that F returns to its initial value when ϕ goes from 0 to 2π, describing a circle of large radius about the origin. An expression involving z^{a-c}, as Eq. (5.2.60) does, cannot be a correct expression for F for $|z|$ large for all phase angles ϕ of z. And we have just seen that it is not; for z real and negative the expression $[\Gamma(c)/\Gamma(c-a)](-z)^{-a}$ is the proper one (and this second expression cannot, in turn, be valid for all ϕ's, for it has a term z^{-a} which would produce multivaluedness).

Just to see how this goes in more detail, let us take the case of $z = |z|e^{i\phi}$ where $|z|$ is very large. We first take ϕ between zero and π (z in the upper half plane). The path of integration in Eq. (5.3.46) is then deformed to that shown by the solid lines in Fig. 5.11, going from $t = 0$ to $t = -\infty e^{-i\phi}$ and from thence back to $t = 1$. Thus again the integral splits into two integrals. For the first let $t = -we^{-i\phi}/|z| = we^{i(\pi-\phi)}/|z|$, and for the second let $t = 1 - ue^{-i\phi}/|z|$, where both u and w are real. The integral representation then becomes

$$F(a|c|z) = \frac{\Gamma(c)}{\Gamma(a)\Gamma(c-a)} \left\{ \frac{e^{|z|e^{i\phi} - i(c-a)\phi}}{|z|^{c-a}} \int_0^\infty e^{-u} u^{c-a-1} \left(1 - \frac{u}{z}\right)^{a-1} du \right.$$
$$\left. + \frac{e^{ia(\pi-\phi)}}{|z|^a} \int_0^\infty e^{-w} w^{a-1} \left(1 + \frac{w}{z}\right)^{c-a-1} dw \right\} \quad (5.3.50)$$

for $0 < \phi < \pi$. Writing only the dominant terms in the expansions of $[1 \pm (w \text{ or } u/z)]$ (*i.e.*, assuming that these quantities are practically equal to 1 over the range of values of w or u for which the integrand is not negligible) we have for the asymptotic formula for $z = |z|e^{i\phi}$, $|z|$ large, $0 < \phi < \pi$

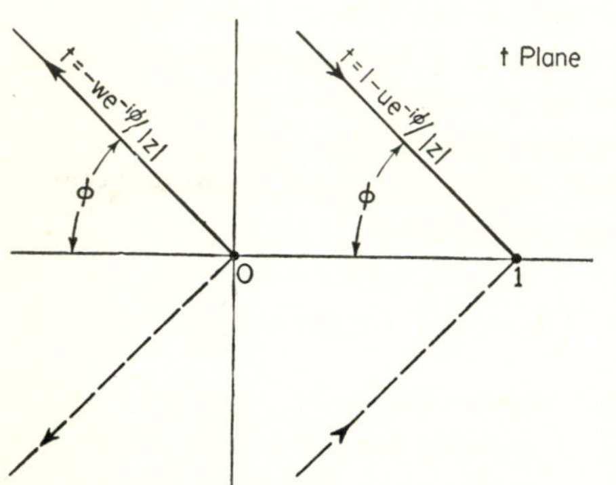

Fig. 5.11 Modification of contour for computing asymptotic behavior of confluent hypergeometric function. Independent variable is $z = |z|e^{i\phi}$, with $|z| \gg 1$.

$$F(a|c|z) \to \frac{\Gamma(c)}{\Gamma(a)} |z|^{a-c} e^{i(a-c)\phi} e^{|z|e^{i\phi}} + \frac{\Gamma(c)}{\Gamma(c-a)} |z|^{-a} e^{ia(\pi-\phi)}$$

Each of these terms has an unavoidable error, inherent in asymptotic series, which is small when z is large but which is zero only when z is infinite. When ϕ is zero (z real and positive), the second term in this expression is smaller than the unavoidable error in the first term (and thus should not be included), and when ϕ is π (z real and negative), the first term is smaller than the unavoidable error in the second (and thus should not be included). When $\phi = \pi/2$ (z positive imaginary), both terms are of about equal size and both must be used to get the correct value.

To find the expression for $0 > \phi > -\pi$ we use the dashed path of integration shown in Fig. 5.11, with $t = -we^{-i\phi}/|z| = we^{-i(\phi+\pi)}/|z|$ and $t = 1 - ue^{-i\phi}/|z|$, respectively. Here we have let $1/-z = e^{-i(\phi+\pi)}/|z|$ in the expression for w, instead of the expression $e^{i(\pi-\phi)}/|z|$ used before, because we want w to be real and positive when t is along the dashed line and because we have taken the cut for the integrand from $t = 0$ to $t = -\infty$. With this change we find that

$$F(a|c|z) = \frac{\Gamma(c)}{\Gamma(a)} |z|^{a-c} e^{i(a-c)\phi} e^{|z|e^{i\phi}} + \frac{\Gamma(c)}{\Gamma(c-a)} |z|^{-a} e^{-ia(\pi+\phi)}$$

when $z = |z|e^{i\phi}$ and $0 > \phi > -\pi$. We note that, as ϕ goes negative from 0 to $-\pi/2$, the second term emerges from its "eclipse" at $\phi = 0$, differing from the corresponding term at $\phi = +\pi/2$ by a factor $e^{-2\pi ia}$. This sudden change in the second term at $\phi = 0$ does not require a discontinuity in F as ϕ goes through zero, for it is just at $\phi = 0$ that the second term is smaller than the inherent error in the first term, so any change in the second term does not count.

However, the change is just what is required to make the second term real for both $\phi = \pi$ and $\phi = -\pi$. If the additional factor $e^{-2\pi ia}$ had not been included in the expression for the range $0 > \phi > -\pi$, the asymptotic formula for $\phi = -\pi$ would have been $e^{2\pi ia}$ times the formula for $\phi = +\pi$, which would have been wrong because F is analytic about $z = 0$. In order to maintain this analyticity the two terms in the asymptotic expansion play "hide and seek" with each other; as each goes into "eclipse," becoming smaller than the error in the other, it changes phase, emerging from obscurity with enough of an additional phase factor to ensure continuity over the range of ϕ where it is large.

This "hide-and-seek" behavior of the terms in an asymptotic expansion is called the *Stokes' phenomenon*, after the person who first called attention to it. It must be displayed by the first term at $\phi = \pi$, where the first term is smaller than the error in the second. For instance, in the range $\pi < \phi < 2\pi$, in order that F for $\phi = 2\pi$ equal F for $\phi = 0$, we must have

$$F(a|c|z) \to \frac{\Gamma(c)}{\Gamma(a)} |z|^{a-c} e^{i(\phi-2\pi)(a-c)} e^z + \frac{\Gamma(c)}{\Gamma(c-a)} |z|^{-a} e^{ia(\pi-\phi)}$$

with an additional factor $e^{2\pi i(c-a)}$ in the first term. This result, of course, comes out from the asymptotic calculation of the integrals, for in the range $\pi < \phi < 2\pi$ we must use $t = 1 - ue^{i(2\pi-\phi)}/|z|$ for the first integral in order to avoid the cut from $t = 0$ to $t = -\infty$.

Collecting all our formulas, we can write, for $z = |z|e^{i\phi}$ with $|z|$ large,

$$\begin{aligned}
\frac{\Gamma(a)\Gamma(c-a)}{\Gamma(c)} F(a|c|z) &\to \Gamma(a)|z|^{-a}; && \phi = -\pi \\
&\to \Gamma(c-a)|z|^{a-c}e^{i(a-c)\phi}e^z \\
&\qquad + \Gamma(a)|z|^{-a}e^{-ia(\pi+\phi)}; && -\pi < \phi < 0 \\
&\to \Gamma(c-a)|z|^{a-c}e^z; && \phi = 0 \\
&\to \Gamma(c-a)|z|^{a-c}e^{i(a-c)\phi}e^z \\
&\qquad + \Gamma(a)|z|^{-a}e^{ia(\pi-\phi)}; && 0 < \phi < \pi \\
&\to \Gamma(a)|z|^{-a}; && \phi = \pi \\
&\to \Gamma(c-a)|z|^{a-c}e^{i(a-c)(\phi-2\pi)}e^z \\
&\qquad + \Gamma(a)|z|^{-a}e^{ia(\pi-\phi)}; && \pi < \phi < 2\pi \\
&\to \Gamma(c-a)|z|^{a-c}e^z; && \phi = 2\pi \\
&\qquad \text{and so on}
\end{aligned} \tag{5.3.51}$$

which shows the Stokes' phenomenon clearly, each term vanishing in turn as ϕ becomes an integral multiple of π, emerging on the other side with just enough phase change to keep the function single-valued as the circuit is made.

Solutions of the Third Kind. For fitting boundary conditions for large values of z it would be more satisfactory to use solutions which tend either to z^{-a} or to $z^{a-c}e^z$ but not both. Reference to Eq. (5.3.50) shows how this can be done. For $z = |z|e^{i\phi}$ we define

$$U_1(a|c|z) = \frac{e^z z^{a-c}}{\Gamma(c-a)} \int_0^\infty e^{-u} u^{c-a-1} \left(1 - \frac{u}{z}\right)^{a-1} du \to z^{a-c}e^z; \quad 0 < \phi < 2\pi$$

$$U_2(a|c|z) = \frac{e^{ia\pi}z^{-a}}{\Gamma(a)} \int_0^\infty e^{-u} u^{a-1} \left(1 + \frac{u}{z}\right)^{c-a-1} du \to e^{ia\pi}z^{-a}; \quad -\pi < \phi < \pi \tag{5.3.52}$$

When Re $(a - c) > 1$, the integral for U_1 is not valid and we must use a contour around the branch point at $u = 0$. The contour which

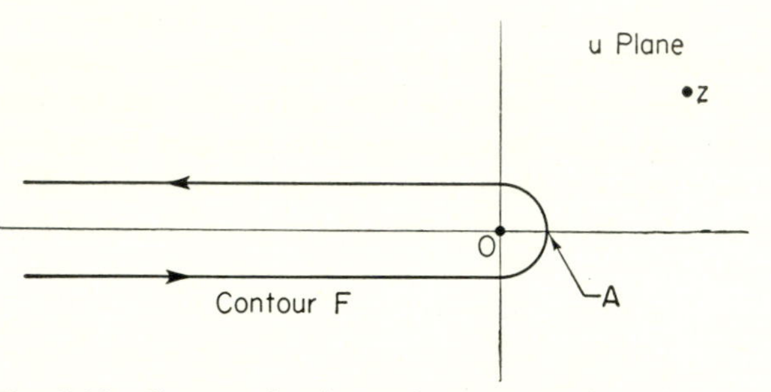

Fig. 5.12 Contour for integral representation of confluent hypergeometric function of the third kind. Integrand is real at A when z is real and positive.

is useful is the one shown in Fig. 5.12, with the integrand real at point A when z is on the negative real axis. The corresponding contour integrals for the U's are

$$U_1(a|c|z) = \frac{\Gamma(a-c+1)}{2\pi i} z^{a-c}e^z \oint_F e^u u^{c-a-1} \left(1 + \frac{u}{z}\right)^{a-1} du$$

$$U_2(a|c|z) = \frac{\Gamma(1-a)}{2\pi i} e^{ia\pi}z^{-a} \oint_F e^u u^{a-1} \left(1 - \frac{u}{z}\right)^{c-a-1} du \tag{5.3.53}$$

which are valid unless $c - a$ (or a, respectively) is a positive integer. Since the point $u = \mp z$ should be outside the contour, we see that the cut for U_1 is the positive real axis, for U_2 the negative real axis. This makes the Stokes' phenomenon for the U's rather different than that for the F's; it will be discussed on page 615.

These solutions may be called the *confluent hypergeometric functions of the third kind*. They are just as good solutions of Eq. (5.3.44) as are $F(a|c|z)$ and $z^{1-c}F(a - c + 1|2 - c|z)$, and any solution may be expressed in their terms. The value of the Wronskian for two independent solutions of Eq. (5.3.44) is [see Eq. (5.2.3)]

$$\Delta = Ae^{-\int (c-z)\, dz/z} = Az^{-c}e^z$$

Use of the asymptotic forms fixes the value of A for the solutions U_1 and U_2 for

$$U_1U_2' - U_2U_1' \to \left[-a\frac{1}{z} - (a-c)\frac{1}{z} - 1\right] z^{-c}e^{z+ia\pi}$$

and since this is an asymptotic formula, we must neglect the terms in $1/z$ compared with the -1. Therefore

$$\Delta(U_1,U_2) = -z^{-c}e^{z+ia\pi} \tag{5.3.54}$$

for all values of z. The usual confluent hypergeometric function can be expressed in terms of them, of course:

$$F(a|c|z) = \frac{\Gamma(c)}{\Gamma(a)} U_1(a|c|z) + \frac{\Gamma(c)}{\Gamma(c-a)} U_2(a|c|z) \tag{5.3.55}$$

for $0 < \phi < \pi$.

Parenthetically it might be pertinent to point out that we do not have to devise new sorts of forms of solutions to express the independent solutions of the ordinary hypergeometric equation about the other singular points. All three of the singular points were regular, and so each pair of solutions could be expressed in terms of the series $F(a,b|c|z)$ with different expressions for a, b, c, and z. In the case of the confluent equation one singular point is regular and the other is irregular, and it is not surprising that the pair of solutions suitable for the irregular point, U_1 and U_2, have a different form from the pair suitable for the regular point, $F(a|c|z)$ and $z^{1-c}F(a-c+1|2-c|z)$.

To finish our discussion we should express $z^{1-c}F(a-c+1|2-c|z)$ in terms of solutions U_1 and U_2. Using Eq. (5.3.46) and making the same separation as was done for Eq. (5.3.50), we obtain

$$\begin{aligned} z^{1-c}F(a-c+1|2-c|z) \\ = \frac{\Gamma(2-c)}{\Gamma(a-c+1)\Gamma(1-a)} \Bigg\{ & \frac{e^{i\pi(a-c+1)}}{z^a} \int_0^\infty e^{-w}w^{a-c}\left(1+\frac{w}{z}\right)^{-a} dw \\ & + z^{a-c}e^z \int_0^\infty e^{-u}u^{-a}\left(1-\frac{u}{z}\right)^{a-c} du \Bigg\} \end{aligned}$$

But this should also be a combination of U_1 and U_2. Comparing asymptotic behavior, we see that the first term should be proportional to U_2 and the second to U_1. In this way we obtain a new set of integral representations for the U's:

$$\begin{aligned} U_1 &= \frac{z^{a-c}e^z}{\Gamma(1-a)} \int_0^\infty e^{-u}u^{-a}\left(1-\frac{u}{z}\right)^{a-c} du = \frac{z^{a-c}e^z}{2\pi i\Gamma(a)} \oint_F e^u u^{-a}\left(1+\frac{u}{z}\right)^{a-c} du \\ U_2 &= \frac{e^{i\pi a}z^{-a}}{\Gamma(a-c+1)} \int_0^\infty e^{-u}u^{a-c}\left(1+\frac{u}{z}\right)^{-a} du \qquad (5.3.56) \\ &= \frac{e^{i\pi a}z^{-a}}{2\pi i\Gamma(c-a)} \oint_F e^u u^{a-c}\left(1-\frac{u}{z}\right)^{-a} du \end{aligned}$$

and also obtain an expression for $z^{1-c}F(a-c+1|2-c|z)$ in terms of U_1 and U_2, for $0 < \phi < \pi$

$$z^{1-c}F(a-c+1|2-c|z) = \frac{\Gamma(2-c)}{\Gamma(a-c+1)} U_1(a|c|z) - \frac{\Gamma(2-c)}{\Gamma(1-a)} e^{-i\pi c} U_2(a|c|z) \quad (5.3.57)$$

which is valid for the phase angle for z over the range $0 < \phi < \pi$. The Stokes' phenomenon for this function must be worked out by using the integral representations (5.3.56) for the asymptotic integrals. The results are

$$\frac{\Gamma(1-a)\Gamma(a-c+1)}{\Gamma(2-c)} z^{1-c}F(a-c+1|2-c|z)$$
$$\to \Gamma(a-c+1)e^{-i\pi(1-c)}|z|^{-a}; \quad \phi = -\pi$$
$$\to \Gamma(1-a)|z|^{a-c}e^{i(a-c)\phi}e^z + \Gamma(a-c+1)|z|^{-a}e^{i\pi(c-1)-ia(\pi+\phi)}; \quad -\pi < \phi < 0$$
$$\to \Gamma(1-a)|z|^{a-c}e^z; \quad \phi = 0$$
$$\to \Gamma(1-a)|z|^{a-c}e^{i(a-c)\phi}e^z + \Gamma(a-c+1)|z|^{-a}e^{i\pi(1-c)+ia(\pi-\phi)}; \quad 0 < \phi < \pi$$
$$\to \Gamma(a-c+1)e^{i\pi(1-c)}|z|^{-a}; \quad \phi = \pi$$
$$\to \Gamma(1-a)|z|^{a-c}e^{-2\pi ia+i(a-c)\phi}e^z + \Gamma(a-c+1)|z|^{-a}e^{i\pi(1-c)+ia(\pi-\phi)}; \quad \pi < \phi < 2\pi$$

and so on. Due to the fact that this second solution is z^{1-c} times an analytic function, the asymptotic value, which is real for $\phi = 0$, must be $e^{-i\pi(1-c)}$ times a real number for $\phi = -\pi$, must be $e^{i\pi(1-c)}$ times a real number for $\phi = \pi$, must be $e^{2\pi i(1-c)}$ times a real number for $\phi = 2\pi$, and so on, and the phase angles generated by the Stokes' phenomenon produce this result.

For completeness we should now express the solutions of the third kind in terms of the solutions of the first kind. These equations are obtained by solving Eqs. (5.3.55) and (5.3.57) simultaneously. The manipulations are somewhat tedious and involve the use of the formula $\sin(\pi x)\,\Gamma(x)\Gamma(1-x) = \pi$. For instance, we have

$$\frac{\Gamma(c-a)}{\Gamma(c)} F(a|c|z) + \frac{\Gamma(1-a)}{\Gamma(2-c)} e^{i\pi c}z^{1-c} F(a-c+1|2-c|z)$$
$$= \left[\frac{\Gamma(c-a)}{\Gamma(a)} + \frac{\Gamma(1-a)e^{i\pi c}}{\Gamma(a-c+1)}\right] U_1 = \Gamma(c-a)\Gamma(1-a)\frac{\sin(\pi c)}{\pi} e^{i\pi(c-a)} U_1$$

Finally

$$U_1(a|c|z) = \frac{\Gamma(1-c)}{\Gamma(1-a)} e^{i\pi(a-c)}F(a|c|z) - \frac{\Gamma(c-1)}{\Gamma(c-a)} e^{i\pi a}z^{1-c}F(a-c+1|2-c|z); \quad 0 < \phi < 2\pi$$
$$U_2(a|c|z) = \frac{\Gamma(1-c)}{\Gamma(a-c+1)} e^{i\pi a}F(a|c|z) + \frac{\Gamma(c-1)}{\Gamma(a)} e^{i\pi a}z^{1-c}F(a-c+1|2-c|z); \quad -\pi < \phi < \pi \quad (5.3.58)$$

The Stokes' phenomenon for the U's (and for the equations relating the U's and F's) can be obtained from these equations, by using the schedules for the phase changes of F and $z^{1-c}F$. If U_1, U_2 are to be the solutions which are asymptotically proportional to $z^{a-c}e^z$ and z^{-a}, respectively, for all values of the phase angle ϕ, and if Eqs. (5.3.58) are supposed to define U_1 and U_2 for $0 < \phi < \pi$, then one can devise rules governing the phase angles in Eqs. (5.3.55), (5.3.57), and (5.3.58) and for the U's. For instance, for $-2\pi < \phi < 0$,

$$U_1 = \frac{\Gamma(1-c)}{\Gamma(1-a)} e^{i\pi(a-c)} F(a|c|z) - \frac{\Gamma(c-1)}{\Gamma(c-a)} e^{i\pi(a-2c)} z^{1-c} F(a-c+1|2-c|z) \to e^{2\pi i(a-c)} z^{a-c} e^z$$

(Since there is a cut for U_1 along the positive real axis, to go from plus to minus values of ϕ we must go from $0 + \epsilon$ to $2\pi - \epsilon$.) A similar jump occurs for U_2 at $\phi = \pi$ or $-\pi$.

The Solution of the Second Kind. So far we have avoided the question of what to do about the solution independent of $F(a|c|z)$ when c is a negative integer. When $c = n + \epsilon$, where n is a positive integer greater than 1 and ϵ vanishes, it turns out that the first $n - 1$ terms of the series for $z^{1-c}F(a - c + 1|2 - c|z)$ stay finite but the remaining terms have a factor $1/(-\epsilon)$ and so go to infinity. The obvious gambit of multiplying through by ϵ before going to the limit does not help, for, as we shall see, the limiting form for $(n - c)z^{1-c}F(a - c + 1|2 - c|z)$, when $c \to n$, turns out to be proportional to the first solution $F(a|c|z)$ and is therefore not an independent solution. This dilemma is also displayed in Eq. (5.3.57), which has a factor $\Gamma(2 - c)$, going to infinity when $c \to 2, 3, 4, \ldots$. The solutions U_1 and U_2 do not become infinite in this way, as the asymptotic expressions show; therefore we should be able to construct another solution out of the U's. A fairly simple way to do this would be to reverse the sign of one of the terms in Eq. (5.3.55), which gives $F(a|c|z)$ in terms of the U's.

We therefore define the *confluent hypergeometric function of the second kind* by the equation

$$G(a|c|z) = \frac{\Gamma(c)}{\Gamma(a)} U_1(a|c|z) - \frac{\Gamma(c)}{\Gamma(c-a)} U_2(a|c|z) \qquad (5.3.59)$$

for $0 < \phi < \pi$. This function is independent of F, as the asymptotic expression shows. It is a useful form, for it produces simple formulas for the U's in terms of F and G; for instance, $U_1 = [\Gamma(a)/\Gamma(c)][F + G]$ and $U_2 = [\Gamma(c - a)/\Gamma(c)][F - G]$. It also stays finite when c is an integer, as will shortly be shown, though in these cases it turns out to have a logarithmic singularity at $z = 0$. But this is just what is to be expected, for when c is an integer, the indices of the confluent hypergeometric equation at $z = 0$ differ by an integer and, as we have seen on pages 533 and 543, we should expect the second solution to have a logarithmic singularity.

To obtain an expansion of G about $z = 0$, we express the U's in Eq. (5.3.59) in terms of F and $z^{1-c}F$, by using Eqs. (5.3.58). After manipulating the gamma functions we finally obtain

$$G(a|c|z) = e^{i\pi a}\frac{\Gamma(c)}{\Gamma(a)}\left\{\frac{\Gamma(1-c)}{\Gamma(1-a)}\left[e^{-i\pi c} + \frac{\sin \pi(a-c)}{\sin(\pi a)}\right]F(a|c|z) - 2\frac{\Gamma(c-1)}{\Gamma(c-a)}z^{1-c}F(a-c+1|2-c|z)\right\} \quad (5.3.60)$$

This does not look too promising, for now *both* terms in the braces go to infinity as $c \to n$, an integer. However, as we mentioned previously, the two solutions equal each other when $c = n$, and the coefficients of the expression above are so designed that the infinite parts just cancel out, leaving a finite expression which is proportional to the *differential of* $F(a|c|z)$ *with respect to c*. This derivative of F with respect to the parameters is a solution of Eq. (5.3.53) when c is an integer, and it is just the second solution we are seeking.

It now behooves us to prove the statements we have been making and actually to compute the expansion of this solution of the second kind about $z = 0$. First, we have to prove that $z^{1-c}F(a - c + 1|2 - c|z)$ is proportional to $F(a|c|z)$ when $c \to n$, an integer. This is simple when $c \to 1$, for $z^{1-c}F(a - c + 1|2 - c|z) \to F(a|1|z)$ when $c \to 1$. Therefore [since $\Gamma(c - 1) \to 1/(c - 1)$ and $\Gamma(1 - c) \to 1/(1 - c)$]

$$G(a|1|z) = e^{i\pi a}\frac{\Gamma(1)}{\Gamma(a)}\lim_{c\to 1}\left\{-\frac{1}{\Gamma(1-a)(c-1)}\cdot\left[\left(e^{-i\pi c} + \frac{\sin \pi(a-c)}{\sin(\pi a)}\right)F(a|c|z) + 2F(a|1|z)\right] - \frac{2}{(c-1)}\left[\frac{1}{\Gamma(c-a)}z^{1-c}F(a-c+1|2-c|z) - \frac{1}{\Gamma(1-a)}F(a|1|z)\right]\right\}$$

Each of the terms in square brackets is of the form $f(c) - f(1)$, so that the limiting form of the quantity, divided by $(c - 1)$, is the derivative of $f(c)$ with respect to c at $c = 1$. Therefore

$$G(a|1|z) = \frac{e^{i\pi a}}{\Gamma(a)}\left\{-\frac{1}{\Gamma(1-a)}\frac{d}{dc}\left[e^{-i\pi c} + \frac{\sin \pi(a-c)}{\sin(\pi a)}\right]F(a|c|z) - 2\frac{d}{dc}\left[\frac{z^{1-c}}{\Gamma(c-a)}F(a-c+1|2-c|z)\right]\right\}_{c=1}$$

But $dz^c/dc = z^c \ln z$ and $d\Gamma(\alpha + c)/dc = \psi(\alpha + c)\Gamma(a + c)$, where $\psi(x)$ is the logarithmic derivative of $\Gamma(x)$,

$$\psi(x) = \frac{1}{\Gamma(x)}\frac{d}{dx}\Gamma(x) = -\gamma + \sum_{r=0}^{\infty}\left[\frac{1}{r+1} - \frac{1}{r+x}\right];$$

$$\gamma = -\psi(1) = 0.5772 \cdots$$

$$\psi(x+n) = \psi(x) + \frac{1}{x} + \frac{1}{x+1} + \frac{1}{x+2} + \cdots + \frac{1}{x+n-1};$$

$$\psi(1-c) = \psi(c) + \pi\cot(\pi c)$$

Therefore

$$G(a|1|z)$$

$$= \frac{e^{i\pi a}}{\Gamma(a)\Gamma(1-a)}\left\{-\frac{d}{dc}\left[e^{-i\pi c} + \frac{\sin\pi(a-c)}{\sin(\pi a)}\right]\sum_{m=0}^{\infty}\frac{\Gamma(a+m)\Gamma(c)z^m}{m!\Gamma(a)\Gamma(c+m)}\right.$$

$$\left.- 2\Gamma(1-a)\frac{d}{dc}\left[\frac{1}{\Gamma(c-a)}\sum_{m=0}^{\infty}\frac{\Gamma(a-c+1+m)\Gamma(2-c)z^{m+1-c}}{m!\Gamma(a-c+1)\Gamma(2-c+m)}\right]\right\}_{c=1}$$

$$= \frac{e^{2\pi ia}-1}{2\pi i}\sum_{m=0}^{\infty}\{[-i\pi - \pi\cot(\pi a) + 2\psi(1) - 2\psi(m+1)]$$

$$- 2[-\psi(1-a) - \ln z + \psi(a)$$

$$+ \psi(m+1) - \psi(a+m) - \psi(1)]\}\frac{\Gamma(a+m)\Gamma(1)}{m!\Gamma(a)\Gamma(m+1)}z^m$$

$$= \frac{e^{2\pi ia}-1}{2\pi i}\left\{[2\ln z + \pi\cot(\pi a) - i\pi + 2\psi(a)]F(a|1|z)\right.$$

$$\left.+2\sum_{m=1}^{\infty}\frac{\Gamma(a+m)}{\Gamma(a)[m!]^2}[\psi(a+m) - \psi(a) + 2\psi(1) - 2\psi(m+1)]z^m\right\} \quad (5.3.61)$$

This solution is independent of $F(a|1|c)$ (the logarithmic term shows this even if the other part is not so obvious about it). The form is interesting, being the logarithm of a quantity proportional to z multiplied by the first solution, plus an infinite series, which is convergent for all finite values of z, which also is independent of $F(a|1|z)$.

For $c = 2, 3, \ldots$, we must work a little harder, but the procedure is quite similar. Expansion of the series shows that for $c - n$ vanishingly small ($n = 2, 3, 4, \ldots$) the function

$$\mathfrak{F}_n(a|c|z) = (c-n)z^{1-c}F(a-c+1|2-c|z)$$

$$= (c-n)\sum_{m=0}^{n-2}\frac{\Gamma(c-a)\Gamma(c-m-1)}{m!\Gamma(c-a-m)\Gamma(c-1)}z^{m-c+1}$$

$$+ (c-n)\frac{\Gamma(a)\Gamma(2-c)}{\Gamma(n)\Gamma(a-c+1)}\sum_{r=0}^{\infty}\frac{\Gamma(a+n-c+r)\Gamma(n)z^{r+n-c}}{\Gamma(r+1+n-c)\Gamma(a)\Gamma(n+r)}$$

becomes the function

$$\frac{\Gamma(a)\Gamma(n-a)\sin(\pi a)}{\pi\Gamma(n)\Gamma(n-1)} F(a|n|z); \quad \text{since } (c-n)\Gamma(2-c) \xrightarrow[c\to n]{} \frac{(-1)^{n-1}}{\Gamma(c-1)}$$

and therefore the first term in Eq. (5.3.60) approaches the second term, but both become infinitely large. Adding and subtracting the term $2[\Gamma(n-1)/\Gamma(n-a)]\mathfrak{F}_n(a|n|z)/(c-n)$ inside the braces, we have

$$\begin{aligned} G(a|n|z) &= e^{i\pi a}\frac{\Gamma(n)}{\Gamma(a)}\lim_{c\to n}\left\{\frac{\Gamma(a)\sin(\pi a)}{\Gamma(c)\pi(c-n)}\left[e^{i\pi(n-c)}+\frac{\sin\pi(a-c)}{\sin\pi(a-n)}\right]F(a|c|z)\right. \\ &\quad - 2\frac{\Gamma(a)}{\Gamma(n)}\frac{\sin(\pi a)}{\pi(c-n)}F(a|n|z) - 2\frac{\Gamma(c-1)}{\Gamma(c-a)}\frac{1}{(c-n)}\mathfrak{F}_n(a|c|z) \\ &\quad \left. + 2\frac{\Gamma(n-1)}{\Gamma(n-a)}\frac{1}{(c-n)}\mathfrak{F}_n(a|n|z)\right\} \\ &= e^{i\pi a}\frac{\Gamma(n)}{\Gamma(a)}\left\{\frac{\sin(\pi a)}{\pi}\frac{d}{dc}\frac{\Gamma(a)}{\Gamma(c)}\left[e^{i\pi(n-c)}+\frac{\sin\pi(a-c)}{\sin\pi(a-n)}\right]F(a|c|z)\right. \\ &\quad \left. - 2\frac{d}{dc}\left[\frac{\Gamma(c-1)}{\Gamma(c-a)}\mathfrak{F}_n(a|c|z)\right]\right\}_{c=n} \end{aligned}$$

Therefore

$$\begin{aligned} \mathcal{G}(a|n|z) &= \frac{e^{2\pi i a}-1}{2\pi i}\left\{[2\ln z + \pi\cot(\pi a) - i\pi]F(a|n|z)\right. \\ &\quad - 2\sum_{m=1}^{\infty}[\psi(m+1)+\psi(n+m) \\ &\quad \left. - \psi(a+m)]\frac{\Gamma(a+m)\Gamma(n)z^m}{m!\Gamma(a)\Gamma(n+m)}\right\} \\ &\quad - 2e^{i\pi a}\frac{\Gamma(n)}{\Gamma(a)}\sum_{r=1}^{n-1}\frac{\Gamma(r)z^{-r}}{\Gamma(n-r)\Gamma(r-a+1)} \end{aligned} \qquad (5.3.62)$$

This expression differs from the general form for $G(a|1|z)$ by the presence of a finite sum of negative powers of z. This, plus the logarithmic term, ensures that G is not analytic at $z = 0$, whereas F is. There is no need to investigate G's for $c = 0$ or for negative integers, for if we have an equation which is satisfied by $z^{1+n}F(a+n+1|2+n|z)$, we use this as a *first solution*, with $c = 2 + n$, and go ahead as outlined above. Since the indices of the confluent hypergeometric equation for $z = 0$ are 0 and $1 - c$, if, at first try, we find c less than 1, we shall find that the equation for $z^{1-c}y$ has a new c larger than 1, and we work with it.

This formula completes our primary task as far as the confluent hypergeometric function is concerned, for we have now written down expressions for both solutions for large and small values of z and for all values of the parameters [the series expressions for the U's may be obtained in terms of the series for F and G by use of Eqs. (5.3.55) and (5.3.59)], including those for which the second solution has a logarithmic singularity. All that remains for the present is a discussion of some of the more useful special cases of these functions.

Bessel Functions. The most useful special function which can be represented in terms of the confluent hypergeometric function is the Bessel function, mentioned on page 550. It is a solution of the equation

$$\frac{d^2\psi}{dz^2} + \frac{1}{z}\frac{d\psi}{dz} + \left(1 - \frac{\nu^2}{z^2}\right)\psi = 0$$

called Bessel's equation. It has a regular singular point at $z = 0$ with indices ν and $-\nu$. The substitution $\psi = z^\nu e^{-iz}F$ yields an equation of confluent hypergeometric form for F

$$zF'' + [(2\nu + 1) - 2iz]F' - i(2\nu + 1)F = 0$$

with solution $F(\nu + \frac{1}{2}|2\nu + 1|2iz)$. The solution of this equation which remains finite at $z = 0$ is called the *Bessel function of the first kind,*

$$\begin{aligned} J_\nu(z) &= \frac{z^\nu e^{-iz}}{2^\nu\Gamma(\nu + 1)} F(\nu + \tfrac{1}{2}|2\nu + 1|2iz) \xrightarrow[z\to 0]{} \frac{1}{\Gamma(\nu + 1)}\left(\frac{z}{2}\right)^\nu \\ &= \frac{z^\nu}{2^\nu\Gamma(\nu + 1)}\left[1 - \frac{1}{1!(\nu + 1)}\left(\frac{z}{2}\right)^2 \right. \\ &\qquad \left. + \frac{1}{2!(\nu + 1)(\nu + 2)}\left(\frac{z}{2}\right)^4 - \cdots\right] \\ &= \frac{(z/2)^\nu}{i\sqrt{\pi}\,\Gamma(\nu + \frac{1}{2})}\int_{-i}^{i} e^{zu}(1 + u^2)^{\nu - \frac{1}{2}}\,du \\ &= \frac{(z/2)^\nu}{\sqrt{\pi}\,\Gamma(\nu + \frac{1}{2})}\int_0^\pi e^{iz\cos\theta}\sin^{2\nu}\theta\,d\theta \end{aligned} \tag{5.3.63}$$

The integral representations may be obtained from that for F [by use of the equation $\Gamma(2\nu + 1) = (2^{2\nu}/\sqrt{\pi})\Gamma(\nu + \frac{1}{2})\Gamma(\nu + 1)$ and by letting $t = \frac{1}{2}(1 - iu)$ or $\frac{1}{2} - \frac{1}{2}\cos\theta$] or by going through the Laplace transform derivation directly. The series may be obtained from the integrals or by multiplying out the series for the exponential and for F (it is interesting to note that the product of a confluent hypergeometric series for imaginary argument and an imaginary exponential can be a real function). When $\operatorname{Re}\nu - \frac{1}{2} < 0$, the line integral may be replaced by a contour integral going around $u = i$ in the positive direction and around $-i$ in the negative direction.

By manipulating the integral representation or the series we can obtain the following recurrence formulas for the Bessel function:

$$\begin{aligned} J_{\nu-1}(z) + J_{\nu+1}(z) &= (2\nu/z)J_\nu(z) \\ J_{\nu-1}(z) - J_{\nu+1}(z) &= 2(d/dz)J_\nu(z) \end{aligned} \tag{5.3.64}$$

Since the formulas can be obtained from the integral representation, they will hold for other solutions of the Bessel equation obtained by changing the limits of integration; *i.e.*, they will be valid for solutions of the second and third kind.

Another form of integral representation, obtained after much manipulation from the last of Eqs. (5.3.63), which will be useful in many applications later, is expressible in various forms (for n an integer)

$$\begin{aligned} J_n(z) &= \frac{1}{2\pi i} \oint u^{-n-1} e^{\frac{1}{2}z(u^2-1)/u}\, du \\ &= \frac{e^{-i\pi n/2}}{\pi} \int_0^\pi e^{iz\cos\phi} \cos(n\phi)\, d\phi \\ &= \frac{1}{2\pi} \int_{-\pi}^{\pi} \cos[n\theta - z\sin\theta]\, d\theta \end{aligned} \tag{5.3.65}$$

where the contour for the first equation is a circle about the origin in a positive direction. That this representation is indeed the Bessel function may be shown by expanding it in series and comparing with the series of Eq. (5.3.63) or by showing that the integral satisfies Bessel's equation and has the correct value at $z = 0$.

This representation, with the aid of Eq. (4.3.4), which defines the Laurent series about an essential singularity at a, can be used to generate a very useful series expansion. The function $e^{z(t^2-1)/2t}$ has an essential singularity at $t = 0$. Therefore its expansion about $t = 0$ is a Laurent series with coefficients

$$a_n = (1/2\pi i) \oint t^{-n-1} e^{z(t^2-1)/2t}\, dt = J_n(z)$$

The series is therefore

$$e^{z(t^2-1)/2t} = \sum_{n=-\infty}^{\infty} t^n J_n(z)$$

which may be used to define the properties of the Bessel functions of integral order. For instance, the recursion formulas (5.3.64) may be obtained by differentiation by z or t, etc. We shall discuss this further in Chap. 11.

This expansion may be used to obtain others. For instance, there is the simple one

$$e^{iz\sin\theta} = \sum_{n=-\infty}^{\infty} e^{in\theta} J_n(z)$$

A more complex but more useful relation may be obtained by expanding the following, where $Z = \sqrt{x^2 + y^2 - 2xy\cos\phi}$; $Z\sin\alpha = x - y\cos\phi$, $Z\cos\alpha = y\sin\phi$:

$$
\begin{aligned}
J_0(Z) &= \frac{1}{2\pi}\int_0^{2\pi} e^{iZ\cos\vartheta}\,d\vartheta = \frac{1}{2\pi}\int_0^{2\pi} e^{iZ\cos(\vartheta-\alpha)}\,d\vartheta \\
&= \frac{1}{2\pi}\int_0^{2\pi} \exp\,[iy\sin\phi\cos\vartheta + i(x - y\cos\phi)\sin\vartheta]\,d\vartheta \\
&= \frac{1}{2\pi}\sum_{n=-\infty}^{\infty} J_n(x)\int_0^{2\pi} e^{iy\sin(\phi-\vartheta)+in\vartheta}\,d\vartheta \\
&= \frac{1}{2\pi}\sum_{n=-\infty}^{\infty} J_n(x)\int_0^{2\pi} e^{iy\cos\theta+in(\theta+\phi-\frac{1}{2}\pi)}\,d\theta \\
&= \sum_{n=-\infty}^{\infty} J_n(x)J_n(y)e^{in\phi} = \sum_{n=-\infty}^{\infty} J_n(x)J_n(y)\cos(n\phi) \qquad (5.3.66)
\end{aligned}
$$

By using other contours of integration for ϑ, we obtain the general formula

$$
Y_0(\sqrt{x^2 + y^2 - 2xy\cos\phi}) = \sum_{m=-\infty}^{\infty} Y_m(x)J_m(y)\cos(m\phi)
$$

where the symbol Y may stand for any of the kinds of Bessel functions, J, N, $H^{(1)}$, or $H^{(2)}$, to be defined shortly. Many other similar expansions will be developed and used later in this book. The present one is worked out here for its usefulness later in this chapter.

Another formula, of use later, relating Bessel functions and Gegenbauer polynomials, can be obtained by using Eq. (5.2.53) and by integrating by parts a modification of the integral representation of Eq. (5.3.63):

$$
\begin{aligned}
J_{\beta+n+\frac{1}{2}}(z) &= \frac{(z/2)^{\beta+n+\frac{1}{2}}}{\sqrt{\pi}\,\Gamma(\beta+n+1)}\int_{-1}^{1} e^{izt}(1-t^2)^{\beta+n}\,dt \\
&= \frac{i^n z^{\beta+\frac{1}{2}}}{2^{n+\beta}\sqrt{2\pi}\,\Gamma(\beta+n+1)}\int_{-1}^{1} e^{izt}\frac{d^n}{dt^n}(1-t^2)^{\beta+n}\,dt \\
&= \frac{n!z^{\beta+\frac{1}{2}}}{i^n\sqrt{2\pi}\,\Gamma(n+2\beta+1)}\int_{-1}^{1} e^{izt}(1-t^2)^{\beta}T_n^{\beta}(t)\,dt \qquad (5.3.67)
\end{aligned}
$$

When ν is a half integer, $n + \frac{1}{2}$, the resulting Bessel functions have a particularly simple form. By comparing the series expansions we can see that

$$
J_{\frac{1}{2}}(z) = \sqrt{2/\pi z}\,\sin z \quad \text{and} \quad J_{-\frac{1}{2}}(z) = \sqrt{2/\pi z}\,\cos z
$$

We shall see later that for the wave equation in spherical coordinates it is convenient to use the *spherical Bessel functions* defined by

$$j_n(z) = \sqrt{\frac{\pi}{2z}}\, J_{n+\frac{1}{2}}(z) \xrightarrow[z\to\infty]{} \frac{1}{z}\cos\left[z - \tfrac{1}{2}\pi(n+1)\right]$$

$$n_n(z) = \sqrt{\frac{\pi}{2z}}\, N_{n+\frac{1}{2}}(z) = (-1)^{n+1}\sqrt{\frac{\pi}{2z}}\, J_{-n-\frac{1}{2}}(z) \rightarrow \frac{1}{z}\sin\left[z - \tfrac{1}{2}\pi(n+1)\right]$$

$$h_n(z) = j_n(z) + i n_n(z) \rightarrow (1/z)\exp\left[iz - \tfrac{1}{2}i\pi(n+1)\right]$$

Some recurrence formulas for these functions are

$$\begin{aligned}
[(2m+1)/z]f_m(z) &= f_{m-1}(z) + f_{m+1}(z) \\
(2m+1)(d/dz)f_m(z) &= m f_{m-1}(z) - (m+1)f_{m+1}(z) \\
(d/dz)[z^{m+1}f_m(z)] &= z^{m+1}f_{m-1}(z) \\
(d/dz)[z^{-m}f_m(z)] &= -z^{-m}f_{m+1}(z)
\end{aligned}$$

where $f_m(z)$ is $j_m(z)$ or $n_m(z)$ or $h_m(z)$. By use of Eq. (5.3.67) we can show that an integral representation for $j_n(z)$ is

$$j_n(z) = \frac{1}{2i^n}\int_0^\pi e^{iz\cos\phi}P_n(\cos\phi)\sin\phi\, d\phi$$

The differential equation which these functions satisfy is

$$\frac{d^2f}{dz^2} + \frac{2}{z}\frac{df}{dz} + \left[1 - \frac{n(n+1)}{z^2}\right]f = 0$$

Bessel's equation has an irregular singular point at $z = \infty$. As is to be expected with such equations, the expansion about this point is an asymptotic type, which diverges for finite values of z when too many terms are taken but which approaches the correct value as z is increased indefinitely if only a finite number of terms are used. The limiting form (the first term in the asymptotic series) can be obtained from Eq. (5.3.51) giving the asymptotic form for $F(a|c|z)$, or it can be obtained directly from the integral representation by deforming the path of integration [going from $-i$ to $-\infty$ and from $-\infty$ to $+1$ for the first representation in Eq. (5.3.63) for instance]. The schedule of asymptotic forms, displaying the Stokes' phenomenon for the Bessel function of the first kind, is (for $z = |z|e^{i\phi}$)

$$\begin{aligned}
J_\nu(z) &\rightarrow e^{-i\pi(\nu+\frac{1}{2})}\sqrt{2/\pi z}\cos\left[z + \tfrac{1}{2}\pi(\nu+\tfrac{1}{2})\right]; & -\tfrac{3}{2}\pi < \phi < -\tfrac{1}{2}\pi \\
&\rightarrow \sqrt{2/\pi z}\cos\left[z - \tfrac{1}{2}\pi(\nu+\tfrac{1}{2})\right]; & -\tfrac{1}{2}\pi < \phi < \tfrac{1}{2}\pi \qquad (5.3.68) \\
&\rightarrow e^{i\pi(\nu+\frac{1}{2})}\sqrt{2/\pi z}\cos\left[z + \tfrac{1}{2}\pi(\nu+\tfrac{1}{2})\right]; & \tfrac{1}{2}\pi < \phi < \tfrac{3}{2}\pi;\ \text{etc.}
\end{aligned}$$

The independent solution about $z = 0$ can be taken to be

$$J_{-\nu}(z) = \frac{2^{3\nu}z^\nu e^{-iz+i\pi\nu}}{\Gamma(1-\nu)}\left[(2iz)^{-2\nu}F(\tfrac{1}{2} - \nu|1 - 2\nu|2iz)\right]$$

except when ν is zero or a positive integer. It goes to infinity as $z^{-\nu}$ when z goes to zero.

Hankel Functions. The functions corresponding to U_1 and U_2 of Eq. (5.3.52), suitable for fitting boundary conditions for large values of z, are likewise obtained by pulling the path of integration in Eq. (5.3.63) out to $-\infty$ and cutting it in two. The upper half gives the function with the positive exponential (we multiply by an extra 2 for reasons apparent later)

$$\begin{aligned} H_\nu^{(1)}(z) &= \frac{2(\frac{1}{2}z)^\nu}{i\sqrt{\pi}\,\Gamma(\nu+\frac{1}{2})} \int_{-\infty}^{i} e^{zt}(1+t^2)^{\nu-\frac{1}{2}}\,dt \\ &= \frac{2^{\nu+1}z^\nu}{\sqrt{\pi}} e^{-iz} U_1(\nu+\tfrac{1}{2}|2\nu+1|2iz) \\ &= \frac{1}{\pi}\int_B e^{iz\cos\phi+i\nu(\phi-\frac{1}{2}\pi)}\,d\phi \\ &\to \sqrt{2/\pi z}\; e^{iz-\frac{1}{2}i\pi(\nu+\frac{1}{2})}; \quad -\tfrac{1}{2}\pi < \phi < \tfrac{3}{2}\pi \end{aligned} \tag{5.3.69}$$

which is called the *Hankel function of the first kind* (or the first Bessel function of the third kind). The path of integration of the second

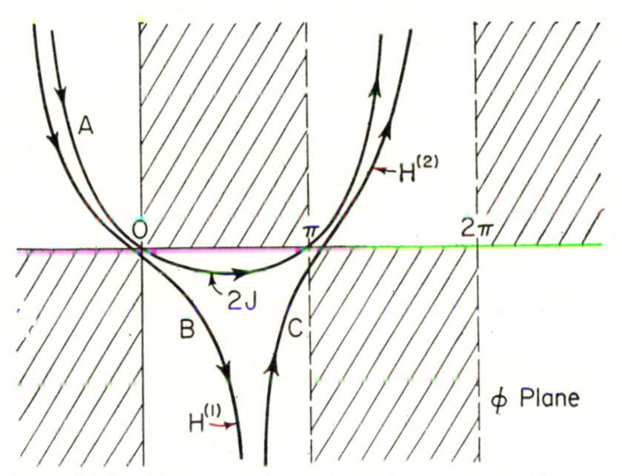

Fig. 5.13 Contours for integral representation of Bessel functions.

integral [which is derived from Eq. (5.3.65)] is shown in Fig. 5.13, going from $+i\infty$ just to the left of the imaginary axis to $-i\infty$ just to the right of the imaginary axis (the reason for such a choice is that in the shaded segments of the ϕ plane the integrand increases without limit as we go away from the real axis, in the unshaded areas the integrand goes to zero, and we thus should terminate our integral in the unshaded regions). This same integral also represents the other functions, for

$$J_\nu(z) = \frac{1}{2\pi}\int_A e^{iz\cos\phi+i\nu(\phi-\frac{1}{2}\pi)}\,d\phi$$

and the contour C represents the other solution for large values of z:

$$\begin{aligned}
H_\nu^{(2)}(z) &= \frac{2(z/2)^\nu}{i\sqrt{\pi}\,\Gamma(\nu+\frac{1}{2})}\int_{-i}^{-\infty} e^{zt}(1+t^2)^{\nu-\frac{1}{2}}\,dt \\
&= \frac{2^{\nu+1}z^\nu e^{-iz}}{\sqrt{\pi}}\,U_2(\nu+\tfrac{1}{2}|2\nu+1|2iz) \\
&= \frac{1}{\pi}\int_C e^{iz\cos\phi+i\nu(\phi-\frac{1}{2}\pi)}\,d\phi \\
&\to \sqrt{2/\pi z}\;e^{-iz+\frac{1}{2}i\pi(\nu+\frac{1}{2})}
\end{aligned} \tag{5.3.70}$$

This function is called the *Hankel function of the second kind.* In many cases, when we are dealing only with the first of these functions, we shall omit the superscript; H_ν always means $H_\nu^{(1)}$ and will be called *the* Hankel function when there is no chance of confusion.

These two solutions are independent, as their asymptotic forms indicate. The Wronskian for independent solutions of the Bessel equation is A/z [by use of Eq. (5.2.3)]. The constant for these two solutions may be fixed by use of their asymptotic forms, and we finally obtain

$$\Delta(H_\nu^{(1)}, H_\nu^{(2)}) = H_\nu^{(1)}\frac{d}{dz}H_\nu^{(2)} - H_\nu^{(2)}\frac{d}{dz}H_\nu^{(1)} = \frac{4}{\pi i z} \tag{5.3.71}$$

Since both $H_\nu^{(1)}$ and $H_\nu^{(2)}$ satisfy the recurrence formulas (5.3.64) (and others which may be found by combining them), we can express the relationship of Eq. (5.3.71) in several ways, such as

$$H_\nu^{(2)}(z)H_{\nu+1}^{(1)}(z) - H_\nu^{(1)}(z)H_{\nu+1}^{(2)}(z) = 4/\pi i z; \quad \text{etc.}$$

There is a close connection between the Hankel functions and the imaginary exponentials on the one hand and the Bessel function J and the cosine function on the other. Just as we have

$$\cos x = \tfrac{1}{2}\,(e^{ix} + e^{-ix})$$

so here we have

$$J_\nu(z) = \tfrac{1}{2}[H_\nu^{(1)}(z) + H_\nu^{(2)}(z)] \tag{5.3.72}$$

as we can see from the definitions above or by simple inspection of Fig. 5.10.

Similarly, from Eqs. (5.3.58) and the definitions of J and the H's in terms of F and the U's, we obtain

$$\begin{aligned}
H_\nu^{(1)}(z) &= \frac{i}{\sin(\pi\nu)}\,[e^{-i\nu\pi}J_\nu(z) - J_{-\nu}(z)]; \\
H_\nu^{(2)}(z) &= \frac{-i}{\sin(\pi\nu)}\,[e^{i\nu\pi}J_\nu(z) - J_{-\nu}(z)]
\end{aligned} \tag{5.3.73}$$

from which we can get the expansion for the Hankel functions near $z = 0$. But before we carry out this calculation we had better examine the behavior of the second solution in case ν is an integer.

Neumann Functions. An examination of the series expansion for $J_\nu(z)$ [Eq. (5.3.63)] shows that, when ν is a negative quantity, $J_\nu(z)$ becomes infinite at $z = 0$ *except when ν is a negative integer.* To examine these special cases, we set $\nu = -n - \epsilon$,

$$J_{-n-\epsilon}(z) = \sum_{m=1}^{\infty} (-1)^{m-1} \frac{(z/2)^{-n-\epsilon-2+2m}}{\Gamma(m)\Gamma(m-n-\epsilon)}$$

$$\xrightarrow[\epsilon \to 0]{} (-1)^n \epsilon \sum_{s=1}^{n} \frac{\Gamma(s+\epsilon)}{\Gamma(n-s+1)} \left(\frac{z}{2}\right)^{n-\epsilon-2s}$$

$$+ (-1)^n \sum_{r=1}^{\infty} (-1)^{r-1} \frac{(z/2)^{n-\epsilon-2+2r}}{\Gamma(r-\epsilon)\Gamma(r+n)}$$

which goes, in the limit of $\epsilon = 0$, to the simple relation

$$J_{-n}(z) = (-1)^n J_n(z); \quad n = 0, 1, 2, \ldots \tag{5.3.74}$$

Therefore when n is zero or an integer, the solution $J_{-n}(z)$ is no longer independent of $J_n(z)$ and we must look further for a second solution.

This situation should be familiar by now; when n is an integer, the indices for the Bessel equation for $z = 0$ differ by an integer and the second solution should have a logarithmic singularity. Neither J_n nor J_{-n} have such a singularity, so we must look elsewhere. Of course, the two Hankel functions do not become proportional to each other when $\nu = n$, so one or both of them must have the logarithmic singularity there (as a matter of fact, both do). But we wish to have a second solution to go along with J_ν. The analogy between J_ν and a cosine supplies the suggestion: we define the Bessel function of the second kind, the *Neumann function*, as being analogous to the sine,

$$\begin{aligned} N_\nu(z) &= (1/2i)[H_\nu^{(1)}(z) - H_\nu^{(2)}(z)] \\ &= \cot(\pi\nu)\, J_\nu(z) - \csc(\pi\nu)\, J_{-\nu}(z) \\ &= \frac{(z/2)^\nu}{i\Gamma(\nu+1)} e^{-iz} G(\nu + \tfrac{1}{2}|2\nu + 1|2iz) \\ &\xrightarrow[z \to \infty]{} \sqrt{2/\pi z}\, \sin[z - \tfrac{1}{2}\pi(\nu + \tfrac{1}{2})]; \quad -\tfrac{1}{2}\pi < \phi < \tfrac{1}{2}\pi \end{aligned} \tag{5.3.75}$$

This function is satisfactory for all values of ν (we need only consider values for Re $\nu > 0$, of course, otherwise we take $J_{-\nu}$ for the first solution). When ν is not zero or an integer, the individual terms are finite, and since the function contains some amount of $J_{-\nu}$ (when $\nu = n + \frac{1}{2}$, the function is all $J_{-\nu}$), it must be independent of J_ν. When ν is zero or an integer, the individual terms are infinite but cancel in such a way as to produce a finite solution having a logarithmic singularity at $z = 0$.

We can utilize Eq. (5.3.61) to obtain this solution, or we can recalculate the series. Since the present case is simpler, we shall perform the calculation again to illustrate again the method. Setting $\nu = n + \epsilon$ and letting ϵ go to zero,

$$\begin{aligned} N_n(z) &= \lim_{\epsilon \to 0} \left[\frac{1}{\pi\epsilon} J_{n+\epsilon}(z) - \frac{(-1)^n}{\pi\epsilon} J_{-n-\epsilon}(z) \right] \\ &= \frac{1}{\pi} \lim_{\epsilon \to 0} \left[\frac{J_{n+\epsilon} - J_n}{\epsilon} - (-1)^n \frac{J_{-n-\epsilon} - J_{-n}}{\epsilon} \right] \\ &= \frac{1}{\pi} \left[\frac{d}{d\nu} J_\nu(z) - (-1)^n \frac{d}{d\nu} J_{-\nu}(z) \right]_{\nu=n} \end{aligned}$$

where we have added and subtracted $J_n[=(-1)^n J_{-n}]$. The derivative of J_ν at $\nu = n$ is

$$\begin{aligned} &\left[\frac{d}{d\nu} \sum_{m=1}^{\infty} \frac{(-1)^{m-1}(z/2)^{\nu+2m-2}}{\Gamma(m)\Gamma(m+\nu)} \right]_{\nu=n} \\ &\qquad = \sum_{m=1}^{\infty} (-1)^{m-1} \left[\ln\left(\frac{z}{2}\right) - \psi(m+n) \right] \frac{(z/2)^{n+2m-2}}{\Gamma(m)\Gamma(m+n)} \\ &\qquad = \left[\ln\left(\frac{z}{2}\right) - \psi(n+1) \right] J_n(z) - \sum_{r=1}^{\infty} (-1)^r \frac{(z/2)^{n+2r}}{r!(n+r)!} \left[\sum_{s=1+n}^{r+n} \frac{1}{s} \right] \end{aligned}$$

where

$$\begin{aligned} \psi(\nu+n) &= -\gamma + \sum_{r=0}^{\infty} \left[\frac{1}{r+1} - \frac{1}{\nu+n+r} \right] \\ &= \psi(\nu) + \frac{1}{\nu} + \frac{1}{\nu+1} + \cdots + \frac{1}{\nu+n-1} \quad \text{and} \quad \gamma = 0.5772 \ldots \end{aligned}$$

Similarly the derivative of $(-1)^n J_{-\nu}$ is

$$\begin{aligned} &\left[\frac{d}{d\epsilon} \sum_{s=1}^{n} \frac{\epsilon(z/2)^{n-\epsilon-2s}\,\Gamma(s+\epsilon)}{\Gamma(n-s+1)} + \frac{d}{d\epsilon} \sum_{r=1}^{\infty} (-1)^{r-1} \frac{(z/2)^{n-\epsilon-2+2r}}{\Gamma(r-\epsilon)\Gamma(r+n)} \right]_{\epsilon=0} \\ &= \sum_{m=0}^{n-1} \frac{(n-m-1)!}{m!(z/2)^{n-2m}} - \sum_{r=1}^{\infty} (-1)^{r-1} \left[\ln\left(\frac{z}{2}\right) - \psi(r) \right] \frac{(z/2)^{n+2r-2}}{\Gamma(r)\Gamma(n+r)} \\ &= -\left[\ln\left(\frac{z}{2}\right) + \gamma \right] J_n(z) + \sum_{m=0}^{n-1} \frac{(n-m-1)!}{m!(z/2)^{n-2m}} \\ &\qquad\qquad + \sum_{m=1}^{\infty} (-1)^m \frac{(z/2)^{n+2m}}{m!(n+m)!} \left[\sum_{s=1}^{m} \frac{1}{s} \right] \end{aligned}$$

Putting these two together we finally arrive at the expansion of the Neumann function about $z = 0$:

$$N_n(z) = \frac{1}{\pi}\left[2\ln\left(\frac{z}{2}\right) + \gamma - \psi(n+1)\right]J_n(z) - \frac{1}{\pi}\sum_{m=0}^{n-1}\frac{(n-m-1)!}{m!(z/2)^{n-2m}}$$
$$- \frac{1}{\pi}\sum_{m=1}^{\infty}(-1)^m\frac{(z/2)^{n+2m}}{m!(n+m)!}\sum_{s=1}^{m}\left[\frac{1}{s}+\frac{1}{s+n}\right] \quad (5.3.76)$$
$$\xrightarrow[z\to 0]{} \frac{2}{\pi}\left[\ln\left(\frac{z}{2}\right)+\gamma\right]; \qquad n = 0$$
$$\xrightarrow[z\to 0]{} -\frac{1}{\pi(n-1)!}\left(\frac{2}{z}\right)^n; \quad n > 0$$

When $n = 0$, the finite sum (from 0 to $n - 1$) is not present. This expansion has the same general form as Eq. (5.3.61) for the confluent hypergeometric function of the second kind; it has a term involving the product of J_n with $\ln z$, a finite series involving negative powers of z (except for $n = 0$ when this is absent), and an additional infinite series which starts with a term in z^{n+2}.

Approximate Formulas for Large Order. The asymptotic formulas given in Eqs. (5.3.68), (5.3.69), (5.3.70), and (5.3.75) assume that the argument z is very much larger than the order ν of the Bessel functions. Useful formulas for $J_\nu(z)$, when *both* z and ν are large, may be obtained by using the method of steepest descents (see Sec. 4.6). When both z and ν are large, the integral $\int_{-\infty}^{i} e^{zt}(1+t^2)^{\nu-\frac{1}{2}}\,dt$, from which we obtained the earlier asymptotic expansions, does not have the largest value of its integrand near $t = i$, as we assumed heretofore. Consequently the first term in the asymptotic expansion is not a good approximation, and we must look for another formula. It turns out that the integral representation

$$(e^{-\frac{1}{2}\pi i\nu}/\pi)\int e^{i(z\cos w+\nu w)}\,dw$$

mentioned on page 623 with the contours shown in Fig. 5.13 is the most suitable for our present purpose. For instance, if the contour is A, the integral is $2J_\nu(z)$; if it is B, the integral is $H_\nu^{(1)}(z)$; and if it is a double contour B and C (with B taken in the reverse order, from top to bottom), then it is $2N_\nu(z)$.

Each of these contours goes from a region where the integrand is vanishingly small (the unshaded areas in Fig. 5.13, far from the real axis) across a region where the integrand is large to another region where the integrand is vanishingly small. If we draw a contour of arbitrary shape between the specified limits, the integrand will vary in a complicated manner, reversing sign often and often having large imaginary parts, but these additional complexities all cancel each other out in the integration, and the result is the same no matter how we have drawn the intermediate part of the contour (for the integrand only has singularities at $w = \infty$). If we are to calculate an approximate value for the integral, we should first pick out the path along which the integrand has the simplest behavior possible. For example, since the integrand is an exponential function, the real part of the exponent controls the magnitude of the integrand and the imaginary part controls the phase. To eliminate undue fluctuations of the integrand we should arrange to perform the integration along a contour for which this phase is constant. We should next try to find a route for which the integrand is large only over one short range of the route, so that calculation in this restricted range is sufficient to give a good approximate answer.

Any one of the contours shown in Fig. 5.13 is a path over a mountain range. We wish to stay in the plains as long as possible in order to cross via the lowest pass and to regain the plains on the other side as fast as possible. It turns out that the path which does this is also the path for which the phase of the integrand stays constant, as was shown in Sec. 4.6.

For our integrand is the exponential of a function of a complex variable $w = u + iv$; the real part of this function controls the magnitude of the integrand and the imaginary part is the phase

$$F(w) = i(z \cos w + \nu w) = f(u,v) + ig(u,v); \quad f = z \sin u \sinh v - \nu v; \quad g = z \cos u \cosh v + \nu u$$

The white areas of Fig. 5.13 are where f becomes large negative, and the shaded areas are where f becomes very large positive. We wish to go over the lowest pass in the $f(u,v)$ surface.

Since f is the real part of a function of a complex variable, it therefore is a solution of Laplace's equation in u and v (see page 353) and it can have no maxima or minima in the finite part of the plane. This means that the top of the lowest pass must be a saddle point, where the surface curves downward in one direction and curves upward in a direction at right angles. In other words the top of the pass is the point where $dF/dw = 0$ and near this point

$$\begin{aligned} F &= F_s + b(w - w_s)^2 + \cdots \\ &= f_s + a[(x^2 - y^2) \cos 2\vartheta - 2xy \sin 2\vartheta + \cdots] \\ &\qquad + ig_s + ia[(x^2 - y^2) \sin 2\vartheta + 2xy \cos 2\vartheta + \cdots] \end{aligned}$$

where $b = ae^{2i\vartheta} = \frac{1}{2}(d^2F/dw^2)_s$ and $x = u - u_s$, $y = v - v_s$. Along the lines $y = -x \tan \vartheta$ and $y = x \cot \vartheta$ (at angles $-\vartheta$ and $\frac{1}{2}\pi - \vartheta$ to the real axis) the imaginary part of F is constant ($g = g_s$) and the real part has its greatest curvature. Along the line $y = -x \tan \vartheta$ the real part of F is

$$\operatorname{Re} F = f = f_s + \tfrac{1}{2}ax^2 \sec^2 \vartheta + \cdots = f_s + \tfrac{1}{2}ad^2 + \cdots$$

where $d = \sqrt{x^2 + y^2}$ is the distance along the line $y = -x \tan \vartheta$. Therefore along the line at an angle $-\vartheta$ with the real axis, f (and therefore the integrand) *rises* as we go away from the saddle point $w = w_s$. Hence this line is along the axis of the mountain range and is *not* the direction to take, for it will lead us among the high peaks.

However, along the line $y = x \cot \vartheta$ the real part of f is

$$f = f_s - \tfrac{1}{2}ax^2 \csc^2 \vartheta + \cdots = f_s - \tfrac{1}{2}as^2 + \cdots$$

where $s = \sqrt{x^2 + y^2}$ is the distance along this line from the saddle point. Therefore along this path we head downward toward the plane in either direction from the top of the pass. If we have no other mountain ranges to cross with our contour, an approximate expression for the integral will then be

$$\int e^{F(w)}\, dw \simeq e^{f_s + ig_s} \int_{-\infty}^{\infty} e^{-\frac{1}{2}as^2 + \frac{1}{2}\pi i - i\vartheta}\, ds = \sqrt{\frac{2\pi}{a}}\, e^{f_s + ig_s - i\vartheta + \frac{1}{2}\pi i}$$

If the contour (in the direction of integration) makes an angle $\frac{1}{2}\pi - \vartheta$ with the real axis near $w = w_s$, the element $dw = e^{\frac{1}{2}\pi i - i\vartheta}\, ds$ where ds is real. [If the integral is in the opposite direction, a factor (-1) must be added to the final result or the term $\frac{1}{2}\pi i$ in the exponent should be changed to $-\frac{1}{2}\pi i$.] This expression will approach the correct value for the integral the more closely the higher and narrower the pass over which we must go (*i.e.*, the larger f_s and a are).

To apply the technique to the Bessel function integral we set $F(w) = iz \cos w + i\nu w$. The saddle points, where dF/dw is zero, are the points where $\sin w_s = \nu/z$. When z is real, there are two cases: one where $z > \nu$ when w_s is along the real axis at the points $\sin^{-1}(\nu/z)$; the other where $z < \nu$ when w_s is complex, at points $(2n + \frac{1}{2})\pi + i \cosh^{-1}(\nu/z)$.

The two cases are shown in Fig. 5.14, the drawing at the right for $z = 0.866\nu$ and that at the left for $z = 1.543\nu$, both plotted on the $w = u + iv$ plane. The light lines are contours for equal magnitudes of $f = \mathrm{Re}\, F$; the heavy lines are the contours which go through one or both of the saddle points S, S' and along which $g = \mathrm{Im}\, F$ is constant. The contour A, coming from $i\infty - \frac{1}{2}\pi$, going through S (or through S and S' on the right) and going to $i\infty + \frac{3}{2}\pi$, gives $2J_\nu(z)$. The contour C, going from $i\infty - \frac{1}{2}\pi$ through S' (on the left) or S (on the right) to $-i\infty + \frac{1}{2}\pi$ gives $H_\nu^{(1)}(z)$, and the contour D gives $H_\nu^{(2)}(z)$. If both C and D are used, starting at $i\infty - \frac{1}{2}\pi$ and $i\infty + \frac{3}{2}\pi$ and ending both at $-i\infty + \frac{1}{2}\pi$, we obtain $2N_\nu(z)$. The contours B and B' are for g constant, but for these cases f increases as we go away from the saddle point. If both ν and z are large, the value of f at the saddle point is much larger than its value anywhere else along A, for instance, and we can neglect all the integrand except that quite close to S.

Taking first the case of $z < \nu$ (we set $z = \nu \operatorname{sech} \alpha$ for convenience), we find that $dF/dw = 0$ at $w = w_s = \frac{1}{2}\pi \pm i\alpha$. The upper point, S, is the one crossed by the contour A, suitable for $2J_\nu(\nu \operatorname{sech} \alpha)$. The value of f at w_s is $f_s = \nu[\tanh\alpha - \alpha]$, and the value of g there is $g_s = \frac{1}{2}\pi\nu$. The second derivative, d^2F/dw^2, at $w = w_s$ is $-\nu\tanh\alpha$, so near the point S an approximate value for F is

$$F \simeq \nu[\tanh\alpha - \alpha] + \tfrac{1}{2}i\nu\pi - \tfrac{1}{2}\nu\tanh\alpha(w - w_s)^2 = F_s + \tfrac{1}{2}ae^{2i\vartheta}(w - w_s)^2;$$
$$a = \nu\tanh\alpha; \quad \vartheta = \tfrac{1}{2}\pi$$

By our earlier discussion the contour which comes down off the pass is at an angle $\frac{1}{2}\pi - \vartheta = 0$ to the real axis and the approximate value

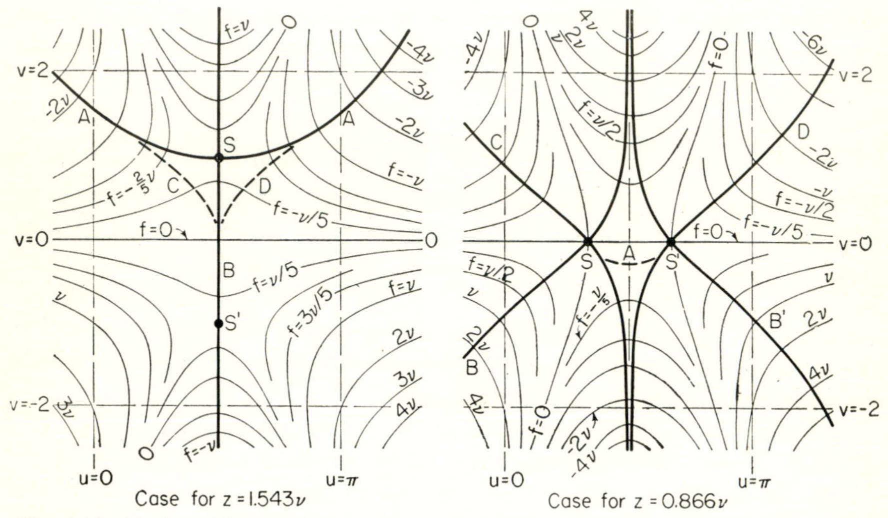

Fig. 5.14 Paths of integration for asymptotic forms of Bessel functions for both z and v large.

of the integral is

$$2J_\nu(\nu\operatorname{sech}\alpha) \simeq \frac{e^{-\frac{1}{2}\pi i\nu}}{\pi}\int_A e^{F(w)}\,dw = \sqrt{\frac{2}{\pi\nu\tanh\alpha}}\,e^{\nu(\tanh\alpha-\alpha)}$$

The contours for the Hankel functions go through S', where

$$f_s = \nu(\alpha - \tanh\alpha); \quad g_s = \tfrac{1}{2}\pi\nu; \quad (d^2F/dw^2)_s = \nu\tanh\alpha; \quad \vartheta = 0;$$
$$a = \nu\tanh\alpha$$

For the first Hankel function the contour goes in the negative direction through S'; for the second function it goes in the positive direction, so

$$H_\nu^{(2)}(\nu\operatorname{sech}\alpha) \simeq -H_\nu^{(1)}(\nu\operatorname{sech}\alpha) = \sqrt{\frac{2}{\pi\nu\tanh\alpha}}\,e^{\nu(\alpha-\tanh\alpha)+\frac{1}{2}\pi i}$$

and therefore the corresponding expression for the Neumann function is

$$-N_\nu(\nu\operatorname{sech}\alpha) \simeq \sqrt{\frac{2}{\pi\nu\tanh\alpha}}\,e^{\nu(\alpha-\tanh\alpha)}$$

When $z > \nu$, we set $z = \nu\sec\beta$, and the saddle points are at $w_s = \frac{1}{2}\pi \pm \beta$. The second derivative $(d^2F/dw^2)_s = \pm i\nu\tan\beta$, so that

$a = \nu \tan \beta$ and $\vartheta = \pm(\pi/4)$; also $f_s = 0$ and $g_s = \frac{1}{2}\pi\nu \mp \nu(\tan \beta - \beta)$. Contour C, for the first Hankel function, crosses $S(w_s = \frac{1}{2}\pi - \beta)$ in the negative direction, so that

$$H_\nu^{(1)}(\nu \sec \beta) \simeq \frac{e^{-\frac{1}{2}\pi i\nu}}{\pi} \sqrt{\frac{2\pi}{\nu \tan \beta}}\, e^{\frac{1}{2}\pi i\nu + i\nu(\tan\beta - \beta) - i\vartheta - \frac{1}{2}\pi i}$$

$$= \sqrt{\frac{2}{\pi\nu \tan \beta}}\, e^{i\nu(\tan\beta - \beta) - \frac{1}{4}\pi i}$$

on the other hand the second function becomes

$$H_\nu^{(2)}(\nu \sec \beta) \simeq \sqrt{\frac{2}{\pi\nu \tan \beta}}\, e^{-i\nu(\tan\beta - \beta) + \frac{1}{4}\pi i}$$

From these two we can compute J_ν and N_ν.

When z is very nearly equal to ν, the two saddle points approach each other and eventually become indistinguishable. For $z = \nu$, at the point $w = \frac{1}{2}\pi$ *both* dF/dz *and* d^2F/dz^2 are zero, so the zero is of higher order and we should expect threefold symmetry of axes. The contours which correspond to $g =$ constant approach $w = \frac{1}{2}\pi$ along the imaginary axis or along the directions $\pm(\pi/6)$ or $\pi \mp (\pi/6)$. The directions $\pi/6$, $\pi - (\pi/6)$, $3\pi/2$ have f decreasing as the distance from the "saddle point" (a saddle for a three-legged rider!) is increased. For $J_\nu(\nu)$ we use the first two directions, the first in reverse. The integral turns out to be

$$J_\nu(\nu) \simeq \frac{\Gamma(\frac{1}{3})}{2^{\frac{2}{3}} 3^{\frac{1}{6}} \pi \nu^{\frac{1}{3}}}$$

When $z - \nu$ is very small compared with ν, we can still use this path of integration and a small correction term can be computed.

We can therefore write the following approximate formulas for large values of ν and of z, for different ranges of z with respect to ν (for ν real):

$$J_\nu(z) \simeq \frac{e^{\nu(\tanh\alpha - \alpha)}}{\sqrt{2\pi\nu \tanh \alpha}}; \quad z < \nu; \quad \tanh \alpha = \sqrt{1 - (z/\nu)^2}$$

$$\simeq \frac{\sin(\pi/3)\,\Gamma(\frac{1}{3})}{3\pi(z/6)^{\frac{1}{3}}} + \frac{\sin(2\pi/3)\,\Gamma(\frac{2}{3})}{3\pi(z/6)^{\frac{2}{3}}}(z - \nu); \quad z \simeq \nu$$

$$\simeq \sqrt{\frac{2}{\pi\nu \tan \beta}} \cos\left[\nu \tan \beta - \nu\beta - \tfrac{1}{4}\pi\right];$$

$$z > \nu; \quad \tan \beta = \sqrt{(z/\nu)^2 - 1} \qquad (5.3.77)$$

$$N_\nu(z) \simeq \frac{-2\, e^{\nu(\alpha - \tanh\alpha)}}{\sqrt{2\pi\nu \tanh \alpha}}; \quad z < \nu$$

$$\simeq -\frac{2\sin(\pi/3)\,\Gamma(\frac{1}{3})}{3\pi(z/6)^{\frac{1}{3}}} + \frac{2\sin(2\pi/3)\,\Gamma(\frac{2}{3})}{3\pi(z/6)^{\frac{2}{3}}}(z - \nu); \quad z \simeq \nu$$

$$\simeq \sqrt{\frac{2}{\pi\nu \tan \beta}} \sin\left[\nu \tan \beta - \nu\beta - \tfrac{1}{4}\pi\right]; \quad z > \nu$$

The Coulomb Wave Function. The radial part of the Schroedinger equation for an electron of mass m, charge $-e$, and total energy E in a centrally symmetric potential field $V(r)$ is

$$\frac{1}{r^2}\frac{d}{dr}\left(r^2\frac{dR}{dr}\right) - \frac{l(l+1)}{r^2}R + \frac{2m}{\hbar^2}[E - V(r)]R = 0$$

where the term $l(l+1)$ is the separation constant coming from the angular factor; if the potential is a function of r alone, then l is an integer. If the potential V is the coulomb field due to a nucleus of charge Ze (heavy enough so that the nucleus is at the center of gravity), V is then equal to $-(e^2Z/r)$. Setting $E = -(me^4Z^2/2\hbar^2)\kappa^2$ and $x = 2\kappa z = (2\kappa me^2Z/\hbar^2)r$ and $R = e^{-\frac{1}{2}x}x^lF(x)$, the equation for F is

$$x\frac{d^2F}{dx^2} + [2(l+1) - x]\frac{dF}{dx} - \left[l + 1 - \frac{1}{\kappa}\right]F = 0$$

which is a confluent hypergeometric equation with $a = [l + 1 - (1/\kappa)]$ and $c = 2l + 2$. The solution of the above equation, which is finite at $z = 0$, is therefore

$$C_l(\kappa,z) = \frac{\sqrt{\pi}\,(2\kappa z)^l e^{-\kappa z}}{2^{2l+1}\Gamma(l+\frac{3}{2})} F\left(l + 1 - \frac{1}{\kappa}\middle|2l+2\middle|2\kappa z\right)$$
$$\xrightarrow[z\to\infty]{} \frac{\Gamma(l+1)}{\Gamma[l+1-(1/\kappa)]}(2\kappa z)^{-1-(1/\kappa)}e^{\kappa z} + \frac{\Gamma(l+1)e^{i\pi[l+1-(1/\kappa)]}}{\Gamma[l+1+(1/\kappa)]}(2\kappa z)^{-1+(1/\kappa)}e^{-\kappa z} \qquad (5.3.78)$$

When κ is real (*i.e.* for energy E negative), this solution, though finite at $z = 0$, becomes infinite at $z \to \infty$ *unless* $\Gamma[l + 1 - (1/\kappa)]$ is infinite. If $1/\kappa$ is a positive integer n larger than l, then $\Gamma[l + 1 - (1/\kappa)]$ is the gamma function of zero or a negative integer, which is infinite. Only in these cases is there a solution which is finite from $r = 0$ to $r = \infty$. Thus the allowed negative values of energy are

$$E_n = -(me^4Z^2/2\hbar^2n^2); \quad n = l+1, l+2, l+3, \ldots$$

which are the values first computed by Bohr.

When κ is imaginary ($\kappa = ik$, for positive energy), C_l is finite for the whole range of z for any real value of k (any positive value of energy). In this case the first (finite) solution takes on the various forms

$$C_l(ik,z) = \frac{\sqrt{\pi}\,(2ikz)^l}{2^{2l+1}\Gamma(l+\frac{3}{2})}\left[1 - \frac{kz}{l+1} + \frac{1-\frac{1}{2}(l+1)}{2!(l+1)(l+\frac{3}{2})}k^2z^2 - \cdots\right]$$
$$= (2ikz)^l\Gamma(l+1)\left[\frac{1}{\Gamma[l+1+(i/k)]}U_1\left(l+1+\frac{i}{k}\middle|2l+2\middle|2ikz\right) + \frac{1}{\Gamma[l+1-(i/k)]}U_2\left(l+1+\frac{i}{k}\middle|2l+2\middle|2ikz\right)\right]e^{-ikz} \qquad (5.3.79)$$
$$\xrightarrow[z\to\infty]{} \frac{\Gamma(l+1)e^{(i\pi l/2)-(\pi/2k)}}{kz|\Gamma[l+1+(i/k)]|}\sin\left[kz + \frac{1}{k}\ln(2kz) - \tfrac{1}{2}\pi l - \Phi_l(k)\right]$$

where $$\left[\Gamma\left(l+1+\frac{i}{k}\right)\Big/\Gamma\left(l+1-\frac{i}{k}\right)\right] = e^{2i\Phi_l(k)}$$

and $$\left|\Gamma\left(l+1+\frac{i}{k}\right)\right|^2 = \left[l^2+\frac{1}{k^2}\right]\left[(l-1)^2+\frac{1}{k^2}\right]\cdots \cdot\left[1+\frac{1}{k^2}\right]\frac{\pi/k}{\sinh(\pi/k)}$$

The second solution, which has a singularity at $z = 0$, can be written

$$D_l(ik,z) = \frac{\sqrt{\pi}\,(2ikz)^l e^{-ikz}}{2^{2l+1}\Gamma(l+\frac{3}{2})} G\left(l+1+\frac{i}{k}\middle|2l+2\middle|2ikz\right)$$
$$\xrightarrow[z\to\infty]{} \frac{\Gamma(l+1)e^{i(\pi l/2)-(\pi/2k)}}{ikz|\Gamma[l+1+(i/k)]|} \cos\left[kz+\frac{1}{k}\ln(2kz)-\tfrac{1}{2}\pi l-\Phi_l(k)\right]$$

The series expansion about $z = 0$, showing the nature of the singularity there, can be obtained from Eq. (5.3.62).

Other solutions, suitable for other equations of interest, will be discussed when the problems are taken up. Thus the wave functions for parabolic and parabolic cylinder coordinates can be expressed in terms of the F's and G's. They will be treated in Chap. 11.

Mathieu Functions. We have intimated on page 587 that the integral representation is most useful for the case of equations with three regular singular points or with one regular and one irregular point. We shall not be disappointed, therefore, to find it not quite so useful with more complex equations. To illustrate the difficulties and to show that a measure of utility still survives, let us apply the Laplace transform to the Mathieu equation (in one of its algebraic forms):

$$(z^2-1)\psi'' + z\psi' + (h^2z^2-b)\psi = 0 \tag{5.3.80}$$

We can see by inspection that the application of the Laplace transform will give an equation for $v(t)$ of the same general form:

$$(t^2+h^2)v'' + 3tv' - (t^2+b-1)v = 0$$

It thus requires as much work to solve the equation for v as it does to solve for ψ; v is also proportional to a Mathieu function, and our integral representation reduces to an integral *equation.*

But an integral equation can be of some value, so it is worth while exploring further. It is more satisfactory to change the scale of t so that the equation for v comes as close as possible to that for ψ. We set $\psi(z) = \int e^{ihtz}\,v(t)\,dt$ and obtain for an equation for v and for the bilinear concomitant

$$(t^2-1)v'' + 3tv' + (h^2t^2-b+1)v = 0$$
$$P(v,e^{ihtz}) = -ihe^{ihzt}\{(t^2-1)v' + [t-ihz(t^2-1)]v\}$$

If v is set equal to $f(t)/\sqrt{1 - t^2}$, the equation for f turns out to be just the same as Eq. (5.3.80) for ψ. Consequently we can say that, if f is a solution of Mathieu's equation (5.3.80), then another solution of the same equation is

$$\psi(z) = A \int e^{ihzt} f(t) \frac{dt}{\sqrt{1 - t^2}} \tag{5.3.81}$$

If the constant A and the limits of integration are chosen properly, then ψ can equal f and we have an *integral equation* for ψ.

But even in its present form we can use the integral to help us in the analytic continuation of the solution, which is why we need an integral representation.

For instance, a solution of Eq. (5.3.80) (see page 565) for a particular value of $b(= be_{2m})$ is

$$Se_{2m}(h,z) = \sum_{n=0}^{\infty} B_{2n} \cos(2n\phi); \quad z = \cos\phi$$

From Eq. (5.3.81), we see that another solution of Eq. (5.3.80) is

$$\psi(z) = A \int_0^{2\pi} e^{ihz\cos\phi} \sum_{n=0}^{\infty} B_{2n} \cos(2n\phi)\, d\phi = 2\pi A \sum_{n=0}^{\infty} (-1)^n B_{2n} J_{2n}(hz)$$

where we have used Eq. (5.3.65) to obtain the Bessel functions. This is an extremely interesting and important relation; that a solution of the Mathieu equation can be expressed in terms of a series of Bessel functions is interesting in itself but that the numerical coefficients for the series are equal (with alternating signs) to the coefficients in the corresponding Fourier series solution is one of those satisfactory denouements which renew one's faith in the essential unity and simplicity of mathematics.

We could, of course, have given ourselves this pleasant surprise in the last section, when we were talking about the series expansion of Mathieu functions. We could have tried using a series of Bessel functions to solve Eq. (5.3.80) (see page 575), and we should have found that the recursion formula for the coefficients is *exactly the same* as the formula (5.2.69) for the Fourier series coefficients (with alternating signs for the a's). Then we should have said that the recurrence formulas (5.3.64) for the Bessel functions, which lead to the recursion formula for the coefficients, are closely related to the recurrence formulas for the trigonometric functions, so that for an equation of the particular symmetry of Mathieu's equation the recursion formulas turn out to be equivalent. Since we waited to demonstrate the property until the present (in order to be better acquainted with Bessel functions), we should now be inclined to say that the close *integral* relationship between trigonometric and

Bessel functions, via the Laplace transform, is the property which is responsible for the coincidence. The two statements are but two aspects of the same general property of course.

As yet we have shown only that the Bessel function series is a solution of Eq. (5.3.80); we have not shown how it is related to $Se_m(h,z)$, which is also a series (the series which is analytic at $z = \pm 1$). To do this we change from $z = \cos\phi$ to $z = \cosh\theta$ ($\theta = i\phi$) in Se and then use Eqs. (5.2.54) relating the hyperbolic functions to the Gegenbauer functions

$$Se_{2m}(h,z) = \sum_{n=0}^{\infty} B_{2n}\cosh(2n\theta) = \sqrt{\frac{\pi}{2}}\sum_{n=0}^{\infty} B_{2n}T_{2n}^{-\frac{1}{2}}(z)$$

By expanding this in powers of z and expanding the Bessel function also in powers of z, we can show that the two series are proportional and are therefore the same solution. Thus we have obtained an analytic continuation of the solution Se, defined for the range $-1 < z < 1$, into the range $1 < z < \infty$.

For the range $1 < z < \infty$ we shall prefer to use a function which has simple asymptotic properties. Using the asymptotic behavior of the Bessel functions [Eq. (5.3.68)], we define

$$Je_{2m}(h,z) = \sqrt{\frac{\pi}{2}}\sum_{n=0}^{\infty}(-1)^{n-m}B_{2n}J_{2n}(hz) \to \frac{1}{\sqrt{hz}}\cos\left[hz - \tfrac{1}{2}\pi(2m + \tfrac{1}{2})\right] \tag{5.3.82}$$

since $\Sigma B_{2n} = 1$. The asymptotic expression is valid for the argument of z between $-\frac{1}{2}\pi$ and $+\frac{1}{2}\pi$. Since this solution is proportional to Se_{2m}, we can find the factor of proportionality by comparing values at any satisfactory values of z. The value $z = 0$ is most satisfactory, for

$$Je_{2m}(h,0) = (-1)^m B_0\sqrt{\frac{\pi}{2}} \quad \text{and} \quad Se_{2m}(h,0) = \sum_{n=0}^{\infty}(-1)^n B_{2n}$$

Therefore
$$\left[\sum_n (-1)^n B_{2n}\right] Je_{2m}(h,z) = (-1)^m B_0\sqrt{\frac{\pi}{2}}\,Se_{2m}(h,z) \tag{5.3.83}$$

We are also now in a position to obtain a second solution for Eq. (5.3.80), by replacing the Bessel by Neumann functions:

$$Ne_{2m}(h,z) = \sqrt{\frac{\pi}{2}}\sum_{n=0}^{\infty}(-1)^{n-m}B_{2n}N_{2n}(hz) \to \frac{1}{\sqrt{hz}}\sin\left[hz - \tfrac{1}{2}\pi(2m + \tfrac{1}{2})\right] \tag{5.3.84}$$

This solution has singularities at $z = \pm 1$ and is, of course, independent of Se_{2m} and Je_{2m}. The series of Eq. (5.3.84) cannot be used for $|z| \le 1$ (as a matter of fact it does not converge well for small values of z just greater than 1), and a different expansion must be used. Since the second solution for the range $|z| \le 1$ is rarely needed in physical problems (since they are not periodic in ϕ), we shall not go into the matter further.

The Laplace Transform and the Separated Wave Equation. Before we continue with our discussion of the Mathieu and spheroidal functions, it will be illuminating to introduce an extremely important technique of solution of the wave equation, which will be used extensively in later chapters. Here we shall use it only to bring out a new point of view with regard to the Laplace transform which will enable us to set up new integral representations practically at will. To do this we return to the discussion of the separation of the wave equation in two dimensions, discussed on pages 498 to 504.

If the coordinates ξ_1 and ξ_2 are separable and represent a conformal transformation from x, y, the Helmholtz equation can be put into the form

$$\frac{\partial^2\psi}{\partial\xi_1^2} + \frac{\partial^2\psi}{\partial\xi_2^2} + k^2[g_1(\xi_1) + g_2(\xi_2)]\psi = 0 \tag{5.3.85}$$

Any solution of the Helmholtz equation, whether it separates into factors in ξ_1, ξ_2 or not, must be a solution of this equation. For instance, e^{ikx} or $J_0(kr)$, when x or r is expressed in terms of ξ_1, ξ_2, satisfies Eq. (5.3.85). If the solution happens to separate in these coordinates, $\psi = X_1(\xi_1)X_2(\xi_2)$, then X_1 and X_2 are solutions of the separated equations

$$(d^2X_1/d\xi_1^2) + k^2[g_1(\xi_1) - \alpha^2]X_1 = 0; \quad (d^2X_2/d\xi_2^2) + k^2[g_2(\xi_2) + \alpha^2]X_2 = 0$$

where α^2 is the separation constant.

The point of reviewing this material here is that we can now show that *any solution of the Helmholtz equation* is a suitable kernel *for the integral representation of one of the separated solutions X_1 in terms of the other X_2.* Suppose that $K(z,t)$ is a solution of the Helmholtz equation $\nabla^2K + k^2K = 0$, expressed in terms of the coordinates ($z = \xi_1$, $t = \xi_2$, to obviate subscripts) in which the solutions X_1, X_2 separate. For instance, K can be e^{ikx}, and the coordinates for separation can be polar coordinates ($z = \ln r$, $t = \phi$); then $K = e^{ikr\cos t}$. Then K satisfies Eq. (5.3.85)

$$\frac{\partial^2K}{\partial z^2} + k^2g_1(z)K = -\frac{\partial^2K}{\partial t^2} - k^2g_2(t)K$$

If now the separated function $X_1(z)$ is to satisfy the equation

$$\mathfrak{L}_z(X_1) = (d^2X_1/dz^2) + k^2(g_1 - \alpha^2)X_1 = 0$$

then the integral representation $X_1(z) = \int K(z,t)v(t)\,dt$ can be used. For

$$\mathfrak{L}_z(K) = \mathfrak{M}_t(K) = -(\partial^2 K/\partial t^2) - k[g_2(t) + \alpha^2]K$$

and because of the absence of the first derivative term and the fact that the second derivative has no factor in t, the adjoint operator $\tilde{\mathfrak{M}}_t = \mathfrak{M}_t$. Therefore the equation for v,

$$(d^2v/dt^2) + k^2[g_2(t) + \alpha^2]v = 0$$

is just the equation satisfied by the separated factor X_2, which goes along with X_1 to form the factored solution of the Helmholtz equation.

What we have just demonstrated is that, if $K(z,t)$ is any solution of $(\nabla^2 + k^2)K$, expressed in terms of separable coordinates z and t, and if a separated solution in the coordinate t is $X_2(t)$, a separated solution in the coordinate z is given by the representation

$$X_1(z) = \int K(z,t)X_2(t)\,dt \tag{5.3.86}$$

and $X_1(z)X_2(t)$ is a separated solution of the equation $(\nabla^2 + k^2)X_1X_2 = 0$ in the coordinates z, t. If X_2 is a simpler function than X_1, we have obtained an integral representation of a complicated function in terms of simpler ones (if K is sufficiently simple).

As a first example of this we can take the polar coordinates $\xi_1 = \ln r$, and $\xi_2 = \phi$, where Eq. (5.3.85) takes the form [see Eq. (5.1.13)]

$$\frac{\partial^2\psi}{\partial\xi_1^2} + \frac{\partial^2\psi}{\partial\xi_2^2} + k^2e^{2\xi_1}\psi = 0$$

A solution of $(\nabla^2 + k^2)K = 0$ is $K = e^{ikx} = \exp(ike^{\xi_1}\cos\phi) = e^{ikr\cos\phi}$ and a solution of the equation for the ϕ factor is $\cos(\alpha\phi)$ (where α is the separation constant). Therefore a solution of the equation for the r factor is

$$X_1(r) = \int e^{ikr\cos\phi}\cos(\alpha\phi)\,d\phi$$

and from Eq. (5.3.65), we see that this is, indeed, an integral representation of the Bessel function $J_\alpha(kr)$ if the integration is between 0 and 2π. A complete separated solution for the separation constant α is $J_\alpha(kr)\cos(\alpha\phi)$.

We could, of course, have used any of the other solutions for X_2, such as $\sin(\alpha\phi)$ or $e^{i\alpha\phi}$, etc., to obtain other representations of Bessel or Neumann functions. Or we could have used other wave solutions for the kernel. Or we could have turned the factors around and expressed the trigonometric functions in terms of an integral of Bessel functions of the sort

$$\int e^{ik\cos\phi\, e^z}J_\alpha(ke^z)\,dz = \int e^{ikr\cos\phi}J_\alpha(kr)\,\frac{dr}{r}$$

which turns out, if the integration is from zero to infinity, to be proportional to $e^{i\alpha\phi}$. To express a simple exponential in terms of a Bessel integral, however, is somewhat of a work of supererogation.

Moving on to the Mathieu functions, we can use the coordinates $i\vartheta = \xi_1 - \beta$, $\phi = \xi_2$ given in Eq. (5.1.16) where $x = d \cos\vartheta \cos\phi$, $y = id \sin\vartheta \sin\phi$. We again use the kernel $e^{ikx} = e^{ih\cos\vartheta\cos\phi}$ $(h = kd)$ and obtain the integral representation

$$X_1(\vartheta) = \int e^{ih\cos\vartheta\cos\phi} X_2(\phi)\, d\phi$$

where both X_1 and X_2 satisfy the equation [see Eq. (5.2.66)]

$$(d^2X/d\phi^2) + [b - h^2\cos^2\phi]X = 0$$

where ϕ is either ϕ or ϑ and X is either X_1 or X_2. Here we have complete symmetry, so that, if limits of integration and constant factors are fixed properly, $X_1 = X_2$ and the equation is an *integral equation* for X instead of an integral representation. From Eqs. (5.3.81) to (5.3.83) we see that the solution $Se_{2m}(h, \cos\vartheta)$ satisfies the integral equation

$$Se_{2m}(h, \cos\vartheta) = \lambda_{2m} \int_0^{2\pi} e^{ih\cos\vartheta\cos\phi}\, Se_{2m}(h, \cos\phi)\, d\phi$$
$$\lambda_{2m} = \frac{1}{2\pi B_0} \sum_{n=0}^{\infty} (-1)^n B_{2n} = \left(\frac{1}{2\pi B_0}\right) Se_{2m}(h,0) \tag{5.3.87}$$

We could just as well have used $e^{iky} = e^{ih\sin\vartheta\sin\phi}$ for our kernel. In fact it is possible to show by this means (see analogous derivation on page 635) that

$$Je_{2m}(h, \cosh\mu) = \frac{\sqrt{\pi/2}}{Se_{2m}(h,0)} (-1)^m \sum_{n=0}^{\infty} B_{2n} J_{2n}(h\sinh\mu) \tag{5.3.88}$$

where $\mu_1 = \xi_1 - \beta = i\vartheta$.

A much more important and useful expansion for Je is obtained by using an integral representation with a Bessel function kernel. A solution of the Helmholtz equation, in polar coordinates, is

$$J_0(kr) = J_0(k\sqrt{x^2 + y^2}) = J_0(h\sqrt{\cos^2\vartheta + \cos^2\phi - 1})$$
$$= J_0(h\sqrt{\tfrac{1}{2}\cos 2\vartheta + \tfrac{1}{2}\cos 2\phi})$$

Therefore another solution of the Mathieu equation (5.2.66) is the integral

$$\int_0^{2\pi} J_0(h\sqrt{\tfrac{1}{2}\cos 2\vartheta + \tfrac{1}{2}\cos 2\phi})\, Se_{2m}(h, \cos\phi)\, d\phi$$

It is finite for all real values of ϑ, is periodic in ϑ, and is even in ϑ (*i.e.*, its value does not change on reversal of the sign of ϑ). Therefore it must be proportional to $Se_{2m}(h, \cos\vartheta)$. That is, we have obtained another integral equation

$$Se_{2m}(h, \cos \vartheta) = \nu_{2m} \int_0^{2\pi} J_0(h \sqrt{\tfrac{1}{2} \cos 2\vartheta + \tfrac{1}{2} \cos 2\phi})\, Se_{2m}(h, \cos \phi)\, d\phi \tag{5.3.89}$$

In order to determine ν_{2m} and, what is more important, to develop a new expansion of Je_{2m}, we utilize Eq. (5.3.66) to expand the kernel. In Eq. (5.3.66) we set $\phi = 2\phi$, $x = \frac{1}{2}he^{i\vartheta}$, $y = -\frac{1}{2}he^{-i\vartheta}$, obtaining

$$J_0(h \sqrt{\tfrac{1}{2} \cos 2\vartheta + \tfrac{1}{2} \cos 2\phi}) = \sum_{n=-\infty}^{\infty} (-1)^n J_n(\tfrac{1}{2}he^{i\vartheta}) J_n(\tfrac{1}{2}he^{-i\vartheta}) \cos(2n\phi)$$

combining this with the Fourier series for Se_{2m} (see page 634) we finally obtain

$$Se_{2m}(h, \cos \vartheta) = 2\pi \nu_{2m} \sum_{n=0}^{\infty} (-1)^n B_{2n} J_n(\tfrac{1}{2}he^{i\vartheta}) J_n(\tfrac{1}{2}he^{-i\vartheta})$$

Setting $\vartheta = 0$ gives us an expression for the constant

$$\nu_{2m} = \frac{1}{2\pi} \left[\sum_{n=0}^{\infty} (-1)^n B_{2n} J_n^2 \left(\frac{h}{2}\right) \right]^{-1}$$

since $Se_{2m}(h,1) = 1$.

It can be shown that $\int_0^{2\pi} J_{2n}(h \cos \vartheta)\, d\vartheta = 2\pi J_n^2 \left(\frac{h}{2}\right)$. From Eqs. (5.3.82) and (5.3.83), with $z = \cos \vartheta$, by integrating over ϑ we have

$$\begin{aligned}\int_0^{2\pi} Se_{2m}(h, \cos \vartheta)\, d\vartheta &= 2\pi B_0 \\ &= \frac{1}{B_0} \left[\sum_{n=0}^{\infty} (-1)^n B_{2n} \right] \sum_{n=0}^{\infty} (-1)^n B_{2n} \int_0^{2\pi} J_{2n}(h \cos \vartheta)\, d\vartheta \\ &= \frac{2\pi}{B_0} \left[\sum_{n=0}^{\infty} (-1)^n B_{2n} \right] \sum_{n=0}^{\infty} (-1)^n B_{2n} J_n^2 \left(\frac{h}{2}\right)\end{aligned}$$

from which we can finally obtain an expression for ν_{2n} in terms of the B_{2n}'s alone:

$$\nu_{2n} = \frac{1}{2\pi B_0^2} \sum_{n=0}^{\infty} (-1)^n B_{2n} = \frac{1}{2\pi B_0^2} Se_{2m}(h,0) Se_{2m}(h,1)$$

More on Mathieu Functions. Collecting all our results for the even periodic functions of even order, we have that the angle function is

$$Se_{2m}(h, \cos \phi) = \sum_{n=0}^{\infty} B_{2n} \cos(2n\phi)$$

where the B's are computed by the methods discussed on pages 564 to 567 and are normalized so that $\Sigma B_{2n} = 1$. The "radial" functions of the first kind are

$$\begin{aligned} Je_{2m}(h, \cosh \mu) &= \frac{\sqrt{\pi/2}}{Se_{2m}(h,1)} \sum_{n=0}^{\infty} (-1)^{n-m} B_{2n} J_{2n}(h \cosh \mu) \\ &= \frac{(-1)^m \sqrt{\pi/2}\, B_0}{Se_{2m}(h,1) Se_{2m}(h,0)} \sum_{n=0}^{\infty} B_{2n} \cosh(2n\mu) \\ &= \frac{(-1)^m \sqrt{\pi/2}}{Se_{2m}(h,0)} \sum_{n=0}^{\infty} B_{2n} J_{2n}(h \sinh \mu) \\ &= \frac{\sqrt{\pi/2}}{B_0} \sum_{n=0}^{\infty} (-1)^{n-m} B_{2n} J_n(\tfrac{1}{2}he^{\mu}) J_n(\tfrac{1}{2}he^{-\mu}) \end{aligned} \tag{5.3.90}$$

Similarly, for the second radial solution we have

$$\begin{aligned} Ne_{2m}(h, \cosh \mu) &= \frac{\sqrt{\pi/2}}{Se_{2m}(h,1)} \sum_{n=0}^{\infty} (-1)^{n-m} B_{2n} N_{2n}(h \cosh \mu) \\ &= \frac{(-1)^m \sqrt{\pi/2}}{Se_{2m}(h,0)} \sum_{n=0}^{\infty} B_{2n} N_{2n}(h \sinh \mu) \\ &= \frac{\sqrt{\pi/2}}{B_0} \sum_{n=0}^{\infty} (-1)^{n-m} B_{2n} N_n(\tfrac{1}{2}he^{\mu}) J_n(\tfrac{1}{2}he^{-\mu}) \end{aligned} \tag{5.3.91}$$

The second series does not converge except for $\sinh \mu > 1$; the third series converges quite satisfactorily for $\mu \geq 0$. For the normalization of Se_{2m} which we have used here, $Se_{2m}(h,1) = \Sigma B_{2n} = 1$ and $Se_{2m}(h,0) = \Sigma(-1)^n B_{2n}$, but the formulas have been written to hold for any normalization.

In order to fit boundary conditions it is often important to know the values of the slope and magnitude of both Je and Ne at $\mu = 0$ (which corresponds to the limiting elliptic cylinder, *i.e.*, a strip of width $d = h/k$). For the first solution we can easily see that for $\mu = 0$

$$Je_{2m}(h, \cosh \mu) = (-1)^m \sqrt{\frac{\pi}{2}} \frac{B_0}{\Sigma(-1)^n B_{2n}}; \quad \frac{d}{d\mu} Je_{2m}(h, \cosh \mu) = 0$$

To obtain the slope of Ne, we utilize the value of the Wronskian

$$\Delta(Je,Ne) = Je_{2m} \frac{d}{d\mu} Ne_{2m} - Ne_{2m} \frac{d}{d\mu} Je_{2m} = 1$$

which can be verified by using the asymptotic forms for Je, Ne. Since the second term is zero at $\mu = 0$, we must have that the slope of Ne must equal the reciprocal of the value of Je at $\mu = 0$.

To find the value of Ne at $\mu = 0$ is more difficult, for we must use the expansion of the Neumann function about the origin. The most straightforward method is to use values of J and N in the expansion

$$Ne_{2m}(h,1) = \frac{\sqrt{\pi/2}}{B_0} \sum_{n=0}^{\infty} (-1)^{n-m} B_{2n} N_n \left(\frac{h}{2}\right) J_n \left(\frac{h}{2}\right)$$

which converges fairly well. Alternatively, according to page 568, Ne_{2m} is a linear combination of Se_{2m} and the second solution Fe_{2m}, defined in Eq. (5.2.79). Comparing slopes and values at $\mu = 0$, we have

$$Ne_{2m}(h, \cosh \mu) = \left[(-1)^m \frac{\sqrt{2/\pi}}{B_0} \sum_n (-1)^n B_{2n} \right] Fe_{2m}(h, \cosh \mu) + [Ne_{2m}(h,1)] Se_{2m}(h, \cosh \mu) \quad (5.3.92)$$

which indicates how the solution behaves near $\mu = 0$.

So far we have discussed only the even Mathieu functions, the Se functions, and their related solutions. There are also odd solutions for different values of the separation constant ($b = bo_m$). The analysis of these solutions goes in an analogous way, with a few modifications due to the change in symmetry. For instance, we cannot use the same integral equation (5.3.89), for the integral over the sine series would be zero. Returning to Eq. (5.3.81), we notice that, if $F(t)$ is a solution of Mathieu's equation, then

$$\Psi(z) = A \sqrt{z^2 - 1} \int e^{ihzt} F(t)\, dt$$

is also a solution. Changing to $z = \cosh \mu$, $t = \cos \phi$ and integrating once by parts we have (for odd values of m, for example)

$$\begin{aligned} Jo_{2m+1}(h, \cosh \mu) &= A \sinh \mu \sum_{n=0}^{\infty} B_{2n+1} \int_0^{2\pi} \sin \phi\, e^{ih \cosh \mu \cos \phi} \sin (2n+1)\phi\, d\phi \\ &= \left(\frac{2\pi}{ih}\right) A \tanh \mu \sum_{n=0}^{\infty} B_{2n+1} i^{2n+1} (2n+1) J_{2n+1}(h \cosh \mu) \end{aligned}$$

The rest of the procedure is carried through with similar changes and results in

$$Jo_{2m+1}(h, \cosh \mu)$$

$$= \frac{\tanh \mu}{So'_{2m+1}(h,1)} \sqrt{\frac{\pi}{2}} \sum_{n=0}^{\infty} (-1)^{n-m}(2n+1)B_{2n+1}J_{2n+1}(h \cosh \mu)$$

$$= \frac{(-1)^m \sqrt{\pi/8}\, B_1 h}{So'_{2m+1}(h,1)So_{2m+1}(h,0)} \sum_{n=0}^{\infty} B_{2n+1} \sinh [(2n+1)\mu]$$

$$= \frac{(-1)^m \sqrt{\pi/2}}{So_{2m+1}(h,0)} \sum_{n=0}^{\infty} B_{2n+1}J_{2n+1}(h \sinh \mu)$$

$$= \frac{\sqrt{\pi/2}}{B_1} \sum_{n=0}^{\infty} (-1)^{n-m}B_{2n+1}[J_n(\tfrac{1}{2}he^{-\mu})J_{n+1}(\tfrac{1}{2}he^{\mu}) - J_{n+1}(\tfrac{1}{2}he^{-\mu})J_n(\tfrac{1}{2}he^{\mu})]$$

$$\to \frac{1}{\sqrt{h \cosh \mu}} \cos [h \cosh \mu - \tfrac{1}{2}\pi(2m + \tfrac{3}{2})] \qquad (5.3.93)$$

where So' is the derivative of $So(h \cos \phi)$ with respect to ϕ.

The second solutions No are built in an analogous manner. Since the first solution Jo has zero value at $\mu = 0$, the magnitude of No there is related to the slope of Jo through the Wronskian. The slope of No at $\mu = 0$ may be calculated by means of the double-Bessel series

$$No_{2m+1}(h, \cosh \mu) = \frac{\sqrt{\pi/2}}{B_1} \sum_{n=0}^{\infty} (-1)^{n-m}B_{2n+1}[J_n(\tfrac{1}{2}he^{-\mu})N_{n+1}(\tfrac{1}{2}he^{\mu}) - J_{n+1}(\tfrac{1}{2}he^{-\mu})N_n(\tfrac{1}{2}he^{\mu})]$$

as was suggested for the values of the functions Ne.

It is also possible to devise Mathieu functions of the third kind, similar to the Hankel functions, by combining the functions of first and second kind. For instance,

$$He_{2n}(h,z) = Je_{2n}(h,z) + i\, Ne_{2n}(h,z) = \sqrt{\frac{\pi}{2}} \sum_{n=0}^{\infty} (-1)^{n-m}B_{2n}H_{2n}(hz)$$

and so on. But further discussion of the special properties of Mathieu functions may better be deferred to a later chapter, when we have a broader background of technique and have a more immediate need for the functions.

Spheroidal Wave Functions. The more general equation (5.2.66)

$$(z^2 - 1)\psi'' + 2(a+1)z\psi' + (h^2z^2 - b)\psi = 0$$

turns up in the case of the spheroidal coordinates. If, for instance, the prolate spheroidal coordinates are μ, ϑ, ϕ, the separated equations are

$$(d^2\Phi/d\phi^2) + m^2\Phi = 0$$

$$\frac{1}{\sin\vartheta}\frac{d}{d\vartheta}\left[\sin\vartheta\frac{dS}{d\vartheta}\right] - \left[b + \frac{m^2}{\sin^2\vartheta} + h^2\cos^2\vartheta\right]S = 0$$

$$\frac{1}{\sinh\mu}\frac{d}{d\mu}\left[\sinh\mu\frac{dR}{d\mu}\right] + \left[b - \frac{m^2}{\sinh^2\mu} + h^2\cosh^2\mu\right]R = 0 \tag{5.3.94}$$

The solution of the first equation requires that m be an integer if the solution is to be periodic. If in the second equation we set $\cos\vartheta = z$, $S = \sin^m\vartheta\,\psi(z)$ or in the third we set $\cosh\mu = z$, $R = \sinh^m\mu\,\psi(z)$, the equation for ψ will be of the form of Eq. (5.2.66), repeated above, with a equal to the integer m.

On pages 573 and 575 we discussed the possibility of solving this equation in terms of a series of Gegenbauer functions. We indicated that a solution could be written in the form

$$S_{ml}(h,z) = (1 - z^2)^{m/2}\sum_{n=0}^{\infty} d_{2n}T_{2n}^m(z) \tag{5.3.95}$$

when l is an even integer. When l is odd the sum is over $d_{2n+1}T_{2n+1}^m(z)$. The corresponding values of the separation constant for which S is finite over the range $-1 \leq z \leq 1$ are called b_{ml}. The normalization of the d's is to be such that S_{ml} has the same value or slope at $z = 1$ as does $T_{l-m}^m(z)$. For instance, when $m = 0$, we require that $S_{0l}(h,1) = T_l^0(1) = P_l(1) = 1$ so that, for $m = 0$, $\sum_n d_{2n} = 1$ or $\sum_n d_{2n+1} = 1$.

The "radial" solutions, analogous to the *Je*, *Jo* functions for the elliptic cylinder case, are obtained by using the integral of (5.3.67) together with the integral equation for these functions (which may be obtained by the Laplace transform or by the method of pages 636 to 639):

$$\psi(h,z) = A(z^2 - 1)^{m/2}\int e^{ihzt}(1 - t^2)^{m/2}f(t)\,dt$$

If $f(t)$ is a solution of Eq. (5.2.66) for $a = m$, then ψ is also a solution of the same equation. From Eq. (5.3.67), which can be rewritten for $a = m$ to be

$$\int_{-1}^{1} e^{ihzt}(1 - t^2)^m T_{2n}^m(t)\,dt = \frac{2(2n + 2m)!(-1)^n}{(2n)!(hz)^m}\,j_{2n+m}(hz)$$

where $j_\nu(x) = \sqrt{\pi/2x}\,J_{\nu+\frac{1}{2}}(x)$ is the spherical Bessel function mentioned on page 622. Combining these two equations, by setting in for $f(t)$ the series for $S_{me}(h,t)$, we can show that a solution of the "radial" equation for $b = b_{ml}$ is

$$je_{ml}(h, \cosh \mu) = \frac{\tanh^m \mu}{\sum_n d_{2n}(2n + 2m)!/(2n)!} \cdot$$

$$\cdot \sum_{n=0}^{\infty} (-1)^{n-l} \frac{(2n + 2m)!}{(2n)!} d_{2n} j_{m+2n}(h \cosh \mu)$$

$$\to \frac{1}{h \cosh \mu} \sin [h \cosh \mu - \tfrac{1}{2}\pi(l + m)] \qquad (5.3.96)$$

A second solution, $ne_{ml}(h, \cosh \mu)$, may be obtained by substituting $n_{m+2n}(h \cosh \mu)$ for $j_{m+2n}(h, \cosh \mu)$, where n is the spherical Neumann function defined on page 622.

Bessel function integral equations may also be obtained for these functions, but we shall find them easier to develop and discuss later in the book (Chap. 11). We shall conclude this section by a discussion of other types of integral representations which occasionally are useful.

Kernels Which Are Functions of zt. We have already discussed the properties of the nucleus e^{zt}, connected with the Laplace transform. However, any nucleus of the form $K(zt)$ has the property that $z(dK/dz) = t(dK/dt)$. This, by itself, is not particularly useful, but if $K(w)$ satisfies the differential equation

$$\mathfrak{N}\left(w \frac{d}{dw}\right) K(w) = w^n \mathfrak{R}\left(w \frac{d}{dw}\right) K(w)$$

where $\mathfrak{R}[w(d/dw)]$ is a comparatively simple differential operator, then we can use K as a kernel for some integral representations. A few examples will show the technique.

The Bessel function $J_\nu(w)$ satisfies the equation

$$\left[w^2 \frac{d^2}{dw^2} + w \frac{d}{dw} - \nu^2\right] J_\nu(w) = -w^2 J_\nu(w)$$

Therefore any equation of the form

$$\mathfrak{L}_z(\psi) = \frac{d^2\psi}{dz^2} + \left[\alpha z + \frac{1}{z}\right] \frac{d\psi}{dz} + \left[\alpha\mu - \frac{\nu^2}{z^2}\right] \psi = 0 \qquad (5.3.97)$$

may be solved by setting $\psi = \int J_\nu(zt)v(t)\, dt$, for $\mathfrak{L}_z$ operating on $J_\nu(zt)$ produces

$$t^2 \left[J_\nu''(w) + \frac{1}{w} J_\nu'(w) - \frac{\nu^2}{w^2} J_\nu(w)\right] + \alpha t \frac{d}{dt} J_\nu(w) + \alpha\mu J_\nu(w) = 0$$

The expression in the brackets is just equal to $-J_\nu(w)$ so that

$$\mathfrak{L}_z\left(\int J_\nu v\, dt\right) = \int \mathfrak{M}_t(J_\nu) v\, dt; \quad \text{where } \mathfrak{M}_t(J_\nu) = \alpha t(d/dt)J_\nu + (\alpha\mu - t^2)J_\nu$$

The adjoint equation is

$$\tilde{\mathfrak{M}}_t(v) = -(d/dt)(\alpha t v) + [\alpha\mu - t^2]v = 0$$

with solution $v(t) = At^{\mu-1}e^{-t^2/2\alpha}$ and bilinear concomitant

$$P = A\alpha t^\mu e^{-t^2/2\alpha}J_\nu(tz)$$

Therefore a solution of Eq. (5.3.97) is

$$\psi(z) = A\int_0^\infty t^{\mu-1}e^{-t^2/2\alpha}J_\nu(tz)\,dt \tag{5.3.98}$$

as long as Re μ + Re $\nu > 0$. This solution is of some interest, for the wave equation for parabolic coordinates, which can be written as

$$F'' + \frac{1}{z}F' + \left(\gamma^2 - \frac{\nu^2}{z^2} - \beta^2 z^2\right)F = 0$$

has a solution

$$F = Ae^{-\beta z^2/2}\int_0^\infty t^{\gamma^2/2\beta}e^{-t^2/4\beta}J_\nu(tz)\,dt \tag{5.3.99}$$

Finally, the equation

$$(z^2 - 1)\psi'' + \left(z - \frac{1}{z}\right)\psi' + \left(-\mu^2 + \frac{\nu^2}{z^2}\right)\psi = 0$$

which has three regular singular points, at $z = 0$ and ± 1, and one irregular point at infinity, also has a solution of the form

$$\psi(z) = \int J_\nu(zt)v(t)\,dt$$

In this case the operator $\mathfrak{M}(J)$ is

$$\mathfrak{M}(J) = t^2\frac{d^2J}{dt} + t\frac{dJ}{dt} + (t^2 - \mu^2)J$$

and the adjoint equation

$$\tilde{\mathfrak{M}}(v) = t^2\left[\frac{d^2v}{dt^2} + \frac{3}{t}\frac{dv}{dt} + \left(1 - \frac{\mu^2 - 1}{t^2}\right)v\right] = 0$$

is related to the Bessel equation again, having a solution $(1/t)J_\mu(t)$. The bilinear concomitant is

$$P(J_\nu,v) = J_\nu(zt)J_\mu(t) + \frac{d}{dt}[tJ_\nu(zt)J_\mu(t)] - 2J_\nu(zt)\frac{d}{dt}[tJ_\mu(t)]$$

and as long as Re μ + Re $\nu > -1$, we can have that

$$\psi(z) = \int_0^\infty J_\nu(zt)J_\mu(t)\frac{dt}{t}$$

This quantity is known to have a discontinuity at $z = 1$.

There are a number of other integral representations which are of some value in quite special cases but are of little use on any other equation. Such solutions usually have to be found by trial and error or by "hunch." In any case there is little point in our spending time cataloguing them or in trying to give recipes for when to use which. We have covered the field of the most useful transforms, and the others can be discovered in the literature.

Problems for Chapter 5

5.1 Set up the Helmholtz equation in conical coordinates and separate it. What are the shapes of the coordinate surfaces? For what physical problems would the equation be useful?

5.2 Set up the Laplace equation in bispherical coordinates and separate it. Show that the constant k_1^2 of Eq. (5.1.47) is $\frac{1}{4}$ and that $R = (x^2 + y^2)^{\frac{1}{4}}$.

5.3 Set up the Schroedinger equation for an electron in a diatomic molecule in the prolate spheroidal coordinates

$$\xi = (r_1 + r_2)/a; \quad \eta = (r_1 - r_2)/a; \quad \varphi = \tan^{-1}(y/x)$$

where r_1 is the distance from one nucleus and r_2 the distance from the other and the nuclei are supposed to lie at the points $z = \pm\frac{1}{2}a$, $x = y = 0$. Express x, y, z in terms of ξ, η, φ, obtain the scale factors; set up the Schroedinger equation and the Stäckel determinant. Show that for the potential function $-(c_1/r_1) - (c_2/r_2)$ the Schroedinger equation separates. Obtain the separated equations.

5.4 The exponential coordinates of Prob. 1.9 are

$$\xi = \ln(x^2 + y^2) - z; \quad \eta = \tfrac{1}{2}(x^2 + y^2) + z; \quad \varphi = \tan^{-1}(y/x)$$

Sketch the surfaces, find the scale factors, set up the wave equation, and show that the equation does not separate.

5.5 Hyperboloidal coordinates are defined by the equations

$$\lambda^4 = z^2(x^2 + y^2); \quad \mu^2 = \tfrac{1}{2}(z^2 - x^2 - y^2); \quad \varphi = \tan^{-1}(y/x)$$

Sketch some of the coordinate surfaces, compute the scale factors, set up the wave equation, and show that it does not separate.

5.6 Rotational coordinates are characterized by having an axis of rotational symmetry (the x axis, for example), the coordinates being $\lambda(r,x)$, $\mu(r,x)$, and $\varphi = \tan^{-1}(z/y)$ $(r^2 = y^2 + z^2)$. The φ factor, when φ is unobstructed, is $\sin(m\varphi)$ or $\cos(m\varphi)$, and if we set the solution ψ of the three-dimensional Laplace equation equal to $e^{\pm im\varphi}\varphi(r,x)/r$, the equation for φ is

$$\frac{\partial^2\varphi}{\partial x^2} + \frac{\partial^2\varphi}{\partial r^2} + \left(\frac{\frac{1}{4} - m^2}{r^2}\right)\varphi = 0$$

Investigate the separability of this equation, for the coordinates λ and μ, as follows: Set $z = x + ir$ and $w = \lambda + i\mu$, so that z is a function of w or vice versa. Use the techniques of Eqs. (5.1.6) *et seq.* to show that the requirement that $|z'|^2/r^2 = -4z'\bar{z}'/(z - \bar{z})^2$ be equal to $f(\lambda) + g(\mu)$ gives rise to the equation

$$F + 2\bar{z}G + \bar{z}^2H = \bar{F} + 2z\bar{G} + z^2\bar{H}$$

where

$$F = (1/z')[z'''z^2 + 6(z')^3 - 6z''z'z]$$
$$G = [3z'' - (z'''z/z')]; \quad H = z'''/z'$$

and $z' = dz/dw$, etc. Show that the most general solution of this equation is for

$$d^2F/dz^2 = c_1; \quad d^2G/dz^2 = c_2; \quad d^2H/dz^2 = c_3$$

which eventually results in the solution

$$(dz/dw)^2 = a_0 + a_1z + a_2z^2 + a_3z^3 + a_4z^4$$

Solve this equation for the various different sorts of coordinate systems $z(w)$ which allow separation of the λ, μ part of the Laplace equation:

$$\frac{\partial^2\varphi}{\partial\lambda^2} + \frac{\partial^2\varphi}{\partial\mu^2} + \left|\frac{dz}{dw}\right|^2\left(\frac{\frac{1}{4} - m^2}{r^2}\right)\varphi = 0$$

Show that the case $z = w$ is the usual cylindrical coordinate system; $(z')^2 = z$, $z = \frac{1}{2}w^2$ is the parabolic system; $z' = z$, $z = e^w$ the spherical system; $(z')^2 = z^2 \pm 1$ the two spheroidal systems; and $z' = 1 \pm z^2$ the bispherical and the toroidal coordinate systems. Sketch the coordinate systems corresponding to $(z')^2 = z^3$ and $(z') = z^2$.

5.7 For the rotational coordinates discussed in Prob. 5.6 discuss the case $(z')^2 = \alpha(1 - z^2)(1 - k^2z^2)$, which results in $z = a\,\mathrm{sn}(w,k)$ [see Eq. (4.5.74)]. Sketch on the $z = x + ir$ plane, for $k = 0.6$, enough of the coordinate lines $\lambda =$ constant, $\mu =$ constant to indicate the nature of the system. Set up the λ, μ part of the Laplace equation and separate it. For what physical problem would this coordinate system be useful?

5.8 Analyze the rotational coordinate system (see Prob. 5.6) corresponding to the relation $z = a\,\mathrm{cn}(w,k)$ $(z = x + ir;\ w = \lambda + i\mu)$. Sketch the shape of the coordinate lines, λ, $\mu =$ constant on the z plane. Separate the Laplace equation in these coordinates. For what physical situation would this system be useful? [See Eq. (4.5.77) for definition of cn].

5.9 Analyze the rotational coordinate system (see Prob. 5.6) corresponding to the relation $z = a\,\mathrm{dn}(w,k)$ $(z = x + ir;\ w = \lambda + i\mu)$ [see Eq. (4.5.77)]. Sketch the coordinate system, separate the Laplace equation, and indicate the physical situation for which this system would be appropriate.

5.10 Set up the Laplace equation in the polar coordinates r, ϕ, and separate it. Find the basic fundamental set of solutions of these two equations at the points $r = a$, $\phi = 0$.

5.11 The Schroedinger equation for an electron in the one-dimensional potential field $V = (\hbar^2/2M)x^2$ is $\psi'' + (k - x^2)\psi = 0$, where $k = 2MW/\hbar^2$. One solution of this equation, for $k = 1$, is $\psi_1 = \exp(-\frac{1}{2}x^2)$. Find the basic fundamental set of solutions at $x = 0$.

5.12 Find the general solution of

$$(d^2\psi/dx^2) - (6/x^2)\psi = x \ln x$$

5.13 A solution of the Legendre equation

$$(1 - x^2)\psi'' - 2x\psi' + 2\psi = 0$$

is $\psi = x$. Find the basic fundamental set at $x = 0$. What is the Wronskian for this set? What solution has the value 2 at $x = 1$? What solutions have the value 2 at $x = 0$? Why is the answer unique in one case and not in the other?

5.14 The Lame equation is

$$\psi'' + \left[\frac{z}{z^2 - a^2} + \frac{z}{z^2 - b^2}\right]\psi' + \frac{k - m(m+1)z^2}{(z^2 - a^2)(z^2 - b^2)}\psi = 0$$

Locate the singular points of this equation, and give the indices of the solutions at each point. What is the basic fundamental set of solutions about $z = 0$? What is the Wronskian for this set?

5.15 Show that the only singular point of

$$\psi'' - 2az\psi' + [E + 2bcz - (a^2 - b^2)z^2]\psi = 0$$

is the irregular point at $z \to \infty$. Show that, by setting $\psi = \exp(\alpha z + \beta z^2)F(z)$ and adjusting the values of α and β, an equation for F will be obtained which will allow a series expansion of F about $z \to \infty$. Write out three terms of this series. Write down an equation which will have only an irregular singular point of infinity, for which the solution will be $\psi = \exp(\alpha z + \beta z^2 + \gamma z^3)G(z)$ where $G(z)$ is a series of the sort $a_0 z^s + a_1 z^{s+1} + \cdots$. Compare this equation with the above and with $\psi'' + k^2\psi = 0$, which also has an irregular singular point at $z \to \infty$. What can you suggest concerning the classification of irregular singular points?

5.16 In Eq. (5.2.26), with one regular singular point, the exponents of the two solutions about this point add up to -1. For the equation with two regular singular points (5.2.28) the sum of the exponents of the solution about one point plus the sum of those for the solutions about the other $(\lambda + \mu - \lambda - \mu)$ add up to zero. What is the corresponding statement concerning Eq. (5.2.36) for three regular singular points? What is the corresponding equation for four regular singular points and no irregular points, and what is the corresponding statement? By induction, what is the sum of the exponents about all singular points for an equation with N regular points and no irregular points?

5.17 Show that a solution of

$$\psi'' + \left\{-\tfrac{1}{4}(c-1)^2 + [ab - \tfrac{1}{2}c(a+b-c+1)]\left(\frac{e^{-x}}{1-e^{-x}}\right)\right.$$
$$\left. - \tfrac{1}{4}(a+b-c+1)(a+b-c-1)\left(\frac{e^{-x}}{1-e^{-x}}\right)^2\right\}\psi = 0$$

is

$$\psi = [1-e^{-x}]^{\frac{1}{2}(a+b-c+1)}e^{-\frac{1}{2}(c+1)x}F(a,b|c|e^{-x})$$

For what values of a, b, and c is the hypergeometric series a finite polynomial and is ψ finite in the range $0 \le x \le \infty$?

5.18 Set up the Schroedinger equation in one dimension for a particle of mass M in a potential field $-(\hbar^2A^2/2M)\,\mathrm{sech}^2(x/d)$. Changing the independent variable into $z = \frac{1}{2} + \frac{1}{2}\tanh(x/d)$, show that the resulting equation in z has three regular singular points. Express the solutions of this equation in terms of hypergeometric functions of z. What solution stays finite when $x \to -\infty$? Find the values of the energy for which this solution is a finite polynomial in z. Is this solution finite at $z \to \infty$?

5.19 Show that the equation for $\psi(x) = [z^{\frac{1}{2}c}e^{-\frac{1}{2}z}/\sqrt{z'}]\,F(a|c|z)$, where z is a function of x, is

$$\psi'' + \left\{-\tfrac{1}{4}(z')^2 - \tfrac{3}{4}\left(\frac{z''}{z'}\right)^2 - c(c-1)\left(\frac{z'}{z}\right)^2 + (c-a)\frac{(z')^2}{z}\right.$$
$$\left. + \tfrac{1}{2}\left(\frac{z'''}{z'}\right)\right\}\psi = 0$$

where $z' = dz/dx$, etc. Show that the equation for

$$\psi(x) = z^{\frac{1}{2}c}(1-z)^{\frac{1}{2}(a+b-c+1)}(z')^{-\frac{1}{2}}F(a,b|c|z)$$

is

$$\psi'' + \left\{\tfrac{1}{4}c(2-c)\left(\frac{z'}{z}\right)^2 - \tfrac{1}{4}(a+b-c+1)(a+b+c-1)\left(\frac{z'}{z-1}\right)^2\right.$$
$$\left. - \tfrac{3}{4}\left(\frac{z''}{z'}\right)^2 + \tfrac{1}{2}\left(\frac{z'''}{z'}\right) + [ab - \tfrac{1}{2}c(a+b-c+1)]\frac{(z')^2}{z(z-1)}\right\}\psi = 0$$

What is the form of these equations for $z = x^n$? For $z = e^{-x}$?

5.20 Separate the Helmholtz equation in spherical coordinates, and show that the radial equation for $x = kr$,

$$\frac{1}{x^2}\frac{d}{dx}\left(x^2\frac{d\psi}{dx}\right) + \left[1 - \frac{n(n+1)}{x^2}\right]\psi = 0$$

has solutions $j_n(x) = \sqrt{\pi/2x}\, J_{n+\frac{1}{2}}(x)$, $n_n(x) = \sqrt{\pi/2x}\, N_{n+\frac{1}{2}}(x)$. Show that a solution of this equation is

$$h_n(x) = \frac{(2n)!}{i2^n n!} \frac{e^{ix}}{x^{n+1}} F(-n|-2n|-2ix)$$

Show that its asymptotic form is $e^{ix}/i^{n+1}x$ and therefore that $h_n = j_n + in_n$.

5.21 Set up the Schroedinger equation in the spherical coordinates r, ϑ, φ for an electron in the potential field $V = -e^2Z/r$. Separate the equation, and show that the solution of the radial equation can be expressed in terms of the confluent hypergeometric function. Find the two asymptotic series for the solution which is independent of ϑ and φ. Find what values of the energy makes the asymptotic series break off and become a finite polynomial (in which case it is not an asymptotic series, it is an exact solution). For what energies is this solution finite for all values of r $(0 \le r \le \infty)$?

5.22 What is the asymptotic series about $z = \infty$ for the spheroidal equation

$$(z^2 - 1)\psi'' + 2(a + 1)\psi' + (h^2z^2 - b)\psi = 0$$

What is the asymptotic series about $z = 0$ for the equation

$$\psi'' + (2/z)\psi' + [(a/z^4) - (b/x^6)]\psi = 0$$

5.23 A very approximate expression for the potential of a conduction electron in a metallic lattice is

$$V = \frac{\hbar^2}{2M} U_0 \left[\cos^2\left(\frac{\pi x}{l_x}\right) + \cos^2\left(\frac{\pi y}{l_y}\right) + \cos^2\left(\frac{\pi z}{l_z}\right)\right]$$

Show that the separated equations all have the form of the Mathieu equation (5.2.67) with $h^2 = l^2U/\pi^2$ and b proportional to the energy of the electron. Using Eq. (5.2.71) compute values of the phase factor s for $h = 1$, for $b = 0.3, 0.469, 1.0, 1.242$, and 1.5. Which of these values of b results in an allowed solution (one which is finite for $-\infty \le x \le \infty$)?

5.24 Using Eqs. (5.2.77) *et seq.*, compute be_0 and the Fourier coefficients of $Se_0(h, \cos\varphi)$ for $h = 2$.

5.25 Set up the Helmholtz equation in prolate spheroidal coordinates (VIII) and show that, if the solution is independent of ξ_3, then the equations for the ξ_1 and ξ_2 factors have the form

$$(x^2 - 1)\psi'' + 2x\psi' + (h^2x^2 - b)\psi = 0$$

Discuss the singular points of this equation, and set up the three-term recursion formula for the coefficients of the series expansion about $x = 0$. Set up the continued fraction equation relating b and a which must hold if the series is to converge at $x = \pm 1$ (no negative powers of x and the series must converge) for the solution having zero slope at $x - 0$ for $h = 1$.

5.26 Expand the solution of the equation of Prob. 5.25 in spherical harmonics $P_n(x)$ (see page 573), and obtain the three-term recursion formula for the coefficients. Obtain from this a continued fraction equation relating h^2 and a. Solve it for b for $h = 1$, and compare with the results of Prob. 5.25.

5.27 Use the Laplace transform to show that a solution of the equation

$$[xf(D) + F(D)]\psi = 0$$

where $D\psi = d\psi/dx$, $D^2\psi = d^2\psi/dx^2$, etc., and where f and F are finite polynomials in powers of D, is

$$\psi = \int_a^b \exp\{xt + \int[F(t)/f(t)]\,dt\}\,[dt/f(t)]$$

where a and b are chosen so that

$$[\exp(xt + \int F\,dt/f)]_a^b = 0$$

for all values of x. Use this formula to compute the integral representation of the confluent hypergeometric function.

5.28 Show that a solution of the equation

$$z(z-1)(z-a)\psi'' - (\alpha - 1)z(2z - a)\psi' + \alpha(\alpha - 1)(z + 1)\psi = 0$$

has the form

$$\psi = A\int(z - t)^{\alpha}(t - 1)^{-\beta-1}(t - a)^{\beta-1}t^{-\alpha}dt$$

where $\beta = (\alpha - 1)/(a - 1)$. What must be the limits of integration in order that ψ be a solution? Which choice of limits and of A yields the solution which is unity at $z = 0$? What is the behavior of this solution near the three other singular points?

5.29 Show that the solution of the equation

$$(d^{n-1}\psi/dz^{n-1}) - z\psi = a$$

is

$$\psi = \sum_{s=0}^{n-1} A_s e^{2\pi is/n} \int_0^\infty \exp[zte^{2\pi is/n} - (t^n/n)]\,dt$$

where

$$\sum_{s=0}^{n-1} A_s = a$$

5.30 Show that a solution of the equation

$$z(d^3\psi/dz^3) - \psi = 0$$

is

$$\psi = \int_0^\infty \sin(z/u)\, e^{-\frac{1}{2}u^2}u\,du$$

5.31 Taking the definition of the gamma function and by changing variables of integration, show that

$$\Gamma(p)\Gamma(q) = \Gamma(p+q)\int_0^1 u^{p-1}(1-u)^{q-1}\,du$$
$$= \frac{-e^{-\pi i(p+q)}\Gamma(p+q)}{\sin(\pi p)\,\sin(\pi q)} \oint u^{p-q}\,(1-u)^{q-1}\,du$$

where contour C is similar to that shown in Fig. 5.7, but about 0 and 1 (where is the integrand real?). Expand $F(a,b|c|z)$ about $z = 0$, and utilize the above equation to substitute for $\Gamma(b + n)/\Gamma(c + n)$ in the series (set $p = b + n$, $q = c - b$), thus obtaining a series in $(uz)^n$ inside the contour integral. Show that this series may be summed, eventually obtaining

$$F(a,b|c|z) = \frac{\Gamma(c)\Gamma(1 - b)}{4\pi \sin \pi(c - b)} e^{-\pi ic} \oint_C u^{b-1}(1 - u)^{c-b-1}(1 - uz)^{-a}\, du$$

Show that this is equivalent to the first part of Eq. (5.3.16).

5.32 Apply the Euler transform to show that a solution of

$$(1 - z^2)\psi'' - 2z\psi' + [n(n + 1) - m^2/(1 - z^2)]\psi = 0$$

is the associated Legendre function (n, m not necessarily integers)

$$P_n^m(z) = \frac{\Gamma(n + m + 1)}{2^n\pi i\Gamma(n + 1)} (1 - z^2)^{\frac{1}{2}m} \oint \frac{(t^2 - 1)^n\, dt}{(t - z)^{n+m+1}}$$

where the contour goes about both $+1$ and z in a positive direction. By modification of the integration variable show that

$$P_n^m(z) = \frac{i^m\Gamma(n + m + 1)}{\pi\Gamma(n + 1)} \int_0^\pi [z + (z^2 - 1)^{\frac{1}{2}} \cos \varphi]^n \cos (m\varphi)\, d\varphi$$

Show that this function is equal to the one defined in Eq. (5.3.36). Show that these functions are involved in all potential and wave problems in spherical coordinates.

5.33 By use of Eqs. (5.3.33) *et seq.*, show that, for n an integer,

$$Q_n(z) = \tfrac{1}{2}P_n(z) \ln [(z + 1)/(z - 1)] - W_{n-1}(z)$$

where $W_{n-1}(z)$ is a polynomial in z of degree $n - 1$. Thus show that, when x is a real quantity between -1 and $+1$,

$$\lim_{\epsilon \to 0} [Q_n(x + i\epsilon) - Q_n(x - i\epsilon)] = -\pi i P_n(x)$$

5.34 Use Cauchy's theorem to show that

$$Q_n(z) = \frac{1}{2\pi i}\left[\oint_{C_2} \frac{Q_n(w)\, dw}{(w - z)} - \oint_{C_1} \frac{Q_n(w)\, dw}{(w - z)}\right]$$

where contour C_1 is around $w = \pm 1$ but is inside $w = z$ and contour C_2 is a circle of radius $R \gg |z|$ and 1. Show that the integral around C_2 is zero for $n = 0$ or a positive integer. Reduce the contour C_1 to a circuit close to the line between ± 1, and by use of the second result of Prob. 5.33, show that

$$Q_n(z) = \tfrac{1}{2} \int_{-1}^1 P_n(w) \frac{dw}{(z - w)}$$

5.35 Prove that polynomial $W_{n-1}(z)$ of Prob. 5.33 is

$$W_{n-1}(z) = \frac{2n-1}{n} P_{n-1}(z) + \frac{2n-5}{3(n-1)} P_{n-3}(z) + \frac{2n-9}{5(n-2)} P_{n-5}(z) + \cdots$$

5.36 We may define the second solution of the hypergeometric equation about $z = 0$ as

$$y_2^0(a,b|c|z) = + \frac{\Gamma(c)\Gamma(c-a-b)}{\Gamma(c-a)\Gamma(c-b)} F(a, b|a+b-c+1|1-z)$$
$$- \frac{\Gamma(c)\Gamma(a+b-c)}{\Gamma(a)\Gamma(b)} (1-z)^{c-a-b} F(c-a, c-b|c-a-b+1|1-z)$$

Show that this is equal to

$$\left[\frac{\sin \pi(c-a) \sin \pi(c-b) + \sin \pi a \sin \pi b}{\sin \pi(c-a) \sin \pi(c-b) - \sin \pi a \sin \pi b}\right] F(a,b|c|z)$$
$$+ \left[\frac{2\pi z^{1-c} \sin \pi c\, \Gamma(c)\, \Gamma(c-1)}{\sin \pi(c-a) \sin \pi(c-b) - \sin \pi a \sin \pi b}\right] \cdot \frac{F(a-c+1, b-c+1|2-c|z)}{\Gamma(a)\Gamma(b)\Gamma(c-a)\Gamma(c-b)}$$

Show that the limiting form of this function, as $c \to 1$, is the series given on page 668.

5.37 Prove that

$$J_m(z) = \frac{1}{2\pi i^n} \int_0^{2\pi} e^{iz \cos u} \cos(mu)\, du$$

5.38 Find the asymptotic series for the Whittaker function $U_2(a|c|z)$. From it, by the use of Eq. (5.3.3), obtain the relation

$$U_2(a|c|z) = \frac{e^{i\pi a} z^{-a}}{2\pi i} \int_{-i\infty}^{i\infty} \frac{\Gamma(t+a-c+1)\Gamma(t+a)}{\Gamma(a-c+1)\Gamma(a)} \Gamma(-t)\, z^{-t}\, dt$$

What is the exact path of the contour? Duplicating the procedure used to obtain Eq. (5.3.5), obtain Eq. (5.3.58) (discuss the limits of convergence at each step).

5.39 Show that the radial factor for solutions of the Helmholtz equation in polar, spherical, and conical coordinates satisfies the Bessel equation

$$\frac{1}{z}\frac{d}{dz}\left(z\frac{dJ}{dz}\right) + \left(1 - \frac{\lambda^2}{z^2}\right) J = 0$$

Locate and describe the singular points; work out the first three terms in the series expansion for solution J_λ, regular at $z = 0$, about each. Use the Laplace transform to derive the integral representation of Eq. (5.3.63). Show that a second solution of this equation is

$$N_\lambda(z) = \cot(\pi\lambda)\, J_\lambda(z) - \csc(\pi\lambda)\, J_{-\lambda}(z)$$

Compute the first three terms in the series expansion of N_λ (for λ not an integer) about $z = 0$ and the first three terms in its asymptotic expansion. Show that, for $\lambda = 0$,

$$N_0(z) = \frac{1}{\pi} \lim_{\lambda\to 0} \left[\frac{d}{d\lambda} J_\lambda(z) - (-1)^\lambda \frac{d}{d\lambda} J_{-\lambda}(z) \right]$$

$$= \frac{2}{\pi} [\ln(\tfrac{1}{2}z) + \gamma] J_0(z) - \frac{2}{\pi} \sum_{m=1}^{\infty} \frac{(-1)^m}{(m!)^2} (\tfrac{1}{2}z)^{2m} \left[\sum_{s=1}^{m} \left(\frac{1}{s}\right) \right]$$

5.40 Show that a solution of the Schroedinger equation in parabolic coordinates $x = \sqrt{\lambda\mu}\cos\varphi$, $y = \sqrt{\lambda\mu}\sin\varphi$, $z = \frac{1}{2}(\lambda - \mu)$, $r = \frac{1}{2}(\lambda + \mu)$ for a particle of mass M in a potential $V = -\eta^2/r$ is

$$\psi = Ne^{im\varphi}(\lambda\mu)^{\frac{1}{2}m} e^{-\frac{1}{2}ik(\lambda+\mu)} \cdot F(\tfrac{1}{2}m + \tfrac{1}{2} - \sigma|m + 1|ik\lambda) F(\tfrac{1}{2}m + \tfrac{1}{2} - \tau|m + 1|ik\mu)$$

where $\sigma + \tau = -i(M\eta^2/\hbar^2 k)$ and where $k^2 = (2ME/\hbar^2) = (Mv/\hbar)^2$. Show that, for $m = 0$, $\sigma = -\frac{1}{2}$, and $N = \Gamma[1 - i(\eta^2/\hbar v)]e^{(\pi\eta^2/2\hbar v)}$, the solution has an asymptotic form

$$\psi \to \exp[ikz - i(\eta^2/hv)\ln k(r - z)] + \left\{ \frac{\eta^2 \exp[i(\eta^2/\hbar v)\ln(1 - z/r) - 2i\delta]}{Mv^2(r - z)} \right\} \exp\left[ikr + \left(\frac{i\eta^2}{\hbar v}\right) \ln(kr) \right]$$

where $\Gamma[1 - i(\eta^2/\hbar v)] = |\Gamma| e^{-i\delta}$. Discuss the physical significance of this result and obtain the Rutherford scattering law.

5.41 In a manner analogous to that indicated in the text for $Je_{2m}(h, \cosh\mu)$, prove the series expansions for the "radial" Mathieu function Je_{2m+1}:

$$Je_{2m+1}(h, \cosh\mu) = \sqrt{\tfrac{1}{2}\pi} \sum_{n=0}^{\infty} (1)^{n-m} B_{2n+1} J_{2n+1}(h\cosh\mu)$$

$$= \frac{\sqrt{\frac{1}{2}\pi}}{B_1} \sum_{n=0}^{\infty} (1)^{n-m} B_{2n+1}[J_n(\tfrac{1}{2}he^{-\mu}) J_{n+1}(\tfrac{1}{2}he^{\mu}) - J_{n+1}(\tfrac{1}{2}he^{-\mu}) J_n(\tfrac{1}{2}he^{\mu})]$$

where B_{2n+1} are the coefficients of the Fourier series for the "angle" function $Se_{2m+1}(h, \cos\vartheta)$ as defined on page 565.

5.42 By the use of the Laplace transform, show that, if $u(s) = \int_0^\infty e^{-st}K(t)\,dt$ and $v(s) = \int_0^\infty e^{-st}\varphi(t)\,dt$, then $f(x) = \int^x K(x - t)\varphi(t)\,dt$ where $u(s)v(s) = \int_0^\infty e^{-st}f(t)\,dt$. From this, prove that

$$n \int_0^x J_m(x - t)\, J_n(t) \left(\frac{dt}{t}\right) = J_{m+n}(x)$$

Table of Separable Coordinates in Three Dimensions

The coordinate system is defined by the interrelations between the rectangular coordinates x, y, z and the curvilinear coordinates ξ_1, ξ_2, ξ_3 or by the scale factors $h_n = \sqrt{(\partial x/\partial \xi_n)^2 + (\partial y/\partial \xi_n)^2 + (\partial z/\partial \xi_n)^2}$, etc., having the property (see page 24)

$$ds^2 = dx^2 + dy^2 + dz^2 = \sum_n h_n^2 (d\xi_n)^2$$

The expressions for the Laplacian, gradient, curl, etc., in terms of the h's are given in the table on page 115. The standard partial differential equation $\nabla^2\psi + k_1^2\psi = 0$ becomes

$$\sum_m \frac{1}{h_1h_2h_3} \frac{\partial}{\partial \xi_m} \left[\frac{h_1h_2h_3}{h_m^2} \frac{\partial \psi}{\partial \xi_m} \right] + k_1^2\psi = 0$$

where $k_1^2 = 0$ for the Laplace equation, $k_1^2 = \text{constant}$ for the wave equation, and $k_1^2 = \epsilon_1 - V(\xi)$ for the Schroedinger equation for a single particle in a potential field V.

For separation the quantity $h_1h_2h_3/h_n^2$ must factor as follows:

$$h_1h_2h_3/h_1^2 = g_1(\xi_2,\xi_3)f_1(\xi_1); \quad \text{etc.}$$

The Stäckel determinant is

$$S = \begin{vmatrix} \Phi_{11}(\xi_1) & \Phi_{12}(\xi_1) & \Phi_{13}(\xi_1) \\ \Phi_{21}(\xi_2) & \Phi_{22}(\xi_2) & \Phi_{23}(\xi_2) \\ \Phi_{31}(\xi_3) & \Phi_{32}(\xi_3) & \Phi_{33}(\xi_3) \end{vmatrix} = \frac{h_1h_2h_3}{f_1(\xi_1)f_2(\xi_2)f_3(\xi_3)}$$

The first minor of S for the element $\Phi_{m1}(\xi_m)$ is related to the scale factors by the equation

$$M_m = \partial S/\partial \Phi_{m1} = S/h_m^2$$

where

$$M_1(\xi_2,\xi_3) = \Phi_{22}\Phi_{33} - \Phi_{23}\Phi_{32}; \quad M_2(\xi_1,\xi_3) = \Phi_{13}\Phi_{32} - \Phi_{12}\Phi_{33};$$
$$M_3(\xi_1,\xi_2) = \Phi_{12}\Phi_{23} - \Phi_{22}\Phi_{13}$$

Therefore
$$\sum_m \frac{\Phi_{mn}(\xi_m)}{h_m^2} = \frac{1}{S}\sum_m \Phi_{mn}M_m = \delta_{n1}$$

and also
$$g_1(\xi_2,\xi_3) = \frac{h_1h_2h_3}{f_1h_1^2} = \left(\frac{M_1}{S}\right)\left(\frac{Sf_1f_2f_3}{f_1}\right) = M_1(\xi_2,\xi_3)f_2(\xi_2)f_3(\xi_3); \quad \text{etc.}$$

The standard partial differential equation then becomes

$$\sum_m \frac{M_m}{S}\left[\frac{1}{f_m}\frac{\partial}{\partial \xi_m}\left(f_m \frac{\partial \psi}{\partial \xi_m}\right)\right] + k_1^2\psi = 0$$

and the three separated equations, for the wave equation (k_1^2 = constant) are, $\psi = X_1(\xi_1)X_2(\xi_2)X_3(\xi_3)$,

$$\frac{1}{f_m(\xi_m)}\frac{d}{d\xi_m}\left[f_m(\xi_m)\frac{dX_m}{d\xi_m}\right] + \sum_n \Phi_{mn}(\xi_m)k_n^2 X_m = 0$$

where k_2^2 and k_3^2 are the separation constants. The partial differential equation is obtained by multiplying the $m = 1$ equation by $(M_1/S)X_2X_3$, and so on, for $m = 2$ and 3, and then summing over m.

For the Schroedinger equation to separate the potential V must have the form

$$V = \sum_m \frac{v_m(\xi_m)}{h_m^2} = \sum_m \left(\frac{M_m}{S}\right) v_m(\xi_m)$$

where v_m is a function of ξ_m alone. The separated equations are, in this case,

$$\frac{1}{f_m}\frac{d}{d\xi_m}\left[f_m\frac{dX_m}{\partial\xi_m}\right] + \left[\sum_n \Phi_{mn}\epsilon_n - v_m\right]X_m = 0$$

where ϵ_2 and ϵ_3 are the separation constants.

The following table lists the scale factors h_m, the related functions f_m, and the Stäckel determinant for the 11 different separable three-dimensional coordinates for the wave equation. The singular points of the three separated equations in their canonical forms are also given. In the cases where alternate scales for the coordinates are sometimes used, these alternate expressions are also given. The general forms for the potential function V for separation of the Schroedinger equation are also given.

I Rectangular Coordinates

$x = \xi_1$; $y = \xi_2$; $z = \xi_3$; $h_1 = h_2 = h_3 = 1$; $f_1 = f_2 = f_3 = 1$

$$S = \begin{vmatrix} 1 & -1 & -1 \\ 0 & 1 & 0 \\ 0 & 0 & 1 \end{vmatrix} = 1$$

Irregular singular point at infinity all three equations.

General form for $V = u(x) + v(y) + w(z)$

II Circular Cylinder Coordinates (rotational) (Fig. 5.15)

$x = \xi_1\xi_2$; $y = \xi_1\sqrt{1 - \xi_2^2}$; $z = \xi_3$; $h_1 = h_3 = 1$; $h_2 = \xi_1/\sqrt{1 - \xi_2^2}$

$f_1 = \xi_1$; $f_2 = \sqrt{1 - \xi_2^2}$; $f_3 = 1$; $\xi_1 = r$; $\xi_2 = \cos\phi$; $\xi_3 = z$

$$S = \begin{vmatrix} 1 & -(1/\xi_1^2) & -1 \\ 0 & 1/(1 - \xi_2^2) & 0 \\ 0 & 0 & 1 \end{vmatrix} = \frac{1}{1 - \xi_2^2}$$

Equation for ξ_1, regular singular point at 0, irregular singular point at ∞.

Equation for ξ_2, regular singular points at -1, $+1$, ∞.

Equation for ξ_3, irregular singular point at ∞.

General form for $V = u(r) + (1/r^2)v(\phi) + w(z)$

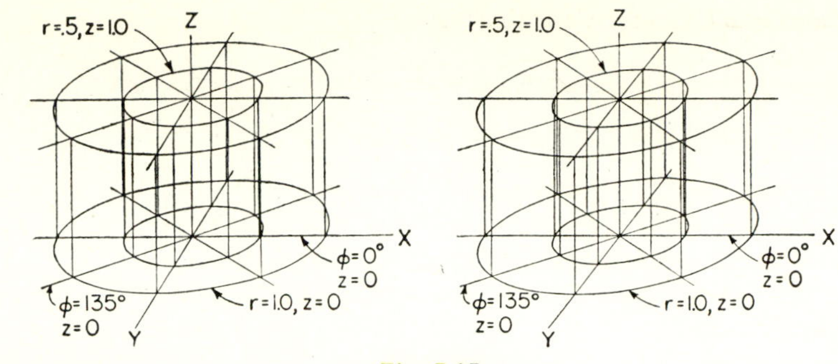

Fig. 5.15.

III Elliptic Cylinder Coordinates

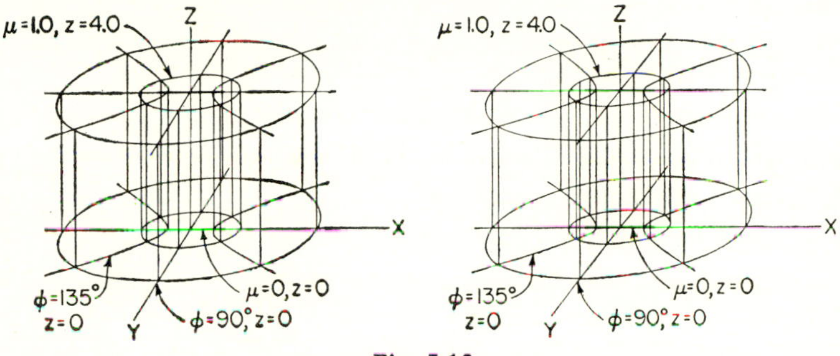

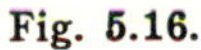
Fig. 5.16.

$x = \xi_1\xi_2;\quad y = \sqrt{(\xi_1^2 - d^2)(1 - \xi_2^2)};\quad z = \xi_3;$

$h_1 = \sqrt{(\xi_1^2 - d^2\xi_2^2)/(\xi_1^2 - d^2)};\quad h_2 = \sqrt{(\xi_1^2 - d^2\xi_2^2)/(1 - \xi_2^2)};\quad h_3 = 1$

$f_1 = \sqrt{\xi_1^2 - d^2};\quad f_2 = \sqrt{1 - \xi_2^2};\quad f_3 = 1$

$\xi_1 = d \cosh \mu = \frac{1}{2}(r_1 + r_2);\quad \xi_3 = z;\quad \xi_2 = \cos \varphi = (1/2d)(r_1 - r_2)$ (see page 556)

$$S = \begin{vmatrix} 1 & -1/(\xi_1^2 - d^2) & -1 \\ d^2 & 1/(1 - \xi_2^2) & -d^2 \\ 0 & 0 & 1 \end{vmatrix} = \frac{\xi_1^2 - d^2\xi_2^2}{(\xi_1^2 - d^2)(1 - \xi_2^2)}$$

Equation for ξ_1 has regular singular points at $-d$, $+d$, irregular singular point at ∞.

Equation for ξ_2 has regular singular points at -1, $+1$, irregular singular point at ∞.

Equation for ξ_3 has irregular singular point at ∞.

General form for $V = \{[u(r_1 + r_2) + v(r_1 - r_2)]/r_1r_2\} + w(z)$.

IV Parabolic Cylinder Coordinates

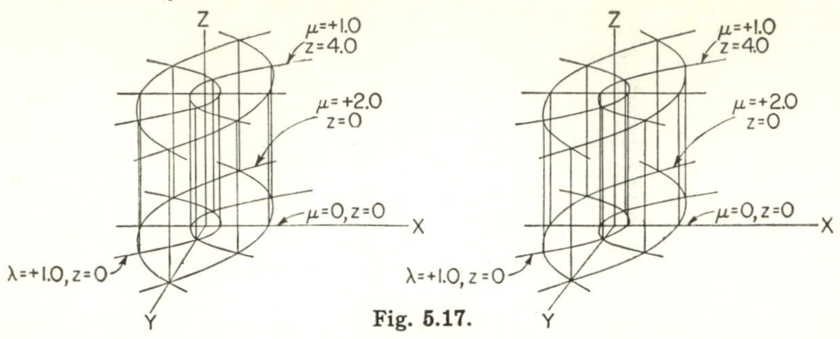

Fig. 5.17.

$x = \frac{1}{2}(\xi_1^2 - \xi_2^2)$; $y = \xi_1\xi_2$; $z = \xi_3$; $h_1 = h_2 = \sqrt{\xi_1^2 + \xi_2^2}$; $h_3 = 1$; $f_1 = f_2 = f_3 = 1$

$$S = \begin{vmatrix} 0 & \xi_1^2 & -1 \\ 0 & \xi_2^2 & 1 \\ 1 & -1 & 0 \end{vmatrix} = \xi_1^2 + \xi_2^2$$

Equations for ξ_1, ξ_2, ξ_3 have irregular singular points at ∞.

General form for $V = \{[u(\xi_1) + v(\xi_2)]/\sqrt{x^2 + y^2}\} + w(z)$

V Spherical Coordinates (rotational)

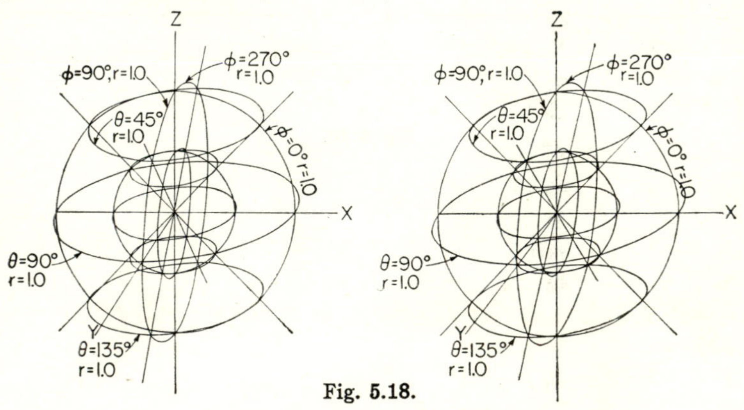

Fig. 5.18.

$x = \xi_1\xi_3\sqrt{1 - \xi_2^2}$; $y = \xi_1\sqrt{(1 - \xi_2^2)(1 - \xi_3^2)}$; $z = \xi_1\xi_2$

$h_1 = 1$; $h_2 = \xi_1/\sqrt{1 - \xi_2^2}$; $h_3 = \xi_1\sqrt{(1 - \xi_2^2)/(1 - \xi_3^2)}$

$f_1 = \xi_1^2$; $f_2 = 1 - \xi_2^2$; $f_3 = \sqrt{1 - \xi_3^2}$

$\xi_1 = r$; $\xi_2 = \cos\vartheta$; $\xi_3 = \cos\varphi$

$$S = \begin{vmatrix} 1 & 1/\xi_1^2 & 0 \\ 0 & 1/(\xi_2^2 - 1) & 1/(\xi_2^2 - 1)^2 \\ 0 & 0 & 1/(\xi_3^2 - 1) \end{vmatrix} = \frac{1}{(1 - \xi_2^2)(1 - \xi_3^2)}$$

Equation for ξ_1 has regular singular point at 0, irregular singular point at ∞.

Equations for ξ_2, ξ_3 have regular singular points at -1, $+1$, ∞.

General form for $V = u(r) + (1/r^2)v(\vartheta) + (1/r^2 \sin^2\vartheta)w(\varphi)$.

VI Conical Coordinates

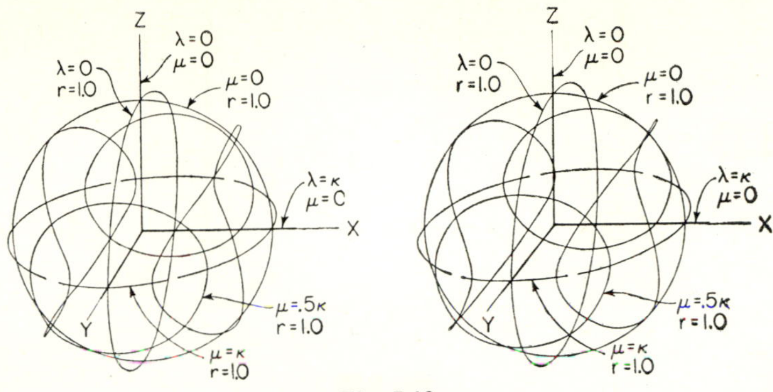

Fig. 5.19.

$$x = (\xi_1/\alpha)\sqrt{(\alpha^2 - \xi_2^2)(\alpha^2 + \xi_3^2)};\quad y = (\xi_1/\beta)\sqrt{(\beta^2 + \xi_2^2)(\beta^2 - \xi_3^2)}$$
$$z = \xi_1\xi_2\xi_3/\alpha\beta;\quad \alpha^2 + \beta^2 = 1$$
$$h_1 = 1;\quad h_2 = \xi_1\sqrt{\frac{\xi_2^2 + \xi_3^2}{(\alpha^2 - \xi_2^2)(\beta^2 + \xi_2^2)}};\quad h_3 = \xi_1\sqrt{\frac{\xi_2^2 + \xi_3^2}{(\alpha^2 + \xi_3^2)(\beta^2 - \xi_3^2)}}$$
$$f_1 = \xi_1^2;\quad f_2 = \sqrt{(\alpha^2 - \xi_2^2)(\beta^2 + \xi_2^2)};\quad f_3 = \sqrt{(\alpha^2 + \xi_3^2)(\beta^2 - \xi_3^2)}$$

$\xi_1 = r;\quad \xi_2 = \alpha\,\mathrm{cn}(\lambda,\alpha);\quad \xi_3 = \beta\,\mathrm{cn}(\mu,\beta);$ cn is one of the elliptic functions so that

$$x = r\,\mathrm{dn}(\lambda,\alpha)\,\mathrm{sn}(\mu,\beta);\quad y = r\,\mathrm{sn}(\lambda,\alpha)\,\mathrm{dn}(\mu,\beta);\quad z = r\,\mathrm{cn}(\lambda,\alpha)\,\mathrm{cn}(\mu,\beta)$$

$$S = \begin{vmatrix} 1 & \dfrac{1}{\xi_1^2} & \dfrac{1}{(\alpha^2+\beta^2)\xi_1^2} \\ 0 & \dfrac{1}{\xi_2^2 - \alpha^2} & \dfrac{1}{(\alpha^2+\beta^2)(\xi_2^2+\beta^2)} \\ 0 & \dfrac{1}{\xi_3^2 + \alpha^2} & \dfrac{1}{(\alpha^2+\beta^2)(\xi_3^2-\beta^2)} \end{vmatrix} = \frac{\xi_2^2 + \xi_3^2}{(\alpha^2 - \xi_2^2)(\beta^2 + \xi_2^2)(\alpha^2 + \xi_3^2)(\beta^2 - \xi_3^2)}$$

Equation for ξ_1 has regular singular point at 0, irregular singular point at ∞.

Equation for ξ_2 has regular singular points at $\pm\alpha$, $\pm i\beta$, ∞.

Equation for ξ_3 has regular singular points at $\pm i\alpha$, $\pm\beta$, ∞.

General form for $V = u(r) + [v(\xi_2) + w(\xi_3)]/(\xi_2^2 + \xi_3^2)$.

VII Parabolic Coordinates (rotational)

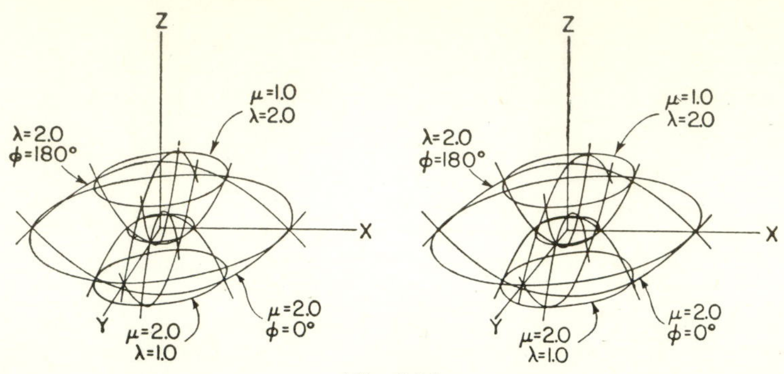

Fig. 5.20.

$$x = \xi_1\xi_2\xi_3;\quad y = \xi_1\xi_2\sqrt{1-\xi_3^2};\quad z = \tfrac{1}{2}(\xi_1^2 - \xi_2^2)$$
$$h_1 = h_2 = \sqrt{\xi_1^2 + \xi_2^2};\quad h_3 = \xi_1\xi_2/\sqrt{1-\xi_3^2}$$
$$f_1 = \xi_1;\quad f_2 = \xi_2;\quad f_3 = \sqrt{1-\xi_3^2}$$
$$\xi_1 = \lambda;\quad \xi_2 = \mu;\quad \xi_3 = \cos\varphi$$

If $r^2 = x^2 + y^2 + z^2$, then $\xi_1^2 = r + z$ and $\xi_2^2 = r - z$.

$$S = \begin{vmatrix} \xi_1^2 & 1 & 1/\xi_1^2 \\ \xi_2^2 & -1 & 1/\xi_2^2 \\ 0 & 0 & 1/(\xi_3^2 - 1) \end{vmatrix} = \frac{\xi_1^2 + \xi_2^2}{1 - \xi_3^2}$$

Equations for ξ_1, ξ_2 have regular singular points at 0, irregular singular points at ∞.

Equation for ξ_3 has regular singular points at $-1, +1, \infty$.

General form for $V = \dfrac{[u(\xi_1) + v(\xi_2)]}{\sqrt{x^2 + y^2 + z^2}} + \dfrac{w(\xi_3)}{x^2 + y^2}$.

VIII Prolate Spheroidal Coordinates (rotational)

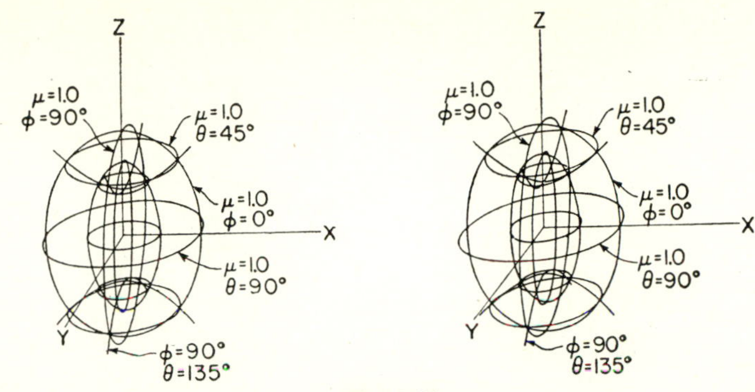

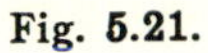

Fig. 5.21.

$$x = \xi_3 \sqrt{(\xi_1^2 - d^2)(1 - \xi_2^2)};\quad y = \sqrt{(\xi_1^2 - d^2)(1 - \xi_2^2)(1 - \xi_3^2)};\quad z = \xi_1 \xi_2$$

$$h_1 = \sqrt{\frac{\xi_1^2 - d^2\xi_2^2}{\xi_1^2 - d^2}};\quad h_2 = \sqrt{\frac{\xi_1^2 - d^2\xi_2^2}{1 - \xi_2^2}};\quad h_3 = \sqrt{\frac{(\xi_1^2 - d^2)(1 - \xi_2^2)}{(1 - \xi_3^2)}}$$

$${}_1 = \xi_1^2 - d^2;\quad f_2 = 1 - \xi_2^2;\quad f_3 = \sqrt{1 - \xi_3^2};$$

$$\xi_1 = d \cosh \mu = \tfrac{1}{2}(r_1 + r_2);\quad \xi_2 = \cos \vartheta = (1/2d)(r_1 - r_2);\quad \xi_3 = \cos \varphi$$

$$S = \begin{vmatrix} 1 & \dfrac{1}{\xi_1^2 - d^2} & \dfrac{d^2}{(\xi_1^2 - d^2)^2} \\ d^2 & \dfrac{1}{\xi_2^2 - 1} & \dfrac{1}{(\xi_2^2 - 1)^2} \\ 0 & 0 & \dfrac{1}{(\xi_3^2 - 1)} \end{vmatrix} = \frac{(\xi_1^2 - d^2\xi_2^2)}{(\xi_1^2 - d^2)(1 - \xi_2^2)(1 - \xi_3^2)}$$

Equation for ξ_1 has regular singular points at $-d$, $+d$, irregular singular point at ∞.

Equation for ξ_2 has regular singular points at -1 and $+1$, irregular singular point at ∞.

Equation for ξ_3 has regular singular points at -1, $+1$, ∞.

$$\text{General form for } V = \frac{u(r_1 + r_2) + v(r_1 - r_2)}{r_1 r_2} + \frac{w(\varphi)}{\sinh \mu \sin \vartheta}.$$

IX Oblate Spheroidal Coordinates (rotational)

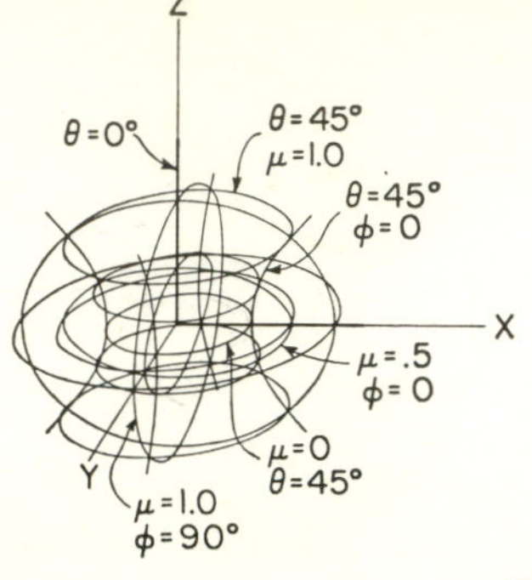

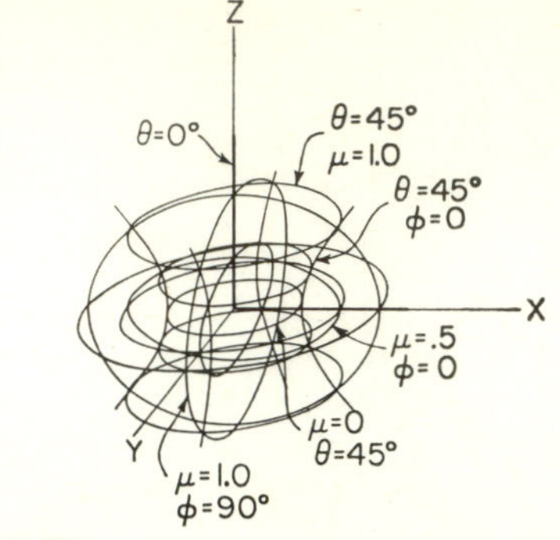

Fig. 5.22.

$$x = \xi_3\sqrt{(\xi_1^2 + d^2)(1 - \xi_2^2)}; \quad y = \sqrt{(\xi_1^2 + d^2)(1 - \xi_2^2)(1 - \xi_3^2)}; \quad z = \xi_1\xi_2$$

$$h_1 = \sqrt{\frac{\xi_1^2 + d^2\xi_2^2}{\xi_1^2 + d^2}}; \quad h_2 = \sqrt{\frac{\xi_1^2 + d^2\xi_2^2}{1 - \xi_2^2}}; \quad h_3 = \sqrt{\frac{(\xi_1^2 + d^2)(1 - \xi_2^2)}{(1 - \xi_3^2)}}$$

$$f_1 = \xi_1^2 + d^2; \quad f_2 = 1 - \xi_2^2; \quad f_3 = \sqrt{1 - \xi_3^2}$$

$$\xi_1 = d \sinh \mu; \quad \xi_2 = \cos \vartheta; \quad \xi_3 = \cos \varphi$$

$$S = \begin{vmatrix} 1 & \dfrac{1}{\xi_1^2 + d^2} & \dfrac{-d^2}{(\xi_1^2 + d^2)^2} \\ -d^2 & \dfrac{1}{\xi_2^2 - 1} & \dfrac{1}{(\xi_2^2 - 1)^2} \\ 0 & 0 & \dfrac{1}{(\xi_3^2 - 1)} \end{vmatrix} = \frac{(\xi_1^2 + d^2\xi_2^2)}{(\xi_1^2 + d^2)(1 - \xi_2^2)(1 - \xi_3^2)}$$

Equation for ξ_1 has regular singular points at $-id$, $+id$, irregular singular point at ∞.

Equation for ξ_2 has regular singular points at -1, $+1$, irregular singular point at ∞.

Equation for ξ_3 has regular singular points at -1, $+1$, ∞.

General form for $V = \dfrac{u(\xi_1) + v(\xi_2)}{\xi_1^2 + d^2\xi_2^2} + \dfrac{w(\xi_3)}{(\xi_1^2 + d^2)(1 - \xi_2^2)}$.

X Ellipsoidal Coordinates

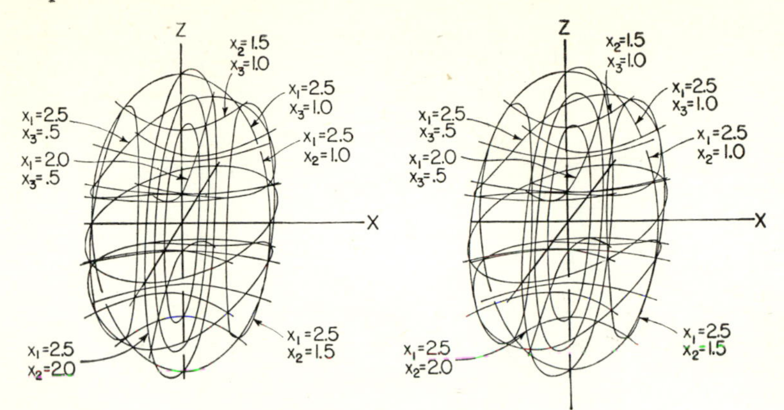

Fig. 5.23 $\xi_n = bx_n,\ a = 2b.$

$$x = \sqrt{\frac{(\xi_1^2 - a^2)(\xi_2^2 - a^2)(\xi_3^2 - a^2)}{a^2(a^2 - b^2)}};\quad y = \sqrt{\frac{(\xi_1^2 - b^2)(\xi_2^2 - b^2)(\xi_3^2 - b^2)}{b^2(b^2 - a^2)}};$$

$$z = \frac{\xi_1\xi_2\xi_3}{ab};\quad h_1 = \sqrt{\frac{(\xi_1^2 - \xi_2^2)(\xi_1^2 - \xi_3^2)}{(\xi_1^2 - a^2)(\xi_1^2 - b^2)}};\quad h_2 = \sqrt{\frac{(\xi_2^2 - \xi_1^2)(\xi_2^2 - \xi_3^2)}{(\xi_2^2 - a^2)(\xi_2^2 - b^2)}};$$

$$h_3 = \sqrt{\frac{(\xi_3^2 - \xi_1^2)(\xi_3^2 - \xi_2^2)}{(\xi_3^2 - a^2)(\xi_3^2 - b^2)}}$$

$$f_1 = \sqrt{(\xi_1^2 - a^2)(\xi_1^2 - b^2)};\quad f_2 = \sqrt{(\xi_2^2 - a^2)(\xi_2^2 - b^2)};$$

$$f_3 = \sqrt{(\xi_3^2 - a^2)(\xi_3^2 - b^2)};\quad \xi_1^2 \geq a^2 \geq \xi_2^2 \geq b^2 \geq \xi_3^2 \geq 0$$

$$S = \begin{vmatrix} 1 & \dfrac{1}{(\xi_1^2 - a^2)} & \dfrac{1}{(\xi_1^2 - b^2)(a^2 - b^2)} \\ 1 & \dfrac{1}{(\xi_2^2 - a^2)} & \dfrac{1}{(\xi_2^2 - b^2)(a^2 - b^2)} \\ 1 & \dfrac{1}{(\xi_3^2 - a^2)} & \dfrac{1}{(\xi_3^2 - b^2)(a^2 - b^2)} \end{vmatrix}$$

Equations for ξ_1, ξ_2, ξ_3 have regular singular points at $-a, -b, +b, +a, \infty$.
General form for $V = \dfrac{(\xi_2^2 - \xi_3^2)u(\xi_1) + (\xi_1^2 - \xi_3^2)v(\xi_2) + (\xi_1^2 - \xi_2^2)w(\xi_3)}{(\xi_1^2 - \xi_2^2)(\xi_1^2 - \xi_3^2)(\xi_2^2 - \xi_3^2)}$.

$$\xi_1 = a\,\frac{\mathrm{dn}(\lambda,k)}{\mathrm{cn}(\lambda,k)};\quad \xi_2 = a\,\mathrm{dn}(\mu,k');\quad \xi_3 = b\,\mathrm{sn}(\nu,k)$$

$$b = ka;\quad \sqrt{a^2 - b^2} = k'a = d$$

$$x = d\,\frac{\mathrm{sn}(\lambda,k)\,\mathrm{sn}(\mu,k')\,\mathrm{dn}(\nu,k)}{\mathrm{cn}(\lambda,k)};\quad y = d\,\frac{\mathrm{cn}(\mu,k')\,\mathrm{cn}(\nu,k)}{\mathrm{cn}(\lambda,k)};$$

$$z = a\,\frac{\mathrm{dn}(\lambda,k)\,\mathrm{dn}(\mu,k')\,\mathrm{sn}(\nu,k)}{\mathrm{cn}(\lambda,k)}$$

XI Paraboloidal Coordinates

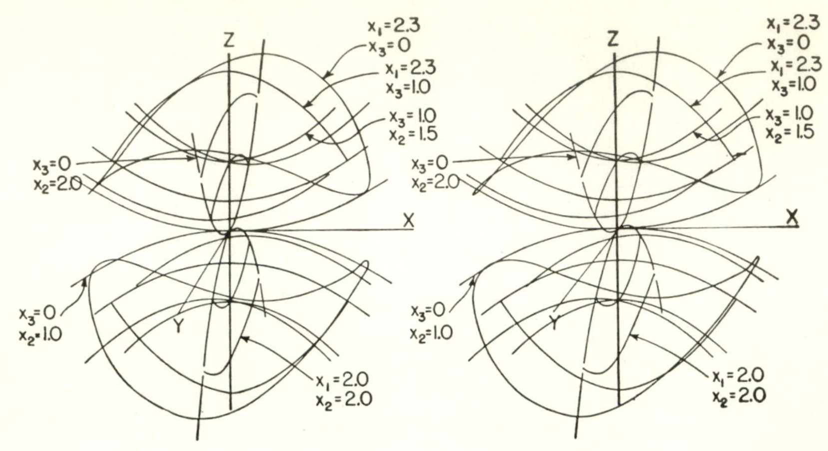

Fig. 5.24 $\xi_n = bx_n,\ a = 2b.$

$$x = \sqrt{\frac{(\xi_1^2 - a^2)(\xi_2^2 - a^2)(\xi_3^2 - a^2)}{a^2 - b^2}};\quad y = \sqrt{\frac{(\xi_1^2 - b^2)(\xi_2^2 - b^2)(\xi_3^2 - b^2)}{b^2 - a^2}};$$

$$z = \tfrac{1}{2}(\xi_1^2 + \xi_2^2 + \xi_3^2 - a^2 - b^2)$$

$$h_1 = \xi_1 \sqrt{\frac{(\xi_1^2 - \xi_2^2)(\xi_1^2 - \xi_3^2)}{(\xi_1^2 - a^2)(\xi_1^2 - b^2)}};\quad h_2 = \xi_2 \sqrt{\frac{(\xi_3^2 - \xi_1^2)(\xi_2^2 - \xi_3^2)}{(\xi_2^2 - a^2)(\xi_2^2 - b^2)}};$$

$$h_3 = \xi_3 \sqrt{\frac{(\xi_3^2 - \xi_1^2)(\xi_3^2 - \xi_2^2)}{(\xi_3^2 - a^2)(\xi_3^2 - b^2)}}$$

$$f_1 = (1/\xi_1)\sqrt{(\xi_1^2 - a^2)(\xi_1^2 - b^2)};\quad f_2 = (1/\xi_2)\sqrt{(\xi_2^2 - a^2)(\xi_2^2 - b^2)};$$

$$f_3 = (1/\xi_3)\sqrt{(\xi_3^2 - a^2)(\xi_3^2 - b^2)}$$

$$S = \begin{vmatrix} \xi_1^2 & \dfrac{\xi_1^2}{(\xi_1^2 - a^2)} & \dfrac{\xi_1^2}{(\xi_1^2 - b^2)(a^2 - b^2)} \\ \xi_2^2 & \dfrac{\xi_2^2}{(\xi_2^2 - a^2)} & \dfrac{\xi_2^2}{(\xi_2^2 - b^2)(a^2 - b^2)} \\ \xi_3^2 & \dfrac{\xi_3^2}{(\xi_3^2 - a^2)} & \dfrac{\xi_3^2}{(\xi_3^2 - b^2)(a^2 - b^2)} \end{vmatrix}$$

Equations for ξ_1, ξ_2, ξ_3 have regular singular points at $-a$, $-b$, 0, $+a$, $+b$.

General form for $V = \dfrac{(\xi_2^2 - \xi_3^2)u(\xi_1) + (\xi_1^2 - \xi_3^2)v(\xi_2) + (\xi_1^2 - \xi_2^2)w(\xi_2)}{(\xi_1^2 - \xi_2^2)(\xi_1^2 - \xi_3^2)(\xi_2^2 - \xi_3^2)}$.

$$\xi_1 = a\,\frac{\mathrm{dn}(\lambda,k)}{\mathrm{cn}(\lambda,k)};\quad \xi_2 = a\,\mathrm{dn}(\nu,k');\quad \xi_3 = a\sqrt{1 - \frac{k'^2}{\mathrm{cn}^2(\mu,k)}}$$

$$b = ka;\quad \sqrt{a^2 - b^2} = k'a = \sqrt{d};\quad x = d\,\frac{\mathrm{sn}(\lambda,k)\ \mathrm{sn}(\nu,k')}{\mathrm{cn}(\lambda,k)\ \mathrm{cn}(\mu,k)};$$

$$y = d\,\frac{\mathrm{sn}(\mu,k)\ \mathrm{cn}(\nu,k')}{\mathrm{cn}(\lambda,k)\ \mathrm{cn}(\mu,k)};\quad z = \frac{d}{2}\left[\frac{\mathrm{sn}^2(\lambda,k)}{\mathrm{cn}^2(\lambda,k)} - \frac{\mathrm{sn}^2(\mu,k)}{\mathrm{cn}^2(\mu,k)} + \frac{\mathrm{dn}^2(\nu,k')}{k'^2}\right]$$

The Laplace equation in two dimensions separates in any coordinate system which is a conformal transform from the rectangular system, x, y.

The Laplace equation in three dimensions separates in all the 11 coordinates, listed above, for which the wave equation separates. In addition, the solution of the Laplace equation can be separated into the following form:

$$\psi = [X_1(\xi_1)X_2(\xi_2)X_3(\xi_3)/R(\xi_1,\xi_2,\xi_3)]$$

and separated equations for the X's can be found if the preceding equations are modified somewhat. We set

$$h_1h_2h_3/h_1^2 = g_1(\xi_2,\xi_3)f_1(\xi_1)R^2$$

The Laplace equation then becomes

$$\sum_n \frac{1}{h_n^2X_n}\left[\frac{1}{f_n}\frac{d}{d\xi_n}\left(f_n\frac{dX_n}{d\xi_n}\right)\right] = \sum_n \frac{1}{h_n^2R}\left[\frac{1}{f_n}\frac{\partial}{\partial\xi_n}\left(f_n\frac{\partial R}{\partial\epsilon_n}\right)\right]$$

If the right-hand side of this equation is equal to $[-k_1^2/u(\xi_1,\xi_2,\xi_3)]$, where k_1 is a constant, and if

$$h_1h_2h_3 = Sf_1f_2f_3R^2u$$

where S is the Stäckel determinant, then the equation reduces to one where the previous techniques can be employed.

Two coordinate systems for which the Laplace equation separates in this sense are the bispherical and the toroidal systems:

Bispherical Coordinates (Fig. 5.25)

$$x = a\xi_3\frac{\sqrt{1-\xi_2^2}}{\xi_1-\xi_2};\quad y = a\frac{\sqrt{(1-\xi_2^2)(1-\xi_3^2)}}{\xi_1-\xi_2};\quad z = a\frac{\sqrt{\xi_1^2-1}}{\xi_1-\xi_2}$$

$$h_1 = \frac{a}{(\xi_1-\xi_2)\sqrt{\xi_1^2-1}};\quad h_2 = \frac{a}{(\xi_1-\xi_2)\sqrt{1-\xi_2^2}};$$

$$h_3 = \frac{a}{\xi_1-\xi_2}\sqrt{\frac{1-\xi_2^2}{1-\xi_3^2}}$$

$$f_1 = \sqrt{\xi_1^2-1};\quad f_2 = 1-\xi_2^2;\quad f_3 = \sqrt{1-\xi_3^2};$$

$$r = \sqrt{x^2+y^2+z^2} = a\sqrt{(\xi_1+\xi_2)/(\xi_1-\xi_2)};\quad k_1^2 = \tfrac{1}{4}$$

$$R^2 = \frac{a}{\xi_1-\xi_2};\quad u = \frac{a^2}{(\xi_1-\xi_2)^2};\quad \xi_1 = \cosh\mu;\quad \xi_2 = \cos\eta;\quad \xi_3 = \cos\varphi$$

$$S = \begin{vmatrix} \dfrac{1}{\xi_1^2-1} & \dfrac{1}{\xi_1^2-1} & 0 \\ 0 & \dfrac{-1}{1-\xi_2^2} & \dfrac{1}{(1-\xi_2^2)^2} \\ 0 & 0 & \dfrac{-1}{1-\xi_3^2} \end{vmatrix} = \frac{1}{(\xi_1^2-1)(1-\xi_2^2)(1-\xi_3^2)}$$

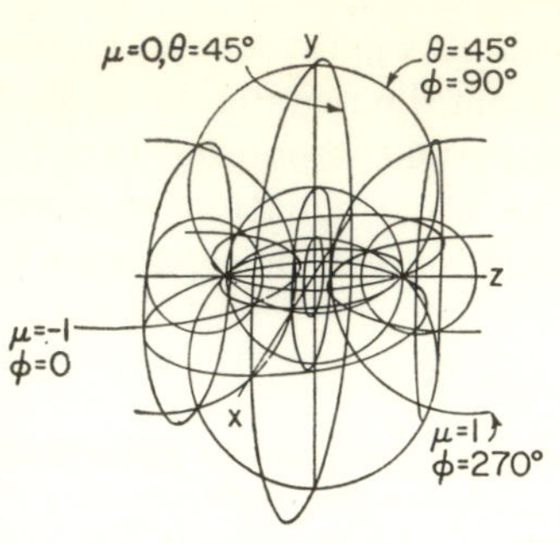

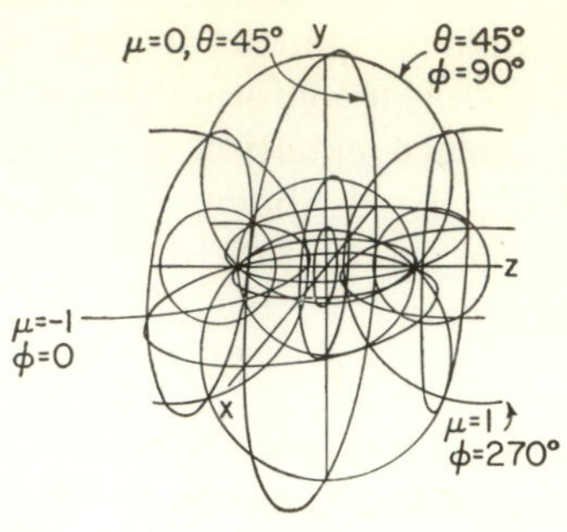

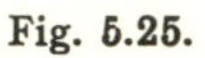
Fig. 5.25.

Toroidal Coordinates

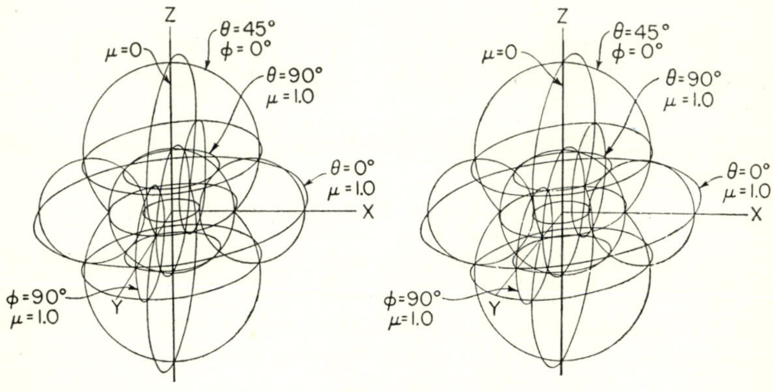

Fig. 5.26.

$$x = a\xi_3 \frac{\sqrt{\xi_1^2 - 1}}{\xi_1 - \xi_2}; \quad y = a\frac{\sqrt{(\xi_1^2 - 1)(1 - \xi_3^2)}}{\xi_1 - \xi_2}; \quad z = a\frac{\sqrt{1 - \xi_2^2}}{\xi_1 - \xi_2}$$

$$h_1 = \frac{a}{(\xi_1 - \xi_2)\sqrt{\xi_1^2 - 1}}; \quad h_2 = \frac{a}{(\xi_1 - \xi_2)\sqrt{1 - \xi_2^2}};$$

$$h_3 = \frac{a}{(\xi_1 - \xi_2)}\sqrt{\frac{\xi_1^2 - 1}{1 - \xi_3^2}}$$

$$f_1 = \xi_1^2 - 1; \quad f_2 = \sqrt{1 - \xi_2^2}; \quad f_3 = \sqrt{1 - \xi_3^2}$$

$$r = \sqrt{x^2 + y^2 + z^2} = a\sqrt{(\xi_1 + \xi_2)/(\xi_1 - \xi_2)}$$

$$R^2 = \frac{a}{\xi_1 - \xi_2}; \quad u = \frac{a^2}{(\xi_1 - \xi_2)^2}; \quad k_1^2 = \tfrac{1}{4}$$

$$\xi_1 = \cosh\mu; \quad \xi_2 = \cos\vartheta; \quad \xi_3 = \cos\varphi$$

$$S = \begin{vmatrix} \dfrac{1}{(\xi_1^2 - 1)} & \dfrac{-1}{(\xi_1^2 - 1)} & \dfrac{-1}{(\xi_1^2 - 1)^2} \\ 0 & \dfrac{1}{(1 - \xi_2^2)} & 0 \\ 0 & 0 & \dfrac{1}{1 - \xi_3^2} \end{vmatrix} = \frac{1}{(\xi_1^2 - 1)(1 - \xi_2^2)(1 - \xi_3^2)}$$

Second-order Differential Equations and Their Solutions

One Regular Singular Point (see page 536):
Canonical form, point at ∞:

$$\frac{d^2y}{dz^2} = 0; \quad \text{solutions } y_1(z) = 1; \quad y_2(z) = z$$

General form, point at a:

$$\frac{d^2y}{dw^2} + \frac{2}{w - a}\frac{dy}{dw} = 0; \quad \text{solutions } y_1 = 1; \quad y_2 = \frac{1}{w - a}$$

One Irregular Singular Point (see page 537):
Canonical form, point at ∞:

$$\frac{d^2y}{dz^2} - k^2y = 0; \quad \text{solutions } y_1 = e^{kz}; \quad y_2 = e^{-kz}$$

General form, point at a:

$$\frac{d^2y}{dw^2} + \frac{2}{w - a}\frac{dy}{dw} - \frac{k^2y}{(w - a)^4} = 0; \quad \text{solutions } y_1 = e^{k/(w-a)};$$
$$y_2 = e^{-k/(w-a)}$$

We do not consider equations with an irregular singular point of higher species.

Two Regular Singular Points (see page 536):
Canonical form, points at $0, \infty$, indices 0 and $-a$ $(a \geq 0)$:

$$z\frac{d^2y}{dz^2} + (1 + a)\frac{dy}{dz} = 0; \quad \text{solutions } y_1 = 1; \quad y_2 = z^{-a}$$

when $a = 0$, $y_2 = \ln z$.
General form, points at a and c, indices λ and μ $(\lambda \neq \mu)$:

$$\frac{d^2y}{dw^2} + \frac{2w + c(\lambda + \mu - 1) - a(\lambda + \mu + 1)}{(w - a)(w - c)}\frac{dy}{dw} + \frac{\lambda\mu(a - c)^2}{(w - a)^2(w - c)^2}y = 0$$

$$\text{Solutions } y_1 = \left(\frac{w - a}{w - c}\right)^\lambda; \quad y_2 = \left(\frac{w - a}{w - c}\right)^\mu$$

when $\mu = \lambda$; $y_2 = \ln (w - a/w - c)$.

Three Regular Singular Points (see pages 538 and 589):
Canonical form; points $0, 1, \infty$; indices $0, 1 - c$ (at 0); $0, c - a - b$ (at 1); and a, b (at ∞). (We can always arrange it so that $\text{Re } c \geq 1$, just as with the confluent hypergeometric function.)

$$z(z-1)\frac{d^2y}{dz^2} + [(a+b+1)z - c]\frac{dy}{dz} + aby = 0$$

(hypergeometric equation)

Solutions $y_1(a,b|c|z)$ and $y_2(a,b|c|z)$ about the singular points
Series expansion for y_1 for $|z| < 1$, valid for Re $c > 0$

$$y_1^0 = F(a,b|c|z) = \frac{\Gamma(c)}{\Gamma(a)\Gamma(b)} \sum_{n=0}^{\infty} \frac{\Gamma(a+n)\Gamma(b+n)z^n}{\Gamma(c+n)n!}$$

(hypergeometric series)

$$F(a,b|c|z) = F(b,a|c|z) = (1-z)^{c-a-b}F(c-a,\, c-b|c|z)$$
$$= (1-z)^{-a}F\left(a,\, c-b|\, c|\, \frac{z}{z-1}\right)$$
$$= (1-z)^{-b}F\left(c-a,\, b|\, c|\, \frac{z}{z-1}\right) \quad (\text{Re } z < \tfrac{1}{2})$$

$$F(a,b|c|1) = [\Gamma(c)\Gamma(c-a-b)]/[\Gamma(c-a)\Gamma(c-b)]$$
$$F(a,b|2b|z) = (1-\tfrac{1}{2}z)^{-a}F[\tfrac{1}{2}a,\, \tfrac{1}{2}a + \tfrac{1}{2}|b + \tfrac{1}{2}|z^2/(2-z)^2]$$
$$F(2\alpha,2\beta|\alpha+\beta+\tfrac{1}{2}|z) = F(\alpha,\beta|\alpha+\beta+\tfrac{1}{2}|4z - 4z^2)$$
$$zF(a,b|c|z) = (c-1)/(a-b)[F(a-1,\, b|\, c-1|\, z) - F(a,b-1|\, c-1|\, z)]$$
$$F(a,b|c|z) = 1/(a-b)[aF(a+1,\, b|\, c|\, z) - bF(a,\, b+1|\, c|\, z)]$$
$$(d/dz)F(a,b|c|z) = (ab/c)F(a+1,\, b+1|\, c+1|\, z)$$

A second solution about $z = 0$ is $z^{1-c}F(a-c+1,\, b-c+1|\, 2-c|\, z)$, but this is not independent of y_1^0 when $c = 1, 2, 3, \ldots$. An independent solution for all values of Re $c \geq 1$ is

$$y_2^0 = G(a,b|c|z) = \left(\frac{\sin\pi(c-a)\sin\pi(c-b) + \sin\pi a \sin\pi b}{\sin\pi(c-a)\sin\pi(c-b) - \sin\pi a\sin\pi b}\right)F(a,b|c|z)$$
$$+ \left(\frac{2\pi z^{1-c}\sin\pi c\,\Gamma(c)\Gamma(c-1)}{\sin\pi(c-a)\sin\pi(c-b) - \sin\pi a\sin\pi b}\right) \cdot \frac{F(a-c+1,\, b-c+1|\, 2-c|z)}{\Gamma(a)\Gamma(b)\Gamma(c-a)\Gamma(c-b)}$$

The series can be constructed from the definition of F except when $c = 1, 2, 3, \ldots$.

When c is an integer, by a limiting process we can show that

$$G(a,b|1|z) = \frac{2\sin(\pi a)\sin(\pi b)}{-\pi\sin\pi(a+b)}\Big\{[\ln z + 2\gamma + \psi(a) + \psi(b)$$
$$+ \tfrac{1}{2}\pi\cot(\pi a) + \tfrac{1}{2}\pi\cot(\pi b)]F(a,b|1|z)$$
$$+ \sum_{n=0}^{\infty}\frac{\Gamma(a+n)\Gamma(b+n)}{\Gamma(a)\Gamma(b)[n!]^2}\left[\sum_{r=0}^{n-1}\left(\frac{1}{a+r} + \frac{1}{b+r} - \frac{2}{r+1}\right)\right]z^n\Big\}$$

For $m = 2, 3, 4, \ldots$

$$G(a,b|m|z) = \frac{2\sin(\pi a)\sin(\pi b)}{-\pi\sin\pi(a+b)}\Big\{[\ln z + \gamma + \psi(a) + \psi(b) - \psi(m) + \tfrac{1}{2}\pi\cot(\pi a) + \tfrac{1}{2}\pi\cot(\pi b)]F(a,b|m|z)$$

$$+ \sum_{n=0}^{\infty} \frac{\Gamma(a+n)\Gamma(b+n)\Gamma(m)}{\Gamma(a)\Gamma(b)\Gamma(n+m)n!} \cdot \left[\sum_{r=0}^{n-1}\left(\frac{1}{a+r} + \frac{1}{b+r} - \frac{1}{r+1} - \frac{1}{r+m}\right)\right] z^n$$

$$- \sum_{n=1}^{m-1} \frac{\Gamma(a-n)\Gamma(b-n)\Gamma(n)\Gamma(m)}{\Gamma(a)\Gamma(b)\Gamma(m-n)} (-z)^{-n}\Big\}$$

Integral representations for the first solution about $z = 0$:

$$y_1^0(a,b|c|z) = \frac{\Gamma(c)}{\Gamma(a)\Gamma(b)} \int_{-i\infty}^{i\infty} \frac{\Gamma(a+t)\Gamma(b+t)}{2\pi i\Gamma(c+t)} \Gamma(-t)(-z)^t\,dt$$

where the contour goes to the left of the points $t = 0, 1, 2, \ldots$ and to the right of the points $-a, -a-1, -a-2, \ldots$, and $-b, -b-1, -b-2, \ldots$ (a, b cannot be negative integers).

$$y_1^0(a,b|c|z) = \frac{\Gamma(c)}{\Gamma(b)\Gamma(c-b)} \int_1^{\infty} (t-z)^{-a}t^{a-c}(t-1)^{c-b-1}\,dt$$

or

$$y_1^0(a,b|c|z) = \frac{\Gamma(c)(1-z)^{c-a-b}}{\Gamma(c-b)\Gamma(b)} \int_1^{\infty} (t-z)^{a-c}t^{-a}(t-1)^{b-1}\,dt$$

when $\operatorname{Re} c > \operatorname{Re} b > 0$ and (for the second representation) when $b+1$ is not a real number larger than unity. The letters a and b may be interchanged in these integrals to give representations of y_1 when $\operatorname{Re} c > \operatorname{Re} a > 0$.

$$y_1^0(a,b|c|z) = \frac{\Gamma(c)z^{1-c}}{\Gamma(b)\Gamma(c-b)} \int_0^z (z-t)^{c-b-1}t^{b-1}(1-t)^{-a}\,dt$$

$$= \frac{e^{-i\pi b}}{4\pi^2}\Gamma(c)\Gamma(1-b)\Gamma(1+b-c)z^{1-c} \oint_C (t-z)^{c-b-1}t^{b-1}(1-t)^{-a}\,dt$$

$$= \frac{e^{i\pi(c-b-1)}}{4\pi^2}\Gamma(c)\Gamma(1-b)\Gamma(1+b-c)z^{1-c}(1-z)^{c-a-b} \oint_C (t-z)^{b-1}t^{c-b-1}(1-t)^{a-c}\,dt$$

where the contours C are folded figure eights, going around $t = z$ in a positive direction (counterclockwise) around $t = 0$ in a positive direction, then around $t = z$ and $t = 0$ in a negative direction. Range of validity for the contour integral is for any values of the parameters except for $-c + 1$, $+b$, and $c - b$ positive integers.

Solutions about $z = 1$ and $z = \infty$:

$$y_1^1(a,b|c|z) = F(a,b|a + b - c + 1|1 - z); \quad \text{series valid for } |1 - z| < 1$$
$$y_2^1(a,b|c|z) = (1 - z)^{c-a-b}F(c - a, c - b|c - a - b + 1|1 - z)$$

though we should have to use $G(a,b|a + b - c + 1|1 - z)$ if $a + b - c + 1 = 1, 2, 3, \ldots$.

$$y_1^\infty(a,b|c|z) = z^{-a}F(a, 1 - c + a|1 - b + a|1/z); \quad \text{series valid for } |z| > 1$$
$$y_2^\infty(a,b|c|z) = z^{-b}F(b, 1 - c + b|1 - a + b|1/z)$$

though we should have to use $z^{-a}G(a, 1 - c + a|1 - b + a|1/z)$ if $a - b = 1, 2, 3, \ldots$, or $z^{-b}G(b, 1 - c + b|1 - a + b|1/z)$ if $a - b = 0, -1, -2, \ldots$.

The primary formulas relating the series solutions about one singularity to those about the other are

$$F(a,b|c|z) = \frac{\Gamma(c)\Gamma(c - a - b)}{\Gamma(c - a)\Gamma(c - b)} F(a,b|a + b - c + 1|1 - z)$$
$$+ \frac{\Gamma(c)\Gamma(a + b - c)}{\Gamma(a)\Gamma(b)} (1 - z)^{c-a-b}F(c - a, c - b|c - a - b + 1|1 - z)$$

$$z^{1-c}F(a - c + 1, b - c + 1|2 - c|z)$$
$$= \frac{\Gamma(2-c)\Gamma(c - a - b)}{\Gamma(1 - a)\Gamma(1 - b)} F(a,b|a + b - c + 1|1 - z)$$
$$+ \frac{\Gamma(2 - c)\Gamma(a + b - c)}{\Gamma(b - c + 1)\Gamma(a - c + 1)} \cdot$$
$$\cdot (1 - z)^{c-a-b}F(c - a, c - b|c - a - b + 1|1 - z)$$

$$F(a,b|c|z) = \frac{\Gamma(c)\Gamma(b - a)}{\Gamma(b)\Gamma(c - a)} (-z)^{-a}F\left(a, 1 - c + a|1 - b + a|\frac{1}{z}\right)$$
$$+ \frac{\Gamma(c)\Gamma(a - b)}{\Gamma(a)\Gamma(c - b)} (-z)^{-b}F\left(b, 1 - c + b|1 - a + b|\frac{1}{z}\right)$$

Consequently, the joining equations, relating solutions for $z = 0$ with those for $z = 1$ and ∞, are

$$y_1^0(a,b|c|z) = \frac{\Gamma(c)\Gamma(a + b - c)}{\Gamma(a)\Gamma(b)} y_2^1(a,b|c|z) + \frac{\Gamma(c)\Gamma(c - a - b)}{\Gamma(c - a)\Gamma(c - b)} y_1^1(a,b|c|z)$$
$$y_2^0(a,b|c|z) = \frac{\Gamma(c)\Gamma(a + b - c)}{\Gamma(a)\Gamma(b)} y_2^1(a,b|c|z) - \frac{\Gamma(c)\Gamma(c - a - b)}{\Gamma(c - a)\Gamma(c - b)} y_1^1(a,b|c|z)$$
$$y_1^0(a,b|c|z) = \frac{\Gamma(c)\Gamma(a - b)}{\Gamma(b)\Gamma(a - c)} e^{-i\pi a}y_1^\infty(a,b|c|z) + \frac{\Gamma(c)\Gamma(b - a)}{\Gamma(a)\Gamma(b - c)} e^{-i\pi b}y_2^\infty(a,b|c|z)$$

$$[\sin \pi(c-a) \sin \pi(c-b) - \sin \pi a \sin \pi b] y_2^0(a,b|c|z)$$
$$= \left\{ \sin \pi(c-a) \sin \pi(c-b) \left[1 + e^{i\pi c} \frac{\Gamma(a-c+1)\Gamma(a-c)}{\Gamma(a)\Gamma(a-1)} \right] + \sin \pi a \sin \pi b \right\} \frac{\Gamma(c)\Gamma(a-b)}{\Gamma(b)\Gamma(a-c)} e^{-i\pi a} y_1^\infty(a,b|c|z)$$
$$+ \left\{ \sin \pi(c-a) \sin \pi(c-b) \left[1 + e^{i\pi c} \frac{\Gamma(b-c+1)\Gamma(b-c)}{\Gamma(b)\Gamma(b-1)} \right] + \sin \pi a \sin \pi b \frac{\Gamma(c)\Gamma(b-a)}{\Gamma(a)\Gamma(b-c)} \right\} e^{-i\pi b} y_2^\infty(a,b|c|z)$$

One Regular, One Irregular Singular Points (see pages 550 and 604):

Canonical form, points 0 (regular), ∞ (irregular), indices at 0 are 0 and $1 - c$ ($c \geq 1$):

$$z \frac{d^2y}{dz^2} + (c - z)\frac{dy}{dz} - ay = 0; \quad \text{confluent hypergeometric equation}$$

If it should not happen that Re $c \geq 1$, we set $y = z^{1-c}F$ and the equation for F will have the same form, with the real part of a new $c \geq 1$.

Solutions $y_1^0(a|c|z)$ and $y_2^0(a|c|z)$ about $z = 0$

Series expansion for y_1 about 0, valid for $|z|$ finite:

$$y_1^0 = F(a|c|z) = \frac{\Gamma(c)}{\Gamma(a)} \sum_{n=0}^{\infty} \frac{\Gamma(a+n)z^n}{\Gamma(c+n)n!}; \quad \text{confluent hypergeometric series}$$
$$= e^z F(c-a|c|-z) = \lim_{b\to\infty} [F(a, b + c|c|z/b)]$$

$$F(a|c|z) = \frac{1}{(c-a-1)} [(c-1)F(a|c-1|z) - aF(a+1|c|z)]$$
$$zF(a|c|z) = (c-1)[F(a|c-1|z) - F(a-1|c-1|z)]$$
$$(d/dz)F(a|c|z) = (a/c)F(a+1|c+1|z)$$

Series expansion for y_2 about 0 (G is the Gordon function):

$$y_2^0 = G(a|c|z) = e^{i\pi a} \frac{\Gamma(c)}{\Gamma(a)} \left\{ \frac{\Gamma(1-c)}{\Gamma(1-a)} \left[e^{-i\pi c} + \frac{\sin \pi(a-c)}{\sin(\pi a)} \right] F(a|c|z) - 2\frac{\Gamma(c-1)}{\Gamma(c-a)} z^{1-c} F(a-c+1|2-c|z) \right\}$$

The series can be constructed from the definition of F above except when $c = 1, 2, 3, \ldots$.

By a limiting process (see page 616) we can show that the formula for G, for c an integer, becomes

$$G(a|1|z) = \frac{e^{i\pi a}}{\pi} \sin(\pi a) \left\{ [2 \ln z + \pi \cot(\pi a) - i\pi + 2\psi(a)]F(a|1|z) + 2 \sum_{m=1}^{\infty} \frac{\Gamma(a+m)}{\Gamma(a)[m!]^2} [\psi(a+m) - \psi(a) + 2\psi(1) - 2\psi(m+1)]z^m \right\}$$

When $c = n = 2, 3, 4, \ldots$

$$G(a|n|z) = \frac{e^{i\pi a}}{\pi}\sin(\pi a)\left\{[2\ln z + \pi\cot(\pi a) - i\pi]F(a|n|z) - 2\sum_{m=1}^{\infty}[\psi(m+1) + \psi(n+m) - \psi(a+m)]\frac{\Gamma(a+m)(n-1)!z^m}{m!\Gamma(a)(n+m-1)!}\right\} - 2e^{i\pi a}\frac{(n-1)!}{\Gamma(a)}\sum_{r=1}^{n-1}\frac{(r-1)!(z)^{-r}}{(n-r-1)!\Gamma(r-a+1)}$$

Integral representations for y_1^0:

$$y_1^0(a|c|z) = \frac{\Gamma(c)}{2\pi i\Gamma(a)}\int_{-i\infty}^{i\infty}\frac{\Gamma(a+t)}{\Gamma(c+t)}\Gamma(-t)(-z)^t\,dt$$

where the contour goes to the left of $t = 0, 1, 2, \ldots$ and to the right of $-a, -a-1, \ldots$. Valid for $a \neq 0, -1, -2, \ldots$

$$y_1^0(a|c|z) = \frac{\Gamma(c)}{\Gamma(a)\Gamma(c-a)}\int_0^1 e^{zt}t^{a-1}(1-t)^{c-a-1}\,dt;\quad \operatorname{Re} c > \operatorname{Re} a > 0$$
$$= \frac{e^{-i\pi a}}{4\pi^2}\Gamma(c)\Gamma(1-a)\Gamma(a-c+1)\oint_c e^{zt}t^{a-1}(t-1)^{c-a-1}\,dt$$

where the contour is a folded figure eight, going in a positive direction about 1 and 0, then in a negative direction about both. The integrand is real for z real and t real and >1. Valid for all values of a and c for which the gamma functions are analytic. Asymptotic expansions for $z = |z|e^{i\phi} \gg 1$, a or c, for $0 < \phi < \pi$:

$$y_1^0(a|c|z) \to \frac{\Gamma(c)}{\Gamma(a)}z^{a-c}e^z + \frac{\Gamma(c)}{\Gamma(c-a)}(-z)^{-a};\quad (-z) = |z|e^{i(\phi-\pi)}$$
$$y_2^0(a|c|z) \to \frac{\Gamma(c)}{\Gamma(a)}z^{a-c}e^z - \frac{\Gamma(c)}{\Gamma(c-a)}(-z)^{-a}$$

For other ranges of ϕ see pages 611 and 614.

Integral representations for solutions about $z = \infty$ (the Whittaker functions):

$$y_1^\infty = U_1(a|c|z) = \frac{z^{a-c}e^z}{\Gamma(c-a)}\int_0^\infty e^{-u}u^{c-a-1}\left(1 - \frac{u}{z}\right)^{a-1}du$$
$$= \frac{z^{a-c}e^z}{\Gamma(1-a)}\int_0^\infty e^{-u}u^{-a}\left(1-\frac{u}{z}\right)^{a-c}du \to z^{a-c}e^z;\quad 0 < \phi < 2\pi$$

$$y_2^\infty = U_2(a|c|z) = \frac{(-z)^{-a}}{\Gamma(a)}\int_0^\infty e^{-u}u^{a-1}\left(1+\frac{u}{z}\right)^{c-a-1}du$$
$$= \frac{(-z)^{-a}}{\Gamma(a-c+1)}\int_0^\infty e^{-u}u^{a-c}\left(1+\frac{u}{z}\right)^{-a}du \to (-z)^{-a};\quad -\pi < \phi < \pi$$

First integral valid for $\operatorname{Re} c > \operatorname{Re} a$, second for $\operatorname{Re} a < 1$, etc. For other ranges see contour integrals of Eqs. (5.3.53) and (5.3.56).

Joining equations connecting solutions about 0 with those about ∞:

$$U_1(a|c|z) = \frac{\Gamma(1-c)}{\Gamma(1-a)} e^{i\pi(a-c)} F(a|c|z) - \frac{\Gamma(c-1)}{\Gamma(c-a)} e^{i\pi a} z^{1-c} F(a-c+1|2-c|z)$$

$$U_2(a|c|z) = \frac{\Gamma(1-c)}{\Gamma(a-c+1)} e^{i\pi a} F(a|c|z) + \frac{\Gamma(c-1)}{\Gamma(a)} e^{i\pi a} z^{1-c} F(a-c+1|2-c|z)$$

$$y_1^0(a|c|z) = \frac{\Gamma(c)}{\Gamma(a)} y_1^\infty(a|c|z) + \frac{\Gamma(c)}{\Gamma(c-a)} y_2^\infty(a|c|z)$$

$$y_2^0(a|c|z) = \frac{\Gamma(c)}{\Gamma(a)} y_1^\infty(a|c|z) - \frac{\Gamma(c)}{\Gamma(c-a)} y_2^\infty(a|c|z)$$

Two Irregular Singular Points (see pages 556 and 638):

Canonical form, points at 0, ∞:

$$z^2 \frac{d^2y}{dz^2} + z\frac{dy}{dz} + \left(\tfrac{1}{4}h^2z^2 - a + \tfrac{1}{4}h^2\frac{1}{z^2}\right) y = 0$$

Setting $z = e^{i\phi}$, we obtain the Mathieu equation

$$\frac{d^2y}{d\phi^2} + [a - \tfrac{1}{2}h^2 \cos(2\phi)]y = 0; \quad a = b - \tfrac{1}{2}h^2$$

(See also the equation for two regular and one irregular point.)

For general values of a (or b), the two solutions are

$$\mathcal{S}(b,h,e^{i\phi}) = e^{is\phi} \sum_{n=-\infty}^{\infty} a_n e^{2in\phi}; \quad \mathcal{S}(b,h,e^{-i\phi}) = e^{-is\phi} \sum_{n=-\infty}^{\infty} a_n e^{-2ni\phi}$$

with the coefficients a_n computed by continued-fraction formulas (see page 559).

When b has the particular values which make s an integer, these two solutions are both periodic and not independent. For these periodic cases we use the functions (called Mathieu functions) defined as follows:

Even angular functions about $\phi = 0$; $b = be_{2m}$ or be_{2m+1}:

$$Se_{2m}(h, \cos\phi) = \sum_{n=0}^{\infty} B_{2n}\cos(2n\phi); \quad \sum_n B_{2n} = 1$$

$$Se_{2m+1}(h, \cos\phi) = \sum_{n=0}^{\infty} B_{2n+1}\cos(2n+1)\phi; \quad \sum_n B_{2n+1} = 1$$

Odd angular functions about $\phi = 0$, $b = bo_{2m}$ or bo_{2m+1}:

$$So_{2m}(h, \cos\phi) = \sum_{n=1}^{\infty} B_{2n} \sin(2n\phi); \quad \sum_n (2n)B_{2n} = 1$$

$$So_{2m+1}(h, \cos\phi) = \sum_{n=0}^{\infty} B_{2n+1} \sin[(2n+1)\phi]; \quad \sum_n (2n+1)B_{2n+1} = 1$$

where the coefficients B are functions of h and are different for each different Se or So.

The second solutions, for the same values of the separation constant, are [see Eq. (5.2.79)]

$$Fe_{2m}(h, \cos\phi) = \gamma^e_{2m} \left\{\phi Se_{2m}(h, \cos\phi) + \sum_{n=1}^{\infty} D_{2n} \sin(2n\phi)\right\};$$

$$\gamma^e_{2m} = \left[1 + \sum_n 2nD_{2n}\right]^{-1}$$

$$Fe_{2m+1}(h, \cos\phi) = \gamma^e_{2m+1} \left\{\phi Se_{2m+1}(h, \cos\phi) + \sum_{n=0}^{\infty} D_{2n+1} \sin[(2n+1)\phi]\right\};$$

$$\gamma^e_{2m+1} = \left[1 + \sum_n (2n+1)D_{2n+1}\right]^{-1}$$

$$Fo_{2m}(h, \cos\phi) = \gamma^o_{2m} \left\{\phi So_{2m}(h, \cos\phi) + \sum_{n=0}^{\infty} D_{2n} \cos(2n\phi)\right\};$$

$$\gamma^o_{2m} = \left[\sum_n D_{2n}\right]^{-1}$$

$$Fo_{2m+1}(h, \cos\phi) = \gamma^o_{2m+1} \left\{\phi So_{2m+1}(h, \cos\phi) + \sum_{n=0}^{\infty} D_{2n+1} \cos[(2n+1)\phi]\right\};$$

$$\gamma^o_{2m+1} = \left[\sum_n D_{2n+1}\right]^{-1}$$

For further details of computation of the coefficients B and D, see pages 565 *et seq.* For behavior of solutions for complex values of ϕ, see pages 635 *et seq.*, and also the tables at the end of Chap. 11.

Bibliography

Pertinent articles and books related to the problem of the separation of variables:

Bocher, M.: "Uber die Reihenentwickelungen der Potentialtheorie," Leipzig, 1894 (dissertation).

Eisenhart, L. P.: Separable Systems of Staeckel, *Ann. Math.*, **35**, 284 (1934).

Eisenhart, L. P.: Separable Systems in Euclidean 3-space, *Phys. Rev.*, **45,** 427 (1934).

Eisenhart, L. P.: Potentials for Which Schroedinger Equations Are Separable, *Phys. Rev.*, **74,** 87 (1948).

Michel: Exhaustion of Neumann's Mode of Solution for the Motion of Solids of Revolution, etc., *Messenger of Mathematics*, **19,** 83 (1890).

Redheffer, R. M.: "Separation of Laplace's Equation," Massachusetts Institute of Technology, Cambridge, 1948 (dissertation).

Robertson, H. P.: Bemerkung uber separierbare Systeme in der Wellenmechanik, *Math. Ann.*, **98,** 749 (1927).

Additional material on the solution of ordinary differential equations:

Bateman, H.: "Partial Differential Equations of Mathematical Physics," Cambridge, New York, 1932.

Forsyth, A. R.: "Theory of Differential Equations," Vol. 4, Cambridge, New York, 1890.

Ince, E. L.: "Ordinary Differential Equations," Longmans, New York, 1927, reprint Dover, New York, 1945.

Riemann-Weber, "Differential- und Integralgleichungen der Mechanik und Physik," Vieweg, Brunswick, 1935.

Schlesinger, L.: "Theorie der Differentialgleichungen," Goschen, Leipzig, 1922.

Whittaker, E. T., and G. N. Watson: "Modern Analysis," Cambridge, New York, 1927.

Works containing further details about the special functions discussed in Secs. 5.2 and 5.3:

Gray, A., G. B. Mathews, and T. M. MacRobert: "Treatise on Bessel Functions," Macmillan, London, 1922.

Hobson, E. W.: "Theory of Spherical and Ellipsoidal Harmonics," Cambridge, New York, 1931.

Klein, F.: "Vorlesungen uber die Hypergeometrische Funktion," Springer, Berlin, 1933.

MacRobert, T. M.: "Spherical Harmonics," Methuen, London, 1927, reprint, Dover, New York, 1948.

McLachlan, N. W.: "Bessel Functions for Engineers," Oxford, New York, 1934.

McLachlan, N. W.: "Theory and Application of Mathieu Functions," Oxford, New York, 1947.

Stratton, J. A., P. M. Morse, L. J. Chu, and R. A. Hutner: "Elliptic Cylinder and Spheroidal Wave Functions," Wiley, New York, 1941.

Strutt, M. J. O.: "Lamesche, Mathieusche und verwandte Funktionen," Springer, Berlin, 1932, reprint Edwards Bros., Inc., Ann Arbor, Mich.

Watson, G. N.: "Treatise on the Theory of Bessel Functions," Cambridge, New York, 1944.

Books containing tabulations of formulas relating the functions of interest, supplementing the tables at the ends of the chapters in the present work:

Jahnke, E., and F. Emde: "Tables of Functions," B. G. Teubner, Leipzig, 1933, reprint Dover, New York, 1945.

Madelung, E.: "Mathematischen Hilfsmittel des Physikers," Springer, Berlin, 1936, reprint Dover, New York, 1943.

Magnus, W., and F. Oberhettinger: "Special Functions of Mathematical Physics," Springer, Berlin, 1943, reprint, Chelsea, New York, 1949.

CHAPTER 6

Boundary Conditions and Eigenfunctions

We have now discussed the methods of solving the ordinary differential equations which will come up in the solution of many problems in the behavior of continuous media. As we have seen, a problem is not uniquely specified if we simply give the differential equation which the solution must satisfy, for there are an infinite number of solutions of every equation of the type we have studied. In order to make the problem a definite one, with a unique answer, we must pick, out of the mass of possible solutions, the one which has certain definite properties along definite boundary surfaces. Any physical problem must state not only the differential equation which is to be solved but also the *boundary conditions* which the solution must satisfy. The satisfying of the boundary conditions is often as difficult a task as the solving of the differential equation.

The first fact which we must notice is that we cannot try to make the solutions of a given equation satisfy *any* sort of boundary conditions; we should not try to "squeeze a right-hand foot into a left-hand shoe," so to speak. For each type of equation which we have discussed in Chap. 2, there is a definite set of boundary conditions which will give unique answers, and any other sort of conditions will give nonunique or impossible answers. Now, of course, an actual physical problem will always have the right sort of boundary conditions to give it a unique answer (or, at least, so we all hope!), and if we make our statement of the problem correspond to the actualities, we shall always have the right boundary conditions for the equations. But it is not always easy to tell just what boundary conditions correspond to "actuality," and it is well for us to know what conditions are suitable for what equations so we can be guided in making our mathematical problems fit the physical problems as closely as possible.

6.1 *Types of Equations and of Boundary Conditions*

Let us first discuss a two-dimensional example in order to bring out the concepts without confusing by complexity. All the two-dimensional partial differential equations for scalar fields, which we discussed in Chaps. 2 and 3, and many of the equations for components of vector fields have the general form

$$A(x,y)\frac{\partial^2\psi}{\partial x^2} + 2B(x,y)\frac{\partial^2\psi}{\partial x\,\partial y} + C(x,y)\frac{\partial^2\psi}{\partial y^2} = F\left(x,y,\psi,\frac{\partial\psi}{\partial x},\frac{\partial\psi}{\partial y}\right) \quad (6.1.1)$$

where, if the equation is linear in ψ, F has the form

$$D(x,y)\frac{\partial\psi}{\partial x} + E(x,y)\frac{\partial\psi}{\partial y} + G(x,y)\psi + H(x,y)$$

This is, of course, the most general linear partial differential equation in the two variables, x and y. These two coordinates may be either two space coordinates or one space coordinate plus time.

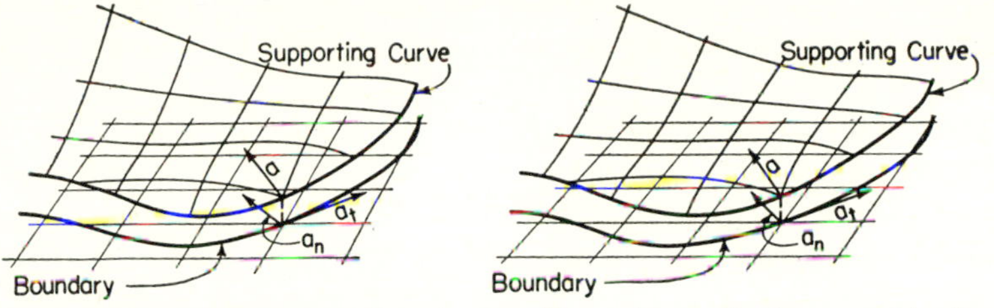

Fig. 6.1 Boundary conditions in two dimensions. Surface $z = \psi(x,y)$, boundary curve $x = \xi(s)$, $y = \eta(s)$, unit vectors $\mathbf{a}_t$ and $\mathbf{a}_n$ in x, y plane; vector $\mathbf{a}$ tangent to surface at boundary.

There is a nondenumerable infinity of solutions of this equation; the additional conditions imposed by the problem, which serve to fix on one particular solution as being appropriate, are called *boundary conditions*. Usually they take the form of the specification of the behavior of the solution on or near some *boundary line* (or surface, in three dimensions). (From this point of view, initial conditions are just boundary conditions in time.) It naturally is of interest to see what sort of curves these boundary curves may be and what sort of specification of the field along the line there must be in order that a unique answer result.

For a two-dimensional problem the solution $\psi(x,y)$ may be represented by the surface $z = \psi(x,y)$. The *boundary* is a specified curve on the (x,y) plane, *not* the edge of the surface $z = \psi(x,y)$ which is above the boundary curve. The *boundary conditions* are then represented by the height of the ψ surface above the boundary curve and/or the slope of the ψ surface normal to the boundary curve (see Fig. 6.1). The edge of the ψ surface just above the boundary curve (which is in general not a plane curve) is sometimes called the *supporting curve* for the boundary.

If the distance along the boundary from some origin is s and the parametric equations for the boundary curve are $x = \xi(s)$, $y = \eta(s)$, then the equation for the supporting curve is $z = \psi(\xi,\eta) = \psi(s)$. The unit vector tangent to the boundary at the point s on the boundary is $\mathbf{a}_t = \mathbf{i}(d\xi/ds) + \mathbf{j}(d\eta/ds)$, and the unit vector normal to the curve is $\mathbf{a}_n = \mathbf{a}_t \times \mathbf{k} = [-\mathbf{j}(d\xi/ds) + \mathbf{i}(d\eta/ds)]$. The expressions for these vectors are particularly simple because we have said that s is the *distance* along the boundary curve, so that $\sqrt{(d\xi/ds)^2 + (d\eta/ds)^2} = 1$ (why?). Since $\mathbf{a}_n$ is an axial vector, having a choice of directions, we choose axes and directions (for this section, not for Chap. 7) so that $\mathbf{a}_n$ points *inward* (on the side of the boundary which contains the solution). In terms of these vectors and derivatives the gradient of ψ normal to the boundary at s is

$$\mathbf{a}_n \cdot \operatorname{grad} \psi = \frac{\partial\psi}{\partial x}\frac{d\eta}{ds} - \frac{\partial\psi}{\partial y}\frac{\partial\xi}{ds} = N(s)$$

where $\partial\psi/\partial y$ and $\partial\psi/\partial x$ are taken at the points $x = \xi(s)$, $y = \eta(s)$. In terms of these definitions we can now visualize the various types of boundary conditions.

Types of Boundary Conditions. In every case, of course, we must specify the shape of the boundary. It may be a closed curve for the Laplace equation in two space dimensions, or it may be an open, U-shaped boundary consisting of a line parallel to the space axis and two lines parallel to the time axis for a string (wave equation in time and one space dimension) fixed at the ends and given a specified start at a given time, and so on. As mentioned on page 690, the boundary is *closed* if it completely surrounds the solution (even if part of the boundary is at infinity); it is *open* if the boundary goes to infinity and no boundary conditions are imposed along the part at infinity.

In a one-dimensional case, the solution of a second-order equation is uniquely specified if we specify its initial value and slope. By analogy we might expect that, if the boundary were parallel to one of the axes, specification of the value of ψ along the boundary [*i.e.*, specifying $\psi(s)$] and of the gradient of ψ normal to the boundary [*i.e.*, specifying $N(s)$, in this case $\partial\psi/\partial y$] will uniquely fix the solution. This is correct, as will be shown later, but it is too special a case to satisfy us. We should sometimes like to have a boundary which is not contiguous with a coordinate line but is free to take any reasonable shape. It is not quite so obvious that specification of value and normal gradient on a boundary of any shape will give a unique result (nor is it true!), and we shall have to determine the sort of boundaries which are satisfactory.

The type of boundary condition mentioned in the last paragraph, the specifying of value *and* normal gradient, is called the *Cauchy boundary condition*, and the problem of determining the shape of boundary and type of equation which yields unique and reasonable solutions for Cauchy conditions is called the *Cauchy problem*, after the investigator who first studied it in detail. Specifying the initial shape and velocity of an infinite flexible string corresponds to Cauchy conditions along the line t = constant. As we know, this uniquely specifies the solution.

On the other hand if the solution is to be set up inside a closed boundary, it might be expected that Cauchy conditions are too much requirement and might rule out all solutions. Perhaps one only needs to specify the value $\psi(s)$ alone or the normal gradient $N(s)$ alone along the boundary in order to obtain a unique answer.

The specifying only of values along the boundary is called *Dirichlet conditions*, and the specifying only of slopes is called *Neumann conditions*. A potential problem, such as the determination of electric potential inside a sequence of conductors at specified potentials, corresponds to Dirichlet conditions. On the other hand the determination of velocity potential around solid bodies, where the fluid must flow tangential to the surface of the solids and the normal potential gradient at the surface is zero, corresponds to Neumann conditions. Alternatively we may, at times, need to specify the value of some linear combination of value and slope, a single boundary condition which is intermediate between Dirichlet and Neumann conditions.

In terms of our supporting line in Fig. 6.1, Cauchy conditions correspond to our specifying not only the line $\psi(s) = z$ but also the normal slope at the edge of the surface $\psi(x,y) = z$. It is as though, instead of a line, we had a thin ribbon as a support to the ψ surface, a twisted ribbon which specified slope perpendicular to its axis as well as height above the z axis (but not higher derivatives). For Dirichlet conditions the supporting line is really a line, not a ribbon. For Neumann conditions the ribbon is free to move up and down, only the "slant" of the ribbon is fixed. Sometimes these two conditions are *homogeneous*, when $\alpha\psi(s) + \beta N(s) = 0$, for α, β specified but independent of s, and sometimes the conditions are *inhomogeneous*, when $\alpha\psi(s) + \beta N(s) = F(s)$. The distinction will be of interest in Sec. 6.3.

But we now must go back to our general equation (6.1.1) and see under what circumstances Cauchy conditions along the curve $x = \xi(s)$, $y = \eta(s)$ will result in a unique solution.

Cauchy's Problem and Characteristic Curves. In order to compute ψ at some distance away from the boundary we can have recourse to some two-dimensional power series, analogous to Taylor's series:

$$\psi(x,y) = \psi(\xi,\eta) + \left[(x-\xi)\frac{\partial\psi}{\partial x} + (y-\eta)\frac{\partial\psi}{\partial y}\right] + \tfrac{1}{2}\left[(x-\xi)^2\frac{\partial^2\psi}{\partial x^2} + 2(x-\xi)(y-\eta)\frac{\partial^2\psi}{\partial x\,\partial y} + (y-\eta)^2\frac{\partial^2\psi}{\partial y^2}\right] + \cdots \tag{6.1.2}$$

where ψ and all its derivatives on the right-hand side of the equation are evaluated at the boundary point (ξ,η). Once these partial derivatives of ψ are all evaluated at the boundary, then ψ is uniquely specified within the radius of convergence of the series, *i.e.*, over all of a strip contiguous to the boundary line, which may be infinite in width depending on the nature of the equation. If we can work out a recipe for computing the partial derivatives, we shall have the Cauchy problem well along toward solution. This is not so straightforward as it may at first seem, for we are given only the equation for ψ, the parametric equations for the boundary, and the values of $\psi(s)$ and $N(s)$ on the boundary and from these data are to compute all the double infinity of values of the partial derivatives for each point (ξ,η) on the boundary.

It is not too difficult to express the first derivatives in terms of known quantities. There are two of them, and there are two equations, one giving the specified normal gradient $N(s)$ and the other the rate of change of the known value $\psi(s)$ along the boundary:

$$N(s) = \left(\frac{d\eta}{ds}\right)\left(\frac{\partial\psi}{\partial x}\right) - \left(\frac{d\xi}{ds}\right)\left(\frac{\partial\psi}{\partial y}\right) = \mathbf{a}_n \cdot \operatorname{grad}\psi; \quad \text{at } x = \xi,\ y = \eta$$

$$\frac{d}{ds}\psi(s) = \left(\frac{d\xi}{ds}\right)\left(\frac{\partial\psi}{dx}\right) + \left(\frac{d\eta}{ds}\right)\left(\frac{\partial\psi}{dy}\right) = \mathbf{a}_t \cdot \operatorname{grad}\psi; \quad \text{at } x = \xi,\ y = \eta$$

Since the determinant of the coefficients $(d\xi/ds)^2 + (d\eta/ds)^2 = 1$, there is *always* a solution for these equations:

$$\left(\frac{\partial\psi}{\partial x}\right)_{\xi,n} = N(s)\left(\frac{d\eta}{ds}\right) + \left(\frac{d\xi}{ds}\right)\left(\frac{d\psi}{ds}\right) = p(s);$$
$$\left(\frac{\partial\psi}{\partial y}\right)_{\xi,n} = \left(\frac{d\eta}{ds}\right)\left(\frac{d\psi}{ds}\right) - \left(\frac{d\xi}{ds}\right)N(s) = q(s) \quad (6.1.3)$$

But the next step, to obtain the second derivatives, is not so simple. It is also the crucial step, for if we can find the three second partials, we shall find that solving for the higher derivatives is simply "more of the same." Now that we have solved for the first derivatives, we know p and q, given in Eq. (6.1.3) as functions of the parameter s. Two of the needed three equations for the second derivatives are obtained by writing down the expression for the known rate of change of p and q with s in terms of these second derivatives; the third equation is the differential equation ψ must satisfy, Eq. (6.1.1) itself:

$$\left(\frac{d\xi}{ds}\right)\left(\frac{\partial^2\psi}{\partial x^2}\right) + \left(\frac{d\eta}{ds}\right)\left(\frac{\partial^2\psi}{\partial x\,\partial y}\right) = \frac{dp}{ds}$$
$$\left(\frac{d\xi}{ds}\right)\left(\frac{\partial^2\psi}{\partial x\,\partial y}\right) + \left(\frac{d\eta}{ds}\right)\left(\frac{\partial^2\psi}{\partial y^2}\right) = \frac{dq}{ds}$$
$$A(s)\left(\frac{\partial^2\psi}{\partial x^2}\right) + 2B(s)\left(\frac{\partial^2\psi}{\partial x\,\partial y}\right) + C(s)\left(\frac{\partial^2\psi}{\partial y^2}\right) = F(s)$$

where $A(s)$, etc., are the known values of the coefficients at the point $\xi(s)$, $\eta(s)$ on the boundary.

These three equations can be solved, to find the three partials, *unless* the determinant of the coefficients

$$\Delta = \begin{vmatrix} \frac{d\xi}{ds} & \frac{d\eta}{ds} & 0 \\ 0 & \frac{d\xi}{ds} & \frac{d\eta}{ds} \\ A & 2B & C \end{vmatrix} = C\left(\frac{d\xi}{ds}\right)^2 - 2B\left(\frac{d\xi}{ds}\right)\left(\frac{d\eta}{ds}\right) + A\left(\frac{d\eta}{ds}\right)^2 \quad (6.1.4)$$

is zero. If the determinant Δ is not zero, all the higher partials can be solved for by successive differentiations of known quantities with respect to s, the distance along the boundary, and the resulting Taylor's series will uniquely specify the resulting solution, within some finite area of convergence. Thus we have shown that Cauchy conditions on a boundary do choose a particular solution *unless the boundary is such that the determinant Δ is zero along it.*

The equation $\Delta = 0$ is the equation of a curve

$$C(x,y)(dx)^2 - 2B(x,y)\,dx\,dy + A(x,y)(dy)^2 = 0 \quad (6.1.5)$$

(where we have changed the differentials $d\xi$, $d\eta$ into the more familiar dx, dy) or, rather, of two families of curves, for this equation may be factored, giving

$$A\,dy = (B + \sqrt{B^2 - AC})\,dx; \quad A\,dy = (B - \sqrt{B^2 - AC})\,dx \quad (6.1.6)$$

These curves are characteristic of the partial differential equation (6.1.1) and are called the *characteristics* of the equation. If the boundary line happens to coincide with one of them, then specifying Cauchy conditions along it *will not* uniquely specify the solution; if the boundary cuts each curve of each family once, then Cauchy conditions along it *will* uniquely specify a solution.

Hyperbolic Equations. In order to have this statement mean anything physically, the two families of characteristics must be real curves. This means that our statement (as it stands) applies only to those partial differential equations for which $B^2(x,y) > A(x,y)C(x,y)$ *everywhere.* Such equations are called *hyperbolic equations.* The wave equation

$$\frac{\partial^2\psi}{\partial x^2} - \frac{1}{c^2}\frac{\partial^2\psi}{\partial t^2} = 0$$

is a hyperbolic equation if t is considered as being the second coordinate y. Equation (2.3.29), for supersonic flow, is also a hyperbolic equation.

For hyperbolic equations the natural coordinate system is formed from the two families of characteristics, which are real. Integration of the first of Eqs. (6.1.6) gives a solution $\lambda(x,y)$ = constant; integration of the second gives $\mu(x,y)$ = constant; and λ and μ are the natural coordinates. Since motion along one of the characteristics λ = constant corresponds to $(\partial\lambda/\partial x)\,dx + (\partial\lambda/\partial y)\,dy = 0$ (the gradient of λ is perpendicular to the vector $\mathbf{i}\,dx + \mathbf{j}\,dy$ for motion along the characteristic), substituting from this equation back into Eq. (6.1.5) shows that

$$A\left(\frac{\partial\lambda}{\partial x}\right)^2 + 2B\left(\frac{\partial\lambda}{\partial x}\right)\left(\frac{\partial\lambda}{\partial y}\right) + C\left(\frac{\partial\lambda}{\partial y}\right)^2 = 0 \tag{6.1.7}$$

with a similar equation for the partials of μ, for the other family.

We now go back to the original equation (6.1.1) and express it in the new coordinates. For instance,

$$\left(\frac{\partial^2\psi}{\partial x^2}\right) = \left(\frac{\partial^2\psi}{\partial\lambda^2}\right)\left(\frac{\partial\lambda}{\partial x}\right)^2 + 2\left(\frac{\partial^2\psi}{\partial\lambda\,\partial\mu}\right)\left(\frac{\partial\lambda}{\partial x}\right)\left(\frac{\partial\mu}{\partial x}\right) + \left(\frac{\partial^2\psi}{\partial\mu^2}\right)\left(\frac{\partial\mu}{\partial x}\right)^2$$

plus terms in $\partial\psi/\partial x$ and $\partial\psi/\partial y$. We finally obtain

$$\begin{aligned}\frac{\partial^2\psi}{\partial\lambda^2}&\left[A\left(\frac{\partial\lambda}{\partial x}\right)^2 + 2B\left(\frac{\partial\lambda}{\partial x}\right)\left(\frac{\partial\lambda}{\partial y}\right) + C\left(\frac{\partial\lambda}{\partial y}\right)^2\right]\\ &+ 2\frac{\partial^2\psi}{\partial\lambda\,\partial\mu}\left[A\left(\frac{\partial\lambda}{\partial x}\right)\left(\frac{\partial\mu}{\partial x}\right) + B\left(\frac{\partial\lambda}{\partial x}\frac{\partial\mu}{\partial y} + \frac{\partial\lambda}{\partial y}\frac{\partial\mu}{\partial x}\right) + C\left(\frac{\partial\lambda}{\partial y}\right)\left(\frac{\partial\mu}{\partial y}\right)\right]\\ &+ \frac{\partial^2\psi}{\partial\mu^2}\left[A\left(\frac{\partial\mu}{\partial x}\right)^2 + 2B\left(\frac{\partial\mu}{\partial x}\right)\left(\frac{\partial\mu}{\partial y}\right) + C\left(\frac{\partial\mu}{\partial\mu}\right)^2\right] = G\left(\frac{\partial\psi}{\partial x}, \frac{\partial\psi}{\partial y}, \psi, x, y\right)\end{aligned}$$

But the first and third expressions in brackets are zero because λ and μ are characteristic functions of the equation. If the equation is homogeneous, G can be put into the form $a(\partial\psi/\partial\lambda) + b(\partial\psi/\partial\mu) + c\psi$, and the second bracket expression (which is *not* zero) and a, b, and c can be made functions of λ and μ.

We thus arrive at the *normal form of the hyperbolic equation*

$$\frac{\partial^2\psi}{\partial\lambda\,\partial\mu} = P\left(\frac{\partial\psi}{\partial\lambda}\right) + Q\left(\frac{\partial\psi}{\partial\mu}\right) + R\psi \tag{6.1.8}$$

where P, Q, and R are functions of λ and μ. If these quantities (P,Q,R) are zero, as they often are (for the wave equation in one space dimension and the equation for supersonic flow, for instance), the solution of Eq. (6.1.8) is

$$\psi = f(\lambda) + g(\mu) \tag{6.1.9}$$

where f can be *any* function of λ and g *any* function of μ. For the wave equation, for example, $\lambda = x - ct$ and $\mu = x + ct$, so that $\psi = f(x - ct) + g(x + ct)$, corresponding to waves of arbitrary shape traveling to the right and to the left with velocity c and $-c$. We shall discuss a case where P and Q are not zero on page 687.

We have thus shown that solutions of at least some hyperbolic equations are similar to traveling waves and that the families of characteristics correspond to the wave fronts. When the normal form of the equation has the particularly simple form

$$\partial^2\psi/\partial\lambda\,\partial\mu = 0 \tag{6.1.10}$$

waves of any shape may be present with fronts along λ and along μ.

When the boundary crosses both families of characteristics (as in the first of Fig. 6.2), then the Cauchy conditions will uniquely determine both $f(\lambda)$ and $g(\mu)$. Each point on the boundary, labeled by the distance s from an origin, corresponds to a given value of λ and μ. Specifying $\psi(s)$ and $N(s)$ at this point gives two simultaneous equations which serve

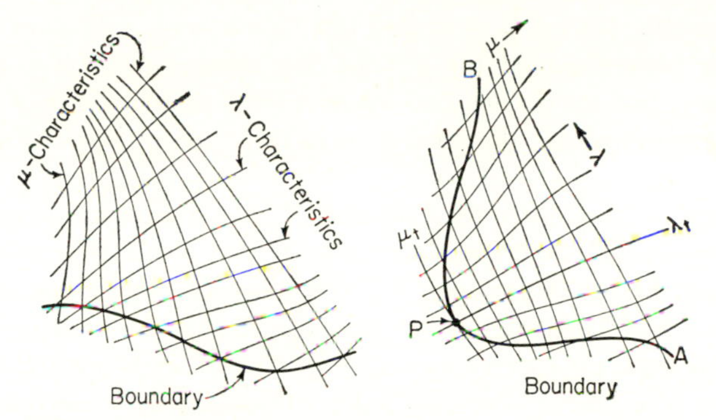

Fig. 6.2 Intersection of families of characteristics, for a hyperbolic equation, with boundary line.

to determine both f and g for this pair of values of λ and μ. If the boundary crosses all of both families of characteristics, then f and g will be specified for all values of λ and μ and the field will be uniquely determined everywhere. When the characteristics are everywhere real, if $\psi(s)$ and $N(s)$ are bounded and continuous, then f and g will also be bounded and continuous and so will $\psi(x,y)$.

Cauchy Conditions and Hyperbolic Equations. We now can see why Cauchy conditions do not specify the solution when the boundary coincides with a characteristic. If the boundary is along $\mu = \mu_0$, say, then Cauchy conditions give data concerning only $g(\mu)$ and the slope of $g(\mu)$ at $\mu = \mu_0$ and *nothing at all* about the behavior of *g for any other* μ. In this case $f(\lambda)$ is determined, because the line $\mu = \mu_0$ cuts all the family of λ characteristics, so the value of $\psi(s)$ [which in this case can be written $\psi(\lambda)$] is $f(\lambda)$. The normal slope $N(s)$ determines $dg/d\mu$ at $\mu = \mu_0$ but no higher derivatives can be determined, and so $g(\mu)$, for any other μ, is wholly undetermined. In general the values of f and g *are determined only* for those values of λ and μ *which are crossed by the boundary line.*

To put it still another way, a boundary which coincides with a characteristic is a boundary which travels along with a wave front. Since it never comes in contact with any of the rest of the wave traveling its direction, it can affect only the wave traveling in the opposite direction (*i.e.*, it can determine only f, not g). The close relation between these statements and the discussion on page 168, of shock waves which appear when a fluid flows past a boundary at a speed greater than the speed of sound, should be apparent by now.

We also can see now what happens when the boundary curves around so that it crosses a family of characteristics *twice*, as in the second part of Fig. 6.2. At the point $P(\lambda_t,\mu_t)$ the boundary is tangent to the characteristic $\mu = \mu_t$; for all values of $\mu > \mu_t$ the boundary cuts the μ characteristics twice and the characteristics for $\mu < \mu_t$ are not cut at all. Suppose that Cauchy conditions are given on the PA part of the boundary. This determines $f(\lambda)$ for $\lambda < \lambda_t$ and $g(\mu)$ for $\mu > \mu_t$. It does not matter that $g(\mu)$ is undetermined for $\mu < \mu_t$, for these values of μ are outside the boundary, but we do need to know values of $f(\lambda)$ for $\lambda > \lambda_t$. These must be determined by boundary conditions along the PB part of the boundary.

If Cauchy conditions [both $\psi(s)$ and $N(s)$] are given along the portion PB, the solution will be "overdetermined," for along this portion $g(\mu)$ is already fixed, being determined by the Cauchy conditions on PA, and only $f(\lambda)$ for $\lambda > \lambda_t$ needs to be determined by the conditions on PB. This may be done by specifying *either* $\psi(s)$ *or* $N(s)$ along PB (or a linear combination of ψ and N) but *not both*. Consequently Dirichlet or Neumann conditions (or the intermediate combination) are sufficient for PB. Of course, we could also obtain a unique answer by setting Cauchy conditions on PB and Dirichlet or Neumann conditions on PA.

In general, we can say that, if the boundary is curved so that it cuts a family of characteristics twice, then Cauchy conditions are needed on the part of the boundary on one side of the point which is tangent to a characteristic and Dirichlet or Neumann conditions on the other side are sufficient. It is not difficult to carry this reasoning on to the case where there is more than one point of tangency. For instance, for a U-shaped boundary, Cauchy conditions are needed along the base of the U and Dirichlet or Neumann conditions are enough along the sides, and for a Z-shaped boundary, Cauchy conditions along the top and bottom of the Z, with Dirichlet or Neumann conditions along the diagonal part, will be proper.

It is also not difficult to see that, when the boundary is closed, so that every interior characteristic crosses the boundary twice, Cauchy conditions on any finite part of the boundary overdetermine the solution. It is not too easy to see, however, whether Dirichlet conditions (or Neumann) all around the boundary are sufficient, and we shall have to postpone our discussion of this until later in the chapter.

It might be well to discuss a simple case of the above problem in order to see how it works out in practice. The case of the flexible string is the simplest hyperbolic equation, where the displacement ψ of the string depends on x and t according to the equation (see page 124)

$$\frac{\partial\psi}{\partial x^2} - \frac{1}{c^2}\frac{\partial^2\psi}{\partial t^2} = 0$$

The characteristic functions are $\lambda = x - ct$ and $\mu = x + ct$, the equation having its normal form

$$\partial^2\psi/\partial\lambda\,\partial\mu = 0$$

with solution $\psi = f(\lambda) + g(\mu)$.

Specifying initial value and slope of an infinite string corresponds to Cauchy conditions on a simple, open boundary cutting the characteristics but once. If the initial shape, at $t = 0$, is $\psi_0(x)$ [$= \psi(s)$] and the

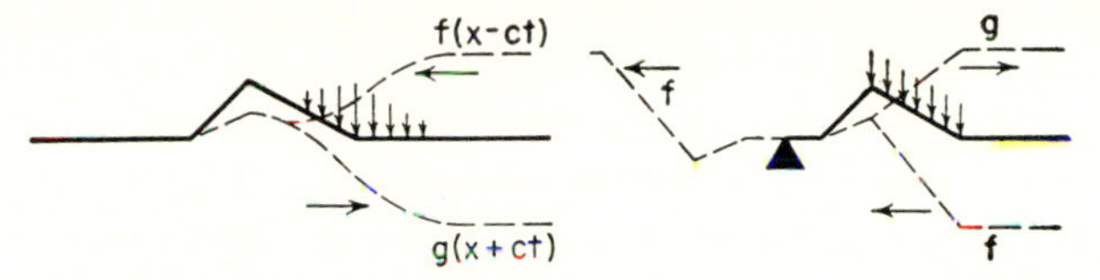

Fig. 6.3 Initial shape (solid line) and initial velocity (arrows) of string. Subsequent motion is given by sum of dotted lines f and g, moving in opposite directions.

initial velocity is $V_0(x)$ [$= N(s)$], we must arrange the functions f and g so that

$$f(x) + g(x) = \psi_0(x) \quad \text{and} \quad -f'(x) + g'(x) = (1/c)V_0(x)$$

where the prime indicates differentiation with respect to the argument. It is not difficult to see that

$$f(\lambda) = \tfrac{1}{2}\psi_0(\lambda) - \frac{1}{2c}\int_0^\lambda V_0(w)\,dw; \quad g(\mu) = \tfrac{1}{2}\psi_0(\mu) + \frac{1}{2c}\int_0^\mu V_0(w)\,dw \tag{6.1.11}$$

The solution is then $\psi(x,t) = f(x - ct) + g(x + ct)$, consisting of the sum of two waves traveling in opposite directions with velocity c and $-c$. This is shown in the first part of Fig. 6.3.

If now the string is clamped at $x = 0$, the boundary is L-shaped and cuts the μ characteristics at two points. The values of $f(\lambda)$ and $g(\mu)$ for λ and μ larger than zero are fixed by the Cauchy conditions of initial value and velocity along the $t = 0$, $x > 0$ part of the boundary, and the value of $f(\lambda)$ for $\lambda < 0$ is determined by the Dirichlet condition that $\psi = 0$ for the $x = 0$, $t > 0$ part of the boundary. Values of g for $\mu < 0$ are not fixed by the boundary conditions, nor are they needed.

The procedure for fitting these boundary conditions is to choose a value for $f(-\lambda)$ which is an "image" of $g(\mu)$, so that for any value of t the two will just cancel at $x = 0$. For initial shape and velocity $\psi_0(x)$ and $V_0(x)$ we can see that the proper solution is $\psi = f(x - ct) + g(x + ct)$, where

$$g(\mu) = \tfrac{1}{2}\psi_0(\mu) + \frac{1}{2c}\int_0^\mu V_0(w)\,dw; \quad \mu > 0$$

$$f(\lambda) = \begin{cases} \tfrac{1}{2}\psi_0(\lambda) - \dfrac{1}{2c}\displaystyle\int_0^\lambda V_0(w)\,dw; & \lambda > 0 \\ -\tfrac{1}{2}\psi_0(-\lambda) - \dfrac{1}{2c}\displaystyle\int_0^{-\lambda} V_0(w)\,dw; & \lambda < 0 \end{cases} \tag{6.1.12}$$

This is illustrated in the second part of Fig. 6.3. We see that the values of $f(\lambda)$ for negative λ produce a reflected wave, the reverse of the μ wave reflected in the end support.

Clamps at both ends of the string correspond to a U-shaped boundary, which causes a periodicity in the reflections, for the initial (Cauchy) conditions along the finite part of the string are reflected first at one end and then at the other. A closed boundary for this case would be to set "final conditions" at time t, as well as initial conditions at $t = 0$. If we specify Cauchy conditions (both value and velocity) at $t = 0$, then the shape and velocity are fixed at $t = t_1$ and our final conditions cannot be arbitrary values of ψ_1 and V_1 or we shall have a contradiction. We might expect that specifying ψ_0 only at $t = 0$ and ψ_1 only at $t = t_1$ (Dirichlet conditions) would uniquely specify the solution, but the periodic motion of the finite string enters to confound us. As we know, such a finite string has free vibrations of period $2l/nc$, where l is the distance between the rigid supports and where n is any integer. If t_1 is equal to any multiple of any of these periods (*i.e.*, is *any rational number* times $2l/c$), then the string could have had a periodic vibration (of *any* amplitude) of the right frequency to go through zero at $t = 0$ and also go through zero at $t = t_1$; and this would not have registered in the value of ψ at either the beginning or the end.

We therefore see that Dirichlet (or Neumann) conditions on a closed boundary do not uniquely specify the solution of this simple hyperbolic equation. In fact it is difficult to see what sort of boundary conditions on a closed boundary do not over- or underdetermine the solution. But we also see that a closed boundary is not a very "natural" boundary for a hyperbolic equation, so the puzzle need not distract us.

Waves in Several Space Dimensions. The extension to several space dimensions of the general concepts discussed above is not very difficult. In an equation of hyperbolic type there is one coordinate which has its second derivative term of opposite sign from the rest. This results in equations for real characteristic surfaces, which often correspond to wave surfaces. The discussion of boundary surfaces and boundary conditions can easily be generalized, though a detailed discussion would necessarily be prolix.

It will be of interest, however, to discuss the generalization of the type of boundary fitting for the wave equation which was exemplified in Eqs. (6.1.11) and (6.1.12). In the two-dimensional case (one space dimension) a disturbance at time t and position x is propagated outward in both directions with finite velocity but *without change in shape* as the wave travels. The wave represented by $f(x - ct)$ moves to the right with unchanged shape as it travels, and similarly for $g(x + ct)$, going to the left. We might ask if waves from a point disturbance travel outward with unchanged shape in any number of dimensions.

Anticipating the discussions of Chap. 7, we state that a disturbance at point $(x_1, x_2, \ldots, x_n)$ in n dimensions, at time t, propagates outward equally in all directions from the point. Therefore we are here interested in a solution of the wave equation depending only on time and on the distance r from one point to another in n dimensions. The hyperspherical coordinates in n dimensions are defined as follows:

$$\begin{aligned} x_1 &= r \cos \theta_1 \\ x_2 &= r \sin \theta_1 \cos \theta_2 \\ &\cdots\cdots\cdots\cdots \\ x_{n-1} &= r \sin \theta_1 \sin \theta_2 \cdots \sin \theta_{n-2} \cos \theta_{n-1} \\ x_n &= r \sin \theta_1 \sin \theta_2 \cdots \sin \theta_{n-2} \sin \theta_{n-1} \end{aligned} \qquad (6.1.13)$$

Discarding derivatives with respect to any of the angles, we find that the wave equation for a "spherically symmetric" wave in n space dimensions is

$$\frac{1}{r^{n-1}} \frac{\partial}{\partial r}\left[r^{n-1} \frac{\partial \psi}{\partial r}\right] - \frac{1}{c^2} \frac{\partial^2 \psi}{\partial t^2} = 0 \quad \text{or} \quad \frac{\partial^2 \psi}{\partial r^2} - \frac{1}{c^2} \frac{\partial^2 \psi}{\partial t^2} = -\frac{n-1}{r} \frac{\partial \psi}{\partial r} \qquad (6.1.14)$$

This last is obviously a hyperbolic equation of the form of Eq. (6.1.1), and it will be of interest to treat it as such. The characteristics, as before, are $\lambda = r - ct$ and $\mu = r + ct$. The normal form of the equation is

$$\frac{\partial^2 \psi}{\partial \lambda \, \partial \mu} = -\frac{1}{2} \frac{n-1}{\lambda + \mu}\left[\frac{\partial \psi}{\partial \lambda} + \frac{\partial \psi}{\partial \mu}\right] \qquad (6.1.15)$$

Only when $n = 1$ is the solution $\psi = f(\lambda) + g(\mu)$; this is the case already discussed. If a solution giving waves which propagate outward and inward without change of shape is impossible except for one space dimension, we might try to see whether we can find a solution which changes only by diminishing its amplitude as it moves away from the point disturbance. In other words we try the solution $\psi = F(\lambda,\mu)(\lambda + \mu)^a = F(\lambda,\mu)(2r)^a$ in the hopes that F will turn out to be $f(\lambda) + g(\mu)$.

Substituting in Eq. (6.1.15) shows that, unless $a = \frac{1}{2}(1 - n)$, the equation for F is worse than that for ψ. When $a = \frac{1}{2}(1 - n)$, in other words when $\psi = F/(2r)^{\frac{1}{2}(n-1)}$, the equation for F is

$$\partial^2 F/\partial\lambda\,\partial\mu = \tfrac{1}{4}(n-1)(n-3)F/(\lambda+\mu)^2$$

Only for $n = 1$ or $n = 3$ will F be simply the sum of a function of λ and a function of μ. We have treated the case of $n = 1$; it now seems that for three space dimensions we can obtain a solution

$$\psi(r,t) = (1/r)[f(r - ct) + g(r + ct)]; \quad r^2 = x^2 + y^2 + z^2 \qquad (6.1.16)$$

which represents outgoing and incoming waves propagated with no change of shape other than diminution of amplitude as r increases. In two space dimensions evidently waves radiating out from a point change shape even more drastically than this.

To go further in this direction will encroach on Chap. 7 overmuch. Suffice it to say here that explosive (single-pulse) waves spread out radially in an odd number of space dimensions ($n = 1, 3, 5, \ldots$) as sharp pulses, diminishing in amplitude but maintaining their sharp fronts and rears, giving no warning of their onset and leaving no "wake" behind them after they have passed. In contrast, waves from a sharp pulse in space of an even number of dimensions ($n = 2, 4, 6, \ldots$) spread out radially and maintain a sharp wave *front*, giving no warning ahead of the time $t = r/c$, but they leave a "wake" behind, the disturbance continuing long after the crest has passed. The reasons for this interesting difference will also be touched upon in Chap. 11.

Elliptic Equations and Complex Variables. We must now return to Eq. (6.1.6) for the characteristics, to see what must be done when the characteristics are not real curves. When $A(x,y)C(x,y) > B^2(x,y)$ for all values of x and y, Eq. (6.1.1) is called an *elliptic equation.* The equations for the characteristics are complex conjugates of each other, and if the characteristic function is $\lambda(x,y) = u(x,y) + iv(x,y)$, then the other function is its complex conjugate, $\mu(x,y) = u(x,y) - iv(x,y)$ (where u and v are real functions of x,y). Changing Eq. (6.1.1) to the form of Eq. (6.1.8),

$$\frac{\partial^2\psi}{\partial\lambda\,\partial\mu} = P\left(\frac{\partial\psi}{\partial\lambda}\right) + Q\left(\frac{\partial\psi}{\partial\mu}\right) + R\psi$$

is not quite so useful here, for λ and μ are complex variables. It is generally more useful to use the real and imaginary parts of λ and μ, u and v, for the natural coordinates.

Consequently the *normal form of the elliptic equation* is

$$\frac{\partial^2\psi}{\partial u^2} + \frac{\partial^2\psi}{\partial v^2} = S\left(\frac{\partial\psi}{\partial u}\right) + T\left(\frac{\partial\psi}{\partial v}\right) + U\psi \qquad (6.1.17)$$

The Laplace equation is an elliptic equation, as is the equation for subsonic (Mach number < 1) flow of a compressible fluid (see page 168) and also the Helmholtz equation. The Poisson equation is an inhomogeneous form of the elliptic equation, with an additional term, $\rho(x,y)$, on the right-hand side.

When P, Q, and R (or S, T, and U) are zero, the solutions can again be expressed in the form

$$\psi = f(\lambda) + g(\mu) = f(u + iv) + g(u - iv) \tag{6.1.18}$$

In other words ψ is any function of the complex variable $u + iv$ plus any function of its conjugate. The applications of the theory of functions to solutions of the two-dimensional Laplace equation have been touched on in Chap. 4 and will be discussed in more detail in Chap. 10.

The connection between analytic functions and solutions of a two-dimensional elliptic equation sheds some light on the relations between these solutions and their boundary conditions. To illustrate, let us take the Laplace equation $\nabla^2\psi = 0$ in two dimensions and try to apply Cauchy conditions along the x axis to determine ψ in the upper half plane. The characteristic functions for the Laplace equation are $\lambda = x + iy$, $\mu = x - iy$, and the general solution is $\psi = f(x + iy) + g(x - iy)$. The boundary conditions to be tried are $\psi = \psi_0(x)$ and $\partial\psi/\partial y = N_0(x)$ when $y = 0$. A solution is

$$\begin{aligned}
f(\lambda) &= \tfrac{1}{2}\psi_0(\lambda) - \tfrac{1}{2}i \int_0^\lambda N_0(z)\, dz \\
g(\mu) &= \tfrac{1}{2}\psi_0(\mu) + \tfrac{1}{2}i \int_0^\mu N_0(z)\, dz \\
\psi &= \mathrm{Re}\,[\psi_0(x + iy)] - \mathrm{Im}\,[\chi_0(x + iy)] \\
\chi_0(z) &= \int_0^z N_0(\xi)\, d\xi
\end{aligned} \tag{6.1.19}$$

Therefore ψ is the real part of the function $\psi_0(z) + i\chi_0(z)$ of the variable $z = x + iy$. For physically sensible boundary conditions ψ_0 and χ_0 are reasonable functions *along the real axis* (*i.e.*, along the boundary). But there are many functions of z which are finite along the real axis yet which have poles and/or essential singularities somewhere off the real axis. In fact we have proved, in Chap. 4, that the only function of z which stays finite *everywhere* is a constant. Therefore unless our boundary conditions are that $\psi_0 =$ constant, $N_0 = 0$, we are sure that $\psi(x,y)$ has an infinity somewhere on the complex plane. Of course all the infinities may be in the lower half of the plane, outside the boundary and therefore harmless, but only a very small fluctuation of the "wrong sort" in ψ_0 or N_0 in some region of the boundary will result in an infinity somewhere in the positive half plane. In other words, a function of $x + iy$ is just *too sensitive* to small fluctuations of its value or slope along the real axis to be amenable to control by boundary conditions of this sort.

The contrast to the hyperbolic equation becomes apparent also. Equation (6.1.11) gives values of $\psi(x,y)$ for the hyperbolic equation in terms of ψ_0 and N_0 *for real values of the argument*, and if ψ_0 and the integral of N_0 are bounded and continuous along the boundary, ψ will also be bounded and continuous everywhere in space; but Eq. (6.1.19) shows that $\psi(x,y)$ for an elliptic equation is expressed in terms of ψ_0 and N_0 *for complex values of the argument*, and just because ψ_0 and the integral of N_0 are bounded and continuous along the real axis is no guarantee that they will be bounded everywhere over the upper half of the z plane.

Another way of looking at this fundamental difference between hyperbolic and elliptic equations lies in the general behavior of the two solutions. A solution of Laplace's equation (which is an elliptic equation), for instance, can have no maxima or minima (see page 7); therefore, if the boundary conditions start the solution increasing in a given direction, the solution has no choice but to keep on increasing until it goes to infinity at some point (this is not a maximum; it is a singularity!) unless the other edge of the boundary intervenes in time. On the other hand, solutions of the wave equation (which is a hyperbolic equation) *can* have maxima and minima; therefore if the boundary conditions start the solution increasing in a given direction, the solution will eventually change its gradient, describing a wave crest in that direction. We shall return later to this oversensitivity of solutions of elliptic equations to Cauchy conditions on an open boundary.

But if Cauchy conditions on an open boundary are too critical, for solutions of elliptic equations, to be used in physical problems, Dirichlet or Neumann conditions on an open boundary are not sufficient to give a unique answer. So it seems that closed boundaries are preferable for elliptic equations, with Dirichlet or Neumann conditions (since Cauchy conditions overspecify the solution for closed boundaries). The difficulty we ran into with the solution of the hyperbolic equation with closed boundaries (due to the possibility of waves being present which are not perceptible at the boundaries) is not present with solutions of elliptic equations, for wave motion is not present. We shall indicate in more detail later that Dirichlet or Neumann conditions on a closed boundary usually give a unique solution for elliptic equations.

It must, of course, be pointed out that the "closed boundary" for these cases may go to infinity for part of the circuit. The point is that ψ_0 (or N_0) must be given even for the part of the boundary at infinity in order that a unique solution be obtained. On the other hand, boundary conditions at infinity for the case of the hyperbolic equation, discussed earlier, are not necessary and in fact are redundant if Cauchy conditions are used along the finite part of the boundary.

Similarly, solutions of the form of Eq. (6.1.16) are not particularly useful for elliptic equations, partly because of the difficulties of the complex characteristics, mentioned above, but also because of the fact that no one coordinate differs from the other in sign of second derivative as does the time term in the wave equation. Accordingly, for the elliptic equation (such as the Laplace equation) it is advisable to consider all coordinates on a par. For a point source in n dimensions, the equation for the "spherically symmetric" solution is

$$\frac{1}{r^{n-1}}\frac{d}{dr}\left[r^{n-1}\frac{d\psi}{dr}\right] = 0; \quad r^2 = x_1^2 + x_2^2 + \cdots + x_n^2$$

The solution is $\psi = a + (b/r^{n-2})$, where a and b are constants (the exception is for $n = 2$, where the solution is $a + b \ln r$). For the elliptic equation this is a satisfactory solution, for it is finite everywhere except at the source, $r = 0$. A similar solution can be set up for the wave equation, with $r^2 = x_1^2 + x_2^2 + \cdots + x_{n-1}^2 - c^2t^2$, but because of the last term the solution is infinite whenever $c^2t^2 = x_1^2 + x_2^2 + \cdots + x_{n-1}^2$; it has not the usefulness which the elliptic equation solution has.

Parabolic Equations. One limiting case of Eq. (6.1.1) should be discussed, that for which $B^2(x,y) = A(x,y)C(x,y)$ everywhere. In this case there is only one family of characteristics, the integrals of the equation $A\,dy = B\,dx$ [see Eqs. (6.1.6)], which may be called $\lambda(x,y)$. Expressing Eq. (6.1.1) in terms of new coordinates λ and x, we have, eventually, for the *normal form of the parabolic equation*

$$\frac{\partial^2\psi}{\partial x^2} = K(x,\lambda)\frac{\partial\psi}{\partial\lambda} + L(x,\lambda)\frac{\partial\psi}{\partial x} + M(x,\lambda)\psi \tag{6.1.20}$$

since both the terms

$$A\left(\frac{\partial\lambda}{\partial x}\right)^2 + 2B\left(\frac{\partial\lambda}{\partial x}\right)\left(\frac{\partial\lambda}{\partial y}\right) + C\left(\frac{\partial\lambda}{\partial y}\right)^2 \quad \text{and} \quad A\left(\frac{\partial\lambda}{\partial x}\right) + B\left(\frac{\partial\lambda}{\partial y}\right)$$

are zero. In this case there is only one second derivative term; only the first derivative with respect to the characteristic function enters.

The diffusion equation is of this form, with t taking the place of λ:

$$\frac{\partial^2\psi}{\partial x^2} = a^2\frac{\partial\psi}{\partial t}$$

This equation is "lopsided" with respect to time; if the sign of t is reversed, we obtain a different sort of solution, whereas the wave equation is symmetric with respect to time. This difference is primarily due to the fact that the diffusion equation is not a "conservative" equation. Entropy increases continually as time goes on (and usually free energy decreases), whereas for the wave equation the energy remains constant (unless friction is included). As one might expect (see page 138), as time increases, all irregularities in the diffusion solution are "ironed out"; for near a maximum in ψ, there the curvature $\partial^2\psi/\partial x^2$ is negative and ψ decreases with time there. Only when all curvature (and therefore irregularity) of ψ is zero can ψ be independent of time.

It should appear reasonable that Dirichlet conditions are satisfactory for this equation and that a boundary should be used which is open in the direction of *increasing t*. Going backward in time tends to increase the irregularities in the solution; the shorter the range of the irregularity the faster the increase; so, though we can predict what a given ψ will become a time t *later*, we cannot be sure what the same ψ was a time t *earlier* (in other words the solution for t negative is oversensitive to boundary values at $t = 0$). This difference between prophecy and history is typical of the parabolic equation and is not typical of the wave equation, for instance. For the wave equation one can look into the past just as easily as into the future.

For the n-dimensional parabolic equation one of the coordinates (call it x_n) enters in the first derivative term only and all the rest enter as second derivatives. The equation for a "spherically symmetric" solution in n space dimensions [compare with Eq. (6.1.14)] is

$$\frac{1}{r^{n-1}}\frac{\partial}{\partial r}\left[r^{n-1}\frac{\partial\psi}{\partial r}\right] - a^2\frac{\partial\psi}{\partial t} = 0 \tag{6.1.21}$$

which is itself a parabolic equation. A solution in an inverse power of r is possible but not particularly useful if there is dependence on t. We shall show, in Sec. 7.4, that the most useful "primitive" solution is one which is very concentrated at $t = 0$ but which diffuses outward as t increases. Choosing a form $\psi = F(t)e^{-r^2/\alpha t}$, which has such properties, we soon find that F is inversely proportional to $t^{n/2}$. So, finally we find that a solution of Eq. (6.1.21) is

$$\psi = [a/2\sqrt{\pi t}]^n e^{-a^2r^2/4t}; \quad t > 0 \tag{6.1.22}$$

where $r^2 = x_1^2 + x_2^2 + \cdots + x_n^2$. This has been so normalized that its integral over all n-dimensional space is unity. The function is very concentrated for small values of t, is quite "spread out" for large values of t, and is not valid for t negative. Its properties will be more thoroughly explored in Sec. 7.4.

6.2 *Difference Equations and Boundary Conditions*

We have several times (pages 132 and 235) utilized the limiting relationship between continuous variables and quantities defined only for integral values of some parameter, between series and integrals, between particles and continuum. There is a corresponding relation between differential equations and difference equations which we touched upon when discussing recursion formulas (see page 540) and which is apparent from the limiting definition of the derivative. Related to the quantity y_n, which is a denumerable infinity of values of y for each integral value of n, is the continuous function $y(x)$ of the continuous variable x. Related to the first, second, etc., differences of y_n

$$\Delta(y_n) = y_{n+1} - y_n; \quad \Delta^2(y_n) = \Delta[\Delta(y_n)] = y_{n+2} - 2y_{n+1} + y_n; \quad \text{etc.} \tag{6.2.1}$$

are the various derivatives of $y(x)$. (The relationship would be somewhat closer if we should discuss $[\Delta(y_n)/h]$, etc., where h is related to the unit change in the index n and goes to zero as the "spacing" in n goes to zero, but we can dispense with h for the preliminary analysis.)

Corresponding to the differential equations we discussed in the previous chapter there are difference equations which are, in principle, easier to analyze but often, in practice, more difficult to solve. Some of the behavior of the solutions, however, is useful to contrast and compare with behavior of the solutions of the corresponding differential equation. This is particularly true in regard to the relationship between solutions and boundary conditions, so it will be profitable to digress here to discuss the general properties of difference equations.

First-order Linear Difference Equations. In the previous chapter we found several recursion formulas (see, for instance, page 542) which involved only two successive coefficients for the series expansion

$$A_n a_n + D_n a_{n+1} = 0$$

from which the coefficients a_n can be determined in terms of a_0, for instance. This is equivalent to the first-order difference equation

$$\Delta(y_n) = C_n y_n; \quad C_n = -1 - (A_n/D_n) \tag{6.2.2}$$

It is obvious that definite solutions for y_n may be obtained once the value of one a_n is fixed. Usually this boundary condition (equivalent to Dirichlet conditions) is expressed by giving the value of y_0. Then the y for larger values of n is the product

$$y_n = y_0 \prod_{r=0}^{n-1} (C_r + 1) = y_0 \exp\left\{\sum_{r=0}^{n-1} \ln (1 + C_r)\right\} \tag{6.2.3}$$

The corresponding differential equation is $dy/dx = f(x)y$, having a solution

$$y(x) = y_0 \exp\left\{\int_0^x f(x)\,dx\right\}$$

The relation between the two is apparent. For the inhomogeneous equation

$$\Delta(y_n) - C_n y_n = B_n \tag{6.2.4}$$

we can try a solution $y_n = A_n \prod_{r=0}^{n-1} (C_r + 1)$. Setting this in Eq. (6.2.4),

$$A_{n+1} \prod_{r=0}^{n} (C_r + 1) - A_n \prod_{r=0}^{n-1} (C_r + 1) - C_n A_n \prod_{r=0}^{n-1} (C_r + 1) = B_n$$

$$\Delta(A_n) \prod_{r=0}^{n} (C_r + 1) = B_n \quad \text{or} \quad \Delta(A_n) = B_n \Big/ \prod_{r=0}^{n} (C_r + 1)$$

Therefore the complete solution of Eq. (6.2.4) is

$$y_n = \left[\prod_{r=0}^{n-1} (1 + C_r)\right] \left\{ y_0 + \sum_{s=0}^{n-1} \frac{B_s}{\prod_{r=0}^{s} (1 + C_r)} \right\} \tag{6.2.5}$$

which is closely related to the solution

$$y(x) = e^{\int F\,dx}[y_0 + \int e^{-\int F\,dx} r\, dx]$$

of the differential equation

$$(dy/dx) + F(x)y = r(x)$$

If $C_0 = 0$, $C_n = c$ and $B_n = b$, independent of n, the solution is

$$\begin{aligned} y_n &= (1 + c)^{n-1} \left\{ y_0 + \sum_{s=0}^{n-1} \frac{b}{(1 + c)^s} \right\} \\ &= (1 + c)^{n-1} \left\{ y_0 + b \left[\frac{1 - (1 + c)^{-n}}{1 - (1 + c)^{-1}} \right] \right\} \\ &= y_0(1 + c)^{n-1} + \frac{b}{c}[(1 + c)^n - 1] \end{aligned}$$

We could go on to second-order difference equations, paralleling the discussion of Chap. 5 in almost complete detail. But we are interested now in using difference equations to give us a better insight into the behavior of partial differential equations.

Difference Equations for Several Dimensions. A very useful method for calculation of approximations to the solution of ordinary and partial differential equations is to replace the continuous independent variables by discontinuous variables, having values only at integral values of some spacing h. For the differential equation

$$\frac{\partial^2 \psi}{\partial x^2} + \frac{\partial^2 \psi}{\partial y^2} = 0$$

for instance, we mark off a rectangular grid or net of spacing h in both directions and compute ψ only at the net points where the grid lines cross. The value of ψ at $x = mh$, $y = nh$ is then labeled $\psi(m,n)$. If h is small enough, a good approximation to the differential equation will be the *difference equation*

$$(1/h^2)[\Delta_m^2\psi + \Delta_n^2\psi] = 0$$

or

$$\tfrac{1}{4}[\psi(m-1,n) + \psi(m+1,n) + \psi(m,n-1) + \psi(m,n+1)] = \psi(m,n) \tag{6.2.6}$$

it is possible to develop methods for solving such equations numerically, for use when the related differential equations cannot be solved exactly; here we wish to study the problem only enough to see the parallelism between differential and difference equations with regard to their reaction to boundary conditions.

What we shall do in this section is to consider the medium, or field, not to be continuous at first, but to be a sort of lattice of spacing h, where h is finite though small. The field ψ will have meaning only at the net points, and the finite differences between the ψ's for neighboring points will be governed by difference equations. We shall derive results about the solutions of boundary-value problems for the difference equations corresponding to the three types of partial differential equations discussed in the preceding section. Having proved the results for a lattice of finite spacing, we can then make the spacing infinitely minute and can be sure that the conclusions will usually hold for the corresponding differential equations. The process of going to the limit is not always so simple as it might seem; it can be shown that for the theorems we are to consider no unforeseen complications arise.

We need use only two dimensions to demonstrate these theorems; more dimensions only increase complexity. And we shall also use only the simplest forms for the three types of equation. For example, the simplest *elliptic equation* is the Laplace equation; its correlated difference equation has been given in Eq. (6.2.6). For a nonhomogeneous elliptic equation we can use Poisson's equation

$$\frac{\partial^2\psi}{\partial x^2} + \frac{\partial^2\psi}{\partial y^2} = -F(x,y)$$

Its difference equation is

$$\tfrac{1}{4}[\psi(m+1,n) + \psi(m-1,n) + \psi(m,n+1) + \psi(m,n-1)] - \psi(m,n) = -(h^2/4)F(m,n) \tag{6.2.7}$$

The simplest *hyperbolic equation* in two dimensions is the wave equation

$$\frac{\partial^2\psi}{\partial x^2} - \frac{\partial^2\psi}{\partial y^2} = 0; \quad y = ct$$

having a difference equation of the form

$$\psi(m+1,n)+\psi(m-1,n)=\psi(m,n+1)+\psi(m,n-1) \quad (6.2.8)$$

Likewise the simplest parabolic equation is the diffusion equation

$$\frac{\partial^2\psi}{\partial x^2}=a^2\frac{\partial\psi}{\partial t}$$

having a related difference equation which may be written

$$\psi(m,n+1)=C[\psi(m+1,n)+\psi(m-1,n)-2\psi(m,n)]+\psi(m,n)$$

or

$$\psi(m,n-1)=-C[\psi(m+1,n)+\psi(m-1,n)-2\psi(m,n)]+\psi(m,n) \quad (6.2.9)$$

depending on whether we wish to work backward or forward in n (the time dimension). We have made $C = 1/ha^2$.

The Elliptic Equation and Dirichlet Conditions. As an example of this type of analysis let us consider the net-point analogue of an elliptic equation with Dirichlet conditions along a closed boundary. We stated on page 690 that these boundary conditions for this equation would result in a unique solution; let us see whether the net-point method can help us prove the statement. The fundamental lattice is shown in Fig. 6.4. The boundary points (open circles) are the rows $m = 0$, $n = 0$, $m = M$, and $n = N$; we specify values for ψ for all these points. Given these boundary values, we wish to determine unique values for ψ for all the interior points (black dots) by use of the difference equation (6.2.6). If a convergent procedure can be developed for computing each ψ in terms of the boundary ψ's, which gives a unique answer for all interior points, we can consider the problem solved and the fact proved that Dirichlet conditions give a unique answer.

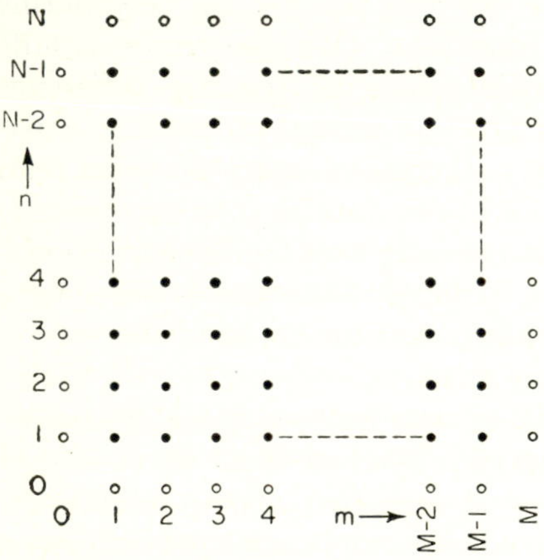

Fig. 6.4 Fundamental lattice for solution of difference equation analogous to Poisson's differential equation.

A physical example of this net problem is a lattice of rubber bands stretched between equally spaced boundary pegs, each extending a different height above a plane. The height of each net point above the plane corresponds to the value of $\psi(n,m)$. The difference equation (and also the rubber-band model) states that the value of ψ at the net point (m,n) *is the average of the values of* ψ *at the four neighboring net points.* We have already (page 7) said that this is the physical significance of the Laplace equation. The statement can be used to compute the values of the interior ψ's.

For instance, the value of ψ at the interior point (2,1) can be computed in terms of the values of its neighbors:

$$\psi(2,1) = \tfrac{1}{4}[\psi(3,1) + \psi(1,1) + \psi(2,2)] + \tfrac{1}{4}\psi(2,0)$$

where $\psi(2,0)$ is a value of ψ at the boundary point (2,0) and is therefore specified. The other three ψ's on the right are for interior points and are not yet known. But we can express them in terms of the average of values at *their* neighbors, giving

$$\tfrac{13}{16}\psi(2,1) = \tfrac{1}{16}[\psi(4,1) + \psi(2,3) + 2\psi(3,2) + 2\psi(1,2)] \\ + \tfrac{1}{4}\psi(2,0) + \tfrac{1}{16}[\psi(0,1) + \psi(1,0) + \psi(3,0)]$$

giving $\psi(2,1)$ in terms of still more interior points and boundary values.

Again expressing the ψ's for interior points in terms of the average value of their neighbors, we obtain for a third equation in the sequence

$$\tfrac{12}{16}\psi(2,1) = \tfrac{1}{64}[\psi(5,1) + \psi(2,4) + 2\psi(3,1) + 3\psi(4,2) + 3\psi(3,3) \\ + 3\psi(1,3) + 4\psi(2,2)] + \tfrac{15}{64}\psi(2,0) \\ + \tfrac{1}{16}[\psi(0,1) + \psi(1,0) + \psi(3,0)] + \tfrac{1}{32}\psi(0,2) + \tfrac{1}{64}\psi(4,0)$$

and so on, for the other equations in the sequence. After each substitution, $\psi(2,1)$ will be expressed in terms of coefficients times other net potentials and other coefficients times boundary values (which have specified values). As we keep on with the substitutions, we notice two important facts: The coefficients of the interior potentials approach zero rapidly (so rapidly that even the sum of all the coefficients of *all* the interior potentials, in any one of the sequence of equations, approaches zero), whereas the coefficients of the boundary potentials approach finite values. For instance, the value of the coefficient of $\psi(3,1)$ is $\frac{1}{4}$ for the first equation, 0 for the second, $\frac{1}{24}$ for the third in the sequence, and so on, whereas the coefficient of the boundary value $\psi(2,0)$ is $\frac{1}{4}$, $\frac{4}{13}$, $\frac{5}{16}$, and so on.

This means that, if we continue with the substitutions long enough, the term, in the eventual equation for $\psi(2,1)$ concerned with the ψ for another interior point, can be made as small as we please compared with the corresponding term for a boundary value. In the limit, therefore, $\psi(2,1)$ can be expressed explicitly in terms of *just* boundary potentials, and if *all* of these are specified, the value of $\psi(2,1)$ will be uniquely specified. The same argument can be used to show that *every* net potential is uniquely determined as soon as boundary potentials are specified all along a closed boundary enclosing the net. If part of the boundary is at infinity, this does not invalidate the argument; as a matter of fact all we usually need to know about the boundary ψ's at infinity to obtain a unique value of $\psi(m,n)$ is that they are all finite in value (*i.e.*, not infinite).

There has been elaborated a procedure, called the *relaxation method*, by which this sequence of calculations can be made in an orderly and quickly convergent manner. All we need to know here, however, is that it *can* be done, which has just been demonstrated. Now, by letting h go to zero, it can be proved that Dirichlet conditions on a closed boundary give unique results for solutions of the Laplace equation. A simple generalization of the same argument will suffice to extend this statement to the general elliptic equation (6.1.17) and to more than two dimensions.

It is not much more difficult to carry through the same argument for Neumann conditions. For a net these are that the differences between each boundary ψ and that for the net point next in from the boundary are all specified. We can show that, as long as these differences satisfy a single, over-all limitation, then the procedure outlined in the preceding pages converges here also and we can obtain unique values for the interior ψ's (except for an arbitrary, additive constant).

The nature of the over-all limitation to the Neumann conditions is indicated by the following examples. Suppose that all the boundary differences are specified to make the solution *increase* toward the interior; how is one to find a solution which has *no maximum* in the interior [for Eq. (6.2.6) allows no maximum]? Obviously not all the boundary gradients can have the same sign. The limitation can be more clearly indicated by remembering that the velocity potential of an incompressible fluid obeys the Laplace equation and that Neumann conditions on a surface then correspond to specifying the fluid flux across the surface. To specify that this flux be all inward across a closed surface would be to specify a net positive rate of increase of fluid inside the surface, an impossible requirement if the fluid is incompressible. In this case, therefore, we should have to require that the integral of the specified gradient, over the whole boundary surface, be zero; aside from this single limitation any values of the normal gradient may be assigned on the boundary.

We shall expect, therefore, that correct boundary conditions for the Laplace equation may be Neumann conditions on a closed boundary, provided that the specified boundary gradients satisfy some integral requirement over the boundary surface. Also, since the boundary conditions are gradients, not values, we shall need to specify the value of ψ at some single point to obtain a unique answer.

Eigenfunctions. Let us discuss a particularly simple case, just to show what can be done by net-point calculations. We take the lattice shown in Fig. 6.5, with four interior points and eight boundary points. We shall now proceed to build up our solution for any boundary conditions in terms of solutions for simple boundary conditions. This can be done because our difference equation is *linear;* if $\psi(m,n)$ is the solution when the boundary values are $\psi_0(0,n)$ or $\psi_0(m,0)$, etc., then $A\psi(m,n)$ is the solution for boundary values $A\psi_0$; and if $\psi'(m,n)$ is the solution for boundary values ψ'_0, then the solution for boundary values $\psi_0 + \psi'_0$ is $\psi(m,n) + \psi'(m,n)$.

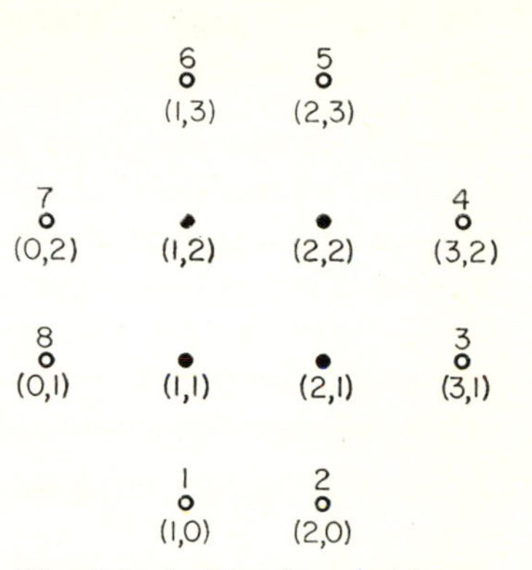

Fig. 6.5 Lattice for simple examples of boundary-value specification.

For instance, we can solve the difference equation for the simple case $\psi_1(1,0) = \psi_2(2,0) = 1$, all other boundary ψ's equal zero. For then, by symmetry, $\psi(1,1) = \psi(2,1)$ and $\psi(1,2) = \psi(2,2)$. Setting down the difference equations for two interior points, we have

$$\psi(1,1) = \tfrac{1}{4} + \tfrac{1}{4}\psi(2,1) + \tfrac{1}{4}\psi(1,2) \quad \text{or} \quad \tfrac{3}{4}\psi(1,1) = \tfrac{1}{4} + \tfrac{1}{4}\psi(1,2)$$
$$\psi(1,2) = \tfrac{1}{4}\psi(1,1) + \tfrac{1}{4}\psi(2,2) \quad \text{or} \quad \tfrac{3}{4}\psi(1,2) = \tfrac{1}{4}\psi(1,1)$$

These can be solved to obtain

$$\psi(1,1) = \psi(2,1) = \tfrac{3}{8}; \quad \psi(1,2) = \psi(2,2) = \tfrac{1}{8} \tag{6.2.10}$$

This matrix of values may be called $\Psi_1(m,n)$.

We can also solve for the case $\psi_1(1,0) = -\psi_2(2,0) = 1$, other boundary ψ's zero. This will give

$$\psi(1,1) = -\psi(2,1) = \tfrac{5}{24}; \quad \psi(1,2) = -\psi(2,2) = \tfrac{1}{24} \tag{6.2.11}$$

which matrix of values may be called $\Psi_2(m,n)$.

It is not difficult now to see that the solution corresponding to $\psi_1(1,0) = a$; $\psi_2(2,0) = b$, all other boundary ψ's $= 0$, is

$$\Psi(m,n) = \frac{a+b}{2}\,\Psi_1(m,n) + \frac{a-b}{2}\,\Psi_2(m,n)$$

Solutions for the other three parts of the boundary may be obtained by rotating these solutions by 90, 180, or 270°. By suitable additions we can obtain a solution for any sorts of assumed values of the boundary potentials.

This method can be extended to regions bounded by a rectangular boundary of any size. We first assume boundary values along one side of a particularly simple type, all other boundary values to be zero. For instance, we set $\psi(m,0) = \phi(m)$ (where ϕ is to be determined), $\psi(0,n) = \psi(m,N) = \psi(M,n) = 0$. We then *separate* the difference equation, by setting $\psi(m,n) = f(n)\phi(m)$, where the difference equations for f and ϕ are

$$\begin{aligned} \phi(m) &= \tfrac{1}{2}[\phi(m+1) + \phi(m-1)] + C\phi(m) \\ \text{and} \quad f(n) &= \tfrac{1}{2}[f(n+1) + f(n-1)] - Cf(n) \end{aligned} \tag{6.2.12}$$

where C is the *separation constant.* The difference equation (6.2.6) is obtained by multiplying the first of these equations by $f(n)$, the second by $\phi(m)$, adding, and dividing by 2. The analogy between this and the process of separation of partial differential equations is obvious.

We see that for this technique to be simple the functions $\phi_\nu(m)$, describing the simple types of boundary values along one side when the other sides are zero, must satisfy the difference equation

$$\phi_\nu(m) = \tfrac{1}{2}\phi_\nu(m-1) + \tfrac{1}{2}\phi_\nu(m+1) + C_\nu\phi_\nu(m)$$

with $\phi_\nu(0)$ and $\phi_\nu(M)$ equal to zero. Reference to page 133 indicates that possible solutions are

$$\phi_\nu(m) = \sin(\pi\nu m/M); \quad C_\nu = 2\sin^2(\pi\nu/2M); \quad \nu = 1, 2, \ldots, M \tag{6.2.13}$$

These M different functions of m have the property that a linear combination can be found to fit any set of values for the boundary potentials we wish along the $n = 0$ part of the boundary, keeping the rest of the boundary zero.

We must next solve for the corresponding functions $f_\nu(n)$, by finding the solutions of

$$f_\nu(n) = \tfrac{1}{2}f_\nu(n-1) + \tfrac{1}{2}f_\nu(n+1) - 2\sin^2(\pi\nu/2M)f_\nu(n) \tag{6.2.14}$$

which is zero at $n = N$ and unity at $n = 0$. The general solution for arbitrary boundary values along the $n = 0$ part of the boundary is a sum of the products $\phi_\nu(m)f_\nu(n)$ over the M allowed values of ν.

Functions of the sort defined by Eq. (6.2.13) are called *eigenfunctions;* they will be discussed later in the chapter in relation to differential equations. By their means we can obtain solutions satisfying any boundary conditions on boundary surfaces of particularly simple shape.

Green's Functions. But we can also rearrange the functions discussed above to give a different aspect to the solutions. The solutions of Eqs. (6.2.10) and (6.2.11) for the simple lattice of Fig. 6.5 can be combined to give the special one

$$\begin{gathered} G_1(m,n) = \tfrac{1}{2}\Psi_1(m,n) + \tfrac{1}{2}\Psi_2(m,n) \\ G_1(1,1) = \tfrac{7}{24}; \quad G_1(2,1) = \tfrac{1}{12} = G_1(1,2); \quad G_1(2,2) = \tfrac{1}{24} \end{gathered} \tag{6.2.15}$$

which is a solution of difference equation (6.2.6) satisfying the simple boundary condition that ψ at boundary point 1 (1,0) is 1 and all others are zero. The solution for boundary point 2 (2,0) to be 1 and all others zero is $G_2(m,n) = \frac{1}{2}\Psi_1(m,n) - \frac{1}{2}\Psi_2(m,n)$, which is the function G_1 reflected in the vertical bisector of the lattice [that is, $G_2(1,1) = \frac{1}{12} = G_2(2,2)$, $G_2(2,1) = \frac{7}{24}$, $G_2(1,2) = \frac{1}{24}$] and the functions for the other boundary points in turn can be obtained by reflecting the first solution in the various axes of symmetry of the lattice.

These quantities, which can be considered to be functions of the boundary point and of the interior point, are called *Green's functions* for the boundary. Since $G_s(m,n)$ corresponds to the boundary condition $\psi_s = 1$ and all other boundary ψ's $= 0$, we can quickly build up the solution for the boundary conditions, ψ at the sth boundary point $= \psi_s$, by the sum

$$\Psi(m,n) = \sum_{s=1} \psi_s G_s(m,n) \qquad (6.2.16)$$

We multiply the Green's function for the sth boundary point by the boundary value at that point and sum over all boundary points. In Chap. 7 we shall discuss the generalization of this technique to differential equation solutions.

The solution of the Poisson difference equation (6.2.7) may be found by a very similar technique. We solve this equation for the boundary potentials all zero and for $F(1,1) = 4/h^2$, all other F's $= 0$.

$$G(1,1|1,1) = \tfrac{7}{6};\quad G(1,1|1,2) = G(1,1|2,1) = \tfrac{1}{3};\quad G(1,1|2,2) = \tfrac{1}{6}$$

The solution for all boundary values zero and for the function $F(m,n)$ any fixed set of values is therefore

$$\Psi(m,n) = \sum_{r,s} F(r,s)G(r,s|m,n) \qquad (6.2.17)$$

a sum of values of F for the different interior net points, each multiplied by the solution for ψ for that net point. If we also have some of the boundary potentials different from zero, we can add to the function of Eq. (6.2.17) enough of Eqs. (6.2.16) to fit these boundary conditions.

The solution $G(r,s|m,n)$ is called the Green's function for the pair of interior points (r,s) and (m,n). It turns out to be symmetric for interchange of the points; *i.e.*, the potential at point (m,n) due to unit "charge" F at point (r,s) is equal to the potential at (r,s) due to unit charge at (m,n). This is an aspect of the *principle of reciprocity* to be discussed several times later.

Incidentally the solution G for a lattice boundary of any shape represents the following, purely imaginary problem: We let the lattice represent the streets of a symmetrically laid-out city. A saloon is located at the corner of r Street and s Avenue. At some early hour in the morning a person is ejected from the saloon and starts at random along either r Street or s Avenue. The streets are somewhat slippery, and each time the person crosses a street or an avenue (*i.e.*, at each net point he encounters), he slips and falls down. Such is the efficacy of his previous potations, he completely forgets the direction he has been traveling by the time he regains his feet, so his next block-long stroll is equally likely to be in any of the four directions away from his last spill. This continues, more and more bruises being collected, several streets and avenues being traversed (some of them several times), until he finally arrives at a boundary point. At each boundary intersection a policeman is stationed, one of whom promptly claps the befuddled wanderer in the nearest calaboose.

Of course there are many possible paths which the person might take between the saloon and his arrest, and we naturally cannot predict exactly which path he will take. We can, however, predict the *probability* that he will fall down at the corner of m Street and n Avenue. [Technically speaking, we can compute the probability density of falls there, the chance that he falls once plus twice the chance that he falls twice there, etc. Put another way, if the stroll is habitual, this probability is the ratio of times he falls at (m,n) to his total falls averaged over all evenings.] This probability *is just the Green's function* $G(r,s|m,n)$ which we have been computing. The example we have discussed here is a special case of the *problem of random flight*, which has engaged the study of several eminent mathematicians and which is of interest in the study of Brownian motion and also of the motion of stars in a galaxy.

The Elliptic Equation and Cauchy Conditions. To complete the story about the elliptic equation we should see what happens when Cauchy conditions are applied along the boundary $n = 0$. This means that $\psi(m,0)$ and $\psi(m,1)$ are specified for all values of m. From Eq. (6.2.6) we can then find all the $\psi(m,2)$'s,

$$\psi(m,2) = 4\psi(m,1) - \psi(m,0) - \psi(m + 1, 1) - \psi(m - 1, 1)$$

and so on for all n's successively. If there are side boundaries (say at $m = 0$ and $m = M$), they cannot have Cauchy conditions along them or some ψ's (the ones for $m = 1$ and $m = M - 1$) would be overspecified; Dirichlet or Neumann conditions are all that are allowed.

Since the ψ's are already specified by setting the Cauchy conditions on $n = 0$, we cannot have an upper boundary at, say, $n = N$, with arbitrarily set boundary conditions of any kind. For an elliptic equation inside a closed boundary Cauchy conditions on any portion of the boundary are too many conditions.

On the other hand, if we do not have an upper boundary, then $\psi(m,n)$ will, in general, increase without limit as n increases. For ψ cannot have a maximum or minimum, and if ψ starts up or down, as n is increased, it must keep on increasing or decreasing indefinitely as n continues to increase. Any infinitesimal change in the Cauchy conditions at $n = 0$

will cause an indefinitely large change in ψ for very large values of n. Such a sensitivity to the boundary conditions is not physically sensible.

All the arguments given above hold no matter what value the net spacing has, so they will be also true when h goes to zero. Consequently *no solution of the Laplace equation can have a maximum or minimum; Dirichlet or Neumann conditions over a closed boundary* are the correct conditions to use; Cauchy conditions over even a portion of a closed boundary are too much and over an open boundary result in too great a sensitivity to small changes in conditions to be physically satisfactory. These conclusions about boundary conditions also apply to the Poisson equation (and, indeed, to any elliptic equation), though the solution of Poisson's equation may have a maximum or minimum.

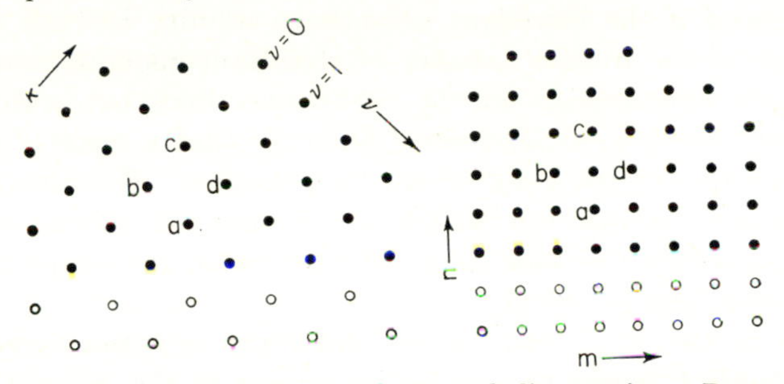

Fig. 6.6 Alternate lattices for hyperbolic equation. Rows $\kappa = \frac{1}{2}(m + n) = \text{constant}$ and $\nu = \frac{1}{2}(m - n) = \text{constant}$ are the characteristics.

The Hyperbolic Difference Equation. For the hyperbolic equation we can trace our net in the x, y directions as given in Eq. (6.2.8) or we can draw the lattice along the characteristics and make the difference equation analogous to the normal form of Eq. (6.1.8), which is $\partial^2\psi/\partial\lambda\,\partial\mu = 0$, for the wave equation

$$\psi(\kappa + 1,\nu + 1) + \psi(\kappa,\nu) = \psi(\kappa + 1,\nu) + \psi(\kappa,\nu + 1) \qquad (6.2.18)$$

where the characteristics are the lattice lines $\kappa = \text{constant}$ or $\nu = \text{constant}$ (see Fig. 6.6). Here potentials at the corners of a diamond-shaped figure are related; Eq. (6.2.18) states that the average value of the potentials at the ends of the horizontal diagonal equals the average of the values at the ends of the vertical diagonal. Cauchy conditions would correspond to the specifying of values at the circled points, a double row along the bottom. With these values specified, Eq. (6.2.18) can be used to compute the potentials at the first row of black points, then the next, and so on. If the function begins to increase at some point then, by successively applying Eq. (6.2.18) one finds that the increase does not go on indefinitely but spreads diagonally outward, thus "diluting" the increase. Consequently the function keeps itself within bounds and cannot increase without limit as does the solution of the Laplace equation.

Therefore slight changes in the boundary conditions produce only slight changes in the function, which become imperceptible as we move away from the boundary; in other words, the solution is *stable* for Cauchy conditions on an open boundary.

The difficulties encountered when the boundary coincides with a characteristic can also be exemplified. Suppose that values are specified along the lines $\nu = 0$ and $\nu = 1$. Applying Eq. (6.2.18) we see that the equations for $\nu = 0$ just check the internal consistency of the boundary values. The next set, of the form

$$\psi(\kappa + 1,2) - \psi(\kappa,2) = \psi(\kappa + 1,1) - \psi(\kappa,1)$$

cannot give us a unique solution because each equation contains *two* unknowns, and if the boundary extends to infinity in both directions, the solution of the infinite number of simultaneous equations does not converge (the determinant of the coefficients does not converge). If the boundary bends around at some point, so that a value of one $\psi(\kappa,2)$ is given, the rest of the $\psi(\kappa,2)$'s can be obtained. But this possibility has come about because the boundary does not everywhere correspond to the characteristic; as long as it coincides with the characteristic, no unique solution can be found.

Return to the (x,y) form of the difference equation, given in Eq. (6.2.8), according to the second lattice shown in Fig. 6.6. Here again the net points related are those in a diamond-shaped figure, the average at the ends of the horizontal diagonal equaling the average at the ends of the vertical diagonal. This persistance of form of the requirements should not be surprising, for the wave equation should represent the same interrelationship no matter what coordinates we express it in. In the horizontal network this means that the equations omit mentioning the net value at the center of the diamond (this was true for the κ, ν representation also, but it was not so noticeable, since the net was so drawn that center points were omitted). The omission of the center point is one aspect of the fundamental difference between the Laplace and the wave equation. It is the basic reason why Dirichlet conditions on a closed boundary do not result in a unique answer for the wave equation but do result in a unique answer for the Laplace equation.

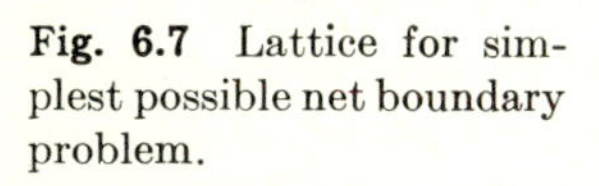

Fig. 6.7 Lattice for simplest possible net boundary problem.

To make this still clearer we can simplify the net-point Dirichlet problem for closed boundaries to the very simplest case shown in Fig. 6.7, for four boundary points and one lattice point. If the equation is the Laplace equation, the value of ψ at the center point is equal to the average of the values at the four boundary points, and if these are all given, a unique solution results. If, however, the equation is the wave equation, this equation relates the values of the boundary points and *says nothing whatever* about the value at the central point. Therefore if the values at the four boundary points are given, they may or may not be consistent with the requirements of the wave equation, but in either case no unique answer for the lattice point is obtained.

Returning to the lattice for m, n at the right in Fig. 6.6, we can fit boundary conditions by separating Eq. (6.2.8) into a difference equation in m and one in n,

$$\begin{aligned} \psi(m,n) &= \phi(m)f(n) \\ \phi(m+1) + \phi(m-1) + (k^2-2)\phi(m) &= \Delta^2\phi + k^2\phi = 0 \qquad (6.2.19) \\ f(n+1) + f(n-1) + (k^2-2)f(n) &= \Delta^2 f + k^2 f = 0 \end{aligned}$$

solve the one for m for particularly simple types of functions [in terms of which we can fit boundary conditions easily, as that given in Eq. (6.2.13) for instance], and then solve for the corresponding f factors; the final solution will be the sum of such products which fit the boundary conditions at $n = 0$ and 1. Or else we can solve the problem for the boundary conditions of unit value or slope at $m = M$ and for zero value and slope elsewhere on the boundary, thus obtaining Green's functions for the boundary. The final solution will again be a sum of these Green's functions times the boundary values along the boundary points.

The Parabolic Difference Equation. Taking up finally Eq. (6.2.9) for the simplest parabolic equation, we see that, if Dirichlet conditions are set along the horizontal boundary $n = 0$, the first equation allows a unique solution for the lattice points $n = 1$, then $n = 2$, and so on. Whether the solution will be stable or not will depend on the value of C. If $\psi(m,n)$ for $m = m_0$ is larger than either $\psi(m_0 + 1, n)$ or $\psi(m_0 - 1, n)$, then $\psi(m, n + 1)$ will have a smaller value than $\psi(m,n)$, since

$$\psi(m, n+1) = \psi(m,n) + 2C\{\tfrac{1}{2}[\psi(m-1, n) + \psi(m+1, n)] - \psi(m,n)\}$$

If C is not so large as to make $\psi(m, n + 1)$ go negative and start unstable oscillations, this will tend to even out irregularities in the solution and eventually, for large enough values of n, tend to bring all the ψ's to the same value (see Prob. 6.3).

If, on the other hand, we use the second form of the equation, to work backward in the n (time) direction, we shall find that irregularities in value of ψ tend to magnify any slight irregularity in the boundary values, resulting eventually in a very large irregularity in the function for large enough negative n's. The solution is therefore unstable for any boundary conditions (it is unstable for Dirichlet and Neumann conditions; Cauchy conditions overspecify the solution). By going to the limit, we can say that solutions of parabolic equations give stable and unique results for Dirichlet conditions on an open boundary if we are working in the positive direction along the characteristic but are unstable in the negative direction.

Physically this comes about because parabolic equations (the diffusion equation is an example) represent situations where the entropy is increasing as time increases. Therefore irregularities in the field ψ will tend to "iron out" as time increases; the sharper the irregularity, the more rapidly will it disappear. If we wish to work backward in time to find out what field, a minute (or an hour) before, eventually diffused into the specified distribution, we shall not be able to tell how many sharp irregularities there had been, at this earlier time, which had practically disappeared by the later time and thus did not show up to an appreciable extent in the specified distribution.

The results of the discussion of this section may be summed up in the following table:

Conditions	Boundary	Hyperbolic equation	Elliptic equation	Parabolic equation
Dirichlet or Neumann (Value or slope specified)	Open	Insufficient	Insufficient	**Unique, stable solution in positive direction, unstable in negative direction**
	Closed	Solution not unique	**Unique, stable solution** (see page 698 for Neumann conditions)	Solution overspecified
Cauchy (Value and slope specified)	Open	**Unique, stable solution**	Solution unstable	Solution overspecified
	Closed	Solution overspecified	Solution overspecified	Solution overspecified

The satisfactory combinations of equations and boundary conditions are those indicated by boldface type. We note again that Dirichlet-Neumann conditions may be *homogeneous* $[\alpha\psi(s) + \beta N(s) = 0]$ or *inhomogeneous* $[\alpha\psi(s) + \beta N(s) = F(s)]$. Homogeneous Dirichlet conditions mean that ψ will be *zero* on the boundary; inhomogeneous Dirichlet conditions mean that ψ will have specified nonzero values on the boundary, and so on.

6.3 *Eigenfunctions and Their Uses*

We have now reached a point where we must cease dealing with generalities and begin to get down to cases. We have spent the first two sections of this chapter showing, in general, what kinds of boundary conditions are suitable to what partial differential equations and under what conditions we might expect a unique solution to correspond to our specifications. We now must study in detail the techniques for generating these unique solutions to fit specific cases.

The process of fitting of boundary conditions is somewhat analogous to the process of solving an ordinary differential equation. In neither case is the procedure straightforward; we must assume some general form, which seems likely to satisfy our requirements, and then adjust things to satisfy them in detail (if, indeed, this turns out to be possible!) Even for the comparatively simple process of integrating a function $f(x)$ with respect to x, for instance, we must, basically, *guess* the form of the integral and then show its correctness by differentiation. Many of our guesses for solutions of differential equations are very general in form, such as power series or integral representations, but we find the specific solution by inserting the assumed form into the differential equation and then trying to make it fit.

The general forms for the fitting of boundary conditions are also either in terms of a *series* of functions or else in terms of an integral of some function over the boundary. As with the solutions of ordinary differential equations we shall start with a discussion of the use of series; the next chapter will be devoted to the fitting of boundary conditions by integration. In practice the application of the series method usually requires the separation of the partial differential equation in coordinates suitable for the boundaries, and before we plunge into technical details, we shall work through a simple example to see how it goes.

Fourier Series. Suppose we wish to solve the two-dimensional Laplace equation

$$\nabla^2\psi = \frac{\partial^2\psi}{\partial x^2} + \frac{\partial^2\psi}{\partial y^2} = 0$$

subject to boundary conditions on a rectangular boundary enclosing the area within the lines $x = 0$, $x = a$, $y = 0$, $y = b$. It is obvious that the rectangular coordinates x, y are suitable for this boundary, so we can separate the equation into the ordinary equations

$$\psi = X(x)Y(y); \quad (d^2X/dx^2) + k^2X = 0; \quad (d^2Y/dy^2) - k^2Y = 0$$

where k^2 is the separation constant. This is an elliptic equation, so Dirichlet or Neumann conditions on a closed boundary are suitable. See the discussion for the lattice case on page 697.

Suppose, to begin with, that our boundary conditions are the particularly simple Dirichlet conditions that ψ is to be zero (homogeneous conditions) along $x = 0$, $x = a$, and $y = b$ and to have an arbitrary sequence of values $\psi_s(x)$ (inhomogeneous conditions) along $y = 0$.

The solution of the x equation which is zero at $x = 0$ and at $x = a$ is $\sin(\pi nx/a)$, where n is integer (we have set $k = \pi n/a$), and the solution of the y equation which is zero at $y = b$ is $\sinh [k(b - y)]$. In other words the only values of k for which the solution can satisfy zero boundary conditions on three sides are the values $k = \pi n/a$ ($n = 1, 2, 3, \ldots$), and the only solutions, for these values, which fit these conditions are the ones mentioned in the previous sentence.

Consequently the most general solution of $\nabla^2\psi = 0$, in two dimensions, which satisfies the homogeneous conditions that ψ be zero along $x = 0$, $x = a$, $y = b$ can be represented by the series

$$\psi = \sum_{n=1}^{\infty} A_n \sinh\left[\left(\frac{\pi n}{a}\right)(b - y)\right] \sin\left(\frac{\pi nx}{a}\right) \tag{6.3.1}$$

To make as categorical a statement as this requires that we be sure that this series form can represent *all possible* solutions satisfying the zero conditions along the three sides, no matter what (inhomogeneous) boundary condition it satisfies along the fourth side, $y = 0$. To do this we shall have to prove that the series of Eq. (6.3.1) can be made to fit *all possible* boundary conditions along $y = 0$. Since Dirichlet conditions along a closed boundary specify a unique solution, we can then be sure that such a series represents the solution corresponding to the specified boundary conditions along $y = 0$. Then if we find another form of function (a closed form or integral form) which also satisfies the same conditions, we can be sure that this new form corresponds to the series and, vice versa, that the series can represent the new form.

Consequently the crucial step in justifying our statement above is to prove that series (6.3.1) can satisfy all possible Dirichlet conditions along $y = 0$. Of course the phrase "all possible" is pretty optimistic in its generality, and we may have to "whittle it down" a bit by means of more exact definitions.

However, for the simple case we are now considering, series (6.3.1) for $y = 0$ is a Fourier series. We shall see later in this chapter that such a series can be fitted to any function which is continuous over the range $0 < x < a$ with the exception of a finite number of discontinuities in value or slope (and which goes to zero at $x = 0$ and $x = a$, of course). Such functions are said to be *piecewise continuous*. The Fourier series will fit all the continuous parts of such an arbitrary function, though if we wish to be captious and specify arbitrary peaks (in both value and slope) of zero breadth at the discontinuities, the series may not be able to fit all these peaks as well. But such arbitrary peaks of zero breadth are mathematical niceties of little physical significance; consequently we can define "all possible" as being "all piecewise continuous" and satisfy all but the most exacting.

Being still more specific, we can say that the statement that the series $\phi(x) = \Sigma B_n \sin(\pi nx/a)$ *can fit* the piecewise continuous function $f(x)$ over the range $0 < x < a$ means that we can choose values of the B_n's such that

$$\int_0^a [\phi(x) - f(x)]^2 \, dx = 0$$

This requirement would, of course, overlook discrepancies in noninfinite peaks of zero width at the discontinuities, since the area under such peaks is zero. From a physical point of view this is satisfactory. We might say that our requirement of fit for the series is a *least-squares* requirement. We shall have to justify our statement concerning the least-squares fit for Fourier series, as well as for other functions, later in this section.

When we are satisfied that the series *can* fit the required boundary conditions, it is not difficult to compute the required values of the coefficients A_n. We set up the equation corresponding to the boundary condition along $y = 0$,

$$\psi_s(x) = \sum_{n=1}^{\infty} A_n \sinh\left(\frac{\pi nb}{a}\right) \sin\left(\frac{\pi nx}{a}\right)$$

multiply both sides by $\sin(\pi mx/a)$, and integrate over x from 0 to a. All but one of the integrals on the right turn out to be zero because of the properties of the sine function when integrated over an integral number of periods. The one left, for $n = m$, is just $\frac{1}{2}aA_m \sinh(\pi mb/a)$. Consequently the coefficients A_m in Eq. (6.3.1) can be easily determined and the solution of $\nabla^2\psi = 0$ which satisfies the boundary conditions $\psi = 0$ at $x = 0$, $x = a$, $y = b$, $\psi = \psi_s(x)$ at $y = 0$, is the series

$$\psi(x,y) = \sum_{n=1}^{\infty} \left[\frac{2}{a}\int_0^a \psi_s(\xi) \sin\left(\frac{\pi n\xi}{a}\right) d\xi\right] \sin\left(\frac{\pi nx}{a}\right) \frac{\sinh[(\pi n/a)(b - y)]}{\sinh(\pi nb/a)} \tag{6.3.2}$$

This series satisfies the boundary conditions and is a solution of the Laplace equation in x and y. Therefore it must be the unique solution we are seeking.

In the limiting case of $b \to \infty$ the region inside the boundary becomes infinite, but according to page 690, the Dirichlet conditions must still be imposed over the whole boundary in order to obtain a unique result. The hyperbolic sine ratio reduces to the simple exponential $\exp[-(\pi ny/a)]$, but otherwise the series remains the same. If now we expand the boundaries in the x direction, we eventually change from a Fourier series to a Fourier integral. Following the discussion in Sec. 4.8, we find the solution of the two-dimensional Laplace equation valid in the upper half plane and satisfying the boundary conditions $\psi = \psi_s(x)$ at $y = 0$ and $\psi = 0$ at infinity to be

$$\psi(x,y) = \frac{1}{2\pi}\left\{\int_{-\infty}^{0} e^{ikx+ky}\,dk \int_{-\infty}^{\infty} \psi_s(\xi)e^{-ik\xi}\,d\xi + \int_{0}^{\infty} e^{ikx-ky}\,dk \int_{-\infty}^{\infty} \psi_s(\xi)e^{-ik\xi}\,d\xi\right\}$$

$$= \frac{1}{\pi}\int_0^\infty e^{-ky}\,dk \int_{-\infty}^{\infty} \psi_s(\xi)\cos[k(x-\xi)]\,d\xi \qquad (6.3.3)$$

where the integration over k is split into two parts so that the dependence on y always vanishes at $y \to \infty$ (that is, we use e^{+ky} for negative k and e^{-ky} for positive k). When $y = 0$, we obtain the usual Fourier integral (4.8.2).

The Green's Function. We notice that, both for finite boundaries [using series (6.3.2)] and for infinite boundaries [using integral (6.3.3)], we can rearrange the solution to be an integral over the boundary $y = 0$:

$$\psi(x,y) = \int_S \psi_s(\xi)G(x,y|\xi)\,d\xi \qquad (6.3.4)$$

where for a finite boundary the integration is between 0 and a, the ends of the boundary, and the function G is the series

$$G(x,y|\xi) = \frac{2}{a}\sum_{n=1}^{\infty} \frac{\sinh[(\pi n/a)(b-y)]}{\sinh(\pi nb/a)} \sin\left(\frac{\pi nx}{a}\right) \sin\left(\frac{\pi n\xi}{a}\right)$$

whereas for an infinite boundary the integration over the boundary S is from $-\infty$ to $+\infty$ and the function G is

$$G(x,y|\xi) = \frac{1}{\pi}\int_0^\infty e^{-ky}\cos[k(x-\xi)]\,dk = \frac{(y/\pi)}{(x-\xi)^2 + y^2}$$

The function G is called the *Green's function* for the boundary surface S ($y = 0$ in this case).

We thus see that our solution of the Dirichlet problem can be expressed in terms of an integral of a Green's function, suitable to the equation and boundary shape, multiplied by the specified boundary-value function, integrated over the boundary. But the ramifications of this idea are to be worked out in the next chapter; we are to concentrate here on the process of determining the functions forming the series and how the series are formed.

We have so far considered only the portion $x = 0$ of the boundary to have boundary values ψ_s different from zero. To fit the conditions where ψ is different from zero along other parts of the rectangular boundary we use obvious modifications of the functions used in series (6.3.2) or integral (6.3.3). For instance, for fitting conditions along $y = 0$ we use the series

$$\sum_{n=1}^{\infty} B_n \sinh\left[\left(\frac{\pi n}{b}\right)(a - x)\right] \sin\left(\frac{\pi n y}{b}\right)$$

and so on, for $x = a$ and $y = b$, adding the individual series to obtain the final solution. Therefore we can set up a Green's function to use in the integrand for any boundary values at any point along the rectangular boundary. See page 700 for the analogous case for a lattice.

Eigenfunctions. The functions $\sin(\pi n x/a)$ for integral values of n are the simplest examples of a sequence of *eigenfunctions*, which form the basis of the series method of fitting boundary conditions. We see that this method involves separating the partial differential equation in terms of coordinates ξ_n such that the boundary corresponds to one or more coordinate surfaces, $\xi_s =$ constant. The factor in the solution giving its dependence along the boundary surface [the factor $\sin(\pi n x/a)$ in the example] depends on a separation constant as well as on position x along the boundary (the boundary $y = 0$ in the example). This factor must also satisfy some simple boundary conditions (at the two ends $x = 0$ and $x = a$ in the example), and we find that only for certain values of the separation constant (n an integer in the example) will this factor fit the boundary conditions. The other factor (sinh $[(\pi n/a)(b - x)]$ in the example) is then adjusted to fit the conditions at the other end of the enclosed region ($y = b$ in the example), and the complete solution is then a sum of these products for all allowed values of the separation constant.

The crucial part of the method, we can see, is the determination of the form of the factor giving dependence along the boundary [the factor $\sin(\pi n x/a)$ in the example] and the corresponding allowed values of the separation constant, which fit the conditions at the ends 0, a of the boundary. Solutions of an ordinary differential equation, containing a separation constant, which satisfy simple boundary conditions at two ends of a range of the independent variable are called *eigenfunctions*, and the values of separation constant which allow it to fit the conditions are called *eigenvalues* (sometimes the terms used are *characteristic functions* and *characteristic values*). In the example above the functions $\sin(\pi n x/a)$ are the eigenfunctions and the values $(\pi n/a)^2$ (n an integer) are the eigenvalues. Presumably a series of these eigenfunctions, for all eigenvalues of the separation constant, should be able to represent any arbitrarily chosen sequence of boundary values. We must show in general that such a series *can* represent any chosen function over the important range of the independent value, and we must also show how we compute the coefficients in the series. (It should be noted that we can have eigenfunctions for more than a single dimension and that we can have eigenfunction solutions for integral, as well as differential, equations.)

In a general two-dimensional example, for instance, we choose the coordinates ξ_1, ξ_2 such that the boundary corresponds to $\xi_1 = a_1$, $\xi_1 = a_2$, $\xi_2 = b_1$ $\xi_2 = b_2$. We next set up the solution corresponding to $\psi = 0$ along three parts of the boundary ($\xi_1 = a_1$, $\xi_1 = a_2$; $\xi_2 = b_2$, for instance) and ψ equal to some arbitrary sequence of boundary values $f(\xi_1)$ along the fourth part $\xi_2 = b_1$. This can be done if we can separate the equation for the coordinates ξ. Solutions of the ξ_1 equation which go to zero at $\xi_1 = a_1$ and $\xi_1 = a_2$ are then the *eigenfunctions* for the problem.

Types of Boundary Conditions. We have so far discussed the case of Dirichlet conditions on a closed boundary, but the same technique can be used for other types of boundary conditions. For instance, for Neumann conditions on a boundary enclosing the rectangle of dimensions a and b, we first fit the conditions along the side $y = 0$, making the conditions along the other three sides the homogeneous ones that the *normal gradient* be zero. The proper eigenfunctions are then $\cos(\pi nx/a)$, with n any integer, and the series which has normal gradient equal to $N_s(x)$ along the x axis from 0 to a and zero normal gradient along $y = b$, $x = 0$, and $x = a$ is

$$\psi(x,y) = \sum_{n=1}^{\infty} \left[\frac{2}{\pi n} \int_0^a N_s(\xi) \cos\left(\frac{\pi n\xi}{a}\right) d\xi\right] \cos\left(\frac{\pi nx}{a}\right) \frac{\cosh[(\pi n/a)(b - y)]}{\sinh(\pi nb/a)}$$

In this case, as we noted in page 698, we have had to require that

$$\int^a N_s(\xi)\, d\xi = 0.$$

If we do not have closed boundaries, we might need to use Cauchy conditions over part of the boundary if the equation were a hyperbolic one. For instance, for a flexible string of length a, clamped at $x = 0$ and $x = a$, the boundary conditions are homogeneous Dirichlet conditions ($\psi = 0$) at $x = 0$ and $x = a$ for all values of time t, but the initial conditions (boundary conditions along $t = 0$) must specify both value, $\psi = \psi_0(x)$, and initial velocity, $\partial\psi/\partial t = U_0(x)$ (see page 586). The proper series would be

$$\psi = \sum_{n=1}^{\infty} \sin\left(\frac{\pi nx}{a}\right)\left[A_n \cos\left(\frac{\pi nct}{a}\right) + B_n \sin\left(\frac{\pi nct}{a}\right)\right]$$

which is a solution of the wave equation $c^2(\partial^2\psi/\partial x^2) = \partial^2\psi/\partial t^2$. The functions $\sin(\pi nx/a)$ are the eigenfunctions, and the values $\pi n/a$ are the eigenvalues. The values of the coefficients A_n, B_n are determined from ψ_0, U_0 in a manner similar to that used on page 709.

Sometimes the boundary conditions determining the eigenfunctions are not the specifying of zero value or slope at the two ends of the range. The requirement may be the general homogeneous one that the ratio of value to slope be constant, independent of the value of the separation constant; for instance,

$$\alpha(\partial\psi/\partial x) = \beta\psi \text{ at } x = 0 \text{ and } x = a$$

When $\alpha = 0$, this reduces to homogeneous Dirichlet conditions; when $\beta = 0$, it reduces to homogeneous Neumann conditions. We note that boundary conditions for eigenfunctions are always *homogeneous* (see page 679) when the boundary point is an ordinary point.

Another type of boundary condition is exemplified by the case of the solution of the Laplace equation inside a circle of radius a, with Dirichlet conditions applied along the circle. The appropriate coordinates are, of course, the polar coordinates r, ϕ, and the Laplace equation, and the separated equations are

$$\frac{1}{r}\frac{\partial}{\partial r}\left(r\frac{\partial\psi}{\partial r}\right) + \frac{1}{r^2}\frac{\partial^2\psi}{\partial\phi^2} = 0; \quad \psi = R(r)\Phi(\phi);$$

$$\frac{d^2\Phi}{d\phi^2} + m^2\Phi = 0; \quad \frac{1}{r}\frac{d}{dr}\left(r\frac{dR}{dr}\right) - \frac{m^2}{r^2}R = 0$$

The solution of the equation in ϕ is $[a\cos(m\phi) + b\sin(m\phi)]$, with arbitrary values for a, b, and m. But if we are to have a physically sensible solution inside the circle $r = a$, we must have the solution continuous and finite in this region. The angle ϕ runs continuously from 0 to 2π, coming back to 0 as the radius vector is swung in a complete revolution. Therefore the function $\Phi(2\pi)$ must equal the function $\Phi(0)$, and in general, $\Phi(\phi) = \Phi(\phi + 2\pi)$. In order that this be possible, the separation constant m *must be an integer*, and the eigenfunctions for ϕ are therefore both $\sin(m\phi)$ and $\cos(m\phi)$, with m an integer. In this case there was no specifying of values of the function at the two ends of the range, $\phi = 0$ and $\phi = 2\pi$. The condition was simply one of continuity, which corresponded to the requirement that Φ be *periodic* in ϕ with period 2π. We shall often encounter the requirement of periodicity as a boundary condition in later chapters.

Formally, the interesting range for r is bounded at both ends, $r = 0$ and $r = a$, although the physical boundary is only at $r = a$. We might begin to worry as to what boundary condition to apply at $r = 0$, where there is no physical boundary but only a concentration point for the coordinate system, if we did not remember that a concentration point corresponds to a singular point for the differential equation, and the simple requirement that ψ be finite or continuous or analytic there is sufficient to limit our choice of function. The general solution of the equation for R is

$$R = \begin{cases} a_0 + b_0 \ln r; & m = 0 \\ a_m r^m + (b_m/r^m); & m > 0 \end{cases}$$

and the second term must be omitted in order that R be finite and continuous inside the boundary. (We note in passing that, if the circle $r = a$ were the inner boundary and if ψ is to be finite at $r = \infty$, then the first term must be omitted for $m > 0$ and the $\ln r$ term must be omitted for $m = 0$.)

The solution of the Laplace equation inside the circle $r = a$, with $\psi = \psi_s(\phi)$ along the circular boundary, is then given by the series

$$\psi = \sum_{n=0}^{\infty} [A_n \cos(n\phi) + B_n \sin(n\phi)] \left(\frac{r}{a}\right)^n$$

where
$$A_0 = \frac{1}{2\pi} \int_0^{2\pi} \psi_s(\alpha)\, d\alpha; \quad B_0 = 0$$

$$A_n = \frac{1}{\pi} \int_0^{2\pi} \psi_s(\alpha) \cos(n\alpha)\, d\alpha; \quad B_n = \frac{1}{\pi} \int_0^{2\pi} \psi_s(\alpha) \sin(n\alpha)\, d\alpha$$

In still other cases the boundary extends from one concentration point to another, so that the two boundary points are adjacent singular points of the differential equation for the eigenfunction, and the boundary conditions may simply be that the solution remain finite at both singular points. The Laplace equation in spherical coordinates r, ϑ, ϕ,

$$\frac{1}{r^2} \frac{\partial}{\partial r}\left(r^2 \frac{\partial \psi}{\partial r}\right) + \frac{1}{r^2 \sin \vartheta} \frac{\partial}{\partial \vartheta}\left(\sin \vartheta \frac{\partial \psi}{\partial \vartheta}\right) + \frac{1}{r^2 \sin^2 \vartheta} \frac{\partial^2 \psi}{\partial \phi^2} = 0$$

separates into
$$\psi = R(r)\Theta(\vartheta)\Phi(\phi)$$

$$\frac{1}{r^2} \frac{d}{dr}\left(r^2 \frac{dR}{dr}\right) - \frac{n(n+1)}{r^2} R = 0$$

$$\frac{d}{dz}\left[(1 - z^2) \frac{d\Theta}{dz}\right] + \left[n(n+1) - \frac{m^2}{1 - z^2}\right]\Theta = 0; \quad z = \cos \vartheta$$

$$\frac{d^2\Phi}{d\phi^2} + m^2\Phi = 0$$

If the boundary is the sphere $r = a$ surrounding the interior space, then the last two equations will serve to determine eigenfunctions and the boundary conditions of continuity and finiteness serve to fix the allowed values of n and m. Since the axial angle ϕ is continuous from 0 to 2π, determination that m should be an integer and that the eigenfunctions are $\sin m\phi$, $\cos m\phi$, follows the same line as that given above for the circular boundary. Solutions of the equation for ϑ are the Legendre functions (see page 548). We saw in Chap. 5 that only when

n is an integer can there be a solution which stays finite at both singular points $z = 1$ and $z = -1$. The restrictions on the R factor that it be finite at the singular point $r = 0$ fix its form also. Therefore the solution must be expressible in terms of the series

$$\psi = \sum_{m=0}^{\infty} \sum_{n=m}^{\infty} [A_{mn} \cos(m\phi) + B_{mn} \sin(m\phi)] \sin^m\vartheta \, T^m_{n-m}(\cos \vartheta) \left(\frac{r}{a}\right)^n$$

where the T's are the Gegenbauer polynomials defined in Eqs. (5.2.53) and (5.3.36). The functions $\sin^m\vartheta \, T^m_{n-m}(\cos \vartheta)$ are sometimes called the *associated Legendre functions*.

The boundary conditions may be Dirichlet ones on the sphere $r = a$. These may be expressed by requiring that $\psi = \psi_S(\vartheta,\phi)$, where ψ_S is a piecewise continuous function of ϑ and ϕ. We shall show later that the integral

$$\int_0^\pi \sin^{2m+1}\vartheta \, T^m_{n-m}(\cos \vartheta) \, T^m_{k-m}(\cos \vartheta) \, d\vartheta$$

is zero if k differs from n and shall later compute the value of the integral when $k = n$. At present all we need to know is that, if we multiply both sides of the series for ψ by

$$\cos(l\phi) \sin^{l+1}\vartheta \, T^l_{k-l}(\cos \vartheta) \, d\phi \, d\vartheta$$

and integrate over the surface of the sphere $r = a$, all terms in the series vanish except the one for which $m = l$ and $n = k$, and this one can be written as $A_{lk}\Lambda_{lk}$, with Λ_{lk} a constant of known value. Therefore the series coefficients can be shown to be

$$A_{mn} = \frac{1}{\Lambda_{mn}} \int_0^{2\pi} \cos(m\phi) \, d\phi \int_0^\pi \psi_S(\vartheta,\phi) \sin^{m+1}\vartheta \, T^m_{n-m}(\cos \vartheta) \, d\vartheta$$
$$B_{mn} = \frac{1}{\Lambda_{mn}} \int_0^{2\pi} \sin(m\phi) \, d\phi \int_0^\pi \psi_S(\vartheta,\phi) \sin^{m+1}\vartheta \, T^m_{n-m}(\cos \vartheta) \, d\vartheta$$

which finally determines the coefficients of the series in terms of the boundary values $\psi_S(\vartheta,\phi)$.

In these examples we have indicated the general properties of some eigenfunctions and some of their uses. They may be solutions of the ordinary differential equations which arise from the separation of one of the partial differential equations derived earlier, corresponding to particular values of the separation constants (eigenvalues). The eigenvalues are then fixed by the requirement that the eigenfunctions satisfy some sort of boundary conditions at the two ends of a given range of values of the independent variable. If these end points are singular points of the differential equation, the boundary conditions may be simply that the solution remain finite there. If the end points are ordinary points, the conditions may be the homogeneous ones that the ratio between value and slope of the function equal a certain constant, independent of the separation constant, or that the solution be periodic, with period equal to some constant value, and so on.

We must find how to determine eigenvalues and eigenfunctions from boundary conditions, but more important still, we must show that the sequence of eigenfunctions so determined can be formed into a series which can fit any piecewise continuous function in the range between the boundary points. The rest of this section will be devoted to a discussion of these general points, with frequent pauses to illustrate concepts by means of specific examples. The properties of eigenfunction solutions of integral equations will be taken up in Chap. 8.

Abstract Vector Space. The eigenfunctions we have so far used as examples, the terms $\sin(\pi nx/a)$ of a Fourier series, for instance, have certain properties distantly analogous to the components of vectors, properties which we have already discussed in Secs. 1.6 and 2.6. The components of a vector $\mathbf{F}_k$ in n-dimensional space can be written F_{km} (that is, if $\mathbf{a}_m$ is a unit vector in the direction of the x_m axis, where $\mathbf{a}_1, \ldots, \mathbf{a}_n$ are mutually orthogonal, then $F_{km} = \mathbf{F}_k \cdot \mathbf{a}_m$). Then the scalar product of $\mathbf{F}_k$ with another vector $\mathbf{F}_l$ is

$$(\mathbf{F}_k \cdot \mathbf{F}_l) = \sum_{m=1}^{n} F_{km} F_{lm}$$

which is zero if $\mathbf{F}_k$ and $\mathbf{F}_l$ are mutually orthogonal (perpendicular to each other). In particular the magnitude of the vector $\mathbf{F}_k$ is equal to the square root of the scalar product of the vector with itself:

$$F_k = \left[\sum_{m=1}^{n} F_{km}^2 \right]^{\frac{1}{2}} = \sqrt{\mathbf{F}_k \cdot \mathbf{F}_k}$$

The function $\sin(\pi mx/a)$ is a function of the integer m and of the continuous variable x. It can be considered as a representation of a vector $\mathbf{S}_m$ in an *abstract vector space* of an infinite number of dimensions. The scalar product of two such vectors, $\mathbf{S}_m$ and $\mathbf{S}_n$, can be defined by the equation

$$(\mathbf{S}_m \cdot \mathbf{S}_n) = \int_0^a \sin\left(\frac{\pi mx}{a}\right) \sin\left(\frac{\pi nx}{a}\right) dx \tag{6.3.5}$$

which is a typical limiting form of the usual sum of component products. It is as though for each value of x in the range $(0 < x < a)$ there corresponded a different direction and a different unit vector $\mathbf{e}(x)$ in function space, each $\mathbf{e}$ for each x being perpendicular to every other $\mathbf{e}$ for every other x, and the quantity $\sin(\pi nx/a)$, for a given x, being somehow the component of $\mathbf{S}_n$ in the direction given by $\mathbf{e}(x)$.

The magnitude of $\mathbf{S}_n$ is then the square root of the scalar product of $\mathbf{S}_n$ with itself:

$$S_n = \left[\int_0^a \sin^2 \left(\frac{\pi n x}{a} \right) dx \right]^{\frac{1}{2}} = \sqrt{\frac{a}{2}} \tag{6.3.6}$$

The eigenfunctions $\sin(\pi n x/a)$ thus define an *eigenvector* $\mathbf{S}_n$. Because of the integral properties of the eigenfunctions, each eigenvector is orthogonal to the other, for the scalar product $(\mathbf{S}_m \cdot \mathbf{S}_n)$ defined in Eq. (6.3.5) is zero unless $m = n$. If we divide each $\mathbf{S}_n$ by $\sqrt{a/2}$, we obtain a set of mutually orthogonal unit vectors

$$\mathbf{e}_n = \sqrt{2/a}\, \mathbf{S}_n$$

which define a set of *normal coordinates* in function space, which are just as useful as the original coordinates and unit vectors $\mathbf{e}(x)$.

In particular, the unit vector $\mathbf{e}(x)$ can be represented by the new unit vectors

$$\mathbf{e}(x) = \sum_n \mathbf{e}_n \sqrt{\frac{2}{a}} \sin \left(\frac{\pi n x}{a} \right) = \frac{2}{a} \sum_n \mathbf{S}_n \sin \left(\frac{\pi n x}{a} \right) \tag{6.3.7}$$

and the *normalized* eigenfunctions $\sqrt{2/a}\, \sin(\pi n x/a)$ are analogous to the direction cosines defining one set of axes in terms of another [see Eq. (1.3.1)].

It may seem rather difficult to "swallow" that a space characterized by the nondenumerable infinity of unit vectors $\mathbf{e}(x)$ should be completely represented by the denumerable infinity of eigenvectors $\mathbf{e}_n$. Actually this is not true, of course, for an arbitrarily chosen vector in $\mathbf{e}(x)$ space would be represented by a "function" of x with an infinite number of discontinuities. Somehow, we have picked out of the nondenumerably infinite-dimensional space a subspace representing all functions which are continuous, except for a finite number of discontinuities (called *piecewise continuous* functions). These drastically specialized choices out of all possible, infinitely discontinuous functions may then be represented by the denumerably infinite set of eigenvectors $\mathbf{e}_n$. This limitation to our transformation should be kept in mind as we go ahead with our investigation.

Within this limitation, however, we should be able to express any arbitrary vector $\mathbf{F}$ in abstract vector space in terms of its components along the normal axes:

$$\mathbf{F} = \sum_n F_n \mathbf{e}_n$$

This same vector can also be represented in terms of the original unit vectors $\mathbf{e}(x)$, with components $F(x)$, which are related to F_n by the equations

$$F_n = [\mathbf{F} \cdot \mathbf{e}_n] = \sqrt{\frac{2}{a}} \int_0^a F(x) \sin\left(\frac{\pi n x}{a}\right) dx;$$
$$F(x) = [\mathbf{F} \cdot \mathbf{e}(x)] = \sqrt{\frac{2}{a}} \sum_{n=1}^{\infty} F_n \sin\left(\frac{\pi n x}{a}\right) \tag{6.3.8}$$

These are just the equations defining the Fourier series expansion of a function $F(x)$.

The directions given by the eigenvectors $\mathbf{S}_n$ (or $\mathbf{e}_n$) are somehow determined by the differential equation

$$d^2\psi/dx^2 = -k^2\psi$$

for the sine functions and by the homogeneous boundary conditions that $\psi = 0$ at $x = 0$ and $x = a$. Therefore, corresponding to this differential equation and these boundary conditions there is an *operator equation* (see Sec. 2.6)

$$\mathfrak{A} \cdot \mathbf{S} = -k^2\mathbf{S}$$

with corresponding boundary conditions, which serves to fix *normal axes* for the operator $\mathfrak{A}$ and eigenvalues for the constant k.

Another differential equation and/or other boundary conditions determine other normal axes, corresponding to new eigenvectors $\mathbf{E}_n$ defined by new eigenfunctions $\psi_n(x)$ (which may be complex). These vectors are also orthogonal, so that (see page 60)

$$(\mathbf{E}_n^* \cdot \mathbf{E}_m) = \int \bar{\psi}_n(x)\psi_m(x)\, dx = \begin{cases} 0; & n \neq m \\ E_n^2; & n = m \end{cases} \tag{6.3.9}$$

where E_n is the magnitude of $\mathbf{E}_n$. Again, any vector $\mathbf{F}$ may be expressed in terms of these eigenvectors, and the relation between the components of $\mathbf{F}$ along the new normal axes and its components $F(x)$ along the directions $\mathbf{e}(x)$ are

$$F(x) = [\mathbf{F}^* \cdot \mathbf{e}(x)] = \sum_n \frac{\bar{F}_n}{E_n} \psi_n(x); \quad \bar{F}_n = \left[\frac{\mathbf{E}_n^*}{E_n} \cdot \mathbf{F}\right] = \frac{1}{E_n} \int F(x)\bar{\psi}_n(x)\, dx \tag{6.3.10}$$

which gives equations by means of which we can expand any function $F(x)$ in terms of the eigenfunctions $\psi_n(x)$.

Finally, we can compute the scalar product of the eigenvector corresponding to $\psi_n(x)$ and the eigenvector corresponding to $\sin(\pi n x/a)$ (we can if the function space for the two is identical, *i.e.*, if the boundaries for both sets of eigenfunctions are 0 and a). This scalar product is

$$(\mathbf{E}_m^* \cdot \mathbf{e}_n) = \sqrt{\frac{2}{a}} \int_0^a \bar{\psi}_m(x) \sin\left(\frac{\pi n x}{a}\right) dx$$

One other property of these eigenvectors will be useful later. Since $\mathbf{e}(x)$ is orthogonal to $\mathbf{e}(x')$ as long as $x' \neq x$, we have, by using an equation analogous to (6.3.7),

$$[\mathbf{e}^*(x) \cdot \mathbf{e}(x')] = \sum_n \frac{1}{E_n^2} \bar{\psi}_n(x)\psi_n(x') = \delta(x - x') \qquad (6.3.11)$$

This quantity is zero when $x' \neq x$ and has such a value at $x = x'$ that

$$\int_{-\infty}^{\infty} \delta(x - x')\, dx = 1$$

In other words it is the Dirac delta function defined on page 22. Naturally there are questions of convergence connected with this series representation, questions we shall take up later.

But we still have not come to grips with details. The vector analogy is fruitful of ideas and helpful in broadening concepts, but it is very abstract. In order fully to understand all of its implications we must return to our differential equations and our boundary conditions.

Sturm-Liouville Problem. The ordinary differential equation arising from the separation of the general partial differential equation $\nabla^2\psi + k^2\psi = 0$ can [see Eq. (5.1.28)] be written in the form

$$\frac{d}{dz}\left[p(z)\frac{d\psi}{dz}\right] + [q(z) + \lambda r(z)]\psi = 0 \qquad (6.3.12)$$

This equation is called the *Liouville equation*. The parameter λ is the separation constant (in some cases more than one separation constant appears; we shall postpone consideration of these cases till later). Each of the functions p, q, r are characteristic of the coordinates used in the separation, and for the separable coordinates we studied in Sec. 5.1, p and r are simple algebraic functions of z, with a finite number of zeros and poles. The function q, particularly for the Schroedinger equation, is a more complicated function, but not even q will have a singularity *inside* the range of z (though it may have a singularity *at* one or both of the boundaries). The points where $p(z)$ is zero are singular points of the equation, and the useful range of z usually extends from one such point to the next; at any rate a singular point may be at the beginning or end of the range, but never in the middle. In other words $p(z)$ does not change sign anywhere in the whole range of z, and therefore p can *always be taken as positive*. It also turns out that r, too, does not change sign (in the cases of interest), so r can be considered as *always positive*.

The Sturm-Liouville problem is essentially the problem of determining the dependence of the general behavior of ψ on the parameter λ and the dependence of the eigenvalues of λ on the homogeneous boundary conditions imposed on ψ.

In part of our discussion we shall be comparing solutions for different values of the separation constant λ; for instance, the solution for $\lambda = \lambda_n$ could be called ψ_n. The procedure which is of greatest use in such a comparison is to multiply the equation for ψ_1 by ψ_2 and the equation for ψ_2 by ψ_1 and subtract:

$$\psi_2 \frac{d}{dz}\left[p\frac{d\psi_1}{dz}\right] + \psi_2[q + r\lambda_1]\psi_1 - \psi_1\frac{d}{dz}\left[p\frac{d\psi_2}{dz}\right] - \psi_1[q + r\lambda_2]\psi_2 = 0$$

or

$$\frac{d}{dz}\left[\psi_2 p\frac{d\psi_1}{dz} - \psi_1 p\frac{d\psi_2}{dz}\right] = (\lambda_2 - \lambda_1)r\psi_1\psi_2 \qquad (6.3.13)$$

The reason we can obtain so simple a relation for comparison is because the Liouville equation is *self-adjoint*. If it were of the more general type

$$\mathfrak{L}_z(\psi) = f(z)\frac{d^2\psi}{dz^2} + g(z)\frac{d\psi}{dz} + [h(z) + \lambda j(z)]\psi = 0$$

for ψ_1, we should have to use the adjoint equation

$$\tilde{\mathfrak{L}}_z(\psi) = \frac{d^2}{dz^2}[f\psi] - \frac{d}{dz}[g\psi] + [h + \lambda j]\psi = 0$$

for ψ_2 in order to express the difference as a total derivative plus $(\lambda_1 - \lambda_2)$ times a function (see page 527). The equation adjoint to the Liouville equation is the same equation (try it!).

Now let us integrate Eq. (6.3.13) over z, beginning at the left-hand boundary point (which we shall call a) and ending at some arbitrary point z, before we reach the right-hand boundary point. We have

$$p\left(\psi_2\frac{d\psi_1}{dz} - \psi_1\frac{d\psi_2}{dz}\right) - \left[p\left(\psi_2\frac{d\psi_1}{dz} - \psi_1\frac{d\psi_2}{dz}\right)\right]_{z=a} = (\lambda_2 - \lambda_1)\int_a^z r\psi_1\psi_2\,dz$$

where the quantity in the parentheses would be the Wronskian for ψ_2 and ψ_1 *if* $\lambda_1 = \lambda_2$ (but for our purposes here $\lambda_1 \neq \lambda_2$).

If a is a singular point of the differential equation, p is zero and the quantity in square brackets is zero as long as ψ_1 and ψ_2 are finite at the singular point. If a is a regular point and the boundary condition is the homogeneous one that $\alpha(d\psi/dz) - \beta\psi = 0$ at $z = a$ (where α and β are independent of λ, see page 713), then $\psi_1(d\psi_2/dz) - \psi_2(d\psi_1/dz) = (\beta/\alpha)(\psi_1\psi_2 - \psi_2\psi_1) = 0$ at $z = a$, and again the quantity in square brackets is zero. Therefore for nearly any of the usual boundary conditions at $z = a$, we have

$$p\left(\psi_2\frac{d\psi_1}{dz} - \psi_1\frac{d\psi_2}{dz}\right) = (\lambda_2 - \lambda_1)\int_a^z r\psi_1\psi_2\,dz \qquad (6.3.14)$$

For some range of values of λ, the solution ψ which satisfies the boundary condition at $z = a$ *oscillates; i.e.*, it rises to a maximum as z is made larger than a, and then it reduces in value, finally going through zero and on to a negative maximum, and so on. Suppose that we choose λ_1 to be in the range of λ for which ψ oscillates, and suppose that we choose ζ to be the smallest value of z (the value closest to $z = a$) for which $\psi_1 = 0$. Then Eq. (6.3.14) becomes

$$\left(p\psi_2 \frac{d\psi_1}{dz}\right)_{z=\zeta} = (\lambda_2 - \lambda_1) \int_a^\zeta r\psi_1\psi_2\, dz; \quad \psi_1(\zeta) = 0 \qquad (6.3.15)$$

Since ζ is the smallest zero of ψ_1, ψ_1 does not change sign between $z = a$ and $z = \zeta$ and therefore ψ_1 can be taken to be positive over the range $a < z < \zeta$. We have already shown, on page 719, that for the equations we are concerned with both p and r are positive over the whole range $a < z < b$, where b is the upper boundary point. We can also say that $d\psi_1/dz$ is negative at $z = \zeta$, since this is the place where ψ_1 goes from positive values through zero to negative values.

Let us choose λ_2 to be greater than λ_1, so that $(\lambda_2 - \lambda_1)$ is a positive quantity. What now does Eq. (6.3.15) tell us of the behavior of ψ_2 (assuming that ψ_2 satisfies the same boundary conditions at $z = a$ and corresponds to λ_2)? Does ψ_2 go to zero in the range $a < z < \zeta$ or not?

Suppose ψ_2 did not go to zero in the range $a < z < \zeta$. In this case we could assume that ψ_2 also is positive in this whole range and is still positive at $z = \zeta$. Then the quantity on the left in Eq. (6.3.15) will be negative (p is positive, $d\psi_1/dx$ is negative, ψ_2 is positive, at $z = \zeta$), whereas the quantity on the right will be positive ($\lambda_2 - \lambda_1$ positive; ψ_1, ψ_2, and r positive). This contradiction indicates that ψ_2 *must* go through zero somewhere in the range $a < z < \zeta$. Further application of the same reasoning for ranges of z from this first zero of ψ_1 to the second zero, and so on, shows that the distance between successive zeros of ψ_2 is smaller than the distance between successive zeros of ψ_1 if $\lambda_2 > \lambda_1$.

In other words, *the larger the value of* λ, *the closer together lie the zeros of* ψ (assuming that p and r are everywhere positive in the range $a < z < b$). Or, vice versa, as the value of λ is decreased, the distance between successive zeros of ψ increases.

For a low enough value of λ (this value may be negative) there will be no zero value of ψ inside the range $a < z < b$. For some value λ_0 of λ the function ψ will have a zero at $z = b$, the other boundary point, and no zero between a and b; for $\lambda < \lambda_0$ there will be no zeros in the range $a < z < b$. If the boundary condition at $z = b$ is that $\psi = 0$, then λ_0 is an eigenvalue of λ and *there are no other eigenvalues less than* λ_0. Similarly, by juggling Eq. (6.3.14) around, we can show that, no matter what the boundary condition is, there will be some *lowest eigenvalue* of λ, which can be called λ_0. *All other eigenvalues of* λ *are larger than* λ_0. It is not difficult to see that the eigenfunction ψ_0 corresponding to λ_0 has the least possible number of zeros between the boundary points a and b;

in fact in the majority of cases ψ_0 has *no* zeros between a and b (though it may be zero *at* a and/or b if the boundary conditions so require).

It is often useful to rearrange the functions q and r in the Liouville equation (6.3.12) so that λ_0 is zero. This can be done, for in the expression $[q(z) + \lambda r(z)]$ we can always add $\lambda_0 r(z)$ to $q(z)$ and subtract it from $\lambda r(z)$, producing a new q, independent of λ, and a new λ, equal to the old λ minus λ_0. Naturally the lowest eigenvalue for the new equation would be zero, and therefore all other eigenvalues are positive.

If we now increase the value of λ from λ_0, the corresponding ψ, if it fits the boundary conditions at $z = a$, will not fit the conditions at b. As we increase λ still further, however, we shall eventually find the next eigenvalue λ_1 corresponding to an eigenfunction ψ_1, which has one more zero than ψ_0 (in most cases this means one zero, since ψ_0 usually has no zeros). As λ is further increased, the distance between nodes (another word for zeros of ψ) grows still smaller until, at the next eigenvalue λ_2, there is one more node inside the range $a - b$. This is shown graphically in Fig. 6.8. Thus it is possible to set up a sequence of eigenfunctions $\psi_0, \psi_1, \psi_2, \psi_3, \ldots$, each satisfying the specified homogeneous boundary conditions at a and b and so ordered that the corresponding eigenvalues $\lambda_0, \lambda_1, \lambda_2, \lambda_3, \ldots$ form a continuously increasing sequence of values, so that $\lambda_{n+1} > \lambda_n$. When this is done, our previous discussion shows that the number of nodes in the eigenfunctions ψ_n between a and b also form a continuously increasing sequence, so that ψ_{n+1} has one more node in the range a, b than does ψ_n. We have tacitly assumed here that all the allowed values of λ are real, so we could order them in a straightforward way. That eigenvalues for Eq. (6.3.12) are all real will be proved on page 728.

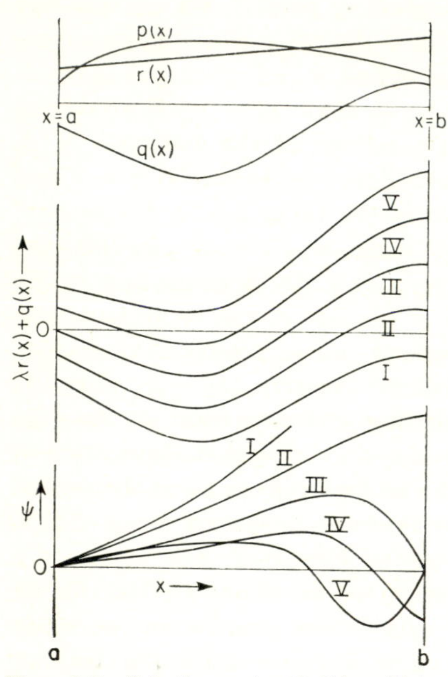

Fig. 6.8 Solutions ψ of Liouville's equation for different values of the separation constant λ. Solutions III and V satisfy the boundary conditions $\psi = 0$ at $x = a$ and b.

A graphical presentation may make this sequence of arguments more understandable. In Fig. 6.8 we show a typical case: The two boundary points are at $x = a$ and $x = b$, and the eigenfunction ψ is to be zero at both these points. Curves for p, r, and q are shown in the top graph; both p and r are positive over the range $a < x < b$, but q is not. The middle graph shows curves for $(\lambda r + q)$ for a sequence of values of λ, that for the lowest value being the lowest curve, and so on. The bottom graph shows curves for the solution of the Liouville equation which is zero at $x = a$ for the values of λ giving the plots in the middle set of curves.

We notice several interesting points about the relation between ψ and the curve for the corresponding $(\lambda r + q)$. *Wherever* $(\lambda r + q)$ *is negative* ψ *curves away from the* $\psi = 0$ *line.* This property, that slope increases with increase of x if ψ is positive and decreases with x if ψ is negative, is characteristic of any combination of the exponentials $e^{\alpha x}$, $e^{-\alpha x}$ and will be called *exponential behavior.* Second, *wherever* $(\lambda r + q)$ *is positive,* ψ *curves toward the axis; i.e.*, if ψ is positive, the slope decreases with x and vice versa. This property is characteristic of trigonometric functions and will be called *sinusoidal behavior.*

These characteristics of ψ can be shown by integrating the Liouville equation once:

$$\left[p\,\frac{d\psi}{dx}\right]_{x_1}^{x_2} = -\int_{x_1}^{x_2} [\lambda r + q]\psi\,dx; \quad x_2 > x_1,\ p > 0$$

If $[\lambda r + q]$ is negative in $x_1 < x < x_2$, then when ψ is positive in that range, $p(d\psi/dx)$ for x_2 is larger than $p(d\psi/dx)$ for x_1, and so on.

We have taken case I in Fig. 6.8 for a value of λ such that $(\lambda r + q)$ is everywhere negative; the corresponding ψ, having exponential behavior, has no chance to "turn back" again and come to zero at $x = b$. In case II $(\lambda r + q)$ is negative over a short range of x, enough to curve the corresponding ψ over somewhat, but not yet enough to bring it to zero. In case III we have hit just the right value of λ (the eigenvalue λ_0) which allows enough region where $(\lambda r + q)$ is positive to curve ψ around so it comes right back to zero at $x = b$. Case IV is for a still higher value of λ, where the oscillation of ψ is more marked. One node has appeared, but it has not "slid over" toward $x = a$ far enough to allow another half cycle of oscillation before x reaches k. Finally, for case V, λ has reached the value for which ψ can go through zero and bend back again so as to become exactly zero at exactly $x = b$.

The continuation of the argument is clear; it is also clear that the same sort of argument can be made for other boundary conditions. It is also not difficult to show that the difference $(\lambda_{n+1} - \lambda_n)$ must always be finite if the distance $(b - a)$ is finite, even in the limit of large n. Utilizing Eq. (6.3.14) again, we set the limits of integration a and b. We set $\lambda_1 = \lambda_n$, an eigenvalue, and ψ_n the corresponding eigenfunction (with $\psi = 0$ boundary conditions, for instance). Suppose that we take λ_2 to be intermediate between λ_n and λ_{n+1}, such that $\psi_2 = 0$ at $x = a$ but $d\psi_2/dx = 0$ at $x = b$. Consideration of the ψ curves of Fig. 6.8 shows that, if there is an even number of nodes of ψ_n between a and b, then

$d\psi_n/dx$ is negative at $x = b$ and ψ_2 (defined in the last sentence) is also negative at $x = b$; if the number of nodes in ψ_n is odd, then both $d\psi_n/dx$ and ψ_2 are positive at $x = b$. Therefore in the equation

$$(\lambda_2 - \lambda_n) \int_a^b \psi_n(x)\psi_2(x)r(x)\,dx = \psi_2(b)\psi_n'(b)p(b)$$

the quantity on the right-hand side is always positive and is not infinitesimal no matter how large n is.

Since λ_2 is, by definition, larger than λ_n (but smaller than λ_{n+1}), it follows that the integral is always positive. If $(b - a)$ is not infinite, the integral cannot be infinite in magnitude, for neither of the ψ's (or r) is infinite anywhere. Consequently the difference $(\lambda_2 - \lambda_n)$ cannot be infinitesimally small unless $(b - a)$ is infinite. Since $\lambda_{n+1} > \lambda_2$, we have the useful result that the difference $(\lambda_{n+1} - \lambda_n)$ cannot be infinitesimal if $(b - a)$ is not infinite, no matter how large n is. Therefore *the sequence of values* $\lambda_0, \lambda_1, \ldots, \lambda_n, \lambda_{n+1}, \ldots,$ *can have no limit point nor any upper bound* but must continue on to $+\infty$. This result will be useful in our later analysis. We thus can have a continuous distribution of eigenvalues *only if* $(b - a)$ *is infinite.*

These conclusions, that there is a lowest eigenvalue λ_0 and that ordering the eigenfunctions in order of increasing eigenvalues also orders them in order of increasing number of nodes in the range a, b, are the useful results of a study of the Sturm-Liouville problem. They depend on our assumption that r is positive everywhere in the range $a < z < b$ and that p is also positive everywhere in this range (this last is equivalent to the statement that there is no singular point of the differential equation *inside* the range a, b). These assumptions are true for all the separated equations obtained in Sec. 5.1. The results have been obtained by the use of the theorem stating that, the larger the value of the separation constant λ, the smaller are the internodal distances of the corresponding solution $\psi(z)$ (this theorem is called *Sturm's first comparison theorem*).

Although there is a lowest eigenvalue, λ_0, we have already seen that *there is no greatest eigenvalue,* that for each eigenvalue λ_n, with eigenfunction λ_n, there can always be a next larger eigenvalue $\lambda_{n+1} > \lambda_n$, with eigenfunction ψ_{n+1}, which has one more node in the range a, b than does ψ_n. Therefore the sequence of eigenvalues is infinite in number, extending from the lowest, λ_0, to infinity.

As an example of this behavior we can consider the case of the equation $(d^2\psi/dz^2) + \lambda\psi = 0$, with the Neumann boundary condition that $d\psi/dz = 0$ at $z = 0$ and the more complicated homogeneous condition that $d\psi/dz = \psi/a$ at $z = a$. Here neither end point is a singular point of the equation.

When λ is negative, the solution which has zero slope at $z = 0$ is $\psi = \cosh(kz)$, where $\lambda = -k^2$. The value of k for which ψ satisfies the condition at $z = a$ is obtained from the solution of transcendental equation

$$(1/w)\coth(w) = 1; \quad w = ka$$

A similar function has been discussed in Sec. 4.4 (see Fig. 4.19). We can easily show that there is only one real root for w; all other roots are imaginary if a is real and positive. The one real solution, k_0, is, approximately, $k_0 \simeq 1.200/a$. Therefore the one negative value of λ (which must be the lowest eigenvalue) is, approximately,

$$\lambda = \lambda_0 = -k_0^2 \simeq -(1.440/a^2)$$

(to four significant figures) and the corresponding eigenfunction is

$$\psi_0(z) \simeq \cosh(1.200z/a)$$

which has no nodes in the range 0, a.

All other eigenvalues must be for positive values of λ, for which we can set $\psi = \cos(kz)$, $\lambda = k^2$. This cosine function satisfies the Neumann conditions at $z = 0$; to satisfy the conditions at $z = a$, we must adjust k so that

$$\left(\frac{d\psi}{dz}\right)_a = -k\sin(ka) = \frac{\psi}{a} = \frac{\cos(ka)}{a} \quad \text{or} \quad \cot(w) = -w; \quad w = ka$$

There is an infinite sequence of solutions of this equation,

$$k_1 \simeq 2.798/a = (\pi - .344)/a$$
$$k_2 \simeq 6.121/a = (2\pi - .162)/a$$
$$\cdots\cdots\cdots\cdots\cdots$$

or to somewhat lesser accuracy,

$$k_n \simeq (n\pi/a) - (1/n\pi a); \quad n = 1, 2, 3, \ldots$$

where

$$\lambda_n = k_n^2 \simeq \left(\frac{n\pi}{a}\right)^2 - \left(\frac{2}{a^2}\right) \quad \text{and} \quad \psi_n = \cos(k_n z) \simeq \cos\left[\left(n\pi - \frac{1}{n\pi}\right)\frac{z}{a}\right]$$

It is not difficult to prove that ψ_n has just n nodes in the range $0 < z < a$. Except for the negative value of the lowest eigenvalue λ_0, all the other eigenvalues are positive, having no upper bound. Of course, we could rearrange our equation, by writing it

$$(d^2\psi/dz^2) + [-k_0^2 + \lambda^0]\psi = 0; \quad p = 1; \quad q = -k_0^2; \quad r = 1$$

where the new eigenfunctions are $\lambda_n^0 = \lambda_n + k_0^2$. By this readjustment we have made the lowest eigenvalue λ_0^0 equal to zero and have ensured that no eigenvalues are negative.

Degeneracy. Some boundary conditions may not be exclusive enough to result in a unique solution for each allowed value of λ. For instance, for the equation $(d^2\psi/d\phi^2) + \lambda\psi = 0$, in some cases (see pages 507 and 713) the sole boundary condition is that ψ be periodic in ϕ with period 2π. In this case the allowed values of λ turn out to be m^2 ($m = 0, 1, 2, \ldots$), but for each value of m any linear combination of $\sin(m\phi)$ and $\cos(m\phi)$ will satisfy the periodicity conditions, so there is an infinity of independent solutions. Such cases, where more than one eigenfunction belongs to a single eigenvalue, are called *degenerate cases.*

In cases of double degeneracy any pair of independent solutions χ_n and Ω_n, corresponding to a given eigenvalue λ_n, can be used to express the most general eigenfunction for this eigenvalue. Usually, however, it is best to choose the two solutions so they are orthogonal (see page 718) to each other, *i.e.*, so that

$$\int_a^b \chi_n(z)\,\Omega_n(z)r\,dz = 0$$

This can usually be done by adding certain symmetry requirements over and above the boundary conditions, which pick either one or the other set of independent solutions.

For instance, for the equation $(d^2\psi/d\phi^2) + m^2\psi = 0$ and for periodic conditions, the functions $\sin(m\phi)$ are odd functions of ϕ and the functions $\cos(m\phi)$ are even functions of ϕ and they are mutually orthogonal, for

$$\int_0^{2\pi} \sin(m\phi)\cos(n\phi)\,d\phi = 0$$

The usefulness of this orthogonality requirement will shortly become evident. At any rate we could label the functions $\cos(n\phi)$ with the notation $\psi_{en}(z)$ (e for even) and the functions $\sin(n\phi)$ with the notation $\psi_{on}(z)$ (o for odd), and we should have the array ψ_{eo} for $\lambda_0 = 0$, ψ_{e1} and ψ_{o1} for $\lambda_1 = 1$, ψ_{e2} and ψ_{o2} for $\lambda_2 = 4$, and so on, two eigenfunctions for each eigenvalue except the lowest.

Series of Eigenfunctions. As we have seen earlier in this section, eigenfunctions enable us to fit boundary conditions by means of series. Once we have obtained our sequence of eigenfunctions $\psi_n(z)$, we should be able to express any piecewise continuous function $F(z)$ in terms of a series

$$F(z) = \sum_{n=0}^{\infty} A_n\psi_n(z)$$

between the boundary points a, b [see Eq. (6.3.8)]. In order that this be *possible,* we must show that the sequence of eigenfunctions is a *complete set.* This somehow is analogous to a requirement that the set of eigenvectors in function space, corresponding to the set of eigenfunctions, cover all the "dimensions" in function space which are required by the arbitrary vector corresponding to $F(z)$ [see discussion following Eq.

(6.3.7)]. If $F(z)$ is analytic in a certain region around $z = 0$, all that is necessary is to show that the eigenfunction series corresponds to a power series.

For instance, if every power of z can be expressed in terms of a convergent series of eigenfunctions, we can transform directly from the power series to the eigenfunction series and be sure that the second series will converge in the region where $F(z)$ is analytic.

But we can often go further than this. For instance, we shall show later that a function with a finite number of discontinuities in the range (a,b) (such piecewise continuous functions are, of course, not analytic at their discontinuities) can be represented by a series of eigenfunctions, giving a "least-squares" fit in the sense of page 709. Such series cannot be differentiated (for obvious reasons), and usually they are not uniformly convergent. But they usually can be integrated, and the resulting integral series will converge uniformly. Therefore eigenfunction series can go further than power series and can represent a certain class of nonanalytic functions, though the resulting series must be dealt with carefully.

In order that least-squares representation of $F(z)$ by a series of eigenfunctions be *easy* to accomplish, we should have that the sequence be *mutually orthogonal, i.e.*, that the corresponding eigenvectors in function space be mutually perpendicular [see Eq. (6.3.5)]. This means that

$$\int_a^b \psi_n(z)\psi_m(z)r(z)\,dz = \begin{cases} 0; & m \neq n \\ E_n^2; & m = n \end{cases} \tag{6.3.16}$$

Equation (6.3.14) will enable us to prove this important property if we set the upper limit of integration equal to the upper boundary point b; for then if ψ_n satisfies reasonable boundary conditions at b (finiteness if b is a singular point; fixed relation between value and slope, independent of λ, if b is an ordinary point), the left-hand side of the equation is zero and

$$(\lambda_n - \lambda_m) \int_a^b \psi_n\psi_m r\,dz = 0$$

showing that, if ψ_n and ψ_m belong to different eigenvalues ($\lambda_n \neq \lambda_m$) they are orthogonal. We note here that in Sec. 11.1 we shall show that, when the boundary conditions vary as λ is varied, the set of eigenfunctions is *not* mutually orthogonal.

Even for the general periodic boundary condition the resulting eigenfunctions, if they belong to a different eigenvalue, are orthogonal, for if we integrate Eq. (6.3.13) between a and b, we obtain

$$\left[\psi_n \frac{d\psi_m}{dz} - \psi_m \frac{d\psi_n}{dz}\right]_a^b = (\lambda_n - \lambda_m) \int_a^b \psi_m\psi_n r\,dz = 0$$

the expression in square brackets being the same at b as it is at a, because of the periodicity, so the two cancel. This does not prove that, in degenerate cases, the several eigenfunctions belonging to the *same* eigenfunctions are orthogonal, but we have indicated on page 726 that we can always choose functions which *are* mutually orthogonal. Consequently our sequence of eigenfunctions ψ_n may be made all mutually orthogonal, whether the case is a degenerate one or not, and any arbitrary, piecewise continuous function $F(z)$ can be represented by the series

$$F(z) = \sum_{n=0}^{\infty} \frac{F_n}{E_n} \psi_n(z); \quad F_n = \frac{1}{E_n} \int_a^b F(x)\psi_n(x)r(x)\,dx \qquad (6.3.17)$$

in the range $a < z < b$ (if the set ψ_n is a complete set).

Incidentally this orthogonality ensures that the eigenvalues of the Liouville equation (6.3.12) are real as long as the functions p, q, r are real along the useful range of z. If there were an allowed value of λ which was complex, then, by symmetry, its complex conjugate would also be an eigenvalue. The two eigenfunctions would also be complex conjugates of each other, so that, if ψ_n, say, were $u + iv$, then ψ_m would be $u - iv$. However, it still would be true that these two eigenfunctions, corresponding to different eigenvalues, would have to be orthogonal, so we should have to have

$$\int_a^b (u^2 + v^2)r\,dz = 0$$

which would be inconsistent with our assumptions about r, u, and v. Consequently we cannot have complex eigenvalues of the Liouville equation, and our worries, expressed on page 722, were groundless.

In accordance with our discussion on page 717, we can say that the density of unit vectors $\mathbf{e}(z)$ between z and $z + dz$ is proportional to $r(z)\,dz$, rather than just dz (for this reason r is sometimes called the *density function*). Thus the scalar product of two vectors $\mathbf{F}$ and $\mathbf{G}$ will be

$$(\mathbf{F}^* \cdot \mathbf{G}) = \int_a^b \bar{F}(z)G(z)r(z)\,dz$$

where $\bar{F}(z)$ and $G(z)$ are the components of $\mathbf{F}^*$ and $\mathbf{G}$ along the various directions given by $\mathbf{e}(z)$. The eigenfunctions ψ_n are components of the mutually orthogonal eigenvectors $\mathbf{E}_n$, of magnitude

$$E_n = \left[\int_a^b |\psi_n|^2 r\,dz \right]^{\frac{1}{2}}$$

called the *normalizing factor* for ψ_n.

The components of the unit eigenvectors $\mathbf{e}_n$ are then $\psi_n(x)/E_n$ which are direction cosines of $\mathbf{e}(x)$ to the normal axes defined by the eigenvectors $\mathbf{e}_n$. Consequently, by a generalization of Eq. (6.3.11)

$$\frac{1}{E_n E_m} \int_a^b \bar{\psi}_n \psi_m r \, dz = \delta_{nm}; \quad \sum_n \frac{r}{E_n^2} \bar{\psi}_n(x) \psi_n(z) = \delta(x - z)$$

where $\delta(x - z)$ is the Dirac delta function (see page 122). The series in the second equation is not absolutely convergent, so that it must be used only in cases where subsequent integration will ensure convergence. For instance, from the equation

$$\int_a^b F(x) \delta(x - z) \, dx = F(z)$$

defining $\delta(x - z)$, by use of the symmetric series of eigenfunctions we obtain Eq. (6.3.17) immediately.

Eigenfunctions which are divided by the amplitude E_n so that they are direction cosines, components of unit vectors, are said to be *normalized.* A set of eigenfunctions which are mutually orthogonal and which are also normalized is called an *orthonormal* set. The property of orthogonality is quite important and useful, but the property of normalization is only of formal utility. We shall nearly always be using orthogonal sets of eigenfunctions, but we shall seldom bother to normalize them, preferring to include the *normalization constants* E_n explicitly in our equations.

Factorization of the Sturm-Liouville Equation. We first encountered the eigenvalue problem in Chap. 1, in connection with the determination of the principal axes of a linear vector operator. The discussion there, particularly the part concerning operators in abstract vector space, corresponds closely with the discussion in this chapter as we have just seen. This correspondence can be brought out still more clearly by a further examination of the operator techniques outlined on pages 89 and 244. There it was shown that a purely operator calculation, using "factored" operators which raised or lowered the eigenvalue, also provided solutions of the corresponding Schroedinger equation, which is a Sturm-Liouville equation. We shall now show that, in a number of instances, the Sturm-Liouville equation may be "factorized" into linear differential operators which can be used to obtain eigenvalues and eigenfunctions by methods analogous to those of Chap. 1.

The process of obtaining these linear differential operators from the differential equation is called *factorization.* Turning back to the derivation of the harmonic oscillator problem, on pages 244 *et seq.*, we see that the equation

$$(d^2\psi_n/dx^2) + (2n + 1 - x^2)\psi_n = 0$$

is equivalent to the pair of first-order differential-recurrence equations

$$\left(\frac{d}{dx} - x\right)\psi_n = \sqrt{2(n + 1)}\,\psi_{n+1}; \quad \left(\frac{d}{dx} + x\right)\psi_n = -\sqrt{2n}\,\psi_{n-1}$$

relating the eigenfunction ψ_n to its "next neighbors" ψ_{n-1} and ψ_{n+1}. Vice versa, from these two first-order equations we can obtain the eigenvalues $(2n + 1)$ for the original second-order equation and can obtain explicit differential expressions for the eigenfunctions, properly normalized.

To generalize this, we change the Sturm-Liouville equation,

$$\frac{d}{dz}\left(p\,\frac{d\psi}{dz}\right) + (q + \lambda r)\psi = 0$$

into what we will call the modified Sturm-Liouville equation,

$$(d^2\Phi/dx^2) + [\lambda - V(x)]\Phi = 0$$

where $\Phi = (pr)^{\frac{1}{4}}\psi;\quad x = \int \sqrt{p/r}\,dz$

$$-V(x) = \left(\frac{q}{r}\right) + \tfrac{3}{16}\left[\left(\frac{p'}{p}\right)^2 + \left(\frac{r'}{r}\right)^2\right] - \tfrac{1}{8}\left(\frac{p'r'}{pr}\right) - \tfrac{1}{4}\left[\left(\frac{p''}{p}\right) + \left(\frac{r''}{r}\right)\right]$$

This has the general form of a one-dimensional Schroedinger equation, with V the potential. We shall consider here the case where the two limits of the range of x (a and b) are contiguous singular points of the equation. In this case neither p nor r is zero in the range $a < x < b$, and the only requirement on Φ is that it be finite at a and b or that $\int_a^b \Phi^2\,dx$ be bounded (in other words that Φ be *quadratically integrable*).

We now inquire whether part of the equation for Φ can be factorized into the pair of differential operators

$$\mathcal{G}^- = [u(x) - (d/dx)];\quad \mathcal{G}^+ = [u(x) + (d/dx)]$$

such that

$$\mathcal{G}^+\mathcal{G}^-\Phi_n = (\lambda_n + a)\Phi_n;\quad \mathcal{G}^-\mathcal{G}^+\Phi_n = (\lambda_n - a)\Phi_n$$

Adding and subtracting the two assumed equations, we see that

$$[u^2(x) - (d^2/dx^2)]\Phi_n = \lambda_n\Phi_n;\quad a = du/dx$$

Consequently if a is to be independent of x, u must be linearly dependent on x and thus u^2, which equals V, must be proportional to x^2, bringing us to the equation for the harmonic oscillator. The differential operators $\mathcal{G}^-$ and $\mathcal{G}^+$ correspond to the abstract vector operators defined in Eq. (2.6.30) and are to be manipulated in the way outlined in Chap. 2 to obtain the eigenvalues and eigenfunctions.

This process can be generalized, to correspond to other forms for V, when V is (or can be formally made) dependent on some parameter m, which may be allowed to take on values differing by unity for the purposes of the analysis. We consider u and a also to be functions of this parameter and set up the two factorized equations

$$\mathcal{G}_{m+1}^-\Phi_n(m|x) \equiv [u_{m+1}(x) - (d/dx)]\Phi_n(m|x) = \sqrt{\lambda_n - a_{m+1}}\,\Phi_n(m+1|x)$$
$$\mathcal{G}_m^+\Phi_n(m|x) \equiv [u_m(x) + (d/dx)]\Phi_n(m|x) = \sqrt{\lambda_n - a_m}\,\Phi_n(m-1|x)$$

or the equivalent pair

$$\mathcal{G}^+_{m+1}\mathcal{G}^-_{m+1}\Phi_n(m|x) = (\lambda_n - a_{m+1})\Phi_n(m|x)$$
$$\mathcal{G}^-_m\mathcal{G}^+_m\Phi_n(m|x) = (\lambda_n - a_m)\Phi_n(m|x) \tag{6.3.18}$$

which are to be adjusted so that they are equivalent to the modified Sturm-Liouville equation

$$[V_m(x) - (d^2/dx^2)]\Phi_n(m|x) = \lambda_n\Phi_n(m|x)$$

To determine the relation between $u_m(x)$ and a_m and the potential function $V_m(x)$, we subtract and add the second pair of equations (reducing m in the first to $m - 1$) and compare with the corresponding difference and sum for the equation with V_m,

$$du_m/dx = \tfrac{1}{2}[V_{m-1}(x) - V_m(x)];\quad u_m^2 + a_m = \tfrac{1}{2}[V_{m-1}(x) + V_m(x)]$$

if λ is considered to be independent of m. Since a_m is supposed to be independent of x, we can differentiate the second equation, substitute the first one, and finally obtain

$$u_m(x) = \tfrac{1}{2}[(V'_{m-1} + V'_m)/(V_{m-1} - V_m)]$$
$$a_m = \tfrac{1}{2}(V_{m-1} + V_m) - u_m^2 \tag{6.3.19}$$

where $V' = dV/dx$. Consequently the modified Sturm-Liouville equation is equivalent to the factorized equations (6.3.18) with the quantities u_m and a_m related to the potential function V_m by Eqs. (6.3.19).

It is obvious that not all forms of V will give a satisfactory factorization. For one thing, we have assumed that the allowed values of λ are independent of m, and this, as we shall see, requires that a_m be independent of x. Consequently $V_m(x)$ must be such a function of x and m that the expression for a_m given in Eq. (6.3.19) is independent of x. This results in a limited number of possibilities, several of which result in useful sequences of eigenfunctions. A number of the possibilities are tabulated at the end of this chapter; here we shall discuss only one example, resulting in the Gegenbauer polynomials (or spherical harmonics) defined in Eq. (5.2.52) or (5.3.35) and at the end of this chapter.

The equation for these polynomials is

$$(z^2 - 1)\frac{d^2T}{dz^2} + 2(m + 1)z\frac{dT}{dz} - \lambda^0 T = 0$$

where, if $T = T_l^m(z)$, the eigenvalue λ^0 will turn out to be $l(l + 2m + 1)$ where $l = n - m$. Let us note here that the parameter m is not necessarily restricted to integral values. Our factorizing relates the eigenfunction for a given value of m to those differing by unity in m, but often we can start from any value of m and travel up or down by unit steps. Other conditions may limit the values of m; this process does not.

The modified form of the equation is obtained by setting

$$z = \cos x; \quad \Phi = \sin^{m+\frac{1}{2}} x \, T$$

and is
$$\frac{d^2\Phi}{dx^2} + \lambda\Phi - \frac{m^2 - \frac{1}{4}}{\sin^2 x}\Phi = 0; \quad \lambda = \lambda^0 + (m + \tfrac{1}{2})^2$$

In this case the potential function is

$$V_m(x) = (m + \tfrac{1}{2})(m - \tfrac{1}{2}) \csc^2 x$$

and consequently $u_m(x) = (m - \frac{1}{2}) \cot(x); \quad a_m = (m - \frac{1}{2})^2$

so that, for this case, a_m is independent of x as required.

Once we have factorized the specific Sturm-Liouville equation under consideration, if it turns out to be factorizable and a_m is independent of x, we can then proceed to solve by methods analogous to those employed in Sec. 2.6. We can show, for example, that the differential operator $\mathcal{G}^+$ is the adjoint of the operator $\mathcal{G}^-$, so that

$$\int_a^b f\mathcal{G}_m^-(\psi)\, dx = \int_a^b \psi\mathcal{G}_m^+(f)\, dx$$

as long as f and ψ are quadratically integrable over the range of x between the limits a and b (which we have assumed are singular points of the equation).

Equations (6.3.18) then indicate that, if $y_n(m|x)$ is an eigenfunction of the modified Sturm-Liouville equation (*i.e.*, is quadratically integrable over $a \leq x \leq b$) for eigenvalue λ_n, then $\mathcal{G}_m^+ y_n(m|x)$ is an eigenfunction (*i.e.*, also quadratically integrable) for the same eigenfunction λ_n but for a value of m reduced by unity, which can be called $y_n(m - 1|x)$. (The functions y are not necessarily normalized; we reserve the symbol Φ for the eigenfunction which is normalized to unity over the range $a \leq x \leq b$). In other words

$$\mathcal{G}_m^+ y_n(m|x) = y_n(m - 1|x); \quad \text{also } \mathcal{G}_{m+1}^- y_n(m|x) = y_n(m + 1|x)$$

unless either of these is identically zero.

We next apply the technique of Sec. 2.6 to obtain the allowed values of λ_n. When a_m is an *increasing* function of m, that is, when $a_{m+1} > a_m$, we set

$$\begin{aligned}
\int_a^b [y_n(m + 1|x)]^2\, dx &= \int^b [\mathcal{G}_{m+1}^- y_n(m|x)][\mathcal{G}_{m+1}^- y_n(m|x)]\, dx \\
&= \int_a^b y_n(m|x)[\mathcal{G}_{m+1}^+ \mathcal{G}_{m+1}^- y_n(m|x)]\, dx \\
&= (\lambda_n - a_{m+1}) \int_a^b [y_n(m|x)]^2\, dx
\end{aligned}$$

so that, if $y_n(m|x)$ is quadratically integrable, so is $y_n(m + 1|x)$ unless $(\lambda_n - a_{m+1})$ is negative, which would be impossible, since the y's are real. Consequently there can be no eigenfunctions for m larger than a certain maximum value n; to ensure this we have only to make $\lambda_n = a_{n+1}$, which fixes the value of the eigenvalue λ_n, since a_{n+1} is known.

To determine the eigenfunctions we note that

$$\mathcal{G}_{n+1}^{-}y_n(n|x) \equiv [u_{n+1}(x) - (d/dx)]y_n(n|x) = 0$$

since $\lambda_n = a_{n+1}$ and thus $y_n(n+1|x)$ is identically zero. Consequently

$$y_n(n|x) = \exp\left[\int u_{n+1}(x)\,dx\right]$$

and

$$y_n(m|x) = \mathcal{G}_{m+1}^{+}\mathcal{G}_{m+2}^{+} \cdots \mathcal{G}_n^{+} \exp\left[\int u_{n+1}(x)\,dx\right]$$

Furthermore, from the y's we can obtain the properly normalized eigenfunctions Φ. We have

$$\Phi_n(n|x) = C_n \exp\left[\int u_{n+1}\,dx\right];\quad C_n = \left[\int_a^b \exp\left(2\int u_{n+1}\,dx\right)dx\right]^{-\frac{1}{2}}$$

and if $\int_a^b [\Phi_n(m|x)]^2\,dx = 1$, we can see that

$$\int_a^b [\mathcal{G}_m^{+}\Phi_n(m|x)]^2\,dx = (\lambda_n - a_m)$$

Thus we have shown that Eqs. (6.3.18) are equivalent to the pair of equations just preceding (6.3.18) if the functions Φ are normalized and that the second of this pair provides a means to obtain the normalized eigenfunction $\Phi_n(m|x)$ from the known function $\Phi_n(n|x)$:

$$\Phi_n(m|x) = \frac{C_n}{\sqrt{(a_{n+1}-a_n)\cdots(a_{n+1}-a_{m+1})}}\,\mathcal{G}_{m+1}^{+}\mathcal{G}_{m+2}^{+}\cdots\mathcal{G}_n^{+}\left(e^{\int u_{n+1}dx}\right);$$
$$m = n,\ n-1,\ n-2,\ \ldots$$

On the other hand, when a_m is a *decreasing* function of m ($a_{m+1} < a_m$), we turn the argument around and show that there must be some *lower* limit n of m, such that $\mathcal{G}_n^{+}y_n(n|x)$ is identically zero. Then $\lambda_n = a_n$ and

$$\Phi_n(m|x) = \frac{C_n}{\sqrt{(a_n-a_{n+1})\cdots(a_n-a_m)}}\,\mathcal{G}_m^{-}\mathcal{G}_{m-1}^{-}\cdots\mathcal{G}_{n+1}^{-}\left(e^{-\int u_n dx}\right);$$
$$m = n,\ n+1,\ n+2,\ \ldots$$

where here

$$C_n = \left[\int^b \exp\left(-2\int u_n dx\right)dx\right]^{-\frac{1}{2}}$$

Thus both of the pair of equations preceding (6.3.18) hold for normalized eigenfunctions, no matter what functions a_m is of m, but we use the first to obtain $\Phi_n(m|x)$ when $a_{m+1} < a_m$, the second when $a_{m+1} > a_m$.

Returning to our example of the associated Legendre functions (or Gegenbauer polynomials) we see that the limits for x are 0 and π and that $a_{m+1} = (m+\frac{1}{2})^2 > a_m$. Therefore the eigenvalues are

$$\lambda_n = a_{n+1} = (n+\tfrac{1}{2})^2;\quad \lambda_{nm}^0 = \lambda_n - (m+\tfrac{1}{2})^2 = l(l+2m+1)$$

where $l = n - m$. Since $m = n, n-1, \ldots, l$ can never be negative. The starting function is

$$\Phi_n(n|x) = C_n \exp\left[(n + \tfrac{1}{2})\int \cot(x)\,dx\right] = C_n \sin^{n+\frac{1}{2}} x$$

$$C_n = \left[\int_0^\pi \sin^{2n+1} x\,dx\right]^{-\frac{1}{2}} = \frac{\sqrt{\frac{1}{2}(2n+1)!}}{2^n n!}$$

and the other eigenfunctions are

$$\Phi_n(m|x) = \frac{\sqrt{(n+\frac{1}{2})(n+m)!(n-m)!}}{2^n n!}\left[(m+\tfrac{1}{2})\cot x + \frac{d}{dx}\right]\cdots$$

$$\cdot\left[(n-\tfrac{1}{2})\cot x + \frac{d}{dx}\right]\sin^{n+\frac{1}{2}} x$$

$$= (-1)^{n-m}\frac{\sqrt{(n+\frac{1}{2})(n+m)!(n-m)!}}{2^n n!(1-z^2)^{\frac{1}{2}m-\frac{1}{4}}}\frac{d^{n-m}}{dz^{n-m}}(1-z^2)^n$$

where $z = \cos x$. Comparing this with the tables at the end of this chapter shows that, for m and n integers and $l = n - m$, the Gegenbauer polynomials and associated Legendre functions are related to the Φ's by the equations

$$\Phi_n(m|x) = (-1)^{n-m}\sqrt{\frac{(n-m)!(n+\frac{1}{2})}{(n+m)!}}\sqrt{\sin x}\,P_n^m(\cos x)$$

$$T_l^m(z) = (-1)^l\sqrt{\frac{(l+2m)!}{l!(l+m+\frac{1}{2})}}(1-z^2)^{-\frac{1}{2}m-\frac{1}{4}}\Phi_{l+m}(m|\cos^{-1} z)$$

In the tables at the end of this chapter we shall enumerate the other forms for $V_m(x)$ and the corresponding $u_m(x)$ which result in a_m independent of x and which consequently allow factorization in the manner outlined above. It is clear that this method of calculating eigenfunctions is, of course, closely related to the abstract vector representation. The operators $\mathfrak{G}$ can also be used to prove the orthogonality of the eigenfunctions. The techniques outlined give us directly various recurrence formulas relating the various eigenfunctions, which are useful for further calculations.

Many equations can be factorized in several different ways. For example, the equation for $P_n^m(\cos x) = \sin^m x T_{n-m}^m(\cos x)$,

$$\frac{1}{\sin x}\frac{d}{dx}\left(\sin x\frac{dP}{dx}\right) + \left[n(n+1) - \frac{m^2}{\sin^2 x}\right]P = 0$$

has already been factorized with respect to the parameter m, by setting $\lambda = n(n+1)$ and $\Phi_n(m|x) = C\sqrt{\sin x}\,P_n^m(\cos x)$. But we can also consider $-m^2$ as our eigenvalue λ and factorize with respect to n; by setting $t = \ln[\tan(\frac{1}{2}x)]$ we obtain

$$\frac{d^2\psi}{dt^2} + \left[\frac{n(n+1)}{\cosh^2 t} + \lambda\right]\psi = 0; \quad \psi = CP; \quad \cosh t = \csc x$$

which may be factorized by setting $u_n(t) = n\tanh t$, $a_n = -n^2$. In this case a_n is a decreasing function of n, so that the second form for

$\psi_m(n|t)$ is used, the values of n never being less than some chosen value m, where $\lambda_m = -m^2$ and the first normalized eigenfunction is

$$\psi_m(m|t) = \sqrt{\frac{m(2m)!}{2^{2m}(m!)^2}} \cosh^{-m} t; \quad \int_{-\infty}^{\infty} |\psi_m(m|t)|^2 \, dt = 1$$

The other eigenfunctions, for $n = m, m + 1, m + 2, \ldots$, are

$$\psi_m(n|t) = \frac{(2m)!}{2^m m!} \sqrt{\frac{m}{(n-m)!(n+m)!}} \left[n \tanh t - \frac{d}{dt} \right] \cdots \cdot \left[(m+1) \tanh t - \frac{d}{dt} \right] \cosh^{-m} t$$

The recurrence relations for the ψ's may be expressed in terms of the variable $x = 2 \tan^{-1} e^t$,

$$\left[-n \cos x - \sin x \frac{d}{dx} \right] \psi_m(n-1|t) = \sqrt{(n-m)(n+m)} \, \psi_m(n|t)$$

$$\left[-n \cos x + \sin x \frac{d}{dx} \right] \psi_m(n|t) = \sqrt{(n-m)(n+m)} \, \psi_m(n-1|t)$$

Since

$$1 = \int_{-\infty}^{\infty} [\psi_m(n|t)]^2 \, dt = C^2 \int_0^{\pi} [P_n^m(\cos x)]^2 \frac{dx}{\sin x} = C^2 \frac{(n+m)!}{m(n-m)!}$$

we see that

$$\psi_m(n|t) = \sqrt{\frac{m(n-m)!}{(n+m)!}} P_n^m(\tanh t) = (-1)^{n-m} \sqrt{\frac{m}{(n+\frac{1}{2}) \sin x}} \Phi_n(m|\cos^{-1} \tanh t)$$

where Φ is the eigenfunction obtained from the first factorization.

We thus have obtained four differential-recurrence relations for the Φ eigenfunctions or the related ψ functions (we prefer to deal with the Φ functions because they are normalized with respect to the angle x, which is the variable in most physical problems). From these can be obtained other useful recurrence relations. For example, from the set

$$\left[(n + \tfrac{1}{2}) \cos x + \sin x \frac{d}{dx} \right] \Phi_n(m|x) = \sqrt{\frac{(n+m+1)(n-m+1)(2n+1)}{(2n+3)}} \, \Phi_{n+1}(m|x)$$

$$\left[(n + \tfrac{1}{2}) \cos x - \sin x \frac{d}{dx} \right] \Phi_n(m|x) = \sqrt{\frac{(n+m)(n-m)(2n+1)}{(2n-1)}} \, \Phi_{n-1}(m|x)$$

which are derived from the equations for the ψ's, by addition we obtain the usual recurrence relation for the normalized Legendre functions

$$\cos x\,\Phi_n(m|x) = \sqrt{\frac{(n+m+1)(n-m+1)}{(2n+1)(2n+3)}}\,\Phi_{n+1}(m|x) + \sqrt{\frac{(n+m)(n-m)}{(2n-1)(2n+1)}}\,\Phi_{n-1}(m|x)$$

and by inserting this into one of the equations derived earlier for the Φ's,

$$[(m+\tfrac{1}{2})\cos x - \sin x(d/dx)]\Phi_n(m|x) = \sqrt{(n+m+1)(n-m)}\,\sin x\,\Phi_n(m+1|x)$$

or

$$\sqrt{(n+m+1)(n-m)}\,\sin x\,\Phi_n(m|x) = (n+m+1)\cos x\,\Phi_n(m+1|x) - [(n+\tfrac{1}{2})\cos x - \sin x(d/dx)]\Phi_n(m+1|x)$$

we obtain

$$\sin x\,\Phi_n(m|x) = \sqrt{\frac{(n+m+1)(n+m+2)}{(2n+1)(2n+3)}}\,\Phi_{n+1}(m+1|x) - \sqrt{\frac{(n-m-1)(n-m)}{(2n-1)(2n+1)}}\,\Phi_{n-1}(m+1|x)$$

These and many other relations of use can be similarly derived. Some of these relations will be derived by other methods later in this section (see page 749). But we must now return to our main investigation, that of the completeness of a set of eigenfunctions.

Eigenfunctions and the Variational Principle. A complete set of eigenfunctions is one which can provide a "least-squares" representation (see page 709) of any piecewise continuous function. To show that the sets of eigenfunctions we have been considering are complete, we turn to the variational techniques developed in Chap. 3. Reference to Eqs. (3.1.7) *et seq.* shows that the solution of the Sturm-Liouville equation (6.3.12) is the same as the minimizing of the following integral:

$$\int_a^b \left[p(z)\left(\frac{d\psi}{dz}\right)^2 - q(z)\psi^2 \right] dz$$

subject to the auxiliary condition that

$$\int_a^b \psi^2 r(z)\,dz = \text{constant}$$

In other words, the Euler equation for this variational problem is just the Liouville equation (6.3.12).

Let us examine this statement in somewhat greater detail. Suppose that we choose for a trial function, $\psi(z)$, one which satisfies the boundary conditions at a and b and which is *normalized* with respect to the density function r,

$$\int^b \psi^2 r\,dz = 1$$

(this is one of the few places where it is definitely advantageous to normalize the functions). We then compute the integral

$$\Omega(\psi) = \int_a^b \left[p\left(\frac{d\psi}{dz}\right)^2 - q\psi^2 \right] dz \tag{6.3.20}$$

and vary the trial function in every possible way (subject to the restriction that ψ is normalized and satisfies the boundary conditions) until we find the ψ which results in the *lowest possible* value of Ω. The fact that the Liouville equation corresponds to the variational equation ensures that the ψ for which Ω is lowest possible is ψ_0, the lowest eigenfunction, and the corresponding value of Ω is

$$\begin{aligned}\Omega(\psi_0) &= \int_a^b \left[p\left(\frac{d\psi_0}{dz}\right)^2 - q\psi_0^2 \right] dz = \left[p\psi_0 \frac{d\psi_0}{dz} \right]_b^a \\ &\qquad - \int_a^b \psi_0 \left[\frac{d}{dz}\left(p\frac{d\psi_0}{dz}\right) + q\psi_0 \right] dz \\ &= \int_a^b \psi_0[\lambda_0 r\psi_0]\,dz = \lambda_0\end{aligned}$$

the lowest eigenvalue. We obtain this result (as shown in the equations) by integrating the term $p(d\psi/dz)^2$ by parts, by using Eq. (6.3.12) to obtain the last integrand and by utilizing the boundary conditions at a and b to show that the quantity in brackets is zero. For the discussion to follow, we assume that we have rearranged q and λ so that $\lambda_0 = 0$.

To find the next eigenfunction and eigenvalue, we minimize Ω for a normalized trial function ψ subject to the boundary conditions at a and b, which, in addition, is orthogonal to ψ_0, found above. This minimum value of Ω is λ_1, and the corresponding trial function is ψ_1. So we can proceed; to find ψ_n and λ_n we compute Ω for a trial function satisfying the boundary conditions at a and b, which is normalized and which is orthogonal to $\psi_0, \psi_1, \ldots, \psi_{n-1}$, which have all been previously computed. The function which then makes Ω a minimum is ψ_n and Ω for ψ_n is equal to λ_n.

For degenerate cases we shall find a whole sequence of functions which will give the same minimum value of Ω. However, we can always pick out, by symmetry or other additional requirements, a set of mutually orthogonal solutions, which can be considered to be the standard solutions for the one degenerate value of λ_n.

The Completeness of a Set of Eigenfunctions. We can reword our conclusions of the previous paragraphs as follows: suppose that F is *any function* satisfying the boundary conditions and normalized in the range $a - b$; then the integral $\Omega(F)$, defined in Eq. (6.3.20), is *never smaller than* λ_0. Since we have adjusted q and λ so that $\lambda_0 = 0$, we have that $\Omega(F)$ is *never negative.* Likewise if F_n is a function such that

$$\int_a^b F_n^2 r\,dz = 1;\quad \int_a^b F_n\psi_m r\,dz = 0;\quad m = 0, 1, 2, \ldots, n-1, n$$

then the integral $\Omega(F_n)$ is *never smaller than* λ_{n+1}.

The series

$$\sum_{m=0}^{\infty} C_m\psi_m(x);\quad C_m = \frac{1}{E_m^2}\int_a^b f(z)\psi_m(z)r(z)\,dz \tag{6.3.21}$$

is supposed to equal (on the average) the function $f(z)$ over the range a, b. If it is to do so, the difference between the function f and the first n terms of the series

$$f_n(z) = f(z) - \sum_{m=0}^{n} C_m\psi_m(z)$$

must approach zero (on the average) as n approaches infinity. In other words, if the series is to be a good least-squares fit, according to page 709, the quantity

$$a_n^2 = \int_a^b f_n^2(z)r(z)\,dz = \int_a^b f^2 r\,dz - \sum_{m=0}^{n} E_m^2 C_m^2$$

should approach zero as n approaches infinity.

We can now apply our variational argument to obtain a measure of the size of a_n, for the function

$$F_n = f_n(z)/a_n$$

has the properties specified on this page:

$$\int_a^b F_n^2 r\,dz = 1;\quad \text{by definition of } a_n$$

$$\int_a^b F_n\psi_m r\,dz = \begin{cases} (E_m^2/a_n)(C_m - C_m) = 0; & m \le n \\ (E_m^2/a_n)C_m \ne 0; & m > n \end{cases}$$

because of the orthogonality of the ψ's and the definition of C_m. Therefore, according to the findings of the previous page,

$$\Omega(F_n) = \frac{1}{a_n^2}\int_a^b \left[p\left(\frac{df}{dz}\right)^2 - qf^2\right]dz - \frac{2}{a_n^2}\int_a^b \sum_{m=0}^{n} C_m\left[p\frac{df}{dz}\frac{d\psi_m}{dz} - qf\psi_m\right]dz$$
$$+ \frac{1}{a_n^2}\sum_{m=0}^{n} C_m \int_a^b \sum_{s=0}^{n} C_s\left[p\frac{d\psi_m}{dz}\frac{d\psi_s}{dz} - q\psi_m\psi_s\right]dz \ge \lambda_{n+1}$$

The first term in this integral is $\Omega(f)/a_n^2$, which, as we have seen above, is never negative. The first terms in the second integral can be integrated by parts, and by using the facts that ψ_m satisfies the Liouville equation and that both f and the ψ's satisfy the boundary conditions, we obtain for the second integral

$$+\frac{2}{a_n^2}\int_a^b f\sum_{m=0}^{n} C_m\left[\frac{d}{dz}\left(p\,\frac{d\psi_m}{dz}\right)+q\psi_m\right]dz-\frac{2}{a_n^2}\left[\sum_{m=0}^{n} C_m pf\,\frac{d\psi_m}{dz}\right]_a^b$$
$$=-\frac{2}{a_n^2}\sum_{m=0}^{n} C_m\lambda_m\int rf\psi_m\,dz=-\frac{2}{a_n^2}\sum_{m=0}^{n} C_m^2\lambda_m E_m^2$$

Similarly the third integral equals $\left(\frac{1}{a_n^2}\right)\sum C_m^2\lambda_m E_m^2$ so that

$$\Omega(F_n)=\frac{1}{a_n^2}\left[\Omega(f)-\sum_{m=0}^{n} E_m^2C_m^2\lambda_m\right]\geq\lambda_{n+1}$$

Since $\Sigma E_m^2C_m^2\lambda_m$ must be positive (none of the λ_m's is negative), we therefore have that

$$\Omega(f)>a_n^2\lambda_{n+1}\quad\text{or}\quad a_n^2<[\Omega(f)/\lambda_{n+1}]$$

But $\Omega(f)$ is a positive quantity, independent of n, and we have proved that λ_{n+1} approaches infinity as n approaches infinity. Therefore *a_n approaches zero as n approaches infinity*, and we have, by this procedure, *proved that the series* (6.3.21) *is a least-squares fit for the function f* in the range $a<z<b$. Therefore if ψ_n is a sequence of eigenfunctions, solutions of Eq. (6.3.12) (which can be adjusted so that $\lambda_0=0$), and satisfying boundary conditions such that $[p\psi(d\psi/dz)]_a^b$ vanishes, then *the set ψ_n is a complete set.* This proof completes our study of the fundamental characteristics of eigenfunctions. We can now turn to special cases and to the discussion of general techniques which will be useful later.

Asymptotic Formulas. It is sometimes useful to obtain approximate expressions for the eigenfunctions and eigenvalues of high order (n large). We return to the Liouville equation

$$\frac{d}{dz}\left(p\,\frac{d\psi}{dz}\right)+(\lambda r+q)\psi=0$$

and make the substitutions

$$y=(pr)^{\frac{1}{4}}\psi;\quad \xi=\frac{1}{J}\int_a^z\sqrt{\frac{r}{p}}\,dz;\quad J=\frac{1}{\pi}\int_a^b\sqrt{\frac{r}{p}}\,dz$$

to obtain the transformed equation

$$(d^2y/d\xi^2)+[k^2-w(\xi)]y=0\tag{6.3.22}$$

where $\quad k^2=J^2\lambda\quad$ and $\quad w=\left[\frac{1}{(pr)^{\frac{1}{4}}}\frac{d^2}{d\xi^2}(pr)^{\frac{1}{4}}-J^2\left(\frac{q}{r}\right)\right]$

when expressed in terms of the new variable ξ. The range of the new independent variable is $0 \leq \xi \leq \pi$.

When λ is large, k^2 is large compared with w, and we would expect y to approximate a sinusoidal dependence on ξ. To make this specific, we can rearrange the equation as though it were an inhomogeneous equation:

$$(d^2y/d\xi^2) + k^2y = wy$$

and solve it as though wy were the inhomogeneous part. Referring to Eq. (5.2.19) we see that an expression for y is

$$y(\xi) = A \sin(k\xi) + B \cos(k\xi) + \frac{1}{k} \int_{\xi_0}^{\xi} \sin[k(\xi - t)]w(t)y(t)\, dt \qquad (6.3.23)$$

where ξ_0 is some suitable lower limit for integration. This is, of course, an integral equation for y and not a solution. In Chap. 8 we shall study its solution in greater detail. Here, however, we are concerned only with the solutions when λ is very large, large enough so that we can neglect the integral compared with the first two terms (if λ is large, k is large, and the integrand, times $1/k$, vanishes). The values of A and B and of k must be adjusted so that the boundary conditions are satisfied.

If neither a nor b is a singular point, the problem of fitting our asymptotic form to the boundary conditions is fairly simple. If the requirements are that ψ be zero at a and b, the first approximation, obtained by neglecting the integral in Eq. (6.3.23), is simply

$$k = n; \quad \psi_n \simeq [1/(rp)^{\frac{1}{4}}] \sin(n\xi); \quad \lambda_n = (n/J)^2 \qquad (6.3.24)$$

which is valid as long as n^2 is larger than w everywhere in the range $0 < \xi < \pi$. A better approximation can then be obtained by setting this back into the integral in Eq. (6.3.23) and computing a correction term, but we shall avoid such details for a moment.

If the boundary conditions are that $\alpha\psi + \beta(d\psi/dz) = 0$ at $z = a$ and a similar relation, with α and β possibly different constants, at $z = b$, the procedure is somewhat more complicated. We set, for our first approximation,

$$\psi \simeq [1/(pr)^{\frac{1}{4}}] \cos(k\xi + \theta)$$

where θ is a phase angle to be determined from the boundary conditions. Since k^2 is to be larger than $w(\xi)$ everywhere in the range $(0 < \xi < \pi)$, we have

$$\frac{d\psi}{dz} \simeq \frac{1}{(pr)^{\frac{1}{4}}} \frac{d\xi}{dz} \frac{d}{d\xi} [\cos(k\xi + \theta)] = - \frac{1}{(pr)^{\frac{1}{4}}} \frac{k}{J} \sqrt{\frac{r}{p}} \sin(k\xi + \theta)$$

since the derivative of the $(pr)^{-\frac{1}{4}}$ factor is small compared with the derivative of the cosine factor for large k. The solution for θ is obtained from the equation

$$\cot\theta \simeq \frac{k}{J}\left[\frac{\beta}{\alpha}\sqrt{\frac{r}{p}}\right]_{z=a}$$

If k is large enough, this reduces to

$$\theta(k) \simeq \frac{J}{k}\left[\frac{\alpha}{\beta}\sqrt{\frac{p}{r}}\right];\quad \text{evaluated at } z = a$$

Similar fitting at $z = b$ determines the allowed values of k and therefore of $\lambda = (k/J)^2$:

$$k_n \simeq n + \frac{J}{\pi n}\left[\frac{\alpha}{\beta}\sqrt{\frac{p}{r}}\right]_a^b;\quad \psi_n = \frac{1}{(pr)^{\frac{1}{4}}}\cos\left[k_n\xi + \theta(n)\right] \tag{6.3.25}$$

where n is an integer.

Thus when a and b are ordinary points, the higher eigenvalues of the Liouville equation are approximately equal to the *square of π times a large integer n divided by the square of the integral of $\sqrt{r/p}$ between a and b*. The corresponding eigenfunction is approximately equal to the cosine of (πn) times the ratio between the integral of $\sqrt{r/p}$ from a to z and the integral of $\sqrt{r/p}$ from a to b [with an additional "amplitude factor" $(pr)^{-\frac{1}{4}}$ modifying the amplitude].

When a or b, or both, are singular points, where p goes to zero, the integrals defining J and ζ [given just above Eq. (6.3.22)] may diverge, and we may have to modify our definitions somewhat. In addition the function q may go to infinity at such points; in any case the function w will become infinite there. We can, of course, solve the Liouville equation near the singular point in terms of a power series and find the nature of the singularities of the two solutions there. If our equation is sensible physically, one of the solutions will be finite there.

To see the way things go, let us take a specific example: that of the Bessel equation, resulting from the separation of the wave equation in polar coordinates (see pages 550 and 619),

$$\frac{d}{dz}\left(z\frac{d\psi}{dz}\right) + \left(\lambda z - \frac{m^2}{z}\right)\psi = 0 \tag{6.3.26}$$

where $z = r = \xi_1$, where m is an integer (determined by periodic boundary conditions on the ξ_2 factor) and where λ is the separation constant. In this case we have $p = z$, $r = z$, $q = -(m^2/z)$. Suppose we take the singular point $z = 0$ as the limit a.

Making the substitution for Eq. (6.3.22) we have

$$\psi = \frac{1}{\sqrt{z}}y;\quad J = \frac{1}{\pi}\int_0^b dz = \frac{b}{\pi};\quad \zeta = \frac{\pi z}{b};\quad k^2 = (b^2/\pi^2)\lambda$$

$(d^2y/d\zeta^2) + [k^2 - (m^2 - \frac{1}{4})/\zeta^2]y = 0$ and the integral equation for y is

$$y(\zeta) = \cos(k\zeta + \theta) - \frac{1}{k}\int_\zeta^\infty \sin[k(\zeta - t)]\, y(t) \left[\frac{m^2 - \frac{1}{4}}{t^2}\right] dt$$

where the limits of the integral were chosen to ensure convergence. For large values of k and z, the first term is sufficiently accurate and we need not worry about the integral except when ζ is near zero. As long as we can find the value of the phase angle θ which corresponds to the requirement that ψ is finite when $z = 0$, we need never use the asymptotic form for ζ small (where it is inaccurate).

We find the value of θ by means of the Stokes' phenomenon (see page 609), by requiring that the change in the asymptotic formula for ψ, as ζ is taken in a circle about $\zeta = 0$, correspond to the change in the series solution about $\zeta = 0$. Examination of the indicial equation (see page 532) for Eq. (6.3.26) shows that the finite solution at $z = 0$ has the form z^m times a Taylor's series in z. Consequently, if ψ is real for z large along the positive real axis (phase angle $\phi = 0$), it is $e^{im\pi}$ times a real function when $\phi = \pi$ (z negative and large), and its leading term should be $e^{im\phi}$ times a real function when ϕ is not an integral multiple of π (z complex and large).

Setting $\psi \simeq (1/2\sqrt{z})[e^{ik\zeta+i\theta} + e^{-ik\zeta-i\theta}]$, which is real for $\phi = 0$, we first examine the behavior for $\phi = \frac{1}{2}\pi$ ($\zeta = i|\zeta|$). Here the first term in the brackets is negligible for ζ large, and the asymptotic formula gives

$$\begin{aligned}\psi &= e^{im\pi/2} \cdot \text{real function} \\ &\simeq (1/2\sqrt{|z|})\, e^{k|\zeta| - i\theta - \frac{1}{4}i\pi}; \quad z = |z|e^{\frac{1}{2}i\pi}\end{aligned}$$

so that $\theta = -\frac{1}{2}\pi(m + \frac{1}{2})$. To check this we also try $\phi = -\frac{1}{2}\pi$ ($\zeta = -i|\zeta|$). Here the second term is negligible and

$$\begin{aligned}\psi &= e^{-im\pi/2} \cdot \text{real function} \\ &\simeq (1/2\sqrt{|z|})\, e^{k|\zeta| + i\theta + \frac{1}{4}i\pi}; \quad z = |z|e^{-\frac{1}{2}i\pi}\end{aligned}$$

so again $\theta = -\frac{1}{2}\pi(m + \frac{1}{2})$, which checks. Therefore the asymptotic form for the eigenfunctions in this case is

$$\psi \simeq \frac{1}{\sqrt{z}} \cos\left[\frac{\pi k z}{b} - \tfrac{1}{2}\pi(m + \tfrac{1}{2})\right]; \quad -\tfrac{1}{2}\pi < \phi < \tfrac{1}{2}\pi$$

which is the asymptotic behavior given in Eq. (5.3.68). If the boundary condition at $z = b$ is that $\psi = 0$, the asymptotic expression for the eigenvalues is

$$k_n \simeq n + \tfrac{1}{2}(m + \tfrac{1}{2}); \quad \lambda_n = (\pi k_n/b)^2; \quad n \text{ large}$$

This same technique can be carried through for other cases where one or both limits are singular points. A more complete discussion of asymptotic formulas will be given in Chap. 9. From our discussion here we can say that any of the usual sorts of boundary conditions (whether at singular or ordinary points) can be fitted to the order of approximation suitable for asymptotic formulas by adjusting the phase angle θ in the expression

$$\psi \simeq \frac{1}{(pr)^{\frac{1}{4}}} \cos\left[\sqrt{\lambda}\int_a^z \sqrt{\frac{r}{p}}\,dz + \theta\right] \tag{6.3.27}$$

Likewise the asymptotic formula for the higher eigenvalues will turn out to be

$$\lambda_n \simeq \left[\int_a^b \sqrt{\frac{r}{p}}\,dz\right]^{-2} [n\pi + \alpha]^2$$

where the value of α depends on the specific kind of boundary conditions at both a and b.

Comparison with Fourier Series. We see from this discussion that every eigenfunction series behaves like a Fourier series for the higher terms in the series. In fact we can prove that the difference between the first n terms of an eigenfunction series and the first n terms in a Fourier series, covering the same range and for the same function, is uniformly convergent as n is increased indefinitely.

To show this we transfer to the new independent variable ζ and new dependent variable y as defined in Eq. (6.3.22). The function we wish to represent is $F(z)$, which is multiplied by $(pr)^{\frac{1}{4}}$, and this function, expressed in terms of ζ, will be called $f(\zeta)$. No new singularities or discontinuities are introduced in f by this procedure. The expansion of f in terms of ζ is given by the formula [see Eqs. (6.3.17) and (6.3.18)]

$$\begin{aligned} f(\zeta) &= [p(z)r(z)]^{\frac{1}{4}} \int_a^b F(t) \sum_{m=0}^{\infty} \frac{1}{E_m^2} \psi_m(z)\psi_m(t)r(t)\,dt \\ &= \int_a^b f(\tau) \sum_{m=0}^{\infty} \frac{1}{E_m^2} y_m(\zeta)y_m(\tau) \sqrt{\frac{r(t)}{p(t)}}\,dt \\ &= \int_0^\pi f(\tau) \sum_{m=0}^{\infty} \frac{1}{N_m^2} y_m(\zeta)y_m(\tau)\,d\tau \end{aligned} \tag{6.3.28}$$

where $\psi_m(z) = y_m(\zeta)/(pr)^{\frac{1}{4}}$; $F(t) = f(\tau)/(pr)^{\frac{1}{4}}$

$$\zeta = \frac{\pi}{J}\int_a^z \sqrt{\frac{r(z)}{p(z)}}\,dz; \quad \tau = \frac{\pi}{J}\int_a^t \sqrt{\frac{r}{p}}\,dz; \quad J = \int_a^b \sqrt{\frac{r}{p}}\,dz$$

$$E_m^2 = \int_a^b \psi_m^2 r\,dz; \quad N_m^2 = \int_0^\pi y_m^2(\zeta)\,d\zeta = \frac{\pi E_m^2}{J}$$

But the function f can be expressed in terms of a Fourier series in the range $0 < \zeta < \pi$. For instance, we can use the cosine series

$$f(\zeta) = \int_0^\pi f(\tau) \sum_{m=0}^{\infty} \frac{\epsilon_m}{\pi} \cos(m\zeta) \cos(m\tau)\, d\tau \tag{6.3.29}$$

where $\epsilon_0 = 1$, $\epsilon_n = 2$ $(n > 0)$ is called the *Neumann factor.*

If the boundary conditions on ψ are such that the asymptotic form for ψ_n is that given by Eq. (6.3.25), then the form of the y terms in the series will approach the form of the cosine terms as n increases. The two series will approach each other, term by term, and eventually the individual terms will be alike, a much more detailed correspondence than one might expect. Many series of eigenfunctions are only conditionally convergent (*i.e.*, converge only because the terms alternate in sign and partially cancel out). Many of the difficulties attending discussions of the representation of peculiar functions by means of eigenfunctions come about due to this weaker convergence. These difficult cases have been worked out in some detail for Fourier series, so if we can show that what holds for Fourier series also holds for other eigenfunction series, we shall have saved ourselves a great deal of work. If it turns out that the difference between the first n terms of series (6.3.28) and the first n terms of series (6.3.29) is *absolutely convergent*, as n goes to infinity (*i.e.*, the *magnitude* of the difference converges), then we shall have established the detailed correspondence we need.

For example, according to Eq. (6.3.28) the series

$$\sum_{m=0}^{\infty} \frac{1}{N_m^2}\, y_m(\zeta) y_m(\tau)$$

somehow represents the delta function $\delta(\zeta - \tau)$. Such a series is far from being absolutely convergent, but we can show that, if the boundary conditions are those resulting in Eq. (6.3.25), then the function

$$\Phi_n(\zeta,\tau) = \sum_{m=0}^{n} \left[\frac{1}{N_m^2}\, y_m(\zeta) y_m(\tau) - \frac{\epsilon_m}{\pi} \cos(m\zeta) \cos(m\tau) \right]$$

is bounded as n goes to infinity. We prove this interesting correspondence by referring once again to the asymptotic form for y_m, obtained from Eq. (6.3.25):

$$y_m(\zeta) \simeq \cos\left[m\zeta + \frac{\zeta}{\pi m}(B - A) + \frac{A}{m} \right];$$

$$A = \frac{J\alpha}{\beta}\sqrt{\frac{p}{r}} \text{ at } z = a,\ B \text{ is same at } z = b$$

$$y_m(\zeta) \simeq \cos\left[\frac{\zeta}{\pi m}(B - A) + \frac{A}{m}\right]\cos(m\zeta) - \sin\frac{1}{m}\left[\left(\frac{\zeta}{\pi}\right)(B - A) + A\right]\sin(m\zeta)$$

$$\underset{m\to\infty}{\longrightarrow} \cos(m\zeta)\left\{1 - \frac{1}{2m^2}\left[\frac{\zeta}{\pi}(B - A) + A\right]^2\right\} - \frac{1}{m}\left[\frac{\zeta}{\pi}(B - A) + A\right]\sin(m\zeta)$$

and the limiting form for the normalizing constant is $N_m^2 \simeq \pi +$ (terms in $1/m^2$ and higher). The difference between the terms in brackets in Φ_n is large for small values of m, but for larger and larger values of m (when n is large enough), the difference approaches the small quantity

$$-\frac{1}{m}\left\{\left[\frac{\zeta}{\pi}(B - A) + A\right]\sin(m\zeta)\cos(m\tau) + \left[\frac{\tau}{\pi}(B - A) + A\right]\sin(m\tau)\cos(m\zeta)\right\} + (\text{terms in } 1/m^2 \text{ and higher})$$

The $1/m^2$ terms converge absolutely, and therefore their sum is bounded; the series $\sum \frac{\sin(m\zeta)}{m}$ converges, and therefore the sum of the terms multiplied by $1/m$ is also bounded, which proves that the function Φ_n is bounded as $n \to \infty$. We can say that it is always less than some finite quantity D, no matter what value n or ζ or τ has (as long as $0 < \zeta < \pi$, $0 < \tau < \pi$).

By the same argument we can show that the difference between the first n terms of the expansion of a continuous function $f(\zeta)$ in terms of the eigenfunctions y_n and the first n terms of the expansion of $f(\zeta)$ in terms of cosines,

$$\int_0^\pi \Phi_n(\zeta,\tau)f(\tau)\,d\tau$$

approaches zero uniformly as $n \to \infty$. We have here compared our eigenfunction series with a cosine series. For some other boundary conditions we should have to compare with a sine series, etc., but it turns out in each case that we can devise a Fourier series having the same close correspondence as that exhibited above.

To state our conclusions in formal language: The expansion of any continuous function in eigenfunctions *converges or diverges at any point as the related Fourier series converges or diverges at that point.* It converges absolutely in any interval when and only when the corresponding Fourier series converges absolutely in the same interval.

The Gibbs' Phenomenon. We have shown that a properly chosen series of eigenfunctions can give a least-squares fit, even for functions which have a finite number of discontinuities. Certain peculiarities of this fit near the discontinuities should be pointed out, however, which indicate the difficulties arising when such series are used to compute values of the function near a discontinuity.

Figure 6.9 shows a function $F(x)$, with a discontinuity at $x = x_0$. The first n terms in the corresponding eigenfunction series

$$S_n(x) = \int_a^b F(u) \sum_{m=0}^{n} \frac{1}{E_m^2} \psi_m(x)\psi_m(u)r(u)\, du$$

cannot fit the discontinuity, for a finite number of terms of a convergent series cannot have the infinite slope required by the discontinuity. As indicated by the lighter line in the figure, the finite series $S_n(x)$ tries to achieve the infinite slope at $x = x_0$, but in the attempt, it *overshoots* the discontinuity by a certain amount. The resulting curve for S_n resembles the intensity curve for light diffracted at the edge of a screen, the wavelength of the light being inversely proportional to n. Even in the limit of n infinite this "overshooting" persists and the complete series has "flanges" on the ends of the discontinuity, as shown by $D+$ and $D-$ in the second part of Fig. 6.9. These additional peaks, being of zero width, make no difference to a least-squares fit, but they do indicate the limitations of the process of representing functions by series of eigenfunctions.

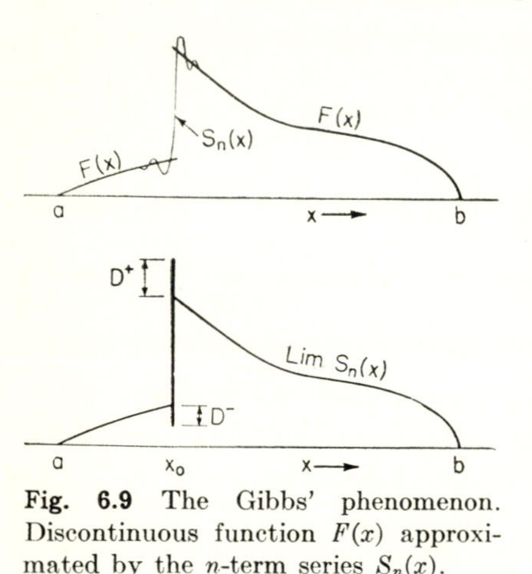

Fig. 6.9 The Gibbs' phenomenon. Discontinuous function $F(x)$ approximated by the n-term series $S_n(x)$.

We can be more specific in our discussion of the Gibbs' phenomenon if we take a Fourier series as an example. The representation of the function

$$F(x) = \begin{cases} +1; & 0 < x < \pi \\ -1; & \pi < x < 2\pi \end{cases}$$

by a series periodic in x with period 2π is

$$S(x) = \frac{1}{\pi}\left[\int_0^\pi dt - \int_\pi^{2\pi} dt\right]\left\{\tfrac{1}{2} + \sum_{m=1}^{\infty} [\cos(mx)\cos(mt) + \sin(mx)\sin(mt)]\right\}$$

$$= \frac{4}{\pi}\sum_{s=0}^{\infty} \frac{1}{(2s+1)} \sin[(2s+1)x]$$

We have here a degenerate case, but the use of the sum of products of all eigenfunctions inside the integral still holds.

In this case, if we wish to compute the sum of all terms between $\sin x$ and $\sin (nx)$ (which is not the first n terms for in this case every other term, for n even, is missing), we can sum inside the integral before integrating. The integrand is

$$\frac{1}{2\pi} + \frac{1}{\pi}\sum_{m=1}^{n} \cos [m(x-t)] = \frac{1}{2\pi}\sum_{m=-n}^{n} e^{im(x-t)} = \frac{1}{2\pi} e^{-in(x-t)} \sum_{s=0}^{2n} [e^{i(x-t)}]^s$$

$$= \frac{e^{-in(x-t)}}{2\pi}\left[\frac{1 - e^{i(2n+1)(x-t)}}{1 - e^{i(x-t)}}\right] = \frac{\sin [(n+\frac{1}{2})(x-t)]}{2\pi \sin [\frac{1}{2}(x-t)]} \tag{6.3.30}$$

so that the sum, up to $\sin(nx)$, is

$$\begin{aligned} S_n(x) &= \frac{1}{2\pi}\left\{\int_0^\pi \frac{\sin [(n+\frac{1}{2})(t-x)]}{\sin [\frac{1}{2}(t-x)]}\,dt - \int_\pi^{2\pi} \frac{\sin [(n+\frac{1}{2})(t-x)]}{\sin [\frac{1}{2}(t-x)]}\,dt\right\} \\ &= \frac{1}{2\pi}\left\{\int_{-x}^{\pi-x} d\theta - \int_x^{\pi+x} d\theta\right\} \frac{\sin (n+\frac{1}{2})\theta}{\sin (\theta/2)} \\ &= \frac{1}{2\pi}\left\{\int_{-x}^{x} \frac{\sin (n+\frac{1}{2})\theta}{\sin (\theta/2)}\,d\theta - \int_{\pi-x}^{\pi+x} \frac{\sin (n+\frac{1}{2})\theta}{\sin (\theta/2)}\,d\theta\right\} \\ &= \frac{1}{\pi}\left\{\int_{-(n+\frac{1}{2})x}^{+(n+\frac{1}{2})x} du - \int_{(n+\frac{1}{2})(\pi-x)}^{(n+\frac{1}{2})(\pi+x)} du\right\} \frac{\sin u}{(2n+1)\sin [u/(2n+1)]} \end{aligned} \tag{6.3.31}$$

When $0 < x < \pi$ and n is large, the first of the two integrals of the last two forms is much larger than the second. The last form shows that, when $n \to \infty$, as long as x does not simultaneously go to zero (that is, x is some finite quantity larger than zero and less than π), the second integral tends to zero and the first integral tends to

$$\frac{1}{\pi}\int_{-\infty}^{\infty} \frac{\sin u}{u}\,du = 1$$

Therefore in the range $\epsilon < x < (\pi - \epsilon)$, where ϵ is small but finite, the sum S_n converges to 1 as $n \to \infty$; likewise for $\pi + \epsilon < x < 2\pi - \epsilon$ (and also for $-\pi + \epsilon < x < -\epsilon$) $S_n \to -1$ as $n \to \infty$.

But suppose we compute the value of S_n for $x = \pi/(2n+1)$ as n goes to infinity. Neglecting again the second integral, which still goes to zero, we have

$$S_n\left(\frac{\pi}{2n+1}\right) \to \frac{1}{\pi}\int_{-\pi/2}^{\pi/2} \frac{\sin u}{u}\,du = 1.179; \quad \text{as } n \to \infty$$

Consequently the series, in trying to follow the discontinuity at $x = 0$, overshoots the mark by about 18 per cent over a region of vanishingly small width, before settling down to the correct value of unity.

We have already shown that other eigenfunction series behave like Fourier series with respect to convergence. We should expect a similar sort of overshooting at discontinuities for other series. Since the region of overshooting is of vanishingly small width, an integral of the series would give correct values, even when the integration was carried through a discontinuity. But we shall need to exercise care in differentiating such series. If a given series converges, its integral will certainly converge, but its derivative may not.

Generating Functions, Legendre Polynomials. Most sets of eigenfunctions have simple equations relating one function with adjacent functions (recurrence formulas), and many have fairly simple expressions for normalization constants. Many of these useful equations can be found most easily by detailed study of the expansion of a fairly simple function in terms of the eigenfunctions. We search for a function of two variables $\Phi(t,z)$, such that the expansion comes out of the form

$$\Phi(t,z) = \sum_{n=0}^{\infty} a_n t^n \psi_n(z) \tag{6.3.32}$$

where ψ_n is one of the set of eigenfunctions under study and the coefficient a_n is independent of z and t. In other words, the function should be so related to the set ψ_n that the terms in the expansion are a power of t, times the corresponding eigenfunction of z, times a coefficient independent of z and t. Such a function Φ is called a *generating function* for the set of eigenfunctions ψ_n. A comparison between Eq. (6.3.32) and Eqs. (4.3.3) and (4.3.1) will indicate that the generating function is related to one or more integral representations of the eigenfunction.

Generating functions are often closely related to the Green's function solutions mentioned on pages 701 and 710 and to be discussed in the next chapter. As an example of the sort of help generating functions can be, we shall derive the properties of the Legendre polynomials (see page 597). These turn up in the solution of the Laplace equation in spherical coordinates and are related to the Green's function $1/R$, where $R^2 = r_1^2 + r_2^2 - 2r_1r_2 \cos\vartheta$, by the equation [see Eq. (5.3.28)]

$$\frac{1}{R} = \sum_{n=0}^{\infty} \left(\frac{r_2^n}{r_1^{n+1}}\right) P_n(\cos\vartheta); \quad r_1 > r_2$$

We simplify this to obtain the generating function for $P_n(z)$

$$\Phi(t,z) = \frac{1}{\sqrt{1 + t^2 - 2tz}} = \sum_{n=0}^{\infty} t^n P_n(z); \quad |t| < 1$$

From this, by differentiation by t, we obtain

$$\partial\Phi/\partial t = \frac{z - t}{(1 + t^2 - 2tz)^{\frac{3}{2}}} \quad \text{or} \quad (1 + t^2 - 2tz)(d\Phi/dt) + (t - z)\Phi = 0$$

Substituting the series expression into this equation, we obtain

$$\sum_{m=0}^{\infty} mt^{m-1}P_m(z) + \sum_{s=0}^{\infty} st^{s+1}P_s(z) - \sum_{n=0}^{\infty} 2nzt^nP_n(z) + \sum_{s=0}^{\infty} t^{s+1}P_s(z) - \sum_{n=0}^{\infty} zt^nP_n(z) = 0$$

Equating the coefficients of t^n, we obtain the recurrence equation

$$(2n + 1)zP_n(z) = (n + 1)P_{n+1}(z) + nP_{n-1}(z); \quad \text{for } n = 0, 1, 2, 3, \ldots \tag{6.3.33}$$

Differentiating Φ by z gives us another relation,

$$\frac{\partial\Phi}{\partial z} = \frac{t}{(1 + t^2 - 2tz)^{\frac{3}{2}}} \quad \text{or} \quad (1 + t^2 - 2tz)\frac{\partial\Phi}{\partial z} = t\Phi$$

from which one obtains

$$\sum_{s=0}^{\infty} t^sP'_s + \sum_{m=0}^{\infty} t^{m+2}P'_m = \sum_{n=0}^{\infty} t^{n+1}[2zP'_n + P_n]; \quad \text{where } P'_n = \frac{d}{dz}P_n(z)$$

or, equating coefficients of powers of t^{n+1}, we obtain

$$P'_0 = 0; \quad P'_1 = P_0; \quad P'_{n+1} + P'_{n-1} = 2zP'_n + P_n; \quad n > 0 \tag{6.3.34}$$

We can also obtain

$$2t^2\frac{\partial\Phi}{\partial t} + t\Phi = (1 - t^2)\frac{\partial\Phi}{\partial z}$$

which results in the recurrence equations

$$P'_0 = 0; \quad P'_1 = P_0; \quad (2n + 1)P_n = P'_{n+1} - P'_{n-1}; \quad n > 0 \tag{6.3.35}$$

By juggling these three equations around or else by further shuffling of the generating function, we can obtain the further formulas

$$P'_{n+1} = (n + 1)P_n + zP'_n; \quad P'_{n-1} = -nP_n + zP'_n$$
$$(1 - z^2)P'_n = nP_{n-1} - nzP_n \tag{6.3.36}$$

and finally

$$(1 - z^2)P''_n - 2zP'_n + n(n + 1)P_n = 0$$

This last is, of course, the differential equation [see Eq. (5.3.23)] for the Legendre functions $P_n(z)$.

By actual expansion of $\Phi(t,z)$, we obtain $P_0(z) = 1$, $P_1(z) = z$; the rest of the P's can be obtained from Eq. (6.3.33). Tedious but straightforward algebra gives us the general formula

$$P_n(z) = \frac{(2n)!}{2^n(n!)^2}\left\{z^n - \frac{n(n-1)}{2(2n-1)}z^{n-2} + \frac{n(n-1)(n-2)(n-3)}{2.4(2n-1)(2n-3)}z^{n-4} - \cdots\right\} \tag{6.3.37}$$

from which we can find that

$$P_n(0) = \begin{cases} 0; & n \text{ odd} \\ (-1)^{\frac{1}{2}n}\dfrac{1\cdot 3\cdot 5\cdots(n-1)}{2\cdot 4\cdot 6\cdots(n)}; & n \text{ even}\end{cases} \tag{6.3.38}$$

From this, or by setting $z = \pm 1$ in the expansion for Φ and comparing coefficients of t^n, we obtain

$$P_n(1) = 1; \quad P_n(-1) = (-1)^n \tag{6.3.39}$$

We see that each of these eigenfunctions is a polynomial in z of degree n, those for even n having only even powers of z and those for odd n only odd powers. Conversely we can express each power of z in terms of a finite number of P_n's, an odd power in terms of P_n's for odd values of n equal to or less than the power of z, an even power for all even n's equal to or less than the power. These expansions may also be obtained by straightforward algebra:

$$1 = P_0(z); \quad z = P_1(z); \quad z^2 = \tfrac{2}{3}P_2(z) + \tfrac{1}{3}P_0(z); \quad z^3 = \tfrac{2}{5}P_3(z) + \tfrac{3}{5}P_1(z)$$

$$z^n = \frac{(2n+1)n(n-2)\cdots 2}{(2n+1)(2n-1)\cdots(n+1)}P_n(z) + \frac{(2n-3)n(n-2)\cdots 4}{(2n-1)(2n-3)\cdots(n+1)}P_{n-2}(z) + \cdots + \frac{1}{n+1}P_0(z); \quad n = 4, 6, 8, \ldots \tag{6.3.40}$$

$$z^n = \frac{(2n+1)(n-1)(n-3)\cdots 2}{(2n+1)(2n-1)\cdots(n+2)}P_n(z) + \frac{(2n-3)(n-1)(n-3)\cdots 4}{(2n-1)(2n-3)\cdots(n+2)}P_{n-2}(z) + \cdots + \frac{3}{n+2}P_1(z); \quad n = 5, 7, 9, \ldots$$

Therefore any function, analytic in the region $|z| < 1$, which can thus be expanded in a convergent power series in z can be expanded in a convergent series of the function $P_n(z)$. The generating function Φ for the P_n's is closely related to the kernel of the integral representation for P_α, given in Eq. (5.3.26) as, of course, it must be.

It might be remarked, in passing, that the Legendre functions can be obtained in the following manner: We wish to represent any analytic function of z in the range $|z| < 1$; we can do this by means of a power series, but the powers of z are not orthogonal to integration from -1

to $+1$ (though they are to integration in a circle around the origin); consequently we start with 1 and z (which are orthogonal), pick a combination of z^2 and 1 which is orthogonal to 1 (and to z), and so on, each function $\psi_n(z)$ being a linear combination of $z^n, z^{n-2}, \ldots$, which is orthogonal to ψ_m $(m < n)$ for integration from -1 to $+1$. The resulting ψ_n will be proportional to the spherical harmonic $P_n(z)$ (see discussion in the table at end of this chapter).

The formula given on page 549, expressing $P_n(z)$ in terms of the nth derivative of $(z^2 - 1)$, can also be obtained from the generating function. The solution of the quadratic equation $y = z + \frac{1}{2}t(y^2 - 1)$ which approaches z as $t \to 0$ is

$$y = \frac{1}{t}[1 - \sqrt{1 - 2zt + t^2}] \quad \text{and} \quad \frac{\partial y}{\partial z} = \Phi(t,z) = P_0 + \sum_{n=1}^{\infty} t^n P_n(z)$$

But the solution of the quadratic equation which approaches z as $t \to 0$ can be given by the Lagrange expansion [see Eq. (4.5.8)]

$$y = z + \sum_{n=1}^{\infty} \frac{t^n}{2^n n!} \frac{d^{n-1}}{dz^{n-1}} (z^2 - 1)^n$$

so that
$$\frac{\partial y}{\partial z} = 1 + \sum_{n=1}^{\infty} \frac{t^n}{2^n n!} \frac{d^n}{dz^n} (z^2 - 1)^n = \Phi(t,z)$$

Equating coefficients of equal powers of t, we have

$$P_0(z) = 1; \quad P_n(z) = \frac{1}{2^n n!} \frac{d^n}{dz^n} (z^2 - 1)^n \qquad (6.3.41)$$

This last equation enables us to compute the normalization integral for the Legendre functions. Suppose first that $n > m$ in the integral

$$I_{mn} = \int_{-1}^{1} P_m(z) P_n(z)\, dz = \frac{(-1)^{m+n}}{2^{m+n} m! n!} \int_{-1}^{1} \frac{d^m}{dz^m} (1 - z^2)^m \frac{d^n}{dz^n} (1 - z^2)^n\, dz$$

We integrate by parts m times, obtaining

$$\begin{aligned} I_{mn} &= \frac{(-1)^{m+n-1}}{2^{m+n} m! n!} \int_{-1}^{1} \frac{d^{m+1}}{dz^{m+1}} (1 - z^2)^m \frac{d^{n-1}}{dz^{n-1}} (1 - z^2)^n\, dz = \cdots \\ &= \frac{(-1)^n}{2^{m+n} m! n!} \int_{-1}^{1} \frac{d^{2m}}{dz^{2m}} (1 - z^2)^m \frac{d^{n-m}}{dz^{n-m}} (1 - z^2)^n\, dz \\ &= \frac{(-1)^{n-m}(2m)!}{2^{m+n} m! n!} \int_{-1}^{1} \frac{d^{n-m}}{dz^{n-m}} (1 - z^2)^n\, dz = 0; \quad n > m \end{aligned}$$

If $n = m$, on the other hand, this last integral is

$$I_{nn} = \frac{(2n)!}{2^{2n}(n!)^2} \int_{-1}^{1} (1 - z^2)^n\, dz = \frac{2}{2n + 1} = E_n^2 \qquad (6.3.42)$$

Thus we have shown that these functions are mutually orthogonal (as, of course, they must be, since they are eigenfunctions), and we have computed the values of the normalizing constants E_n [see Eq. (6.3.16)]. We can apply our general variational argument to show that the set is a complete one and that the series

$$S(x) = \int_{-1}^{1} f(\zeta) \left[\sum_{n=0}^{\infty} \left(\frac{2n+1}{2} \right) P_n(x) P_n(\zeta) \right] d\zeta \tag{6.3.43}$$

gives a good least-squares fit to an arbitrary, piecewise continuous function $f(x)$. We should expect, of course, that Gibbs' phenomenon would be observed next to each discontinuity but that we could integrate the series and be sure that such integrals would everywhere equal the corresponding integrals of the function f. (We might have trouble with differentials of the series, however.) For instance, as long as it is used for integration with continuous functions, the series in the square brackets in Eq. (6.3.43) can be considered to be equivalent to the delta function $\delta(x - \zeta)$ in the range $-1 < (x,\zeta) < 1$.

In addition to the delta function, another extremely useful discontinuous function may be represented by a series of Legendre polynomials

$$\frac{1}{z - \zeta} = \sum_{n=0}^{\infty} \frac{2n+1}{2} \left[\int_{-1}^{1} \frac{P_n(w)}{z - w} dw \right] P_n(\zeta)$$

If z is not on the real axis between -1 and $+1$, the quantity in square brackets converges and may be computed by integration by parts:

$$\begin{aligned}(-1)^n \frac{2n+1}{2^{n+1} n!} \int_{-1}^{1} \frac{1}{z - w} \frac{d^n}{dw^n} (1 - w^2)^n \, dw \\ = \left(\frac{2n+1}{2^{n+1}} \right) \int_{-1}^{1} \frac{(1 - w^2)^n}{(z - w)^{n+1}} dw \\ = (2n + 1) Q_n(z)\end{aligned}$$

Where $Q_n(z)$ [see Eq. (5.3.29)] is the Legendre function of the second kind. Consequently we obtain the useful series

$$\frac{1}{z - \zeta} = \sum_{n=0}^{\infty} (2n + 1) Q_n(z) P_n(\zeta) \tag{6.3.44}$$

This converges only when $|z| > 1$ and $|\zeta| < 1$, but integrals of the series multiplied by analytic functions can be carried out over still wider ranges of the variables, as with the series for the delta function. It is important, however, to remember that the function $Q_n(z)$ has a branch cut along the real axis of z, between -1 and $+1$.

This last equation has several interesting and useful consequences. In the first place the expansion for a function $f(z)$ which is analytic within and on an ellipse C with foci at ± 1 is

$$f(z) = \sum_{n=0}^{\infty} a_n P_n(z)$$

where the alternate forms of the equation determining the coefficients are

$$\begin{aligned} a_n &= \frac{2n+1}{2\pi i} \oint_C f(t) Q_n(t)\, dt = (n + \tfrac{1}{2}) \int_{-1}^{1} f(x) P_n(x)\, dx \\ &= \frac{n + \frac{1}{2}}{2^n n!} \int_{-1}^{1} (1 - x^2)^n \frac{d^n}{dx^n} f(x)\, dx \end{aligned} \tag{6.3.45}$$

The first form of the integral is useful with functions analytic for large values of $|z|$, for the asymptotic form (5.3.30) of Q_n may be used along a large ellipse for the contour C.

In the second place, by using the generating function for P_n, we find that

$$(1 + t^2 - 2tz)^{-\frac{1}{2}} \cosh^{-1}\left[\frac{t - z}{\sqrt{z^2 - 1}}\right] = \sum_{n=0}^{\infty} t^n Q_n(z) \tag{6.3.46}$$

which defines the generating function for Q_n (though Q_n is not an eigenfunction).

Many other useful sets of one-dimensional eigenfunctions can be worked out from suitable generating functions. Some of them will be listed in the table at the end of this chapter, and some will be the subjects of problems.

Eigenfunctions in Several Dimensions. So far we have been discussing eigenfunctions in one dimension, solutions of a Liouville equation in one independent variable satisfying boundary conditions of fairly general type at the two ends of the range. Partial differential equations in three or more variables give rise to eigenfunctions in more than one variable. These are solutions of a partial differential equation, with one of the coordinates separated off, leaving a separation constant behind. The solution of this equation must satisfy some simple boundary condition along some boundary line or surface (usually either homogeneous Dirichlet or Neumann conditions or a homogeneous combination). This is usually possible only for certain discrete values (eigenvalues) of the separation constant.

Many of the arguments of the Sturm-Liouville theory can be carried over to the multidimensional case. There is a lowest eigenvalue for most forms of the equation, and the eigenfunction corresponding to this lowest value usually has no nodal lines (or surfaces) within the boundary. The eigenfunctions cannot be put in any simple, one-dimensional array, however, as was the case in one dimension. The difficulty is that there is more than one solution with one node (even when there is no degeneracy). One can place the solutions in some sort of a two-dimensional array, however, for it usually turns out that the nodes for a two-dimensional case fall into two families (see discussion on page 497), and an increase in the number of nodes of one kind always results in an increase in the corresponding eigenvalue.

In those cases where there is a lowest eigenvalue, where there is no upper bound to the eigenvalues, and where there is a variational equation equivalent to the differential equation for the eigenfunctions, the arguments set forth in the previous pages can be utilized to prove that these more general eigenfunctions are a complete, orthogonal set (or if there are degenerate cases, they can be made orthogonal) which can represent any piecewise continuous function inside the boundary in terms of a series.

When the partial differential equation for the eigenfunctions itself separates, these matters are easy to prove. Each eigenfunction is a simple product of eigenfunctions of the separated coordinates (except for degenerate cases, where the solutions may be a finite combination of products for each of the degenerate states). The orthogonality and completeness may be derived from the orthogonality and completeness of the one-dimensional factors.

For instance, we can consider the case of a uniform, flexible membrane, of mass ρ per unit area, stretched under uniform tension T per unit length between rigid supports. The equation of motion is the wave equation

$$\nabla^2\Psi - \frac{1}{c^2}\frac{\partial^2\Psi}{\partial t^2} = 0; \quad c^2 = \frac{T}{\rho}$$

where Ψ is the displacement of the membrane from its equilibrium position. By requiring simple harmonic dependence on time, we can separate off the time factor: $\Psi = \psi(x,y)e^{-i\omega t}$, where

$$\frac{\partial^2\psi}{\partial x^2} + \frac{\partial^2\psi}{\partial y^2} + k^2\psi = 0; \quad k = \frac{\omega}{c} \tag{6.3.47}$$

which is a Helmholtz equation in two variables x and y (or, by transformation, r and ϕ, etc.) with separation constant k^2.

If the supports are rectangular ones, at $x = 0$ and a, $y = 0$ and b, the eigenfunctions and eigenvalues are

$$\psi_{mn}(x,y) = \sin(\pi mx/a)\sin(\pi ny/b)$$
$$k^2_{mn} = \pi^2[(m/a)^2 + (n/b)^2]; \quad m, n = 1, 2, 3, 4, \ldots \tag{6.3.48}$$

Since the individual factors are mutually orthogonal, the products also are, and by the same reasoning as before, the ψ_{mn}'s are a complete set. The nodal lines (lines where $\psi = 0$ inside the boundary) are either perpendicular to the x axis or else perpendicular to the y axis. The number of the first equals $(m - 1)$, and the number of the second is $(n - 1)$. Therefore, although the arranging of the k_{mn}^2's in order of size bears no simple relation to the sequence of numbers m and n, nevertheless an increase in m by 1 increases the nodes perpendicular to x by 1 and increases the corresponding k^2, and an increase of n by 1 increases the nodes perpendicular to y and also increases k^2.

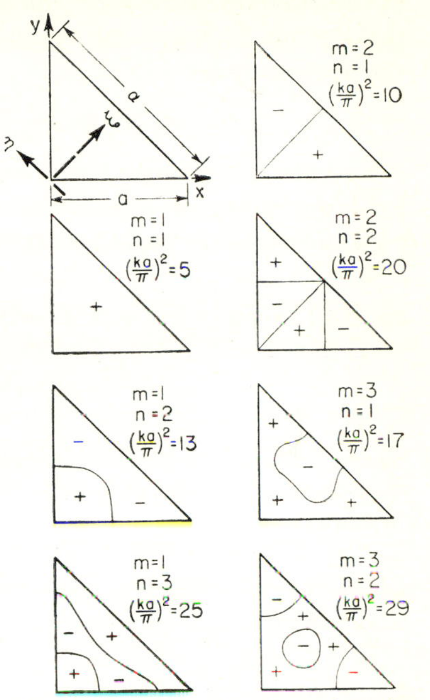

Fig. 6.10 Eigenfunctions for the triangular membrane. Lines inside the boundaries indicate position of nodal lines, where $\psi = 0$, for a few values of the quantum numbers m and n.

If $a = b$, we have a condition of degeneracy, for $k_{mn} = k_{nm}$, and there are at least two eigenfunctions for the same eigenvalue of k. In this case, we could use any linear combination of ψ_{mn} and ψ_{nm} for our eigenfunction. Since the ψ_{mn}'s are already mutually orthogonal, it is advisable to use them as they stand.

We can then expand any piecewise continuous function $f(x,y)$ in a series

$$f(x,y) = \sum_{m,n} \left[\frac{4}{ab} \int_0^a du \int_0^b dv\, f(u,v)\psi_{mn}(u,v) \right] \psi_{mn}(x,y) \quad (6.3.49)$$

If f is an analytic function of x and of y in the ranges $(0,a)$ and $(0,b)$, respectively, then the series will be uniformly convergent in this range; if f has discontinuities, the series may exhibit Gibbs' phenomenon at the discontinuities, but the integral of the series will be a convergent series.

The analysis is rather more involved when the equation is not separable in coordinates suitable for the boundary. Unfortunately only two nonseparable cases have been solved in detail, one for a boundary which is an isosceles right triangle. This is probably too simple to bring out all the complexities; nevertheless it is worth some discussion (see Sec. 11.2).

To make the discussion specific, let us consider the membrane again, only this time we choose the boundary to be the three straight lines, from the origin to $(0,a)$, from $(0,a)$ to $(a,0)$, and from $(a,0)$ to the origin again. The equations are the same as before, and for simple harmonic time dependence, the Helmholtz equation (6.3.47) applies. A solution is the product $\sin(\mu\pi x/a)\cdot\sin(\nu\pi y/a)$, which is zero along the x and y parts of the boundary but is not zero along the diagonal part. However, the combination

$$\sin(\mu\pi x/a)\,\sin(\nu\pi y/a) \mp \sin(\mu\pi y/a)\,\sin(\nu\pi x/a)$$

is zero along the diagonal if μ and ν are integers (the + or − sign being used, depending on whether $|\mu - \nu|$ is an even or an odd integer). The eigenfunctions for this case are therefore

$$\psi_{mn}(x,y) = \sin\left[\frac{\pi}{a}(m+n)x\right]\sin\left[\frac{\pi}{a}ny\right] - (-1)^m\sin\left[\frac{\pi}{a}(m+n)y\right]\sin\left[\frac{\pi}{a}nx\right] \quad (6.3.50)$$

where m and n are positive integers. To show how this fits the boundary conditions along the diagonal, we rotate our axes by $\pi/4$, setting

$$x = (1/\sqrt{2})(\xi - \eta);\quad y = (1/\sqrt{2})(\xi + \eta);\quad \alpha = \sqrt{2}\,a$$

where the coordinates ξ and η are shown in Fig. 6.10. Performing a few trigonometric manipulations we obtain

$$\begin{aligned}\psi_{mn} &= \sin\left[\frac{\pi}{\alpha}(m+2n)\xi\right]\sin\left[\frac{\pi}{\alpha}m\eta\right] - \sin\left[\frac{\pi}{\alpha}(m+2n)\eta\right]\sin\left[\frac{\pi}{\alpha}m\xi\right]; \\ &\qquad m = 2, 4, \ldots \\ &= \cos\left[\frac{\pi}{\alpha}(m+2n)\eta\right]\cos\left[\frac{\pi}{\alpha}m\xi\right] - \cos\left[\frac{\pi}{\alpha}(m+2n)\xi\right]\cos\left[\frac{\pi}{\alpha}m\eta\right]; \\ &\qquad m = 1, 3, \ldots\end{aligned}$$

which shows that ψ vanishes for $\xi = \alpha/2$, which corresponds to the diagonal boundary.

The eigenvalues for this case are

$$k^2_{mn} = \left(\frac{\pi}{a}\right)^2[(m+n)^2 + n^2] = \left(\frac{\pi}{a}\right)^2[m^2 + 2mn + 2n^2] \quad (6.3.51)$$

which have a lowest value ($m = n = 1$) and no upper bound. The eigenfunctions are mutually orthogonal, and they are solutions of the variational equation

$$\delta\iint\left[\left(\frac{\partial\psi}{\partial x}\right)^2 + \left(\frac{\partial\psi}{\partial y}\right)^2\right]dx\,dy = 0;\quad \iint\psi^2\,dx\,dy = E^2_{nm} = \tfrac{1}{8}a^2$$

Therefore the functions form a complete set. The interrelation between the values of k^2 and the number and disposition of the nodal lines is not obvious, however.

Nodal lines for some of the eigenfunctions are shown in Fig. 6.10.

We notice that they do not fall into two mutually orthogonal families as do the ones for the rectangular boundary (and, indeed, for all separable cases). All we can say is that, if either m or n is increased, more nodes are introduced and k^2 is increased (which is, after all, the essential result of the Sturm-Liouville analysis).

The expansion of a function $f(x,y)$ in terms of these eigenfunctions has the same form as that given in Eq. (6.3.49), except that the integrals in the brackets are over the triangular area within the new boundaries instead of over the rectangular area and that the factor $4/a^2$ in front of the integrals is changed to $8/a^2$. The same remarks can be made concerning convergence as have been made several times before.

Separability of Separation Constants. The complications arising when no suitable separable coordinates exist for a given boundary are usually so severe that the set of eigenfunctions cannot be computed and the boundary-value problem cannot be solved. But even when separation can be achieved, certain complications, hitherto glossed over, can arise which make solution of the boundary-value problem quite difficult in practice. These difficulties arise in the cases, discussed on pages 516 to 518, where there is not complete separation of the *separation constants*.

When the separation constants separate [case A, given in Eq. (5.1.41)], the equations for the factors for ξ_2 and ξ_3 contain only one separation constant each and thus correspond to the simple form of the Liouville equation, which we have already discussed in detail. The boundary conditions for the factor X_2 fix the eigenvalues of k_2, those for X_3 the eigenvalues of k_3, and these values can then be inserted, *as known constants*, into the equation for X_1 to determine the eigenvalues of k_1. The series of eigenfunctions to fit a boundary condition along the surface ξ_1 = constant is then a simple double series over all eigenvalues of k_2 and all eigenvalues of k_3.

In case B_2, given in Eqs. (5.1.42), it is still possible to obtain the eigenvalues for one factor, X_3, without knowing the values of the other separation constants.

But in the other cases the separation constants do not separate, and the setting up of a series of eigenfunctions to satisfy boundary conditions on a surface will involve a simultaneous solution for two eigenvalues for each term in the series. As an example of the complications which arise and also as an indication as to the method of solution, we consider the case of the vibrations of an elliptic membrane. We separate the Helmholtz equation $\nabla^2\psi + k^2\psi = 0$ in the elliptic coordinates

$$x = d \cos \phi \cosh \mu; \quad y = d \sin \phi \sinh \mu; \quad \psi = M(\mu)\Phi(\phi)$$

obtaining

$$(d^2M/d\mu^2) + [h^2 \cosh^2 \mu - b]M = 0; \quad (d^2\Phi/d\phi^2) + [b - h^2 \cos^2 \phi]\Phi = 0$$

where $h = kd$ and k is related to the vibration frequency $\nu = \omega/2\pi$ (for use of the Helmholtz equation corresponds to an assumption that the motion is simple harmonic, with time factor $e^{-i\omega t}$) by the relation $k = \omega/c = 2\pi\nu/c$. The constant b is the separation constant, but of course, $h = 2\pi\nu d/c$ is also a separation constant, so both equations contain both constants. Referring to Sec. 5.2, we see that the second of these equations is the same as Eq. (5.2.67).

The boundary condition for Φ is that it be periodic with ϕ, and the boundary condition for M is that it be finite and continuous at $\mu = 0$ and that it be zero at $\mu = \mu_0$, the ellipse corresponding to the boundary. We must simultaneously adjust b and h so that both sets of conditions are fulfilled.

To do this we solve the equation for ϕ, assuming that h is known. As shown on page 562, the periodic solutions of Mathieu's equation are expressible as Fourier series. There are two sorts: one set, $Se_m(h, \cos\phi)$, even in ϕ, and the other set, $So_m(h, \cos\phi)$, odd in ϕ; corresponding to the first set of eigenfunctions are the eigenvalues $be_m(h)$ and to the second set are $bo_m(h)$. These eigenvalues are "interleaved," so that, for a given value of h,

$$be_0 < bo_1 < be_1 < \cdots bo_m < be_m < bo_{m+1} < \cdots$$

Therefore, *if h were fixed arbitrarily*, the sequence Se, So, forms a complete, orthogonal set of eigenfunctions.

But h is not fixed, so all we can say is that we have obtained a relationship between h and a sequence of values of b which will allow the boundary conditions on Φ to be satisfied.

The solutions for M are those given in Eq. (5.3.90) or (5.3.91), and we must first decide whether Je or Ne fits the requirements of continuity at $\mu = 0$. Examination of the coordinates themselves shows that ($\mu = 0$, $\phi = \alpha$) and ($\mu = 0$, $\phi = -\alpha$) are really the same point ($x = d\cos\alpha$, $y = 0$), and a few minutes cogitation will show that, in order that ψ be continuous in value and slope across the line $\mu = 0$, when factor Φ is even in ϕ, then factor M must have zero slope at $\mu = 0$; when factor Φ is odd in ϕ, then M must go to zero at $\mu = 0$. Reference to page 635 shows that the proper solution to go with $Se_m(h, \cos\phi)$ is $Je_m(h, \cos\phi)$ for the same value of h and of m (*i.e.*, the same value of b) and the proper one to go with So_m is Jo_m with corresponding h and m. The functions Ne, No are therefore not used in this problem.

We finally are supposed to set $M = 0$ at $\mu = \mu_0$. For any given value of b, it is possible to adjust h so that this is obtained, but of course, b also depends on h, so h and b must be determined simultaneously. In practice, we would compute a series of curves giving the sequence of values of μ for which $Je_m(h, \cosh\mu)$, for instance, went to zero for each different value of h. These values could be called $\mu^e_{mn}(h)$, the subscript m

corresponding to that of Je_m and the subscript n labeling which in the sequence of zero points we mean (μ_{m1} can be the smallest value, μ_{m2} next, and so on).

We thus have a twofold array of values of μ, onefold for different values of m and the other for different values of n. Each of these values is a function of h. We now invert the problem by solving for the value of h to make each one of the roots $\mu_{mn}(h)$ equal to μ_0, the coordinate of the boundary. The root of the equation $\mu^e_{mn}(h) = \mu_0$ will be called he_{mn}, and that of the equation $\mu^0_{mn}(h) = \mu_0$ will be called ho_{mn}. From these values we can then compute the allowed values of frequency ν of free vibration of the elliptic membrane. Corresponding to the particular frequency $\omega^e_{mn}/2\pi = (c/2\pi d)he_{mn}$ we shall have the two-dimensional eigenfunction $Se_m(he_{mn}, \cos\phi)Je_m(he_{mn}, \cos h\mu)$, and corresponding to the eigenvalue $\omega^0_{mn}/2\pi = (c/2\pi d)ho_{mn}$ we have the other eigenfunction $So_m(ho_{mn}, \cos\phi)Jo_m(ho_{mn}, \cosh\mu)$. The complete solution will thus be

$$\sum_{m,n} [A_{mn}Se_mJe_m \cos(\omega^e_{mn}t + \alpha_{mn}) + B_{mn}So_mJo_m \cos(\omega^0_{mn}t + \beta_{mn})]$$

with the A's, B's, α's, and β's determined by the initial displacement and velocity of the membrane.

It is apparent, of course, that the factors Se, So are mutually orthogonal *if they all correspond to the same value of h.* But the terms in the above series are each for a *different* value of h, so the factors Se_m are not all mutually orthogonal, nor are the factors So_m (though all the Se's are still orthogonal to all the So's). However, the general arguments which were outlined on page 727 show us that our present two-dimensional eigenfunctions are mutually orthogonal, so it must be that the functions $Je_m(he_{mn}, \cosh\mu)$ must be mutually orthogonal over the range $0 < \mu < \mu_0$ and likewise the functions $Jo_m(ho_{mn}, \cosh\mu)$, for different values of n, whereas the functions $Se_m(he_{mn}, \cos\phi)$ must be mutually orthogonal for different values of m. Therefore all terms in the series are orthogonal to all others, after integrating over the area of the membrane, and the coefficients A, B and phase angles α, β can be determined in the usual manner.

By this example we have shown that the eigenfunction technique of fitting boundary conditions can always be used when the equation separates in the coordinates suitable for the boundary, even though the separation constants may not be separated in the resulting ordinary equations. Such cases require a great deal more computation to obtain a solution than do the cases where one of the separated equations contains only one separation constant, but the calculations are straightforward, and no new principles need to be invoked.

Density of Eigenvalues. The rather haphazard distribution of eigenvalues exhibited by even the simple two-dimensional cases brings up a question we have hitherto neglected: Is there anything we can say about the number of eigenvalues between λ and $\lambda + \epsilon$? This number is a discontinuous function of λ and ϵ (since the eigenvalues are a discrete set), but we might try to compute the *average* number of eigenvalues between λ and $\lambda + \epsilon$. Presumably this quantity would be a "smoothed-out" approximation to the actual number and would be a continuous function of λ and ϵ.

Such an *approximate density function* for eigenvalues can be obtained for the one-dimensional case by using the asymptotic formula following Eq. (6.3.27). We shall actually calculate the density of k_n's, where $k_n^2 = \lambda_n$. This is generally more useful, for if the equation is the wave equation, k is proportional to the allowed frequency of free vibration of the system. The asymptotic formula for the eigenvalues of k is

$$k_n = \sqrt{\lambda_n} \simeq (n\pi + \alpha) \Big/ \left(\int_a^b \sqrt{\frac{r}{p}}\, dz \right)$$

This formula means that the higher values of k are distributed uniformly along the real k axis. The asymptotic spacing between successive values is $\pi/(\int \sqrt{r/p}\, dz)$, the number of eigenvalues less than k is

$$n(k) \simeq \frac{k}{\pi} \int_a^b \sqrt{\frac{r}{p}}\, dz$$

and the average number of eigenvalues between k and $k + dk$ is therefore

$$dn \simeq \left[\frac{1}{\pi} \int_a^b \sqrt{\frac{r}{p}}\, dz \right] dk \tag{6.3.52}$$

where the quantity in brackets can be called the *average density* of eigenvalues of k, for large values of k.

We notice that this density is proportional to an integral of a function $\sqrt{r/p}$, integrated over the interval a, b. If r and p have the same dimensions (in which case $1/k$ has dimensions of length), then the integral of $\sqrt{r/p}$ has the dimensions of a length. For the wave equation in one dimension, r and p are unity and the average density of eigenvalues is just $1/\pi$ times the distance from a to b. The integrand $\sqrt{r/p}$ is related to the scale factors of the curvilinear coordinates, so that a deviation of $\sqrt{r/p}$ from unity is somehow related to the curvature of the coordinate under consideration. The integral $\int \sqrt{r/p}\, dz$ can be called the *effective interval length* for the coordinate and the boundaries in question.

Turning now to the two-dimensional case we find here that the density of eigenvalues of k is not approximately independent of k. For example, plotting the values given in Eq. (6.3.48), for a rectangular membrane, we see that the density seems to *increase* with increasing k.

That this is correct can be shown by utilizing the particularly simple relationship between k and the integers m and n. Equation (6.3.48) is analogous to the equation relating the distance from the origin to a point specified by the cartesian coordinates $\pi m/a$ and $\pi n/b$. These points, for integral values of m and n, are plotted in Fig. 6.11 as the intersections of a rectangular grid. An allowed value of k corresponds to the distance of any one of these grid points to the origin.

We can then speak of "the density of allowed points in k space." Since the spacing of the grid in the two directions is π/a and π/b, the average density of points is ab/π^2, where $ab = A$, the *area enclosed in the boundaries.* Consequently for this simple case the average number of eigenvalues of k less than a value k is

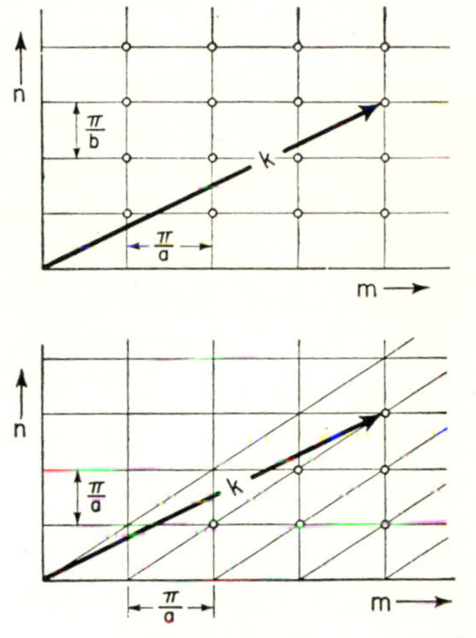

Fig. 6.11 Distribution of eigenvalues for a rectangular and triangular membrane. Length of vector k is value $\sqrt{\lambda}$.

$$n(k) \simeq \frac{A}{\pi^2}\left(\tfrac{1}{4}\pi k^2\right) = \frac{A}{4\pi}k^2$$

where the quantity in parenthesis is the area between the lines $r = k$, $y = 0$, and $x = 0$. The differential of this

$$dn \simeq [Ak/2\pi]\,dk \qquad (6.3.53)$$

gives the density of eigenvalues of k for this simple case (the quantity in brackets). We see that it is proportional to k (which is what we wanted to prove) and also is proportional to the area A inside the rectangular boundary.

We can go further than this, however, for we note that the area covered by the dots, in Fig. 6.11, is not quite all of the first quadrant. Since the points for $m = 0$ or for $n = 0$ are missing (for the boundary condition $\psi = 0$), we should remove half a grid strip adjacent to the two axes, and a more accurate formula would be

$$n(k) \simeq (ab/\pi^2)[\tfrac{1}{4}\pi k^2 - (k\pi/2a) - (k\pi/2b)] = (A/4\pi)k^2 - (L/4\pi)k$$

or

$$dn \simeq [(Ak/2\pi) - (L/4\pi)]\,dk \qquad (6.3.54)$$

where $L = 2a + 2b$ is the perimeter of the boundary.

This second term in the expression for the density of eigenvalues depends on the boundary conditions. For instance, if the boundary conditions at the inner surface of the rectangle were that the normal gradient of ψ be zero instead of ψ, then the expression for the eigenvalues of k would be the same but the points for $m = 0$ and $n = 0$ would now be allowed. Consequently the density function would be $[(Ak/2\pi) + (L/4\pi)]$.

It appears that the first term in the average density is independent of the precise kind of boundary conditions applied as long as *some* sort of condition is applied; it depends only on the area inside the boundary. The second term depends on the particular boundary condition imposed.

Other cases where exact solutions of the Helmholtz equation in two dimensions are known (for instance, the case of a circular membrane) can be worked out. It turns out that, if the shape of the boundary is changed but the area A is not changed, the points representing allowed values of k are moved around on the "k plane" but *the average density of points is not changed.* In addition one can show that, if the boundary conditions are not changed as the boundary shape is varied, the correction term (in terms of L, the perimeter of the boundary) is not changed in form. One can verify both of these statements by taking the case of the triangular membrane. Here the reduction to half the area seems to eliminate half of the points in the first quadrant. Detailed counting shows that, for the boundary condition $\psi = 0$, Eq. (6.3.54) holds, with $A = \frac{1}{2}a^2$ and $L = 2a + \sqrt{2a^2}$.

Equation (6.3.54) presumably holds for boundaries of any shape, for conditions $\psi = 0$ at the boundaries. It has been proved that the first term (which is the preponderating term for large k) is valid in general. The proof has not been extended to the second term, though there have not been found any contradictory cases among those which can be worked out in detail. In any case, if we are dealing with large values of k, we can neglect the second term in the density, using just the first term, which is known to be valid for all boundaries and all reasonable boundary conditions.

Thus we have shown that for the higher eigenfunctions the density of eigenvalues is independent of k for one-dimensional cases, is proportional to k for two-dimensional cases, and, by extension, is proportional to k^2 for three-dimensional cases. In each case the density is also proportional to the "size" of the space within the boundary; length for one dimension, area for two, and so on. If the equation is the Helmholtz equation, and if the interval is not curved, these "sizes" are the actual lengths, areas, etc., but if the coordinates are curved and the line or surface bounded by the boundary is curved, the "sizes" are effective lengths, areas, etc., being the integral of certain combinations of scale factors, similar to the integral $\int \sqrt{r/p}\, dz$ for the one-dimensional case.

Continuous Distribution of Eigenvalues. We see, from the discussion above, that the spacing between eigenvalues diminishes as the size of the boundaries increases. For instance, for the one-dimensional case, the average difference between successive eigenvalues, according to Eq. (6.3.52), is π divided by the effective interval length $\int \sqrt{r/p}\, dz$. As this length goes to infinity, the separation between eigenvalues goes to zero until *all values* of λ (or k) *larger than the lowest value are eigenvalues.*

In this limiting case we have a continuous distribution of eigenvalues, and our series representations turn into integral representations.

Our first example, on page 709, can be extended to show the transition. Suppose that the value of a, the distance between the two ends of the range of x, is increased indefinitely. The eigenfunction $\sin(\pi n x/a)$ for a given n will have a longer and longer wavelength, and its normalization constant $E_n = \sqrt{a/2}$ will become larger and larger. The Fourier series

$$f(x) = \sum_{n=1}^{\infty} \sin\left(\frac{\pi n x}{a}\right)\left[\frac{2}{a}\int_0^a f(\zeta)\sin\left(\frac{\pi n \zeta}{a}\right)d\zeta\right]$$

will conserve its general form, but each individual term will reduce in size and (for a given finite value of x) will change less and less rapidly from n to successive n until, near the limit, each term is vanishingly small and the rate of change with n of the coefficients in the brackets is vanishingly slow. In fact, in the limit, it is not advisable to use n as the variable of summation, but $k = n\pi/a$, the square root of the separation constant λ. The spacing between allowed values of this variable becomes smaller and smaller, until eventually k is a continuous variable and the summation over n becomes an integral over k from zero to infinity.

As a is made ever larger, the average number of eigenvalues of k between k and $k + dk$, $(a/\pi)\,dk$, given by Eq. (6.3.52), becomes more and more nearly equal to the actual number, as more and more allowed values of k are found in any finite segment dk. At the same time the successive terms in the sum, for the eigenfunctions for those eigenvalues within dk, differ less and less from each other (at least for finite values of x), so that we can finally represent the sum of all the terms between $n = ak/\pi$ and $n + (a\,dk/\pi)$ by the expression

$$\left(\frac{a\,dk}{\pi}\right)\sin(kx)\left[\frac{2}{a}\int_0^a f(\zeta)\sin(k\zeta)\,d\zeta\right]$$

and when a goes to the limit, the sum finally becomes the integral

$$f(x) = \frac{2}{\pi}\int_0^\infty \sin(kx)\int_0^\infty f(\zeta)\sin(k\zeta)\,d\zeta\,dk \tag{6.3.55}$$

which can be used to express any piecewise continuous function in the range $0 < x < \infty$ which goes to zero at the ends of the range. This is one form of the Fourier integral (see page 454). The more general form

$$f(x) = \frac{1}{2\pi}\int_{-\infty}^{\infty} e^{ikx}\,dk\int_{-\infty}^{\infty} f(\zeta)e^{-ik\zeta}\,d\zeta \tag{6.3.56}$$

can be obtained by using the more general boundary conditions that the functions be periodic in x with period a.

But after all, it should not be necessary to obtain the properties of eigenfunctions for continuous eigenvalues by a clumsy limiting process. From the point of view of function space, we can express a vector **F** representing any function F in terms of its components $F(x)$ along the unit vectors $\mathbf{e}(x)$ (corresponding to each value of x in the allowed range) or in terms of its components $f(k)$ along the alternative set of unit vectors $\mathbf{e}(k)$ (corresponding to each allowed value of k). Instead of having one continuous set and one discrete set, we now have two continuous sets, a more symmetric arrangement.

The eigenfunctions are still the projections of the vectors $\mathbf{e}(k)$ on the vectors $\mathbf{e}(x)$, but since both sets of vectors are preferably unit vectors, these eigenfunctions are now direction cosines and the whole formalism gains considerably in symmetry. The eigenfunctions $\psi(k,x)$ are at the same time components of the vectors $\mathbf{e}(k)$ on the vectors $\mathbf{e}(x)$ but also components of the vector $\mathbf{e}(x)$ on the vectors $\mathbf{e}(k)$. Extending Eq. (6.3.17), the component $F(x)$ of an arbitrary vector **F** in the direction given by $\mathbf{e}(x)$ is related to the component $f(k)$ of **F** in the direction specified by $\mathbf{e}(k)$ by the equations

$$F(z) = \int f(k)\psi(k,z)r(k)\,dk; \quad f(k) = \int F(z)\bar{\psi}(k,z)r(z)\,dz \qquad (6.3.57)$$

where the function $r(k)$ is related to the density of eigenvalues of k along k, just as $r(x)$ is related to the density of eigenvalues along x.

The range of integration for these integrals is over the allowed ranges of k and x, sometimes from 0 to ∞, but more often from $-\infty$ to ∞. Letting k go to $-\infty$ does not mean that there is no lower bound to the eigenvalues of λ, for $\lambda = k^2$ and, as long as k is real, $\lambda > 0$. Furthermore when we include negative values of k, we can arrange to have ψ a complex quantity (as e^{ikx} in the example) if we use the complex conjugate $\bar{\psi}$ in the second integral.

The normalization and orthogonality properties of these eigenfunctions are obtained by extending Eqs. (6.3.18):

$$\begin{aligned} r(k)\int\psi(k,z)\bar{\psi}(\kappa,z)r(z)\,dz &= \delta(k-\kappa) \\ r(z)\int\psi(k,z)\bar{\psi}(k,\zeta)r(k)\,dk &= \delta(z-\zeta) \end{aligned} \qquad (6.3.58)$$

both of the integrals corresponding to delta functions. The functions ψ are direction cosines, since both $\mathbf{e}(x)$ and $\mathbf{e}(k)$ are unit vectors, so the factors E_m are unity and do not explicitly appear in the formulas. As always with expressions corresponding to delta functions, they are to be used in integrals, not in differentials or by themselves. The most useful form of the normalization integral corresponds to the usual definition of the delta function

$$\int_{-\Delta}^{\Delta} \delta(z-\zeta)\,d\zeta = \begin{cases} 0; & |z| > \Delta \\ 1; & |z| < \Delta \end{cases}$$

In other words, in the limit $\Delta \to 0$

$$\int_{k_0-\Delta}^{k_0+\Delta} r(\kappa)\, d\kappa \int \psi(k,z)\bar{\psi}(\kappa,z) r(z)\, dz = \begin{cases} 0; & |k - k_0| > \Delta \\ 1; & |k - k_0| < \Delta \end{cases} \tag{6.3.59}$$

with a similar equation, reversing z and k, corresponding to $\delta(z - \zeta)$. Usually the limits of the second integral are first set finite but large (0 to R, or $-R$ to R, R large) and then later extended to infinity, for ease in calculation.

As an example of this we consider the Bessel equation resulting from the separation of the Helmholtz equation in polar coordinates

$$\frac{d}{dr}\left(r\frac{dR}{dr}\right) + \left(k^2 r - \frac{m^2}{r}\right) R = 0 \tag{6.3.60}$$

where m is the separation constant coming from the ϕ factor and is an integer if ϕ can go from 0 to 2π (in other words, if Φ is periodic in ϕ).

If the boundary conditions for R are that it be finite at the singular point $r = 0$ and that $R = 0$ at $r = a$, the eigenfunctions are the Bessel functions [see Eq. (5.3.63)]

$$\begin{aligned} \psi_n(r) &= J_m(\alpha_n r/a); \quad J_m(\alpha_n) = 0 \\ k_n &= (\alpha_n/a); \quad n = 0, 1, 2, \ldots \end{aligned} \tag{6.3.61}$$

These functions are orthogonal (with density function r)

$$\int_0^a J_m\left(\frac{\alpha_n r}{a}\right) J_m\left(\frac{\alpha_l r}{a}\right) r\, dr = \begin{cases} 0; & l \neq n \\ -\dfrac{a^2}{2} J_{m+1}(\alpha_n) J_{m-1}(\alpha_n); & l = n \end{cases}$$

so that we have a complete set of eigenfunctions ($n = 0, 1, 2, \ldots$) for each value of m.

If a goes to infinity, the eigenvalues of k are a continuous sequence from $k = 0$ to infinity. Therefore the function $J_m(kr)$ is proportional to an eigenfunction for a continuous k.

To normalize, we set $\psi(k,z) = AJ_m(kz)$ and determine A by using Eq. (6.3.59). As mentioned previously, to simplify the calculations we make the upper limit of the second integral R and let R go to ∞ later. Using the asymptotic expression for the J's and doing the calculations for $m > 0$, we have

$$\begin{aligned} A^2 &\int_{k_0-\Delta}^{k_0+\Delta} \kappa\, d\kappa \int_0^R J_m(kz) J_m(\kappa z) z\, dz \\ &= A^2 \int_{k_0-\Delta}^{k_0+\Delta} \kappa\, d\kappa \left[\frac{\kappa R J_m(kR) J_{m-1}(\kappa R) - kR J_m(\kappa R) J_{m-1}(kR)}{k^2 - \kappa^2}\right] \\ &\simeq A^2 \int_{k_0-\Delta}^{k_0+\Delta} \kappa\, d\kappa \left\{\frac{2/\pi}{k^2 - \kappa^2} \sin\,[(k - \kappa)R]\right\} \xrightarrow[R\to\infty]{} \begin{cases} A^2; & |k - k_0| < \Delta \\ 0 & |k - k_0| > \Delta \end{cases} \end{aligned}$$

This final result also holds for $m = 0$, though some of the intermediate steps are different in form. Therefore the normalizing constant A is equal to unity, the normalized eigenfunction is just $J_m(kz)$, and the expression corresponding to the Fourier integral formula (6.3.55) is just

$$f(z) = \int_0^\infty J_m(kz)k\,dk \int_0^\infty f(\zeta)J_m(k\zeta)\,\zeta\,d\zeta \qquad (6.3.62)$$

which is called the *Fourier-Bessel integral.*

Eigenvalues for the Schroedinger Equation. In a number of solutions of the Schroedinger equation, cases are encountered where the eigenvalues are discrete for a certain range of values and are continuous for the rest of the range. A consideration of the relationship between the Schroedinger equation [see Eq. (2.6.28)] and the Sturm-Liouville problem will indicate how these come about and will perhaps shed further light on the Sturm-Liouville results [see also the discussion of Eqs. (12.3.25) and (12.3.27)].

The Schroedinger equation in one dimension is

$$\frac{d^2\psi}{dx^2} + \frac{2m}{\hbar^2}[E - V(x)]\psi = 0 \qquad (6.3.63)$$

where m is the mass of the particle under consideration, E is its total energy in the state considered, V is the potential energy, and $\hbar = h/2\pi$ is the modified Planck's constant. The probability that the particle is between x and $x + dx$ is proportional to $|\psi|^2\,dx$, and the average "current density" of the particle is proportional to the imaginary part of $\bar{\psi}(d\psi/dx)$ (that is, if ψ is a real function for x real, then the net current is zero).

According to classical mechanics the particle would be only where the particle energy E is larger than the potential energy V; in such regions the greater the value of $E - V$ (= kinetic energy), the greater the current density and the smaller the particle density. In fact the probable density of the particle would be proportional to (1/velocity) $\propto 1/\sqrt{E - V}$, and the probable current density would be proportional to $\sqrt{E - V}$. No energy E would be possible which would be less than V everywhere; all energies E would be possible for E larger than V somewhere. If a particle is in one minimum of potential separated from another minimum by a peak higher than E, then the particle cannot go from one minimum to the other.

In contrast, the Schroedinger equation is more restrictive in regard to allowed values of energy but less restrictive on location of the particle, as the derivation of the equation (pages 243 and 314) has suggested. Let us take a case where $V(x)$ has a minimum (we can place the minimum at $x = 0$ for convenience) and where the asymptotic value of V is larger than this minimum and analyze the equation from the point of view of the Sturm-Liouville problem (see page 722).

When E is less than V everywhere, the solution of Eq. (6.3.63) is not oscillatory; it behaves more like the real exponentials e^x and e^{-x}, and no combination of the two independent solutions will produce a solution which is finite both at $-\infty$ and at $+\infty$. As indicated in Fig. 6.8, if the solution goes to zero at $-\infty$, it will go to infinity at $x = +\infty$. Therefore no energy is allowed which is less than $V(x)$ everywhere along the real axis of x.

When E is less than $V(-\infty)$ and $V(+\infty)$ but larger than $V(0)$, the minimum value of V, ψ still behaves like the real exponentials in those regions where $E < V$, curving away from the x axis, but where $E > V$, it curves toward the axis, behaving like a trigonometric function. Starting ψ out at zero at $-\infty$, it will rise exponentially until x reaches the value where $V(x) = E$, then for a time ψ will curve back toward the axis. Beyond the second point where $V = E$, it will again have an exponential behavior. For some value of E, larger than $V(0)$, the curving back will be just sufficient so that ψ will fit smoothly onto a function similar to $e^{-\alpha x}$ in the right-hand region, so that it curves upward just enough to go to zero at $x = +\infty$.

This value of E is the *lowest eigenvalue* of the energy. The corresponding eigenfunction has a maximum value somewhere near $x = 0$, in the region where $E > V$. This is the region within which classic physics would expect the particle to be confined. But the quantity $|\psi|^2$ is not suddenly zero when $V > E$; it drops down exponentially to zero on both sides if E is an eigenvalue; so that wave mechanics predicts a small but finite probability that the particle be in a region where potential energy is larger than total energy (*i.e.*, where kinetic energy is negative).

If E is increased further, the corresponding ψ again becomes infinite at $+\infty$ or $-\infty$ until, at the next eigenvalue, ψ can again go to zero at both ends of the range. This eigenfunction has a node somewhere in the region where $E > V$.

And so it goes, only a discrete set of energies is allowed, the values determined by the requirement that ψ be finite, until we reach the energy equal to the lowest asymptotic value of V [we assume, to avoid complicated wording, that $V(-\infty) < V(+\infty)$]. Above this energy E is larger than V from $-\infty$ to some finite positive value of x and is oscillatory over all this range of x. Therefore neither solution of the differential equation goes to infinity at $x = -\infty$, and we can choose, for any value of E, the proper linear combination which will go to zero exponentially at $+\infty$. Thus *any value of the energy above* $V(-\infty)$ *is allowed.*

Since the vanishing exponential solutions are real functions, for x real, the eigenfunctions for the discrete energies [less than either $V(-\infty)$ or $V(\infty)$] are real everywhere (to be more exact, the phase angle of these eigenfunctions is independent of x and might as well be set zero). Since they are real and their derivatives are real, the average current density [proportional to the imaginary part of $\bar{\psi}(d\psi/dx)$] is zero. This is explained classically by saying that the particle reverses its direction of motion at the points where V rises above E, so that at each point the particle is equally likely to be found going in either direction and the *net* current is zero.

This is still true for the continuous range of eigenvalues of E between $V(-\infty)$ and $V(+\infty)$, for the particle here comes from $-\infty$, rebounds from the potential "hill," and returns to $-\infty$.

When E is larger than both $V(-\infty)$ and $V(+\infty)$, both solutions of the equation are finite everywhere, so that it is possible to choose linear combinations at will. Some of these combinations will give nonzero values for the average current density, corresponding to the classic statement that, for E larger than $V(-\infty)$ and $V(+\infty)$, a particle can go from $-\infty$ to $+\infty$ without being reflected. This whole question will be discussed further in a later chapter.

Discrete and Continuous Eigenvalues. As an example of this sort of behavior we take the symmetric case of a potential function $V = -V_0 \operatorname{sech}^2(x/d)$ [see Eq. (12.3.22)]. This potential has a minimum $-V_0$ at $x = 0$ and rises asymptotically to zero at $\pm\infty$. By suitable choice of scale factors and constants, the Schroedinger equation (6.3.63) comes to have the form

$$(d^2\psi/dw^2) + (\lambda + Q \operatorname{sech}^2 w)\psi = 0; \quad w = x/d \tag{6.3.64}$$

Further transformations, such as letting $\lambda = -K^2$, $z = \frac{1}{2}[1 + \tanh w] = e^w/(e^w + e^{-w})$, and $\psi = \cosh^{-K} w\, F(z)$, result in the equation

$$z(z-1)\frac{d^2F}{dz^2} + (K+1)(2z-1)\frac{dF}{dz} + [K(K+1) - Q]F = 0$$

which is just Eq. (5.2.42) for the hypergeometric function

$$F(K + \tfrac{1}{2} + P, K + \tfrac{1}{2} - P | K + 1 | z)$$

where $P = \sqrt{Q + \frac{1}{4}}$. There is a second solution, but it becomes infinite at $z = 0$. Therefore the solution which remains finite at $x = -\infty$ is

$$\psi = \cosh^{-K}(w) F\left(K + \tfrac{1}{2} + P, K + \tfrac{1}{2} - P | K + 1 | \frac{e^w}{e^w + e^{-w}}\right) \tag{6.3.65}$$

Examination of Eq. (5.2.49) shows that this function becomes infinite at $z = 1$ ($w \to \infty$) unless the quantity $K + \frac{1}{2} - P$ is a negative integer, *i.e.*, unless $K = \sqrt{Q + \frac{1}{4}} - \frac{1}{2} - n$, where n can be zero, or any positive integer less than $\sqrt{Q + \frac{1}{4}} - \frac{1}{2}$. The discrete eigenvalues of λ, which is proportional to the energy, are therefore

$$\lambda_n = -[(n + \tfrac{1}{2}) - \sqrt{Q + \tfrac{1}{4}}]^2; \quad n = 0, 1, 2, \ldots, n < \sqrt{Q + \tfrac{1}{4}} - \tfrac{1}{2} \tag{6.3.66}$$

For such values of λ, the eigenfunction ψ is a polynomial in z. We note that, as long as Q is positive (*i.e.*, as long as there is a potential "valley"), there will be at least one discrete state ($n = 0$), though there may not be more than one if Q is small.

For positive values of the energy, K is imaginary, and we can set $K = ik$, where $\lambda = k^2$. In this case the two independent solutions near $z = 0$ ($w \to -\infty$) are

$$[2 \cosh w]^{-ik} F\left(P + \tfrac{1}{2} + ik, -P + \tfrac{1}{2} + ik \middle| 1 + ik \middle| \frac{e^w}{e^w + e^{-w}}\right)$$

which might be called $\psi(k)$, and

$$e^{-ikw} F\left(P + \tfrac{1}{2}, -P + \tfrac{1}{2} \middle| 1 - ik \middle| \frac{e^w}{e^w + e^{-w}}\right)$$

which might be called $\psi(-k)$. Thus we have two eigenfunctions for each value of λ. For w very large negative, these two solutions reduce to $e^{ikw} = e^{ikx/d}$ and $e^{-ikw} = e^{-ikx/d}$, both of which are finite at $w = -\infty$. One represents a wave in the positive direction (which has a correspondingly positive current density), and the other a wave in the negative direction (with negative current density). Similarly both solutions near $w = +\infty$ are finite for all values of k, so every positive value of the energy parameter λ is allowed.

In this case we have both discrete and continuous eigenvalues. Our expansion in terms of the eigenfunctions must include a series over the few discrete values (sum over n for n less than $\sqrt{Q + \frac{1}{4}} - \frac{1}{2}$) and an integral over k from $-\infty$ to $+\infty$ (thus including both eigenfunctions for each positive eigenvalue of λ). To obtain the exact form of the sum and integral, with the proper values of the normalizing factors, would require more detailed analysis of the problem than is worth while here. It will be taken up again in Sec. 12.3.

Differentiation and Integration as Operators. This discussion of the Schroedinger equation (together with the treatment of factorization on page 729) may serve to remind us that it is occasionally useful to consider differentiation and multiplication by a constant (and also integration) as operators. We consider the vector $\mathbf{f}$ to have components $f(x)$, the vector $a\mathbf{f}$ to have components $af(x)$, the vector $\mathfrak{D}\mathbf{f}$ to have components $[df(x)/dx]$, and the vector $\mathfrak{Q}\mathbf{f}$ to have components

$$\int_0^x f(w)\, dw$$

These operators can be added (*i.e.*, they satisfy the distributive and commutative laws for addition); the vector $[\mathcal{P} + \mathcal{Q}]\mathbf{f} = [\mathcal{Q} + \mathcal{P}]\mathbf{f}$ has components

$$\frac{d}{dx} f(x) + \int_0^x f(w)\, dw = \int_0^x f(w)\, dw + \frac{d}{dx} f(x)$$

for instance.

Both $\mathcal{P}$ and $\mathcal{Q}$ commute with the scalar operator a, but $\mathcal{P}$ does not commute with $\mathcal{Q}$, for

$$\frac{d}{dx} \int_0^x f(w)\, dw = f(x) \quad \text{but} \quad \int_0^x \frac{d}{dw} f(w)\, dw = f(x) - f(0)$$

so that

$$\mathcal{P}\mathcal{Q} = 1 \quad \text{but} \quad \mathcal{Q}\mathcal{P} \neq 1 \tag{6.3.67}$$

where 1 is the unity operator. However if order of multiplication is watched, we can consider the integration operator $\mathcal{Q}$ to be the inverse of the differentiation operator $\mathcal{P}$ and it can be written $\mathcal{P}^{-1}$.

Both $\mathcal{P}$ and $(\mathcal{P})^{-1}$ may be iterated. The components of $\mathcal{P}^n\mathbf{f}$ are $(d^n/dx^n)f(x)$, and the components of $\mathcal{P}^{-n}\mathbf{f}$ are

$$\int_0^x du_1 \int^{u_1} \cdots \int_0^{u_{n-1}} du_n\, f(u_n)$$

The positive powers of $\mathcal{P}$ operating on a constant vector $\mathbf{A}$ give zero, but the negative powers give a nonzero result. For instance, the components of $\mathcal{P}^{-n}\mathbf{A}$ are $A(x^n/n!)$. By integration by parts we can see that the components of $\mathcal{P}^{-2}\mathbf{f}$ are

$$\int_0^x du \int_0^u f(w)\, dw = \int_0^x (x - w) f(w)\, dw$$

Continued application of this procedure shows that the components of $\mathcal{P}^{-n}\mathbf{f}$ are

$$\int_0^x \frac{(x - w)^{n-1}}{(n - 1)!} f(w)\, dw \tag{6.3.68}$$

Series of operators can be handled (subject to the usual requirements of convergence) just as ordinary series. For instance, the components of the vector

$$\left\{\sum_n a_n \mathcal{P}^{-n}\right\} \mathbf{f} \quad \text{are} \quad \int_0^x \left[\sum_n a_n \frac{(x - w)^{n-1}}{(n - 1)!}\right] f(w)\, dw$$

provided the series of integrals converges. By this means we can determine the meaning of reciprocal expressions, such as $[\mathcal{P}^{-1}/(1 - \alpha\mathcal{P}^{-1})]\mathbf{f}$. Going at this formally, we compute by expanding on a series of powers of $\mathcal{P}^{-1}$. The components of the resulting vector, $[\mathcal{P}^{-1} + \alpha\mathcal{P}^{-2} + \alpha^2\mathcal{P}^{-3} + \cdots]\mathbf{f}$, are

$$\int_0^x \left[1 + \alpha \frac{(x-w)}{1!} + \alpha^2 \frac{(x-w)^2}{2!} + \cdots \right] f(w)\, dw = e^{\alpha x} \int_0^x e^{-\alpha w} f(w)\, dw \tag{6.3.69}$$

a surprisingly compact result.

To see how this technique can be used and, incidentally, to validate somewhat this rather reckless manipulation of symbols, we compute the solution of the first-order differential equation

$$(dy/dx) - \alpha y = f(x) \tag{6.3.70}$$

By usual means, we can find that the solution of this equation is

$$y(x) = y(0)e^{\alpha x} + e^{\alpha x} \int_0^x e^{-\alpha w} f(w)\, dw$$

To solve Eq. (6.3.70) by symbolic means we note that the equation is equivalent to the operator equation $(\mathcal{P} - \alpha)\mathbf{y} = \mathbf{f}$, or if we had integrated first with respect to x, the equation would be

$$\mathcal{P}^{-1}\mathcal{P}\mathbf{y} - \alpha\mathcal{P}^{-1}\mathbf{y} = \mathcal{P}^{-1}\mathbf{f}$$

Referring to Eq. (6.3.67) and dividing both sides by $(1 - \alpha\mathcal{P}^{-1})$, we find that $y(x) - y(0)$ is just equal to the components of the vector $[\mathcal{P}^{-1}\mathbf{f}/(1 - \alpha\mathcal{P}^{-1})]$, which by Eq. (6.3.69) is

$$e^{\alpha x} \int_0^x e^{-\alpha w} f(w)\, dw$$

which checks with the solution obtained in the usual way.

We should note that $[\mathcal{P}^{-1}/(1 - \alpha\mathcal{P}^{-1})]$ is not necessarily equal to $[1/(\mathcal{P} - \alpha)]$. If we had assumed these were equal, we might have tried to solve Eq. (6.3.70) by assuming that y was equal to the components of $[\mathcal{P} - \alpha]^{-1}\mathbf{f}$ and the result would not have included the necessary constant $y(0)$.

This technique, sketched here, will be discussed in more detail later. All we wish to point out at present is that differential equations may be considered as representations of *operational equations*, with operators $\mathcal{P}$ and $\mathcal{P}^{-1}$ corresponding to differentiation and integration, and that such operational equations are amenable to a certain amount of algebraic manipulation as long as we follow known rules for commutation.

The Eigenvalue Problem in Abstract Vector Space. The question of discrete eigenvalues and allowed solutions has a much wider application than just to differential equations, as a reperusal of Sec. 2.6 will indicate and as the abstract vector space analogy suggests. In many cases in theoretical physics we have situations which can be represented by some sort of operator operating on a vector. In the case of elasticity it is a strain (or stress) operator operating on a usual three-vector to give the resulting displacement; in the case of the ordinary differential equation it is a combination of the operators $\mathcal{P}$ just defined, operating on the "vector" corresponding to the function $\psi(x)$ [which can be considered as the component of the vector along the direction given by the unit vector $\mathbf{e}(x)$]; in the case of the Dirac equation the operator is a combination of operators which interchange the four components $\psi_1(x) \cdot \cdot \cdot \psi_4(x)$ and of differential operators which act on the x dependence of the four ψ's. In any of these cases we can talk about a vector **F** and an operator $\mathfrak{A}$ which, in general, changes **F** into another vector **E**, as was discussed in Sec. 1.6.

The vector **F** can be described in terms of its components along suitable coordinates [x, y, z components for strain displacement, $F(x)$ for each x for differential equations, different spin states for the Dirac equation, etc.]. The operator must correspondingly be given by a *matrix* of components

$$\mathbf{F} = \sum_n F_n \mathbf{e}_n; \quad \mathfrak{A} = \sum_{mn} \mathbf{e}_m A_{mn} \mathbf{e}_n^*$$

(see Eq. 1.6.35)

$$\mathfrak{A} \cdot \mathbf{F} = \sum_m \left[\sum_n A_{mn} F_n \right] \mathbf{e}_m = \mathbf{E}$$

the matrix (A_{mn}) representing $\mathfrak{A}$. When the axes of reference are rotated, the components of **F** and of $\mathfrak{A}$ are modified according to the usual rules of transformation given in Chap. 1.

The fiction of a discrete set of axes, represented by the unit vectors $\mathbf{e}_n$, can be used even when the "subscripts" are continuously variable [as with $\mathbf{e}(x)$] and the scalar product has to be represented as an integral over x, instead of a sum over the subscript n. In the following it should not be difficult to make the necessary change from sum to integral when this is necessary. We should recollect the discussion following Eq. (6.3.7) in this connection.

For instance, corresponding to the operator equation $\mathfrak{A} \cdot \mathbf{F} = \mathbf{E}$ is the differential equation $\mathfrak{a}_x F(x) = E(x)$, where $\mathfrak{a}_x$ is an ordinary differential operator of the sort

$$\mathfrak{a}_x = f(x) \frac{d^2}{dx^2} + g(x) \frac{d}{dx} + h(x)$$

which we have been discussing in this chapter. Corresponding to the scalar product $(\mathbf{G}^* \cdot \mathfrak{A} \cdot \mathbf{F})$ we have the integral

$$\int_a^b \bar{G}(x) \mathfrak{a}_x F(x)\, dx$$

($\bar{G}$ is the complex conjugate of G) and so on.

In nearly all cases of interest the operator $\mathfrak{A}$ is *Hermitian*, which means that its Hermitian *adjoint*, $\mathfrak{A}^*$, obtained by interchanging rows and columns in the matrix and then taking the complex conjugate, is equal to

$\mathfrak{A}$ itself. We worked out some of the consequences of this on pages 83 to 87. Let us see what it means for a differential operator having components corresponding to the continuum of values of x.

The Hermitian adjoint of an operator $\mathfrak{A}$ is the operator $\mathfrak{A}^*$, such that if

$$\mathfrak{A} \cdot \mathbf{F} = \mathbf{E} \quad \text{then} \quad \mathbf{E}^* = \mathbf{F}^* \cdot \mathfrak{A}^*$$

in other words

$$\mathbf{G}^* \cdot (\mathfrak{A} \cdot \mathbf{F}) = (\mathfrak{A}^* \cdot \mathbf{G})^* \cdot \mathbf{F} \tag{6.3.71}$$

When the operator is real, then the concept of Hermitian adjointness corresponds to the concept of adjointness for differential operators given on page 527. For instance, in integral form, Eq. (6.3.71) becomes

$$\int_a^b \bar{G}(x)\mathfrak{a}_x F(x)\,dx = \int_a^b \bar{G}(x)\left[f\frac{d^2F}{dx^2} + g\frac{dF}{dx} + hF\right]dx$$

But according to Eq. (5.2.10) this last integral is equal to

$$\int_a^b \overline{[\tilde{\mathfrak{a}}_x G]}\,F\,dx + [P(\bar{G},F)]_a^b$$

where $\tilde{\mathfrak{a}}_x$ is the *adjoint differential operator* given by

$$\frac{d^2}{dx^2}(fG) - \frac{d}{dx}(gG) + hG = \tilde{\mathfrak{a}}_x G$$

and discussed on page 584. If both $F(x)$ and $G(x)$ satisfy suitable boundary conditions at a and b, the bilinear concomitant $P(\bar{G},F)$ is zero at a and b and we have

$$\int_a^b \bar{G}\mathfrak{a}_x F\,dx = \int_a^b \overline{[\tilde{\mathfrak{a}}_x G]}F\,dx$$

which corresponds to Eq. (6.3.71), defining the adjoint of an operator $\mathfrak{a}_x$. This means that our use of the word *adjoint* in relation to a generalized operator $\mathfrak{A}$ corresponds to our use of adjoint in relation to an ordinary differential operator $\mathfrak{a}_x$ if $\mathfrak{a}_x$ is real. If $\mathfrak{a}_x$ is not real, then its Hermitian adjoint is the complex conjugate of its adjoint, $\bar{\mathfrak{a}}^* = \tilde{\mathfrak{a}}$. Consequently an ordinary differential operator which is *self-adjoint*, as defined on page 720, corresponds to a real *Hermitian operator*, which is correspondingly self-adjoint in the operator sense. (We have already seen that the Liouville equation is self-adjoint.) This will be discussed more fully in Sec. 7.5.

Whether it be a differential operator or a more general kind of operator, if $\mathfrak{A}$ is Hermitian, then

$$(\mathbf{G}^* \cdot \mathfrak{A} \cdot \mathbf{F}) = (\mathbf{G}^* \cdot \mathfrak{A}^* \cdot \mathbf{F}) \tag{6.3.72}$$

This means that the quantity $(\mathbf{F}^* \cdot \mathfrak{A} \cdot \mathbf{F})$ is a real quantity, no matter what the vector $\mathbf{F}$, if $\mathfrak{A}$ is Hermitian. In quantum mechanics this quantity would be called the *average value* of $\mathfrak{A}$ in the state characterized by the vector $\mathbf{F}$. If $\mathfrak{A}$ corresponds to a physical quantity (position, momentum, etc.), its average value should be real, of course.

In a great number of cases this average value $(\mathbf{F}^* \cdot \mathfrak{A} \cdot \mathbf{F})$ *is always positive* (as well as being real) for all vectors $\mathbf{F}$. In such cases the operator is said to be *positive definite* as well as Hermitian.

Corresponding to each operator $\mathfrak{A}$ is a sequence of *eigenvectors* $\mathbf{E}_n$, such that

$$\mathfrak{A} \cdot \mathbf{E}_n = a_n \mathbf{E}_n$$

where a_n is the eigenvalue of $\mathfrak{A}$ corresponding to $\mathbf{E}_n$. From what we have said above, it is easy to prove that, if $\mathfrak{A}$ is Hermitian and positive definite, then *all its eigenvalues will be real* and *all of them will be greater than zero*. As we have seen earlier, the eigenvalues may be a series of discrete values or a continuous range or a combination of both.

We have discussed eigenvectors several times already; our purpose here is to link up our earlier findings with the findings of the present section on eigenfunctions. For instance, in connection with Eq. (1.6.9), we showed that the eigenvectors for a specific operator were mutually orthogonal, which corresponds to the finding that eigenfunctions, for a given differential equation and boundary conditions, are mutually orthogonal. We can, of course, normalize our eigenvectors to obtain the mutually orthogonal set of unit vectors $\mathbf{e}_n$. There are as many of them as there are "dimensions" in the abstract vector space suitable for the operator $\mathfrak{A}$. Consequently any vector in the same abstract space can be expressed in terms of its components along the principal axes for $\mathfrak{A}$:

$$\mathbf{F} = \sum_n F_n \mathbf{e}_n; \quad F_n = (\mathbf{F}^* \cdot \mathbf{e}_n)$$

This seemingly obvious statement corresponds to the fundamental expansion theorem for eigenfunctions. To show the generality of this theorem, proved earlier for differential operators, we shall here sketch the argument for a general operator $\mathfrak{A}$. In order that the eigenvectors for a given operator $\mathfrak{A}$ form a *complete set*, it is sufficient that:

1. $\mathfrak{A}$ is *self-adjoint* (or *Hermitian); i.e.*, for any vector $\mathbf{F}$ $(\mathbf{F}^* \cdot \mathfrak{A} \cdot \mathbf{F})$ is real.

2. $\mathfrak{A}$ is *positive definite; i.e.*, for any vector $\mathbf{F}$ $(\mathbf{F}^* \cdot \mathfrak{A} \cdot \mathbf{F})$ is greater than zero.

3. That the equation for the eigenvectors

$$\mathfrak{A} \cdot \mathbf{E}_n = a_n \mathbf{E}_n \tag{6.3.73}$$

corresponds to some variational principle. Such a variational principle can be quite general. For instance, we can compute the scalar (real and positive) quantity

$$D(\mathbf{F}) = (\mathbf{F}^* \cdot \mathfrak{A} \cdot \mathbf{F})/(\mathbf{F}^* \cdot \mathbf{F}) \tag{6.3.74}$$

for an arbitrary vector **F**. The variational requirement that **F** be the vector for which *D is a minimum* gives the eigenvector $\mathbf{E}_0$ for Eq. (6.3.74), and the value of D for $\mathbf{F} = \mathbf{E}_0$ is just the eigenvalue a_0.

To prove this we consider the variation of D, as **F** and $\mathbf{F}^*$ are varied by arbitrary, small amounts $\delta\mathbf{F}$ and $\delta\mathbf{F}^*$, and require that $\delta D = 0$. Multiplying across by $(\mathbf{F}^* \cdot \mathbf{F})$, we require that

$$D[(\delta\mathbf{F}^* \cdot \mathbf{F}) + (\mathbf{F}^* \cdot \delta F)] = [(\delta\mathbf{F}^* \cdot \mathfrak{A} \cdot \mathbf{F}) + (\mathbf{F}^* \cdot \mathfrak{A} \cdot \delta\mathbf{F})]$$

or

$$[\delta\mathbf{F}^* \cdot (\mathfrak{A} \cdot \mathbf{F} - D\mathbf{F})] + [(\mathbf{F}^* \cdot \mathfrak{A} - D\mathbf{F}^*) \cdot \delta\mathbf{F}] = 0$$

Since $\delta\mathbf{F}^*$ and $\delta\mathbf{F}$ are arbitrary and independent variations, the variational requirement is thus equivalent to Eq. (6.3.73) and its conjugate, with D equal to a.

As we mentioned above, the vector which gives the minimum value of D is $\mathbf{E}_0$ and the corresponding value of D is a_0, the lowest eigenvalue (which is greater than zero, since $\mathfrak{A}$ is positive definite).

The vector which gives a minimum value of D, subject to the additional requirement that it be orthogonal to $\mathbf{E}_0$, is $\mathbf{E}_1$, and the corresponding value of D is a_1, the next eigenvalue. To prove this statement we have to consider that the operator $\mathfrak{A}$ can be dealt with formally as if it were an analytic function, *i.e.*, that the operators $\mathfrak{A}^{-1}$, $\sqrt{\mathfrak{A}}$, etc., have the same meaning as do their algebraic counterparts. We have indicated that this is possible for differential and integral operators.

If this is so, we can show that the eigenvectors of any function $\mathfrak{H}(\mathfrak{A})$ which can be expressed as a series of powers of $\mathfrak{A}$ are the corresponding eigenvectors of $\mathfrak{A}$, and the eigenvalues are the equivalent functions $H(a_n)$ of the corresponding eigenvalue a_n. From this it follows that the solutions of the variational problem $\delta B = 0$, where

$$B = [\mathbf{F}^* \cdot \mathfrak{H}(\mathfrak{A}) \cdot \mathbf{F}]/[\mathbf{F}^* \cdot \mathbf{F}]$$

are just the eigenvectors $\mathbf{E}_n$, which are solutions of Eq. (6.3.73) *if* the related algebraic function $H(a)$ is always real and positive when a is real and positive. The resulting stationary values of B are the values $H(a_n)$. (The proof of this statement can be left to the reader; it is, of course, true only if $\mathfrak{A}$ is positive definite.)

The product $K = [\mathbf{F}^* \cdot (\mathfrak{A} - a_0) \cdot \mathbf{F}]$, where a_0 is the lowest eigenfunction of $\mathfrak{A}$, is never negative, no matter what vector **F** we choose, as a consideration of the variational equation will show. Nor is the product $J = [\mathbf{F}^* \cdot (\mathfrak{A} - a_0)(\mathfrak{A} - a_1)\mathbf{F}]$, where a_1 is the next eigenvalue, for the minimum values of

$$\frac{J}{\mathbf{F}^* \cdot \mathbf{F}} = \frac{\mathbf{F}^*[\mathfrak{A} - \frac{1}{2}(a_0 + a_1)]^2\mathbf{F}}{[\mathbf{F}^* \cdot \mathbf{F}]} - \tfrac{1}{4}(a_0 - a_1)^2$$

are for **F** equal to either $\mathbf{E}_0$ or $\mathbf{E}_1$. For these two vectors J is zero, whereas K is only zero for $\mathbf{F} = \mathbf{E}_0$. Consequently *the quantity* $[J/K]$ *is* *never negative, and it is zero only when* $\mathbf{F} = \mathbf{E}_1$. A rewording of this last sentence will show us that we have proved the statement, made above, which we wished to prove. Suppose we set up a vector $\mathbf{G} = \sqrt{\mathfrak{A} - a_0} \cdot \mathbf{F}$ which is automatically orthogonal to the lowest eigenvector $\mathbf{E}_0$ ($\mathbf{G}^* \cdot \mathbf{E}_0 = 0$) but which is otherwise quite arbitrary. We then set up the variational ratio

$$D(\mathbf{G}) = \frac{(\mathbf{G}^* \cdot \mathfrak{A} \cdot \mathbf{G})}{(\mathbf{G}^* \cdot \mathbf{G})} = \frac{J}{K} + a_1$$

The sentence in italics shows that the minimum value of D is for $\mathbf{G} = \sqrt{a_1 - a_0}\,\mathbf{E}_1$, and its value for that $\mathbf{G}$ is a_1, which is what we set out to prove.

Since we have shown that the minimum value of D for a vector orthogonal to $\mathbf{E}_0$ is a_1 and the corresponding vector is $\mathbf{E}_1$, we can extend the argument and arrive at the statement

$$D(\mathbf{G}) \geq a_{s+1} \quad \text{if} \quad (\mathbf{G}^* \cdot \mathbf{E}_m) = 0; \quad \text{for } m = 0, 1, \ldots, s \qquad (6.3.75)$$

which is equivalent to the statement preceding Eq. (6.3.21).

For the general case of an arbitrary operator, there may not be an infinite sequence of eigenvalues. If the vector space has only n dimensions (*i.e.*, if all vectors in the space can be expressed as a linear combination of only n mutually orthogonal vectors), then there are just n eigenvectors and n eigenvalues (if there are degenerate states, some of the eigenvalues will be equal but the n eigenvectors will still be mutually orthogonal), for as long as we can make up a vector orthogonal to the first s eigenvectors, we can find still another eigenvector and eigenvalue. Only when $s = n$ will it be impossible to find another orthogonal vector, and at that point the sequence will stop. Therefore for vector spaces with a finite number of dimensions, the number of mutually orthogonal eigenvectors will equal the number of dimensions and this set of eigenvectors will be complete, for by definition any vector in this space can be expressed as a linear sum of this number of orthogonal vectors.

For vector spaces with an infinite number of dimensions the proof of completeness is not so simple [see comments following Eq. (6.3.7)]. To make it watertight we should prove that the sequence of eigenvalues tends to infinity as n goes to infinity. But a proof of this theorem for the most general sort of positive-definite, Hermitian operator would carry us too far afield into the intricacies of modern algebra. We indicated that it is true for differential operators of the Liouville type on page 724. It is also true for quantum mechanical operators having an infinite number of possible states. *If* we assume that it *is* true in general, then our proof of completeness will follow the same lines as the proof given on pages 736 to 739. Because the arguments using abstract operators are "cleaner" than those for differential equations, we shall run through the proof again.

We wish to express an arbitrary vector **G** in terms of the unit eigenvectors $\mathbf{e}_m$, so we set up the finite sum

$$\mathbf{S}_n = \sum_{m=0}^{n} C_m \mathbf{e}_m; \quad C_m = (\mathbf{e}_m^* \cdot \mathbf{G}) \tag{6.3.76}$$

where $$\mathfrak{A} \cdot \mathbf{e}_m = a_m \mathbf{e}_m \quad \text{and} \quad (\mathbf{e}_m^* \cdot \mathbf{e}_s) = \delta_{ms}$$

The vector

$$\mathbf{J}_n = \mathbf{G} - \mathbf{S}_n$$

is orthogonal to the first n eigenvectors of $\mathfrak{A}$, since $(\mathbf{F}_m^* \cdot \mathbf{J}_n) = 0$ for $m \leq n$. Therefore by Eq. (6.3.75) we have that

$$D(J_n) = \frac{1}{(\mathbf{J}_n^* \cdot \mathbf{J}_n)} [\mathbf{J}_n^* \cdot \mathfrak{A} \cdot \mathbf{J}_n] \geq a_{n+1}$$

or $$(\mathbf{J}_n^* \cdot \mathbf{J}_n) \leq \frac{1}{a_{n+1}} \left[\left(\mathbf{G}^* - \sum_{m=0}^{n} \bar{C}_m \mathbf{e}_m^* \right) \mathfrak{A} \left(\mathbf{G} - \sum_{m=0}^{n} C_m \mathbf{e}_m \right) \right]$$

$$= \frac{1}{a_{n+1}} \left[(\mathbf{G}^* \cdot \mathfrak{A} \cdot \mathbf{G}) - \sum_{m=0}^{n} a_m |C_m|^2 \right]$$

But the series which is the second term in the brackets is positive, being the sum of products of squares of quantities times eigenvalues (which are all positive). Therefore

$$(\mathbf{J}_m^* \cdot \mathbf{J}_n) < \frac{1}{a_{n+1}} (\mathbf{G}^* \cdot \mathfrak{A} \cdot \mathbf{G})$$

Since neither of the quantities in parentheses can be negative (nor can a_{n+1}), since $(\mathbf{G}^* \cdot \mathfrak{A} \cdot \mathbf{G})$ is independent of n, and since $a_{n+1} \to \infty$ as $n \to \infty$, therefore $(\mathbf{J}_n^* \cdot \mathbf{J}_n) \to 0$ as $n \to \infty$ and therefore $\mathbf{J}_n$, which is the difference between the arbitrary vector **G** and the first n terms of its series expansion in terms of the eigenvectors $\mathbf{e}_m$, goes to zero as n goes to infinity. Therefore the complete series $(n \to \infty)$ equals the vector **G**, and we have proved again that the $\mathbf{e}_n$'s are a complete set of eigenvectors.

Similarly any operator (in the same abstract space) can be expressed in terms of its components along the principal axes for $\mathfrak{A}$. In particular, the operator $\mathfrak{A}$ itself has the particularly simple form

$$\mathfrak{A} = \sum_n \mathbf{e}_n a_n \mathbf{e}_n^*; \quad \text{that is, } A_{mn} = a_n \delta_{mn}$$

where a_n is the eigenvalue of $\mathfrak{A}$ corresponding to the eigenvector $\mathbf{e}_n$. In other words, when expressed in terms of its own principal axes, the matrix for an operator is a *diagonal matrix*.

Other general properties, applicable alike to abstract vector operators and to ordinary differential operators, have been already discussed in

Secs. 1.6 and 2.6, and still others will be derived later. It should be apparent by now that the abstract vector representation has many advantages of simplicity, because of our familiarity with the simple geometrical analogue, which makes it invaluable as an alternative point of view for nearly all our problems.

Problems for Chapter 6

6.1 A net-point potential $\varphi(m,n)$ satisfies the difference equation (6.2.6) and is to satisfy boundary conditions on the boundary lines $n = 0$, $n = 5$, $m = 0$, $m = 5$. Show that the solution satisfying the requirement that φ have the value φ_ν at the νth boundary point is given by

$$\varphi(m,n) = \sum_\nu G(m,n|\nu)\varphi_\nu$$

where $G(m,n|\nu)$ is the solution of Eq. (6.2.6) which is zero at all boundary points except the νth and has unit value at the νth boundary point. Show that all the G's can be obtained from the ones connected with the boundary points (0,1) and (0,2). Compute these two G's, correct to three places of decimals, for each interior net point.

6.2 Show that a solution of the Poisson difference equation

$$[\psi(m+1, n) + \psi(m-1, n) + \psi(m, n+1) + \psi(m, n-1) - 4\psi(m,n)] = F(m,n)$$

with $F(m,n)$ specified, and with ψ zero at all boundary points, is

$$\psi(m,n) = \sum_{\mu\nu} G(m,n|\mu,\nu)F(\mu,\nu)$$

where $G(m,n|\mu,\nu)$ is a solution of the Poisson difference equation for $F(\mu,\nu) = 1$, all other F's $= 0$, and for G zero at all boundary points. What are the values of G for the 4×4 net of Prob. 6.1? How can these results be combined with those of Prob. 6.1 to obtain a general solution of the Poisson difference equation satisfying general boundary conditions?

6.3 The differential equation is the simple parabolic one:

$$\partial^2\psi/\partial x^2 = \partial\psi/\partial t$$

and the boundaries are $x = 0$, $x = \pi$, and $t = 0$. For the boundary condition $\psi = 0$ at $x = 0$ and $x = \pi$ show that the solution for $t \geq 0$ is

$$\psi(x,t) = \sum_{\nu=1}^{\infty} A_\nu \sin(\nu x) \exp[-\nu^2 t]$$

where the A's are chosen to fit the initial value of ψ at $t = 0$. Now consider the net-point approximation to this equation, obtained by dividing the range $0 \le x \le \pi$ into N equal parts with spacing $h = \pi/N$ and dividing t into intervals of length $k = \pi/M$. Show that the corresponding solution of the equation

$$(1/h^2)[\psi(m+1, n) + \psi(m-1, n) - 2\psi(m,n)] = (1/k)[\psi(m, n+1) - \psi(m,n)]$$

$$\psi(0,n) = \psi(N,n) = 0; \quad n \ge 0$$

is

$$\psi(m,n) = \sum_{\nu=1}^{N-1} B_\nu \sin(\nu mh) \exp\left\{n \ln\left[1 - \frac{4k}{h^2}\sin^2(\tfrac{1}{2}h\nu)\right]\right\}$$

What happens to this solution if k has been chosen larger than h^2? What limitation must be imposed on the size of k in order that the solution be stable? Suppose the initial conditions are such that the coefficients A_ν in the exact solution are negligible for $\nu > \nu_{\max}$. What can you say concerning the choice of h and k which will result in a net-point solution which is reasonably accurate (to 1 per cent, say) in the range $0 \le t \le \pi$ and which is not, on the other hand, so "fine-grained" (N and M too large) as to make the numerical computations too laborious?

6.4 The initial values of $\psi(x,t)$ of Prob. 6.3 is

$$\psi(x,0) = \begin{cases} \tfrac{1}{4}\pi x; & (0 \le x \le \tfrac{1}{2}\pi) \\ \tfrac{1}{4}\pi(\pi - x); & (\tfrac{1}{2}\pi \le x \le \pi) \end{cases} = \sum_{\sigma=0}^{\infty} \frac{1}{(2\sigma+1)^2}\sin[(2\sigma+1)x]$$

Compute values of $\psi(x,t)$ for $x = \tfrac{1}{4}\pi, \tfrac{1}{2}\pi$; $t = \tfrac{1}{4}\pi, \tfrac{1}{2}\pi$ for the exact solution. Then compute values of ψ, by use of the difference equation

$$\psi(m, n+1) = [1 - (2k/h^2)]\psi(m,n) + (k/h^2)[\psi(m+1, n) + \psi(m-1, n)]$$

starting at $n = 0$ and working upward (for increasing n) for the same initial conditions. Take $h = \tfrac{1}{4}\pi$ ($N = 4$) and make the computations for two choices of k: $k = \pi/4$ and $k = \pi/16$. Compare with the four exact values already computed.

6.5 Is the partial differential equation

$$(\partial^2\psi/\partial x^2) - y^2(\partial^2\psi/\partial y^2) - y(\partial\psi/\partial y) = 0$$

elliptic or hyperbolic? What are the equations for the characteristics? Sketch a few of them. Show that, if Cauchy conditions are applied at the boundary $y = y_0 > 0$, the solution for $y > y_0$ is

$$\psi = \tfrac{1}{2}\psi_0\left(x + \ln\frac{y}{y_0}\right) + \tfrac{1}{2}\psi_0\left(x - \ln\frac{y}{y_0}\right) + \tfrac{1}{2}y_0\varphi_0\left(x + \ln\frac{y}{y_0}\right) - \tfrac{1}{2}y_0\varphi_0\left(x - \ln\frac{y}{y_0}\right)$$

where the value of ψ at $y = y_0$ is $\psi_0(x)$ and where $\varphi_0(z) = \int v_0(z)\, dz$, $v_0(x)$ being the initial value of $\partial\psi/\partial y$ at $y = y_0$. Why does this solution fail at $y_0 = 0$?

6.6 Build up a sequence of mutually orthogonal polynomials in x for the range $-1 \leq x \leq 1$. Start with $y_0 = 1$, $y_1 = x$, . . . with y_n a polynomial of degree n such that

$$\int_{-1}^{1} y_n(x) y_m(x)\, dx = 0; \quad m = 0, 1, \ldots, n-1$$

Obtain the first four such polynomials. Show that, in general, those polynomials for even n's have no odd powers of x, those for odd n have no even powers. Show that the polynomials obtained are proportional to the Legendre polynomials $P_n(x)$. Is this process of building up a set of orthogonal polynomials a unique one? If not, what restrictions must be added to make the process unique? Is the resulting set of functions a complete set? How can you be sure?

6.7 Repeat the process of Prob. 6.6 for the range $0 \leq x \leq 1$ and the orthogonality requirement

$$\int_0^1 y_n(x) y_m(x) x\, dx = 0; \quad m = 0, 1, \ldots, n-1$$

Start again with $y_0 = 1$ and obtain the first four polynomials. Compare this with the set $\varphi_n(x) = J_0(\pi\alpha_{an} x)$, where α_{0n} is the nth root of the equation $[dJ_0(\pi\alpha)/d\alpha] = 0$. Show that this is also a mutually orthogonal set of functions for the same range of x and for the same density function x. For what problems would each be useful?

6.8 The polynomials $\Upsilon_n(x)$ (related to the Tschebyscheff polynomials of page 782) are defined by the generating function

$$\frac{1 - t^2}{1 - 2tx + t^2} = \sum_{n=0}^{\infty} \Upsilon_n(x) t^n$$

Obtain the first four polynomials and, by manipulation of the generating function, show that

$$\Upsilon_1(x) - 2x\Upsilon_0(x) = 0; \quad 2\Upsilon_0(x) - 2x\Upsilon_1(x) + \Upsilon_2(x) = 0$$
$$\Upsilon_{n+1}(x) - 2x\Upsilon_n(x) + \Upsilon_{n-1}(x) = 0; \quad n > 1$$

and consequently that $\Upsilon_n(x) = \epsilon_n \cos [n \cos^{-1} x]$. Show that

$$\int_{-1}^{1} \Upsilon_m(x)\Upsilon_n(x) \frac{dx}{\sqrt{1 - x^2}} = \epsilon_n \pi \delta_{mn}$$

6.9 The Jacobi polynomials are defined as

$$J_n(a,c|x) = F(a + n, -n|\, c|\, x)$$

Write out the first four polynomials and show that the set is complete (for what range?). By use of the contour integral obtained from Eq. (5.3.21) and subsequent use of Eq. (4.3.1) show that

$$J_n(a,c|x) = \frac{x^{1-c}(1-x)^{c-a}\Gamma(c)}{\Gamma(c+n)} \frac{d^n}{dx^n} [x^{c+n-1}(1-x)^{a+n-c}]$$

Show that $\dfrac{d}{dx} J_n(a,c|x) = -\dfrac{n(n+a)}{c} J_{n-1}(a+2,\, c+1|\, x)$

$$xJ_n(a,c|x) = \frac{c-1}{2n+a} [J_n(a-1,\, c-1|\, x) - J_{n+1}(a-1,\, c-1|\, x)]$$

and that

$$\int_0^1 x^{c-1}(1-x)^{a-c} J_n(a,c|x) J_m(a,c|x)\, dx = \frac{n![\Gamma(c)]^2\Gamma(n+a-c+1)}{(a+2n)\Gamma(a+n)\Gamma(c+n)} \delta_{mn}$$

Express $P_n(x)$ and $T_n^\beta(x)$ in terms of the J's.

6.10 The radial function for the Helmholtz equation in spherical coordinates is $j_n(kr) = \sqrt{\pi/2kr}\, J_{n+\frac{1}{2}}(kr)$. Show that the eigenfunctions for a standing acoustical wave inside a rigid spherical shell are $j_n(\pi\beta_{nm}r/a)$, where β_{nm} is the mth root of the equation $[dj_n(\pi\beta)/d\beta] = 0$. Show that these form a complete, orthogonal set for the range $0 \leq r \leq a$. By letting $a \to \infty$, show that

$$f(z) = \frac{2}{\pi} \int_0^\infty j_n(zu)u^2\, du \int_0^\infty f(v) j_n(uv) v^2\, dv$$

6.11 Show that

$$\int_0^\infty e^{-zt} t^a L_n^a(t)\, dt = [\Gamma(a+n+1)]^2 \frac{(z-1)^n}{n!z^{a+n+1}}$$

6.12 Prove that

$$\sum_{n=0}^\infty H_n(x)H_n(y) \frac{t^n}{2^n n!} = \frac{1}{\sqrt{1-t^2}} \exp\left[\frac{2xyt - t^2(x^2+y^2)}{1-t^2}\right]$$

Table of Useful Eigenfunctions and Their Properties

We choose a range of the variable z and a density function $r(z)$ such that $r(z)$, times any positive power of z, integrated over the chosen range of z, is finite. We then choose the lowest eigenfunction to be $\psi_0(z) = 1$. The next eigenfunction, $\psi_1(z)$, is chosen to be that combination of 1 and z which is orthogonal to ψ_0 in the chosen range and for the chosen density r. Then $\psi_2(z)$ is that combination of z^2, z, and 1 which is orthogonal to ψ_0

and ψ_1, and so on. We can thus set up a set of eigenfunctions by a purely mechanical procedure, called the *Schmidt method*, which will serve as a basis for expanding any piecewise continuous function of z in the chosen range. It is usually found that the eigenfunctions thus obtained are ones which also arise from the solution of some Liouville equation, with boundary conditions, or which are obtained from some generating function. Three useful cases will be summarized here, for three ranges of z and for different density functions $r(z)$. See also the table of Jacobi polynomials at the end of Chap. 12.

I Range $-1 \le z \le +1$; **Density Function** $(1 - z^2)^\beta$: **Gegenbauer polynomials:** $T_n^\beta(z)$ (see page 748).

Generating function:

$$\frac{2^\beta}{(1 + t^2 - 2tz)^{\beta+\frac{1}{2}}} = \frac{\sqrt{\pi}}{\Gamma(\beta + \frac{1}{2})} \sum_{n=0}^{\infty} t^n T_n^\beta(z)$$

Special cases: $T_a^0(z) = P_n(z)$, Legendre polynomials (see Eq. 5.3.24)

$(1 - z^2)^{m/2} T_{n-m}^m(z) = P_n^m(z)$, associated Legendre functions [see Eq. (5.3.38)]

$nT_n^{-\frac{1}{2}}(z) = \sqrt{\frac{2}{\pi}} \cosh\,[n \cosh^{-1} z]$, Tschebyscheff polynomials [see Eq. (5.3.43)]

$\sqrt{z^2 - 1}\, T_{n-1}^{\frac{1}{2}}(z) = \sqrt{\frac{2}{\pi}} \sinh\,[n \cosh^{-1} z]$, Tschebyscheff polynomials [see Eq. (5.3.43)]

$$T_0^\beta(z) = \frac{2^\beta}{\sqrt{\pi}} \Gamma(\beta + \tfrac{1}{2}) = 1 \cdot 1 \cdot 3 \cdot 5 \cdots (2\beta - 1);\ \text{when } \beta = 0, 1, 2, \ldots$$

$$T_1^\beta(z) = \frac{2^{\beta+1}}{\sqrt{\pi}} \Gamma(\beta + \tfrac{3}{2})z = 1 \cdot 3 \cdot 5 \cdots (2\beta + 1)z;\ \text{when } \beta = 0, 1, 2, \ldots$$

Recurrence formulas, relating these polynomials, obtained from the generating function

$$\frac{d}{dz} T_n^\beta(z) = T_{n-1}^{\beta+1}(z);$$

$$\frac{d}{dz}[(z^2 - 1)^\beta T_n^\beta(z)] = (n + 1)(n + 2\beta)(z^2 - 1)^{\beta-1} T_{n+1}^{\beta-1}(z)$$

$$(2\beta + 2n + 1)zT_n^\beta(z) = (n + 1)T_{n+1}^\beta(z) + (2\beta + n)T_{n-1}^\beta(z)$$

$$(2\beta + 2n + 1)T_n^\beta(z) = \frac{d}{dz}[T_{n+1}^\beta(z) - T_{n-1}^\beta(z)] = T_n^{\beta+1}(z) - T_{n-2}^{\beta+1}(z)$$

$$(2\beta + 1)T_n^\beta(z) + 2z\frac{d}{dz} T_n^\beta(z) = T_n^{\beta+1}(z) + T_{n-2}^{\beta+1}(z)$$

$$(n + 2\beta + 1)T_n^\beta(z) = T_n^{\beta+1}(z) - zT_{n-1}^{\beta+1}(z)$$

$$nT_n^\beta(z) = zT_{n-1}^{\beta+1}(z) - T_{n-2}^{\beta+1}(z)$$

$$(2\beta + 2n + 1)(z^2 - 1)T_{n-1}^{\beta+1}(z) = n(n+1)T_{n+1}^{\beta}(z) - (n+2\beta)(n+2\beta+1)T_{n-1}^{\beta}(z)$$

$$T_n^\beta(z) = \frac{2^\beta \Gamma(\beta + \frac{1}{2})}{n!\sqrt{\pi}} \cdot \frac{(2\beta+1)(2\beta+3)\cdots(2\beta+2n-3)(2\beta+2n-1)}{(2\beta+n+1)(2\beta+n+2)\cdots(2\beta+2n-1)(2\beta+2n)} \cdot \frac{1}{(z^2-1)^\beta}\frac{d^n}{dz^n}(z^2-1)^{n+\beta}$$

$$(z^2-1)\frac{d^2}{dz^2}T_n^\beta(z) + 2(\beta+1)z\frac{d}{dz}T_n^\beta(z) - n(n+2\beta+1)T_n^\beta(z) = 0$$

which is related to the hypergeometric equation, having three regular singularities. $\beta = m$, an integer, results in *associated Legendre* polynomials,

$$T_n^m = \frac{P_{n+m}^m}{(1-z^2)^{\frac{1}{2}m}} = \frac{d^m}{dz^m}(P_{n+m})$$

$$T_0^0 = 1;\quad T_0^1 = 1;\quad T_0^2 = 3;\quad T_0^3 = 15;\quad \ldots$$

$$T_1^0 = z;\quad T_1^1 = 3z;\quad T_1^2 = 15z;\quad T_1^3 = 105z;\quad \ldots$$

$$T_2^0 = \tfrac{1}{2}(3z^2-1);\quad T_2^1 = \tfrac{3}{2}(5z^2-1);\quad T_2^2 = \tfrac{15}{2}(7z^2-1);\quad \ldots$$

$$T_3^0 = \tfrac{1}{2}(5z^3-3z);\quad T_3^1 = \tfrac{5}{2}(7z^3-3z);\quad T_3^2 = \tfrac{105}{2}(3z^3-z);\quad \ldots$$

$$T_4^0 = \tfrac{1}{8}(35z^4-30z^2+3);\quad T_4^1 = \tfrac{15}{8}(21z^4-14z^2+1);\quad \ldots$$

. . . .

$$T_n^m = \frac{(2n+2m)!}{2^{n+m}n!(n+m)!}\left[z^n - \frac{n(n-1)}{2(2n+2m-1)}z^{n-2} + \frac{n(n-1)(n-2)(n-3)}{2.4(2n+2m-1)(2n+2m-3)}z^{n-4}\cdots\right]$$

see also Eqs. (6.3.37) and (6.3.40).

Normalization integral:

$$\int_{-1}^{1}(1-z^2)^\beta T_m^\beta(z)T_n^\beta(z)\,dz = \delta_{mn}\frac{2\Gamma(n+2\beta+1)}{(2n+2\beta+1)\Gamma(n+1)}$$

Special values:

$$T_n^\beta(z) = (-1)^n T_n^\beta(-z)$$

$$T_n^\beta(1) = \frac{\Gamma(n+2\beta+1)}{2^\beta n!\Gamma(\beta+1)};\quad T_n^\beta(0) = 0;\quad n = 1, 3, 5, \ldots$$

$$T_n^\beta(0) = (-1)^{\frac{1}{2}n}\frac{2^\beta\Gamma(\beta+\frac{1}{2}n+\frac{1}{2})}{\sqrt{\pi}\,(\frac{1}{2}n)!};\quad \text{when } n = 0, 2, 4, \ldots$$

Relation with hyyergeometric function:

$$T_n^\beta(z) = \frac{\Gamma(n+2\beta+1)}{2^\beta n!\Gamma(\beta+1)}F\left(-n, n+2\beta+1|1+\beta|\frac{1-z}{2}\right)$$

$$= (-1)^n\frac{\Gamma(n+2\beta+1)}{2^\beta n!\Gamma(\beta+1)}F\left(-n, n+2\beta+1|1+\beta|\frac{1+z}{2}\right)$$

Addition formula:

$$T_n^\beta(\cos\vartheta\cos\vartheta_0 + \sin\vartheta\sin\vartheta_0\cos\varphi)$$
$$= \sqrt{2\pi}\sum_{m=0}^{n}\frac{(\beta+m)(n-m)!}{\Gamma(2\beta+n+m+1)}[\sin\vartheta\sin\vartheta_0]^m \cdot$$
$$\cdot\, T_{n-m}^{\beta+m}(\cos\vartheta)\, T_{n-m}^{\beta+m}(\cos\vartheta_0)\, T_m^{\beta-\frac{1}{2}}(\cos\varphi)$$

II Range $0 \le z < \infty$; Density Function $z^a e^{-z}$: Laguerre polynomials: $L_n^a(z)$

Generating function:

$$\frac{e^{-zt/(1-t)}}{(1-t)^{a+1}} = \sum_{n=0}^{\infty}\frac{t^n}{\Gamma(n+a+1)}L_n^a(z)$$

Special cases: $L_n^0(z) = e^z\dfrac{d^n}{dz^n}(z^n e^{-z})$

$$L_0^a(z) = \Gamma(a+1);\quad L_1^a(z) = \Gamma(a+2)[(a+1)-z]$$

Recurrence formulas:

$$\frac{d}{dz}L_n^a(z) = -L_{n-1}^{a+1}(z);\quad \text{except for } n = 0$$

$$\frac{d}{dz}[z^a e^{-z}L_n^a(z)] = (n+1)z^{a-1}e^{-z}L_{n+1}^{a-1}(z)$$

$$L_n^a(z) = \frac{d}{dz}\left[L_n^a(z) - \frac{1}{n+a+1}L_{n+1}^a(z)\right]$$

$$zL_n^a(z) = (a+2n+1)L_n^a(z) - \frac{n+1}{a+n+1}L_{n+1}^a(z) - (a+n)^2L_{n-1}^a(z)$$

$$z\frac{d}{dz}L_n^a(z) = (z-a)L_n^a(z) + (n+1)L_{n+1}^{a-1}(z)$$

$$L_n^a(z) = \frac{\Gamma(a+n+1)}{\Gamma(n+1)}\frac{e^z}{z^a}\frac{d^n}{dz^n}[z^{a+n}e^{-z}]$$

$$z\frac{d^2}{dz^2}L_n^a(z) + (a+1-z)\frac{d}{dz}L_n^a(z) + nL_n^a(z) = 0$$

which is the confluent hypergeometric equation, with regular singular point at $z = 0$ and an irregular point at ∞.

$a = m$, an integer, results in *associated Laguerre polynomials.*

$$L_n^m(z) = (-1)^m\frac{d^m}{dz^m}[L_{n+m}^0(z)]$$

$L_0^0 = 1;\quad L_0^1 = 1;\quad L_0^2 = 2;\quad L_0^3 = 6;\quad L_0^m = m!;\quad \ldots$

$L_1^0 = 1 - z;\quad L_1^1 = 4 - 2z;\quad L_1^2 = 18 - 6z;\quad L_1^3 = 96 - 24z;\quad \ldots$

$L_2^0 = 2 - 4z + z^2;\quad L_2^1 = 18 - 18z + 3z^2;\quad L_2^2 = 144 - 96z + 12z^2;\quad \ldots$

$L_3^0 = 6 - 18z + 9z^2 - z^3$; $L_3^1 = 96 - 144z + 48z^2 - 4z^3$; . . .

. . . .

$$L_n^m = \frac{[(m+n)!]^2}{n!m!} F(-n|m+1|z)$$

where F is a confluent hypergeometric series with $n + 1$ terms.

Normalization integrals:

$$\int_0^\infty z^a e^{-z} L_m^a(z) L_n^a(z)\, dz = \delta_{mn} \frac{[\Gamma(a+n+1)]^3}{\Gamma(n+1)}$$

$$\int_0^\infty z^p e^{-z} L_m^{p-\mu}(z) L_n^{p-\nu}(z)\, dz = (-1)^{m+n}\Gamma(p+m-\mu+1)\Gamma(p+n-\nu+1) \cdot \\ \cdot \mu!\nu! \sum_\sigma \frac{\Gamma(p+\sigma+1)}{\sigma!(m-\sigma)!(n-\sigma)!(\sigma+\mu-m)!(\sigma+\nu-n)!}$$

where m, n, μ, ν are integers or zero and σ takes on all integral values larger than $m - \mu - 1$ or $n - \nu - 1$ and smaller than $m + 1$ or $n + 1$ (if these requirements cannot be met the integral is zero).

Relation with confluent hypergeometric series for general values of a:

$$L_n^a(z) = \frac{[\Gamma(a+n+1)]^2}{n!\Gamma(a+1)} F(-n|a+1|z) \\ = \frac{[\Gamma(a+n+1)]^2}{n!\Gamma(a+1)} e^z F(a+n+1|a+1|-z) \\ \xrightarrow[z\to\infty]{} \frac{\Gamma(a+n+1)}{n!} (-z)^n$$

Addition formula and other equations:

$$L_n^a(x+y) = e^y \sum_{m=0}^\infty \frac{(-1)^m}{m!} \frac{\Gamma(n+a+1)}{\Gamma(m+n+a+1)} y^m L_n^{a+m}(x)$$

$$(1+t)^a e^{-xt} = \sum_{n=0}^\infty \frac{t^n}{\Gamma(a+1)} L_n^{a-n}(x)$$

$$x^m = \sum_{n=0}^\infty \frac{m!\Gamma(a+m+1)(-1)^n}{(m-n)![(\Gamma(a+n+1)]^2} L_n^a(x)$$

$$e^t(xt)^{-\frac{1}{2}a} J_a(2\sqrt{xt}) = \sum_{n=0}^\infty \frac{t^n}{[\Gamma(a+n+1)]^2} L_n^a(x)$$

$$\int_0^\infty e^{-\frac{1}{2}u} u^{\nu+1} J_\nu(\tfrac{1}{2}uz) L_n^{2\nu}(u)\, du \\ = 8(n+\nu+\tfrac{1}{2})\Gamma(n+2\nu+1) \frac{(2z)^\nu}{(z^2+1)^{\nu+\frac{3}{2}}} T_n^\nu\left(\frac{z^2-1}{z^2+1}\right)$$

where m and n are integers but a and ν need not be integers.

III Range $-\infty < z < \infty$; Density Function e^{-z^2}: Hermite polynomials: $H_n(z)$

Generating function: $e^{-t^2+2tz} = \sum_{n=0}^{\infty} \frac{t^n}{n!} H_n(z)$

Recurrence formulas:

$$\frac{d}{dz} H_n(z) = 2nH_{n-1}(z)$$

$$zH_n(z) = nH_{n-1}(z) + \tfrac{1}{2}H_{n+1}(z)$$

$$\frac{d}{dz}[e^{-z^2}H_n(z)] = -2e^{-z^2}H_{n+1}(z)$$

$$H_n(z) = (-1)^n e^{z^2} \frac{d^n}{dz^n} e^{-z^2} = \frac{2^n}{\sqrt{\pi}} \int_{-\infty}^{\infty} (z + it)^n e^{-t^2}\, dt$$

$$\frac{d^2}{dz^2} H_n(z) - 2z \frac{d}{dz} H_n(z) + 2nH_n(z) = 0$$

which is the equation for the confluent hypergeometric functions

$$F(-\tfrac{1}{2}n|\tfrac{1}{2}|z^2) \quad \text{and} \quad zF(-\tfrac{1}{2}n + \tfrac{1}{2}|\tfrac{1}{2}|z^2)$$

$$H_0 = 1; \quad H_1 = 2z; \quad H_2 = 4z^2 - 2$$

$$H_3 = 8z^3 - 12z; \quad H_4 = 16z^4 - 48z^2 + 12$$

$$H_n = (-1)^{\frac{1}{2}n} \frac{n!}{(\frac{1}{2}n)!} F(-\tfrac{1}{2}n|\tfrac{1}{2}|z^2); \quad n = 0, 2, 4, \ldots$$

$$= 2(-1)^{\frac{1}{2}(n-1)} \frac{n!}{(\frac{1}{2}n - \frac{1}{2})!} zF(-\tfrac{1}{2}n + \tfrac{1}{2}|\tfrac{1}{2}|z^2); \quad n = 1, 3, 5, \ldots$$

$$\xrightarrow[z \to \infty]{} 2^n z^n$$

$$x^m = \sum_{s=0}^{s \le \frac{1}{2}m} \frac{m!}{2^m s!(m - 2s)!} H_{m-2s}(x)$$

Normalization integral and other formulas:

$$\int_{-\infty}^{\infty} H_m(z)H_n(z)e^{-z^2}\, dz = \delta_{mn}2^n n! \sqrt{\pi}$$

$$H_{2n}(z) = \frac{(-4^n)n!}{\Gamma(n + \frac{1}{2})} L_n^{-\frac{1}{2}}(z^2); \quad H_{2n+1}(z) = \frac{2(-4)^n n!}{\Gamma(n + \frac{3}{2})} zL_n^{\frac{1}{2}}(z^2)$$

$$[e^{-(x^2+y^2-2xyz)/(1-z^2)}]/\sqrt{1 + z^2} = e^{-x^2-y^2} \sum_{m=0}^{\infty} \left(\frac{z^m}{2^m m!}\right) H_m(x)H_m(y)$$

$$\exp\{-\tfrac{1}{2}[(u^2 + v^2)\cos\varphi + 2uv\sin\varphi]\}$$

$$= \sec(\tfrac{1}{2}\varphi) \sum_{m=0}^{\infty} \frac{i^m \tan^m(\frac{1}{2}\varphi)}{m!} e^{\frac{1}{2}(v^2-u^2)} H_m(u)H_m(iv)$$

A variation in these polynomials, similar to the superscripts β and a in cases I and II, may be introduced by choosing the density function to be e^{-z^2+2az} instead of e^{-z^2}. All this does, however, is to shift the center of the polynomials from $z = 0$ to $z = a$. The new generating function is

$$e^{-t^2+2t(z-a)} = \sum_{n=0}^{\infty} \frac{t^n}{n!} H_n^a(z) = e^{-2ta} \sum_{m=0} \frac{t^m}{m!} H_m(z)$$

where

$$H_n^a(z) = H_n(z - a) = \sum_{m=0}^{n} \frac{(-2a)^{n-m} n!}{m!(n-m)!} H_m(z)$$

$$= 2^{-\frac{1}{2}n} \sum_{m=0}^{n} \frac{n!}{m!(n-m)!} H_m(z\sqrt{2}) H_{n-m}(-a\sqrt{2})$$

or

$$e^{-\frac{1}{2}(z-a)^2} H_n(z - a)$$
$$= e^{\frac{1}{4}a^2 - \frac{1}{2}z^2} \left\{ \sum_{m=0}^{n} H_m(z) \frac{n!(-a)^{n-m}}{m!(n-m)!} F(n+1 | n-m+1 | -\tfrac{1}{2}a^2) \right.$$
$$\left. + \sum_{m=n+1}^{\infty} H_m(z) \frac{(a/2)^{m-n}}{(m-n)!} F(m+1 | m-n+1 | -\tfrac{1}{2}a^2) \right\}$$

The recurrence formulas are all the same, about the new origin. The normalization integral is

$$\int_{-\infty}^{\infty} e^{-z^2+2az} [H_n^a(z)]^2 \, dz = e^{a^2} \int_{-\infty}^{\infty} e^{-(z-a)^2} [H_n(z-a)]^2 \, dz = 2^n n! \sqrt{\pi}\, e^{a^2}$$

With these three sets of eigenfunctions we have covered the various possibilities of singularities of the density function at the end points: The Gegenbauer polynomials correspond to a density having a branch point at both ends of the interval, the Laguerre polynomials to a density function with a branch point at one end and an essential singularity at the other, and the Hermite polynomials to a density function with essential singularities at both ends. The values of the independent variable at the two end points may be changed from the standard values given above by obvious transformations. For instance, for a range from $z = -a$ to $z = \infty$ with density function having branch points at both ends, we use the set of eigenfunctions $T_n^\beta[z/(z + 2a)]$, with density function $r = 2^{2\beta+1}a^{\beta+1}(z + a)^\beta/(z + 2a)^{2\beta+2}$, and so on.

Eigenfunctions by the Factorization Method

The fundamental equation is the Schroedinger type

$$(d^2\Phi/dx^2) + [\lambda - V_m(x)]\Phi = 0$$

where λ is the eigenvalue and the corresponding eigenfunction Φ is required to be quadratically integrable over the range $a \leq x \leq b$, where a and b are contiguous singular points of the equation. Parameter m may be continuously variable or may only have discrete values (in which case the scale is adjusted so that these values are integral values). This equation is sometimes equivalent to the following operator equations:

$$\mathcal{G}^+_{m+1}\mathcal{G}^-_{m+1}\Phi_n(m|x) = (\lambda_n - a_{m+1})\Phi_n(m|x);$$
$$\mathcal{G}^-_m\mathcal{G}^+_m\Phi_n(m|x) = (\lambda_n - a_m)\Phi_n(m|x)$$

where $\quad \mathcal{G}^+_m = [u_m(x) + (d/dx)]; \quad \mathcal{G}^-_m = [u_m(x) - (d/dx)]$

are mutually adjoint operators. When $V_m(x)$ is such that the factorization can be made, with a_m a function of m but not of x; then the eigenvalues are independent of m and

when $\quad a_{m+1} > a_m; \quad \lambda_n = a_{n+1}; \quad n = m, m + 1, m + 2, \dots$

$$\Phi_n(n|x) = \int_a^b [\exp(2\textstyle\int u_{n+1}\,dx)\,dx]^{-\frac{1}{2}} e^{\int u_{n+1}(x)\,dx}$$

$$\Phi_n(m|x) = \frac{1}{\sqrt{a_{n+1} - a_{m+1}}}\left[u_{m+1}(x) + \frac{d}{dx}\right]\Phi_n(m+1|x)$$

$$\int_a^b \Phi_n(m|x)\Phi_{n'}(m|x)\,dx = \delta_{nn'}$$

when $\quad a_{m+1} < a_m; \quad \lambda_n = a_n; \quad n = m, m - 1, m - 2, \dots$

$$\Phi_n(n|x) = \left[\int_a^b \exp\left(-2\int u_n\,dx\right)dx\right]^{-\frac{1}{2}} e^{-\int u_n(x)\,dx}$$

$$\Phi_n(m|x) = \frac{1}{\sqrt{a_n - a_m}}[u_m(x) - (d/dx)]\Phi_n(m-1|x)$$

where Φ_n is again an orthonormal set.

The various sorts of functions V which will allow factorization can be obtained by determining those functions u_m which satisfy

$$u^2_{m+1} - u^2_m + \frac{d}{dx}u_{m+1} + \frac{d}{dx}u_m = a_m - a_{m+1}$$

for which a_m is independent of x. Then the corresponding function V for the original equation is

$$V_m(x) = u^2_m(x) - \frac{d}{dx}u_m(x) + a_m = u^2_{m+1}(x) + \frac{d}{dx}u_{m+1}(x) + a_{m+1}$$

The trivially simple case is for u_m to be independent of x; then $a_m = -u_m^2$ and $V_m = 0$, the eigenfunctions being the trigonometric functions. Other possibilities are

$$u_m = v(x) + mw(x)$$

where, in order that a_m be independent of x we must have that $w^2 + w' =$ constant; $v' + vw =$ constant.

$$u_m = (1/m)y(x) + mw(x)$$

where we must have $y =$ constant and $w^2 + w' =$ constant. Any other choice of dependence on m and x does not allow a_m to be independent of x.

Solving these equations for v, w, and y, for various values of the constants (including zero values), we obtain the following specific forms for $u_m(x)$, a_m, and $V_m(x)$, which include all the possibilities for the outlined method of factorization:

$$(A)\quad u_m = (m + c)b \cot [b(x + p)] + d \csc [b(x + p)]; \quad a_m = b^2(m + c)^2$$
$$V_m = \{b^2(m + c)(m + c + 1) + d^2 + 2bd(m + c + \tfrac{1}{2}) \cos [b(x + p)]\} \csc^2 [b(x + p)]$$

from which, by transformation of variables and by choice of values of the constants b, c, d, and p, one obtains the spherical harmonic functions and other eigenfunctions related to the hypergeometric function.

$$(B)\quad u_m = de^{bx} - m - c; \quad a_m = b^2(m + c)^2$$
$$V_m = -d^2e^{2bx} + 2bd(m + c + \tfrac{1}{2})e^{bx}$$

from which, by transformation, one obtains the Laguerre functions and other eigenfunctions related to the confluent hypergeometric function.

$$(C)\quad u_m = (m + c)(1/x) + \tfrac{1}{2}bx; \quad a_m = -2bm + \tfrac{1}{2}b$$
$$V_m = -(m + c)(m + c + 1)(1/x)^2 - \tfrac{1}{4}b^2x^2 + b(m - c)$$

also giving confluent hypergeometric functions.

$$(D)\quad u_m = bx + d; \quad a_m = -2bm$$
$$V_m = -(bx + d)^2 + b(2m + 1)$$

giving a generalization of the Hermite polynomials.

$$(E)\quad u_m = ma \cot [b(x + p)] + (q/m); \quad a_m = b^2m^2 - (q^2/m^2)$$
$$V_m = -m(m + 1)b^2 \csc^2 [b(x + p)] - 2bq \cot [b(x + p)]$$

related to the hypergeometric function [see Eq. (12.3.22)].

$$(F)\quad u_m = (m/x) + (q/m); \quad a_m = -(q/m)^2$$
$$V_m = -(2q/x) - m(m + 1)/x^2$$

resulting in Laguerre polynomials [see Eq. (12.3.38)].

Bibliography

References on types of partial differential equations, on types of boundary conditions and on difference equations:

Bateman, H.: "Partial Differential Equations of Mathematical Physics," Cambridge, New York, 1932.

Courant, R., and D. Hilbert: "Methoden der Mathematischen Physik," Vol. 2, Springer, Berlin, 1937, reprint Interscience, 1943.

Courant, R., K. Friedrichs, and H. Lewy: "Über die partiellen Differenzengleichungen der mathematischen Physik," *Math. Ann.*, **100,** 32 (1928).

Hadamard, J. S.: "Lectures on Cauchy's Problem in Linear Partial Differential Equations," Yale University Press, New Haven, 1923.

Phillips, H. B., and N. Wiener: Nets and the Dirichlet Problem, *J. Math. Phys.*, **2,** 105 (March, 1923).

Poeckels, F.: "Uber die Partielledifferentialgleichung $\nabla^2 u + k^2 u = 0$," B. G. Teubner, Leipzig, 1891.

Sommerfeld, A.: "Partial Differential Equations in Physics," Academic Press, New York, 1949.

Webster, A. G.: "Partial Differential Equations of Mathematical Physics," Stechert, New York, 1933.

Books giving fairly complete discussions of eigenfunctions, their properties and uses:

Bateman, H.: "Partial Differential Equations of Mathematical Physics," Cambridge, New York, 1932.

Bibliography of Orthogonal Polynomials, National Research Council, Washington, 1940.

Courant, R., and D. Hilbert: "Methoden der Mathematischen. Physik," Vol. 1, Springer, Berlin, 1937.

Ince, E. L.: "Ordinary Differential Equations," Chaps. 10 and 11, Longmans, New York, 1927, reprint Dover, New York, 1945.

Infeld, L., and T. E. Hull: "Factorization Method," *Rev. Modern Phys.*, **23,** 21 (1951).

Kemble, E. C.: "Fundamental Principles of Quantum Mechanics," Chaps. 3 and 4, McGraw-Hill, New York, 1937.

Magnus, W., and F. Oberhettinger: "Special Functions of Mathematical Physics," Springer, Berlin, 1943, reprint Chelsea, New York, 1949.

Riemann-Weber: "Differential- und Integralgleichungen der Mechanik und Physik," ed. by P. Frank and R. von Mises, Vol. 1, Chaps. 7 and 8, Vieweg, Brunswick, 1935, reprint Rosenberg, New York, 1943.

Sommerfeld, A.: "Wellenmechanik," Vieweg, Brunswick, 1937, reprint Ungar, New York, 1948.

Szego, G.: "Orthogonal Polynomials," American Mathematical Society, New York, 1939.

CHAPTER 7

Green's Functions

In the last chapter we began the study of the central problem of field theory, that of fitting the solution of a given partial differential equation to suitable boundary conditions. There we explored the technique of expansion in eigenfunctions, a method which can be used in a perfectly straightforward way whenever we can find a coordinate system, suitable to the boundaries, in which the given partial differential equation will separate. But the result usually comes out in terms of an infinite series, which often converges rather slowly, thus making it difficult to obtain a general insight into the over-all behavior of the solution, its peculiarities near edges, etc. For some aspects of problems it would be more desirable to have a closed function represent the solution, even if it were an integral representation involving closed functions. The Green's function technique is just such an approach.

The method is obvious enough physically. To obtain the field caused by a distributed source (or charge or heat generator or whatever it is that causes the field) we calculate the effects of each elementary portion of source and add them all. If $G(\mathbf{r}|\mathbf{r}_0)$ is the field at the *observer's point* $\mathbf{r}$ caused by a unit point source at the *source point* $\mathbf{r}_0$, then the field at $\mathbf{r}$ caused by a source distribution $\rho(\mathbf{r}_0)$ is the integral of $G\rho$ over the whole range of $\mathbf{r}_0$ occupied by the source. The function G is called the *Green's function.*

Boundary conditions can also be satisfied in the same way. We compute the field at $\mathbf{r}$ for the boundary value (or normal gradient, depending on whether Dirichlet or Neumann conditions are pertinent) zero at every point on the surface *except* for $\mathbf{r}_0^s$ (which is on the surface). At $\mathbf{r}_0^s$ the boundary value has a delta function behavior, so that its integral over a small surface area near $\mathbf{r}_0^s$ is unity. This field at $\mathbf{r}$ (*not* on the boundary) we can call $G(\mathbf{r}|\mathbf{r}_0^s)$; then the general solution, for an arbitrary choice of boundary values $\psi_0(\mathbf{r}_0^s)$ (or else gradients N_0) is equal to the integral of $G\psi_0$ (or else GN_0) over the boundary surface. These functions G are also called Green's functions.

It is not particularly surprising that one can solve the inhomogeneous equation for a field caused by a source distribution, by means of a product of the source density with a Green's function integrated over space, and that the solution of the homogeneous equation having specified values on a surface can be obtained in terms of a product of these values with another Green's function, integrated over the boundary surface. What is so useful and (perhaps) surprising is that these two functions are *not* different; they are *essentially the same function.* For each of the linear, partial differential equations of Chaps. 1 to 3, one can obtain a function which, when integrated over volume, represents a distributed source. When it (or its gradient) is integrated over a surface, it represents the field caused by specified boundary conditions on the surface.

Physically this means that boundary conditions on a surface can be thought of as being equivalent to source distributions on this surface. For the electrostatic case this is, perhaps, not a new concept. The boundary condition on a grounded conductor is that the potential be zero at its surface. Placing a surface dipole distribution just outside the conductor (a surface charge $+\sigma$ just outside the conductor and another surface charge $-\sigma$ just outside that) will result in values of the potential just outside the dipole layer which differ from zero by an amount proportional to the dipole density (to σ times the spacing between $+\sigma$ and $-\sigma$). Nor is it so new in the case of the flow of an incompressible fluid. The boundary condition at a rigid surface is that the normal gradient of the velocity potential be zero at the surface. Placing a single layer of sources next to this rigid surface will result in values of normal gradient of the velocity potential which are proportional to the surface density of the source layer. As we shall see, the fact that boundary conditions can be satisfied by surface integrals of source functions makes the use of source (Green's) functions particularly useful.

It is desirable that we underscore this dualism between sources and boundary conditions by our choice of vocabulary. The equation for a field in the presence of sources is an *inhomogeneous* partial differential equation (*e.g.*, the Poisson equation $\nabla^2\psi = -4\pi\rho$). The inhomogeneous term, which does not contain ψ, contains the source density ρ. Conversely the equation for a field with no sources present is a *homogeneous* equation (*e.g.*, the Laplace equation $\nabla^2\psi = 0$).

Analogously we can say (in fact we have said) that boundary conditions requiring that the field be zero at the surface are *homogeneous boundary conditions* (zero values are homogeneous Dirichlet condition; zero normal gradients are homogeneous Neumann conditions; the requirements that $a\psi + b(\partial\psi/\partial n)$ be zero at the surface are homogeneous mixed conditions).

Conversely the requirements that ψ have specified values ψ_0 (not everywhere zero) at the surface are called *inhomogeneous* Dirichlet conditions; in this case the boundary values may be said to be "caused" by a surface dipole layer of sources, corresponding to an inhomogeneous equation. Likewise the requirements that $\partial\psi/\partial n = N_0$ (not everywhere zero) on the surface are called inhomogeneous Neumann conditions, and the requirements that $a\psi + b(\partial\psi/\partial n) = F_0$ on the surface can be called inhomogeneous mixed conditions. When either equation or boundary conditions are inhomogeneous, sources may be said to be present; when both are homogeneous, no sources are present.

Of course there is another, more obvious, reason why both are given the same descriptive adjective. Solutions of homogeneous equations or for homogeneous boundary conditions may be multiplied by an arbitrary constant factor and still be solutions; solutions of inhomogeneous equations or for inhomogeneous boundary conditions may not be so modified.

The Green's function is therefore a solution for a case which is *homogeneous everywhere except at one point.* When the point is on the boundary, the Green's function may be used to satisfy inhomogeneous boundary conditions; when it is out in space, it may be used to satisfy the inhomogeneous equation. Thus, by our choice of vocabulary, we are able to make statements which hold both for boundary conditions and source distributions; we have made our words conform to our equations.

7.1 *Source Points and Boundary Points*

In the previous chapter we used the concepts of abstract vector space to "geometrize" our ideas of functions. A function $F(x,y,z)$ was considered as being a handy notation for writing down the components of the vector $\mathbf{F}$ along all the nondenumerably infinite directions, corresponding to all the points (x,y,z) inside the boundaries. The delta function $\delta(\mathbf{r} - \mathbf{r}_0)$ represented a unit vector $\mathbf{e}(\mathbf{r}_0)$ in the direction corresponding to the point (x_0,y_0,z_0) (where $\mathbf{r} = x\mathbf{i} + y\mathbf{j} + z\mathbf{k}$; we should note that $\mathbf{r}$ is a vector in three space whereas $\mathbf{e}$ and $\mathbf{F}$ are vectors in abstract vector space).

Formulation in Abstract Vector Space. In Chap. 6 and Sec. 1.6 we discussed the transformation of coordinates from directions given by the unit vectors $\mathbf{e}(\mathbf{r})$ to those for unit vectors $\mathbf{e}_n$, which latter correspond to eigenfunction solutions ψ_n of certain differential equations

$$\mathcal{L}(\psi_n) = \lambda_n\psi_n$$

The corresponding vectors $\mathbf{e}_n$ are eigenvectors for the abstract vector operator $\mathfrak{L}$, corresponding to the differential operator $\mathcal{L}$, such that

$$\mathfrak{L}(\mathbf{e}_n) = \lambda_n\mathbf{e}_n \tag{7.1.1}$$

We showed that the new unit vectors $\mathbf{e}_n$ were mutually orthogonal and that the vector corresponding to the required solution, fitting the specified boundary conditions, could be built up in a unique manner by summing the individual eigenvectors

$$\mathbf{F} = \Sigma A_n \mathbf{e}_n \quad \text{or} \quad F(x,y,z) = \Sigma A_n \psi_n(x,y,z)$$

Since the differential operators $\mathfrak{L}$ and the corresponding vector operators $\mathfrak{L}$ are *linear*, solutions may be added and series expansions are possible. A straightforward method of calculating the components A_n was developed, which allowed our whole abstract picture to become a powerful, practical technique for solving boundary-value problems.

But it should be obvious, by now, that other useful resolutions for **F** are possible. Another possibility is demonstrated by a study of the inhomogeneous equation

$$\mathfrak{L}(F) = -4\pi\rho(x,y,z) \tag{7.1.2}$$

To solve this equation by means of eigenfunctions we expand both ρ and F in eigenfunctions. If the vector corresponding to ρ is $\mathbf{P} = \Sigma B_n \mathbf{e}_n$ and if we assume that $\mathbf{F} = \Sigma A_n \mathbf{e}_n$, then the unknown coefficients A_n may be determined by insertion in the equation

$$\mathfrak{L}(\mathbf{F}) = -4\pi\mathbf{P}; \quad \Sigma\lambda_n A_n \mathbf{e}_n = -4\pi\Sigma B_n \mathbf{e}_n$$

However, we could have resolved the inhomogeneous vector **P** in terms of the unit vectors $\mathbf{e}(x_0,y_0,z_0)$ instead of the $\mathbf{e}_n$'s,

$$\mathbf{P} = \sum_{x_0 y_0 z_0} \rho(x_0,y_0,z_0)\mathbf{e}(x_0,y_0,z_0)$$

corresponding to the equation (which is one definition of the delta function)

$$\rho(x,y,z) = \iiint \rho(\mathbf{r}_0)\delta(\mathbf{r} - \mathbf{r}_0)\, dx_0\, dy_0\, dz_0$$

We then solve the simpler inhomogeneous equation

$$\mathfrak{L}(\mathbf{G}) = -4\pi\mathbf{e}(x_0,y_0,z_0) \tag{7.1.3}$$

(if we can). The components of the solution **G** in the (x,y,z) system are solutions of the simpler inhomogeneous differential equation

$$\mathfrak{L}(G) = -4\pi\delta(\mathbf{r} - \mathbf{r}_0) \tag{7.1.4}$$

The components G, obtained from the solution of (7.1.4), are functions, both of the coordinates x, y, z (the independent variables of the differential operator $\mathfrak{L}$) and also of x_0, y_0, z_0 (the position of the delta function "source") corresponding to the unit vector $\mathbf{e}(x_0,y_0,z_0)$ chosen for Eq. (7.1.3). We shall show, at the end of Sec. 7.5, that the functions $\mathbf{G}(x,y,z|x_0,y_0,z_0) = \mathbf{G}(\mathbf{r}|\mathbf{r}_0)$, for different values of x_0, y_0, z_0, are components, along the directions $\mathbf{e}(\mathbf{r})$, of an *operator*, rather than a vector.

This operator changes the vector **P**, for the inhomogeneous part, into vector **F**, the solution. Because of linearity we expect that a solution of

$$\mathfrak{L}(\mathbf{F}) = -4\pi \sum_{x_0y_0z_0} \rho(\mathbf{r}_0)\mathbf{e}(\mathbf{r}_0) \quad \text{is} \quad \mathbf{F} = \sum_{x_0y_0z_0} \rho(\mathbf{r}_0)\mathbf{G}(\mathbf{r}|\mathbf{r}_0) \qquad (7.1.5)$$

a sum of all the individual solutions for unit vectors on the right-hand side, multiplied each by the appropriate amplitude $\rho(\mathbf{r}_0)$. Consequently, we should expect that a solution of the inhomogeneous differential equation (7.1.2) would be

$$F(x,y,z) = \iiint G(x,y,z|x_0,y_0,z_0)\rho(x_0,y_0,z_0)\, dx_0\, dy_0\, dz_0 \qquad (7.1.6)$$

where G is the solution of Eq. (7.1.4) and is called a *Green's function*. It is thus apparent that, in terms of abstract vectors, a solution by Green's functions is a representation in terms of the unit vectors $\mathbf{e}(x,y,z)$ whereas a solution by eigenfunctions is a representation in terms of the unit vectors $\mathbf{e}_n$. A much more complete discussion of this representation will be given at the end of this chapter.

Of course, it will take the rest of this chapter to outline how the unit solutions G are determined, when the representation converges, and all the other precautionary details analogous to the ones we had to learn, in the last chapter, before we could use the eigenfunction technique with confidence.

Boundary Conditions and Surface Charges. We have not yet shown how the solution of an inhomogeneous *equation* (with homogeneous boundary conditions) will help us to solve a homogeneous equation with inhomogeneous *boundary conditions*. Perhaps a simple example will clarify the principle before we immerse ourselves in details. It will be shown later that a solution of the Poisson equation

$$\nabla^2 G = -4\pi\delta(\mathbf{r} - \mathbf{r}_0)$$

with homogeneous Dirichlet conditions ($G = 0$ on the surface S in Fig. 7.1) is a function which goes to infinity as $1/|\mathbf{r} - \mathbf{r}_0|$ for $\mathbf{r}$ near the source point $\mathbf{r}_0$. What we wish to point out here is that this same solution G can be used to build up a solution for an arbitrary charge distribution inside the surface S *and also for an arbitrary set of Dirichlet boundary conditions* on S (that is, for $\psi = \psi_s$ on the surface).

What is done is to replace the inhomogeneous boundary conditions on S by homogeneous conditions, together with a *surface distribution of charge* just inside the surface. Let us magnify up the situation near the surface, as shown in Fig. 7.2. The charge distribution to replace the inhomogeneous boundary conditions is a surface layer of density σ/ϵ spaced a very small distance ϵ out from the surface S. We make ϵ much smaller than the radius of curvature of the surface and also smaller than the distances over which σ varies appreciably. We have replaced the inhomogeneous conditions by homogeneous conditions plus this charge layer, so we now require that the potential be zero at the surface. For distances of the order of ϵ, the surface is plane (and may be taken as the y, z plane) and the charge density σ may be considered to be uniform.

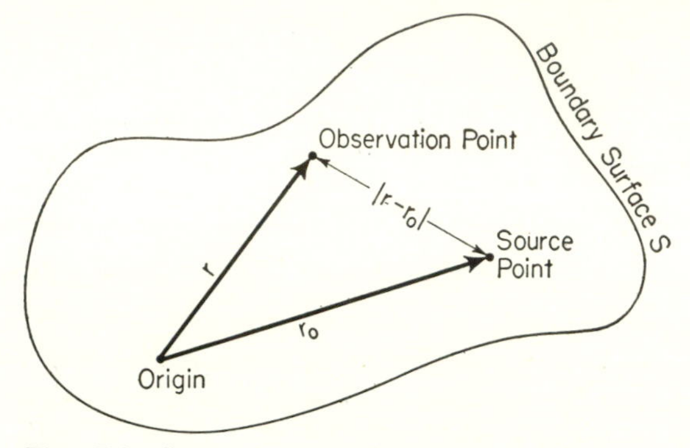

Fig. 7.1 Source point, observation point, and boundary surface for Green's function.

We thus have the situation of a uniform charge sheet of surface density σ/ϵ a distance ϵ in front of a grounded plane conductor at $x = 0$. From elementary electrostatics we remember that the normal gradient of the potential changes by an amount $4\pi\sigma/\epsilon$ when going through such a sheet of charge. Because ϵ is so small, the gradient between charge and bound-

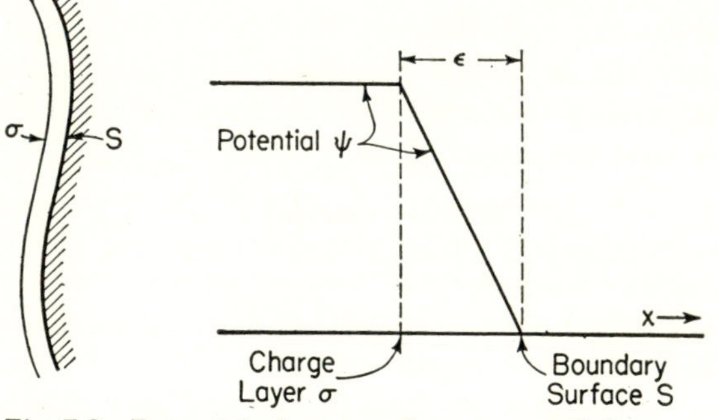

Fig. 7.2 Potential of a source layer σ a small distance ϵ outside a grounded surface S.

ary must be very much larger than the gradient outside the sheet; in fact we can neglect the latter with respect to the former.

Therefore, the gradient between $x = -\epsilon$ and $x = 0$ must equal $-(4\pi\sigma/\epsilon)$, and the potential in this region must be

$$\psi = -(4\pi\sigma/\epsilon)x; \quad -\epsilon < x < 0$$

and thus the potential just outside the charge sheet at $x = -\epsilon$ must be ψ, which is $(4\pi\sigma)$. Therefore, if we made this surface density σ, which is infinitesimally close to the grounded surface, equal to $\psi_s/4\pi$, then *the potential just outside the charge sheet would be just* ψ_s, the boundary value we wished to satisfy.

We have hence made plausible the idea that a solution of a homogeneous equation satisfying inhomogeneous boundary conditions is equivalent to a solution of an inhomogeneous equation satisfying homogeneous boundary conditions, with the inhomogeneous part being a surface layer of charge, proportional to the inhomogeneous boundary values and infinitesimally close to the boundary surface. Of course, we have not *proved* this equivalence; we have only made it plausible. Nor have we seen where the equivalence breaks down (it might be expected, for instance, that the equivalence breaks down between the charge layer and the surface). But having pointed out the goal, we should be able more easily to recognize progress toward it.

Incidentally, it should not be difficult to make plausible the satisfying of inhomogeneous Neumann conditions by a similar replacement with homogeneous conditions ($\partial\psi/\partial n$ zero on S) plus a surface layer, on the surface, of surface density proportional to the specified boundary gradients. Thus we begin to see why solutions of inhomogeneous equations are related to solutions for inhomogeneous conditions and how surface layers can be substituted for boundary conditions.

A Simple Example. Before dealing with the problem in full generality we shall go into some detail with a simple example. We take the two-dimensional Poisson equation

$$\nabla^2\psi = \left(\frac{\partial^2\psi}{\partial x^2} + \frac{\partial^2\psi}{\partial y^2}\right) = -4\pi\rho(x,y) \tag{7.1.7}$$

inside the rectangular boundaries, $x = 0$, $x = a$, $y = 0$, $y = b$. First, we recapitulate the eigenfunction method of solving the case for $\rho = 0$ (homogeneous equation) for homogeneous boundary conditions ($\psi = 0$) on the three sides $x = 0$, $x = a$, $y = 0$, but with the inhomogeneous boundary conditions $\psi = \psi_b(x)$ along $y = b$. Equation (6.3.2) shows that this solution is

$$\psi(x,y) = \int^b \psi_b(\xi)G^b(x,y|\xi)\, d\xi$$

where

$$G^b(x,y|\xi) = \left[\frac{2}{a}\sum_{n=0}^{\infty} \frac{\sinh(\pi ny/a)}{\sinh(\pi nb/a)} \sin\left(\frac{\pi nx}{a}\right) \sin\left(\frac{\pi n\xi}{a}\right)\right] \tag{7.1.8}$$

The quantity G^b, in the brackets, is a function of the coordinates x and y and also of the position ξ on the surface $y = b$. It may be called the *Green's function* for boundary conditions on the surface $y = b$. We multiply it by the specified boundary value for ψ and integrate over the boundary to obtain the solution.

To show how this function is related to the solution of Poisson's equation for a point charge (a point charge in two dimensions is a line charge in three dimensions), we next investigate two forms for the solution of the inhomogeneous equation [which, according to Eq. (7.1.4), should be the equation for the Green's function for a point source at x_0, y_0],

$$\nabla^2\psi = -4\pi\delta(x - x_0)\delta(y - y_0) \tag{7.1.9}$$

with the homogeneous boundary conditions $\psi = 0$ on all four boundary lines.

Very close to the source $[R^2 = (x - x_0)^2 + (y - y_0)^2 \ll a^2,b^2]$ the solution must behave as though the boundaries were not present. The solution of Eq. (7.1.9) for boundaries at infinity is $\psi = -2 \ln R = -\ln [(x - x_0)^2 + (y - y_0)^2]$, so that one would expect the solution of Eq. (7.1.9) with finite boundaries to go to infinity as $-2 \ln R$ as the distance R becomes very much smaller than the distance between the source point x_0, y_0 and the nearest boundary.

There are two ways of solving Eq. (7.1.9) [or Eq. (7.1.7) for that matter]. One, the simplest analytically and the nastiest computationally, is to expand ψ in a double Fourier series

$$\psi = \sum_{m,n} A_{mn} \sin\left(\frac{\pi mx}{a}\right) \sin\left(\frac{\pi ny}{b}\right)$$

This series is not a solution of Laplace's equation $\nabla^2\psi = 0$, but we shall shortly show that it *can* be a solution of Poisson's equation (7.1.7); in fact, to be a solution of (7.1.9), it must be a solution of Laplace's equation *except for one point*, at (x_0,y_0). To solve (7.1.7) we expand $\rho(x,y)$ in Fourier series

$$\rho(x,y) = \sum_{n,m} P_{mn} \sin\left(\frac{\pi mx}{a}\right) \sin\left(\frac{\pi ny}{b}\right)$$

$$P_{mn} = \left(\frac{4}{ab}\right) \int_0^a d\xi \int_0^b d\eta \sin\left(\frac{\pi m\xi}{a}\right) \sin\left(\frac{\pi n\eta}{b}\right) \rho(\xi,\eta)$$

and insert both series in Eq. (7.1.7). The coefficients A_{mn} can then be determined, and we finally obtain for our solution

$$\psi(x,y) = \frac{4}{\pi} \sum_{m,n} \frac{P_{mn}}{(m/a)^2 + (n/b)^2} \sin\left(\frac{\pi mx}{a}\right) \sin\left(\frac{\pi ny}{b}\right)$$

$$= \int_0^a d\xi \int_0^b d\eta\, G(x,y|\xi,\eta)\rho(\xi,\eta) \tag{7.1.10}$$

where

$$G(x,y|\xi,\eta) = \frac{16}{\pi ab} \sum_{m,n} \left[\frac{\sin(m\pi x/a) \sin(n\pi y/b) \sin(n\pi\xi/a) \sin(n\pi\eta/b)}{(m/a)^2 + (n/b)^2}\right]$$

is the Green's function for the inhomogeneous equation. It is not difficult to see that G is the solution of Eq. (7.1.9) for homogeneous boundary conditions. It is the potential at point x, y of a unit point charge at ξ, η. Its poor convergence arises from the fact that it must approach infinity logarithmically as (x,y) approaches (ξ,η) (as we mentioned a few pages ago and as we shall prove in the next section). For more reasonable distributions of ρ series (7.1.10) converges more rapidly.

One other property of G, which will be generally true of these Green's functions: *It is symmetric in interchange of* (x,y) *and* (ξ,η). In other words, the potential at (x,y) caused by a charge at (ξ,η) equals the potential at (ξ,η) caused by the same charge at (x,y), boundary conditions remaining equal. Exchange of source and observer does not change G, a result sometimes called the *principle of reciprocity.*

Relation between Volume and Surface Green's Functions. But series (7.1.10) is still far from the form of the quantity in brackets in Eq. (7.1.8). In the first place, the earlier quantity is a single series, whereas Eq. (7.1.10) is a double series. Let us try to solve Eq. (7.1.7) in terms of a single series. We set

$$\psi(x,y) = \sum_m F_m(y) \sin\left(\frac{\pi m x}{a}\right); \quad \rho(x,y) = \sum_m \rho_m(y) \sin\left(\frac{\pi m x}{a}\right) \qquad (7.1.11)$$

where F_m is to be computed and

$$\rho_m(y) = \frac{2}{a}\int_0^a \rho(\xi,y)\sin\left(\frac{\pi m \xi}{a}\right) d\xi$$

Inserting this in Eq. (7.1.7) we obtain an ordinary, inhomogeneous equation for $F_m(y)$:

$$\frac{d^2}{dy^2}F_m - \left(\frac{\pi m}{a}\right)^2 F_m = -4\pi\rho_m(y)$$

Referring to Eq. (5.2.19) for the solution of such an equation, we note that two independent solutions of the homogeneous part are

$$y_1 = \sinh(\pi m y/a); \quad y_2 = \sinh[(\pi m/a)(b - y)]$$

These are independent, for their Wronskian

$$\begin{aligned}\Delta(y_1,y_2) &= -\frac{\pi m}{a}\left\{\sinh\left(\frac{\pi m y}{a}\right)\cosh\left[\frac{\pi m}{a}(b-y)\right]\right. \\ &\qquad\qquad \left. + \cosh\left(\frac{\pi m y}{a}\right)\sinh\left[\frac{\pi m}{a}(b-y)\right]\right\} \\ &= -(\pi m/a)\sinh(\pi m b/a) \neq 0\end{aligned}$$

is a constant (as it must be). Inserting all this in Eq. (5.2.19) and setting the limits of integration so that F_m is zero at $y = 0$ and $y = b$, we have

$$F_m(y) = \frac{4a}{m\sinh(\pi mb/a)}\left\{\sinh\left(\frac{\pi my}{a}\right)\int_y^b \rho_m(\eta)\sinh\left[\frac{\pi m}{a}(b-\eta)\right]d\eta + \sinh\left[\frac{\pi m}{a}(b-y)\right]\int_0^y \rho_m(\eta)\sinh\left(\frac{\pi m\eta}{a}\right)d\eta\right\} = \int_0^b g_m(y|\eta)\rho_m(\eta)\,d\eta$$

where

$$g_m(y|\eta) = \frac{4a}{m\sinh(\pi mb/a)}\begin{cases}\sinh(\pi my/a)\sinh[(\pi m/a)(b-\eta)]; & \eta > y\\ \sinh(\pi m\eta/a)\sinh[(\pi m/a)(b-y)]; & \eta < y\end{cases}$$

The function $g_m(y|\eta)$ goes to zero for either y or η either zero or b, and it has a discontinuity in slope (see page 123) of an amount (-4π) at $y = \eta$.

Finally, inserting all these solutions back in Eq. (7.1.11), we have the simple form for the solution

$$\psi(x,y) = \int_0^a d\xi \int_0^b d\eta\, G(x,y|\xi,\eta)\rho(\xi,\eta) \tag{7.1.12}$$

where

$$G(x,y|\xi,\eta) = \sum_{m=0}^{\infty} \frac{8\sin\left(\frac{\pi mx}{a}\right)\sin\left(\frac{\pi m\xi}{a}\right)}{m\sinh(\pi mb/a)} \cdot \begin{cases}\sinh(\pi my/a)\sinh[(\pi m/a)(b-\eta)]; & \eta > y\\ \sinh(\pi m\eta/a)\sinh[(\pi m/a)(b-y)]; & \eta < y\end{cases}$$

Since this integral for ψ has the same form as the integral in Eq. (7.1.10), the G given here must equal the G given there. A simple (but tedious) process of expansion of g_m into a Fourier series in y will show that the two G's are indeed identical.

This latest expression for G, however, is best arranged to display the relation between the solution of an inhomogeneous equation with homogeneous boundary conditions and the solution, given in (7.1.8), of the homogeneous equation with inhomogeneous boundary conditions; for if the only charge inside the boundaries is a sheet of charge, of surface density $(1/4\pi\epsilon)\psi_b(\xi)$ a vanishingly small distance ϵ away from the surface $y = b$, then at $\eta = b - \epsilon$, which is the only place where ρ differs from zero,

$$G(x,y|\xi,b-\epsilon) \xrightarrow[\epsilon\to 0]{} \sum_m \frac{8\pi\epsilon}{a}\sin\left(\frac{\pi mx}{a}\right) \cdot \sin\left(\frac{\pi m\xi}{a}\right)\frac{\sinh(\pi my/a)}{\sinh(\pi mb/a)};\quad y < b - \epsilon$$

Since the range $b - \epsilon < y < b$ is to become vanishingly small, it will not enter our discussions, though we must remember that the slope of G suffers a discontinuity in this infinitesimal range, going to zero at $y = b$. Inserting this value of G in Eq. (7.1.12) we can see that the potential from a thin layer of charge, of surface density $(1/4\pi\epsilon)\psi_b(x)$ an infinitesimal distance ϵ from the grounded plate $y = b$, is represented by exactly the same integral as is the potential arising when the surface $y = b$ is given the inhomogeneous boundary condition $\psi_b(x)$, given in (7.1.8).

The General Solution. We can arrive at the same equality of the two solutions by a somewhat different, and more generally useful, route. We notice that

$$\left[\frac{\partial}{\partial\eta} G(x,y|\xi,\eta)\right]_{y<\eta} \xrightarrow[y\to b]{} -4\pi G_b(x,y|\xi) \qquad (7.1.13)$$

where G_b is defined in Eq. (7.1.8), so that we can rewrite Eq. (7.1.8) as follows:

$$\psi(x,y) = -\frac{1}{4\pi}\int_{S_0} \psi_0(S_0)\left[\frac{\partial}{\partial n_0} G(x,y|S_0)\right] dS_0 \qquad (7.1.14)$$

where the quantity in square brackets is the gradient of $G(x,y|x_0,y_0)$, in the x_0, y_0 coordinates, normal to the boundary surface S_0 (in this case S_0 is $y_0 = b$, so the gradient is the derivative with respect to y_0 or η) with the coordinates x_0, y_0 taken to be on this surface. The integration of this, times the specified boundary value ψ_0, is taken over the surface (in this case the integration is over x_0 or ξ).

This shorthand notation does bring out the relation between a Green's function for the inhomogeneous equation and the Green's function for inhomogeneous boundary conditions, but it also leaves several things unsaid. For instance, it neglects to point out the nature of the discontinuity in G produced by the limiting process of setting x_0, y_0 on the surface S_0 (since G has a discontinuity in slope at $y = y_0 = \eta$, the gradient of G has a discontinuity in *value*, which discontinuity is brought directly to the surface when x_0, y_0 is brought to the surface). There is no discontinuity in the resulting solution ψ *within* the region enclosed by the boundary surface, but there must be a discontinuity just at the boundary [as a matter of fact this discontinuity must be such that ψ is zero just beyond the surface and is $\psi_0(S_0)$ just inside the surface, in the region enclosed by the boundary].

In future use of formula (7.1.14) we must be careful that, in obtaining the Green's function for boundary conditions from $G(\mathbf{r}|\mathbf{r}_0)$, we first let the "source point" x_0, y_0 approach the surface S_0 and only afterward allow the point x, y to approach the corresponding boundary surface S for the "observation point" space. Going to the limit in this order, we see, from Eqs. (7.1.13) and (7.1.8), that the limit, as η first approaches b and then y approaches b, is

$$+\frac{\partial}{\partial\eta} G(x,y|\xi,\eta) \to -\frac{8\pi}{a}\sum_n \sin\left(\frac{n\pi x}{a}\right)\sin\left(\frac{n\pi\xi}{a}\right) = -4\pi\delta(x-\xi)$$

Therefore, the Green's function for the surface (evaluated on the surface) is a delta function, corresponding to boundary conditions zero except at the point $x = \xi$. This, of course, corresponds to our intuitive feeling as to the nature of the Green's function.

To recapitulate: An alternative to the eigenfunction technique for solving boundary-value problems involves solving the inhomogeneous equation for homogeneous boundary conditions for a "point source" at some point $\mathbf{r}_0$ inside the surface. The resulting function $G(\mathbf{r}|\mathbf{r}_0)$, the Green's function for the interior volume, is symmetric to interchange of $\mathbf{r}$ and $\mathbf{r}_0$, has a discontinuity (in value or slope) at $\mathbf{r} = \mathbf{r}_0$, and satisfies homogeneous boundary conditions for $\mathbf{r}$ on the boundary surface S and for $\mathbf{r}_0$ on the similar surface S_0. The solution of the general inhomogeneous equation then has the form given in Eq. (7.1.12), $\psi(\mathbf{r})$ being obtained by integrating $G\rho$ over the $\mathbf{r}_0$ space inside S_0.

If the homogeneous boundary conditions satisfied by G are that $G = 0$ on S and S_0, then the solution for inhomogeneous boundary conditions $[\psi = \psi(S)$ on $S]$ has the form given in Eq. (7.1.14), where the gradient of G is in the $\mathbf{r}_0$ coordinates, normal to S_0, for $\mathbf{r}_0$ vanishingly close to S_0. Therefore, the point $\mathbf{r}_0$ is now "on" S_0, and the integration, times $\psi(S_0)$ is taken over the boundary *surface*.

If the homogeneous boundary conditions on G are Neumann conditions (normal gradient of G zero on S), then the solution for inhomogeneous conditions $[\partial\psi/\partial n = N(S)$ on $S]$ has the form

$$\psi(x,y) = 4\pi \int_{S_0} N(S_0)G(x,y|S_0)\, dS_0 \tag{7.1.15}$$

This will be proved later. In fact, the discussion of this section has not finished our "proof" of Eq. (7.1.14); it has just begun the investigation.

Green's Functions and Generating Functions. Formulas (7.1.8) to (7.1.12) show that Green's functions and eigenfunctions are closely related. Later in this chapter we shall obtain a general formula for the expansion of a Green's function in eigenfunctions. It certainly is not difficult to see that such expansion formulas can prove to be a prolific source of generating functions for eigenfunctions [see Eq. (6.3.32)]. In the case discussed in this section, we obtain a generating function by integrating $G_b(x,y|\xi)$, defined in Eq. (7.1.8), over ξ from zero to a. Only the terms for n an odd integer give nonzero integrals, so that, by expanding the hyperbolic sine into its constituent exponentials, we have

$$\int_0^a G_b(x,y|\xi)\, d\xi = \frac{4}{\pi} \sum_{n=0}^{\infty} \frac{\sinh[(2n+1)(\pi y/a)]}{\sinh[(2n+1)(\pi b/a)]} \frac{\sin[(2n+1)(\pi x/a)]}{(2n+1)}$$

$$= \frac{2}{\pi} \sum_{n=-\infty}^{\infty} \frac{\gamma^{2n+1}}{2n+1} \frac{\sin[(2n+1)(\pi x/a)]}{\sinh[(2n+1)(\pi b/a)]};\quad \gamma = e^{\pi y/a} \tag{7.1.16}$$

This relationship is not particularly useful, since G_b is not given in closed form. It is easy to see, however, that, when it is possible to obtain a closed form for G, we can then produce generating functions for the related eigenfunctions.

7.2 *Green's Functions for Steady Waves*

Before we deal with more complicated cases, it is advisable to simplify our notation by reverting to vector symbolism. The point x, y, z in three dimensions is represented by the vector $\mathbf{r} = x\mathbf{i} + y\mathbf{j} + z\mathbf{k}$. The Green's function $G(\mathbf{r}|\mathbf{r}_0)$ depends on the position of two points: the *observation point,* where the field is measured, represented by $\mathbf{r}$ and given in terms of the *observational coordinates* x, y, z, and the source point, where the unit source is placed, represented by $\mathbf{r}_0$ and given in terms of the *source coordinates* x_0, y_0, z_0. The boundary surface, in the observational coordinates, is labeled S, in the source coordinates S_0. A Green's function with observational point at $\mathbf{r}$ and with source on the boundary surface is labeled $G(\mathbf{r}|\mathbf{r}_0^s)$. The element of volume in observer's space is labeled dv ($= dx\,dy\,dz$), and the axial vector representing an element of boundary surface is $d\mathbf{A}$ for the observational coordinates, $d\mathbf{A}_0$ for the source coordinates. These vectors both point *outward,* away from the volume inside the boundary.

The element of normal gradient of G at the surface, in the source coordinates, would then be written $\text{grad}_0\,[G(\mathbf{r}|\mathbf{r}_0^s)] \cdot d\mathbf{A}_0$, where the zero subscript after grad indicates that the derivative is to be taken with respect to x_0, y_0, z_0. The delta function for three dimensions can be written $\delta(\mathbf{r} - \mathbf{r}_0)$; it has the integral properties

$$\iiint F(\mathbf{r})\delta(\mathbf{r} - \mathbf{r}_0)\,dv = F(\mathbf{r}_0) \tag{7.2.1}$$

Finally the Laplacian in the source coordinates is ∇_0^2, having a zero subscript, whereas in observer's space it is simply ∇^2.

Green's Theorem. In order to obtain a fairly rigorous derivation of the properties of the Green's function, we shall find it advantageous to use a variant of Gauss' theorem. For any closed surface S, Gauss' theorem given in Eq. (1.4.7), states that the normal outflow integral over the surface S of any "reasonable" vector is equal to the integral of the divergence of this vector over the whole volume inside the surface S. By "reasonable," we mean that the vector should not have discontinuities in value or slope at S and that its divergence is integrable.

We next consider two "reasonable" scalar functions of position $\mathbf{r}$, $U(\mathbf{r})$ and $V(\mathbf{r})$. A vector may be constructed from them by using the gradient, U grad V. Gauss' theorem states that the normal outflow integral of this vector, over S, equals the integral of the divergence of

U grad V over the volume inside S. Since

$$\operatorname{div}(U \operatorname{grad} V) = (\operatorname{grad} U) \cdot (\operatorname{grad} V) + U\nabla^2 V$$

we have, in our agreed-upon notation (with $d\mathbf{A}$ pointing away from the enclosed region),

$$\oint U \operatorname{grad} V \cdot d\mathbf{A} = \iiint (\operatorname{grad} U \cdot \operatorname{grad} V)\, dv + \iiint U\nabla^2 V\, dv$$

But we can also obtain a similar equation for the other vector V grad U. Subtracting one from the other we obtain

$$\oint [U \operatorname{grad} V - V \operatorname{grad} U] \cdot d\mathbf{A} = \iiint [U\nabla^2 V - V\nabla^2 U]\, dv \quad (7.2.2)$$

where the surface integral is of the outward normal component of the vector in brackets over the closed surface S and the volume integral is of the scalar quantity in brackets over the whole volume inside S. The boundary surface may be finite in extent or infinite. For instance, it may be the surface of a finite sphere, with the interior of the sphere being the volume integrated over and the direction of $d\mathbf{A}$ pointing away from the center of the sphere. Or it may be the surface of this finite sphere, plus the infinite sphere, with the "inside" being the infinite volume between these spheres, the $d\mathbf{A}$ on the finite sphere pointing toward its center (away from the "inside") and the $d\mathbf{A}$ on the infinite sphere pointing outward. Other special examples can be worked out from the same principles.

This relation between surface and volume integrals is called *Green's theorem*. As we have seen, it is a special case of Gauss' theorem. Its use in deriving the properties of the Green's function will shortly be apparent.

Green's Function for the Helmholtz Equation. For a beginning, let us study the Green's function for the Helmholtz operator

$$\mathfrak{L}\psi \equiv \nabla^2\psi + k^2\psi = 0 \quad (7.2.3)$$

for some boundary conditions on a closed surface S. According to our hasty preview in the first section we should now prove the following properties of the Green's function.

1. The Green's function will be a symmetric function of two sets of coordinates, those of the observation point and those of the source point:

$$G_k(\mathbf{r}|\mathbf{r}_0) = G_k(\mathbf{r}_0|\mathbf{r}); \quad \text{reciprocity relation} \quad (7.2.4)$$

This function will satisfy some homogeneous boundary conditions on both S and S_0 and will have a discontinuity (of a sort to be determined) at $\mathbf{r} = \mathbf{r}_0$.

2. Employing this function, it is possible to obtain a solution of the inhomogeneous equation with the given homogeneous boundary conditions or else the solution of the homogeneous equation with inhomogene-

ous boundary conditions. Because of the linearity of the equation, we can also solve the inhomogeneous equation with inhomogeneous boundary conditions by superposition of both individual solutions.

3. The solutions for inhomogeneous boundary conditions will have a discontinuity at the boundary. For instance, if ψ is specified on the surface (Dirichlet conditions), the solution will have the specified value ψ just inside the boundary and will be zero just outside. For Neumann conditions, where normal gradient is specified, the discontinuity is in the gradient.

The required Green's function is the solution of the inhomogeneous Helmholtz equation

$$\nabla^2 G_k(\mathbf{r}|\mathbf{r}_0) + k^2 G_k(\mathbf{r}|\mathbf{r}_0) = -4\pi\delta(\mathbf{r} - \mathbf{r}_0) \tag{7.2.5}$$

for a unit point source at $\mathbf{r}_0$, which satisfies homogeneous boundary conditions (either zero value or zero normal gradient of G) on the boundary surface S (and also, in the source coordinates, on S_0). The delta function is defined in Eq. (7.2.1) and is symmetric with respect to rotation about $\mathbf{r}_0$. More complex sources, such as dipoles, etc., may be handled by superposition of solutions for this simple source. Incidentally, ∇^2 in (7.2.5) operates on the observational coordinates.

We now wish to show that the inhomogeneous equation

$$\nabla^2\psi + k^2\psi = -4\pi\rho(\mathbf{r}) \tag{7.2.6}$$

subject to arbitrary Dirichlet (or Neumann) conditions on the closed boundary surface S may be expressed in terms of G. To do this we multiply Eq. (7.2.5) by ψ and Eq. (7.2.6) by G and subtract, exchanging $\mathbf{r}$ for $\mathbf{r}_0$ at the same time:

$$G_k(\mathbf{r}_0|\mathbf{r})\nabla_0^2\psi(\mathbf{r}_0) - \psi(\mathbf{r}_0)\nabla_0^2 G_k(\mathbf{r}_0|\mathbf{r}) = 4\pi[\psi(\mathbf{r}_0)\delta(\mathbf{r} - \mathbf{r}_0) - G_k(\mathbf{r}_0|\mathbf{r})\rho(\mathbf{r}_0)]$$

Integrating this over all source coordinates, x_0, y_0, z_0 inside S_0, we obtain, because of the properties of the delta function and Eq. (7.2.4),

$$\frac{1}{4\pi}\iiint [G_k(\mathbf{r}|\mathbf{r}_0)\nabla_0^2\psi(\mathbf{r}_0) - \psi(\mathbf{r}_0)\nabla_0^2 G_k(\mathbf{r}|\mathbf{r}_0)]\,dv_0 + \iiint \rho(\mathbf{r}_0)G_k(\mathbf{r}|\mathbf{r}_0)\,dv_0 = \begin{cases} \psi(\mathbf{r}); & \text{if } \mathbf{r} \text{ is inside } S \\ 0; & \text{if } \mathbf{r} \text{ is outside } S \end{cases}$$

It will be of interest to indicate what value the right-hand side has when the observation point is exactly on S. This is to some extent a matter of convention, though once we decide on the convention, we must be consistent about it. At any rate, unless we state otherwise, we shall in the future assume that this discontinuous function is $\psi(\mathbf{r})$ within *and on* S and is zero outside S.

We now use Green's theorem (7.2.2) to simplify the left-hand side. According to our convention we measure the gradient pointing outward from the surface (or, if you like, the surface element $d\mathbf{A}$ points away from the interior of the volume, where the field is to be evaluated), so that the surface integrals are the usual normal outflow integrals. The use of Eq. (7.2.2) gives

$$\frac{1}{4\pi}\oint [G_k(\mathbf{r}|\mathbf{r}_0^s)\,\mathrm{grad}_0\,\psi(\mathbf{r}_0^s) - \psi(\mathbf{r}_0^s)\,\mathrm{grad}_0\,G_k(\mathbf{r}|\mathbf{r}_0^s)]\cdot d\mathbf{A}_0 + \iiint \rho(\mathbf{r}_0)G_k(\mathbf{r}|\mathbf{r}_0)\,dv_0 = \begin{cases}\psi(\mathbf{r}); & \mathbf{r}\text{ within and on }S\\ 0; & \mathbf{r}\text{ outside }S\end{cases} \tag{7.2.7}$$

from which we can obtain our solutions for either inhomogeneous equation or inhomogeneous boundary condition.

Solution of the Inhomogeneous Equation. For instance, for the inhomogeneous equation ($\rho \neq 0$), if the boundary condition is homogeneous Dirichlet ($\psi = 0$ at S), we choose G also to be zero at both S and S_0. Then the surface integral over S_0 is zero and

$$\psi(\mathbf{r}) = \iiint \rho(\mathbf{r}_0)G_k(\mathbf{r}|\mathbf{r}_0)\,dv_0 \tag{7.2.8}$$

for $\mathbf{r}$ inside or on S. This function automatically satisfies the homogeneous boundary conditions ($\psi = 0$ at S) and is a solution of Eq. (7.2.6). If the boundary conditions are homogeneous Neumann (normal gradient of ψ zero on S), then we adjust G_k so that it satisfies the same conditions for both sets of coordinates, observational and source. Then again the surface integral is zero and Eq. (7.2.8) holds. Therefore (7.2.8) is the solution of the inhomogeneous equation (7.2.6) for homogeneous boundary conditions when G_k satisfies the same conditions as does ψ.

As a matter of fact, this equation still holds if the boundary conditions are that ψ on S equals a function $f(\mathbf{r}_0^s)$ times the normal gradient on S as long as G_k satisfies these same conditions. The most general homogeneous boundary conditions are

$$\psi(\mathbf{r}^s) = f(\mathbf{r}^s)\frac{\partial}{\partial n}\psi;\quad G_k(\mathbf{r}^s|\mathbf{r}_0) = f(\mathbf{r}^s)\frac{\partial}{\partial n}G_k;\quad G_k(\mathbf{r}|\mathbf{r}_0^s) = f(\mathbf{r}_0^s)\frac{\partial}{\partial n_0}G_k$$

When $f = 0$, the conditions are homogeneous Dirichlet; when $1/f = 0$, they are homogeneous Neumann. No matter what the values of f, the surface integral of Eq. (7.2.7) will vanish and leave us Eq. (7.2.8) for the solution. Of course, not all conditions are allowed in physical problems. For instance, if there is a net charge inside S, we could not expect to have the normal gradient of ψ be zero all over the boundary, so we would not expect $\partial G/\partial n$ to be zero on the boundary for the Laplace equation solution. Consideration of the physics involved usually indicates what is possible.

For *in*homogeneous boundary conditions, if they are to be Dirichlet conditions [$\psi = \psi_0(\mathbf{r}_0^s)$ on S_0], we make G_k zero at S and also at S_0. Then we obtain the solution we already discussed on page 801.

$$\psi(\mathbf{r}) = -(1/4\pi)\oint \psi_0(\mathbf{r}_0^s)[\text{grad}_0\, G_k(\mathbf{r}|\mathbf{r}_0^s)] \cdot d\mathbf{A}_0 \tag{7.2.9}$$

for $\mathbf{r}$ inside and on S, for inhomogeneous Dirichlet conditions on ψ, if G_k is zero on S and S_0. We note again that the vector $d\mathbf{A}_0$ points *outward*, away from the inside region where the field is measured, so that the gradient integrated is the *outward* normal gradient. Here we obtain our solution in terms of an integration over the boundary surface only; the inhomogeneous boundary conditions are satisfied by *putting a layer of charge on the surface.* For Dirichlet conditions this is a doublet layer, as evidenced by the use of grad G.

If ψ is to satisfy the inhomogeneous Neumann boundary conditions that the normal outward gradient of ψ on S be equal to $N(\mathbf{r}^s)$, then we take the normal gradient to be zero for G_k on S and obtain the solution

$$\psi(\mathbf{r}) = +(1/4\pi)\oint G_k(\mathbf{r}|\mathbf{r}_0^s)\mathbf{N}(\mathbf{r}_0^s) \cdot d\mathbf{A}_0 \tag{7.2.10}$$

when $\mathbf{r}$ is within or on S, where $\mathbf{N}$ is the specified vector gradient on S and the normal gradient of G is zero at S and S_0. In this case the surface charge to force the normal gradient to be N is just $N/4\pi$, a single charge layer, not a doublet layer.

Finally, if the boundary conditions are the general inhomogeneous form

$$(\partial/\partial n_0)\psi + f(\mathbf{r}_0^s)\psi = F_0(\mathbf{r}_0^s) \tag{7.2.11}$$

then we require that G satisfy the corresponding homogeneous form

$$(\partial/\partial n_0)G_k + f(\mathbf{r}_0^s)G_k = 0$$

where both normal gradients are outward and the solution for the homogeneous equation with requirement (7.2.11) is

$$\psi(\mathbf{r}) = \begin{cases} +(1/4\pi)\oint G_k(\mathbf{r}|\mathbf{r}_0^s)F_0(\mathbf{r}_0^s)\, dA_0 \\ -\dfrac{1}{4\pi}\oint \dfrac{F_0(\mathbf{r}_0^s)}{f(\mathbf{r}_0^s)}\, \text{grad}_0\,[G_k(\mathbf{r}|\mathbf{r}_0^s)] \cdot d\mathbf{A}_0 \end{cases} \tag{7.2.12}$$

Either form of the surface integral may be used. The first is more useful if f is small or zero (Neumann conditions); the second when $1/f$ is small or zero (Dirichlet conditions) but F/f is bounded.

The solutions given in Eqs. (7.2.9), (7.2.10), and (7.2.12) are for the homogeneous equation ($\rho = 0$ inside the boundary). If we wish to solve the inhomogeneous equation (7.2.6) for inhomogeneous boundary conditions, we add a volume integral of the type of Eq. (7.2.8) to the appropriate surface integral, with the G in the volume integral satisfying the same homogeneous boundary conditions as the G in the surface integral.

General Properties of the Green's Function. We have now proved in a fairly rigorous manner most of the statements made in Sec. 7.1 and on page 802. The boundary conditions are satisfied by putting on the surface a distribution of dipoles (for Dirichlet conditions) or a simple charge distribution (for Neumann conditions) of an amount proportional to the required value or normal gradient of ψ times the normal gradient or value of the appropriate Green's function G. We also note the discontinuity in the solution, the integral giving ψ on and inside S, but being zero outside S.

We have not as yet proved directly that G is a symmetric function of $\mathbf{r}$ and $\mathbf{r}_0$, as stated in Eq. (7.2.4). We have that, as a function of $\mathbf{r}$, G satisfies Eq. (7.2.5) and that, for a source at $\mathbf{r}_1$, it satisfies

$$\nabla^2 G_k(\mathbf{r}|\mathbf{r}_1) + k^2 G_k(\mathbf{r}|\mathbf{r}_1) = -4\pi\delta(\mathbf{r} - \mathbf{r}_1)$$

Multiplying (7.2.5) by $G(\mathbf{r}|\mathbf{r}_1)$ and this new equation by $G(\mathbf{r}|\mathbf{r}_0)$, subtracting, and using Green's theorem (7.2.2), we have

$$\begin{aligned} &-\oint \{G(\mathbf{r}|\mathbf{r}_1)\operatorname{grad}[G(\mathbf{r}|\mathbf{r}_0)] - G(\mathbf{r}|\mathbf{r}_0)\operatorname{grad}[G(\mathbf{r}|\mathbf{r}_1)]\} \cdot d\mathbf{A} \\ &\quad = 4\pi\iiint[G(\mathbf{r}|\mathbf{r}_1)\delta(\mathbf{r} - \mathbf{r}_0) - G(\mathbf{r}|\mathbf{r}_0)\delta(\mathbf{r} - \mathbf{r}_1)]\,dv = 4\pi[G(\mathbf{r}_0|\mathbf{r}_1) - G(\mathbf{r}_1|\mathbf{r}_0)] \end{aligned}$$

Since both G's satisfy the same homogeneous boundary conditions, the surface integral vanishes and therefore we have the reciprocity condition $G(\mathbf{r}_0|\mathbf{r}_1) = G(\mathbf{r}_1|\mathbf{r}_0)$ as long as both $\mathbf{r}_0$ and $\mathbf{r}_1$ are inside or on the surface.

But still more important, it remains to investigate the behavior of $G(\mathbf{r}|\mathbf{r}_0)$ for the observation point near the source point, where the magnitude of $\mathbf{R} = \mathbf{r} - \mathbf{r}_0$ is small compared with the distance of either from the nearest point on the boundary surface. From the nature of Eq. (7.2.5) we see that, when $R = \sqrt{(x - x_0)^2 + (y - y_0)^2 + (z - z_0)^2}$ is small compared with the distance to S or S_0, the function G is just a function of R. In other words, the source is completely symmetrical, so that G cannot depend on the direction of R, only on its magnitude. We have also noted earlier that we should expect that G would have a singularity at $R = 0$.

To restate this in more mathematical language, we should expect that $G(\mathbf{r}|\mathbf{r}_0)$ could be separated into two parts: a part which is everywhere regular and continuous inside S and which depends on the boundary conditions imposed on G at S and a part which is regular and continuous inside S *except* at $\mathbf{r} = \mathbf{r}_0$, which is a function of R alone, having a singularity at $R = 0$. This last part we can call $g_k(R)$. Therefore we can say that, if observation point and/or source point are not infinitesimally close to S,

$$G_k(\mathbf{r}|\mathbf{r}_0) \xrightarrow[R\to 0]{} g_k(R); \quad \mathbf{R} = \mathbf{r} - \mathbf{r}_0 \tag{7.2.13}$$

To find the behavior of g_k for R small, we integrate both sides of Eq. (7.2.5) over a small sphere of radius ϵ about $\mathbf{r}_0$ (the integration is in observer's space, x, y, z). This gives us

$$\iiint \nabla^2 G_k(\mathbf{r}|\mathbf{r}_0)\, dv + k^2 \iiint G_k(\mathbf{r}|\mathbf{r}_0)\, dv = -4\pi$$

The integral on the right-hand side equals -4π because of the properties of the delta function and because the sphere integrated over includes the point $\mathbf{r} = \mathbf{r}_0$.

We now make ϵ small enough so that Eq. (7.2.13) holds and substitute g_k for G_k in the remaining integrals. Moreover we presume that the Laplacian of a singularity is more "singular" than the singularity itself, so that, in the limit, the first integral predominates. We therefore obtain

$$\iiint \nabla^2 g_k(R)\, dv \rightarrow -4\pi \quad \text{as} \quad \epsilon \rightarrow 0$$

where the coordinates for integration over the spherical volume are R, ϑ, ϕ and the volume element is $dv = R^2\, dR \sin\vartheta\, d\vartheta\, d\phi$.

Using Gauss' theorem (1.4.7) we have that the net outflow of grad g over the surface of a sphere about $\mathbf{r}_0$ of radius ϵ is equal to -4π:

$$\oint \operatorname{grad}(g_k) \cdot d\mathbf{A} \rightarrow -4\pi$$

Since g_k depends only on R, the radial coordinate of the little sphere, grad(g_k) is everywhere in the radial direction, parallel to $d\mathbf{A}$, and its magnitude is the same everywhere on the surface. Therefore the surface integral is equal, in the limit, to dg/dR, at $R = \epsilon$, times the area of the sphere, $4\pi\epsilon^2$, and we arrive at the formula

$$[(dg/dR)_{R=\epsilon}](4\pi\epsilon^2) \rightarrow -4\pi; \quad \epsilon \rightarrow 0$$

or, what is the same thing,

$$(d/dR)g_k(R) \rightarrow -(1/R^2); \quad (R \rightarrow 0)$$

Thus finally, we obtain $g_k(R) \rightarrow (1/R)$, or

$$G_k(\mathbf{r}|\mathbf{r}_0) \rightarrow (1/R); \quad \text{as } R = |\mathbf{r} - \mathbf{r}_0| \rightarrow 0 \tag{7.2.14}$$

when neither $\mathbf{r}$ nor $\mathbf{r}_0$ is very close to S or S_0. Having obtained this, it is not hard to see that we were correct in discarding the volume integral of G above.

Thus we can say that, viewed as a function of the coordinates (x,y,z) of the observation point, the Green's function is a regular continuous solution of the homogeneous equation $\nabla^2 G + k^2 G = 0$, inside S, *except* at the point $\mathbf{r} = \mathbf{r}_0$, where it has a singularity as specified in Eq. (7.2.14). This singularity is a consequence of the presence of the unit point source, Eq. (7.2.5) being inhomogeneous *only* at this point. We note that the limiting form $1/R$ is independent of k, so that the result holds also for the Laplace equation.

Equation (7.2.14) holds for three dimensions. A similar discussion can be carried through for the two-dimensional case, in which it can be shown that for two dimensions

$$G(\mathbf{r}|\mathbf{r}_0) \to -2 \ln R; \quad R \to 0 \tag{7.2.15}$$

For the one-dimensional case, Eq. (7.2.5) becomes

$$(d^2/dx^2)G + k^2G = -4\pi\delta(x - x_0)$$

As indicated on page 800, we have here a discontinuity in slope, so that, when we integrate over x, from $x_0 - \epsilon$ to $x_0 + \epsilon$, the integral of k^2G goes to zero whereas the integral of the second derivative

$$\int_{x_0-\epsilon}^{x_0+\epsilon} \left(\frac{d^2G}{dx^2}\right) dx = \left[\frac{dG}{dx}\right]_{x_0-\epsilon}^{x_0+\epsilon} \to -4\pi; \quad \epsilon \to 0 \tag{7.2.16}$$

Therefore, for one dimension, the Green's function G has a discontinuity in slope equal to -4π at $x = x_0$.

Referring back to pages 123 and 206, we see that the Green's functions we mentioned earlier, for the Helmholtz equation, satisfied the requirements listed in Eq. (7.2.14), (7.2.15), or (7.2.16), depending on the number of dimensions involved. For instance, subsequent to Eq. (2.1.10), we showed that the Green's function for the string is $(2\pi i/k)e^{ik|x-x_0|}$, which satisfies Eq. (7.2.16). Likewise we know that the Green's function for a point source in three dimensions is e^{ikR}/R.

The Effect of Boundary Conditions. Next we inquire how the boundary conditions affect the Green's function, so that we can see how to build up a usable function for each case of interest. Presumably the simplest case is one where the boundary is at infinity, for here its effects should be least. For three dimensions we wish a solution of $(\nabla^2 + k^2)G = 0$, except for $\mathbf{r} = \mathbf{r}_0$, which goes to zero (and/or its slope is zero) at $R \to \infty$ and which has a singularity $1/R$ as $R \to 0$. Two such solutions are available (plus any linear combination of the two):

$$e^{ikR}/R,\ e^{-ikR}/R,\ \cos(kR)/R,\ \text{etc.}; \quad R = |\mathbf{r} - \mathbf{r}_0|$$

Obviously, even for a boundary at infinity, the boundary conditions must choose between these.

To make the choice we go back to the original, time-dependent solution of the wave equation, $\nabla^2\Psi = (1/c^2)(\partial^2\Psi/\partial t^2)$. We obtained the Helmholtz equation by specifying a simple-harmonic dependence on time of the solution of the wave equation. On pages 125 and 754 we set this solution equal to $\psi e^{-i\omega t} = \psi e^{-ikct}$, whence ψ is a solution of the Helmholtz equation. Consequently if G is to be used for a simple-harmonic solution of the wave equation, the complete solution is Ge^{-ikct}. Hence, if we wish to have waves go *outward* from the source point, we use

$$G_k(\mathbf{r}|\mathbf{r}_0) = g_k(R) = e^{ikR}/R; \quad \text{boundary at infinity} \tag{7.2.17}$$

for then the complete solution is $(1/R)e^{ik(R-ct)}$, representing an outgoing wave. On the rare chance that we wish *incoming* waves, we should choose the e^{-ikR}/R function. For boundaries everywhere a finite distance from $\mathbf{r}_0$, the energy might not be completely dissipated at the surface. Some might be reflected directly back to the source, and we might have to use a combination of outgoing and incoming waves. In most cases, however, we use the outgoing wave solution given in Eq. (7.2.17).

Solutions of the Helmholtz equation are sometimes involved in solutions of the diffusion equation $\nabla^2\Psi = (1/a^2)(\partial\Psi/\partial t)$. Here, to obtain a diffusion outward of the material measured by Ψ, we set $\Psi = \psi e^{-k^2a^2t}$, whence ψ is again a solution of $\nabla^2\psi + k^2\psi = 0$. We usually wish ψ to be real everywhere; hence

$$G_k(\mathbf{r}|\mathbf{r}_0) = [\cos(kR)]/R$$

so that a possible Green's function for the diffusion equation is

$$(1/R)\cos(kR)e^{-a^2k^2t}$$

suitable for outgoing diffusion, when the boundary is at infinity and when the diffusion "source" has an exponentially decaying dependence on time. Actually, as will be shown later, a quite different solution of the diffusion equation, suitable for much more "normal" sorts of "sources," is usually used for a Green's function for the diffusion equation.

In the case of two dimensions, the solutions of the Helmholtz equation symmetric about $R = 0$ are solutions of the equation

$$\frac{1}{R}\frac{d}{dR}\left(R\frac{dg}{dR}\right) + k^2g = 0$$

which is the equation for the Bessel functions of order zero [see Eq. (5.3.63)]. We need solutions which have a singularity at $R = 0$; in fact we desire the singularity given in Eq. (7.2.15). The possible solutions are the Hankel functions [see Eq. (5.3.69)]; in particular

$$\begin{aligned} G_k(\mathbf{r}|\mathbf{r}_0) = g_k(R) = i\pi H_0^{(1)}(kR) &\to -2\ln(kR); \qquad R \to 0 \\ &\to \sqrt{\frac{2\pi}{kR}}\,e^{i(kR+\frac{1}{4}\pi)}; \quad R \to \infty \end{aligned} \tag{7.2.18}$$

is the correct Green's function for two dimensions for boundaries at infinity for *outgoing* waves. If incoming waves are desired, the second Hankel function is used, or if a real function be required (as for special use for the diffusion equation), one can use π times the Neumann function $N_0(kR)$ [see Eq. (5.3.75)].

Finally, for one dimension the Green's function for outgoing waves, extending to infinity in both directions, has already been given in Eq. (7.2.16). It is

$$G_k(x|x_0) = g_k(R) = \frac{2\pi i}{k}\,e^{ik|x-x_0|} \tag{7.2.19}$$

Method of Images. We now turn to the problem of obtaining Green's functions for bounded domains. One method, *the method of images*, makes heavy use of the function $g_k(R)$ which has been obtained above; we consider it first. What is the physical effect of introducing boundaries? Without boundaries, $g_k(R)$ is the appropriate Green's function. When a surface is introduced, in electrostatics, a potential is developed which is due to the induced charge on the bounding surface, which in turn is caused by the applied electric field $g_k(R)$. In acoustics, the effect of the boundary is to cause reflections which then must be added to the wave developed by the source to give the total pressure. We may therefore expect that

$$G_k(\mathbf{r}|\mathbf{r}_0) = g_k(R) + F_k(\mathbf{r}|\mathbf{r}_0) \tag{7.2.20}$$

where $F_k(\mathbf{r}|\mathbf{r}_0)$ represents the boundary effects. $F_k(\mathbf{r}|\mathbf{r}_0)$ cannot have a singularity within the region, so that, as $r \to r_0$, $G_k(\mathbf{r}|\mathbf{r}_0) \to g_k(R)$. The

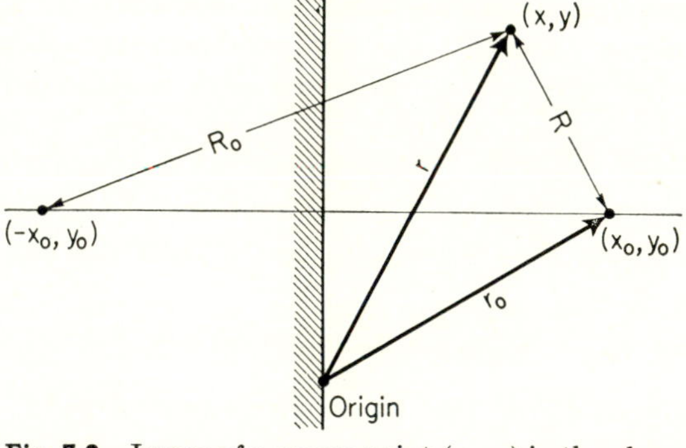

Fig. 7.3 Image of a source point (x_0,y_0) in the plane $x = 0$.

method of images, since it describes the *reflected* wave in acoustics or the induced charge in electrostatics, may be employed to determine $F_k(\mathbf{r}|\mathbf{r}_0)$.

Consider the simple case illustrated in Figure 7.3. A unit line charge is placed at (x_0,y_0) in front of an infinitely conducting metal plane. The potential on the surface of the plane at $x = 0$ must be zero, so the Green's function must satisfy Dirichlet boundary conditions. In the method of images, a line charge of opposite sign is introduced at the image point $(-x_0,y_0)$ as illustrated. The total potential for $x > 0$ is then

$$G_0(\mathbf{r}|\mathbf{r}_0) = -2\ln(R/R_0); \quad x \geq 0 \tag{7.2.21}$$

On the equipotential plane $R = R_0$ and $G_0 = 0$, so that the boundary conditions are satisfied. Since $g_k(R) = -2 \ln R$, we see that $F_0 = 2 \ln R_0$. The only singularity exhibited in the region of physical interest $(x > 0)$ occurs at the position of the charge, $R = 0$. The other singularity present occurs at $R_0 = 0(x < 0)$, outside the region where the expression (7.2.21) no longer applies.

It is easy to generalize (7.2.21) so that G_k may be found as well. Introducing the image as before, we find

$$G_k(\mathbf{r}|\mathbf{r}_0) = \pi i[H_0^{(1)}(kR) - H_0^{(1)}(kR_0)] \tag{7.2.22}$$

The reflection of the incident waves by the mirror, in this case requiring G to be zero, is given by $H_0^{(1)}(kR_0)$. (From here on, in this section, we are going to omit the superscript (1), since we will always use the Hankel function of the first kind.)

If, on the other hand, the surface were rigid or if, in the Laplace case, it was desired to solve problems involving known charge distributions rather than potentials, the appropriate boundary conditions should be Neumann, $\partial\psi/\partial n = 0$. The method of images may be applied also here; the charge or source at the image is taken to be of the same sign (or phase) so that

$$G_0(\mathbf{r}|\mathbf{r}_0) = -2\ln(RR_0); \quad G_k(\mathbf{r}|\mathbf{r}_0) = \pi i[H_0(kR) + H_0(kR_0)]$$

The boundary conditions are satisfied, as may be verified by evaluating the derivative with respect to x at $x = 0$.

As an illustration let us compute the effect of having a potential $f(y)$ on the plane, rather than zero potential. Then from Eq. (7.2.9) we have

$$\psi(\mathbf{r}) = -(1/4\pi)\int \psi(\mathbf{r}_0^s)(\partial G_0/\partial n_0)\,dS_0 = +(1/4\pi)\int_{-\infty}^{\infty} f(y_0)(\partial G_0/\partial x_0)_{x_0=0}\,dy_0$$

Introducing Green's function (7.2.21) and evaluating the derivative, we find

$$\psi(\mathbf{r}) = \frac{x}{\pi}\int_{-\infty}^{\infty}\left[\frac{f(y')}{x^2 + (y - y')^2}\right]dy' \tag{7.2.23}$$

We have seen this formula before. It was derived by means of complex variable theory in Chap. 4 [Eq. (4.2.13)], the only requirement being that ψ satisfy the Laplace equation. It is of interest to check directly that the solution satisfies the boundary conditions.

As mentioned on page 801 this should yield a representation of the δ function. For this purpose let us consider the function

$$\Delta(x,\eta) = \frac{1}{\pi}\left[\frac{x}{x^2 + \eta^2}\right]$$

We note that $\int_{-\infty}^{\infty} \Delta(x,\eta)\,d\eta = 1$. To investigate its properties as $x \to 0$, $\eta \to 0$ rewrite $\Delta(x,\eta) = (1/\pi x)[1 + (\eta/x)^2]^{-1}$. If we set $\eta = 0$; then as $x \to 0$, $\Delta(x,0)(=1/\pi x)$ increases strongly. On the other hand, for $\eta \neq 0$, as $x \to 0$, $\Delta(x,\eta)(\simeq x/\pi\eta^2)$ goes to zero. We see that $\lim_{x\to 0}[\Delta(x,\eta)]$ is a function whose integral is always unity, whose value at $\eta = 0$ increases toward infinity in the limit $x \to 0$, and whose value, for $\eta \neq 0$, goes to zero. Therefore

$$\lim_{x \to 0} [\Delta(x,\eta)] = \delta(\eta)$$

Hence in (7.2.23)

$$\lim_{x \to 0} [\psi(\mathbf{r})] = \int \delta(y - y')f(y')\, dy' = f(y)$$

as it should. The function $\Delta(x,y)$ is particularly useful for the Laplace equation problem because it is a solution of the Laplace equation, being the real part of the analytic function $1/\pi z = 1/\pi(x + iy)$.

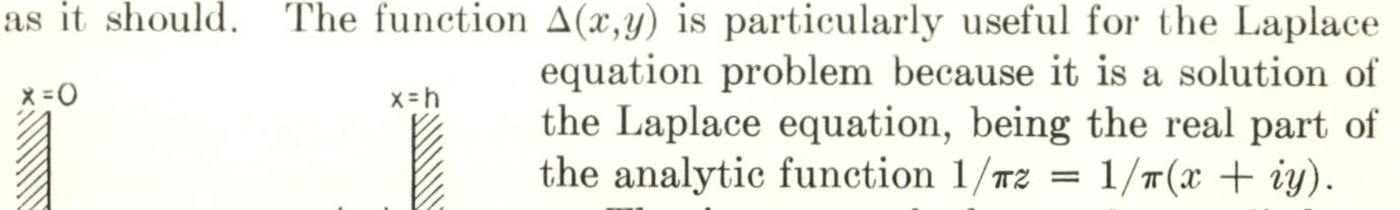

Fig. 7.4 Field between parallel planes from source at (x_0,y_0).

The image method may be applied to other geometries. For example, if the bounding surface is a circle, the image method may again be applied. The application of the Green's function theory yields the Poisson integral formula (4.2.24). Further details will be relegated to the problems and to Chap. 10, for the image method for a circle will operate for only the Laplace equation.

Certain other symmetrically shaped boundaries permit the use of the image method. Generally speaking, however, these lead to infinite series with their usual limitations of usefulness. As an example consider the arrangement illustrated in Figure 7.4 where a source is placed between two infinite planes $x = 0$, $x = h$ upon which the Green's function is required to satisfy the Neumann boundary conditions $\partial\psi/\partial n = 0$.

Series of Images. We apply the image method in a stepwise fashion, by disregarding each boundary $x = 0$ and $x = h$ in turn. Disregarding the $x = h$ boundary, we introduce an image at $x = -x_0$, and disregarding $x = 0$, we introduce an image at $x = 2h - x_0$. These are labeled 1 and 2 in Fig. 7.5. However, although 1 and 0 together give rise to potentials which satisfy the Neumann conditions at $x = 0$, it is necessary to add the effect of source 2, which, of course, does not satisfy $\partial\psi/\partial n = 0$ at $x = 0$. To offset the effect of 2, an image (3) of 2 with respect to $x = 0$ is now introduced at $(x_0 - 2h, y_0)$. Similarly to offset the effect of 1 at $x = h$ a source 4 is introduced at $x = (2h + x_0)$. But now it becomes necessary to offset the effect of 3 at $x = h$; it is necessary to add an image (6) at $(3h - x_0, y_0)$. The process is continued indefinitely, leading to an infinite number of images. This should occasion no surprise, since there are an infinite number of reflections for any ray started at the source. Each image corresponds to one of these reflections.

Let us now write down the consequent Green's function. The sources are located at $(x_0 + 2nh, y_0)$ and at $(2mh - x_0, y_0)$, m, n integers. Therefore

$$G_k = \pi i \sum_{n=-\infty}^{\infty} \{H_0[k\sqrt{(x - x_0 - 2nh)^2 + (y - y_0)^2}] + H_0[k\sqrt{(x + x_0 - 2nh)^2 + (y - y_0)^2}]\}$$

(where we mean the Hankel function of the first kind, though we have not bothered to write out the superscript). Writing this more succinctly

$$G_k = \pi i \sum_{n=-\infty}^{\infty} [H_0(k|\mathbf{r} - \mathbf{r}'_n|) + H_0(k|\mathbf{r} - \mathbf{r}''_n|)] \qquad (7.2.24)$$

where $\mathbf{r}'_n = \mathbf{a}_x[2nh + x_0] + \mathbf{a}_y y_0$; $\mathbf{r}''_n = \mathbf{a}_x[2nh - x_0] + \mathbf{a}_y y_0$

Series (7.2.24) is useful when only the direct source and perhaps the first few reflections are important. This will occur whenever the observation

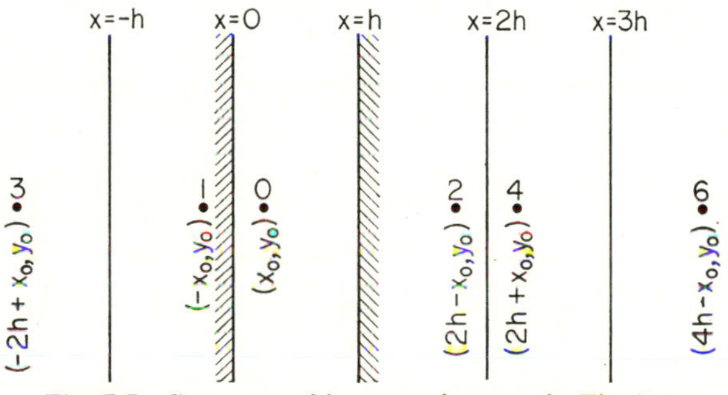

Fig. 7.5 Sequence of images of source in Fig. 7.4.

point $\mathbf{r}$ is very close to the source point, that is $\mathbf{r} \to \mathbf{r}_0$, because of the singularity of $H_0(k|\mathbf{r} - \mathbf{r}'_0|)$. The remaining reflections form a background correction to the direct effect of the source. Unfortunately series (7.2.24) does not converge very rapidly, so computing the correction is a fairly tedious matter. To compute its order of magnitude let us examine the behavior of the individual terms as $|n| \to \infty$. Then

$$|\mathbf{r} - \mathbf{r}'_n| \xrightarrow[|n|\to\infty]{} 2|n|h - (x - x_0); \quad |\mathbf{r} - \mathbf{r}''_n| \xrightarrow[|n|\to\infty]{} 2|n|h - (x + x_0) \qquad (7.2.25)$$

Note that these approximations are valid only if

$$2|n|h \gg \sqrt{(x - x')^2 + (y - y')^2}.$$

Clearly this asymptotic form is achieved more quickly, *i.e.*, for smaller values of n, the closer the observation point to the source. For large values of $2nhk$ the Hankel functions may be replaced by their asymptotic form, so that

$$H_0(k|\mathbf{r} - \mathbf{r}'_n|) + H_0(k|\mathbf{r} - \mathbf{r}''_n|) \xrightarrow[k|n|h\to\infty]{} \sqrt{\frac{1}{i\pi k|n|h}}\,[e^{ik(2|n|h-x+x_0)} + e^{ik(2|n|h-x-x_0)}] = \sqrt{\frac{4}{i\pi k|n|h}}\, e^{ik(2|n|h-x)}\cos(kx_0)$$

The higher terms in the series for G_k, therefore, approach the series

$$\Sigma = \frac{4}{\sqrt{i\pi kh}}\, e^{-ikx}\cos(kx_0)\sum_{n=N}^{\infty}\frac{e^{2iknh}}{\sqrt{n}} \tag{7.2.26}$$

where N is chosen sufficiently large to ensure the validity of the approximations made so far. We estimate the infinite sum by replacing it by the corresponding integral

$$\sum_{n=N}^{\infty}\frac{e^{2iknh}}{\sqrt{n}} \simeq \int_N^\infty \frac{e^{2ikh\nu}}{\sqrt{\nu}}\,d\nu \xrightarrow[2kh\sqrt{N}\to\infty]{} \frac{e^{2ikhN}}{2ikh\sqrt{N}} \tag{7.2.27}$$

The integral may also be evaluated exactly in terms of the Fresnel integrals. If

$$C(u) = \int_0^u \cos\left(\frac{\pi}{2}t^2\right)dt; \quad S(u) = \int_0^u \sin\left(\frac{\pi}{2}t^2\right)dt \tag{7.2.28}$$

then $$\int_N^\infty \frac{e^{2ikh\nu}}{\sqrt{\nu}}\,d\nu = \sqrt{\frac{\pi}{kn}}\left[\tfrac{1}{2} - C\left(\sqrt{\frac{4khN}{\pi}}\right) + \tfrac{1}{2}i - iS\left(\sqrt{\frac{4khN}{\pi}}\right)\right]$$

Employing the simpler expression (7.2.27), Σ becomes

$$\Sigma \simeq \frac{2}{\sqrt{k^3h^3\pi N}}\, e^{-\frac{3}{4}\pi i}e^{2ikh\sqrt{N}-ikx}\cos(kx_0) \tag{7.2.29}$$

We thus see that, when kh is much larger than unity, the whole of series (7.2.24) can be expressed in terms of the simple expression (7.2.29) with $N = 1$. For the wave equation, $k = 2\pi/\lambda$ where λ is the wavelength, so that the simple expression may be used for the whole sum when λ is much smaller than h, the spacing between the plates. The only term not included in Σ is then the one for $n = 0$, the direct effect of the source on the observer. To repeat, when $h \gg \lambda$ and when $|x - x_0| \ll h$, then the value of ψ at the observation point (x,y) equals the direct term $\pi i H_0(k|\mathbf{r} - \mathbf{r}_0|)$ plus a small correction proportional to Σ.

Other Expansions. In the event that either or both of these conditions are not satisfied, more drastic treatment must be applied. If $\mathbf{r}$ is at some distance from the source and kh is neither large nor small, the image expansion may be rearranged to obtain a more convergent series. This may be performed with the aid of the Poisson sum formula [Eq. (4.8.28)];

$$\sum_{-\infty}^{\infty} f(2\pi n) = \frac{1}{2\pi} \sum_{\nu=-\infty}^{\infty} F(\nu)$$

where

$$F(\nu) = \int_{-\infty}^{\infty} f(\tau) e^{-i\nu\tau}\, d\tau \tag{7.2.30}$$

To apply the Poisson sum formula to this problem, the Fourier transform of $H_0(k|\mathbf{r} - \mathbf{r}_0|)$ is required. We shall show later on in this chapter (see page 823) that

$$H_0(k|\mathbf{r} - \mathbf{r}_0|) = \frac{i}{\pi^2} \int_{-\infty}^{\infty} dK_x \int_{-\infty}^{\infty} dK_y \left[\frac{e^{i\mathbf{K}\cdot(\mathbf{r}-\mathbf{r}_0)}}{k^2 - K^2}\right] \tag{7.2.31}$$

The integral is not completely defined without specifying the manner in which the pole at $K = k$ is to be circumvented. This will be done in the course of the calculation.

We must now evaluate

$$I = \int_{-\infty}^{\infty} e^{-i\nu\tau} \left\{H_0\left[k\sqrt{\left(x - x_0 - \frac{\tau}{\pi}h\right)^2 + (y - y_0)^2}\right] + H_0\left[k\sqrt{\left(x + x_0 - \frac{\tau}{\pi}h\right)^2 + (y - y_0)^2}\right]\right\} d\tau$$

Introducing (7.2.31) into the integrand, I becomes

$$I = \frac{i}{\pi^2} \int_{-\infty}^{\infty} dK_y \int_{-\infty}^{\infty} dK_x \int_{-\infty}^{\infty} d\tau\, e^{-i\nu\tau} \left\{\frac{e^{i\left[K_x\left(x - x_0 - \frac{\tau h}{\pi} + K_y(y - y_0)\right)\right]}}{k^2 - K^2} + \frac{e^{i\left[K_x\left(x + x_0 - \frac{\tau h}{\pi}\right) + K_y(y - y_0)\right]}}{k^2 - K^2}\right\}$$

$$= \frac{2i}{\pi^2} \int_{-\infty}^{\infty} dK_y\, e^{iK_y(y-y_0)} \int_{-\infty}^{\infty} \frac{dK_x e^{iK_x x}}{k^2 - K^2} \cos(K_x x_0) \int_{-\infty}^{\infty} d\tau\, e^{-i\nu\tau} e^{-iK_x\tau h/\pi}$$

The τ integral may be immediately performed in terms of the delta function (actually we are just using the Fourier integral theorem):

$$I = \frac{4i}{\pi} \int_{-\infty}^{\infty} dK_y e^{iK_y(y-y_0)} \int_{-\infty}^{\infty} dK_x \frac{e^{iK_x x} \cos(K_x x_0)\, \delta[\nu + (K_x h/\pi)]}{k^2 - K_x^2 - K_y^2}$$

The K_x integral is easily evaluated by employing the δ function property $\int_{-\infty}^{\infty} f(z)\delta(z - a)\, dz = f(a)$. Then

$$I = \frac{4i}{h} e^{-(i\pi\nu x/h)} \cos\left(\frac{\pi\nu x_0}{h}\right) \int_{-\infty}^{\infty} \frac{e^{iK_y(y-y_0)}\, dK_y}{[k^2 - (\pi^2\nu^2/h^2)] - K_y^2}$$

The final integral can be performed only after the path of integration on the K_y plane is specified. The particular path C, illustrated in Fig. 7.6, is chosen so that I satisfies the boundary condition that the point $x = x_0$, $y = y_0$ be a source only, rather than a sink or both source and sink.

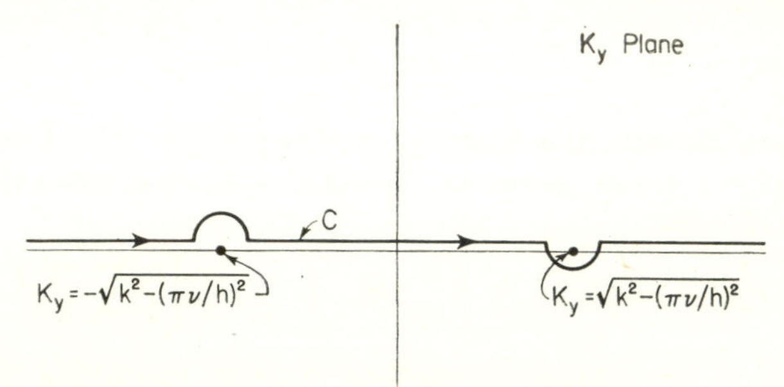

Fig. 7.6 Contour C for integration of Eq. (7.2.31).

The evaluation of this integral by means of Cauchy's integral formula [Eq. (4.2.8)] is discussed in Chap. 4, page 415. We find that

$$I = \left(\frac{4\pi}{h}\right) e^{-i\pi\nu x/h} \cos\left(\frac{\pi\nu x_0}{h}\right) \frac{e^{i|y-y_0|\sqrt{k^2-(\pi\nu/h)^2}}}{\sqrt{k^2 - (\pi\nu/h)^2}}$$

The final expansion for the Green's function becomes

$$G_k(\mathbf{r}|\mathbf{r}_0) = \left(\frac{2\pi i}{h}\right) \sum_{\nu=-\infty}^{\infty} e^{-i\pi\nu x/h} \cos\left(\frac{\pi\nu x_0}{h}\right) \frac{e^{i|y-y_0|\sqrt{k^2-(\pi\nu/h)^2}}}{\sqrt{k^2 - (\pi\nu/h)^2}}$$

$$= \left(\frac{2\pi i}{h}\right) \sum_{\nu=0}^{\infty} \epsilon_\nu \cos\left(\frac{\pi\nu x}{h}\right) \cos\left(\frac{\pi\nu x_0}{h}\right) \frac{e^{i|y-y_0|\sqrt{k^2-(\pi\nu/h)^2}}}{\sqrt{k^2 - (\pi\nu/h)^2}} \tag{7.2.32}$$

This result is particularly useful whenever $|y - y_0| \gg 1$, for as soon as $(\pi\nu/h) > k$, terms in the series decrease exponentially. The number of terms required to obtain an accurate value for the series is thus of the order of $hk/\pi = 2h/\lambda$. [Note that, when ν becomes large, the corresponding terms in (7.2.32) become independent of k.] We see that the expansion given above complements that for Σ. The expansion in images (7.2.24) is feasible if $hk \gg 1$; expansion (7.2.32) applies when $hk \ll 1$. The expansion in images is appropriate for short wavelengths and close to the source, for then the effects of the boundary are less important; expansion (7.2.32) is appropriate for long wavelengths and considerable distance from the source. Expansion (7.2.32) is a Fourier expansion quite similar to Eq. (7.1.12) and may be obtained more directly than by the roundabout mannér via images and Poisson's sum rule, which we have employed here. The importance of the derivation we have given lies in its exhibition of the connection between the two types of expansions.

Equation (7.2.32), unlike Eq. (7.2.29), is exactly equal to series (7.2.24), and it always converges. Unless $hk/\pi = 2h/\lambda$ is of the order of unity, the series does not converge very rapidly and we must look around for means of improving its convergence. To this end we return to the parenthetic remark of the last paragraph, that the terms in the expansion (7.2.32) for ν large are independent of k. This suggests examination of $G_0(\mathbf{r}|\mathbf{r}_0)$, the Green's function for the Laplace equation, which may often be exhibited in closed form. If we write

$$G_k = G_0 + (G_k - G_0)$$

then the expansion for $(G_k - G_0)$ will converge more rapidly than that for G_k alone.

If we had chosen Dirichlet conditions at $x = 0$ and $x = h$, the corresponding static G_0 would come out in closed form; $G_0(\mathbf{r}|\mathbf{r}_0)$ would then be the static potential for a unit charge at x_0, y_0, between two grounded plates. To bring out more of the difficulties, we have picked Neumann conditions, where G_0 corresponds to the steady flow of a fluid out of a unit source at x_0, y_0. But steady flow requires a sink (in this case at infinity) as well as a source; this has not been included. A slight modification is thus required to take the sink into account, as we shall proceed to show.

We start with the series

$$\Gamma_0(\mathbf{r}|\mathbf{r}_0) = 4\sum_{\nu=1}^{\infty}\left(\frac{1}{\nu}\right)\cos\left(\frac{\pi\nu x}{h}\right)\cos\left(\frac{\pi\nu x_0}{h}\right)e^{-(\pi\nu/h)|y-y_0|} \qquad (7.2.33)$$

which is what Eq. (7.2.32) reduces to when $k = 0$ (except for the omission of the $\nu = 0$ term). By repeated use of the relation

$$\sum_{\nu=1}^{\infty}\left(\frac{1}{\nu}\right)e^{-\nu b} = -\ln(1 - e^{-b})$$

we obtain

$$\Gamma_0 = R(x + x_0|y - y_0) + R(x - x_0|y - y_0) \qquad (7.2.34)$$

where $\quad R(a|b) = -\ln[1 - 2e^{-\pi|b|/h}\cos(\pi a/h) + e^{-2\pi|b|/h}]$

It is not difficult to show (by computing $\nabla^2\Gamma_0$ if necessary) that Γ_0 is a solution of the Poisson equation

$$\nabla^2\Gamma_0 = -4\pi[\delta(\mathbf{r} - \mathbf{r}_0) - (1/h)\delta(y - y_0)] \qquad (7.2.35)$$

which corresponds to a unit positive charge at (x_0,y_0) and a unit negative charge spread uniformly along the line $y = y_0$, perpendicular to the two boundary lines $x = 0$ and $x = h$. Since the entire charge distribution averages to zero between the two boundary lines we may fit Neumann conditions without having the static solution approach infinite values at infinity. It may also be seen directly that $\partial\Gamma/\partial x$ may equal zero for $x = 0$ and $x = h$, by differentiating (7.2.35).

The final expression for G_k is therefore

$$G_k(r|r_0) = R(x + x_0|y - y_0) + R(x - x_0|y - y_0) + \left(\frac{2\pi i}{hk}\right) e^{ik|y-y_0|} + 4\sum_{\nu=1}^{\infty} \cos\left(\frac{\pi\nu x}{h}\right)\cos\left(\frac{\pi\nu x_0}{h}\right)\left\{\frac{e^{-\sqrt{(\pi\nu/h)^2 - k^2}|y-y_0|}}{\sqrt{\nu^2 - (kh/\pi)^2}} - \left(\frac{1}{\nu}\right) e^{-(\pi\nu/h)|y-y_0|}\right\} \quad (7.2.36)$$

This series converges quite rapidly. Other cases, where the static Green's function turns out to be a closed expression, may be worked out from the results of Chap. 10.

The method of images is restricted to boundaries which are composed of straight lines in two dimensions or planes in three dimensions. There is one exception to this rule. In the case of the Laplace equation (with Dirichlet conditions) the method of images may be applied to the circle in two dimensions and to the sphere in three dimensions. This limitation on the image method may be expected from an elementary knowledge of geometrical optics, for it is well known that the only mirror for which the image of a point source is itself a point is the plane mirror. Of course this does not mean that the image method cannot be applied to other shapes, rather that it can be applied only approximately. We therefore turn to a more general representation of Green's functions, by eigenfunctions.

Expansion of Green's Function in Eigenfunctions. The method of eigenfunctions discussed in Chap. 6 is limited only by the ease with which the requisite eigenfunctions can be determined. Since precise solutions are available for only the separable coordinate systems, the expansion of Green's functions in eigenfunctions is practical for only these cases.

Let the eigenfunctions be ψ_n and the corresponding eigenvalues k_n^2, that is,

$$\nabla^2\psi_n + k_n^2\psi_n = 0 \quad (7.2.37)$$

Here n represents all the required indices defining the particular ψ_n under discussion. Moreover, as was shown in the preceding chapter, the functions ψ_n form an orthonormal set:

$$\int \bar{\psi}_n\psi_m \, dV = \delta_{nm} \quad (7.2.38)$$

where the region of integration R is bounded by the surface upon which ψ_n satisfies homogeneous boundary conditions. The Green's function $G_k(\mathbf{r}|\mathbf{r}_0)$ satisfies the same conditions. In addition it is assumed that the functions ψ_n form a complete set so that it is possible to expand $G_k(\mathbf{r}|\mathbf{r}_0)$ in a series of ψ_n:

$$G_k(\mathbf{r}|\mathbf{r}_0) = \sum_n A_m\psi_m(\mathbf{r})$$

Introducing this expansion into the partial differential equation satisfied by G_k,

$$\nabla^2 G_k + k^2 G_k = -4\pi\delta(\mathbf{r} - \mathbf{r}_0)$$

we find that

$$\Sigma A_m(k^2 - k_m^2)\psi_m(\mathbf{r}) = -4\pi\delta(\mathbf{r} - \mathbf{r}_0)$$

Employing Eq. (7.2.38), we multiply both sides of the above equation by $\bar{\psi}_n(\mathbf{r})$ and integrate over the volume R. Then

$$A_n = \frac{4\pi\bar{\psi}_n(\mathbf{r}_0)}{k_n^2 - k^2}$$

so that

$$G_k(\mathbf{r}|\mathbf{r}_0) = 4\pi \sum_n \frac{\bar{\psi}_n(\mathbf{r}_0)\psi_n(\mathbf{r})}{k_n^2 - k^2} \tag{7.2.39}$$

the desired expansion. An example of such an expansion was given in Eq. (7.1.10).

One unexpected feature of Eq. (7.2.39) is its unsymmetrical dependence on $\mathbf{r}$ and $\mathbf{r}_0$ for complex ψ_n in face of the proof given earlier that G_k must depend symmetrically on these variables. This is, of course, not a real dilemma. The solution lies in recognizing that, since the scalar Helmholtz equation does not involve any complex numbers explicitly, $\bar{\psi}_n$ is also a solution of (7.2.38) and therefore will be included as one of the orthonormal set ψ_n. We have here a simple case of a degeneracy, for to one eigenvalue k_n^2 there belong two eigenfunctions, $\bar{\psi}_n$ as well as ψ_n. Thus included in the sum (7.2.39) there will be the term

$$\frac{\bar{\psi}_m(\mathbf{r}_0)\psi_m(\mathbf{r})}{k_m^2 - k^2} \quad \text{and also the term} \quad \frac{\bar{\psi}_m(\mathbf{r})\psi_m(\mathbf{r}_0)}{k_m^2 - k^2}$$

so that actually (7.2.39) is symmetric and real.

Another matter of interest is the behavior of G_k as $k \to k_n$. We see that, as a function of k, G_k is analytic except for simple poles at $k = \pm k_n$, with residues $\mp 2\pi\bar{\psi}_n(r')\psi_n(r)/k_n$. Thus if it should happen that a Green's function is known in a closed form, the eigenfunctions ψ_n and the eigenvalues k_n may be found by investigating G_k at its poles.

The singularities have a simple physical interpretation, for they are just the infinities which occur when a nondissipative vibrating system is driven at one of its resonant frequencies. To make this correspondence more explicit, we recall that the partial differential equation satisfied by the velocity potential set up by a point source at $\mathbf{r}_0$, with angular frequency ω, is

$$\nabla^2\psi - \frac{1}{c^2}\frac{\partial^2\psi}{\partial t^2} = -4\pi\delta(\mathbf{r} - \mathbf{r}_0)e^{-i\omega t}$$

But $\psi = e^{-i\omega t}G_k$, $k = \omega/c$. Hence if $k = k_n$, the system is being driven at one of its resonant frequencies and gives an infinite response if the system has no friction. There is one situation for which the response will not be infinite. This occurs if the source space-dependence is orthogonal to $\bar{\psi}_n$. For if ψ satisfies

$$(\nabla^2\psi + k^2\psi) = -4\pi\rho$$

then

$$\psi = \int \rho G_k\, dV_0 = 4\pi \sum_m \frac{\int \bar{\psi}_m(\mathbf{r}_0)\rho(\mathbf{r}_0)\, dV_0}{k_m - k^2}\psi_m(\mathbf{r}) \tag{7.2.40}$$

The nth term vanishes if $\int \bar{\psi}_n(r)\rho(r)\, dV = 0$. Then the nth term in series (7.2.39) is missing, and k can equal k_n without G_k becoming infinite. The Green's function for such problems (we shall use here the term "modified Green's function" and the notation Γ_{k_n}) satisfies

$$(\nabla^2 + k_n^2)\Gamma_{k_n} = -4\pi[\delta(\mathbf{r} - \mathbf{r}_0) - \bar{\psi}_n(\mathbf{r}_0)\psi_n(\mathbf{r}) - \bar{\psi}_n(\mathbf{r})\psi_n(\mathbf{r}_0)]$$

Then

$$\Gamma_{k_n} = 4\pi \sum_{m \neq n} \frac{\bar{\psi}_m(\mathbf{r}_0)\psi_m(\mathbf{r})}{k_m^2 - k_n^2} \tag{7.2.41}$$

where by $m \neq n$ we mean that all terms for which $k_n = \pm k_n$ are to be left out. We have already discussed one such case for the Green's function for the Laplace equation when the boundary conditions were homogeneous Neumann. In that case one of the eigenvalues was $k = 0$, corresponding to a constant for an eigenfunction. The k for the Laplace equation was also zero, and we found it advisable to use the modified Green's function Γ_0.

We shall now give some examples of the application of (7.2.39). These are fairly simple when completely enclosed regions are under discussion, for when the eigenfunctions and the corresponding eigenvalues are known, it is only necessary to normalize the ψ's in order to be able to fill in the formula. An example of the expansion under these circumstances is given in Eq. (7.1.10).

Expansions for the Infinite Domain. We therefore turn to other types of regions, of which the simplest is unbounded and infinite. We showed earlier in this section [Eq. (7.2.18)] that the two-dimensional Green's function for this case, for a source (as opposed to a sink) is

$\pi i H_0(k|\mathbf{r} - \mathbf{r}_0|)$. A possible complete orthonormal set in which to expand this is furnished by the plane wave

$$[1/(2\pi)]e^{i\mathbf{K}\cdot\mathbf{r}}$$

where $\mathbf{K} \cdot \mathbf{r} = K_x x + K_y y$ where K_x and K_y may assume any numerical value. To obtain a complete set it is necessary, according to the Fourier integral theorem, to have the range in K extend from $-\infty$ to $+\infty$ following a route in the complex planes of K_x and K_y joining these two points.

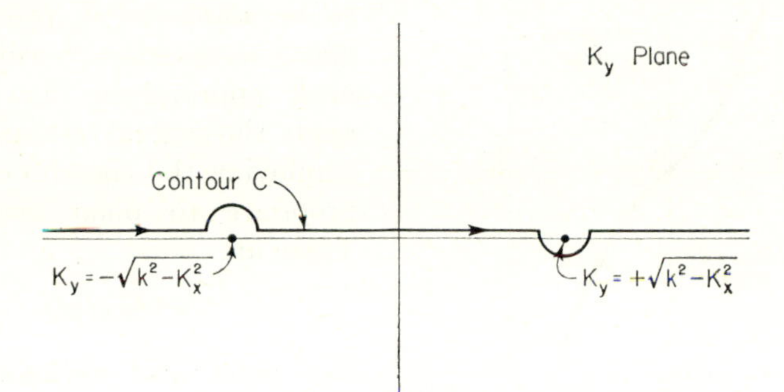

Fig. 7.7 Contour C for integration of Eq. (7.2.42).

Since K_x and K_y are continuous variables (see page 762 for the discrete-continuous transition), the sum in (7.2.39) must be replaced by an integral:

$$g_k(R) = i\pi H_0(kR) = \frac{1}{\pi}\int_{-\infty}^{\infty} dK_x \int_{-\infty}^{\infty} dK_y \frac{e^{i\mathbf{K}\cdot\mathbf{R}}}{K^2 - k^2} \qquad (7.2.42)$$

This representation was utilized in Eq. (7.2.31) ($K^2 = K_x^2 + K_y^2$). Again it is important to note that the integral is undefined unless the path of integration about the poles of the integrand is specified. The K_y path of integration is given in Fig. 7.7. It is chosen so as to lead to an outgoing wave from the source point $R = 0$. The K_y integrations may then be performed to yield, for $Y > 0$,

$$i\pi H_0(kR) = i\int_{-\infty}^{\infty} dK_x \frac{e^{\,i(K_xX + \sqrt{k^2 - K_x^2}\,Y)}}{\sqrt{k^2 - K_x^2}}$$

Let $K_x = k\cos(\vartheta + \phi)$ where $\phi = \tan^{-1}(Y/X)$

$$\pi H_0(kR) = \int_{i\infty}^{-i\infty} e^{ikR\cos\vartheta}\, d\vartheta$$

The contour of integration for ϑ must, of course, be such as to yield a convergent integral. In view of the original limits, it must go from $-i\infty$ to $+i\infty$. Convergence is obtained by running the contour somewhat to the left of the imaginary axis in the upper half plane and to the right of it in the lower half plane as is illustrated in Fig. 7.8. Our final result is the well-known integral representation for the Hankel function [see Eq. (5.3.69)]:

$$H_0(kR) = \left(\frac{1}{\pi}\right) \int_{-\pi/2+i\infty}^{\pi/2-i\infty} e^{ikR\cos\vartheta}\, d\vartheta \tag{7.2.43}$$

Polar Coordinates. Equation (7.2.42) is the proper representation of the Green's function $g_k(\mathbf{r})$ for a two-dimensional infinite region, to be employed in problems for which rectangular coordinates are most appropriate. Let us now apply the general formula (7.2.39) employing the eigenfunctions appropriate to polar coordinates. These are

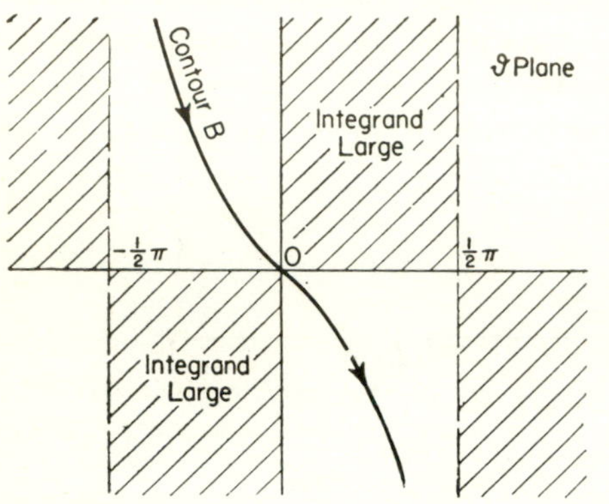

Fig. 7.8 Contour B for integral representation of $H_0(kR)$.

$$e^{im\phi}J_m(kr)$$

One must also normalize. The normalization factor for the ϕ dependence is $1/\sqrt{2\pi}$. The normalization factor N_m for the radial dependence is obtained from the equation appropriate to continuous eigenvalues (in this case k):

$$\lim_{\Delta k\to 0, R\to\infty} \left\{ N_m^2 \int_{k-\Delta k}^{k+\Delta k} dk' \int_0^R J_m(kr)J_m(k'r)r\, dr \right\} = 1$$

The value of the indefinite integral over r is (see formulas at end of Chap. 11)

$$\int J_m(kr)J_m(k'r)r\, dr = \frac{k'rJ_m(kr)J_{m-1}(k'r) - krJ_m(kr')J_{m+1}(kr)}{k^2 - k'^2}$$

Since R is large, the asymptotic behavior, $J_m(x) \simeq \sqrt{2/\pi x}\cos[x - \frac{1}{2}\pi(m + \frac{1}{2})]$, given in Eq. (5.3.68) may be used. We find that N_m is independent of m and is equal to $\sqrt{k}$. Hence the normalized eigenfunctions are

$$\sqrt{k/2\pi}\; e^{im\phi}J_m(kr) \tag{7.2.44}$$

Substituting (7.2.44) into (7.2.39) yields

$$i\pi H_0(kR) = \left(\frac{1}{2\pi}\right) \sum_{m=-\infty}^{\infty} e^{im(\phi-\phi_0)} \int_{-\infty}^{\infty} \frac{J_m(Kr)J_m(Kr_0)}{K^2 - k^2} K\, dK$$

where the contour of integration is still to be specified. Actual evaluation demonstrates that, for outgoing waves, the required contour is just the one illustrated in Fig. 7.7 with the poles at $\pm k$. The above expansion may be rewritten so as to involve only positive m, as follows:

$$i\pi H_0(kR) = \left(\frac{1}{2\pi}\right) \sum_{m=0}^{\infty} \epsilon_m \cos[m(\phi - \phi_0)] \int_{-\infty}^{\infty} \frac{J_m(Kr)J_m(Kr_0)}{K^2 - k^2} K\, dK \tag{7.2.45}$$

It is possible to evaluate the K integral by methods of function theory. However, the procedure is just a difficult way to do a simple calculation. It is preferable to derive expansion (7.2.45) by another procedure, which may be extended to other coordinate systems and boundary surfaces.

A General Technique. The method is just that employed in Sec. 7.1 to establish a connection between the surface Green's function $G(x,y|\xi)$ [Eq. (7.1.8)] and the volume Green's function [Eq. (7.1.10)]. We expand the volume Green's function in terms of a complete set of functions involving all but one of the coordinates (in the present case there are only two coordinates, r and ϕ) with coefficients which are undetermined functions of the uninvolved coordinate. Thus let

$$g_k(R) = \sum_{-\infty}^{\infty} \left(\frac{1}{2\pi}\right) e^{im(\phi-\phi_0)} p_m(r|r_0) \tag{7.2.46}$$

Comparing with (7.2.45) we note that

$$p_m(r|r_0) = \int_{-\infty}^{\infty} \frac{J_m(Kr)J_m(Kr_0)}{K^2 - k^2} K\, dK$$

Introduce (7.2.46) into the equation for $g_k(\mathbf{r}|\mathbf{r}_0)$:

$$\nabla^2 g_k + k^2 g_k = -4\pi\delta(\mathbf{r} - \mathbf{r}_0)$$

In polar coordinates this becomes

$$\frac{1}{r}\frac{\partial}{\partial r}\left[r\frac{\partial g_k}{\partial r}\right] + \frac{1}{r^2}\frac{\partial^2 g_k}{\partial\phi^2} + k^2 g_k = -\frac{4\pi\delta(r - r_0)\delta(\phi - \phi_0)}{r} \tag{7.2.47}$$

[The right-hand side of this equation involves the expression of the δ function in polar coordinates. This expression must satisfy the requirement that $\delta(\mathbf{r} - \mathbf{r}_0)$ must vanish unless $r = r_0$ and $\phi = \phi_0$ and must integrate to unity over all space, $\int\int \delta(\mathbf{r} - \mathbf{r}_0) r\, dr\, d\phi = 1$. It is easy to verify that these requirements are satisfied.]

Inserting expansion (7.2.46) into Eq. (7.2.47) one obtains

$$\sum_{-\infty}^{\infty} \left(\frac{1}{2\pi}\right) e^{im(\phi-\phi_0)} \left[\frac{1}{r}\frac{d}{dr}\left(r\frac{dp_m}{dr}\right) + \left(k^2 - \frac{m^2}{r^2}\right) p_m\right] = -\frac{4\pi\delta(r - r_0)\delta(\phi - \phi_0)}{r}$$

Multiply both sides of this equation by $e^{-in\phi}$ and integrate over ϕ from 0 to 2π. The integration on the left-hand side involves the orthogonal properties of the set $e^{im\phi}$. We obtain

$$\frac{1}{r}\frac{d}{dr}\left(r\frac{dp_m}{dr}\right) + \left(k^2 - \frac{m^2}{r^2}\right)p_m = -\frac{4\pi\delta(r - r_0)}{r} \tag{7.2.48}$$

We see that $p_m(r|r_0)$ is a *one-dimensional Green's function* for a Sturm-Liouville operator $(d/dr)r(d/dr) + r[k^2 - (m^2/r^2)]$ [see Eq. (6.3.12)]. The solution of the inhomogeneous linear second-order differential equation $\mathfrak{L}(\psi) = v$ is given in Eq. (5.2.19) as

$$\psi = y_2 \int^z \frac{vy_1\,dz}{\Delta(y_1,y_2)} + y_1 \int_z \frac{vy_2\,dz}{\Delta(y_1,y_2)}$$

where z is the independent variable, y_1 and y_2 are the two independent solutions of the homogeneous equation

$$\frac{d}{dz}z\frac{dy}{dz} + z\left(k^2 - \frac{m^2}{z^2}\right)y = 0$$

and $\Delta(y_1,y_2)$ is the Wronskian

$$\Delta(y_1,y_2) = \begin{vmatrix} y_1 & y_1' \\ y_2 & y_2' \end{vmatrix}$$

The function v is the inhomogeneous term, in this case $-4\pi\delta(r - r_0)/r$.

The limits of integration in the expression for ψ depend upon the particular choice of independent functions y_1 and y_2, and the boundary conditions on p_m. We shall take the limits (this is permitted since we have not chosen y_1 and y_2) to be less than $z(=r)$ in the first integral and greater than z in the second. Thus

$$p_m = y_2(r)\int_a^r \frac{-4\pi\delta(u - r_0)}{r_0\Delta(y_1,y_2)}y_1(u)\,du + y_1(r)\int_r^b \frac{-4\pi\delta(u - r_0)y_2(u)\,du}{r_0\Delta(y_1,y_2)}$$

For $r < r_0$, the first integral vanishes, while for $r > r_0$, the second integral vanishes. Hence

$$p_m = \frac{-4\pi}{r_0\Delta(y_1,y_2)}\begin{cases} y_1(r)y_2(r_0); & r \le r_0 \\ y_2(r)y_1(r_0); & r \ge r_0 \end{cases} \tag{7.2.49}$$

The Wronskian is evaluated at r_0. Boundary conditions determine which of the solutions of the homogeneous equation are to be employed. In the case under discussion the solutions are the Bessel functions $J_m(kr)$, $N_m(kr)$ or any linear combination. The boundary conditions are (1) p_m is to be finite at $r = 0$, since the only singularity of g_k occurs at $r = r_0$, and (2) the point $r = r_0$ must be a source, since $g_k(R)$ has been taken to be a diverging wave. Hence $y_1 = J_m(kr)$ and $y_2 = H_m(kr)$.

Finally we must evaluate $\Delta(y_1,y_2)$ at $r = r_0$. It is useful to use the relation Eq. (5.2.3), giving the space dependence of the Wronskian

$$\Delta(z) \doteq \Delta(z_0)[f(z_0)/f(z)]$$

where the differential equation satisfied by y_1 (or y_2) is

$$\frac{d}{dz}\left(f\frac{dy}{dz}\right) + qy = 0$$

In the case under discussion $f = r$, so that $\Delta(y_1,y_2) = \text{constant}/r$. To determine the constant, one may employ the first terms of either the power series about the origin ($r = 0$) or the asymptotic series about $r = \infty$, since the relation $\Delta(y_1,y_2) = \text{constant}/r$ must be satisfied for each term of the power or asymptotic series for $\Delta(y_1,y_2)$. To illustrate, we utilize

$$J_m(kr) \xrightarrow[kr\to\infty]{} \sqrt{2/\pi kr}\cos\left[kr - \tfrac{1}{2}\pi(m + \tfrac{1}{2})\right]$$

$$H_m(kr) \xrightarrow[kr\to\infty]{} \sqrt{2/\pi kr}\, e^{i[kr - \frac{1}{2}\pi(m+\frac{1}{2})]}$$

The Wronskian is asymptotically

$$\left(\frac{2}{\pi kr}\right)\left\{ik\cos\left[kr - \tfrac{1}{2}\pi(m + \tfrac{1}{2})\right]e^{i[kr-\frac{1}{2}\pi(m+\frac{1}{2})]} + k\sin\left[kr - \tfrac{1}{2}\pi(m + \tfrac{1}{2})\right]e^{i[kr-\frac{1}{2}\pi(m+\frac{1}{2})]}\right\}$$

This equals $2i/\pi r$ so that $\Delta(y_1,y_2) = 2i/\pi r$. Finally, collecting all our results,

$$p_m = 2\pi^2 i\begin{cases} J_m(kr)H_m(kr_0); & r \le r_0 \\ J_m(kr_0)H_m(kr); & r \ge r_0 \end{cases} \tag{7.2.50}$$

Note that p_m is the value of the integral occurring in (7.2.26). Introducing (7.2.50) into the expansion for $g_k(k) = i\pi H_0(kR)$, an expansion for $H_0(kR)$ is obtained:

$$H_0(kR) = \sum_{-\infty}^{\infty} e^{im(\phi-\phi_0)}\begin{cases} J_m(kr)H_m(kr_0); & r \le r_0 \\ J_m(kr_0)H_m(kr); & r \ge r_0 \end{cases}$$

$$H_0(kR) = \sum_{m=0}^{\infty} \epsilon_m \cos[m(\phi - \phi_0)]\begin{cases} J_m(kr)H_m(kr_0); & r \le r_0 \\ J_m(kr_0)H_m(kr); & r \ge r_0 \end{cases} \tag{7.2.51}$$

We have given the derivation of (7.2.51) in detail because it will serve as the prototype for the calculation of expansions of other Green's functions. These expansions are of considerable use, as the following calculation will show. We shall derive the expansion of e^{ikx} (a plane wave traveling from left to right) in polar coordinates and then the integral representation of the Bessel function J_m [described in Eq. (5.3.65)]. We note that $H_0(kR)$ represents a wave traveling from the source at r_0. To obtain a plane wave traveling from left to right it is necessary to place the source at $-\infty$, *i.e.*, let $r_0 \to \infty$ and $\phi_0 \to \pi$. Then

$$R = \sqrt{r^2 - 2rr_0 \cos(\phi - \phi_0) + r_0^2} \xrightarrow[\substack{r_0' \to \infty \\ \phi_0' = \pi}]{} r_0\left(1 + \frac{r}{r_0}\cos\phi\right) = r_0 + x$$

Hence
$$H_0(kR) \xrightarrow[\substack{r_0 \to \infty \\ \phi_0 = \pi}]{} \sqrt{\frac{2}{\pi k r_0}}\, e^{i[k(r_0+x) - \frac{1}{4}\pi]}$$

$$= \sum_{m=0}^{\infty} \epsilon_m (-1)^m \cos(m\phi)\, J_m(kr) \sqrt{\frac{2}{\pi k r_0}}\, e^{i[kr_0 - \frac{1}{2}\pi(m+\frac{1}{2})]}$$

or
$$e^{ikx} = \sum_{m=0}^{\infty} \epsilon_m i^m \cos(m\phi)\, J_m(kr) \tag{7.2.52}$$

the desired expansion. This series has been given in Chap. 5 in another form [see Eq. (5.3.65)].

Finally, by employing the orthogonality properties of the set $\cos(m\phi)$ it is possible to derive an integral representation for $J_m(kr)$. Multiply both sides of Eq. (7.2.52) by $\cos(\nu\phi)$, and integrate from 0 to π. Then

$$J_\nu(kr) = \frac{i^{-\nu}}{\pi} \int_0^\pi e^{ikr\cos\phi} \cos(\nu\phi)\, d\phi \tag{7.2.53}$$

This relation was derived in another fashion in Chap. 5 [see Eq. (5.3.65)].

A General Formula. Let us now turn to the problem of deriving the expansion of the Green's function for any of the generalized coordinate systems for which the scalar Helmholtz equation is separable. We need to review some of the results of the discussion of separation (see pages 655 *et seq.*). If ξ_1, ξ_2, and ξ_3 are three orthogonal, generalized coordinates, with scale factors h_1, h_2, and h_3, then the Laplacian is

$$\nabla^2\psi = \sum_{n=1}^{3} \frac{M_n}{Sf_n} \frac{\partial}{\partial \xi_n}\left(f_n \frac{\partial\psi}{\partial \xi_n}\right) \tag{7.2.54}$$

The quantities f_n are functions of ξ_n only (that is, f_1 is a function of ξ_1 only); S is the Stäckel determinant [Eq. (5.1.25)] whose elements Φ_{nm} are functions of ξ_n only (that is, Φ_{1m} is a function ξ_1 only); M_n is the minor of S which multiplies Φ_{n1} in the expansion of S [Eq. (5.1.26)]. M_1 is a function of ξ_2 and ξ_3 but not of ξ_1. The scalar Helmholtz equation separates in coordinates ξ_1, ξ_2, ξ_3, so that

$$\psi = X_1(\xi_1)X_2(\xi_2)X_3(\xi_3)$$

where
$$\frac{1}{f_n X_n} \frac{d}{d\xi_n}\left[f_n \frac{dX_n}{d\xi_n}\right] + \sum_{m=1}^{3} \Phi_{nm}(\xi_n) k_m^2 = 0 \tag{7.2.55}$$

where $k_1^2 = k^2$ and k_2^2 and k_3^2 are two separation constants. The factors f_n and the elements of S are given in the table at the end of Chap. 5. We shall also need the Robertson condition (5.1.32), $h_1h_2h_3 = S f_1f_2f_3$.

To see how we expand a Green's function in these general coordinates it is well to return to the derivation of Eq. (7.2.51), the expansion of a Green's function in polar coordinates. There the ϕ factors turned out to be eigenfunctions, independent of the constant k in the Helmholtz equation. The r factors, on the other hand, depended on k and on the eigenvalues m of the ϕ factors and for these and other reasons could not be made into eigenfunctions.

The function G was then expanded into an eigenfunction series in the ϕ factors. The r factors for each term in the series then satisfied an inhomogeneous equation which could be solved in terms of the two solutions of the homogeneous equation, and the expansion was achieved.

We try the same procedure with the three general coordinates ξ_1, ξ_2, ξ_3. Of the three separation constants, $k_1 = k$, k_2, k_3, the first one, k, is fixed in value by the Helmholtz equation which we are solving. The other two, k_2 and k_3, are available to become eigenvalues for a set of two-dimensional eigenfunctions, in terms of which we are to expand the Green's function G. Usually the choice of which two of the three coordinate factors are to be eigenfunctions is an obvious one. In spherical coordinates, for instance, two of the three coordinates, ϕ and ϑ, are angles, with finite range of values and simple boundary conditions (periodicity and finiteness) which may be imposed to obtain eigenfunctions. In other cases (such as the circular cylinder coordinates, r, ϕ, z) only one of the coordinates (ϕ for the circular cylinder) has a finite range of values, and one of the other two (z, for instance) must produce a set of eigenfunctions for an infinite domain, having a continuous range of eigenvalues for one of the separation constants k_2 or k_3.

Suppose we find that the ξ_2 and ξ_3 factors may be made into eigenfunctions, with corresponding pairs of eigenvalues for k_2 and k_3. We order these eigenfunctions in some manner, with respect to the allowed values of k_2 and k_3; for instance, the lowest value of k_2 may be labeled k_{20}, the next k_{21}, and the mth k_{2m}, whereas the allowed values of k_3 are k_{30}, k_{31}, . . . , k_{3n} . . . ; the eigenfunction corresponding to k_{2m}, k_{3n} being $X_{2m}(\xi_2)X_{3n}(\xi_3)$ though X_2 may also depend on n and X_3 on m. To simplify the notation we label the pairs of integers (m,n) by a single letter, p or q, and express the eigenfunction product by a single letter W. Then the pth eigenfunction for the coordinates ξ_2, ξ_3 is $W_p(\xi_2,\xi_3)$ with eigenvalues k_{2p}, k_{3p}. In what follows we shall assume that both ξ_2 and ξ_3 have finite ranges of values, so that both k_{2p} and k_{3p} have a discrete set of values and the eigenfunction expansion is a series over p (m and n). The extension to cases where one or both of the eigenvalue sequences is a continuum, so that the expansion is an integral (like a Fourier integral, instead of a Fourier series), is one which is not difficult in any particular case.

Therefore we assume the existence of a complete set of eigenfunctions $W_q(\xi_2,\xi_3)$ (we chose ξ_2, and ξ_3 as examples, any pair for which the necessary conditions are satisfied) satisfying the orthonormal condition

$$\iint \bar{W}_q(\xi_2,\xi_3) W_p(\xi_2,\xi_3)\rho(\xi_2,\xi_3)\, d\xi_2\, d\xi_3 = \delta_{qp} \tag{7.2.56}$$

where ρ is a weight function. (See page 781 for a discussion of weight functions in one dimension.) We shall assume that this set of eigenfunctions exists for arbitrary k for the range in ξ_2 and ξ_3 in the domain of interest. The function W_q may depend upon k, of course. Then in analogy to (7.2.46), we write

$$G_k(\mathbf{r}|\mathbf{r}') = \sum_q X_{1q}(\xi_1|\xi_1')B_q(\xi_2',\xi_3')W_q(\xi_2,\xi_3) \tag{7.2.57}$$

where the functions X_q and B_q are to be determined.

In generalized coordinates the equation determining G_k is

$$\sum_{n=1}^{3} \frac{M_n}{Sf_n}\frac{\partial}{\partial \xi_n}\left[f_n \frac{\partial G}{\partial \xi_n}\right] + k_1^2 G = -4\pi \frac{\delta(\xi_1 - \xi_1')\delta(\xi_2 - \xi_2')\delta(\xi_3 - \xi_3')}{h_1 h_2 h_3} \tag{7.2.58}$$

Note that the representation chosen for $\delta(\mathbf{r} - \mathbf{r}')$ as given by the coefficient of -4π is such as to vanish unless all three coordinates ξ_1, ξ_2, ξ_3 equal the corresponding primed coordinates, respectively, and so that its integral over all of space is unity. We now proceed to introduce series (7.2.57) into Eq. (7.2.58). We require the result of applying $(\nabla^2 + k_1^2)$ to $W_q(\xi_2,\xi_3)$. Only two of the terms in the sum in (7.2.58) involve derivatives of W_q. Since W_q is a product of $X_2(\xi_2)$ and $X_3(\xi_3)$, solutions of Eqs. (7.2.55), it follows that

$$\sum_{n=2}^{3} \frac{M_n}{Sf_n}\frac{\partial}{\partial \xi_n}\left[f_n \frac{\partial W_q}{\partial \xi_n}\right] = -\sum_{n=2}^{3} \frac{M_n}{S}\left\{\sum_{m=1}^{3} \Phi_{nm}(\xi_n)k_{mq}^2\right\} W_q$$

where the separation constants have been given the additional label q to indicate their correspondence to function W_q.

The sum over n may be simplified by utilizing the properties of the determinant S [see Eq. (5.1.27)].

$$\sum_n M_n \Phi_{nm} = S\delta_{1m}$$

Then

$$\sum_{n=2,3} \frac{M_n}{Sf_n}\frac{\partial}{\partial \xi_n}\left[f_n \frac{\partial W_q}{\partial \xi_n}\right] = -k_1^2 W_q + \frac{M_1}{S}\left[\sum_m k_{mq}^2 \Phi_{1m}\right] W_q$$

Hence, upon substitution of the series (7.2.57) into (7.2.58) one obtains

$$\sum_q B_q(\xi_2',\xi_3')W_q(\xi_2,\xi_3)\frac{M_1}{S}\left\{\frac{1}{f_1}\frac{d}{d\xi_1}\left[f_1\frac{dX_{1q}}{d\xi_1}\right]+\sum_{m=1}^{3}k_{mq}^2\Phi_{1m}(\xi_1)X_{1q}\right\}$$
$$= -4\pi\frac{\delta(\xi_1-\xi_1')\delta(\xi_2-\xi_2')\delta(\xi_3-\xi_3')}{h_1h_2h_3}$$

By employing the orthonormality condition of (7.2.56) and the Robertson condition (5.1.32) it is found that

$$B_q(\xi_2',\xi_3') = \frac{\rho(\xi_2',\xi_3')\bar{W}_q(\xi_2',\xi_3')}{M_1(\xi_2',\xi_3')f_2(\xi_2')f_3(\xi_3')} \tag{7.2.59}$$

and

$$\frac{1}{f_1}\frac{d}{d\xi_1}\left[f_1\frac{dX_{1q}}{d\xi_1}\right]+\left[\sum_{m=1}^{3}k_{mq}^2\Phi_{1m}\right]X_{1q} = -\left(\frac{4\pi}{f_1}\right)\delta(\xi_1-\xi_1') \tag{7.2.60}$$

Thus, as in the example in polar coordinates discussed above, X_{1q} is a one-dimensional Green's function. Following the procedure employed in solving (7.2.48), it is possible to express X_{1q} in terms of two independent solutions (y_1 and y_2) of the homogeneous equation

$$\frac{1}{f_1}\frac{d}{d\xi_1}\left[f_1\frac{dy}{d\xi_1}\right]+\sum k_{mp}^2\Phi_{1m}\,y = 0$$

We obtain

$$X_{1q}(\xi_1|\xi_1') = -\frac{4\pi}{\Delta(y_{1q},y_{2q})f_1(\xi_1')}\begin{cases} y_{1q}(\xi_1)y_{2q}(\xi_1'); & \xi_1 \le \xi_1' \\ y_{1q}(\xi_1')y_{2q}(\xi_1); & \xi_1 \ge \xi_1' \end{cases} \tag{7.2.61}$$

where Δ is the Wronskian evaluated at ξ_1'. As in the discussion on page 826, which solutions y_1 and y_2 are used depends upon the boundary conditions of the problem. The form of the result for the Wronskian is

$$\Delta(y_1,y_2) = \text{constant}/f_1 \tag{7.2.62}$$

so that the factor $\Delta(y_1,y_2)f_1$ is a constant.

The expansion for $G_k(\mathbf{r}|\mathbf{r}')$ is then

$$G_k(\mathbf{r}|\mathbf{r}') = -4\pi\left(\frac{h_1}{h_2h_3}\right)\rho(\xi_2',\xi_3')\sum_q \bar{W}_q(\xi_2',\xi_3')W_q(\xi_2,\xi_3)\,\cdot$$
$$\cdot\,\frac{1}{\Delta(y_{1q},y_{2q})}\begin{cases} y_{1q}(\xi_1)y_{2q}(\xi_1'); & \xi_1 \le \xi_1' \\ y_{1q}(\xi_1')y_{2q}(\xi_1); & \xi_1 \ge \xi_1' \end{cases} \tag{7.2.63}$$

where the scale factors h and the Wronskian Δ are functions of the primed coordinates and where ρ is the density function defined in Eq. (7.2.56). From the expansion it is generally possible, by the procedure employed in the polar coordinate case discussed above, to obtain plane wave expansions and integral representations of the functions in the plane wave expansion. Finally, since it is now possible to express plane waves and Green's function in different coordinate systems, it becomes possible to express the solutions of the Helmholtz equation in one coordinate system in terms of those solutions appropriate to another solution.

Green's Functions and Eigenfunctions. The Green's function for the Sturm-Liouville problem (see page 719)

$$\frac{d}{dx}\left[p\,\frac{dG}{dx}\right] + [q + \lambda r]G = -4\pi\delta(x - x_0) \tag{7.2.64}$$

may be expressed in an infinite series (7.2.39) in the eigenfunctions of the homogeneous differential equation corresponding to (7.2.64):

$$\frac{d}{dx}\left[p\,\frac{d\psi_n}{dx}\right] + [q + \lambda_n r]\psi_n = 0 \tag{7.2.65}$$

with homogeneous boundary conditions at $x = a$ and $x = b$. These eigenfunctions may be arranged to form a complete orthonormal series so that

$$\int_a^b \bar{\psi}_n(x)\psi_m(x)r\,dx = \delta_{nm} \tag{7.2.66}$$

The infinite series for $G_\lambda(x|x_0)$ is

$$G_\lambda(x|x_0) = 4\pi \sum_m \frac{\bar{\psi}_n(x)\psi_n(x_0)}{\lambda_n - \lambda} \tag{7.2.67}$$

On the other hand, the function G may be expressed in terms of two independent solutions y_1 and y_2 of (7.2.64) by the method following Eq. (7.2.49). One obtains

$$G_\lambda(x|x_0) = -\frac{4\pi}{p(x_0)\Delta(y_1,y_2)} \begin{cases} y_1(x)y_2(x_0); & x \le x_0 \\ y_2(x)y_1(x_0); & x \ge x_0 \end{cases} \tag{7.2.68}$$

This alternative expression may be employed by comparison with (7.2.67) to obtain information on the functions ψ_n and the corresponding eigenvalues λ_n. It forms a very powerful method for investigation of the properties of eigenfunctions and indeed has been employed at times as the basis for the entire theory outlined in Chap. 6. Here we shall be content with the discussion of those results which seem of practical importance.

The central idea consists in noting *that, as a function of* λ, $G_\lambda(x|x_0)$ has simple poles at $\lambda = \lambda_n$ with residues $-4\pi\bar{\psi}_n(x)\psi_n(x_0)$. These very singularities must be present in the closed form (7.2.68). Hence by examining (7.2.68) it is (in principle) possible to obtain the eigenvalues λ_n and, in addition, the corresponding eigenfunctions ψ_n already normalized.

Let us clarify the suggested procedure by means of a simple example. Let (7.2.65) be

$$(d^2\psi/dx^2) + \lambda\psi = 0$$

and the boundary conditions $\psi = 0$ at $x = a$, $x = b$ $(a < b)$. Then the appropriate y_1 is $\sin[\sqrt{\lambda}\,(x - a)]$ and $y_2 = \sin[\sqrt{\lambda}\,(x - b)]$. The value of $\Delta(y_1,y_2)$ is

$$\sqrt{\lambda}\sin[\sqrt{\lambda}\,(x-a)]\cos[\sqrt{\lambda}\,(x-b)] - \sqrt{\lambda}\cos[\sqrt{\lambda}\,(x-a)]\sin[\sqrt{\lambda}\,(x-b)] = \sqrt{\lambda}\sin[\sqrt{\lambda}\,(b-a)]$$

Therefore

$$G_\lambda(x|x_0) = \frac{-4\pi}{\sqrt{\lambda}\sin[\sqrt{\lambda}\,(b-a)]}\begin{cases}\sin[\sqrt{\lambda}\,(x-a)]\sin[\sqrt{\lambda}\,(x_0-b)]; & x \le x_0 \\ \sin[\sqrt{\lambda}\,(x_0-a)]\sin[\sqrt{\lambda}\,(x-b)]; & x \ge x_0\end{cases}$$

The eigenvalues occur at the zeros of $\sin[\sqrt{\lambda}\,(b - a)]$, so that one obtains the familiar result $\sqrt{\lambda_n} = [n\pi/(b - a)]$. The residue here is

$$-[8\pi/(b-a)](-1)^n \sin[n\pi(x-a)/(b-a)]\sin[n\pi(x_0-b)/(b-a)],$$

or $-[8\pi/(b - a)]\sin[n\pi(b - x)/(b - a)]\sin[n\pi(b - x_0)/(b - a)]$: Hence $\psi_n(x)\psi_n(x_0)$ is $[2/(b - a)]\sin[n\pi(b - x)/(b - a)]\sin[\pi n(b - x_0)/(b - a)]$ so that the normalized eigenfunction is

$$\psi_n = \sqrt{\frac{2}{b-a}}\sin\left[\frac{n\pi(b-x)}{b-a}\right]$$

These eigenfunctions satisfy orthogonality condition (7.2.66) with $r = 1$ and form an orthonormal complete set. We thus see that, if any two solutions whose combination satisfies the proper boundary conditions (actually y_1 satisfies those which exist at one boundary point, y_2 at the other) can be obtained, and if the evaluation of the Wronskian can be obtained, then the normalized orthogonal eigenfunctions and eigenvalues may be found.

Comparing this method with the more usual methods discussed in Chap. 6, we find that the amount of labor is the same for all. However, the method just described also yields the normalization which often involves a difficult integral in the other methods.

We shall have occasion to employ this method for more complex functions in the problems for this chapter and for the determination of eigenfunctions and their normalizations which arise in problems of two and higher dimensions. We shall also encounter the same procedures under a somewhat different guise in Sec. 11.1.

7.3 *Green's Function for the Scalar Wave Equation*

The Green's function for the scalar Helmholtz equation, just discussed in Sec. 7.2, is particularly useful in solving inhomogeneous problems, *i.e.* problems which arise whenever sources are present within the volume or on the bounding surface. The Green's function for the scalar wave equation must perform a similar function; thus it should be possible to solve the scalar wave equation, with sources present, in terms of a Green's function. To obtain some notion as to the equation this function must satisfy let us consider a typical inhomogeneous problem. Let ψ satisfy

$$\nabla^2\psi - \frac{1}{c^2}\frac{\partial^2\psi}{\partial t^2} = -4\pi q(r,t) \tag{7.3.1}$$

The function $q(r,t)$ describes the source density, giving not only the distribution of sources in space but also the time dependence of the sources at each point in space. In addition to Eq. (7.3.1) it is necessary to state boundary and initial conditions in order to obtain a unique solution (7.3.1). The condition on the boundary surface may be either Dirichlet or Neumann or a linear combination of both. The conditions in time dimension must be Cauchy (see page 685, Chap. 6). Hence it is necessary to specify the value of ψ and $(\partial\psi/\partial t)$ at $t = t_0$ for every point of the region under consideration. Let these values be $\psi_0(r)$ and $v_0(r)$, respectively.

Inspection of (7.3.1) suggests that the equation determining the Green's function $G(\mathbf{r},t|\mathbf{r}_0,t_0)$ is

$$\nabla^2 G - \frac{1}{c^2}\frac{\partial^2 G}{\partial t^2} = -4\pi\delta(\mathbf{r} - \mathbf{r}_0)\delta(t - t_0) \tag{7.3.2}$$

We see that the source is an impulse at $t = t_0$, located at $\mathbf{r} = \mathbf{r}_0$. G then gives the description of the effect of this impulse as it propagates away from $\mathbf{r} = \mathbf{r}_0$ in the course of time. As in the scalar Helmholtz case, G satisfies the homogeneous form of the boundary conditions satisfied by ψ on the boundary. For initial conditions, it seems reasonable to assume that G and $\partial G/\partial t$ should be zero for $t < t_0$; that is if an impulse occurs at t_0, *no effects of the impulse should be present at an earlier time.*

It should not be thought that this cause-and-effect relation, employed here, is obvious. The unidirectionality of the flow of time is apparent for macroscopic events, but it is not clear that one can extrapolate this experience to microscopic phenomena. Indeed the equations of motion in mechanics and the Maxwell equations, both of which may lead to a wave equation, do not have any asymmetry in time. It may thus be possible, for microscopic events, for "effects" to propagate backward in time; theories have been formulated in recent years which employ such solutions of the wave equation. It would take us too far afield, however, to discuss how such solutions can still lead to a cause-effect time relation for macroscopic events.

For the present we shall be primarily concerned with the initial conditions $G(\mathbf{r},t|\mathbf{r}_0,t_0)$ and $\partial G/\partial t$ zero for $t < t_0$, though the existence of other possibilities should not be forgotten.

The Reciprocity Relation. The directionality in time imposed by the Cauchy conditions, as noted above, means that the generalization of the reciprocity relation $G_k(\mathbf{r}|\mathbf{r}_0) = G_k(\mathbf{r}_0|\mathbf{r})$ to include time is not $G(\mathbf{r},t|\mathbf{r}_0,t_0) = G(\mathbf{r}_0 t_0|\mathbf{r},t)$. Indeed if $t > t_0$, the second of these is zero. In order to obtain a reciprocity relation it is necessary to reverse the direction of the flow of time, so that the reciprocity relation becomes

$$G(\mathbf{r},t|\mathbf{r}_0,t_0) = G(\mathbf{r}_0,-t_0|\mathbf{r},-t) \tag{7.3.3}$$

To interpret (7.3.3) it is convenient to place $t_0 = 0$. Then $G(\mathbf{r},t|\mathbf{r}_0,0) = G(\mathbf{r}_0,0|\mathbf{r},-t)$. We see that the effect, at $\mathbf{r}$ at a time t later than an impulse started at $\mathbf{r}_0$, equals the effect, at $\mathbf{r}_0$ at a time 0, of an impulse started at $\mathbf{r}$ at a time $-t$, that is t earlier.

To prove (7.3.3) let us write the equations satisfied by both of the Green's functions:

$$\nabla^2 G(\mathbf{r},t|\mathbf{r}_0,t_0) - \frac{1}{c^2}\left[\frac{\partial^2 G(\mathbf{r},t|\mathbf{r}_0,t_0)}{\partial t^2}\right] = -4\pi\delta(\mathbf{r}-\mathbf{r}_0)\delta(t-t_0)$$

$$\nabla^2 G(\mathbf{r},-t|\mathbf{r}_1,-t_1) - \frac{1}{c^2}\left[\frac{\partial^2 G(\mathbf{r},-t|\mathbf{r}_1,-t_1)}{\partial t^2}\right] = -4\pi\delta(\mathbf{r}-\mathbf{r}_1)\delta(t-t_1)$$

Multiplying the first of these by $G(\mathbf{r},-t|\mathbf{r}_1,-t_1)$ and the second by $G(\mathbf{r},t|\mathbf{r}_0,t_0)$, subtracting, and integrating over the region under investigation and over time t from $-\infty$ to t' where $t' > t_0$ and $t' > t_1$, then

$$\int_{-\infty}^{t'} dt \int dV \left\{G(\mathbf{r},t|\mathbf{r}_0,t_0)\nabla^2 G(\mathbf{r},-t|\mathbf{r}_1,-t_1) - G(\mathbf{r},-t|\mathbf{r}_1,-t_1)\nabla^2 G(\mathbf{r},t|\mathbf{r}_0,t_0)\right.$$
$$\left. + \frac{1}{c^2} G(\mathbf{r},t|\mathbf{r}_0,t_0)\frac{\partial^2}{\partial t^2} G(\mathbf{r},-t|\mathbf{r}_1,-t_1) - \frac{1}{c^2} G(\mathbf{r},-t|\mathbf{r}_1,-t_1)\frac{\partial^2}{\partial t^2} G(\mathbf{r},t|\mathbf{r}_0,t_0)\right\}$$
$$= 4\pi\{G(\mathbf{r}_0,-t_0|\mathbf{r}_1,-t_1) - G(\mathbf{r}_1,t_1|\mathbf{r}_0,t_0)\} \tag{7.3.4}$$

The left-hand side of the above equation may be transformed by use of Green's theorem and by the identity

$$\frac{\partial}{\partial t}\left[G(\mathbf{r},t|\mathbf{r}_0,t_0)\frac{\partial}{\partial t}G(\mathbf{r},-t|\mathbf{r}_1,-t_1) - G(\mathbf{r},-t|\mathbf{r}_1,-t_1)\frac{\partial}{\partial t}G(\mathbf{r},t|\mathbf{r}_0,t_0)\right]$$
$$= G(\mathbf{r},t|\mathbf{r}_0,t_0)\frac{\partial^2}{\partial t^2}G(\mathbf{r},-t|\mathbf{r}_1,-t_1) - G(\mathbf{r},-t|\mathbf{r}_1,-t_1)\frac{\partial^2}{\partial t^2}G(\mathbf{r},t|\mathbf{r}_0,t_0)$$

We obtain for the left-hand side

$$\int_{-\infty}^{t'} dt \int d\mathbf{S} \cdot [G(\mathbf{r},t|\mathbf{r}_0,t_0) \operatorname{grad} G(\mathbf{r},-t|\mathbf{r}_1,-t_1) - G(\mathbf{r},-t|\mathbf{r}_1,-t_1) \operatorname{grad} G(\mathbf{r},t|\mathbf{r}_0,t_0)]$$
$$+\frac{1}{c^2}\int dV \left[G(\mathbf{r},t|\mathbf{r}_0,t_0)\frac{\partial G(\mathbf{r},-t|\mathbf{r}_1,-t_1)}{\partial t} - G(\mathbf{r},-t|\mathbf{r}_1,-t_1)\frac{\partial G(\mathbf{r},t|\mathbf{r}_0,t_0)}{\partial t}\right]_{t=-\infty}^{t=t'}$$

The first of these integrals vanishes, for both Green's functions satisfy the same homogeneous boundary conditions on S. The second also vanishes, as we shall now see. At the lower limit both $G(\mathbf{r},-\infty|\mathbf{r}_0,t_0)$ and its time derivative vanish in virtue of the causality condition. At the time $t = t'$, $G(\mathbf{r},-t'|\mathbf{r}_1,-t_1)$ and its time derivative vanish, since $-t'$ is earlier than $-t_1$. Thus the left-hand side of (7.3.4) is zero, yielding reciprocity theorem (7.3.3).

We shall demonstrate that it is possible to express the solution (including initial conditions) of the inhomogeneous problem for the scalar wave equation in terms of known inhomogeneities in the Green's function. We shall need Eq. (7.3.1):

$$\nabla_0^2\psi(\mathbf{r}_0,t_0) - \frac{1}{c^2}\frac{\partial^2\psi}{\partial t_0^2} = -4\pi q(\mathbf{r}_0,t_0)$$

also
$$\nabla_0^2 G(\mathbf{r},t|\mathbf{r}_0,t_0) - \frac{1}{c^2}\frac{\partial^2 G(\mathbf{r},t|\mathbf{r}_0,t_0)}{\partial t_0^2} = -4\pi\delta(\mathbf{r}-\mathbf{r}_0)\delta(t-t_0)$$

This last equation may be obtained from (7.3.2) by the use of the reciprocity relation. As is usual, multiply the first equation by G and the second by ψ and subtract. Integrate over the volume of interest and over t_0 from 0 to t^+. By the symbol t^+ we shall mean $t + \epsilon$ where ϵ is arbitrarily small. This limit is employed in order to avoid ending the integration exactly at the peak of a delta function. When employing the final formulas, it is important to keep in mind the fact that the limit is t^+ rather than just t. One obtains

$$\int_0^{t^+} dt_0 \int dV_0 \left\{G\nabla_0^2\psi - \psi\nabla_0^2 G + \frac{1}{c^2}\left(\frac{\partial^2 G}{\partial t_0^2}\psi - G\frac{\partial^2\psi}{\partial t_0^2}\right)\right\} = 4\pi\left\{\psi(\mathbf{r},t) - \int_0^{t^+} dt_0 \int dV_0 q(\mathbf{r}_0,t_0)G\right\}$$

Again employing Green's theorem, etc., we obtain

$$\int_0^{t^+} dt_0 \oint d\mathbf{S}_0 \cdot (G \operatorname{grad}_0 \psi - \psi \operatorname{grad}_0 G) + \frac{1}{c^2}\int dV_0 \left[\frac{\partial G}{\partial t_0}\psi - G\frac{\partial\psi}{\partial t_0}\right]_0^{t^+} + 4\pi\int_0^{t^+} dt_0 \int dV_0 q(\mathbf{r}_0,t_0)G = 4\pi\psi(\mathbf{r},t)$$

The integrand in the first integral is specified by boundary conditions. In the second integral, the integrand vanishes when $t = t^+$ is introduced by virtue of the initial conditions on G. The remaining limit involves only initial conditions. Hence,

$$4\pi\psi(\mathbf{r},t) = 4\pi \int_0^{t^+} dt_0 \int dV_0 G(\mathbf{r},t|\mathbf{r}_0,t_0)q(\mathbf{r}_0,t_0) + \int_0^{t^+} dt_0 \oint d\mathbf{S}_0 \cdot (G \operatorname{grad}_0 \psi - \psi \operatorname{grad}_0 G) - \frac{1}{c^2}\int dV_0 \left[\left(\frac{\partial G}{\partial t_0}\right)_{t_0=0} \psi_0(\mathbf{r}_0) - G_{t_0=0}v_0(\mathbf{r}_0)\right] \quad (7.3.5)$$

where $\psi_0(\mathbf{r}_0)$ and $v_0(\mathbf{r}_0)$ are the initial values of ψ and $\partial\psi/\partial t$.

Equation (7.3.5) gives the complete solution of the inhomogeneous problem including the satisfaction of initial conditions. The surface integrals, as in the Helmholtz case, must be carefully defined. As in that case we shall take a surface value to be the limit of the value of the function as the surface is approached from the interior.

The first two integrals on the right side of the above Eq. (7.3.5) are much the same sort as those appearing in the analogous equation for the case of the Helmholtz equation. The first represents the effect of sources; the second the effect of the boundary conditions on the space boundaries. The last term involves the initial conditions. We may interpret it by asking what sort of source q is needed in order to start the function ψ at $t = 0$ in the manner desired. We may expect that this will require an impulsive type force at a time $t = 0^+$. From (7.3.5) we can show that the source term required to duplicate the initial conditions is

$$(1/c^2)[\psi_0(\mathbf{r}_0)\delta'(t_0) + v_0(\mathbf{r}_0)\delta(t_0)]$$

where by $\delta'(t_0)$ we mean the derivative of the δ function. It has the property

$$\int_a^b f(x)\delta'(x)\,dx = \begin{cases} -f'(0); & \text{if } x = 0 \text{ is within interval } (a,b) \\ 0; & \text{if } x = 0 \text{ is outside interval } (a,b) \end{cases}$$

The physical significance of these terms may be understood. A term of type $v_0\delta(t_0)$ is required to represent an impulsive force, which gives each point of the medium an initial velocity $v_0(\mathbf{r}_0)$. To obtain an initial displacement, an impulse delivered at $t_0 = 0$ must be allowed to develop for a short time until the required displacement is achieved. At this time a second impulse is applied to reduce the velocity to zero but leave the displacement unchanged. It may be seen that the first term $\psi(\mathbf{r}_0,t_0)\delta'(t_0)$ has this form if it is written

$$\lim_{\epsilon\to 0}\left\{\psi(\mathbf{r}_0,t_0)\left[\frac{\delta(t_0+\epsilon) - \delta(t_0-\epsilon)}{2\epsilon}\right]\right\}$$

Form of the Green's Function. Knowledge of G is necessary to make (7.3.5) usable. As in the case of the scalar Helmholtz equation we shall first find G for the infinite domain. Let us call this function g. The method employed in the scalar Helmholtz case involves assessing the relative strength of the singularities in the functions $\nabla^2 g$ and $\partial^2 g/\partial t^2$ in the equation

$$\nabla^2 g - \frac{1}{c^2}\left(\frac{\partial^2 g}{\partial t^2}\right) = -4\pi\delta(\mathbf{r} - \mathbf{r}_0)\delta(t - t_0)$$

It may be argued that $\nabla^2 g$ is the more singular, since it involves the second derivative of a three-dimensional δ function $\delta(\mathbf{r} - \mathbf{r}') = \delta(x - x') \cdot \delta(y - y')\delta(z - z')$. Such an argument is not very satisfying. However, for the moment, let us assume it to be true. We shall return to the above equation later and derive the result we shall obtain in a more rigorous manner.

Integrating both sides of the equation over a small spherical volume surrounding the point $\mathbf{r} = \mathbf{r}_0$, that is, $R = 0$, and neglecting the time derivative term, one obtains as in the previous section

$$g \xrightarrow[R\to 0]{} \delta(t - t_0)/R \tag{7.3.6}$$

As before we now proceed to find a solution of the homogeneous equation satisfying this condition, for it is clear that g satisfies the equation

$$\nabla^2 g - \frac{1}{c^2}\frac{\partial^2 g}{\partial t^2} = 0; \quad R \text{ and } t - t_0 \text{ not equal to zero}$$

At $R = 0$ condition (7.3.6) must be employed. Since we are dealing with point sources in an infinite medium g is a function of R rather than of $\mathbf{r}$ and $\mathbf{r}_0$ separately. Hence

$$\frac{1}{R^2}\frac{\partial}{\partial R}\left(R^2\frac{\partial g}{\partial R}\right) - \frac{1}{c^2}\left(\frac{\partial^2 g}{\partial t^2}\right) = 0 \quad \text{or} \quad \frac{\partial^2(gR)}{\partial R^2} - \frac{1}{c^2}\frac{\partial^2(gR)}{\partial t^2} = 0 \tag{7.3.7}$$

The solutions of this equation are

$$g = \frac{h[(R/c) - (t - t_0)] + k[(R/c) + (t - t_0)]}{R}$$

where h and k are any functions. Comparing with condition (7.3.6) we see that two possibilities (or any linear combination of these) occur, $\delta[(R/c) - (t - t_0)]/R$ or $\delta[(R/c) + (t - t_0)]/R$. The second of these must be eliminated, for it does not satisfy the condition imposed earlier, which requires that the effect of an impulse at a time t_0 be felt at a distance R away at a time $t > t_0$. Therefore

$$g = \frac{\delta[(R/c) - (t - t_0)]}{R}; \quad R, t - t_0 > 0 \tag{7.3.8}$$

representing a spherical shell about the source, expanding with a radial velocity c.

We may now make an a posteriori check of our initial assumption, that the singularity of $\nabla^2 g$ was greater than that for $\partial^2 g/\partial t^2$. This is indicated by the presence of the $1/R$ factor, but to prove it requires a rather nice balancing of infinities. We shall therefore stop to put (7.3.8) on a more firm footing and only then return to discuss the deductions which follow from this formula. Using spherical coordinates for $\delta(\mathbf{R}) = \delta(\mathbf{r} - \mathbf{r}_0)$ and defining

$$\tau = t - t_0$$

it is immediately possible to retrace the steps leading to (7.3.7) and obtain the more general equation, valid also for R and τ equal to zero,

$$\frac{\partial^2(Rg)}{\partial R^2} - \frac{1}{c^2}\frac{\partial^2(Rg)}{\partial \tau^2} = -\frac{2\delta(R)}{R}\,\delta(\tau)$$

The numerical factor 2 enters because the variable R can never be negative. Hence $\int_0^\infty \delta(R)\,dR = \frac{1}{2}$.

To proceed further it is desirable to employ the relation

$$\delta(R)/R = -\delta'(R) \tag{7.3.9}$$

To demonstrate this multiply $\delta(R)/R$ by a differentiable function $f(R)$ and integrate over R. Let $f(R) = f(0) + f'(0)R + f''(0)(R^2/2!) + \cdots$. Then

$$\int_{-\infty}^{\infty} \frac{f(R)\delta(R)}{R}\,dR = f(0)\int_{-\infty}^{\infty}\frac{\delta(R)}{R}\,dR + f'(0)\int_{-\infty}^{\infty}\delta(R)\,dR + \frac{f''(0)}{2!}\int_{-\infty}^{\infty} R\delta(R)\,dR + \cdots$$

The first of these terms is an integral over an odd function, so that it has a Cauchy principal value of zero; the second one gives $f'(0)$; the third and all higher terms give zero. Hence

$$\int_{-\infty}^{\infty}\frac{f(R)\delta(R)}{R}\,dR = f'(0) = \int_{-\infty}^{\infty} f'(R)\delta(R)\,dR = -\int_{-\infty}^{\infty} f(R)\delta'(R)\,dR$$

This equation may also be derived more directly from the definition of derivative as follows:

$$\delta'(R) = \lim_{\epsilon\to 0}\left[\frac{\delta(R+\epsilon) - \delta(R-\epsilon)}{2\epsilon}\right] = \lim_{\epsilon\to 0}\tfrac{1}{2}\left[\frac{\delta(R+\epsilon)}{-R} - \frac{\delta(R-\epsilon)}{R}\right] = -\frac{\delta(R)}{R}$$

Returning to the equation, we may now write

$$\frac{\partial^2}{\partial R^2}(Rg) - \frac{1}{c^2}\frac{\partial^2}{\partial \tau^2}(Rg) = 2\delta'(R)\,\delta(\tau)$$

It is clearly appropriate to introduce the variables

$$\xi = R - c\tau; \quad \eta = R + c\tau \tag{7.3.10}$$

The meaning of $\delta'(R)\,\delta(\tau)$ in the new variables must also be determined. To do this note that

$$\int_{-\infty}^{\infty}\int_{-\infty}^{\infty} d\tau\,dR\,f(\tau,R)\delta'(R)\,\delta(\tau) = -\left(\frac{\partial f}{\partial R}\right)_{R,\tau=0}$$

This may be rewritten $-[(\partial f/\partial\xi) + (\partial f/\partial\eta)]_{\xi,\eta=0}$. Hence under transformation (7.3.10)

$$\delta'(R)\,\delta(\tau) = 2c[\delta'(\xi)\,\delta(\eta) + \delta'(\eta)\,\delta(\xi)] \tag{7.3.11}$$

The factor $2c$ is just the inverse of the change of volume element under the transformation from variables (R,τ) to (ξ,η).

In the new variables ξ and η the equation satisfied by Rg becomes, therefore,

$$\partial^2(Rg)/\partial\xi\,\partial\eta = c[\delta'(\xi)\,\delta(\eta) + \delta'(\eta)\,\delta(\xi)]$$

or $$Rg = c\int_{-\infty}^{\xi} d\xi \int_{-\infty}^{\eta} d\eta\,\delta'(\xi)\,\delta(\eta) + c\int_{\xi}^{\infty} d\xi \int_{\eta}^{\infty} d\eta\,\delta(\xi)\,\delta'(\eta)$$

The limits of integration have been chosen in such a fashion as to yield a solution satisfying the required initial conditions. Upon integration,

$$Rg = c\,\delta(\xi)u(\eta) - c\,\delta(\eta)[1 - u(\xi)] \tag{7.3.12}$$

where $u(\eta) = \begin{cases} 0; & \eta < 0 \\ 1; & \eta > 0 \end{cases}$; therefore $1 - u(\xi) = \begin{cases} 1; & \xi < 0 \\ 0; & \xi > 0 \end{cases}$

The second term of (7.3.12) may be dropped; we can show that one or both of its factors are zero everywhere. The function $\delta(\eta)$ differs from zero only when $\eta = 0$ (*i.e.*, when $c\tau = -R$), but at that point $\xi = 2R$ so that $1 - u(\xi) = 0$. On the other hand, in the first term $\delta(\xi)$ differs from zero when $c\tau = R$, where $\eta = 2R$ and $u(\eta) = 1$. Solving for g in (7.3.12), replacing ξ by $[R - c(t - t_0)]$ and $u(\eta)$ by 1 (for R, $t - t_0 > 1$), Eq. (7.3.12) becomes (7.3.8) as desired.

In order to obtain some concept as to the meaning of (7.3.8), consider the infinite-domain case with initial conditions $\psi = \partial\psi/\partial t = 0$ at $t = 0$. Then

$$\psi(\mathbf{r},t) = \frac{1}{4\pi}\int_0^{t^+} dt_0 \int dV_0 \left\{\frac{\delta[(R/c) - (t - t_0)]}{R}\right\} q(\mathbf{r}_0,t_0); \quad t^+ = t + \epsilon$$

and finally, $$\psi(\mathbf{r},t) = \frac{1}{4\pi}\int dV_0 \left[\frac{q(\mathbf{r}_0, t - R/c)}{R}\right] \tag{7.3.13}$$

We see that the effect at $\mathbf{r}$ at a time t is "caused" by the value of the source function q at $\mathbf{r}_0$ at a time $t - (1/c)|\mathbf{r} - \mathbf{r}_0|$. This is just the statement that the velocity of propagation of the disturbance is c. When the velocity of propagation becomes infinite, the solution reduces to the familiar solution of the Poisson equation, a potential as it should, since the inhomogeneous scalar wave equation becomes the Poisson equation in this limit. As a consequence (7.3.13) is often referred to as the *retarded potential* solution.

Field of a Moving Source. As a simple example consider a point source traveling in an infinite medium with a velocity $\mathbf{v}$. Then $q = q_0\,\delta(\mathbf{r} - \mathbf{v}t)$, where q_0 is related to the strength of the source. From (7.3.5)

$$\psi(\mathbf{r},t) = \frac{q_0}{4\pi}\int_0^{t^+} dt_0 \int dV_0 \left\{\frac{\delta[(R/c) - (t - t_0)]\,\delta(\mathbf{r}_0 - \mathbf{v}t_0)}{R}\right\}$$

$$= \frac{q_0}{4\pi}\int_0^{t^+} dt_0 \left\{\frac{\delta[(1/c)|\mathbf{r} - \mathbf{v}t_0| - (t - t_0)]}{|\mathbf{r} - \mathbf{v}t_0|}\right\}$$

Let
$$p = (1/c)|\mathbf{r} - \mathbf{v}t_0| + t_0$$

Then
$$dp = dt_0\left[\frac{v^2t_0 - \mathbf{v}\cdot\mathbf{r}}{c|\mathbf{r} - \mathbf{v}t_0|} + 1\right]$$

so that
$$\psi(\mathbf{r},t) = \frac{q_0}{4\pi}\int_{r/c}^{t^+ + |r - vt^+|/c} \frac{dp\,\delta(p - t)}{(1/c)(v^2t_0 - \mathbf{v}\cdot\mathbf{r}) + |\mathbf{r} - \mathbf{v}t_0|}$$

The singularity of the δ function occurs when $p = t$. This must define a time t_0 such that a signal emitted by the particle at a time t_0 will arrive at $\mathbf{r}$ at a time t (see Fig. 7.9). The time $(t - t_0)$ equals therefore the distance traveled, $|\mathbf{r} - \mathbf{v}t_0|$ divided by c; $t - t_0 = |\mathbf{r} - \mathbf{v}t_0|/c$, which is just $p = t$. The quantity $\mathbf{r} - \mathbf{v}t_0 = \boldsymbol{\rho}$ is thus the distance of the source from the observation point $\mathbf{r}$, at a time ρ/c earlier than time t. The time t_0, called the *retarded time*, is the solution of the equation $p = t$, giving the time and therefore the source position $\mathbf{v}t_0$ at the retarded time.

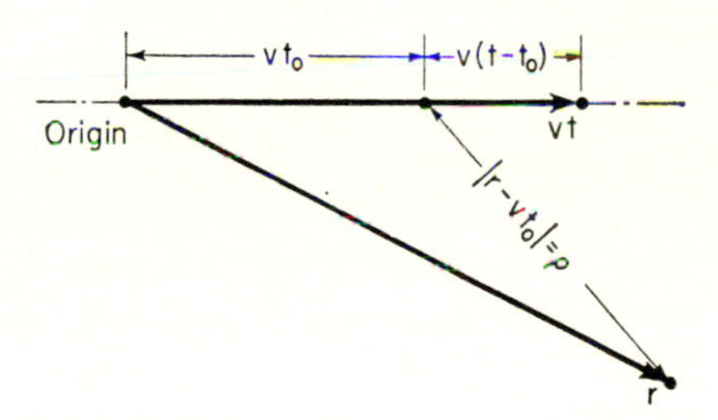

Fig. 7.9 Retarded potential of a moving source.

Integrating over p

$$\psi(r,t) = \frac{q_0}{4\pi}\left[\frac{1}{(1/c)(v^2t_0 - \mathbf{v}\cdot\mathbf{r}) + |\mathbf{r} - \mathbf{v}t_0|}\right]_{p=t}$$

Introducing $\boldsymbol{\rho} = \mathbf{r} - \mathbf{v}t_0$, we have

$$\psi(\mathbf{r},t) = \frac{q_0}{4\pi}\frac{1}{\rho - (\mathbf{v}\cdot\boldsymbol{\rho}/c)} \qquad (7.3.14)$$

where $\boldsymbol{\rho}$ is a vector drawn from the position of the source to the observation point at a time $t_0 = t - \rho/c$. We have obtained this solution in Chap. 2 by another method, where it was shown that the factor $1 - (\mathbf{v} \cdot \boldsymbol{\rho}/c\rho)$ had to be introduced in order to take into account the fact that the source moves in the course of the integration, the factor being required for normalization of the source (see page 214).

Two-dimensional Solution. If the source distribution q is independent of z, we have a problem in which the dependence on space is only two-dimensional; that is, ψ depends only upon x and y. The "two-dimensional point source" for such a problem is a line source, a uniform source extending from $z_0 = -\infty$ to $z_0 = +\infty$ along a line parallel to the z axis passing through (x_0,y_0). The Green's function for two-dimensional problems may be therefore found by integrating the three-dimensional point source from $z_0 = -\infty$ to $z_0 = +\infty$:

$$g(\boldsymbol{\rho},t)|\boldsymbol{\rho}_0,t_0) = \int_{-\infty}^{\infty} \frac{\delta[(R/c) - (t - t_0)]}{R}\, dz_0$$

where $\boldsymbol{\rho} = x\mathbf{i} + y\mathbf{j}$ is the radius vector in the x, y plane. We may expect that g is a function of $|\boldsymbol{\rho} - \boldsymbol{\rho}_0| = P$ and of $t - t_0 = \tau$. Indeed, the above equation may be rewritten

$$g(P,\tau) = \int_{-\infty}^{\infty} \frac{\delta[(R/c) - \tau]}{R}\, d\zeta$$

where $\zeta = z_0 - z$ and where

$$R^2 = \zeta^2 + P^2; \quad d\zeta/dR = R/\zeta$$

Hence

$$g(P,\tau) = 2\int_0^{\infty} \frac{\delta[(R/c) - \tau]}{\sqrt{R^2 - P^2}}\, dR = \frac{2}{\sqrt{c^2\tau^2 - P^2}} \int_0^{\infty} \delta[(R/c) - \tau]\, dR$$

or finally

$$g(P,\tau) = \begin{cases} \dfrac{2c}{\sqrt{c^2\tau^2 - P^2}}; & P < c\tau \\ 0; & P > c\tau \end{cases} \tag{7.3.15}$$

We see, from Eq. (7.3.15), a striking difference between the two- and the three-dimensional cases. In three dimensions, the effect of an impulse, after a time τ has elapsed, will be found concentrated on a sphere of radius $R = c\tau$ whose center is at the source point. This is a virtue of the function $\delta[(R/c) - \tau]$ which occurs in (7.3.8). In two dimensions, the effect at a time τ due to an impulsive source is spread over the entire region $P < c\tau$. To be sure, there is a singularity at $P = c\tau$, but this is very weak indeed when compared with the δ function singularity in the three-dimensional case. The explanation of the difference may be readily seen by examining the line source in three dimensions. The effects, after a time τ has elapsed, of each point source constituting the line source will be found in a different region of the xy plane. Thus we infer that an impulsive line source does not emit a cylindrical shell wave with the disturbance present only at the wave surface $P = c\tau$. Rather there is a "wake," which trails off behind this wave surface. This wake is characteristic of two-dimensional problems and is not encountered in either the three- or one-dimensional problems as we shall see. This has been mentioned in pages 145 and 687.

One-dimensional Solutions. The three-dimensional source distribution corresponding to a Green's function which depends only upon $(x - x_0)$ (not upon $y - y_0$ and $z - z_0$) is the plane source upon which point sources (or equivalent line sources) are uniformly distributed. Such Green's functions are useful for problems in which there is no space dependence upon y or z. It may be obtained from (7.3.15) by integrating $g(P,\tau)$ over y_0 keeping x and x_0 fixed. Let $\xi = x - x_0$, $\eta = y - y_0$; then

$$g(\xi,\tau) = 2c \int_{-\gamma}^{\gamma} \frac{d\eta}{\sqrt{c^2\tau^2 - \xi^2 - \eta^2}}; \quad \gamma = \sqrt{c^2\tau^2 - \xi^2}; \quad |\xi| < c\tau$$
$$= 0; \quad |\xi| > c\tau$$

The integration over η may be easily performed to yield $2c\pi$, so that

$$g(\xi,\tau) = 2c\pi \left[1 - u\left(\frac{|\xi|}{c} - \tau\right)\right] \tag{7.3.16}$$

Again we note that the effect of an impulse delivered at a time t_0, at the point x_0, for one-dimensional situations (or on the entire plane $x = x_0$ in three dimensions) is not concentrated at the point $|x - x_0| = \pm c(t - t_0)$ but rather exists throughout the region of extent $2c(t - t_0)$ with the source point x_0 at the mid-point.

Initial Conditions. To obtain a better grasp of the significance of the various expressions (7.3.8), (7.3.15), and (7.3.16), let us discuss the initial-value problem. Suppose that v_0 and ψ_0, the initial velocity and displacement, are known at every point in space; what are the velocity and displacement at a time t, assuming no sources present, that is $q = 0$? The solution of this problem may be obtained from (7.3.5):

$$\psi(r,t) = \frac{1}{4\pi c^2} \int dV_0 \left\{ g_{t_0=0} v_0(\mathbf{r}_0) - \left(\frac{\partial g}{\partial t_0}\right)_{t_0=0} \psi_0(\mathbf{r}_0) \right\} \tag{7.3.17}$$

The integration extends over all of space [we have also assumed that the surface integral in (7.3.5) vanishes at infinity].

Let us consider the one-dimensional case first, where the evaluation of (7.3.17) is simple and the interpretation of the result is straightforward. For one dimension, (7.3.17) becomes

$$\psi(x,t) = \frac{1}{4\pi c^2} \int dx_0 \left\{ g_{t_0=0} v_0(x_0) - \left(\frac{\partial g}{\partial t_0}\right)_{t_0=0} \psi_0(x_0) \right\}$$

where g is given by Eq. (7.3.16). The functions $g_{t_0=0}$ and $(\partial g/\partial t_0)_{t_0=0}$ have the values $2c\pi\{1 - u[(|\xi|/c) - t]\}$ and $-2c\pi\delta[(|\xi|/c) - t]$, respectively, where $|\xi| = |x - x_0|$. The integration may be easily performed to yield

$$\psi(x,t) = \tfrac{1}{2}\left\{\frac{1}{c}\int_{x-ct}^{x+ct} v_0(x_0)\,dx_0 + \psi_0(x + ct) + \psi_0(x - ct)\right\} \quad (7.3.18)$$

This is the familiar d'Alembert solution of the initial problem in one dimension [see Eq. (11.1.58)]. It may be also obtained directly from the differential equation itself (see page 685).

From Eq. (7.3.18) we see that, if the medium, say a string, is given an initial displacement with no velocity, the initial pulse breaks up into

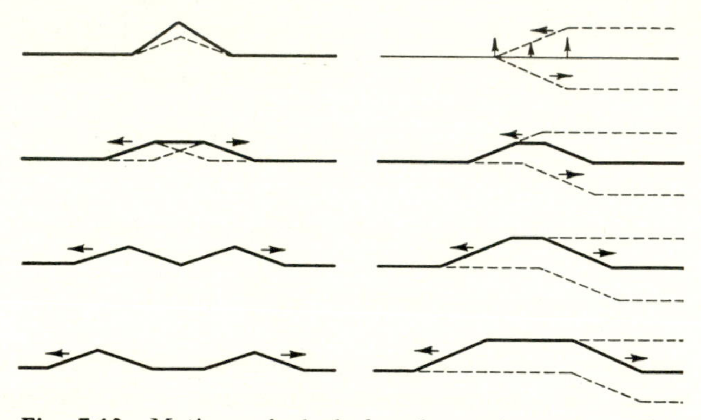

Fig. 7.10 Motions of plucked and struck strings. The solid lines give the shapes of the strings at successive times and the dotted lines give the shapes of the two "partial waves," traveling in opposite directions, whose sum is the shape of the string.

two identical waves, one traveling in the positive x direction and the other in the negative x direction. The sum of the two waves at $t = 0$ adds up to the initial displacement ψ_0. For some time thereafter, while they partially overlap, the composite shape will be rather complicated, until they finally separate. This behavior is illustrated in Fig. 7.10 above. Note that there is no wake trailing off behind each wave. For two-dimensional problems we shall find that such a wake is developed.

The initial problem in two dimensions is solved by (7.3.17). Here it is convenient to place the origin at the point of observation. Then

$$\psi(0,t) = \frac{1}{4\pi c^2}\int dS_0 \left\{g_{t_0=0}v_0(\boldsymbol{\varrho}_0) - \left(\frac{\partial g}{\partial t_0}\right)_{t_0=0}\psi_0(\boldsymbol{\varrho}_0)\right\}$$

where g is given by Eq. (7.3.15). Introducing polar coordinates and observing that $\partial g/\partial t_0 = -(\partial g/\partial t)$, then

$$\psi(0,t) = \frac{1}{2\pi c}\left\{\int_0^{2\pi} d\phi_0 \int_0^{ct} \rho_0\, d\rho_0 \frac{v_0(\boldsymbol{\varrho}_0)}{\sqrt{c^2t^2 - \rho_0^2}} + \frac{\partial}{\partial t}\left[\int_0^{2\pi} d\phi_0 \int_0^{ct} \rho_0\, d\rho_0 \frac{\psi_0(\boldsymbol{\varrho}_0)}{\sqrt{c^2t^2 - \rho_0^2}}\right]\right\} \tag{7.3.19}$$

This equation shows that the value of ψ at a point depends upon the original values of $\partial\psi/\partial t$ and ψ within a circle of radius ct about the observation point.

As in the discussion of the one-dimensional case, consider the initial condition $v_0 = 0$. Moreover, let $\psi_0(\boldsymbol{\varrho}_0) = \delta(\boldsymbol{\varrho}_0 - \boldsymbol{\varrho})$; that is, let the initial motion be an impulse delivered at point $\boldsymbol{\varrho}$. Then

$$\psi(0,t) = \frac{1}{2\pi c}\frac{\partial}{\partial t}\left[\int_0^{2\pi} d\phi_0 \int_0^{ct} \frac{\rho_0\, d\rho_0 \delta(\boldsymbol{\varrho} - \boldsymbol{\varrho}_0)}{\sqrt{c^2t^2 - \rho_0^2}}\right] = \begin{cases} \dfrac{1}{2\pi c}\dfrac{\partial}{\partial t}\left[\dfrac{1}{\sqrt{c^2t^2 - \rho^2}}\right]; & \text{if } \rho < ct \\ 0; & \text{if } \rho > ct \end{cases}$$

Combining and differentiating result in

$$\psi(0,t) = \frac{1}{2\pi c}\frac{\partial}{\partial t}\left[\frac{u(ct - \rho)}{\sqrt{c^2t^2 - \rho^2}}\right] = \frac{1}{2\pi}\left\{-\frac{ct\, u(ct - \rho)}{(c^2t^2 - \rho^2)^{\frac{3}{2}}} + \frac{\delta(ct - \rho)}{\sqrt{c^2t^2 - \rho^2}}\right\}$$

Thus the signal at the origin is zero until $ct = \rho$. At this time, a pulse (the second term in the above expression) arrives at the origin. This is, however, followed by a wake trailing off behind the pulse as described by the first term, which, for $ct \gg \rho$, decreases with time as $1/c^2t^2$. Contrast this to the results obtained for a similar pulse at x_0 in the one-dimensional case. There the signal will arrive at a point x at a time $|x - x_0|/c$. It will be an exact duplicate of the original signal except for a reduction in amplitude by $\frac{1}{2}$; there will be no wake. This difference between one- and two-dimensional cases is illustrated in Fig. 7.11.

The presence of a wake is characteristic of propagation in uniform media in two dimensions. In three as well as in one dimension, the shape of pulse remains unchanged in propagating away from its initial position. This is an immediate consequence of the form of the Green's function for three dimensions as given by Eq. (7.3.8). The presence of the δ function prevents the formation of a wake. It does not necessarily guarantee that the pulse is unchanged in shape. In one-dimensional problems the shape of a pulse is preserved in which a given *displacement* (with zero velocity) is initially established. In three dimensions, the reverse is true. The shape of a pulse is maintained in which the initial displacement is zero and the *velocity* is given initially.

We may see this by deriving the analogue of (7.3.19) and (7.3.18). Introducing the three-dimensional Green's function into (7.3.17) and taking the origin at the observation point, we have

$$\psi(0,t) = \frac{1}{4\pi c^2}\int d\Omega_0 \int dr_0 \left[r_0\delta\left(\frac{r_0}{c} - t\right) v_0(\mathbf{r}_0) - r_0\delta'\left(\frac{r_0}{c} - t\right)\psi_0(\mathbf{r}_0)\right]$$

where $d\Omega_0$ is the element of solid angle of the sphere in the "subzero" coordinates (that is, $\sin\theta_0\, d\theta_0\, d\phi_0$). It is immediately clear that the

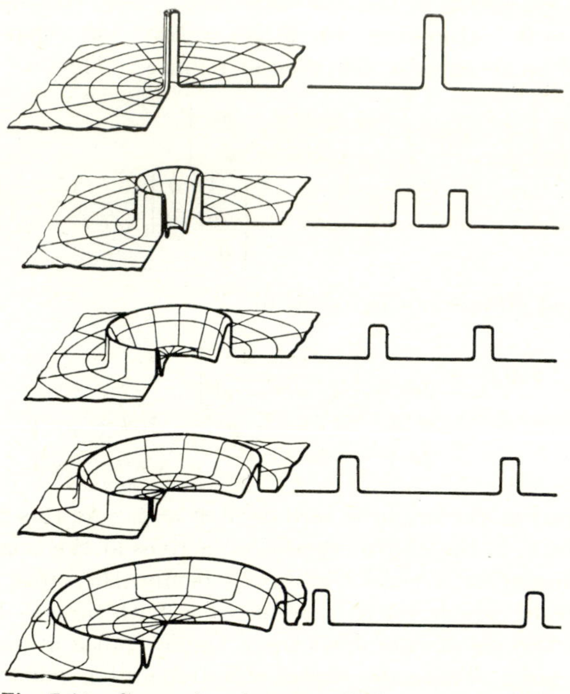

Fig. 7.11 Comparison between the behavior of string and membrane. First sketches show initial shapes, the lower ones the shapes at successive instants later. One-quarter of the membrane has been cut away to show the shape of the membrane, with the "wake" following the initial sharp wave front.

effects which occur at the observation point at a time t have their origin in the conditions initially present on the surface of a sphere of radius ct centered about the observation point. The integration over r_0 may now be performed. We need to write in the dependence of ψ_0 and v_0 on the coordinates θ_0, ϕ_0 on the spherical surface so that $\psi_0(\mathbf{r}_0) = \psi_0(r_0,\theta_0,\phi_0)$. The first integral is evaluated by employing the properties of the δ function. We obtain $[c^2tv_0(ct,\theta_0,\phi_0)]$.

To evaluate the second term, one may integrate by parts or one may employ the following property of δ' (see page 837):

$$\int_{-\infty}^{\infty} f(x')\delta'(x' - x)\, dx' = -f'(x)$$

Thus the second term is $-c^2\{\partial[t\psi_0(ct,\theta_0,\phi_0)]/\partial t\}$. Collecting these results one finally obtains

$$\psi(0,t) = \frac{1}{4\pi}\int d\Omega_0 \left\{tv_0(ct,\theta_0,\phi_0) + \frac{\partial}{\partial t}[t\psi_0(ct,\theta_0,\phi_0)]\right\} \tag{7.3.20}$$

The direct dependence of ψ on v_0 and the appearance of the derivative acting upon ψ_0 are in accordance with our introductory remarks. Equation (7.3.20) is known as *Poisson's solution.*

Huygens' Principle. The Green's function for the infinite region may also be used to obtain the mathematical expression of *Huygens' principle.* From an elementary point of view Huygens' principle postulates that each point on a wave front acts as a point source emitting a spherical wave which travels with a velocity c. The field at a given point some time later is then the sum of the fields of each of these point sources; the envelope of the wavelets from all the points is the next wave front.

To derive this principle, we turn to the general equation (7.3.5) and consider the situation where there are no sources (that is, $q = 0$ within a surface S) and, in addition, the initial values of ψ and $\partial\psi/\partial t$ are zero. We see that the volume integral in Eq. (7.3.5), involving the initial condition, vanishes, so that all that is left is

$$\psi(\mathbf{r},t) = \frac{1}{4\pi}\int_0^{t^+} dt_0 \oint d\mathbf{S}_0 \cdot \left\{\left[\frac{\delta(t_0 - t + R/c)}{R}\right]\operatorname{grad}_0\psi - \psi\operatorname{grad}_0\left[\frac{\delta(t_0 - t + R/c}{R}\right]\right\}$$

where we have inserted Eq. (7.3.8) into what is left of (7.3.5). Integrating over t_0 is not too difficult for the first term:

$$\int_0^{t^+}\left(\frac{1}{R}\right)\delta\left(t_0 - t + \frac{R}{c}\right)\operatorname{grad}_0\psi(\mathbf{r}_0,t_0)\,dt_0 = \left(\frac{1}{R}\right)\operatorname{grad}_0\psi\left(\mathbf{r}_0,t - \frac{R}{c}\right)$$

The second term is not much more difficult as long as we watch our δ's and (δ')'s. We have

$$\begin{aligned}
\int_0^{t^+}\psi\operatorname{grad}_0&\left[\frac{\delta(t_0 - t + R/c)}{R}\right]dt_0 \\
&= \int_0^{t^+}\psi\frac{\partial}{\partial R}\left[\frac{\delta(t_0 - t + R/c)}{R}\right]\operatorname{grad}_0 R\,dt_0 \\
&= \int_0^{t^+}\psi(\mathbf{r}_0,t_0)(\mathbf{R}/R^3)\left[-\delta\left(t_0 - t + \frac{R}{c}\right) + \frac{R}{c}\delta'\left(t_0 - t + \frac{R}{c}\right)\right]dt_0 \\
&= -\frac{\mathbf{R}}{R^3}\left\{\psi\left(\mathbf{r}_0, t - \frac{R}{c}\right) + \frac{R}{c}\left[\frac{\partial}{\partial t_0}\psi(\mathbf{r}_0,t_0)\right]_{t_0=t-(R/c)}\right\}
\end{aligned}$$

Consequently the potential at a point $\mathbf{r}$ and time t, inside a surface S containing no sources and having null initial values internally, is given entirely by the integral of surface values on S:

$$\psi(\mathbf{r},t) = \frac{1}{4\pi} \oint d\mathbf{S}_0 \cdot \left[\left(\frac{1}{R}\right) \operatorname{grad}_0 \psi(\mathbf{r}_0,t_0) + \left(\frac{\mathbf{R}}{R^3}\right) \psi(\mathbf{r}_0,t_0) - (\mathbf{R}/cR^2) \frac{\partial}{\partial t_0} \psi(r_0,t_0) \right]_{t_0=t-(R/c)} \quad (7.3.21)$$

If now part of the surface S_0 is along a wave front and the rest is at infinity or where ψ is zero, we can say that the field value ψ at (r,t) is "caused" by the field ψ in the wave front at a time $t - (R/c)$ earlier. From another point of view, the effect of the wave front on the field in front of it at a later time is equivalent to that of a distribution of sources over the surface of the wave front: an ordinary surface charge proportional to the gradient of ψ normal to the wave front, a double layer proportional to ψ itself, and a curious single sheet, proportional to the time rate of change of ψ on the surface, which is strongest straight ahead of the surface but drops off proportional to the cosine of the angle between the normal to the surface and the direction of propagation (*i.e.*, the direction of $\mathbf{R}$).

In most cases (except for simple and trivial ones) the exact values of ψ, grad ψ, and $\partial\psi/\partial t$ along a whole wave front are not known exactly. But in a great many cases of interest these quantities are approximately known, so that Eq. (7.3.21) may be used to compute approximate values of ψ at a later time. This will be discussed in detail in Sec. 11.4.

Boundaries in the Finite Region. We turn next to the effects of introducing boundaries upon which Green's functions must satisfy specific boundary conditions. The techniques which may be employed here are very similar to those which we discussed in the steady-state case in the preceding Sec. 7.2. As in that case, there are two methods: the method of images and the method of eigenfunctions. Let us discuss the image method first, in which we shall utilize our knowledge of the Green's function for the infinite domain. The only singularity which occurs is naturally at the source point at the time the impulse is initiated, so that generally

$$G(\mathbf{r},t|\mathbf{r}_0,t_0) = \frac{\delta[t_0 - t - (R/c)]}{R} + F(\mathbf{r},t|\mathbf{r}_0,t_0)$$

where F is a solution of the homogeneous wave equation which is free of singularities in the region under discussion.

A simple example is useful here. In Fig. 7.12 the source is at Q; it is initiated at t_0. An infinite rigid plane is at $x = 0$. To satisfy the boundary conditions on the plane we start an image pulse at I, the image of Q at the same time t_0. To obtain the proper cancellation at $x = 0$, the effects of these must be added so that

$$G(\mathbf{r},t|\mathbf{r}_0,t_0) = \frac{\delta[t_0 - t + (R/c)]}{R} + \frac{\delta[t_0 - t + (R'/c)]}{R'}$$

It may be easily verified that $(\partial G/\partial x)_{x=0}$ is zero for all time t. The effect of the second term is to give a reflection which occurs at $x = 0$ at the correct time. This is the only reflection which takes place, and thus only one term in addition to the Green's function for the infinite domain is required. We note that in the case of the wave equation it is necessary to specify not only the position of the image but also the time t' at which the image pulse is started. Fortunately, in most problems this has a simple solution, in that all the images are started at the same time t' as the source pulse itself. For sufficiently regular geometries, it is possible to employ the image method much as in the case of the Helmholtz equation.

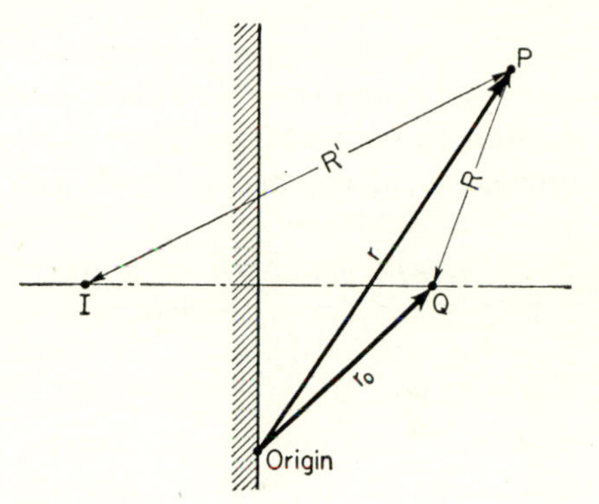

Fig. 7.12 Image at I, of impulsive wave source at Q.

Eigenfunction Expansions. We utilize the Green's function for the Helmholtz equation. The expression $G_k(\mathbf{r}|\mathbf{r}_0)e^{-i\omega t}$ is a solution of the wave equation with a simple-harmonic point source located at $\mathbf{r}_0$. We have, for $\omega = kc$,

$$\nabla^2[G_k e^{-i\omega(t-t_0)}] - \frac{1}{c^2}\frac{\partial^2}{\partial t^2}[G_k e^{-i\omega(t-t_0)}] = -4\pi\delta(\mathbf{r} - \mathbf{r}_0)e^{-i\omega(t-t_0)}$$

By properly superposing these simple harmonic solutions it is possible to obtain the Green's function for a pulse at a point in space, corresponding to Eq. (7.3.2). For this purpose we employ the integral representation of the δ function:

$$\delta(t - t_0) = \left(\frac{1}{2\pi}\right)\int_{-\infty}^{\infty} e^{-i\omega(t-t_0)}\,d\omega$$

From linearity we expect that the Green's function for the pulse is related to the Helmholtz equation solution by the equation

$$G(\mathbf{r},t|\mathbf{r}_0,t_0) = \frac{1}{2\pi}\int_{-\infty}^{\infty} G(\mathbf{r}|\mathbf{r}_0|k)e^{-i\omega(t-t_0)}\,d\omega \qquad (7.3.22)$$

where $\omega = kc$ and $G(\mathbf{r}|\mathbf{r}_0|k) = G_k(\mathbf{r}|\mathbf{r}_0)$. This relation will be derived in a more painstaking manner in Sec. 11.1. The simplicity of Eq. (7.3.22) is misleading. It may be recalled that for finite regions G_k has singularities whenever $k = k_n$, where k_n is an eigenvalue of the scalar Helmholtz equation for a solution ψ_n satisfying the boundary condition satisfied by G_k. More explicitly, if the ψ_n's are normalized, then from (7.2.39),

$$G_k = 4\pi \sum_n \frac{\bar{\psi}_n(\mathbf{r}_0)\psi_n(\mathbf{r})}{k_n^2 - k^2}$$

Thus integration (7.3.22) cannot proceed along the real axis of ω (or k) but must avoid these singularities in some fashion. *The contour choice is dictated by the postulate of causality* discussed earlier. To see this let us introduce the expansion for G_k into integral (7.3.22):

$$G(\mathbf{r},t|\mathbf{r}_0,t_0) = 2c^2 \sum_n \bar{\psi}_n(\mathbf{r}_0)\psi_n(\mathbf{r}) \int_{-\infty}^{\infty} \frac{e^{-i\omega(t-t_0)}}{\omega_n^2 - \omega^2}\,d\omega; \quad \omega_n = ck_n$$

The contour must be chosen so that $G(\mathbf{r},t|\mathbf{r}_0,t_0) = 0$ when $t < t_0$. The appropriate contour is shown in Fig. 7.13. It is parallel to the real axis and just *above* it. When $t > t_0$, the contour may be closed in the lower half plane by a semicircle of large ($\to \infty$) radius without changing the value of the integral. We may now employ the Cauchy integral formula (4.2.9) to obtain $(2\pi/\omega_n) \sin[\omega_n(t - t_0)]$. When $t < t_0$, the contour may be closed by a semicircle in the upper half plane. Since there are no poles in the upper half plane, the value of the integral for $t < t_0$ is zero. Hence

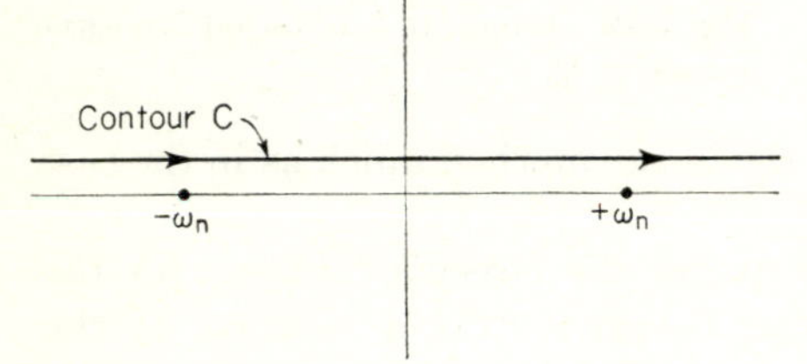

Fig. 7.13 Contour for integral of Eq. (7.3.22).

$$G(\mathbf{r},t|\mathbf{r}_0,t_0) = 4\pi c^2 \sum_n \frac{\sin[\omega_n(t - t_0)]}{\omega_n}\, u(t - t_0)\bar{\psi}_n(\mathbf{r}_0)\psi_n(\mathbf{r}) \quad (7.3.23)$$

where $u(t - t_0)$ is the unit function vanishing for $t < t_0$ and equaling unity for $t > t_0$.

Employing (7.3.23) we may now obtain an explicit evaluation of the initial- and boundary-value problem as given in (7.3.5). Let us consider each term in (7.3.5) separately. The first term ψ_1 gives the effect of sources distributed throughout the volume.

$$\begin{aligned}\psi_1 &= \frac{1}{4\pi}\int_0^{t^+} dt_0 \int dV_0\, G(\mathbf{r},t|\mathbf{r}_0,t_0)q(\mathbf{r}_0,t_0) \\ &= c^2 \sum_n \frac{1}{\omega_n}\int_0^{t^+} dt_0 \int dV_0 \sin[\omega_n(t - t_0)]\, \bar{\psi}_n(\mathbf{r}_0)\psi_n(\mathbf{r})q(\mathbf{r}_0,t_0)\end{aligned}$$

Let $$\phi_n(\mathbf{r},t) = \psi_n(\mathbf{r})e^{-i\omega_n t}$$

Then $$\psi_1 = -c^2 \operatorname{Im}\left\{\sum_n \frac{\phi_n(\mathbf{r},t)}{\omega_n}\int_0^{t^+} dt_0 \int dV_0\, q(\mathbf{r}_0,t_0)\bar{\phi}_n(\mathbf{r}_0,t_0)\right\} \qquad (7.3.24)$$

We see that the amplitude of excitation of the nth mode is proportional to the multiple integral in (7.3.24). The excited amplitude is large when the space dependence of q is very much like that of ϕ_n and if the time dependence is close to $e^{-i\omega_n t}$, a result which agrees with expectations. In the case of exact resonance $q \simeq e^{-i\omega_n t}$ we note that ψ_1 increases linearly with time and no longer oscillates.

The second term in Eq. (7.3.5) gives the effect of sources distributed on the boundary. The results are rather similar to those obtained from the first term. The third term involves the satisfaction of initial conditions. We require $(\partial G/\partial t_0)_{t_0=0}$ and $G_{t_0=0}$ before proceeding.

$$G(\mathbf{r},t|\mathbf{r}_0,0) = 4\pi c^2 \sum_n \frac{\sin(\omega_n t)}{\omega_n} u(t)\bar{\psi}_n(\mathbf{r})\psi_n(\mathbf{r}_0) \qquad (7.3.25)$$

$$\left[\frac{\partial}{\partial t_0} G(\mathbf{r},t|\mathbf{r}_0,t_0)\right]_{t_0=0} = -4\pi c^2 \sum_n \cos(\omega_n t)\, u(t)\bar{\psi}_n(\mathbf{r})\psi_n(\mathbf{r}_0) \qquad (7.3.26)$$

where we have placed $[\sin(\omega_n t)/\omega_n]\delta(t) = 0$. The third term ψ_3 in (7.3.5) becomes

$$\psi_3 = \sum_n \left\{\frac{\sin \omega_n t}{\omega_n} u(t)\psi_n(\mathbf{r}) \int \bar{\psi}_n(\mathbf{r}_0)v_0(\mathbf{r}_0)\, dV_0 + \cos \omega_n t\, \psi_n(\mathbf{r}) \int \bar{\psi}_n(\mathbf{r})\psi_0(\mathbf{r}_0)\, dV_0\right\} \qquad (7.3.27)$$

We may verify at once that $\psi_3(t = 0)$ is just ψ_0 and $(\partial\psi_3/\partial t)_{t=0} = v_0$ as it should. This also shows that we could have obtained Eq. (7.3.27) directly without going through the intermediary of the Green's function. On the other hand it verifies the validity of fundamental formula (7.3.5) for the case of finite regions.

Transient Motion of Circular Membrane. An example will be useful at this point to illustrate the sort of results one obtains for a time-dependent problem. A circular membrane of radius a under a tension T and of mass σ per unit area is given an initial displacement in a small region about its center. The edges of the membrane are fixed, so that the boundary conditions are $\psi(r) = 0$ at $r = a$. We shall represent the initial conditions by means of a δ function:

$$\psi_0(\mathbf{r}) = A\,\delta(\mathbf{r}); \quad v_0(r) = 0 \qquad (7.3.28)$$

Here A is a constant. Introducing (7.3.28) into (7.3.27), the solution of the initial-value problem yields

$$\psi(r,t) = A \sum_n \cos(\omega_n t)\, \psi_n(\mathbf{r})\bar{\psi}_n(0) \tag{7.3.29}$$

To proceed further it is necessary to obtain the eigenfunction $\psi_n(\mathbf{r})$. This will be discussed in great detail in Sec. 11.2. For the present we note that the Helmholtz equation separates in polar coordinates (r,ϕ) and that a general solution, which is finite and single-valued for $r < a$, is a sum of terms $e^{\pm im\phi} J_m(kr)$, where J_m is the Bessel function of the first kind, of order m.

It is now necessary to introduce the boundary conditions at $r = a$. This leads to an equation determining k:

$$J_m(ka) = 0 \tag{7.3.30}$$

Let the values of k satisfying this equation be k_{mp}, the subscript m indicating the order of the Bessel function, the letter p indicating a particular root of (7.3.30). For the purposes of this illustration it will suffice to employ the asymptotic form for $J_m(ka)$ [Eq. (5.3.68)]:

$$J_m(ka) \xrightarrow[ka\to\infty]{} \sqrt{\frac{2}{\pi ka}} \cos\left[ka - \frac{2m+1}{4}\pi\right]$$

Thus $J_m(ka)$ is zero whenever the argument of the cosine is an odd number of $\pi/2$:

$$k_{mp}a \simeq \tfrac{1}{4}(2m+1)\pi + \tfrac{1}{2}(2p+1)\pi; \quad p \text{ integer}$$

We may now return to expression (7.3.29), giving the response of the membrane to an impulse at $r = 0$, $t = 0$. The functions ψ_n are thus

$$\psi_n = N_{mp} J_m(k_{mp}r) e^{\pm im\phi}$$

where to each n we associate a particular couple (m,p) and a particular sign of the exponential. The factor N_{mp} is chosen so that

$$\int \bar{\psi}_n \psi_n \, dA = 1$$

where the area of integration is the membrane. For (7.3.29) we require $J_m(0)$. Since $J_m(z) \xrightarrow[z\to 0]{} 0(z^m)$, we see that $J_m(0) = \delta_{0m}$ [see Eq. (5.3.63)]. Hence the summation (7.3.29) reduces to a sum over zero-order Bessel functions. The absence of any angular dependence is not surprising in view of the circular symmetry of the initiating pulse (7.3.28). The response at any subsequent time t and at a position r is given by

$$\psi(r,t) = A \sum_p \cos(k_{0p}ct) N_{0p}^2 J_0(k_{0p}r) \tag{7.3.31}$$

Equation (7.3.31) is exact. Note that the set $N_{0p}\cos(k_{0p}ct)\,J_0(k_{0p}r)$ describes the free radial vibrations of the membrane. Generally the response to an initial impulse may be expressed in terms of a superposition of free vibrations, each mode vibrating with its own frequency. This is to be contrasted to the response of the system to a steady driving force of a given frequency. In that case, the response has the same frequency as the driving force and the space dependence involves a superposition of the $\psi_n(r)$'s, all of them vibrating with the frequency of the driving force.

Let us consider the response back at the starting point, $r = 0$. Then (7.3.31) becomes

$$\psi(0,t) = A \sum_p \cos(k_{0p}ct)\,N_{0p}^2$$

We introduce the approximate value of the zeros:

$$\psi(0,t) \simeq A \sum_p \cos\left[\left(\frac{2p+3}{4}\right)\frac{\pi ct}{a}\right] N_{0p}^2 \qquad (7.3.32)$$

When will the original pulse refocus at $r = 0$? On first consideration we might think this would occur when $t = 2a/c$, the time for the pulse to go to the edge of the membrane and back. This is, however, not the case. As may be seen from the asymptotic behavior of $J_0(z) \simeq \sqrt{2/\pi z}\cdot\cdot\cos(z - \frac{1}{4}\pi)$, a phase change of $\pi/4$ occurs in passing from the region $r \simeq 0$ to $r \simeq a$. This is characteristic of propagation in two dimensions. No such phase change occurs in either one or three dimensions.

Because of this phase shift, it is necessary for two traversals from the center out to the edge to occur before a final phase shift of π occurs and the pulse is refocused. Hence we may expect that, when $ct = 4a$, the pulse will reform itself at $r = 0$. This may be readily verified by substitution in (7.3.32) for $\psi(0,4a) \simeq -A \sum_p N_{0p}^2$. (The initial pulse $\psi(0,0)$ is $A\Sigma N_{0p}^2$.) We should like to emphasize again that this phenomenon occurs only in two dimensions; in one and three dimensions it does not occur. The pulse from the center of a sphere of radius a re-forms at the center at a time $t = 2a/c$.

There is one final point which also shows the striking difference in wave propagation in two as compared with one or three dimensions. In the latter the initial pulse re-forms exactly at the proper time. In two dimensions this is not so because of the wake developed as the wave progresses. This may be seen in the present instance as follows. Expression (7.3.32) is approximate, for the approximate values of the roots of the Bessel function J_0 were utilized. If the precise values of the roots had been employed, there would have been no value of ct at which the phase $(k_{0p}ct)$ would be exactly the same for all p. In other words, there would be no value of ct for which the free vibration initiated by the pulse would have all returned to their initial phase. Thus the free vibrations would never interfere in the proper fashion to re-form the initial situation exactly.

As another example of the construction of Green's function for the scalar wave equation let us derive expression (7.3.8), the infinite space Green's function, by direct utilization of the superposition method. In that case

$$G_k(\mathbf{r}|\mathbf{r}_0) = e^{ikR}/R; \quad k = \omega/c$$

so that

$$g(R,\tau) = \frac{1}{2\pi R}\int_{-\infty}^{\infty} e^{i(kR-\omega\tau)}\, d\omega = \frac{1}{2\pi R}\int_{-\infty}^{\infty} e^{i\omega[(R/c)-\tau]}\, d\omega$$

It should be noted that we have carefully chosen the relative sign between the factor kR and ωt to be such that $e^{i(kR-\omega\tau)}/R$ represents a wave *diverging* from the source as time progresses, *i.e.*, as τ increases. This is the manner in which we satisfy the causality principle. We now make use of the integral representation for the δ function, Eq. (7.3.22), to obtain

$$g(R,\tau) = \delta[(R/c) - \tau]/R$$

Klein-Gordon Equation. The Green's function for the time-dependent Klein-Gordon equation satisfies the equation

$$\nabla^2 G - \frac{1}{c^2}\frac{\partial^2 G}{\partial t^2} - \kappa^2 G = -4\pi\delta(t - t_0)\delta(\mathbf{r} - \mathbf{r}_0) \tag{7.3.33}$$

It is easy to verify that the Green's function for the Klein-Gordon equation may be employed in much the same way as the Green's function for the scalar wave equation. For example, the reciprocity condition (7.3.3) and the general solution (7.3.5) apply as well here. There are, however, important physical differences between the two. These may be best illustrated by considering the Klein-Gordon Green's function for the infinite domain, thus obtaining the analogue of Eq. (7.3.8). The function $g(\mathbf{r},t|\mathbf{r}_0,t_0)$ may be obtained by superposition of the solutions obtained for a simple harmonic time dependence $e^{-i\omega(t-t_0)}$ rather than the impulsive one given by $\delta(t - t_0)$. The necessary superposition is given by Eq. (7.3.22). The individual solutions may be then given by $g(R|\sqrt{\omega^2 - c^2k^2})$, where

$$[\nabla^2 + (\omega/c)^2 - \kappa^2]\, g[R|\sqrt{\omega^2 - (c\kappa)^2}] = -4\pi\delta(\mathbf{r} - \mathbf{r}_0)$$

The solution of this equation is

$$g = \frac{\exp\,[i\sqrt{(\omega/c)^2 - \kappa^2}\,R]}{R}; \quad R = |\mathbf{r} - \mathbf{r}_0| \tag{7.3.34}$$

In the limit $\omega/c \gg \kappa$ Eq. (7.3.34) becomes $g = e^{i(\omega/c)R}/R$ as it should. For the opposite case $\omega/c \ll \kappa$

$$g \xrightarrow[(\omega/c)\ll\kappa]{} e^{-\kappa R}/R$$

giving a characteristically "damped" space dependence. This is, of course, not related to any dissipation. From the one-dimensional mechanical analogue (Chap. 2, pages 138 *et seq.*), a string embedded in an elastic medium, we see that it is a consequence of the stiffness of the medium.

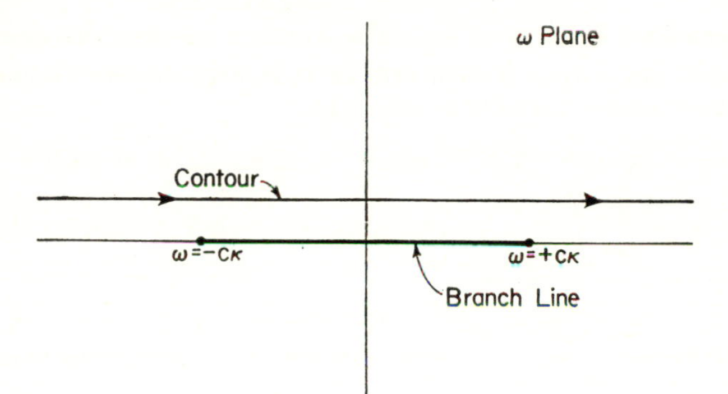

Fig. 7.14 Contour for integral of Eq. (7.3.37) for $R > ct$.

Employing (7.3.22) we may now write as the solution of (7.3.33) valid for an infinite medium

$$g(\mathbf{r},t|\mathbf{r}_0,t_0) = \frac{1}{2\pi R}\int_{-\infty}^{\infty} \exp i[\sqrt{(\omega/c)^2 - \kappa^2}\,R - \omega\tau]\,d\omega \tag{7.3.35}$$

where $\tau = t - t_0$. Function g is a function of R and τ only, as expected. We must now specify the path of integration. Before doing so it is convenient, for convergence questions, to introduce the function $h(R,\tau)$ such that

$$\partial h(R,\tau)/\partial R = Rg(\mathbf{r},t|\mathbf{r}_0,t_0) \tag{7.3.36}$$

Hence

$$h(R,\tau) = \frac{1}{2\pi i}\int_{-\infty}^{\infty} \frac{\exp i[\sqrt{(\omega/c)^2 - \kappa^2}\,R - \omega\tau]}{\sqrt{(\omega/c)^2 - \kappa^2}}\,d\omega \tag{7.3.37}$$

The integrand has branch points at $\omega = \pm c\kappa$. The relation of the path of integration relative to these branch points is determined by the causality condition. We choose the path and branch line shown in Fig. 7.14. First note that $h = 0$ if $R > c\tau$, as the causality postulate would demand for this case. In the limit of large ω the exponent in (7.3.37) approaches $i\omega[(R/c) - \tau] = i\omega|(R/c) - \tau|$. The path of integration may then be closed in the upper half of the ω plane without changing the value of the integral. Since the integrand has no singularities in the upper half plane, the integral is zero.

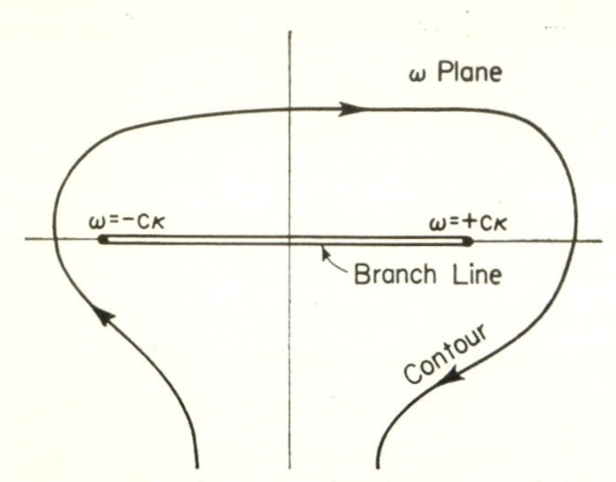

Fig. 7.15 Contour for integral of Eq. (7.3.37) for $R < ct$.

Now consider h for $R < c\tau$. The contour is then deformed to the one shown in Fig. 7.15. It may now be reduced to a more familiar form. We introduce a new variable ϑ, such that

$$\tau = |\sqrt{\tau^2 - (R/c)^2}| \cosh \vartheta \quad \text{and let} \quad \omega = c\kappa \cosh x$$

Then $h(R,\tau) = \dfrac{c}{2\pi i} \displaystyle\int_{\infty + \frac{3}{2}\pi i}^{\infty - \frac{1}{2}\pi i} \exp\left[-i\kappa c|\sqrt{\tau^2 - (R/c)^2}| \cosh (x - \vartheta)\right] dx$

Finally let $(x - \vartheta) = i\xi$; then

$$h(R,\tau) = \frac{c}{2\pi} \int_{+\frac{3}{2}\pi - i\infty}^{-\frac{1}{2}\pi - i\infty} \exp\left[-i\kappa c|\sqrt{\tau^2 - (R/c)^2}| \cos \xi\right] d\xi$$

This is just the integral representation of the Bessel function of zero order [see Eq. (5.3.65)] so that

$$h(R,\tau) = -cJ_0[\kappa c \sqrt{\tau^2 - (R/c)^2}]; \quad R < c\tau$$

Combining this with the expression for $c\tau < R$ we finally obtain

$$h(R,\tau) = -cJ_0[\kappa c \sqrt{\tau^2 - (R/c)^2}]\, u[\tau - (R/c)]$$

The Green's function g is then

$$g(R,\tau) = \frac{1}{R}\frac{\partial h}{\partial R} = \frac{\delta[\tau - (R/c)]}{R} J_0[\kappa c \sqrt{\tau^2 - (R/c)^2}] - \frac{\kappa}{\sqrt{\tau^2 - (R/c)^2}} J_1[\kappa c \sqrt{\tau^2 - (R/c)^2}]\, u[\tau - (R/c)]$$

or

$$g(R,\tau) = \frac{\delta[\tau - (R/c)]}{R} - \frac{\kappa}{\sqrt{\tau^2 - (R/c)^2}} J_1[\kappa c \sqrt{\tau^2 - (R/c)^2}] \cdot u[\tau - (R/c)] \quad (7.3.38)$$

We observe that in the limit $\kappa \to 0$, $g(R,\tau)$ reduces to the Green's function for the scalar wave equation (7.3.8) as it must.

The effect of a pulse at a distance R from its position and at a time τ later vanishes as long as $R > c\tau$, that is as long as the wave initiated by the pulse has not had sufficient time to reach the observation point R. At $R = c\tau$, the original pulse arrives, diminished in amplitude, as it must be, by the geometrical factor $1/R$. It is then followed by a wake which is given by the second term in (7.3.38). This wake, for large values of the time, will decrease in amplitude by the factor $[\tau^2 - (R/c)^2]^{-\frac{3}{4}}$. We may understand this phenomenon by noting that the phase velocity of a plane wave satisfying the Klein-Gordon equation is a function of ω:

$$v = \frac{\omega}{\sqrt{(\omega/c)^2 - \kappa^2}} \quad \text{or} \quad \frac{v}{c} = \frac{\omega}{\sqrt{\omega^2 - (\kappa c)^2}}$$

Since a pulse is made up of many frequencies combined, it is not surprising that the associated plane waves do not arrive at the observation point in the same relative phases as they had at the start. An equivalent description may be obtained from the mechanical realization of the Klein-Gordon equation, as given in Sec. 2.1.

7.4 *Green's Function for Diffusion*

The diffusion equation differs in many qualitative aspects from the scalar wave equation, and of course, the Green's functions will exhibit these differences. The most important single feature is the asymmetry of the diffusion equation with respect to the time variable. For example, if $\psi(\mathbf{r},t)$ is a solution of the scalar wave equation, so is $\psi(\mathbf{r},-t)$. However, if $\psi(\mathbf{r},t)$ is a solution of the diffusion equation

$$\nabla^2\psi = a^2(\partial\psi/\partial t) \tag{7.4.1}$$

the function $\psi(\mathbf{r},-t)$ is not; it is a solution of a quite different equation

$$\nabla^2\psi(\mathbf{r},-t) = -a^2(\partial\psi/\partial t)$$

Thus the equation carries with it a directionality in time; *i.e.*, it differentiates between past and future. The scalar wave equation and indeed all equations which are applicable to microscopic (*e.g.*, atomic) phenomena are symmetric in time. The directionality in time of the diffusion equation is a consequence of the fact that the field which does the diffusing represents the behavior of some average property of an ensemble of many particles. As can be inferred from the theorems of thermodynamics, irregularities in such averages, which may initially exist, will smooth out as time progresses. Looking to the future, entropy increases; looking to the past, entropy was smaller.

Causality and Reciprocity. As in the case of the scalar wave equation, it is possible to solve the various inhomogeneous problems and the initial-value problem in terms of a Green's function which satisfies homogeneous boundary conditions and a causality condition:

$$G(\mathbf{r},t|\mathbf{r}_0,t_0) = 0 \quad \text{if } t < t_0 \tag{7.4.2}$$

The equation satisfied by G involves an impulsive point source:

$$\nabla^2 G - a^2(\partial G/\partial t) = -4\pi\delta(\mathbf{r} - \mathbf{r}_0)\delta(t - t_0) \tag{7.4.3}$$

To interpret (7.4.3) let G be the temperature of a medium. Then the impulsive point source involves introducing a unit of heat at $\mathbf{r}_0$, at a time t_0. G then gives the temperature at a future time for any other point of the medium and thus describes the manner in which heat diffuses away from its initial position.

The function G satisfies a reciprocity condition where, as in the scalar wave equation, a time reversal is involved by virtue of the time sequence (7.4.2) demanded by causality. We shall show that

$$G(\mathbf{r},t|\mathbf{r}_0,t_0) = G(\mathbf{r}_0,-t_0|\mathbf{r},-t) \tag{7.4.4}$$

The function $G(\mathbf{r}_0,-t_0|\mathbf{r},-t)$ gives the effect at $\mathbf{r}_0$ at a time $-t_0$ of a heat source introduced into the medium at $\mathbf{r}$ at a time $-t$. Since $t_0 < t$, the time sequence is still properly ordered. Another interpretation may be obtained by considering the *adjoint* function $\tilde{G}(\mathbf{r},t|\mathbf{r}_0,t_0)$ defined by the relation

$$G(\mathbf{r},-t|\mathbf{r}_0,-t_0) = \tilde{G}(\mathbf{r},t|\mathbf{r}_0,t_0) \tag{7.4.5}$$

$\tilde{G}$ satisfies the time-reversed equation.

$$\nabla^2\tilde{G} + a^2(\partial\tilde{G}/\partial t) = -4\pi\delta(\mathbf{r} - \mathbf{r}_0)\delta(t - t_0)$$

Condition (7.4.2) is replaced by $\tilde{G}(\mathbf{r},t|\mathbf{r}_0,t_0) = 0$ if $t > t_0$. In other words $\tilde{G}$ gives the development backward in time of a source placed at $\mathbf{r}_0$ at a time t_0. The reciprocity condition now reads

$$G(\mathbf{r},t|\mathbf{r}_0,t_0) = \tilde{G}(\mathbf{r}_0,t_0|\mathbf{r},t) \tag{7.4.6}$$

Function G describes development as time increases, leading from the initial source to the final distribution. Function $\tilde{G}$ describes the same process in reverse time order, beginning with the final distribution and going backward in time to the initial source. The question of functions and their adjoints will be discussed later in this chapter.

The proof of (7.4.4) or (7.4.6) follows the pattern developed in the preceding section. The two equations to be considered are

$$\nabla^2 G(\mathbf{r},t|\mathbf{r}_0,t_0) - a^2\frac{\partial}{\partial t}G(\mathbf{r},t|\mathbf{r}_0,t_0) = -4\pi\delta(\mathbf{r} - \mathbf{r}_0)\delta(t - t_0)$$

$$\nabla^2 G(\mathbf{r},-t|\mathbf{r}_1,-t_1) + a^2\frac{\partial}{\partial t}G(\mathbf{r},-t|\mathbf{r}_1,-t_1) = -4\pi\delta(\mathbf{r} - \mathbf{r}_1)\delta(t - t_1)$$

Multiply the first of these by $G(\mathbf{r},-t|\mathbf{r}_1,-t_1)$ and the second by $G(\mathbf{r},t|\mathbf{r}_0,t_0)$, subtract, and integrate over the region of interest and over t from $-\infty$ to t_0^+. Then using Green's theorem one obtains

$$\int_{-\infty}^{t_0^+} dt \int \Big\{G(\mathbf{r},-t|\mathbf{r}_1,-t_1)\ \mathrm{grad}[G(\mathbf{r},t|\mathbf{r}_0,t_0] - G(\mathbf{r},t|\mathbf{r}_0,t_0)\ \mathrm{grad}[G(\mathbf{r}\ -t|\mathbf{r}_1,-t_1]\Big\} \cdot d\mathbf{S}$$
$$-a^2 \int dV \int_0^{t_0^+} \Big\{G(\mathbf{r},-t|\mathbf{r}_1,-t_1)\ \frac{\partial}{\partial t}\ [G(\mathbf{r},t|\mathbf{r}_0,t_0)] + G(\mathbf{r},t|\mathbf{r}_0,t_0)\ \frac{\partial}{\partial t}\ [G(\mathbf{r},-t|\mathbf{r}_1,-t_1)]\Big\}\ dt$$
$$= 4\pi[G(\mathbf{r}_1,t_1|\mathbf{r}_0,t_0) - G(\mathbf{r}_0,-t_0|\mathbf{r}_1,-t_1)]$$

The first of the integrals vanishes by virtue of the homogeneous boundary conditions satisfied by G. In the second, the time integration may be performed to obtain

$$[G(\mathbf{r},-t|\mathbf{r}_1,-t_1)G(\mathbf{r},t|\mathbf{r}_0,t_0)]_{t=-\infty}^{t=t_0^+}$$

At the lower limit the second of the two factors vanishes because of (7.4.2). At the upper limit the first factor vanishes again because of (7.4.2), and it is recognized that we have tacitly assumed in all of this that t_1 is within the region of integration.

The reciprocity condition now follows immediately. We may also obtain the equations satisfied by G and $\tilde{G}$ as functions of t_0. For example, from (7.4.6)

$$\begin{aligned} \nabla_0^2 G + a^2(\partial G/\partial t_0) &= -4\pi\delta(\mathbf{r} - \mathbf{r}_0)\delta(t - t_0) \\ \nabla_0^2 \tilde{G} - a^2(\partial \tilde{G}/\partial t_0) &= -4\pi\delta(\mathbf{r} - \mathbf{r}_0)\delta(t - t_0) \end{aligned} \tag{7.4.7}$$

Inhomogeneous Boundary Conditions. We shall now obtain the solution of the inhomogeneous diffusion equation, with inhomogeneous boundary conditions and given initial conditions, in terms of G. The equation to be solved is

$$\nabla_0^2\psi - a^2(\partial\psi/\partial t_0) = -4\pi\rho(\mathbf{r}_0,t_0) \tag{7.4.8}$$

where ρ, the source function, is a known function of the space and time coordinates. Multiply this equation by G and the first of Eqs. (7.4.7) by ψ; subtract the two equations; integrate over space and over time from 0 to t^+:

$$\int_0^{t^+} dt_0 \int dV_0\,[\psi\nabla_0^2 G - G\nabla_0^2\psi] + a^2 \int dV_0 \int_0^{t^+} dt_0 \left[\psi\left(\frac{\partial G}{\partial t_0}\right) + G\left(\frac{\partial \psi}{\partial t_0}\right)\right]$$
$$= 4\pi \int_0^{t^+} dt_0 \int dV_0 \rho G - 4\pi\psi(\mathbf{r},t)$$

We may apply Green's theorem to the first of these integrals. In the case of the second, the time integration may be performed. Note that $G(\mathbf{r},t|\mathbf{r}_0,t^+) = 0$. Finally

$$\psi(\mathbf{r},t) = \int_0^{t^+} dt_0 \int dV_0\, \rho(\mathbf{r}_0,t_0)G(\mathbf{r},t|\mathbf{r}_0,t_0)$$
$$+ \frac{1}{4\pi}\int_0^{t^+} dt_0 \int d\mathbf{S}_0 \cdot [G\,\mathrm{grad}_0\,\psi - \psi\,\mathrm{grad}_0\,G] + \frac{a^2}{4\pi}\int dV_0[\psi G]_{t_0=0} \quad (7.4.9)$$

G is chosen so as to satisfy homogeneous boundary conditions corresponding to the boundary conditions satisfied by ψ. For example, if ψ satisfies homogeneous or inhomogeneous Dirichlet condition, G is chosen to satisfy homogeneous Dirichlet conditions. The first two terms of (7.4.9) represent the familiar effects of volume sources and boundary conditions, while the third term includes the effects of the initial value ψ_0 of ψ. If the initial value of $\partial\psi/\partial t$ should be given, it is necessary to consider the equation satisfied by $\partial\psi/\partial t$ rather than that satisfied by ψ. Let $v = \partial\psi/\partial t$. Then from (7.4.8) we obtain

$$\nabla^2 v - a^2(\partial v/\partial t) = -4\pi(\partial\rho/\partial t)$$

an equation of the same form as (7.4.8) and to which the same analysis may be applied. As a consequence, either type of initial condition may be discussed by means of (7.4.9). As we saw in Chap. 6, we should not specify *both* initial value and slope for the diffusion equation.

Green's Function for Infinite Domain. We now go on to construct specific examples of Green's functions for this case. As usual the Green's function $g(R,\tau)$, $R = |\mathbf{r} - \mathbf{r}_0|$, $\tau = t - t_0$ for the infinite medium is the first to be discussed. It is possible to derive the expression for one, two, or three dimensions simultaneously. Let g be a one-, two-, or three-dimensional Fourier integral:

$$g(R,\tau) = \frac{1}{(2\pi)^n}\int e^{i\mathbf{p}\cdot\mathbf{R}}\gamma(\mathbf{p},\tau)\,dV_p$$

where n is 1, 2, or 3 depending on the number of dimensions and the dimensionality of the integration variable dV_p is the same. Since

$$\nabla^2 g - a^2\left(\frac{\partial g}{\partial\tau}\right) = \frac{1}{(2\pi)^n}\int e^{i\mathbf{p}\cdot\mathbf{R}}\left[-p^2\gamma - a^2\left(\frac{\partial\gamma}{\partial\tau}\right)\right]dV_p$$

and
$$\delta(\mathbf{R}) = \frac{1}{(2\pi)^n}\int e^{i\mathbf{p}\cdot\mathbf{R}}\,dV_p$$

we finally obtain an equation for γ:

$$a^2(d\gamma/d\tau) + p^2\gamma = 4\pi\delta(\tau)$$

with a solution
$$\gamma = (4\pi/a^2)e^{-(p^2/a^2)\tau}u(\tau)$$

where we have picked that solution which conforms with the causality requirement. Hence

$$g(R,\tau) = \left[\frac{4\pi}{(2\pi)^n a^2}\right] u(\tau) \int e^{i\mathbf{p}\cdot\mathbf{R}} e^{-(p^2/a^2)\tau}\, dV_p$$

or
$$g(R,\tau) = \left[\frac{4\pi}{(2\pi)^n a^2}\right] u(\tau) \left[\int_{-\infty}^{\infty} e^{ip_xR_x} e^{-(p_x^2/a^2)\tau}\, dp_x\right] \cdots$$

By rearranging the exponent in the integral it becomes possible to evaluate each term of the product exactly. In the first term

$$ip_xR_x - \left(\frac{p_x^2}{a^2}\right)\tau = -\left(\frac{p_x\sqrt{\tau}}{a} - \frac{iaR_x}{2\sqrt{\tau}}\right)^2 - \left(\frac{a^2R_x^2}{4\tau}\right) = -\frac{\tau}{a^2}\xi^2 - \left(\frac{a^2R_x^2}{4\tau}\right)$$

Hence the integral may be written

$$\int_{-\infty}^{\infty} e^{-(\tau\xi^2/a^2)-(a^2R_x/4\tau)}\, dp_x; \quad \xi = p_x - \frac{ia^2R_x}{2\tau}$$

and by an appropriate change of variable this equals

$$e^{-(a^2R_x^2/4\tau)} \int_{-\infty}^{\infty} e^{-(\tau\xi^2/a^2)}\, d\xi = a\sqrt{\frac{\pi}{\tau}}\, e^{-(a^2R_x^2/4\tau)}$$

Introducing this result into the expression for g we obtain

$$g(R,\tau) = \frac{4\pi}{a^2}\left(\frac{a}{2\sqrt{\pi\tau}}\right)^n e^{-(a^2R^2/4\tau)}u(\tau) \tag{7.4.10}$$

The function g satisfies an important integral property which is valid for all values of n:

$$\int g(R,\tau)\, dV = \frac{4\pi}{a^2}; \quad \tau > 0 \tag{7.4.11}$$

This equation is an expression of the conservation of heat energy. At a time t_0 and at $\mathbf{r}_0$, a source of heat is introduced. The heat diffuses out through the medium, but in such a fashion that the total heat energy is unchanged.

The function $g(R,\tau)$ is plotted in Fig. 7.16 for one dimension, $n = 1$, for several values of τ. We note that the curve has a sharp maximum at $R = 0$, that the width of the curve increases with increasing τ. The quantity $\sqrt{(4\tau/a^2)}$ is a measure of this width. At $\tau = 0$ there is zero width, since the heat has just been added and is all concentrated at $R = 0$. Since relation (7.4.11) still holds, we see that $g(R,\tau) \xrightarrow[\tau\to 0]{} (4\pi/a^2)\delta(R)$, which is often employed as an example of a δ function. As τ changes from zero, the temperature immediately rises everywhere, the most pronounced rise occurring, of course, near $R = 0$, that is, for $R < \sqrt{(4\tau/a^2)}$.

We shall later consider cases wherein inertia is properly accounted for so that there is a finite velocity of propagation.

Let us now introduce (7.4.10) into Eq. (7.4.9) giving $\psi(\mathbf{r},t)$ in terms of the Green's function and initial value, source distribution, and boundary conditions. Consider the initial-value problem. Let ρ be zero and the

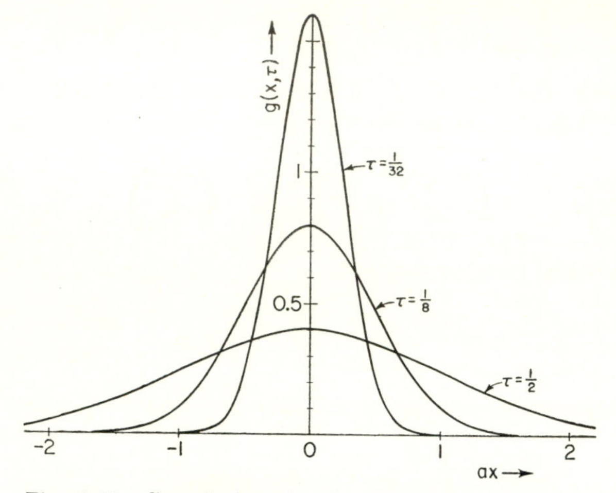

Fig. 7.16 Green's function for diffusion equation for one dimension, for unit source introduced at $x = 0$, $t = 0$, as function of x for different times t.

volume of interest be the infinite domain, so that in the second term G and its derivatives are zero. Then

$$\psi(\mathbf{r},t) = \frac{a^2}{4\pi} \int dV_0 \psi_0(r_0) g(R,t) \tag{7.4.12}$$

In one dimension, this reduces to

$$\psi(x,t) = \frac{a}{2\sqrt{\pi t}} \int_{-\infty}^{\infty} e^{-a^2(x-x_0)^2/4t} \psi_0(x_0)\, dx_0 \tag{7.4.12a}$$

Equation (7.4.12) has a simple interpretation in terms of initial source density. At $t_0 = 0$, each element of space was given an amount of heat equal to $\psi_0(x_0)\, dx_0$ for the one-dimensional case. The effect of this source at a later time is obtained by multiplying by the function $(a/2\sqrt{\pi t})\, e^{-a^2(x-x_0)^2/4t}$, which describes how this heat diffuses away from x_0. The final ψ at x is obtained by a linear superposition of the effects at x of each of the sources at different x_0's.

Finite Boundaries. The proper G for bounded domains may be obtained by the method of images or by the method of eigenfunctions. The method of images is completely similar to that already described for the scalar Helmholtz equation. As a simple case consider a semi-infinite bar starting at $x = 0$ and extending to $x = \infty$. We assume that the temperature ψ is a function of x and t only. The temperature at the end of the bar is varied with a simple harmonic time dependence

$$\psi(0,t) = T_0 \cos \omega t$$

To employ Eq. (7.4.8) we require a Green's function whose value at the end of the bar is 0. This is provided by the method of images:

$$G(x,t|x_0,t_0) = \frac{4\pi}{a^2}\left(\frac{a}{2\sqrt{\pi(t-t_0)}}\right)[e^{-a^2(x-x_0)^2/4(t-t_0)} - e^{-a^2(x+x_0)^2/4(t-t_0)}] \tag{7.4.13}$$

Green's function (7.4.13) is now introduced into (7.4.8). In order to obtain a steady-state solution we shift the "beginning of time" to $t = -\infty$ at which time initial values are zero. In the absence of sources,

$$\psi(x,t) = \frac{1}{4\pi}\int_{-\infty}^{t^+} dt_0 \psi(0,t_0)\left(\frac{\partial G}{\partial x_0}\right)_{x_0=0} dt_0$$

or

$$\psi(x,t) = \frac{xaT_0}{2\sqrt{\pi}}\int_{-\infty}^{t} \cos(\omega t_0)\, e^{-a^2x^2/4(t-t_0)}\left[\frac{dt_0}{(t-t_0)^{\frac{3}{2}}}\right]$$

A more convenient integration variable is given by

$$\xi^2 = [a^2x^2/4(t-t_0)]$$

Then

$$\begin{aligned}
\psi(x,t) &= \frac{2T_0}{\sqrt{\pi}}\int_0^\infty d\xi\, e^{-\xi^2}\cos\omega\left(t - \frac{a^2x^2}{4\xi^2}\right) \\
&= \frac{2T_0}{\sqrt{\pi}}\,\mathrm{Re}\left\{\int_0^\infty d\xi\, e^{i\omega t - \xi^2 - i\omega(a^2x^2/4\xi^2)}\right\} \\
&= \frac{2T_0}{\sqrt{\pi}}\,\mathrm{Re}\left\{e^{i\omega t - \sqrt{i\omega}\,ax}\int_0^\infty d\xi\, e^{-[\xi - (\sqrt{i\omega}\,ax/2\xi)]^2}\right\}
\end{aligned}$$

We can show that this integral is a constant, independent of x. We set

$$J(\alpha) = \int_0^\infty d\xi\, e^{-[\xi - (\alpha^2/\xi)]^2}$$

The integral J may be written in another form by the substitution $\eta = \alpha^2/\xi$:

$$J(\alpha) = \int_0^\infty \left(\frac{\alpha^2}{\eta^2}\right) d\eta\, e^{-[\eta - (\alpha^2/\eta)]^2}$$

Hence

$$0 = \int_0^\infty d\xi\left(1 - \frac{\alpha^2}{\xi^2}\right) e^{-[\xi - (\alpha^2/\xi)]^2}$$

By differentiating the first form for J we find

$$\begin{aligned}
J'(\alpha) &= 2\int_0^\infty d\xi\left[\xi - \left(\frac{\alpha^2}{\xi}\right)\right]\left(\frac{2\alpha}{\xi}\right) e^{-[\xi - (\alpha^2/\xi)]^2} \\
&= 4\alpha\int_0^\infty d\xi\left[1 - \left(\frac{\alpha^2}{\xi^2}\right)\right] e^{-[\xi - (\alpha^2/\xi)]^2} = 0
\end{aligned}$$

Therefore J is independent of α and is equal to its value for $\alpha = 0$, which is just $\frac{1}{2}\sqrt{\pi}$. Hence

$$\psi(x,t) = \mathrm{Re}[T_0\, e^{i\omega t - \sqrt{i\omega}\, ax}] = T_0\, e^{-\sqrt{(\omega/2)}ax} \cos[\omega t - \sqrt{(\omega/2)}\, ax] \quad (7.4.14)$$

This obviously satisfies the boundary conditions at $x = 0$. The function $\psi(x,t)$ represents a temperature wave moving with a velocity $\sqrt{(2\omega/a)}$ but attenuating as x increases. The wave velocity is, as we see, dependent on the frequency of the temperature variation, being greater the higher the frequency and rising to infinity as ω is increased indefinitely, an apparent paradox which we shall discuss shortly.

From the response to a harmonic variation of the boundary value of ψ one may obtain, by Fourier superposition, the response to any sort of time dependence of the boundary value.

Let
$$\psi(0,t) = T(t) = \frac{1}{2\pi}\,\mathrm{Re}\left\{\int_{-\infty}^{\infty} T(\omega)\, e^{-i\omega t}\, d\omega\right\}$$

Then
$$\psi(x,t) = \frac{1}{2\pi}\,\mathrm{Re}\left\{\int_{-\infty}^{\infty} T(\omega)\, e^{i\omega t - \sqrt{i\omega}\, ax}\, d\omega\right\} \quad (7.4.15)$$

This expression may be employed in place of (7.4.9) for the semi-infinite one-dimensional case. The technique of obtaining the Fourier transforms is discussed in Sec. 11.1, and applications of the technique to diffusion problems are given in Sec. 12.1.

Eigenfunction Solutions. The Green's function may also be expanded in terms of eigenfunctions. Let u_n be a solution of a scalar Helmholtz equation in a region bounded by a surface S, upon which u_n satisfies homogeneous boundary conditions. Then

$$\nabla^2 u_n + k_n^2 u_n = 0 \quad \text{and} \quad \int \bar{u}_n u_m\, dV = \delta_{nm}$$

Since the functions u_n form a complete set, we may expand G in terms of them; the coefficients of the expansion will be time-dependent of course. Let

$$G(\mathbf{r},t|\mathbf{r}_0,t_0) = \sum_n C_n(t,t_0) u_n(\mathbf{r}) \bar{u}_n(\mathbf{r}_0)$$

Introducing this expansion into Eq. (7.4.3) satisfied by G and noting that $\Sigma u_n(\mathbf{r})\bar{u}_n(\mathbf{r}_0) = \delta(\mathbf{r} - \mathbf{r}_0)$, one obtains a simple first-order differential equation for C_n:

$$a^2\left(\frac{dC_n}{dt}\right) + k_n^2 C_n = 4\pi\delta(t - t_0)$$

or
$$C_n = (4\pi/a^2)\, e^{-(k_n^2/a^2)(t-t_0)} u(t - t_0)$$

Hence
$$G(\mathbf{r},t|\mathbf{r}_0,t_0) = \left(\frac{4\pi}{a^2}\right) u(t - t_0) \sum_n e^{-(k_n^2/a^2)(t-t_0)} u_n(\mathbf{r}) \bar{u}_n(\mathbf{r}_0) \quad (7.4.16)$$

This expression may now be utilized in Eq. (7.4.9) to solve problems involving source distributions, inhomogeneous boundary conditions, and initial values. For example, suppose that there are no sources, and suppose that the boundary conditions are homogeneous. Choose eigenfunctions u_n which satisfy the same homogeneous boundary conditions. Then from (7.4.9)

$$\psi(\mathbf{r},t) = \frac{a^2}{4\pi}\int dV_0\psi_0 G_{t_0=0}$$

$$\psi(\mathbf{r},t) = \sum_n e^{-(k_n^2/a^2)t}u_n(\mathbf{r})\int \bar{u}_n(\mathbf{r}_0)\psi_0(\mathbf{r}_0)\,dV_0;\quad \text{for } t \geq 0 \qquad (7.4.17)$$

At $t = 0$, (7.4.17) reduces to an expansion of ψ_0 in terms of u_n. Indeed expression (7.4.17) is easily derived when the usual eigenfunction expansion is required to satisfy initial conditions. Let

$$\psi(\mathbf{r},t) = \sum_n A_n e^{-(k_n^2/a^2)t}u_n(\mathbf{r})$$

Then placing $t = 0$, one obtains

$$\psi_0(\mathbf{r}) = \sum_n u_n(\mathbf{r})A_n$$

from which expansion (7.4.17) follows.

Maximum Velocity of Heat Transmission. As we have mentioned in Sec. 2.4, the diffusion equation governing the transmission of heat in a gas is an approximation to the rather complicated motion of the gas molecules. One of the immediately obvious shortcomings of the diffusion approximation is its prediction that the temperature of a body will rise instantaneously everywhere (though not equally) if heat is introduced at some point in the body. For example, the point source function $G(R,\tau)$, Eq. (7.4.10), differs from zero for all values of R just as soon as τ is greater than zero. As such instantaneous propagation of heat is impossible, we must assume that the diffusion equation is correct only after a sufficiently long time has elapsed. This time depends naturally upon the velocity of propagation of the heat, which in turn depends upon the mean free path λ of the gas molecules. The velocity of propagation of a disturbance in a gas is, of course, the velocity of sound, c. Once the time required for the temperature to get to the point in question is exceeded, we may presume that then the diffusion equations apply. The partial differential equation which includes this effect is

$$\nabla^2\psi = a^2(\partial\psi/\partial t) + (1/c^2)(\partial^2\psi/\partial t^2) \qquad (7.4.18)$$

We may also arrive at this equation from another point of view, by considering the change in the sound wave equation due to absorption. We shall encounter this equation when we consider the effects of resistive losses in the vibration of a string and also the propagation of electromagnetic waves in conducting media.

An appropriate Green's function for (7.4.18) satisfies

$$\nabla^2 G - a^2(\partial G/\partial t) - (1/c^2)(\partial^2 G/\partial t^2) = -4\pi\delta(\mathbf{r} - \mathbf{r}_0)\delta(t - t_0) \qquad (7.4.19)$$

We shall again assume the causality principle. The reciprocity condition is

$$G(\mathbf{r},t|\mathbf{r}_0,t_0) = G(\mathbf{r}_0,-t_0|\mathbf{r},-t)$$

The analogue of Eq. (7.4.9) is

$$\psi(\mathbf{r},t) = \int_0^{t^+} dt_0 \int dV_0(\rho G) + \frac{a^2}{4\pi}\int dV_0[\psi G]_{t_0=0} + \frac{1}{4\pi}\int_0^{t^+} dt_0 \oint d\mathbf{S}_0 \cdot [G \operatorname{grad}_0 \psi - \psi \operatorname{grad}_0 G] + \frac{1}{4\pi c^2}\int dV_0 \left[G\frac{\partial\psi}{\partial t_0} - \psi\frac{\partial G}{\partial t_0}\right]_{t_0=0} \quad (7.4.20)$$

We shall now develop the Green's function $g(R,\tau)$, appropriate for Eq. (7.4.20), for the infinite, unbounded domain. Let

$$g(R,\tau) = \int dV_p e^{i\mathbf{p}\cdot\mathbf{R}}\gamma(\mathbf{p},\tau) \quad (7.4.21)$$

Substituting this expression into (7.4.19), we obtain a differential equation determining $\gamma(\mathbf{p},\tau)$:

$$\left(\frac{1}{c^2}\right)\frac{d^2\gamma}{d\tau^2} + a^2\frac{d\gamma}{d\tau} + p^2\gamma = \frac{4\pi}{(2\pi)^n}\,\delta(\tau) \quad (7.4.22)$$

where $n = 1, 2, 3$ according to the number of spatial dimensions involved. Note that γ is a function of p^2 and τ only, so that the angular integrations over the direction of p required by (7.4.21) can be immediately performed. Particular paths of integration are chosen so that the required boundary conditions are satisfied:

$$\begin{aligned} g(R,\tau) &= \frac{2\pi}{iR}\int_{-\infty}^{\infty} e^{ipR}\gamma(p,\tau)p\,dp; && n = 3 \\ &= \pi\int_{-\infty}^{\infty} H_0(pR)\gamma(p,\tau)p\,dp; && n = 2 \\ &= \int_{-\infty}^{\infty} e^{ipR}\gamma(p,\tau)\,dp; && n = 1 \end{aligned} \quad (7.4.23)$$

We note that the three-dimensional Green's function and the one-dimensional Green's function are connected. Let $g_3(R,\tau)$ be the three-dimensional Green's function and g_1 the one-dimensional case. Then

$$g_3(R,\tau) = -\frac{1}{2\pi R}\frac{\partial}{\partial R}[g_1(R,\tau)] \quad (7.4.24)$$

where we have taken into account the difference in the γ's as given by the right-hand side of Eq. (7.4.22).

To determine γ, we must first consider the solutions of the homogeneous form of Eq. (7.4.22). These are

$$e^{-i\omega^+ t} \quad\text{and}\quad e^{-i\omega^- t}$$

where ω^+ and ω^- are solutions of $\omega^2 + i\omega a^2c^2 - p^2c^2 = 0$:

$$\omega^+ = \tfrac{1}{2}[-ia^2c^2 + \sqrt{4p^2c^2 - a^4c^4}]; \quad \omega^- = \tfrac{1}{2}[-ia^2c^2 - \sqrt{4p^2c^2 - a^4c^4}]$$

The proper linear combination of these satisfying the condition that $\gamma(\tau) = 0$ for $\tau < 0$ is

$$\gamma(\tau) = \frac{4\pi c^2 i}{(2\pi)^n} \left\{ \frac{e^{-i\omega^+\tau} - e^{-i\omega^-\tau}}{\omega^+ - \omega^-} \right\} u(\tau)$$

The integral determining g_1 becomes

$$g_1(R,\tau) = 2c^2 iu(\tau) \int_{-\infty}^{\infty} \frac{e^{-i\omega^+\tau} - e^{-i\omega^-\tau}}{\omega^+ - \omega^-} e^{ipR}\, dp$$

where as contour we take a line in the upper half plane of p parallel to the real axis of p. The branch line is chosen to run from $p = a^2c/2$ to $p = -(a^2c/2)$ along the real axis. Let us just consider the integral involving $e^{-i\omega^+\tau}$:

$$\left(\frac{1}{2c}\right) e^{-\frac{1}{2}a^2c^2\tau} \int_C \frac{\exp\{i[pR - \sqrt{p^2 - (a^4c^2/4)}\, c\tau]\}}{\sqrt{p^2 - \frac{1}{4}(a^4c^2)}}\, dp$$

When $R > c\tau$, the contour may be closed by a semicircle in the upper half plane. The integral is then zero, since there are no singularities within the contour. When $R < c\tau$, the contour is deformed so as to extend along the negative imaginary axis. We must then evaluate an integral which is very similar to the integral involved in the calculation of the Green's function for the Klein-Gordon equation. We obtain (see table of Laplace transforms at the end of Chap. 11)

$$-\left(\frac{\pi i}{c}\right) e^{-\frac{1}{2}a^2c^2\tau} J_0[\tfrac{1}{2}a^2c\sqrt{R^2 - c^2\tau^2}]u(c\tau - R)$$

Consider now the contribution coming from the integral involving $e^{-i\omega^-\tau}$:

$$\frac{1}{2c} e^{-\frac{1}{2}a^2c^2\tau} \int_C \frac{\exp[ipR + i\sqrt{p^2 - (a^4c^2/4)}\, c\tau]}{\sqrt{p^2 - \frac{1}{4}(a^4c^2)}}\, dp$$

This integral is zero when $(R + c\tau) > 0$ (recall that in one dimension R can be negative) but is not zero when $(R + c\tau) < 0$. Then we obtain

$$-\left(\frac{\pi i}{c}\right) e^{-\frac{1}{2}a^2c^2\tau} J_0[\tfrac{1}{2}a^2c\sqrt{R^2 - c^2\tau^2}][1 - u(R + c\tau)]$$

Combining these two expressions yields

$$g_1(R,\tau) = 2\pi c\, e^{-\frac{1}{2}a^2c^2\tau} J_0\left[\frac{a^2c}{2}\sqrt{R^2 - c^2\tau^2}\right] u(c\tau - |R|) \quad (7.4.25)$$

The reader should be able to verify that this result tends to the correct limiting forms given by Eqs. (7.4.10) and (7.3.16) as $c \to \infty$ or as $a \to 0$, respectively.

We may now obtain the three-dimensional g from the differential equation (7.4.24):

$$g_3(R,\tau) = \frac{c}{R} e^{-\frac{1}{2}a^2c^2\tau} \left\{ \delta(c\tau - R) + \frac{a^2cR}{2\sqrt{R^2 - c^2\tau^2}} J_1[\tfrac{1}{2}a^2c\sqrt{R^2 - c^2\tau^2}]u(c\tau - R) \right\} \tag{7.4.26}$$

We shall obtain the Green's function for two-dimensional problems by integrating $g_3(r,\tau)$ over the z component of R rather than by direct consideration of (7.4.23). Let $R^2 = \xi^2 + \rho^2$. Then

$$g_2(R,\tau) = \int_{-\infty}^{\infty} d\xi \, g_3(R,\tau)$$

or

$$g_2(R,\tau) = \frac{2c\, e^{-\frac{1}{2}a^2c^2\tau}}{\sqrt{c^2\tau^2 - \rho^2}} u(c\tau - \rho) \left\{ 1 + 2\sinh^2\left[\frac{a^2c}{4}\sqrt{c^2\tau^2 - \rho^2}\right] \right\} \tag{7.4.27}$$

Here we have employed the formula

$$\int_0^\pi I_1(2z\sin\theta)\, d\theta = \frac{2\sinh^2 z}{z}$$

where $I_1(x) = -iJ_1(ix)$ (see tables at end of Chaps. 10 and 11).

The three-dimensional case exhibits the physical phenomena which occur as a consequence of the inclusion of a velocity of propagation into the diffusion equation or the inclusion of a dissipative term into the wave equation. Both terms in (7.4.26) vanish when $R > c\tau$, as is to be expected whenever effects propagate with a finite velocity. The first term is a reproduction of the initial pulse, reduced, however, by two factors. The first, $1/R$, is the geometrical factor which appeared in the solution of the simple wave equation. The second is the factor $e^{-\frac{1}{2}a^2c^2\tau}$ which tells us that this part of the wave, generated by the point source, decays with time as it moves through the medium. The second term in (7.4.26) constitutes the wake. For sufficiently long times $c\tau \gg R$, it is the term which yields the usual diffusion approximation.

These differences may be exhibited in another fashion. Let us solve the one-dimensional initial-value problem. From (7.4.20) we find

$$\psi = \frac{a^2}{4\pi}\int dx_0[\psi g_1]_{t_0=0} + \frac{1}{4\pi c^2}\int dx_0 \left[g_1 \frac{\partial\psi}{\partial t_0} - \psi\frac{\partial g_1}{\partial t_0}\right]_{t_0=0}$$

Then

$$\begin{aligned}\psi = {} & \tfrac{1}{2}e^{-\frac{1}{2}a^2c^2t}[\psi_0(x + ct) + \psi_0(x - ct)] \\ & + e^{-\frac{1}{2}a^2c^2t}\int_{x-ct}^{x+ct}\left\{\frac{a^2c}{4} I_0[\tfrac{1}{2}a^2c\sqrt{c^2t^2 - (x_0 - x)^2}] \right. \\ & \left. + \frac{1}{2c}\frac{\partial}{\partial t} I_0[\tfrac{1}{2}a^2c\sqrt{c^2t - (x_0 - x)^2}]\right\}\psi_0(x_0)\, dx_0 \\ & + \frac{1}{2c}e^{-\frac{1}{2}a^2c^2t}\int_{x-ct}^{x+ct} I_0[\tfrac{1}{2}a^2c\sqrt{c^2t^2 - (x_0 - x)^2}]v_0(x_0)\, dx_0\end{aligned} \tag{7.4.28}$$

where $\psi_0(x_0)$ and $v_0(x_0)$ are the initial values of ψ and $\partial\psi/\partial t$, respectively. Note that this becomes d'Alembert's solution, Eq. (7.3.18), if $a \to 0$. The first term is the same as that given by d'Alembert except for the decay in time as given by $e^{-\frac{1}{2}a^2c^2t}$. The second term is new and represents the effect of diffusion. The third term reduces to the d'Alembert term when $a \to 0$.

7.5 *Green's Function in Abstract Operator Form*

So far our discussion has been limited to a particular type of partial differential equation. The space operator has been ∇^2, and the time operators have been absent or $\partial/\partial t$ or $\partial^2/\partial t^2$ in the Helmholtz, diffusion, and wave equations, respectively. In the present section we shall generalize these considerations so that they apply to any operator, permitting the application of the theory to any of the equations of physics as long as they are linear. Our plan will be to emphasize the important elements in the previous discussion and then to see how they are most appropriately generalized.

It is natural that we shall have to be somewhat abstract. For example, instead of writing out a specific form for the homogeneous equation to be considered, we shall state it in operator form:

$$\mathfrak{A}\psi = 0 \tag{7.5.1}$$

where $\mathfrak{A}$ operates on the coordinates giving the dependence of ψ. For example, in the diffusion equation $\mathfrak{A} = \nabla^2 - (a^2)(\partial/\partial t)$ and is a function of $\mathbf{r}$ and t. Another linear type, the integral equation, is mentioned in Chap. 2 (see page 180) and will be discussed more fully in Chap. 8. Such an operator $\mathfrak{A}$ is

$$\mathfrak{A} = 1 - \int^b K(x,x_0) \cdot \cdot \cdot \, dx_0$$

and the equation $\mathfrak{A}\psi = 0$ reads

$$\psi(x) - \int_a^b K(x,x_0)\psi(x_0)\, dx_0 = 0$$

The variables may include more than just space and time dependence. In transport problems (Sec. 2.4) the distribution function f depends not only on $\mathbf{r}$ and t but also on the momentum $\mathbf{p}$ and energy E.

In the same notation, the equation for the Green's function G is

$$\mathfrak{A}G(\mathbf{x}|\mathbf{x}_0) = -4\pi\delta(\mathbf{x} - \mathbf{x}_0) \tag{7.5.2}$$

where $\mathbf{x}$ is a generalized vector representing all the independent variables which are relevant; $\mathfrak{A}$ operates on $\mathbf{x}$. For the wave equation $\mathbf{x} = \mathbf{a}_x x + \mathbf{a}_y y + \mathbf{a}_z z + \mathbf{a}_t t$, where $\mathbf{a}_x$, etc., are unit, mutually orthogonal vectors.

Then $\delta(\mathbf{x} - \mathbf{x}_0)$ becomes the product of the δ functions for each coordinate; *e.g.*, for the wave equation

$$\delta(\mathbf{x} - \mathbf{x}_0) = \delta(x - x_0)\delta(y - y_0)\delta(z - z_0)\delta(t - t_0)$$

Generalization of Green's Theorem, Adjoint Operators. The most important mathematical tool employed in the analysis of the preceding sections was Green's theorem; our first task will be to generalize it. In differential form Green's theorem states that

$$u\nabla^2 v - v\nabla^2 u = \nabla \cdot (u\nabla v - v\nabla u)$$

A generalization of this equation in terms of $\mathfrak{L}$ which immediately suggests itself is

$$u\mathfrak{L}v - v\mathfrak{L}u = \nabla \cdot \mathbf{P}(u,v) \tag{7.5.3}$$

where $\mathbf{P}$ is a generalized vector in terms of the same unit vectors as that describing $\mathbf{x}$ while ∇ is the corresponding gradient operator. Hence $\nabla \cdot \mathbf{P} = (\partial P_x/\partial x) + (\partial P_y/\partial y) + (\partial P_z/\partial z) + (\partial P_t/\partial t) + \cdots$. For example, in the case of the wave equation, $\mathfrak{L} = \nabla^2 - (1/c^2)(\partial^2/\partial t^2)$, we find from Eq. (7.5.3) that

$$u\left[\nabla^2 - \frac{1}{c^2}\frac{\partial^2}{\partial t^2}\right]v - v\left[\nabla^2 - \frac{1}{c^2}\frac{\partial^2}{\partial t^2}\right]u = \frac{\partial}{\partial x}\left[u\frac{\partial v}{\partial x} - v\frac{\partial u}{\partial x}\right] + \frac{\partial}{\partial y}\left[u\frac{\partial v}{\partial y} - v\frac{\partial u}{\partial y}\right] + \frac{\partial}{\partial z}\left[u\frac{\partial v}{\partial z} - v\frac{\partial u}{\partial z}\right] - \frac{1}{c^2}\frac{\partial}{\partial t}\left[u\frac{\partial v}{\partial t} - v\frac{\partial u}{\partial t}\right]$$

Here $\mathbf{P} = u\nabla v - v\nabla u$, where ∇ is the general gradient operator.

Relation (7.5.3) is not satisfied by all operators $\mathfrak{L}$. For example, in the case of the diffusion equation for one dimension, $\mathfrak{L} = (\partial^2/\partial x^2) - a^2(\partial/\partial t)$, we find

$$u\mathfrak{L}v - v\mathfrak{L}u = \frac{\partial}{\partial x}\left[u\frac{\partial v}{\partial x} - v\frac{\partial u}{\partial x}\right] - a^2\left[u\frac{\partial v}{\partial t} - v\frac{\partial u}{\partial t}\right]$$

The first pair of terms on the left side is in the proper form. However, the second pair cannot be written as the time derivative of a function of u and v. We must therefore generalize Green's theorem beyond (7.5.3):

$$u\mathfrak{L}v - v\tilde{\mathfrak{L}}u = \nabla \cdot \mathbf{P}(u,v) \tag{7.5.4}$$

where $\tilde{\mathfrak{L}}$ is an operator which is called the *adjoint* of $\mathfrak{L}$. When $\tilde{\mathfrak{L}} = \mathfrak{L}$, that is, when (7.5.3) applies, the operator $\mathfrak{L}$ is said to be *self-adjoint*. In the case of the diffusion equation, $\tilde{\mathfrak{L}} = (\partial^2/\partial x^2) + a^2(\partial/\partial t)$. Definition (7.5.4) is just a generalization of a definition of the adjoint operator employed in Chap. 5 (see page 526). In Eq. (5.2.10) $\tilde{\mathfrak{L}}$ was defined by

$$v(z)\mathfrak{L}[y(z)] - y(z)\tilde{\mathfrak{L}}[v(z)] = (d/dz)P(v,y)$$

where $P(v,y)$ is the bilinear concomitant. This is the statement for one-dimensional problems of Eq. (7.5.4).

Looking back at the manner in which Green's theorem is utilized, we see that we shall be interested in the solutions of the equation

$$\tilde{\mathfrak{A}}\tilde{\psi} = 0 \tag{7.5.5}$$

and the corresponding Green's function for the adjoint operator

$$\tilde{\mathfrak{A}}\tilde{G}(\mathbf{x}|\mathbf{x}_0) = -4\pi\delta(\mathbf{x} - \mathbf{x}_0) \tag{7.5.6}$$

Equation (7.5.5) is known as the *adjoint* of (7.5.1), the equation involving $\mathfrak{A}$, while $\tilde{\psi}$ is referred to as the adjoint of ψ. In the case of the one-dimensional diffusion equation, $\tilde{\mathfrak{A}}\tilde{\psi} = 0$ reads

$$\frac{\partial^2\tilde{\psi}}{\partial x^2} + a^2\frac{\partial\tilde{\psi}}{\partial t} = 0$$

We see that $\tilde{\psi}$ satisfies the diffusion equation with the time variable reversed. Hence if $\psi(t)$ is a solution of (7.5.1), then $\tilde{\psi}(t) = \psi(-t)$ is a solution of (7.5.5).

Once the generalization of Green's theorem is available, it becomes possible to solve the inhomogeneous problem

$$\mathfrak{A}\psi = -4\pi\rho(\mathbf{x}) \tag{7.5.7}$$

with inhomogeneous boundary conditions. Since the adjoint operator $\tilde{\mathfrak{A}}$ is involved in Green's theorem (7.5.4), it is clear that we must compare (7.5.7) and (7.5.6). Multiply the latter by $\psi(\mathbf{x})$ and the former by $\tilde{G}(\mathbf{x}|\mathbf{x}_0)$, and subtract:

$$\tilde{G}(\mathbf{x}|\mathbf{x}_0)\mathfrak{A}\psi(\mathbf{x}) - \psi(\mathbf{x})\tilde{\mathfrak{A}}\tilde{G}(\mathbf{x}|\mathbf{x}_0) = 4\pi\psi(\mathbf{x})\delta(\mathbf{x} - \mathbf{x}_0) - 4\pi\rho(\mathbf{x})\tilde{G}(\mathbf{x}|\mathbf{x}_0)$$

Employ (7.5.4), and integrate over a volume in $\mathbf{x}$ space (which includes the entire range of each component of $\mathbf{x}$ which is of physical interest). For example, in the case of the wave equation, it includes integration over time from 0 to t_0^+ and an integration over x, y, and z coordinates within the surface upon which boundary conditions are to be satisfied. Then

$$\psi(\mathbf{x}_0) = \int\rho(\mathbf{x})\tilde{G}(\mathbf{x}|\mathbf{x}_0)\,dv + (1/4\pi)\int\nabla\cdot\mathbf{P}[\tilde{G}(\mathbf{x}|\mathbf{x}_0), \psi(\mathbf{x})]\,dv$$

or $$\psi(\mathbf{x}_0) = \int\rho(\mathbf{x})\tilde{G}(\mathbf{x}|\mathbf{x}_0)\,dv + (1/4\pi)\oint\mathbf{n}\cdot\mathbf{P}[\tilde{G}(\mathbf{x}^s|\mathbf{x}_0), \psi(\mathbf{x}^s)]\,dS \tag{7.5.8}$$

where $\mathbf{n}$ is an outward-pointing unit vector orthogonal to the surface S bounding the volume in $\mathbf{x}$-space. In the scalar wave equation this term is

$$\frac{1}{4\pi}\int_0^{t_0^+} dt \int \mathbf{n}\cdot[\tilde{G}\nabla\psi - \psi\nabla\tilde{G}]\,dS - \frac{1}{4\pi c^2}\int dv\left[\tilde{G}\frac{\partial\psi}{\partial t} - \psi\frac{\partial\tilde{G}}{\partial t}\right]_{t=0}^{t=t_0^+}$$

Effect of Boundary Conditions. To proceed any further it becomes necessary to consider the boundary conditions satisfied by ψ. Consider the case in which ψ satisfies homogeneous boundary conditions on S;

that is, there are no sources of the field ψ on the surface S. By the principle of superposition it must then be possible to obtain the solution as an integral over the volume source distribution $\rho(\mathbf{x})$ multiplied by the effect due to a source at $\mathbf{x}$. This involves us in two considerations. In the first place, we must relate G and $\tilde{G}$. As we shall see, this will lead to a generalized reciprocity condition. We shall postpone the proof of this theorem for a short time. In the second place, in order to obtain a solution of the proper form it is necessary for the surface term in (7.5.8) to vanish. The Green's function $\tilde{G}$ and ψ must satisfy homogeneous boundary conditions which must be so correlated that

$$\oint \mathbf{n} \cdot \mathbf{P}\,[\tilde{G}(\mathbf{x}^s|\mathbf{x}_0), \psi(\mathbf{x}^s)]\,dS = 0 \tag{7.5.9}$$

In the simplest case we have considered, the scalar Helmholtz equation, the surface term vanishes if the Green's function and the function ψ satisfy the same homogeneous boundary condition. In the scalar wave equation, we employed initial values for $(\partial\psi/\partial t)$ and ψ, the Cauchy conditions; for the Green's function we employed the causality condition (see page 834).

It is also possible (as we pointed out earlier, in Sec. 7.2) to determine the proper boundary conditions ψ must satisfy. For example, in the Helmholtz equation, the surface term involves the surface values of ψ and $\partial\psi/\partial n$. Placing both of them equal to zero is manifestly improper, for in that event the surface integral automatically vanishes, the boundary condition on G is left arbitrary, and the solution of the inhomogeneous equation becomes nonunique. Since the solution is in fact unique, the initial assumption about the surface values of ψ and $\partial\psi/\partial n$ is incorrect and we are led to relax the boundary conditions to be either homogeneous Dirichlet or homogeneous Neumann or some linear combination of the two. Then in order for the surface term to vanish, G (in the case of the Helmholtz equation) must satisfy the same boundary condition as ψ. In a similar fashion, the examination of Eq. (7.5.9) will lead to a determination of the proper boundary conditions for ψ and the corresponding ones for $\tilde{G}$.

Having determined in an implicit fashion the boundary conditions on ψ and $\tilde{G}(\mathbf{x}|\mathbf{x}_0)$, we may now turn to the reciprocity condition. We compare the equations satisfied by G and $\tilde{G}$:

$$\mathfrak{L}G(\mathbf{x}|\mathbf{x}_0) = -4\pi\delta(\mathbf{x} - \mathbf{x}_0);\quad \tilde{\mathfrak{L}}\tilde{G}(\mathbf{x}|\mathbf{x}_1) = -4\pi\delta(\mathbf{x} - \mathbf{x}_1)$$

Multiply the first of these by $\tilde{G}$ and the second by G, subtract, and integrate over the relevant volume in $\mathbf{x}$ space. Employing the generalized Green's theorem (7.5.4), we obtain

$$4\pi[\tilde{G}(\mathbf{x}_0|\mathbf{x}_1) - G(\mathbf{x}_1|\mathbf{x}_0)] = \int \mathbf{n} \cdot \mathbf{P}[\tilde{G}(\mathbf{x}^s|\mathbf{x}_0), G(\mathbf{x}^s|\mathbf{x}_1)]\,dS$$

In order that the solution of the inhomogeneous source problem with homogeneous boundary conditions be expressible in terms of G and not $\tilde{G}$, it is necessary for a simple algebraic relation between them to exist, which in turn requires the surface term in the above equation to vanish. Comparing this surface term with (7.5.9), we see that $G(\mathbf{x}|\mathbf{x}_0)$ *satisfies the same conditions on* S *as* ψ, a result which may be expected from our intuitive idea of G and its relation to ψ.

We finally obtain

$$G(\mathbf{x}|\mathbf{x}_0) = \tilde{G}(\mathbf{x}_0|\mathbf{x}) \qquad (7.5.10)$$

In words, the left-hand side of the equation describes the effect at $\mathbf{x}$ of a point source at $\mathbf{x}_0$, the propagation being governed by the operator $\mathfrak{A}$ and the boundary conditions. On the right-hand side, the source is placed at $\mathbf{x}$; the effect is measured at $\mathbf{x}_0$, the propagation from $\mathbf{x}$ to $\mathbf{x}_0$ now being governed by $\tilde{\mathfrak{A}}$ and the corresponding boundary condition for $\tilde{G}$. If $\tilde{G}$ is not the same function as G, a directionality in $\mathbf{x}$ space must exist, for reversing the direction of propagation changes the consequent observations. This irreversibility must be apparent in the operator $\mathfrak{A}$ or in the boundary conditions. For example, the operator $\mathfrak{A}$ for the diffusion equation $\nabla^2 - a^2(\partial/\partial t)$ is not invariant against a change in the sense of time, *i.e.*, against the substitution of $-t$ for $+t$. The operator for the wave equation $\mathfrak{A} = \nabla^2 - (1/c^2)(\partial^2/\partial t^2)$ is self-adjoint ($\tilde{\mathfrak{A}} = \mathfrak{A}$), so that a directionality, for example in the time coordinate, cannot arise from it. However, a directionality can perfectly well arise from the boundary conditions imposed. For example, the application of the causality condition imposes a definite asymmetry with respect to past and future. As a consequence, the reciprocity principle for the Green's function for the wave equation for this initial condition reads

$$G(\mathbf{r},t|\mathbf{r}_0,t_0) = G(\mathbf{r}_0,-t_0|\mathbf{r},-t)$$

so that

$$\tilde{G}(\mathbf{r},t|\mathbf{r}_0 t_0) = G(\mathbf{r},-t|\mathbf{r}_0,-t_0) \qquad (7.5.11)$$

We see that $\tilde{G}$ describes the propagation from a source point $\mathbf{r}_0$ to one at $\mathbf{r}$ with, however, the sense of time reversed, so that the event at t occurs at some time *earlier* than the impulse causing it at a time t_0 (note that $t < t_0$). For example, in the case of the Green's function for the infinite domain

$$\tilde{G}(\mathbf{r},t|\mathbf{r}_0,t_0) = (1/R)\delta[(R/c) + (t - t_0)]; \quad R = |\mathbf{r} - \mathbf{r}_0|$$

At a given R, an effect is felt at a time $t = t_0 - R/c$, that is, at a time R/c earlier than the initiation of the motion at t_0. For this reason $\tilde{G}$ is often referred to as the *advanced potential* while $G = (1/R)\delta[R/c - (t - t_0)]$ is the *retarded potential.* Both are solutions of the source problem with differing initial conditions.

Because of the effect of boundary conditions it is useful to generalize the idea of adjoint. We introduce two terms: *adjoint boundary conditions* and *adjoint problem.* An adjoint problem will be satisfied by $\tilde{\psi}$ if $\tilde{\psi}$ is a solution of

$$\tilde{\mathfrak{L}}\tilde{\psi} = 0$$

and if it satisfies adjoint boundary conditions. We shall define the latter by the requirement

$$\mathbf{n} \cdot \mathbf{P}[\psi,\tilde{\psi}] = 0; \quad \text{on the boundary surface} \tag{7.5.12}$$

Hence if ψ satisfies a certain boundary condition, $\tilde{\psi}$ will satisfy a corresponding boundary condition which we shall call the adjoint boundary condition. A problem is considered self-adjoint when $\tilde{\mathfrak{L}} = \mathfrak{L}$ and the boundary conditions for ψ and $\tilde{\psi}$ are the same. For self-adjoint problems $\tilde{G}(\mathbf{x}|\mathbf{x}_0) = G(\mathbf{x}|\mathbf{x}_0)$.

More on Adjoint Differential Operators. Let us now become more definite and consider some operators and their adjoints. As a first example consider the one-dimensional situation. Here we shall generally be interested in second-order operators, so that we may specialize to

$$\mathfrak{L}v = p\frac{d^2v}{dz^2} + q\frac{dv}{dz} + rv \tag{7.5.13}$$

The adjoint is [see Eqs. (5.2.10) *et seq.*]

$$\tilde{\mathfrak{L}}u = \frac{d^2}{dz^2}(pu) - \frac{d}{dz}(qu) + ru \tag{7.5.14}$$

The bilinear concomitant $P(u,v)$

$$P(u,v) = \left[pu\frac{dv}{dz} - v\frac{d(pu)}{dz} + quv\right] \tag{7.5.15}$$

Under what conditions will $\mathfrak{L}$ be self-adjoint? Upon placing $\tilde{\mathfrak{L}} = \mathfrak{L}$ we find that dp/dz must equal q. Under these circumstances,

$$\mathfrak{L}v = \frac{d}{dz}\left(p\frac{dv}{dz}\right) + rv$$

The equation $\mathfrak{L}v = 0$ is just the Sturm-Liouville one discussed in Sec. 6.3. We see that it corresponds to the only linear self-adjoint operator containing at most second-order differential operators.

When $dp/dz = q$, the bilinear concomitant P is

$$P = p\left(u\frac{dv}{dz} - v\frac{du}{dz}\right)$$

The requirement $\mathbf{n}\cdot\mathbf{P}(\psi,\tilde{\psi}) = 0$ is just that $0 = p\left(\psi\frac{d\tilde{\psi}}{dz} - \tilde{\psi}\frac{d\psi}{dz}\right)_{z=a} = p\left(\psi\frac{\partial\tilde{\psi}}{\partial z} - \tilde{\psi}\frac{\partial\psi}{\partial z}\right)_{z=b}$, where b and a are the points at which boundary conditions are satisfied. If p is finite at the end points, then possible boundary conditions are Dirichlet, $\psi = 0$ at a and b; Neumann, $d\psi/dz = 0$ at a and b; or mixed, $\partial\psi/\partial z = \beta\psi$ at a and b. All of these are self-adjoint boundary conditions, for $\tilde{\psi}$ must satisfy the same boundary conditions as ψ. Periodic boundary conditions $\psi(a) = \psi(b)$ and $(d\psi/dz)_{z=a} = (d\psi/dz)_{z=b}$ are also self-adjoint. Another type of boundary condition occurs when p has a zero at either a or b. In that event $\mathbf{P}$ is zero at the point only if the functions ψ and $\tilde{\psi}$ are bounded. Again this boundary condition is self-adjoint. We have, of course, considered these very same conditions in Chap. 6. For all of them, the Green's function must be symmetric.

Expressions (7.5.13) to (7.5.15) may be generalized to include operators involving higher order differentials and more than one dimension. Consider first the operator

$$\mathfrak{A}_n v = p(d^n v/dz^n) \tag{7.5.16}$$

Any one-dimensional operator is, of course, a linear combination of operators of type $\mathfrak{A}_n$. The adjoint is

$$\tilde{\mathfrak{A}}_n u = (-1)^n (d^n/dz^n)(pu) \tag{7.5.17}$$

The bilinear concomitant is

$$P(u,v) = pu\left(\frac{d^{n-1}v}{dz^{n-1}}\right) - \left(\frac{d(pu)}{dz}\right)\left(\frac{d^{n-2}v}{dz^{n-2}}\right) + \left(\frac{d^2(pu)}{dz^2}\right)\left(\frac{d^{n-3}v}{dz^{n-3}}\right) - \cdots (-1)^{n-1}\left(\frac{d^{n-1}(pu)}{dz^{n-1}}\right)v \tag{7.5.18}$$

In several dimensions, the most general differential operator would be of the form

$$\mathfrak{A} = p(x_1,x_2, \ldots)\frac{\partial^n}{\partial x_1^a \partial x_2^b \cdots \partial x_s^k}; \quad a + b + \cdots + k = n \tag{7.5.19}$$

where $x_1,x_2, \ldots, x_s$ are generalized coordinates. The adjoint $\tilde{\mathfrak{A}}$ is

$$\tilde{\mathfrak{A}} = (-1)^n \frac{\partial^n}{\partial x_1^a \partial x_2^b \cdots}[p(x_1,x_2, \ldots)] \tag{7.5.20}$$

The bilinear concomitant $P(u,v)$ is

$$\begin{aligned}\mathbf{P}(u,v) = \mathbf{a}_1 \bigg[pu \left(\frac{\partial^{(n-1)} v}{\partial x_1^{a-1} \partial x_1^b \cdots \partial x_s^k} \right) - \frac{\partial(pu)}{\partial x_1} \left(\frac{\partial^{(n-2)} v}{\partial x_1^{a-2} \partial x_2^n \cdots \partial x_s^k} \right) \\ + \cdots (-)^{a-1} \frac{\partial^{(a-1)}(pu)}{\partial x_1^{a-1}} \left(\frac{\partial^{(n-a)} v}{\partial x_2^b \cdots \partial x_s^k} \right) \bigg] \\ +(-1)^a \mathbf{a}_2 \bigg[\left(\frac{\partial^a (pu)}{\partial x_1^a} \right) \left(\frac{\partial^{(n-a)} v}{\partial x_2^{b-1} \cdots \partial x_s^k} \right) - \left(\frac{\partial^a (pu)}{\partial x_1^a \partial x_2} \right) \left(\frac{\partial^{(n-a-2)} v}{\partial x_2^{b-2} \cdots \partial x_s^k} \right) \\ + \cdots (-)^{b-1} \left(\frac{\partial^{a+b-1}(pu)}{\partial x_1^a \partial x_2^{b-1}} \right) \left(\frac{\partial^{(n-a-b)} v}{\cdots \partial x_s^k} \right) \bigg] \\ + \cdots +(-1)^{n-k} \mathbf{a}_s \bigg[\left(\frac{\partial^{n-s}(pu)}{\partial x_1^a \partial x_2^b \cdots} \right) \left(\frac{\partial^{k-1} v}{\partial x_s^{k-1}} \right) - \left(\frac{\partial^{n-s+1}(pu)}{\partial x_1^a \partial x_2^b \cdots \partial x_s} \right) \left(\frac{\partial^{k-2} v}{\partial x_s^{k-2}} \right) \\ + \cdots (-1)^{k-1} \left(\frac{\partial^{n-1}(pu)}{\partial x_1^a \partial x_2^b \cdots \partial x_s^{k-1}} \right) v \bigg] \qquad (7.5.21)\end{aligned}$$

where $\mathbf{a}_n$ is the unit vector corresponding to the coordinate x_n.

As a simple example, consider the operator

$$\mathfrak{A} v = p \frac{\partial^2 v}{\partial x^2} + q \frac{\partial^2 v}{\partial y^2} + 2r \frac{\partial^2 v}{\partial x \, \partial y} + t \frac{\partial v}{\partial x} + m \frac{\partial v}{\partial y}$$

Then $$\tilde{\mathfrak{A}} u = \frac{\partial^2 (pu)}{\partial x^2} + \frac{\partial^2 (qu)}{\partial y^2} + 2 \frac{\partial^2 (ru)}{\partial x \, \partial y} - \frac{\partial (tu)}{\partial x} - \frac{\partial (mu)}{\partial y}$$

and

$$\begin{aligned}\mathbf{P}(u,v) = \mathbf{a}_x \left[pu \frac{\partial v}{\partial x} - \frac{\partial (pu)}{\partial x} v + 2\, ru \frac{\partial v}{\partial y} + tuv \right] \\ + \mathbf{a}_y \left[qu \frac{\partial v}{\partial y} - \frac{\partial (qu)}{\partial y} v - 2 \frac{\partial (ru)}{\partial x} v + muv \right]\end{aligned}$$

Because of the symmetrical dependence upon x and y of the r term, it is possible to obtain another expression for $P(u,v)$ [which may be obtained from general formula (7.5.21) by writing an alternate expression for P in which $x_1 = y$ and $x_2 = x$ and averaging with the above in which $x_1 = x$ and $x_2 = y$]:

$$\begin{aligned}\mathbf{P}(u,v) = \mathbf{a}_x \left[pu \frac{\partial v}{\partial x} - \frac{\partial (pu)}{\partial x} v + ru \frac{\partial v}{\partial y} - \frac{\partial (ru)}{\partial y} v + tuv \right] \\ + \mathbf{a}_y \left[qu \frac{\partial v}{\partial y} - \frac{\partial (qu)}{\partial y} v + ru \frac{\partial v}{\partial x} - \frac{\partial (ru)}{\partial x} v + muv \right]\end{aligned}$$

The conditions for the operator $\mathfrak{A}$ being self-adjoint are found to be

$$\frac{\partial p}{\partial x} + \frac{\partial r}{\partial y} - t = 0; \quad \frac{\partial q}{\partial y} + \frac{\partial r}{\partial x} - m = 0$$

For consistency of these two equations we must also require that

$$\frac{\partial}{\partial x} \left(t - \frac{\partial p}{\partial x} \right) = \frac{\partial}{\partial y} \left(m - \frac{\partial q}{\partial y} \right)$$

Then $\mathbf{P}(u,v)$ simplifies considerably to

$$\mathbf{P}(u,v) = \mathbf{a}_x\left[p\left(u\frac{\partial v}{\partial x} - v\frac{\partial u}{\partial x}\right) + r\left(u\frac{\partial v}{\partial y} - v\frac{\partial u}{\partial y}\right)\right] + \mathbf{a}_y\left[q\left(u\frac{\partial v}{\partial y} - v\frac{\partial u}{\partial y}\right) + r\left(u\frac{\partial v}{\partial x} - v\frac{\partial u}{\partial x}\right)\right]$$

Adjoint Integral Operators. The definition of the adjoint operator given in (7.5.4) is not particularly appropriate for an integral operator (7.5.1), and an attempt to employ it without modification is not fruitful. We shall employ an integral definition which is in many ways weaker than (7.5.4) but nevertheless will leave most of our results unchanged. The particular choice we are about to make will remove all surface terms involving $\mathbf{P}$ from the result summarized in (7.5.8); Eq. (7.5.10), the reciprocity condition, will still hold. For integral operators we define the adjoint operator $\tilde{\mathfrak{A}}$ by the equation

$$\int_a^b u\mathfrak{A}v\,dx - \int_a^b v\tilde{\mathfrak{A}}u\,dx = 0 \tag{7.5.22}$$

where a and b are the limits of the region of interest.

This definition follows from the one-dimensional form of (7.5.4) by first integrating (7.5.4) from a to b:

$$\int_a^b [u\mathfrak{A}v - v\tilde{\mathfrak{A}}u]\,dx = [\mathbf{P}(u,v)]_{x=a}^{x=b}$$

We see that (7.5.22) follows if $[\mathbf{P}(u,v)]_{x=a}^{x=b}$ is zero. We recall that this occurs for differential operators when the boundary conditions satisfied by u and v are homogeneous and are adjoint to each other as defined by Eq. (7.5.12). We now consider the consequence of definition (7.5.22).

For example let us examine the operator

$$\mathfrak{A}v = \int_a^b K(x,x_0)v(x_0)\,dx_0$$

Then if $\tilde{\mathfrak{A}}$ is defined by

$$\tilde{\mathfrak{A}}u = \int_a^b \tilde{K}(x,x_0)u(x_0)\,dx_0$$

the defining equation (7.5.22) becomes

$$\int_a^b dx \int_a^b dx_0[v(x)K(x,x_0)u(x_0) - u(x)\tilde{K}(x,x_0)v(x_0)] = 0$$

Interchanging the variables of integration in the first integral this equation becomes

$$\int_a^b dx \int_a^b dx_0\{v(x_0)u(x)[K(x_0,x) - \tilde{K}(x,x_0)]\} = 0$$

Since this is to hold for arbitrary u and v we define $\tilde{K}$ as follows:

$$K(x_0,x) = \tilde{K}(x,x_0) \tag{7.5.23}$$

which is reminiscent of (7.5.10) where a similar statement is made for the Green's function. As we shall see in the following chapter, K is very often a Green's function or related to one, so that the correspondence appearing here is not too surprising. From (7.5.23) we may now write

$$\tilde{\mathfrak{A}}v = \int_a^b \tilde{K}(x_0,x)v(x_0)\,dx$$

The condition for $\tilde{\mathfrak{A}} = \mathfrak{A}$, that is, for the operators to be self-adjoint, is

$$K(x_0,x) = K(x,x_0) \tag{7.5.24}$$

that is, the function K is to be symmetric in the two variables x_0 and x.

Another type of integral operator involves an indefinite integral

$$\mathfrak{A}v = \int_a^x K(x,x_0)v(x_0)\,dx_0$$

We may reduce this operator to one involving definite limits by introducing the unit function so that

$$\mathfrak{A}v = \int_a^b u(x - x_0)K(x,x_0)v(x_0)\,dx_0$$

or if we write

$$M(x,x_0) = u(x - x_0)K(x,x_0)$$

Then the adjoint operator $\tilde{\mathfrak{A}}$ involves

$$\tilde{M}(x,x_0) = M(x_0,x) = u(x_0 - x)K(x_0,x)$$

Hence $\quad \tilde{\mathfrak{A}}\omega = \int_a^b u(x_0 - x)K(x_0,x)\omega(x_0)\,dx_0 = \int_x^b K(x_0,x)\omega(x_0)\,dx_0$

For $\mathfrak{A}$ to be self-adjoint $M(x,x_0) = M(x_0,x)$ or

$$u(x - x_0)K(x,x_0) = u(x_0 - x)K(x_0,x)$$

This equation can never be satisfied, since $u(x - x_0)$ is zero whenever $u(x_0 - x)$ is unity and vice versa. Hence $\mathfrak{A}$ for indefinite integrals is not self-adjoint.

Generalization of these definitions to more than one dimension does not involve any new principles and we shall therefore relegate their discussion to the problems at the end of the chapter.

Generalization to Abstract Vector Space. As with the treatment of eigenfunctions, it is both useful and instructive to extend our discussion now to the representation of the results of this chapter in the abstract symbolism of vector space and vector operators. This symbolism was first treated in Sec. 1.6 and was used extensively in Secs. 2.6 and 6.3. There we showed that any vector $\mathbf{F}$ in abstract space could be expressed in terms of its components with respect to a mutually orthogonal set of eigenvectors:

$$\mathbf{F} = \sum_n F_n\mathbf{e}_n; \quad F_n = \mathbf{F}^* \cdot \mathbf{e}_n$$

for a denumerable infinity of eigenvectors $\mathbf{e}_n$; or

$$\mathbf{F} = \int F(x)\mathbf{e}(x)\,dx; \quad F(x) = \mathbf{F}^* \cdot \mathbf{e}(x)$$

for the nondenumerable set $\mathbf{e}(x)$ (incidentally x here stands for one or more coordinates, such as x, y, t, and the integration is understood to be carried out over the range of these variables inside the boundary). Vectors are represented by a one-dimensional array of components, F_n or $F(x)$.

In addition, we dealt with operators $\mathfrak{A}$, generalizations in abstract vector space of the differential operators discussed in the first part of this section. Here the representation must be a two-dimensional matrix of components:

$$\mathfrak{A} = \sum_{mn} \mathbf{e}_m A_{mn} \mathbf{e}_n^* = \int dx \int \mathbf{e}(x) A(x|x_0) \mathbf{e}^*(x_0)\,dx_0 \qquad (7.5.25)$$

where we could have used $\mathbf{r}$, $\mathbf{r}_0$ instead of x, x_0 and dV instead of dx to emphasize that more than one dimension may be involved. If the matrix A_{mn} is a diagonal one ($A_{mn} = a_m\delta_{mn}$), the unit vectors $\mathbf{e}_n$ are then eigenvectors for $\mathfrak{A}$, satisfying the equation $\mathfrak{A} \cdot \mathbf{e}_n = a_n\mathbf{e}_n$; if A_{mn} is not a diagonal matrix, the set $\mathbf{e}_n$ is an eigenvector set for some other operator, not $\mathfrak{A}$.

Corresponding to the operator $\mathfrak{A}$ operating on the vector $\mathbf{F}$ is the differential operator $\mathcal{A}_r$, operating on the function $F(\mathbf{r})$:

$$\mathfrak{A} \cdot \mathbf{F} = \int [\mathcal{A}_r F(\mathbf{r})]\mathbf{e}(\mathbf{r})\,dV = \int dV \int A(\mathbf{r}|\mathbf{r}_0)F(\mathbf{r}_0)\mathbf{e}(\mathbf{r})\,dV_0$$

where $\mathcal{A}_r$ is the differential (or integral) operator discussed in the first part of this section. The second form illustrates the fact that any operator may be expressed in integral form. This is most easily seen, for differential operators, by the use of the delta function and its derivatives. For instance, if $\mathcal{A}_x = g(x)(d/dx) + r(x)$, the matrix for $\mathfrak{A}$ in terms of the x's is

$$A(x|x_0) = -g(x)\delta'(x_0 - x) + r(x)\delta(x_0 - x)$$

for then (see page 837)

$$\mathcal{A}_x F(x) = \int A(x|x_0)F(x_0)\,dx_0 = g(x)(d/dx)F(x) + r(x)F(x) \quad (7.5.26)$$

Higher derivatives may be expressed in terms of higher "derivatives" of the delta function; it is not difficult to see that all operators dealt with so far in this section may be expressed in terms of an equivalent $A(x|x_0)$, involving delta functions and their derivatives, functions of x and x_0, and, occasionally, the step function $u(x_0 - x)$ (see page 840).

To generalize the Green's function procedure we must first find the operator generalization of Green's theorem, then find the generalization of the delta function corresponding to the unit source on the right-hand side of the inhomogeneous equation, and finally find the actual solution of the generalized operator equation.

We have already generalized Green's theorem by Eq. (7.5.22). This at the same time defines what is meant by an adjoint operator and what is needed for adjoint boundary conditions and also is the basis for the reciprocity relation. All these matters must now be translated to vector-operator language.

Adjoint, Conjugate, and Hermitian Operators. The equation which is equivalent to Green's theorem is Eq. (7.5.22). Written in slightly different form it is

$$\int[\bar{u}(x)\mathfrak{A}_x v(x) - v(x)\tilde{\mathfrak{A}}_x \bar{u}(x)]\,dx = 0 \tag{7.5.27}$$

This is somewhat like the equation defining the Hermitian adjoint of an operator [see Eq. (1.6.34) *et supra* and also Eq. (6.3.72)]

$$\mathbf{U}^* \cdot \mathfrak{A} \cdot \mathbf{V} - (\mathfrak{A}^* \cdot \mathbf{U})^* \cdot \mathbf{V} = 0$$

which becomes

$$\int\{\bar{u}(x)\mathfrak{A}_x v(x) - v(x)\overline{[\mathfrak{A}_x^* u(x)]}\}\,dx = 0$$

or

$$\iint dx\{\bar{u}(x)A(x|x_0)v(x_0) - v(x)\overline{A^*(x|x_0)u}(x_0)\}\,dx_0 = 0$$

in terms of components along the eigenvectors $\mathbf{e}(x)$ if $u(x)$ and $v(x)$ are the corresponding components of the vectors $\mathbf{U}$, $\mathbf{V}$. Since the components of a Hermitian adjoint $A^*(x|x_0)$ are the complex conjugates of the components of $\mathfrak{A}$ itself with row and column reversed, $\bar{A}(x_0|x)$, this last integral automatically is zero.

But Eq. (7.5.27) is not exactly this, and we have signified the difference by using the tilde sign $\tilde{\mathfrak{A}}$ instead of the star $\mathfrak{A}^*$. We must go back to the definition of the adjoint differential operator to find the nature of this difference. What becomes apparent then is that, if $A(x|x_0)$ are the components of the operator $\mathfrak{A}$ along the $\mathbf{e}(x)$ axes, then the components of $\tilde{\mathfrak{A}}$ are $A(x_0|x)$ whereas the components of $\mathfrak{A}^*$ are $\bar{A}(x_0|x)$. In one case we reverse the rows and columns (x and x_0), and in the other case we both reverse rows and columns *and also* take the complex conjugate.

To show that the ordinary adjoint of an operator corresponds to reversal of x and x_0 in the $\mathbf{e}(x)$ components, we can take the case given in Eq. (7.5.27) for the differential operator $g(d/dx) + r$. Here the components turned out to be

$$A(x|x_0) = -g(x)\delta'(x_0 - x) + r(x)\delta(x_0 - x)$$

Simple reversal of x and x_0 results in

$$\tilde{A}(x|x_0) = A(x_0|x) = -g(x_0)\delta'(x - x_0) + r(x_0)\delta(x - x_0)$$

which corresponds to the differential operator

$$\tilde{\mathfrak{a}}_x F(x) = \int \tilde{A}(x|x_0)F(x_0)\,dx_0 = -\frac{d}{dx}(gF) + r(x)F(x)$$

which *is* the adjoint operator as defined in Eqs. (7.5.14).

We can now sum up our conventions regarding Hermitian adjoint, adjoint, and conjugate. The components of $\mathbf{F}^*$, the vector *conjugate* to $\mathbf{F}$, are the complex conjugates of the components of $\mathbf{F}$;

$$\mathbf{F} = \sum_n F_n \mathbf{e}_n = \int F(x)\mathbf{e}(x)\,dx; \quad \mathbf{F}^* = \sum_n \mathbf{e}_n^* \bar{F}_n = \int \mathbf{e}^*(x)\bar{F}(x)\,dx \qquad (7.5.28)$$

(we can also call $\mathbf{F}^*$ adjoint to $\mathbf{F}$). The components of $\tilde{\mathfrak{A}}$, the operator *adjoint* to $\mathfrak{A}$, involve the inversion of row and column; the components of $\bar{\mathfrak{A}}$, the *conjugate* to $\mathfrak{A}$, are just the complex conjugates of those of $\mathfrak{A}$, whereas the *Hermitian adjoint* $\mathfrak{A}^*$ has elements which are reversed, rows and columns, and also are complex conjugates:

$$\begin{aligned} \mathfrak{A} &= \sum_{nm} \mathbf{e}_n A_{nm} \mathbf{e}_m^*; \quad \tilde{\mathfrak{A}} = \sum_{nm} \mathbf{e}_n A_{mn} \mathbf{e}_m^* \\ \bar{\mathfrak{A}} &= \sum_{nm} \mathbf{e}_n \bar{A}_{nm} \mathbf{e}_m^*; \quad \mathfrak{A}^* = \tilde{\bar{\mathfrak{A}}} = \sum_{nm} \mathbf{e}_n \bar{A}_{mn} \mathbf{e}_m^* \\ \tilde{\mathfrak{A}} &= \bar{\mathfrak{A}}^*; \quad \bar{\mathfrak{A}} = \tilde{\mathfrak{A}}^* \end{aligned} \qquad (7.5.29)$$

If $\mathfrak{A}$ is Hermitian, $\mathfrak{A}^* = \mathfrak{A}$, but $\tilde{\mathfrak{A}} \neq \mathfrak{A}$ unless all the elements of $\mathfrak{A}$ are real. If $\mathfrak{A}$ is real, $\bar{\mathfrak{A}} = \mathfrak{A}$ but $\tilde{\mathfrak{A}} \neq \mathfrak{A}$ unless $\mathfrak{A}$ is Hermitian, and so on.

We note that the property of being Hermitian is invariant under a rotation of axes in vector space whereas the property of being real (or of being self-adjoint) is not an invariant. For instance, the matrix given by the components of $\mathfrak{A}$ along $\mathbf{a}_1$ and $\mathbf{a}_2$

$$\mathfrak{A} = \begin{pmatrix} 1 & 1+i \\ 1-i & 2 \end{pmatrix}$$

is Hermitian but is not real or self-adjoint. This matrix has eigenvector $\mathbf{e}_1 = \sqrt{\tfrac{2}{3}}\,[\mathbf{a}_1 - \mathbf{a}_2/(1-i)]$ with eigenvalue 0 and also eigenvector $\mathbf{e}_2 = \sqrt{\tfrac{2}{3}}[\mathbf{a}_2 + \mathbf{a}_1/(1-i)]$ with eigenvalue 2. With respect to these new axes $\mathfrak{A}$, of course, is diagonal:

$$\mathfrak{A} = \begin{pmatrix} 0 & 0 \\ 0 & 2 \end{pmatrix}$$

which is still Hermitian but is also self-adjoint and real.

Green's Function and Green's Operator. The analogue of the inhomogeneous equation $\mathfrak{a}\psi(\mathbf{r}) = -4\pi\rho(\mathbf{r})$ is thus the operator equation $\mathfrak{A}\cdot\mathbf{F} = -4\pi\mathbf{P}$, where $\mathbf{F}$ and $\mathbf{P}$ are abstract vectors and $\mathfrak{A}$ is one of the operators we have been describing. What is the analogue of the equation $\mathfrak{A}_r G(\mathbf{r}|\mathbf{r}_0) = -4\pi\delta(\mathbf{r} - \mathbf{r}_0)$ for the Green's function? We have seen earlier that the components of the idemfactor

$$\mathfrak{J} = \int \mathbf{e}(\mathbf{r})\mathbf{e}^*(\mathbf{r})\, dV; \quad \mathfrak{J} \cdot \mathbf{F} = \mathbf{F}; \quad \text{for any } \mathbf{F} \tag{7.5.30}$$

along the $\mathbf{e}(r)$ axes is just the delta function $\delta(\mathbf{r} - \mathbf{r}_0)$, so the analogue of the right-hand side of the Green's function equation could be $-4\pi\mathfrak{J}$.

This would make the Green's function also analogous to an abstract vector *operator* rather than an abstract vector, for if the right-hand side of the equation is to be an operator, so must the left-hand side. The result is not surprising, for if the analogue of the source-density function ρ is the abstract vector **P**, the analogue of G must be an operator which transforms **P** into the solution **F**. Therefore the analogue of the inhomogeneous equation

$$\mathfrak{A}_r G(\mathbf{r}|\mathbf{r}_0) = -4\pi\delta(\mathbf{r} - \mathbf{r}_0)$$

is

$$\mathfrak{A} \cdot \mathfrak{G} = -4\pi\mathfrak{J} \tag{7.5.31}$$

or, in components along $\mathbf{e}(\mathbf{r})$,

$$\int A(\mathbf{r}|\mathbf{r}')G(\mathbf{r}'|\mathbf{r}_0)\, dV' = -4\pi\delta(\mathbf{r} - \mathbf{r}_0)$$

which is equivalent to the differential equation for G. Thus the generalization of the Green's function is the *Green's operator* $\mathfrak{G}$.

Looking at Eq. (7.5.31), we quickly see that the proper form for the Green's operator is

$$\mathfrak{G} = -4\pi\mathfrak{A}^{-1} \tag{7.5.32}$$

In other words, the Green's function is the $\mathbf{e}(x)$ component representation of (-4π) times the inverse of the operator $\mathfrak{A}$, corresponding to the homogeneous equation. The solution of the general inhomogeneous equation $\mathfrak{A} \cdot \mathbf{F} = -4\pi\mathbf{P}$ is then

$$\mathbf{F} = \mathfrak{G} \cdot \mathbf{P} = -4\pi\mathfrak{A}^{-1} \cdot \mathbf{P} \tag{7.5.33}$$

a very simple and obvious answer when expressed in these general terms. The process of multiplying by a Green's function and integrating is equivalent to multiplying by an inverse operator; $\int G(x|x_0)\rho(x_0)\, dx_0$ are the components of vector $-4\pi\mathfrak{A}^{-1} \cdot \mathbf{P}$.

Reciprocity Relation. We have seen that the generalization of Green's theorem led to the condition

$$\int [\bar{u}\mathfrak{A}_x v - v\tilde{\mathfrak{A}}_x \bar{u}]\, dV = 0$$

where u and v satisfy "self-adjoint" boundary conditions on the boundary surface. The operator $\tilde{\mathfrak{A}}$ is the operator adjoint to $\mathfrak{A}$ (if $\mathfrak{A}$ is Hermitian, it is also the complex conjugate of $\mathfrak{A}$). Its Green's function is $\mathfrak{H}$, where $\tilde{\mathfrak{A}} \cdot \mathfrak{H} = -4\pi\mathfrak{J}$. But, by taking the transpose of Eq. (7.5.31), we obtain

$\tilde{\mathfrak{G}} \cdot \tilde{\mathfrak{A}} = -4\pi\mathfrak{I}$; consequently $\mathfrak{H} = \tilde{\mathfrak{G}}$. In other words the Green's operator for the adjoint operator $\tilde{\mathfrak{A}}$ is the adjoint of the Green's operator $\mathfrak{G}$:

$$\tilde{G}(\mathbf{r}|\mathbf{r}_0) = G(\mathbf{r}_0|\mathbf{r}) \tag{7.5.34}$$

which is the generalization of the reciprocity theorem (7.5.10). In the language of vector space this result is somewhat of a tautology. If we are convinced that the generalization of Green's function is an operator, then (7.5.34) is just the definition of adjoint; on the other hand Eq. (7.5.34) might be taken as a further piece of evidence that $\mathfrak{G}$ really is a perfectly well-behaved operator.

We now see why the discussion at the beginning of this section was not complete and could not be complete at that point. We had to understand the concept of "adjointness" for differential (or integral) operators, for Green's functions, and for boundary conditions before we could see that (7.5.27) really could contain all the essentials of the boundary conditions and properties of differential operators to enable us to determine Green's operators. And we had to explore the behavior of actual Green's functions before we could understand their relation to operators.

Expansion of Green's Operator for Hermitian Case. It is sometimes useful to expand operators and vectors in terms of some other set of eigenvectors than the nondenumerably infinite set $\mathbf{e}(r)$. For instance, if the operator $\mathfrak{A}$ were equal to $\mathfrak{L} - \lambda$, where λ is a multiplicative constant, one might be tempted to expand the quantities involved in terms of the eigenvectors $\mathbf{e}_n$ of the equation

$$\mathfrak{L} \cdot \mathbf{e}_n = \lambda_n \mathbf{e}_n; \quad \mathbf{e}_n^* \cdot \mathbf{e}_m = \delta_{nm}$$

where the operator equation has inherent in it the boundary conditions as well as the differential operator. For instance, the idemfactor $\mathfrak{I}$ is $\Sigma \mathbf{e}_n \mathbf{e}_n^*$, and we could express the Green's operator in terms of its matrix components with respect to these axes:

$$\mathfrak{G} = \sum_{m,n} \mathbf{e}_m G_{mn} \mathbf{e}_n^* \tag{7.5.35}$$

If $\mathfrak{L}$ is a Hermitian operator (and if the boundary conditions are Hermitian), the eigenvalues λ_n are all real and the adjoint operator is the conjugate of $\mathfrak{L}$. The transformation functions, coming from the relation between $\mathbf{e}_n$ and $\mathbf{e}(\mathbf{r})$,

$$\mathbf{e}_n = \int \psi_n(\mathbf{r})\mathbf{e}(\mathbf{r})\, dV; \quad \mathbf{e}(\mathbf{r}) = \sum_m \bar{\psi}_m(\mathbf{r})\mathbf{e}_m \tag{7.5.36}$$

are the complex conjugates of the ψ's for the adjoint operator $\tilde{\mathfrak{L}}$. If $\mathfrak{L}$ is self-adjoint, either the ψ's are real functions of $\mathbf{r}(x,y,z,t)$, or if they are complex, their complex conjugates are also part of the set of eigenfunctions, so that, if we wished (and if the boundary conditions allowed), we could make all the ψ's real (for instance, for the angle ϕ we can use $e^{im\phi}$ with m positive or negative or else we can use $\cos m\phi$ and $\sin m\phi$).

To solve the inhomogeneous equation $[\mathfrak{L} - \lambda] \cdot \mathbf{F} = -4\pi\mathbf{P}$, we have first to solve the operator equation

$$[\mathfrak{L} - \lambda] \cdot \mathfrak{G} = -4\pi\mathfrak{I} \tag{7.5.37}$$

Then operating by both sides of the equation on the vector $\mathbf{P}$, corresponding to the density function, we see that the solution of $(\mathfrak{L} - \lambda) \cdot \mathbf{F} = -4\pi\mathbf{P}$ is $\mathbf{F} = \mathfrak{G} \cdot \mathbf{P}$.

We know that $\mathfrak{G}$ is equal to $-4\pi[\mathfrak{L} - \lambda]^{-1}$, but this formal solution is not of much use. It would be more fruitful to obtain an expansion of $\mathfrak{G}$ in terms of the eigenvectors $\mathbf{e}_n$. Setting (7.5.35) into Eq. (7.5.37) we have that the matrix components for $\mathfrak{G}$ are

$$G_{mn} = \frac{4\pi\delta_{mn}}{\lambda - \lambda_m}; \quad \mathfrak{G} = \sum_n \left(\frac{4\pi}{\lambda - \lambda_n}\right) \mathbf{e}_n\mathbf{e}_n^* \tag{7.5.38}$$

The Green's function is the matrix component of $\mathfrak{G}$ in terms of the eigenvectors $\mathbf{e}(r)$:

$$G(\mathbf{r}|\mathbf{r}_0) = \mathbf{e}^*(\mathbf{r}) \cdot \mathfrak{G} \cdot \mathbf{e}(\mathbf{r}_0) = 4\pi \sum_n \frac{\psi_n(\mathbf{r})\bar{\psi}_n(\mathbf{r}_0)}{\lambda - \lambda_n} \tag{7.5.39}$$

The adjoint to G is $G(\mathbf{r}_0|\mathbf{r})$, which is the complex conjugate of $G(\mathbf{r}|\mathbf{r}_0)$. This, of course, is a consequence of the fact that, if $\mathfrak{L}$ is Hermitian, then G is also [see Eq. (7.5.29)]. If $\mathfrak{L}$ is also self-adjoint, so that its elements in this expansion are all real, then G is also self-adjoint and $G(\mathbf{r}|\mathbf{r}_0) = G(\mathbf{r}_0|\mathbf{r})$. In this case all ψ's are real, or if there are complex ψ's, their conjugates are also eigenfunctions, so the sum of (7.5.39) is symmetric.

Therefore, in general, for a Hermitian operator, the Green's function at $\mathbf{r}$ for a source at $\mathbf{r}_0$ is the complex conjugate of the Green's function at $\mathbf{r}_0$ for a source at $\mathbf{r}$, whereas for a self-adjoint operator (with self-adjoint boundary conditions), source and observation point may be interchanged without changing G. This is the final generalization of the reciprocity relation.

Non-Hermitian Operators: Biorthogonal Functions. At times we are forced to consider differential equations or boundary conditions which do not correspond to Hermitian operator equations. In this case the operator $\mathfrak{L}^*$ differs from the operator $\mathfrak{L}$ and the eigenvectors must also differ. We define the two sets by the usual equations

$$\mathfrak{L} \cdot \mathbf{e}_m = \lambda_m\mathbf{e}_m; \quad \mathfrak{L}^* \cdot \mathbf{f}_n = \mu_n\mathbf{f}_n \tag{7.5.40}$$

where $\mathbf{e}_n$ and $\mathbf{f}_m$ have their conjugate vectors $\mathbf{e}_n^*$ and $\mathbf{f}_n^*$, of course.

The eigenvectors λ_m for $\mathfrak{L}$ may, in principle, be obtained if we know the effect of $\mathfrak{L}$ on some standard set of mutually orthogonal unit vectors $\mathbf{a}_n$ [which may, of course, be $\mathbf{e}(x)$]:

$$\mathfrak{L} = \sum_{m,n} \mathbf{a}_m L_{mn} \mathbf{a}_n^*; \quad \mathfrak{L}^* = \sum_{m,n} \mathbf{a}_m \bar{L}_{nm} \mathbf{a}_n^* \tag{7.5.41}$$

from Eq. (1.6.35). As shown in Sec. 1.6 Eqs. (7.5.40) correspond to the set of simultaneous equations

$$\sum_n L_{mn}(\mathbf{a}_n^* \cdot \mathbf{e}_\nu) = \lambda_\nu(\mathbf{a}_m^* \cdot \mathbf{e}_\nu); \quad \sum_n \bar{L}_{nm}(\mathbf{a}_n^* \cdot \mathbf{f}_\nu) = \mu_\nu(\mathbf{a}_m^* \cdot \mathbf{f}_\nu)$$

which are solved to obtain the eigenvalues λ_ν and μ_ν, by finding the roots of the equations obtained by setting the determinant of the coefficients of $(\mathbf{a}_n^* \cdot \mathbf{e}_\nu)$ and $(\mathbf{a}_n^* \cdot \mathbf{f}_\nu)$ equal to zero.

$$|L_{mn} - \lambda_\nu \delta_{mn}| = 0; \quad |\bar{L}_{mn} - \mu_\nu \delta_{mn}| = 0 \tag{7.5.42}$$

The set of roots λ_ν is closely related to the set of roots μ_ν, as will be apparent by taking the conjugate of the first of Eqs. (7.5.40),

$$(\mathfrak{L} \cdot \mathbf{e}_\nu)^* = \mathbf{e}_\nu^* \cdot \mathfrak{L}^* = \bar{\lambda}_\nu \mathbf{e}_\nu^*$$

which becomes

$$\sum_m (\mathbf{e}_\nu^* \cdot \mathbf{a}_m)\bar{L}_{mn} = \bar{\lambda}_\nu(\mathbf{e}_\nu^* \cdot \mathbf{a}_n)$$

with secular determinant

$$|\bar{L}_{mn} - \bar{\lambda}_\nu \delta_{nm}| = 0$$

which is identical with the second one of Eq. (7.5.42). Consequently the set of roots μ_ν is identical with the set of roots $\bar{\lambda}_\nu$, and we can arrange the subscripts so that $\mu_\nu = \bar{\lambda}_\nu$. This does *not* mean that $\mathbf{e}_m = \mathbf{f}_m^*$ or $\mathbf{e}_n^* = \mathbf{f}_n$, however.

When $\mathfrak{L}$ is not Hermitian, the set of eigenvectors $\mathbf{e}_n$ is not mutually orthogonal but the vectors $\mathbf{f}_n$ have orthogonal relationships with the $\mathbf{e}$'s; for

$$[\mathbf{f}_n^* \cdot \mathfrak{L} \cdot \mathbf{e}_m] = \lambda_m[\mathbf{f}_n^* \cdot \mathbf{e}_m] = [(\mathfrak{L}^* \cdot \mathbf{f}_n)^* \cdot \mathbf{e}_m] = \lambda_n[\mathbf{f}_n^* \cdot \mathbf{e}_m]$$

so that

$$(\lambda_m - \lambda_n)(\mathbf{f}_n^* \cdot \mathbf{e}_m) = 0$$

Consequently the scalar products $(\mathbf{f}_n^* \cdot \mathbf{e}_m)$ are zero except for $m = n$. The expansion of any vector $\mathbf{F}$ is then

$$\mathbf{F} = \Sigma F_n \mathbf{e}_n; \quad \text{where } F_n = (\mathbf{f}_n^* \cdot \mathbf{F}) \tag{7.5.43}$$

The dual set of eigenvectors $\mathbf{e}_m$ and $\mathbf{f}_n$ are called a *biorthogonal set* of eigenvectors. Their representatives in terms of $\mathbf{e}(x)$,

$$\psi_n(\mathbf{r}) = \mathbf{e}^*(\mathbf{r}) \cdot \mathbf{e}_n; \quad \bar{\phi}_n(\mathbf{r}) = \mathbf{f}_n^* \cdot \mathbf{e}(\mathbf{r})$$

are said to form a biorthogonal set of eigenfunctions.

The idemfactor $\mathfrak{J}$ is then $\Sigma \mathbf{e}_n \mathbf{f}_n^*$, and the expansion of the Green's operator is

$$\mathfrak{G} = \sum_{n,m} \mathbf{e}_n G_{nm} \mathbf{f}_m^* = 4\pi \sum_n \frac{\mathbf{e}_n \mathbf{f}_n^*}{\lambda - \lambda_n}$$

and the corresponding Green's function is

$$G(\mathbf{r}|\mathbf{r}_0) = 4\pi \sum_n \frac{\psi_n(\mathbf{r})\bar{\phi}_n(\mathbf{r}_0)}{\lambda - \lambda_n} \tag{7.5.44}$$

This function is symmetrical only when $\phi_n = \bar{\psi}_n$ (as it is for some cases).

In Sec. 11.1 [Eqs. (11.1.21) *et seq.*] we consider the case of a flexible string with homogeneous boundary conditions which depend on the frequency (the slope of ψ at the boundary depends on the velocity of ψ as well as its value). These boundary conditions are not self-adjoint, and the corresponding operator is non-Hermitian, so that biorthogonal eigenfunctions must be used. We solve the problem for the given conditions and also solve the problem for adjoint boundary conditions, the complex conjugate of Eq. (11.1.22). It turns out, in this case, that $\phi_n = \bar{\psi}_n$, so that the series given in Eq. (11.2.25) corresponds to Eq. (7.5.44).

Problems for Chapter 7

7.1 A circular conducting disk of radius a, at a constant potential V, is set flush in a plane conductor of infinite extent (coincident with the plane $z = 0$) held at zero potential. Show that the Green's function suitable for this problem is

$$[(x - x_0)^2 + (y - y_0)^2 + (z - z_0)^2]^{-\frac{1}{2}} - [(x - x_0)^2 + (y - y_0)^2 + (z + z_0)^2]^{-\frac{1}{2}}$$

Show that the potential at the point (x,y,z) caused by this combination of conductors is

$$\psi(r,\vartheta) = \frac{zV}{2\pi} \int^{2\pi} d\varphi \int^{a} y\, dy[r^2 + y^2 - 2ry \sin\vartheta \cos\varphi]^{-\frac{3}{2}}$$

where $r^2 = x^2 + y^2 + z^2$ and $\tan\vartheta = (1/z)\sqrt{x^2 + y^2}$. Find the charge density on the disk and on the infinite conductor in the form of definite integrals. Find the value of ψ for r large compared with a, for r small compared with a.

7.2 Suppose the boundary condition on the plane $z = 0$ is that $\partial\psi/\partial z = V$ on the disk of radius a, is zero elsewhere on the plane. Show that the Green's function is

$$[(x - x_0)^2 + (y - y_0)^2 + (z - z_0)^2]^{-\frac{1}{2}} + [(x - x_0)^2 + (y - y_0)^2 + (z + z_0)^2]^{-\frac{1}{2}}$$

Compute the potential ψ on the surface of the disk and its gradient for $r \gg a$.

7.3 Suppose that the disk of Probs. 7.1 and 7.2 were oscillating normal to its plane with a velocity $Ve^{-i\omega t}$ radiating sound into the region $z > 0$. Show that the proper Green's function is $(e^{ikR}/R) + (e^{ikR'}/R')$, where $k = \omega/c$, and

$$R^2 = (x - x_0)^2 + (y - y_0)^2 + (z - z_0)^2$$
$$(R')^2 = (x - x_0)^2 + (y - y_0)^2 + (z + z_0)^2$$

Show that, when $r^2 = x^2 + y^2 + z^2 \gg a^2$, the asymptotic expression for ψ is

$$\psi \to (Va^2/r)e^{ikr-i\omega t}[J_1(ka \sin \vartheta)/ka \sin \vartheta]$$

Use this result to discuss the Fraunhofer diffraction of waves from a circular orifice.

7.4 The inner surface of a sphere of radius a is kept at potential $\psi_a(\vartheta,\varphi)$, where ϑ and φ are the angle coordinates of a spherical system concentric with the sphere. Show that the Green's function appropriate for this problem is

$$[r^2 + r_0^2 - 2rr_0 \cos \theta]^{-\frac{1}{2}} - [(rr_0/a)^2 + a^2 - 2rr_0 \cos \theta]^{-\frac{1}{2}}$$

where θ is the angle between $\mathbf{r}$ the radius vector to the observation point and $\mathbf{r}_0$ the radius vector to the source point $[\cos \theta = \cos \vartheta \cos \vartheta_0 + \sin \vartheta \sin \vartheta_0 \cos (\varphi - \varphi_0)]$. Show that the interior potential is

$$\psi(r,\vartheta,\varphi) = \frac{a}{4\pi}\left[1 - \left(\frac{r}{a}\right)^2\right] \int_0^{2\pi} d\varphi_0 \int_0^{\pi} \frac{\psi_a(\vartheta_0,\varphi_0) \sin \vartheta_0 \, d\vartheta_0}{[a^2 + r^2 - 2ar \cos \theta]^{\frac{3}{2}}}$$

Find a series expansion of ψ in terms of powers of r/a useful for points near the origin.

7.5 Show that the Green's function in spherical coordinates for the Laplace equation is

$$\frac{1}{R} = \sum_{n,m} \epsilon_m \frac{(n - m)!}{(n + m)!} \cos[m(\varphi - \varphi_0)] P_n^m(\cos \vartheta) P_n^m(\cos \vartheta_0) \begin{cases} (r^n/r_0^{n+1}); & r \leq r_0 \\ (r_0^n/r^{n+1}); & r \geq r_0 \end{cases}$$

and that for the Helmholtz equation is

$$\frac{e^{ikR}}{R} = ik \sum_{n,m} \epsilon_m (2n + 1) \frac{(n - m)!}{(n + m)!} \cos[m(\varphi - \varphi_0)] P_n^m(\cos \vartheta) \cdot$$
$$\cdot P_n^m(\cos \vartheta_0) \begin{cases} j_n(kr)h_n(kr_0); & r \leq r_0 \\ j_n(kr_0)h_n(kr); & r \geq r_0 \end{cases}$$

where j_n and h_n are the spherical Bessel functions (see Prob. 5.20 and the tables at the end of Chap. 11).

7.6 A solution of the Helmholtz equation, originally of the form

$$\psi_0(r) = \Sigma A_{mn} \cos(m\varphi + \alpha_m) P_n^m(\cos \vartheta) j_n(kr)$$

extending throughout all space, is perturbed by the presence of a sphere of radius a with center at the origin, at the surface of which ψ must now satisfy the boundary condition

$$(\partial\psi/\partial r) = \eta\psi; \quad r = a$$

Show that the new ψ solution of the equation $\nabla^2\psi + k^2\psi = 0$ outside the sphere, satisfying the above boundary condition at $r = a$ and the condition at $r \to \infty$ that ψ must equal ψ_0 plus an outgoing wave, is a solution of the following integral equation:

$$\psi(\mathbf{r}) = \psi_0(\mathbf{r}) + \frac{1}{4\pi} \oint \psi(\mathbf{r}_0^s) \left[\frac{\partial}{\partial r_0} G(\mathbf{r}|\mathbf{r}_0^s) - \eta G(\mathbf{r}|\mathbf{r}_0^s) \right] dA_0; \quad r \geq a$$

where G is the second series of Prob. 7.5 and the integration is over the surface of the sphere.

7.7 A wire of radius b is immersed in an oil bath of infinite volume. The heat-diffusion coefficient of both oil and wire is a. Both are originally at zero temperature. A pulse of current is sent through the wire, heating it momentarily up to temperature T_0. Show that the temperature a distance r from the wire axis a time t later is

$$T = \left(\frac{T_0}{2a^2t}\right) e^{-r^2/4a^2t} \int_0^b e^{-y^2/4a^2t} J_0\left(\frac{iyr}{2a^2t}\right) y\, dy$$

By use of the series expansion and asymptotic expression for J_0, compute T for the two limiting cases, one where $2a^2t/r$ is much smaller than b, the other where it is much larger.

7.8 Determine the one-dimensional Green's function $G_k(\mathbf{r}|\mathbf{r}_0)$ for the Bessel differential operator

$$\frac{d}{dr}\left(r \frac{dG_k}{dr}\right) + k^2 r G_k = -\delta(r - r_0); \quad r \leq a$$

where $G_k(a|\mathbf{r}_0) = 0$. Show that G_k is singular whenever $k = k_r$, where $J_0(k_n a) = 0$. From the behavior of G_k at this singularity determine the normalization integral

$$\int_0^a r J_0^2(k_n r)\, dr$$

7.9 Show that, in cylindrical coordinates,

$$\frac{e^{ikR}}{iR} = \sum_m (2 - \delta_{0m}) \cos[m(\varphi - \varphi_0)] \int_0^\infty J_m(\lambda\rho) J_m(\lambda\rho_0) \frac{e^{i\sqrt{k^2 - \lambda^2}|z - z_0|}}{\sqrt{k^2 - \lambda^2}} \lambda\, d\lambda$$

7.10 Let $u = E_x$ and $v = -E_y$, where **E** is a two-dimensional electric field. Show that the equations satisfied by u and v may be summarized as follows:

$$\begin{pmatrix} \partial/\partial x & -\partial/\partial y \\ \partial/\partial y & \partial/\partial x \end{pmatrix} \begin{pmatrix} u \\ v \end{pmatrix} = 0$$

Define a Green's dyadic

$$\mathfrak{G} = \begin{pmatrix} G_{11} & G_{12} \\ G_{21} & G_{22} \end{pmatrix}$$

satisfying the equation

$$\begin{pmatrix} \partial/\partial x & -\partial/\partial y \\ \partial/\partial y & \partial/\partial x \end{pmatrix} \mathfrak{G} = -4\pi\delta(x - x_0)\delta(y - y_0) \begin{pmatrix} 1 & 0 \\ 0 & 1 \end{pmatrix}$$

Show that

$$\mathfrak{G} = \begin{pmatrix} \partial/\partial x & \partial/\partial y \\ -\partial/\partial y & \partial/\partial x \end{pmatrix} G$$

where G is the Green's function for the two-dimensional Laplace equation. Discuss the meaning of $\mathfrak{G}$, and obtain the solution in terms of $\mathfrak{G}$ of the inhomogeneous form of the equations for u and v.

7.11 Let ψ satisfy the following equation:

$$(d^2\psi/dx^2) + k^2\psi = 0; \quad 0 \le x \le l$$

and the boundary condition $\psi(0) = 0$ and $\psi(l) = f\psi'(l)$ where f is a complex constant. Show that the eigenfunctions are $\sin(k_n x)$ where $\tan(k_n) = fk_n$. Show that the adjoint solution satisfies the same equation as ψ but with the boundary conditions $\tilde{\psi}(0) = 0$, $\tilde{\psi}(l) = \bar{f}\tilde{\psi}'(l)$. Show that $\tilde{\psi}_n = \bar{\psi}_n$. Show that

$$\int_0^l \bar{\tilde{\psi}}_n \psi_m \, dx = \int_0^l \psi_n \psi_m \, dx = 0; \quad n \neq m$$

Discuss the normalization for ψ_n, and verify by examining the Green's function for the problem, which may be obtained in closed form.

7.12 A self-adjoint operator $\mathfrak{L}$ may be broken up into two self-adjoint parts $\mathfrak{L}_r$ and $\mathfrak{L}_\rho$, where $\mathfrak{L}_r$ operates on variable r only and $\mathfrak{L}_\rho$ operates on ρ only:

$$\mathfrak{L} = \mathfrak{L}_r + \mathfrak{L}_\rho$$

Let the orthogonal and normalized eigenfunctions of $\mathfrak{L}_\rho$ be $\varphi_n(\rho)$:

$$\mathfrak{L}_\rho \varphi_n(\rho) = \lambda_n \varphi_n(\rho)$$

Show that the Green's function $G_\lambda(r,\rho|r_0,\rho_0)$ which satisfies the equation

$$[\mathfrak{L}(r,\rho) - \lambda]G_\lambda = -\delta(r - r_0)\delta(\rho - \rho_0)$$

is given by

$$G_\lambda = \sum_n g_{\lambda-\lambda_n}(r|r_0)\varphi_n(\rho)\varphi_n(\rho_0)$$

where

$$[\mathfrak{L}_r - (\lambda - \lambda_n)]g_{\lambda-\lambda_n} = -\delta(r - r_0)$$

7.13 If

$$(\mathfrak{L} - \lambda)G_\lambda = -4\pi\delta(\mathbf{r} - \mathbf{r}_0)$$

Show that

$$G_\lambda(\mathbf{r}|\mathbf{r}_0) = G_0(\mathbf{r}|\mathbf{r}_0) - \left(\frac{\lambda}{4\pi}\right)\int G_\lambda(\mathbf{r}|\mathbf{r}_1)G_0(\mathbf{r}_1|\mathbf{r}_0)\,dV_1$$

and

$$G_0(\mathbf{r}|\mathbf{r}_0) = G_\lambda(\mathbf{r}|\mathbf{r}_0) + \left(\frac{\lambda}{4\pi}\right)\int G_0(\mathbf{r}|\mathbf{r}_1)G_\lambda(\mathbf{r}_1|\mathbf{r}_0)\,dV_1$$

7.14 If G is the Green's function for the scalar Helmholtz equation for the semi-infinite region $x > 0$, satisfying mixed boundary conditions

$$\partial\psi/\partial x = F\psi; \quad \text{at } x = 0$$

show that

$$\left(\frac{\partial G}{\partial x}\right) - FG = -\left(\frac{\partial}{\partial x} + F\right)\left(\frac{e^{ikR}}{R} - \frac{e^{ikR'}}{R'}\right)$$

where R is $|\mathbf{r} - \mathbf{r}_0|$ and $R' = |\mathbf{r} + \mathbf{r}_0|$. By integration show that

$$G = (e^{ikR}/R) + T$$

Determine T.

Table of Green's Functions

General Properties. The Green's function $G_\lambda(\mathbf{r}|\mathbf{r}_0)$ satisfies the equation

$$\mathfrak{L}(G) - \lambda G = -4\pi\delta(\mathbf{r} - \mathbf{r}_0)$$

with certain homogeneous boundary conditions on the boundary surface S. Its adjoint [see Eq. (7.5.4)] $\tilde{G}_\lambda(\mathbf{r}|\mathbf{r}_0)$ satisfies the equation

$$\tilde{\mathfrak{L}}(\tilde{G}) - \lambda\tilde{G} = -4\pi\delta(\mathbf{r} - \mathbf{r}_0)$$

satisfying adjoint boundary conditions [see Eq. (7.5.9)] on the boundary surface S: The *reciprocity principle* is that

$$G_\lambda(\mathbf{r}|\mathbf{r}_0) = \tilde{G}_\lambda(\mathbf{r}_0|\mathbf{r})$$

If $\mathfrak{L}$ is Hermitian (if its adjoint equals its conjugate), then G_λ is also Hermitian. In this case the eigenvalues λ_n for $\mathfrak{L}$,

$$\mathfrak{L}(\psi_n) - \lambda_n\psi_n = 0$$

are real, the set of eigenfunctions ψ_n is mutually orthogonal, and

$$G_\lambda(\mathbf{r}|\mathbf{r}_0) = 4\pi \sum_n \frac{\bar{\psi}_n(\mathbf{r}_0)\psi_n(\mathbf{r})}{N_n(\lambda - \lambda_n)}$$

where $N_n = \int |\psi_n|^2 \, dv$. If $\mathfrak{L}$ is not Hermitian, then the eigenfunctions Φ_n of the Hermitian conjugate equation

$$\mathfrak{L}^*(\Phi_n) - \mu_n \Phi_n = 0; \quad \bar{\mathfrak{L}}^* = \tilde{\mathfrak{L}}; \quad \mu_n = \bar{\lambda}_n$$

may not equal the eigenfunctions ψ_n and neither set may be mutually orthogonal. The double set ψ, Φ then constitutes a biorthogonal set, however, and

$$G_\lambda(\mathbf{r}|\mathbf{r}_0) = 4\pi \sum_n \frac{\bar{\Phi}_n(\mathbf{r}_0)\psi_n(\mathbf{r})}{N_n(\lambda - \lambda_n)}$$

and the adjoint $\tilde{G}$ is not necessarily equal to the conjugate $\bar{G}$ [see Eq. (7.5.44); $N_n = \int \bar{\Phi}_n \psi_n \, dv$].

Green's Function for the Helmholtz Equation. G is a solution of

$$\nabla^2 G_k(\mathbf{r}|\mathbf{r}_0) + k^2 G_k(\mathbf{r}|\mathbf{r}_0) = -4\pi\delta(\mathbf{r} - \mathbf{r}_0)$$

satisfying homogeneous boundary conditions on some surface S. Then the reciprocity relation is $G_k(\mathbf{r}|\mathbf{r}_0) = G_k(\mathbf{r}_0|\mathbf{r})$, since the equation is self-adjoint. If ψ is a solution of $(\nabla^2 + k^2)\psi = -4\pi\rho$, having value $\psi_0(\mathbf{r}^s)$ on the surface S and having outward-pointing normal gradient $N_0(\mathbf{r}^s) = (\partial\psi/\partial n)_S$ on S, then within and on S

$$\psi(\mathbf{r}) = \int \rho(\mathbf{r}_0) G_k(\mathbf{r}|\mathbf{r}_0) \, dv_0 + \frac{1}{4\pi} \oint [G_k(\mathbf{r}|\mathbf{r}_0^s) N_0(\mathbf{r}_0^s) - \psi_0(\mathbf{r}_0^s) \frac{\partial}{\partial n_0} G_k(\mathbf{r}|\mathbf{r}_0^s)] \, dA_0$$

where the first integral is over the volume enclosed by S and the second is a normal outflow integral over all of S. The normal gradients are taken in the outward direction, away from the interior where ψ is measured.

If the surface S is at infinity and if outgoing waves are specified [causality condition, Eq. (7.2.17)], then G takes on the simple form $g_k(\mathbf{r}|\mathbf{r}_0)$ for the *infinite domain:*

$$\begin{aligned} g_k(\mathbf{r}|\mathbf{r}_0) &= e^{ikR}/R; && \text{3 dimensions;} \\ & && R^2 = (x - x_0)^2 + (y - y_0)^2 + (z - z_0)^2 \\ &= i\pi H_0^{(1)}(kP); && \text{2 dimensions;} \quad P^2 = (x - x_0)^2 + (y - y_0)^2 \\ &= \left(\frac{2\pi i}{k}\right) e^{ik|x - x_0|}; && \text{1 dimension} \end{aligned}$$

The Green's function for the Poisson equation $\nabla^2\psi = -4\pi\rho$ is $G_0(\mathbf{r}|\mathbf{r}_0)$, for $\lambda = 0$. The corresponding forms for the infinite domain are

$$\begin{aligned} g_0(\mathbf{r}|\mathbf{r}_0) &= (1/R); && \text{3 dimensions} \\ &= -2 \ln R; && \text{2 dimensions} \end{aligned}$$

When surface S coincides, all or in part, with one of the set of separable coordinate surfaces discussed in Sec. 5.1, we can expand G in series of separated solutions. Suppose the boundary conditions (finiteness, periodicity, or homogeneous conditions on the boundary) are such that two of the factors can be eigenfunctions, say the ξ_2 and ξ_3 factors. The ξ_1 factor must also satisfy homogeneous conditions at the surface, which we assume corresponds to the surfaces $\xi_1 = a$, $\xi_1 = b$, $b > a$. The coordinates have scale factors h_1, h_2, h_3; a Stäckel determinant S with elements $\Phi_{mn}(\xi_m)$; and minors $M_m = \partial S/\partial \Phi_{m1}$ [see Eqs. (5.1.25) *et seq.*]. The Helmholtz equation

$$\sum_n \frac{M_n}{Sf_n} \frac{\partial}{\partial \xi_n} \left(f_n \frac{\partial \psi}{\partial \xi_n} \right) + k_1^2 \psi = 0; \quad k_1^2 = k^2$$

separates into $\psi = X_1(\xi_1)X_2(\xi_2)X_3(\xi_3)$, where

$$\frac{1}{f_m} \frac{d}{d\xi_m} \left(f \frac{dX_m}{md\xi_m} \right) + \sum_{n=1}^{3} k_n^2 \Phi_{mn} X_m = 0$$

and eigenfunction solutions satisfying appropriate boundary conditions are chosen for X_3 and X_2:

$$W_q(\xi_2,\xi_3) = \vartheta_{\nu_q}(\xi_2) X_{\mu_q}(\xi_3); \quad \nu, \mu = 0, 1, 2, \ldots$$

These are orthogonal with respect to a density function ρ (often $\rho = h_2h_3$) so that

$$\iint W_p W_q \rho(\xi_2,\xi_3)\, d\xi_2\, d\xi_3 = \begin{cases} 0; & p \neq q \\ N_q; & p = q \end{cases}$$

and the W's constitute a complete set for the coordinates ξ_2, ξ_3 within the surface S.

Two independent solutions, $y_{1q}(\xi_1)$ and $y_{2q}(\xi_1)$, are chosen for the ξ_1 factor, each corresponding to the separation constants of W_q and arranged so that y_1 satisfies the required boundary condition at $\xi_1 = a$ and y_2 the required condition at $\xi_1 = b (b > a)$. Then

$$G_k(\mathbf{r}|\mathbf{r}') = -\left(\frac{4\pi h_1}{h_2 h_3} \right) \sum_q \frac{\rho(\xi_2', \xi_3')}{N_q} \bar{W}_q(\xi_2', \xi_3') W(\xi_2, \xi_3) \cdot$$

$$\cdot \left(\frac{1}{\Delta} \right) \begin{cases} y_{1q}(\xi_1) y_{2q}(\xi_1'); & \xi_1 < \xi_1' \\ y_{2q}(\xi_1) y_{1q}(\xi_1'); & \xi_1 > \xi_1' \end{cases}$$

where the scale factors have functions of the primed coordinates and where Δ is the Wronskian for the two ξ_1' solutions:

$$\Delta = \Delta(y_{1q}, y_{2q}) = y_{1q} y_{2q}' - y_{1q}' y_{2q} = (\text{constant}/f_1), \text{ a function of } \xi_1'$$

Expansion of the Green's function for the infinite domain, of the sort generalized above, is given for two-dimensional polar coordinates in Eqs. (7.2.51) and (11.2.23), for rectangular coordinates in Eq. (11.2.11), for parabolic coordinates in Eq. (11.2.67), and for elliptic coordinates in Eq. (11.2.93). The expansions for three-dimensional systems for rectangular coordinates are in Eq. (11.3.10), for spherical coordinates in Eq. (11.3.44), and for spheroidal coordinates in Eq. (11.3.91). Similar expansions for vector solutions are given in Eqs. (13.3.15) and (13.3.79).

Green's Function for the Wave Equation. G is a solution of

$$\nabla^2 G(\mathbf{r},t|\mathbf{r}_0,t_0) - \frac{1}{c^2}\frac{\partial^2}{\partial t^2}G(\mathbf{r},t|\mathbf{r}_0,t_0) = -4\pi\delta(\mathbf{r}-\mathbf{r}_0)\delta(t-t_0)$$

satisfying homogeneous boundary conditions on surface S and obeying the "causality" requirement that G and $\partial G/\partial t = 0$ everywhere for $t < t_0$. The reciprocity relation is then

$$G(\mathbf{r},t|\mathbf{r}_0,t_0) = G(\mathbf{r}_0,-t_0|\mathbf{r},-t)$$

If $\psi(\mathbf{r},t)$ is a solution of $\nabla^2\psi - (1/c^2)(\partial^2\psi/\partial t^2) = -4\pi\rho(\mathbf{r},t)$ having value $\psi_s(\mathbf{r}^s)$ and outward normal gradient $N_s(\mathbf{r}^s)$ on the surface S and having initial value $\psi_0(\mathbf{r})$ and initial time derivative $v_0(\mathbf{r}) = \partial\psi/\partial t_{t=0}$ within S at $t = 0$, then for $t > 0$ and within and on S

$$\begin{aligned}\psi(\mathbf{r},t) = &\int^{t+\epsilon} dt_0 \int dV_0\, G(\mathbf{r},t|\mathbf{r}_0,t_0)\rho(\mathbf{r}_0,t_0) \\ &+ \frac{1}{4\pi}\int_0^{t+\epsilon} dt_0 \oint dA_0 \left[G(\mathbf{r},t|\mathbf{r}_0^s,t_0)N_s(\mathbf{r}_0^s) - \psi(\mathbf{r}_0^s)\frac{\partial}{\partial n_0}G(\mathbf{r},t|\mathbf{r}_0^s,t_0)\right] \\ &- \frac{1}{4\pi c^2}\int dV_0 \left[\left(\frac{\partial G}{\partial t_0}\right)_{t_0=0}\psi_0(\mathbf{r}_0) - G_{t_0=0}v_0(\mathbf{r}_0)\right]; \quad \epsilon \to 0^+\end{aligned}$$

The closed forms for the Green's function for the infinite domain are

$$\begin{aligned}g(\mathbf{r},t|\mathbf{r}_0,t_0) &= (1/R)\delta[(R/c) - (t-t_0)]; \quad \text{for 3 dimensions;} \\ &\qquad R^2 = (x-x_0)^2 + (y-y_0)^2 + (z-z_0)^2 \\ &= [2c/\sqrt{c^2(t-t_0)^2 - P^2}]u[(t-t_0) - (P/c)]; \\ &\qquad \text{for 2 dimensions;} \quad P^2 = (x-x_0)^2 + (y-y_0)^2 \\ &= 2c\pi u[(t-t_0) - (|x-x_0|/c)]; \quad \text{for 1 dimension}\end{aligned}$$

where $u(x) = 0;\ x < 0;\ u(x) = 1;\ x > 0;\ \delta(x) = u'(x)$

$$\int_{-\infty}^{\infty}\delta(x)f(x+a)\,dx = f(a); \quad \int_{-\infty}^{\infty}\delta'(x)f(x+a)\,dx = -f'(a)$$

The Green's function for the wave equation is related to the Green's function for the Helmholtz equation by the Fourier integral relationship

$$G(\mathbf{r},t|\mathbf{r}_0,t_0) = (c/2\pi)\int_{-\infty}^{\infty} G_k(\mathbf{r}|\mathbf{r}_0)e^{-ikc(t-t_0)}\,dk$$
$$= 4\pi c^2 u(t-t_0)\sum_n \left(\frac{1}{\omega_n}\right)\bar{\psi}_n(\mathbf{r}_0)\psi(\mathbf{r})\sin[\omega_n(t-t_0)]$$

where ψ_n is the eigenfunction solution of $\nabla^2\psi + k_n^2\psi_n = 0$ within S and where $\omega_n = k_n c$. The contour for the integration over k is just above the real axis.

Green's Function for the Diffusion Equation. G is a solution of

$$\nabla^2 G_a(\mathbf{r},t|r_0,t_0) - a^2(\partial/\partial t)G_a(\mathbf{r},t|\mathbf{r}_0,t_0) = -4\pi\delta(\mathbf{r}-\mathbf{r}_0)\delta(t-t_0)$$

satisfying homogeneous boundary conditions on surface S and obeying the causality requirement that G be zero for $t < t_0$. The adjoint function $\tilde{G}_a(\mathbf{r},t|\mathbf{r}_0,t_0) = G_a(\mathbf{r},-t|\mathbf{r}_0,-t_0)$ satisfies the adjoint equation $\nabla^2\tilde{G} + a^2(\partial\tilde{G}/\partial t) = -4\pi\delta(r-r_0)\delta(t-t_0)$. The reciprocity relationship is

$$G_a(\mathbf{r},t|\mathbf{r}_0,t_0) = \tilde{G}_a(\mathbf{r}_0,t_0|\mathbf{r},t) = G_a(\mathbf{r}_0,-t_0|\mathbf{r},-t)$$

If $\psi(\mathbf{r})$ is a solution of $\nabla^2\psi - a^2(\partial\psi/\partial t) = -4\pi\rho(r)$, having value $\psi_s(\mathbf{r}^s)$ and outward normal gradient $N_s(\mathbf{r}^s) = \partial\psi/\partial n$ on surface S and having initial value $\psi_0(\mathbf{r})$ within S at $t = 0$, then for $t > 0$ within and on S

$$\psi(\mathbf{r},t) = \int_0^{t+\epsilon} dt_0 \int dV_0\, \rho(\mathbf{r}_0,t)\, G(\mathbf{r},t|\mathbf{r}_0,t_0)$$
$$+ \frac{1}{4\pi}\int_0^{t+\epsilon} dt_0 \int dA_0 \left[G(\mathbf{r},t|\mathbf{r}_0^s,t_0)\, N_s(\mathbf{r}_0^s) - \psi_s(\mathbf{r}_0^s)\,\frac{\partial}{\partial n_0}G(\mathbf{r},t|\mathbf{r}_0^s,t_0)\right]$$
$$+ \frac{a^2}{4\pi}\int dV_0\, \psi_0(\mathbf{r}_0)\, G(\mathbf{r},t|\mathbf{r}_0,0); \quad \epsilon \to 0^+$$

The form of the Green's function for the infinite domain, for n dimensions, is

$$g_a(\mathbf{r},t|\mathbf{r}_0,t_0) = \frac{4\pi}{a^2}\left(\frac{a}{2\sqrt{\pi\tau}}\right)^n e^{-a^2R^2/4\tau}\, u(\tau)$$

where $\tau = t - t_0$ and $R = |\mathbf{r} - \mathbf{r}_0|$. The Green's function for the diffusion equation is related to the eigenfunctions ψ_n for the related Helmholtz equation, $(\nabla^2 + k_n^2)\psi_n = 0$ for the domain within S, by the equation

$$G_a(\mathbf{r},t|\mathbf{r}_0,t_0) = \frac{4\pi}{a^2}u(t-t_0)\sum_n \frac{e^{-(k_n^2/a^2)(t-t_0)}}{N_n}\,\bar{\psi}_n(\mathbf{r}_0)\,\psi_n(\mathbf{r})$$

where $\int\bar{\psi}_n\psi_m\,dV = \delta_{nm}N_n$.

Bibliography

The literature on Green's functions and their applications is rather spotty. Satisfactory accounts of various aspects of the theory:

Bateman, H.: "Partial Differential Equations of Mathematical Physics," Chap. 2, Cambridge, New York, 1932.

Carslaw, H. S.: "Mathematical Theory of the Conduction of Heat in Solids," Macmillan & Co., Ltd., London, 1921, reprint Dover, New York, 1945.

Courant, R., and D. Hilbert: "Methoden der Mathematischen Physik," Vol. 1, Springer, Berlin, 1937.

Kellogg, O. D.: "Foundations of Potential Theory," Springer, Berlin, 1939, reprint, Ungar, New York, 1944.

Murnaghan, F. D., "Introduction to Applied Mathematics," Wiley, New York, 1948.

Riemann-Weber: "Differential- und Integralgleichungen der Mechanik und Physik," Vieweg, Brunswick, 1935.

Sommerfeld, A.: "Partial Differential Equations in Physics," Academic Press, New York, 1949.

Webster, A. G.: "Partial Differential Equations of Mathematical Physics," Chap. 5, Stechert, New York, 1933.

CHAPTER 8

Integral Equations

In the preceding chapters, we have relied mainly upon the differential equation to describe the propagation of a field ψ. Boundary conditions have to be specified in addition, for the differential equation describes only the local behavior of ψ, relating ψ at a point $\mathbf{r}$ to ψ at $\mathbf{r} + d\mathbf{r}$. Starting from a given point $\mathbf{r}$, the differential equation permits the construction of the many possible solutions in a stepwise fashion. The boundary conditions are then invoked to choose the solution which is appropriate to the physical situation of interest.

Inasmuch as the boundary values are a determining feature, it would be useful to formulate the determining equation for ψ in such a manner as to include the boundary conditions explicitly. Such a formulation must relate $\psi(\mathbf{r})$ not only to the values of ψ at neighboring points but to its values at all points in the region including the boundary points. The integral equation is an equation of this form. Since it contains the boundary conditions, it represents the entire physics of the problem in a very compact form and, as we shall see in many instances, a more convenient form than the more conventional differential equation.

This is not the only reason for studying integral equations. We have already seen, for example, in the discussion of diffusion and transport phenomena, that there are many situations which cannot be represented in terms of differential equations. In other words, there are problems in which the behavior of ψ at $\mathbf{r}$ depends on the values of ψ at some distance from $\mathbf{r}$, not just on the values at neighboring points.

In the first section of this chapter we shall display some of the integral equations which arise in physics and shall discuss their classification into various types, each having different properties and techniques for solution. Then after a discussion of general mathematical properties of these types, we shall devote the rest of this chapter to discussing techniques of solution.

8.1 *Integral Equations of Physics, Their Classification*

We consider first an example from transport theory. Here, as the consequence of a collision, a particle which was originally traveling in a given direction with a certain energy, as specified by its momentum $\mathbf{p}_0$, may acquire a new momentum $\mathbf{p}$ which is very different both in direction and in size from the original $\mathbf{p}_0$. More explicitly, let $P(\mathbf{p}|\mathbf{p}_0)\,d\mathbf{p}\,dt$ be the probability that a particle having a momentum $\mathbf{p}_0$ be scattered into momentum between $\mathbf{p}$ and $\mathbf{p} + d\mathbf{p}$ in a time dt. If the original distribution function $f(\mathbf{r},\mathbf{p}_0,t)\,d\mathbf{p}_0$ gives the relative number of particles having a momentum between $\mathbf{p}_0$ and $\mathbf{p}_0 + d\mathbf{p}_0$ at a point $\mathbf{r}$, then the collisions which occur in a time dt contribute the following to $f(\mathbf{r},\mathbf{p},t)$:

$$[\textstyle\int P(\mathbf{p}|\mathbf{p}_0)\,f(\mathbf{r},\mathbf{p}_0,t)\,d\mathbf{p}_0]\,dt$$

We see immediately that the value of $f(\mathbf{r},\mathbf{p},t)$ at $\mathbf{p}$ is related to *all* the values of $f(\mathbf{r},\mathbf{p}_0,t)$ for all values of $\mathbf{p}_0$ consistent with conservation of momentum and energy. To complete the picture let us obtain the complete equation for f by considering the other changes in $f(\mathbf{r},\mathbf{p},t)$ in a time dt. The above term gives the number of particles scattered into the volume of configuration space at $\mathbf{r}$ and $\mathbf{p}$. A number of particles leave this region because of scattering or because of their complete disappearance by absorption. Let the probability per unit time for scattering or absorption out of $\mathbf{p}$ be $P_T(\mathbf{p})$. If there is no absorption, $P_T(\mathbf{p}) = \int P(\mathbf{p}_0|\mathbf{p})\,d\mathbf{p}_0$. The number of particles leaving in a time dt is

$$P_T(\mathbf{p})\,f(\mathbf{r},\mathbf{p},t)\,dt$$

Finally, even if there are no collisions, a change in f occurs simply because the particles are moving. The particles at $\mathbf{r}$ were, at a time dt earlier, at $\mathbf{r} - (\mathbf{p}/m)\,dt$. Thus

$$f(\mathbf{r},\,\mathbf{p},\,t+dt) = f[\mathbf{r} - (\mathbf{p}/m)\,dt,\,\mathbf{p},\,t] - P_T(\mathbf{p})f(\mathbf{r},\mathbf{p},t)\,dt + [\textstyle\int P(\mathbf{p}|\mathbf{p}_0)f(\mathbf{r},\mathbf{p}_0,t)\,d\mathbf{p}_0]\,dt$$

The equation states that the particles at $\mathbf{r}$ at a time $t + dt$ consist (1) of those which arrived there as a consequence of their motion (2) less those which scattered or absorbed out of the momentum range $d\mathbf{p}$ (3) plus those which were scattered into the range $d\mathbf{p}$ by collision.

By expansion we finally obtain a differential-integral equation (see Secs. 2.4 and 12.2):

$$\frac{\partial f}{\partial t} = -\left(\frac{\mathbf{p}}{m}\cdot\nabla\right)f - P_T f + \int P(\mathbf{p}|\mathbf{p}_0)f(\mathbf{r},\mathbf{p}_0,t)\,d\mathbf{p}_0 \qquad (8.1.1)$$

Under steady-state conditions, f is independent of t and

$$\left(\frac{\mathbf{p}}{m}\cdot\nabla\right)f = -P_T f + \int P(\mathbf{p}|\mathbf{p}_0)f(\mathbf{r},\mathbf{p}_0)\,d\mathbf{p}_0;\quad \text{steady state} \qquad (8.1.2)$$

We emphasize again the dependence of f on the complete functional dependence of f on $\mathbf{p}$ rather than just on the relation of f for neighboring values of **p**. In Sec. 2.4 the probabilities P_T and P are directly related to the cross sections and the above equation is converted by direct integration from a differential-integral equation into an integral equation. We shall again discuss transport equations in Chap. 12.

Example from Acoustics. It should not be thought that this type of equation occurs only for transport problems, where the collision process is a rather obvious source of discontinuous changes in the momentum p. We may, for example, consider a case in acoustics. As we shall see, differential-integral equations arise whenever two systems with distributed mass or other relevant parameters are coupled. For example, consider the vibrations of a membrane set in a rigid plate as illustrated in Fig. 8.1. The vibrations of the membrane give rise to sound waves which in turn react back upon the membrane, influencing its vibration, etc. Suppose that the displacement of the membrane is given by $\psi(y,z)$; the corresponding velocity in the x direction is $\partial\psi/\partial t = -i\omega\psi$, assuming simple harmonic time dependence. The resulting velocity potential in the medium to the right is, from Eq. (7.2.10),

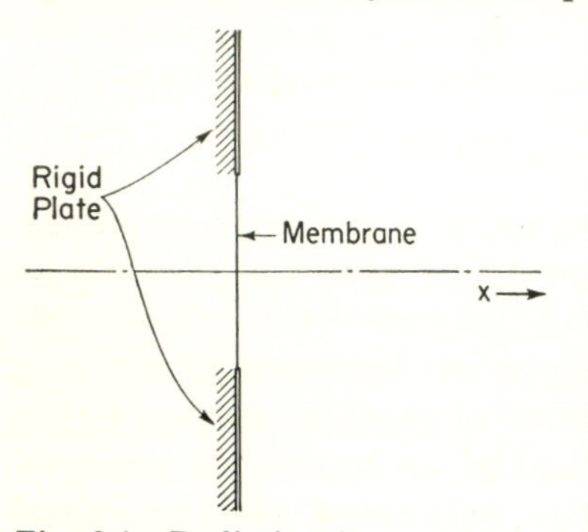

Fig. 8.1 Radiation from a membrane in a rigid plate.

$$\varphi(x,y,z) = -\frac{1}{4\pi}\int G_k(x,y,z|0,y_0,z_0)v_n(y_0,z_0)\,dS_0$$

where G_k is a Green's function satisfying the condition

$$(\partial G_k/\partial n) = 0; \quad \text{at } x = 0$$

where $k = \omega/c$, c = velocity of propagation of sound and where v_n is the normal component of the velocity, *i.e.*, the velocity in the negative x direction. Hence $v_n = i\omega\psi$. The function G_k is provided by the method of images (see page 812):

$$\begin{aligned} G_k(x,y,z|x_0,y_0,z_0) &= (e^{ikR}/R) + (e^{ikR'}/R') \\ (R)^2 &= (x - x_0)^2 + (y - y_0)^2 + (z - z_0)^2 \\ (R')^2 &= (x + x_0)^2 + (y - y_0)^2 + (z - z_0)^2 \end{aligned}$$

The sound thus generated in the region $x > 0$ gives rise to a pressure which now becomes a forcing term in equation of vibration of the membrane. The pressure is related to the velocity potential by the equation

$$p = \rho_0(\partial\varphi/\partial t) = -i\omega\rho_0\varphi$$

where ρ_0 is the mean density of the medium in which the sound propagates. The equation of motion of the membrane is

$$\nabla^2\psi + \kappa^2\psi = -(p/T); \quad \kappa = \omega/V; \quad V = \sqrt{T/\mu}$$

where $\quad T = \text{tension}; \quad \mu = \text{mass/area}$

Substituting for p we obtain

$$\nabla^2\psi + \kappa^2\psi = \frac{\omega^2\rho_0}{2\pi} \int \frac{\exp[ik\sqrt{(y - y_0)^2 + (z - z_0)^2}]}{\sqrt{(y - y_0)^2 + (z - z_0)^2}} \psi(y_0,z_0)\, dS_0 \quad (8.1.3)$$

In this equation we again see that the behavior of ψ depends not only upon the value of ψ at a point on the membrane and neighboring points but also on its values at every point on the membrane in the manner dictated by the integral on the right. The above equation is a differential-integral equation, but by employing the Green's function for the membrane it may be reduced to an integral equation.

It should be clear from the above example that an integral equation will result when an impulse at one point in a medium may be transmitted to a point some distance away through another medium coupled to the first. The equation for describing the vibrations of the first medium will contain a term arising from the propagation in the second medium. This term will involve the values of ψ at all points at which the two media are in contact; the integral in the case discussed above is just such a term. Radiation problems, in which the reactive effects of the emitted radiation back on the source cannot be neglected, will naturally lead to integral equations. Solution of the integral equation will give rise to precise evaluation of radiation resistance or, more general, radiation impedance. We shall consider such problems in more detail in Chaps. 11 and 13.

An Example from Wave Mechanics. As a final example let us turn to quantum mechanics. The Schroedinger equation must be written as an integral equation whenever the potential energy is velocity dependent. Let the potential energy V be

$$V(\mathbf{r},\hbar\nabla/i)$$

where we have already replaced the momentum operator by $(\hbar/i)\nabla$. The Schroedinger equation in differential form is

$$\nabla^2\psi + \frac{2m}{h^2}\{E - V[\mathbf{r},(\hbar/i)\nabla]\}\psi = 0$$

If the dependence of V on ∇ is not that of a simple polynomial, this equation is not one of finite order. To obtain the equivalent integral equation, introduce the Fourier transform of ψ:

$$\psi(\mathbf{r}) = \frac{1}{(2\pi\hbar)^{\frac{3}{2}}} \int_{-\infty}^{\infty} \varphi(\mathbf{p}) e^{(i/\hbar)\mathbf{p}\cdot\mathbf{r}}\, d\mathbf{p}$$

Substituting in the Schroedinger equation, multiplying through by $[1/(2\pi\hbar)^{\frac{3}{2}}]e^{-(i/\hbar)\mathbf{q}\cdot\mathbf{r}}$, and integrating on $\mathbf{r}$ yields

$$\left(\frac{q^2}{2m}\right)\varphi(\mathbf{q}) + \int_{-\infty}^{\infty} \varphi(\mathbf{p})V(\mathbf{p}-\mathbf{q},\mathbf{p})\,d\mathbf{p} = E\varphi(\mathbf{q}) \tag{8.1.4}$$

$$V(\mathbf{p}-\mathbf{q},\mathbf{p}) = \frac{1}{(2\pi\hbar)^{\frac{3}{2}}}\int_{-\infty}^{\infty} e^{i(\mathbf{p}-\mathbf{q})\cdot\mathbf{r}/\hbar}\,V(\mathbf{r},\mathbf{p})\,dV$$

This integral equation determining $\varphi(\mathbf{q})$ was given earlier in Sec. 2.6. The meaning of the integral can be most easily seen if we consider a scattering problem in which the scattering is caused by the potential V. If a plane wave of amplitude $\varphi(\mathbf{p})$ is incident upon the region where the potential exists, then it is scattered. In other words, some of the incident wave is diverted into other directions, possibly with loss of momenta. We now ask for the contribution from a variety of plane waves with different momenta, to a given momentum $\mathbf{q}$ upon scattering by V. This is given by the integral term. Here is a correspondence with transport phenomena, discussed earlier in this chapter, which may be employed to obtain a graphic understanding of quantum mechanics.

We have devoted some space to discussing problems which require integral operators. We have also pointed out that even those problems which are expressible in terms of a differential equation may be reformulated in terms of an integral equation. Several examples of this sort will now be given.

Boundary Conditions and Integral Equations. The integral equation formulation is of particular advantage in the case of boundary-value problems associated with partial differential equations. In the example which follows, a second-order partial differential equation in *two dimensions* will be restated as an integral equation in only *one dimension*. A reduction in the dimensionality is, of course, extremely worth while from the point of view of obtaining both exact and approximate solutions. Consider the Helmholtz equation

$$\nabla^2\psi + k^2\psi = 0$$

Place a barrier along the negative x axis as shown in Fig. 8.2. A plane wave $e^{i\mathbf{k}\cdot\mathbf{r}}$ traveling in the $\mathbf{k}$ direction is incident upon the barrier. We shall be interested in the effect of the barrier on this wave in the case that the solution ψ satisfies the boundary condition $\partial\psi/\partial y = 0$ on the barrier. This solution must satisfy the following conditions at large distance from the origin. In the lower half plane, $y < 0$,

$$\psi \xrightarrow[r\to\infty]{} 2\cos(k_y y)\,e^{ik_x x} + \text{a diverging cylindrical wave} \tag{8.1.5}$$

In the upper half plane, $y > 0$, there will be no reflected wave, so that

$$\psi \xrightarrow[r\to\infty]{} e^{i\mathbf{k}\cdot\mathbf{r}} + \text{a diverging cylindrical wave} \tag{8.1.6}$$

Because of the asymmetry of these boundary conditions we shall have to treat ψ for $y > 0$ and $y < 0$ somewhat differently, so that it will be necessary to verify that ψ and $(\partial\psi/\partial y)$ are continuous at the interface of these two regions, *i.e.*, for $y = 0$, $x > 0$. From the theory of Green's functions given in Chap. 7 we have the general equation for $y \leq 0$:

$$\psi(\mathbf{r}) = 2\cos(k_y y)\, e^{ik_x x} + \frac{1}{4\pi}\oint \left[G_k(\mathbf{r}|\mathbf{r}_0)\,\frac{\partial\psi}{\partial n_0} - \psi\,\frac{\partial G_k(\mathbf{r}|\mathbf{r}_0)}{\partial n_0} \right] dS_0$$

The path of integration is shown in Fig. 8.2. The proper function G_k is chosen by the requirement that for large $\mathbf{r}$ the proper boundary conditions be satisfied. It is clear that G_k must be a diverging source, in which

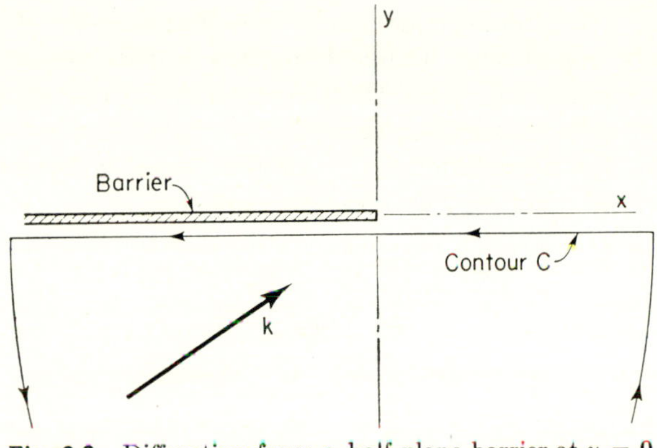

Fig. 8.2 Diffraction from a half-plane barrier at $y = 0$, $x < 0$. Contour for integral equation.

event the integral around the large semicircle vanishes. The integral along the x axis is then simplified by choosing $[\partial G_k(\mathbf{r}|x_0,y_0)/\partial y_0] = 0$ at $y_0 = 0$. Then

$$\psi(\mathbf{r}) = 2\cos(k_y y)\, e^{ik_x x} + \frac{1}{4\pi}\int_0^\infty G_k(\mathbf{r}|x_0,0)\left(\frac{\partial\psi}{\partial y_0}\right)_{y_0=0^-} dx_0; \quad y < 0 \qquad (8.1.7)$$

$$G_k(\mathbf{r}|\mathbf{r}_0) = \pi i[H_0(kR) + H_0(kR')]$$

$$R = \sqrt{(x-x_0)^2 + (y-y_0)^2}; \quad R' = \sqrt{(x-x_0)^2 + (y+y_0)^2}$$

In the region $y > 0$ employing the same G_k one finds

$$\psi(\mathbf{r}) = -\frac{1}{4\pi}\int_0^\infty G_k(\mathbf{r}|x_0,0)\left(\frac{\partial\psi}{\partial y_0}\right)_{y_0=0^+} dx_0; \quad y > 0 \qquad (8.1.8)$$

The boundary condition (8.1.6) for the region $y > 0$ is clearly satisfied.

It is now necessary to introduce continuity conditions. At $y = 0$, the function $\psi(x_0,0)$ computed from (8.1.7) and from (8.1.8) should agree. Hence, for $x > 0$,

$$2e^{ik_x x} = -\frac{1}{4\pi}\int_0^\infty G_k(x,0|x_0,0)\left[\left(\frac{\partial\psi}{\partial y_0}\right)_{y_0=0^+} + \left(\frac{\partial\psi}{\partial y_0}\right)_{y_0=0^-}\right] dx_0$$

Moreover the slopes must be continuous; *i.e.*,

$$\left(\frac{\partial\psi}{\partial y_0}\right)_{y_0=0^+} = \left(\frac{\partial\psi}{\partial y_0}\right)_{y_0=0^-}$$

Hence
$$2e^{ik_xx} = -\frac{1}{2\pi}\int_0^\infty G_k(x,0|x_0,0)\left(\frac{\partial\psi}{\partial y_0}\right)_{y_0=0} dx_0; \quad x > 0$$

or
$$2e^{ik_xx} = -i\int_0^\infty H_0^{(1)}[k|x - x_0|]\left(\frac{\partial\psi}{\partial y_0}\right)_{y_0=0} dx_0 \tag{8.1.9}$$

This is an integral equation for $(\partial\psi/\partial y_0)_{y_0=0}$. Once $(\partial\psi/\partial y_0)_{y_0=0}$ is known it can be substituted into (8.1.7) and (8.1.8) to obtain the diffracted as well as reflected wave. One needs another equation for $x < 0$.

Note that the integral equation is one-dimensional. It includes all the boundary conditions. Other boundary conditions on the barrier or on the aperture (*i.e.*, other than continuity on the aperture) would have led to another integral equation.

Equations for Eigenfunctions. Another type of integral equation may be obtained from the Schroedinger equation:

$$\left[\nabla^2 + \frac{2m}{\hbar^2}(E - V)\right]\psi = 0$$

or
$$[\nabla^2 + k^2 - U]\psi = 0 \tag{8.1.10}$$
$$k^2 = (2m/\hbar^2)E; \quad U = (2m/\hbar^2)V$$

Rewriting (8.1.10) as $(\nabla^2 + k^2)\psi = U\psi$, we see that a solution of (8.1.10) is given by

$$\psi(\mathbf{r}) = -\frac{1}{4\pi}\int G_k(\mathbf{r}|\mathbf{r}_0)U(\mathbf{r}_0)\psi(\mathbf{r}_0)\,dV_0 \tag{8.1.11}$$

where G_k has been chosen so as to ensure satisfaction of boundary conditions for ψ. Equation (8.1.11) is an integral equation for ψ. It differs from (8.1.9) in that ψ appears both within and outside the integral.

To obtain a better understanding of (8.1.11) it is convenient to turn to a one-dimensional example. We consider the Sturm-Liouville problem discussed in Sec. 6.3. The unknown function ψ satisfies

$$\frac{d}{dz}\left[p\frac{d\psi}{dz}\right] + [q(z) + \lambda r(z)]\psi = 0 \tag{8.1.12}$$

To perform the reduction from Eq. (8.1.10) to Eq. (8.1.11) here, the Green's function is introduced, which satisfies

$$\frac{d}{dz}\left[p\frac{dG(z|z_0)}{dz}\right] + qG(z|z_0) = -\delta(z - z_0) \tag{8.1.13}$$

We now must add the boundary conditions to be satisfied by G and ψ. For definiteness one assumes that $\psi(l)$ and $\psi(0)$ are known, *i.e.*, Dirichlet conditions on ψ. The corresponding conditions on G are $G(0|z_0) = 0 = G(l|z_0)$. Then, transposing $(\lambda r\psi)$ in (8.1.12) to the right-hand side of the equation and considering it to be an inhomogeneous term, one finds

$$\psi(z) = \lambda \int_0^l G(z|z_0)r(z_0)\psi(z_0)\, dz_0 + \left[\psi(0)p(0)\,\frac{dG(z_0|z)}{dz_0}\right]_{z_0=0} - \left[\psi(l)p(l)\left(\frac{dG(z_0|z)}{dz_0}\right)\right]_{z_0=l} \tag{8.1.14}$$

This is an integral equation for ψ. The boundary conditions on ψ enter explicitly, indicating again how the integral equation includes all the pertinent data on the problem. There are no additional conditions on ψ to be satisfied. If the boundary conditions satisfied by ψ are homogeneous Dirichlet, $\psi(0) = 0$, $\psi(l) = 0$, then the integral equation for ψ becomes

$$\psi(z) = \lambda \int_0^l G(z|z_0)r(z_0)\psi(z_0)\, dz_0 \tag{8.1.15}$$

Eigenfunctions and Their Integral Equations. Let us illustrate this last equation by giving the integral equation satisfied by the classical orthogonal functions:

(*a*)

$$\frac{d^2\psi}{dz} + \lambda\psi = 0; \quad \psi(l) = \psi(0) = 0$$

$$\psi(z) = \lambda \int_0^l G(z|z_0)\psi(z_0)\, dz_0$$

$$G(z|z_0) = \frac{1}{l}\begin{cases} z(l - z_0); & z < z_0 \\ z_0(l - z); & z > z_0 \end{cases}$$

Solution: $\sin(n\pi z/l)$; $\lambda = (n\pi/l)^2$; n integer

(*b*)

$$\frac{d}{dz}\left[(1 - z^2)\,\frac{d\psi}{dz}\right] + \lambda\psi = 0; \quad \psi \text{ finite at } z = \pm 1$$

$$\psi(z) = \lambda \int_{-1}^1 G(z|z_0)\psi(z_0)\, dz_0 - \tfrac{1}{2}\int_{-1}^1 \psi(z_0)dz_0$$

$$G(z|z_0) = \tfrac{1}{2}\begin{cases} \ln\,[(1 + z)/(1 - z_0)]; & z < z_0 \\ \ln\,[(1 + z_0)/(1 - z)]; & z > z_0 \end{cases}$$

Solution: *Legendre polynomials* $P_n(z)$; $\lambda = (n)(n + 1)$; n integer

(*c*)

$$\frac{1}{z}\frac{d}{dz}\left[z\,\frac{d\psi}{dz}\right] + \left[\lambda - \frac{n^2}{z^2}\right]\psi = 0; \quad \psi \text{ finite at } z = 0,\ \infty$$

$$\psi(z) = \lambda \int_0^\infty G(z|z_0)\psi(z_0)z_0\, dz_0; \quad G(z|z_0) = \left(\frac{1}{2n}\right)\begin{cases} (z/z_0)^n; & z < z_0 \\ (z_0/z)^n; & z > z_0 \end{cases}$$

Solution: *Bessel functions* $J_n(\sqrt{\lambda}\, z)$

(*d*) $$\frac{d^2\psi}{dz^2} + [\beta^2 - \alpha^2 z^2]\psi = 0; \quad \psi(\infty),\ \psi(-\infty) \text{ finite}$$

or $$\frac{d^2\psi}{dz^2} + [\lambda - \alpha - \alpha^2 z^2]\psi = 0; \quad \lambda = \beta^2 + \alpha$$

$$\psi(z) = \lambda \int_{-\infty}^{\infty} G(z|z_0)\psi(z_0)\,dz_0$$

$$G(z|z_0) = \sqrt{\frac{\alpha}{\pi}} \begin{cases} e^{\frac{1}{2}\alpha z^2} \int_{-\infty}^{z} e^{-\alpha\xi^2}\,d\xi\, e^{\frac{1}{2}\alpha z_0^2} \int_{z_0}^{\infty} e^{-\alpha\xi^2}\,d\xi; & z < z_0 \\ e^{\frac{1}{2}\alpha z_0^2} \int_{-\infty}^{z_0} e^{-\alpha\xi^2}\,d\xi\, e^{\frac{1}{2}\alpha z^2} \int_{z}^{\infty} e^{-\alpha\xi^2}\,d\xi; & z > z_0 \end{cases}$$

Solution: *Hermite functions* $e^{-\frac{1}{2}\alpha z^2} H_n(\sqrt{\alpha}\, z)$; $\lambda = 2(n+1)\alpha$; n integer

(*e*) $$\frac{1}{z^2}\frac{d}{dz}\left[z^2 \frac{d\psi}{dz}\right] + \left[-\beta^2 + \frac{2\alpha}{z}\right]\psi = 0; \quad \psi(0),\ \psi(\infty) \text{ finite}$$

We may identify λ with either 2α or $\alpha^2 - \beta^2$. In the first case,

$$\psi(z) = \lambda \int_0^{\infty} G_0(z|z_0) z_0 \psi(z_0)\,dz_0; \quad \lambda = 2\alpha; \quad G(z|z_0) = \frac{e^{-\beta|z-z_0|}}{2\beta z z_0}$$

Solutions: $e^{-\beta z} L_n^1(2\beta z)$ where $L_n(2\beta z)$ are the

Laguerre polynomials; $(\alpha/\beta) - 1 = n$; n integer

In the second case an equivalent integral equation is

(*f*) $$\frac{1}{z^2}\frac{d}{dz}\left[z^2 \frac{d\psi}{dz}\right] + \left[\lambda - \alpha^2 + \frac{2\alpha}{z}\right]\psi = 0; \quad \lambda = \alpha^2 - \beta^2$$

$$\psi(z) = \lambda \int_0^{\infty} G(z|z_0) z_0^2 \psi(z_0)\,dz_0$$

$$G(z|z_0) = e^{-\alpha(z+z_0)} \begin{cases} \int_{z_0}^{\infty} \frac{e^{2\alpha\xi}}{\xi^2}\,d\xi; & z < z_0 \\ \int_{z}^{\infty} \frac{e^{2\alpha\xi}}{\xi^2}\,d\xi; & z > z_0 \end{cases}$$

Types of Integral Equations; Fredholm Equations. We may now proceed to classify the integral equations already discussed and to generalize somewhat. Turning to Eq. (8.1.14), we see that it may be written

$$\psi(z) = \lambda \int_a^b K(z|z_0)\psi(z_0)\,dz_0 + \varphi(z) \tag{8.1.16}$$

where in that case $K(z|z_0) = r(z_0)G(z|z_0)$ and $\varphi(z)$ is a known function; a and b are the fixed points at which ψ satisfies boundary conditions. An integral equation for ψ of the form of (8.1.16) is called an *inhomogeneous Fredholm equation of the second kind*. The quantity $K(z|z_0)$ is called the *kernel* of the integral equation. A kernel is *symmetric* if $K(z|z_0) = K(z_0|z)$. In (8.1.14) the kernel is not symmetric if $r(z_0) \neq 1$.

A homogeneous Fredholm equation of the second kind may be obtained by omitting $\varphi(z)$:

$$\psi(z) = \lambda \int_a^b K(z|z_0)\psi(z_0)\, dz_0 \tag{8.1.17}$$

Equation (8.1.15) and the cases (*a*) to (*e*) tabulated immediately below it are examples of homogeneous Fredholm equations of the second kind. In cases (*a*), (*b*), (*d*), the kernel is symmetric. Cases (*e*) and (*f*) and the kernel in (8.1.14) are examples of a *polar kernel:*

$$K(z|z_0) = G(z|z_0)r(z_0); \quad \text{where } G(z|z_0) = G(z_0|z) \tag{8.1.18}$$

The kernels in every one of the cases are *definite* in the region of interest, *i.e.*, in the range $0 \leq z \leq l$ for case (*a*), $0 \leq z < \infty$ for case (*c*), etc. A *positive definite* kernel satisfies the inequality

$$\int_a^b dz_0 \int_a^b dz[K(z|z_0)\bar{\psi}(z)\psi(z_0)] > 0$$

where K is presumed real and ψ is arbitrary. For a *negative definite* kernel the above integral is always less than zero. In either case the kernel is definite. In the event that neither inequality is true for an arbitrary ψ, the kernel is said to be *indefinite*.

Equation (8.1.9) is an example of a *Fredholm equation of the first kind* which has the general form

$$\varphi(z) = \int_a^b K(z|z_0)\psi(z_0)\, dz_0 \tag{8.1.19}$$

Here ψ is the unknown function, φ is known.

Volterra Equations. Fredholm integral equations (8.1.16), (8.1.17), and (8.1.19) involve definite integrals. When the limits are variable, the corresponding equations are *Volterra equations*. An *inhomogeneous Volterra equation of the second kind* is, corresponding to (8.1.16),

$$\psi(z) = \int_a^z K(z|z_0)\psi(z_0)\, dz_0 + \varphi(z) \tag{8.1.20}$$

In the homogeneous variety $\varphi = 0$. A *Volterra equation of the first kind,* corresponding to (8.1.19), is

$$\varphi(z) = \int_a^z K(z|z_0)\psi(z_0)\, dz_0 \tag{8.1.21}$$

The Volterra equation, when convenient, may be considered as a special case of the corresponding Fredholm equation where the kernel employed in the Fredholm equation is

$$M(z|z_0) = \begin{cases} K(z|z_0); & z_0 < z \\ 0; & z_0 > z \end{cases} \tag{8.1.22}$$

In the preceding chapter, we have met Green's functions which satisfy Eq. (8.1.22). It will be recalled that, in discussing time dependent problems, it turned out that $G(\mathbf{r},t|\mathbf{r}_0,t_0) = 0$ for $t < t_0$, as a consequence of the principle of causality, which requires that an event at a time t_0 cannot cause any effects at a time t earlier than t_0. We may expect that an integral equation involving G as the kernel will be of the Volterra type.

To illustrate the origin of the Volterra equation, consider the motion of a simple harmonic oscillator:

$$(d^2\psi/dt_0^2) + k^2\psi = 0$$

Define a Green's function $G(t|t_0)$ for the impulse function $\delta(t - t_0)$ by

$$d^2G(t|t_0)/dt_0^2 = -\delta(t - t_0) \quad \text{and} \quad G(t|t_0) = 0; \quad \text{if } t < t_0$$

Multiplying the first of these equations by G and the second by ψ, subtracting the two, and integrating on t_0 from $t_0 = 0$ to $t_0 = t^+$ (where t^+ signifies taking the limit of the integral as $t_0 \to t$ from the side for which $t_0 > t$), we have

$$\int_0^{t^+} dt_0 \left\{G(t|t_0)\frac{d^2\psi}{dt_0^2} - \frac{d^2G(t|t_0)}{dt_0^2}\psi\right\} + k^2\int_0^{t^+} G(t|t_0)\psi(t_0)\,dt_0 = \psi(t)$$

or

$$\psi_0\left[\frac{dG\,(t|t_0)}{dt_0}\right]_{t_0=0} - G(t|0)v_0 + k^2\int_0^{t^+} G(t|t_0)\psi(t_0)\,dt_0 = \psi(t) \quad (8.1.23)$$

where ψ_0 and v_0 are the initial values of the displacement ψ and the velocity $d\psi/dt$. Equation (8.1.23) is an inhomogeneous Volterra equation of the second kind. The initial conditions on ψ are explicitly contained in it.

From the example it is clear that Volterra equations will result whenever there is a preferred direction for the independent variable; in the above example this is one of increasing time. Another case of a similar kind occurs in transport theory whenever collisions are made with massive scattering centers. Then the energy of the scattered particles cannot be greater than their energy prior to the scattering. As a consequence there is a degradation in energy giving a preferred direction to the energy variable.

As an example of this sort of equation consider a beam of X rays traversing a material in the direction of increasing x. We shall neglect the change in direction in scattering, presuming that the X rays all propagate directly forward after scattering. In passing through a thickness dx, the number of X rays of a given wavelength are depleted by absorption and scattering out of that wavelength range and are increased by scattering from those X rays whose energy (energy $\propto 1/\lambda$) is greater (in other words, whose wavelength is shorter). Hence if

$f(\lambda,x)\, d\lambda$ is the relative number of X rays with wavelength between λ and $\lambda + d\lambda$, then

$$\frac{\partial f(\lambda,x)}{\partial x} = -\mu f(\lambda,x) + \int_0^\lambda P(\lambda|\lambda_0) f(\lambda_0,x)\, d\lambda_0$$

where μ represents the absorption coefficient and $P(\lambda|\lambda_0)\, d\lambda$ is the probability, per unit thickness, that an X ray having a wavelength λ_0 is scattered into a wavelength region $d\lambda$ at λ. This is an integral-differential equation; to reduce it to a pure integral equation let

$$f(\lambda,x) = \int_0^\infty e^{-px} \psi(\lambda,p)\, dp$$

Then $\psi(\lambda,p)$ satisfies the homogeneous Volterra equation of the second kind:

$$(\mu - p)\psi(\lambda,p) = \int_0^\lambda P(\lambda|\lambda_0)\psi(\lambda_0,p)\, d\lambda_0$$

8.2 *General Properties of Integral Equations*

In discussing the general properties of integral equations it will be convenient to utilize the results obtained for operator equations in abstract vector space. As we shall now see, the Fredholm integral equation is just a transcription of an operator equation to ordinary space. Consider the inhomogeneous equation in vector space:

$$\mathfrak{A} \cdot \mathbf{e} = \lambda \mathbf{e} + \mathbf{f} \tag{8.2.1}$$

Since the Green's function which occurs so frequently in the formulation of integral equations is intimately connected with the inverse operator (see page 882), it will be profitable to rewrite (8.2.1) as

$$\mathbf{e} = \lambda \mathfrak{A}^{-1} \cdot \mathbf{e} + \mathbf{g}; \quad \mathbf{g} = \mathfrak{A}^{-1} \cdot \mathbf{f} \tag{8.2.2}$$

To transcribe this equation to ordinary space let us expand the vectors in terms of the eigenvectors of the operator z, $\mathbf{e}(z_0)$. Let

$$\begin{aligned} \mathbf{e} &= \int \mathbf{e}(z_0)\psi(z_0)\, dz_0 \\ \mathbf{g} &= \int \mathbf{e}(z_0)\varphi(z_0)\, dz_0 \\ \mathfrak{A}^{-1} \cdot \mathbf{e}(z_0) &= \int \mathbf{e}(z_1) K(z_1|z_0)\, dz_1 \\ \mathfrak{A}^{-1} \cdot \mathbf{e} &= \int \mathbf{e}(z_0)\, dz_0 \int K(z_0|z_1)\psi(z_1)\, dz_1 \end{aligned} \tag{8.2.3}$$

Introducing these definitions into (8.2.2) yields an inhomogeneous Fredholm equation of the second kind:

$$\psi(z_0) = \lambda \int K(z_0|z_1)\, \psi(z_1)\, dz_1 + \varphi(z_0) \tag{8.2.4}$$

The limits of integration are included in the definition of K.

The integral equation is thus often equivalent to a differential equation "handled in reverse." Instead of the differential operator, equivalent to $\mathfrak{A}$, which is studied, it is the integral operator, symbolized by $\mathfrak{A}^{-1}$, which is examined. The Fredholm equation of the first kind

$$-\varphi(z_0) = \int K(z_0|z_1)\, \psi(z_1)\, dz_1$$

is equivalent to the operator equation

$$-\mathbf{g} = \mathfrak{A}^{-1} \cdot \mathbf{e} \tag{8.2.5}$$

which is to be solved for $\mathbf{e}$ (or ψ, in coordinate components). The homogeneous Fredholm equation of the second kind

$$\psi(z_0) = \lambda \int K(z_0|z_1)\, \psi(z_1)\, dz_1$$

is equivalent to $\mathbf{e} = \lambda\mathfrak{A}^{-1} \cdot \mathbf{e}$, and the inhomogeneous form (8.2.4) is equivalent to Eq. (8.2.2).

The kernel $K(x_0|x_1)$ enters into all three forms of the Fredholm equation; it is represented by the operator $\mathfrak{A}^{-1}$. Thus it behooves us to examine the properties of K and to deduce from it the properties of the related operator $\mathfrak{A}^{-1}$. Not all operators have an inverse; if there is an operator $\mathfrak{A}^{-1}$, representing a kernel K of an integration, there may not be an operator $\mathfrak{A}$, corresponding to a differential operation (for example, if $\mathfrak{A}^{-1} \cdot \mathbf{f} = 0$ for some $\mathbf{f}$, there is no unique inverse). It is useful to see what types of kernels correspond to operators which do have an inverse $\mathfrak{A}$ and, if they do, whether the differential operator related to $\mathfrak{A}$ is self-adjoint, and so on.

The homogeneous Fredholm equation of the second kind [Eq. (8.2.4) with $\varphi = 0$], if $\mathfrak{A}^{-1}$ has an inverse, is equivalent to the eigenvalue equation $\mathfrak{A} \cdot \mathbf{e} = \lambda\mathbf{e}$ or its differential representation; therefore we should expect λ to take on a sequence of allowed values $\lambda = \lambda_0, \lambda_1, \lambda_2, \ldots$ $(\lambda_{n+1} > \lambda_n)$ with corresponding eigenfunction solutions ψ_n. The solutions of the inhomogeneous equation may be expected to be related to the Green's function solutions of $\mathfrak{A} \cdot \mathbf{e} = \lambda\mathbf{e} + \mathbf{f}$ discussed in Chap. 7. Although we may not wish to transform the integral equation into a differential one before solving, we shall nevertheless wish to keep in mind the reciprocal relationship symbolized by the operators $\mathfrak{A}^{-1}$ and $\mathfrak{A}$ as we study the properties of the various kernels encountered in integral equations.

Kernels of Integral Equations. An integral equation is more or less amenable to solution depending on two properties of its kernel: symmetry and the property which is symbolized by saying that the corresponding operator $\mathfrak{A}^{-1}$ has an inverse $\mathfrak{A}$. A *symmetric* kernel corresponds to an operator which is self-adjoint. If there is an $\mathfrak{A}$, then

$$\mathfrak{A} = \tilde{\mathfrak{A}} \tag{8.2.6}$$

when $K(x_0|x_1) = K(x_1|x_0)$.

A kernel is called *definite* if $\int \bar{f}(x_0)\,dx_0 \int K(x_0|x_1) f(x_1)\,dx_1$ is either always positive for any f or else always negative (either of which implies that the integral is *real*). This corresponds to the operator statement that, for any e, either

$$\begin{aligned} &(\mathbf{e}^* \cdot \mathfrak{A}^{-1} \cdot \mathbf{e}) > 0; \quad \text{positive definite} \\ \text{or} \quad &(\mathbf{e}^* \cdot \mathfrak{A}^{-1} \cdot \mathbf{e}) < 0; \quad \text{negative definite} \end{aligned} \tag{8.2.7}$$

In either case we can be sure that the operator $\mathfrak{A}$, inverse to $\mathfrak{A}^{-1}$, exists. (We note that, if $\mathfrak{A}^{-1}$ is definite, then $\mathfrak{A}$ is definite and both are Hermitian.) Unfortunately not all kernels are definite; we can often find a function, corresponding to an abstract vector $\mathbf{e}$, such that $(\mathbf{e}^* \cdot \mathfrak{A}^{-1} \cdot \mathbf{e}) = 0$, in which case the inverse $\mathfrak{A}$ is not unique. In some cases $(\mathbf{e}^* \cdot \mathfrak{A}^{-1} \cdot \mathbf{e}) \geq 0$ or $(\mathbf{e}^* \cdot \mathfrak{A}^{-1} \cdot \mathbf{e}) \leq 0$, the zero value being included. Such kernels and operators are called *semidefinite* (though the term might be misleading, for a semidefinite kernel is not much better than any other nondefinite kernel; in this case "a miss is as good as a mile").

Some integral equations have a nonsymmetric and/or a nondefinite kernel in their original form; it is useful to know whether we can modify the equation to produce one with a symmetric, definite kernel.

A *polar* kernel [see Eq. (8.1.18)] has the form

$$K(x|x_0) = r(z_0) G(x|x_0)$$

where G is symmetric in x and x_0. This may be transformed into a symmetric kernel by changing the dependent variable, letting $\psi(z) = \varphi(z)/\sqrt{r(z)}$. For example, if $\psi = \varphi/\sqrt{r}$ is inserted into the homogeneous equation of the second kind,

$$\begin{aligned} \psi(z) &= \lambda \int G(z|z_0) r(z_0) \psi(z_0)\,dz_0 \\ \text{becomes} \quad \varphi(z) &= \lambda \int \sqrt{r(z)}\, G(z|z_0) \sqrt{r(z_0)}\, \varphi(z_0)\,dz_0 \end{aligned}$$

The new kernel $\sqrt{r(z)}\, G(z|z_0) \sqrt{r(z_0)}$ is symmetric. In spite of the direct reduction given here, it is nevertheless useful to introduce an operator equation related to the polar one

$$\mathfrak{A} \cdot \mathbf{e} = \lambda \mathfrak{B} \cdot \mathbf{e} \tag{8.2.8}$$

where $\mathfrak{A}$ is symmetric. The analogue to the transformation from ψ to φ may be followed here. Let

$$\begin{aligned} \mathbf{f} &= \sqrt{\mathfrak{B}} \cdot \mathbf{e} \\ \text{then} \quad \mathbf{f} &= \lambda\, (\sqrt{\mathfrak{B}} \cdot \mathfrak{A}^{-1} \cdot \sqrt{\mathfrak{B}}) \cdot \mathbf{f} \end{aligned} \tag{8.2.9}$$

Actually the polar equation is a very special case of the more general form (8.2.8). To see this, transform $\mathbf{e} = \lambda \mathfrak{A}^{-1} \mathfrak{B} \cdot \mathbf{e}$ to ordinary space. Then

$$\begin{aligned} \psi(z) &= \lambda \int M(z|z_0) \psi(z_0)\,dz_0 \\ \text{where} \quad M(z|z_0) &= \int K(z|z_1) L(z_1|z_0)\,dz_1 \end{aligned}$$

and where

$$\mathfrak{A}^{-1} \cdot \mathbf{e}(z) = \int K(z|z_0)\mathbf{e}(z_0)\, dz_0; \quad \mathfrak{B} \cdot \mathbf{e}(z) = \int L(z|z_0)\mathbf{e}(z_0)\, dz_0$$

The special case of the polar kernel occurs when

$$L(z_1|z_0) = \delta(z_1 - z_0)r(z_0)$$

where δ is the Dirac δ function. Computing the integral form yields

$$M(z|z_0) = K(z|z_0)r(z_0)$$

Transformation to Definite Kernels. We wish, if possible, to transform our equation to one with a *symmetric, definite* kernel. For convenience we can choose the sign of K so that the kernel is positive definite. By making K definite, we have automatically limited our problem to one with real kernels and the operators in abstract vector space to Hermitian operators. Moreover, no eigenvector will then exist for which $\mathfrak{A} \cdot \mathbf{e} = 0$. Before starting the discussion it is important to list those operators which may, by some transformation, be reduced to an operator of the above type. Generally speaking *these transformed operators will be semidefinite rather than definite,* for if a vector $\mathbf{e}$ exists such that $\mathfrak{A} \cdot \mathbf{e} = 0$, then $\mathbf{e}$ operated upon by the transformed vector will also be zero.

Any Hermitian real operator may be transformed into a definite operator by iteration. Let

$$\mathfrak{A} \cdot \mathbf{e} = \lambda \mathbf{e}$$

Operating on this equation with $\mathfrak{A}$, that is, *iterating,* we obtain

$$\mathfrak{A}^2 \cdot \mathbf{e} = \lambda^2 \mathbf{e}$$

The operator $\mathfrak{A}^2$ is at least semidefinite, for

$$\mathbf{e}^* \cdot \mathfrak{A}^2 \cdot \mathbf{e} = (\mathfrak{A} \cdot \mathbf{e})^* \cdot (\mathfrak{A} \cdot \mathbf{e}) \geq 0$$

The kernel corresponding to $\mathfrak{A}^{-2}$ is

$$K_2(x|x_0) = \int K(x|x_1)K(x_1|x_0)\, dx_1 \tag{8.2.10}$$

and

$$\psi(x) = \lambda^2 \int K_2(x|x_0)\psi(x_0)\, dx_0; \quad \text{if } \psi(x) = \lambda \int K(x|x_0)\psi(x_0)\, dx_0$$

As a first corollary of this theorem, we note that a skew-Hermitian operator defined by $\mathfrak{A}^* = -\mathfrak{A}$ may also be transformed into a definite operator. A skew-Hermitian operator corresponds to an "antisymmetric" kernel defined by

$$K(x|x_0) = -K(x_0|x) \tag{8.2.11}$$

if K is real; if K is complex, its real part changes sign, its imaginary part does not change, when x and x_0 are interchanged. To prove the corollary, note that a skew-Hermitian operator $\mathfrak{A}$ may be written as

$$\mathfrak{A} = i\mathfrak{B}$$

where $\mathfrak{B}$ is Hermitian. Then $\mathfrak{A} \cdot \mathbf{e} = \lambda\mathbf{e}$ becomes

$$\mathfrak{B} \cdot \mathbf{e} = -i\lambda\mathbf{e}$$

Operating on this equation with $\mathfrak{B}$, one obtains

$$\mathfrak{B}^2 \cdot \mathbf{e} = (-\lambda^2)\mathbf{e} \tag{8.2.12}$$

so that $\mathfrak{B}^2$ is at least semidefinite.

A second corollary is that a polar integral equation may be reduced to an integral equation in which the kernel is real, symmetric, and definite. Corresponding to this corollary, we can show that, if in the equation

$$\mathfrak{A} \cdot \mathbf{e} = \lambda\mathfrak{B} \cdot \mathbf{e}$$

where *either* $\mathfrak{A}$ or $\mathfrak{B}$ (but not both) is not definite (though $\mathfrak{A}$ and $\mathfrak{B}$ are either Hermitian or skew-Hermitian), then an eigenvalue equation for $\mathbf{e}$ can be found in which all the operators are definite. Let us prove this theorem for the case $\mathfrak{A}$ and $\mathfrak{B}$ Hermitian, $\mathfrak{B}$ positive definite, $\mathfrak{A}$ indefinite. If $\mathfrak{B}$ is positive definite, an inverse of $\mathfrak{B}$ exists. Then

$$(\mathfrak{B}^{-1} \cdot \mathfrak{A}) \cdot \mathbf{e} = \lambda\mathbf{e}$$
$$(\mathfrak{B}^{-1}\mathfrak{A}\mathfrak{B}^{-1}\mathfrak{A}) \cdot \mathbf{e} = \lambda^2\mathbf{e}$$

or

$$(\mathfrak{A}\mathfrak{B}^{-1}\mathfrak{A}) \cdot \mathbf{e} = \lambda^2\mathfrak{B}\mathbf{e}$$

The new operator $(\mathfrak{A}\mathfrak{B}^{-1}\mathfrak{A})$ is at least semidefinite.

It is characteristic of the above reductions that the final kernel or operator is independent of λ. A reduction to a Hermitian operator may be made for any operator, but for most cases the final operator will be dependent on λ. To see this, consider

$$(\mathfrak{A} - \lambda) \cdot \mathbf{e} = 0$$

Operate on this equation with $(\mathfrak{A}^* - \bar{\lambda})$ to obtain

$$(\mathfrak{A}^* - \bar{\lambda})(\mathfrak{A} - \lambda) \cdot \mathbf{e} = 0 \quad \text{or} \quad [\lambda\mathfrak{A}^* + \bar{\lambda}\mathfrak{A} - \mathfrak{A}^*\mathfrak{A}] \cdot \mathbf{e} = |\lambda|^2\mathbf{e}$$

The operator $\lambda\mathfrak{A}^* + \bar{\lambda}\mathfrak{A} - \mathfrak{A}^*\mathfrak{A}$ is, of course, Hermitian, and from its derivation it is also definite. However, its use in place of the simpler equation $\mathfrak{A} \cdot \mathbf{e} = \lambda\mathbf{e}$ would be difficult because of the presence of the λ and $\bar{\lambda}$ in the operator itself. We shall therefore not pursue this particular avenue any further. Later on in this section we shall discuss an alternate and more practical procedure.

To summarize, the following types of eigenvalue problems may be transformed into problems having an operator which is at least semidefinite and independent of the eigenvalue λ. If $\mathfrak{A} \cdot \mathbf{e} = \lambda\mathbf{e}$, then the transformation is possible if $\mathfrak{A}$ is Hermitian or skew-Hermitian. If $\mathfrak{A} \cdot \mathbf{e} = \lambda\mathfrak{B} \cdot \mathbf{e}$, then $\mathfrak{A}$ and $\mathfrak{B}$ may be either Hermitian or skew-Hermitian but one of the two must be definite.

Properties of the Symmetric, Definite Kernel. We turn now to a real, positive definite, symmetric kernel. The corresponding operator is Hermitian. From Sec. 6.3, we may extract the following theorems. The homogeneous equation $\mathfrak{A} \cdot \mathbf{e} = \lambda \mathbf{e}$ has nonzero solutions for only particular values of λ, λ_m, the *eigenvalues.* The corresponding *eigenvectors* $\mathbf{e}_m$ form a mutually orthogonal set of vectors, possibly finite in number, which we are permitted to normalize:

$$\mathbf{e}_n^* \cdot \mathbf{e}_m = \delta_{nm} \tag{8.2.13}$$

We assume here that there are no degenerate eigenvalues. The eigenvalues are real and may be arranged in an ascending sequence. There is a smallest eigenvalue λ_0 (greater than zero, if $\mathfrak{A}$ is positive definite); the next largest is λ_1; etc. These results may be based on a *variational* principle for λ:

$$\lambda = \min[\mathbf{e}^* \cdot \mathfrak{A} \cdot \mathbf{e}/\mathbf{e}^* \cdot \mathbf{e}] \tag{8.2.14}$$

A second variational principle, based on $\mathbf{e} = \lambda \mathfrak{A}^{-1} \cdot \mathbf{e}$, is

$$\lambda = \min[(\mathbf{e}^* \cdot \mathbf{e})/(\mathbf{e}^* \cdot \mathfrak{A}^{-1} \cdot \mathbf{e})] \tag{8.2.15}$$

The inhomogeneous problem corresponding to the inhomogeneous Fredholm equation of the second kind has solutions for all values of λ. It may be solved by application of the inverse of $\mathfrak{A} - \lambda$: If

$$\mathfrak{A} \cdot \mathbf{e} = \lambda \mathbf{e} + \mathbf{f}$$

then

$$\mathbf{e} = (\mathfrak{A} - \lambda)^{-1}\mathbf{f} = \mathfrak{G}_\lambda \cdot \mathbf{f} \tag{8.2.16}$$

where $\mathfrak{G}_\lambda$ is the Hermitian Green's operator [see Eq. (7.5.35)].

We now consider some properties of $\mathfrak{G}_\lambda$. First consider its relation to $\mathfrak{A}^{-1}$. From (8.2.16), it immediately follows that

$$\mathfrak{G}_0 = \mathfrak{A}^{-1}$$

the operator corresponding to the kernel K. Moreover,

$$\mathfrak{G}_\lambda = (\mathfrak{A} - \lambda)^{-1} = \mathfrak{A}^{-1}(1 - \lambda\mathfrak{A}^{-1})^{-1}$$

or

$$\mathfrak{G}_\lambda = \sum_m \mathfrak{A}^{-(n+1)}\lambda^n \tag{8.2.17}$$

if the series so represented converges. Corresponding to Eq. (8.2.16), we have the general relation obtained from (8.2.17) that

$$\left[\frac{1}{n!}\frac{\partial^n \mathfrak{G}_\lambda}{\partial \lambda^n}\right]_{\lambda=0} = \mathfrak{A}^{-(n+1)}$$

It is also true that $\mathfrak{G}_\lambda$ may be expanded in terms of the eigenvector $\mathbf{e}_m$, corresponding to the "inverse" equation, $(\mathfrak{A} - \lambda) \cdot \mathbf{e} = 0$:

$$\mathfrak{G}_\lambda = \sum_m \frac{\mathbf{e}_m \mathbf{e}_m^*}{\lambda_m - \lambda} \tag{8.2.18}$$

An expansion for $\mathfrak{A}^{-1}$ may be obtained by placing $\lambda = 0$:

$$\mathfrak{A}^{-1} = \mathfrak{G}_0 = \sum_m \frac{\mathbf{e}_m \mathbf{e}_m^*}{\lambda_m} \tag{8.2.19}$$

The corresponding expansion for $\mathfrak{A}^{-p}$ is

$$\mathfrak{A}^{-p} = \sum_m \frac{\mathbf{e}_m \mathbf{e}_m^*}{\lambda_m^p} \tag{8.2.20}$$

Turning back to the expansion for $\mathfrak{G}_\lambda$ (8.2.18) we note that, as a function of λ, $\mathfrak{G}_\lambda$ is singular at $\lambda = \lambda_m$, that is, at each of the eigenvalues of $\mathfrak{A}$. This is a perfectly general property, independent of the Hermitian or of the definite nature of $\mathfrak{A}$; for if $\mathfrak{A} \cdot \mathbf{e}_m = \lambda_m \mathbf{e}_m$, then

$$\mathfrak{G}_\lambda \cdot \mathbf{e}_m = (\mathfrak{A} - \lambda)^{-1} \cdot \mathbf{e}_m = (\lambda_m - \lambda)^{-1} \mathbf{e}_m$$

We see that, as $\lambda \to \lambda_m$, $\mathfrak{G}_\lambda \cdot \mathbf{e}_m \to \infty$.

The specific nature of the singularities with respect to λ may be most easily expressed by considering the *Spur* [we have also referred to this sum as *expansion factor* in Chaps. 1 and 3, see Eq. (1.6.3)] of $\mathfrak{G}_\lambda$. From Eq. (8.2.18), it equals

$$|\mathfrak{G}_\lambda| = \sum_m \frac{1}{\lambda_m - \lambda} \tag{8.2.21}$$

We note that $|\mathfrak{G}_\lambda|$ is a meromorphic function of λ. It has simple poles at $\lambda = \lambda_m$, each with residue -1. The scalar $|\mathfrak{G}_\lambda|$ may be expressed in terms of $|\mathfrak{A}^{-p}|$ from Eq. (8.2.17)

$$|\mathfrak{G}_\lambda| = \sum_n \lambda_n |\mathfrak{A}^{-(n+1)}|$$

where from (8.2.20)

$$|\mathfrak{A}^{-p}| = \sum_n \frac{1}{\lambda_m^p} \tag{8.2.22}$$

Let us now paraphrase these results in terms of the theory of integral equations. The eigenvalue equation, with K a *real, symmetric, positive-definite* kernel, the homogeneous equation of the second kind,

$$\psi(z) = \lambda \int K(z|z_0)\psi(z_0)\, dz_0$$

has nonzero solutions for only special values of λ, λ_m. Corresponding to these, there are particular ψ's, ψ_m, the eigenfunctions. These eigenfunctions form an orthonormal set, possibly finite in number:

$$\int \bar{\psi}_n(z)\psi_m(z)\, dz = \delta_{nm} \tag{8.2.23}$$

The set $\{\lambda_m\}$ forms a monotonically increasing sequence with a lower bound, λ_0 (> 0 if K is positive definite). A variational principal for λ, obtained from Eq. (8.2.15), is

$$\lambda = \min \left\{ \frac{\int \bar{\psi}(z)\psi(z)\, dz}{\int \bar{\psi}(z) K(z|z_0)\psi(z_0)\, dz\, dz_0} \right\} \tag{8.2.24}$$

Kernels and Green's Functions for the Inhomogeneous Equation. We turn next to the inhomogeneous equation

$$\psi(z) = \lambda \int K(z|z_0)\psi(z_0)\, dz_0 + \chi(z) \tag{8.2.25}$$

This is to be solved in terms of a Green's function. We shall choose one to correspond to the $\mathfrak{G}_\lambda$ of (8.2.16). In abstract vector space Eq. (8.2.25) reads

$$\mathbf{e} = \lambda \mathfrak{A}^{-1} \cdot \mathbf{e} + \mathbf{q}$$

The solution for $\mathbf{e}$ is obtained as follows:

$$(1 - \lambda \mathfrak{A}^{-1}) \cdot \mathbf{e} = \mathbf{q} \quad \text{or} \quad (\mathfrak{A} - \lambda) \cdot \mathbf{e} = \mathfrak{A} \cdot \mathbf{q}$$

Hence

$$\mathbf{e} = (\mathfrak{G}_\lambda \mathfrak{A}) \cdot \mathbf{q}; \quad \mathfrak{G} = (\mathfrak{A} - \lambda)^{-1}$$

This solution is not convenient, inasmuch as it involves the product of two operators. However, a little rearrangement of factors circumvents this difficulty. Write

$$\mathbf{e} = [\mathfrak{G}_\lambda(\mathfrak{A} - \lambda)] \cdot \mathbf{q} + \lambda \mathfrak{G}_\lambda \cdot \mathbf{q}$$

so that

$$\mathbf{e} = \mathbf{q} + \lambda \mathfrak{G}_\lambda \cdot \mathbf{q} \tag{8.2.26}$$

This is the solution of Eq. (8.2.25) in terms of $\mathfrak{G}_\lambda$. Expressed in ordinary space it reads

$$\psi(z) = \chi(z) + \lambda \int G_\lambda(z|z_0)\chi(z_0)\, dz_0 \tag{8.2.27}$$

The Green's function given here is referred to, in the theory of integral equations, as the *solving kernel.*

To obtain Eq. (8.2.27) directly without the intermediary of vector space it is necessary to formulate the integral equation for G_λ. This may be obtained from the defining equation (8.2.16) for $\mathfrak{G}_\lambda$, as follows:

$$(\mathfrak{A} - \lambda)\mathfrak{G}_\lambda = \mathfrak{I}$$

or

$$\mathfrak{A}^{-1}(\mathfrak{A} - \lambda)\mathfrak{G}_\lambda = \mathfrak{A}^{-1} = \mathfrak{G}_\lambda - \lambda \mathfrak{A}^{-1}\mathfrak{G}_\lambda$$

or

$$\mathfrak{G}_\lambda = \mathfrak{A}^{-1} + \lambda \mathfrak{A}^{-1}\mathfrak{G}_\lambda \tag{8.2.28}$$

Translated into ordinary space this becomes an integral equation for G_λ:

$$G_\lambda(z|z_0) = K(z|z_0) + \lambda \int K(z|z_1) G_\lambda(z_1|z_0)\, dz_1 \tag{8.2.29}$$

From this integral equation it is apparent that

$$G_0(z|z_0) = K(z|z_0)$$

Combining Eqs. (8.2.28) and the integral equation for ψ, (8.2.25), it is possible, by employing the symmetry of G_λ and K, to derive solution

(8.2.27). However, it should be emphasized that this symmetry involves the possibility of interchanging limits of integration. This is, of course, contained in the definition of symmetry.

It will be useful, for later discussion, to obtain an integral equation for K in terms of G_λ. Again turning to vector space, note that

$$(\mathfrak{A} - \lambda)\mathfrak{A}^{-1} = \mathfrak{J} - \mathfrak{A}^{-1}\lambda \quad \text{or} \quad \mathfrak{A}^{-1} = \mathfrak{G} - \lambda\mathfrak{G}\mathfrak{A}^{-1}$$

In ordinary space this is

$$K(z|z_0) = G_\lambda(z|z_0) - \lambda\int G_\lambda(z|z_1)K(z_1|z_0)\,dz_1 \tag{8.2.30}$$

The difference between this integral equation and the one for G_λ, Eq. (8.2.29), is only apparent. In Eq. (8.2.30) let z and z_0 be interchanged, so that

$$K(z_0|z) = G_\lambda(z_0|z) - \lambda\int G_\lambda(z_0|z_1)K(z_1|z)\,dz_1$$

Employing the symmetry properties of K and G_λ we are led to Eq. (8.2.29) from (8.2.30). This relation between K and G_λ led Volterra to introduce the term *reciprocal functions* to describe K and $-G_\lambda$.

Expansions corresponding to Eqs. (8.2.18) and (8.2.19) for $\mathfrak{G}_\lambda$ and $\mathfrak{A}^{-1}$ may be obtained for G_λ and K (see discussion on page 908):

$$G_\lambda(z|z_0) = \sum_m \frac{\psi_m(z)\bar{\psi}_m(z_0)}{\lambda_m - \lambda} \tag{8.2.31}$$

and

$$K(z|z_0) = \sum_m \frac{\psi_m(z)\bar{\psi}_m(z_0)}{\lambda_m} \tag{8.2.32}$$

To give the analogues for Eqs. (8.2.17), (8.2.20), and (8.2.22), it is first necessary to obtain the transcription of $\mathfrak{A}^{-p}$ to ordinary space. Corresponding to the expression $\mathfrak{A}^{-p}\cdot\mathbf{e}$ we have

$$\int K_p(z|z_0)\psi(z_0)\,dz_0$$

where it now becomes necessary to find $K_p(z|z_0)$. Consider first $\mathfrak{A}^{-1}\cdot\mathbf{e}$. In ordinary space this is written as

$$\int K(z|z_0)\psi(z_0)\,dz_0$$

The effect of $\mathfrak{A}^{-2}$ on $\mathbf{e}$ may be obtained as $\mathfrak{A}^{-1}\cdot(\mathfrak{A}^{-1}\cdot\mathbf{e})$ or

$$\iint K(z|z_1)K(z_1|z_0)\psi(z_0)\,dz_0\,dz_1$$

Hence

$$K_2(z|z_0) = \int K(z|z_1)K(z_1|z_0)\,dz_1$$

The effect of $\mathfrak{A}^{-3}$ on $\mathbf{e}$ may be obtained by operating with $\mathfrak{A}^{-1}$ on $\mathfrak{A}^{-2}\cdot\mathbf{e}$; hence

$$K_3(z|z_0) = \int K(z|z_1)K_2(z_1|z_0)\,dz_1$$

It is clear from the general relation $\mathfrak{A}^{-(p+q)} = \mathfrak{A}^{-p}\mathfrak{A}^{-q}$ that

$$K_{p+q}(z|z_0) = \int K_p(z|z_1)K_q(z_1|z_0)\,dz_1 = \int K_q(z|z_1)K_p(z_1|z_0)\,dz_1 \tag{8.2.33}$$

We may now write the equivalents of Eqs. (8.2.17), (8.2.20), and (8.2.22). The analogue of (8.2.17) is

$$G_\lambda(z|z_0) = \sum_{n=0}^{\infty} K_{n+1}(z|z_0)\lambda^n \tag{8.2.34}$$

The expansion for K_p is the analogue of that for $\mathfrak{A}^{-p}$ as given in (8.2.20):

$$K_p(z|z_0) = \sum_m \frac{\psi_m(z)\bar{\psi}_m(z_0)}{\lambda_m^p} \tag{8.2.35}$$

The statements as to the scalar values of $\mathfrak{G}_\lambda$ and $\mathfrak{A}^{-p}$ may be converted into statements about G_λ and K_p once the equivalent process to that of obtaining the scalar value $|\mathfrak{G}|$ is known. This consists of placing $z = z_0$ in the kernel, *i.e.*, obtaining the diagonal element of the matrix representing $\mathfrak{G}_\lambda$, and then integrating over z, the equivalent of summing over these diagonal elements. Hence

$$\int G_\lambda(z|z)\, dz = \sum_m \frac{1}{\lambda_m - \lambda} \tag{8.2.36}$$

while

$$\int K_p(z|z)\, dz = \sum_m \frac{1}{\lambda_m^p} = C_p \tag{8.2.37}$$

and

$$\int G_\lambda(z|z)\, dz = \sum_n C_{n+1}\lambda^n \tag{8.2.38}$$

Semidefinite and Indefinite Kernels. In many cases the kernels are not definite and the corresponding operators not Hermitian, so that the discussion above does not apply. Even after iteration, for some cases, the final operator is only semidefinite rather than definite, so that the above theorems are still not appropriate.

What may be said about kernels which are not definite? First of all it is no longer necessarily true that the eigenvalues are real. The eigenvalues again are in some cases finite in number. In the case of the Volterra equation, for example, there are no eigenvalues; no solutions of the homogeneous equation (see page 920). As an example consider the following simple, nondefinite kernel and the corresponding Fredholm integral equation:

$$\psi(z) = \lambda \int_0^1 (z - 2z_0)\psi(z_0)\, dz_0$$

It is clear that $\psi(z)$ is a linear function of z:

$$\psi(z) = \alpha z + \beta$$

The constants may be determined by introduction of the linear function into the integral equation

$$\alpha z + \beta = \lambda \int_0^1 (z - 2z_0)(\alpha z_0 + \beta)\, dz_0$$

Equating like powers of z, one obtains

$$\alpha = \lambda(\tfrac{1}{2}\alpha + \beta); \quad \beta = 2\lambda(-\tfrac{1}{3}\alpha - \tfrac{1}{2}\beta)$$

These are a pair of homogeneous linear simultaneous equations for α and β. For a nonzero solution the determinant of the coefficients must vanish. Hence

$$\begin{vmatrix} 1 - \frac{1}{2}\lambda & -\lambda \\ \frac{2}{3}\lambda & 1 + \lambda \end{vmatrix} = 0$$

There are two roots:

$$\lambda_1 = \tfrac{1}{2}(-3 + i\sqrt{15}); \quad \lambda_2 = \bar{\lambda}_1 = -\tfrac{1}{2}(3 + i\sqrt{15})$$

The corresponding solutions are

$$\psi_1 = z - \tfrac{1}{2} + (1/\lambda_1); \quad \psi_2 = z - \tfrac{1}{2} + (1/\lambda_2)$$

In this example there are only two eigenvalues and there are only two solutions; the two eigenvalues are complex conjugates.

Because the eigenvalues and eigenfunctions *may* be finite in number, it may be, of course, no longer possible to expand an arbitrary function in terms of these eigenfunctions. The statement of completeness must therefore be made only with respect to a certain class of functions. For example, the eigenfunctions for the problem considered just above can be employed to express any *linear* function of z. More important, the eigenfunctions are sufficiently complete to permit the expansion of the kernel $K(z|z_0)$ and the Green's function $G_\lambda(z|z_0)$. Hence it still remains possible to solve this inhomogeneous Fredholm equation of the second kind in terms of its eigenfunctions.

The necessary formalism for this situation has already been considered in Chap. 7. The eigenvalue problem in abstract vector space is

$$\mathfrak{A} \cdot \mathbf{e} = \lambda \mathbf{e} \tag{8.2.39}$$

We define the Hermitian *adjoint eigenvalue problem* [see Eq. (7.5.40)]:

$$\mathfrak{A}^* \cdot \mathbf{f} = \bar{\lambda} \mathbf{f} \tag{8.2.40}$$

The solutions of (8.2.39) are orthogonal to the solution of (8.2.40), so that we may set

$$\mathbf{f}_m^* \cdot \mathbf{e}_n = \delta_{nm} \tag{8.2.41}$$

This relation may be employed to evaluate the coefficients in the expansion of those vectors which may be expressed in terms of $\mathbf{e}_n$ [see Eq. (7.5.43)]:

$$\mathbf{g} = \sum_n g_n \mathbf{e}_n; \quad g_n = (\mathbf{f}_n^* \cdot \mathbf{g}) \tag{8.2.42}$$

Expansions of the Green's operator are also available:

$$\mathfrak{G}_\lambda = \sum_m \frac{\mathbf{e}_m \mathbf{f}_m^*}{\lambda_m - \lambda} \tag{8.2.43}$$

where $\mathfrak{G}_\lambda = (\mathfrak{A} - \lambda)^{-1}$

The Spur of an operator, when expanded in terms of $\mathbf{e}_m$ and $\mathbf{f}_m^*$, is defined by

$$|\mathfrak{A}| = \sum_m \mathbf{f}_m^* \cdot \mathfrak{A} \cdot \mathbf{e}_m \tag{8.2.44}$$

The expansion factors (Spurs) $|\mathfrak{A}^{-1}|$, $|\mathfrak{A}^{-p}|$, $|\mathfrak{G}_\lambda|$ and the expansion of $|\mathfrak{G}_\lambda|$ in terms of $|\mathfrak{A}^{-p}|$ are given in Eqs. (8.2.19) to (8.2.22), respectively. If there are only a finite number of eigenvalues, relations must exist between the Spurs of various powers of $|\mathfrak{A}^{-1}|$. If there are, say, q eigenvalues, it is possible to express these eigenvalues in terms of the Spurs of the first q powers of $|\mathfrak{A}^{-1}|$. Hence $|\mathfrak{A}^{-(q+1)}|$ may be expressed in terms of the scalars $|\mathfrak{A}^{-1}|, |\mathfrak{A}^{-2}|, \ldots, |\mathfrak{A}^{-q}|$.

The variational problem satisfied by the eigenvector solutions of Eqs. (8.2.39) and (8.2.40) is

$$\lambda = \text{stationary value of } (\mathbf{f}^* \cdot \mathfrak{A} \cdot \mathbf{e})/(\mathbf{f}^* \cdot \mathbf{e}) \tag{8.2.45}$$

By varying $\mathbf{f}^*$ in the above equation, we find [see Eq. (6.3.74)] that $\delta\lambda = 0$ yields Eq. (8.2.39), while by varying $\mathbf{e}$, we find that $\delta\lambda = 0$ yields Eq. (8.2.40) for $\mathbf{f}$. Equation (8.2.45) is similar to (8.2.14). A variational problem similar to (8.2.15):

$$\lambda = \text{stationary value of } (\mathbf{f}^* \cdot \mathbf{e})/(\mathbf{f}^* \cdot \mathfrak{A}^{-1} \cdot \mathbf{e}) \tag{8.2.46}$$

The case of the skew-Hermitian operator (corresponding to an antisymmetric real kernel) deserves special attention. In that case $\mathfrak{A}^* = -\mathfrak{A}$. Moreover, if

$$\mathfrak{A} \cdot \mathbf{e} = \lambda \mathbf{e}$$

then $(\mathfrak{A}^*\mathfrak{A}) \cdot \mathbf{e} = -\lambda^2 \mathbf{e}$

From the fact that $\mathfrak{A}^*\mathfrak{A}$ is definite, that is, $(\mathbf{e}^* \cdot \mathfrak{A}^*\mathfrak{A} \cdot \mathbf{e}) \geq 0$, it follows that $-\lambda^2 \geq 0$ or that λ *is a pure imaginary*. A second point of interest concerns the solutions of the adjoint eigenvalue problem. Whenever $\mathfrak{A}$ is skew-Hermitian, it follows that $\mathbf{e}_n = \mathbf{f}_n$; for if

$$\mathfrak{A} \cdot \mathbf{e}_n = \lambda_n \mathbf{e}_n$$

then $\mathfrak{A}^* \cdot \mathbf{e}_n = -\mathfrak{A} \cdot \mathbf{e}_n = -\lambda_n \mathbf{e}_n$

Turning to orthogonality relation (8.2.41), we see that the $\mathbf{e}_n$ vectors for skew-Hermitian operators are mutually orthogonal, as in the case of solutions of eigenvalue problems involving Hermitian operators. The expansion of the Green's operator can be carried through for the latter case, and hence the formulas for positive definite operators may be taken over bodily.

Kernel not Real or Definite. Let us now consider the application of Eqs. (8.2.39) to (8.2.46) to integral equations. Because of the greater complication of the present case we shall illustrate some of the results with the specific integral equation with the kernel $(z - 2z_0)$ discussed above.

The eigenvalue equation with a kernel which is *not* real, positive definite,

$$\psi(z) = \lambda \int K(z|z_0)\psi(z_0)\, dz_0 \tag{8.2.47}$$

has nonzero solutions for only special values of λ; call them λ_m. Corresponding to these values λ_m there are corresponding ψ's, ψ_m, the eigenfunctions. The λ_m's are not necessarily infinite in number; they are not generally real. The eigenfunction set, besides being not necessarily complete, is not mutually orthogonal. For this reason, we must introduce the Hermitian adjoint problem [corresponding to (8.2.40)]

$$\varphi(z) = \bar{\lambda} \int K^*(z_0|z)\varphi(z_0)\, dz_0 \tag{8.2.48}$$

As noted earlier, the eigenvalues of the adjoint problem are complex conjugates of those of (8.2.47). Moreover,

$$\int \bar{\varphi}_p \psi_q\, dz = \delta_{pq} \tag{8.2.49}$$

where φ_p and ψ_q are individual eigenfunctions.

The eigenfunctions and eigenvalues for the kernel $z - 2z_0$ are given above (page 916). The solutions φ_p satisfy the adjoint problem:

$$\varphi(z) = \mu \int_0^1 (z_0 - 2z)\varphi(z_0)\, dz_0$$

(This kernel K^* was obtained by exchanging z and z_0 in K and taking the complex conjugate of the result.)

Again note that $\varphi(z)$ is linear in z, so that $\varphi = az + b$. Then

$$-(a + 2b)\mu = a; \quad (\tfrac{1}{3}a + \tfrac{1}{2}b)\mu = b$$

Hence
$$\begin{vmatrix} \mu + 1 & 2\mu \\ \frac{1}{3}\mu & \frac{1}{2}\mu - 1 \end{vmatrix} = 0$$

This equation, determining μ, is precisely the same as that determining λ. Indeed, we may place

$$\mu_1 = \bar{\lambda}_1; \quad \mu_2 = \bar{\lambda}_2 = \lambda_1$$

However, φ_1 and ψ_1 are not complex conjugates of each other, for

$$\varphi_1 = z - (\tfrac{1}{2}) - (1/2\bar{\lambda}_1); \quad \varphi_2 = z - (\tfrac{1}{2}) - (1/2\bar{\lambda}_2)$$

which is to be compared with the earlier expressions for ψ_1 and ψ_2. It is now very easy to show that, in agreement with Eq. (8.2.49),

$$\int_0^1 \bar{\varphi}_1\psi_2\,dz = 0$$

The functions φ_i and ψ_i have not as yet been normalized in the manner determined by (8.2.49). The normalization integral is

$$\int_0^1 \bar{\varphi}_1\psi_1\,dz = \frac{1}{12} - \frac{1}{2\lambda_1^2}$$

and similarly for the second pair, φ_2 and ψ_2.

Let us now return to the general discussion, to the Green's function and the solution of the inhomogeneous Fredholm equation of the second kind. The solution of the inhomogeneous equation may again be expressed in terms of the Green's function. As in (8.2.27):

$$\psi(z) = \chi(z) + \lambda\int G_\lambda(z|z_0)\chi(z_0)\,dz_0$$

The integral equations for G_λ are again (8.2.28) and (8.2.30). The expansion for $G_\lambda(z|z_0)$ may be obtained from (8.2.43):

$$G_\lambda(z|z_0) = \sum_m \frac{\psi_m(z)\bar{\varphi}_m(z_0)}{\lambda_m - \lambda} \tag{8.2.50}$$

where, of course, ψ_m and φ_m are normalized according to Eq. (8.2.49). In the examples we have been examining,

$$G_\lambda(z|z_0) = \left(\frac{1}{12} - \frac{1}{2\lambda_1^2}\right)^{-1}\frac{\psi_1(z)\bar{\varphi}_1(z_0)}{\lambda_1 - \lambda} + \left(\frac{1}{12} - \frac{1}{2\lambda_2^2}\right)^{-1}\frac{\psi_2(z)\bar{\varphi}_2(z_0)}{\lambda_2 - \lambda}$$

where the normalization has been written in explicitly. Since $\psi_1 = \bar{\psi}_2$, $\varphi_1 = \bar{\varphi}_2$, and $\lambda_1 = \bar{\lambda}_2$, G_λ is real when λ is real. However, we note that $G_\lambda(z|z_0) \neq G_\lambda(z_0|z)$, so that the reciprocity principle is not satisfied. The reader will find no difficulty in proving that (8.2.27) is indeed a solution for arbitrary χ when the kernel is $(z - 2z_0)$ and the above Green's function is employed.

Volterra Integral Equation. The Volterra integral equation [see Eqs. (8.1.21) *et seq.*] furnishes us with an example of a kernel with no eigenvalues. This statement holds for any Volterra equation as long as the kernel is bounded. The Volterra integral equation may be cast in the Fredholm form by introducing the kernel $M(z|z_0)$ [Eq. (8.1.22)]:

$$M(z|z_0) = \begin{cases} K(z|z_0); & z_0 < z \\ 0; & z_0 > z \end{cases}$$

We prove the statement by considering the integral

$$\psi(z) = \lambda\int_0^l M(z|z_0)\psi(z_0)\,dz_0$$

which has the Fredholm form. Assuming K to be bounded, we note that the absence of eigenvalues will be indicated by proving that the series

$$G_\lambda(z|z_0) = \sum_n M_{n+1}(z|z_0)\lambda^n; \quad M_n = n\text{th iteration of } M$$

converges for every finite value of λ. If an eigenvalue in the finite λ plane existed, it would be necessary for $G_\lambda(z|z_0)$ to be singular at this value of λ. [Compare Eqs. (8.2.34) and (8.2.31).] To examine this series it is necessary to evaluate the iterates of M. For example,

$$M_2(z|z_0) = \int_0^l M(z|z_1)M(z_1|z_0)\,dz_1$$

For $z > z_0$ the integration may be broken up into ranges 0 to z_0, z_0 to z, and z to l. In the first the integral vanishes, for here $z_1 < z_0$ and $M(z_1|z_0) = 0$ under these circumstances. The integral from z to l also vanishes, for here $M(z|z_1)$ is zero. Hence

$$M_2(z|z_0) = \int_{z_0}^z K(z|z_1)K(z_1|z_0)\,dz_1; \quad z > z_0$$

For $z < z_0$ a similar argument shows that

$$M_2(z|z_0) = 0; \quad z < z_0$$

Similarly

$$M_3(z|z_0) = \int_0^l M_2(z|z_1)M(z_1|z_0)\,dz_1$$

$$M_3(z|z_0) = \begin{cases} \int_{z_0}^z M_2(z|z_1)K(z_1|z_0)\,dz_1; & z > z_0 \\ 0; & z < z_0 \end{cases}$$

Let us now introduce the assumption that M is bounded. Let $|K| \le m$. Then

$$\begin{aligned} |M| &\le m \\ |M_2| &\le m^2|z - z_0| \\ |M_3| &\le (m^3|z - z_0|^2/2!) \\ |M_n| &\le [m^n|z - z_0|^{n-1}/(n-1)!] \end{aligned}$$

Hence

$$G_\lambda(z|z_0) = \sum_{n=0} \lambda^n M_{n+1}(z|z_0) \le m \sum_n \frac{\lambda^n m^n |z - z_0|^n}{n!} = me^{m\lambda|z-z_0|}$$

This function, as a function of λ, has no singularities for finite values of λ, proving the theorem that a Volterra equation has no eigenvalues as long as its kernel is bounded.

Singular Kernels. The nature of the singularities of a kernel has a strong effect upon the distribution of eigenvalues of the Fredholm equation of the second kind. A kernel is said to be singular if (1) it has discontinuities or (2) it has a singularity within the range of integration or

(3) the limits of integration extend to infinity. Singularities of type (3) are transformed into type (1) by a change of variable. For example, if the limits of integration are zero and infinity, we may transform to an integral with finite limits by means of the substitution $\zeta = 1/(z + 1)$, but this brings in a pole to the finite part of the ζ plane.

Integral equations involving the Green's functions for the Helmholtz and Laplace equation commonly involve all these singularity types. For example, the integral equation arising in the discussion of diffraction by a straight edge, (8.1.9), involves a kernel $H_0^{(1)}[k|x - x_0|]$ which has a logarithmic singularity at $x = x_0$. Moreover, the limits of integration extend to infinity. The kernels of the integral equations listed below, (8.1.15), all have discontinuities in slope.

Kernels which are quadratically integrable may be reduced to a bounded kernel by iteration. For example, if $K(z|z_0)$ has a finite number of discontinuities, the first iterate $K_2(z|z_0)$ has none. This result is essentially a consequence of the theorem that an integral of an integrable function is continuous. Let us illustrate this theorem. Suppose that

$$K(z|z_0) = \begin{cases} k(z|z_0); & \text{for } z < z_0 \\ h(z|z_0); & \text{for } z > z_0 \end{cases}$$

where $k(z|z_0)$ and $h(z|z_0)$ are continuous but do not necessarily have equal values at $z = z_0$. The iterated kernel $K_2(z|z_0)$ is given by the integral

$$K_2(z|z_0) = \int_a^b K(z|z_1)K(z_1|z_0)\,dz_1$$

Introducing the definition of K we can evaluate K_2:

$$\begin{aligned} K_2(z|z_0) &= \int_a^{z_0} h(z|z_1)k(z_1|z_0)\,dz_1 + \int_{z_0}^{z} h(z|z_1)h(z_1|z_0)\,dz_1 \\ &\qquad + \int_z^b k(z|z_1)h(z_1|z_0)\,dz_1; \quad z \geq z_0 \\ &= \int_a^z h(z|z_1)k(z_1|z_0)\,dz_1 + \int_z^{z_0} k(z|z_1)k(z_1|z_0)\,dz_1 \\ &\qquad + \int_{z_0}^b k(z|z_1)h(z_1|z_0)\,dz_1; \quad z \leq z_0 \end{aligned}$$

The function $K_2(z|z_0)$ is continuous for $z \neq z_0$, for by placing $z = z_0$, we find that the expressions valid for $z \geq z_0$ and $z \leq z_0$ join on continuously.

A singular kernel of the type

$$K(z|z_0) = H(z|z_0)/[z - z_0|^\alpha; \quad |H(z|z_0)| \leq M \tag{8.2.51}$$

may be reduced to a nonsingular one by a sufficient number of iterations if $\alpha < 1$, that is, if the kernel is quadratically integrable. Let us compute a few of these:

$$K_2(z|z_0) = \int_a^b \frac{H(z|z_1)H(z_1|z_0)}{|z - z_1|^\alpha |z_1 - z_0|^\alpha}\,dz_1$$

We have

$$|K_2(z|z_0)| \leq M^2 \int_a^b \frac{dz_1}{|z - z_1|^\alpha |z_1 - z_0|^\alpha}$$

$$\leq \frac{M^2}{|z - z_0|^{2\alpha-1}} \int_\gamma^\beta \frac{d\zeta}{|1 - \zeta|^\alpha |\zeta^\alpha|} \leq \frac{M^2}{|z - z_0|^{2\alpha-1}} \int_{-\infty}^\infty \frac{d\zeta}{|1 - \zeta|^\alpha |\zeta^\alpha|}$$

where $\beta = (b - z_0)/(z - z_0)$ and $\gamma = (a - z_0)/(z - z_0)$. The value of the final integral may be easily expressed in terms of beta functions. These are finite *as long as α is not a positive integer or zero.* We assume this to be true. Then

$$|K_2(z|z_0)| \leq \frac{C_0 M^2}{|z - z_0|^{2\alpha-1}}$$

where C_0 is the value of the integral.

It is clear that K_2 is bounded as long as $2\alpha - 1 \leq 0$, that is, as long as $\alpha \leq \frac{1}{2}$. If this condition is not satisfied, we may continue the iterative process. However, as we shall now see, *the iterative process will yield a bounded kernel only if $\alpha < 1$.* The basis of the proof lies in the fact that

$$|K_p(z|z_0)| \leq \int_a^b |K(z|z_1)||K_{p-1}(z_1|z_0)|\, dz_1$$

Hence

$$|K_3(z|z_0)| \leq \int_a^b |K(z|z_1)||K_2(z_1|z_0)|\, dz_1$$

$$\leq C_0 M^3 \int_a^b \frac{dz_1}{|z - z_1|^\alpha |z_1 - z_0|^{2\alpha-1}}$$

$$\leq \frac{C_0 M^3}{|z - z_0|^{3\alpha-2}} \int_{-\infty}^\infty \frac{d\zeta}{|1 - \zeta|^\alpha |\zeta|^{2\alpha-1}}$$

so that

$$|K_3(z|z_0)| \leq C_1 M^3/|z - z_0|^{3\alpha-2}$$

where C_1 is a new constant. Generally

$$K_n(z|z_0) \leq CM^n/|z - z_0|^{n\alpha-(n-1)} \qquad (8.2.52)$$

The nth iterate will be bounded if $n\alpha - (n - 1) < 0$, that is, if

$$\alpha < (n - 1/n) \qquad (8.2.53)$$

Hence for a given $\alpha < 1$ an n can be found such that K_n is bounded.

The Green's function for the Laplace equation gives rise to kernels whose singularities are very much like the ones just being considered, that is, similar to (8.2.51). The three-dimensional Green's function is proportional to

$$[(x - x_0)^2 + (y - y_0)^2 + (z - z_0)^2]^{-\frac{1}{2}} \qquad (8.2.54)$$

The two-dimensional Green's function is proportional to the logarithm of the similar two-dimensional quantity. By considering the more singular kernel

$$[(x - x_0)^2 + (y - y_0)^2]^{-\frac{1}{2}}$$

the two- and three-dimensional cases may be considered together. Consider the three-dimensional case.

Let
$$|K(\mathbf{r}|\mathbf{r}_0)| \leq M/|\mathbf{r} - \mathbf{r}_0|$$

Then

$$K_2(\mathbf{r}|\mathbf{r}_0) \leq M^2 \cdot \int \frac{dx_1\, dy_1\, dz_1}{\sqrt{[(x - x_1)^2 + (y - y_1)^2 + (z - z_1)^2][(x_1 - x_0)^2 + (y - y_0)^2 + (z - z_0)^2]}}$$

We may reduce this case to that of (8.2.51) by employing the inequality

$$(x - x_1)^2 + (y - y_1)^2 + (z_1 - z_1)^2 \geq 3[(x - x_1)^2(y - y_1)^2(z - z_1)^2]^{\frac{1}{3}}$$

Hence

$$K_2(r|r_0) \leq \frac{M^2}{3}\left[\int \frac{dx_1}{(x - x_1)^{\frac{1}{3}}(x_1 - x_0)^{\frac{1}{3}}}\right]\left[\int \frac{dy_1}{(y - y_1)^{\frac{1}{3}}(y_1 - y_0)^{\frac{1}{3}}}\right]\left[\int \frac{dz_1}{(z - z_1)^{\frac{1}{3}}(z_1 - z_0)^{\frac{1}{3}}}\right]$$
$$\leq (CM^2/3)[(x - x_0)(y - y_0)(z - z_0)]^{\frac{1}{3}}$$

Therefore K_2 is bounded. A similar proof holds for the two-dimensional case.

Green's functions appropriate to the Helmholtz equation do not have just the branch point or pole type of singularity as in Eq. (8.2.54) but contain an essential singularity as $|r - r_0| \to \infty$ [see integral equation (8.1.9)]. Kernels of this type cannot be made nonsingular by iteration. We shall call such kernels *intrinsically singular*.

The major difference between an intrinsically singular kernel and a bounded kernel, which is of interest to us, is the nature of the eigenvalue spectrum associated with the homogeneous Fredholm equation of the second kind. It may be shown that for bounded kernels the eigenvalues are denumerable even if there are an infinite number of eigenvalues. This is not necessarily true for intrinsically singular kernels. In that case the eigenvalue spectrum may be continuous; *i.e.*, for a range of λ, *all* values of λ have nonzero solutions ψ_λ.

We may understand this difference as follows. If K_n is quadratically integrable, then there exist meaningful expansions of K_n in terms of a denumerable set of eigenfunctions, expansions which converge in the mean (see page 739). For example, it is possible to expand such a kernel in a double Fourier series. This is, however, not possible for a kernel which is not quadratically integrable. Generally speaking, in addition to a Fourier series, a Fourier integral over a properly chosen path of integration (avoiding the singularity) is required to represent a function which is not quadratically integrable. Indeed we have seen in our discussions of Green's functions that eigenfunction expansions for Green's functions (for both Laplace and Helmholtz equations) in the infinite domain must involve Fourier-type integrals, corresponding to the fact that these Green's functions are not quadratically integrable.

We shall close this section with an example of an integral equation with an intrinsically singular kernel and a continuous eigenvalue spectrum. The equation is

$$\psi(z) = \lambda \int_{-\infty}^{\infty} e^{-|z-z_0|}\psi(z_0)\, dz_0 \tag{8.2.55}$$

The singularity of the kernel occurs at the infinite limits. This may be reduced to a differential equation by differentiating twice. Then

$$(d^2\psi/dz^2) + (2\lambda - 1)\psi = 0 \quad \text{or} \quad \psi_\lambda = Ae^{\sqrt{1-2\lambda}\,z} + Be^{-\sqrt{1-2\lambda}\,z}$$

However, the integral in (8.2.55) exists only if $\mathrm{Re}[\sqrt{1-2\lambda}] < 1$. All values of λ which satisfy this restriction are eigenvalues of (8.2.55). All corresponding ψ_λ are eigenfunction solutions. The eigenvalue spectrum is clearly continuous. Another example is given by case (*c*) following (8.1.15).

8.3 *Solution of Fredholm Equations of the First Kind*

We shall limit ourselves to a discussion of those cases in which an exact solution may be obtained. Approximate techniques will be discussed in Chap. 9. It should be emphasized that approximate methods for the solution of physical problems are most conveniently based on an integral equation formulation. Thus although we shall not be able to solve a great many integral equations exactly, the approximate treatment of such equations will occupy a special position and will be discussed in detail in Chap. 9.

The general methods to be described here are similar to those employed in the solution of differential equations discussed in Chap. 5. The principal feature of the procedure is the expansion of the unknown function in terms of a complete set of functions. This expansion would be given in the form of a sum or integral over the set, with undetermined coefficients. Upon introduction of the expansion into the differential equation a relation among the unknown coefficients could be determined. In other words, *the differential equation is transformed* into an equation or set of equations determining the unknown coefficients. The complete set is chosen, if possible, so that the new equations are readily solvable. For example, if the expansion is a power series $\psi = \Sigma a_n z^{n+s}$, the transformed equation is a difference equation for the coefficients a_n. For certain types discussed in Chap. 5, this difference equation involved only two different values of n and could be easily solved.

Series Solutions for Fredholm Equations. We shall now apply this method to Fredholm integral equations of the first kind, for which it is particularly appropriate. This integral equation, as given in Eq. (8.1.19), is

$$\varphi(z) = \int_a^b K(z|z_0)\psi(z_0)\,dz_0 \tag{8.3.1}$$

Following the suggestion above, let

$$\psi(z) = \sum_n a_n g_n(z) w(z) \tag{8.3.2}$$

where the g_n's form a complete set in the interval (a,b); $w(z)$ is a density function which may be chosen so as to approximate $\psi(z)$ and therefore to improve the convergence of the series (8.3.2). Then

$$\varphi(z) = \sum_n a_n \int_a^b K(z|z_0) g_n(z_0) w(z_0)\,dz_0 \tag{8.3.3}$$

$$\varphi(z) = \sum_n a_n h_n(z); \quad h_n(z) = \int^b K(z|z_0) g_n(z_0) w(z_0)\,dz_0$$

The h_n's are known functions.

The solution of the integral equation is thus reduced to finding the a_n's from the known properties of φ and h_n. This is particularly simple in two cases: If the h_n's are just proportional to powers of z, the series in (8.3.3) is just a power series, so that the unknown coefficients may be determined by comparison with the power series for φ; or second, the h_n's may form an orthogonal set:

$$\int_a^b \bar{h}_n(z) h_m(z) \rho(z)\,dz = N_n \delta_{nm}$$

Then the coefficients a_n may be determined by quadrature:

$$a_n = \frac{1}{N_n}\int_a^b \bar{h}_n(z)\varphi(z)\rho(z)\,dz$$

As we shall see later, there is an important class of cases in which one or the other of these two special circumstances will occur. Unfortunately the h_n's are more usually neither powers of z nor an orthogonal set, in which event it is necessary to reorganize an expansion of φ, in some series of complete functions, into a series in h_n. More explicitly, let φ be expanded in terms of a complete set χ_q:

$$\varphi = \sum_q f_q \chi_q(z) \tag{8.3.4}$$

Then it becomes necessary to express χ_q in terms of h_n:

$$\chi_q = \sum_n \alpha_{qn} h_n \tag{8.3.5}$$

Substituting in Eq. (8.3.4) we obtain

$$\varphi = \sum_{q,n} f_q \alpha_{qn} h_n = \sum_n \left(\sum_q f_q \alpha_{qn} \right) h_n$$

Comparing with (8.3.3) yields

$$a_n = \sum_q f_q \alpha_{qn} \tag{8.3.6}$$

In this fashion the integral equation is solved once the coefficients α_{qn} in (8.3.5) are evaluated.

Determining the Coefficients. Three methods for this evaluation will be discussed. In the first it is assumed that the χ_q's are an arbitrary set of functions not related to h_n in any particular fashion. In the second the χ_q's are constructed from the h_n by the method of orthogonalization; *i.e.*, linear combinations of the h_n's are found which are mutually orthogonal and complete. Then φ may be readily expanded in terms of χ_q, the coefficients being evaluated by integration.

For the first method we expand the functions h_n in Eq. (8.3.5) in terms of χ_q:

$$h_n = \sum_p h_{np} \chi_p$$

Then Eq. (8.3.5) becomes

$$\chi_q = \sum_{n,p} \alpha_{qn} h_{np} \chi_p$$

Hence

$$\sum_n \alpha_{qn} h_{np} = \delta_{qp} \tag{8.3.7}$$

The set of Eqs. (8.3.7) must now be solved for α_{qn}. Let the determinant consisting of elements h_{np} be called H:

$$H = \begin{vmatrix} h_{00} & h_{01} & h_{02} & \cdots \\ h_{10} & h_{11} & h_{12} & \cdots \\ h_{20} & h_{21} & h_{22} & \cdots \\ \cdots & \cdots & \cdots & \cdots \\ \cdots & \cdots & \cdots & \cdots \\ \cdots & \cdots & \cdots & \cdots \end{vmatrix} \tag{8.3.8}$$

Let the minor corresponding to h_{np} be M_{np}. Since $\sum_n M_{nq} h_{np} = H\delta_{qp}$, we have

$$\alpha_{qn} = M_{nq}/H \tag{8.3.9}$$

It is clear from the result that this method will be practical only when each h_n is a combination of just a few of the functions χ_p, for then it would be an easy matter to evaluate the determinant and its minors.

Orthogonalization. The Schmidt method was mentioned in the table after Chapter 6, and will now be discussed at greater length. The problem of interest is as follows: Given a complete set of nonorthogonal functions h_n, to form by linear combination of these functions a new set χ_q which is complete and orthogonal. The Schmidt procedure involves building the χ_q's up in a stepwise fashion. The function χ_0 is taken to be equal to h_0. The function χ_1 is taken to be a linear combination of h_0 and h_1 adjusted to be orthogonal to χ_0; χ_2 a linear combination of h_0, h_1, h_2 orthogonal to χ_1 and χ_0.

It is possible to develop a recurrence relation expressing χ_q in terms of h_q and χ_p, $p < q$. From the method of formation, we see that h_q must be a linear combination of χ_p, $p \leq q$. Employing the orthogonality condition

$$\int_a^b \bar{\chi}_p \chi_q \rho \, dz = \delta_{pq} N_q; \quad N_q = \int_a^b |\chi_p|^2 \rho \, dz$$

the function h_q is given by the sum

$$h_q = \sum_{p=0}^{q} \frac{\int_a^b \bar{\chi}_p h_q \rho \, dz}{N_p} \chi_p$$

where ρ is a density factor. Solving for χ_q one finds

$$\chi_q = \frac{N_q}{\int_a^b \bar{\chi}_q h_q \rho \, dz} \left\{ h_q - \sum_{p=0}^{q-1} \frac{\int_a^b \bar{\chi}_p h_q \rho \, dz}{N_p} \chi_p \right\}$$

We may now choose the normalization of χ_q so that

$$N_q = \int_a^b \bar{\chi}_q h_q \rho \, dz \tag{8.3.10}$$

The expression for χ_q becomes

$$\chi_q = h_q - \sum_{p=0}^{q-1} \frac{\int_a^b \bar{\chi}_p h_q \rho \, dz}{N_p} \chi_p \tag{8.3.11}$$

and

$$N_q = \int_a^b |h_q|^2 \rho \, dz - \sum_{p=0}^{q-1} \frac{\left| \int_a^b \bar{\chi}_p h_q \rho \, dz \right|^2}{N_p} \tag{8.3.12}$$

In more detail:

$$\chi_0 = h_0; \quad N_0 = \int_a^b [h_0|^2 \rho \, dx$$

$$\chi_1 = h_1 - \left(\int_a^b \bar{\chi}_0 h_1 \rho \, dz / N_0\right) \chi_0$$

$$\chi_2 = h_2 - \left(\int_a^b \bar{\chi}_1 h_2 \rho \, dz / N_1\right) \chi_1 - \left(\int_a^b \bar{\chi}_0 h_2 \rho \, dz / N_0\right) \chi_0; \quad \text{etc.}$$

$$N_1 = \int_a^b |h_1|^2 \rho \, dz - \frac{\left|\int_a^b \bar{\chi}_0 h_1 \rho \, dz\right|^2}{N_0}$$

$$N_2 = \int_a^b |h_2|^2 \rho \, dz - \frac{\left|\int_a^b \bar{\chi}_1 h_2 \rho \, dz\right|^2}{N_1} - \frac{\left|\int_a^b \bar{\chi}_0 h_2 \rho \, dz\right|^2}{N_0}; \quad \text{etc.}$$

As an example let us orthogonalize the power series z^n in the interval $(-1,1)$; $\rho = 1$. This should lead to the Legendre functions. We shall show that it does so by computing χ_1 and χ_2, which should be proportional to z and $3z^2 - 1$, respectively. The integrals required for χ_1 are

$$N_0 = 2; \quad \int_a^b \bar{\chi}_0 h_1 \rho \, dz = \int_{-1}^1 z \, dz = 0$$

From this it follows that

$$\chi_1 = z; \quad N_1 = \tfrac{2}{3}$$

For χ_2, the additional integrals needed are

$$\int_a^b \bar{\chi}_1 h_2 \rho \, dz = \int_{-1}^1 z^3 \, dz = 0; \quad \int_a^b \bar{\chi}_0 h_2 \rho \, dz = \int_{-1}^1 z^2 \, dz = \tfrac{2}{3}$$

Hence

$$\chi_2 = z^2 - \tfrac{2}{3} \cdot \tfrac{1}{2} = z^2 - \tfrac{1}{3}$$

which is indeed proportional to P_2. The normalization integral is

$$N_2 = \tfrac{2}{5} - \frac{\tfrac{4}{9}}{2} = \tfrac{8}{45}$$

By continuing this procedure the various Legendre polynomials are developed, together with their normalization integrals. In the table at the end of Chap. 6, other orthogonal polynomials are shown to result when other intervals of integration and other weight factors are employed. For example, the Hermite polynomials correspond to an integration region $(-\infty, \infty)$ for a weight factor e^{-z^2}, while the Laguerre polynomials L_n^α correspond to an integration region $(0, \infty)$ for a weight factor $x^\alpha e^{-x}$.

The Schmidt method, however, is not sufficient for the discussion at hand, for it does not give the coefficients α_{qn} in Eq. (8.3.5) directly. We shall now obtain explicit formulas for α_{qn}. From Eq. (8.3.5) and from the procedure for obtaining the χ's,

$$\chi_q = \sum_{n=0}^{q} \alpha_{qn} h_n$$

Since the χ's are orthogonal, we have

$$\int_a^b \bar{\chi}_p \chi_q \rho \, dz = 0; \quad \text{if } p \neq q$$

The orthogonality condition is met for all p and q if, for $p < q$, one requires

$$\int_a^b \bar{h}_p \chi_q \rho \, dz = 0; \quad p < q \tag{8.3.13}$$

Condition (8.3.13) may now be employed to determine α_{qn}, for upon substituting the expansion for χ_q in (8.3.13) one obtains

$$\sum_{n=0}^{q} \alpha_{qn} d_{np} = 0; \quad \text{if } p < q$$

where

$$d_{np} = \int^b h_n \bar{h}_p \rho \, dz$$

For $p = q$, the sum is equal to N_q. The coefficients α_{qn} may now be expressed in terms of the minors of the elements d_{np} in the determinant of these elements.

Consider the determinants D_q:

$$\begin{aligned} D_0 &= d_{00} \\ D_1 &= \begin{vmatrix} d_{00} & d_{01} \\ d_{10} & d_{11} \end{vmatrix} \\ D_2 &= \begin{vmatrix} d_{00} & d_{01} & d_{02} \\ d_{10} & d_{11} & d_{12} \\ d_{20} & d_{21} & d_{22} \end{vmatrix}; \quad \text{etc.} \end{aligned} \tag{8.3.14}$$

It appears that α_{qn} is proportional to the minor M_{nq} of the element d_{nq} in D_q. This is a consequence of the condition satisfied by M_{nq}:

$$\sum_{n=0}^{q} M_{nq} d_{np} = \delta_{pq} D_q; \quad p \leq q$$

Since we wish the coefficient of h_q in the expression for χ_q to be unity, we shall set $\alpha_{qn} = M_{nq}/M_{qq}$, and the expansion for χ_q now reads

$$\chi_q = \sum_{n=0}^{q} h_n \left(\frac{M_{nq}}{M_{qq}} \right) \tag{8.3.15}$$

This series is just another representation of the Schmidt formula, Eq. (8.3.11), but one which gives the coefficients of the expansion of χ_q explicitly.

As an example let us consider the Legendre functions again; *i.e.*, let $\chi_q = P_q$, $h_n = z^n$, $\rho = 1$, and the region of integration $(-1,1)$. For

P_2, we need to look at D_2:

$$D_2 = \begin{vmatrix} 2 & 0 & \frac{2}{3} \\ 0 & \frac{2}{3} & 0 \\ \frac{2}{3} & 0 & \frac{2}{5} \end{vmatrix}$$

Now from (8.3.15), $P_2 \sim (M_{02}/M_{22}) + z(M_{12}/M_{22}) + z^2 = -\frac{1}{3} + z^2$, which is indeed just proportional to P_2.

It is also possible to express the normalization integrals in terms of the determinants D_q. The integral

$$N_q = \int_a^b \bar{\chi}_q \chi_q \rho \, dz = \sum_{n=0}^{q} \left(\frac{M_{nq}}{M_{qq}}\right) \int_a^b \bar{h}_n \chi_q \rho \, dz$$

From Eq. (8.3.13) N_q becomes

$$N_q = \int_a^b \bar{h}_q \chi_q \rho \, dz = \left(\frac{1}{M_{qq}}\right) \sum_{p=0}^{q} M_{pq} \int_a^b \bar{h}_q h_p \rho \, dz$$

or

$$N_q = \left(\frac{1}{M_{qq}}\right) \sum_{p=0}^{q} M_{pq} d_{pq}$$

This last sum is just the value of the determinant D_q, while M_{qq} equals D_{q-1}. Hence

$$N_q = D_q/D_{q-1} \tag{8.3.16}$$

In the example discussed just above

$$D_2 = \tfrac{32}{135}; \quad D_1 = \tfrac{4}{3}; \quad N_2 = \tfrac{8}{45}$$

Biorthogonal Series. We return to the original series (8.3.3) in which φ is expanded in a set of known but nonorthogonal functions h_n:

$$\varphi(z) = \Sigma a_n h_n(z)$$

The a_n's may be evaluated if it is practical to find a set of functions w_n which satisfy the condition:

$$\int w_n(z) h_m(z) \, dz = N_n \delta_{nm} \tag{8.3.17}$$

Then the coefficients a_n may be determined by quadratures:

$$a_n = \left(\frac{1}{N_n}\right) \int w_n \varphi(z) \, dz \tag{8.3.18}$$

In the above we have not specified the limits of integration or, more generally, the path of integration in the complex plane of z. All that is necessary is that the region of validity of (8.3.18) should include as a subdomain the region in which the integral equation is to hold.

A direct and frontal attack on the problem of determining the $w_n(z)$ is not usually fruitful. Instead we shall discuss a method which is practical whenever the successive h_n's start off with increasingly greater powers of z (such as the Bessel functions):

$$h_n(z) = \sum_{p=n}^{\infty} A_{pn} z^p; \quad A_{nn} = 1 \tag{8.3.19}$$

A_{nn} may always be placed equal to 1, though historically this has not always been done. Hence h_n is regular at the origin. We shall now show that in this case the functions w_n are obtained from the expansion

$$\frac{1}{t-z} = \sum_n w_n(t) h_n(z) \tag{8.3.20}$$

To prove this, evaluate

$$\oint \frac{h_q(t)}{t-z}\, dt = \sum_n \left[\oint w_n(t) h_q(t)\, dt \right] h_n(z)$$

around any closed contour surrounding the point z and the origin in a region in which $h_n(z)$ is analytic. Then according to Cauchy's integral formula the left-hand side of the above equation is just $2\pi i h_n(z)$. Comparing with the right side we see that is possible only if

$$\oint w_n(t) h_q(t)\, dt = 2\pi i \delta_{nq} \tag{8.3.21}$$

Relation (8.3.21) does not depend upon the assumed behavior of h_n as given in (8.3.19). It is general and applies as long as h_n is analytic in the region enclosed by the contour. This is a limitation only in the event that h_n has an essential singularity at a point z, and in that event the expansion of φ in terms of h_n must be carefully examined.

The determination of the functions w depends upon being able to carry out expansion (8.3.20). In the event that (8.3.19) holds, however, it is possible to give a general algorithm. Take the qth derivative of (8.3.20) with respect to z, and then place $z = 0$. We find

$$\frac{q!}{t^{q+1}} = \sum_n w_n(t) \left\{ \frac{d^q}{dz^q} [h_n(z)] \right\}_{z=0}$$

Introducing assumption (8.3.19), we find that

$$\left\{ \frac{d^q}{dz^q} [h_n(z)] \right\}_{z=0} = \begin{cases} q! A_{qn}; & q \geq n \\ 0; & q < n \end{cases}$$

Hence

$$\frac{1}{t^{q+1}} = \sum_{n=0}^{q} w_n(t) A_{qn} \tag{8.3.22}$$

These equations consist of a set of recurrence relations which may be solved in turn to obtain $w_n(t)$. This is most conveniently done by rewriting Eq. (8.3.22) as follows:

$$w_q = \frac{1}{t^{q+1}} - \sum_{n=0}^{q-1} w_n(t) A_{qn}$$

First let $q = 0$. Then

$$w_0(t) = 1/t$$

Let $q = 1$. Then

$$w_1(t) = (1/t^2) - A_{10}w_0 = (1/t^2) - (A_{10}/t)$$

We give here the first four w_n's obtained in this fashion.

$$\begin{aligned}
w_0 &= \frac{1}{t}; \quad w_1 = \frac{1}{t^2} - \frac{A_{10}}{t} \\
w_2 &= \frac{1}{t^3} - \frac{A_{21}}{t^2} + \frac{A_{21}A_{10} - A_{20}}{t} \\
w_3 &= \frac{1}{t^4} - \frac{A_{32}}{t^3} + \frac{A_{32}A_{21} - A_{31}}{t^2} + \frac{-A_{30} + A_{31}A_{10} + A_{32}A_{20} - A_{32}A_{21}A_{10}}{t}
\end{aligned} \tag{8.3.23}$$

From these the general formula may be induced.

It is clear that the w_n's are polynomials in $1/t$, that the maximum power of $1/t$ present in w_n is $(n + 1)$.

Let us now return to the original question of obtaining an expansion of an arbitrary function φ in terms of h_n. This now reduces to the evaluation of integral (8.3.18). The contour has now been chosen to be about the origin. If we write

$$w_n = \sum_{p=0}^{n} \left(\frac{b_{np}}{t^{p+1}}\right) \tag{8.3.24}$$

then Eq. (8.3.18) becomes

$$a_n = \frac{1}{2\pi i} \oint w_n(z)\varphi(z)\, dz = \frac{1}{2\pi i} \sum_{p=0}^{n} b_{np} \oint \frac{\varphi(z)}{z^{p+1}}\, dz$$

From Cauchy's integral formula we know that

$$\frac{1}{2\pi i} \oint \frac{\varphi(z)}{z^{p+1}}\, dz = \frac{\varphi^{(p)}(0)}{p!}$$

where $\varphi^{(p)}(0)$ is the pth derivative of φ with respect to its argument, taken at the origin of the argument. Hence

$$a_n = \sum_{p=0}^{n} \frac{b_{np}}{p!} \varphi^{(p)}(0) \tag{8.3.25}$$

This is the expression determining a_n. It is clear that, in the event that h_n satisfies condition (8.3.19) (series commencing with the nth power of z), it is possible to carry through the determination of a_n without at any time becoming involved in infinite processes. However, if condition (8.3.19) is not satisfied, the determination of $w_n(t)$ is no longer so simple. The method of orthogonalization discussed earlier is then more efficient. Actually it is possible to set up an analogue of the Schmidt procedure.

We conclude the discussion of this method with an example. Suppose that the functions h_n are $J_n(z)$, the Bessel functions. Then we require the value of the coefficients in the series

$$\varphi(z) = \Sigma a_n J_n(z)$$

This type of series is called a *Neumann* series. The corresponding biorthogonal functions are called the *Neumann polynomials* $O_n(t)$. The relation between J_n and O_n is given by

$$\frac{1}{t-z} = \sum_n \epsilon_n O_n(t) J_n(z); \quad \epsilon_n = \begin{cases} 1; & n = 0 \\ 2; & n \neq 0 \end{cases}$$

The functions O_n are not exactly equal to w_n, since J_n does not begin as z^n but rather as $z^n/2^n n!$, and definition (8.3.20) does not involve ϵ_n. However, these are not differences in principle but rather in detail; one must simply solve Eq. (8.3.22) without assuming $A_{qq} = 1$. The Neumann polynomials are given by

$$\begin{aligned} \epsilon_n O_n(t) &= \frac{2^n n!}{t^{n+1}} \left\{ 1 + \frac{t^2}{2(2n-2)} + \frac{t^4}{2 \cdot 4 \cdot (2n-2)(2n-4)} + \cdots \right\} \\ &= \tfrac{1}{2} \sum_{m=0}^{\leq \frac{1}{2}n} \left[\frac{n(n-m-1)!}{m!(t/2)^{n-2m+1}} \right]; \quad n \neq 0 \end{aligned} \tag{8.3.26}$$

The first few are

$$\begin{aligned} O_0(t) &= \frac{1}{t}; \quad O_1(t) = \frac{1}{t^2}; \quad O_2(t) = \frac{1}{t} + \frac{4}{t^3} \\ O_3(t) &= \frac{3}{t^2} + \frac{24}{t^4}; \quad O_4(t) = \frac{1}{t} + \frac{16}{t^3} + \frac{192}{t^5} \\ O_5(t) &= \frac{5}{t^2} + \frac{120}{t^4} + \frac{1920}{t^6} \end{aligned} \tag{8.3.27}$$

The coefficients a_n may be determined just as in Eq. (8.3.25)

$$\begin{aligned} a_n &= {\sum_{s=0,1}^{n}}' \left(\frac{2^s n}{n+s} \right) \binom{\frac{n+s}{2}}{s} \varphi^{(s)}(0); \quad n \neq 0 \\ a_0 &= \varphi(0); \qquad\qquad n = 0 \end{aligned} \tag{8.3.28}$$

Here the prime on the summation signifies that only odd or even s is to be taken in the sum if n is odd or even, respectively. The coefficients $\binom{a}{s}$ are the binomial expansion coefficients (see glossary in the Appendix).

Integral Equations of the First Kind and Generating Functions. We shall now illustrate the general discussion given above by several examples. Consider first the class for which $h_n = z^n$. This occurs whenever the kernel of the integral equation is also the generating function for a set of orthogonal polynomials.

According to page 786, the generating function for the Hermite polynomials $H_n(z_0)$ may be written as

$$e^{-(z-z_0)^2} = \sum_n \frac{e^{-z_0^2} H_n(z_0) z^n}{n!}$$

This expansion may be used to solve the integral equation

$$\varphi(z) = \int_{-\infty}^{\infty} e^{-(z-z_0)^2} \psi(z_0)\, dz_0 \tag{8.3.29}$$

A problem of this type could occur in heat-conduction problems in which the unknown is the necessary initial source distribution required to give a certain temperature distribution. In (8.3.29) let

$$\psi(z_0) = \Sigma a_n H_n(z_0)$$

Employing the expansion of the kernel and the normalization of the Hermite functions (page 786),

$$\int_{-\infty}^{\infty} H_n^2 e^{-z^2}\, dz = 2^n (n!) \sqrt{\pi}$$

Equation (8.3.29) reduces to

$$\varphi(z) = \sqrt{\pi} \sum_n a_n 2^n z^n$$

Hence

$$a_n = \varphi^{(n)}(0)/[2^n n! \sqrt{\pi}]; \quad \text{where } \varphi^{(n)}(0) = [d^n\varphi/dz^n]_{z=0}$$

The solution $\psi(z)$ is then

$$\psi(z) = \frac{1}{\sqrt{\pi}} \sum_n \frac{\varphi^{(n)}(0)}{2^n n!} H_n(z) \tag{8.3.30}$$

Once an integral equation of the first kind has been solved for a given kernel, one may find further solvable examples by applying operators on z to both sides of, say, (8.3.29); the solution remains as (8.3.30). In the example under discussion it is clear that by differentiating the generating function further kernels expandable in terms of Hermite functions and power series may be obtained. Employing the relation

$$H_n(z) = (-)^n e^{z^2} \frac{d^n}{dz^n} e^{-z^2}$$

we have

$$e^{-(z-z_0)^2}H_p(z - z_0) = (-1)^p \sum_{s=0}^{\infty} \frac{e^{-z_0^2}H_{p+s}(z_0)}{s!} z^s$$

Hence the equation

$$\varphi(z) = \int_{-\infty}^{\infty} e^{-(z-z_0)^2}H_p(z - z_0)\psi(z_0)\, dz_0 \tag{8.3.31}$$

has the solution

$$\psi(z) = \sum_{q=0}^{p-1} \alpha_q H_q(z) + \frac{(-1)^p}{\sqrt{\pi}} \sum_{s=0}^{\infty} \frac{\varphi^{(s)}(0)}{2^s s!} H_{p+s}(z) \tag{8.3.32}$$

Here the coefficients α_q, $q \leq p - 1$ are arbitrary, for they cannot be determined from integral equation (8.3.31) inasmuch as H_q is orthogonal to $e^{-(z-z_0)^2}H_p(z - z_0)$.

Finally we may take a linear combination of these various kernels to obtain a new kernel:

$$K(z|z_0) = e^{-(z-z_0)^2} \sum_p a_p H_p(z - z_0) \tag{8.3.33}$$

Clearly any kernel which is a function of $(z - z_0)$ may be expanded in such a series. Solving the corresponding integral equation of the first kind would therefore be a solution for any integral equation involving such a kernel. Later on in this chapter we shall find that a method involving Fourier integrals may also be applied to these kernels.

We now write (8.3.33) as a power series in z:

$$e^{-(z-z_0)^2} \sum_{p=0}^{\infty} a_p H_p(z - z_0) = \sum_{s=0}^{\infty} \frac{z^s}{s!} \left[\sum_{p=0}^{\infty} (-1)^p a_p e^{-z_0^2} H_{p+s}(z_0) \right]$$

Introducing this expansion into the integral equation one obtains

$$\varphi^{(s)}(0) = \sum_{p=0}^{\infty} (-1)^p a_p \int_{-\infty}^{\infty} e^{-z_0^2} H_{p+s}(z_0)\psi(z_0)\, dz_0$$

Let

$$\psi(z) = \sum \frac{C_q H_q(z)}{2^q \sqrt{\pi}\, q!}$$

Then

$$\varphi^{(s)}(0) = \sum_{p=0}^{\infty} (-1)^p a_p C_{p+s}; \quad \text{for each } s \tag{8.3.34}$$

Equations (8.3.34) are a set of equations determining C_p. We write down a few of these equations

$$\varphi(0) = C_0 - a_1C_1 + a_2C_2 - a_3C_3 + a_4C_4 \cdots$$
$$\varphi^{(1)}(0) = C_1 - a_1C_2 + a_2C_3 - a_3C_4 \cdots$$
$$\varphi^{(2)}(0) = C_2 - a_1C_3 + a^2C_4 \cdots$$

where a_0 has been put equal to unity.

The solution to these equations may be obtained by an iterative scheme. We only present the results here. The answer is a linear combination of $\varphi^{(s)}(0)$:

$$C_p = \sum_{s=0}^{\infty} \varphi^{(p+s)}(0) T_s \tag{8.3.35}$$

The coefficients T_s are

$$T_s = \sum (-1)^{r_2+r_4+\cdots} \frac{(r_1 + r_2 + r_3 + \cdots)!}{(r_1)!(r_2)!(r_3)! \cdots} (a_1)^{r_1}(a_2)^{r_2}(a_3)^{r_3} \cdots$$

where all combinations are taken so that

$$r_1 + 2r_2 + 3r_3 + 4r_4 + \cdots = s \tag{8.3.36}$$

We write out the first few T_s:

$$\begin{aligned} &T_0 = 1; \quad T_1 = a_1; \quad T_2 = a_1^2 - a_2; \quad T_3 = a_1^3 - 2a_1a_2 + a_3 \\ &T_4 = a_1^4 - 3a_1^2a_2 + a_2^2 + 2a_3a_1 - a_4 \\ &T_5 = a_1^5 - 4a_1^3a_2 + 3a_1^2a_3 + 3a_1^2a_3 + 3a_2^2a_1 - 2a_3a_2 - 2a_4a_1 + a_5 \end{aligned} \tag{8.3.37}$$

In a similar manner one may solve integral equations involving the generating functions for other orthogonal polynomials. The pertinent details are listed below for Laguerre polynomials:

Generating function:

$$\frac{J_a(2\sqrt{zz_0})}{(zz_0)^{\frac{1}{2}a}} = e^{-z} \sum_{n=0}^{\infty} \frac{L_n^{(a)}(z_0)}{[\Gamma(n + 1 + a)]^2} z^n$$

Normalization:

$$\int_0^{\infty} e^{-z_0} z_0^a [L_n^{(a)}(z_0)]^2 \, dz_0 = \frac{[\Gamma(n + a + 1)]^3}{n!}$$

Then the solution of

$$\varphi(z) = \int_0^{\infty} \frac{J_a(2\sqrt{zz_0})}{(zz_0)^{\frac{1}{2}a}} \psi(z_0) \, dz_0; \quad 0 \le z \le \infty \tag{8.3.38}$$

is

$$\psi(z) = e^{-z} z^a \sum_{n=0}^{\infty} \frac{\chi^{(n)}(0) L_n^{(a)}(z)}{\Gamma(n + a + 1)}; \quad \chi(z) = e^z \varphi(z) \tag{8.3.39}$$

We shall also mention one other expansion from which a solvable integral equation may be obtained. There is, based upon the more common generating function for $L_n^{(a)}$ (see page 784),

$$\left(\frac{z_0}{z}\right)^{a+1} e^{-(z_0/z)} = e^{-z_0} z_0^{a+1} \sum_{n=0}^{\infty} \frac{L_n^{(a)}(z_0)}{\Gamma(n + a + 1)} (1 - z)^n$$

Employing this function as the kernel, the solution to the corresponding integral equation of the first kind is

$$\psi = \frac{1}{z}\sum_{n=0}^{\infty}\frac{(-1)^n}{[\Gamma(n+\alpha+1)]^2}\varphi^{(n)}(1)L_n^{(\alpha)}(z) \tag{8.3.40}$$

Use of Gegenbauer Polynomials. The generating functions for the Legendre polynomials and more generally for the Gegenbauer functions also yield integral equations of the first kind which may be solved. For example, the following kernel:

$$K(z|z_0) = \frac{1}{\sqrt{1-2zz_0+z^2}} = \sum_{n=0}^{\infty} z^n P_n(z_0)$$

The integral equation reads

$$\varphi(z) = \int_{-1}^{1}\frac{\psi(z_0)}{\sqrt{1-2zz_0+z^2}}\,dz_0; \quad -1 \le z \le 1 \tag{8.3.41}$$

Introducing the expansion for K and writing $\psi = \Sigma a_n P_n$ one obtains

$$\varphi(z) = \sum_{n=0}^{\infty}\frac{2a_n}{2n+1}z^n; \quad \text{employing } \int_{-1}^{1}[P_n(z)]^2\,dz = \frac{2}{2n+1}$$

Hence

$$a_n = \frac{2n+1}{2}\left[\frac{\varphi^{(n)}(0)}{n!}\right] \quad \text{and} \quad \psi(z) = \sum_{n=0}^{\infty}P_n(z)\,\frac{2n+1}{2}\left[\frac{\varphi^{(n)}(0)}{n!}\right]$$

Similar results may be obtained for any Gegenbauer function $T_n^\nu(z)$, for the corresponding generating function is (see table at end of Chap. 6).

$$(1-2zz_0+z^2)^{-(\nu+\frac{1}{2})} = \frac{\sqrt{\pi}}{2^\nu\Gamma(\nu+\frac{1}{2})}\sum_{n=0}^{\infty}T_n^\nu(z_0)z^n \tag{8.3.42}$$

The normalization integral is

$$\int_{-1}^{1}(1-z^2)^\nu[T_n^\nu(z)]^2\,dz = \frac{2\Gamma(2\nu+n+1)}{(2\nu+2n+1)n!} \tag{8.3.43}$$

Employing the generating function as a kernel and restricting the range of z from -1 to 1, the solution of the integral equation of the first kind, (8.3.41), is

$$\psi = \frac{2^{\nu-1}\Gamma(\nu+\frac{1}{2})(1-z^2)^\nu}{\sqrt{\pi}}\sum_{n=0}^{\infty}\frac{2\nu+2n+1}{\Gamma(2\nu+n+1)}\varphi^{(n)}(0)T_n^\nu(z) \tag{8.3.44}$$

By giving appropriate values to ν various particular cases of importance are obtained: for $\nu = 0$ the Legendre polynomials, for $\nu =$ integer the associated Legendre function. For $\nu = \frac{1}{2}$

$$\frac{1}{1 - 2zz_0 + z^2} = \sqrt{\frac{\pi}{2}} \sum_{n=0}^{\infty} T_n^{\frac{1}{2}}(z_0) z^n$$

Here $T_n^{\frac{1}{2}}$ are the Tschebyscheff polynomials. The normalization condition is

$$\int_{-1}^{1} [T_n^{\frac{1}{2}}]^2 (1 - z^2)^{\frac{1}{2}}\, dz = 1$$

The solution of the integral equation

$$\varphi(z) = \int_{-1}^{1} \frac{\psi(z_0)}{(1 - 2zz_0 + z^2)}\, dz_0$$

is given by

$$\psi = \sqrt{\frac{2}{\pi}} (1 - z^2)^{\frac{1}{2}} \sum_{n=0}^{\infty} \frac{\varphi^{(n)}(0)}{n!} T_n^{\frac{1}{2}}(z) \tag{8.3.45}$$

In all the cases discussed above other solvable integral equations may be obtained by integrating or differentiating both sides with respect to z.

We have so far considered generating functions in which the associated functions form an orthogonal set. This favorable situation does not always prevail. It would thus be of interest to see how the attack on the problem then proceeds. Consider, for example, the integral equation formed when the kernel is the generating function for the Bessel functions:

$$\varphi(z) = \int_0^{2\pi} e^{iz\cos u}\, \psi(u)\, du; \quad 0 \le z \le \infty \tag{8.3.46}$$

An integral equation of this kind occurs in two-dimensional wave propagation in which the amplitude of a wave is specified along a half plane. The first step in solving (8.3.46) consists in expanding both the kernel and $\psi(u)$ in a Fourier series in $\cos(nu)$:

$$\psi = \Sigma a_n \cos(nu); \quad e^{iz\cos u} = \Sigma \epsilon_n J_n(z) \cos(nu)$$

Then

$$\varphi(z) = \pi \sum_n a_n J_n(z) \tag{8.3.47}$$

The coefficients a_n may be directly expressed in terms of the value of φ and its derivatives at $z = 0$, as given in Eq. (8.3.28).

Integral Equations of the First Kind and Green's Functions. If the kernel should happen to be a Green's function, the solution may be easily obtained. This is a consequence of the fact that any Green's function may be expanded in an eigenfunction series [Eq. (7.2.39)]. For a symmetrical G:

$$G_k(x|x_0) = 4\pi \sum_n \frac{\bar{\chi}_n(x_0)\chi_n(x)}{k_n^2 - k^2} \tag{8.3.48}$$

where the functions χ_n form an orthogonal and normalized set in region $a \le x \le b$. The solution of the problem

$$\varphi(x) = \int_a^b G(x|x_0)\psi(x_0)\,dx_0$$

may then be easily obtained. Introduce the expansion for ψ:

$$\psi(x) = \sum_n a_n\chi_n(x)$$

Then

$$\varphi(x) = 4\pi \sum \frac{a_n}{k_n^2 - k^2}\,\chi_n(x)$$

The coefficients a_n may now be obtained by quadratures:

$$a_n = \left(\frac{1}{4\pi}\right)(k_n^2 - k^2)\int_a^b \bar{\chi}_n(x)\varphi(x)\,dx \tag{8.3.49}$$

A similar procedure may be carried out with Green's functions in two and three dimensions for which the equation satisfied by G_k is separable, for suppose that

$$G_k(\mathbf{r}|\mathbf{r}_0) = 4\pi \sum_n \frac{\bar{\chi}_n(\mathbf{r}_0)\chi_n(\mathbf{r})}{k_n^2 - k^2} \tag{8.3.50}$$

The functions occurring in the sum are functions of two or three variables. However, assuming separability, each χ_n may be written as a product of functions each of which depend on only one variable; moreover each of these forms an orthonormal set in each variable:

$$\chi_n(r) = X_{n1}(x_1)X_{n2}(x_2)X_{n3}(x_3);\quad \int \bar{X}_{n1}X_{m1}\rho(x_1)\,dx_1 = \delta_{nm}$$

If then two of the variables are given specific values, say x_2 and x_3 for both the $\mathbf{r}$ and $\mathbf{r}_0$ coordinates, expansion (8.3.50) for G_k is of the form

$$K(x|x_0) = \sum_k \alpha_k\bar{X}_{k1}(x_0)X_{k1}(x) \tag{8.3.51}$$

An integral equation of the first kind with $K(x|x_0)$ as kernel is readily solvable. The solution is

$$\psi(x) = \rho(x)\sum_k \left[\frac{\int \bar{X}_{k1}(x_0)\varphi(x_0)\rho(x_0)\,dx_0}{\alpha_k}\right]X_{k1}(x) \tag{8.3.52}$$

Many kernels of type (8.3.51) may be formed from one Green's function by taking different values for x_2 and x_3 and finally by taking linear combinations of these.

We shall now illustrate these remarks. A Green's function for the two-dimensional Laplace equation is proportional to $\ln R$, where $R = \sqrt{r^2 - 2rr_0\cos(\phi - \phi_0) + r_0^2} = |\mathbf{r} - \mathbf{r}_0|$. Employing the general theory for expansion of Green's functions as given in the preceding chapter (see also Chap. 10), one finds that in polar coordinates

$$\ln R = \ln r - \sum_{n=1}^{\infty} \left(\frac{1}{n}\right)\left(\frac{r_0}{r}\right)^n \cos[n(\phi - \phi_0)]; \quad r > r_0 \qquad (8.3.53)$$

Let us now take the special values $r = r_0 = 1$. Then the expansion becomes

$$\ln \sqrt{2[1 - \cos(\phi - \phi_0)]} = \ln\left|2 \sin\left(\frac{\phi - \phi_0}{2}\right)\right| = -\sum_{n=1}^{\infty}\left(\frac{1}{n}\right)\cos[n(\phi - \phi_0)]$$

or

$$\ln\left|2 \sin\left(\frac{\phi - \phi_0}{2}\right)\right| = -\sum_{n=1}^{\infty}\left(\frac{1}{n}\right)[\cos(n\phi)\cos(n\phi_0) + \sin(n\phi)\sin(n\phi_0)] \qquad (8.3.54)$$

An integral equation of the first kind with $K(\phi|\phi_0) = \ln|2 \sin \frac{1}{2}(\phi - \phi_0)|$ where $0 \le \phi \le 2\pi$ may then be readily solved.

We can construct another kernel with a simple bilinear expansion of type (8.3.51) by addition:

$$\ln\left|2 \sin\left(\frac{\phi - \phi_0}{2}\right)\right| + \ln\left|2 \sin\left(\frac{\phi + \phi_0}{2}\right)\right| = \ln[2(\cos\phi - \cos\phi_0)]$$

Then

$$\ln[2\,|\cos\phi - \cos\phi_0|] = -\sum_{n=1}^{\infty}\left(\frac{2}{n}\right)\cos(n\phi)\cos(n\phi_0) \qquad (8.3.55)$$

Hence the integral equation

$$\psi(\phi) = \int_0^{\pi} \ln[2\,|\cos\phi - \cos\phi_0|]\psi(\phi_0)\,d\phi_0 \qquad (8.3.56)$$

has the solution

$$\psi(\phi) = a_0 - \frac{2}{\pi^2}\sum_n n\left[\int_0^{\pi}\Psi(\phi_0)\cos(n\phi_0)\,d\phi_0\right]\cos(n\phi) \qquad (8.3.57)$$

where a_0 is arbitrary.

Further useful kernels may be obtained by differentiating the kernels we have just discussed. Differentiating $\ln |2 \sin(\phi - \phi_0)/2|$ yields

$$\tfrac{1}{2} \cot\left(\frac{\phi - \phi_0}{2}\right) = \sum_n [\sin(n\phi) \cos(n\phi_0) - \cos(n\phi) \sin(n\phi_0)]$$

Differentiating $\ln[2 |\cos \phi - \cos \phi_0|]$ yields

$$\frac{1}{\cos \phi_0 - \cos \phi} = - \sum_n 2 \left[\frac{\sin(n\phi)}{\sin(\phi)}\right] \cos(n\phi_0)$$

The integral equation employing this function as a kernel may be transformed into another of some interest in hydrodynamics. The integral equation is

$$\Phi(\varphi) = \int_0^\pi \frac{\psi(\varphi_0)}{\cos \varphi_0 - \cos \varphi} d\varphi_0$$

Let $\cos \varphi = x$, $\Phi(\varphi) = \chi(x)$, $-[\psi(\varphi_0)/\sin \varphi_0] = \Psi(x_0)$. Then

$$\chi(x) = \int_{-1}^1 \frac{\Psi(x_0)}{x - x_0} dx_0 \tag{8.3.58}$$

The manipulations by which we developed a number of kernels from the Green's function $\ln R$ may, of course, be repeated for other Green's functions. The example we have chosen is particularly simple. Indeed the expansions we have listed may be obtained directly rather than from a prior knowledge of a Green's function expansion, and in principle the direct procedure is always possible. However, in the following chapters (Chaps. 10 *et seq.*) we shall list many Green's function expansions for separable coordinate systems and thus shall have automatically classified a number of integral equations of the first kind which are solvable by means of an eigenfunction expansion.

Transforms and Integral Equations of the First Kind. In the relation

$$\Psi(z) = \int K(z|z_0)\psi(z_0)\, dz_0$$

$\Psi(z)$ is often called the *transform* of $\psi(z)$. The solution of the integral equation in which ψ is expressed in terms of Ψ is called an *inversion* of the transform. Transforms in which the relation between Ψ and ψ are particularly simple have been investigated in great detail. We shall consider some examples here.

The most familiar and the most important case is that of the Fourier transform (see Sec. 4.7):

$$\Psi(k) = \frac{1}{\sqrt{2\pi}} \int_{-\infty}^\infty e^{ikz}\, \psi(z)\, dz \tag{8.3.59}$$

The inversion is given by

$$\psi(z) = \frac{1}{\sqrt{2\pi}} \int_{-\infty}^\infty e^{-ikz}\, \Psi(k)\, dk \tag{8.3.60}$$

so that the integral equation (8.3.59) for ψ with the kernel $K(k|z) = e^{ikz}$ has its solution (8.3.60). We refer the reader to Sec. 4.7 for a discussion of the conditions under which the inversion is possible.

It is possible to generalize the Fourier transform to a wide class of functions. As was shown in Chap. 6, the Fourier transform follows the completeness relation for a set of eigenfunctions with a continuous eigenvalue spectrum. Let us first recall the orthogonality and normalization condition appropriate to the continuous spectrum. If k is the eigenvalue, then

$$\int \bar{\varphi}(k|x)\varphi(k_0|x)\,dx = \delta(k - k_0) \tag{8.3.61}$$

where the region of integration extends out to infinity in at least one direction and δ is the Dirac delta function. An arbitrary function ψ may be expanded in terms of φ subject only to the requirement of the existence of the integral:

$$\psi(x) = \int \Psi(k)\varphi(k|x)\,dk \tag{8.3.62}$$

To obtain $\Psi(k_0)$ multiply through both sides of the above equation by $\bar{\varphi}(k_0|x)$ and integrate over x:

$$\int \bar{\varphi}(k_0|x)\psi(x)\,dx = \int\int \Psi(k)\bar{\varphi}(k_0|x)\varphi(k|x)\,dk\,dx$$

Blithely interchanging integrals (a procedure which must be justified in each particular case) and employing Eq. (8.3.61), one obtains

$$\int \bar{\varphi}(k_0|x)\psi(x)\,dx = \int \Psi(k)\delta(k - k_0)\,dk$$

Hence

$$\Psi(k) = \int \bar{\varphi}(k|x)\psi(x)\,dx \tag{8.3.63}$$

We see that, if $\Psi(k)$ is called the *transform* of $\psi(x)$ by the function $\bar{\varphi}$, then ψ is the transform of Ψ by the function φ. In the Fourier case, discussed just above, $\bar{\varphi}(k|x) = (1/\sqrt{2\pi})\exp(ikx)$, and the Fourier integral relations (8.3.59) and (8.3.60) are just special cases of Eqs. (8.3.63) and (8.3.62), respectively.

It is thus clear that any function of two variables which may be normalized according to (8.3.61) may be employed to set up a transform and its inversion. As a consequence those integral equations of the first kind which employ such functions as kernels may be solved by inversion.

An example of a transform of this kind, known as the *Hankel transform*, may be obtained from the orthonormal properties of the Bessel function $J_m(kr)$ where m is arbitrary. In Chap. 7 [see also Eq. (6.3.62)] it was found that

$$\sqrt{kk_0}\int_0^\infty J_m(kr)J_m(k_0r)r\,dr = \delta(k - k_0)$$

Hence in the above discussion we can place

$$\varphi(k|x) = \sqrt{k}\,J_m(kr)$$

Following custom, we define the Hankel transform without the $\sqrt{k}$ factor. Thus

$$\Psi(k) = \int_0^\infty J_m(kr)\psi(r)r\,dr \tag{8.3.64}$$

The inversion is given by

$$\psi(r) = \int_0^\infty J_m(kr)\Psi(k)k\,dk \tag{8.3.65}$$

We may combine Eqs. (8.3.64) and (8.3.65) to obtain

$$\psi(r) = \int_0^\infty k\,dk \int_0^\infty \rho\,d\rho[J_m(kr)J_m(k\rho)\psi(\rho)]$$

The solution of integral equation (8.3.64) is thus given by Eq. (8.3.65).

A number of transforms which are closely related to the Fourier transform (see also Sec. 4.7) are of great importance in the treatment of some integral equations. The *Laplace transform*, which we shall later apply to the Volterra equation, is

$$\Psi(p) = \int_0^\infty e^{-pz}\psi(z)\,dz \tag{8.3.66}$$

where Ψ is the transform of ψ. The inversion is

$$\psi(z) = \frac{1}{2\pi i}\int_{c-i\infty}^{c+i\infty} e^{pz}\Psi(p)\,dp \tag{8.3.67}$$

The Mellin transform is defined by

$$\Psi(s) = \int_0^\infty z^{s-1}\psi(z)\,dz \tag{8.3.68}$$

Its inversion is given by

$$\psi(z) = \frac{1}{2\pi i}\int_{c-i\infty}^{c+i\infty} z^{-s}\Psi(s)\,ds \tag{8.3.69}$$

Both Eqs. (8.3.67) and (8.3.69) may be regarded as solutions of the integral equations of the first kind given by (8.3.66) and (8.3.68), respectively.

Finally we mention a pair of transforms obtained from function theory which are essentially derived from the Cauchy integral formula. These are called *Hilbert transforms*. From Eq. (4.2.18) we have that, if

$$\Psi(z) = \left(\frac{1}{\pi}\right)P\int_{-\infty}^{\infty}\frac{\psi(z_0)}{z_0 - z}\,dz_0 \tag{8.3.70}$$

then

$$\psi(z) = -\left(\frac{1}{\pi}\right)P\int_{-\infty}^{\infty}\frac{\Psi(z_0)}{z_0 - z}\,dz_0 \tag{8.3.71}$$

Here P means principal part of the integral. Equation (8.3.71) is a solution of integral equation (8.3.70) for $\psi(z_0)$.

From (4.2.28) a similar pair are obtained:

$$\begin{aligned}\Psi(\varphi) &= \left(\frac{1}{2\pi}\right)P\int_0^{2\pi}\left[1 + \cot\left(\frac{\varphi_0 - \varphi}{2}\right)\right]\psi(\varphi_0)\,d\varphi_0\\ \psi(\varphi) &= \left(\frac{1}{2\pi}\right)P\int_0^{2\pi}\left[1 - \cot\left(\frac{\varphi_0 - \varphi}{2}\right)\right]\Psi(\varphi_0)\,d\varphi_0\end{aligned} \tag{8.3.72}$$

We have already discussed this particular kernel in the section on Green's functions, following Eq. (8.3.57) where a solution was found in terms of a Fourier series. This connection is not too surprising, since (4.2.28) is based on a Green's function type relation (4.2.25).

In this subsection we have emphasized the relation of the transform to integral equations of the first kind. Later in this chapter we shall employ the transform to change integral equations into forms which are more amenable for solution.

Differential Equations and Integral Equations of the First Kind. We have now seen how to solve a good many sorts of equations of the first kind. Many more can be devised by operating on both sides by some operator, either differential or integral (or both). For example, suppose we know the solution ψ of the equation

$$\varphi(z) = \int K(z|z_0)\psi(z_0)\, dz_0$$

where K and φ are known. We can then find the solution of

$$\chi(z) = \int \mathfrak{L}_z[K(z|z_0)]\psi(z_0)\, dz_0 \tag{8.3.73}$$

provided we can solve the equation

$$\mathfrak{L}_z[\varphi(z)] = \chi(z) \tag{8.3.74}$$

to determine φ from χ. Vice versa, if we know the solution of the equation with χ, we can obtain the solution for φ. Thus, from a solution of a simple integral equation we can go on to solve more complex ones if we can solve the equation obtaining φ from χ. If the operator changing K into $\mathfrak{L}(K)$ is a differential operator, then the equation which must be solved to obtain φ is a differential equation. Thus we can solve an integral equation if we can solve a differential equation, or vice versa, we can solve a differential equation if we can solve an integral equation.

For example, let $K(z|z_0) = e^{-zz_0}$ so that

$$\varphi(z) = \int_0^\infty e^{-zz_0}\psi(z_0)\, dz_0$$

This is just the Laplace transform; the solution ψ is given by the inversion equation (8.3.67). A rather general class of kernels may be obtained from this:

$$\mathcal{K}(z|z_0) = \sum_n z_0^n g_n(z) e^{-zz_0}$$

or

$$\mathcal{K}(z|z_0) = \sum_n (-1)^n g_n(z) \frac{d^n}{dz^n}(e^{-zz_0}) = \mathfrak{L}_z[K(z|z_0)]$$

where

$$\mathfrak{L} = \sum_n (-1)^n g_n(z) \frac{d^n}{dz^n} \tag{8.3.75}$$

Equation (8.3.74) is in this case a differential equation for φ, an equation discussed at some length in Chap. 5:

$$\sum_n (-1)^n g_n(z) \frac{d^n\varphi}{dz^n} = \chi(z)$$

The solution of this differential equation may not be easy. It is necessary, for most available techniques, that the sum be finite; in fact, unless the differential equation is of second or first order, it is not likely to be soluble.

Another example is furnished by the Hermite functions. Let $K(z|z_0) = e^{-(z-z_0)^2}$. If $\mathfrak{L} = \sum_n (-1)^n a_n \frac{d^n}{dz^n}$, then

$$\mathfrak{L}[K(z|z_0)] = \sum_n a_n e^{-(z-z_0)^2} H_n(z - z_0).$$

which allows a very wide range of choice of kernel for the new integral equation. If the a_n's are constants, the kernel $\mathfrak{L}(K)$ is a general function of $(z - z_0)$ in the range $-\infty$ to $+\infty$. If the a's are functions of z, the kernel is even more general. The corresponding differential equation

$$\sum_n (-1)^n a_n \left(\frac{d^n\varphi}{dz^n}\right) = \chi(z)$$

is particularly easy to solve if the a's are all constants; consequently we have here a means of solving any integral equation of the first kind, having a kernel which is a function of $(z - z_0)$ alone and having limits $-\infty$ and $+\infty$.

There is another associated differential equation which may be obtained from Eq. (8.3.73) in a manner closely analogous to the method in which integral representations for differential equations are obtained (see Sec. 5.3). Suppose that

$$\mathfrak{L}_z K(z|z_0) = \mathfrak{M}_{z_0} K(z|z_0)$$

where the subscript indicates the variable on which the operators act. For example, for the $\mathfrak{L}$ in Eq. (8.3.75)

$$\mathfrak{M}_{z_0} = \sum_n (-1)^n z_0^m \, g_n\left(-\frac{d}{dz_0}\right)$$

(For further examples see Sec. 5.3.) We then introduce the adjoint operator $\tilde{\mathfrak{M}}$ defined by

$$u\mathfrak{M}[v] - v\tilde{\mathfrak{M}}[u] = (d/dz)P[u,v]$$

Employing these relations in Eq. (8.3.73) we obtain

$$\chi(z) + P[K(z|a), \psi(a)] - P[K(z|b), \psi(b)] = \int_a^b K(z|z_0)\tilde{\mathfrak{M}}[\psi(z_0)]\, dz_0$$

According to our assumption this integral equation may be solved for $\tilde{\mathfrak{M}}[\psi]$ so that

$$\tilde{\mathfrak{M}}[\psi(z)] = r(z) \tag{8.3.76}$$

This may be a simpler differential equation to solve than (8.3.74). Methods and examples for obtaining $\tilde{\mathfrak{M}}$ from $\mathfrak{M}$ are given in Sec. 5.3, and indeed many examples of $\mathfrak{L}$.

The operator $\mathfrak{L}$ does not have to be restricted to differential operators. For example, the operation involved may be to take a transform of both sides, and problem (8.3.73) would then be the problem of inverting the transform. This powerful method deserves a section to itself, however, and we shall take it up in Sec. 8.4.

The Moment Problem. The nth moment M_n of a function (or more commonly in physics, a distribution) ψ is defined by

$$M_n = \int_a^b z_0^n \psi(z_0)\rho(z_0)\, dz_0 \tag{8.3.77}$$

where $\rho(x)$ is a weighting function. In many cases, particularly in transport phenomena (see Secs. 2.4 and 12.2), it is possible to evaluate the series of moments M_n; then ψ is required.

Before discussing the methods that are available, let us first note that it is possible, by means of a linear change in variable, to shift the limits of integration. This will have the effect of changing the definition of ψ and in shuffling some of the moments. For example, let $\zeta_0 = 2\pi[(z_0 - a)/(b - a)]$. Then

$$M_n = \frac{2\pi}{b-a}\int_0^{2\pi}\left[\frac{(b-a)\zeta_0}{2\pi} + a\right]^n \psi\left(\frac{b-a}{2\pi}\zeta_0 + a\right)\rho\left(\frac{b-a}{2\pi}\zeta_0 + a\right) d\zeta_0$$

Let $\psi(z_0) = \chi(\zeta_0)$ and $\rho(z_0) = \omega(\zeta_0)$ and

$$\mu_n = \int_0^{2\pi} \zeta_0^n \chi(\zeta_0)\omega(\zeta_0)\, d\zeta_0$$

Then $$\mu_n = \left(\frac{2\pi}{b-a}\right)^{n+1}\int_a^b (z_0 - a)^n \psi(z_0)\rho(z_0)\, dz_0$$

Hence $$\mu_n = \left(\frac{2\pi}{b-a}\right)^{n+1}\sum_{s=0}^{n}\binom{n}{s}(-a)^{n-s}M_s \tag{8.3.78}$$

It is therefore clear that, as long as the transformation is linear, it is possible to choose the limits of integration a and b in (8.3.77) without becoming involved in any infinite processes. The limits which are commonly employed are $(-\infty, \infty)$, $(0, \infty)$, and $(-1, 1)$.

The principal technique available for the solution of the moment problem involves relating the moments to the coefficients of the expansion of ψ in polynomials orthogonal in the region (a,b), with the weight factor ρ. For example, suppose that the region of integration is $(-\infty, \infty)$ and the weight factor $e^{-z_0^2}$. [One may, of course, factor $e^{-z_0^2}$ from $\psi(z_0)$ and define a new unknown function $\psi(z_0) = e^{-z_0^2}\varphi(z_0)$.] The corresponding orthogonal polynomials are $H_n(z_0)$. If we now expand $\psi(z_0)$ as follows:

$$\psi(z_0) = \sum a_n H_n(z_0); \quad a_n = \frac{1}{n!2^n \sqrt{\pi}} \int_{-\infty}^{\infty} \psi(z_0) H_n(z_0) e^{-z_0^2}\, dz_0$$

Since the H_n's are polynomials, the integral in a_n may be expressed directly in terms of M_n. For example, $H_0(z_0) = 1$, so that

$$a_0 = \frac{1}{\pi} \int_{-\infty}^{\infty} \psi(z_0) e^{-z_0^2}\, dz_0 = \frac{M_0}{\pi}$$

More generally

$$\begin{aligned} H_n &= (2x)^n - \binom{n}{2} \frac{2!}{1!} (2x)^{n-2} + \binom{n}{4} \frac{4!}{2!} (2x)^{n-4} + \cdots \\ &= \sum_k \frac{(-1)^k n!}{(k!)(n-2k)!} (2x)^{n-2k} \end{aligned}$$

where the last term in the expansion is a constant or the first power of x according as n is even or odd. Then

$$a_n = \frac{1}{\pi} \sum_k \frac{(-1)^k}{(k!)(n-2k!)} \left(\frac{M_{n-2k}}{2^{2k}} \right) \tag{8.3.79}$$

From (8.3.79), the function ψ may be readily evaluated.

There is another way to view the moment problem which is more general in principle though not necessarily more advantageous in practice. We may show that this problem is equivalent to an integral equation of the first kind. For example, return to the illustration above. Multiply both sides of the corresponding equation (8.3.77) by $e^{-z^2}[(2z)^n/n!]$

$$\left[\frac{(2z)^n}{n!}\right] e^{-z^2} M_n = \int_{-\infty}^{\infty} \left[\frac{(2zz_0)^n}{n!}\right] e^{-(z^2+z_0^2)} \psi(z_0)\, dz_0$$

Now by summing over n on both sides one obtains

$$e^{-z^2} \left\{ \sum_{n=0}^{\infty} \left[\frac{(2z)^n}{n!}\right] M_n \right\} = \int_{-\infty}^{\infty} e^{-(z-z_0)^2} \psi(z_0)\, dz_0 \tag{8.3.80}$$

Assuming that the series on the left converges, the moment problem has now been reduced to an integral equation of the first kind. Of course, it may be solved by an expansion of ψ in the Hermite polynomials. However, there are many approximate ways (see Chap. 9) of solving integral equations which may be more practical than (8.3.79) in many cases.

Similar analyses may be set up for the other intervals of integration. For the interval $(0, \infty)$, the weight factor e^{-z_0} and the Laguerre polynomials are appropriate. In the region $(-1, 1)$, for $\rho = 1$, the proper polynomials are the Legendre polynomials. If $\rho = \sqrt{1 - z_0^2}$, the proper polynomials are the Tschebyscheff polynomials. It is quite important to note that the expansion of ψ in terms of the orthogonal polynomials depends on the weight factor ρ; indeed convergence may be quite different for different weight factors. However, as indicated above, by removing a factor from $\psi(z_0)$ and designating the remainder as the unknown function, it is possible arbitrarily to change the weight factor. Let us call this factor $\omega(z)$, so that

$$M_n = \int_a^b z_0^n \rho(z_0)\omega(z_0)\varphi(z_0)\, dz_0$$

where $\psi = \omega\varphi$. The most appropriate choice for ω, that is, the one which will yield the most rapid convergence, is a function which is as close to the unknown ψ as is possible. Of course if it is exactly ψ, there is only one term in the polynomial expansion for φ, namely $\varphi = 1$. It is clearly advantageous to employ whatever information is available to make as close a guess to ψ as is possible and then to use this guess for ω.

Recapitulation. We have investigated those solutions of Fredholm integral equations of the first kind which are obtained by expanding the unknown function in a complete set. The most general kernels for which solutions could be exactly obtained are characterized by an expansion of the form

$$K(z|z_0) = \sum_n a_n \psi_n(z)\varphi_n(z_0)$$

where both ψ_n and φ_n form complete and orthogonal sets. Orthogonality is not in principle essential but in practice of great convenience. However, procedures which may be adopted when one of the sets is not orthogonal have been discussed.

It might be pointed out that Volterra equations of the first kind may be solved by some of the methods outlined in this section, though in general the procedure is more difficult and the full machinery of the Schmidt method or the biorthogonal series must be used. Luckily many kernels of Volterra equations are of the form $v(z - z_0)$, a case which is discussed in some detail in Sec. 8.5.

8.4 *Solution of Integral Equations of the Second Kind*

The methods which may be employed to solve the Fredholm integral equation of the second kind may also be classified according to the manner in which the kernel may be expanded. This classification is, of course, very similar to that given in Sec. 8.3, though the manipulation of the equation is quite different. It is thus advisable to approach the classification in a different manner from that used in Sec. 8.3, though, as we shall see, the same kinds of kernels will be discussed here as formerly. As in the previous section, we confine ourselves to methods of exact solution; approximate solutions will receive their full attention in Chap. 9. Indeed one method of series solution, called the Fredholm solution, is so important to the study of perturbation theory that its exposition will be put off till the next chapter, though it could just as well have been inserted here.

The equation under consideration is

$$\psi(z) = \lambda \int_a^b K(z|z_0)\psi(z_0)\, dz_0; \quad a \leq z \leq b \tag{8.4.1}$$

Solutions to this problem exist for only special values of λ, labeled λ_n, with corresponding solutions ψ_n. If the kernel K is symmetric and nonsingular, the eigenvalues are real, the eigenvalue spectrum is discrete. If it is symmetric but singular, part of the spectrum may be continuous; if it is not symmetric, the eigenvalues are not necessarily real.

We may expand $K(z|z_0)$ in a complete set $h_n(z)$; the coefficients will be functions of z_0, $g_n(z_0)$:

$$K(z|z_0) = \sum_n h_n(z) g_n(z_0) \tag{8.4.2}$$

Then

$$\psi(z) = \lambda \sum_n h_n(z) \int_a^b g_n(z_0)\psi(z_0)\, dz_0$$

This suggests that we place

$$\psi(z) = \sum_n A_n h_n(z) \tag{8.4.3}$$

(It should be emphasized that this expansion will prove meaningful in the following development only if it converges.) Hence

$$A_n = \lambda \int_a^b g_n(z_0)\psi(z_0)\, dz_0; \quad \text{for each } n \tag{8.4.4}$$

$$A_n = \lambda \sum_p A_p \int_a^b g_n(z_0) h_p(z_0)\, dz_0$$

If we let

$$\int_a^b g_n(z_0) h_p(z_0)\, dz_0 = \alpha_{pn}$$

Then

$$A_n = \lambda \sum_p A_p \alpha_{pn}$$

or

$$\sum_p A_p(\lambda\alpha_{pn} - \delta_{pn}) = 0 \tag{8.4.5}$$

This is a set of linear *homogeneous* simultaneous equations for A_p. Nonzero solutions for A_p exist only if the determinant of the coefficients vanishes:

$$\begin{vmatrix} (\lambda\alpha_{00} - 1) & \lambda\alpha_{10} & \lambda\alpha_{20} & \lambda\alpha_{30} & \cdots \\ \lambda\alpha_{01} & (\lambda\alpha_{11} - 1) & \lambda\alpha_{21} & \lambda\alpha_{31} & \cdots \\ \lambda\alpha_{02} & \lambda\alpha_{12} & (\lambda\alpha_{22} - 1) & \lambda\alpha_{32} & \cdots \\ \lambda\alpha_{03} & \lambda\alpha_{13} & \lambda\alpha_{23} & (\lambda\alpha_{33} - 1) & \cdots \\ \cdot & \cdot & \cdot & \cdot & \\ \cdot & \cdot & \cdot & \cdot & \\ \cdot & \cdot & \cdot & \cdot & \\ \cdot & \cdot & \cdot & \cdot & \end{vmatrix} = 0$$

or more succinctly:

$$|\lambda\alpha_{pn} - \delta_{pn}| = 0 \tag{8.4.6}$$

This determinantal equation may be solved for the possible values of λ while Eq. (8.4.5) yields the corresponding values of A_p and therefore by virtue of (8.4.3), determines ψ. For the purpose of making a connection with the problem of solving differential equations, note that (8.4.5) forms a set of *recursion relations* for A_p much like those which occur when a power series or any other complete set of functions is substituted in a differential equation.

There is no "royal road" to the discovery of the "best" type of expansion for the kernel K; there is no infallible rule which tells which expansion will make the solution easiest. Indeed it is usually true that several alternative expansions are possible for each type of kernel, and often each must be tried in order to discover the most useful one for the purpose at hand. Accordingly our classification of methods of solution will be according to types of expansions rather than according to kernels, according to the nature of the coefficients α_{pn} which result when one has decided on a set of g's and h's for a given K. Often expansions of different classes may be applied to a given K, though usually one expansion appears more appropriate. We shall now review the more useful classes of expansions of K, giving examples of the kinds of kernels for which the expansion is appropriate and some details as to techniques of solution.

Expansions of the First Class. Our classification will be dependent upon the properties of the matrix α_{pn} and therefore upon the relation of the set g_n to h_p. For example, the *first class* is the one in which α_{pn} is *diagonal;* that is, $\alpha_{pn} = \alpha_n\delta_{pn}$. This means that g_n is a set which is orthogonal to all h_p except for the $p = n$ term. The eigenvalues λ_n are then

$$\lambda_n = 1/\alpha_n \tag{8.4.7}$$

with

$$\psi_n = h_n$$

One may verify this solution directly by substitution in the integral equation (8.4.1).

Green's function kernels furnish examples of the *diagonal* type of expansion. In the preceding section, from the expansion for $\ln R$ we obtained Eq. (8.3.54):

$$K(\varphi|\varphi_0) = \ln|2\sin(\varphi - \varphi_0)/2|$$
$$= -\sum_{n=1}^{\infty} \frac{1}{n} [\cos(n\varphi)\cos(n\varphi_0) + \sin(n\varphi)\sin(n\varphi_0)]$$

In the region $0 \le \varphi \le 2\pi$, the eigenfunctions of the corresponding integral equations are $\cos(n\varphi)$, and $\sin(n\varphi)$; both have eigenvalues $\lambda_n = -n/\pi$.

A second example is furnished by combining $\ln|2\sin(\varphi - \varphi_0)/2|$ and $\ln|2\sin(\varphi + \varphi_0)/2|$ to obtain (8.3.55):

$$K(\varphi|\varphi_0) = \ln[2|\cos\varphi - \cos\varphi_0|] = -\sum_{n=1}^{\infty} \left(\frac{2}{n}\right) \cos(n\varphi)\cos(n\varphi_0)$$

In the region $0 \le \varphi \le \pi$, the corresponding integral equation has eigenfunctions $\cos(n\varphi)$ and eigenvalues $\lambda_n = -(n/\pi)$.

By considering the kernel obtained by differentiating this last $K(\varphi|\varphi_0)$ [Eq. (8.3.55)], we shall examine a case which illustrates an essential difference between an equation of the first kind and one of the second kind. After differentiation the kernel becomes

$$\frac{1}{\cos\varphi - \cos\varphi_0} = -2\sum_{n=1}^{\infty} \frac{\sin(n\varphi)}{\sin\varphi} \cos(n\varphi_0)$$

We note that, although $\sin(n\varphi)$ and $\cos(n\varphi)$ form complete sets, they are not mutually orthogonal in the region $0 \le \varphi \le \pi$. In the solution of the equation of the first kind, this does not create a difficulty, since the inhomogeneous term in the equation does not bear any explicit relation to ψ whereas this is not at all true for equations of the second kind.

Expansions of the Second Class. As the second class of expansions we consider those for which

$$\begin{aligned} &\alpha_{pn} = 0; \quad \text{if } p < n; \quad \text{type } a \\ \text{or} \qquad &\alpha_{pn} = 0; \quad \text{if } p > n; \quad \text{type } b \end{aligned} \tag{8.4.8}$$

We shall call such expansions *semidiagonal*. The function g_n may be expanded in terms of h_p as follows:

$$
\begin{aligned}
g_n(z) &= \sum_{p=n}^{\infty} \alpha_{pn} h_p(z); \quad \text{type } a \\
g_n(z) &= \sum_{p=0}^{n} \alpha_{pn} h_p(z); \quad \text{type } b
\end{aligned}
\tag{8.4.9}
$$

The recursion relation between A_p becomes for case a

$$A_n = \lambda \sum_{p=n}^{\infty} A_p \alpha_{pn}$$

This relation includes as a special case (when only two α_{pn} exist for a given n) the two-term recursion relations which are of such fundamental importance in differential equations.

The determinantal equations determining λ now take on a particularly simple form. For case a:

$$\begin{vmatrix} \lambda\alpha_{00} - 1 & \lambda\alpha_{10} & \lambda\alpha_{20} & \lambda\alpha_{30} & \cdots \\ 0 & \lambda\alpha_{11} - 1 & \lambda\alpha_{21} & \lambda\alpha_{31} & \cdots \\ 0 & 0 & \lambda\alpha_{22} - 1 & \lambda\alpha_{32} & \cdots \\ 0 & 0 & 0 & \lambda\alpha_{33} - 1 & \cdots \\ \cdot & \cdot & \cdot & \cdot & \\ \cdot & \cdot & \cdot & \cdot & \\ \cdot & \cdot & \cdot & \cdot & \end{vmatrix} = 0$$

The expansion of the determinant is just $(\lambda\alpha_{00} - 1)(\lambda\alpha_{11} - 1) \cdots$, so that the eigenvalues are $\lambda_n = 1/\alpha_{nn}$. The corresponding eigenfunctions involve h_p, $p \leq n$. Hence

$$\psi_0 = h_0; \quad \psi_1 = h_1 + [\lambda_1\alpha_{10}/(1 - \lambda_1\alpha_{00})]h_0; \quad \text{etc.}$$

These are obtained from the recursion relations for A_p.

It is instructive to consider a special case of case a in which α_{pn} differs from zero only for $p = n$ and $p = n + 1$. Hence the recursion relation becomes

$$A_n(\lambda\alpha_{nn} - 1) + A_{n+1}\lambda\alpha_{n+1,\,n} = 0$$

This is a two-term recursion formula and can be easily solved:

$$A_p = A_0 \prod_{j=0}^{p-1} (-1)^{j+1} \left[\frac{\lambda\alpha_{j,j} - 1}{\lambda\alpha_{j+1,\,j}} \right]$$

We see that choosing $\lambda = \lambda_n = 1/\alpha_{nn}$ results in $A_p = 0$ if $p > n$; in other words the series for ψ_n in terms of h_p breaks off at $p = n$. However, what prevents *any* λ from being employed in evaluating A_p and therefore obtaining solutions with a continuous spectrum for λ? The only limiting condition is the convergence of the series for ψ. Only those values of λ for which the series for ψ does converge are possible. The situation here is completely similar to that which obtains in differential equations. For example, in solving the Hermite differential equation by power series, a two-term recursion formula results. Convergent power series cannot be obtained without picking special values corresponding to the special values of λ in the case under discussion (see 768). Coming back then to the integral equation, it is important to realize that in addition to the solutions $\lambda_n = 1/\alpha_{nn}$, *there may also be solutions for which the spectrum for* λ *is continuous.*

Consider type b next. The recursion relation is

$$A_n = \lambda \sum_{p=0}^{n} A_p \alpha_{pn}$$

The determinantal equation is now

$$\begin{vmatrix} \lambda\alpha_{00} - 1 & 0 & 0 & 0 & \cdots \\ \lambda\alpha_{01} & \lambda\alpha_{11} - 1 & 0 & 0 & \cdots \\ \lambda\alpha_{02} & \lambda\alpha_{12} & \lambda\alpha_{22} - 1 & 0 & \cdots \\ \lambda\alpha_{03} & \lambda\alpha_{13} & \lambda\alpha_{23} & \lambda\alpha_{33} - 1 & \cdots \\ \cdot & \cdot & \cdot & \cdot & \\ \cdot & \cdot & \cdot & \cdot & \\ \cdot & \cdot & \cdot & \cdot & \end{vmatrix} = 0$$

As in type a, the eigenvalues are $\lambda_n = 1/\alpha_{nn}$. The corresponding eigenfunctions are not finite combinations of h_p as they were in case a. For example, for ψ_0 we have the determining equations

$$A_n(1 - \lambda_0\alpha_{nn}) = \lambda_0 \sum_{p=0}^{n-1} A_p \alpha_{pn}$$

Hence

$$A_1 = [(\lambda_0 A_0 \alpha_{01})/(1 - \lambda_0\alpha_{11})]; \quad A_2(1 - \lambda_0\alpha_{22}) = \lambda_0 A_1 \alpha_{12} + \lambda_0 A_0 \alpha_{02}$$

or

$$A_2 = \frac{\lambda_0 A_0}{1 - \lambda_0\alpha_{22}} \left[\frac{\alpha_{01}}{1 - \lambda_0\alpha_{11}} + \alpha_{02} \right]$$

and so on. The words of caution concerning the values of λ, added after the discussion of type a, apply here as well.

Examples of kernels of the semidiagonal type are furnished by generating functions. Consider

$$K(z|z_0) = \frac{1}{\sqrt{1 - 2zz_0 + z_0^2}} = \sum_n P_n(z) z_0^n$$

The independent variable ranges from -1 to 1. Upon substituting $\psi = \Sigma A_p P_p(z)$, the integral equation becomes

$$A_n = \lambda \sum_p A_p \int_{-1}^{1} z_0^n P_p(z_0)\, dz_0 \tag{8.4.10}$$

The integral vanishes as soon as $p > n$, so that the kernel is semidiagonal and of type b. The eigenvalues are given by

$$\lambda_n = 1 \Big/ \int_{-1}^{1} z_0^n P_n(z_0)\, dz_0$$

The integral may be evaluated by noting that z_0^n may be expressed as a linear combination of P_p, $p \leq n$ and that only the term in P_n will survive after the integration. Hence in expressing z_0^n in terms of P_p only the z_0^n term in P_n need be considered. Now

$$P_n = \frac{(2n)!}{2^n(n!)^2}[z^n - \cdots]$$

Accordingly
$$z^n = \frac{2^n(n!)^2}{(2n)!}[P_n + \cdots]$$

Finally
$$\lambda_n = \frac{(2n+1)!}{2^{n+1}(n!)^2} = \frac{2^n\Gamma(n+\frac{3}{2})}{n!\Gamma(\frac{1}{2})}$$

$$\lambda_0 = \tfrac{1}{2};\quad \lambda_1 = \tfrac{3}{2};\quad \lambda_2 = \tfrac{15}{4};\ \ldots$$

The coefficients A_n for each particular λ may be obtained directly from recursion relations (8.4.10). By interchanging the roles of z and P_n in the above discussion, we may obtain a semi-infinite kernel of type a. The kernel is now

$$K(z|z_0) = \frac{1}{\sqrt{1 - 2z_0z + z^2}} = \sum_n P_n(z_0)z^n$$

Then
$$\psi = \lambda \int_{-1}^{1} K(z|z_0)\psi(z_0)\, dz_0$$

becomes
$$\psi = \sum_n z^n \int_{-1}^{1} P_n\psi\, dz_0$$

Let
$$\psi = \sum_n A_n z^n$$

Then
$$A_n = \sum_p A_p \int_{-1}^{1} P_n z^p\, dz$$

Since $\int_{-1}^{1} P_n z^p\, dz = 0$ if $p < n$, we may write

$$A_n = \sum_{p=n}^{\infty} A_p\alpha_{pn};\quad \alpha_{pn} = \int_{-1}^{1} P_n z^p\, dz = \frac{p!}{2^n(p-n)!}\frac{\Gamma(p-n+1/2)}{\Gamma(p+n+3/2)}$$

The eigenvalues are given by $\lambda_n = 1/\alpha_{nn}$ and are identical with those of the problem discussed immediately above. The eigenfunctions are, however, more easily obtained in this case.

$$\psi_0 = 1$$
$$\psi_1 = z$$
$$\psi_2 = z^2 + \frac{(\alpha_{20}/\alpha_{22})}{1 - (\alpha_{00}/\alpha_{22})}; \quad \text{etc.}$$

where the α's are given above.

Expansions of the Third Class. We now turn to a third class of expansions which we shall call *finite*. These are defined by the condition

$$\alpha_{pn} = 0; \quad \text{if either } p \text{ or } n \text{ is bigger than } r \tag{8.4.11}$$

For this class, the determinantal equation (8.4.6) evolves a finite determinant of r rows and columns. Upon expansion, the determinant becomes a polynomial of rth degree. The equation thus has r roots which may be determined by standard methods. Of course both diagonal and semidiagonal kernels may be finite.

In the simplest example of this class, $K(z|z_0)$ is a factorable function:

$$K(z|z_0) = h(z)g(z_0)$$

The integral equation is

$$\psi(z) = \lambda h(z) \int g(z_0)\psi(z_0)\, dz_0$$

Since the integral is just a constant, we see immediately that $\psi(z) = h(z)$. Hence

$$\lambda = 1/\int g(z_0)h(z_0)\, dz_0$$

This result may also be obtained from the determinantal equation (8.4.6), for in this case only α_{00} is different from zero. Accordingly (8.4.6) becomes

$$\lambda\alpha_{00} - 1 = 0; \quad \lambda = 1/\alpha_{00}$$

in agreement with the result just obtained directly from the integral equation.

Solutions may also be explicitly obtained if only two terms occur in the expansion of K. We give these results for reference. The determinantal equation

$$\begin{vmatrix} (\lambda\alpha_{00} - 1) & \lambda\alpha_{10} \\ \lambda\alpha_{01} & (\lambda\alpha_{11} - 1) \end{vmatrix} = 0$$

has two solutions:

$$\lambda_{\pm} = \frac{(\alpha_{00} + \alpha_{11}) \pm \sqrt{(\alpha_{00} - \alpha_{11})^2 + 4\alpha_{10}\alpha_{01}}}{2(\alpha_{00}\alpha_{11} - \alpha_{10}\alpha_{01})} \tag{8.4.12}$$

where λ_+ refers to the solution with the square root added to $\alpha_{00} + \alpha_{11}$. The corresponding eigenfunctions are

$$\begin{aligned}\psi_+ &= h_0 + (1 - \lambda_+\alpha_{00})/(\lambda_+\alpha_{10})h_1 \\ \psi_- &= h_0 + (1 - \lambda_-\alpha_{00})/(\lambda_-\alpha_{10})h_1\end{aligned} \tag{8.4.13}$$

Other Classes. Finally we turn to the case for which the choice of h_n has been so unfortunate as to yield an expansion for K which does not fall into any of the classes, diagonal, semidiagonal, or finite, discussed above. It is then generally impossible to obtain an explicit expression for the eigenvalue λ, and one must have recourse to approximate or numerical techniques. These will be discussed in Chap. 9. There is, however, one situation for which the equation for λ is sufficiently simple to warrant discussion. Suppose that

$$\alpha_{pn} = 0; \quad \text{if } p \neq n - 1,\ n,\ n + 1 \tag{8.4.14}$$

In other words the expansion of g_n in terms of h_n involves but three terms. Assuming the h_n's to be normalized, this expansion would be

$$g_n = \alpha_{n-1,n}\, h_{n-1} + \alpha_{n,n}\, h_n + \alpha_{n+1,n}\, h_{n+1}$$

The determinantal equation (8.4.6) becomes

$$\begin{vmatrix} (\lambda\alpha_{00} - 1) & \lambda\alpha_{10} & 0 & 0 & \cdot & \cdot & \cdot & \cdot \\ \lambda\alpha_{01} & (\lambda\alpha_{11} - 1) & \lambda\alpha_{21} & 0 & \cdot & \cdot & \cdot & \cdot \\ 0 & \lambda\alpha_{12} & (\lambda\alpha_{22} - 1) & \lambda\alpha_{32} & 0 & \cdot & \cdot & \cdot \\ 0 & 0 & \lambda\alpha_{23} & (\lambda\alpha_{33} - 1) & \lambda\alpha_{43} & 0 & \cdot & \cdot \\ \cdot & \cdot & \cdot & \cdot & \cdot & \cdot & \cdot & \cdot \\ \cdot & \cdot & \cdot & \cdot & \cdot & \cdot & \cdot & \cdot \\ \cdot & \cdot & \cdot & \cdot & \cdot & \cdot & \cdot & \cdot \end{vmatrix} = 0$$

The recursion relation connecting the unknown coefficients A_p in the expansion of ψ, as given by (8.4.5), is

$$\lambda\alpha_{n-1,\,n}A_{n-1} + (\lambda\alpha_{n,n} - 1)A_n + \lambda\alpha_{n+1,\,n}A_{n+1} = 0 \tag{8.4.15}$$

This is a three-term recursion formula. We have discussed the solution of such difference equations in Chap. 5, page 565, where a similar problem turned up in the solution of the Mathieu differential equation.

The solution proceeds by introducing the new dependent variable

$$G_n = A_n/A_{n-1}; \quad G_0 = \infty$$

Then

$$\lambda\alpha_{n-1,\,n} + (\lambda\alpha_{n,n} - 1)G_n + \lambda\alpha_{n+1,\,n}G_{n+1}G_n = 0$$

To simplify the notation, let

$$\begin{gathered}\alpha_{n,n}/\alpha_{n-1,\,n} = -p_n; \quad \alpha_{n+1,\,n}/\alpha_{n-1,\,n} = -q_n \\ 1/\lambda = \mu; \quad 1/\alpha_{n-1,\,n} = -r_n\end{gathered}$$

Then (8.4.15) may be written

$$-1 + (p_n - \mu r_n)G_n + q_n G_n G_{n+1} = 0 \tag{8.4.16}$$

Solving this equation for G_n in terms of G_{n+1}, we obtain

$$G_n = 1/(p_n - \mu r_n + q_n G_{n+1})$$

By introducing into this expression the corresponding expression for G_{n+1}, one finds

$$G_n = \cfrac{1}{p_n - \mu r_n + \cfrac{q_n}{p_{n+1} - \mu r_{n+1} + q_{n+1}G_{n+2}}}$$

Continuing this process, a continued fraction is obtained for G_n:

$$G_n = \cfrac{1}{p_n - \mu r_n + \cfrac{q_n}{p_{n+1} - \mu r_{n+1} + \cfrac{q_{n+1}}{p_{n+2} - \mu r_{n+2} + \cfrac{q_{n+2}}{p_{n+3} - \mu r_{n+3} + \cdots}}}}$$

For G_1:

$$C_1 = \cfrac{1}{p_1 - \mu r_1 + \cfrac{q_1}{p_2 - \mu r_2 + \cfrac{q_2}{p_3 - \mu r_3 + \cdots}}} \tag{8.4.17}$$

We may also solve (8.4.16) for G_{n+1}:

$$G_{n+1} = -\frac{(p_n - \mu r_n)}{q_n} + \frac{1}{q_n G_n}$$

From this a finite continued fraction is obtained:

$$G_{n+1} = -\frac{(p_n - \mu r_n)}{q_n} \tag{8.4.18}$$
$$+ \cfrac{1}{-q_n \left\{ \cfrac{-(p_{n-1} - \mu r_{n-1})}{q_{n-1}} + \cfrac{1}{q_{n-1}\left[\cfrac{-(\cdots)}{q_{n-2}} + \cdots \cfrac{-q_1(p_0 - \mu)}{q_0}\right]} \right\}}$$

For $n = 0$, (8.4.18) becomes

$$G_1 = -(p_0 - \mu r_0)/q_0$$

Equating this expression and (8.4.17) yields an equation determining $\mu = 1/\lambda$:

$$-\frac{p_0 - \mu r_0}{q_0} = \cfrac{1}{p_1 - \mu r_1 + \cfrac{q_1}{p_2 - \mu r_2 + \cfrac{q_2}{p_3 - \mu r_3 + \cdots}}} \tag{8.4.19}$$

The numerical methods which must be employed to solve this equation are described in Chap. 5, page 566, and need not be reviewed here. In any event an expansion of a kernel of type (8.4.14) may be reduced to solving Eq. (8.4.19). Once μ is determined, the corresponding G_{n+1} may be evaluated from (8.4.18), and consequently the A_n's may be computed from $G_n = A_n/A_{n-1}$.

Inhomogeneous Fredholm Integral Equation of the Second Kind. The solution of the inhomogeneous equation of the second kind has been considered in Sec. 8.2. It is instructive to rederive the result obtained there by direct considerations. The equation is

$$\psi = \lambda \int_a^b K(z|z_0)\psi(z_0)\, dz_0 + \chi(z) \tag{8.4.20}$$

Let

$$\psi = \sum_n A_n\psi_n$$

where

$$\psi_n = \lambda_n \int^b K(z|z_0)\psi_n(z_0)\, dz_0$$

Moreover, K may be expanded in a series of orthonormal eigenfunctions:

$$K(z|z_0) = \sum_n (1/\lambda_n)\psi_n(z)\varphi_n(z_0)$$

where

$$\int_a^b \psi_n\varphi_m\, dz = \delta_{nm}$$

Then in (8.4.20)

$$\sum_n A_n\psi_n = \lambda \sum_{n,p} \left(\frac{\psi_n A_p}{\lambda_n}\right) \int_a^b \varphi_n\psi_p\, dz_0 + \chi = \sum_n \left(\frac{\lambda A_n}{\lambda_n}\right)\psi_n + \chi$$

or

$$\sum_n A_n\left(1 - \frac{\lambda}{\lambda_n}\right)\psi_n = \chi$$

Hence

$$A_n = \left(\frac{\lambda_n}{\lambda_n - \lambda}\right) \int_a^b \chi\varphi_n\, dz$$

and

$$\psi = \sum_n \psi_n \left(\frac{\lambda_n}{\lambda_n - \lambda}\right) \int_a^b \chi\varphi_n\, dz \tag{8.4.21}$$

Solution (8.4.21) implies the availability of the solutions ψ_n in at least the number required to attain convergence in the series over n. When these are not available, it is sometimes convenient to rewrite integral equation (8.4.20) as an integral equation of the first kind. We may then apply the techniques of Sec. 8.3. Equation (8.4.20) may be written

$$\chi(z) = \int_a^b [\delta(z - z_0) - \lambda K(z|z_0)]\psi(z_0)\, dz_0 \tag{8.4.22}$$

where $\delta(z - z_0)$ is the Dirac δ function. This is an equation of the first kind with the kernel $[\delta(z - z_0) - \lambda K(z|z_0)]$. From solution (8.4.21) it is apparent that those values of λ for which ψ, considered as a function of λ, has a pole are the eigenvalues λ_n.

8.5 *Fourier Transforms and Integral Equations*

In Sec. 5.3 we investigated a technique of transforming from one differential equation to another, at times simpler than the first. Instead of trying to solve the specified differential equation directly, we looked at the equation for the corresponding Fourier (or Laplace), Mellin, or Euler transform. In the case that one of the equations, for a transform, was simpler than the original equation, we then obtained an integral representation for the solution of the original equation, a procedure which we found was quite powerful.

The same technique can, at times, lighten our work in solving integral equations. We determine the integral equation governing a transform of the original unknown; if this is simpler than the original equation, we have then obtained an integral representation of our original unknown.

Of course this procedure is but a special case of the techniques of Sec. 8.3, of expansion of the unknown in terms of a complete set of eigenfunctions. If the eigenvalue has a continuous range of allowed values, then the expansion is an integral over the continuous allowed range of the eigenvalue, an integral representation instead of a series expansion. For example, the coefficient of the expansion of $\psi(x)$ in terms of the eigenfunctions $\sqrt{1/2\pi}\, e^{ikx}$ over the continuous range $-\infty < k < \infty$ is just the Fourier transform of ψ.

This technique is particularly valuable when the transformed integral equation reduces to just an algebraic equation, which occurs whenever the transformed kernel turns out to be diagonal. But a few examples will illustrate the point better than further general discussion. At first we shall not worry too much about fine points (such as whether the transform exists!); we shall return to these after the salient features of the technique are exemplified.

The Fourier Transform and Kernels of Form $v(x - x_0)$. In the first place the range of values appropriate for Fourier transforms are the complete range $-\infty$ to $+\infty$, so our investigation will first center on equations of the form (Fredholm equation of the second kind over $-\infty < x < \infty$)

$$\psi(x) = \varphi(x) + \lambda \int_{-\infty}^{\infty} w(x|x_0)\psi(x_0)\, dx_0 \tag{8.5.1}$$

We now take the transform of both sides. Assuming that the transforms exist (we shall be more careful later; the present discussion will be simplified in order to avoid beclouding the central idea with details), we find

$$\Psi(k) = \Phi(k) + \lambda\mathfrak{F}\left[\int_{-\infty}^{\infty} w(x|x_0)\psi(x_0)\,dx_0\right]$$

To evaluate the Fourier transform of the integral

$$\mathfrak{F}\left[\int_{-\infty}^{\infty} w(x|x_0)\psi(x_0)\,dx_0\right] = \frac{1}{\sqrt{2\pi}}\int_{-\infty}^{\infty} e^{ikx}\,dx \int_{-\infty}^{\infty} w(x|x_0)\psi(x_0)\,dx_0$$

we express $\psi(x_0)$ in terms of $\Psi(k_0)$ by means of the inversion formula

$$\psi(x_0) = \frac{1}{\sqrt{2\pi}}\int_{-\infty}^{\infty} e^{-ik_0x_0}\Psi(k_0)\,dk_0$$

We may then write

$$\mathfrak{F}\left[\int_{-\infty}^{\infty} w(x|x_0)\psi(x_0)\,dx_0\right] = \int_{-\infty}^{\infty} W(k|k_0)\Psi(k_0)\,dk_0$$

where

$$W(k|k_0) = \frac{1}{2\pi}\int_{-\infty}^{\infty} dx_0 \int_{-\infty}^{\infty} dx\,[e^{-ik_0x_0}w(x|x_0)e^{ikx}] \tag{8.5.2}$$

The transformed integral equation is

$$\Psi(k) = \Phi(k) + \lambda\int_{-\infty}^{\infty} W(k|k_0)\Psi(k_0)\,dk_0 \tag{8.5.3}$$

This transformation is of value if the new kernel is simpler than the old. This is often the case, for a rather complex function can often be represented as the Fourier integral of a comparatively simple functional form.

It is apparent from Eq. (8.5.3) that a form of $W(k|k_0)$ for which this integral equation is immediately solvable is

$$W(k|k_0) = \sqrt{2\pi}\,V(k)\delta(k - k_0) \tag{8.5.4}$$

where δ is the Dirac δ function. Then Eq. (8.5.3) becomes

$$\Psi(k) = \Phi(k) + \sqrt{2\pi}\,\lambda V(k)\Psi(k)$$

This is now a simple algebraic equation for $\Psi(k)$:

$$\Psi(k) = \Phi(k)/[1 - \sqrt{2\pi}\,\lambda V(k)] \tag{8.5.5}$$

We may now obtain $\psi(x)$ by employing the Fourier inversion formula.

The requirement that the transformed kernel be diagonal (*i.e.*, be proportional to a delta function in $(k - k_0)$ imposes a certain restriction on the form of the original kernel $w(x|x_0)$. If we can learn to recognize kernels satisfying this restriction, we shall be able to recognize which integral equations are amenable to treatment by the Fourier transform. Equation (8.5.2) for the transformed kernel is a double Fourier transform, applied to both coordinates x and x_0; its inversion is

$$w(x|x_0) = \frac{1}{2\pi} \int_{-\infty}^{\infty} \int_{-\infty}^{\infty} e^{-ikx} W(k|k_0) e^{ik_0x_0} \, dk \, dk_0$$

Introducing the special form (8.5.4) into this, we obtain

$$\begin{aligned} w(x|x_0) &= \frac{1}{\sqrt{2\pi}} \int_{-\infty}^{\infty} V(k) e^{-ikx} \, dk \int_{-\infty}^{\infty} e^{ik_0x_0} \delta(k - k_0) \, dk_0 \\ &= \frac{1}{\sqrt{2\pi}} \int_{-\infty}^{\infty} V(k) e^{ik(x_0 - x)} \, dk = v(x - x_0) \end{aligned} \tag{8.5.6}$$

if

$$v(x) = \frac{1}{\sqrt{2\pi}} \int_{-\infty}^{\infty} V(k) e^{-ikx} \, dk$$

where v is the Fourier transform of V. Consequently, *any Fredholm equation of the second kind* (for the variable range $-\infty$ to ∞) *may be simplified by use of the Fourier transform if the kernel is a function of the difference* $(x - x_0)$, as stated in Eq. (8.5.6).

The Hankel Transform. If the range of the variables is from zero to infinity, we might try the Hankel transform [see Eq. (6.3.62)]:

$$\mathcal{H}(f) = F(k) = \int_0^\infty f(x) J_0(kx) x \, dx \tag{8.5.7}$$

with its inversion formula

$$f(x) = \int_0^\infty F(k) J_0(kx) k \, dk = \mathcal{H}(F) \tag{8.5.8}$$

If we apply the Hankel transform to the equation

$$\psi(x) = \varphi(x) + \lambda \int_0^\infty w(x|x_0) \psi(x_0) x_0 \, dx_0; \quad 0 \le x < \infty$$

then

$$\Psi(k) = \Phi(k) + \lambda \mathcal{H} \left[\int_0^\infty w(x|x_0) \psi(x_0) x_0 \, dx_0 \right]$$

$$\Psi(k) = \Phi(k) + \lambda \int_0^\infty W(k|k_0) \Psi(k_0) k_0 \, dk_0$$

where

$$W(k|k_0) = \int_0^\infty x \, dx \int_0^\infty x_0 \, dx_0 [J_0(kx) w(x|x_0) J_0(k_0x_0)] \tag{8.5.9}$$

The inversion of this equation gives

$$w(x|x_0) = \int_0^\infty k \, dk \int_0^\infty k_0 \, dk_0 [J_0(kx) W(k|k_0) J_0(k_0x_0)]$$

As before, for this transform to be of decided aid to us, the transformed kernel should be diagonal:

$$W(k|k_0) = (1/k) V(k) \delta(k - k_0)$$

which results in a relationship limiting the form of the original kernel w:

$$w(x|x_0) = \int_0^\infty J_0(kx)V(k)J_0(kx_0)k\,dk \qquad (8.5.10)$$

This restriction on w cannot be expressed in so simple a form as was the analogue for the Fourier transform, so it is harder to recognize in a particular kernel. A couple of examples might be of interest:

(*a*) $$V(k) = e^{-p^2k^2}; \quad w(x|x_0) = \left(\frac{1}{2p^2}\right) e^{-(x^2+x_0^2)/4p^2} I_0\left(\frac{xx_0}{2p^2}\right)$$

where I_0 is a Bessel function of imaginary argument.

(*b*) $$V(k) = e^{-ak}; \quad w(x|x_0) = (1/\pi\sqrt{xx_0})Q_{-\frac{1}{2}}[(x^2 + x_0^2 + a^2)/2xx_0]$$

where Q_n is the second solution of the Legendre equation [see Eq. (5.3.29)].

Turning to more general considerations, we note that the $w(x|x_0)$ satisfying Eq. (8.5.10) is symmetric; *i.e.*,

$$w(x|x_0) = w(x_0|x)$$

To obtain more definite information we expand $J_0(kx)$ in a power series:

$$w(x|x_0) = \sum_m \frac{(-1)^m}{(m!)^2}\left(\frac{x}{2}\right)^{2m} \int_0^\infty k^{2m+1}V(k)J_0(kx_0)\,dk$$

Now $V(k)$ is the Hankel transform of $v(x)$:

$$v(x_0) = \int_0^\infty kV(k)J_0(kx_0)\,dk$$

We may evaluate the more general integral in terms of $v(x)$ by noting that

$$\left[\frac{d^2}{dx_0^2} + \frac{1}{x_0}\frac{d}{dx_0}\right] v = -\int_0^\infty k^3V(k)J_0(kx_0)\,dk$$

Hence

$$\left[\frac{d^2}{dx_0^2} + \frac{1}{x_0}\frac{d}{dx_0}\right]^m v = \int_0^\infty k^{2m+1}V(k)J_0(kx_0)\,dk$$

and

$$w(x|x_0) = \sum_{m=0}^\infty \frac{(-1)^m}{(m!)^2}\left(\frac{x}{2}\right)^{2m}\left[\frac{d^2}{dx_0^2} + \frac{1}{x_0}\frac{d}{dx_0}\right]^m v(x_0) \qquad (8.5.11)$$

We may therefore verify whether a given kernel is of the type whose Hankel transform is diagonal by seeing if w is symmetric in x and x_0 and if, upon expansion in a power series in x, the coefficients follow the form (8.5.11).

The Kernel $v(x - x_0)$ in the Infinite Domain. The Fourier transform of the solution of the Fredholm equation of the second kind (8.5.1) with $w(x|x_0) = v(x - x_0)$, as given by Eq. (8.5.5), may be obtained more directly by the faltung theorem [Eq. (4.8.25)],

$$\mathfrak{F}\left[\int_{-\infty}^{\infty} v(x - x_0)\psi(x_0)\,dx_0\right] = \sqrt{2\pi}\,V(k)\Psi(k)$$

The Fourier transform of Eq. (8.5.1) is then

$$\Psi(k) = \Phi(k) + \sqrt{2\pi}\,\lambda V(k)\Psi(k) \tag{8.5.12}$$

which may be solved for $\Psi(k)$ to obtain Eq. (8.5.5). This treatment is possible only if there is a region of the complex plane of k in which Eq. (8.5.12) holds.

In Sec. 4.8 we saw that the integrals representing Fourier transforms of most functions do not converge over the whole complex plane of k. In most cases the transform is represented by the integral only in a band of the k plane parallel to the real k axis, extended from $-\infty$ to $+\infty$ with respect to the real part of k but bounded on the upper or lower side, or both, with respect to the imaginary part of k. Within these bands the transform is analytic everywhere; the presence of a singularity sets the upper or lower bound of the band. Consequently we must take care, when we deal with an equation relating the Fourier transforms of several functions, that their bands of analyticity overlap.

As we have also seen in Sec. 4.8, the Fourier transform of a function $\varphi(x)$ may have two different forms in two different bands of the k plane. For example, if the integral $\int_0^\infty \varphi(x)e^{-\tau x}\,dx$ converges only for $\tau \geq \tau_0'$, and if the integral $\int_{-\infty}^0 \varphi(x)e^{-\tau x}\,dx$ converges only for $\tau \leq \tau_1'$, then we found we could express φ as [see Eq. (4.8.19)]

$$\varphi(x) = \frac{1}{\sqrt{2\pi}}\left\{\int_{-\infty+i\tau_0'}^{\infty+i\tau_0'} \Phi_+(k)e^{-ikx}\,dk + \int_{-\infty+i\tau_1'}^{\infty+i\tau_1'} \Phi_-(k)e^{-ikx}\,dk\right\} \tag{8.5.13}$$

where Φ_+ is analytic throughout the region $\text{Im}\,k > \tau_0'$ and Φ_- is analytic throughout the band $\text{Im}\,k < \tau_1'$. In such a case it is desirable to have the band in which transform V is analytic overlap the region where both Φ_+ and Φ_- are analytic, so it is possible to extend Eq. (8.5.12) by analytic continuation.

Likewise the Fourier transform of the unknown ψ may have to be split into two functions: one, Ψ_+, analytic throughout the band $\text{Im}\,k > \tau_0''$ and the other, Ψ_-, analytic for $\text{Im}\,k < \tau_1''$. What we have said in the preceding paragraph is that, if the transform $V(k)$ is analytic throughout the band

$$\tau_1''' < \text{Im}\,k < \tau_0''' \tag{8.5.14}$$

then it is necessary that

$$\tau_1''' < \tau_1' \text{ and } \tau_1''; \quad \tau_0''' > \tau_0' \text{ and } \tau_0''$$

Under these conditions we may now apply inversion formula (8.5.13) to the entire integral equation, obtaining

$$\int_{-\infty+i\tau_0}^{\infty+i\tau_0} \Psi_+(k)e^{-ikx}\,dk + \int_{-\infty+i\tau_1}^{\infty+i\tau_1} \Psi_-(k)e^{-ikx}\,dk = \int_{-\infty+i\tau_0}^{\infty+i\tau_0} \Phi_+(k)e^{-ikx}\,dx$$
$$+ \int_{-\infty+i\tau_1}^{\infty+i\tau_1} \Phi_-(k)e^{-ikx}\,dk + \int_{-\infty+i\tau_0}^{\infty+i\tau_0} \Psi_+(k)[\sqrt{2\pi}\,\lambda V(k)]e^{-ikx}\,dk$$
$$+ \int_{-\infty+i\tau_1}^{\infty+i\tau_1} \Psi_-(k)[\sqrt{2\pi}\,\lambda V(k)]e^{-ikx}\,dk$$

or

$$\int_{-\infty+i\tau_0}^{\infty+i\tau_0} [(1 - \sqrt{2\pi}\,\lambda V)\Psi_+ - \Phi_+]e^{-ikx}\,dk$$
$$+ \int_{-\infty+i\tau_1}^{\infty+i\tau_1} [(1 - \sqrt{2\pi}\,\lambda V)\Psi_- - \Phi_-]e^{-ikx}\,dk = 0 \quad (8.5.15)$$

Here τ_0 is greater than the maximum of (τ_0',τ_0'') but is less than τ_0''' while τ_1 is less than the minimum of (τ_1',τ_1'') but greater than τ_1'''. Then the conditions of the theorem following Eq. (4.8.20) are met, and we may conclude that

$$[(1 - \sqrt{2\pi}\,\lambda V)\Psi_+ - \Phi_+] + [(1 - \sqrt{2\pi}\,\lambda V)\Psi_- - \Phi_-] = 0$$

Returning to this theorem, we note that not only is the sum of the two integrands equal to zero but each integrand separately must be analytic over the whole strip of analyticity. Thus integrand $[(1 - \sqrt{2\pi}\,\lambda V)\Psi_+ - \Phi_+]$ must be equal to some function $S_+(k)$ which is analytic in the strip $\tau_0 \le \text{Im}\,k \le \tau_1$ and goes to zero for $|\text{Re}\,k| \to \infty$ sufficiently rapidly for the integrals to converge. In order that the sum of the integrands be zero, we must then have $[(1 - \sqrt{2\pi}\,\lambda V)\Psi_- - \Phi_-] = S_-(k) = -S_+(k)$.

Consequently the equivalent of Eq. (8.5.15) is the pair of equations

$$\Psi_+(k) = \frac{\Phi_+(k) + S_+(k)}{1 - \sqrt{2\pi}\,\lambda V(k)}; \quad \Psi_-(k) = \frac{\Phi_-(k) + S_-(k)}{1 - \sqrt{2\pi}\,\lambda V(k)} \quad (8.5.16)$$

where $S_+ = -S_-$ is analytic throughout the strip $\tau_0 \le \text{Im}\,k \le \tau_1$. These reduce to Eq. (8.5.5) when τ_0' and τ_1' can be placed equal to zero, so that $\Phi_+ = \Phi_-$ (in which case S_+ must equal S_- and therefore both must be zero). We can now apply inversion formula (8.5.13) to obtain our solution

$$\psi(x) = \frac{1}{\sqrt{2\pi}} \left\{ \int_{-\infty+i\tau_0}^{\infty+i\tau_0} \frac{\Phi_+ e^{-ikx}}{1 - \sqrt{2\pi}\,\lambda V}\,dk + \int_{-\infty+i\tau_1}^{\infty+i\tau_1} \frac{\Phi_- e^{-ikx}}{1 - \sqrt{2\pi}\,\lambda V}\,dk \right.$$
$$\left. + \oint \frac{S_+ e^{-ikx}}{1 - \sqrt{2\pi}\,\lambda V}\,dk \right\} \quad (8.5.17)$$

The first two integrals are equivalent to the steady-state solution (particular solution) of an inhomogeneous differential equation, and the last integral, which is along a closed contour within the strip of analyticity, is equivalent to the transient solution (complementary function) which remains when $\varphi = 0$.

The Homogeneous Equation. Let us first examine the complementary function, the solution of the homogeneous equation

$$\psi(x) = \lambda \int_{-\infty}^{\infty} v(x - x_0)\psi(x_0)\, dx_0 \tag{8.5.18}$$

The solution is, from (8.5.17),

$$\psi_c(x) = \oint \frac{S_+(k)e^{-ikx}}{1 - \sqrt{2\pi}\,\lambda V(k)}\, dk \tag{8.5.19}$$

where $V(k)$ is the Fourier transform of $v(z)$ and where the contour is a closed one within the strip of analyticity of $V(k)$, $\tau_1''' < \text{Im}\, k < \tau_0'''$, with S analytic everywhere within this strip. The integrand is thus zero unless $[1 - \sqrt{2\pi}\,\lambda V(k)]$ has zeros or branch points within the strip.

Such behavior is not surprising, for we learned, in Sec. 8.2, that nonzero solutions of the homogeneous equation occur only for certain values of λ, the eigenvalues. In the present case, as a matter of fact, continuous bands of values of λ are allowed because the infinite range of the integration makes this type of integral equation a singular case [see Eq. (8.2.55)]. All values of λ are allowed which cause one or more zeros of $(1 - \sqrt{2\pi}\,\lambda V)$ to occur within the strip of analyticity. For most of the values of λ which do result in nonzero solutions these zeros of $(1 - \sqrt{2\pi}\,\lambda V)$ will be simple ones (going to zero linearly with $k - k_r$), so that the singularities of the integrand of Eq. (8.5.19) will usually be simple poles. Suppose that the poles within the strip are at $k_0, k_1, \ldots, k_r, \ldots, k_n$ (each of these are functions of λ), and suppose that the residue of the factor $S_+/(1 - \sqrt{2\pi}\,\lambda V)$ at k_r is $A_r/2\pi i$; then

$$\psi_c(x) = \sum_{r=0}^{n} A_r e^{-ik_r x} \tag{8.5.20}$$

(There may be, for some special values of λ, zeros of $1 - \sqrt{2\pi}\,\lambda V$ of a higher order than the first, in which case terms of the sort $B_s x^{t-1} e^{-ik_s x}$ will occur in the expression for ψ.) Actually the constants A_r are not yet determined, for S_+ is so far just any analytic function of k. Their values are determined by initial or boundary conditions, just as with the usual complementary function for a differential equation. In some cases, ratios between some of the coefficients must be chosen so that the combination is a solution of the original integral equation; this can be done easily by substituting (8.5.20) into (8.5.18).

It might be instructive to verify that $A_r e^{-ik_r x}$ is a solution of the homogeneous equation (8.5.18), by direct substitution

$$\begin{aligned} A_r e^{-ik_r x} &= \lambda A_r \int_{-\infty}^{\infty} v(x - x_0) e^{ik_r x_0}\, dx_0 \\ &= \lambda A_r \int_{-\infty}^{\infty} v(y) e^{-ik_r(x-y)}\, dy \\ &= A_r \sqrt{2\pi}\, \lambda V(k_r) e^{-ik_r x} \end{aligned}$$

or

$$1 - \sqrt{2\pi}\, \lambda V(k_r) = 0 \tag{8.5.21}$$

if we remember that $V(k)$ is the Fourier transform of $v(z)$. The last equation is, of course, satisfied; k_r was originally defined as being a root of this equation.

An Example. Consider the integral equation

$$\psi(x) = A e^{\alpha|x|} + \lambda \int_{-\infty}^{\infty} e^{-|x-x_0|} \psi(x_0)\, dx_0 \tag{8.5.22}$$

We first consider the solution of the corresponding homogeneous equation

$$\chi(x) = \lambda \int_{-\infty}^{\infty} e^{-|x-x_0|} \chi(x_0)\, dx_0 \tag{8.5.23}$$

[as given by Eq. (8.5.20)]. For this purpose we need to find the zeros of $1 - \lambda \sqrt{2\pi}\, V$, where V is the transform of $e^{-|x|}$. This may be easily obtained:

$$V(k) = \frac{1}{\sqrt{2\pi}} \int_{-\infty}^{\infty} e^{-|x|} e^{ikx}\, dx \quad \text{or} \quad V(k) = \frac{\sqrt{2/\pi}}{1 + k^2}$$

where $V(k)$ is analytic for the region $|\text{Im } k| < 1$. We then obtain

$$1 - \sqrt{2\pi}\, \lambda V = [k^2 - (2\lambda - 1)]/(1 + k^2)$$

The zeros occur at $k = \pm k_0$ where $k_0 = \sqrt{2\lambda - 1}$. These zeros are simple. Hence the solutions of the homogeneous equation are

$$\chi = e^{\pm ik_0 x} \tag{8.5.24}$$

These are solutions only if $\pm k_0$ is within the domain of regularity of V, $|\text{Im } k_0| < 1$, which corresponds to the requirement that $\text{Im } \lambda < 2 \text{ Re } \lambda$. These solutions may be verified directly by substitution in the original integral equation for x or by reducing the latter to a differential equation by differentiating the entire equation twice. We then obtain

$$(d^2\chi/dx^2) + k_0^2 \chi = 0$$

We consider next the solution of the inhomogeneous equation as given by the first two terms of Eq. (8.5.17). For this purpose, Φ_+ and Φ_- are required:

$$\Phi_+ = \frac{A}{\sqrt{2\pi}} \int_0^{\infty} e^{\alpha x} e^{ikx}\, dx = -\frac{A}{\sqrt{2\pi}\,(\alpha + ik)}; \quad \text{Im } k > \alpha$$

$$\Phi_- = \frac{A}{\sqrt{2\pi}} \int_{-\infty}^{0} e^{-\alpha x} e^{ikx}\, dx = \frac{A}{\sqrt{2\pi}\,(ik - \alpha)}; \quad \text{Im } k < -\alpha$$

The solution (8.5.17) is valid for only $\alpha < 1$. Then

$$\psi(x) = \frac{A}{2\pi i}\left\{-\int_{-\infty+i\tau_0''}^{\infty+i\tau_0''} \frac{(1+k^2)e^{-ikx}}{(k^2-k_0^2)(k-i\alpha)}\,dk + \int_{-\infty+i\tau_1''}^{\infty+i\tau_1''} \frac{(1+k^2)e^{-ikx}}{(k^2-k_0^2)(k+i\alpha)}\,dk\right\} \tag{8.5.25}$$

where we may add any solution of the homogeneous equation, *i.e.*, any linear combination of the two solutions given in Eq. (8.5.24). The integration limits satisfy the conditions $\tau_0'' < 1$, $\tau_1'' > -1$ as shown in Fig. 8.3. Consider two cases, $x > 0$, and $x < 0$. In the first case we can add on a semicircle in the lower half plane to the contour of each integral and so obtain a closed contour. The integrals may then be

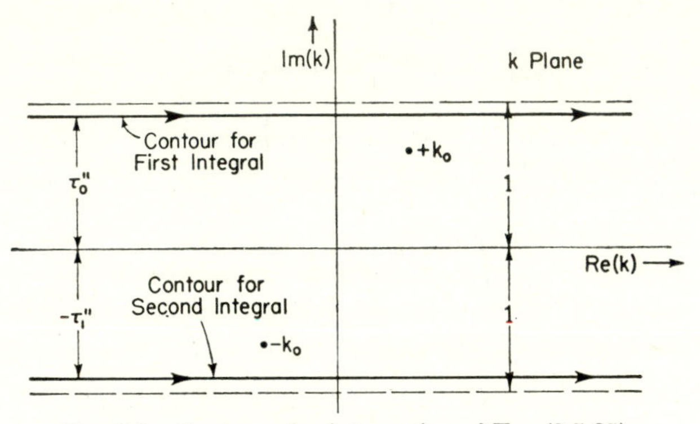

Fig. 8.3 Contours for integration of Eq. (8.5.25).

evaluated by an application of Cauchy's integral formula. The second integral vanishes since all the singularities of its integrand occur outside the contour. The first term may be readily evaluated to yield

$$\psi_+ = A\left\{\frac{\alpha^2-1}{\alpha^2+k_0^2}e^{\alpha x} + \frac{(k_0^2+1)e^{-ik_0x}}{2k_0(k_0-i\alpha)} + \frac{(k_0^2+1)e^{ik_0x}}{2k_0(k_0+i\alpha)}\right\}$$

$$= A\left\{\frac{\alpha^2-1}{\alpha^2+k_0^2}e^{\alpha x} + \frac{k_0^2+1}{k_0\sqrt{\alpha^2+k_0^2}}\cos\left[k_0x - \tan^{-1}\left(\frac{\alpha}{k_0}\right)\right]\right\};\quad x>0$$

For $x < 0$, the semicircle now runs in the upper half plane, so that only the second integral contributes:

$$\psi_- = A\left\{\frac{\alpha^2-1}{\alpha^2+k_0^2}e^{-\alpha x} + \frac{k_0^2+1}{k_0\sqrt{\alpha^2+k_0^2}}\cos\left[k_0x + \tan^{-1}\left(\frac{\alpha}{k_0}\right)\right]\right\};\quad x<0$$

We may combine these two forms into

$$\psi = A\left\{\frac{\alpha^2-1}{\alpha^2+k_0^2}e^{\alpha|x|} + \frac{k_0^2+1}{k_0\sqrt{\alpha^2+k_0^2}}\cos\left[k_0|x| - \tan^{-1}\left(\frac{\alpha}{k_0}\right)\right]\right\} \tag{8.5.26}$$

Function ψ is clearly continuous at $x = 0$, having the value A there; but it has a discontinuity in the first derivative. The discontinuity at $x = 0$ cannot be modified by adding solutions of the homogeneous equation. These can only modify the value of ψ at $x = 0$. We note that the particular solution integrals of (8.5.25) bring in a certain amount of the complementary function χ, enough to make $\psi(0)$ equal A.

The above analysis may be carried through as long as $V(k)$ is analytic in a strip including the real axis of k. The only possible obstacles are the evaluation of the transforms Φ_+, Φ_-, and V.

Branch Points. There is no great difficulty in extending these results to cases in which $V(k)$ has branch points present in the strip between the real axis of k. However, it is essential that a strip exists in which $V(k)$ and Φ_+ are analytic, similarly for $V(k)$ and Φ_-. If these conditions are satisfied it is only necessary to introduce branch lines and to make certain that in carrying out the details we do not integrate along contours which cross branch lines. This may cause some hardship in the evaluation of $\psi(x)$ from Ψ_+ and Ψ_-, since it is no longer possible to perform the entire inversion by means of the Cauchy integral formula. One always remains with integrals along the branch lines which may or may not be readily expressible in terms of elementary functions.

In this connection, it is important to remember that the Fourier transform contains information on the asymptotic behavior of $\psi(x)$. From the results on page 462 we know that

$$\Psi_+(k) \xrightarrow[\sigma\to\infty]{} \frac{\psi_+(0)}{\sqrt{2\pi}\,(-ik)}; \quad \text{if } \psi_+(0) \neq 0; \quad k = \sigma + i\tau$$

or

$$-i\sqrt{2\pi}\, k\Psi_+(k) \to \psi_+(0)$$

In the example discussed above

$$\Psi_+(k) \xrightarrow[\sigma\to\infty]{} -(A/ik\sqrt{2\pi})$$

Hence $\psi(0) = A$, in agreement with the result obtained from the complete solution (8.5.26).

The Kernel $v(x + x_0)$ in the Infinite Domain. The faltung theorem may be applied here also. We may again represent $\psi(x)$ and $\varphi(x)$ through inversion formula (8.5.13). The representation of

$$\int_{-\infty}^{\infty} v(x + x_0)\psi(x_0)\, dx_0$$

is obtained by changing the variable of integration from x_0 to $-\xi_0$ and then applying the faltung theorem. We find that

$$\int_{-\infty}^{\infty} v(x + x_0)\psi(x_0)\, dx_0 = \frac{1}{\sqrt{2\pi}} \int_{-\infty+i\tau_0}^{\infty+i\tau_0} \Psi(-k)[\sqrt{2\pi}\, V(k)]e^{-ikx}\, dk$$
$$+ \frac{1}{\sqrt{2\pi}} \int_{-\infty+i\tau_1}^{\infty+i\tau_1} \Psi_+(-k)[\sqrt{2\pi}\, V(k)]e^{-ikx}\, dk$$

Here the integrand of the first term, as well as $\Phi_+(k)$ and $\Psi_+(k)$, is presumed analytic for $\text{Im } k > \tau_0$ while the second integrand, Φ_- and Ψ_-, are analytic for $\text{Im } k < \tau_1$. Note that, if $\Phi_+(k)$ is analytic for $\text{Im } k > \tau_0'$, then $\Phi_+(-k)$ is analytic for $\text{Im } k < \tau_0'$. The equation corresponding to (8.5.15) is therefore

$$\{\Psi_+(k) - \sqrt{2\pi}\,\lambda V(k)\Psi_-(-k) - \Phi_+(k)\} + \{\Psi_-(k) - \sqrt{2\pi}\,\lambda V(k)\Psi_+(-k) - \Phi_-(k)\} = 0$$

Each term enclosed by the braces is analytic in a strip parallel to the real axis, as in (8.5.15). Dropping the terms S_+ and S_-, which occurred in Eq. (8.5.16) and which we later realized involved solutions of the homogeneous equation, we may obtain particular solutions. Hence

$$\begin{aligned}\Psi_+(k) - \sqrt{2\pi}\,\lambda V(k)\Psi_-(-k) &= \Phi_+(k)\\ \Psi_-(k) - \sqrt{2\pi}\,\lambda V(k)\Psi_+(-k) &= \Phi_-(k)\end{aligned}$$

In the second of these equations we replace k by $(-k)$. This can be done only if $\Phi_-(-k)$ and $V(-k)$ are analytic in the region where $\Phi_+(k)$ and $V(k)$ are both analytic. The second equation now reads

$$\Psi_-(-k) - \sqrt{2\pi}\,\lambda V(-k)\Psi_+(k) = \Phi_-(-k)$$

We may then solve for $\Psi_+(k)$:

$$\Psi_+(k) = \frac{\Phi_+(k) + \sqrt{2\pi}\,\lambda V(k)\Phi_-(-k)}{1 - 2\pi\lambda^2 V(k)V(-k)} \tag{8.5.27}$$

Similarly

$$\Psi_-(k) = \frac{\Phi_-(k) + \sqrt{2\pi}\,\lambda V(k)\Phi_+(-k)}{1 - 2\pi\lambda^2 V(k)V(-k)} \tag{8.5.28}$$

Formula (8.5.17) obtained for the kernel $v(x - x_0)$ is now replaced by

$$\begin{aligned}\psi(x) = {} & \frac{1}{\sqrt{2\pi}} \int_{-\infty+i\tau_0}^{\infty+i\tau_0} \left[\frac{\Phi_+(k) + \sqrt{2\pi}\,\lambda V(k)\Phi_-(-k)}{1 - 2\pi\lambda^2 V(k)V(-k)}\right] e^{-ikx}\,dk \\ & + \frac{1}{\sqrt{2\pi}} \int_{+\infty+i\tau_1}^{\infty+i\tau_1} \left[\frac{\Phi_-(k) + \sqrt{2\pi}\,\lambda V(k)\Phi_+(-k)}{1 - 2\pi\lambda^2 V(k)V(-k)}\right] e^{-ikx}\,dk \end{aligned} \tag{8.5.29}$$

The integrand of the first integral is analytic for $\text{Im } k > \tau_0$; the second for $\text{Im } k < \tau_1$.

The solution of the *homogeneous equation* may be obtained from (8.5.29) by replacing Φ_+ by S_+, Φ_- by S_- and finally by placing $S_- = -S_+$ so that

$$\psi = \frac{1}{\sqrt{2\pi}} \oint \left[\frac{S_+(k) - \sqrt{2\pi}\,\lambda V(k)S_+(-k)}{1 - 2\pi\lambda^2 V(k)V(-k)}\right] e^{-ikx}\,dk$$

where the contour is within the region in which $V(k)$ and $V(-k)$ are analytic. By reversing the direction of integration on the second term ψ may be written as

$$\psi = \frac{1}{\sqrt{2\pi}} \oint S_+(k) \left[\frac{1 + \sqrt{2\pi}\,\lambda V(-k)e^{2ikx}}{1 - 2\pi\lambda^2 V(k)V(-k)} \right] e^{-ikx}\, dk \qquad (8.5.30)$$

Since S_+ and $1 + \sqrt{2\pi}\,\lambda V(-k)$ are both analytic in the region within the contour, we may replace their product by one analytic function of k. Hence the homogeneous solutions are given by

$$\psi = \sum_{r,s} A_{rs} x^{s-1} e^{-ik_r x} \qquad (8.5.31)$$

where it has been assumed that

$$\frac{1}{1 - 2\pi\lambda^2 V(k)V(-k)} = \frac{b_s}{(k - k_r)^s} + \frac{b_{s-1}}{(k - k_r)^{s-1}} + \cdots$$

The coefficients A_{rs} are to be determined by initial or boundary conditions, though ratios between some of them must be adjusted to be consistent with the original integral equation.

An Example. To illustrate, consider the integral equation

$$\psi(x) = Ae^{\alpha|x|} + \lambda \int_{-\infty}^{\infty} e^{-|x+x_0|} \psi(x_0)\, dx_0$$

From the results obtained in the discussion of integral equation (8.5.22) we have

$$\begin{aligned} V(k) &= \sqrt{2/\pi}/(1 + k^2); & &|\mathrm{Im}\, k| < 1 \\ \Phi_+ &= -A/\sqrt{2\pi}\,(\alpha + ik); & &\mathrm{Im}\, k > \alpha \\ \Phi_- &= A/\sqrt{2\pi}\,(ik - \alpha); & &\mathrm{Im}\, k < -\alpha \end{aligned}$$

Consider the solutions of the homogeneous equations first. We look then at the roots of $1 - 2\pi\lambda^2 V(k)V(-k) = 0$ which for this case are given by

$$\begin{aligned} k &= \pm k_0; & k_0 &= \sqrt{2\lambda - 1} \\ k &= \pm ik_1; & k_1 &= \sqrt{2\lambda + 1} \end{aligned}$$

From (8.5.31) we may write the solution of the homogeneous equation as

$$\psi = a_1 e^{-ik_0 x} + a_2 e^{ik_0 x} + b_1 e^{-k_1 x} + b_2 e^{-k_1 x}$$

It is now necessary to evaluate the ratios between coefficients of this expression by direct substitution in the integral equation. One obtains

$$\psi(x) = \lambda \sqrt{2\pi}\,[a_1 V(k_0) e^{ik_0 x} + a_2 V(-k_0) e^{-ik_0 x} + b_1 V(-ik_1) e^{k_1 x} + b_2 V(ik_1) e^{-k_1 x}]$$

Equating coefficients of like exponentials yields the independent pair of simultaneous equations

$$a_1 - \lambda \sqrt{2\pi}\, V(-k_0)\, a_2 = 0; \quad \lambda \sqrt{2\pi}\, V(k_0)\, a_1 - a_2 = 0$$

and

$$b_1 - \lambda \sqrt{2\pi}\, V(ik_1)\, b_2 = 0; \quad \lambda \sqrt{2\pi}\, V(-ik_1)\, b_1 - b_2 = 0$$

The requirement for nonzero solutions, that the determinant of these equations vanish, leads to the equation employed above to determine k_0 and k_1 and is therefore automatically satisfied.

The ratio of the coefficients may now be found. For the k_0 solutions it is

$$a_2/a_1 = \lambda \sqrt{2\pi}\, V(k_0) = 1$$

Hence $\psi = \cos k_0x$ is a solution of the homogeneous equation. Similarly $\cosh k_1x$ is another independent solution.

Turning now to the inhomogeneous equation (8.5.29), we see that there is no essential difficulty in evaluating the integrals by applying Cauchy's integral formula with a procedure completely analogous to that employed in example (8.5.22).

Applications of the Laplace Transform. As may be predicted from the above discussion, the Laplace transform may be most gainfully employed for integral equations whose kernels permit the application of the faltung theorem for the Laplace transformation:

$$\mathfrak{L}\left[\int^x v(x - x_0) f(x_0)\, dx_0\right] = V(p)F(p)$$

where $V(p)$ and $F(p)$ are the Laplace transforms of $v(x)$ and $f(x)$, respectively. This suggests that we should consider Volterra integral equations

$$f(x) = \varphi(x) + \int_0^x v(x - x_0) f(x_0)\, dx_0; \quad x > 0 \qquad (8.5.32)$$

Examples of Volterra equations arising from the solution of vibration problems and from energy absorption of X rays in matter have been discussed on pages 905 to 907.

For the present let us consider the general situation. Take the Laplace transform of both sides of (8.5.32). Assume that the transform of φ is analytic for $\text{Re}\, p > \tau_0$ and that the region of analyticity of the transform of v has at least a strip, parallel to the imaginary axis of p, in common with the band for φ. In that strip,

$$F(p) = \Phi(p) + V(p)F(p)$$

Solving for $F(p)$, we have

$$F(p) = \Phi(p)/[1 - V(p)] \qquad (8.5.33)$$

Hence a particular solution of Eq. (8.5.32) is obtained by inverting the Laplace transform [Eq. (4.8.32)]

$$f(x) = \frac{1}{2\pi i}\int_{-i\infty+\tau_0}^{i\infty+\tau_0} \left[\frac{\Phi(p)}{1 - V(p)}\right] e^{px}\, dp; \quad \text{Re}\, x > 0 \qquad (8.5.34)$$

There are no nonzero solutions of the homogeneous Volterra equation, so that formula (8.5.34) gives the unique solution of Eq. (8.5.33).

As an example of this, we shall consider the Volterra integral equation resulting from the differential equation

$$(d^2\psi/dt^2) + k^2\psi = 0$$

initial conditions at $t = 0$, $\psi = \psi_0$, $\partial\psi/\partial t = v_0$. The equivalent integral equation is given in Eq. (8.1.23):

$$\psi(t) = \psi_0 \left[\frac{\partial G(t|t_0)}{\partial t_0}\right]_{t_0=0} - v_0 G(t|0) + k^2 \int_0^{t^+} G(t|t_0)\psi(t_0)\,dt_0$$

The Green's function G is determined by

$$\partial^2 G/\partial t^2 = -\delta(t - t_0);$$
$$G(t|t_0) = 0; \quad \text{if } t < t_0$$

The equations may be readily solved:

$$G(t|t_0) = (t_0 - t)u(t - t_0)$$

where $u(t)$ is the unit function (see page 840). Substituting this result for G, the integral equation becomes

$$\psi(t) = \psi_0 + v_0 t + k^2 \int_0^{t^+} (t_0 - t) \cdot \psi(t_0)\,dt_0; \quad t > 0$$

This is now in proper form to apply the preceding discussion:

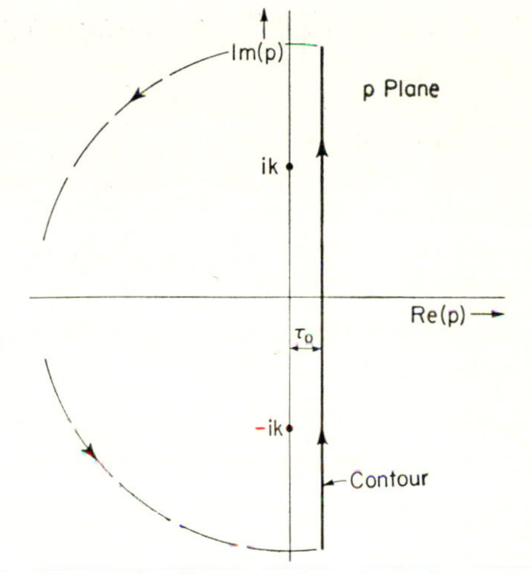

Fig. 8.4 Contour for inversion of Laplace transform.

$$\varphi(t) = \psi_0 + v_0 t; \quad \Phi(p) = (\psi_0/p) + (v_0/p^2); \quad \operatorname{Re} p > 0$$
$$v(t - t_0) = k^2(t_0 - t); \quad V(p) = -k^2/p^2; \quad \operatorname{Re} p > 0$$

Substituting in solution (8.5.34) one obtains

$$\psi(t) = \frac{1}{2\pi i}\int_{-i\infty+\tau_0}^{i\infty+\tau_0} \left[\frac{v_0 + p\psi_0}{p^2 + k^2}\right] e^{px}\,dp; \quad \tau_0 > 0$$

We may evaluate this integral by adding on an infinite semicircle, as indicated in Fig. 8.4, and then applying Cauchy's integral formula. We, of course, obtain the familiar result

$$\psi = \psi_0 \cos(kt) + (v_0/k)\sin(kt)$$

Volterra Integral Equation, Limits (x, ∞). The Laplace transform may also be applied to integral equations of the following form:

$$\psi(x) = \varphi(x) + \int_x^\infty v(x - x_0)\psi(x_0)\,dx_0 \tag{8.5.35}$$

Integral equations of this type occur in transport problems where x may be the energy after a collision and x_0 the energy before collision (see Secs. 2.4 and 12.2). For collisions with fixed systems with no internal degree of freedom, $x_0 \geq x$; that is, the collisions always result in a loss of energy of the incident particle.

To solve Eq. (8.5.35) by application of the Laplace transform it is necessary to develop a faltung theorem for the form

$$\int_x^\infty v(x - x_0)\psi(x_0)\, dx_0$$

We start from the faltung theorem for Fourier transforms:

$$\mathfrak{F}\left\{\int_{-\infty}^\infty g(x - x_0)\phi(x_0)\, dx_0\right\} = \sqrt{2\pi}\, G_f(k)\Phi_f(k)$$

Now let $g(x) = v_-(x)$, that is, equal $v(x)$ for $x < 0$ and equal zero for $x > 0$; similarly let $\phi(x) = \psi_+(x)$, that is, $\psi(x)$ for $x > 0$ and zero for $x < 0$. Then the above equation becomes

$$\mathfrak{F}\left\{\int_x^\infty v(x - x_0)\psi(x_0)\, dx_0\right\} = \sqrt{2\pi}\, [V_-(k)]_f[\Psi_+(k)]_f$$

To convert the Fourier into a Laplace transform, we recall that $F_l(p) = \sqrt{2\pi}\, [F_+(ip)]_f$. Hence

$$\mathfrak{L}\left\{\int_x^\infty v(x - x_0)\psi(x_0)\, dx_0\right\} = \sqrt{2\pi}\, [V_-(ip)]_f[\Psi_+(p)]_l$$

We can express $[\sqrt{2\pi}\, V_-(ip)]_f$ in terms of a Laplace transform:

$$[\sqrt{2\pi}\, V_-(ip)]_f = \int_{-\infty}^0 v(x)e^{-px}\, dx; \quad [\sqrt{2\pi}\, V_-(ip)]_f = \int_0^\infty v(-x)e^{px}\, dx$$

If we therefore let $v(-x) = w(x)$, then

$$[\sqrt{2\pi}\, V_-(ip)]_f = W_l(-p)$$

Finally

$$\mathfrak{L}\left\{\int_x^\infty v(x - x_0)\psi(x_0)\, dx_0\right\} = W_l(-p)\Psi_l(p) \tag{8.5.36}$$

We may now return to integral equation (8.5.35). Taking the Laplace transform of both sides (we shall dispense with subscript l from here on, since we shall be dealing with Laplace transforms only), one obtains

$$\Psi(p) = \Phi(p) + W(-p)\Psi(p)$$

or

$$\Psi(p) = \Phi(p)/[1 - W(-p)] \tag{8.5.37}$$

Finally

$$\psi(x) = \frac{1}{2\pi i}\int_{-i\infty+\tau_0}^{i\infty+\tau_0} \left[\frac{\Phi(p)}{1 - W(-p)}\right] e^{px}\, dp \tag{8.5.38}$$

is a particular solution of integral equation (8.5.35). It should be emphasized that, for solution (8.5.37) or (8.5.38) to be meaningful, it is necessary for the regions of analyticity of $W(-p)$ and $\Phi(p)$ to overlap. As has been mentioned before, if this occurs for only certain range of the parameters in either Φ or W, it may be possible to extend the range by analytic continuation.

As an example for this case let

$$\varphi(x) = C; \quad v(x) = Ae^{\alpha x}; \quad A, \alpha \text{ real and positive}$$

so that (8.4.35) is

$$\psi(x) = C + A \int_x^\infty e^{\alpha(x-x_0)}\psi(x_0)\, dx_0$$

We now apply the Laplace transform [though before we carry through the analysis we should point out that for this particular $v(x)$, the integral equation may be reduced to a first-order differential equation which can be easily solved]. To fill in formula (8.5.38), $\Phi(p)$ and $W(-p)$ are required:

$$\begin{aligned} \Phi(p) &= C/p; \quad \operatorname{Re} p > 0 \\ W(-p) &= \int_0^\infty e^{px} v(-x)\, dx = A \int_0^\infty e^{(p-\alpha)x}\, dx \\ &= A/(\alpha - p); \quad \operatorname{Re} p < \alpha \end{aligned}$$

Note that $W(-p)$ and $\Phi(p)$ are analytic in a common strip only if $\alpha > 0$. Equation (8.5.38) becomes

$$\psi(x) = \frac{1}{2\pi i} \int_{-i\infty+\tau_0}^{i\infty+\tau_0} \left\{ \frac{C(p-\alpha)}{p[p-(\alpha-A)]} \right\} e^{px}\, dp$$

where $0 < \tau_0 < \alpha$. We again close the contour by adding a semicircle extending around the left-hand half plane of p as in Fig. 8.4 above. Then the integral for ψ may be evaluated by the Cauchy integral formula, there being simple poles at $p = 0$ and at $p = \alpha - A$ (if τ_0 is taken greater than $\alpha - A$). This arbitrariness concerning the residue at $\alpha - A$, which may be included or not as desired, corresponds to the arbitrariness of the complementary function, a solution of the homogeneous equation, which in this case is proportional to the difference of two particular solutions of the inhomogeneous linear equation. We obtain

$$\psi(x) = \frac{C\alpha}{\alpha - A} - \frac{CA}{\alpha - A} e^{(\alpha-A)x}$$

The first term is a particular solution which represents the "steady state" induced by the "source" term C if $\alpha < A$. The second term indicates that the solution of the homogeneous equation, obtained by placing $C = 0$, is proportional to $e^{(\alpha-A)x}$. We note the homogeneous term again filling the familiar role of the transient.

Mellin Transform. We commence this subsection by recalling the definition and inversion formula as given in Chap. 4 and Prob. (4.48). The Mellin transform of $f(x)$ is given by

$$F(s) = \int_0^\infty f(x)x^{s-1}\,dx$$

If this should not exist, then it is often possible to introduce the "half-plane" transforms equivalent to those introduced in Fourier transform theory:

$$F_-(s) = \int_0^1 f(x)x^{s-1}\,dx; \quad F_+(s) = \int_1^\infty f(x)x^{s-1}\,dx \tag{8.5.39}$$

F_- exists for Re $s > \sigma_0$ whereas F_+ exists for Re $s < \sigma_1$. In the event that $F(s)$ does not exist, $\sigma_0 > \sigma_1$, while the opposite inequality holds if f does exist.

The inversion formula may accordingly be written

$$f(x) = \frac{1}{2\pi i}\left\{\int_{-i\infty+\sigma_0'}^{i\infty+\sigma_0'}\left[\frac{F_-(s)}{x^s}\right]ds + \int_{-i\infty+\sigma_1'}^{i\infty+\sigma_1'}\left[\frac{F_+(s)}{x^s}\right]ds\right\} \tag{8.5.40}$$

where $\sigma_0' > \sigma_0$ and $\sigma_1' < \sigma_1$. The faltung theorem is

$$\mathfrak{M}\left\{\int_0^\infty \psi(x_0)v\left(\frac{x}{x_0}\right)\left(\frac{dx_0}{x_0}\right)\right]\right\} = V(s)\Psi(s) \tag{8.5.41}$$

suggesting that the Mellin transform may be gainfully employed in solving integral equations of the following type:

$$\psi(x) = \varphi(x) + \int_0^\infty v\left(\frac{x}{x_0}\right)\psi(x_0)\left(\frac{dx_0}{x_0}\right) \tag{8.5.42}$$

The analysis leading to the solution of this equation is so completely similar to the corresponding Fourier integral treatment [leading to solution (8.5.17)] that we give only the results here:

$$\psi(x) = \frac{1}{2\pi i}\left\{\int_{-i\infty+\sigma_0'}^{i\infty+\sigma_0'}\left[\frac{\Phi_-}{1-V}\right]\frac{ds}{x^s} + \int_{-i\infty+\sigma_1'}^{i\infty+\sigma_1'}\left[\frac{\Phi_+}{1-V}\right]\frac{ds}{x^s} + \oint\left[\frac{S}{1-V}\right]\frac{ds}{x^s}\right\} \tag{8.5.43}$$

where the contour integral on S is within the region $\sigma_0' <$ Re $p < \sigma_1'$, within which S must be analytic. The solution χ of the homogeneous counterpart of (8.5.42) is given just by the contour integral. If the zeros of $1 - V$ occur at s_r, and if these zeros are of order t, then

$$\chi = \Sigma B_{rt}(\ln x)^{t-1}x^{-s_r} \tag{8.5.44}$$

where B_{rt} are arbitrary constants.

As an example of the use of the Mellin transform, we set

$$\varphi(x) = Ae^{-\alpha x}; \quad v(x/x_0) = Ce^{-(x/x_0)}$$

so that integral equation (8.5.42) becomes

$$\psi(x) = Ae^{-\alpha x} + C\int_0^\infty e^{-(x/x_0)}\psi(x_0)\left(\frac{dx_0}{x_0}\right)$$

The necessary Mellin transforms are

$$\Phi(s) = A[\Gamma(s)/\alpha^s]; \quad \text{Re}\, s > 0$$
$$V(s) = C\Gamma(s); \quad \text{Re}\, s > 0$$

We have not made any decomposition of Φ into Φ_+ and Φ_-, since the regions of regularity of Φ and V are identical.

Consider first the solutions of the homogeneous equation. Following the recipe given by Eq. (8.5.44) we discover that

$$\chi = \sum_r B_r x^{-s_r} \tag{8.5.45}$$

where s_r are the roots of the equation

$$\Gamma(s_r) = 1/C$$

The zeros of $\Gamma(s) - (1/C)$ are of order 1; there are an infinite number of solutions.

The inhomogeneous equation has the particular solution

$$\psi(x) = \frac{A}{2\pi i}\int_{-i\infty+\sigma_0'}^{i\infty+\sigma_0'}\left[\frac{\Gamma(s)}{1 - C\Gamma(s)}\right]\frac{ds}{(\alpha x)^s}; \quad \sigma_0' > 0$$

This integral may be evaluated by use of the Cauchy integral formula. For $\alpha x > 1$, the contour is closed by a semicircle in the right-hand half plane. In that event, the only singularity of the integrand occurs at s_0 for which $1 - C\Gamma(s_0) = 0$. Then

$$\psi = \frac{A}{C(\alpha x)^{s_0}\psi_1(s_0)}; \quad \alpha x > 1 \tag{8.5.46}$$

where $\psi_1(s_0)$ is the logarithmic derivative of the gamma function at s_0. For $\alpha x < 0$ the singularities occur at all the negative roots of $1 - C\Gamma(s) = 0$, so that

$$\psi = -\frac{A}{C}\sum_{r=1}^{\infty}\frac{1}{(\alpha x)^{s_r}\psi_1(s_r)}; \quad \alpha x < 1 \tag{8.5.47}$$

[Note that these results for ψ are not solutions of the homogeneous equation as given in Eq. (8.5.45).] The series for ψ in the range $(\alpha x) < 1$

converges very well, since s_r is a sequence of negative numbers whose absolute value increases with r.

The Method of Wiener and Hopf. It is possible to extend the class of integral equations which can be solved through the use of Fourier transforms so as to include the following type:

$$\psi(x) = \lambda \int_0^\infty v(x - x_0)\psi(x_0)\, dx_0 \tag{8.5.48}$$

as well as its inhomogeneous counterparts of both the first and second kind. It is important to realize that the above equation is presumed to hold for all *real* values of x, both positive and negative. To make this point more obvious in the writing of the integral equation, let us introduce the functions ψ_+ and ψ_- with the usual definitions:

$$\psi_+(x) = \psi(x); \quad x > 0$$
$$\psi_-(x) = \psi(x); \quad x < 0$$

Then Eq. (8.4.48) is

$$\psi_+(x) + \psi_-(x) = \lambda \int_0^\infty v(x - x_0)\psi_+(x_0)\, dx_0$$
$$= \lambda \int_{-\infty}^\infty v(x - x_0)\psi_+(x_0)\, dx_0 \tag{8.5.49}$$

In other words, it is possible to express both ψ_- and ψ_+ in terms of ψ_+:

$$\begin{aligned} \psi_-(x) &= \lambda \int_0^\infty v(x - x_0)\psi_+(x_0)\, dx_0; \quad x < 0 \\ \psi_+(x) &= \lambda \int_0^\infty v(x - x_0)\psi_+(x_0)\, dx_0; \quad x > 0 \end{aligned} \tag{8.5.50}$$

Integral equations of this type, called *the Wiener-Hopf type*, occur whenever we are dealing with boundary value problems where the boundaries are semi-infinite rather than infinite, the latter giving rise to integral equations discussed earlier in this section. The problem of diffraction of waves by a half plane furnishes us with an example; others will be considered in Chaps. 11 and 12.

In our discussion of the Wiener-Hopf solution, we shall concentrate on the formal technique employed and the conditions under which these manipulations are valid, concluding with illustrations. We shall do this without detailed, rigorous proofs; most of the elements of the proofs have been discussed in Chap. 4. The additional few required are perforce established in the course of solving the problem.

As in the discussion given for the problems treated earlier in this section, we assume that the Fourier transform $v(x) \to V(k)$ is regular in the region

$$-\tau_1 < \operatorname{Im} k < \tau_0$$

a condition employed in Eq. (8.5.14). This corresponds to the asymptotic dependence of $v(x)$ given by

$$v(x) \xrightarrow[x\to\infty]{} e^{-\tau_1 x}; \quad v(x) \xrightarrow[x\to-\infty]{} e^{\tau_0 x}$$

We look for solutions of Eq. (8.5.48) which behave like $e^{\mu x}$ for $x \to \infty$ (where $\mu < -\tau_1$). This condition in parentheses is necessary for the

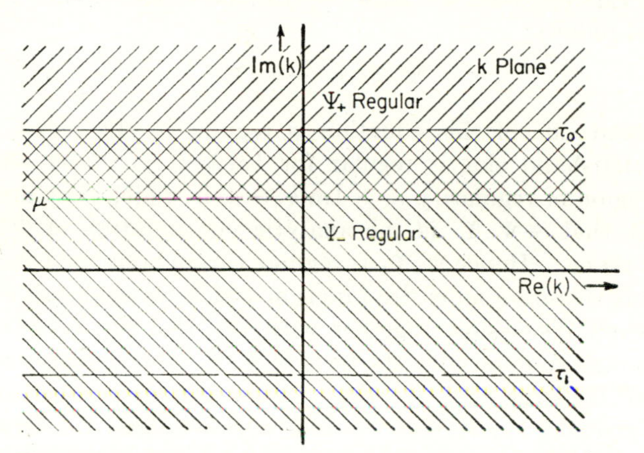

Fig. 8.5 Regions of regularity for Ψ_+ and Ψ_- in the k plane.

convergence of the integral in the integral equation. We may determine the asymptotic dependence of ψ_- directly from Eq. (8.5.50).

$$\psi_- \xrightarrow[x\to-\infty]{} \lambda \int_0^\infty e^{\tau_0(x-x_0)}\psi_+(x_0)\, dx_0 = \lambda e^{\tau_0 x}\left[\int_0^\infty e^{-\tau_0 x_0}\psi_+(x_0)\, dx_0\right]$$

Consequently it is necessary that $\mu < \tau_0$ and that ψ_- behave like $e^{\tau_0 x}$ for $x \to -\infty$. The band of regularity of Ψ_+ is therefore for $\text{Im}\, k > \mu$, while for Ψ_- the band is for $\text{Im}\, k < \tau_0$. These are illustrated in Fig. 8.5, where it is seen that there is a strip $\mu < \text{Im}\, k < \tau_0$ wherein all the relevant transforms, V, Ψ_+, and Ψ_-, are regular. This result is, as we shall see, fundamental in the Wiener-Hopf technique.

The transform of expression (8.5.49) may now be taken, employing the faltung theorem:

$$\Psi_+ + \Psi_- = \sqrt{2\pi}\,\lambda V \Psi_+$$

or

$$\Psi_+(1 - \sqrt{2\pi}\,\lambda V) + \Psi_- = 0 \tag{8.5.51}$$

It is clear that some added information must be brought to bear on this equation before Ψ_+ and Ψ_- can be independently determined. This is provided by the method of *factorization* as applied to the function $(1 - \sqrt{2\pi}\,\lambda V)$. This quantity is regular in the strip $-\tau_1 < \text{Im}\, k < \tau_0$. We now seek to break it up into factors Υ_+ and Υ_- such that

$$1 - \sqrt{2\pi}\,\lambda V = \Upsilon_+(k)/\Upsilon_-(k) \tag{8.5.52}$$

These factors are to be regular and free of zeros in the half planes $\operatorname{Im} k > \mu$ and $\operatorname{Im} k < \tau_0$, respectively. It is usual in addition to require that Υ_+ and Υ_- have algebraic growth, as compared with exponential growth. That this is possible is shown by Wiener and Hopf in their original memoir.

In any given problem this factorization must be carried out explicitly. Assuming Eq. (8.5.52), we may rewrite the equation for Ψ_+ and Ψ_-, (8.5.51), as follows:

$$\Psi_+ \Upsilon_+ = -\Psi_- \Upsilon_- \tag{8.5.53}$$

The left-hand side of this equation is regular in the region $\operatorname{Im} k > \mu$, while the left-hand side is regular for $\operatorname{Im} k < \tau_0$. Since they have a common region of regularity $\tau_0 > \operatorname{Im} k > \mu$, in which they are equal, we may assert that $-\Psi_-\Upsilon_-$ is the analytic continuation of $\Psi_+\Upsilon_+$ in the lower half plane. Hence $\Psi_+\Upsilon_+$ is regular throughout the entire complex plane and is therefore an entire function which we shall call $P(k)$.

This whole discussion, together with Eq. (8.5.52), is, of course, not definite enough to determine the form of $\Psi_+\Upsilon_+$; this reveals itself through its behavior for large k. Note that Υ_+ has already been chosen to have algebraic growth; *i.e.* it behaves like a polynomial for large k. The behavior of Ψ_+ for large k is determined by the behavior of $\psi_+(x)$ as $x \to 0^+$. The condition that $\psi_+(x)$ be integrable at the origin, necessary for the existence of Ψ_+, leads to the asymptotic dependence

$$\Psi_+(k) \xrightarrow[|k| \to \infty]{} 0$$

We therefore observe that $P(k)$ is a polynomial of degree less than Υ_+ (since P/Υ_+ approaches zero for large $|k|$). This fixes the form of $P(k)$; the undetermined constants may be obtained by substitution into the original equation (8.5.48).

Equation (8.5.53) may now be solved for Ψ_+ and Ψ_-:

$$\Psi_+ = P(k)/\Upsilon_+(k); \quad \Psi_- = -P(k)/\Upsilon_-(k) \tag{8.5.54}$$

The inversion formula is

$$\psi_+(x) = \frac{1}{\sqrt{2\pi}} \int_{-\infty+i\tau}^{\infty+i\tau} \left[\frac{P(k)}{\Upsilon_+(k)}\right] e^{-ikx}\, dk; \quad \mu < \tau < \tau_0 \tag{8.5.55}$$

From ψ_+ one may determine ψ_- by substitution in (8.5.50) or more directly

$$\psi_-(x) = -\frac{1}{\sqrt{2\pi}} \int_{-\infty+i\tau}^{\infty+i\tau} \left[\frac{P(k)}{\Upsilon_-(k)}\right] e^{-ikx}\, dk; \quad \mu < \tau < \tau_0 \tag{8.5.56}$$

This completes the solution of the Wiener-Hopf integral equation (8.5.48). We shall now illustrate this discussion with some examples.

Illustrations of the Method. A very simple example is furnished by the integral equation

$$\psi(x) = \lambda \int_0^\infty e^{-|x-x_0|}\psi(x_0)\,dx_0$$

The transform of $e^{-|x|}$, $V(k)$ is

$$V(k) = \frac{2}{\sqrt{2\pi}\,(1+k^2)};\quad \tau_0 = \tau_1 = 1$$

The quantity which must be factored into Υ_+ and Υ_- of Eq. (8.5.52) is

$$(1 - \sqrt{2\pi}\,\lambda V) = \left[\frac{k^2 - (2\lambda - 1)}{k^2 + 1}\right] = \frac{\Upsilon_+}{\Upsilon_-}$$

The factorization may be performed by inspection:

$$1 - \sqrt{2\pi}\,\lambda V = \left[\frac{k^2 - (2\lambda - 1)}{k + i}\right]\left[\frac{1}{k - i}\right]$$

so that $\qquad \Upsilon_+ = [k^2 - (2\lambda - 1)]/(k + i);\quad \Upsilon_- = (k - i)$

The first of these, Υ_+, is clearly regular and free of zeros for $\mathrm{Im}\,k > \mu$, where μ is less than $\tau_0 = 1$, while Υ_- is regular and free of zeros for $\mathrm{Im}\,k < \tau_0$, as long as $\mathrm{Re}\,\lambda > 0$. Hence

$$P(k) = \Psi_+(k)[k^2 - (2\lambda - 1)]/(k + i) = -(k - i)\Psi_-(k)$$

The function $P(k)$ is determined from the condition that it must be regular in the finite complex plane of k, while $\Psi_+(k) \to 0$ as $|k| \to \infty$. From this it follows that $P(k)$ for this example must be a constant C. It cannot increase as rapidly as k, for this would imply $\Psi_+(k) \to 1$. It cannot decrease more rapidly than a constant, for this would imply a singularity (pole or branch point) in the finite complex plane. We may now solve for Ψ_+ and Ψ_-:

$$\Psi_+ = C(k + i)/[k^2 - (2\lambda - 1)];\quad \Psi_- = -C/(k - i)$$

and $$\psi_+ = C\int_{-\infty+i\tau}^{\infty+i\tau} \{(k + i)/[k^2 - (2\lambda - 1)]\}e^{-ikx}\,dx$$

Since, in this expression, $x > 0$, we may close the contour with a semicircle in the lower half plane. Employing Cauchy's formula one obtains

$$\begin{aligned}\psi_+ &= -\frac{C}{4\pi i}\left\{\left[\frac{\sqrt{2\lambda - 1} + i}{\sqrt{2\lambda - 1}}\right]e^{-i\sqrt{2\lambda-1}\,x} + \left[\frac{\sqrt{2\lambda - 1} - i}{\sqrt{2\lambda - 1}}\right]e^{i\sqrt{2\lambda-1}\,x}\right\}\\ &= D\{\cos(\sqrt{2\lambda - 1}\,x) + [\sin(\sqrt{2\lambda - 1}\,x)/(\sqrt{2\lambda - 1})]\};\quad x > 0\end{aligned} \tag{8.5.57}$$

where D is a new constant. In the same manner ψ_- may be evaluated:

$$\psi_- = De^x;\quad x < 0 \tag{8.5.58}$$

It was, of course, not necessary to employ the Wiener-Hopf technique to solve this integral equation. One may easily verify from the integral equation directly that, for $x > 0$, $\psi = \psi_+$ satisfies the differential equation

$$\psi'' + (2\lambda - 1)\psi = 0; \quad x > 0$$

For $x < 0$, $\psi = \psi_-$ is given by

$$\psi_- = \lambda \int_0^\infty e^{-|x-x_0|}\psi_+(x_0)\,dx_0 \quad \text{or} \quad \psi_- = \lambda e^x \int_0^\infty e^{-x_0}\psi_+(x_0)\,dx_0$$

This is obviously in the form given by (8.5.58). The solution given in Eqs. (8.5.57) and (8.5.58) is that one which is continuous and has a continuous slope at $x = 0$.

This factorization is not so mysterious as it might seem. In the first place it is not necessarily unique, since the requirements on $P = \Psi_+\Upsilon_+ = -\Psi_-\Upsilon_-$ and on the asymptotic form of Ψ are not completely rigid. It turns out, however, that the interrelations between P and the Υ's and the Ψ's are such that the final solutions come out the same, no matter which choice is made. In many cases the factorization *is* unique. It is in the example just given. For instance, we might have tried

$$\Upsilon_+ = \{[k^2 - (2\lambda - 1)]/(k + i)\}(k - \beta); \quad \Upsilon_- = (k - i)(k - \beta)$$

If Υ_+ is to have no zeros in the range $\operatorname{Im} k > \mu < 1$, we must have $\operatorname{Im} \beta < \mu < 1$; but in this case Υ_- would be zero at $k = \beta$, which would be in the region where Υ_- is not supposed to have zeros. Or we might try

$$\Upsilon_+ = [(k^2 - 2\lambda + 1)/(k + i)(k - \beta)]; \quad \Upsilon_- = (k - i)/(k - \beta)$$

but this would put a pole in an undesirable region of the k plane. Consequently the only choice which keeps the zeros and poles of Υ_+ below i and those for Υ_- above i is the one given, if we are to restrict our choice to functions which go to infinity with a finite power of k as $|k| \to \infty$.

As a second example, we turn to a problem considered by A. E. Heins:

$$\psi(x) = \lambda \int_0^\infty \frac{\psi(x_0)\,dx_0}{\cosh[\frac{1}{2}(x - x_0)]}$$

The Fourier transform of $\operatorname{sech}[\frac{1}{2}(x - x_0)]$ may be readily obtained by contour integration.

The function to be factored is

$$1 - \sqrt{2\pi}\,\lambda V = 1 - \frac{\lambda\pi}{\cosh(\pi k)} = \frac{\cosh(\pi k) - \lambda\pi}{\cosh(\pi k)}$$

Let $\cos(\pi\alpha) = \lambda\pi$ where $|\alpha| < \frac{1}{2}$; then

$$\cosh(\pi k) - \cos(\pi\alpha) = 2\sin[\tfrac{1}{2}\pi(\alpha + ik)]\sin[\tfrac{1}{2}\pi(\alpha - ik)]$$

These factors may in turn be expressed directly in terms of their zeros by means of their infinite product representations [Eq. (4.3.8)] or equivalently in terms of Γ functions from the relation [Eq. (4.5.33)]

$$\Gamma(z)\Gamma(1 - z) = \pi \csc(\pi z)$$

so that
$$\sin\left[\pi\left(\frac{\alpha + ik}{2}\right)\right] = \frac{\pi}{\Gamma(\frac{1}{2}\alpha + \frac{1}{2}ik)\Gamma(1 - \frac{1}{2}\alpha - \frac{1}{2}ik)}$$

and similarly

$$\sin\left[\pi\left(\frac{\alpha - ik}{2}\right)\right] = \frac{\pi}{\Gamma(\frac{1}{2}\alpha - \frac{1}{2}ik)\Gamma(1 - \frac{1}{2}\alpha + \frac{1}{2}ik)}$$

Also
$$\cosh(\pi k) = \frac{\pi}{\Gamma(\frac{1}{2} - ik)\Gamma(\frac{1}{2} + ik)}$$

Consequently

$$[1 - \sqrt{2\pi}\,\lambda V] = \frac{2\pi\Gamma(\frac{1}{2} - ik)\Gamma(\frac{1}{2} + ik)}{\Gamma\left(\frac{\alpha + ik}{2}\right)\Gamma\left(\frac{\alpha - ik}{2}\right)\Gamma\left(1 - \frac{\alpha + ik}{2}\right)\Gamma\left(1 - \frac{\alpha - ik}{2}\right)}$$

From this we can write down, with some arbitrariness,

$$\Upsilon_+ = \frac{\pi\Gamma(\frac{1}{2} - ik)(\alpha + ik)e^{\chi(k)}}{\Gamma\left(\frac{\alpha - ik}{2}\right)\Gamma\left(1 - \frac{\alpha + ik}{2}\right)};$$

$$\Upsilon_- = \frac{\Gamma\left(1 + \frac{\alpha + ik}{2}\right)\Gamma\left(1 - \frac{\alpha - ik}{2}\right)e^{\chi(k)}}{\Gamma(\frac{1}{2} + ik)}$$

The function $\chi(k)$ is determined by the requirement that Υ_+ and Υ_- be of algebraic growth for large values of k. To examine the behavior for large k we employ Stirling's theorem:

$$\ln[\Gamma(z)] \xrightarrow[z \to \infty]{} [z - \tfrac{1}{2}] \ln z - z + \tfrac{1}{2}\ln(2\pi)$$

Then
$$\ln(\Upsilon_+) \xrightarrow[|k| \to \infty]{} \chi - ik \ln 2 + \ln(ik) + \cdots$$

In order that Υ_+ behave like a polynomial for large $|k|$, we must choose $\chi = ik \ln 2$, in which case we have

$$\Upsilon_+ \xrightarrow[|k| \to \infty]{} ik$$

Of course this holds only where Υ_+ is regular.

We may now determine $P(k)$. Since $\Psi_+ \xrightarrow[|k| \to \infty]{} 0$, it follows from the regularity of $P(k) = -\Psi_-\Upsilon_- = \Psi_+\Upsilon_+$ that $P(k)$ is a constant which we shall call C, which thus determines function Ψ_+:

$$\Psi_+ = \frac{C}{\Upsilon_+} = \frac{C}{\Upsilon_-}\left[\frac{\cosh(\pi k)}{\cosh(\pi k) - \cos(\pi\alpha)}\right]$$

The latter form will prove to be more useful in the present problem. The function $\psi_+(x)$ is

$$\psi_+(x) = \frac{C}{\sqrt{2\pi}} \int_{-\infty+i\tau}^{\infty+i\tau} e^{-ikx} \frac{\cosh(\pi k)}{[\cosh(\pi k) - \cos(\pi\alpha)]} \frac{dk}{\Upsilon_-}$$

For $x > 0$, we may close the contour in the lower half plane where Υ_- is regular. The poles of the integrand occur then only at the zeros of $[\cosh(\pi k) - \cos(\pi\alpha)]$ which are at $-ik = -2n \pm \alpha$, $n = 0, 1, 2, \ldots$. Therefore

$$\psi_+(x) = C' \left[\frac{\cot(\pi\alpha)}{\pi}\right] \sum_{n=0}^{\infty} \left\{\frac{e^{-(2n+\alpha)x}}{\Upsilon_-[-(2n+\alpha)i]} - \frac{e^{-(2n-\alpha)x}}{\Upsilon_-[-(2n-\alpha)i]}\right\}$$

where C' is a new constant. Substituting for Υ_-, ψ_+ becomes

$$\psi_+ = C' \left[\frac{\cot(\pi\alpha)}{\pi}\right] \sum_{n=0}^{\infty} \left\{\frac{\Gamma(\frac{1}{2} + \alpha + 2n)}{\Gamma(1 + \alpha + n)} \frac{(2e^x)^{-(2n+\alpha)}}{n!} - \frac{\Gamma(\frac{1}{2} - \alpha + 2n)}{\Gamma(1 - \alpha + n)} \frac{(2e^x)^{-(2n-\alpha)}}{n!}\right\}$$

This may be reduced to a linear combination of hypergeometric functions of e^{2x} by use of the duplication formula for the gamma function

$$\sqrt{\pi}\,\Gamma(2z) = 2^{2z-1}\Gamma(z)\Gamma(z + \tfrac{1}{2})$$

Hence

$$\psi_+ = C'' \left[\frac{\cot(\pi\alpha)}{\pi}\right] \sum_{n=0}^{\infty} \left\{\frac{\Gamma(\frac{1}{4} + \frac{1}{2}\alpha + n)\Gamma(\frac{3}{4} + \frac{1}{2}\alpha + n)}{\Gamma(1 + \alpha + n)} \frac{e^{-(2n+\alpha)x}}{n!} - \frac{\Gamma(\frac{1}{4} - \frac{1}{2}\alpha + n)\Gamma(\frac{3}{4} - \frac{1}{2}\alpha + n)}{\Gamma(1 - \alpha + n)} \frac{e^{-(2n-\alpha)x}}{n!}\right\}$$

where C'' is still another constant. In terms of the hypergeometric function

$$\psi_+ = C'' \left[\frac{\cot(\pi\alpha)}{\pi}\right] \cdot \left\{\frac{\Gamma(\frac{1}{4} + \frac{1}{2}\alpha)\Gamma(\frac{3}{4} + \frac{1}{2}\alpha)}{\Gamma(1 + \alpha)} e^{-\alpha x} F(\tfrac{1}{4} + \tfrac{1}{2}\alpha, \tfrac{3}{4} + \tfrac{1}{2}\alpha | 1 + \alpha | e^{-2x}) - \frac{\Gamma(\frac{1}{4} - \frac{1}{2}\alpha)\Gamma(\frac{3}{4} - \frac{1}{2}\alpha)}{\Gamma(1 - \alpha)} e^{\alpha x} F(\tfrac{1}{4} - \tfrac{1}{2}\alpha, \tfrac{3}{4} - \tfrac{1}{2}\alpha | 1 - \alpha | e^{-2x})\right\}$$

This may in turn be expressed in terms of $Q_{\alpha-\frac{1}{2}}$, $Q_{-\alpha-\frac{1}{2}}$ the second solution of the Legendre differential equation. We have

$$Q_{\alpha-\frac{1}{2}}(e^x) = \frac{\sqrt{\pi}\,\Gamma(\frac{1}{2} + \alpha)}{2^{\alpha+\frac{1}{2}}\Gamma(1 + \alpha)} e^{-x(\frac{1}{2}+\alpha)} F(\tfrac{1}{4} + \tfrac{1}{2}\alpha, \tfrac{3}{4} + \tfrac{1}{2}\alpha | 1 + \alpha | e^{-2x})$$

Using the duplication formula for gamma functions,

$$\psi_+ = C''' \left[\frac{\cot(\pi\alpha)}{\pi}\right] e^{\frac{1}{2}x}[Q_{\alpha-\frac{1}{2}}(e^x) - Q_{-\alpha-\frac{1}{2}}(e^x)] \tag{8.5.59}$$

Finally from the relation between Legendre functions of the first and second kind,

$$P_\beta = \left[\frac{\tan(\pi\beta)}{\pi}\right][Q_\beta - Q_{-\beta-1}]$$

we have

$$\psi_+ = -C'''e^{\frac{1}{2}x}P_{\alpha-\frac{1}{2}}(e^x); \quad x > 0 \tag{8.5.60}$$

We have thus been able to express the solution of the integral equation for $x > 0$ in terms of hypergeometric functions or Legendre functions which when combined with connection formulas (see tables at the end of Chap. 5) are sufficient for the representation of ψ for all values of x.

The Milne Problem. From the preceding examples, it should be apparent that the factorization of the expression $[1 - \sqrt{2\pi}\,\lambda V]$ is the only step in the Wiener-Hopf procedure which may offer an essential difficulty. In the above problems, this factorization was performed by inspection (after some juggling!). The problem we are to discuss now involves a kernel whose Fourier transform V has a branch point, making factorization by inspection well nigh impossible. We shall now see how this difficulty can be surmounted in a manner which is completely general but consequently cumbersome. The problem, the Milne problem, arises in the passage of radiation (or any other velocity-independent set of particles such as slow neutrons) through a nonabsorbing homogeneous medium in which the radiation is scattered isotropically (see Sec. 2.4). The medium is taken to be semi-infinite in extent, and we consider the situation in which the distribution function f [see Eqs. (8.1.2)] is independent of (x,y), as well as of $|p|$, the momentum.

Under these assumptions, Eq. (8.1.2) becomes

$$v \cos\vartheta \left(\frac{\partial f}{\partial z}\right) = -P_T f + P_T \int f(z,\vartheta_0)\left(\frac{d\Omega_0}{4\pi}\right)$$

where ϑ is the angle between the direction of motion of the particle and the z axis. We may perform the integration over the spherical angle φ_0. In addition, it is convenient to introduce the variables

$$\zeta = (z/vP_T); \quad \mu = \cos\vartheta$$

Note that $P_T = N\sigma$ where N equals the (number of atoms/volume) while σ is the total scattering cross section. The equation is then [see Eq. (2.4.16)]

$$\mu\left[\frac{\partial f(\zeta,\mu)}{\partial \zeta}\right] = -f(\zeta,\mu) + \tfrac{1}{2}\int_{-1}^{1} f(\zeta,\mu_0)\,d\mu_0 \tag{8.5.61}$$

To convert this equation into an integral equation we consider it as a first-order equation in ζ and integrate accordingly. The "solution" is

$$f(\zeta,\mu) = Ae^{-(\zeta/\mu)} + \left(\frac{1}{2\mu}\right)\int_a^\zeta e^{-(\zeta-\zeta_0)/\mu}\,d\zeta_0 \int_{-1}^1 f(\zeta_0,\mu_0)\,d\mu_0$$

The constant of integration in the second term and the constant coefficient A are set by the boundary conditions. Consider the two angular regions $0 < \mu < 1$ and $-1 < \mu < 0$ (the first for particles going to the right, the second for z components of velocity negative). Let us call the corresponding f's, f_a and f_b; that is,

$$\begin{aligned} f_b &= 0; \quad f_a = f; \quad 0 < \mu < 1 \\ f_b &= f; \quad f_a = 0; \quad -1 < \mu < 0 \end{aligned}$$

Then the boundary conditions are that

$$\left.\begin{aligned} &\text{At } \zeta = 0; \quad f_a = I_0(\mu) \\ &\text{At } \zeta = \infty; \quad f_b = 0 \end{aligned}\right\}; \quad f_a, f_b > 0 \text{ for all finite } \zeta \qquad (8.5.62)$$

Here I_0 is the distribution function for the incident beam of particles.

These boundary conditions are met by the following expressions [see Eq. (2.4.19)]:

$$\begin{aligned} f_a &= I_0 e^{-(\zeta/\mu)} + (1/2\mu)\int_0^\zeta e^{-(\zeta-\zeta_0)/\mu}\,d\zeta_0 \int_{-1}^1 f(\zeta_0,\mu_0)\,d\mu_0 \\ f_b &= (1/2|\mu|)\int_\zeta^\infty e^{(\zeta-\zeta_0)/|\mu|}\,d\zeta_0 \int_{-1}^1 f(\zeta_0,\mu_0)\,d\mu_0 \end{aligned} \qquad (8.5.63)$$

To reduce this to a single integral equation, it is useful to change the dependent variable form f to

$$\rho(\zeta) = \int_{-1}^1 f(\zeta,\mu)\,d\mu = \int_{-1}^1 (f_a + f_b)\,d\mu \qquad (8.5.64)$$

Introducing (f_a and f_b) from (8.5.63) into (8.5.64) yields

$$\rho(\zeta) = \int_0^1 I_0(\mu)e^{-(\zeta/\mu)}\,d\mu + \int_0^1 \frac{d\mu}{2\mu}\int_0^\infty e^{-|\zeta-\zeta_0|/\mu}\rho(\zeta_0)\,d\zeta_0$$

Interchanging μ and ζ_0 integrations yields [see Eq. (2.4.20)]

$$\begin{aligned} \rho(\zeta) &= \int_0^1 I_0(\mu)e^{-(\zeta/\mu)}\,d\mu + \int_0^\infty v(\zeta - \zeta_0)\rho(\zeta_0)\,d\zeta_0 \qquad (8.5.65) \\ v(\zeta - \zeta_0) &= \tfrac{1}{2}\int_0^1 e^{-|\zeta-\zeta_0|/\mu}\left(\frac{d\mu}{\mu}\right) \end{aligned}$$

We recognize Eq. (8.5.65) as an inhomogeneous equation of the Wiener-Hopf type. The Milne equation is obtained by dropping the source term to obtain the homogeneous equation

$$\rho(\zeta) = \int_0^\infty v(\zeta - \zeta_0)\rho(\zeta_0)\,d\zeta_0 \qquad (8.5.66)$$

The comparatively trivial change in the Wiener-Hopf procedure which must be included for inhomogeneous equations will be discussed later. The solutions of Eq. (8.5.66) will give the asymptotic (ζ large) form of the solutions of the inhomogeneous equation.

Once ρ is known, the complete distribution may be obtained by integration of expression (8.5.63). Of particular interest is the emergent distribution at the lower surface, $\zeta = 0$. Here

$$f_b(0) = \frac{1}{2|\mu|} \int_0^\infty e^{-\zeta/|\mu|}\rho(\zeta)\, d\zeta \quad \text{or} \quad f_b(0) = \frac{\sqrt{2\pi}}{2|\mu|} R_+\left(\frac{i}{|\mu|}\right)$$

where $R_+(k)$ is the Fourier transform of $\rho_+(\zeta)$. This result is particularly interesting, since it shows that one need not invert the Fourier transform solution of Eq. (8.5.66) in order to obtain the emergent distribution at the surface of the material.

We now return to (8.5.66) and consider its solution via the Wiener-Hopf method. The transform of v is

$$V(k) = \frac{1}{\sqrt{2\pi}}\left(\frac{\tan^{-1} k}{k}\right) = \frac{1}{\sqrt{2\pi}\,(2ik)} \ln\left(\frac{1 + ik}{1 - ik}\right)$$

so that the function Υ which is to be factored into Υ_+/Υ_- is:

$$\Upsilon = (1 - \sqrt{2\pi}\,\lambda V) = [k - \tan^{-1} k]/k \tag{8.5.67}$$

Υ is analytic in the strip $|\text{Im}\, k| < 1$. However, it has branch points at $k = \pm i$, making it impossible to use the simple inspection technique discussed in the two preceding examples.

A General Method for Factorization. The method to be discussed is based on the theorem that any function $f(k)$ analytic in a strip $|\text{Im}\, k| < \alpha$ is the *sum* of two functions, one analytic for $\text{Im}\, k > -\alpha$, the other for $\text{Im}\, k < \alpha$. It will be recognized that this is just a special case of the derivation leading to the Laurent expansion [Eq. (4.3.4) and discussion following]. The procedure adopted there is to apply Cauchy's integral formula to the boundary of the region in which $f(k)$ is analytic, in that case a circular annulus (for which the strip is a special case in which the center of the annulus is placed at infinity). The Cauchy integral breaks up into two integrals: one along the outer circle in a positive direction, the other about the inner circle in the negative direction (the two circles corresponding to the two lines bounding the strip). The function represented by the integral along the outer circle is analytic within the outer circle, *i.e.*, within the annulus and also the inner circle, while the function represented by the integral along the inner circle is analytic outside the inner circle. Placing the center of the annulus at $k = -i\infty$, we have

$$q(k) = q_-(k) - q_+(k)$$

where
$$\left.\begin{aligned} q_-(k) &= \frac{1}{2\pi i}\int_{-\infty+\beta i}^{\infty+\beta i}\frac{q(\eta)}{\eta-k}\,d\eta \\ q_+(k) &= \frac{1}{2\pi i}\int_{-\infty-\beta i}^{\infty-\beta i}\frac{q(\eta)}{\eta-k}\,d\eta \end{aligned}\right\};\quad \beta < \alpha \tag{8.5.68}$$

The component q_- is analytic for $\text{Im}\,k < \beta$, while q_+ is analytic for $\text{Im}\,k > -\beta$, where $\beta < \alpha$, α being the limit of convergence of the transform V of the kernel.

We are now ready to employ Eqs. (8.5.68) for the purpose of writing $\Upsilon = \Upsilon_+/\Upsilon_-$ as required in Eq. (8.5.52). This desideratum is equivalent to the breaking up of $\ln\Upsilon$ into $\ln\Upsilon_+ - \ln\Upsilon_-$, which may be done with the aid of Eq. (8.5.68) if $\ln\Upsilon$ is free of singularities in the strip. Since Υ is analytic there, difficulties can occur only where Υ has zeros, say at k_r. Moreover, to ensure convergence of the integrals in (8.5.68) we should require that $q(\eta)\to 0$ as $|\eta|\to\infty$ within the strip. If $q(k)$ is to be the logarithm of Υ, then Υ must go to unity at $|k|\to\infty$.

Function q cannot be the logarithm of just Υ; we must introduce enough factors to cancel out the zeros of Υ inside the region of analyticity $|\text{Im}\,k| < \alpha$ of V and then to bring the whole argument of the logarithm to unity for $|k|\to\infty$. We have said that the zeros of Υ inside the region of analyticity are at k_r ($r = 1, 2, \ldots, N$). The ratio $\Upsilon/\Pi(k-k_r)$ therefore has no zeros in the strip [sometimes the zeros are of higher order, in which case $(k-k_r)$ must enter with a higher power]. This ratio, however, will not go to unity at $|k|\to\infty$; we need to multiply it by some polynomial in k (which has no zeros in the strip). If, for example, $[\Upsilon/\Pi(k-k_r)]\to(k^{-M}/C)$ as $|k|\to\infty$, we could multiply by $C(k^2+\alpha^2)^{\frac{1}{2}M}$, taking

$$q = \ln\left[C\,\Upsilon(k)(k^2+\alpha^2)^{\frac{1}{2}M}\Big/\prod_r(k-k_r)\right] \tag{8.5.69}$$

The reasons for using the factor $(k^2+\alpha^2)^{\frac{1}{2}M}$ are fairly obvious. We cannot have any zeros of this factor inside the range of analyticity of $\Upsilon(|\text{Im}\,k| < \alpha)$, so we put them just at the edge of the strip. We want them in pairs, one at the top and another at the bottom, so that f_+ can have all of one and f_- all of the other. The constant C is chosen just to make the quantity in the brackets go to unity at $|k|\to\infty$.

Writing $q = -q_+ + q_-$, after computing Eqs. (8.5.68), we have

$$\Upsilon = [\Pi(k-k_r)/C(k^2+\alpha^2)^{\frac{1}{2}M}]e^{(q_--q_+)}$$

We may now factor Υ into Υ_+ and Υ_- by inspection:

$$\Upsilon_+ = [\Pi(k-k_r)e^{-q_+}/C(k+i\alpha)^{\frac{1}{2}M}];\quad \Upsilon_- = (k-i\alpha)^{\frac{1}{2}M}e^{-q_-} \tag{8.5.70}$$

Thus the necessary factorization is formally accomplished and a general procedure is available for the Wiener-Hopf method (if we can calculate the q's!).

Milne Problem, Continued. In the Milne problem, Υ is given by Eq. (8.5.67) as $[1 - (\tan^{-1} k/k)]$. The only zero of this function occurs at $k = 0$, where the zero is of double order. Hence

$$q = \ln\left[\left(\frac{k^2+1}{k^2}\right)\left(1 - \frac{\tan^{-1} k}{k}\right)\right]; \quad \alpha = 1$$

so that

$$q_+ = \frac{1}{2\pi i}\int_{-\infty-\beta i}^{\infty-\beta i} \ln\left[\left(\frac{\eta^2+1}{\eta^2}\right)\left(1 - \frac{\tan^{-1}\eta}{\eta}\right)\right]\frac{d\eta}{\eta - k}; \quad \beta < 1 \qquad (8.5.71)$$

while q_- involves the same integrand with limits $\infty + \beta i$, $-\infty + \beta i$. As a final step we must determine $P(k) = \Psi_+\Upsilon_+ = -\Psi_-\Upsilon_-$ where, according to (8.5.70),

$$\Upsilon_+ = k^2 e^{-q_+}/(k+i); \quad \Upsilon_- = (k-i)e^{-q_-} \qquad (8.5.72)$$

From the convergence of the integrals for q_+ and q_- it follows that both are bounded for large k in their respective regions of regularity. Hence e^{-q_+} approaches a constant, so that $\Upsilon_+ \to k$. Again since $\Psi_+(k) \to 0$, and since P is an entire function, P must equal a constant, say A. Therefore the Fourier transform of ρ_+ is

$$R_+ = \frac{A}{\Upsilon_+} = \left[\frac{A(k+i)}{k^2}\right]e^{q_+(k)} \qquad (8.5.73)$$

where q_+ is given by Eq. (8.5.71).

To obtain the angular distribution it is now necessary [according to (8.5.67)] to obtain $R_+(i/|\mu|)$. The chief difficulty here is in the evaluation of the integral q_+. We shall reduce it to a form suitable for numerical evaluation. For this purpose, we may take $\beta = 0$ and integrate along the real axis of η (as long as $k \neq 0$). Then

$$q_+(k) = \frac{k}{\pi i}\int_0^\infty \ln\left[\left(\frac{\eta^2+1}{\eta^2}\right)\left(1 - \frac{\tan^{-1}\eta}{\eta}\right)\right]\frac{d\eta}{(\eta^2 - k^2)} \qquad (8.5.74)$$

which is of suitable form for $k \neq 0$.

To determine the asymptotic form of $\rho(\zeta)$, it is necessary to have the power series expansion of R_+ in powers of k. We evaluate $q_+(0)$ and then higher terms from Eq. (8.5.74). Integral (8.5.71) for $k = 0$ may be evaluated by considering a contour consisting of the real axis except for a small semicircle about the point $\eta = 0$ (below it). The integrals along the real axis add up to zero because of the oddness of the integrand.

The integral about the circle gives

$$\tfrac{1}{2}\ln\left[\left(\frac{\eta^2+1}{\eta^2}\right)\left(1-\frac{\tan^{-1}\eta}{\eta}\right)\right] \quad \text{at } \eta = 0$$

which is $\frac{1}{2}\ln\frac{1}{3}$ so that

$$q_+(0) = -\ln\sqrt{3}$$

To obtain the next term we employ Eq. (8.5.74) where we make the integrand regular at $\eta = 0$:

$$q_+(k) = \frac{k}{\pi i}\int_0^\infty \frac{\ln\left[\dfrac{3(\eta^2+1)}{\eta^2}\left(1-\dfrac{\tan^{-1}\eta}{\eta}\right)\right]}{(\eta^2-k^2)}\,d\eta$$

Consequently

$$q'_+(0) = \frac{1}{\pi i}\int_0^\infty \frac{\ln\left[\dfrac{3(\eta^2+1)}{\eta^2}\left(1-\dfrac{\tan^{-1}\eta}{\eta}\right)\right]}{\eta^2}\,d\eta \qquad (8.5.75)$$

Returning to R_+ we see that its power series expansion is

$$R_+ \xrightarrow[k\to 0]{} \frac{A}{\sqrt{3}}\,\frac{(k+i)}{k^2}\,[1+kq'_+(0)] \to \frac{Ai}{\sqrt{3}}\left[\frac{1}{k^2}+\frac{1+iq'_+(0)}{ik}\right]$$

Hence

$$\rho_+(\zeta) \xrightarrow[\zeta\to 0]{} \text{constant } [1+iq'_+(0)+\zeta]$$

The constant $1 + iq'_+(0)$ may be evaluated by integrating Eq. (8.5.75) by parts. We obtain $1 + iq'_+(0) \simeq 0.7104$ . . . . This equation and its solution will be discussed in more detail in Sec. 12.2.

Inhomogeneous Wiener-Hopf Equation. The equation

$$\psi(x) = \varphi(x) + \lambda\int_0^\infty v(x-x_0)\psi(x_0)\,dx_0 \qquad (8.5.76)$$

may be solved by the same technique employed for the homogeneous case. Again take Fourier transforms:

$$\Psi_+ + \Psi_- = \Phi_+ + \Phi_- + \sqrt{2\pi}\,\lambda V\Psi_+ \qquad (8.5.77)$$

Introducing the factorization $(1 - \sqrt{2\pi}\,\lambda V) = (\Upsilon_+/\Upsilon_-)$, this equation may be written

$$\Psi_+\Upsilon_+ + \Upsilon_-(\Psi_- - \Phi_-) - \Upsilon_-\Phi_+ = 0 \qquad (8.5.78)$$

The first two terms are in the desired form in that they are analytic in overlapping half planes. The third term is not and must therefore be broken up into a *sum* of two terms, one analytic in the upper half plane, the other analytic in the lower. It is, of course, necessary that there be a strip in which both Υ_- and Υ_+ are regular. If this is so, we may employ the separation given by Eq. (8.5.68) where q is now $\Upsilon_-\Phi_+$. We therefore write

$$\Upsilon_-\Phi_+ = q_- - q_+$$

Equation (8.5.78) may now be written

$$\Psi_+\Upsilon_+ + q_+ = q_- + \Upsilon_-(\Phi_- - \Psi_-) \equiv P \tag{8.5.79}$$

The left-hand side of this equation is analytic in the upper half plane, the right-hand side in the lower half plane. There is a common strip of regularity, so that, as in the case of the homogeneous equation, the right-hand side is the analytic continuation of the left into the lower half plane. The function so defined, P, is regular everywhere in the finite plane and is therefore an entire function. As in the case of the homogeneous equation, the character of P is determined by the asymptotic behavior of one of its defining expressions. Once P is determined,

$$\Psi_+ = (P - q_+)/\Upsilon_+ \tag{8.5.80}$$

which now may be introduced into the inversion formula to obtain $\psi_+(x)$.

If the equation is of the first kind rather than of the second kind, *i.e.*,

$$0 = \varphi(x) + \lambda \int_0^\infty v(x - x_0)\psi(x_0)\, dx_0$$

the solution may be expressed in the form (8.5.79) and (8.5.80) if Υ_+ and Υ_- are now defined by

$$-\lambda \sqrt{2\pi}\, V = \Upsilon_+/\Upsilon_-$$

This completes our general discussion of integral equations. We shall apply many of the techniques touched on here, in later chapters. It must have been noticeable throughout this chapter that the subject of integral equations is not at all so well organized as the subject of ordinary differential equations. There are no rules, so simple as the rules for finding singular points, which enable us to classify kernels of an integral equation and to recognize easily what representation, integral or series, for the unknown function will lead most easily to the solution. This situation is in part due to the fact that integral equations generally represent a much more complex physical and mathematical situation. Only rarely is an integral equation equivalent to a second-order differential equation. More often it corresponds to a differential equation of infinite order.

However, there are some cases for which we may obtain a straightforward solution. For example, kernels of the form $v(x - x_0)$ may be discussed with the air of the Fourier transform. For many cases, however, suitable algorithms for general solution have not been discovered. In spite of this, the integral equation formulation of a problem will prove to be useful, for as we shall see in the next chapter, it forms the basis for the development of many of the approximate solutions to the equations of physics.

Table of Integral Equations and Their Solutions

Types of Equations. A Fredholm equation involves integrals between fixed limits; a Volterra equation involves one variable limit; an equation of the first kind has the unknown function ψ only inside the integral; an equation of the second kind has ψ outside the integral as well:

Fredholm equation of the first kind:

$$\varphi(z) = \int_a^b K(z|z_0)\psi(z_0)\,dz_0; \quad \varphi \text{ and } K \text{ known}$$

Fredholm equation of the second kind:

$$\psi(z) = \varphi(z) + \lambda \int_a^b K(z|z_0)\psi(z_0)\,dz_0; \quad \varphi \text{ and } K \text{ known}$$

When $\varphi = 0$ this equation is homogeneous.

Volterra equation of the first kind:

$$\varphi(z) = \int_a^z K(z|z_0)\psi(z_0)\,dz_0; \quad \varphi \text{ and } K \text{ known}$$

Volterra equation of the second kind:

$$\psi(z) = \varphi(z) + \int_a^z K(z|z_0)\psi(z_0)\,dz_0; \quad \varphi \text{ and } K \text{ known}$$

There are no nonzero solutions of the homogeneous equation, for $\varphi = 0$. The function $K(z|z_0)$ is called the *kernel* of the equation.

Types of Kernels. A *symmetric* kernel satisfies the equation $K(z|z_0) = K(z_0|z)$. A *polar* kernel is of the form $r(z_0)G(z|z_0)$, where G is symmetric. A Fredholm equation with symmetric kernel is *self-adjoint;* the similar Volterra equation is not. A *definite* kernel is one for which either

$$\begin{aligned} &\textstyle\int f(z)\,dz \int K(z|z_0)f(z_0)\,dz_0 > 0; \quad \text{positive definite} \\ &\textstyle\int f(z)\,dz \int K(z|z_0)f(z_0)\,dz_0 < 0; \quad \text{negative definite} \end{aligned}$$

for any function f finite in the range of integration appropriate for the equation using K. A *Hermitian* kernel is one for which $K(z_0|z) = \bar{K}(z|z_0)$; a *skew-Hermitian* one has $K(z_0|z) = -\bar{K}(z|z_0)$. Both of these can be converted into definite or semidefinite (kernels for which $\geq$ or $\leq$ is substituted for $>$ or $<$ in the above definitions) kernels by iteration [see Eqs. (8.2.10) *et seq.*]. An equation with a polar kernel may be transformed into one with a symmetric kernel [see Eq. (8.2.8)].

A *real, positive-definite, symmetric* kernel in a Fredholm equation has the following properties: The homogeneous equation of the second kind has nonzero solutions for a sequence of *eigenvalues* λ_n of λ ($0 < \lambda_0 < \lambda_1 < \lambda_2 \cdots$); the corresponding solutions $\psi_n(z)$ are eigenfunctions, constituting an orthogonal set for $a \le x \le b$ (see page 913). They satisfy the variational principle that

$$\delta \left\{ \frac{\int \bar{\psi}(z)\psi(z)\, dz}{\int\int \bar{\psi}(z) K(z|z_0)\psi(z_0)\, dz_0\, dz} \right\} = 0$$

[see Eq. (6.3.19) for the analogous equation for a differential equation]. The stationary values of the quantity in braces are the eigenvalues λ_n [see Eq. (8.2.24)].

A *singular* kernel has discontinuities or singularities within the range of integration, or else the limits of integration extend to infinity. Some of these kernels may be reduced to nonsingular kernels by iteration; those which cannot are called *intrinsically singular* (see page 924). Homogeneous Fredholm equations with intrinsically singular kernels may have a nondenumerable infinity of eigenvalues; all values of λ within certain ranges may be allowed.

Green's Function for the Inhomogeneous Equation. The inhomogeneous Fredholm equation of the second kind may be solved in terms of the Green's function $G_\lambda(z|z_0)$, a solution of

$$G_\lambda(z|z_0) = K(z|z_0) + \lambda \int K(z|z_1) G_\lambda(z_1|z_0)\, dz_1$$

The solution is

$$\psi(z) = \varphi(z) + \lambda \int G_\lambda(z|z_0)\varphi(z_0)\, dz_0$$

The Green's function G_λ may be expanded in various series involving the eigenfunction solutions ψ_n of the homogeneous equation, the eigenvalues λ_n, or the iterated kernel K_n:

$$G_\lambda(z|z_0) = \sum_m \frac{\psi_m(z)\bar{\psi}_m(z_0)}{\lambda_m - \lambda}; \quad \psi_m \text{ normalized to unity}$$

$$= \sum_{n=0}^{\infty} K_{n+1}(z|z_0)\lambda^n$$

$$K(z|z_0) = K_1(z|z_0) = G_0(z|z_0)$$

$$K_{n+1}(z|z_0) = \int K_n(z|z_1) K(z_1|z_0)\, dz_1 = \sum_m \frac{\psi_m(z)\bar{\psi}_m(z_0)}{(\lambda_m)^{n+1}}$$

The Spur of the Green's function is

$$\int G_\lambda(z|z)\,dz = \sum_m \frac{1}{\lambda_m - \lambda} = \sum_{n=0}^{\infty} C_{n+1}\lambda^n$$

$$C_n = \int K_n(z|z)\,dz = \sum_m \left(\frac{1}{\lambda_m}\right)^n$$

Solutions of Fredholm Equations of the First Kind. This equation is usually solved by expanding φ, K, and the unknown ψ in some series and equating coefficients. The type of series used depends on the type of kernel K.

a. K is a *generating function* for some set of eigenfunctions χ_n appropriate for the limits of integration,

$$K(z|z_0) = \Sigma b_n z^n \bar{\chi}_n(z_0)$$

assuming $\psi = \Sigma a_n \chi_n$ we have $\varphi(z) = \Sigma a_n b_n z^n$ and the unknown coefficients a_n may be determined by comparison with the series expansion of φ.

b. K is a *Green's function* expressible in appropriate eigenfunctions χ_n,

$$K(z|z_0) = \Sigma b_n \bar{\chi}_n(z_0)\chi_n(z)$$

assuming $\psi = \Sigma a_n \chi_n$ we have $a_n = (1/b_n)\int \varphi(z)\bar{\chi}_n(z)\,dz$ which evaluates the coefficients a_n.

If neither of these possibilities can be exploited, the coefficients a_n may be found by the Schmidt method [see Eqs. (8.3.10) *et seq.*] or by biorthogonal functions [see Eqs. (8.3.19) *et seq.*] or by numerical methods (see Chap. 9).

For infinite range of integration, where the kernel is singular, one can sometimes use the fact that the Fredholm equation of the first kind defines an integral transform relation between the known φ and the unknown ψ, through the kernel K:

c. $K(z|z_0) = e^{izz_0}$, integration $-\infty$ to ∞, Fourier transform.

Solution: $$\psi(z) = \frac{1}{2\pi}\int_{-\infty}^{\infty} \varphi(k)e^{-izk}\,dk$$ [Eq. (8.3.59)]

d. $K = e^{-zz_0}$, integration 0 to ∞, Laplace transform.

Solution: $$\psi(z) = \frac{1}{2\pi i}\int_{-i\infty+\epsilon}^{i\infty+\epsilon} \varphi(p)e^{pz}\,dp$$ [Eq. (8.3.66)]

e. $K = J_m(zz_0)z_0$, integration 0 to ∞, Hankel transform.

Solution: $$\psi(z) = \int_0^{\infty} J_m(zr)\varphi(r)r\,dr$$ [Eq. (8.3.64) *f.* $K = (z_0)^{z-1}$, integration 0 to ∞, Mellin transform.

Solution: $$\psi(z) = \frac{1}{2\pi i}\int_{-i\infty+\epsilon}^{i\infty+\epsilon} z^{-s}\varphi(s)\,ds \qquad \text{[Eq. (8.3.68)]}$$

Solutions of Volterra Equations of the First Kind. Types *a* and *b* are not appropriate here; usually the Schmidt method of orthogonalization or the biorthogonal series must be resorted to. For a kernel of the form $K = v(z - z_0)$ the Laplace transform [Eqs. (4.8.30) *et seq.*] may be utilized:

a. $K = v(z - z_0)$, integration 0 to z.

Solution: $$\psi(z) = \frac{1}{2\pi i}\int_{-i\infty+\epsilon}^{i\infty+\epsilon}\left[\frac{\Phi(p)}{V(p)}\right]e^{pz}\,dp; \quad \operatorname{Re} z > 0 \qquad \text{[Eq. (8.5.34)]}$$

where $$V(p) = \int_0^\infty v(x)e^{-px}\,dx; \quad \Phi(p) = \int_0^\infty \varphi(x)e^{-px}\,dx$$

b. $K = v(z - z_0)$, integration z to ∞.

Solution: $$\psi(z) = \frac{1}{2\pi i}\int_{-i\infty+\epsilon}^{i\infty+\epsilon}\left[\frac{\Phi(p)}{V(-p)}\right]e^{pz}\,dp$$

where $$V(-p) = \int_0^\infty v(-x)e^{px}\,dx$$

Solutions of Fredholm Equations of the Second Kind. The kernel may be expressed in terms of an appropriate, orthonormal set of eigenfunctions χ_n; $K(z|z_0) = \Sigma g_n(z_0)\chi_n(z)$. Then assuming $\psi = \Sigma a_n\chi_n$, the homogeneous equation reduces to the set of simultaneous equations

$$\sum_n a_n[\alpha_{mn} - (1/\lambda)\delta_{mn}] = 0; \quad \alpha_{mn} = \int \chi_m(z_0)g_n(z_0)\,dz_0$$

The secular determinant $|\alpha_{mn} - (1/\lambda)\delta_{mn}|$ must be zero. Roots of this equation give the eigenvalues λ_n, and the corresponding series for ψ are the eigenfunctions. Several cases simplify:

a. K a Green's function, then g_n is proportional to $\bar{\chi}_n$, $\alpha_{mn} = \alpha_n\delta_{mn}$. The secular determinant is diagonal, the eigenvalues are $1/\alpha_n$, and the eigenfunctions are χ_n.

b. K a generating function or other form making the secular determinant *semidiagonal* [see Eq. (8.4.8)].

c. K such as to make the secular determinant *finite* [see Eq. (8.4.11)].

The inhomogeneous equation is then solved in terms of a series of the eigenfunctions ψ_n of the homogeneous equation and of the biorthogonal set φ_n, solutions of the adjoint equation (if K is symmetric, $\varphi_n = \bar{\psi}_n$). Then $K(z|z_0) = \Sigma(1/\lambda_n)\psi_n(z)\varphi_n(z_0)$, and the particular integral is

$$\psi(z) = \sum_n \left[\left(\frac{\lambda_n}{\lambda_n - \lambda} \right) \int \varphi(x) \varphi_n(x) \, dx \right] \psi_n(z)$$

where $\varphi(z)$ is the inhomogeneous term in the integral equation.

For infinite range of integration a continuum of eigenvalues of λ is allowed, the secular determinant becomes an integral, and integral transform techniques must be used:

d. $K = v(z - z_0)$, integration $-\infty$ to ∞, Fourier transform. If both transforms $\Phi(k)$ and $V(k)$ are regular for $\text{Im}\, k = \tau$, $-\infty \leq \text{Re}\, k \leq \infty$, then the solution is

$$\psi(z) = \frac{1}{\sqrt{2\pi}} \int_{-\infty + i\tau}^{\infty + i\tau} \frac{\Phi(k) e^{-ikz} \, dk}{1 - \sqrt{2\pi}\, \lambda V(k)}$$

If separate transforms must be established for $z < 0$ and for $z > 0$, then Eq. (8.5.17) holds.

e. $K = v(z - z_0)$, integration 0 to ∞, Wiener-Hopf type. For solution see Eqs. (8.5.55), (8.5.56), and (8.5.76).

f. $K = v(z + z_0)$, integration $-\infty$ to ∞, Fourier transform. For solution see Eq. (8.5.29).

g. $K = (1/z_0)v(z/z_0)$, integration 0 to ∞, Mellin transform. For solution see Eq. (8.5.43).

Solutions of Volterra Equation of the Second Kind. There are no nonzero solutions of the homogeneous equation. In the cases $K = v(z - z_0)$ integration 0 to z or z to ∞, the Laplace transform can be invoked, as with the Volterra equation of the first kind. Solutions of the inhomogeneous equation of the second kind are given in Eqs. (8.5.34) and (8.5.38).

Bibliography

General texts on integral equations:

Courant, R., and D. Hilbert: "Methoden der Mathematischen Physik," Vol. 1, Chap. 3, Springer, Berlin, 1937, reprint Interscience, 1943.

Hamel, G.: "Integralgleichungen," Springer, Berlin, 1937.

Kneser, A.: "Integralgleichungen und ihre Anwendung in der Mathematischen Physik," Vieweg, Brunswick, 1911.

Kowalewski, G. W. H.: "Integralgleichungen," De Gruyter, Berlin, 1930.

Lovitt, W. V.: "Linear Integral Equations," McGraw-Hill, New York, 1924, reprint Dover, New York, 1950.

Shohat, J. A., and J. D. Tamarkin: "The Problem of Moments," American Mathematical Society, New York, 1943.

Vivanti, G.: "Elemente der Theorie der linearen Integralgleichungen," Helwinsche, Hanover, 1929.

Whittaker, E. T., and G. N. Watson: "Modern Analysis," Chap. 11, Cambridge, New York, 1927.

References on the use of Fourier and Laplace transforms to solve integral equations:

Doetsch, G.: "Theorie und Anwendung der Laplace-Transformation," Springer, Berlin, 1937, reprint Dover, New York, 1943.

Hopf, E.: "Mathematical Problems of Radiative Equilibrium," Cambridge, New York, 1934.

Paley, R. E. A. C., and N. Wiener: "Fourier Transforms in the Complex Plane," American Mathematical Society, New York, 1934.

Smithies, F.: Singular Integral Equations, *Proc. London Math. Soc.*, **46,** 409 (1939).

Titchmarsh, E. C.: "Introduction to the Theory of Fourier Integrals," Oxford, New York, 1937.

Index

B

G

N

S

T

Y

Z